T0383009

21st Century Guidebook to Fungi

Second Edition

The mysterious world of fungi is once again unearthed in this expansive second edition. This textbook provides readers with an all-embracing view of the Kingdom *Fungi*, ranging in scope from ecology and evolution, diversity and taxonomy, cell biology and biochemistry, to genetics and genomics, biotechnology and bioinformatics. Adopting a unique systems biology approach – and using explanatory figures and colour illustrations – the authors emphasise the diverse interactions between fungi and other organisms. They outline how recent advances in molecular techniques and computational biology have fundamentally changed our understanding of fungal biology, and have updated chapters and references throughout the book in light of this. This is a fascinating and accessible guide, which will appeal to a broad readership – from aspiring mycologists at undergraduate and graduate level to those studying related disciplines. Online resources are hosted on a complementary website.

DAVID MOORE served as Reader in Genetics in the Faculty of Life Sciences, University of Manchester, UK until his retirement in 2009. He was elected President of the British Mycological Society in 1997, and was Membership Secretary from 2000 to 2003. He has served as Executive Editor of the journal *Fungal Biology* (formerly *Mycological Research*).

GEOFFREY D. ROBSON was Senior Lecturer, University of Manchester, UK. He served as General Secretary of the British Mycological Society for many years and was President of the Society from 2013 to 2014.

ANTHONY P. J. TRINCI is Emeritus Professor in the Faculty of Biology, Medicine and Health, University of Manchester, UK, and was previously Barker Professor of Cryptogamic Botany and Dean of the School of Biological Sciences. He is past President of the Society for General Microbiology and the British Mycological Society.

Frontispiece: A thousand mushrooms crowd to a keyhole … . They lift frail heads in gravity and good faith … .They are begging us, you see, in their wordless way … .To do something, to speak on their behalf. Or at least not to close the door again. (Lines from Derek Mahon's poem 'A Disused Shed in Co. Wexford'. Source: New Collected Poems (The Gallery Press, 2011): https://www.poetryfoundation.org/poems/92154/a-disused-shed-in-co-wexford). *Psathyrella multipedata* (crowded brittlestem) photographed by David Moore in the RHS Garden Harlow Carr, Harrogate, Yorkshire.

21st Century Guidebook to Fungi

Second Edition

David Moore
University of Manchester

Geoffrey D. Robson
University of Manchester

Anthony P. J. Trinci
University of Manchester

CAMBRIDGE
UNIVERSITY PRESS

CAMBRIDGE
UNIVERSITY PRESS

Shaftesbury Road, Cambridge CB2 8EA, United Kingdom

One Liberty Plaza, 20th Floor, New York, NY 10006, USA

477 Williamstown Road, Port Melbourne, VIC 3207, Australia

314–321, 3rd Floor, Plot 3, Splendor Forum, Jasola District Centre, New Delhi – 110025, India

103 Penang Road, #05–06/07, Visioncrest Commercial, Singapore 238467

Cambridge University Press is part of Cambridge University Press & Assessment, a department of the University of Cambridge.

We share the University's mission to contribute to society through the pursuit of education, learning and research at the highest international levels of excellence.

www.cambridge.org
Information on this title: www.cambridge.org/9781108745680

DOI: 10.1017/9781108776387

First published 2011
Second edition 2020 (version 2, January 2025)

Printed in Great Britain by CPI Group (UK) Ltd, Croydon CR0 4YY, January 2025

A catalogue record for this publication is available from the British Library

ISBN 978-1-108-74568-0 Paperback

CONTENTS

PREFACE TO THE SECOND EDITION

Ten years ago, when we first set out to write this book, we aimed to provide a broad understanding of the biology of fungi and the biological systems to which fungi contribute. We hoped the book that would emerge would provide an all-round view of fungal biology, with a scope ranging over ecology, evolution, diversity, cell biology, genetics, biochemistry, molecular biology, biotechnology, genomics and bioinformatics. We were very pleased, and proud, of the book that was published in 2011. That satisfaction was increased enormously when the book was acclaimed around the world by our peers. Reviewers and purchasers alike were appreciative of the broad scope of our text; their reviews included the phrases: 'The *21st Century Guidebook to Fungi* is a game changer ... '; 'Moore, Robson and Trinci have now made fungi accessible to everyone'; 'All you ever wanted to know about fungi: at last: a university level book on fungi to sit next to the classics of cell biology and biochemistry'; 'This remarkably comprehensive volume will be useful to every scientist and educator'.

As well as liking the content, our reviewers also approved of the way the book was organised and written. They commented: 'This is an innovative text, both in its presentation and its organisation'; 'The authors' clear, comprehensible and accurate writing style is just perfect for a textbook'; 'written in a delightful prose, integrating concepts and interdisciplinary knowledge'; 'very user friendly ... making this an outstanding resource'. Overall, the consensus was that the *21st Century Guidebook to Fungi* was: 'a valuable contribution to modernising fungal science education'; 'the best comprehensive mycology text available'. Finally, we were particularly pleased with the *Journal of the American Library Association*'s review, which started: 'The content and quality of this book is simply breathtaking!'

But, you will have noticed that time passes, even breathtaking books age and go out of print; and that's what happened to the *21st Century Guidebook to Fungi*. The third printing went out of print in early 2017. In many respects, authors regard their books in much the same way as they view their children, and we couldn't allow our literary offspring to wither and die on the publisher's vine. In the belief that it remains essential to maintain the availability of a textbook of such user-perceived value, we decided to produce this thoroughly updated **second edition** of the *21st Century Guidebook to Fungi*. As the original text was so well received, this second edition is not a major reorganisation. We have taken the opportunity to rearrange a few parts of the text, but the key difference is a **thorough** section-by-section update from mycology of the first decade to represent mycology of the second decade of the twenty-first century and beyond.

All sections have been revised, but several sections deserved more extensive treatment. This applied particularly to the variety of applications of fungal genome analyses developed in the past 10 years: genomics, proteomics, transcriptomics, metabolomics, metagenomics and several other areas of integrative biology, including but not limited to bioinformatics and computational biology, though we have tried to remain succinct even with these. There were also a few of our own recent publications that warranted proper inclusion; including the mycological alternative interpretation of eukaryote evolution from the book *Fungal Biology in the Origin and Emergence of Life* (Moore, 2013), and some aspects from two other books: *The Algorithmic Fungus* (Moore & Meškauskas, 2017) and *Fungiflex: The Untold Story* (Moore & Novak Frazer, 2017).

One intended omission from the second edition is the compact disc (CD) that was included with the first edition. Although this was widely appreciated as a valuable accessory at the time, we believe that a CD format is no longer appropriate, simply because CD players are so rarely encountered today. Instead, with the second edition, we are expanding the online representation of the *21st Century Guidebook to Fungi*.

In writing the second edition, we have continued our intention to show Kingdom *Fungi* as a **biological system** with its own intrinsic interest rather than as a diverse group of individually fascinating, but still separate, organisms. In addition, we have maintained our model of a tourist guide to a holiday destination. Tourist guides do not attempt a comprehensive depiction of a location, but they bring attention to a broad range of places you might visit, describe enough for you to decide if you **are** interested, and tell you how to get there. We called this a *Guidebook* because we have always been aware of the impossibility of writing a comprehensive, monographic treatment of an entire kingdom, so each section of **your** *Guidebook to Fungi* directs **you** to an interesting aspect of fungal biology and, perhaps unusually for a textbook, provides references to external resources that will provide more information; and there are more references in the second edition.

These references include internet URLs or DOI URLs. The acronym DOI stands for Digital Object Identifier, which uniquely identifies where an electronic document (or another electronic object) can be found on the internet and remains fixed. Other information about a document may change over time, including where to find it, but its DOI will not change and will always direct you to the original electronic document. To access one of these references, enter the DOI URL into your browser and you will be taken to the document on the website of the original publisher. Almost always you will have free access to the abstract or summary of the article, and an increasing number of publishers provide free access to the full text. But even for those that do not provide free access, if your institution maintains a subscription to the products of that publisher you may be able to download the complete text of the article. Save the downloaded document to your hard disk to build your own reprint collection.

We want to restate our sincere thanks to our families for their help and understanding while we produced this text, and all those students of ours who have made constructive comments on this *Guidebook* as it developed over the years. We also thank **Dr Dominic Lewis**, our Commissioning Editor, and the rest of the Cambridge University Press staff and contractors (especially Charlie Howell in Cambridge, and Rajeswari Azayecoche at Integra Software Services) for all their efforts on our behalf in achieving the publication of this book. We extend our appreciation to those anonymous reviewers of the first draft, and Professor David L. Hawksworth CBE for his comments on the proof, all of whose suggestions have so improved the final text.

Special thanks go to **Professor Paul S. Dyer**, School of Life Sciences, University of Nottingham, who was so insistent that we do something about the *21st Century Guidebook to Fungi* going out of print that we were driven to write this second edition. We also restate our gratitude to the many other friends and colleagues who supplied illustrations used in this book: **Professor M. Catherine Aime**, Louisiana State University; **Dr G. W. Beakes**, Newcastle University; **Professor Meredith Blackwell**, Louisiana State University; **Dr Manfred Binder**, Clark University; **Professor C. Kevin Boyce**, University of Chicago; **Professor Jacques Brodeur**, Université de Montréal; **Professor Mark Brundrett**, University of Western Australia; **James Burn**, emapsite.com sales team, Reading; **Sheila** and **Jack Fisher**, Chichester; **Forestry Images**, http://www.forestryimages.org; **Dr Elizabeth Frieders**, University of Wisconsin–Platteville; **Professor G.M. Gadd FRSE**, University of Dundee; **Dr Daniel Henk**, Imperial College; **Professor David S. Hibbett**, Clark University; **Dr Kentaro Hosaka**, National Museum of Nature and Science, Japan; **Dr Carol Hotton**, National Museum of Natural History, Washington, DC; **Dr F. M. Hueber**, National Museum of Natural History, Washington, DC; **Dr Timothy Y. James**, University of Michigan; **Dr P. R. Johnston**, Landcare Research, New Zealand; **Tom Jorstad**, Smithsonian Institution; **Pamela Kaminski**, http://pkaminski.homestead.com; **Dr Bryce Kendrick**, http://www.mycolog.com; **Geoffrey Kibby**, *Field Mycology*; **Dr Cletus P. Kurtzman**, USDA/ARS Peoria; **Dr Roselyne Labbé**, Agriculture and Agri-Food Canada; **Dr Marc-André Lachance**, Western Ontario University; **Professor Karl-Henrik Larsson**, Göteborg University; **Dr Heino Lepp**, Australian National Botanical Gardens; **Dr Peter M. Letcher**, University of Alabama; **Professor Xingzhong Liu**, Chinese Academy of Sciences, Beijing; **Dr Mark Loftus**, Sylvan Inc, PA, USA; **Dr Joyce E. Longcore**, University of Maine; **Dr P. Brandon Matheny**, University of Tennessee; **Dr Audrius Meškauskas**, Switzerland; **Professor Steven L. Miller**, University of Wyoming; **Dr Randy Molina**, *Mycorrhiza* and USDA Forest Service; **Dr Jean-Marc Moncalvo**, Royal Ontario Museum and University of Toronto; **Elizabeth Moore**, Stockport; **NASA**'s Space Telescope Science Institute; **Dr Stephen F. Nelsen**, University of Wisconsin–Madison; **Professor Birgit Nordbring-Hertz**, Lund University; **Dr Lily Novak Frazer**, University Hospital of South Manchester; **Dr Ingo Nuss**, Mintraching-Sengkofen, Germany; **Dr Kerry O'Donnell**, USDA/ARS Peoria; **Dr Fritz Oehl**, ART Zürich; **Dr Lise Øvreås**, University of Bergen; **Mary Parrish**, Smithsonian Institution; **Dr Jens H. Petersen**, University of Aarhus; **Professor Nick D. Read**, Institute of Cell Biology, University of Edinburgh; **Professor Dirk Redecker**, INRA/Université de Bourgogne; **Professor Karl Ritz**, Cranfield University; **Dr Carmen Sánchez**, Universidad Autónoma de Tlaxcala México; **Professor Marc-André Selosse**, Université Montpellier II; **Dr Sabrina Setaro**, Wake Forest University; **Dr Karen Snetselaar**, *Mycologia*, Saint Joseph's University, Philadelphia; **Malcolm Storey**, http://www.bioimages.org.uk; **Professor Junta Sugiyama**, TechnoSuruga Co. Ltd, Tokyo; **Dr Sung-Oui Suh**, American Type Culture Collection; **Mr John L. Taylor**, Manchester; **Professor Vigdis Torsvik**, University of Bergen; **Professor John Webster**, University of Exeter; **Dr Alexander Weir**, SUNY–ESF, New York; **Professor Merlin M. White**, Boise State University; **Alex Wild** Photography, Illinois; **Ence Yang**, Chinese Academy of Sciences, Beijing.

We must end this Preface on an unwelcome sad note. As the text of this second edition was reaching completion, our kind and gentle friend, greatly valued colleague and co-author, **Geoffrey David Robson**, died suddenly on 15 May 2018. This volume is dedicated to his memory as a mycologist of distinction.

David Moore and Tony Trinci
Stockport, Cheshire

CHAPTER 1

Twenty-First Century Fungal Communities

This second edition of our book continues to emphasise interactions between fungi and other organisms to bring out the functions and behaviours of biological systems; but this edition features a thorough section-by-section update from the mycology of the years 2008/10 to the mycology of 2018/20. However, we continue to maintain our aim to:

- concentrate on integration rather than reduction, to satisfy those who would see systems biology as a paradigm of scientific method;
- include computational modelling and bioinformatics for those who view systems biology in terms of operational research protocols;
- bring together data about biological systems from diverse interdisciplinary sources;
- offer our readers a comprehensive, all-embracing view of today's fungal biology.

In this chapter, we examine present-day communities, starting with the essential terrestrial habitat and the nature and formation of soil. We emphasise the contributions made by fungi to soil structure and chemistry, particularly what has come to be called geomycology. We also discuss the diversity of organisms in soil and illustrate interactions between bacteria, amoebae (including slime moulds), fungi, nematodes, microarthropods and larger animals. The origins of agriculture are briefly mentioned and our dependence on fungi briefly illustrated.

1.1 WHAT AND WHERE ARE FUNGI?

'How many of you think that fungi are bacteria?' is a question asked at a summer school for Year 10 pupils (4th year in secondary school, 14 years of age at entry), by one of the pupils who had attended a workshop session of ours. When all attendees (approximately 170 pupils) were asked 'Hands up all those who think fungi are plants,' about 15 hands went up, but when asked 'Hands up all those who think fungi are bacteria,' at least 150 hands went up!

When teaching mycology, we were used to battling against the mistaken idea that fungi are plants, but it was a shock to find that so many pupils believed that fungi are bacteria so close to the end of their statutory education. After all, it's a bigger error than for them to think that whales are fish; at least whales and fish are in the same biological kingdom. Does such ignorance matter? We say it does. The practical reason it matters is because the activities of fungi are crucially important in our everyday lives. The educational reason it matters is that fungi form what is arguably the largest kingdom of higher organisms on the planet. Ignorance of this kingdom is a major blot on our personal education.

Fungi are not bacteria, because fungi are eukaryotes and they have the complex cell structures and abilities to make tissues and organs that we expect of higher organisms. Unfortunately, even though fungi make up such a large group of higher organisms, most current biology teaching, from school-level upwards, concentrates on animals, with a trickle of information about plants. The result is that the majority of school and college students (and, since they've been through the same system, current university academics) are ignorant of fungal biology and therefore of their own dependence on fungi in everyday life. This institutional ignorance about fungi, generated by the lack of an appropriate treatment of fungal biology in national school curricula, seems to apply throughout Europe, North, South and Central America, and Australasia; indeed, most of the world. We have a history of writing about this institutional ignorance of fungi, in several formats, in several arenas and at several academic levels. We stated in the first edition of this book that although fungi comprise what is arguably the most pivotal kingdom of organisms on the planet, these organisms are often bypassed and ignored by the majority of biologists. We say 'pivotal' here because molecular phylogenies place animals and fungi together at the root of evolutionary trees. It is likely that the first eukaryotes would have been recognised as 'fungal in nature' by features currently associated with that kingdom. So, in a sense, those primitive 'fungi' effectively invented the eukaryotic lifestyle.

We will not tax our readers further, except to say that this tragic situation clearly persists. To quote Lozano Garza and Reynaga Peña (2017): 'Even in specialised Biology books, the issue of fungi, as compared to plants or animals, is not given equal importance. The results of this analysis partially explain the reasons for the common but scientifically incorrect belief that fungi are bacteria or plants.' This institutional ignorance is the feature which figured most in our decision to write the first edition of this textbook and its persistence fuelled the desire to write the second edition.

It is also worth emphasising that the contribution that fungi make to human existence is close to crucial, too. Imagine life without bread, without alcohol, without soft drinks, without cheese, coffee or chocolate, without cholesterol-controlling drugs (the 'statins') or without antibiotics, and you are imagining a much less satisfactory existence than we currently enjoy. As we will show in later chapters:

- Fungi (known as anaerobic chytrids) help to digest the grass eaten by cows (and other domesticated grazing animals) and by so doing indirectly provide the milk for our breakfast, the steak for dinner and the leather for shoes.
- Fungi make plant roots work more effectively (more than 95% of all terrestrial plants depend on mycorrhizal fungi) and, even leaving aside the effect of this on the evolution of land plants, by so doing mycorrhizal fungi help provide the corn for our cornflakes, oats for our porridge, potatoes, lettuce, cabbage, peas, celery, herbs, spices, cotton, flax, timber, etc. And even oxygen for our daily breath.
- The characteristic fungal lifestyle is the secretion of enzymes into their environment to digest nutrients externally; we harness this feature in our biotechnology to produce enzymes to start our cheese making, clarify our fruit juices, distress denim for 'stone-washed' jeans and, conversely, provide fabric conditioners to repair day-to-day damage to our clothes in the weekly wash.
- Fungi also produce a range of compounds that enable them to compete with other organisms in their ecosystem; when we harness these for our own purposes we create products like:
  - cyclosporine, which suppresses the immune response in transplant patients and prevents organ rejection;
  - the statins, which help increase the lifespan of so many people these days by controlling cholesterol levels;
  - and even today's most widely used agricultural fungicides, the strobilurins.

But fungi are not always benevolent. There are fungal diseases of all our crops that we need to understand and control. In many cases crop losses of 20 to 50% are *expected* by the agricultural industry today. As the human population increases such losses in primary production cannot be sustained. And there is more to fungal infection of humans than athlete's foot and a disfigured toenail. Opportunistic fungal infections of patients are an increasing clinical challenge, as the majority of patients with chronic immunodeficiency now die of fungal infections; and yet we lack a sufficient range of good drugs to treat fungal infections.

Our answers to the questions in the title of this section 'What and where are fungi?' are that fungi comprise the most crucial kingdom of eukaryotic organisms on the planet, and that they exist everywhere on Earth. Remember this: 'when looking for nature-based solutions to some of our most critical global challenges, fungi could provide many of the answers' (Willis, 2018).

1.2 SOIL, THE ESSENTIAL TERRESTRIAL HABITAT

The conventional estimate is that 75% of the Earth is covered with water; oceans, lakes, rivers, streams. However, less than 1% of the known species of fungi have been found in marine habitats (see pp. 346–351 in Carlile *et al.*, 2001). Fresh water is inhabited by many water moulds (an informal grouping that includes the most ancient fungi and fungus-like organisms, which we will discuss in more detail in Chapter 3), but the overwhelming majority of fungi occur in association with soil; where 'in association with' means in or on the soil, or in or on some live or dead plant or animal that is in or on the soil.

The soil environment is the most complex on Earth and provides a range of habitats that support an enormous population of soil organisms. Soil is characterised by a heterogeneity, which is measured across physical scales varying from nanometres to kilometres, and differs in chemical, physical and biological characteristics in both space and time. The nature of soil is determined by the interaction of geology, climate and vegetation, and is a biochemical product of the organisms participating in its formation (Voroney & Heck, 2015).

As Wikipedia points out (https://en.wikipedia.org/wiki/Soil) 'Soil is commonly referred to as earth,' so it is the substance from which our planet takes its name. Soil is, therefore, essential to terrestrial habitats.

But they are categories: grassland, forest, coastal, desert, tundra and even cities and suburbs, and ultimately all these environments depend on their soil. Without soil, no grass; so no grassland habitat. Without soil, no trees; so no forest habitat. Few, if any, organisms can be found on bare rock, wind-blown sand or ice. Fundamentally, terrestrial life on Earth depends upon 'earth', and showing how fungi contribute to the formation of soil is where we choose to start our story.

1.3 HOW MUCH SOIL IS THERE AND WHERE IS IT?

Only about 7.5% of the Earth's surface provides the agricultural soil on which we depend for the world's food supply (Table 1.1). This fragment competes, sometimes unsuccessfully, with all other needs: housing, cities, schools, hospitals, shopping centres, land fills, etc.

Indeed, there may not be enough soil in the first place. A subsistence diet requires about 180 kg of grain per person per year. This can be produced on 0.045 ha of land. In contrast, an affluent high-meat diet requires at least four times more grain (and four times more land, 0.18 ha) because the animals are fed on grain and conversion of grain to meat is very inefficient.

The Earth has about 0.25 ha of farmland per person, but only about 0.12 ha per person of farmland is suitable for producing grain crops. As it stands, the Earth does not have enough land for all inhabitants to enjoy an affluent diet as that is presently defined (see Table 1–2 in Miller & Gardiner, 2004; and see Figure 12.12, in section 12.4 below, for a potential alternative).

This chapter describes the key physical and chemical features of the soil habitat that govern the biodiversity and activity of soil organisms.

1.4 THE NATURE OF SOIL AND WHO MADE IT

Soil is that part of the Earth's surface comprised of fragmented rock and humus. It is made up of solid, liquid and gaseous phases (Needelman, 2013).

- The solid phase is mineral and organic matter and includes many living organisms.
- The liquid phase is the 'soil solution', from which plants and other organisms take up nutrients and water.
- The gaseous phase is the soil atmosphere, supplying oxygen to plant roots and other organisms for respiration.

The solid phase is made up of minerals and organic matter. Minerals may be either primary or secondary. Primary minerals are those that cooled from a molten mass, and are chemically unchanged from the day they came into existence. Secondary minerals form by chemical modification, precipitation or recrystallisation of chemicals released by the weathering of parental rocks. Rocks are mixtures of minerals. Igneous rock forms from molten magma. If the magma cools slowly it forms a network of large crystals as in granite, if it cools rapidly it forms small crystals as in basalt. Sedimentary rocks are cemented accumulations of minerals: common sedimentary rocks include limestone, sandstone, quartzite, and shale. Metamorphic rocks, which arise when an existing rock is transformed by exposure to high temperatures and pressures, include slate (hardened shale) and marble (hardened limestone).

Table 1.1. How much soil is there? Broad estimates of the coverage of the Earth's surface by different features

Surface feature	Percentage coverage
Aquatic: oceans, seas, rivers and lakes	75.0
Deserts: polar and mountain regions unsuitable for agriculture	12.5
Rocky and other poor-quality terrestrial regions unsuitable for agriculture	5.0
Terrestrial regions suitable for agriculture	7.5

Weathering is the term applied to the processes that cause rocks to disintegrate into smaller parts. Loose or unconsolidated products of weathering are called soil minerals. These may be fragmented versions of primary minerals (e.g. sand is fragmented quartz rock) or may be secondary minerals, like clays, slowly formed through chemical interactions in the soil, then further chemically modified with time. The elements most commonly found in soil minerals are silicon, oxygen and aluminium.

Physical and chemical processes contribute to weathering. The main physical weathering effect is the force exerted by the expansion of water as it freezes, so physical weathering is most pronounced in cold climates. In dry climates, abrasion by materials suspended in the wind causes weathering (a similar effect occurs in flowing water). Chemical weathering predominates in warm and/or moist climates. It is generally more important for soil formation than physical. Chemical processes include:

- oxidation and reduction (of great importance for iron-containing minerals);
- carbonation (dissolution of minerals in water made acidic by carbon dioxide);
- hydrolysis (when water splits into hydrogen and hydroxide, and one or both components participate directly in the chemical process); and
- hydration (when water is incorporated into the crystal structure of a mineral, changing the properties of that mineral) (Miller & Gardiner, 2004).

Soils are highly dynamic environments. They change over time, as their particles are moved downward by the leaching effect of rainwater and laterally by wind, water and ice.

The most potent soil-forming factor is often considered to be the climate, mainly temperature and rainfall. Temperature affects the rates of chemical reactions, so that soils of warmer climates tend to mature more rapidly. However, living organisms (the soil biota) both affect, and are affected by, soil formation. First thoughts tend to be about the profound effects of vegetation on soil formation. For one thing, the extent of vegetation cover influences water runoff and erosion. Fairly obviously, the vegetation type and amount directly affect the type and amount of organic matter that accumulates on and in the soil. Grasslands and forests form different soils, as there is more rapid nutrient cycling in grassland.

Organic matter deposited on the surface contributes to soil solids. It is moved downward physically through rainwater leaching and influences soil chemistry, pH and nutrient supply as it goes. This organic matter is the food source for most microorganisms in the soil, so the vegetation influences soil microbial populations by providing their nutrients. Old soils can lose their ability to produce vegetation fast enough to keep up with microbial decomposition. In healthy agricultural soils, organic material is initially decomposed rapidly, but within about a year, organic materials like crop residues 'stabilise' and the remaining residues decay very slowly. This slowly decomposing material is comprised of 'humic substances' (commonly called humus). Humic substances are natural non-living organic substances that occur in all aquatic and terrestrial environments. They are found in sediments, peat, sewage, composts and other deposits. This soil organic matter represents the main carbon reservoir in the biosphere, estimated at a grand total of 1,600 × 10^{15} g C (Grinhut *et al.*, 2007). The organic matter of soil is crucial to its agricultural value because it aids structure, nutrition and water relations, everything that contributes to soil tilth (*tilth* is an Old English word that describes the structure and quality of cultivated soil in the sense that good tilth corresponds to potentially good crop growth).

Decomposing organic matter provides nutrients to other soil organisms (including, but not exclusively crop plants). Stable organic matter does not do this, but it improves the ability of the soil to hold nutrients and water. An organic soil is dominated by organic matter, rather than minerals. Such soils are found in wetlands, especially cold wetlands, where the primary production of organic materials by the plants exceeds the rates of decomposition in the soil. Ultimately, this equation results in peat formation.

The spaces between soil particles form the pore space, which contains air and water. The water, called the soil solution, contains soluble salts, organic solutes and some suspended colloids. The amount and behaviour of soil water is controlled by pore size, influenced by proportions of coarse material (like sand) and fine minerals (such as clays). Small pores have a greater affinity for water and hold it very tightly. Larger pores allow water to escape easily, by drainage or into the atmosphere by evaporation. Soil 'air' has more CO_2, but less O_2, than the open atmosphere. This is because organisms in the soil consume O_2 and produce CO_2, producing corresponding concentration gradients between the soil and the atmosphere. Similarly, soil air always has a relative humidity near 100%. Respiration releases water vapour, which evaporates only slowly into the atmosphere above the soil.

Soil is a dynamic matrix of organic and mineral constituents enclosing a network of voids and pores, which contain liquids and gases. It is also a living system. Soil organic matter includes living organisms: bacteria, fungi, algae, protozoa and multicellular animals, from rotifers and microarthropods to worms and small mammals (Haynes, 2014). The soil biota is extremely important to soil processes. Although living macroorganisms are usually not considered part of the soil, they can have considerable effect on soil; remember Darwin's experiments on earthworms, animals that excrete more bacteria into the soil than they consume (Darwin, 1881). This is true even if we leave aside human activities like ploughing, irrigating, mining, clearing, waste-disposing, excavating, levelling, building, draining, flooding, etc.

1.5 SOIL BIOTA ARE EXTREMELY VARIED AND NUMEROUS

In about 5 cm^3 of agricultural soil you are likely to find:

- at least 5 billion bacteria;
- 5 million protozoa;
- 5,000 nematodes (about 0.3–1.5 mm long); the most common multicellular animals in soil;
- about 6 mites and other microarthropods; this equates to up to 600,000 m^{-2}.

For larger organisms we must look at quadrats of about 1 m^2:

- Earthworms – maybe 300 per square metre. Earthworm casts add more bacteria back to the soil than the worm eats. More bacteria mean healthier soil.
- There may be around 20,000 km of hyphae per square metre. Above ground, a meadow may look like separate plants. Underground, the plants are interconnected by their fungal associates (mycorrhizas) so they all belong to a single web of living things.
- Small mammals (mice, voles, shrews and moles) that depend on the earthworms, arthropods and fungi for their nutrition, and in their turn feed predators; owls, foxes, etc., so the food web extends from microbes to large animals.

RESOURCES BOX 1.1 Life in the Soil

Deep Down & Dirty: the Science of Soil. A close-up of creatures living beneath the soil, made by the British Broadcasting Corporation (BBC): https://www.youtube.com/watch?v=gYXoXiQ3vC0

The Living Soil Beneath Our Feet, made by the California Academy of Sciences: https://www.youtube.com/watch?v=MIREaT9hFCw

The development of communities of soil biota is characterised by progressive addition with many pioneer species remaining throughout soil development. Size and diversity of soil biota communities increases rapidly during the first 20 to 50 years and then more or less stabilises after hundreds of years. Plant biomass and soil organic matter content do not reach a peak for many hundreds or even thousands of years (Haynes, 2014). Development of communities of soil fauna is less rapid than that of the microbial communities because dispersal is slower, and some faunal species require a certain depth of organic topsoil and/or litter layer before high populations can develop. With increasing time, the food web, which is based on organic detritus, becomes increasingly complex (Haynes, 2014). Organism abundance, diversity and activity vary in a patchy fashion both horizontally across a landscape and vertically through the soil profile. Different groups of soil organisms exhibit different spatial patterns, with the spatial heterogeneity in microbial properties being an inherent feature of soils (Frey, 2015).

1.6 MICROBIAL DIVERSITY IN SOIL

The word 'diversity' when used in relation to organisms in a habitat describes complexity and variability at different levels of biological organisation:

- genetic variability within taxa (which may be species);
- the number (also called richness) of taxa;
- relative abundance (or evenness) of taxa; and
- number and abundance of functional groups.

Important aspects of diversity at the ecosystem level are:

- the range of processes;
- complexity of interactions;
- number of trophic levels.

Thus, measurements of microbial diversity must include multiple methods, integrating measures at the total community level and partial approaches that target subsets of the community with specific structural or functional attributes. For example, you might be trying to assess all decomposers, or all leaf-eaters, all root diseases, etc., each of which will give you just a partial view of the community in the habitat.

Simply attempting to count the number of microorganisms in soil raises difficulties. Because they are microscopic counting and identifying them with conventional techniques requires them to be cultivated. Yet not all can be cultivated; some have growth requirements that are so fastidious they may be difficult or impossible to provide. In many other cases, the growth requirements are simply unknown. The filamentous nature of most fungi creates the additional difficulty of recognising an individual fungus, and disentangling an extensive mycelial network from the substratum it is exploring and penetrating. Techniques based on chemical analysis to quantify some characteristic component of the fungal cell have been successfully used to quantify fungal biomass in soils, composts (in mushroom farming) and timber. Measurement of chitin (as amino sugar) can be used where confusion with arthropod exoskeletons can be excluded, but measurement of ergosterol, which is a characteristic component of fungal membranes, is more generally applicable. Methods based on RNA and DNA probes and polymerase chain reaction (PCR) were initially developed to *identify* organisms, but they revealed an *immense diversity* of microbes in natural habitats (Prosser, 2002; Torsvik & Øvreås, 2002; Wellington *et al.*, 2003; Anderson & Parkin, 2007). Frequently, less than 1% of the microorganisms detected this way can be cultivated and characterised as live cultures (Figure 1.1).

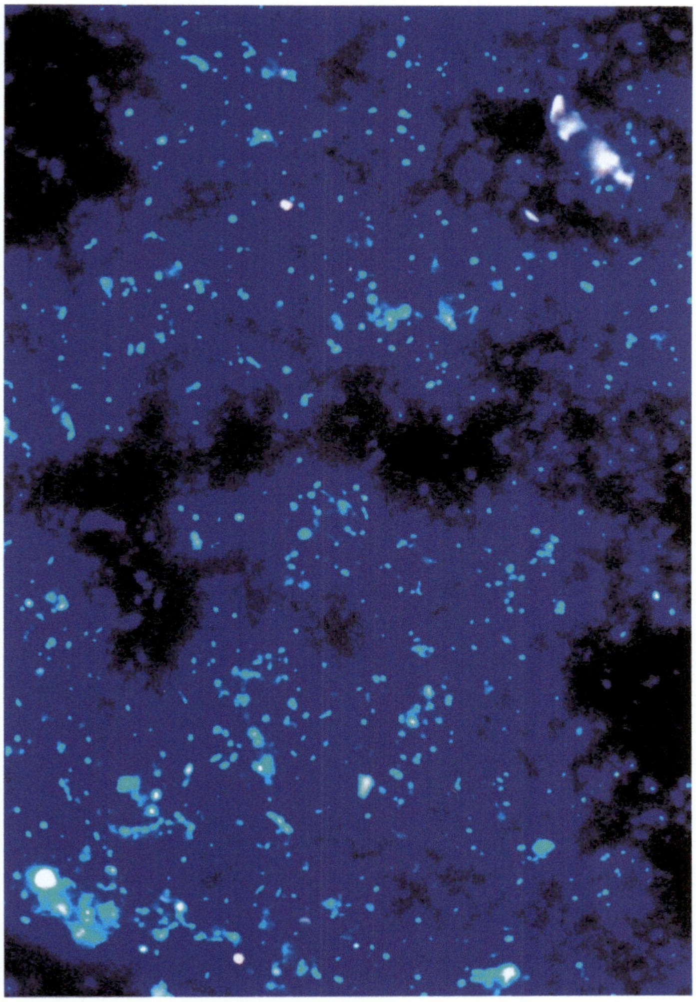

Figure 1.1. This picture (from Torsvik & Øvreås, 2002) shows an epifluorescence micrograph of soil microorganisms stained with the fluorochrome DAPI (4′-6-diamidino-2-phenylindole), which detects intact DNA. This sample had a visible count of 4×10^{10} cells g^{-1} soil (dry weight); but the viable count was 4×10^6 colony-forming units g^{-1} soil (dry weight) when estimated by plating on agar media. Reproduced with permission from Elsevier.

This contrast between the numbers of microbes (of all sorts) that are known to exist and the numbers that can be cultivated is not unusual, and certainly applies to fungi (Prosser, 2002; Mitchell & Zuccaro, 2006; Anderson & Parkin, 2007). There may be several reasons for this discrepancy, including the unknown growth requirements of the uncultured organisms and the intractable dormancy of their resistant structures. For example, it has been demonstrated that *arbuscular mycorrhizal fungi are fatty acid auxotrophs*; that is, they depend on their host plants to synthesise and transfer, from the plant to the mycorrhizal fungus, essential substrates for the production of fungal lipids that the fungus cannot make for itself (Luginbuehl *et al.*, 2017). Also, there are many obligately biotrophic fungi that cause serious and widespread diseases of crop plants, but which cannot be cultured *in vitro* (Tang *et al.*, 2018). Today, the numerical discrepancy between living presence and culturability (as illustrated in Figure 1.1) is most likely to be detected by environmental DNA testing (Taberlet *et al.*, 2018), although the use of environmental DNA samples to define sequence based taxa without physical specimens and formal nomenclature is hotly debated (Hongsanan *et al.*, 2018).

1.7 MICROBIAL DIVERSITY IN GENERAL

Microorganisms exist in every conceivable place on Earth, even in extreme environments. The tropics are richer in microbial species diversity than the temperate zones, but deserts may feature an equal amount, if not more, microbial diversity, and microbial communities can be found on rocks and within deep rock crevices (e.g. Staley *et al.*, 1982). Temperature may be the only limitation as to where they can and cannot exist (Hunter-Cevera, 1998).

The conservative estimate is that there are 1.5 million species of fungi on Earth, of which only 120,000 species have been isolated or described (Hawksworth, 1997, 2001; Hawksworth & Lücking, 2017). The former estimate was made by comparing the number of species of fungi and vascular plants described for geographic regions. For example, in the British Isles there are about six times more species of fungi than species of vascular plants. Extrapolating this ratio to the 270,000 species of vascular plants in the world gives an estimate of 1,620,000 fungi. Now, this figure needs to be corrected (to 1,504,800) to account for the double counting of fungal species resulting from the practice of giving separate specific names to the asexual and sexual stages of some fungi (because it may not be known that the two reproductive stages belong to the same fungus). In recent years molecular approaches to species recognition, together with the recognition of new habitats, hyperdiverse environments and unstudied collections have prompted revision of this conservative estimate of fungal diversity. Hawksworth and Lücking (2017) conclude that the range 2.2–3.8 million species of fungi on Earth is a better estimate. They point out that this means that 'at best just 8%, and in the worst case scenario just 3%' of the world's fungi have been formally described and named so far.

You might ask: 'Where are the other 92 to 97% of undescribed fungi?' In fact, recent developments of molecular phylogeny have revealed an unexpected diversity, many times greater than this, among fungi, and indeed, most other organisms. We mentioned in the previous Section the numerical discrepancy between living presence detected by environmental DNA testing and culturability (as illustrated in Fig. 1.1). Applying molecular phylogenetics to the DNA testing identifies many cryptic species (species recognised only by analysis of DNA sequences). Based on such data, the number of fungal species on the Earth has been estimated to be **12 (11.7–13.2) million** compared to the estimate of **2.2–3.8 million** species indicated above (Wu *et al.*, 2019).

Intensive studies of specific fungal genera and families have demonstrated that in countries and areas that were hitherto neglected by mycological taxonomists, up to over 90% of the collected specimens may constitute undescribed species (Hyde *et al.*, 2018). But the novelty goes beyond recognising more species; a number of novel taxa including new divisions, classes, orders and new families have been established in the last decade by molecular phylogenetics, and many of these ***dark matter fungi*** belong to early diverging branches of the fungal evolutionary tree (Grossart *et al.*, 2016).

The range and prevalence of the genetic material being detected has led to it being called **biological dark matter** [https://en.wikipedia.org/wiki/Biological_dark_matter] by (informal) analogy with cosmological dark matter, which is a form of invisible matter thought to account for approximately 85% of the matter in the universe because of its gravitational influences. Biological dark matter is an extremely active and important research topic, which we cannot pursue further here but we do suggest a few representative references: Wu *et al.*, 2011; Carey, 2015; Lok, 2015; Wu *et al.*, 2019.

Part of the answer to our question about the location of all those undescribed fungi is that there are not many mycologists in the world today and not much work has been done in several unique geographical regions or habitats. Many 'missing fungi' may be associated with tropical forests, for example. Insects may be another large source of missing fungi as many fungi are already known to be associated with insects. Finally,

RESOURCES BOX 1.2 About the Diversity of Fungi

We do not intend to expand further on the topic of fungal biodiversity here because we prefer to concentrate on other aspects of the fungal contribution to the soil community, but if you wish to investigate fungal diversity further we recommend the following literature references (and the papers referenced in them). Many of these topics will be discussed later in this book.

Aptroot, A., Cáceres, M.E.S., Johnston, M.K. & Lücking R. (2016). How diverse is the lichenized fungal Family Trypetheliaceae (Ascomycota: Dothideomycetes): a quantitative prediction of global species richness. *Lichenologist*, 48: 983–1011. DOI: https://doi.org/10.1017/S0024282916000463.

Bass, D. & Richards, T.A. (2011). Three reasons to re-evaluate fungal diversity 'on Earth and in the ocean'. *Fungal Biology Reviews*, 25: 159–164. DOI: https://doi.org/10.1016/j.fbr.2011.10.003.

Hawksworth, D.L., Hibbett, D.S., Kirk, P.M. & Lücking, R. (2016). (308–310) Proposals to permit DNA sequence data to serve as types of names of fungi. *Taxon*, 65: 899–900. DOI: https://doi.org/10.12705/654.31.

Hawksworth, D.L. & Lücking, R. (2017). Fungal diversity revisited: 2.2 to 3.8 million species. In *The Fungal Kingdom*, ed. J. Heitman, B. Howlett, P. Crous *et al.* Washington, DC: ASM Press, pp. 79–95. DOI: https://doi.org/10.1128/microbiolspec.FUNK-0052-2016.

Hibbett, D.S. (2016). The invisible dimension of fungal diversity. *Science*, 351: 1150–1151. DOI: https://doi.org/10.1126/science.aae0380.

Hyde, K.D., Bussaban, B., Paulus, B. *et al.* (2007). Diversity of saprobic microfungi. *Biodiversity and Conservation*, 16: 7–35. DOI: https://doi.org/10.1007/s10531-006-9119-5.

box continues

Leavitt, S. D., Divakar, P. K, Crespo, A. & Lumbsch, H. T. (2016). A matter of time: understanding the limits of the power of molecular data for delimiting species boundaries. *Herzogia*, 29: 479–492. DOI: https://doi.org/10.13158/heia.29.2.2016.479.
Mora, C., Tittensor, D.P., Adl, S., Simpson, A.G.B. & Worm, B. (2011). How many species are there on Earth and in the ocean? *PLoS Biology*, 9: article e1001127. DOI: https://doi.org/10.1371/journal.pbio.1001127.
Mueller, G.M. & Schmit, J.P. (2007). Fungal biodiversity: what do we know? What can we predict? *Biodiversity and Conservation*, 16: 1–5. DOI: https://doi.org/10.1007/s10531-006-9117-7.
Mueller, G.M., Schmit, J.P., Leacock, P.R. *et al.* (2007). Global diversity and distribution of macrofungi. *Biodiversity and Conservation*, 16: 37–48. DOI: https://doi.org/10.1007/s10531-006-9108-8.
Schmit, J.P. & Mueller, G.M. (2007). An estimate of the lower limit of global fungal diversity. *Biodiversity and Conservation*, 16: 99–111. DOI: https://doi.org/10.1007/s10531-006-9129-3.
Schoch, C.L., Seifert, K.A., Huhndorf, S *et al.* (2012). Nuclear ribosomal internal transcribed spacer (ITS) region as a universal DNA barcode marker for Fungi. *Proceedings of the National Academy of Sciences of the United States of America*, 109: 6241–6246. DOI: https://doi.org/10.1073/pnas.1117018109.
Shearer, C.A., Descals, E. Kohlmeyer, B *et al.* (2007). Fungal biodiversity in aquatic habitats. *Biodiversity and Conservation*, 16: 49–67. DOI: https://doi.org/10.1007/s10531-006-9120-z.
Taberlet, P., Bonin, A., Zinger, L. & Coissac, E. (2018). *Environmental DNA For Biodiversity Research and Monitoring*. Oxford, UK: Oxford University Press. ISBN-10: 0198767285, ISBN-13: 978-0198767282.
Tedersoo, L., Bahram, M., Põlme, S. *et al.* (2014). Global diversity and geography of soil fungi. *Science*, 346: article 1256688. DOI: http://dx.doi.org/10.1126/science.1256688.
Tedersoo, L., Bahram, M., Puusepp, R., Nilsson, R.H., James, T.Y. (2017). Novel soil-inhabiting clades fill gaps in the fungal tree of life. *Microbiome*, 5: article 42 (10 pp.). DOI: https://doi.org/10.1186/s40168-017-0259-5.
Truong, C., Mujic, A.B., Healy, R *et al.* (2017). How to know the fungi: combining field inventories and DNA-barcoding to document fungal diversity. *New Phytologist*, 214: 913–919. DOI: https://doi.org/10.1111/nph.14509.

A greater understanding of Earth's biodiversity, coupled with the responsible stewarding of its resources, are among the most crucial aims and challenges of the Earth BioGenome Project as described in the following:

Lewin, H.A., Robinson, G.E., Kress, W.J *et al.* (2018). Earth BioGenome Project: sequencing life for the future of life. *Proceedings of the National Academy of Sciences of the United States of America*, 115: 4325–4333. DOI: https://doi.org/10.1073/pnas.1720115115.

See also the *Earth BioGenome Project* website at https://www.earthbiogenome.org/ and the *Darwin Tree of Life Project* at https://www.sanger.ac.uk/news/view/genetic-code-66000-uk-species-be-sequenced.

many missing fungi may be discovered in specialised habitats, which have not yet been explored at all or have been only poorly investigated. The rumen and hindguts of herbivorous animals and the inner surfaces of Antarctic rocks do not sound like very promising habitats, but they are examples of habitats that have, unexpectedly, already yielded novel fungi.

Wherever they occur, fungal communities are very diverse metabolically, physiologically and taxonomically (Bass & Richards, 2011; Money, 2014). Given the benefits that man has derived from the fungi we know about, it is surprising, and disappointing, that more efforts have not been made to seek out these still unknown fungi (see section 3.9 for discussion of the species concept in fungi).

1.8 GEOMYCOLOGY

The fungal contribution to the soil community is usually seen as some aspect of their involvement in biomass recycling (releasing nutrients for plants), or direct involvement as components of food webs (as part of the nutrition of some animal, large or small). These aspects of fungal biology are undeniably extremely important and will be discussed in some detail in later chapters in this book. Here, we will only mention these points because we want to emphasise something that usually gets much less attention, which is fungal involvement in the geological transformations that produce and modify soils (Gadd, 2016, 2017; Robson, 2017).

Fungi are intimately involved in biogeochemical transformations on large and small scales, and although such transformations occur in both aquatic and terrestrial habitats, the terrestrial environment is where fungi have the greatest influence. The areas in which fungi have fundamental importance include:

- organic and inorganic transformations and element cycling (e.g. Lepp *et al.*, 1987);
- rock and mineral transformations;
- bioweathering (Kirtzel *et al.*, 2017, 2018);
- mineral formation (Robson, 2017);
- fungal–clay interactions; and
- metal–fungal interactions (Robson, 2017; and see Figure 1.2).

Many of these processes are relevant to the potential use of fungi in environmental biotechnology such as bioremediation (Burford *et al.*, 2003; Conceição *et al.*, 2019; Gadd, 2004, 2007, 2016, 2017; Robson, 2017) and recovery of high-value rare earth elements or other precious metals from wastes (Boczonádi *et al.*, 2019).

Fungi affect the physical structure of soils at a variety of spatial scales via electrostatic charge, and adhesive and enmeshment mechanisms. They also produce large quantities of extracellular polysaccharides and hydrophobic compounds that affect water infiltration properties of soils. Fungal decomposition of

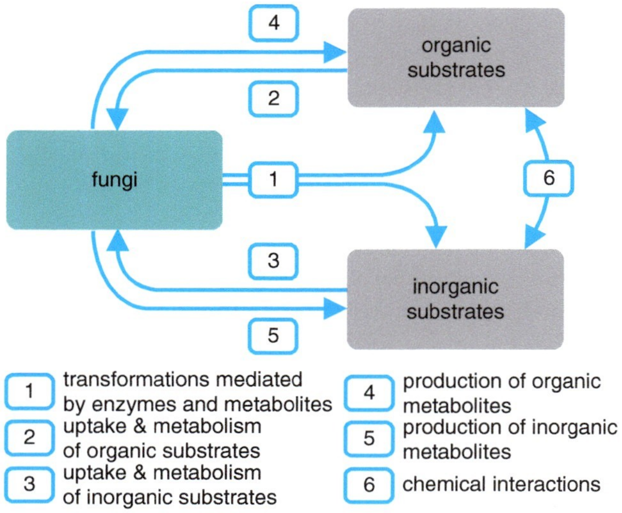

Figure 1.2. Diagrammatic representation of fungal action on organic and inorganic substrates which may be naturally occurring and/or synthetic. Full key: (1) organic and inorganic transformations mediated by enzymes and metabolites, e.g. H-ions, carbon dioxide, and organic acids, and physicochemical changes occurring as a result of metabolism; (2) uptake, metabolism or degradation of organic substrates; (3) uptake, accumulation, sorption, metabolism of inorganic substrates; (4) production of organic metabolites, exopolymers, and biomass; (5) production of inorganic metabolites, secondary minerals and transformed metal(loid)s; (6) chemical interactions between organic and inorganic substances, e.g. complexation and chelation, which can modify bioavailability, toxicity and mobility. Organisms in this model may also translocate nutrients. Modified from Gadd (2004; and see Gadd, 2016, 2017).

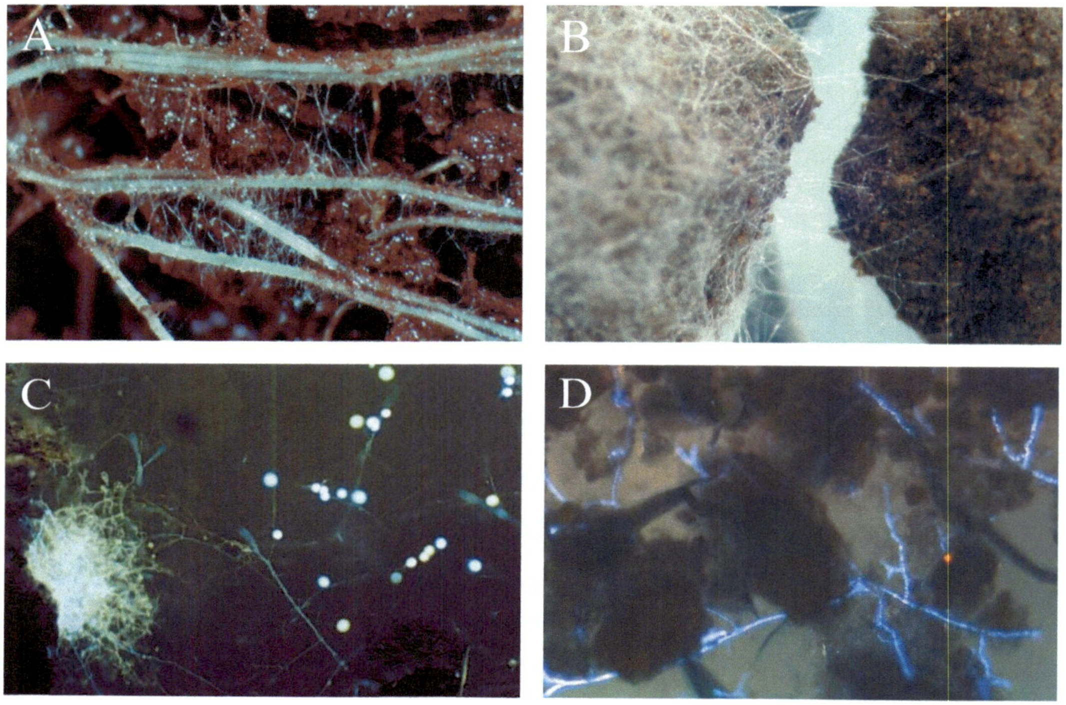

Figure 1.3. Fungal mycelia visualised in the soil environment. (A) Unidentified hyphae bridging roots of *Plantago lanceolata* growing in non-sterile field soil. Note the shiny films of mucilage; image width, 2 cm. (B) Hyphae of *Fusarium oxysporum* f. sp. *raphani* colonising a pair of adjacent soil aggregates. Aggregate on left is sterile, hence extensive mycelial development. Aggregate on right is non-sterile; reduced mycelial growth is due to competition from other microorganisms and reduced nutrients; image width, 1 cm. (C) Unidentified mycelium growing in soil pore, visualised in a thin-section of undisturbed pasture soil with a fluorescent stain. Note proliferation of hyphae on pore wall in left of image. Bright spherical objects are sporangia; image width, 150 μm. (D) Mycelium of *Rhizoctonia solani* growing in sterilised arable soil, visualised in thin-section with a fluorescent stain; image width, 150 μm. Modified from Ritz and Young (2004) using images kindly supplied by Professor Karl Ritz, Cranfield University, UK. Reproduced with permission from Elsevier.

organic matter can also destroy soil structure through effects on soil aggregation. In turn, soil structure affects fungi. The filamentous growth form of fungi is an efficient adaptation for life in a heterogeneous environment like soil, but the labyrinthine pore network will itself determine how fungal mycelia can grow through and function within the soil (Figure 1.3).

The distribution of water within soils plays a crucial role in governing fungal development and activity, as does the spatial distribution of nutrient resources (Ritz & Young, 2004). In aerobic environments fungi are of great importance on rock surfaces, in soil and at the plant root–soil interface (Table 1.2).

Table 1.2. Roles and activities of fungi in biogeochemical processes

Fungal role and/or activity	Biogeochemical consequences
Growth and mycelium development	Stabilisation of soil structure; soil particulate aggregation; penetration of pores, fissures and grain boundaries in rocks and minerals; mineral tunnelling; biomechanical disruption of solid substrates; plant colonisation and/or infection (mycorrhizas, pathogens, parasites); animal colonisation and/or infection (symbiotic, pathogens, parasites); translocation of inorganic and organic nutrients; assisted redistribution of bacteria; production of exopolymeric substances (serve as nutrient resource for other organisms); water retention and translocation; surfaces for bacterial growth, transport and migration; cord formation (enhanced nutrient translocation); mycelium acting as a reservoir of nitrogen and/or other elements (e.g. wood-decay fungi).
Metabolism: carbon and energy metabolism	Organic matter decomposition; cycling and/or transformations of component elements of organic compounds and biomass: carbon, hydrogen, oxygen, nitrogen, phosphorus, sulfur, metals, metalloids, radionuclides (natural and accumulated from anthropogenic sources); breakdown of polymers; altered geochemistry of local environment, e.g. changes in redox, oxygen, pH; production of inorganic and organic metabolites, e.g. protons, carbon dioxide, organic acids, with resultant effects on the substrate; extracellular enzyme production; fossil fuel degradation; oxalate formation; metalloid methylation (e.g. arsenic, selenium); xenobiotic degradation (e.g. polynuclear aromatic hydrocarbons); organometal formation and/or degradation (note: lack of fungal decomposition in anaerobic conditions caused by waterlogging can lead to organic soil formation, e.g. peat).
Inorganic nutrition	Altered distribution and cycling of inorganic nutrient species, e.g. nitrogen, sulfur, phosphorus, essential and inessential metals, by transport and accumulation; transformation and incorporation of inorganic elements into macromolecules; alterations in oxidation state; metal(loid) oxido-reductions; heterotrophic nitrification; siderophore production for iron(III) capture; translocation of nitrogen, phosphorus, calcium, magnesium, sodium, potassium through mycelium and/or to plant hosts; water transport to and from plant hosts; metalloid oxyanion transport and accumulation; degradation of organic and inorganic sulfur compounds.
Mineral dissolution	Rock and mineral deterioration and bioweathering including carbonates, silicates, phosphates and sulfides; bioleaching of metals and other components; manganese dioxide reduction; element redistributions including transfer from terrestrial to aquatic systems; altered bioavailability of e.g. metals, phosphorus, sulfur, silicon, aluminium; altered plant and microbial nutrition or toxicity; early stages of mineral soil formation; deterioration of building stone, cement, plaster, concrete, etc.
Mineral formation	Element immobilisation including metals, radionuclides, carbon, phosphorus, and sulfur; mycogenic carbonate formation; limestone calcrete cementation; mycogenic metal oxalate formation; metal detoxification; contribution to patinas on rocks (e.g. 'desert varnish'); soil storage of carbon and other elements.
Physicochemical properties Sorption of soluble and particulate metal species Exopolysaccharide production	Altered metal distribution and bioavailability; metal detoxification; metal-loaded food source for invertebrates; prelude to secondary mineral formation. Complexation of cations; provision of hydrated matrix for mineral formation; enhanced adherence to substrate; clay mineral binding; stabilisation of soil aggregates; matrix for bacterial growth; chemical interactions of exopolysaccharide with mineral substrates.
Mutualistic symbiotic associations: mycorrhizas, lichens, insects and other invertebrates	Altered mobility and bioavailability of nutrient and inessential metals, nitrogen, phosphorus, sulfur, etc.; altered carbon flow and transfer between plant, fungus and rhizosphere organisms; altered plant productivity; mineral dissolution and metal and nutrient release from bound and mineral sources; altered biogeochemistry in soil–plant root region; altered microbial activity in plant root region; altered metal distributions between plant and fungus; water transport to and from the plant.

table continues

Fungal role and/or activity	Biogeochemical consequences
	Pioneer colonisation of rocks and minerals; bioweathering; mineral dissolution and/or formation; metal accumulation and redistribution; metal accumulation by dry or wet deposition, particulate entrapment; metal sorption; enrichment of carbon, nitrogen, etc.; early stages of mineral soil formation; development of geochemically-active microbial populations; mineral dissolution by metabolites including 'lichen acids'; biophysical disruption of substrate. Fungal populations in gut aid degradation of plant material; invertebrates mechanically render plant residues more amenable for decomposition; cultivation of fungal gardens by certain insects (organic matter decomposition and recycling); transfer of fungi between plant hosts by insects (aiding infection and disease).
Pathogenic effects: plant and animal pathogenicity	Plant infection and colonisation; animal predation (e.g. nematodes) and infection (e.g. insects, etc.); redistribution of elements and nutrients; increased supply of organic material for decomposition; stimulation of other geochemically-active microbial populations.

Such activities take place in aquatic and terrestrial ecosystems, as well as in artificial and anthropogenic systems, their relative importance depending on the species present and physicochemical factors that affect activity. The terrestrial environment is the main locale of fungal-mediated biogeochemical change, especially in mineral soils and the plant root zone, and on exposed rocks and mineral surfaces. There is rather a limited amount of knowledge on fungal biogeochemistry in freshwater and marine systems, sediments, and the deep subsurface. Fungal roles have been arbitrarily split into categories based on growth, organic and inorganic metabolism, physicochemical attributes and symbiotic relationships. However, it should be noted that many, if not all, of these are interlinked, and almost all directly or indirectly depend on the mode of fungal growth (including symbiotic relationships) and accompanying heterotrophic metabolism. This, in turn, is dependent on a utilisable carbon source for biosynthesis and energy, and other essential elements, such as nitrogen, oxygen, phosphorus, sulfur and many metals, for structural and cellular components. Mineral dissolution and formation are outlined separately although these processes clearly depend on metabolic activity and growth form (modified from Table 1 in Gadd, 2007; and see Gadd, 2016, 2017).

Many fungi can grow oligotrophically, which means they can thrive in environments that are low in food sources. They do this by scavenging nutrients from the air and rainwater. This ability enables them to survive on stone and rock surfaces. Fungi can cause weathering of a wide range of rocks. In Iceland and other subpolar regions, bioweathering of basalt outcrops by fungal communities is believed to be the first weathering process. Lichens are important at early stages of rock colonisation and mineral soil formation, while free-living fungi are also major biodeterioration agents of stone, wood, plaster, cement and other building materials. There is increasing evidence that fungi are important components of rock-inhabiting microbial communities, with significant roles in mineral dissolution and secondary mineral formation.

Several fungi can dissolve minerals and mobilise metals more efficiently than bacteria. Mycorrhizal fungi are involved in mineral transformations and redistributions of inorganic nutrients (e.g. essential metal ions and phosphate; Figures 1.4–1.6). Proteomics analysis has demonstrated that bioweathering of black slate, which can contain up to 20% organic carbon, is caused by the laccase enzymes of the white-rot basidiomycete fungus, *Schizophyllum commune* (Kirtzel *et al.*, 2017, 2018).

These roles of fungi in soil geochemistry, especially metal cycling, have been included under the term 'geomycology', defined as 'the study of the role fungi have played and are

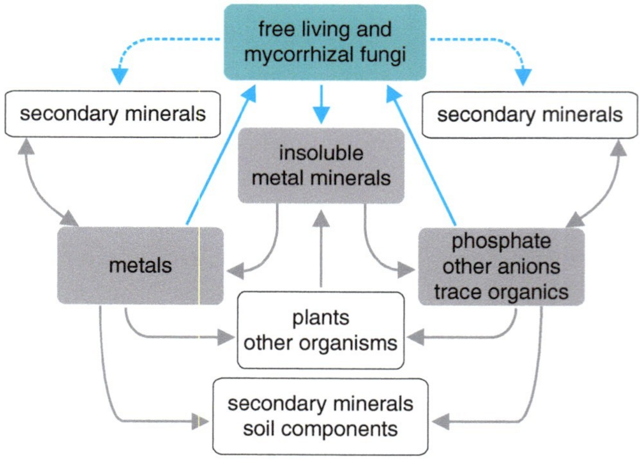

Figure 1.4. Action of free-living and mycorrhizal fungi on insoluble metal minerals in the terrestrial environment resulting in release of mineral components; metal(s), anionic substances, trace organics and other impurities. These can be taken up by living organisms (biota) as well as forming secondary minerals with soil components or fungal metabolites and/or biomass. Released minerals can also be absorbed or adsorbed or otherwise removed by organic and inorganic soil components. The dashed arrows imply secondary mineral formation because of secreted metabolites as well as fungal action on non-biogenic minerals. Possible losses to groundwater are not shown. Blue arrows indicate processes driven by the fungi. Modified from Gadd (2004; and see Gadd, 2016, 2017).

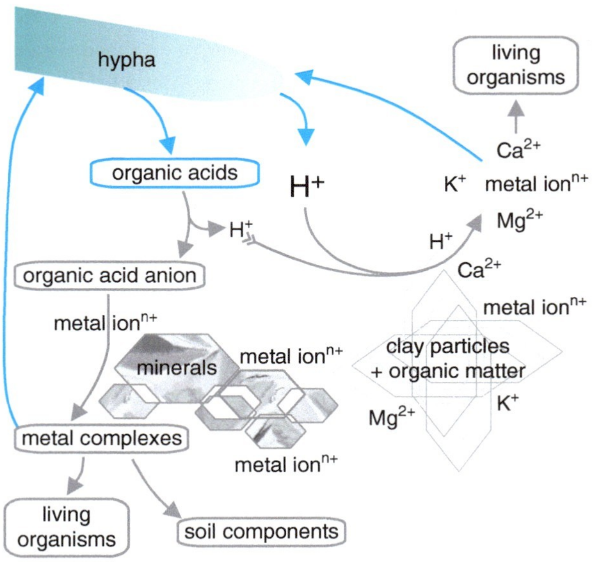

Figure 1.5. Proton- and organic acid-mediated dissolution of metals from soil components and minerals. Proton release from the hypha results in cation exchange with metal ions on clay particles, colloids etc. and metal displacement from mineral surfaces. Released metals can interact with biomass, can be taken up by other organisms, and can react with other environmental components. Organic acid anions, e.g. citrate, may cause mineral dissolution or removal by complex formation. Metal complexes can interact with live organisms as well as environmental constituents: in some circumstances complex formation may be followed by crystallisation, e.g. metal oxalate formation. Blue arrows indicate processes driven by the fungi. Modified from Gadd (2004; and see Gadd, 2016, 2017).

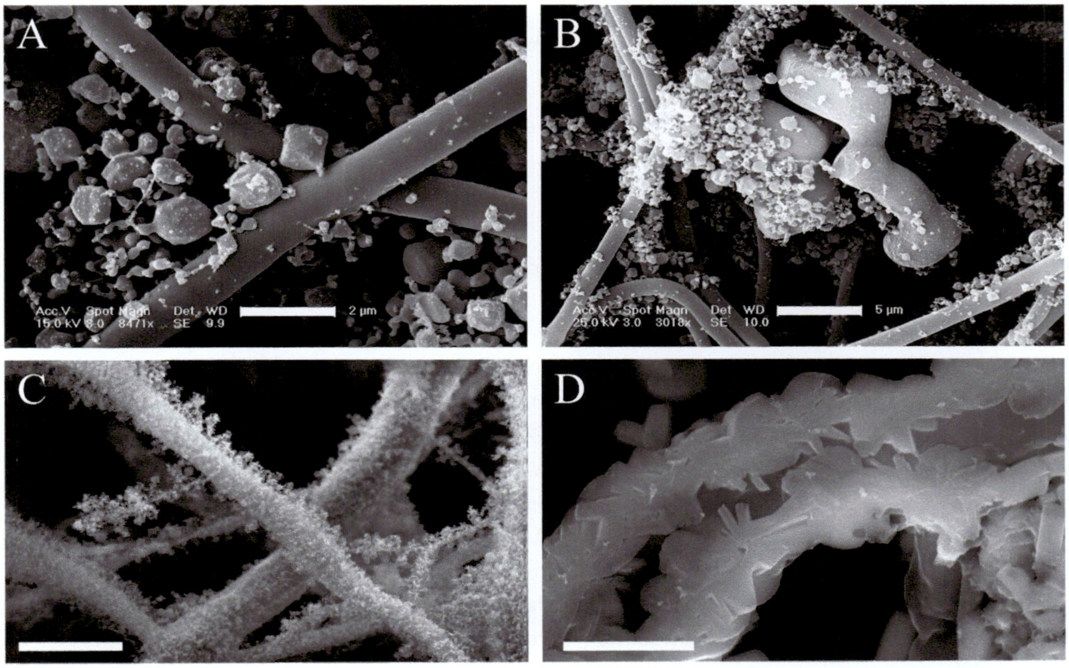

Figure 1.6. Photomicrographs of fungal hyphae showing examples of minerals formed by hyphal growth; in these cases, formation of uranium-containing biominerals following growth on medium containing uranium salts or uranium ore. Scale bars: (A) 2 µm; (B) 5 µm; (C) 20 µm; (D) 5 µm. Modified from Gadd (2007) using graphic files kindly supplied by Professor G. M. Gadd, University of Dundee, UK. Reproduced with permission from Elsevier. See Gadd (2016, 2017).

playing in fundamental geological processes' (Burford *et al.*, 2003; Gadd, 2004, 2007, 2016, 2017).

1.9 THE ORIGINS OF AGRICULTURE AND OUR DEPENDENCE ON FUNGI

As the last ice age came to an end, the consequential climatic and environmental changes forced humans to utilise an ever-wider variety of food resources. Although hunting and gathering persisted (and still exists in certain regions of the world), new food production techniques gained importance. The controlled cultivation of plants, what we might now call agriculture, began to be practised in different parts of the world between 11,000 and 14,000 years ago. This was soon followed by the close management and eventual domestication of the animals that are common on farms around the world today. The four major centres from which agriculture evolved were the Middle East and Europe, Africa, the Americas, and China and South-East Asia.

European agriculture originated in the 'Fertile Crescent', centred on the Tigris and Euphrates rivers. The region is also known as Mesopotamia, which refers to an area now occupied by modern Iraq, eastern Syria, south-east Turkey and south-west Iran (Riehl *et al*., 2013). Farmers in Mesopotamia were using irrigation to improve crop yields 8,000 years ago. This region saw the domestication of wild cereals, such as wheat (*Triticum*) and barley (*Hordeum*), as well as several legumes and fruit, including grapes, melons, almonds and dates. The region also saw the first domestication of many of the animals with which we are familiar today: dogs, goats, sheep, pigs, cattle, horses and camels were domesticated in succession from wild relatives indigenous to the region.

Plants and animals were domesticated through human-controlled selection (an unconscious use of applied genetics). Animals provided food resources, in the form of meat, and a variety of secondary products, including milk, dairy products, hides and wool. Animals also provided traction and power, more extensive travel and new forms of energy. Improved crops gave greater surpluses and provided a source of wealth for economic exchange and trade, providing some release from the daily search for food and the opportunity to develop a civilised way of life. Within just a few thousand years, farming lifestyles became a global phenomenon. The shift from hunting and gathering to agriculture spread quite rapidly from the various originating centres by both migration with colonisation and adoption of new technologies.

Agriculture reached the central Mediterranean region about 8,000 years ago, most of Western Europe about 7,500 years ago and the Iberian Peninsula and British Isles about 7,000 years ago (Whittle, 2001; Renfrew & Bahn, 2004).

With the spread of agricultural civilisation went the spread of agricultural fungi, good and bad. Fungi have always accompanied the steady march of civilisation across human settlements; the animals and plants were accompanied by their fungal parasites and commensalisms, and fungi accompanied technologies such as baking, brewing and cheese making. We have been dependent on fungi since we became human. That observation raises the questions of how long the fungi have been on Earth and where they came from. Those topics are dealt with in Chapter 2.

1.10 REFERENCES

Anderson, I.C. & Parkin, P.I. (2007). Detection of active soil fungi by RT-PCR amplification of precursor rRNA molecules. *Journal of Microbiological Methods*, **68**: 248–253. DOI: https://doi.org/10.1016/j.mimet.2006.08.005.

Aptroot, A., Cáceres, M.E.S., Johnston, M.K. & Lücking, R. (2016). How diverse is the lichenized fungal Family Trypetheliaceae (Ascomycota: Dothideomycetes): a quantitative prediction of global species richness. *Lichenologist* **48**:983–1011. DOI: https://doi.org/10.1017/S0024282916000463.

Bass, D. & Richards, T.A. (2011). Three reasons to re-evaluate fungal diversity 'on Earth and in the ocean'. *Fungal Biology Reviews*, **25**: 159–164. DOI: https://doi.org/10.1016/j.fbr.2011.10.003.

Boczonádi, I., Jakab, Á., Baranyai, E. *et al.* (2019). The potential application of *Aspergillus oryzae* in the biosorption of rare earth element ions present in seepage waters from a post-uranium-mining area. In *Abstracts Book, 30th Asilomar Fungal Genetics Conference.* Pacific Grove, CA: FGC, abstract 317T, p. 79. http://conferences.genetics-gsa.org/fungal/2019/program-book.

Burford, E.P., Kierans, M. & Gadd, G.M. (2003). Geomycology: fungi in mineral substrata. *Mycologist*, **17**: 98–107. DOI: https://doi.org/10.1017/S0269915X03003112.

Carlile, M. J., Watkinson, S. C. & Gooday, G. W. (2001). *The Fungi*, 2nd edition. London: Academic Press, Chapters 2, 4, 6 & 7. ISBN: 0127384464.

Carey, N. (2015). *Junk DNA: A Journey Through the Dark Matter of the Genome.* New York, USA: Columbia University Press. Pp. 360. ISBN-10: 0231170858; ISBN-13: 978-0231170857.

Conceição, A.A., Barbosa Cunha, J.R., Oliveira Vieira, V. *et al.* (2019). Bioconversion and biotransformation efficiencies of wild macrofungi. In *Biology of Macrofungi. Fungal Biology*, ed. B.P. Singh Lallawmsanga & A.K. Passari. Springer Nature: Cham, Switzerland, pp. 361–377. ISBN: 9783030026219. DOI: https://doi.org/10.1007/978-3-030-02622-6_18.

Darwin, C. (2017[1881]). *The Formation of Vegetable Mould Through the Action of Worms with Observations of Their Habits.* Scotts Valley, CA: CreateSpace Independent Publishing Platform. ISBN-10: 1978460171, ISBN-13: 978-1978460171. http://www.gutenberg.org/ebooks/2355

Frey, S.D. (2015). The spatial distribution of soil biota. In *Soil Microbiology, Ecology and Biochemistry*, 4th edition, ed. E.A. Paul. Boston, MA: Academic Press, pp. 223–244. DOI: https://doi.org/10.1016/B978-0-12-415955-6.00008-6.

Gadd, G.M. (2004). Mycotransformation of organic and inorganic substrates. *Mycologist*, **18**: 60–70. DOI: https://doi.org/10.1017/S0269915XO4002022.

Gadd, G.M. (2007). Geomycology: biogeochemical transformations of rocks, minerals, metals and radionuclides by fungi, bioweathering and bioremediation. *Mycological Research*, **111**: 3–49. DOI: https://doi.org/10.1016/j.mycres.2006.12.001.

Gadd, G.M. (2016). Geomycology. In *Fungal Applications in Sustainable Environmental Biotechnology*, ed. D. Purchase. *Fungal Biology* book series. Cham, Switzerland: Springer International, pp. 371–401. DOI: https://doi.org/10.1007/978-3-319-42852-9_15.

Gadd, G.M. (2017). The geomycology of elemental cycling and transformations in the environment. In *The Fungal Kingdom*, ed. J. Heitman, B. Howlett, P. Crous *et al.* Washington, DC: ASM Press, pp. 371–386. DOI: https://doi.org/10.1128/microbiolspec.FUNK-0010-2016.

Grinhut, T., Hadar, Y. & Chen, Y. (2007). Degradation and transformation of humic substances by saprotrophic fungi: processes and mechanisms. *Fungal Biology Reviews* **21**: 179–189. DOI: https://doi.org/10.1016/j.fbr.2007.09.003

Grossart, H.-P., Wurzbacher, C., James, T.Y. & Kagami, M. (2016). Discovery of dark matter fungi in aquatic ecosystems demands a reappraisal of the phylogeny and ecology of zoosporic fungi. *Fungal Ecology*, **19**: 28–38. DOI: https://doi.org/10.1016/j.funeco.2015.06.004.

Hawksworth, D.L. (1997). The fascination of fungi: exploring fungal diversity. *Mycologist* **11**: 18–22. DOI: https://doi.org/10.1016/S0269-915X(97)80062-6.

Hawksworth, D.L. (2001). The magnitude of fungal diversity: the 1.5 million species estimate revisited. *Mycological Research*, **105**: 1422–1432. DOI: https://doi.org/10.1017/S0953756201004725.

Hawksworth, D.L., Hibbett, D.S., Kirk, P.M. & Lücking, R. (2016). (308–310) Proposals to permit DNA sequence data to serve as types of names of fungi. *Taxon*, **65**: 899–900. DOI: https://doi.org/10.12705/654.31.

Hawksworth, D. L. & Lücking, R. (2017). Fungal diversity revisited: 2.2 to 3.8 million species. In *The Fungal Kingdom*, ed. J. Heitman, B. Howlett, P. Crous *et al.* Washington, DC: ASM Press, pp. 79–95. DOI: https://doi.org/10.1128/microbiolspec.FUNK-0052-2016.

Haynes, R.J. (2014). Nature of the belowground ecosystem and its development during pedogenesis. *Advances in Agronomy*, **127**: 43–109. DOI: https://doi.org/10.1016/B978-0-12-800131-8.00002-9.

Hibbett, D.S. (2016). The invisible dimension of fungal diversity. *Science*, **351**: 1150–1151. DOI: https://doi.org/10.1126/science.aae0380.

Hongsanan, S., Jeewon, R., Purahong, W. *et al.* (2018). Can we use environmental DNA as holotypes? *Fungal Diversity*, **92**: 1–30. DOI: https://doi.org/10.1007/s13225-018-0404-x.

Hunter-Cevera, J. C. (1998). The value of microbial diversity. *Current Opinion in Microbiology*, **1**: 278–285. DOI: https://doi.org/10.1016/S1369-5274(98)80030-1.

Hyde, K.D., Norphanphoun, C., Chen, J., Dissanayake, A.J., Doilom, M., Hongsanan, S., Jayawardena, R.S., Jeewon, R., Perera, R.H., Thongbai, B., Wanasinghe, D.N., Wisitrassameewong, K., Tibpromma, S. & Stadler, M. (2018). Thailand's amazing diversity – an estimated 55–96% of fungi in northern Thailand are novel. *Fungal Diversity*, **93**: 215–239. DOI: https://doi.org/10.1007/s13225-018-0415-7.

Kirtzel, J., Siegel, D., Krause, K. & Kothe, E. (2017). Stone-eating fungi: mechanisms in bioweathering and the potential role of laccases in black slate degradation with the basidiomycete Schizophyllum commune. *Advances in Applied Microbiology*, **99**: 83–101. DOI: https://doi.org/10.1016/bs.aambs.2017.01.002.

Kirtzel, J., Madhavan, S., Wielsch, N. *et al.* (2018). Enzymatic bioweathering and metal mobilization from black slate by the basidiomycete Schizophyllum commune. *Frontiers in Microbiology*, **9**: article 2545. DOI: https://doi.org/10.3389/fmicb.2018.02545.

Lepp, N.W., Harrison, S.C.S. & Morrell, B.G. (1987). A role for *Amanita muscaria* L. in the circulation of cadmium and vanadium in a non-polluted woodland. *Environmental Geochemistry and Health*, **9**: 61–64. DOI: https://doi.org/10.1007/BF02057276.

Lok, C. (2015). Mining the microbial dark matter. *Nature*, **522**: 270–273. DOI: https://doi.org/10.1038/522270a.

Lozano Garza, O.A. & Reynaga Peña, C.G. (2017). Los hongos desde la perspectiva educativa en México [Fungi from the educational perspective in Mexico]. In *Enfoques en Investigación e Innovación en Educación: Vol. 2. Prácticas educativas digitales, pedagogía y curriculum*, ed. A. Domínguez, J. Sánchez & T. Guerra. Monterrey, Mexico: REDIIEN, pp. 281–303. ISBN: 978-607-96725-2-2.

Luginbuehl, L.H., Menard, G.N., Kurup, S., *et al.* (2017). Fatty acids in arbuscular mycorrhizal fungi are synthesized by the host plant. *Science*, **2017**: article aan0081. DOI: https://doi.org/10.1126/science.aan0081.

Miller, R.W. & Gardiner, D.T. (2004). *Soils in our Environment*, 10th edition. Upper Saddle River, NJ: Pearson/Prentice Hall Publishers. ISBN: 0130481955.

Mitchell, J.I. & Zuccaro, A. (2006). Sequences, the environment and fungi. *Mycologist*, **20**: 62–74. DOI: https://doi.org/10.1016/j.mycol.2005.11.004.

Money, N.P. (2014). *Microbiology: A Very Short Introduction*. Oxford: Oxford University Press. ISBN-13: 9780199681686. DOI: https://doi.org/10.1093/actrade/9780199681686.003.0001.

Needelman, B.A. (2013). What are soils? *Nature Education Knowledge*, **4**(3): 2–10. URL: https://www.nature.com/scitable/knowledge/library/what-are-soils-67647639.

Prosser, J.I. (2002). Molecular and functional diversity in soil microorganisms. *Plant and Soil*, **244**: 9–17. DOI: https://doi.org/10.1023/A:1020208100281.

Renfrew, C. & Bahn, P. G. (2004). *Archaeology: Theories, Methods, and Practice*, 4th edition. London: Thames & Hudson Ltd. (See Chapter 7.) ISBN: 0500284415.

Riehl, S., Zeidi, M. & Conard, N.J. (2013). Emergence of agriculture in the foothills of the Zagros Mountains of Iran. *Science*, **341**: 65–67. DOI: https://doi.org/10.1126/science.1236743.

Ritz, K. & Young, I.M. (2004). Interactions between soil structure and fungi. *Mycologist*, **18**: 52–59. DOI: https://doi.org/10.1017/S0269915XO4002010.

Robson, G.D. (2017). Fungi: geoactive agents of metal and mineral transformations. *Environmental Microbiology*, **19**: 2533–2536. DOI: https://doi.org/10.1111/1462-2920.13807.

Staley, J.T., Palmer, F. & Adams, J.B. (1982). Microcolonial fungi: common inhabitants on desert rocks? *Science*, **215**: 1093–1095. DOI: https://doi.org/10.1126/science.215.4536.1093.

Taberlet, P., Bonin, A., Zinger, L. & Coissac, E. (2018). *Environmental DNA For Biodiversity Research and Monitoring*. Oxford, UK: Oxford University Press. ISBN-10: 0198767285, ISBN-13: 978-0198767282.

Tang, C., Xu, Q., Zhao, M., Wang, X. & Kang, Z. (2018). Understanding the lifestyles and pathogenicity mechanisms of obligate biotrophic fungi in wheat: the emerging genomics era. *The Crop Journal*, **6**: 60–67. DOI: https://doi.org/10.1016/j.cj.2017.11.003.

Torsvik, V. & Øvreås, L. (2002). Microbial diversity and function in soil: from genes to ecosystems. *Current Opinion in Microbiology*, **5**: 240–245. DOI: https://doi.org/10.1016/S1369-5274(02)00324-7.

Voroney, R.P. & Heck, R.J. (2015). The soil habitat. In *Soil Microbiology, Ecology and Biochemistry* (4th edition), ed. E.A. Paul. Boston, USA: Academic Press, pp. 15–39. DOI: https://doi.org/10.1016/B978-0-12-415955-6.00002-5.

Wellington, E.M., Berry, A. & Krsek, M. (2003). Resolving functional diversity in relation to microbial community structure in soil: exploiting genomics and stable isotope probing. *Current Opinion in Microbiology*, **6**: 295–301. DOI: https://doi.org/10.1016/S1369-5274(03)00066-3.

Whittle, A. (2001). The first farmers. In *The Oxford Illustrated History of Prehistoric Europe*, ed. B.W. Cunliffe. Oxford, UK: Oxford University Press, pp. 136–166. ISBN: 0192854410.

Willis, K.J. (ed.) (2018). *State of the World's Fungi 2018.* Richmond, UK: Royal Botanic Gardens, Kew. ISBN: 978-1-84246-678-0. https://stateoftheworldsfungi.org/2018/

Wu, B., Hussain, M., Zhang, W., Stadler, M., Liu, X. & Xiang, M. (2019). Current insights into fungal species diversity and perspective on naming the environmental DNA sequences of fungi. *Mycology*, **10**: 127–140. DOI: https://doi.org/10.1080/21501203.2019.1614106.

Wu, D., Wu, M., Halpern, A., Rusch, D.B., Yooseph, S., Frazier, M., Venter, J.C. & Eisen, J.A. (2011). Stalking the fourth domain in metagenomic data: searching for, discovering, and interpreting novel, deep branches in marker gene phylogenetic trees. *PLoS ONE*, **6**: article number e18011. DOI: https://doi.org/10.1371/journal.pone.0018011.

CHAPTER 2

Evolutionary Origins

This chapter considers the evolutionary origins and phylogenetics of fungi. We present these against the background of **global evolution** in the hope of improving appreciation of the time scales involved.

Reading this chapter could be the longest task you'll have in your life as a student, because we plan to deal with all the time that has ever existed. We are doing this to provide some context to the enormous lengths of time that we must think about when discussing the origins of one of the major eukaryotic kingdoms. Using some of the most recent molecular phylogenetic analyses, we describe fungi as an ancient and successful lineage, arguably the first terrestrial eukaryotes. To do this we have to talk in billions (10^9) of years, and we find it difficult to envisage a billion years. It's even more difficult, given the amazing changes witnessed during a pitifully short human lifetime, to imagine the sorts of changes that can occur in a billion years.

It's slightly easier to think in terms of fractions and percentages and to take in the lifetime of the universe by equating a billion years to a little over 7% of the total age of the universe and extending that to realise that the lifetime of our Sun is about one-third of the age of the universe.

Well, we did say only ***slightly*** easier!

From there we go on to consider planet Earth as a habitat (our habitat, ***your*** habitat) and the unique series of events that make this planet (the 'Goldilocks planet') so suitable for the sustained evolution of life. Following discussion of the three domains that make up the tree of life, we outline the origin of Kingdom *Fungi* from the opisthokonts, discuss some fossil fungi and introduce the fungal phylogeny.

2.1 LIFE, THE UNIVERSE AND EVERYTHING

The ultimate product of the long pathway of cosmic evolution is ***your*** habitat. This little bit of rock with some unique characteristics is where you were born. It keeps you alive right now and it will deal with your remains after you die. It is important that you understand how this pretty blue planet got here and how the unique set of circumstances arose that make your existence possible.

Enough preamble, let's sweep back to the beginning of the story: 13.77 billion years ago (see NASA, 2012), the universe, and all time and space, began with the **Big Bang**.

- Three hundred thousand years after the Big Bang hydrogen nuclei captured electrons, forming the first atoms.
- Six hundred million years after the Big Bang the first galaxies were formed.

Most cosmologists believe that present-day galaxies resulted from the gravitational consequences of small variations in the density of matter in the universe. For example, when the universe was one-thousandth its present size (about 500,000 years after the Big Bang), the density of matter in the region of space that now contains, our home galaxy (the Milky Way) might have been 0.5% higher than in adjacent regions. The higher density caused this region of space to expand more slowly than surrounding regions, because of gravitational interactions; and that, in turn, increased the relative over-density.

Later, when the universe was one-hundredth its present size (around 15 million years after the Big Bang), our region of space was probably 5% denser than the surrounding regions. This evolution continued as the universe expanded and aged, and when the universe was one-fifth its present size (roughly 12.5 billion years ago), our locality was probably twice as dense as neighbouring regions.

The inner portions of our galaxy (and similar galaxies) were assembled about 12 billion years ago. Stars in the outer regions of our galaxy were probably assembled in the more recent past. Our Sun was formed about 5 billion years ago (that is, when the universe was 64% of its current age).

Of course, we can't see into the ultra-deep time at the birth of the universe, but the Hubble Space Telescope got pretty close. In March 2004, astronomers at the Space Telescope Science Institute released the deepest view of the visible universe achieved at that time. Called the Hubble Ultra-Deep Field (HUDF), the million-second-long exposure (that's about 16 days) revealed the first galaxies to emerge from the time shortly after the Big Bang when the first stars started to reheat the cold, dark universe (Figures 2.1 and 2.2).

The HUDF image was an astonishing portrait of the origin of our habitat. Importantly, it shows that galaxies evolved quickly.

The HUDF showed numerous galaxies. Galaxies mean stars. Stars mean nuclear reactions producing an ever-widening range of elements. When the universe was created it was composed of hydrogen and helium, with some lithium, boron and beryllium. Stars emit radiation, including visible light, as a by-product of the **nuclear fusion** reactions in their core, which create the more massive elements, such as carbon, oxygen, nitrogen, iron, calcium, etc., by fusing together atoms of the lighter ones. When the star's nuclear fuel is exhausted the star explodes, scattering its outer layers back into space, adding to the gas clouds within

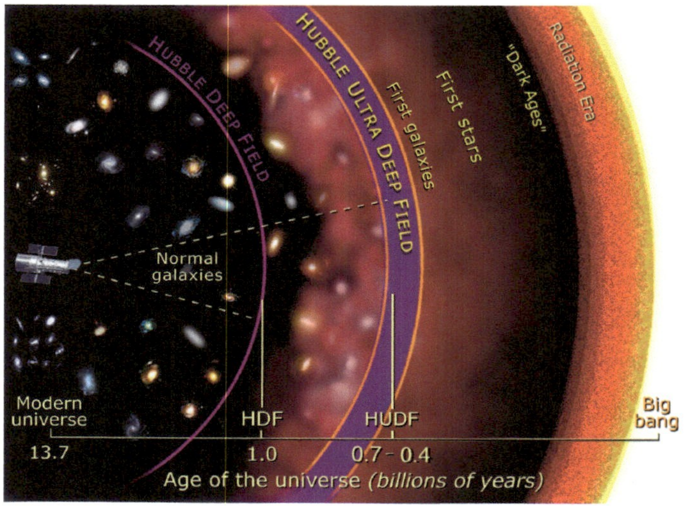

Figure 2.2. This image shows the place of the HUDF data within NASA's original interpretation of the history of the universe (note the time axis is logarithmic; image credit: NASA and A. Field, STScI (Space Telescope Science Institute, 3700 San Martin Drive, Baltimore, MD 21218, USA)). In 2013, measurements made by the Wilkinson Microwave Anisotropy Probe (WMAP satellite) enabled the precision of the estimated age of the universe to be improved to 13.77 ± 0.059 billion years. See https://map.gsfc.nasa.gov/universe/uni_age.html.

Figure 2.1. Ultra-Deep Field of the Hubble Space Telescope (credit: NASA, ESA, S. Beckwith (STScI) and the HUDF Team). The HUDF is actually two separate images (assembled by Anton Koekemoer) taken by Hubble's Advanced Camera for Surveys (ACS; an 11.3-day exposure) plus the near infrared camera and multi-object spectrometer (NICMOS; a further 4.5-day exposure). The HUDF field contains an estimated 10,000 galaxies that existed between 400 and 800 million years after the Big Bang. In ground-based images, the patch of sky in which the HUDF is located (in the constellation Fornax, just below the constellation Orion) is empty to the eye and is even empty to the best ground-based optical telescopes. Image produced by STScI (Space Telescope Science Institute, 3700 San Martin Drive, Baltimore, MD 21218, USA). For further information and images, visit http://hubblesite.org. For more information about cosmology, visit Cosmology: The Study of the Universe at https://map.gsfc.nasa.gov/universe/index.html and view the 2012 Hubble Extreme Deep Field at https://apod.nasa.gov/apod/ap121014.html.

which later stars form. Those later stars will contain more of the heavier elements and will be able to create even more massive elements in their nuclear fusion reactions. Repetitive cycling within stars produces all of the elements of the periodic table, but 'All of the atoms on the Earth except hydrogen and most of the helium are recycled material; they were not created on the Earth. They were created in the stars' (Strobel, 2019).

The HUDF shows that some of the most important developmental changes happened in the universe well within 1 billion years of the Big Bang. In the present universe, star lifetimes vary with the mass of the star from about 3×10^6 years for stars with masses around 60 times the mass of the Sun, to about 32×10^6 years for those with masses about 10 times that of the Sun, and on to 10^9 years for stars of about the same mass as the Sun. The first billion years (which is a little over 7% of the total age of the universe) would have provided sufficient time for several-to-many star lifetimes. These stellar life cycles would have created, even that distance of time ago, all the elements which became the building blocks of chemistry. And chemistry means evolution. Chemical evolution provides the potential for life.

2.2 PLANET EARTH: YOUR HABITAT

The Earth was formed 4.5×10^9 years ago by accretion within the cloud of dust and gases left over from the formation of the Sun. Its main ingredients were iron and silicates, with smaller quantities of many other elements, some of them radioactive. Energy released by radioactive decay (mostly of uranium, thorium and potassium) heated the planet to high enough temperatures to melt some of the constituents. The iron melted before the silicates and, being denser, sank towards the centre of the aggregate, where it accumulated to form our liquid **iron core**.

At the time the Earth formed, other planets were also being formed, of course. These eventually became our present-day neighbours in the solar system. But in addition, another planetary body formed in the same orbit as the young Earth. It was roughly the size of present-day Mars (a third or half the size of Earth), sometimes called **Theia**, and it hit the Earth late in Earth's growth process. The heat energy from the impact melted the two planets; the iron Theia contained slurped into the Earth's liquid iron core, and rocky debris from the impact exploded into orbit around the Earth and gradually aggregated into our Moon (Figure 2.3).

Around the molten iron core (now itself the size of present-day Mars), a thin but stable crust of solid rock formed as Earth cooled. Water, rising from the interior of the planet through volcanoes and fissures and arriving from space through comet impacts, collected in natural depressions in the crust to form the oceans. This highlights another important point: the Earth is comfortably located in what astronomers have called the **'Goldilocks orbit'**, which is at just the right distance from the Sun to be in a temperature range in which water can exist as a liquid ('Goldilocks' because, in the fairy tale, Goldilocks found a breakfast that was 'just right, not too hot and not too cold'). Astronomers are simple, gentle, people, but we would recommend that you read Andy Gardner's paper about the anthropic principle (essentially that life must be possible for there to be a living entity to observe it) and cosmological natural selection (Gardner, 2014).

Figure 2.3. The Earth and Moon as imaged by the Mariner 10 spacecraft in 1973 from 2.6 million km distance. These images have been combined to illustrate the relative sizes of the two bodies. Credits: NASA/JPL/Northwestern University.

2.3 THE GOLDILOCKS PLANET

So why are we describing all this cosmology in a chapter about the evolutionary origins of fungi? Our reason is that all this cosmology was ***essential*** to make the Earth–Moon binary system such a very special habitat for life. A habitat that is unique in its:

- orbital position: not too hot and not too cold;
- molten iron core;
- nature as, essentially, a binary planetary system.

As mentioned, our orbital position is crucial to maintaining liquid water on the surface of the planet. You only need to think of our seasonal changes to realise how exact that positioning must be. The Earth's axis is tipped over about 23.5° from the vertical and it is our annual orbital motion around that which causes our seasons. We authors are European northern hemisphere people, and we are used to wintery ice and snow from December to February and warm summers from June to September. But these seasons only result from the old planet being angled slightly towards the Sun in summer, or away from the Sun in winter. Just that small rotational shift towards or away from the Sun is sufficient to generate an enormous temperature differential. At the time of writing, the highest recorded temperature in Europe (in Athens, Greece) is 48.0°C; compared to the lowest recorded European temperature of minus 58.1°C (at Ust' Shchugor, Komi Republic, Russia) (source: https://en.wikipedia.org/wiki/List_of_weather_records). You will appreciate that if Earth's orbit was very slightly closer to the Sun, let's say by about one Earth-***radius***, surface temperatures would be intolerably high **all the time**. Alternatively, if the planet was a similar distance further from the Sun, it would be in **permanent deep freeze**.

That highly improbable impact between proto-Earth and Theia not only topped up the Earth's liquid iron core but increased the rotation of both planet and core. The molten core spins now, and the still-spinning iron generates our magnetic field, the **magnetosphere**, producing a magnetic field strong enough to protect Earth from the solar wind.

The **solar wind** is a stream of charged particles ejected from the upper atmosphere of the Sun and consists mostly of high-energy electrons and protons. Mars is unprotected, and the solar wind has stripped away up to a third of its original atmosphere. Even the dense atmosphere of Venus is being eroded by the solar wind, so much so that space probes have discovered a comet-like tail of drifting atmosphere that stretches back from Venus to the orbit of the Earth.

Earth is safe, protected from the solar wind by its magnetic field, which deflects charged particles but also serves as an electromagnetic energy transmission line to the Earth's upper atmosphere and ionosphere through the **aurora borealis** in the north and **aurora australis** in the south. By preventing the solar wind stripping away the ozone layer our magnetosphere also protects the surface of Earth from biologically harmful **ultraviolet** (UV) radiation emitted by the Sun, as well as the high-energy particles of the solar wind. This is crucially important to biology. UV wavelengths shorter than 280 nm (also known as **'germicidal UV'** or UV-C) are entirely screened out by ozone at around 35 km altitude. For radiation with a wavelength of 290 nm (UV-B, which is most damaging to DNA), the intensity at Earth's surface is only 10^{-8} that at the top of the atmosphere, due mainly to high altitude **ozone**.

And that leaves the contribution of our enormous **Moon**, by far the **largest satellite** in the solar system ***in proportion to its planet*** and constructed from the rocky remnants of the Theia–Earth impact (Figure 2.3). The presence of the Moon gradually reduced the rotation rate to a level that reduces temperature variations on the Earth's surface to life-supporting limits. It also stabilises our **axial tilt**, which ensures annual seasons and challenging environments to drive evolution. Similarly, the Moon generates **tidal effects** in both rocks and water, and the latter also produces variable shoreline environments that spur chemical and biological evolution (Zalasiewicz & Williams, 2013).

RESOURCES BOX 2.1 The Story of Planet Earth

For more information about the origins of the Earth we recommend the following collection of publications and websites.

Burger, W.C. (2002). *Perfect Planet, Clever Species: How Unique Are We?* Amherst, NY: Prometheus Books. ISBN-10: 1591020166, ISBN-13: 978-1591020165.

Davies, P. (2008). *The Goldilocks Enigma. Why Is the Universe Just Right For Life?* London: Mariner Books. ISBN-10: 0547053584, ISBN-13: 978-0547053585.

Lamb, S. & Sington, D. (1998). *Earth Story. The Forces that Have Shaped Our Planet.* Princeton, NJ: Princeton University Press. ISBN-10: 0691002290, ISBN-13: 978-0691002293.

Zalasiewicz, J. & Williams, M. (2013). *The Goldilocks Planet: The 4 Billion Year Story of Earth's Climate.* Oxford, UK: Oxford University Press. ISBN-10: 0199683506, ISBN-13: 978-0199683505.

Visit the following web pages

http://www.psi.edu/epo/moon/moon.html
https://en.wikipedia.org/wiki/Giant-impact_hypothesis
https://en.wikipedia.org/wiki/Impact_event

2.4 THE TREE OF LIFE HAS THREE DOMAINS

There is evidence for the activity of living organisms in terrestrial rocks that are 3.5×10^9 years old. These oldest known fossils are the remains of bacteria-like organisms. Over the last 3.5 billion years, which is about a quarter of the age of the universe, living organisms on Earth have diversified and adapted to almost every environment imaginable.

For a rather lesser period of time, something like a few hundred years, humanity has worked at systematically finding, naming and classifying all those organisms. In 1866, Ernst Haeckel made one of the first attempts at constructing a phylogenetic **classification** by dividing organisms into three categories: the Plantae, the Animalia and the Protista (microorganisms). For the best part of a century, biologists were content to classify living things as either a plant or animal. In this classification, fungi and lichens were classified with bacteria and algae in a group called the Thallophyta, which formed part of the Kingdom *Plantae*. Because of this, mycology developed as a branch of botany. This is somewhat ironic since we now know that the true fungi are more closely related to animals than they are to plants.

Towards the middle of the twentieth century the nature of bacteria became clearer, but fungi were still classified in the plant kingdom (strictly speaking into Subkingdom Cryptogamia, Division Fungi, Subdivision Eumycotina) and were separated into four classes: the Phycomycetes, Ascomycetes, Basidiomycetes and Deuteromycetes (the latter were also known as Fungi Imperfecti and were separated off because they lacked a sexual cycle). These traditional groups of 'fungi' were identified by the morphology of their sexual organs, whether their hyphae had septa (cross-walls), and the ploidy (degree of repetition of the basic number of chromosomes) of nuclei in their vegetative mycelium. The slime moulds, all grouped at that time in Subdivision Myxomycotina, were also included in Division Fungi.

By the 1950s and 1960s, it had become very clear that this system failed to deal properly with the fungi, protists and bacteria and in 1969 **Robert Harding Whittaker** published a classification scheme in which he divided organisms into five kingdoms: Animalia, Plantae, Fungi, Protista (eukaryotic microorganisms and a mixed grouping of protozoa and algae) and Monera (prokaryotic microorganisms, bacteria and archaea). By the 1970s, a system of five kingdoms had come to be accepted as the best way to classify living organisms. Fundamentally, a distinction is made between prokaryotes and the **four eukaryotic kingdoms** (plants, animals, fungi and protists). The prokaryote–eukaryote distinction recognises the 'higher organism' traits that eukaryotic organisms share, such as nuclei, cytoskeletons, internal membranes, and mitotic and meiotic division cycles.

The **endosymbiosis theory** of **Lynn Margulis** (Margulis, 2004) accounts for the origin of the set of features that characterise the modern eukaryote through a sequence of symbiotic relationships being established between 'prokaryotic' partners. The **mitochondria** of eukaryotes evolving from aerobic 'bacteria' living within a host cell; **chloroplasts** of eukaryotes evolving from endosymbiotic 'cyanobacteria'; eukaryotic **cilia and flagella** arising from endosymbiotic 'spirochetes', the basal bodies from which eukaryotic cilia and flagella develop creating the mitotic spindle and thus contributing to the **cytoskeleton**.

The early eukaryotes were **anaerobic**, sometimes aerotolerant, organisms that lacked mitochondria and **peroxisomes**. It is presumed that these groups diversified prior to the endosymbiotic events which gave rise to mitochondria. Today, most of these 'primitive' eukaryotes are parasites of other eukaryotes, from microorganisms to humans, but free-living relatives of these parasites branch deep in the evolutionary tree.

These kinds of classification schemes of organisms were historically based on taxonomy according to similarity, but this approach has been replaced by one firmly based on **phylogenetics**. The term comes from the Greek words *phyle*, meaning tribe or race, and *genetikos*, meaning from birth, so it is applied to the study of evolutionary relatedness within and between populations. Phylogenetic taxonomy is a 'natural' taxonomy because the classification is based on grouping by ancestral traits. However, before the development of molecular techniques, it really was not possible to determine evolutionary relationships of present-day organisms on a sufficiently comprehensive scale to construct a meaningful ***Tree of Life***.

The breakthrough was made by **Carl Woese** who concluded that, because all organisms possessed **small subunit rRNA** (SSU rRNA, so called because they form part of the small subunit of a ribosome), the SSU rRNA gene would be a perfect candidate for the universal chronometer of all life.

SSU rRNA genes (16S rRNA in prokaryotes and mitochondria, 18S rRNA in eukaryotes) display a mosaic of **conservation** patterns, with rapidly evolving regions interspersed among moderately or nearly invariant regions (it has been estimated that about 56% of the nucleotide positions in 18S rRNA data sets are not free to vary and have not undergone substitutions useful in phylogenetic reconstructions). This variation in conservation permits SSU rRNA gene sequences to be used as sophisticated chronometers of evolution with the slowly evolving regions recording events that occurred many millions of years ago, and the rapidly evolving regions chronicling more recent events.

In Woese's procedure, pairs of SSU rRNA gene sequences from different organisms were aligned, and the differences counted and considered to be some measure of 'evolutionary distance' between the organisms. Pairwise differences between many organisms were then used to infer phylogenetic trees, maps that represent the evolutionary paths leading to the SSU rRNA gene sequences of present-day organisms. Of course, such trees rely on many assumptions, among which are assumptions about the rate of mutational change (the '**evolutionary clock**') and that rRNA genes are free from artefacts generated by convergent evolution or lateral gene transfer (Woese, 1987; Woese *et al.*, 1990).

Woese's studies called into question many beliefs about evolutionary relationships between organisms and brought order to biological diversity. Most importantly, the tree of life constructed from SSU rRNA gene sequences led Woese to recognise three primary lines of evolutionary descent (first called kingdoms, but subsequently renamed domains; a new taxon ***above*** the level of kingdom): the Bacteria (now called ***Eubacteria***), the ***Archaea*** and the Eucarya (now called ***Eukaryota***). These three domains were thought to have diverged from some '**universal ancestor**' (which was the first chemical aggregate to be truly ***alive***). The first two domains contain prokaryotic microorganisms and the third domain contains all eukaryotic organisms.

There has been a good deal of speculation about the universal ancestor, which, rather than being a primitive 'prokaryote', might have had a complex cell like a eukaryote, with archaea and eubacteria evolving from it by reduction and simplification (Doolittle, 2000). Penny and Poole (1999) pointed out that modern eukaryotes use RNAs to catalyse intron splicing and stable RNA processing, and suggested that these processes could be 'molecular fossils' from the RNA world that may have been the first step in the origin of life, before the evolution of protein catalysis. The universal ancestor might have possessed some extremely primitive features that are now considered to be characteristic of present-day eukaryotes. If these relics of the RNA world were present in the universal ancestor, it does not mean that the ancestor was eukaryotic. Rather, the ancestor contained a mix of features that were selected and combined in different ways during evolution of the present-day archaea, eubacteria and eukaryotes.

Unfortunately, because the evolutionary clock is not constant in different lineages, the time of occurrence of evolutionary events in the tree of life cannot be extracted reliably from SSU rRNA gene sequences alone, so other sequences have to be brought into the analysis. All molecules chosen for phylogenetic studies must:

- be universally distributed across the group chosen for study;
- be functionally homologous;
- change in sequence at a rate proportionate with the evolutionary distance to be measured (the broader the phylogenetic distance being measured, the slower must be the rate at which the sequence changes).

Although SSU rRNA genes satisfy these criteria, it is important not to allow phylogenetic trees based on a single gene to dominate evolutionary and systematic conclusions. Instead, account needs to be taken of phylogenetic trees based on a range of conserved molecules with a judgement being made about the weight to be given to the different lines of evidence so obtained.

Among the sequences that have been used successfully in multigene phylogenies are ATPase genes. The ATPase enzymes are composed of several different kinds of subunits, each related among themselves. The F-, V- and A-enzymes have catalytic and non-catalytic subunits, which are thought to have arisen during early **gene duplications** prior to the separation of the three domains. Because of this, the phylogenetic tree inferred from catalytic subunits can be rooted with the non-catalytic subunits. Using ATPase subunits, the root of the tree of life was placed on the *Eubacteria* branch, making the *Eukaryota* and *Archaea* sister taxa. This conclusion was supported by studies using aminoacyl-tRNA synthetase gene sequences. These genes form a series of 20 enzyme families, with each family the result of gene duplication again argued to predate the origin of the three domains. Thus, aminoacyl-tRNA synthase phylogenetic trees can be rooted in a fashion like that for the ATPase trees.

Table 2.1. Ribosomal rRNA sequences used to identify and classify fungi

rRNA molecule	Structural rRNA?	Transcribed?	Level of conservation	Taxa which can be distinguished
18S rRNA (small subunit RNA)	Yes	Yes	Highly conserved domains interspersed with conserved domains	From domains to classes
5S rRNA	Yes	Yes	Conserved domains	Classes and orders
28S rRNA (large subunit RNA)	Yes	Yes	Conserved domains interspersed with variable domains	From phyla to species
Internally transcribed spacers (ITS) 1 and 2	No	Yes	Variable domains	Species and closely related genera
Intergenic spacer (IGS)	No	No	Highly variable domains	Strains and races

The **ribosomal gene cluster** is another regular contributor to construction of phylogenetic trees. Ribosomal genes are usually present in genomes in large numbers (100–200 tandem repeats) but they evolve as a single unit. The large quantities of rRNAs expressed by cells make their isolation and purification relatively easy despite the nearly ubiquitous occurrence of stable RNAses. Additionally, rDNA can be sequenced using specific oligonucleotide primers and the PCR. Because each genome contains many identical copies of the ribosomal genes, at least one copy for molecular analysis can usually be recovered, even from low-quality DNA preparations. Importantly, rDNA sequencing results in fewer artefacts than rRNA sequencing, and offers the opportunity of sequencing both strands of the rRNA sequences, providing a further check against sequencing errors.

Three regions of the ribosomal gene cluster in eukaryotes code for rRNA genes, which are transcribed into 5.8S, 18S and 28S RNA molecules that form part of the ribosome structure. Of the approximately 9,000 nucleotides in a ribosomal repeat, the 18S gene accounts for about 1,800 base pairs (bp), the 5.8S gene for about 120 bp and the 28S gene for about 3,200 bp. Interspersed between the rRNA genes are **spacer regions**. The areas that lie between the 18S and 5.8S and between the 5.8S and 28S genes are called internally transcribed spacers (ITS1 and ITS2).

The ITS1 and ITS2 regions are transcribed together as a single unit, then cleaved into separate RNA products, and finally eliminated in the process that yields rRNA. The region that separates one ribosomal gene cluster from the next is called the **intergenic spacer region** (IGS), which is made up of the non-transcribed spacer region (NTS) and the externally transcribed spacer region (ETS). The rRNA product from the ETS region suffers a similar fate to the internally transcribed spacers. The rRNA genes, the transcribed spacers (ITS and ETS) and the non-transcribed spacer (NTS) evolve at different rates and because of this their sequences have become widely used to discriminate between fungal taxa at levels from the kingdom to the intraspecific strains and races (Table 2.1).

Sequences other than the ribosomal gene cluster and ATPases that have been used in phylogenetic studies of eukaryotes include: ribosomal protein factors, α-tubulin, β-tubulin, actins and cytochromes. Protein sequences, which are made up of 20 amino acids, offer several advantages over DNA sequences (made up of four nucleotides) for some phylogenetic studies because **homology** (similarity due to shared ancestry) is more easily distinguished from **analogy** (similarity in structure or function that evolved through different pathways and different ancestries, a process known as convergent evolution). Further, length changes are infrequent in protein-coding genes because insertions and deletions often lead to such large shifts in the reading frame that they are fatal, and so fail to persist in the lineage. Mitochondria and their protein-coding genes are present in multiple copies in most cells and mitochondrial genes are easy to amplify, even from starting DNA of low quality. However, relatively few phylogenetic studies of fungi have used mitochondrial protein-coding genes.

It has been estimated that the domain *Eubacteria* and the common ancestor to the domains *Archaea* and *Eukaryota* diverged about 4 billion years ago, with the *Eukaryota* arising on the *Archaea* lineage about 2 billion years ago (Figure 2.4).

The root of the universal tree of life remains controversial, though. Evidence for ancient lateral transfers of genes and uncertainty over the ancestry of *Archaea* make the distant origins of the major lineages of Life uncertain. Inferring **ancient relationships** is full of difficulties. If the data contain too little information, random errors can swamp the truth, and random errors can be introduced by the mathematical model used to interpret the data. The method used to establish the phylogenetic trees can also introduce systematic errors if it is too simplistic (explanations and further references in Keeling *et al.*, 2005). Improvements in sensitivity of detection methods have

Figure 2.4. One view of the most ancient relationships of the major lineages of the Domains of Life, from the Tree of Life Web Project (http://tolweb.org/Life_on_Earth/1/1997.01.01). *Eocytes* are a group of sulfur-dependent bacteria with a unique pattern of organisation of ribosomal large and small subunits. They are closely related to eukaryotes. Most of the best-investigated species of *Archaea* are members of two main phyla: the *Euryarchaeota*, composed of methanogens and extreme halophiles, and *Crenarchaeota*, composed of the extreme thermophiles Other groups have been tentatively created, but the classification of archaea, and of prokaryotes in general, is a rapidly changing arena into which we do not wish to venture.

demonstrated that archaea, and other prokaryotes, are common and widely distributed in all habitats, and have allowed the discovery and identification of many organisms that have not been cultivated in the laboratory. Consequently, the classification of archaea, and of prokaryotes in general, is a rapidly changing field. As we do not intend to give any further description of these organisms, we recommend that you refer to the regularly updated Wikipedia pages that give reliable information:

- https://en.wikipedia.org/wiki/Archaea;
- https://en.wikipedia.org/wiki/Bacteria;
- https://en.wikipedia.org/wiki/Prokaryote;

and view the *Taxonomic Outline of Bacteria and Archaea* at:

- http://www.taxonomicoutline.org/.

Errors amplify as you attempt to reach back further in time, in what is called **deep time** (hundreds of millions to billions of years; Table 2.2) where you are looking for deep divergences between major groups of organisms. Fossils are necessary to calibrate phylogenetic trees to a real timeline, but there is a very patchy fossil record and the older the fossil the greater the debate about its nature. The result of such errors and uncertainties is that the timings inferred for major events (like the divergences of major eukaryote groups) in different studies can differ by several hundred million years, or more. For example, one study claims the common ancestor of living eukaryotes existed 2.3 billion years ago. Another puts the time of eukaryote divergence at 0.95 to 1.26 billion years ago. The two studies used different methods and different phylogenetic models; their different dates could mean that the common ancestor existed for a billion years before evolving into the plant, animal and fungal lines or it could mean that the best we can say is that the divergence event occurred sometime between 0.95 and 2.3 billion years ago.

Archaea and *Eukaryota* are sister groups with a common ancestor (Figure 2.4), so their modern representatives share many properties that differ from those found in the *Eubacteria*. For example, the RNA polymerase of *Archaea* and *Eukaryota* resemble each other in subunit composition and sequence far more closely than either resembles the type of polymerase found in *Eubacteria*. Also, the *Archaea* and *Eukaryota* use TATA-binding

Table 2.2. Some terminology explained (with particular reference to fungi)

Classification: the assignment of objects to defined categories. Biologists classify species of organisms. Classification places organisms within groups that show their relationships to other organisms. Modern classification was started by Carolus Linnaeus in the eighteenth century. He grouped species according to shared physical characteristics. Linnaean groupings were subsequently revised to make them consistent with the Darwinian principle of common descent. See https://en.wikipedia.org/wiki/Scientific_classification.
Systematics: the study of relationships and classification of organisms and the processes by which it has evolved and by which it is maintained. Systematics deals specifically with relationships through time. Used to understand the evolutionary history of life on Earth and uses taxonomy as a primary tool in understanding organisms and existing classification systems.
Nomenclature: the allocation of scientific names to the taxons a systematist considers as meriting formal recognition (nomenclature of fungi is governed by the International Code of Nomenclature for algae, fungi, and plants).
Taxonomy: the describing, identifying, classifying and naming of organisms. Organisms are grouped according to their morphological and/or molecular characteristics into taxa of particular ranks: species, which are grouped into genera (singular: genus), and then higher taxa, like families, orders, classes, phyla (singular: phylum) or divisions, kingdoms, domains. In phylogenetic taxonomy (or cladistic taxonomy), organisms are classified by clades, which are based on evolutionary grouping by ancestral traits. By using clades as the criteria for separation, cladistic taxonomy, using cladograms, can categorise taxa into unranked groups. Visit the following web pages: https://en.wikipedia.org/wiki/Taxonomy http://taxonomicon.taxonomy.nl/ http://sn2000.taxonomy.nl/

table continues

Taxonomy of fungi: at the time of writing, the taxonomy of fungi is governed by the International Code of Nomenclature for algae, fungi, and plants. Although the rank of species is basic, there is no universally applicable definition! Corrections are afoot, however. Key changes were made in the rules relating to the nomenclature of fungi at the *XIX International Botanical Congress* in Shenzhen, China, in July 2017. The most significant being the transfer of decision-making on matters related only to the naming of fungi from *International Botanical* to ***International Mycological Congresses*** (Hawksworth *et al.*, 2017). This was detailed during the Fungal Nomenclature Session on 18 July 2018 at ***IMC11*** in Puerto Rico. Also, a contribution to the definition of fungal species would be to allow DNA sequence data to serve as ***types*** for the names of fungi (Hawksworth *et al.*, 2016).

MycoBank [http://www.mycobank.org/] is an online database designed as a service that documents new fungal names and combinations, and associated data such as descriptions and illustrations, sequence alignments and polyphasic identifications of fungi and yeasts against curated reference databases. The polyphasic approach uses genomic, chemotaxonomic, cultural, pathogenic and phenotypic methods of characterising the fungus, in combination, to establish its taxonomic position (Robert *et al.*, 2013). Since 2013, all new fungal names have to be registered with MycoBank in order to be valid.

The **Index Fungorum** database and website has moved and is now based at the Mycology Section, Royal Botanic Gardens Kew in association with Landcare Research-NZ (the New Zealand Crown Research Institute for terrestrial biodiversity and land resources, managing the national fungal collection) and the Institute of Microbiology, Chinese Academy of Science. The Index can be accessed at http://www.indexfungorum.org/.

Taxon (plural = taxa): a taxonomic group of any rank.

Cladistics: a method of systematics which aims to reconstruct the genealogical descent of organisms through objective and repeatable analysis, leading to a natural classification or phylogeny.

Cladistics is based on three basic assumptions:

- that taxa are united into natural groups on the basis of shared derived characters;
- that all groups recognised must be descended from a single ancestor (i.e. be monophyletic);
- that the most parsimonious pattern (= the one requiring the fewest steps to account for relationships) is the one most likely to be correct.

The product of **cladistic analysis** is a tree-like branching diagram (called a **cladogram** or **phylogenetic tree**) which shows the pattern of relationships between the organisms based on the characters used.

In a cladogram, organisms form the leaves (extreme ends of branches), and each branching node (divergence) is ideally binary (two-way). The two taxa on either side of a divergence are called **sister taxa** or sister groups. Each subtree, whether it contains one item or thousands of items, is called a clade.

A **natural group** has all its organisms sharing a unique ancestor (one which they do not share with any other organisms on the diagram) for that clade. See http://en.wikipedia.org/wiki/Cladistics.

Monophyletic groups, also called **clades**, are composed of a single ancestor together with all of its descendants and are generally considered as the only 'natural' grouping. They are very important in phylogenetic classification. In contrast, a paraphyletic group contains some, but not all, of the descendants of a common ancestor. The members included are those that have changed little from the ancestral state; those that have changed more are excluded. **Polyphyletic groups** are formed when two lineages convergently evolve similar characters. Organisms classified into the same polyphyletic group share observable similarities rather than phylogenetic or evolutionary relationships.

Common ancestor: a group of organisms is said to have common descent if they have a common ancestor. In biology, the theory of universal common descent proposes that all organisms on Earth are descended from a common ancestor or ancestral gene pool. See https://en.wikipedia.org/wiki/Common_descent.

Deep time: the concept of **geological time**. See http://www.pbs.org/wgbh/evolution/change/deeptime/index.html.

Deep divergences: divergence between taxons inferred to have occurred in deep geological time, typically at least 1 billion years ago.

DNA barcoding: a taxonomic method that uses a short (about 700 nucleotides in length) genetic marker in the DNA of an organism to identify it as belonging to a particular species. The DNA marker chosen is amplified by PCR in DNA extracted from a tissue sample and the amplified sequence is submitted for sequencing. Capillary sequencers have now largely displaced slab gel instruments, and short DNA barcodes can be quickly processed from thousands of specimens and unambiguously analysed by computer programs. The sequencer generates a 'barcode' because it detects the four bases that make up the sequence using fluorescent dyes and the sequencer's colorimeter reports the sequence by printing appropriately coloured peaks (or bars) that designate the

table continues

sequence. If the sequence contains a run of the same base, then the peak (or bar) is broad; where the sequence has a single base, the peak (or bar) is narrow. So, just as the unique pattern of bars in a universal product code (UPC) in commerce identifies each consumer product, a 'DNA barcode' is a unique pattern of DNA sequence that identifies each living thing (providing the right DNA marker is chosen!). Botanists tend to use a portion of the chloroplast DNA of plants, those interested in animals can use mitochondrial cytochrome *c* oxidase subunit I or 16S rRNA. Recent research is demonstrating that the nuclear ribosomal internal transcribed spacer (ITS) region could serve as a universal DNA barcode marker for fungi (Schoch *et al.*, 2012; Truong *et al.*, 2017).

A little fungal taxonomy:
Principal taxonomic ranks (and the characteristic word-endings) in use for fungi today are as follows (the current recommendation by the International Code is to print all of these taxon names in italics, or underline in typescript):

Domain *Eukarya*
Kingdom *Fungi*
Phylum ... mycota
Subphylum ... mycotina
Class ... mycetes
Order ... ales
Family ... aceae
Genus
Species

Refer to Hyde *et al.* (2017) and He *et al.* (2019) for information about current ideas for the development of classification schemes and ranking of taxa in fungi, and the most likely historical divergence times of those taxa.

proteins to regulate the initiation of transcription, while the *Eubacteria* use sigma transcription factors.

Overall, the *Archaea* and *Eukaryota* are more closely related to each other than either is to the *Eubacteria* and it follows from this that, despite their prokaryotic nature, members of the *Archaea* should not be regarded as bacteria, a conclusion that some bacteriologists still find difficult to accept (similarly, some mycologists find it difficult to accept that the *Oomycota* are not true fungi).

A more complete fossil record with improved molecular dating methods and a better understanding of molecular evolution will be needed before the true ages of the eukaryote lines of evolution can be determined with any certainty.

As we have explained above, phylogenetic trees are currently built using a wide variety of data, which are largely, but not entirely, molecular in nature. It is now accepted that molecular sequences are generally more revealing of evolutionary relationships than are classical phenotypes and this is particularly true for microorganisms. Some of the terminology that has grown up in this area of research is explained in Table 2.2.

The opinions presented above are largely based on studies that built on the established schemes of evolutionary relationships developed using morphology and biochemistry that were available when the first edition of this book was written (for example: Doolittle *et al.*, 1996; Embley & Hirt, 1998; Aravind & Subramanian, 1999; Philippe *et al.*, 2000; Keeling *et al.*, 2005). More recent work has featured enormously improved methods of analysis (and by that we mean both improved sequence analysis of biopolymers *and* improved computational software and firmware) that allow phylogenetic comparisons to be made between multiple single copy ***genes***, whole ***genomes***, all the RNA sequences produced by the studied organisms (their ***transcriptomes***), all their polypeptide sequences (their ***proteomes***) or profiles of all the small-molecule metabolite profiles in cells, tissues or organisms (their ***metabolomes***), or any combination of these. We will consider these recent analyses as they apply to fungi as appropriate throughout this text (with the overall summary description brought together in Chapter 6). For the moment, we must be satisfied with generalisations about the phylogeny of the domains of organisms; and the generalisations we favour are that eukaryotes and prokaryotes diverged 2 billion years ago, and plants, animals and fungi diverged from one another 1 billion years ago.

Today microbes often live in mixed communities that are capable of rapid attachment to surfaces as a **biofilm**. Biofilms are an important component of our present environment, being found essentially everywhere on Earth, including in extreme environments (https://en.wikipedia.org/wiki/Biofilm; Flemming & Wingender, 2010). Most are beneficial or harmless but those that form on clinical equipment and many household surfaces can cause harm. Biofilm formation is so obviously beneficial to the organisms in the community that it must have arisen at an early stage in evolution. So perhaps we should think in terms of the microorganisms that existed for the first 3 billion years of life on Earth forming extensive biofilms over moist surfaces (and that would include the surfaces of bodies of shallow water). Some of those biofilms would have contained photosynthetic microbes, cyanobacterial as well as, eventually, algal.

Analyses of carbon isotope ratios in ancient carbonate rocks suggest that the '***greening of Earth***' started about 850 million years ago in coastal regions and resulted in an extensive spread of photosynthetic microbes as 'an explosion of photosynthesising communities on late Precambrian land surfaces' (Hand, 2009; Knauth & Kennedy, 2009; for counterarguments see Arthur, 2009). Geological evidence indicates that oxygenic photosynthesis (which uses water as an electron donor and produces molecular oxygen) by cyanobacteria became important around 2 billion years ago (Cardona *et al.*, 2015; https://en.wikipedia.org/wiki/Evolution_of_photosynthesis). By 1,000 million years ago this 'greening' was sufficient to increase atmospheric oxygen, alter the chemical breakdown of rocks on the Earth's surface and increase nutrient flux and organic matter in ancient soils and sediments (Sage *et al.*, 2012; Williams *et al.*, 2013).

Primitive biofilms would have also contained fungi (and even more primitive filamentous organisms that had not yet evolved into fungi, because an important ability of hyphal growth is that filamentous hyphae can escape from the biofilm). Even more importantly, ***filamentous hyphae***, by secreting their digestive enzymes into their environment, can exploit the biofilm, digesting the adhesives, gums and other polymers that make up the biofilm matrix and parasitising the photosynthetic microbes to recruit photobionts into primitive lichen-like arrangements.

2.5 THE KINGDOM *FUNGI*

As we have seen, around the middle of the twentieth century the three major kingdoms of eukaryotes were finally recognised. One of the crucial character differences was the mode of nutrition:

- Animals engulf.
- Plants photosynthesise.
- Fungi absorb externally digested nutrients.

To these can be added many other differences. For example: in their cell membranes animals use cholesterol, fungi use ergosterol; in their cell walls, plants use cellulose (a glucose polymer), fungi use chitin (a glucosamine polymer); recent genomic surveys show that plant genomes lack gene sequences that are crucial in animal development, and vice versa, and fungal genomes have none of the sequences that are important in controlling multicellular development in animals or plants. This latter point implies that animals, plants and fungi separated at a unicellular grade of organisation.

The fungal kingdom is now recognised as one of the oldest and largest clades of living organisms on Earth. Kingdom *Fungi* is a monophyletic group which diverged from a common ancestor with the animals about 800–900 million years ago. We continue to follow the phylogenetic classification scheme suggested by Hibbett *et al.* (2007), as this appears to be a well-corroborated phylogenetic framework but modified in accordance with McLaughlin *et al.* (2009), Jones *et al.* (2011), Powell and Letcher (2014), Spatafora *et al.* (2016) and McCarthy and Fitzpatrick (2017). In this summarised arrangement, the *Eumycota* (or true fungi), that make up the monophyletic clade called Kingdom *Fungi*, is comprised of the 10 phyla (the taxon 'phylum' has been borrowed from animal taxonomy) listed below:

- *Cryptomycota;*
- *Microsporidia;*
- *Chytridiomycota;*
- *Monoblepharidomycota;*
- *Neocallimastigomycota;*
- *Blastocladiomycota;*
- ***Zoopagomycota*** (comprises subphyla *Entomophthoromycotina*, *Kickxellomycotina* and *Zoopagomycotina*);
- ***Mucoromycota*** (comprises subphyla *Glomeromycotina*, *Mortierellomycotina* and *Mucoromycotina*);
- ***Ascomycota*** (about 65,000 species in 6,355 genera);
- ***Basidiomycota*** (about 32,000 species in 1,589 genera).

The last two phyla are combined in Subkingdom ***Dikarya*** by Hibbett *et al.* (2007). These taxa will be described in more detail in Chapter 3 and we say something about Zygomycota and *Microsporidia* below. **The Mycota** (2nd edition), **Vol. VII** (in two parts **A** and **B**) (McLaughlin & Spatafora, 2014, 2015) includes treatments of the systematics and related topics for fungi and fungus-like organisms in four eukaryotic ***supergroups*** (the informal rank above kingdom), three of fungus-like organisms: **Amoebozoa**, **Excavata** and **SAR** (*Straminipila*, *Alveolata* and *Rhizaria*) and the true fungi in supergroup **Opisthokonta** that comprises the phyla mentioned above. It also includesd specialised chapters on nomenclature, techniques and evolution. The proteomic fungal tree of life is similar to, but not identical to, the outline shown above (Choi & Kim, 2017). The classification of Kingdom *Fungi* and fungus-like organisms is outlined (and fully referenced) in **Appendix 1** at the end of this book and this includes an explanation of 'supergroups' (see Figure 2.5).

For the moment, remember that when fungi were still classified in the plant kingdom (Subkingdom Cryptogamia, Division Fungi, Subdivision Eumycotina) they were separated into four classes:

- Phycomycetes;
- Ascomycetes;
- Basidiomycetes;
- Deuteromycetes (the latter also known as Fungi Imperfecti because they lacked a sexual cycle).

You may still encounter these traditional names for groups of fungi, but if they are used today, you must appreciate that they can only be used informally. Many organisms included in these groups (particularly among the phycomycetes and the slime moulds) are no longer considered to be true fungi, even though mycologists might study them. This applies to many of the water moulds, like the ***Oomycota*** (which include the plant pathogen *Phytophthora*), and ***Hyphochytriomycota***, all of which have been removed from the fungi, and are now classified with brown algae and diatoms in the **Kingdom *Chromista*** or **Kingdom *Straminipila*** (see section 3.10). Similarly, the Amoebidales, which are parasites or commensals of living arthropods and were previously considered to be trichomycete fungi within the 'Zygomycota' are now considered to be protozoan animals. None of the slime moulds are now considered to belong to Kingdom *Fungi* and their relationship to other organisms, especially animals, is still in dispute (see Appendix 1).

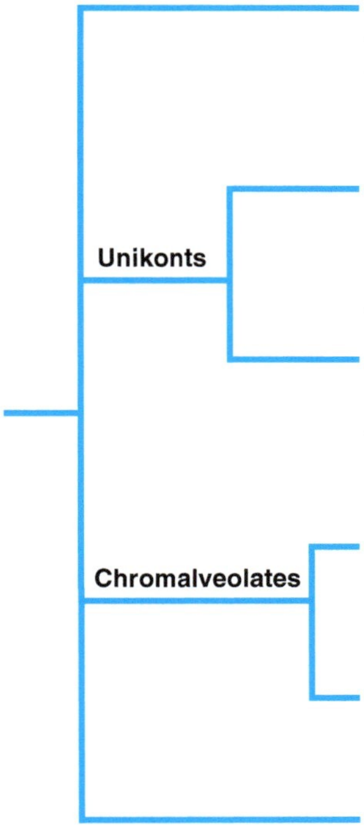

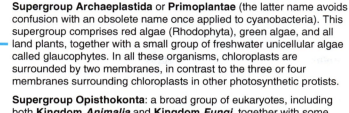

Supergroup Archaeplastida or **Primoplantae** (the latter name avoids confusion with an obsolete name once applied to cyanobacteria). This supergroup comprises red algae (Rhodophyta), green algae, and all land plants, together with a small group of freshwater unicellular algae called glaucophytes. In all these organisms, chloroplasts are surrounded by two membranes, in contrast to the three or four membranes surrounding chloroplasts in other photosynthetic protists.

Supergroup Opisthokonta: a broad group of eukaryotes, including both **Kingdom *Animalia*** and **Kingdom *Fungi***, together with some eukaryotic microorganisms (choanoflagellates) that are sometimes grouped in the phylum *Choanozoa*

Supergroup Amoebozoa: amoeba-like protists, many with blunt, fingerlike, lobed pseudopods and tubular mitochondrial cristae. Contains many of the best-known amoeboid organisms, including the genus *Amoeba* itself, as well as several varieties of slime moulds. Amoebozoa is a monophyletic clade, often shown as the sister group to Opisthokonta.

Supergroup SAR, comprising ***Straminipila***, the ***Alveolata***, and the ***Rhizaria*** (Cercozoa, amoebae and flagellates common in soil; Foraminifera (amoeboids with reticulose pseudopods, common as marine benthos); and Radiolaria (amoeboids with axopods, common as marine plankton). Many produce shells or skeletons of complex structure, which form the bulk of protozoan fossils.

Supergroup Chromalveolata: a varied assemblage including *Cryptophyta* (cryptomonad algae) and *Haptophyta* (coccoliths and other phytoplankton, some of which form toxic algal blooms.

Supergroup Excavata: flagellate protozoa; containing free-living, symbiotic forms, and some important parasites of humans. Classified on the basis of flagellar structures and considered to be the basal lineage of flagellated organisms.

Figure 2.5. A cladogram (phylogenetic tree) showing the relationships and contents of the five to seven 'Supergroups' into which the global tree of eukaryotes is currently subdivided (Adl *et al.*, 2005, 2012; Burki, 2014). The taxonomically formal kingdoms have been assigned to the deliberately informal supergroups by phylogenomic analysis, a method that uses large alignments of tens to hundreds of genes to reconstruct evolutionary histories. The intention of 'supergrouping' is to summarise the relationships arrived at so far while maintaining flexibility as further work on phylogenomics resolves contentious relationships (see https://www.sciencenews.org/article/tree-life-gets-makeover). See Chapter 3 and, especially, Appendix 1 for further details.

Molecular analyses have led to dramatic changes in our understanding of relationships of fungi placed in the traditional phyla *Chytridiomycota* and 'Zygomycota'. The ***Chytridiomycota*** is retained in the 2007 scheme, but in a much more restricted sense. For one thing, one of its traditional orders, the *Blastocladiales*, has been raised to phylum status as the **Blastocladiomycota**. Similarly, the group of anaerobic rumen chytrids previously known as Order *Neocallimastigales* has also been recognised as a distinct phylum, the *Neocallimastigomycota*.

In contrast, the phylum 'Zygomycota' is *not* accepted in the most recent classification because of remaining doubts about relationships between the groups that have traditionally been placed in this phylum (Benny *et al.*, 2014). Benny *et al.* (2014) recognised the phylum 'Glomeromycota' and left four subphyla (*Entomophthoromycotina*, *Kickxellomycotina*, *Mucoromycotina* and *Zoopagomycotina*) as *'incertae sedis'*, which means 'of uncertain placement'. Later, Spatafora *et al.* (2016) clarified the situation by proposing the two phyla **Zoopagomycota** (comprising subphyla *Entomophthoromycotina*, *Kickxellomycotina* and *Zoopagomycotina*) and **Mucoromycota** (comprising subphyla *Glomeromycotina*, *Mortierellomycotina* and *Mucoromycotina*). The name 'Zygomycota' may be reinstated at some time in the future, but at the time of writing, 'Zygomycota' has not been given a proper diagnosis and can only be used informally. We continue to use the old term 'zygomycetes' in Chapter 3 and elsewhere as a convenient 'container' for the subphyla mentioned above and, more importantly, because you are bound to encounter it in other, older, books.

Kingdom *Fungi* has also gained a few recruits on the basis of recent molecular phylogenetic analysis, notably *Pneumocystis*, the *Microsporidia* and *Hyaloraphidium*. *Pneumocystis carinii* is a pathogen that causes pneumonia in mammals, including humans with weakened immune systems (the human pathogen is more correctly called *Pneumocystis jirovecii*). ***Pneumocystis* pneumonia** (or **PCP**) is the most common opportunistic infection in people with human immunodeficiency virus (HIV) and has been a major cause of death of people infected with HIV. *Pneumocystis* was initially described as a trypanosome, but evidence from sequence analyses of several genes places it in the *Taphrinomycotina* in the *Ascomycota*.

The **Microsporidia** (Didier *et al.*, 2014) are obligate intracellular parasites of animals. They are extremely reduced organisms, without mitochondria. Most infect insects, but they are also responsible for common diseases of crustaceans and fish, and have been found in most other animal groups, including humans (probably transmitted through contaminated food and/or water). They were thought to be a unique phylum of protozoa for many years. Molecular studies show that these organisms are related to the zygomycetes indicating that they should be classified in Kingdom *Fungi* or at least as a sister Kingdom to *Fungi*.

Hyaloraphidium curvatum, an organism previously classified as a colourless green alga is now recognised as a fungus based on molecular sequence data, which show it is a member of the *Monoblepharidales* in the *Chytridiomycota*.

2.6 THE OPISTHOKONTS

The **opisthokont** clade is a distinct lineage of eukaryote groups which share an ultrastructural identity and have no evident sister group; it comprises all the true fungi including chytrids, microsporidia, collar-flagellate protists (*Choanozoa*) and Kingdom *Animalia* (which includes us in the metazoa) (Figure 2.5).

All molecular and ultrastructural studies have strongly supported the idea that opisthokonts form a monophyletic group (it is informally called **supergroup Opisthokonta**; Figure 2.5). The name 'opisthokont' comes from the Greek and means 'posterior flagellum', so the common characteristic that gives them their name is that flagellate cells, when they occur, are propelled by a ***single*** posterior flagellum, and this applies as much to chytrid zoospores as to the sperm of mammals. In contrast, other eukaryote groups that have motile cells propel them with one or more ***anterior*** flagella, and that 'one or more' phrase results in the name **heterokont**.

Heterokonts are organisms in which the motile cells have unequal flagella. This is the most nutritionally diverse eukaryote supergroup and includes several ecologically important groupings. The formal name of the assemblage is the **Heterokonta** and it is placed within the Kingdom *Chromista* alongside haptophytes and cryptomonads by Cavalier-Smith and Chao (2006). The *Chromista* seems to represent an independent evolutionary line that diverged from the same common ancestor as plants, fungi and animals. Heterokonta include:

- Multicellular brown seaweeds, which are the most common type of seaweed on rocky beaches, some forming kelp forests with fronds up to 50 m long.
- The usually parasitic oomycetes that include *Phytophthora*, the pseudofungus that caused the great Irish potato famine of 1845 and *Pythium*, the cause of seed rot and damping-off in seedlings, which has been confused with fungi because of its filamentous growth form. The inclusion of several groups of 'pseudofungi' and 'water moulds', especially genera that have been studied long and hard by mycologists over the years (names to look out for are genera like *Saprolegnia*, *Achlya*, *Albugo*, *Bremia*, *Plasmopara*), is why the Heterokonta is important to us.
- Numerous protists of major importance, such as photosynthetic diatoms, which are a primary component of plankton.
- Numerous groups of chlorophyll *c*-containing algae (chlorophyll *c* and a number of other pigments found in the *Chromista* are not found in any group of true plants).
- Several non-photosynthetic groups that feed phagotrophically or absorptively.

This is one of the most actively researched groups of eukaryotes, partly because some biologists doubt that the group is monophyletic, so the components of the group, and its name, are frequently revised (refer to the Tree of Life project website, Keeling *et al.*, 2009). A few people treat the *Chromista* as identical in composition with the heterokonts, describe them as straminopiles, or seek to change the name of the kingdom to ***Straminipila***. However, although the name *Chromista* has nomenclatural precedence (see discussion in Cavalier-Smith & Chao, 2006), we will use the name ***Straminipila***, following Beakes *et al.* (2014).

2.7 FOSSIL FUNGI

Most fungal structures are very poor candidates for preservation over long periods of time as fossils. Fungal hyphae have so few unique morphological features that it has been difficult to establish much of a fossil record for fungi (Taylor *et al.*, 2015; Edwards *et al.*, 2018; Krings *et al.*, 2018).

Some of the oldest terrestrial fossils of any sort are large fibrous things called **nematophytes**. They are part of what is known as 'phytodebris' and provide the earliest evidence for land organisms. They have been found from the mid-Ordovician (460 million years ago; see the geological time scale in Figure 2.6) to the early Devonian, suggesting that they lasted a period of at least 40 million years. This phytodebris certainly contains fossils of bryophyte-like plants, but it has been suggested that some of the nematophytes (particularly the genus ***Prototaxites***) were terrestrial fungi (Hueber, 2001).

Two remarkable things about these fossils are that they were extremely large (Figure 2.7), and so ***common*** that they were a major component of these early terrestrial ecosystems, both in terms of abundance and diversity. They included by far the ***largest organisms*** present in their ancient environments. Indeed, 'specimens of *Prototaxites* over a metre across have been reported' (Wellman & Gray, 2000). They are described as 'Devonian fossil logs' and have been conjectured to be kelp-like aquatic algae, rolled up carpets of liverworts, giant lichens or enormous saprotrophic fungal sporophores.

Boyce *et al.* (2007) measured the ratio between different carbon isotopes to determine the nature of the fossils. The argument runs that a photosynthetic primary producer will have a relatively constant ratio of carbon isotopes between individuals because they are all using atmospheric carbon. On the other hand, organisms (like fungi) that are classified as consumers will take on the isotope ratios of whatever they are digesting locally, and therefore individual specimens end up with widely differing isotope ratios. Boyce *et al.* (2007) found too much isotopic variance between individual *Prototaxites* fossils for them to be photosynthetic primary producers. Instead, *Prototaxites* was a consumer and, taken together with direct microscopic observation of their anatomy (Hueber, 2001), this demonstrates that these enormous fossils, the largest land organisms to have lived up to their point in time, were giant fungi. Another factor is that

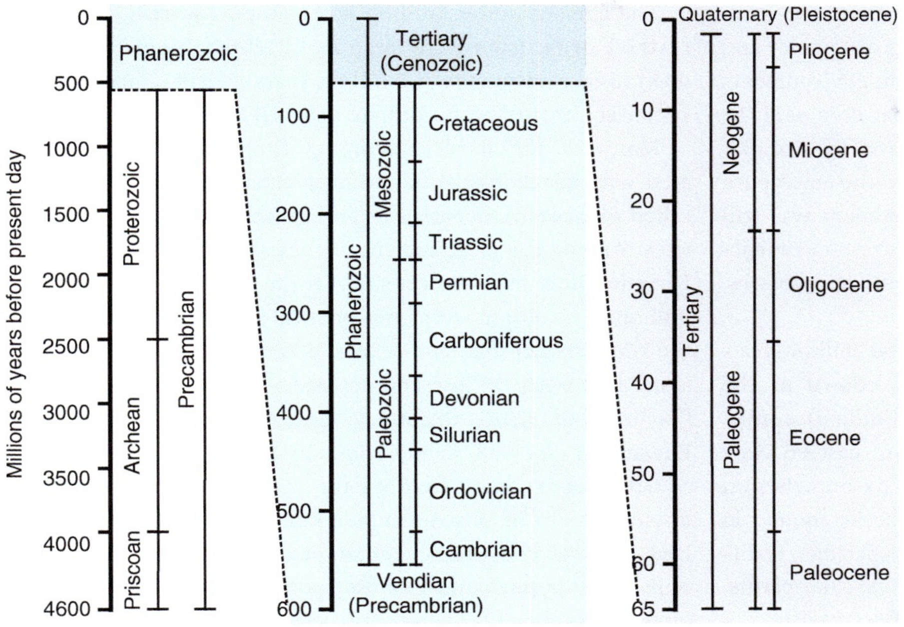

Figure 2.6. The geological time scales. The figure shows how the entire span of time during which the Earth has existed can be classified into eras and epochs. Time axes on the central and right-hand panels are magnified views of the upper section of the previous panel. The last 65 million years is called the Cenozoic era, and is divided into several epochs, which began with the Palaeocene approximately 65 million years ago and runs to the present day. The Palaeocene, which lasted from 54–65 million years ago is followed by the Eocene (34–54 million years ago), the Oligocene (24–34 million years ago), the Miocene (5–24 million years ago), the Pliocene (1.8–5 million years ago) and the Pleistocene or Ice Ages (1.8 million years ago to 10,000 years ago). The period since the last retreat of the glaciers and present glacial warming is called the Holocene, or Recent (10,000 years ago to present day). Source (and see for more details): https://en.wikipedia.org/wiki/Geologic_time_scale (page edit of January 2018).

Figure 2.7. Lower Devonian *Prototaxites* compression fossils, approximately 2 m tall, *in situ* in the Bordeaux Quarry, near Cross Point on the Restigouche River in the Gaspésie region of eastern Quebec, Canada. What you are looking at here is a stream bed that is turned more or less on end (that is, at right angles to its original position), so that you are now looking down on it. You can see fossilised impressions of at least three large *Prototaxites* specimens that must have formed something like a log jam in the stream. Dr Francis Hueber, who first suggested *Prototaxites* fossils are fungal in origin (Hueber, 2001), is posing alongside as a scale marker.
Part of this image, kindly supplied by Dr Carol Hotton of the Smithsonian Institution, appears as Figure 1A in Boyce *et al.* (2007).

the isotope ratios of *Prototaxites* fossils can indicate the sorts of primary producers available for its mycelium to digest, and several *Prototaxites* individuals tested come from environments that just did not include the vascular plants that dominate the modern world. Even though conquest of the land by vascular plants had begun about 40 million years prior to the emergence of *Prototaxites*. Instead, the *Prototaxites* environment was still dependent on the more ancient primary producers: cyanobacteria (blue-green algae), eukaryotic algae, lichens and mosses, liverworts and their relatives (bryophytes).

Retallack and Landing (2014) examined a 386 million-year-old fossil of *Prototaxites loganii*, a specimen known as the 'Schunnemunk tree' (as it comes from the Middle Devonian Bellvale Sandstone on Schunnemunk Mountain, eastern New York). This specimen was 8.83 m long and had six branches on the upper 1.2 m of the main axis, each branch being about 1 m long and 9 cm in diameter. The histology of this specimen is different from that expected of algae, liverworts or vascular plants. The authors claim it best resembles a lichen tissue with coccoid phycobionts, and though *Prototaxites* has previously been placed within *Basidiomycota* (as a sort of giant club fungus), they consider the organisms may belong within *Mucoromycotina* or *Glomeromycotina*.

The current understanding is that the **first truly large terrestrial organisms** were large multicellular fungi that developed to take advantage of 2 billion years' worth of accumulated bacterial, protist and bryophyte debris (Figure 2.8).

Many of the most convincing fungal fossils are associated with plant fossils, including glomeromycotan (used to be called glomalean) mycorrhizas and ascomycete or chytrid parasites. Among the most ancient of these are from the **Devonian Rhynie chert** of Aberdeenshire in the north of Scotland (400 million years old) in which mycorrhizal fungi (recognisably from the *Glomeromycotina*) and several other fungi have been found associated with the preserved tissues of early vascular plants (Taylor *et al.*, 2015, Chapter 7, Glomeromycota, pp. 103–128; Edwards *et al.*, 2018; Krings *et al.*, 2018).

Glomeromycotan fossils have also been found in mid-Ordovician rocks of Wisconsin (460 million years old). The fossilised material consisted of entangled, occasionally branching, nonseptate hyphae and globose spores. The age of these fossil glomeromycotan fungi indicates that such fungi were present before the first vascular plants arose, when the land flora most likely only consisted of bryophytes, lichens and cyanobacteria. Today, the *Glomeromycotina* form the **arbuscular mycorrhizal symbiosis**, which is ubiquitous in modern vascular plants and

Figure 2.8. Two artistic impressions of the Lower Devonian landscape of some 400 million years ago, dominated by specimens of *Prototaxites* up to 9 m tall. (A) A painting by Mary Parrish of the Smithsonian Institution, Washington, which was prepared for the publication about *Prototaxites* fossils by Hueber (2001). (B) A painting by Geoffrey Kibby that appeared under the title 'an artist's impression of the landscape of the Devonian period' as a back-cover image on the magazine *Field Mycology* in April 2008. In the landscape portrayed in these paintings the fungus *Prototaxites* dominates as the largest terrestrial organism to have lived up to this point in the history of the Earth. Although vascular plants had emerged by this time, these landscapes were still dependent on the more ancient primary producers: cyanobacteria (blue-green algae), eukaryotic algae, lichens and mosses, liverworts, and their bryophytes relatives. What you are seeing here is the physical expression of the dominance of fungi in the Earth's biosphere. This physical dominance of *Prototaxites* lasted at least 40 million years (about 20 times longer than the genus *Homo* has so far existed on Earth). Images kindly supplied by Tom Jorstad of the Smithsonian Institution and Geoffrey Kibby, senior editor of *Field Mycology*. The painting by Mary Parrish courtesy of and © Smithsonian Institution. Reproduced with permission from Elsevier.

has also been reported in modern hepatics and hornworts. As the ancient fungi were present both prior to the emergence of vascular plants and in the tissues of early vascular plant fossils, it seems reasonable to suppose that arbuscular mycorrhizas played an important role in the success of early terrestrial plants (Redecker *et al.*, 2000).

Other convincing fossils are more recent. Microfossils of hyphae with clamp connections are known from the Pennsylvanian/Carboniferous (300 million years ago; see Figure 2.6). However, the only convincing mushroom fossils found so far are preserved in amber dating from the Cretaceous (about 90 million years ago). These '**mushrooms in amber**' (Hibbett *et al.*, 1995) bear a strong resemblance to the existing genera *Marasmius* and *Marasmiellus*, which are quite common in modern woodlands; yet, when they were preserved, the dinosaurs still ruled the Earth. In other words, the mushrooms **you** see when you trek through the forest are almost identical to those seen by dinosaurs in their forests, although, of course, the forest plants are very different.

Amber dated to the Eocene (54–34 million years ago; see the geological time scale in Figure 2.6) has been found that contains the remains of several **filamentous mould fungi**. Among the finds are sooty moulds in European amber dating back to 22–54 million years ago. Present-day sooty moulds are a mixed group of saprotrophs, usually with dark-coloured hyphae, which produce colonies superficially on living plants (as harmless epiphytes). Most present-day sooty moulds use arthropod excretions for nutrition and live closely associated with aphids, scale insects and other producers of honeydew. All the fossils are composed of darkly coloured hyphae with features identical to the present-day genus *Metacapnodium*, suggesting that *Metacapnodium* **hyphae have remained unchanged for tens of millions of years** (Rikkinen *et al.*, 2003). Possibly the most impressive fossil is a piece of amber from the Baltic region that contains an inclusion of a springtail (a collembolan arthropod) which is **overgrown by an** *Aspergillus* **species** (Dörfelt & Schmidt, 2005). The surface of the springtail is densely covered in places by excellently preserved hyphae and conidiophores. Numerous **sporulating conidiophores** can be seen easily, and conidial heads with radial chains of conidia are clearly visible. As well as superficial hyphae at the cuticle, the springtail is loosely penetrated by branched substrate hyphae, so the authors suggest that the fungus may be parasitic and describe it as a new species, *Aspergillus collembolorum* (Dörfelt & Schmidt, 2005).

A survey of fleshy fungi from amber deposits around the world found representatives of gilled fungi, puffballs, bird's nest fungi, gasteroid fungi, xylaroid fungi, hymenomycetes, polypores and the first fossil morel from amber collected in the Dominican Republic (Poinar, 2016). So fossil evidence shows the fungi to be important members of terrestrial ecosystems up to 500 million years ago. Molecular phylogenetic evidence suggests that fungi are much older.

To begin with, molecular phylogenetic studies tended to use single gene sequences. The most popular in fungi have been the nuclear ribosomal DNA (rDNA) locus, particularly that encoding small subunit (18S) ribosomal RNA but including the nuclear large ribosomal RNA subunit (nucLSU), mitochondrial rDNAs, and complete or near complete mitochondrial genomes. Single gene phylogenies may not provide sufficient information nor be truly representative to resolve a fungal phylogeny with sufficient confidence, so broader studies provide better information. Sequences of protein-coding genes can be a problem because it is difficult to design primers for PCR amplification that can be reliably applied to a wide range of taxa. Also, heterozygous loci in heterokaryons can complicate interpretations.

2.8 THE FUNGAL PHYLOGENY

A global phylogeny of fungi first emerged using data from several gene regions: 18S rRNA, 28S rRNA, 5.8S rRNA, elongation factor-1 (EF1) and RNA polymerase II subunits (RPB1 and RPB2). Data for all these gene regions were combined (a total number of 6,436 aligned nucleotides) for 199 fungi (James *et al.*, 2006), and there have been several more genome-level studies since this work was published, many of which were combined by McCarthy and Fitzpatrick (2017) (and see Appendix 1). The outcome is an enormous cladogram which we cannot reproduce here, but a highly simplified evolutionary tree is shown in Figure 2.9.

Significantly, this extensive analysis generally supports the more traditional arrangement into: *Ascomycota*, *Basidiomycota*, various zygomycetes and *Chytridiomycota*, but it obviously adds new detail to that traditional structure. *Ascomycota* and *Basidiomycota* are united as the *Dikarya*, fungi in which at least part of the life cycle is characterised by cells with paired nuclei. The closest relatives of these two sister groups are the *Glomeromycotina* (which was for a long time included as the Glomales within the 'Zygomycota'). Neither the 'Zygomycota' nor the *Chytridiomycota* are monophyletic groups; they have representatives in different clades or branches of the tree that are grouped into those phyla by their shared primitive morphologies (such groups are called paraphyletic). This is why, in the latest classification (see Chapter 3), the *Chytridiomycota* is redefined and, as mentioned above, the 'Zygomycota' is demoted from rank as a formal taxon and becomes an informal name, at least for the time being.

Note the microsporidia and *Rozella* (**Cryptomycota**) branches in Figure 2.9, which come out as basal to all other fungi in this analysis. *Rozella*, a genus of chytrid that is parasitic on other *Chytridiomycota*, seems to be one of the most primitive fungi. ***Microsporidia***, which are parasites of animals, seem to be derived from an endoparasitic chytrid ancestor similar to *Rozella*, on the earliest diverging branch of the fungal phylogenetic tree.

The fungi, animals and plants are the only three eukaryotic kingdoms of life that developed multicellular tissues in

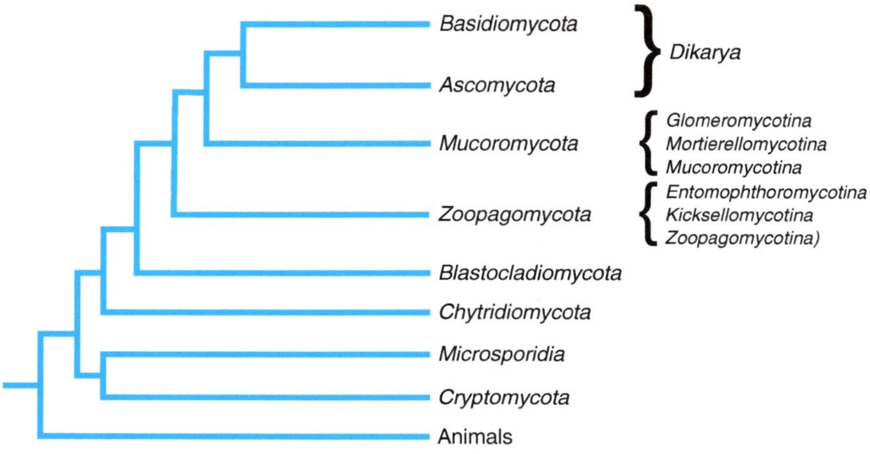

Figure 2.9. The main branches of the tree of life for Kingdom *Fungi*, as established by the AFTOL project and subsequent publications (referenced in the text). All the traditional phyla are represented: *Ascomycota*, *Basidiomycota*, zygomycetes (as *Mucoromycota*, which includes *Glomeromycotina*, and *Zoopagomycota*), and *Chytridiomycota* (see discussion in the text of this chapter and Chapter 3 for *Blastocladiomycota* and *Neocallimastigomycota*). *Ascomycota* and *Basidiomycota* are united as the *Dikarya*, fungi in which at least part of the life cycle is characterised by cells with paired nuclei ('dikaryons'). The closest relatives of these two sister groups are the *Glomeromycotina*. The multigene phylogeny of fungi established by the AFTOL project placed the *Glomeromycotina* (using the name Glomeromycota) in a basal position as a sister group of *Asco-* and *Basidiomycota* (together called Subkingdom *Dikarya*); this relationship to *Dikarya* is only slightly changed by the realignment of the group as a subphylum. Although, in formal terms, it is now Phylum *Mucoromycota* that shared a most recent common ancestor with *Dikarya*, the *Glomeromycotina* was the first lineage of *Mucoromycota* to diverge (Spatafora *et al.*, 2016). View Appendix 1 for further details (and references).

terrestrial environments. They are thought to have diverged from each other roughly 1 billion years ago. This study continues to support the view that the ancestors of fungi were simple aquatic cells with flagellated spores, like current chytrids. What it changes is the idea that there was a single loss of the chytrid flagellum as terrestrial fungi diversified. Rather, the study argues for at least four independent losses of the flagellum during early evolution of Kingdom *Fungi*, coinciding with the evolution of new mechanisms of spore dispersal.

Estimating the historical time of appearance of the major fungal groups remains a major problem, though whole-genome analyses are beginning to aid understanding (Reynolds *et al.*, 2017). A significant contribution to the problem is that horizontal gene transfer has been a significant factor in fungal evolution (Richards *et al.*, 2011; see sections 6.5 and 19.15) and, of course, fungi have been so successful in the billion years or so of their existence.

Today's fungal kingdom is arguably the most abundant and diverse group of organisms on Earth. Fungi are found in every terrestrial ecosystem as mutualist partners, pathogens, parasites or saprotrophs. As we mentioned in section 1.7, it is estimated that the kingdom contains 2.2 to 3.8 million species (Hawksworth & Lücking, 2017), but only about 3–8% of these have been described. If most of the unknowns are members of the traditional taxa, then current phylogenetic inferences will be unchallenged by additional discoveries. However, novel fungal groups could be awaiting discovery by DNA-based environmental sampling, which is already starting to reveal microscopic, undescribed and unculturable fungi. Because unknowns are unknowns, we cannot predict how such discoveries might affect our understanding of fungal origins and evolution. An old quotation we'd like to associate with this summary is: 'evidence accumulates to support the long-held view that the history of fungi is not marked by change and extinctions but by conservatism and continuity' (Pyrozynski, 1976). In other words, fungal evolution is based on the principle: if it works ... don't fix it. Our current understanding of the broad sweep of fungal evolution is summarised in Figure 2.10.

An interesting development in the first few years of the twenty-first century was a trend to suggest that the first ***terrestrial*** eukaryotes might have been fungal (rather than animal or plant). A few titles will illustrate this: 'Terrestrial life – fungal from the start?' (Blackwell, 2000); 'Early cell evolution, eukaryotes, anoxia, sulfide, oxygen, fungi first (?), and a Tree of Genomes revisited' (Martin *et al.*, 2003); and *'Devonian landscape heterogeneity recorded by a giant fungus'* (Boyce *et al.*, 2007). For mycologists this was certainly a refreshing development because, prior to this, fungi have always been ignored when theorists pondered the origin and early emergence of life on this planet. The complete version of what we like to call 'The Mycologists' Tale' appeared in the book entitled ***Fungal Biology in the Origin and Emergence of Life*** (Moore, 2013).

This book offers a new interpretation (illustrated here in Figure 2.11) of the early radiation of eukaryotes based on the emergence of major innovations in cell biology that apply uniquely to present-day fungi. These emphasised increasingly detailed management of the positioning and distribution of membrane-bound compartments (vacuoles, vesicles and

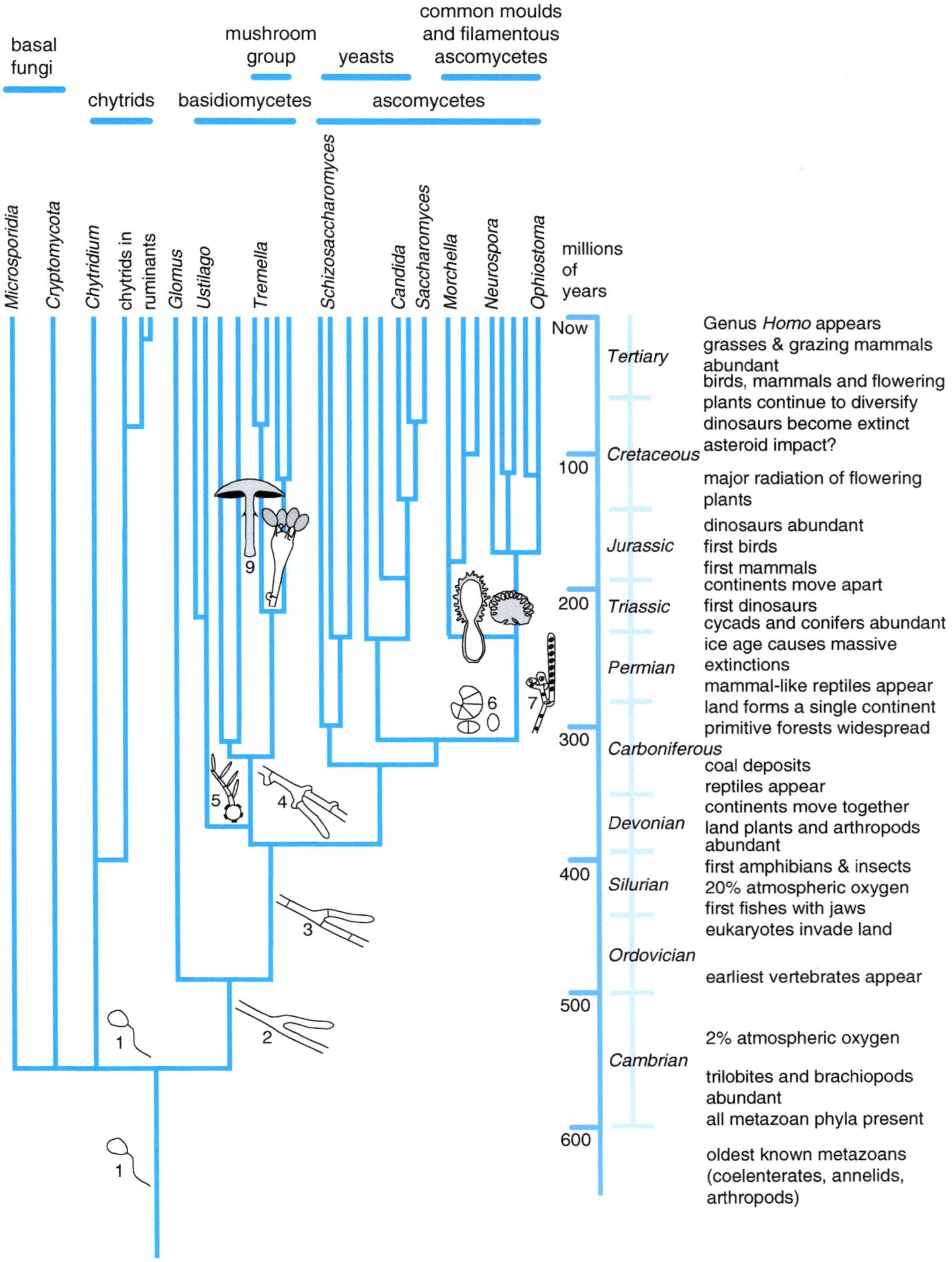

Figure 2.10. A summary of fungal evolution. This phylogenetic tree summarises current ideas about the broad sweep of fungal evolution and puts it into context by showing some markers of geological time and animal and plant evolutionary features. This is a cladogram showing phylogeny of the true fungi largely based on the 18 S rDNA gene sequence but modified by phylogenomic analysis. Branch lengths in the cladogram are proportional to the average rate of nucleotide substitution (1% per 100 million years), so the cladogram becomes an evolutionary tree, which has been calibrated using fossil fungi, fungal hosts and/or symbionts. The time scale on the right shows the context of other major evolutionary events in geological time. The numerals and cartoons on the cladogram illustrate major milestones of fungal morphological evolution: terrestrial higher fungi diverged from water moulds (1) as branching filaments without septa (2) about 550 million years ago (Mya); the *Glomeromycotina* diverged from the progenitor of ascomycetes and basidiomycetes about 490 Mya, and the latter lineage evolved septate filaments (3); clamp connections mark early basidiomycetes (4); basidia (smut-like, 5), asexual spores (6) and asci (7) probably evolved early in the major radiations of basidiomycetes and ascomycetes; filamentous ascomycetes diverged from the yeast lineage about 310 Mya, and sporophores (8) presumably evolved before the Permian divergences because they are present in all the lineages today; mushroom fungi (9), with their characteristic holobasidium (10) probably radiated 130–200 Mya, soon after flowering plants became an important part of the flora. It is interesting to note that coals deposited in the Cretaceous and Tertiary periods show much more evidence of fungal decay than the much older Carboniferous coals, reflecting the radiation of aggressive wood-decay basidiomycetes from the Triassic onwards. Also note the relatively recent radiation of (anaerobic) chytrids as grasses and grazing mammals became more abundant. Revised from Moore & Novak Frazer (2002).

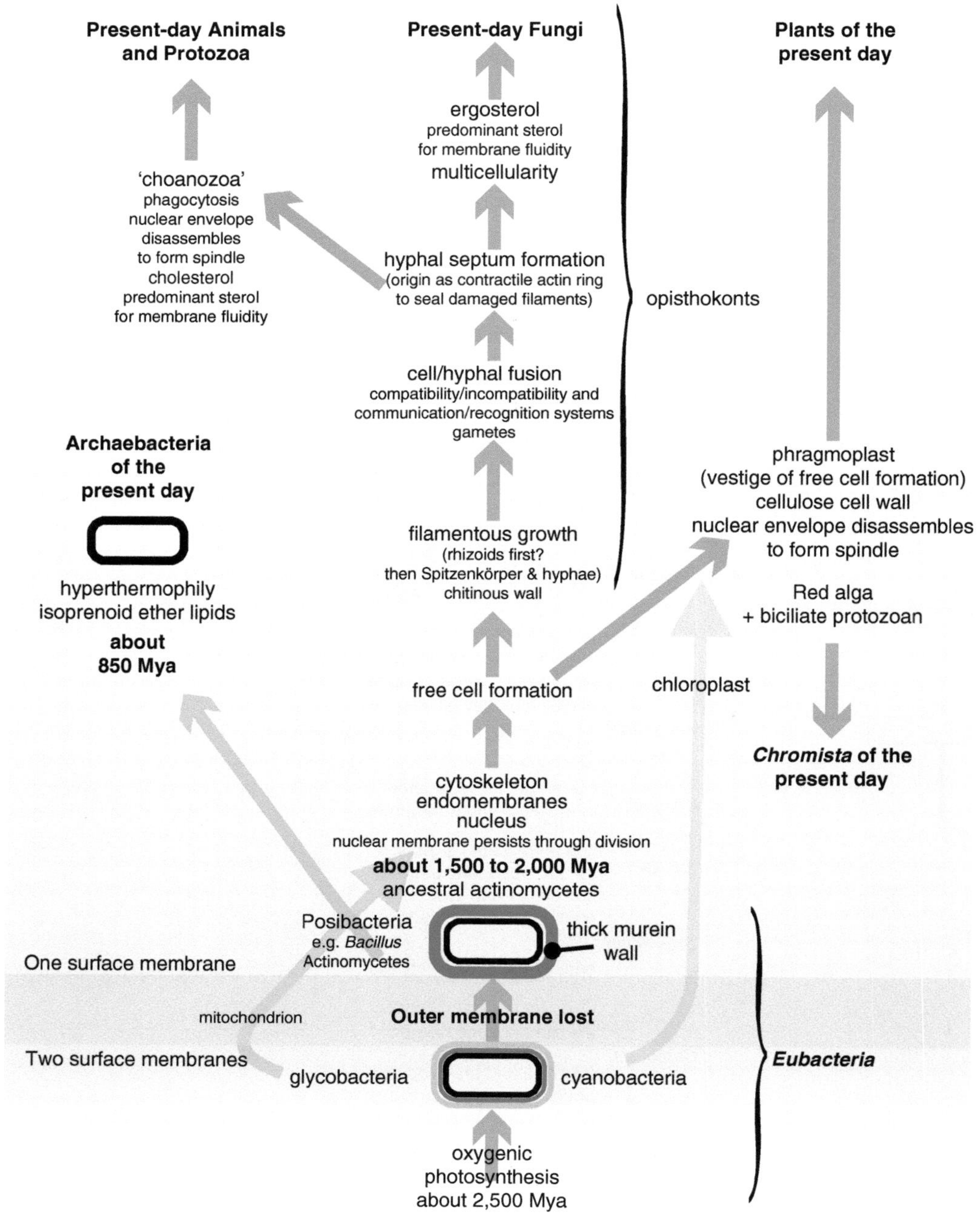

Figure 2.11. The mycologist's tree of life. The lower part of this diagram is based on Cavalier-Smith's tree of life (Cavalier-Smith, 2010a; his Figure 6), which emphasises major evolutionary changes in membrane topology and chemistry, except that the most ancient bacteria are shown here to be heterotrophic descendants of LUCA (the last universal common ancestor of all current life on Earth). Eukaryotes diverge from actinobacterial ancestors about 1,500 Mya (million years ago) and the bulk of this illustration deals with eukaryote evolution. The most ancient stem eukaryotes are considered to exhibit characteristics that would today be thought of as applying to primitive fungi. Their evolution emphasises increasingly detailed management of the positioning and distribution of membrane-bound compartments (vacuoles, vesicles and microvesicles) by the filamentous components of the cytoskeleton (microfilaments, intermediate filaments and microtubules); culminating, as far as filamentous fungi are concerned, with emergence of the Spitzenkörper and apical hyphal extension. Uniquely among present-day eukaryotes, the fungi maintain their nuclear membrane intact as the nuclear division progresses. The subsequent evolution of Kingdom *Fungi* is outlined in Figure 2.10.

microvesicles) by the filamentous components of the cytoskeleton (microfilaments, intermediate filaments and microtubules), culminating, as far as filamentous fungi are concerned, with emergence of the Spitzenkörper and apical hyphal extension. These features of present-day fungi are described in detail elsewhere in this book (Chapter 5, sections 5.9, 5.10 and 5.14) and we will reference where we describe other features as we mention them.

The interpretation suggests that the last universal common ancestor of all current life on Earth (usually known by the acronym LUCA) was a heterotrophic, mesophilic prokaryote, essentially a 'bacterial' cell with the cell enveloped by two distinct lipid bilayer membranes. The earliest prokaryotes used prebiotically synthesised organic carbon compounds as nutrients but, as these supplies diminished, they were outstripped by the anoxygenically photosynthetic Chlorobacteria as the most primitive surviving prokaryotic phylum. This interpretation follows the deep phylogeny of the tree of life published by Cavalier-Smith (2006, 2010a & b), who argued that thermophiles evolved late, making Archaebacteria the youngest bacterial phylum and the sisters, rather than ancestors of eukaryotes (in contrast to the interpretations described in section 2.4 above), which themselves diverged from actinobacterial ancestors. Eukaryotes are generally thought to have appeared about 1.5 billion years ago. So, for a total of about 2 billion years the only living organisms on the planet were prokaryotes together, presumably, with their associated viruses.

The primitive eukaryotic stem featured primitive nuclear structures (including the nuclear membrane remaining ***intact*** as the nuclear division progresses, which is a unique characteristic of present-day fungi) (Chapter 5, section 5.6), added the mitochondrion by enslavement of a bacterium (Chapter 5, section 5.9) and evolved those aspects of the endomembrane system and cytoskeletal architecture that are also unique characteristics of present-day fungi, in the following probable temporal sequence:

- Free cell formation (Chapter 3), by managing positioning of wall- and membrane-forming vesicles to enclose volumes of cytoplasm to subdivide sporangia into spores, with adoption of a chitinous cell wall, possibly as an adaptation of muramopeptide oligosaccharide synthesis from the wall of an actinobacterial ancestor. This is a possible branch point to plants if the phragmoplast is assumed to be a vestige of free cell formation and the cell wall was adapted to be a polymer of glucose rather than N-acetylglucosamine, possibly for economy in usage of reduced nitrogen in organisms abandoning heterotrophy. Plants also evolved a means to disassemble the nuclear envelope to form the division spindle.
- Filamentous growth (Chapter 4), first to make rhizoids then apically extending with the Spitzenkörper as the organising centre for hyphal extension and morphogenesis to make nucleated hyphae to explore and exploit the then extant biofilm and terrestrial debris that had accumulated during 2 billion years of prokaryote growth, life and death.
- Hyphal/cell fusion (Chapters 4 and 5, especially section 5.15), with associated cytoplasmic (vegetative) and nuclear (sexual) compatibility/incompatibility systems, hypha-to-hypha communication/recognition systems, autotropism, gravitropism and intrahyphal communication using secondary metabolites, including the evolution of gametes.
- Hyphal septum formation (Chapter 4, section 4.12; and Chapter 5, section 5.16), initially dependent on a contractile ring of actin to seal the membrane of damaged filaments rapidly, later developing ingressive wall synthesis to strengthen the seal, and ultimately cross-wall formation at regular intervals to initiate multicellular development. Possibly combined with the (accidental?) fixation on ergosterol as the quantitatively predominant sterol involved with controlling membrane fluidity in fungi.

This last is a possible branch point from chytrid-level fungi to animals (of the sort presently called choanozoa). The animal stem gradually lost its wall and adapted cytoskeletal organisation/vesicle trafficking originally used in wall synthesis to the new function of phagocytosis. Animals also developed disassembly of the nuclear envelope as the division spindle forms and adopted cholesterol as the predominant sterol for membrane fluidity, as well as equatorially contractile cell division.

Through this sequence of events filamentous fungi emerged 1.5 billion years ago as the first crown group of eukaryotes. They emerged to exploit the components of the biofilms in which they lived and the debris left by 2 billion years of prokaryote growth, and they have been cleaning up the planet ever since (Moore, 2013).

Finally, to illustrate the ancient importance of fungi, and maybe suggest something that accounts for their success through the rest of geological time we offer a few quotations, which relate to the Permian–Triassic (P–Tr) extinction event that occurred approximately 251 million years ago. The evolution of life on Earth has been interrupted by several mass extinction events. The P–Tr event, informally known as the Great Dying, was the Earth's most severe extinction event (so far!), with about 96% of all marine species and 70% of terrestrial vertebrates becoming extinct. This catastrophic ecological crisis was triggered by the effects of severe changes in atmospheric chemistry arising from the largest volcanic eruption in the past 500 million years of Earth's geological history, which formed what are now known as the Siberian Traps flood basalts. When first formed, these are thought to have covered an area in Siberia about the size of Australia.

Plants suffered massive extinctions as well as animals: 'excessive dieback of arboreous vegetation, effecting destabilisation and subsequent collapse of terrestrial ecosystems with concomitant loss of standing biomass' occurred 'throughout the world'.

However, the result of all this death and destruction is that 'sedimentary organic matter preserved in latest Permian deposits is characterised by unparalleled abundances of fungal remains, irrespective of depositional environment (marine, lacustrine [= lake sediments], fluviatile [= river/stream deposits]), floral provinciality, and climatic zonation'. The quotations were taken from Visscher *et al.* (1996).

The Cretaceous–Tertiary (K–T) extinction of 65 million years ago is another one that we all know a little bit about, because it was caused by a meteor collision that caused the **Chicxulub** crater in Mexico and is blamed for the extinction of the dinosaurs.

The K–T boundary is characterised by high concentration of the element iridium, which is rare on Earth but common in space debris, such as asteroids and meteors. Current understanding is that a meteor hit the Earth at the end of the Cretaceous and the iridium-rich layer resulted as the worldwide dust cloud produced by the impact settled to the ground. As the Cretaceous is the last geological period in which dinosaur fossils are found, the belief is that the meteor collision at Chicxulub caused the extinction of the dinosaurs. There was also widespread deforestation right at the end of the Cretaceous, which is assumed to be due to post-impact conditions of high humidity (caused by widespread rain), decreased sunlight and cooler global temperatures resulting from increased atmospheric sulfur aerosols and dust.

However, coincident with all this death and destruction of animal and plant life at the K–T boundary there is a massive ***proliferation*** of fungal fossils. Vajda and McLoughlin (2004) put it like this: 'This fungi-rich interval implies wholesale dieback of photosynthetic vegetation at the K-T boundary in this region. The fungal peak is interpreted to represent a dramatic increase in the available substrates for [saprotrophic] organisms (which are not dependent on photosynthesis) provided by global forest dieback after the Chixculub impact.'

It is the same story as at the P–Tr extinction boundary: while the rest of the world was dying, the fungi were having a party! But that might not be the full significance of this anecdote, because Casadevall (2005) suggests that the massive increase in the number of fungal spores in the atmosphere of the time caused fungal diseases that 'could have contributed to the demise of dinosaurs and the flourishing of mammalian species'. The impact of fungi on our own origins is as great as their impact on the world habitat.

2.9 REFERENCES

Adl, S.M., Simpson, A.G.B., Farmer, M.A. *et al.* (2005). The new higher level classification of Eukaryotes with emphasis on the taxonomy of protists. *Journal of Eukaryotic Microbiology*, 52: 399–451. DOI: https://doi.org/10.1111/j.1550-7408.2005.00053.x.

Adl, S.M., Simpson, A.G.B., Lane, C.E. *et al.* (2012). The revised classification of Eukaryotes. *Journal of Eukaryotic Microbiology*, **59**: 429–493. DOI: https://doi.org/10.1111/j.1550-7408.2012.00644.x.

Aravind, L. & Subramanian, G. (1999). Origin of multicellular eukaryotes: insights from proteome comparisons. *Current Opinion in Genetics & Development*, **9**: 688–694. DOI: https://doi.org/10.1016/S0959-437X(99)00028-3.

Arthur, M.A. (2009). Biogeochemistry: carbonate rocks deconstructed. *Nature*, **460**: 698–699. DOI: https://doi.org/10.1038/460698a.

Beakes, G.W., Honda, D. & Thines, M. (2014). Systematics of the Straminipila: Labyrinthulomycota, Hyphochytriomycota, and Oomycota. In *The Mycota Systematics and Evolution*, VII part A, 2nd edition, ed. D.J. McLaughlin & J.W. Spatafora. Berlin: Springer-Verlag, pp. 39–97. ISBN: 978-3-642-55317-2. DOI: https://doi.org/10.1007/978-3-642-55318-9_3.

Benny, G.L., Humber, R.A. & Voigt, K. (2014). Zygomycetous fungi: Phylum Entomophthoromycota and Subphyla Kickxellomycotina, Mortierellomycotina, Mucoromycotina, and Zoopagomycotina. In *The Mycota Systematics and Evolution*, VII part A, 2nd edition, ed. D.J. McLaughlin & J.W. Spatafora. Berlin: Springer-Verlag, pp. 209–250. ISBN: 978-3-642-55317-2. DOI: https://doi.org/10.1007/978-3-642-55318-9_8.

Blackwell, M. (2000). Terrestrial life: fungal from the start? *Science*, **289**: 1884–1885. DOI: https://doi.org/10.1126/science.289.5486.1884.

Boyce, C.K., Hotton, C.L., Fogel, M.L. *et al.* (2007). Devonian landscape heterogeneity recorded by a giant fungus. *Geology*, **35**: 399–402. DOI: https://doi.org/10.1130/G23384A.1.

Burki, F. (2014). The eukaryotic tree of life from a global phylogenomic perspective. *Cold Spring Harbor Perspectives in Biology*, **6**: a016147. DOI: https://doi.org/10.1101/cshperspect.a016147.

Cardona, T., Murray, J.W. & Rutherford, A.W. (2015). Origin and evolution of water oxidation before the last common ancestor of the Cyanobacteria. *Molecular Biology and Evolution*, **32**: 1310–1328. DOI: https://doi.org/10.1093/molbev/msv024.

Casadevall, A. (2005). Fungal virulence, vertebrate endothermy, and dinosaur extinction: is there a connection? *Fungal Genetics and Biology*, **42**: 98–106. DOI: https://doi.org/10.1016/j.fgb.2004.11.008.

Cavalier-Smith, T. (2006). Cell evolution and Earth history: stasis and revolution. *Philosophical Transactions of the Royal Society of London, Series B*, **361**: 969–1006. DOI: https://doi.org/10.1098/rstb.2006.1842.

Cavalier-Smith, T. (2010a). Deep phylogeny, ancestral groups and the four ages of life. *Philosophical Transactions of the Royal Society of London, Series B*, **365**: 111–132. DOI: https://doi.org/10.1098/rstb.2009.0161.

Cavalier-Smith, T. (2010b). Kingdoms Protozoa and Chromista and the eozoan root of the eukaryotic tree. *Biology Letters*, 6: 342–345. DOI: https://doi.org/10.1098/rsbl.2009.0948.

Cavalier-Smith, T. & Chao, E.E-Y. (2006). Phylogeny and megasystematics of phagotrophic heterokonts (Kingdom Chromista). *Journal of Molecular Evolution*, **62**: 388–420. DOI: https://doi.org/10.1007/s00239-004-0353-8.

Choi, J.-J. & Kim, S.-H. (2017). A genome Tree of Life for the fungi kingdom. *Proceedings of the National Academy of Sciences of the United States of America*, **114**: 9391–9396. DOI: https://doi.org/10.1073/pnas.1711939114.

Didier, E.S., Becnel, J.J., Kent, M.L., Sanders, J.L. & Weiss, L.M. (2014). Microsporidia. In *The Mycota Systematics and Evolution*, VII part A, 2nd edition, ed. D.J. McLaughlin & J.W. Spatafora. Berlin: Springer-Verlag, pp. 115–140. ISBN: 978-3-642-55317-2. DOI: https://doi.org/10.1007/978-3-642-55318-9_5.

Doolittle, R.F., Feng, D.F., Tsang, S., Cho, G. & Little, E. (1996). Determining divergence times of the major kingdoms of living organisms with a protein clock. *Science*, **271**: 470–477. DOI: https://doi.org/10.1126/science.271.5248.470.

Doolittle, W.F. (2000). The nature of the universal ancestor and the evolution of the proteome. *Current Opinion in Structural Biology*, **10**: 355–358. DOI: https://doi.org/10.1016/S0959-440X(00)00096-8.

Dörfelt, H. & Schmidt, A.R. (2005). A fossil *Aspergillus* from Baltic amber. *Mycological Research*, 109: 956–960. DOI: https://doi.org/10.1017/S0953756205003497.

Edwards, D., Kenrick, P. & Dolan, L. (2018). History and contemporary significance of the Rhynie cherts – our earliest preserved terrestrial ecosystem. *Philosophical Transactions of the Royal Society of London, Series B*, **373**: article number 20160489. DOI: https://doi.org/10.1098/rstb.2016.0489.

Embley, T.M. & Hirt, R.P. (1998). Early branching eukaryotes? *Current Opinion in Genetics & Development*, **8**: 624–629. DOI: https://doi.org/10.1016/S0959-437X(98)80029-4.

Flemming, H-C. & Wingender, J. (2010). The biofilm matrix. *Nature Reviews Microbiology*, **8**: 623–633. DOI: https://doi.org/10.1038/nrmicro2415.

Gardner, A. (2014). Life, the universe and everything. *Biology & Philosophy*, **29**: 207–215. DOI: https://doi.org/10.1007/s10539-013-9417-8.

Hand, E. (2009). When Earth greened over. *Nature*, **460**: 161. DOI: https://doi.org/10.1038/460161a.

Hawksworth, D.L., Hibbett, D.S., Kirk, P.M. & Lücking, R. (2016). Proposals to permit DNA sequence data to serve as types of names of fungi. *Taxon*, **65**: 899–900. DOI: https://doi.org/10.12705/654.31.

Hawksworth, D.L. & Lücking, R. (2017). Fungal diversity revisited: 2.2 to 3.8 million species. In *The Fungal Kingdom*, ed. J. Heitman, B. Howlett, P. Crous, E. Stukenbrock, T. James & N. Gow. Washington, DC: ASM Press, pp. 79–95. DOI: https://doi.org/10.1128/microbiolspec.FUNK-0052-2016.

Hawksworth, D.L., May, T.W. & Redhead, S.A. (2017). Fungal nomenclature evolving: changes adopted by the 19th International Botanical Congress in Shenzhen 2017, and procedures for the Fungal Nomenclature Session at the 11th International Mycological Congress in Puerto Rico 2018. *IMA Fungus*, **8**(2): 211–218. DOI: https://doi.org/10.5598/imafungus.2017.08.02.01.

He, M.-Q., Zhao, R.-L., Hyde, K.D., Begerow, D., Kemler, M. and 65 others. (2019). Notes, outline and divergence times of Basidiomycota. *Fungal Diversity*, **99**: 105–367. DOI: https://doi.org/10.1007/s13225-019-00435-4.

Hibbett, D.S., Binder, M., Bischoff, J.F. *et al.* (2007). A higher-level phylogenetic classification of the Fungi. *Mycological Research*, 111: 509–547. DOI: https://doi.org/10.1016/j.mycres.2007.03.004.

Hibbett, D.S., Grimaldi, D. & Donoghue, M.J. (1995). Cretaceous mushrooms in amber. *Nature*, **377**: 487. DOI: https://doi.org/10.1038/377487a0.

Hueber, F.M. (2001). Rotted wood–alga–fungus: the history and life of *Prototaxites* Dawson 1859. *Review of Paleobotany and Palynology*, **116**: 123–148. DOI: https://doi.org/10.1016/S0034-6667(01)00058-6.

Hyde, K.D., Maharachchikumbura, S.S.N., Hongsanan, S. *et al.* (2017). The ranking of fungi: a tribute to David L. Hawksworth on his 70th birthday. *Fungal Diversity*, **84**: 1–23. DOI: https://doi.org/10.1007/s13225-017-0383-3.

James, T.Y., Kauff, F., Schoch, C.L. *et al.* (2006). Reconstructing the early evolution of Fungi using a six-gene phylogeny. *Nature*, **443**: 818–822. DOI: https://doi.org/10.1038/nature05110.

Jones, M.D.M., Richards, T.A., Hawksworth, D.L. & Bass, D. (2011). Validation and justification of the phylum name *Cryptomycota* phyl. nov. *IMA Fungus*, **2**: 173–175. DOI: https://doi.org/10.5598/imafungus.2011.02.02.08.

Keeling, P.J., Burger, G., Durnford *et al.* (2005). The tree of eukaryotes. *Trends in Ecology and Evolution*, **20**: 670–676. DOI: https://doi.org/10.1016/j.tree.2005.09.005.

Keeling, P.J, Leander, B.S. & Simpson, A. (2009). Eukaryotes. Eukaryota, organisms with nucleated cells. Version 28 October 2009. URL: http://tolweb.org/Eukaryotes/3/2009.10.28.

Knauth, L.P. & Kennedy, M.J. (2009). The late Precambrian greening of the Earth. *Nature*, **460**: 728–732. DOI: https://doi.org/10.1038/nature08213.

Krings, M., Harper, C.J. & Taylor, E.L. (2018). Fungi and fungal interactions in the Rhynie chert: a review of the evidence, with the description of *Perexiflasca tayloriana* gen. et sp. nov. *Philosophical Transactions of the Royal Society of London, Series B*, **373**: article number 20160500. DOI: https://doi.org/10.1098/rstb.2016.0500.

Margulis, L. (2004). Serial endosymbiotic theory (SET) and composite individuality. Transition from bacterial to eukaryotic genomes. *Microbiology Today*, **31**: 172–174. URL: http://www.socgenmicrobiol.org.uk/pubs/micro_today/pdf/110406.pdf.

Martin, W., Rotte, C., Hoffmeister, M. *et al.* (2003). Early cell evolution, eukaryotes, anoxia, sulfide, oxygen, fungi first (?), and a Tree of Genomes revisited. *International Union of Biochemistry and Molecular Biology: Life*, **55**: 193–204. DOI: https://doi.org/10.1080/1521654031000141231.

McCarthy, C.G.P. & Fitzpatrick, D.A. (2017). Multiple approaches to phylogenomic reconstruction of the fungal kingdom. *Advances in Genetics*, 100: 211–266. DOI: https://doi.org/10.1016/bs.adgen.2017.09.006.

McLaughlin, D.J., Hibbett, D.S., Lutzoni, F., Spatafora, J.W. & Vilgalys, R. (2009). The search for the fungal tree of life. *Trends in Microbiology*, **17**: 488–497. DOI: https://doi.org/10.1016/j.tim.2009.08.001.

McLaughlin, D.J. & Spatafora, J.W. (ed.) (2014). *The Mycota Systematics and Evolution*, VII part A, 2nd edition. Berlin: Springer-Verlag. ISBN: 978-3-642-55317-2. DOI: https://doi.org/10.1007/978-3-642-55318-9.

McLaughlin, D.J. & Spatafora, J.W. (ed.) (2015). *The Mycota Systematics and Evolution*, VII part B, 2nd edition. Berlin: Springer-Verlag. ISBN: 978-3-662-46010-8. DOI: https://doi.org/10.1007/978-3-662-46011-5.

Moore, D. (2013). *Fungal Biology in the Origin and Emergence of Life*. Cambridge, UK: Cambridge University Press. 230 pp. ISBN-10: 1107652774, ISBN-13: 978-1107652774.

Moore, D. & Novak Frazer, L. (2002). *Essential Fungal Genetics*. New York: Springer-Verlag. ISBN-10: 0387953671, ISBN-13: 978-0387953670.

NASA (2012). *Universe 101: Our Universe*. Website. https://map.gsfc.nasa.gov/universe/uni_age.html.

Penny, D. & Poole, A. (1999). The nature of the last universal common ancestor. *Current Opinion in Genetics & Development*, **9**: 672–677. DOI: https://doi.org/10.1016/S0959-437X(99)00020-9.

Philippe, H., Germot, A. & Moreira, D. (2000). The new phylogeny of eukaryotes. *Current Opinion in Genetics & Development*, **10**: 596–601. DOI: https://doi.org/10.1016/S0959-437X(00)00137-4.

Pirozynski, K.A. (1976). Fossil fungi. *Annual Review of Phytopathology*, **14**: 237–246. DOI: https://doi.org/10.1146/annurev.py.14.090176.001321.

Poinar, G. Jr (2016). Fossil fleshy fungi ('mushrooms') in amber. *Fungal Genomics & Biology*, **6**: 142–147. DOI: https://doi.org/10.4172/2165-8056.1000142.

Powell, M.J. & Letcher, P.M. (2014). Chytridiomycota, Monoblepharidomycota, and Neocallimastigomycota. In *The Mycota Systematics and Evolution*, VII part A, 2nd edition, ed. D.J. McLaughlin & J.W. Spatafora. Berlin: Springer-Verlag, pp. 141–176. ISBN: 978-3-642-55317-2. DOI: https://doi.org/10.1007/978-3-642-55318-9_6.

Redecker, D., Kodner, R. & Graham, L.E. (2000). Glomalean fungi from the Ordovician. *Science*, **289**: 1920–1921. DOI: https://doi.org/10.1126/science.289.5486.1920.

Retallack, G.J. & Landing, E. (2014). Affinities and architecture of Devonian trunks of *Prototaxites loganii*. *Mycologia*, **106**: 1143–1158. DOI: https://doi.org/10.3852/13-390.

Reynolds, N.K., Smith, M.E., Tretter, E.D. *et al.* (2017). Resolving relationships at the animal-fungal divergence: a molecular phylogenetic study of the protist trichomycetes (Ichthyosporea, Eccrinida). *Molecular Phylogenetics and Evolution*,**109**: 447–464. DOI: https://doi.org/10.1016/j.ympev.2017.02.007.

Richards, T.A., Leonard, G., Soanes, D.M. & Talbot, N.J. (2011). Gene transfer into the fungi. *Fungal Biology Reviews*, **25**: 98–110. DOI: https://doi.org/10.1016/j.fbr.2011.04.003.

Rikkinen, J., Dörfelt, H., Schmidt, A.R. & Wunderlich, J. (2003). Sooty moulds from European Tertiary amber, with notes on the systematic position of *Rosaria* ('Cyanobacteria'). *Mycological Research*, **107**: 251–256. DOI: https://doi.org/10.1017/S0953756203007330.

Robert, V., Vu, D., Amor, A. B., van de Wiele, N., Brouwer, C. and 31 others (2013). MycoBank gearing up for new horizons. *IMA fungus*, **4**: 371-379. DOI: https://doi.org/10.5598/imafungus.2013.04.02.16.

Sage, R.F., Sage, T.L. & Kocacinar, F. (2012). Photorespiration and the evolution of C4 photosynthesis. *Annual Review of Plant Biology*, **63**:19–47. DOI: https://doi.org/10.1146/annurev-arplant-042811-105511.

Schoch, C.L., Seifert, K.A., Huhndorf, S. *et al.* (2012). Nuclear ribosomal internal transcribed spacer (ITS) region as a universal DNA barcode marker for Fungi. *Proceedings of the National Academy of Sciences of the United States of America*, **109**: 6241–6246. DOI: https://doi.org/10.1073/pnas.1117018109.

Spatafora, J.W., Chang, Y., Benny, G.L. *et al.* (2016). A phylum-level phylogenetic classification of zygomycete fungi based on genome-scale data. *Mycologia*, **108**: 1028–1046. DOI: https://doi.org/10.3852/16-042.

Strobel, N. (2019). *Astronomy Notes*. Website. http://www.astronomynotes.com/index.html.

Taylor, T.N., Krings M. & Taylor, E.L. (2015). *Fossil Fungi*. San Diego, CA: Academic Press. ISBN-10: 0123877318, ISBN-13: 978-0123877314. DOI: https://doi.org/10.1016/B978-0-12-387731-4.00014-1.

Truong, C., Mujic, A.B., Healy, R. *et al.* (2017). How to know the fungi: combining field inventories and DNA-barcoding to document fungal diversity. *New Phytologist*, **214**: 913–919. DOI: https://doi.org/10.1111/nph.14509.

Vajda, V. & McLoughlin, S. (2004). Fungal proliferation at the Cretaceous–Tertiary boundary. *Science*, **303**: 1489. DOI: https://doi.org/10.1126/science.1093807.

Visscher, H., Brinkuis, H., Dilcher, D.L. *et al.*(1996). The terminal Paleozoic fungal event: evidence of terrestrial ecosystem destabilization and collapse. *Proceedings of the National Academy of Sciences of the United States of America*, **93**: 2155–2158. URL: http://www.jstor.org/stable/38482.

Wellman, C.H. & Gray, J. (2000). The microfossil record of early land plants. *Philosophical Transactions of the Royal Society of London, Series B*, **355**: 717–732. URL: http://www.jstor.org/stable/3066802.

Williams, B.P., Johnston, I.G., Covshoff, S. & Hibberd, J.M. (2013). Phenotypic landscape inference reveals multiple evolutionary paths to C_4 photosynthesis. *eLife*, **2**: e00961. DOI: https://doi.org/10.7554/eLife.00961.

Woese, C.R. (1987). Bacterial evolution. *Microbiological Reviews*, **51**: 221–271. URL: https://www.ncbi.nlm.nih.gov/pmc/articles/PMC373105/pdf/microrev00049-0051.pdf.

Woese, C.R., Kandler, O. & Wheels, M.L. (1990). Towards a natural system of organisms: proposal for the domains Archaea, Bacteria and Eucarya. *Proceedings of the National Academy of Sciences of the United States of America*, **87**: 4576–4579. URL: http://www.jstor.org/stable/2354364.

Zalasiewicz, J. & Williams, M. (2013). *The Goldilocks Planet: The 4 Billion Year Story of Earth's Climate*. Oxford UK: Oxford University Press. ISBN-10: 0199683506, ISBN-13: 978-0199683505.

CHAPTER 3

Natural Classification of Fungi

In this chapter, we will give you an overview of the organisms that make up Kingdom *Fungi*. We'll try to emphasise a more ecosystem-oriented approach because we want to avoid rambling taxonomy-driven species lists. However, there may be as many as 3.8 million species currently present on Earth (see section 1.7 above), so we need to know ***some*** by name and understand the **natural classification of fungi**. A natural classification is the arrangement of organisms into groups based on their evolutionary relationships.

To begin with, we will describe enough of the taxonomic structure to provide a foundation for understanding the breadth of the group we are studying, and the major phyla that make up the kingdom, already mentioned in section 2.5: *Cryptomycota*, *Microsporidia* (see section 17.2 and Didier *et al.*, 2014), *Chytridiomycota*, *Monoblepharidomycota*, *Neocallimastigomycota*, *Blastocladiomycota*, *Zoopagomycota*, *Mucoromycota*, *Ascomycota* and *Basidiomycota*.

We will then discuss what the word (or concept) 'species' means in fungi and the ways in which a fungal species might be defined. In the final sections of the chapter, we will examine those fungus-like organisms, which we call 'the untrue fungi', and which have some of the characteristics of fungi without being closely related in an evolutionary sense. In our final section, we draw a few key aspects about fungi in the natural environment and in their natural communities from this overview of Kingdom *Fungi* that we will develop more fully in later chapters about ecosystem mycology.

3.1 THE MEMBERS OF THE KINGDOM

Approximately 120,000 fungal species have been described to date, the majority being members of the *Ascomycota* (over 64,000 known species) and the *Basidiomycota* (over 32,000 known species). These numbers will always be approximate because about 1,000 new species are described each year (and have been for several decades), and because not all 'new' descriptions are genuinely new (some have been described before under different names). In these cases, the names are described as **synonyms** and one of them (the first published name that follows the international rules of nomenclature) is chosen as the true name for the species. The rest are reduced to synonymy, but they stay on the list and can never be used for any other species.

Because there are so many species of fungi for which no sexual stage is known, at least at the time when they are first described, a particular peculiarity of fungal taxonomy in the past has been the practice of assigning different generic and specific names to the asexual (called the **anamorph**) and sexual (the **teleomorph**) stages. When subsequent work established the connection between the two stages, two names applied to the one organism and in this case the name that was defined first took precedence. An example is that the well-known *Aspergillus nidulans* is an anamorph that has a teleomorph which was named *Emericella*. Strictly speaking, the generic name *Emericella* should have taken precedence, but the geneticists and molecular biologists who had spent their lives working with *Aspergillus nidulans* were unwilling to call their strains *Emericella nidulans*. This sort of thing can be a confusing complication to students of mycology, so you must be aware of this historical peculiarity and take it in your stride. One thing this peculiarity of giving different names to sexual and asexual stages did, of course, is increase the number of taxonomic ***names*** recorded for fungi. Fortunately, this practice ended in 2011; so, *Aspergillus* is the right name for *A. nidulans* and not *Emericella*.

In fact, around 300,000 species ***names*** exist, and this provides another opportunity for estimating how many real species we know about because major 'monographs' (the research studies in which taxonomists critically review all the species in a genus or family) suggest an average rate of synonymy of about 2.5 invalid names for each valid name. If this ratio is applied to all 300,000 names, we can estimate the upper limit of accepted known species to be about 120,000. This compares with an estimate of up to 3.8 million species currently present on Earth (Hawksworth & Lücking, 2017; see section 1.7), which leaves up to 3.6 million species of fungi still to be found and described.

Our ***Outline Classification of Fungi*** (and related organisms) is shown in Appendix 1 of this text; in this chapter, we will concentrate on the biology of the organisms classified as fungi.

3.2 THE CHYTRIDS

Members of the Phylum ***Chytridiomycota***, often referred to as **chytrid fungi** or **chytrids**, are morphologically simple organisms with a global distribution and approximately 700 described species that can be found from the tropics to the Arctic regions. Chytrids occur in aquatic environments such as streams, ponds, estuaries and marine systems, living as parasites of algae and planktonic organisms. Many chytrids, perhaps the majority, occur in terrestrial forest, agricultural and desert soils, and in acidic bogs as saprotrophs on difficult-to-digest substrata, such as pollen grains, chitin, keratin and cellulose. Some soil chytrids are obligate parasites of vascular plants.

Chytrids reproduce through the production of motile spores (**zoospores**), which are typically propelled by a single flagellum directed towards the rear. Chytrid morphology is very simple (Figure 3.1).

Eucarpic chytrids are those that consist of a sporangium and filamentous **rhizoids**, and contrast with **holocarpic** chytrids that produce thalli that are entirely converted to **sporangia** during reproduction. Zoospore-producing sporangia (zoosporangia; always the result of **asexual reproduction**) have thin walls. Resting spores may be formed sexually or asexually. They have thick cell walls and may germinate to produce a sporangium after a dormant period.

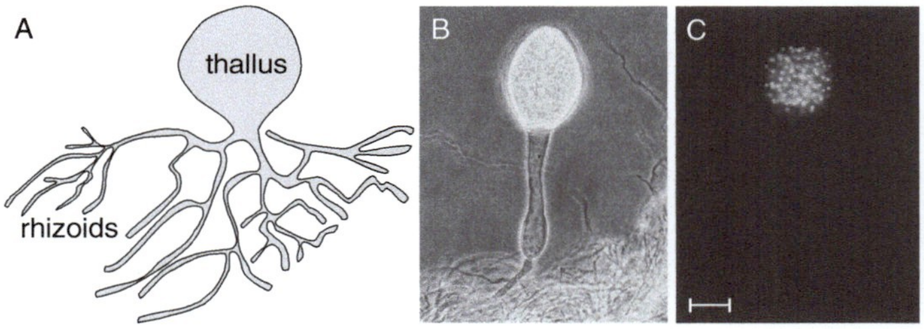

Figure 3.1. General morphology of chytrids. (A) Sketch diagram of a eucarpic chytrid thallus. The most obvious morphological feature is the thallus, the main body of the chytrid, perhaps 10 μm diameter, in which most of the cytoplasm resides and from which a system of branching elements emerges. The latter is called the rhizoidal system; the rhizoids anchor the fungus in its substratum and secrete digestive enzymes. The thallus is converted into a sporangium during reproduction, so the sporangium is sac-like and its protoplasm becomes internally divided to produce zoospores. (B) Phase contrast microphotograph of a thallus of the rumen chytrid, *Neocallimastix* sp., showing the single sporangium of the monocentric thallus and its rhizoidal system. (C) The same field of view as in B, but this time with DAPI (4′-6-diamidino-2-phenylindole) fluorescence staining. DAPI forms fluorescent complexes with natural double-stranded DNA, so very specifically stains nuclei. In image C the fluorescent staining is limited to the thallus/sporangium, showing that the rhizoids do not contain nuclei. Scale bar, 40 μm. Images B and C modified from Trinci *et al.* (1994); reproduced with permission of Elsevier. See Figure 3.2 for more images illustrating chytrid diversity.

In **monocentric chytrids**, the thallus produces a single sporangium, whereas those described as **polycentric** are individual chytrids in which several sporangia form on a network of rhizoids termed a **rhizomycelium**. Other features used for taxonomy include whether the sporangium has a lid-like operculum that opens to release zoospores, and whether there is (subsporangial) swelling just below the sporangium (called the apophysis). Chytrids have also been distinguished on the basis of whether they grow on (**epibiotic**) or within (**endobiotic**) their substratum.

Zoospores are usually 2–10 μm diameter, lack a wall, contain a single nucleus and are propelled by a single posterior whiplash flagellum (though some anaerobic **rumen chytrids**, closely related but classified in a separate phylum, the *Neocallimastigomycota*, have multiflagellate zoospores) (Figure 3.2).

Names to look out for among the *Chytridiomycota* include *Rhizophydium*, which is the largest genus in the *Chytridiales*, its members having one reproductive centre per thallus (= monocentric), have rhizoids (= eucarpic) and develop endogenously (the zoospore cyst enlarges to form the zoosporangium). Species of *Rhizophydium* release zoospores through one or more pores or, in some species, a large part of the sporangial wall melts away to release zoospores. *Rhizophydium sphaerotheca* develops on pollen grains and *Rhizophydium globosum* develops on algae. **Plant parasites** include *Synchytrium* on potatoes and *Olpidium* on cucurbits (including cucumber, squashes, pumpkins, gherkins, melons of all kinds and various gourds).

The only known **chytrid parasite** of vertebrates is *Batrachochytrium dendrobatidis*, which causes a fatal epidermal infection (chytridiomycosis) of amphibian species and has been responsible for mass mortality and major population declines of amphibians around the world (see section 17.7). *Batrachochytrium* grows on the keratinised skin of adult frogs and keratinised mouth parts of tadpoles. The resulting epidermal hyperplasia (increase in the number of the cells as a chronic inflammatory response) impairs cutaneous respiration and osmoregulation causing widespread fatalities in adults. Toxin production has not been implicated.

Although most chytrids are infrequently encountered and scarce to rare, a few species are relatively common, even abundant, in most freshwater and soil habitats (e.g. species of *Rhizophydium, Rhizophlyctis, Phlyctochytrium* and *Chytriomyces* are readily isolated by flooding soil samples with water and 'baiting' with cellulose, chitin, pollen grains or hemp seeds). In fact, chytrids are very important components of freshwater ecosystems, and in their review of environmental factors that affect growth and population composition of chytrids in aquatic

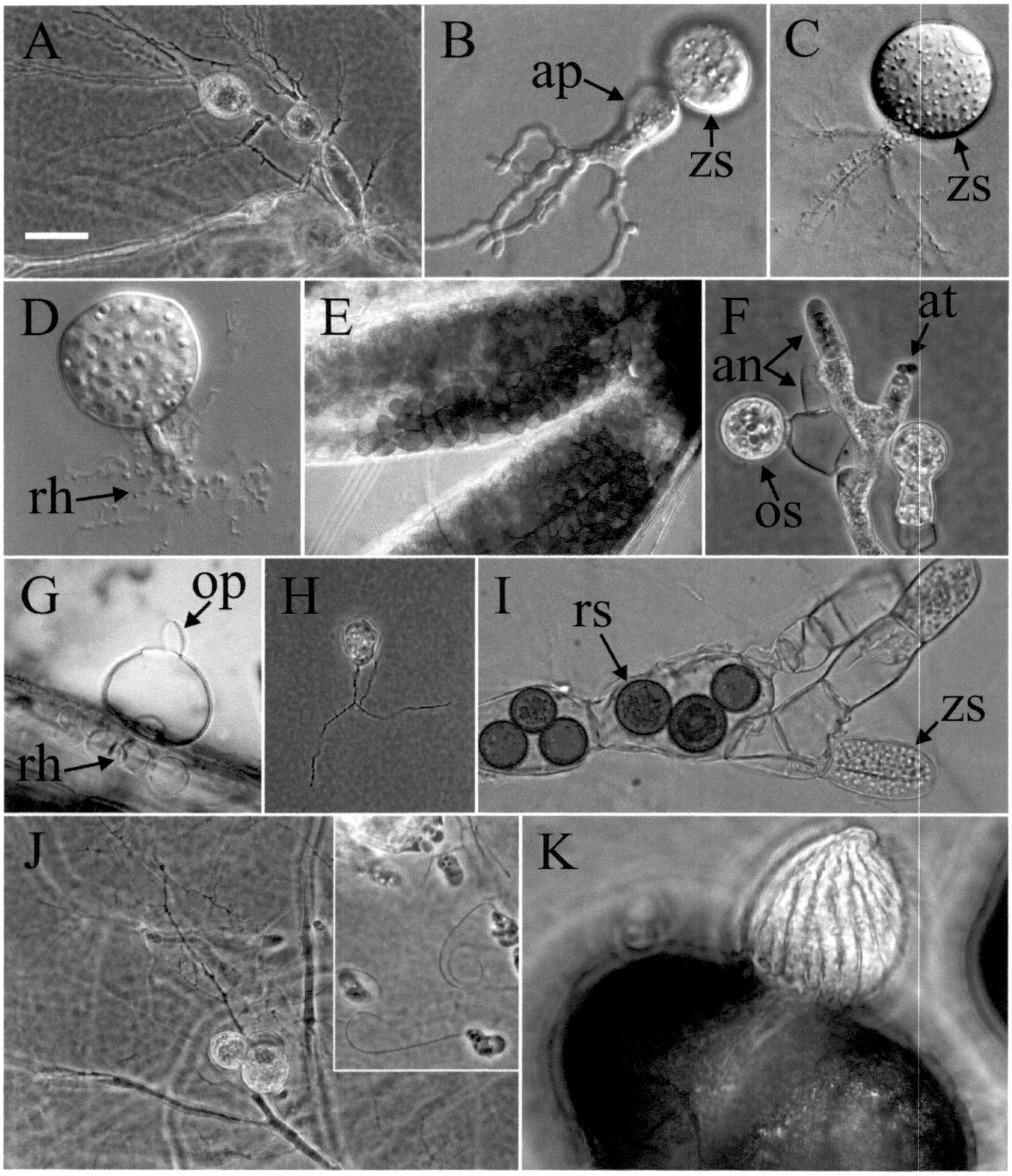

Figure 3.2. Light micrographs of representative chytrids to give some indication of their biodiversity. (A) *Catenomyces persicinus*, polycentric thallus with apophysate (swollen) rhizoidal axis and intercalary zoosporangia. (B) *Spizellomyces plurigibbosus*, monocentric, inoperculate zoosporangium (zs), with swollen, apophysate rhizoidal axis (ap) and branched rhizoids that are blunt at the tips. (C) *Chytriomyces hyalinus*, monocentric, operculate zoosporangium (zs) with long branched rhizoids that taper to less than 0.5 μm at the tips. (D) *Terramyces subangulosum*, monocentric, inoperculate sporangium with a thick rhizoidal axis, and densely branched rhizoids (rh) that taper to less than 0.5 μm at tips. (E) *Coelomomyces stegomyiae*, elliptical resting spores inside the anal gills of its mosquito host. (F) *Monoblepharis polymorpha*, mature zygote or oospore (os), empty and mature antheridia (an) and male gametes (antherozoids, at) emerging from antheridium. (G) *Catenochytridium* sp., monocentric, operculate (op) zoosporangium with catenulate (chain-like) rhizoids (rh). (H) *Lobulomyces angularis*, monocentric, operculate sporangium with threadlike rhizoids that branch several μm from the sporangial base. (I) *Rozella allomycis* parasitising hyphae of *Allomyces* (another chytrid, though now placed in the phylum *Cryptomycota* (Jones *et al*., 2011), is the earliest diverging lineage of Fungi followed in phylogenetic time by *Chytridiomycota* and *Blastocladiomycota*). The parasite grows inside the host and causes it to produce hypertrophied, highly septate cells within which the parasite forms thick-walled resting spores (rs) or unwalled zoosporangia (zs) that use the host's cell wall as its own. (J) The polycentric thallus of *Polychytrium aggregatum* with at least two sporangia; the inset shows the zoospores of *Polychytrium aggregatum*. (K) *Blyttiomyces helicus* growing on pollen grain and forming an epibiotic, inoperculate sporangium with distinct helical pattern. Scale: approximate sizes indicated by scale bar in A, 10 μm. Images by Dr Timothy Y. James, Department of Ecology and Evolutionary Biology, University of Michigan, Dr Joyce E. Longcore, School of Biology and Ecology, University of Maine and Dr Peter M. Letcher, Department of Biological Sciences, University of Alabama. Figure modified from James *et al*. (2006) using graphic files kindly supplied by Dr T. Y. James. Reprinted with permission from Mycologia. ©The Mycological Society of America.

habitats, Gleason *et al.* (2008) identify five roles for chytrids in food web dynamics:

- chytrid zoospores are a good food source for zooplankton;
- chytrids decompose particulate organic matter;
- chytrids are parasites of aquatic plants;
- chytrids are parasites of aquatic animals;
- chytrids convert inorganic compounds into organic compounds.

Anaerobic chytrids are probably the most numerous and the most economically important members of the group, because they occur in the rumen and hindgut of many larger mammalian herbivores, including all farmed animals (see the section on ***Neocallimastigomycota*** below).

There has always been debate about whether chytrids are 'true' fungi, but they have chitin in their cell walls, they use the α-aminoadipic acid pathway to synthesise lysine and use glycogen as storage carbohydrates, which are all characteristics of true fungi. For many years the chytrids have been treated as **true fungi**, but very primitive ones because of their simple morphology and their dependence on zoospores for sexual reproduction. Recent phylogenetic evidence using mitochondrial genes, nuclear rDNA sequences and whole genomes conclusively demonstrated that chytrids are true fungi and that they occupy a basal position in the fungal phylogenetic tree.

We are basing our descriptions of the group on the Assembling the Fungal Tree of Life (AFTOL) summary of chytrid phylogeny (James *et al.*, 2006) and the subsequent overall classification of fungi published by Hibbett *et al.* (2007). Prior to AFTOL, and based on life cycle and mode of reproduction, together with gross morphology and zoospore ultrastructure, the *Chytridiomycota* were divided into five orders:

- *Chytridiales*: characterised by zygotic meiosis in which haploid individuals fuse to form a diploid zygote which immediately undergoes meiosis;
- *Spizellomycetales*: separated from the *Chytridiales* on the basis of distinctive ultrastructure;
- *Monoblepharidales*: which are oogamous;
- *Neocallimastigales*: anaerobic rumen symbionts; now raised to phylum status as the *Neocallimastigomycota* (see section 3.3);
- *Blastocladiales*: which have sporic meiosis, in which meiosis results in the production of haploid spores, and an alternation of 'sporophytic' and 'gametophytic' generations; now raised to phylum status as the *Blastocladiomycota* (see section 3.4).

The AFTOL study identified four major lineages of chytrid fungi (that is, the *Chytridiomycota* as described above is not monophyletic). The *Blastocladiales* and *Neocallimastigales* are both phylogenetically distinct from other chytrids and were elevated to the level of phylum (***Blastocladiomycota*** and ***Neocallimastigomycota***).

The *Neocallimastigomycota* emerges as the **earliest diverging lineage**, originating long, long before its current hosts, ruminant herbivores, appeared on the scene. Removing the *Neocallimastigomycota* and *Blastocladiomycota* leaves four orders within *Chytridiomycota*: *Chytridiales*, *Spizellomycetales*, *Rhizophydiales* and *Monoblepharidales*. This summary may need revision as further molecular analyses resolve other phylogenetically distinct clades of flagellated fungi (James *et al.*, 2006; Powell & Letcher, 2014).

3.3 *NEOCALLIMASTIGOMYCOTA*

The **anaerobic chytrids** occur in terrestrial and aquatic anaerobic environments and are outstandingly important economically because they are crucial to the microbiome within the rumen and hindgut of most large mammalian herbivores, including all farmed animals (Trinci *et al.*, 1994). Right up to the 1980s, it was believed that obligately anaerobic fungi did not exist but these chytrids are truly **obligate anaerobes** that lack mitochondria but contain hydrogenosomes instead.

Obligately anaerobic chytrids have a crucial role in the primary colonisation and enzymic degradation of lignocellulose in plant materials eaten by herbivores and have therefore been integral to the evolution of herbivores, and to the prosperity of animal husbandry since humans first domesticated animals (see section 15.5). Although they are morphologically similar to other chytrids, differences are sufficient for them to be placed in their own phylum, called *Neocallimastigomycota*, which is a focus for extensive research (Hibbett *et al.*, 2007; Gruninger *et al.*, 2014; Powell & Letcher, 2014; Edwards *et al.*, 2017; Wang *et al.*, 2017).

These chytrids are potent producers of the enzymes needed to degrade cellulose. Their own carbon metabolism relies on fermentation of glucose to acetate, lactate, ethanol and hydrogen. They possess an organelle called a **hydrogenosome** that generates adenosine triphosphate (ATP) and appears to be a degenerate mitochondrion lacking a genome (Trinci *et al.*, 1994; van der Giezen, 2002). Figure 3.3 gives a little further information about the metabolism of anaerobic chytrids.

Anaerobic chytrids may be monocentric or polycentric, and zoospores may be multiflagellate or uniflagellate. About eight genera (comprising about 20 species) have been described in the single order, *Neocallimastigales*. Examples are: *Neocallimastix frontalis*, the original isolate from the domestic cow and notable for multiflagellate zoospores; *Orpinomyces*, which occurs widely in cattle; and *Piromyces*, which has been isolated from horses and elephants.

Rumen chytrid zoospores encyst on plant material in the animal's rumen and intestine, forming a thallus with a well-developed rhizoidal system that penetrates the plant material. These chytrids are potent producers of enzymes needed to degrade cellulose (animals do not produce their own cellulose-degrading enzymes) so they are crucial to the herbivore's ability to digest

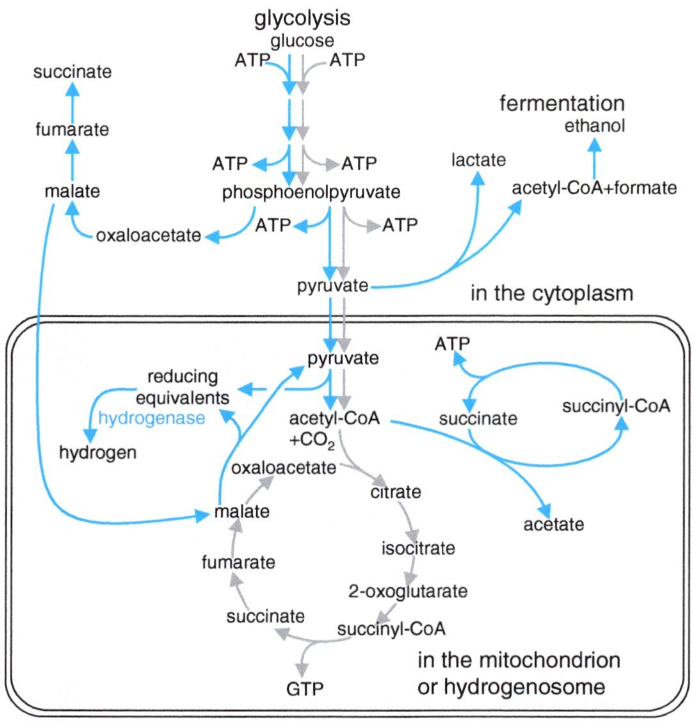

Figure 3.3. Generalised pathway of aerobic carbohydrate degradation in eukaryotes compared to that of the anaerobic fungus *Neocallimastix frontalis*. Aerobic degradation is indicated by grey arrows and the *Neocallimastix* anaerobic route by blue arrows. In the mitochondrial pathway pyruvate is reductively decarboxylated to acetyl-CoA by the pyruvate dehydrogenase complex. In the hydrogenosome pyruvate: ferredoxin oxidoreductase catalyses the oxidative decarboxylation of pyruvate to acetyl-CoA and CO_2 and generates reducing equivalents (chemicals that transfer the equivalent of one electron in redox reactions). See Figures 11.7 for more details of glycolysis and 11.9 for more details of the tricarboxylic acid (TCA) cycle. Modified from van der Giezen (2002).

its food. The chytrids pass from mother to offspring, probably through licking or faecal contamination of feed. No sexual stage is known.

3.4 *BLASTOCLADIOMYCOTA*

Members of this phylum, which you will find called *Blastocladiales* in older textbooks, are saprotrophs as well as parasites of fungi, algae, plants and invertebrates. They may be facultatively anerobic in oxygen-depleted environments. The saprotrophs are easily found on decaying fruits and plant litter. All members of this phylum (14 genera) have zoospores with a distinct ribosome-filled cap around the nucleus (James *et al.*, 2014).

The thallus may be monocentric or polycentric and becomes mycelial in *Allomyces*. Other representative genera are: *Physoderma, Blastocladiella* and *Coelomomyces*. *Physoderma* spp. are parasitic on higher plants. *Coelomomyces* is an obligate endoparasite of insects with alternating sporangia and gametangia stages in mosquito larvae and copepod (fish lice) hosts, respectively.

Characteristically, the *Blastocladiomycota* have life cycles with what is described as a **sporic meiosis**; that is, meiosis results in the production of haploid spores that can develop directly into a new, but now haploid, individual. This results in a regular **alternation of generations** between haploid gametothallus and diploid sporothallus individuals (Figure 3.4). In general terms, a multicellular diploid adult organism (the sporothallus) produces a sporangium within which meiosis occurs. Meiosis typically produces four haploid meiotic products, which are zoospores. Under proper conditions these germinate and develop into a multicellular haploid gametothallus organism. This differentiates gametangia that produce gametes by mitosis. Gametangia and gametes are both haploid. Gametes find each other, unite, and produce a diploid zygote that matures into a young diploid sporothallus to complete the life history.

Allomyces is **anisogamous**; female gametes are colourless and sluggish, male gametes are orange (contain α-carotene) and very active, swimming in arcs interspersed with a jerky tumbling movement. This activity is an aspect of the mechanism that allows male gametes to find female gametes, which they do because female gametes produce a **chemical attractant**.

This is a hormone called **sirenin** (Figure 3.5), which is a sesquiterpene that consists of a cyclopropyl ring attached to an isohexenyl side chain ($C_{15}H_{24}O_2$ with a molecular mass of 236).

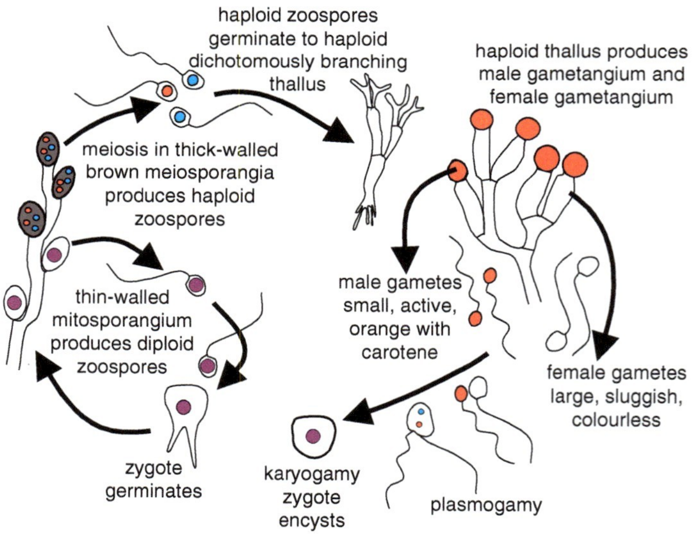

Figure 3.4. The *Allomyces* life cycle. *Allomyces* produces a branched thallus with marked polarity (basal rhizoids and apical sporangia). Sexual reproduction by anisoplanogametes; female gametes are colourless (produced in oogonia) and twice the size of the orange male gametes (produced in antheridia). Two different thalli are formed: haploid gametothallus and diploid sporothallus (the diploid phase is prolonged).

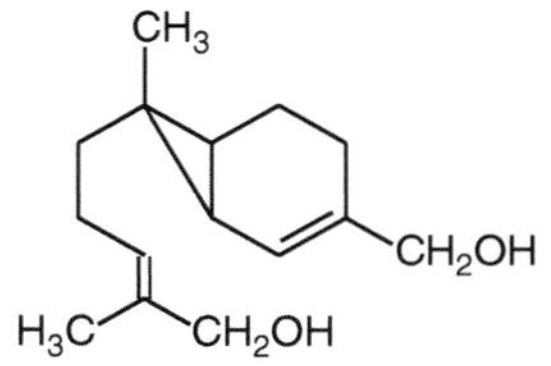

Figure 3.5. Molecular structure of the *Allomyces* male-attracting pheromone sirenin. Biological activity requires a terminal hydroxymethyl group on the side chain and a hydrophobic group must be present at the other end of the sirenin molecule (Pommerville *et al.*, 1990). The chemical gets its name from the Sirens of Greek mythology.

To synthesise this molecule the female gamete converts **acetyl-CoA** to **farnesyl pyrophosphate**, which in turn is converted to sirenin. So much acetyl-CoA is diverted to form sirenin that there is less ATP available in the mitochondrion for flagella motion, which is why the female gametes are not active.

However, the male gametes are active, and they swim in arcs; they have membrane receptors that respond to sirenin concentration. Sirenin stimulates the influx of calcium ions (Ca^{2+}) into the sperm cytoplasm and the physiological response is to reduce the length of the arc in the swimming of the male gametes; that is, the pheromone influences the frequency of directional changes and the duration of the chemotactic run, resulting in movement towards the source of the **pheromone** (the female).

As the male gamete nears the highest concentration of sirenin, the arcs disappear; however, the tumbling motion becomes exaggerated. Thus, the male gametes are very erratic and active near the female gametes and this response ensures **syngamy**. The resultant zygote is a motile zoospore (with tinsel-type flagellum) that settles down in the environment to grow into a diploid thallus.

Sirenin is, therefore, a **sex pheromone** (a hormone produced by one partner to elicit a sexual response in the other). Only male gametes respond to sirenin, to which they are highly sensitive (sensitivity threshold about 1×10^{-10} M). In fact, their sensitivity of response to sirenin (they react to as little as 20 pg ml^{-1}) is 20 million times greater than their response to nutrients (400 μg ml^{-1}) (pg = 10^{-12} g, μg = 10^{-6} g). The importance of this very sensitive hormonal system in *Allomyces* is that it enables gametes to find each other in an aquatic ecosystem (preventing gamete loss or wastage) and by so doing increases the chance of successful sexual reproduction.

Besides sirenin, the sperm cells of *Allomyces macrogynus* produce a female attractant, called **parisin**. Some general features of this molecule suggest it may be similar structurally to sirenin in being a terpene, but the molecular nature of parisin and its effect on female gametes have not been completely resolved.

We bring this story to your attention now to make an important point about the general biology of fungi, which is that even these 'primitive' organisms have evolved a precise and efficient **cell targeting system**. This clearly comprises a very specific chemical attractant produced by one cell and a very exact receptor of that hormone in the other cell, which is linked to an intracellular signalling cascade that amplifies the signal to an extent that makes the reception process exquisitely sensitive to the hormone. Further discussion of pheromones in fungi can be found in Sections 6.1.4–6.1.6 in the book *Fungal Morphogenesis* (Moore, 1998). The point to note from this discussion is that fungi produce a full chemical spectrum of hormones: terpenoid, sterol and peptide hormones. Just like animals.

Another example that reveals an important truth about fungal biology is found in the way organisms like *Blastocladiella* make their zoospores. *Blastocladiella* has been used for extensive research on reproductive physiology, biochemistry and cell biology, and use of the electron microscope to examine the ultrastructure of zoospore formation revealed a unique feature of fungal biology.

To emphasise the significance of this, let's carry out the thought experiment of working out what would happen if these fungi were either animals or plants. The situation is that we are converting the chytrid thallus, a single sac-like cell, into a sporangium. Initially there is a single nucleus, but this will undergo several mitotic divisions so that the volume of the sporangium can be subdivided into many zoospores, each of which will have a single mitotically produced nucleus. How will that subdivision be managed?

If *Blastocladiella* was an animal, then at each division the dividing cell would become constricted at the equator of the mitotic spindle and two daughter cells would be produced as a result of the cleavage of the mother cell. Through successive rounds of mitosis, more and more cells would be produced; just like a developing animal embryo.

If *Blastocladiella* was a plant, then at each nuclear division a daughter cell wall would be formed across the equator of the mitotic division spindle. Daughter cells would then be successively halved in size (but doubled in number) as each round of mitosis occurred.

But *Blastocladiella* is neither animal nor plant, and it does neither of these things. Instead, *Blastocladiella* uses a uniquely fungal mechanism. Its zoospores are formed by cleavage of the multinucleate protoplasm in the zoosporangium, yes, but this happens as masses of **cytoplasmic vesicles** fuse to one another to create the borders between adjacent zoospores (Figure 3.6). We can do no better than quote the original description (Lessie & Lovett, 1968):

> Soon after the beginning of flagella formation it is possible to find early stages of 'cleavage furrow' formation … . This process … involves the fusion of many small vesicles … cleavage vesicle fusion results in progressive expansion of the primary cleavage furrows and it appears that this activity is simultaneously initiated at many points. Occasionally vesicles can be found in somewhat linear arrangements

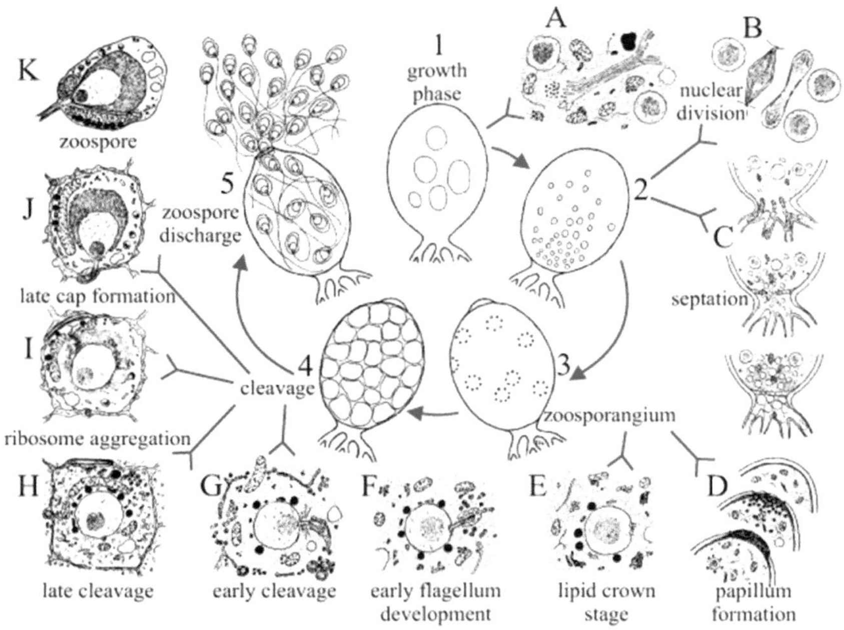

Figure 3.6. The process of sporangium formation and zoospore differentiation in *Blastocladiella*. The inner diagrams illustrate the changing appearance of sporangia during differentiation, as seen in the light microscope. The outer series summarises the intracellular changes observed in electronmicrographs. Development progresses clockwise in this figure and the experiments used synchronous cultures to determine the timing of these events as follows: (1) 15 h 30 min after the start of cultivation (zoospores germinate synchronously on transfer to a medium containing 10 mM $MgCl_2$ + 50 mM KCl; this transfer = time zero); (2) 17 h 30 min; (3) 18 h; (4) 18 h 40 min; (5) 19 h to 19 h 13 min. Intracellular events were timed as follows: (A) 15 h 30 min; (B) 16 h 30 min to 17h 30 min; (C) 16 h 30 min to 17 h 30 min; (D) 17 to 18 h; (E) 18 h; (F) 18 h 10 min; (G) 18 h 20 min; (H) 18 h 40 min; (I) 18 h 50 min; (J) 18 h 50 min to 19 h; (K) 19 h 15 min. Modified from the original drawing by James S. Lovett in Lessie and Lovett (1968), which illustrates all of these features with electronmicrographs, though it does refer to the *Blastocladiella* thalli as both 'fungus' and 'plant', something which would not be done now. Reproduced with permission of the Botanical Society of America.

> over a short distance. They more often occur in less orderly clusters and fuse in irregular ring-shaped patterns lying roughly in the plane of the developing cleavage furrow. The frequent occurrence of cytoplasmic peninsulas surrounded by U-shaped areas of cleavage vesicle suggests that many of the rings may in fact be short cylinders; if so, the closure and interconnection of the rings may be irregular and only gradually assume the form of a regular furrow. .... The cleavage furrows also fuse with the earlier formed vesicles surrounding the flagella with the result that these finally lie within the cleavage furrows and outside of the uninucleate blocks of cytoplasm delineated by the [newly formed] membrane system.

This remarkably precise zoospore generating pattern is repeated throughout the chytrids, and indeed ***throughout the fungi***. Compare the description quoted above with this description of sporogenesis in the **mucoraceous** (***terrestrial***) fungus *Gilbertella persicaria* (Bracker, 1968):

> During cleavage, the principal structural changes involve pattern transformations of protoplasmic membranes ... small vesicles are formed, apparently from special cisternae [of the endoplasmic reticulum]. The disappearance of these initial vesicles coincides with the appearance of cleavage vesicles ... distinguished by the presence of granules on the inner surface of the vesicle membrane ... Cleavage is initiated endogenously by the coalescence of cleavage vesicles to form a ramifying tubular cleavage apparatus. The cleavage apparatus demarcates the boundaries of potential spore initials. Lateral expansion of elements of the cleavage apparatus results in furrow-like configurations which converge to cut out spore initials as independent cells. The cleavage membrane is transformed to the plasma membrane of spore initials during late cleavage ... . The marker granules that were present around the periphery of the cleavage vesicles are found on the outer surfaces of spore plasma membranes after cleavage. The granules fuse to form a continuous spore envelope, and subsequently the spore wall is laid down centripetally. Thus, the envelope becomes the outermost spore wall layer ...

The process described here has been called **'free cell formation'** and for **ascospore** formation has also been summarised in a similar way (Reeves, 1967):

> A summary of the main points of free cell formation is as follows: 1. in the 8-nucleate ascus each of the haploid

nuclei forms a beak with a persistent central-body and astral rays at the tip of the beak; 2. the astral rays swing outward and down and form a thin membrane which cuts out the young spore; 3. the membrane around each spore separates the sporoplasm and included nucleus, leaving the epiplasm in the ascus ...

We are emphasising this point because it makes the general rule that where a volume of cytoplasm needs to be subdivided in fungi, **the mechanism depends on the organised distribution of cytoplasmic microvesicles**; the microvesicles then fuse together to create the separation of the cytoplasm. ***This is the way the fungi do it*** (and a similar cleavage system produces zoospores in sporangia of the fungus-like *Oomycota*), so note well this major difference from plants (no cross-walls are formed) and animals (there is no constrictive cell cleavage).

We find this mechanism to be remarkable and worthy of emphasis because it raises so many questions about the molecular mechanism(s) involved in determining ***how*** the cytoplasmic domains contributing to each individual spore are defined. We have chosen to illustrate the point with quotations from papers published at about the same time, in the late 1960s, to illustrate another point that we find remarkable: ***which is that we cannot describe the mechanism(s) in much more detail 50 years later, as this quotation reveals*** (Tehler *et al.*, 2003):

> Free cell formation is generally considered a specific feature of the Ascomycotina although it is evidently shared or partly shared with the Basidiomycotina. Early stages of basidiospore development follow the same general pattern as that of the free cell formation process in the Ascomycotina: the haploid nuclei become free in the cytoplasm and develop into individual cells together with part of the plasma from the mother cell ... The Basidiomycotina are specialised by way of their nuclei and part of the plasma, which are forced to migrate by a vacuolation process, through a sterigma into a special structure formed by the sporangium wall, which will be cut off from the basidium and in which the spore formation is completed.

It is a major cause of dissatisfaction with twenty-first century mycology that the molecular mechanism(s) involved in determining such crucial aspects of the unique cell biology of fungi are still unknown.

3.5 THE TRADITIONAL ZYGOMYCETES

The thousand or so species in this group generally have a **multinucleate mycelium** with little or no septation (i.e. coenocytic) that produces asexual spores in a mitosporangium (a sporangium in which spores are produced by mitosis). But the characteristic feature of the informal 'Phylum' **zygomycetes** is that they form sexual spores in a zygosporangium.

The name is derived from the way in which they reproduce sexually. Zygos is Greek for 'joining' or 'a yoke' (the wooden crosspiece bound to the necks of a pair of oxen or horses that is then connected to the load they are pulling). So zygomycetes reproduce through the fusion or conjugation of two hyphal branches (= **gametangia**) to form a zygote which eventually gives rise to the **zygosporangium** when it produces spores.

The gametangia arise from hyphae of a single mycelium in **homothallic** species, or from different but sexually compatible mycelia in heterothallic species. Zygosporangia usually have thickened walls and act as resting spores; they are notoriously difficult to germinate in the laboratory. The **zygospore** wall contains **sporopollenin**, which is a complex, highly cross-linked polymer also found in the exine (outer wall layer) of pollen grains. Its presence, together with melanin, helps to explain the long-lived nature of zygospores. Long-lived resistant spores spread the organism in time, to survive periodic adverse environmental conditions; while non-dormant spores (dispersal spores) spread the organism in space, to colonise new territory.

These fungi are ecologically very diverse, very widely distributed and very common, though frequently overlooked. Some of this diversity is illustrated in Figure 3.7.

Clearly, this group of zygomycetes is a primitive and early diverging lineage of Kingdom *Fungi*. Primitive features are that they have coenocytic aseptate hyphae for all or part of their life cycle and they are unable to make tissues or complex *sporophore* structures. They may be primitive, but they are certainly successful, and you are never far away from a member of this group. In the traditional classification, some organisms were included that are more properly placed with the slime moulds (e.g. Amoebidiales and Eccrinales), protists that separated from the true fungi when animals and fungi diverged from the opisthokont lineage. The problem with the **traditional 'Phylum Zygomycota'** is that it is **polyphyletic** and cannot retain rank as a natural phylum. In the 2007 classification (Hibbett *et al.*, 2007), the name remains undefined leaving the following subphyla in an uncertain position (Benny *et al.*, 2014):

- ***Mucoromycotina*** (containing Orders *Mucorales*, *Endogonales* and *Mortierellales*);
- ***Entomophthoromycotina*** (Order *Entomophthorales*);
- ***Zoopagomycotina*** (Order *Zoopagales*);
- ***Kickxellomycotina*** (Orders *Kickxellales*, *Dimargaritales*, *Harpellales* and *Asellariales*).

Spatafora *et al.* (2016) have formally redefined the group using phylogenetic analyses of a genome-scale data set for 46 taxa, including 25 zygomycetes and 192 proteins. They abandon use of the name 'Zygomycota' to avoid any further confusion and instead recognise that zygomycetes comprise ***two major clades***, leading them to propose a formal phylogenetic classification comprising the 2 phyla ***Zoopagomycota*** and ***Mucoromycota***, 6 subphyla, 4 classes and 16 orders.

Figure 3.7. Scanning electron micrographs of zygomycete fungi illustrating morphological diversity of sexual and asexual reproductive structures. (A–E) *Mucorales* (*Mucoromycota*). Zygospores of (A) *Cokeromyces recurvatus*, (B) *Cunninghamella homothallicus*, (C) *Radiomyces spectabilis* and (D) *Absidia spinosa*. (E) *Hesseltinella vesiculosa* sporangia. (F–G) *Mortierellales* (*Mucoromycota*). (F) *Mortierella* (*Gamsiella*) *multidivaricata* chlamydospore. (G) *Lobosporangium transversalis* sporangia borne on arachnoid mycelium. (H–J) *Zoopagales* (*Zoopagomycota*). (H) *Piptocephalis cormbifera* immature sporangia. (I) *Syncephalis cornu* sporophore bearing senescent sporangia with uniseriate sporangiospores. (J) *Rhopalomyces elegans* fertile vesicle with monosporous sporangia. (K) *Basidiobolus ranarum* monosporous sporangium. (L) *Dimargaritales* (*Zoopagomycota*), *Dispira cornuta* two-spored sporangia. (M–O) *Kickxellales* (*Zoopagomycota*), monosporous sporangia of (M) *Spiromyces minutus*, (N) *Linderina pennispora* and (O) *Kickxella alabastrina*. Scale bars: A–F, I–O, 10 μm; G, H, 20 μm. Photographs by Dr Kerry O'Donnell, United States Department of Agriculture, Agricultural Research Service, Peoria, Illinois. Plate modified from White *et al.* (2006) using graphic files kindly supplied by Dr Merlin White, Boise State University, Idaho, USA. Reprinted with permission from Mycologia. ©The Mycological Society of America.

Zoopagomycota comprises the subphyla ***Entomophthoromycotina***, ***Kickxellomycotina*** and ***Zoopagomycotina***. It constitutes the earliest diverging lineage of zygomycetes and contains species that are primarily parasites and pathogens of small animals (e.g. amoeba, insects, etc.) and other fungi, that is, mycoparasites.

Mucoromycota comprises subphyla ***Glomeromycotina*** (see section 3.6), ***Mortierellomycotina*** and ***Mucoromycotina*** and is a sister phylum to the *Dikarya* (*Ascomycota* and *Basidiomycota*). It is the more derived clade of zygomycetes and mainly consists of mycorrhizal fungi, root endophytes and decomposers of plant material.

What follows are edited and summarised notes from the phylogenetic commentaries given by Spatafora *et al.* (2016); you should refer to the original for the formal referencing of synonyms, authorities and typifications.

In Phylum ***Zoopagomycota***, sexual reproduction, where known, involves the production of zygospores by gametangial conjugation. Morphologies associated with asexual reproductive states include sporangia, merosporangia, conidia and chlamydospores. *Zoopagomycota* is the earliest diverging clade of zygomycetous fungi. It comprises three subphyla in which associations with animals (e.g. pathogens, commensals, mutualists) form a common ecological theme, although species from several lineages are mycoparasites (e.g. *Syncephalis, Piptocephalis* and *Dimargaritales*, which are obligate mycoparasites). Although some of the fungi in *Zoopagomycota* can be maintained in axenic culture, most species of *Zoopagomycota* are most frequently observed growing in association with a host organism and are more difficult to maintain in pure culture than species of *Mucoromycota*. Haustoria are produced by some of the animal pathogens and mycoparasites. *Zoopagomycota* hyphae may be compartmentalised by septa that may be complete or uniperforate with electron-opaque lenticular plugs in the latter. Zygospore formation typically involves modified hyphal tips, thallus cells or hyphal bodies (yeast-like cells) that function as gametangia.

- Subphylum: ***Entomophthoromycotina***, includes three classes and three orders of saprobic and insect pathogenic fungi. The thallus may consist of coenocytic or septate hyphae, which may fragment to form hyphal bodies, or it may comprise only hyphal bodies. Asexual reproduction is by conidiogenesis from branched or unbranched conidiophores; primary conidia are forcibly discharged, and secondary conidia are either forcibly or passively released. Sexual reproduction involves the formation of either zygospores by gametangial copulation, involving hyphal compartments or hyphal bodies. Classes: *Basidiobolomycetes, Entomophthoromycetes, Neozygitomycetes*. Orders: *Basidiobolales, Entomophthorales* (parasites of small animals, including arthropods, rotifers and even amoebae); *Neozygitales*. The words 'entomo' + 'phthora' mean 'insect destroyer' and *Entomophthora* is the generic name for fungi that attack and kill flies and many other insects with two wings (= *Diptera*). *Entomophthora muscae*, attacks houseflies. The fungus grows inside the body of the insect. The mycelium of the fungus grows into the region of the fly's brain that controls behaviour, inducing the fly to land on a nearby surface and crawl up as high as possible. Eventually the hyphae of the fungus grow throughout the body of the fly and it dies. The mycelium then proliferates to the point where it pushes the abdominal segments apart and bursts through to give the fly a banded appearance; the bands being tight columns of cells that produce and shoot off the spores that eventually form a white halo encircling the body of the dead fly, waiting to infect the next victim.
- Subphylum: ***Kickxellomycotina***, species may be saprobes, mycoparasites and symbionts of insects; the latter includes ***Harpellales*** that are typically found within the hindguts of aquatic life history stages. Mycelium is regularly divided into compartments by bifurcate septa that often have lenticular occlusions. Sexual reproduction involves the formation of variously shaped zygospores by gametangial conjugation of relatively undifferentiated sexual hyphal compartments. Sporophores may be produced from septate, simple or branched somatic hyphae. Asexual reproduction involves the production of single- or multispored merosporangia arising from a specialised vesicle (sporocladium), sporiferous branchlets or an undifferentiated sporophore apex. Orders: *Asellariales, Dimargaritales, Harpellales, Kickxellales*. The *Harpellales* and *Asellariales* are specialised for attachment to the gut wall of arthropods and are obligate endosymbionts. They are so widespread that just about every arthropod that crawls past you will carry these fungi within (see Figure 3.8).
- Subphylum: ***Zoopagomycotina***, includes mycoparasites and predators or parasites of small invertebrates and amoebae. The hyphal diameter is characteristically narrow in thalli that are branched or unbranched; sometimes specialised haustoria are produced in association with hosts. Only a handful of species have been successfully maintained in axenic culture. Sexual reproduction, where known, is by gametangial conjugation, forming globose zygospores on suspensor cells. Asexual reproduction is by arthrospores, chlamydospores, conidia or multispored merosporangia that may be simple or branched.

Phylum ***Mucoromycota*** of Spatafora *et al.* (2016) contains *Mucoromycotina, Mortierellomycotina* and *Glomeromycotina*. Characters associated with sexual reproductive states, where known, include zygospore production by gametangial conjugation. Asexual reproductive states can involve chlamydospores and spores produced in sporangia and sporangioles. *Mucoromycota* shares a most recent common ancestor with *Dikarya* and is characterised by plant symbionts, decomposers of plant debris, plant pathogens and only rare ecological interactions with animals (primarily opportunistic infections). Zygospores tend to be globose, smooth or ornamented, and produced on opposed or apposed suspensor cells with or without appendages. Asexual reproduction typically involves the production of sporangiospores in sporangia or sporangioles or chlamydospores. Hyphae tend to be large diameter and coenocytic except for the delimitation of reproductive structures by adventitious septa.

- Subphylum: ***Glomeromycotina*** contains *Archaeosporales, Diversisporales, Glomerales* and *Paraglomerales* (Redecker & Schüßler, 2014). Sexual reproduction is unknown and asexual reproduction is by specialised spores that resemble azygospores or chlamydospores. Class *Glomeromycetes*. Orders: *Archaeosporales, Diversisporales, Glomerales, Paraglomerales* (see section 3.6 below).
- Subphylum: ***Mortierellomycotina***. Order: *Mortierellales*. *Mortierellomycotina* reproduce asexually by sporangia. Molecular phylogenetic analyses reveal considerable diversity within *Mortierellomycotina* and environmental

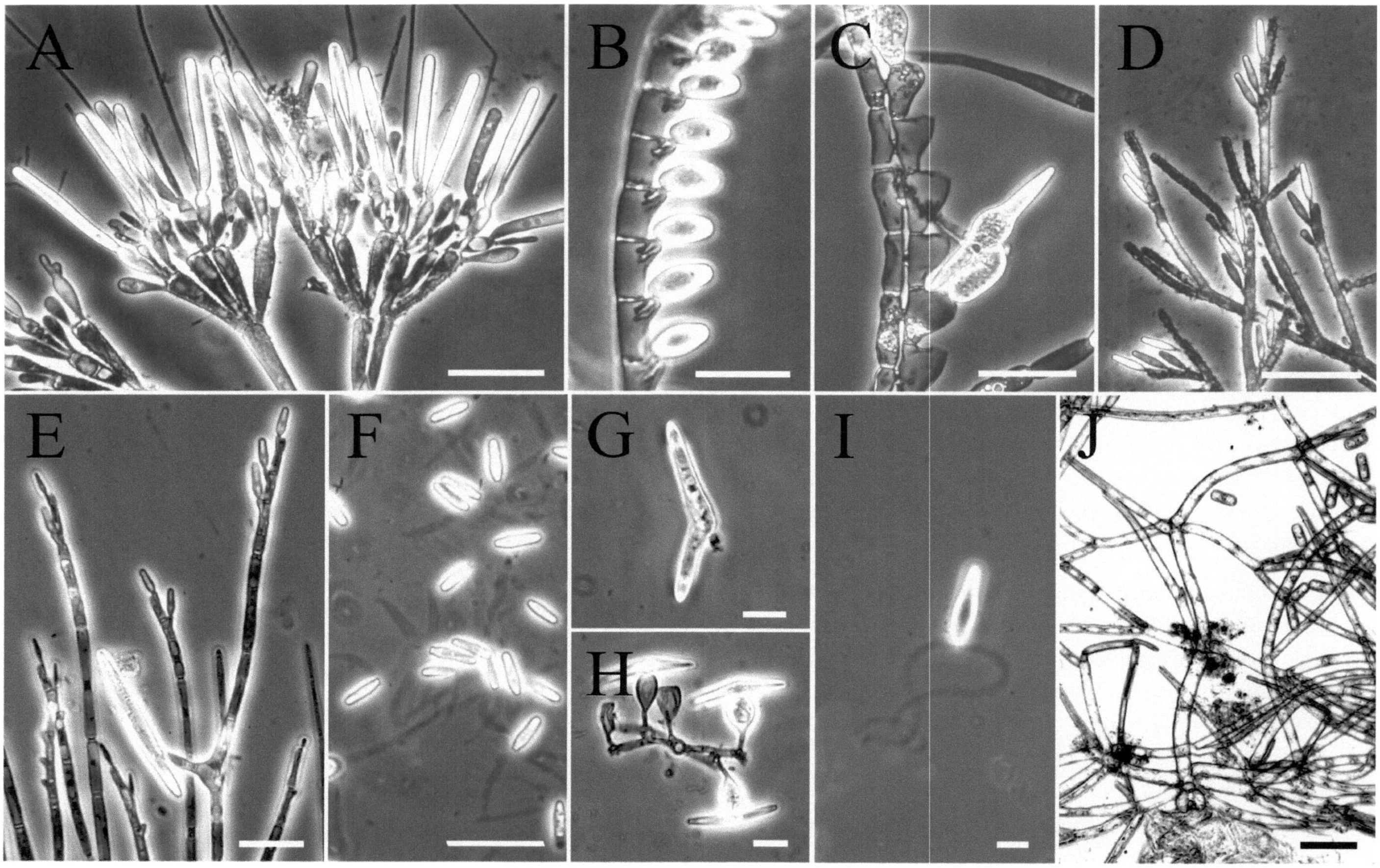

Figure 3.8. Phase contrast photomicrographs illustrating sexual and asexual reproductive structures in trichomycetes of Phylum *Zoopagomycota*. (A–I) *Harpellales* from aquatic insect larvae. (A) *Orphella catalaunica*, sporulating heads with attached cylindrical asexual spores. (B, C) *Harpellomyces eccentricus*, trichospores in (B) with their appendages visible within generative cells; (C) shows a zygospore arising from conjugated cells. (D) *Smittium culisetae*, trichospores attached to fertile branchlets. (E, F) *Capniomyces stellatus*, (E) attached, biconical zygospore and immature trichospores; (F) released trichospores with multiple appendages. (G) *Furculomyces boomerangus*, released bent zygospore with a short collar. (H, I) *Genistelloides hibernus*, (H) biconical zygospores with swollen zygosporophores (similar to suspensors in zygomycetes) attached to two conjugated branches; (I) released trichospore with two appendages. (J) *Asellariales* from a marine isopod, *Asellaria ligiae*, thallus attached by a holdfast cell to the hindgut cuticle and releasing small cylindrical arthrospores. Scale bars: A–I, 20 μm; J, 100 μm. Modified from White *et al.* (2006) using graphic files kindly supplied by Dr Merlin White, Boise State University, Idaho, USA. Reprinted with permission from Mycologia. ©The Mycological Society of America.

sampling shows a diversity of taxa associated with soils, rhizosphere and plant roots. *Mortierella* species are known as prolific producers of fatty acids, especially arachidonic acid, and they frequently harbour bacterial endosymbionts. Most species of *Mortierellomycotina* only form microscopic colonies, but at least two species in the genus *Modicella* make multicellular sporocarps.

- Subphylum: ***Mucoromycotina***; Orders: *Endogonales*, *Mucorales* and Umbelopsidales. This subphylum features the largest number of described species of *Mucoromycota* and includes the well-known model species *Mucor mucedo* and *Phycomyces blakesleeanus*. It also includes industrially important species of *Rhizopus* and other genera. Where known, sexual reproduction within *Mucoromycotina* is by typical zygospore formation and asexual reproduction typically involves the copious production of sporangia and/or sporangioles. Species are frequently isolated from soil, dung, plant debris and sugar-rich plant parts (like fruits). Fungi in the *Mucoromycotina* represent the majority of zygomycetous fungi in pure culture. Some ***Mucorales*** are extremely fast growing in culture and can often be found on mouldy bread or fruit. ***Endogonales*** are either **ecto**mycorrhizal on plant roots (remember the *Glomeromycotina* are **endo**mycorrhizal) or are saprotrophic. Sexual reproduction involves the production of zygospores by apposed gametangia within a simple sporocarp that may be hypogeous, embedded in heavily decayed wood, or produced among foliage of mosses or liverworts.

Ectomycorrhizas have probably evolved twice within *Endogonales*. *Endogonales* represents an independent origin of mycorrhizas relative to the arbuscular mycorrhizas of *Glomeromycotina* and ectomycorrhizas of *Dikarya* and, like many of *Mucoromycota*, they harbour endohyphal bacteria. Order **Umbelopsidales** contains the genus *Umbelopsis*. Asexual reproduction is by sporangia and chlamydospores. Sporangiophores may be branched. Sporangia are typically pigmented red or ochre, multi- or single-spored and with or without conspicuous columella. Sporangiospores are globose, ellipsoidal or polyhedral and pigmented like sporangia. Chlamydospores are filled with oil globules and often abundant in culture. Sexual reproduction is unknown. Species in the Umbelopsidales were previously classified in *Mucorales* or *Mortierellales*. Umbelopsidales is a distant sister group to *Mucorales*. Like *Mortierellales*, species of Umbelopsidales are frequently isolated from rhizosphere soils, with increasing evidence that these fungi occur as root endophytes.

Asexual spores in the traditional zygomycetes are single celled and include mitosporic sporangiospores (made within the sporangium by internal cleavage of sporangial cytoplasm, as described above for *Blastocladiella*; see Figure 3.6) and true **conidia** (which are borne on specialised hyphae called **conidiophores**). Sporangiospores are dispersed by wind or animals after rupture of the sporangium wall. Conidia are formed by the *Entomophthorales* and are forcibly discharged.

Names to look out for include *Mucor*, a typical filamentous mould that occurs as a saprotroph in soil, and on decaying fruits and vegetables. It is found everywhere in nature. Some *Mucor* spp. cause diseases (called zygomycosis) in humans, frogs and other amphibians, cattle and pigs. Strains isolated from human infections are unusually thermotolerant, most other strains being unable to grow at 37°C. Species names to look out for are *Mucor hiemalis, Mucor amphibiorum, Mucor circinelloides* and *Mucor racemosus*.

The genus *Rhizopus* is characterised by very fast-growing and coarse hyphae and colonies spread by means of creeping aerial hyphae, known as stolons, that branch into pigmented root-like rhizoids when they contact the substratum. Sporangiophores, singly or in groups, arise from nodes directly above the rhizoids. The fungus is common on decaying fruits, and in soil and house dust, and can be a pest in microbiology laboratories because its rapid growth and dry, easily air-distributed spores enable it to take over culture media in a few days.

Rhizopus oligosporus is used in some Asian countries to transform soy bean paste into the fermented food called tempeh. However, it also produces the alkaloid agroclavine, which is toxic to humans. *Rhizopus oryzae* (also known as *Rhizopus arrhizus*) is the most common causative agent of zygomycosis, accounting for some 60% of cases; it is a serious (and often fatal) opportunistic infection in immunosuppressed patients (in an **opportunistic infection** a vigorous saprotrophic organism becomes a disease organism by taking advantage of a weakened host).

Rhizopus rot is a soft rot of harvested or over-ripe stone fruits, such as peaches, nectarines, sweet cherries and plums; it can be caused by several species, including *Rhizopus nigricans* and *Rhizopus stolonifer*.

Phycomyces blakesleeanus is another filamentous fungus in the order *Mucorales*; this one gets its fame from the amount of basic research done with it over the years. *Phycomyces* produces numerous tall sporangiophores that are very sensitive to light. An enormous amount of research has been directed to investigating phototropism, and other environmental sensitivities, such as gravitropism. Spore germination, carotene biosynthesis and sexual development have all been studied in depth and genetic analysis carried out. It may be worth mentioning that the nonseptate, and therefore unicellular, sporangiophores of *Phycomyces blakesleeanus* are more than 100 μm wide (0.1 mm) and can grow to a height of 60 cm or more. Another remarkable fungus!

Basidiobolus is usually considered to be a member of the *Entomophthorales* (although some phylogenetic studies have grouped *Basidiobolus* with *Chytridiomycota*), but in the AFTOL study *Basidiobolus* spp. emerged in a novel position, separate from the majority of chytrids and *Entomophthorales* (White *et al.*, 2006). *Basidiobolus* is regularly isolated from the dung of amphibians, reptiles and insectivorous bats, as well as woodlice, plant debris and soil. It is cosmopolitan, but human infections due to *Basidiobolus* are reported mostly from Africa, South America and tropical Asia. Isolates pathogenic to humans belong to a unique species, *Basidiobolus ranarum*. The real claims to fame, though, come from the size of the mycelial cells, several hundred micrometres long (the apical compartment can be 300–400 μm long, and the fact that the cells have a single large nucleus which may be more than 25 μm in length (a yeast cell is only about 5 μm long). The nucleus undergoes a mitosis that is readily visible by conventional light microscopy and this feature has been exploited to study how inhibitors of mitosis work. This is a good example of using a very unusual fungus to study a biological problem. The huge nucleus contains an equally huge nucleolus (about 15 μm in length) (Jordan *et al.*, 1980). In addition, *Basidiobolus* has remarkable conidia which are shot explosively from the mycelium using what has been described as a rocket mechanism; the conidium remains intact and attached to the upper part of the conidiophore as the apex of the conidiophore bursts explosively. *Basidiobolus* appears ideal for studies of the cell cycle, mitosis and explosive spore dispersal.

The genus *Cunninghamella* is also commonly found in soil and plant material, particularly in Mediterranean and subtropical zones. It has also been recovered from animal material, cheese and Brazil nuts. *Cunninghamella bertholletiae, Cunninghamella elegans* and *Cunninghamella echinulata* are the most common species. *Cunninghamella bertholletiae* is the only known human and animal pathogen, another opportunistic fungus that may cause infections in immunocompromised hosts.

Mortierella species are also common soil fungi and the genus contains over 70 recognised species, however, *Mortierella wolfii* is probably the only (again, opportunistic) pathogen of humans and other animals.

A peculiarity of *Mucoromycota* is the occurrence of ***endosymbiotic bacteria*** (often with greatly reduced genomes) dwelling ***within the fungal cells***. Like the endobacteria of insects, those of fungi show a range of behaviours from mutualism to antagonism. The data suggest that some of the benefits derived by mycorrhizal fungi from their endobacteria may be shared with the plant host, giving rise to a three-level inter-domain/inter-kingdom interaction (Bonfante & Desirò, 2017). Here we refer you to section 14.10 for an account of the way the *Rhizobium*–legume symbioses form the bacterium–plant symbiotic interface by making use of molecular components derived from the arbuscular mycorrhizal fungal partner of the legume.

3.6 *GLOMEROMYCOTINA*

Until recently, **arbuscular mycorrhizal** (AM) fungi have generally been classified in the 'Zygomycota' (in Order Glomales), but they do not form the zygospores characteristic of zygomycetes, and ***all 'glomalean' fungi form mutualistic symbioses with plants***, known as **arbuscular mycorrhizas**; there is one exception, the genus *Geosiphon*, which is a symbiont of cyanobacteria in the genus *Nostoc*. The fungi are dependent on carbohydrates derived from their photoautotrophic symbiotic partners and represent a significant carbon sink in terrestrial ecosystems. In return, AM fungi provide inorganic nutrients (mainly phosphorus) to the plants and are therefore essential for plant nutrition. Because most land plants form AM, these fungi must be considered as vital components of terrestrial ecosystems. Initial molecular studies suggested a separate phylum was appropriate for the AM fungi, the 'Glomeromycota' (Redecker & Raab, 2006), consisting of the four orders *Diversisporales*, *Glomerales*, *Archaeosporales* and *Paraglomerales* (Redecker *et al.*, 2013; Redecker & Schüßler, 2014). More recently, based on genome-scale data, these four orders have been placed in Class *Glomeromycetes* in a newly erected subphylum ***Glomeromycotina***, which replaces the name 'Glomeromycota'. *Glomeromycotina* being a subphylum of Phylum *Mucoromycota* (Spatafora *et al.*, 2016; and see section 3.5).

Traditionally, taxonomy of AM fungi has been based on characteristics of the relatively large (80 to 500 μm diameter) **multinucleate spores**. There is no evidence that the *Glomeromycotina* reproduce sexually. Studies using molecular marker genes have detected little or no genetic recombination, although it has been shown that *Glomus* species contain 51 genes encoding all the tools necessary for meiosis (Halary *et al.*, 2011), which might imply that *Glomus* species have a cryptic sexual cycle (Kuck & Poggeler, 2009; Halary *et al.*, 2013). Clearly, there is a severe limit to the morphology that can be used in taxonomy and fewer than 200 species (or ***morphotypes***) are recognisable; however, combined morphological and molecular analyses have led to the description of 3 classes, 5 orders, 14 families and 29 genera (Oehl *et al.*, 2011). The spores have a layered wall and features of this can be used to describe morphospecies. Similarly, spores may be formed singly, in clusters or aggregated in so-called sporocarps, and the mode of spore formation has been important in describing genera and families (Figure 3.9).

This is a very challenging group to study. AM fungi are obligate symbionts and none of them has been cultivated without their plant hosts; most probably because AM fungi depend on their host plants to synthesise lipids that they cannot make for themselves (Luginbuehl *et al.*, 2017). This is one of several metabolic abilities that are not found in the AM genome, though there are many transporters and secreted proteins, implying evolutionary adaptation to mutualistic obligate biotrophy (Tang *et al.*, 2016). Genomic data from AM fungi have now revealed indirect evidence of sexual reproduction, such as intact sets of meiotic genes, and homologues of the mating type genes found in other fungi (Corradi & Brachmann, 2017). Nevertheless, pure AM biomass can be obtained only from cultures in transformed plant roots that can be cultivated in tissue culture, but only a small number of AM species are available in this form. Most samples from nature are DNA samples of 'cryptic species' that cannot be cultivated even in this way. Consequently, there is an ongoing controversy about the definition, dispersal and distribution of species in this clade. The controversy is far from being settled, but it has been well described by Bruns *et al.* (2017).

Yet, in ecological terms, this is possibly the most important group of fungi because AM fungi form **endomycorrhizal associations** with a majority of land plants, improving their uptake of nutrients and resistance to stresses. The association is essential for plant ecosystem function because the plants depend on it for their mineral nutrient uptake, which is efficiently performed by the mycelium of the fungal symbionts that extends outside the roots. Within root cells, AM fungi form hyphal coils or typical tree-like structures, the **arbuscules**.

Some also produce storage organs, termed **vesicles** (hence, another frequently used name for them: vesicular-arbuscular mycorrhizas or VAM fungi). In phylogenetic terms, they are important because they are the oldest unambiguous filamentous fungi known from the fossil record (see section 2.7). Put these two facts together and you get the suggestion that early colonisation of the land surface on Earth was promoted by the success of this plant–fungal symbiosis.

Ten genera are recognised currently in the *Glomeromycotina*. Names to look out for include *Glomus*, which is the largest genus in the phylum, with more than 70 morphospecies (spores, typically with layered wall structure, are formed by budding from

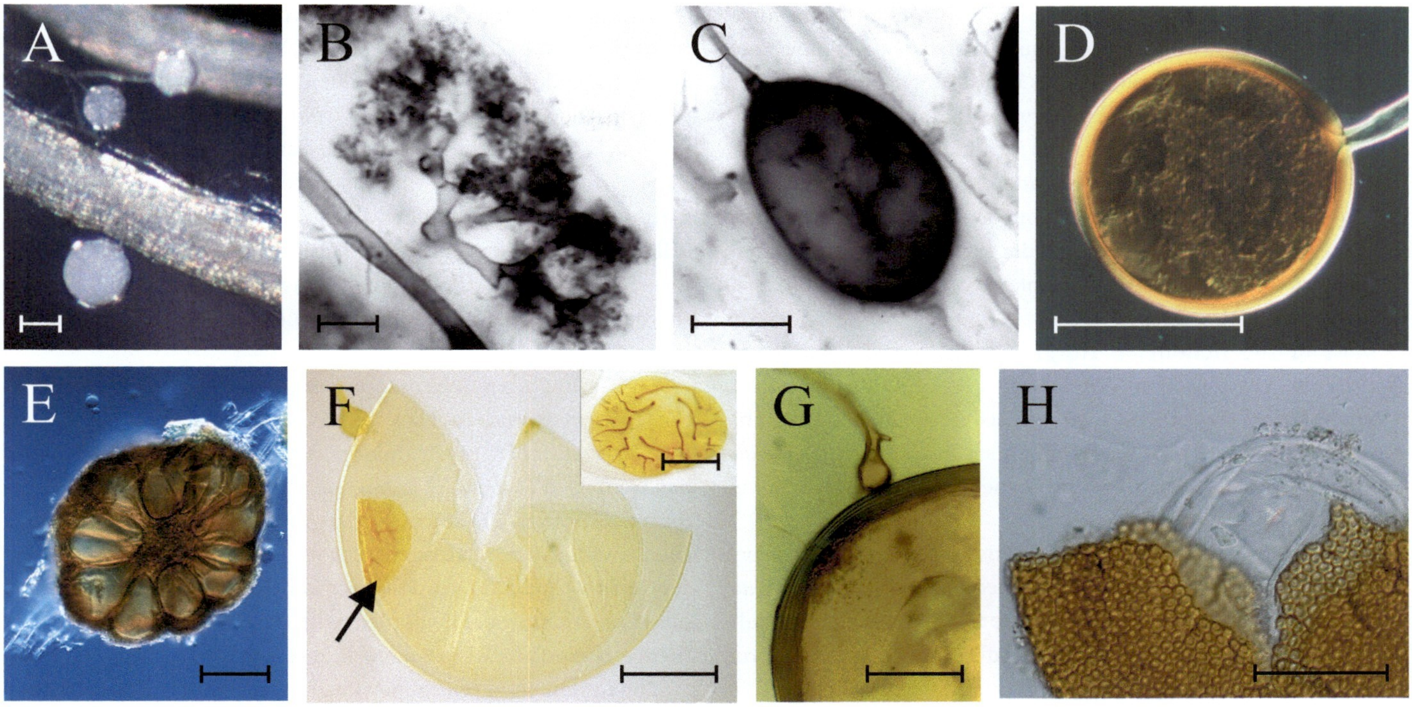

Figure 3.9. Some characteristic morphological features of fungi belonging to Subphylum *Glomeromycotina*. (A) Roots of *Plantago media* colonised with hyphae and spores of *Glomus clarum*. (B) Arbuscule of *Glomus mosseae* stained with chlorazol black. (C) Vesicle of *Glomus mosseae*. (D) Spore of *Glomus* sp. showing the hyphal attachment. (E) Section of a sporocarp of *Glomus sinuosum* with spores grouped around a hyphal plexus and covered by a layer of hyphae. (F) Spore of *Scutellospora cerradensis*, showing bulbous sporogenous cell and inner flexible walls with germination shield (arrow); inset: germination shield of *Scutellospora scutata* in face view (photomicrographs by Fritz Oehl, Agroscope Reckenholz-Tänikon, Switzerland). (G) Germinating spore of *Gigaspora decipiens* with sporogenous cell, warty germination layer and germination hypha. (H) Spore of *Acaulospora denticulata* with tooth-like wall ornamentations and inner germinal walls. Scale bars, A, E, F, G, 100 μm; D, H, 50 μm; B, C, 5 μm. Modified from Redecker and Raab (2006) using graphic files kindly supplied by Professor Dirk Redecker, Université de Bourgogne, France. Reprinted with permission from Mycologia. ©The Mycological Society of America.

a hyphal tip), placed in Family Glomeraceae, Order *Glomerales*. For a reminder of the characteristic word-endings used for the principal taxonomic ranks of fungi today see Table 2.2.

Gigaspora and *Scutellospora* are closely related genera in Family Gigasporaceae; their spores form on a bulbous sporogenous cell and germinate through a newly formed opening in the spore wall. These two genera do not form vesicles within roots, and the mycelium external to the root bears 'auxiliary cells' of unknown function. For the Family Acaulosporaceae, the diagnostic feature is the formation of spores next to a 'saccule' that collapses during spore maturation and eventually disappears; the genera included are *Acaulospora* and *Entrophospora*.

Geosiphon pyriformis (Family Geosiphonaceae) is the only member of the phylum that is symbiotic with a cyanobacterium. It forms an endosymbiosis with *Nostoc punctiforme*, the photobionts being harboured in fungal bladders up to 2 mm in size (Redecker & Raab, 2006).

Because of the difficulties (outlined above) resulting from their obligate symbiosis and our inability to culture them, there is no clear notion of what constitutes a species in the *Glomeromycotina* at the moment. Environmental studies using molecular markers have revealed a great deal of diversity suggesting that the number of 200 or so described morphospecies might considerably underestimate the true diversity of the *Glomeromycotina*.

3.7 ASCOMYCOTA

With at least 64,000 species known, *Ascomycota* (informally, **ascomycetes**) is the largest group in Kingdom *Fungi*. Like the fungi that belong to its sister group, the *Basidiomycota*, most species within the *Ascomycota* are **filamentous fungi** that produce a mycelium in which the hyphae are **regularly septate**. What characterised the *Ascomycota* is that their sexual spores (**ascospores**) are formed within a sac-like structure hyphal cell which is called an **ascus** (from the Greek askos which means 'bag').

Ascomycota includes many species of great importance. **Names to look out for** include the plant pathogens *Fusarium, Magnaporthe* and *Cryphonectria*, as well as medically important genera like *Candida* and *Pneumocystis* that cause human disease; *Penicillium chrysogenum*, the producer of penicillin, the first antibiotic to be discovered; and *Penicillium citrinum* and *Aspergillus terreus*, which were among the first moulds shown to produce precursors of today's 'wonder drugs', the statins, which are crucial to the management cholesterol in many millions of patients (see section 11.13).

Most of the **lichenised fungi** belong to the *Ascomycota*, as does the yeast used in our baking and brewing industries, *Saccharomyces cerevisiae*. As we will see later (section 5.2), this yeast has been used as a model organism in science experimentation for well over a century and is the best characterised of all eukaryotic cells.

Ascomycota contributes to other foods: *Penicillium camemberti* and *Penicillium roqueforti* condition cheese; *Fusarium venenatum* provides the mycoprotein used to make the vegetarian meat substitute Quorn®; and two celebrated edible fungi are the morel, *Morchella esculenta*, and truffle, *Tuber magnatum* (white truffle of northern Italy) and *Tuber melanosporum* (black truffle of the Périgord region in France) (see Chapter 11). Some of these moulds (especially *Aspergillus* spp.) produce metabolites that spoil food and, like aflatoxins, can be extremely toxic (see section 14.4).

Ascomycota also includes the majority of fungi for which sexual reproduction has not been observed. In traditional classifications (based largely on morphological features) these asexual fungi have been placed in a separate group, variously called Deuteromycetes, Deuteromycotina or Deuteromycota (according to whether the taxonomist concerned thought they deserved the status of class, subphylum or phylum). Molecular techniques now permit them to be classified among their sexually reproducing relatives from comparison of nucleic acid sequences, which makes this separation redundant.

The AFTOL review of the phylogeny of Phylum *Ascomycota* (Blackwell *et al.*, 2006) remains current, with just some minor changes to reflect more recent improvements in understanding. AFTOL divided the phylum into three major evolutionary lineages, given the rank of subphylum:

- ***Taphrinomycotina***, also known, informally, as the archiascomycetes; for example: *Protomyces, Taphrina, Pneumocystis*.
- ***Saccharomycotina***, also known, informally, as the hemiascomycetes; for example: *Saccharomyces, Pichia, Candida*.
- ***Pezizomycotina***, also known, informally, as the euascomycetes; for example: *Aspergillus, Neurospora, Peziza*.

Kurtzman and Sugiyama (2015) discuss the ecology, physiology, molecular biology, biotechnology, phylogeny and systematics of *Saccharomycotina* and *Taphrinomycotina*, and focus on changes in knowledge resulting from molecular studies.

Members of the ***Taphrinomycotina*** do not make sporophores (called **ascomata** in this phylum). The group includes a varied collection of organisms, including the filamentous plant pathogen *Taphrina*. *Taphrina deformans* causes peach leaf curl, and witches' broom disease of birch is caused by *Taphrina betulina*; other species attack oak, poplar, maple and many others). *Taphrina* normally grows in a yeast form until it infects plant tissues in which typical filamentous hyphae are formed. Ultimately the fungus forms a naked layer of asci on the deformed, frequently brightly pigmented surfaces of their hosts (Figure 3.10).

The fission yeast *Schizosaccharomyces pombe* is also placed in the *Taphrinomycotina*. This species, first isolated in 1893 from East African millet beer ('pombe' is the Swahili word for beer), has been used as a model organism in molecular and cell biology for over 50 years. These yeast cells grow by apical extension, and then divide by medial fission through a new, centrally placed, septum to produce two daughter cells of equal sizes. The regularity of this process, coupled with ease of cultivation, makes them a powerful tool in cell cycle research.

The fission yeast researcher **Paul Nurse** won the **2001 Nobel Prize in Physiology or Medicine** for his work on cyclin-dependent kinases in cell cycle regulation, together with Lee Hartwell (who developed the genetic analysis of the cell cycle in budding yeast, *Saccharomyces cerevisiae*, and introduced the concept of checkpoints) and **Tim Hunt** (who discovered cyclins in sea urchins) (see section 5.2).

Other, even more enigmatic inclusions in the *Taphrinomycotina* are the human pathogen *Pneumocystis* and the widely distributed Genus *Neolecta*. Long considered to be a protozoan, *Pneumocystis* is now clearly accepted as a yeast-like fungus. *Pneumocystis* pneumonia (PCP) is caused by *Pneumocystis jirovecii* (named in honour of the Czech parasitologist Otto Jirovec). Pneumocystosis is one of the most common infections in immunosuppressed patients because acquired immune deficiency syndrome (AIDS) and other immunity impairments such as immunodeficiencies, steroid treatment, organ transplantation medication and cancers predispose the patient to *Pneumocyctis* infection.

Neolecta has been found in Asia, North and South America, and northern Europe, in association with trees. The fungus produces club-shaped and brightly coloured sporophores, known as 'earth tongues', which are a few to several centimetres tall. *Neolecta vitellina* grows from rootlets of its host, but it is not known whether the fungus is parasitic, saprotrophic or symbiotic.

The ***Saccharomycotina*** contains a single order, the *Saccharomycetales* that includes the majority of the ascomycetous yeasts, including the economically important genera *Saccharomyces* and *Candida*. Yeasts usually grow as single cells that reproduce by budding or less frequently by fission. Asci and ascospores are not enclosed in the sporophores (ascomata) commonly found in the filamentous ascomycetes.

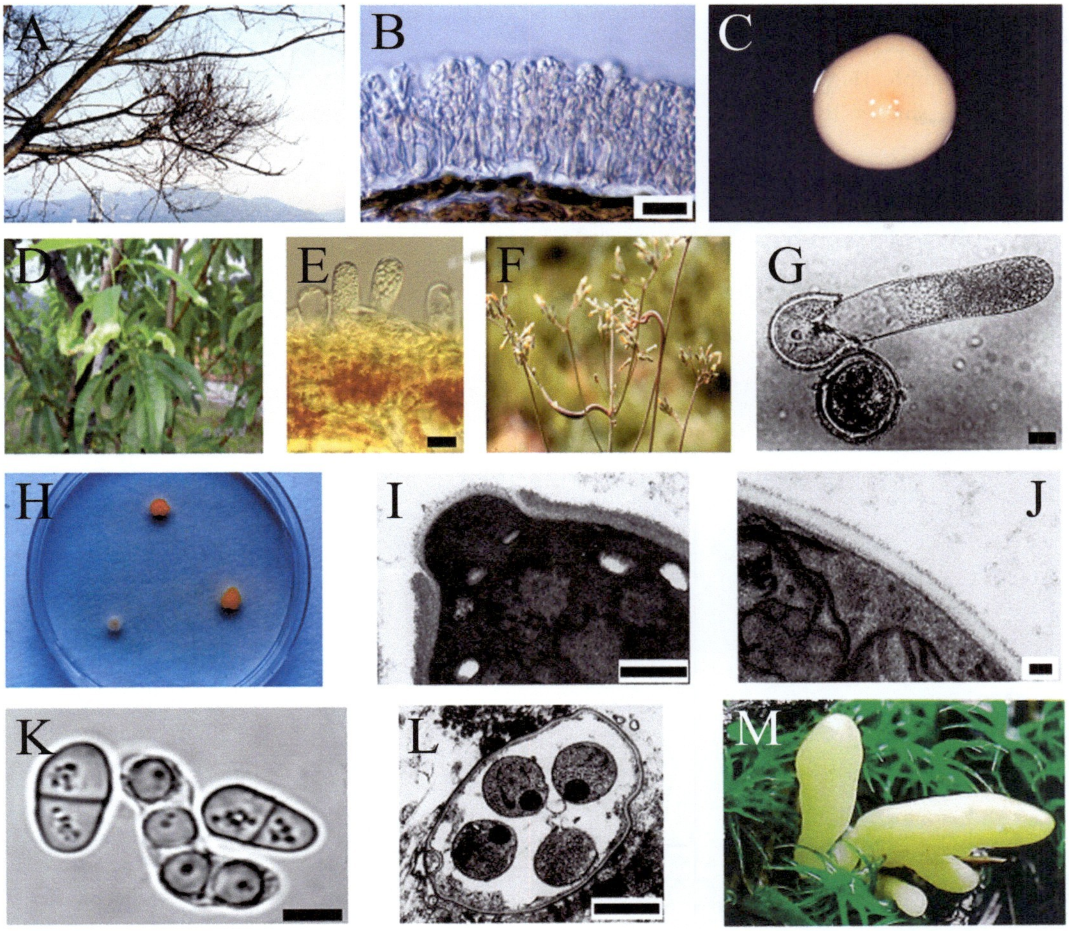

Figure 3.10. Macro- and micromorphology showing the diversity of representative taxa in the *Taphrinomycotina*. (A) Witches' broom disease of a Japanese cherry tree (*Cerasus yedoensis*) caused by *Taphrina wiesneri*. (B) Mature asci of *Taphrina wiesneri* on leaf tissue of a Japanese cherry tree showing ascospore budding (blastoconidium formation). (C) A colony on potato dextrose agar. (D) Symptoms of peach leaf curl disease caused by *Taphrina deformans*. (E) Hymenium of *Taphrina caerulescens*, causal agent of oak leaf curl. (F) Galls induced by *Protomyces inouyei* on stem of *Youngia japonica*. (G) Germination of a thick-walled resting spore of *Protomyces inouyei* in water. (H) Colonies of *Rhodosporidium toruloides* (anamorph: *Rhodotorula glutinis*) (top), *Saitoella complicata* (right) and *Taphrina wiesneri* (left) on potato dextrose agar. (I) *Saitoella complicata*: TEM showing enteroblastic budding, characteristic of basidiomycetous yeast. (J) *Saitoella complicata*. TEM showing the cell wall ultrastructure of the ascomycete type composed of one thin, dark layer and a broad, light inner layer. (K) *Schizosaccharomyces pombe*. Fission and an ascus containing four ascospores. (L) *Pneumocystis jirovecii*. Mature cyst containing intracystic bodies (endospores). (M) *Neolecta vitellina*. Bright yellow sporophores that can grow several cm tall. Scale bars: B, E, G, 20 µm; I 0.5 µm; J, 0.1 µm; K, 5 µm; L, 1 µm. Modified from Sugiyama *et al.* (2006) using graphic files kindly supplied by Professor Junta Sugiyama, TechnoSuruga Co. Ltd, Tokyo, Japan. Reprinted with permission from Mycologia. ©The Mycological Society of America.

Louis Pasteur, in 1857, demonstrated that yeasts ***caused*** the fermentation of grape juice to wine and eventually, **yeasts** were recognised as fungi. The name *Saccharomyces* means 'sugar fungus'. However, historical records from ancient Egypt and China depict brewing and baking 8,000 to 10,000 years ago, and analysis of whole-genome sequences of over a thousand different *Saccharomyces cerevisiae* yeast strains isolated around the world from varied ecological niches in nature, as well as baking and brewing industries, demonstrated that *Saccharomyces cerevisiae* originated in East Asia about 15,000 years ago (Peter *et al.*, 2018). The authors showed that while domesticated isolates exhibit high variation in ploidy, aneuploidy and genome content, genome evolution in wild isolates was mainly driven by the accumulation of single nucleotide polymorphisms, most of which are present at very low frequencies. Several independent domestication events spread budding yeast around the globe; genomic markers of domestication appeared about 4,000 years ago in sake yeast isolates, though such markers appeared in wine yeast isolates only 1,500 years ago. Why the same organism was domesticated for brewing and baking is probably due to the rare ability of *Saccharomyces cerevisiae* to ferment high external glucose concentrations, producing ethanol and CO_2, ***under aerobic conditions***. In most other organisms, fermentation metabolism requires anaerobic conditions. This unusual ability of yeast is known as the ***Crabtree effect***.

Yeast cellular morphology is rather simple (Figures 3.11 and 3.12), and yeast genomes tend to be smaller than those of filamentous fungi, but it has a highly adapted growth form. Not all yeasts are ascomycetes; there are members of the *Basidiomycota* that adopt a yeast form and the term 'yeast-like' also has been applied to dimorphic members of the zygomycete

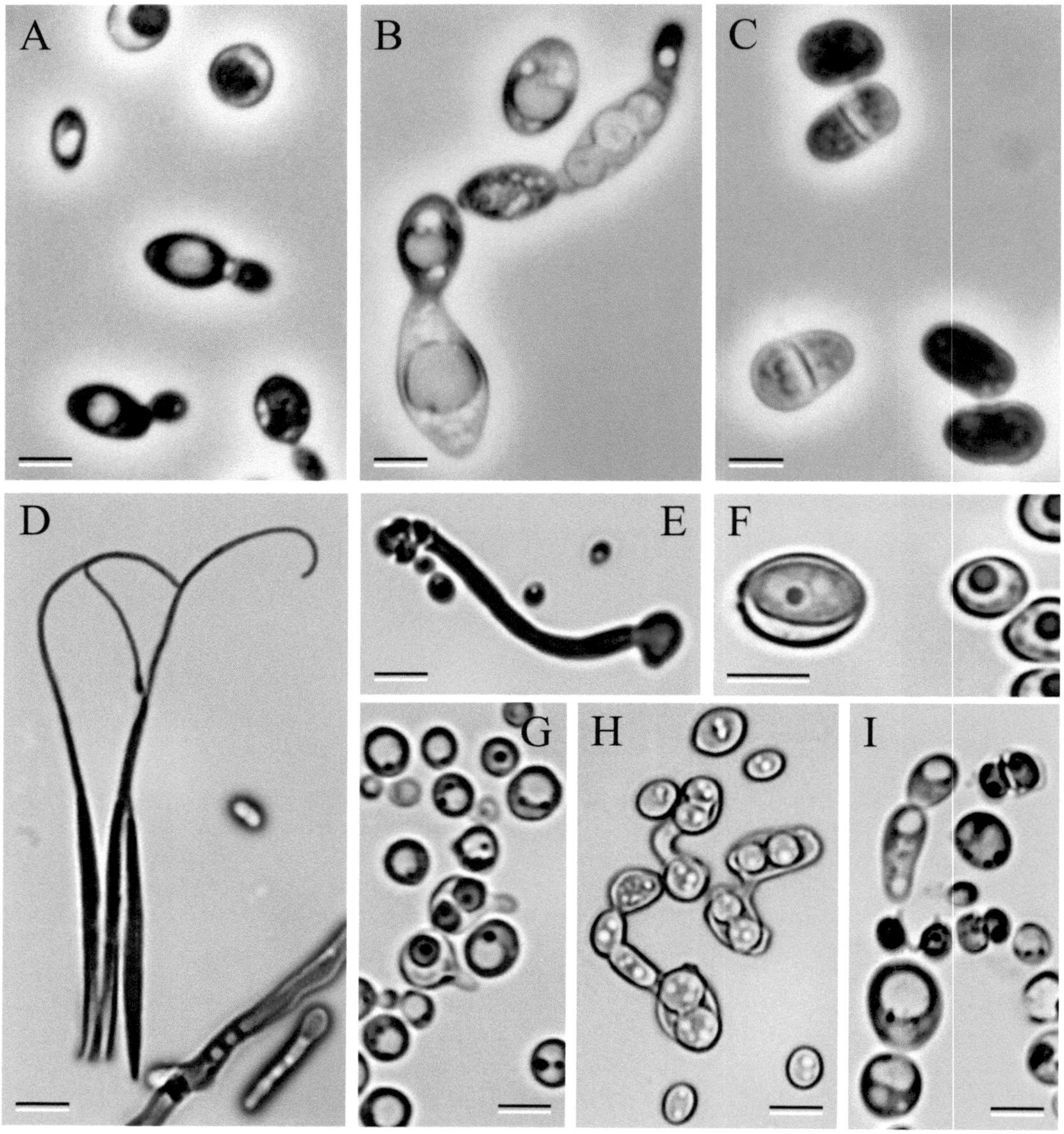

Figure 3.11. Ascomycete yeasts in pure culture. (A) *Saccharomyces cerevisiae*. Cells dividing by multilateral budding. (B) *Saccharomycodes ludwigii*. Cell division by bipolar budding. Note the ascus with four spherical ascospores above the dividing cell. (C) *Schizosaccharomyces pombe*. Cell division by fission. Once divided, the newly formed cells often will be morphologically indistinguishable from cells formed by budding. (D) *Eremothecium* (*Nematospora*) *coryli*. Free, needle-shaped ascospores with whip-like tails of extended wall material. Members of this group are some of the few yeasts that are plant pathogens. (E) *Pachysolen tannophilus*. A single ascus forms on the tip of an elongated refractile tube. The ascus wall becomes deliquescent and releases four hat-shaped ascospores. *Pachysolen tannophilus* was the first yeast discovered to ferment the pentose sugar D-xylose, a major component of hemicellulose from biomass. (F) *Lodderomyces elongisporus*. Persistent ascus with a single ellipsoidal ascospore. This is the only species of the clade, that includes *Candida albicans* and *Candida tropicalis*, which is known to form ascospores. (G) *Torulaspora delbrueckii*. Asci with 1–2 spherical ascospores. Asci often form an elongated extension that may function as a bud conjugant. (H) *Zygosaccharomyces bailii*. Asci with spherical ascospores. Often there are two ascospores per conjugant giving rise to the term 'dumbbell-shaped' asci. This species is one of the most aggressive food spoilage yeasts known. (I) *Pichia bispora*. Ascospores are hat-shaped and released from the asci at maturity. Hat-shaped ascospores are produced by species in a variety of different genera (see Figure 3.12A, for example). A–C, phase contrast. D–I, bright field; all photographs by Dr C. P. Kurtzman, ARS, USDA. Scale bars, 5 μm. Modified from Suh *et al.* (2006) using graphic files kindly supplied by Dr Sung-Oui Suh, American Type Culture Collection, Mycology Collection. Reprinted with permission from Mycologia. ©The Mycological Society of America.

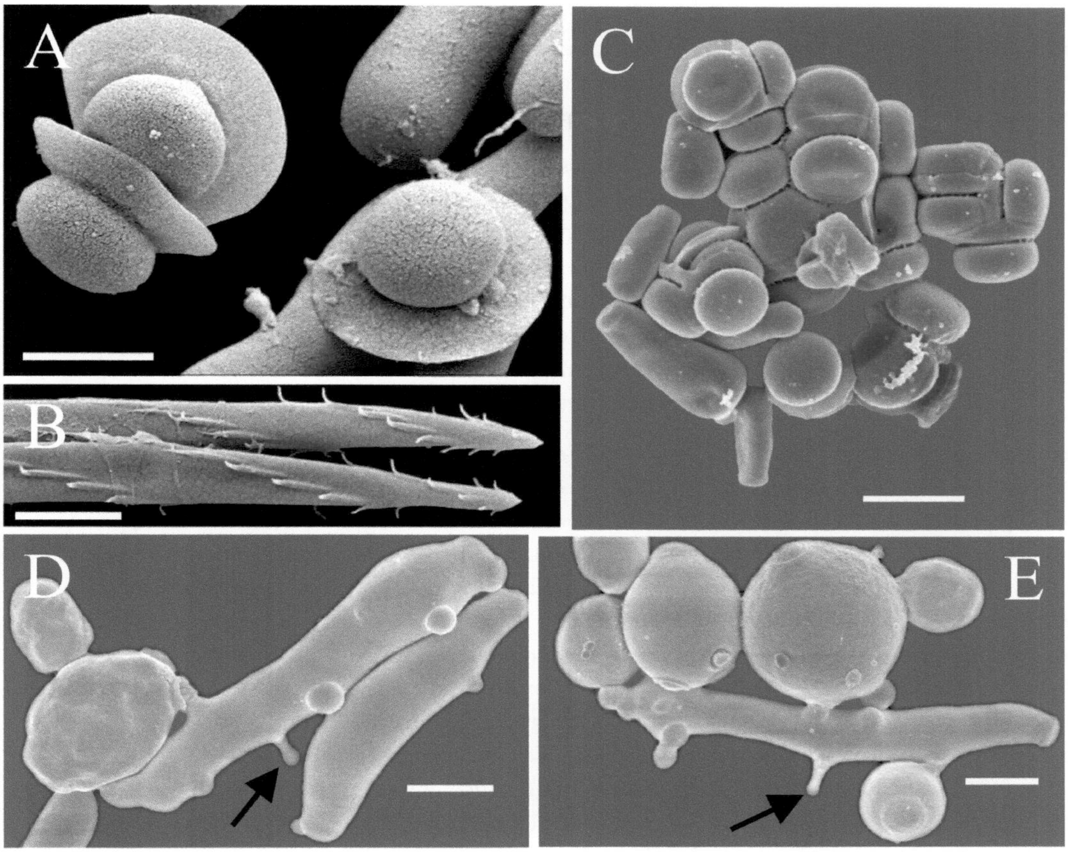

Figure 3.12. Scanning electron micrographs of yeasts. (A) Ascospores of *Kodamaea anthophila*. (B) End of ascospores of *Metschnikowia borealis* released by treatment of the ascus with Mureinase (a commercial multi-component enzyme preparation). (C) Agglutinated ascospores of *Saccharomycopsis synnaedendrus*. (D, E). Elongated predaceous cells of *Arthroascus schoenii* penetrating ovoid cells of *Saccharomyces cerevisiae* by means of narrow infection pegs (arrows). Scale bars, 2 μm. Photographs by Dr M.-A. Lachance, Western Ontario University, Canada. Modified from Suh *et al.* (2006) using graphic files kindly supplied by Dr Sung-Oui Suh, American Type Culture Collection, Mycology Collection. Reprinted with permission from Mycologia. ©The Mycological Society of America.

genus *Mucor* and a few other fungi, particularly pathogenic species. Ascomycete yeasts are usually found in specialised habitats, which tend to be small volumes of liquid rich in organic carbon (for example, flower nectaries). Basidiomycete yeasts, in contrast, seem to be adapted to colonising solid surfaces that are poor in nutrients.

Candida albicans is commensal on humans, its success apparently being the combination of an extracellular lipase activity, the ability to form invasive hyphae and the ability to grow at 37°C. *Candida albicans* lives on the skin, in the mouth, gut and other mucous membranes of about 80% of the human population, and for most of the time with no harmful effect. When the balance between the normal microorganisms is lost, for example because of antibiotic treatment, hormonal disturbance or immunocompromise, overgrowth of *Candida albicans* results in candidiasis, or 'thrush'. This common condition is usually easily cured, but in immunocompromised patients, such as HIV-positive individuals, the yeast form of *Candida* reacts to environmental cues by switching into an invasive filamentous growth form and a systemic and very serious infection can result.

The ***Pezizomycotina*** or **euascomycetes** ('true ascomycetes') make up the majority (about 90%) of the *Ascomycota* and includes most lichen fungi. These are the filamentous ascomycetes and their characteristic feature is that sexually reproducing species produce ascomata within which their sexual spores are formed in a sac-like ascus. Inevitably with such a large range of organisms, its members can be found in all aquatic and terrestrial habitats and participating in all ecosystems, including wood and litter decay, animal and plant pathogens, mycorrhizas and lichens (with only a few exceptions, all lichenised fungi belong to this group).

Prior to the application of molecular phylogenetics, classification of *Pezizomycotina* was based on the morphology and development of ascomata and asci. The four main **ascoma morphologies** are:

- Apothecia
- Perithecia
- Cleistothecia
- Ascostromata.

Organisms that produced these sporophores are said to be apothecial, perithecial, cleistothecial or ascostromatic, as appropriate.

Apothecia are typically disc- to cup-shaped to spoon-shaped (spathulate) and produce their asci in a well-defined tissue layer, a **hymenium**, which is exposed to the air.

Perithecia and cleistothecia are, respectively, partially or completely closed ascomata; their asci are formed in the central cavity ('centrum') of the ascoma. In perithecia, which are considered 'true' ascomata, the inner wall of the perithecium forms at the same time as the ascogenous hyphae develop. Asci are formed in a defined hymenium and are frequently mixed with **sterile paraphyses** arising from the subhymenial tissue although paraphyses are absent in some lineages (e.g. *Hypocreales*). The term **hamathecium** or hamathecial tissue is a general term applied to whatever tissue separates asci within the ascoma (it is derived from the Greek 'háma', meaning 'all together'). It may originate from different parts of the sporophore, be composed of paraphyses or other sterile cells, be generally distributed or localised; it may even be absent (e.g. *Dothidea*).

In ascostromata, the asci develop in preformed spaces, called **locules**, and the stroma often forms a flask-shaped (pseudothecia) or open, cup-shaped (hysterothecia and thyriothecia) structure that resembles the gross morphology of perithecia or apothecia.

Ascus walls appear to be multilayered in transmission electron micrographs, so a classification of asci has developed based on the number and thickness of wall layers as well as on the mechanism by which the ascospores are released (= **dehiscence**). The descriptive names (illustrated in Appendix 2) for the major ascus types include:

- **Unitunicate**, ascus with relatively thin walls; encompassing operculate, inoperculate and prototunicate asci:
  - **prototunicate**, produced by apothecial, cleistothecial and perithecial fungi; thin walled, globose to broadly club-shaped, ascospores released passively by disintegration of the ascus wall
  - **operculate**, found in apothecial fungi; release ascospores through a defined opening with a 'lid' (operculum) that is formed either at the ascus apex or just below it
  - **inoperculate**, produced by apothecial, cleistothecial and perithecial fungi; typically thin walled, the tip of the ascus usually has a small pore filled with loose wall material; the spores are discharged through this pore, or if there is no pore, dehiscence by rupture of the ascus apex.
- **Bitunicate**, conspicuously thick walled with two walls, called the exotunica and endotunica; produced by ascostromatic lichenised and non-lichenised species and ascohymenial lichens. In the traditional definition of bitunicate asci, fissitunicate dehiscence occurs (a fissitunicate ascus is a double-walled ascus where the inner wall pops completely out of the outer wall during dehiscence in a jack-in-the-box manner; it happens when the endotunica ruptures through the exotunica). Other dehiscence mechanisms exist among 'bitunicate' ascus morphologies involving little to no wall separation; these occur especially in lichenised taxa.

Some of these morphologies (photographs in Figure 3.13 and diagrams in Figure 9.15) most likely represent ancestral traits for the *Pezizomycotina* (e.g. apothecium), while others have occurred several times through convergent evolution (e.g. cleistothecium, prototunicate asci).

The current classification, based on multigene phylogenies of *Ascomycota*, divides *Pezizomycotina* into at least 10 major clades (Spatafora *et al.*, 2006), though some uncertainties remain:

- ***Arthoniomycetes*** (apothecia; bitunicate asci); includes the lichen *Lecanactis abietina* (Schoch & Grube, 2015).
- ***Dothideomycetes*** (ascostromata; bitunicate asci) (used to be included in loculoascomycetes); contains *Dothidia, Aureobasidium, Pleospora, Tyrannosorus* (yes, honestly) and *Tubeufia* (Schoch & Grube, 2015).
- ***Eurotiomycetes*** (perithecia, cleistothecia, ascostromata; bitunicate or prototunicate asci) (used to be called loculoascomycetes); contains *Aspergillus, Penicillium, Histoplasma* and *Coccidioides* (Geiser *et al.*, 2015).
- ***Laboulbeniomycetes*** (perithecia; prototunicate asci); comprises ectoparasites of insects and other arthropods (*Laboulbeniales*), e.g. *Herpomyces*, and mycoparasites and coprophiles (*Pyxidiophorales*), e.g. *Pyxidiophora* and *Rhynchonectria*; ascospores characterised by holdfasts.
- ***Sordariomycetes*** (perithecia, cleistothecia; inoperculate, prototunicate asci); contains the bulk of the traditional 'pyrenomycetes', including *Sordaria, Cordyceps, Neurospora, Hypocrea, Verticillium, Bombardia, Xylaria* and *Diaporthe* (Zhang & Wang, 2015; Hyde *et al.*, 2017).
- ***Lecanoromycetes*** (apothecia, perithecia; bitunicate, inoperculate, prototunicate asci); composed exclusively of lichen-forming ascomycetes like *Lecanora, Cladonia, Usnea, Peltigera* and *Lobaria* (Gueidan *et al.*, 2015).
- ***Leotiomycetes*** (apothecia, cleistothecia; inoperculate, prototunicate asci) (the inoperculate discomycetes); contains *Leotia, Sclerotinia, Monilinia, Mitrula, Hymenoscyphus, Microglossum* and *Cudonia* (Zhang & Wang, 2015).
- ***Lichinomycetes*** (apothecia; bitunicate, inoperculate, prototunicate asci); includes the lichen *Lempholemma, Peltula*.
- ***Orbiliomycetes*** (apothecia; inoperculate asci); contains *Orbilia* (Pfister, 2015).
- ***Pezizomycetes*** (apothecia; operculate asci); includes the operculate 'discomycetes' *Peziza, Aleuria, Morchella, Gyromitra, Tuber* and *Pyronema* (Pfister, 2015).

To these may be added:

- ***Geoglossomycetes***, *Geoglossum* and *Trichoglossum*, commonly called earth tongues, were previously included in the *Leotiomycetes* but rejected as members

Figure 3.13. Examples of the diversity of ascomata in the *Pezizomycotina*. (A) Apothecia of *Aleuria aurantia* (*Pezizomycetes*); *Aleuria* sporophores are whitish, yellowish to orange (hence the common name orange peel fungus) and may be 1–30 mm tall and 1–160 mm wide. (B) Apothecia of the cup fungus, *Peziza howsei* (*Pezizomycetes*); *Peziza* produces soft and fragile, disc- to cup-shaped sporophore, 2–120 mm high and 5–150 mm across, may have a stem up to 30 mm long, colour whitish, or yellow, pink, blue, buff, brown, or grey to blackish. (C) Yellow apothecia of the glasscup, *Orbilia delicatula* (Orbiliomycetes); soft and fragile and only 0.05–0.3 mm high, 0.1–2 mm wide, on a white mycelium. (D) Apothecia of *Trichophaea hybrida* (*Pezizomycetes*); 1–5 mm high and 1–15 mm wide, buff to dark grey and the outer surface downy to hairy. (E) Club-shaped or spatulate apothecium of the Earthtongue, *Geoglossum cookeanum* (*Leotiomycetes*); generally soft and fragile, 10–100 mm tall, 3–25 mm wide, purple to brownish black. (F) Cinnamon jellybaby, *Cudonia confusa* (*Leotiomycetes*); rather mushroom-like, club-shaped sporophore (though the hymenium covers the outer surface of the 'cap' or head), 20–80 mm tall, 5–20 mm across the head. (G) Saddle fungus, *Helvella crispa* (*Pezizomycetes*); with a saddle-shaped, folded and stipitate apothecium, up to 50 mm high and 30 mm wide. (H) Sporophore of the morel, *Morchella esculenta* (*Pezizomycetes*); the sporophore on the left has been sliced open vertically to show the hollow stem and head, 30–300 mm high and 15–160 mm wide. (I) The summer truffle, *Tuber aestivum* (*Pezizomycetes*); sporophore is tuberoid to spherical, longest dimension up to 70 mm, this specimen has been sliced open to show the massively folded spore-bearing inner mass of flesh (the gleba). See http://www.mycokey.com/ for more details. Photographs A–H by Jens H. Petersen, University of Aarhus, Denmark; I by Jan Vesterholt, University of Copenhagen, Denmark.

of that class by DNA analyses (Spatafora, 2007, in the Tree of Life Web Project at http://tolweb.org/Pezizomycotina/29296/2007.12.19).

- ***Xylonomycetes***, this group is an example of the discovery of new fungi by multigene phylogenetic analyses of 'exotic' habitats; this new major lineage of *Ascomycota* was discovered in a survey of endophytic fungi cultured from living sapwood and leaves of rubber trees (*Hevea* spp.) in remote forests of Peru (Gazis *et al.*, 2012).
- ***Coniocybomycetes***, a heterogeneous assemblage of fungi sharing the presence of a mazaedium (a sporophore of some lichens in which the ascospores lie freely in a powdery mass that is enclosed in an ascome wall or peridium) separated from within the *Arthoniomycetes*, *Eurotiomycetes*, *Lecanoromycetes* and *Leotiomycetes* by multigene phylogenetic analyses (Prieto *et al.*, 2013).

The AFTOL data (Spatafora *et al.*, 2006) strongly indicated *Orbiliomycetes* and *Pezizomycetes* as the most basal classes of *Pezizomycotina*, both consist of species that produce apothecia. The fossil genus *Paleopyrenomycites* from the Early Devonian Rhynie chert (about 400 million years old) is the oldest ***accepted*** fossil member of *Pezizomycotina*, although its position within this subphylum is unclear (Beimforde *et al.*, 2014).

3.8 *BASIDIOMYCOTA*

Basidiomycota (informally, **basidiomycetes**) is another large group, comprising about 32,000 known species. Morphologically, ecologically and taxonomically this is a very diverse group, but its members share the feature that their sexual spores are **exospores**, formed on a **basidium** (and are therefore called **basidiospores**).

Basidiomycota includes the plant pathogens that cause **smut** and **rust** diseases, ectomycorrhizal species which are of key significance to forest ecosystems, saprotrophic species that can decay the lignin (**'white-rot fungi'**) as well the cellulose of plant litter, and the most noticeable and frequently encountered **mushroom** fungi. Several basidiomycetes have mutualistic associations with insects (e.g. leaf-cutter ants, termites, ambrosia beetles) that are based on the ability of the fungus to digest plant litter efficiently.

All the fungi whose sporophores are **commercially farmed** as food are basidiomycetes (truffle 'cultivation' is different, being more a matter of the silviculture of the truffle groves; see section 12.6). On the other hand, some members of this group produce toxins that can be hallucinogenic (e.g. *Psilocybe cubensis*) or deadly poisonous (e.g. the 'Destroying Angel', *Amanita virosa*). The species that causes cryptococcal meningitis in persons suffering compromised immunity (due to HIV infection, cancer chemotherapy, metabolic immunosuppression to maintain a transplanted organ) is also a basidiomycete; the disease is caused by the asexual form (called the anamorph), which grows as a yeast. This form is given the generic name *Cryptococcus*. The sexual form or teleomorph is called *Filobasidiella*.

The *Basidiomycota* divides into three distinct evolutionary groups which are given subphylum status in the AFTOL study (Blackwell *et al.*, 2006; Hibbett *et al.*, 2007; 2014) and this still represents the most widely accepted classification scheme (and see He *et al.*, 2019, for the most recent, and extensive, phylogenetic treatment):

- ***Ustilaginomycotina*** (smuts and allies; the traditional ***Ustilaginomycetes***)
- ***Pucciniomycotina*** (rusts and allies; the traditional ***Urediniomycetes***)
- ***Agaricomycotina*** (the traditional **hymenomycetes** or basidiomycetes and allies: producing a macroscopic sporophore with a hymenium spread over gills (as in mushrooms), within pores (boletes and bracket fungi), or over a toothed (hedgehog fungi), coralloid (coral fungi), labyrinthoid (daedaleoid fungi), wrinkled (merulioid fungi), or smooth to diffusely encrusted (corticioid fungi with resupinate sporophores) tissue structure; or completely enclosed (gasteroid fungi)).

To these may be added:

- The Class ***Wallemiomycetes*** (*incertae sedis*, which is Latin for '*of uncertain placement*'), which comprises three species of fungi in Genus *Wallemia* and is a 500 million-year-old sister group of *Agaricomycotina* (Padamsee *et al.*, 2012; Zajc *et al.*, 2013). These are the most halophilic fungi known; *Wallemia ichthyophaga* requires at least 1.5 M NaCl for growth *in vitro* and thrives even in saturated NaCl solution. A limited number of strains of *Wallemia* have been isolated so far; from salt pans used to crystallise salt from sea water ('solar salterns') and salted meat (Zajc *et al.*, 2016).

Ustilaginomycotina is a group of about 1,200 fungi in approximately 62 genera, most of which parasitise plants, mainly non-woody angiosperm herbs, especially grasses (*Poaceae*) and sedges (*Cyperaceae*). The best-known members, *Ustilago* and *Tilletia*, cause diseases of cereals known as smut or **stinking smut** that can severely reduce crop yield. *Ustilago maydis*, which causes smut disease on maize (*Zea mays*), is used widely as a model organism for plant pathogenesis and was the first basidiomycete plant pathogen to have its complete genome worked out (Begerow *et al.*, 2014).

In these fungi, the basidiospores establish a yeast-like budding haploid phase (the cells are called **sporidia**) that grows saprotrophically; eventually **conjugation** of compatible haploid cells produces a dikaryotic, parasitic mycelium that infects the host plant (Figure 3.14).

Mating (controlled by **mating type factors**, see section 9.5) is essential for infection of host plants, and the dikaryotic plant-parasitic phase ends with the production of teliospores, which, in most of these fungi, develop thick walls and are the dispersal spore.

Teliospores are formed in a **sorus** (a sorus is a mass of spores that bursts through the host epidermis; the word comes from the Greek soros, heap) which is on or in the parenchyma tissue of the host. The location depends on species; sori may appear on roots, stems, leaves, flowers, seeds, etc. As the teliospores are usually powdery and dark brown or black, the plant looks as though it is covered in particles of dirt or soot, so giving rise to the common name **'smut disease'**.

Other names to look out for are *Graphiola, Exobasidium* and *Microstroma*. The group also includes some animal pathogens in Genus *Malassezia*, which are lipophilic yeasts (no dikaryophase known) isolated from the skin of warm-blooded animals. These fungi require fat to grow and are found in areas of the skin with many sebaceous glands: scalp, face and upper part of the body. In humans *Malassezia globosa* causes dandruff and seborrhoeic

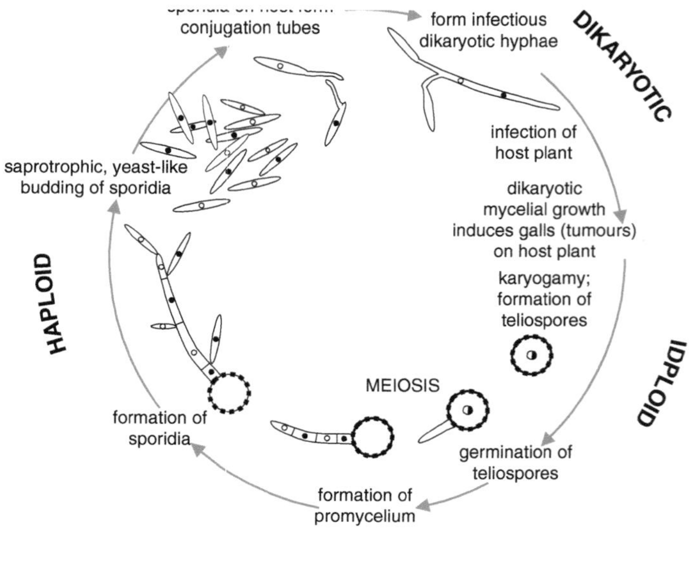

Figure 3.14. Diagram of the life cycle of *Ustilago maydis*. Modified from Chapter 2 in Moore & Novak Frazer (2002).

dermatitis. It is possible that *Malassezia* species originated from a dikaryophase parasitic on plants.

Roughly 8,400 of the species assigned to *Basidiomycota* belong to Subphylum ***Pucciniomycotina***. The *Pucciniomycotina* differs from the other two subphyla by the possession of simple septal pores lacking membrane-bound caps, and by the sugar composition of the cell wall. Most members of the *Pucciniomycotina* are parasitic, and about 90% belong to a single order of plant parasites, *Pucciniales*, known as **rust fungi**, that cause some of the most devastating diseases of crops (Figure 3.15); but the group also includes pathogens that attack other fungi and those that attack insects (Aime *et al*., 2014).

The rusts form a natural (monophyletic) group. **Names to look out for** include *Puccinia* (*Puccinia graminis* causes stem rust of cereals), *Uromyces* (*Uromyces appendiculatus* causes a rust disease in both temperate and tropical crops of the common bean, *Phaseolus vulgaris*). *Endophyllum* has only been recorded as a rust disease of the perennial woody shrub *Chrysanthemoides monilifera*; both host and rust originate in South Africa, but *Chrysanthemoides* is an aggressive weed in southern Australia and *Endophyllum* is showing promise as a biocontrol agent.

Basidiomycetous yeasts include *Rhodosporidium* (and its anamorph *Rhodotorula*), and *Sporidiobolus* (Figure 3.15D and E) and its anamorph *Sporobolomyces*, which can be commonly isolated from surfaces in the domestic environment.

The *Agaricomycotina* (traditionally known as **hymenomycetes**) is a diverse group that includes about 21,400 described species, which is about 65% of known *Basidiomycota* (or about **a fifth of all fungi**). The *Agaricomycotina* probably originated between 380 and 960 million years ago. Today the group includes many **wood decayers, litter decomposers** and **ectomycorrhizal fungi**. It also includes a lesser number of important plant pathogens, one of which is the **largest organism on the planet**. The AFTOL classification scheme for Subphylum *Agaricomycotina* is listed below. Here we will briefly describe the major groups.

The largest clade is Order ***Agaricales*** or **euagarics clade** (in the subclass *Agaricomycetidae*), which is made up of the mushroom-forming fungi and includes more than half of all known species of the homobasidiomycetes, like the type genus *Agaricus* (Kerrigan, 2016). More than 13,200 species and roughly 410 genera have been ascribed to the 33 families of the order. Traditionally, the shape and morphology of the mature

Figure 3.15. Representatives of the *Pucciniomycotina*. (A) *Jola javensis* (*Platygloeales*) sporophores on *Sematophyllum swartzii* (photograph by Dr Elizabeth Frieders, University of Wisconsin-Platteville). (B) *Septobasidium burtii* fungal mat completely covering scale insects (photograph by Dr Daniel Henk, Imperial College London). (C) *Eocronartium muscicola* sporophores on a moss (photograph by Dr Stephen F. Nelsen, University of Wisconsin-Madison). (D) Yeast and filamentous cells of *Sporidiobolus pararoseus*. (E) Cultures of two *Sporidiobolus* species. (F) *Phragmidium* sp. (*Pucciniales*) on *Rosa rubiginosa*. Modified from Aime *et al.* (2006) using graphic files kindly supplied by Dr M. Catherine Aime, Louisiana State University. Reprinted with permission from Mycologia. ©The Mycological Society of America.

sporophore, the colour of the spore deposit, and various anatomical and cytological features have figured largely in the taxonomy; but, because apparently similar structures and shapes can evolve in different ways, this approach led to the establishment of many artificial groups. For example, there are several ways of creating the folded surface that we call **gills**. Since they arise in different ways, such features are only superficially similar (like, for example, the wings of butterflies and wings of birds); they are described as **analogous organs** (as distinct from **homologous organs**, which are those that have a shared evolutionary ancestry). Shape and morphology of mature sporophores, even though these are the easiest to find in the field, can be misleading because they do not contribute to a natural classification. This is why introduction of molecular sequence comparisons is prompting major, and sometimes surprising, revisions in classification of these organisms. Phylogenies based on analysis of ribosomal RNA sequences has transformed the classification of the *Agaricales*, and has shown that ecological traits have been underused in the past in diagnosing natural groups (Figure 3.16).

Another major group of mushroom-forming fungi that has representatives in most forest ecosystems around the world is Order ***Boletales***. This order contains around 1,320 described species although, as it is most diverse in the still under-researched tropics, this probably underestimates the real biodiversity. The *Boletales* includes conspicuous stipitate-pileate forms ('mushrooms' with stems and caps) that mainly have tubular, but sometimes lamellate (gilled) hymenophores, as well as intermediates that show transitions between the two types of hymenophore structures. The *Boletales* also includes puffball-like forms (that used to be called gasteromycetes), resupinate or crust-like fungi that produce smooth, merulioid (wrinkled to warted) or hydnoid (toothed) hymenophores. As well as diverse morphologies, the *Boletales* exploit diverse habitats, including wood decay but they differ from their sister clades (*Agaricales* and *Atheliales*) by not having white-rot fungi in the group. Instead, saprotrophic *Boletales* have developed a unique brown rot that is especially adapted to decaying conifer timber. The majority of *Boletales* form mycorrhizal associations and some are mycoparasites (Figure 3.17).

Some of the most eccentric sporophore morphologies occur in the ***Hysterangiales***, ***Geastrales***, ***Gomphales*** and ***Phallales*** (combined in a grouping known as **gomphoid-phalloid fungi**, meaning 'like *Gomphus* and/or like *Phallus*'). This is where we find the **cage fungi**, the **stinkhorns** and **puffballs** (Figure 3.18). The *Gomphales* contains both gasteroid (sporophores 'stomach-like' with completely enclosed hymenium) and nongasteroid taxa, with gasteroid morphology being derived from **epigeous** taxa (epigeous means occurring above the surface of the ground; contrasts with **hypogeous**, which describes underground or semi-underground sporophores). On the other hand, in the *Phallales* the truffle-like hypogeous form is an ancestral morphology of the stinkhorn sporophores. Although basidiospore maturation occurs within the enclosed young primordial sporophores of the stinkhorn (the '**eggs**'), lifting the mature spore producing tissue into the air where flies can easily find it represents an independent instance of the origin of stipes within the *Basidiomycota*.

The ***Cantharellales***, which includes mushroom fungi, such as *Cantharellus* and *Craterellus* (Figure 3.19), was set apart from other gilled fungi very early in the history of mycology on the grounds that the gills are formed when the hymenophore folds into pleats like a fan, so these were called 'false gills' in comparison with the allegedly 'true gills' of most other mushrooms (understood to be individual, plate-like or blade-like things, structurally separate from one another and from the flesh of the cap). In today's phylogeny the difference is still important but is less surprising because we recognise the lamellate (gilled) hymenophore as a simple strategy to increase the area available for spore formation, and, more importantly, it is a strategy that can be arrived at through several different evolutionary routes.

The ***Hymenochaetales***, which is dominated by wood-decaying species, also includes many variations of sporophore types. Many species have basidiomata that are effused (stretched out flat) or effused-reflexed (with the edge turned up or turned back); a few form stipitate mushroom-like (agaricoid), coral-like (clavarioid) and spathulate (shaped like a spoon) to rosette-like basidiomata (Figure 3.20).

Most species have vegetative (sterile) cells in the sporophore tissue, often accompanying the basidia in the hymenium. Although these could collectively be called **cystidia**, many unique terms have been introduced for them to describe their distinctive morphologies; a simple example is that most species in Family Hymenochaetaceae have characteristic cystidia called **setae**. Because microscopic distinctions like these, and sporophore shapes, formed such an important basis for the traditional classification of fungi, what is now known as the hymenochaetoid clade draws its members from several traditional families: Agaricaceae, Polyporaceae, Corticiaceae, Stereaceae and Hymenochaetaceae, but includes only the type genus for the last family mentioned.

The ***Russulales*** is, probably, the most morphologically diverse group, because it contains resupinate, discoid, effused-reflexed, clavarioid, pileate and gasteroid sporophores, and hymenophores that may be smooth, poroid, hydnoid, lamellate or labyrinthoid. Members of the *Russulales* are primarily saprotrophs but others are **ectomycorrhizal**, root parasites and **insect symbionts** (Figure 3.21).

The AFTOL classification scheme for Subphylum *Agaricomycotina* looks like this:

- ***Agaricomycetes*** (Hibbett *et al.*, 2014)
  - *Agaricomycetidae*
    - *Agaricales* (Figure 3.16)
    - *Atheliales* (Figure 3.22)
    - *Boletales* (Figure 3.17)

Figure 3.16. Photographs depicting some of the diversity of the *Agaricales*. (A) *Plicaturopsis crispa*. (B) *Podoserpula pusio* (photo by Heino Lepp). (C) *Pterula echo* (photo by Dave McLaughlin). (D) *Camarophyllus borealis*. (E) *Ampulloclitocybe clavipes*. (F) *Resupinatus applicatus*. (G) *Mycena* aff. *pura*. (H) *Crucibulum laeve* (photo by Mark Steinmetz, Mykoweb). (I) *Nolanea* sp. (J) *Volvariella gloiocephala*. (K) *Crepidotus fimbriatus*. (L) Basidiospores with germ pore of *Psilocybe squamosa* (photo by Roy Halling). (M) *Camarophyllopsis hymenocephala* (photo by D. Jean Lodge). (N) inverse lamellar trama and pleurocystidia of *Pluteus* (photo from D. E. Stuntz slide collection). (O) *Clitocybe subditopoda*. (P) *Cortinarius bolaris*. (Q) *Cylindrobasidium evolvens*. (R) *Tricholoma columbetta*. Modified from Figure 2 in Matheny *et al.* (2006) using graphic files kindly supplied by Dr P. Brandon Matheny, University of Tennessee, USA. Reprinted with permission from Mycologia. ©The Mycological Society of America.

Figure 3.17. Morphological diversity in *Boletales*. (A) *Bondarcevomyces taxi*. (B) *Bondarcevomyces taxi*, pores. (C) *Coniophora puteana*. (D) *Leucogyrophana mollusca*. (E) *Hygrophoropsis aurantiaca*. (F) *Suillus granulatus*. (G) *Chroogomphus vinicolor*. (H) *Boletinellus merulioides*, hymenophore; (I) *Calostoma cinnabarinum*; (J) *Scleroderma septentrionale*. (K) *Meiorganum neocaledonicum*, young hymenophore. (L) '*Tylopilus*' *chromapes*. (M) *Phylloporus centroamericanus*. (N) *Xerocomus* sp. Modified from Figure 1 in Binder and Hibbett (2006) using graphic files kindly supplied by Dr Manfred Binder, Biology Department, Clark University, USA. Reprinted with permission from Mycologia. ©The Mycological Society of America.

Figure 3.18. Macro- and microscopic characters of the gomphoid-phalloid fungi. (A–E) represent the *Hysterangiales* clade: (A) *Hysterangium setchellii*. (B) *Mesophellia castanea*. (C) *Gallacea scleroderma*. (D) Basidiospores of *Hysterangium inflatum* enclosed in utricle. (E) Basidiospores of *Austrogautieria rodwayi* (photograph by L. Rodway). (F–L) represent the *Phallales* clade: (F) *Phallus impudicus* (photograph by Koukichi Maruyama). (G) *Aseroë rubra*. (H) *Ileodictyon cibarium*. (I) *Lysurus mokusin* (photograph by Ikuo Asai). (J) *Claustula fischeri*. (K) *Phallobata alba* (photograph by Peter Johnston). (L) Basidiospores of *Ileodictyon cibarium*. (M–T) represent the *Gomphales* clade: (M) *Ramaria fennica* (photograph by Ikuo Asai). (N) *Turbinellus* (*Gomphus*) *floccosus* (photograph by Ikuo Asai). (O) *Gautieria* sp. (P) *Kavinia* sp. (photograph by Patrick Leacock). (Q) Basidiospores of *Ramaria botrytis* with cyanophilic ornamentation (photograph by Koukichi Maruyama). (R) Ampullate hypha of *Ramaria eumorpha*. (S) Acanthohypha of *Ramaria cystidiophora* stained with cotton blue (photograph by Efren Cazares). (T) Hyphal mat formed by *Ramaria* sp.; note sharp contrast in colour between the white soil of the mycelial mat and the black soil elsewhere. (U–Z) represent the *Geastrales* clade: (U) *Geastrum fornicatum*. (V) *Pyrenogaster pityophilus*. (W) *Sphaerobolus stellatus*. (X) Basidiospore of *Geastrum coronatum*. (Y) Basidiospore of *Myriostoma coliforme*. (Z) Basidiospores and germinating gemmae of *Sphaerobolus stellatus*. Scale bars: A, B, V, 5 mm; C, G, I–K, O, P, U, 1 cm; D, E, L, Q–S, Z, 5 μm; F, H, M, N, 5 cm; T, 10 cm; W, 1 mm; X, Y, 1 μm. Modified from Figure 1 in Hosaka *et al*. (2006) using graphic files kindly supplied.by Dr Kentaro Hosaka, National Museum of Nature and Science, Japan. Reprinted with permission from Mycologia. ©The Mycological Society of America.

Figure 3.19. Morphological diversity in the cantharelloid clade. Basidiomata of: (A) *Cantharellus cibarius* (image by J.-M. Moncalvo). (B) *Craterellus tubaeformis* (image by M. Wood). (C) *Sistotrema confluens* (image by R. Halling). (D) *Multiclavula mucida* (image by M. Wood). (E) *Botryobasidium subcoronatum*, sporophores on an old polypore (image by E. Langer). (F) *Sistotrema coroniferum* (image by K.-H. Larsson). (G) *Clavulina cinerea* (image by E. Langer). Modified from Figure 2 in Moncalvo *et al.* (2006) using graphic files kindly supplied by Dr J.-M. Moncalvo, Royal Ontario Museum and Department of Botany, University of Toronto, Canada.

- *Phallomycetidae*
  - *Geastrales* (Figure 3.18)
  - *Gomphales* (Figure 3.18)
  - *Hysterangiales* (Figure 3.18)
  - *Phallales* (Figure 3.18)
- **Dacrymycetes** (Oberwinkler, 2014)
  - *Dacrymycetales* (Figure 3.22)
- **Tremellomycetes** (Weiss *et al.*, 2014)
  - *Cystofilobasidiales*
  - *Filobasidiales*
  - *Tremellales* (Figure 3.22)
- **Agaricomycetes** [*incertae sedis* (position still uncertain)]
  - *Auriculariales* (Figure 3.22)
  - *Cantharellales* (Figure 3.19)
  - *Corticiales*
  - *Gloeophyllales* (Figure 3.22)
  - *Hymenochaetales* (Figure 3.20)
  - *Polyporales* (Figure 3.22)
  - *Russulales* (Figure 3.21)
  - *Sebacinales* (Figure 3.22)
  - *Thelephorales* (Figure 3.22)
  - *Trechisporales* (Figure 3.22).

Names to look out for include *Phellinus weirii* ('laminated root rot' of many fir trees, and a root and butt rot of western red cedar), *Heterobasidion annosum* (the most damaging root pathogen of coniferous trees in the northern hemisphere), *Thanatephorus cucumeris*, which is the teleomorph of *Rhizoctonia solani* and

Figure 3.20. Macro- and micro-characters in *Hymenochaetales*. (A–I) Basidiome and hymenophore types. (A) *Cotylidia pannosa*, stipitate with smooth hymenophore (photo David Mitchel, http://www.nifg.org.uk/photos.htm). (B) *Coltricia perennis*, stipitate with poroid hymenophore. (C) *Contumyces rosella*, stipitate with lamellae (gills). (D) *Clavariachaete rubiginosa* (photo Roy Halling). (E) *Phellinus robustus*, sessile to effuse-reflexed with poroid hymenophore (photo Andrej Kunca, Forest Research Institute, Slovakia, http://www.forestryimages.org). (F) *Coltricia montagnei*, stipitate with contrical (concentric) lamellae (photo Dianna Smith, http://www.mushroo-mexpert.com). (G) *Hydnochaete olivacea*, resupinate to effuse-reflexed with coarse, compressed spines (called aculei). (H) *Resinicium bicolor*, resupinate with small, rounded aculei (spines). (I) *Hyphodontia arguta*, resupinate with acute aculei. (J) Setal cystidium of *Hymenochaete cinnamomea*. Scale bars: A, D, 10 mm; B, 2 mm; G–I, 1 mm, J, 10 μm. Modified from Larsson *et al.* (2006) using graphic files kindly supplied by Professor Karl-Henrik Larsson, Göteborg University, Sweden. Reprinted with permission from Mycologia. ©The Mycological Society of America.

Figure 3.21. Sporophore morphology and hymenophore types in the *Russulales*. (A) Pileate hydnoid sporophores of *Auriscalpium vulgare*, ×2. (B) Clavarioid smooth sporophore of *Artomyces pyxidata*, ×0.5. (C) Effused-reflexed smooth sporophores of *Stereum ostrea*, ×0.3. (D) Discoid smooth sporophores of *Peniophora rufa*, ×0.5. (E) Resupinate (effused or corticioid) smooth sporophores of *Variaria investiens*, ×0.2. (F) Pileate lamellate (agaricoid) sporophores of *Russula discopus*, ×0.5. Modified from Miller *et al.* (2006) using images kindly supplied by Dr Steven L. Miller, University of Wyoming, USA. Reprinted with permission from Mycologia. ©The Mycological Society of America.

a very common soil-borne pathogen with a great diversity of host plants. Humans don't escape, either: *Filobasidiella neoformans* is the teleomorph of *Cryptococcus*, referred to above as the cause of cryptococcal meningitis in immunocompromised patients.

We turn the tables by eating enormous amounts of hymenomycete sporophores: all commercially farmed fungi are hymenomycetes; these include the **button mushroom** (*Agaricus bisporus*), **oyster mushroom** (*Pleurotus* spp.), **shiitake** (*Lentinula edodes*), **paddy straw mushroom** (*Volvariella volvacea*), **enokitake** (*Flammulina velutipes*) and **shimejitake** (*Hypsizygus tessulatus*). Other edible mushrooms with massive worldwide markets (amounting to billions of dollars annually) are field-collected mycorrhizal species like *Cantharellus cibarius* (the **chanterelle**), *Boletus edulis* (**cep, penny bun** or **porcini**) and *Tricholoma matsutake* (**matsutake**).

For a long time, the Guinness Book of World Records gave the **largest known sporophore** of a fungus as a 160 kg specimen of the polypore *Bridgeoporus nobilissimus*. This has since been displaced by a very large specimen of *Rigidoporus ulmarius* (a bracket fungus that grows at the base of the trunks of deciduous trees, usually elm) growing in the Royal Botanic Gardens at Kew which, in the 1999 edition, was recorded as 163 × 140 cm with a circumference of 480 cm. Sporophores of this size put the **giant puffball** (*Calvatia gigantea*) in the shade, as these (edible) hymenomycete sporophores average around 10 to 70 cm diameter, although they have been known to reach diameters up to 150 cm and weights of 20 kg. The largest, most extensive, and longest-lived mycelium currently known is a mycelium of *Armillaria gallica* in the Blue Mountains/Malheur National Forest in Eastern Oregon, USA. This one individual fungus covers an area of nearly 900 ha (3.4 square miles) and is estimated to be more than 2,400 years old.

Figure 3.22. Sporophores of jelly fungi and crust fungi (resupinate or inverted forms) of the *Agaricomycotina*. (A) *Tremella mesenterica* (*Tremellales*). (B) *Tremella fuciformis* (*Tremellales*). (C) *Dacryopinax spathularia* (*Dacrymycetales*). (D) *Tremellodendron pallidum* (*Sebacinales*). (E) *Auricularia auricula-judae* (*Auriculariales*). (F) *Exidiopsis* sp. (*Auriculariales*). (G) *Trechispora* sp. (*Trechisporales*). (H) *Tomentella* sp. (*Thelephorales*). (I) *Athelia* sp. (probably *Atheliales*). (J) *Veluticeps* sp. (*Gloeophyllales*). (K) *Phlebia* sp. (*Polyporales*). (L) *Ganoderma australe* (*Polyporales*). (M) *Hydnellum* sp. (*Thelephorales*). (N) *Neolentinus lepideus* (*Gloeophyllales*). Images A–C and F–L by Heino Lepp, Australian National Botanical Gardens (http://www.anbg.gov.au/index.html); images D and E by Pamela Kaminski; used with permission and modified from Hibbett (2006) using images kindly supplied by Dr David S. Hibbett, Clark University, USA. Reprinted with permission from Mycologia. ©The Mycological Society of America.

3.9 THE SPECIES CONCEPT IN FUNGI

Most of what we have discussed so far depends absolutely on the ability to identify each organism, so that you can state confidently the exact name of the species you are dealing with. That, of course, applies to everything you might want to do with a fungus; whether you want to image it or sequence its genome, you must give it an accurate name. The identification side of this depends on accurate descriptions, and the knowledge and skill of the person making the identification. Accurate descriptions should be assured by the international rules of taxonomy and nomenclature; knowledge and skill can be learned. As more use is made of molecular detection methods a rapidly increasing number of **cryptic species** are being discovered. This is where two or more species are hidden under one species name because the boundaries between them are not evident in the existing species definition. When a new method of analysis is applied the boundaries between closely related sibling species become evident. Cryptic species may be unrecognised pathogens, or toxin producers or spoilage organisms. Global transport and trade make it imperative that we have effective and reliable methods to identify these hidden hazards (Cai *et al.*, 2011). What is worrying is that we are not very clear about the 'unit' you might need to identify.

We know you need to identify a particular species; but what is 'a species'? The debate over how species should be defined (the 'species concept') has been going on for many years but we don't have a universal definition of 'species' that meets with widespread agreement. It is all a matter of your notion of how a 'species' might be described, your ***species concept***; and there might be more than one way of defining the species in which you have an interest. Mayden (1997) discussed 22 different species concepts. We do not intend to go to those lengths, but we will consider the most commonly applied species concepts in fungal taxonomy.

- **The morphological species concept**. Historically, biologists, including mycologists, first recognised species based on morphological similarity; this creates what are known as morphospecies. The difficulty with this is in finding characters that convincingly define the boundaries of a species (you need to know the boundaries so that the next specimen can either be placed within or outside that species). If you doubt the difficulty, look around at other people and think what morphological features you might use for describing the human species. Body morphology? Skin colour? Eye colour? Hair distribution? Face shape? You use all of these to identify individuals, but they are poor candidates to circumscribe the species. In fungi, there are not only fewer morphological characters to use, but they may also be produced infrequently and/or temporarily and/or very variable. In addition, the natural variation of characters may be influenced by environmental variation in ways that are difficult to measure. Add to these issues the fact that similar morphologies can be arrived at by very different evolutionary routes (convergent evolution) and you'll not be surprised to learn that most biologists, and mycologists in particular, believe that the morphological species concept is the least satisfactory one to use. Unfortunately, the foundation for fungal (particularly mushroom) classification was established by Elias Fries between about 1820 and 1875, with emphasis on morphological features, such as gills, pores and spore colour. This was practical for identification, but it obscured the phylogenetic origins of many of these features.
- **The biological species concept**. The dominant idea in biology (especially animal biology) is the biological species concept in which a species is defined as an interbreeding population that is somehow reproductively isolated from other populations. This looks simple and obvious but is also flawed; and especially so for fungi. The first problem is to identify the reproductive barrier(s) between populations made up of morphologically similar individuals. The concept is not even applicable to homothallic and asexual organisms, which excludes about 20% of fungi; nor is it applicable to the large number of organisms that cannot be cultivated, because it depends on mating tests being carried out on specimens raised in the laboratory. The biological species concept also has difficulty coping with populations that are geographically isolated (or rather, the meaning of geographic isolation is difficult to apply consistently). Populations that do not interbreed because of geographic isolation will evolve independently and may well diverge sufficiently to become different species on other criteria, and yet still interbreed successfully when brought together artificially if they retain common ancestral sexual characteristics. The underlying question, though, is what constitutes an effective reproductive barrier. Geography, even global geography, does not help with fungi because within the group there are many organisms that produce spores that can glide through the atmosphere across and between continents, ignoring oceans and mountain ranges. At the other extreme, there are microscopic fungi that may be located in such restricted habitats that they are effectively isolated from their relatives just a few metres away. Overall, there are too many severe restrictions on application of the biological species concept widely in fungi for it to be a serious contender for that elusive universal definition.
- **Ecological and physiological species concepts**. It seems a reasonable assumption that parasitic or symbiotic fungi have at least some measure of host specificity and that such a fungal species might be defined on the basis of habitat and/or host relationships. Equally reasonable are the expectations that ecological adaptations influence fungal speciation, and that physiological features contributing to adaptation to habitat and/or host could characterise a fungal

species. In effect, the habitat and/or host relationship is being cast in the role of reproductive isolation mechanism. An ecological and/or physiological species concept has been used for a long time with plant pathogenic fungi. The concept chiefly differentiates species by their ecological niche and the constraints on their evolution that determine their maintenance and reproduction in that niche (and the niche may be a specific species of host plant, or even a specific cultivar of a host species). Again, it sounds reasonable, but there are problems. We are largely ignorant of the exact physiological, biochemical and/or genetic nature of whatever it is that determines substrate specificity and host specificity, so it is little better than using morphological characters. Furthermore, application of the concept is severely limited in practice. A plant pathologist might be able to characterise a fungus of interest from its host spectrum, or a medical mycologist from its serotype, but there is a vast range of other fungi for which this approach simply fails. It is another concept that cannot offer the universal definition that would be most valuable.

- **Evolutionary/phylogenetic species concept**. Definitions based on molecular analyses seem likely to be the most promising. Their foundation is the ancestry of the species. They envisage a species as being a monophyletic group of organisms sharing molecular characters that derive from a common ancestor. This, and the fact that the operation is done with the DNA sequence itself, is inherently satisfying. Furthermore, the concept does not have any obvious built-in exclusions or limitations. Particularly important to fungi is the fact that the analysis can be applied to asexual organisms; anamorphic and teleomorphic stages can be covered by a single species concept. Nor is there any shortage of characters; fungal genomes provide more than enough sequences for this to be a definition of universal applicability. Clearly, if molecular analyses could be found that complement the historically established morphological species definitions there is a promise that species could be identified rapidly, and biological diversity evaluated reliably. We have already referred to several studies in which sequence analyses of multiple genes have been used in phylogenetic studies and such markers have been applied to the ***identification*** of fungal species since the 1990s. However, a new strategy based on the sequencing of standardised genomic fragments (**DNA barcoding**) was established early in the twenty-first century (Costa & Carvalho, 2007; Hollingsworth, 2007). The **DNA barcode** approach involves the use of a standard DNA region that is specific for a taxonomic group. The DNA marker you choose is amplified and sequenced. The sequencer generates a 'barcode' because it detects the four bases that make up the sequence using fluorescent dyes and the sequencer's colorimeter reports the sequence by printing appropriately coloured peaks (or bars) that designate the sequence. If the sequence contains a run of the same base, then the peak (or bar) is broad; where the sequence has a single base, the peak (or bar) is narrow. Just as the unique pattern of bars in a universal product code (UPC) in the supermarket identifies each consumer product, a 'DNA barcode' is a unique pattern of DNA sequence that identifies each living thing (providing you've made the right choice of DNA marker!). Botanists tend to use a portion of the chloroplast DNA of plants, those interested in animals can use mitochondrial cytochrome *c* oxidase subunit I or 16S rRNA. Schoch *et al.* (2012) identified several DNA regions as promising universal barcodes for fungi and found that the **nuclear ribosomal RNA internal transcribed spacer** (ITS) region exhibited the highest probability of correct identification for a range of fungal lineages. Since then, the **ITS region** has been accepted as the standard barcode marker for fungi, although there are indications that this sequence is not always useful in identifying genera and a careful, informed selection must be made (Badotti *et al.*, 2017; Truong *et al.*, 2017). Another word of warning is that different analytical platforms yield different outcomes from the same data sets. The time required for computation differs, but more importantly, the quality of error filtering and hence the quality of the output largely depends on the platform used. Therefore, taxonomic assignments derived from ITS data need to be validated by other methodologies (Anslan *et al.*, 2018). However, as techniques improve it may be only a matter of effort and time for this approach to be applied effectively to a good range of fungal species.

Different species concepts give different outcomes. Use of different methods of defining species inevitably results in recognition of different things. The indications are that application of the phylogenetic species concept will lead to the recognition of far more species than have already been recognised using morphological, biological or physiological species concepts.

A study across all major groups of eukaryotes examining the effect of applying the phylogenetic species concept to organisms previously well classified using other criteria showed an average 48% increase in the number of species documented. The increase tends to be greater in fungi. In most traditional fungal groups, a reclassification using molecular methods together with the phylogenetic species concept leads to a 2–4 times increase in the number of species recognised.

This means more names and more taxonomy, which is not greeted very sympathetically by those who are not taxonomists. But just think what biological truth is being revealed here. If molecular sequence analysis characteristically reveals 2–4 times more species than we know about using non-molecular methods, then the extent of biodiversity and the species richness of every ecosystem are 2–4 times greater than we currently imagine. And that is just for 'species' we know about!

3.10 THE UNTRUE FUNGI

In our discussion so far, we have sometimes used the description **'true fungi'** to refer to those organisms that belong to Kingdom *Fungi*. This implies that there are some organisms that might be called **'untrue fungi'** because they look like fungi but are not fungi. And, well, yes this is the case (Rossman & Palm, 2006).

Water moulds. This is an informal grouping that includes the most ancient fungi and fungus-like organisms:

- The ***Chytridiomycota***, water moulds that are the ancestral group of the true fungi (Kingdom *Fungi*) and which we have already discussed (section 3.2 above).
- ***Oomycota*** and ***Hyphochytriomycota***, water moulds that are not true fungi but have affinities to the algae. They are now placed in **Kingdom *Chromista*** (as Subphylum ***Pseudofungi***; Cavalier-Smith, 2018), or (our preference) **Kingdom *Straminipila*** (Beakes *et al.*, 2014).

Phylum *Oomycota* (Beakes *et al.*, 2014) consists of about 600 species in 90 genera, placed in the following orders:

- *Leptomitales*; example genera: *Apodachlyella, Ducellieria, Leptolegniella, Leptomitus*
- *Myzocytiopsidales*; example genus: *Crypticola*
- *Olpidiopsidales*; example genus: *Olpidiopsis*
- *Peronosporales*; example genera: *Albugo, Peronospora, Bremia, Plasmopara*
- *Pythiales*; example genera *Pythium, Phytophthora, Pythiogeton*
- *Rhipidiales*; example genus: *Rhipidium*
- *Salilagenidiales*; example genus: *Haliphthoros*
- *Saprolegniales*; example genera: *Leptolegnia, Achlya, Saprolegnia*
- *Sclerosporales*; example genera: *Sclerospora, Verrucalvus*
- *Anisolpidiales*; example genus: *Anisolpidium*
- *Lagenismatales*; example genus: *Lagenisma*
- *Rozellopsidales*; example genera: *Pseudosphaerita, Rozellopsis*
- *Haptoglossales*; example genera: *Haptoglossa, Lagena, Electrogella, Eurychasma, Pontisma, Sirolpidium*.

The *Oomycota* produce **biflagellate zoospores** bearing one whiplash and one tinsel-type flagellum on each zoospore (Table 3.1; see also Chapter 2 in Carlile *et al.*, 2001). The genera *Saprolegnia, Achlya, Phytophthora* and *Pythium* are prime examples.

Although these are primitive organisms, at least in terms of their evolutionary position with respect to Fungi, they are highly adapted to their lifestyle. This is illustrated by the behaviour of **zoospores** of *Oomycota*:

- Temperature influences the fate of the zoosporangium in *Phytophthora infestans* (cause of potato blight): below 15°C the zoosporangium forms zoospores, but above 20°C it forms

Table 3.1. Some differences between *Oomycota* and true fungi

Character	*Oomycota*, Kingdom *Straminipila*	*Eumycota*, Kingdom *Fungi*
Flagella, when present	Biflagellated cells with anterior tinsel and posterior whiplash flagella	Uniflagellate or multiflagellate cells with posterior whiplash flagella
Microfibrils in cell walls	Cellulose (also present in algal and plant cell walls)	Chitin or chitosan
Protein in cell wall characterised by	Hydroxyproline	Proline
Sterol in membrane	Cholesterol, demosterol	Ergosterol
Ploidy throughout most of life cycle	Diploid	Usually haploid or dikaryotic, rarely diploid
Cristae in mitochondria	Tubular	Plate-like (like most animals)
Golgi	Stacks of cisternae, similar to those in plants and algae	Simple cisternae which are not stacked
Intermediate in lysine biosynthesis	α-ε-diaminopimelic acid (DAP) (as in algae and plants)	α-aminoadipic acid (AAA) (as in euglenoid protozoa)
Nicotinamide adenine dinucleotide (NAD)-linked isocitric dehydrogenase	Absent	Present
Base pair at the very base of helix 47 in the variable regions of the 18S rRNA sequences	AU	UA

a germ tube. So, in the cold water of the soil, the zoospores will swim to find new hosts; in the warm sunshine, the sporangium will infect the plant.

- After release, zoospores typically swim for many hours; zoospores of *Phytophthora megasperma* swim at 88 μm s^{-1} at 15°C; it only takes them 11 s to swim a distance of 1 mm.
- Zoospores may show amoeboid movement when in contact with a solid substratum; allowing slow-speed targeting of the host.
- Zoospores show tactic movements. A 'taxis' is a ***movement*** towards or away from a stimulus (tropism is ***growth*** towards or away from a stimulus).
- Zoospores of *Phytophthora palmivora* are negatively geotactic (they swim upwards; which is where the nice newly formed host leaves and buds will be found).
- Zoospores of *Pythium aphanidermatum* have a positive chemotaxis to roots.

Hyphae of *Oomycota* are also positively chemotropic (hyphae of true fungi are not chemotropic, although they do exhibit other tropisms, particularly autotropism).

Saprolegnia is another important genus. *Saprolegnia* species are parasites of freshwater fish and fish eggs and can cause economic damage on fish farms. Reproduction is mainly asexual, but the life cycle includes a sexual phase (Figure 3.23).

Hyphal branches become modified into long **zoosporangia** separated from the hypha by septa. Biflagellate zoospores released from a zoosporangium swim for a while and then encyst. Each eventually gives rise to a secondary zoospore, which also encysts and then germinates to produce a new mycelium.

For sexual reproduction, compatible **oogonia** and **antheridia** develop on the same diploid mycelium. Meiosis occurs within these gametangia. In mating, antheridia grow towards the oogonia and develop tubular processes called 'fertilisation tubes', which penetrate the oogonia. Male nuclei travel through the fertilisation tubes to fuse with the female nuclei within (karyogamy). Following karyogamy, a thick-walled zygote, called the oospore, is produced. The oospore germinates into hyphae, which then produce a zoosporangium.

Pheromones in *Achlya*:

- *Achlya bisexualis* is heterothallic. Sterols are involved in the hormonal mechanisms regulating sexual reproduction in *Achlya* (as they are in animals).
- The female mycelium produces antheridiol, which induces male hyphae to make antheridial branches.
- Antheridial hyphae are attracted by antheridiol (this is a chemotropism) to the female mycelium.
- Male mycelium produces the hormone oogoniol (another sterol), which induces female mycelium to produce oogonia.
- Antheridia grow around the oogonium, producing fertilisation tubes (plasmogamy) and enable karyogamy which results in oospore formation. So even these primitive organisms have a sterol hormone mechanism for cell targeting.

Other members of **Kingdom *Straminipila*** are:

- **Phylum *Hyphochytriomycota*** (Beakes *et al.*, 2014), which are microscopic organisms that form a small thallus, often with branched rhizoids, occurring as parasites or saprotrophs on algae and fungi in freshwater and in soil. The whole of the thallus is eventually converted into a reproductive structure. Only 23 species (in six genera) are known and are placed in Order Hyphochytriales (example genera are: *Hyphochytrium, Rhizidiomyces*).
- **Phylum *Labyrinthulomycota*** (Beakes *et al.*, 2014), in which the feeding stage comprises an ectoplasmic network and spindle-shaped or spherical cells that move within the network by gliding over one another. They occur in both salt- and freshwater in association with algae and other

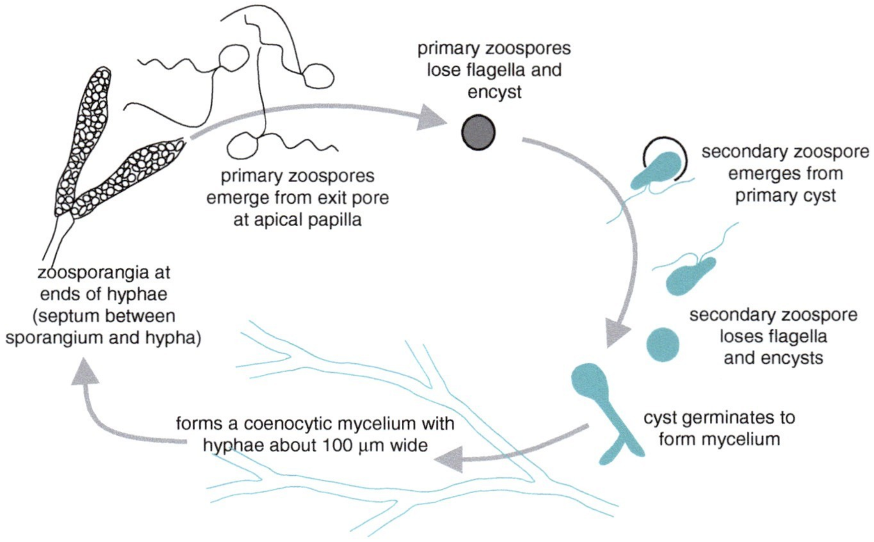

Figure 3.23. Diagrammatic life cycle of *Saprolegnia*.

chromists. There are about 45 species in 10 genera, placed in the two orders Labyrinthulales (e.g. *Labyrinthula*) and Thraustochytriales (e.g. *Thraustochytrium*).

Slime moulds. The organisms known as slime moulds are now all placed in **Kingdom *Protozoa***. They do not form hyphae, and they generally lack cell walls, being capable of ingesting food particles by phagocytosis. The slime moulds fail to meet normal definitions of fungi, but they produce sporophores which have a superficial resemblance to those of fungi, and this is why they have been called 'moulds' and have been studied by mycologists and included in most textbooks on mycology. They are placed in three phyla in the ***Protozoa***:

- **Order *Plasmodiophoromycota*** (included in the **Phylum *Phytomyxea***) are obligate intracellular symbionts or parasites of plant, algal or fungal cells living in freshwater or soil habitats. They have multinucleate, unwalled plasmodia. There are about 15 genera with 50 species in Order Plasmodiophorales; example genera are *Plasmodiophora, Polymyxa* and *Spongospora*. *Plasmodiophora* and *Spongospora* cause serious plant diseases (Bulman & Braselton, 2014).
- **Phylum *Myxomycota*** (in the **Supergroup Amoebozoa**) are free-living unicellular or plasmodial amoeboid slime moulds. A total of 900 species assigned to 80 genera, and seven orders, among which are the *Dictyosteliales* (e.g. *Dictyostelium*), *Physarales* (e.g. *Didymium, Physarum, Fuligo*) and *Stemonitales* (e.g. *Stemonitis*) (Stephenson, 2014).
- **Phylum *Acrasiomycota*** (in the **Supergroup Excavata**) are generally saprotrophic amoeboid slime moulds, found on a wide range of decaying plant material. A total of 12 species assigned to six genera in the single order *Acrasiales* (e.g. *Acrasis, Copromyxa*) (Stephenson, 2014).

Finally, in the **Phylum *Choanozoa*** (Kingdom *Protozoa*) are placed two groups of organisms previously misclassified as trichomycete fungi. They are now placed in **Class *Mesomycetozoea*** making up the orders ***Amoebidiales*** and ***Eccrinales***. They are intimately associated with arthropods, insects, millipedes and crustaceans and have a coenocytic thallus attached to the host by a holdfast.

3.11 ECOSYSTEM MYCOLOGY

We want to draw from this lengthy discussion of Kingdom *Fungi* a few key aspects about fungi in the natural environment and in their natural communities that we will develop more fully in later chapters (references to the relevant sections are included below).

The chytrids are the only true fungi that are **aquatic** and have actively motile spores. All other fungi are terrestrial. In fact, **fungi were among the first organisms to venture onto the land**, and they are now found in all terrestrial habitats. Often in association with other organisms. However, the statement that 'chytrids are the only true fungi that are aquatic' does not mean that other groups of fungi are excluded from aquatic habitats. Far from it. There are many fungi associated with freshwater (and which show extreme adaptations of their spores to a (passive) planktonic distribution), and with marine environments, especially mangroves (and spores of these species often produce enormously strong adhesives that work under water). However, it remains the case that less than 1% of the known species of fungi have been found in aquatic habitats (see pp. 346–351 in Carlile *et al.*, 2001; Landy & Jones, 2006).

A kingdom-specific characteristic of fungi is that they obtain their nutrients by **external digestion of substrates**. 'Wood' (that is, plant secondary cell wall) is the most widespread substrate on the planet. This is **lignocellulose**; an intimate mixture of lignin, hemicellulose and cellulose. About 95% of the Earth's terrestrial biomass is lignocellulose. Ability to degrade **lignin** is restricted to fungi – *Basidiomycota* and a few *Ascomycota* (see section 11.7).

Wood-degrading fungi are divided into those producing **white rots** and those causing **brown rots**. With a white rot, wood becomes markedly paler as the organism digests the (coloured) phenolics in lignin. This mainly affects hardwoods. In a white rot, hemicellulose, cellulose and lignin are degraded simultaneously. *Trametes versicolor, Phanerochaete chrysosporium* and *Xylaria polymorpha* (an ascomycete) are examples.

With a brown-rot fungus, wood becomes darker brown as the digestion proceeds. In this case hemicellulose and cellulose are preferentially removed, because the fungus does not degrade the lignin. This particularly affects softwoods. *Piptoporus betulinus, Serpula lacrymans* and *Coniophora puteana* are examples.

A **mycorrhiza** is a symbiotic interaction between fungi and plant roots that developed very early in the process of colonisation of the terrestrial environment. Mycorrhizas originated over 450 million years ago. More than 6,000 fungi are capable of forming mycorrhizas and at least 95% of vascular plants today have mycorrhizas associated with their roots. There are several types (see sections 14.8 to 14.17):

- **Endomycorrhizas**. Here the fungal structure is almost entirely within the host root, so that the root looks normal. AM (arbuscular mycorrhizas) are the commonest of all mycorrhizas, being associated with the roots of about 80% of plant species, including many crop plants. AM mycorrhizas have an ancient origin, fungi are assigned to Subphylum *Glomeromycotina* (section 3.6), e.g. *Glomus* spp. Enhanced growth of the plant host occurs mainly because the fungus improves phosphate availability to plant, but plant-to-plant transfer of nutrients can occur via the fungus.
- **Ericoid endomycorrhizas**. These are mycorrhizas of heather (*Erica*), ling (*Calluna*) and bilberry (*Vaccinium*); plants of mountain moorland and lowland heaths. The fungi involved here belong to the *Ascomycota*, e.g. *Hymenoscyphus ericae*. These mycorrhizas improve nitrogen and phosphorus uptake by the plant; the nitrogen being derived from the fungus breaking down polypeptides in the soil. In

extremely harsh conditions (e.g. winter in the Pennines), the mycorrhiza may support the host with carbon nutrients (again from polypeptide digestion). Normally, though, the fungus takes photosynthetically produced carbohydrates ***from*** the plant host.

- **Orchidaceous endomycorrhizas**. These are like ericoid mycorrhizas but have a carbon nutrition more dedicated to supporting the host. Orchids (Family Orchidaceae) form the largest and most diverse group of flowering plants, with over 800 described genera and 25,000 to 30,000 species (with another 100,000+ hybrids and cultivars produced by horticulturists since the introduction of tropical species in the nineteenth century). Because of their intimate relationships with pollinators and their symbiosis with orchid mycorrhizal fungi, orchids are considered, along with the grasses, to exhibit the most advanced flowering plant evolution. The orchidaceous endomycorrhizal fungus utilises complex carbon sources in soil and the products of digestion are made available to the orchid. At the seedling stage the orchid is dependent on the fungus and can be interpreted as parasitising the fungus as a result. An example is the fungus *Rhizoctonia* (basidiomycetous, anamorphic fungus which is a widespread pathogen of non-orchidaceous crop plants).
- **Ectomycorrhizas**. These are the most advanced symbiotic association between higher plants and fungi. The root system is surrounded by a sheath of fungal tissue up to several millimetres thick from which hyphae penetrate between the outermost cell layers of the root and from which a network of hyphal elements (hyphae, strands and rhizomorphs) extend out to explore the soil. About 3% of seed plants, including most forest trees (temperate and tropical), have ectomycorrhizas. Fungi involved are mostly *Basidiomycota* (also some *Ascomycota*). *Basidiomycota* include most of the common woodland mushrooms; for example, species of *Amanita, Boletus* and *Tricholoma*. There are both highly specific (*Boletus elegans* and larch) and nonspecific associations (*Amanita muscaria* with 20 or more trees). In the other 'specificity direction', 40 different fungal species can form mycorrhizas with pine. Ectomycorrhizal fungi depend on the host for the bulk of their carbon. Only a few of them can utilise cellulose and lignin as saprotrophs. The fungus provides enhanced mineral ion uptake to the plant, particularly phosphate and ammonium ions. Fungi efficiently utilise organic compounds in the soil containing nitrogen (polypeptides) and phosphate (nucleic acids); the plant cannot do this for itself. Most plants, pines especially, fail to grow or grow only poorly when they lack their ectomycorrhizas.

Lichens. These are usually associations between a fungus and a green alga, although the fungus can survive independently in nature (see section 14.18). Some lichens contain both a cyanobacterium and a green alga (so they are tripartite associations), and other bacteria may ***also*** contribute to the community. There are about 13,500 species of fungi involved in lichens, representing about 20% of all known fungi. They are mainly *Ascomycota* with a few *Basidiomycota*. Lichens are extremely resistant to environmental extremes and are pioneer colonisers on rock faces, tree bark, stone walls, roofing tiles, etc., as well as early colonisers of terrestrial habitats (there are fossils from the Triassic period).

Endophytes. Aside from pathogens or mycorrhizas some, maybe many, plants harbour other fungi that can affect their growth. These fungi are called 'endophytes' because they exist within plants (see section 14.19). Endophytes are at least harmless and may be beneficial. A wide range of plants have now been examined and endophytes have been found in most of them, including aquatic plants, and red and brown algae; indeed, it has been said that: 'All we know for certain is that endophytes are present in any healthy plant tissue!' (Sieber, 2007).

Many different fungi can be isolated from plants growing in their native habitat, representing all taxonomic groups of fungi. They can be present in most plant parts, but especially the leaves, which may be fully colonised by a variety of fungi within a few weeks of emergence. Endophytes remain within the plant tissue, except that sporophore structures may emerge through the surface of the plant. Most endophytes are transmitted horizontally; that is, each plant is colonised by fungal propagules that arrive from the environment. The source of transmission has been determined in only a few cases. Propagules of endophytes have been found in the body of insect pests of the plant host, and at least two insect pathogens have been identified as endophytes, so insects may disperse some fungi from plant to plant.

Endophytes became a hot research topic when it was found that some which live entirely within grasses are responsible for the toxicity of some grasses to livestock. It has become evident that there are numerous endophytic fungi. A functional relation to the host is not always obvious. Some may be simple passengers; living in the inner spaces of the plant in much the same way as they would live in any other moist, secluded place. But there are some intriguing stories, like the endophyte in oak leaves that remains dormant until an insect activates it by chewing on the leaf. The fungus responds to the insect attack by becoming a pathogen, killing a zone of the leaf surrounding the insect so the insect dies for lack of live leaf tissue to feed on. With the insect pest dead, the fungus returns to being harmless and the oak's leaves can photosynthesise in peace!

Epiphytes. Fungi that live on the surfaces of plants are called epiphytes (see section 14.20). Some show special adaptations to the plant surface, which is a challenging environment, being dry, waxy and exposed to direct sunlight. Epiphytes are often coloured (particularly melanised) to protect them from ultraviolet radiation, and some can digest lipids sufficiently to use the waxy layer covering the leaf epidermis. The yeast form usually has a short life cycle, which enables yeast epiphytes to multiply even if favourable conditions last only a short time.

3.12 REFERENCES

Aime, M.C., Matheny, P.B., Henk, D.A. *et al.* (2006). An overview of the higher level classification of Pucciniomycotina based on combined analyses of nuclear large and small subunit rDNA sequences. *Mycologia*, 98: 896–905. DOI: https://doi.org/10.3852/mycologia.98.6.896.

Aime, M.C., Toome, M. & Mclaughlin, D.J. (2014). Pucciniomycotina. In *The Mycota Systematics and Evolution*, VII part A, 2nd edition, ed. D.J. McLaughlin & J.W. Spatafora. Berlin: Springer-Verlag, pp. 271–294. ISBN: 978-3-642-55317-2. DOI: https://doi.org/10.1007/978-3-642-55318-9_10.

Anslan, S., Nilsson, R.H., Wurzbacher, C. *et al.* (2018). Great differences in performance and outcome of high-throughput sequencing data analysis platforms for fungal metabarcoding. *PeerJ Preprints*, 6: article e27019v2. DOI: https://doi.org/10.7287/peerj.preprints.27019v2.

Badotti, F., de Oliveira, F.S., Garcia, C.F. *et al.* (2017). Effectiveness of ITS and sub-regions as DNA barcode markers for the identification of Basidiomycota (Fungi). *BMC Microbiology*, 17: 42 (12 pp.). DOI: https://doi.org/10.1186/s12866-017-0958-x.

Beakes, G.W., Honda, D. & Thines, M. (2014). Systematics of the Straminipila: Labyrinthulomycota, Hyphochytriomycota, and Oomycota. In *The Mycota Systematics and Evolution*, VII part A, 2nd edition, ed. D.J. McLaughlin & J.W. Spatafora. Berlin: Springer-Verlag, pp. 39–97. ISBN: 978-3-642-55317-2. DOI: https://doi.org/10.1007/978-3-642-55318-9_3.

Begerow, D., Schäfer, A.M., Kellner, R. *et al.* (2014). Ustilaginomycotina. In *The Mycota Systematics and Evolution*, VII part A, 2nd edition, ed. D.J. McLaughlin & J.W. Spatafora. Berlin: Springer-Verlag, pp. 295–329. ISBN: 978-3-642-55317-2. DOI: https://doi.org/10.1007/978-3-642-55318-9_11.

Beimforde, C., Feldberg, K., Nylinder, S. *et al.* (2014). Estimating the Phanerozoic history of the Ascomycota lineages: combining fossil and molecular data. *Molecular Phylogenetics and Evolution*, 78: 386–398. DOI: https://doi.org/10.1016/j.ympev.2014.04.024.

Benny, G.L., Humber, R.A. & Voigt, K. (2014). Zygomycetous fungi: phylum Entomophthoromycota and subphyla Kickxellomycotina, Mortierellomycotina, Mucoromycotina, and Zoopagomycotina. In *The Mycota Systematics and Evolution*, VII part A, 2nd edition, ed. D.J. McLaughlin & J.W. Spatafora. Berlin: Springer-Verlag, pp. 209–250. ISBN: 978-3-642-55317-2. DOI: https://doi.org/10.1007/978-3-642-55318-9_8.

Binder, M. & Hibbett, D.S. (2006). Molecular systematics and biological diversification of Boletales. *Mycologia*, 98: 971–981. DOI: https://doi.org/10.3852/mycologia.98.6.971.

Blackwell, M., Hibbett, D.S., Taylor, J.W. & Spatafora, J.W. (2006). Research coordination networks: a phylogeny for Kingdom Fungi (deep hypha). *Mycologia*, 98: 829–837. DOI: https://doi.org/10.3852/mycologia.98.6.829.

Bonfante, P. & Desirò, A. (2017). Who lives in a fungus? The diversity, origins and functions of fungal endobacteria living in Mucoromycota. *The ISME Journal*, 11: 1727–1735. DOI: https://doi.org/10.1038/ismej.2017.21.

Bracker, C.E. (1968). The ultrastructure and development of sporangia in *Gilbertella persicaria*. *Mycologia*, 60: 1016–1067. DOI: https://doi.org/10.2307/3757290.

Bruns, T.D., Corradi, N., Redecker, D., Taylor, J.W. & Öpik, M. (2017). Glomeromycotina: what is a species and why should we care? *New Phytologist*, 2017: Viewpoint Early View (online version of record before inclusion in an issue), article 0028-646X. DOI: https://doi.org/10.1111/nph.14913.

Bulman, S. & Braselton, J.P. (2014). Rhizaria: Phytomyxea. In *The Mycota Systematics and Evolution*, VII part A, 2nd edition, ed. D.J. McLaughlin & J.W. Spatafora. Berlin: Springer-Verlag, pp. 99–112. ISBN: 978-3-642-55317-2. DOI: https://doi.org/10.1007/978-3-642-55318-9_4.

Cai, L., Giraud, T., Zhang, N. Begerow, D., Cai, G. & Shivas, R.G. (2011). The evolution of species concepts and species recognition criteria in plant pathogenic fungi. *Fungal Diversity*, 50: 121. DOI: https://doi.org/10.1007/s13225-011-0127-8.

Carlile, M.J., Watkinson, S.C. & Gooday, G.W. (2001). *The Fungi*, 2nd edition. London: Academic Press. ISBN: 0127384464.

Cavalier-Smith, T. (2018). Kingdom Chromista and its eight phyla: a new synthesis emphasising periplastid protein targeting, cytoskeletal and periplastid evolution, and ancient divergences. *Protoplasma*, 255: 297–357. DOI: https://doi.org/10.1007/s00709-017-1147-3.

Corradi, N. & Brachmann, A. (2017). Fungal mating in the most widespread plant symbionts? *Trends in Plant Science*, 22: 175–183. DOI: https://doi.org/10.1016/j.tplants.2016.10.010.

Costa, F.O. & Carvalho, G.R. (2007). The Barcode of Life Initiative: synopsis and prospective societal impacts of DNA barcoding of fish. *Genomics, Society and Policy*, 3: 29–40. DOI: https://doi.org/10.1186/1746-5354-3-2-29.

Didier, E.S., Becnel, J.J., Kent, M.L., Sanders, J.L. & Weiss, L.M. (2014). Microsporidia. In *The Mycota Systematics and Evolution*, VII part A, 2nd edition, ed. D.J. McLaughlin & J.W. Spatafora. Berlin: Springer-Verlag, pp. 115–140. ISBN: 978-3-642-55317-2. DOI: https://doi.org/10.1007/978-3-642-55318-9_5.

Edwards, J.E., Forster, R.J., Callaghan, T.M. *et al.* (2017). PCR and omics based techniques to study the diversity, ecology and biology of anaerobic fungi: insights, challenges and opportunities. *Frontiers in Microbiology*, 8: 1657. DOI: https://doi.org/10.3389/fmicb.2017.01657.

Gazis, R., Miądlikowska, J., Lutzoni, F., Arnold, A.E. & Chaverri, P. (2012). Culture-based study of endophytes associated with rubber trees in Peru reveals a new class of pezizomycotina: Xylonomycetes. *Molecular Phylogenetics and Evolution*, 65: 294–304. DOI: https://doi.org/10.1016/j.ympev.2012.06.019.

Geiser, D.M., Lobuglio, K.F. & Gueidan, C. (2015). Pezizomycotina: Eurotiomycetes. In *The Mycota Systematics and Evolution*, VII part B, 2nd edition, ed. D.J. McLaughlin & J.W. Spatafora. Berlin: Springer-Verlag, pp. 121–141. ISBN: 978-3-662-46010-8. DOI: https://doi.org/10.1007/978-3-662-46011-5_5.

Gleason, F.H., Kagami, M., LeFèvre, E. & Sime-Ngando, T. (2008). The ecology of chytrids in aquatic ecosystems: roles in food web dynamics. *Fungal Biology Reviews*, 22: 17–25. DOI: https://doi.org/10.1016/j.fbr.2008.02.001.

Gruninger, R.J., Puniya, A.K., Callaghan, A.M. *et al.* (2014). Anaerobic fungi (phylum Neocallimastigomycota): advances in understanding their taxonomy, life cycle, ecology, role and biotechnological potential. *FEMS Microbiology Ecology*, 90: 1–17. DOI: https://doi.org/10.1111/1574-6941.12383.

Gueidan, C., Hill, D.J., Miądlikowska, J. & Lutzoni, F. (2015). Pezizomycotina: Lecanoromycetes. In *The Mycota Systematics and Evolution*, VII part B, 2nd edition, ed. D.J. McLaughlin & J.W. Spatafora. Berlin: Springer-Verlag, pp. 89–120. ISBN: 978-3-662-46010-8. DOI: https://doi.org/10.1007/978-3-662-46011-5_4.

Halary, S., Daubois, L., Terrat, Y., Ellenberger, S., Wöstemeyer, J. & Hijri, M. (2013). Mating type gene homologues and putative sex pheromone-sensing pathway in arbuscular mycorrhizal fungi, a presumably asexual plant root symbiont. *PLoS ONE*, 8(11): article number e80729. DOI: https://doi.org/10.1371/journal.pone.0080729.

Halary, S., Malik, S.B., Lildhar, L., Slamovits, C.H., Hijri, M. & Corradi, N. (2011). Conserved meiotic machinery in *Glomus* spp., a putatively ancient asexual fungal lineage. *Genome Biology and Evolution*, 3: 950–958. DOI: https://doi.org/10.1093/gbe/evr089.

Hawksworth, D.L. & Lücking, R. (2017). Fungal diversity revisited: 2.2 to 3.8 million species. In *The Fungal Kingdom*, ed. J. Heitman, B. Howlett, P. Crous *et al.* Washington, DC: ASM Press, pp. 79–95. DOI: https://doi.org/10.1128/microbiolspec.FUNK-0052-2016.

He, M.-Q., Zhao, R.-L., Hyde, K.D., Begerow, D., Kemler, M. and 65 others. (2019). Notes, outline and divergence times of Basidiomycota. *Fungal Diversity*, **99**: 105–367. DOI: https://doi.org/10.1007/s13225-019-00435-4.

Hibbett, D.S. (2006). A phylogenetic overview of the Agaricomycotina. *Mycologia*, 98: 917–925. DOI: https://doi.org/10.3852/mycologia.98.6.917.

Hibbett, D.S., Bauer, R., Binder, M. *et al.* (2014). Agaricomycetes. In *The Mycota Systematics and Evolution*, VII part A, 2nd edition, ed. D.J. McLaughlin & J.W. Spatafora. Berlin: Springer-Verlag, pp. 373–429. ISBN: 978-3-642-55317-2. DOI: https://doi.org/10.1007/978-3-642-55318-9_14.

Hibbett, D.S., Binder, M., Bischoff, J.F. *et al.* (2007). A higher-level phylogenetic classification of the Fungi. *Mycological Research*, 111: 509–547. DOI: https://doi.org/10.1016/j.mycres.2007.03.004.

Hollingsworth, P.M. (2007). DNA barcoding: potential users. *Genomics, Society and Policy*, 3: 44–47. DOI: https://doi.org/10.1186/1746-5354-3-2-44.

Hosaka, K., Bates, S.T., Beever, R.E. *et al.* (2006). Molecular phylogenetics of the gomphoid-phalloid fungi with an establishment of the new subclass Phallomycetidae and two new orders. *Mycologia*, 98: 949–959. DOI: https://doi.org/10.3852/mycologia.98.6.949.

Hyde, K.D., Maharachchikumbura, S.S.N., Hongsanan, S. *et al.* (2017). The ranking of fungi: a tribute to David L. Hawksworth on his 70th birthday. *Fungal Diversity*, 84: 1–23. DOI: https://doi.org/10.1007/s13225-017-0383-3.

James, T.Y., Letcher, P.M., Longcore, J.E. *et al.* (2006). A molecular phylogeny of the flagellated fungi (Chytridiomycota) and description of a new phylum (Blastocladiomycota). *Mycologia*, 98: 860–871. DOI: https://doi.org/10.3852/mycologia.98.6.860.

James, T.Y., Porter, T.M. & Martin, W.W. (2014). Blastocladiomycota. In *The Mycota Systematics and Evolution*, VII part A, 2nd edition, ed. D.J. McLaughlin & J.W. Spatafora. Berlin: Springer-Verlag, pp. 177–208. ISBN: 978-3-642-55317-2. DOI: https://doi.org/10.1007/978-3-642-55318-9_7.

Jones, M.D.M., Richards, T.A., Hawksworth, D.L. & Bass, D. (2011). Validation and justification of the phylum name Cryptomycota phyl. nov. *IMA Fungus*, 2: 173–175. DOI: https://doi.org/10.5598/imafungus.2011.02.02.08.

Jordan, E.G., Birkett, J.A. & Trinci, A.P.J. (1980). Effect of temperature on nucleolar size in *Basidiobolus ranarum*. *Transactions of the British Mycological Society*, 74: 214–217. DOI: https://doi.org/10.1016/S0007-1536(80)80035-0.

Kerrigan, R.W. (2016). *Agaricus* of North America. *Memoirs of the New York Botanical Garden*, ***114***. New York: NYBG Press. 592 pp. ISBN: 978-0-89327-536-5.

Kück, U. & Pöggeler, S. (2009). Cryptic sex in fungi. *Fungal Biology Reviews*, 23: 86–90. DOI: https://doi.org/10.1016/j.fbr.2009.10.004.

Kurtzman, C.P. & Sugiyama, J. (2015). Saccharomycotina and Taphrinomycotina: the yeasts and yeastlike fungi of the Ascomycota. In *The Mycota Systematics and Evolution*, VII part B, 2nd edition, ed. D.J. McLaughlin & J.W. Spatafora. Berlin: Springer-Verlag, pp. 3–33. ISBN: 978-3-662-46010-8. DOI: https://doi.org/10.1007/978-3-662-46011-5_1.

Landy, E.T. & Jones, G.M. (2006). What is the fungal diversity of marine ecosystems in Europe? *Mycologist*, 20: 15–21. DOI: https://doi.org/10.1016/j.mycol.2005.11.010.

Larsson, K.-H., Parmasto, E., Fischer, M. *et al.* (2006). Hymenochaetales: a molecular phylogeny for the hymenochaetoid clade. *Mycologia*, 98: 926–936. DOI: https://doi.org/10.3852/mycologia.98.6.926.

Lessie, P.E. & Lovett, J.S. (1968). Ultrastructural changes during sporangium formation and zoospore differentiation in *Blastocladiella emersonii*. *American Journal of Botany*, 55: 220–236. URL: https://www.jstor.org/stable/2440456?seq=1#page_scan_tab_contents.

Luginbuehl, L.H., Menard, G.N., Kurup, S. *et al.* (2017). Fatty acids in arbuscular mycorrhizal fungi are synthesized by the host plant. *Science*, 2017: article aan0081. DOI: https://doi.org/10.1126/science.aan0081.

Matheny, P.B., Curtis, J.M., Hofstetter, V. *et al.* (2006). Major clades of Agaricales: a multilocus phylogenetic overview. *Mycologia*, 98: 982–995. DOI: https://doi.org/10.3852/mycologia.98.6.982.

Mayden, R.L. (1997). A hierarchy of species concepts: the denouement in the saga of the species problem. In *Species: The Units of Biodiversity*, ed. M.F. Claridge, H.A. Dawah & M.R. Wilson. London: Chapman and Hall, pp. 381–424. ISBN-10: 0412631202, ISBN-13: 978-0412631207.

Miller, S.L., Larsson, E., Larsson, K.-H., Verbeken, A. & Nuytinck, J. (2006). Perspectives in the new Russulales. *Mycologia*, 98: 960–970. DOI: https://doi.org/10.3852/mycologia.98.6.960.

Moncalvo, J.-M., Nilsson, R.H., Koster, B. *et al.* (2006). The cantharelloid clade: dealing with incongruent gene trees and phylogenetic reconstruction methods. *Mycologia*, 98: 937–948. DOI: https://doi.org/10.1080/15572536.2006.11832623.

Moore, D. (1998). *Fungal Morphogenesis*. New York: Cambridge University Press. ISBN-10: 0521552958, ISBN-13: 978-0521552950.

Moore, D. & Novak Frazer, L. (2002). *Essential Fungal Genetics*. New York: Springer-Verlag. ISBN-10: 0387953671, ISBN-13: 978-0387953670.

Oberwinkler, F. (2014). Dacrymycetes. In *The Mycota Systematics and Evolution*, VII part A, 2nd edition, ed. D.J. McLaughlin & J.W. Spatafora. Berlin: Springer-Verlag, pp. 357–372. ISBN: 978-3-642-55317-2. DOI: https://doi.org/10.1007/978-3-642-55318-9_13.

Oehl, F., Sieverding, E., Palenzuela, J., Ineichen, K. & Alves da Silva, G. (2011). Advances in Glomeromycota taxonomy and classification. *IMA Fungus*, 2: 191–199. DOI: https://doi.org/10.5598/imafungus.2011.02.02.10.

Padamsee, M. Kumar, T.K.A., Riley, R. *et al.* (2012). The genome of the xerotolerant mold *Wallemia sebi* reveals adaptations to osmotic stress and suggests cryptic sexual reproduction. *Fungal Genetics and Biology*, 49: 217–226. DOI: https://doi.org/10.1016/j.fgb.2012.01.007.

Peter, J., De Chiara, M., Friedrich, A. *et al.* (2018). Genome evolution across 1,011 *Saccharomyces cerevisiae* isolates. *Nature*, 556: 339–344. DOI: https://doi.org/10.1038/s41586-018-0030-5.

Pfister, D.H. (2015). Pezizomycotina: Pezizomycetes, Orbiliomycetes. In *The Mycota Systematics and Evolution*, VII part B, 2nd edition, ed. D.J. McLaughlin & J.W. Spatafora. Berlin: Springer-Verlag, pp. 35–55. ISBN: 978-3-662-46010-8. DOI: https://doi.org/10.1007/978-3-662-46011-5_2.

Pommerville, J.C., Strickland, J.B. & Harding, K.E. (1990). Pheromone interactions and ionic communication in gametes of aquatic fungus *Allomyces macrogynus*. *Journal of Chemical Ecology*, 16: 121–131. DOI: https://doi.org/10.1007/BF01021274.

Powell, M.J. & Letcher, P.M. (2014). Chytridiomycota, Monoblepharidomycota, and Neocallimastigomycota. In *The Mycota Systematics and Evolution*, VII part A, 2nd edition, ed. D.J. McLaughlin & J.W. Spatafora. Berlin: Springer-Verlag, pp. 141–176. ISBN: 978-3-642-55317-2. DOI: https://doi.org/10.1007/978-3-642-55318-9_6.

Prieto, M., Baloch, E., Tehler, A. & Wedin, M. (2013), Mazaedium evolution in the Ascomycota (Fungi) and the classification of mazaediate groups of formerly unclear relationship. *Cladistics*, 29: 296–308. DOI: https://doi.org/10.1111/j.1096-0031.2012.00429.x.

Redecker, D. & Raab, P. (2006). Phylogeny of the Glomeromycota (arbuscular mycorrhizal fungi): recent developments and new gene markers. *Mycologia*, 98: 885–895. DOI: https://doi.org/10.3852/mycologia.98.6.885.

Redecker, D. & Schüßler, A. (2014). Glomeromycota. In *The Mycota Systematics and Evolution*, VII part A, 2nd edition, ed. D.J. McLaughlin & J.W. Spatafora. Berlin: Springer-Verlag, pp. 251–270. ISBN: 978-3-642-55317-2. DOI: https://doi.org/10.1007/978-3-642-55318-9_9.

Redecker, D., Schüßler, A., Stockinger, H. *et al.* (2013). An evidence-based consensus for the classification of arbuscular mycorrhizal fungi (Glomeromycota). *Mycorrhiza*, 23: 515–531. DOI: https://doi.org/10.1007/s00572-013-0486-y.

Reeves, F. Jr (1967). The fine structure of ascospore formation in *Pyronema domesticum*. *Mycologia*, 59: 1018–1033. DOI: https://doi.org/10.2307/3757272.

Rossman, A.Y. & Palm, M.E. (2006). Why are *Phytophthora* and other Oomycota not true fungi? *Outlooks on Pest Management*, 17: 217–219. URL: https://www.apsnet.org/edcenter/intropp/PathogenGroups/Pages/Oomycetes.aspx.

Schoch, C. & Grube, M. (2015). Pezizomycotina: Dothideomycetes and Arthoniomycetes. In *The Mycota Systematics and Evolution*, VII part B, 2nd edition, ed. D.J. McLaughlin & J.W. Spatafora. Berlin: Springer-Verlag, pp. 143–176. ISBN: 978-3-662-46010-8. DOI: https://doi.org/10.1007/978-3-662-46011-5_6.

Schoch, C.L., Seifert, K.A., Huhndorf, S. *et al.* (2012). Nuclear ribosomal internal transcribed spacer (ITS) region as a universal DNA barcode marker for Fungi. *Proceedings of the National Academy of Sciences of the United States of America*, 109: 6241–6246. DOI: https://doi.org/10.1073/pnas.1117018109.

Sieber, T.N. (2007). Endophytic fungi in forest trees: are they mutualists? *Fungal Biology Reviews*, 21: 75–89. DOI: https://doi.org/10.1016/j.fbr.2007.05.004.

Spatafora, J.W., Chang, Y., Benny, G.L. *et al.* (2016). A phylum-level phylogenetic classification of zygomycete fungi based on genome-scale data. *Mycologia*, 108: 1028–1046. DOI: https://doi.org/10.3852/16-042.

Spatafora, J.W., Sung, G.-H., Johnson, D. *et al.* (2006). A five-gene phylogeny of Pezizomycotina. *Mycologia*, 98: 1018–1028. DOI: https://doi.org/10.3852/mycologia.98.6.1018.

Stephenson, S.L. (2014). Excavata: Acrasiomycota; Amoebozoa: Dictyosteliomycota, Myxomycota. In *The Mycota Systematics and Evolution*, VII part A, 2nd edition, ed. D.J. McLaughlin & J.W. Spatafora. Berlin: Springer-Verlag, pp. 21–38. ISBN: 978-3-642-55317-2. DOI: https://doi.org/10.1007/978-3-642-55318-9_2.

Sugiyama, J., Hosaka, K. & Suh, S.-O. (2006). Early diverging Ascomycota: phylogenetic divergence and related evolutionary enigmas. *Mycologia*, 98: 996–1005. DOI: https://doi.org/10.3852/mycologia.98.6.996.

Suh, S.-O., Blackwell, M., Kurtzman, C.P. & Lachance, M.-A. (2006). Phylogenetics of Saccharomycetales, the ascomycete yeasts. *Mycologia*, 98: 1006–1017. DOI: https://doi.org/10.3852/mycologia.98.6.1006.

Tang, N., San Clemente, H., Roy, S. *et al.* (2016). A survey of the gene repertoire of *Gigaspora rosea* unravels conserved features among Glomeromycota for obligate biotrophy. *Frontiers in Microbiology*, 7: 233. DOI: https://doi.org/10.3389/fmicb.2016.00233.

Tehler, A., Little, D.P. & Farris, J.S. (2003). The full-length phylogenetic tree from 1551 ribosomal sequences of chitinous fungi, *Fungi*. *Mycological Research*, 107: 901–916. DOI: https://doi.org/10.1017/S0953756203008128.

Trinci, A.P.J., Davies, D.R., Gull, K. *et al.* (1994). Anaerobic fungi in herbivorous animals. *Mycological Research*, 98: 129–152. DOI: https://doi.org/10.1016/S0953-7562(09)80178-0.

Truong, C., Mujic, A.B., Healy, R. *et al.* (2017). How to know the fungi: combining field inventories and DNA-barcoding to document fungal diversity. *New Phytologist*, 214: 913–919. DOI: https://doi.org/10.1111/nph.14509.

van der Giezen, M. (2002). Strange fungi with even stranger insides. *Mycologist*, 16: 129–131. DOI: https://doi.org/10.1017/s0269915x02003051.

Wang, X., Liu, X. & Groenewald, J.Z. (2017). Phylogeny of anaerobic fungi (phylum Neocallimastigomycota), with contributions from yak in China. *Antonie van Leeuwenhoek*, 110: 87–103. DOI: https://doi.org/10.1007/s10482-016-0779-1.

Weiss, M., Bauer, R., Sampaio, J.P. & Oberwinkler, F. (2014). Tremellomycetes and related groups. In *The Mycota Systematics and Evolution*, VII part A, 2nd edition, ed. D.J. McLaughlin & J.W. Spatafora. Berlin: Springer-Verlag, pp. 331–355. ISBN: 978-3-642-55317-2. DOI: https://doi.org/10.1007/978-3-642-55318-9_12.

White, M.M., James, T.Y., O'Donnell, K., Cafaro, M.J., Tanabe, Y. & Sugiyama, J. (2006). Phylogeny of the Zygomycota based on nuclear ribosomal sequence data. *Mycologia*, 98: 872–884. DOI: https://doi.org/10.3852/mycologia.98.6.872.

Zajc, J., Jančič, S., Zalar, P. & Gunde-Cimerman, N. (2016). *Wallemia*. In *Molecular Biology of Food and Water Borne Mycotoxigenic and Mycotic Fungi*, ed. R.R.M. Paterson & N. Lima. Boca Raton, FL CRC Press, pp. 569–581. ISBN-10: 1466559861, ISBN-13: 978-1466559868.

Zajc, J., Liu, Y., Dai, W. *et al.* (2013). Genome and transcriptome sequencing of the halophilic fungus *Wallemia ichthyophaga*: haloadaptations present and absent. *BMC Genomics*, 14: 617. DOI: https://doi.org/10.1186/1471-2164-14-617.

Zhang, N. & Wang, Z. (2015). Pezizomycotina: Sordariomycetes and Leotiomycetes. In *The Mycota Systematics and Evolution*, VII part B, 2nd edition, ed. D.J. McLaughlin & J.W. Spatafora. Berlin: Springer-Verlag, pp. 57–88. ISBN: 978-3-662-46010-8. DOI: https://doi.org/10.1007/978-3-662-46011-5_3.

CHAPTER 4

Hyphal Cell Biology and Growth on Solid Substrates

Although their mode of nutrition is important in defining members of Kingdom *Fungi*, the fundamental aspect of cell biology which sets most fungi off from most members of the other major kingdoms is the **apical extension** of their tubular hyphae. These possess controls which ensure that hyphae normally grow away from one another to form the typical 'colony' with an outwardly migrating growing front. Extension growth of the hypha is limited to the apex and this pattern of growth makes the vegetative fungal mycelium an exploratory, invasive organism; and **exploration and invasion is the fundamental lifestyle of fungi**. This lifestyle allows filamentous fungi to dominate their ecosystems because it gives them the tools they need to find and colonise new substrates rapidly. The success of this growth habit can be judged from the extraordinary diversity of fungal species, their distribution in virtually every habitat on the planet and the parallel evolution of a similar growth strategy by other important soil microorganisms, the prokaryotic streptomycetes and some of the *Oomycota* in Kingdom *Straminipila* (e.g. *Saprolegnia* and *Achlya*; see section 3.10). A general introduction to the fungal lifestyle can be found in Chapter 3 of the book *Slayers, Saviors, Servants, and Sex: An Exposé of Kingdom Fungi* (Moore, 2001).

In this chapter, we will discuss the hyphal mode of growth in some detail, explaining how hyphae emerge during spore germination and how hyphae contribute to colony formation. Mycelium growth kinetics is a key topic in understanding the nature of fungi; here we show how that understanding has been built from experiments with living fungi. We consider how fungal colonies grow to maturity and the morphological differentiation that can be seen in fungal colonies. At the cellular, or hyphal, level we show the meaning of 'duplication cycle' in moulds, how this depends on regulation of nuclear migration, and how it contributes to hyphal growth kinetics. Turning then to communities of hyphae, we explain autotropic reactions, and consider hyphal branching and septation. Finally, we draw it all together by discussing the ecological advantage of mycelial growth in colonising solid substrates.

4.1 MYCELIUM: THE HYPHAL MODE OF GROWTH

Members of Kingdom *Fungi* have made a major success of filamentous extension, even though the mycelial growth strategy is not unique to them. In all organisms in which it is used, high rates of filament extension are achieved by generating biomass, most importantly membrane and cell wall precursors, over a long length of filament behind the tip. This biomass is transported to the tip to extend the filament, and as long as the food-gathering activities of the rest of the mycelium can supply the nutrients, tip extension can continue.

The strategy has arisen, by convergent evolution, in *Oomycota* and streptomycete bacteria but there are major differences between these organisms in the mechanisms used to put the strategy into effect, so don't be misled (as early biologists were) into confusing similar morphology with phylogenetic relationship. Apical extension of pollen tubes in flowering plants and several systems in developing animals (neurons, blood vessels, insect tracheary systems, ducts in lungs, kidneys and glands) are also based upon a similar approach of **apically extending branching filaments**, but for different purposes (Davies, 2006). Indeed, although the fundamental kinetics of all these different filamentous systems are similar, they are 'fine-tuned' in their extension rates, branching frequencies and tropisms to suit their specific biological function(s).

As far as members of Kingdom *Fungi* are concerned, hyphal **tip growth** is usually thought of as the primary mode of growth in filamentous fungi but though apical extension is the crucial way in which the hyphal structure is established, it is only the start of a multifaceted chain of events that extend over the full length of the hypha (Read, 2011). Overall, the characteristic behaviour pattern of fungi is to **explore** the habitat with rapidly growing, sparsely branched hyphae, then, when some of those hyphae find a nutrient resource, the extension rate declines, rate of branching increases, and the mycelium **captures and exploits** the resource, from which it subsequently sends out a new generation of exploratory hyphae and/or populations of spores. This pattern of behaviour can be recognised from the microscopic scale as tiny saprotrophic colonies find minute fragments of nutrients on plant surfaces, to the landscape scale as pathogens of trees and wood-decay saprotrophs search across the forest floor for new hosts or freshly felled timber (Carlile, 1995; Lindahl & Olsson, 2004; Money, 2004, 2008; Watkinson *et al.*, 2005).

The polarised growth of the hyphal tip requires a progressive supply of proteins, lipids and cell wall precursors to the hyphal tip. This transport is managed by vesicle trafficking by way of the actin and microtubule cytoskeleton and polarity marker proteins; the arrangement, and ***rearrangement*** (tip polarity is maintained by repeated transient assembly and disassembly of polarity sites), of the cytoskeleton is crucial to maintain hyphal polarity (Takeshita, 2016). In the next few chapters, we will describe and explain the aspects of fungal cell biology that enable this pattern of behaviour. In this chapter, we will concentrate on the macroscopic aspects of hyphal extension and what can be deduced and established from observation of whole mycelia. In the next chapter, our attention will turn to the microscopic and molecular cell biology that characterises fungal hyphae and yeast cells, before turning to the 'population aspects' of the meaning of fungal individuality and the consequences of interactions between hyphae from the same and from different fungal individuals, culminating in the outcome of genetic crosses between individuals.

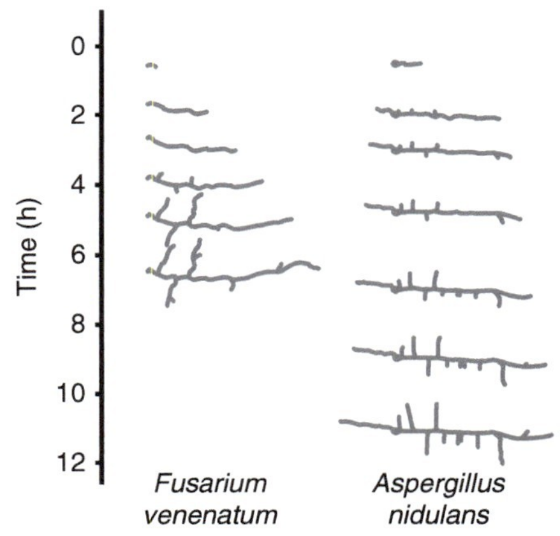

Figure 4.1. Time lapse drawings showing development of young germlings of *Fusarium venenatum* and *Aspergillus nidulans* during the initial 12 hours of growth at 25°C on agar-solidified medium. Note that most of the first-formed branches are oriented at close to 90° to the long axis of the main germ tube hypha, as new hyphal tips grow directly away from their parent hypha to explore the substratum.

4.2 SPORE GERMINATION AND DORMANCY

Spores are products of both sexual and asexual reproduction and act as the prime **units of dispersal** in fungi. Most spores that settle on an appropriate substrate under favourable environmental conditions germinate to produce one or more germ tubes and a new fungal mycelium. Except for sterile species, the spores of all fungi can enter into a dormant phase, during which metabolism is reduced by about 50%. Several types of dormancy exist in fungi; classed as either **exogenous** or **endogenous** (Feofilova *et al.*, 2012). If a spore is faced with unfavourable conditions, such as lack of nutrients, low temperature, an unfavourable pH or the presence of an inhibitor (for example, on the surface of a plant), the spore remains dormant and delays germination. Spores under these conditions are **exogenously dormant** and will only germinate when environmental conditions become favourable.

Some fungi produce spores that fail to germinate immediately, even under favourable conditions because of factors within the spore such as nutrient impermeability or the presence of endogenous inhibitors. Spores of this sort are said to be **endogenously dormant**. Dormancy of these spores is usually broken by ageing or by some physiological shock permitting nutrients to begin to enter, or the endogenous inhibitors to leach out of the spore. A classic example is that dormancy of the ascospores of *Neurospora crassa* is broken by a 30-minute heat shock at 60°C or exposure to 0.12 mM furfural ($C_5H_4O_2$). This relates to natural physiology because *Neurospora crassa* is one of the first moulds to emerge on organic remains after bush fires. Clearly, the heat of the fire will activate dormant spores; furfural is prepared by acid distillation of the sugar xylose and, as many plant hemicelluloses contain xylans, bush fires are likely also to produce furfural.

Prior to the emergence of a **germ tube**, fungal spores undergo a process of swelling (**spherical growth**) during which spores increase in diameter up to four times due mainly to uptake of water. During this phase, the metabolic activity of the spore increases greatly and protein, DNA and RNA production all increase rapidly. This is followed by the emergence of one or more germ tubes (young hyphal tips) that extend away from the spore in typical apically polarised manner (Figure 4.1).

4.3 THE FUNGAL LIFESTYLE: COLONY FORMATION

Following germination, the **extension rate** of the germ tube **increases towards a maximum** linear rate, at which point the hypha attains a linear extension rate. The maximum rate of hyphal extension varies greatly between different fungal species and is also dependent on environmental conditions such as temperature, pH and nutrient availability. For example, the extension rate of leading hyphae of *Penicillium chrysogenum* is about 75 μm h^{-1} at 25°C, while those of *Neurospora crassa* grown at 37°C can extend at up to about 6,000 μm h^{-1}.

Before the maximum rate of extension is attained, a **lateral branch** is formed to produce a new growing hypha the extension rate of which also accelerates towards the maximum. As the germling continues to grow, new lateral branches form at an exponential rate (Figure 4.2).

Although **individual hyphae** in the developing mycelium eventually attain a maximum **linear rate**, the overall **growth of the mycelium is exponential** (we will illustrate these features later, and also show how the measurements can be generalised algebraically). During early growth, nutrients surrounding the young mycelium are in excess and the mycelium is unrestricted and undifferentiated. During undifferentiated growth, the mean rate of hyphal extension is dependent on the **specific growth rate** of the organism (the maximum rate of growth in biomass per unit time) and the **manner and degree of branching**.

In the older mycelium (Figure 4.3), **hyphal fusions** are evident at the colony centre, and hyphal **avoidance reactions** at the colony margin. Between them, the sketches in Figures 4.2 and 4.3 portray the main growth processes that influence the distribution of hyphae in a mycelium, which are:

- polarised hyphal growth;
- branching frequency;
- **autotropism** (the 'self-avoidance' reaction that makes vegetative hyphae grow away from the already existing mycelium).

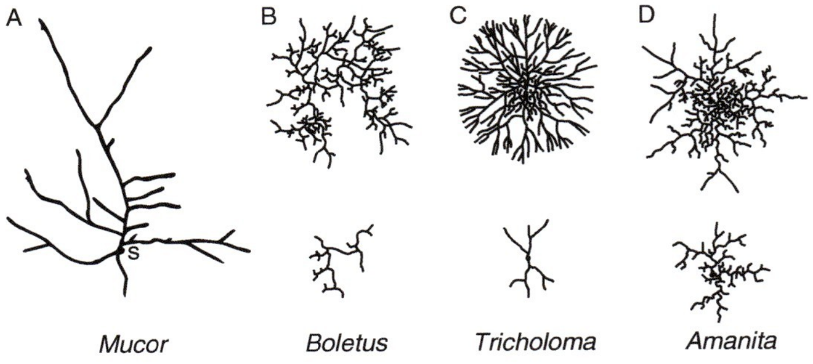

Figure 4.2. These hand-drawn pictures illustrate the well-ordered, young mycelium which colonises solid substrates effectively and efficiently. (A) A young germling of *Mucor* (S, position of the germinated spore). (B–D) show a very young germling below and a slightly more mature colony above. (B) *Boletus*. (C) *Tricholoma*. (D) *Amanita*. (A is redrawn from Trinci, 1974; B–D redrawn after Fries, 1943.)

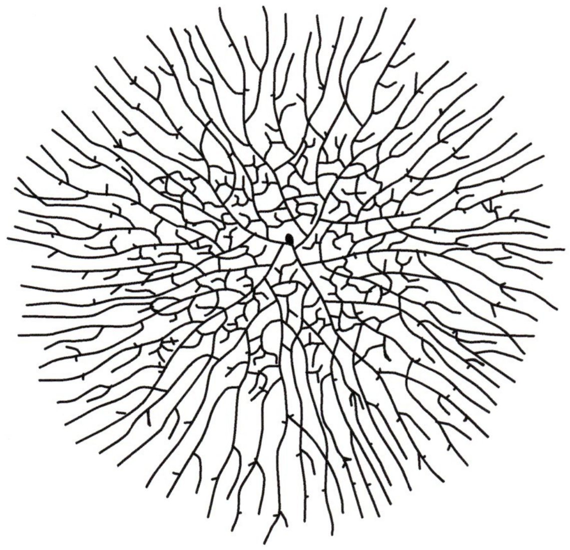

Figure 4.3. A maturing fungal colony. Notice how the growing hyphae are oriented outward into uncolonised regions while the production of branches and hyphal fusions centrally ensures the mycelium becomes a network that efficiently exploits available substrate. This hand-drawn sketch of *Coprinus sterquilinus* comes from one volume of A. H. R. Buller's epic series *Researches on Fungi* (Buller, 1909–1934).

It is important to recognise that these are **dynamic relationships**. They change with the age and with the developmental state of the mycelium as its biological functions change, indeed the distribution of biomass in a fungal colony varies with the age of the hyphae. One part of a mycelium may be growing as a rapidly extending, sparsely branched exploratory sector, another part may be a highly branched and interconnected network exploiting a nutrient resource, while a third region reverses the autotropism so that hyphal tips congregate and cooperate in formation of a sporophore.

4.4 MYCELIUM GROWTH KINETICS

One lesson to draw from the morphological description in section 4.3 is that the gross morphology can be very misleading. Systems that depend on outwardly directed extension, regular branching and avoidance reactions (autotropism) can end up looking almost identical though they may not be related in any biological sense (for example, a fungal mycelium compared with blood vasculature or with the growth of nerve fibres).

Evidently, from the description so far, in fungi there is an intimate relationship between hyphal extension rate, branch initiation and growth rate. Because of its importance to the understanding of the growth of filamentous fungi, it is essential to distinguish between 'growth' and 'extension':

- **growth** (biosynthesis, or biomass production) occurs **throughout a hypha**;
- **extension** only occurs at the **hyphal tip**.

Thus, at the tip it is best to refer to hyphal ***extension*** rather than hyphal ***growth***.

Eukaryotic hyphal organisms (members of Kingdom *Fungi* and some *Oomycota*) can extend at very rapid rates because they have exploited the tactic of producing biomass, including membrane and cell wall precursors, over a long length of hypha subtending the extending tip (a 5–6 mm length in the case of *Neurospora crassa*) and rapidly transporting this biomass to the tip where the membrane and cell wall precursors are rapidly converted to new protoplasmic membrane and new cell wall. The rate at which new membrane and cell wall is added to the extension zone of a hypha and is then transformed into rigidified wall in *Neurospora crassa* is quite remarkable considering it supports the extension growth mentioned above of 100 µm min^{-1} when grown at 37°C.

The basic features of fungal growth kinetics can be established with relatively straightforward experiments. A classic paper by Steele and Trinci (1975) showed that for a strain of *Neurospora crassa*:

- there was a direct relationship between extension zone length and the rate of hyphal extension;
- extension zone length and hyphal diameter were not affected by temperature;
- at a specific temperature the extension zone expansion time (the time taken for a tip to expand from its minimum to its maximum diameter) was a constant (less than 60 s).

These experiments indicated that, at 25°C, only seconds elapse between cell wall precursors of *Neurospora crassa* being added to the **plastic extension zone** of the wall at the tip and these materials being transformed into **rigidified hyphal wall**. Not all fungal species achieve such rapid rates of wall rigidification; hence the wide variation in the maximum rates of extension of fungal hyphae.

When growth rate is unaltered but extension rate is decreased (for example, by inclusion of inhibitory metabolite analogues such as L-sorbose or validamycin A in the medium), newly synthesised biomass must be redirected towards increased branch formation. The important point here is that we are distinguishing between growth rate (as the rate in increase of biomass, which should strictly be called the **specific growth rate**) and the **extension rate** (the rate at which the colony margin marches across the substratum).

It is fairly obvious that if the mycelium makes a lot of biomass but the hyphal tips do not extend by an equivalent amount then the biomass must go somewhere and the only place it can go is into more branches. This is the verbal description of the kinetic equation that describes a mycelium that is exploiting a resource (it also, incidentally, describes the 'make a sporophore' situation if you also add a reversal of the normal autotropism so that the branches can cluster together rather than grow away from each other).

The contrasting situation consists of a rapidly extending mycelial margin, in which the majority of the biomass is devoted to driving the hyphal tip extension with very little branching. Here again, you can readily understand that if the mycelium is making more biomass but is not making many branches then the extension rate of the mycelial margin must increase. For the moment these are theoretical interpretations based on visual observation. What we have done so far is create a model of hyphal growth in words. It's pretty powerful, in that it seems capable of explaining quite a lot, but it is just a word model, so how can we establish whether it represents the truth? To enable statistical testing we need some numbers so we can convert this into a mathematical model.

We need to grow some mycelia and instead of simply observing, we need to measure the length of hypha formed, its rate of formation and how many branches arise. Let's say we germinate a spore on a medium providing all nutrients in excess and in sufficient volume to prevent any inhibitors accumulating. Then we count the **number of hyphal tips** and measure the **total length of hypha** produced. Some representative data are shown in Figures 4.4, 4.5 and 4.6.

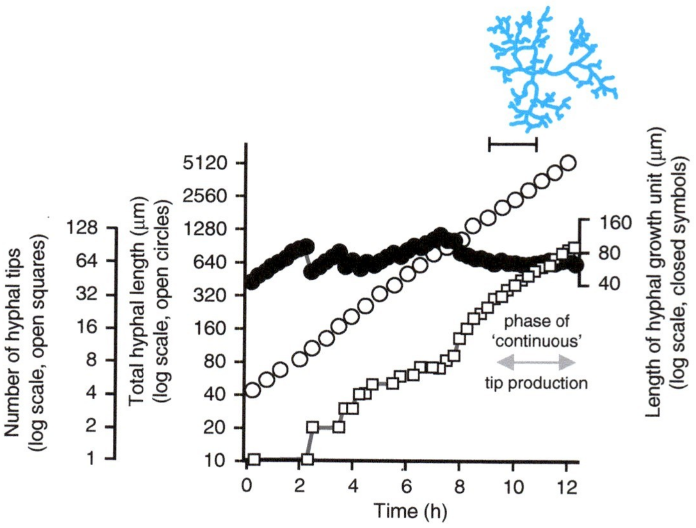

Figure 4.4. Initial growth of a mycelium of a spreading colonial mutant of *Neurospora crassa* on solid medium at 25°C. Open circles, total hyphal length produced by the germinating spore in μm; open squares, number of hyphal tips; solid circles, length of the hyphal growth unit in μm. The final appearance of the mycelium whose growth is recorded graphically is shown in the sketch at top right of the panel (scale bar, 250 μm).

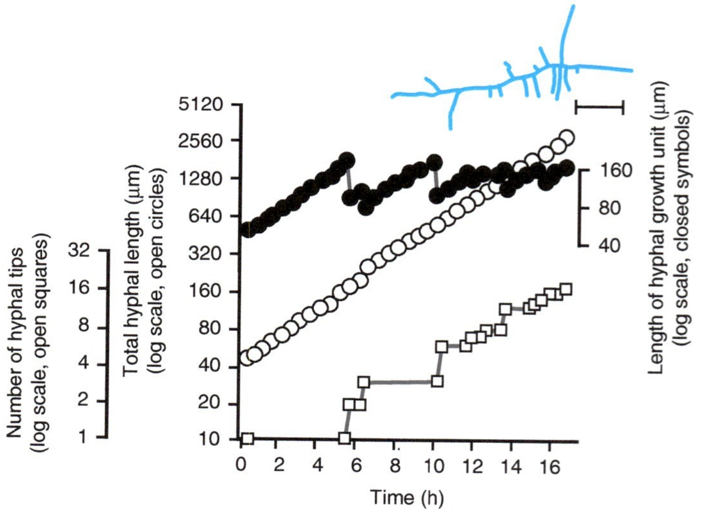

Figure 4.5. Initial growth of a mycelium of *Aspergillus nidulans* on solid medium at 25°C. Open circles, total hyphal length produced by the germinating spore in μm; open squares, number of hyphal tips; solid circles, length of the hyphal growth unit in μm. The final appearance of the mycelium whose growth is recorded graphically is shown in the sketch at top right of the panel (scale bar, 250 μm).

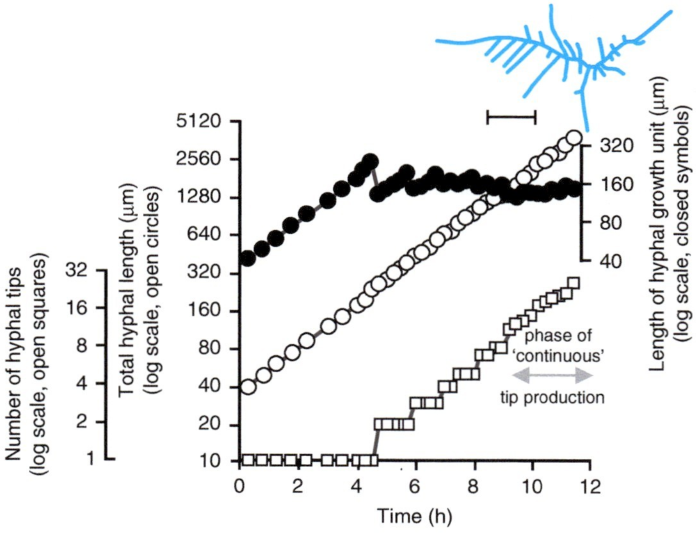

Figure 4.6. Initial growth of a mycelium of *Geotrichum candidum* on solid medium at 25°C. Open circles, total hyphal length produced by the germinating spore in µm; open squares, number of hyphal tips; solid circles, length of the hyphal growth unit in µm. The final appearance of the mycelium whose growth is recorded graphically is shown in the sketch at top right of the panel (scale bar, 250 µm).

In all these examples, note that the graphs for both number of hyphal tips and total hyphal length become straight lines. As the vertical axis is logarithmic in these plots, this shows that these variables increase exponentially. Further, these two lines are eventually parallel, so they must be increasing exponentially at the same rate. Evidently, 'growth', when measured in terms of total mycelial length, is exponential because of exponential branch formation. Total mycelial length and the total number of hyphal apices (all lead hyphae and branches) increase exponentially at the same specific rate. This rate is equivalent to the specific growth rate of the organism under the same growth conditions in liquid medium, where biomass can be measured directly as dry weight. We can refine our word model now to this:

- During extension of a hypha, a new branch is initiated when the mean volume of cytoplasm per hyphal tip (we will call this **the hyphal growth unit**) exceeds a particular critical value.
- If the hyphal growth unit is 'the mean volume of cytoplasm per hyphal tip' it can be calculated as the ratio of total mycelial length to total number of hyphal tips.

Biologically, the hyphal growth unit can be interpreted as the average volume of cytoplasm (equivalent to average length of hypha of uniform diameter) necessary to support the extension growth of a single average hyphal apex.

For a range of fungi, the hyphal growth unit increases following spore germination but then exhibits a series of damped oscillations tending towards a constant value (see top line in Figures 4.4–4.6). Such constancy demonstrates that over the mycelium as a whole, and not just in single hyphae, the number of branches is regulated in accord with increasing cytoplasmic volume. The hyphal growth unit is the population parameter which describes this. The hyphal growth unit is a length, measured in µm. Hyphal growth unit values have been measured for 21 fungal species, including both zygomycetes and *Ascomycota* that were cultured on a defined medium at 25°C; the values observed ranged from 35 to 682 µm with an average of 182 µm (Table 4.1).

Table 4.1. Lengths of hyphal growth units of a range of fungi grown on semi-solid glucose + mineral salts + vitamins medium at 25°C

Zygomycetes	Hyphal growth unit, G (µm)	*Ascomycota*	Hyphal growth unit, G (µm)
Actinomucor repens	352	*Botrytis fabae*	125
Cunninghamella sp.	35	*Cladosporium* sp.	59
Mucor hiemalis	95	*Fusarium avenaceum*	620
Mucor rammanicinus	37	*Fusarium vaucerium*	682
Rhizopus stolonifer	124	*Fusarium venenatum*	266
Ascomycota		*Geotrichum candidum*	110
Aspergillus nidulans	130	*Neurospora crassa*	402
Aspergillus giganteus	77	*Penicillium chrysogenum*	48
Aspergillus oryzae	167	*Penicillium claviforme*	104
Aspergillus niger	77	*Verticillium* sp.	82
Aspergillus wentii	66	*Trichoderma viride*	160

Data from Bull and Trinci (1977).

On average, therefore, across this range of fungi a 182 μm length of hypha is required to support extension of the hyphal apex. As we will see later, provision of resources for hyphal tip extension is visible cytologically as the flow of many **vesicles** (small vacuoles) towards the hyphal tip. A mathematical model has been created that produced numerical predictions of changes in total mycelial length, number of branches and distances between branches that compared well with observations of live fungi. This model assumes:

- vesicles are produced at a constant rate in distal hyphal regions;
- they are transported to the tip at a constant rate;
- at the hyphal tip vesicles accumulate and fuse with existing wall and membrane to give hyphal extension.

Obviously, a very close correspondence between theoretical predictions and real life observations encourages the belief that the description we have been giving is correct (Trinci *et al.*, 1994, 2001).

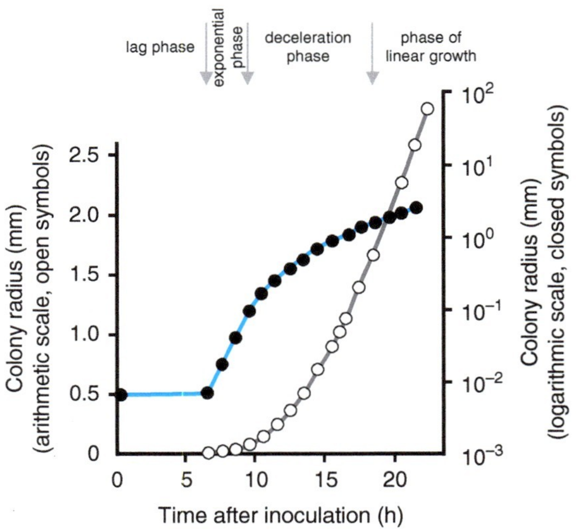

Figure 4.7. Growth phases of *Aspergillus nidulans* colonies. Modified from Trinci (1969).

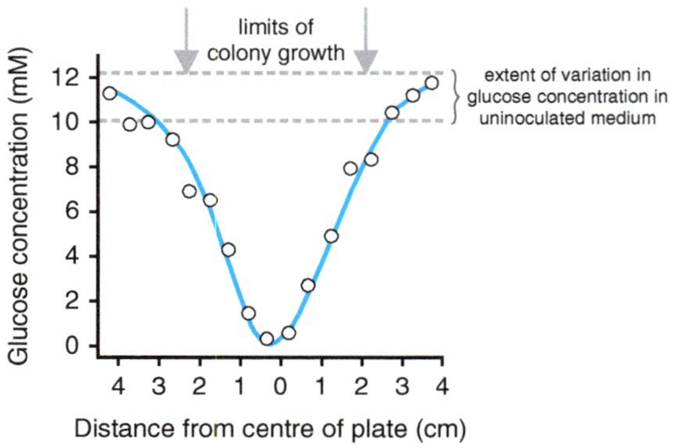

Figure 4.8. Glucose concentration in the medium below a colony of *Rhizoctonia cerealis* grown at 25°C on 20 ml of agar medium in a 9 cm Petri dish.

4.5 COLONY GROWTH TO MATURITY

A limitation of the observations we have shown you so far is that they deal with very young mycelia growing under unrestricted conditions. In the real world, this situation can last for only a short time. Before too long, growth of maturing mycelia is affected by nutrient limitation, changes in pH and growth inhibitors (metabolic waste products and secondary metabolites that the mycelium itself leaks into the medium). This is described as heterogeneous growth under restricted conditions.

If the growth of the culture is followed by measuring biomass, such a growth pattern can be shown to occur in the conventional sequence of phases of growth of a culture of a unicellular microorganism: namely **lag phase, exponential phase** (growth occurring under unrestricted conditions), **linear** (= constant growth rate) and **deceleration** phases of growth, which occur as conditions become restricting. The linear phase can be the predominant phase in growth of filamentous fungi in nature, but the important point is that all phases can be demonstrated during filamentous fungal growth on a solid medium (an agar-solidified medium in the laboratory; Figure 4.7; but see Chapter 19 for detailed discussion of growth kinetics in fermenters).

Growth of a fungus on a solid medium will eventually result in the establishment of conditions below the centre of the colony which are less favourable for growth than was initially the case (as described above). Figure 4.8 shows the gradient in glucose concentration established in the medium beneath and around a colony of *Rhizoctonia* grown on solid medium. The gradient results from glucose uptake by the fungus and diffusion of glucose from uncolonised to colonised parts of the substrate. Similar gradients are established for other nutrients, oxygen and pH.

Development of less favourable conditions for growth than those present at the start causes a deceleration from the maximum rate; this is known as restricted growth, and growth may cease eventually. Restricted growth is also brought about in some fungi by genetically programmed senescence.

4.6 MORPHOLOGICAL DIFFERENTIATION OF FUNGAL COLONIES

Changes in the medium of this sort mean that there are consequential changes in the conditions experienced by hyphae at different positions in the colony and this results in **hyphal differentiation**. Hyphal differentiation is very evident in the morphological differentiation of fungal colonies (Figures 4.9 and 4.10); as nutrients become depleted beneath the centre of the colony and metabolic products accumulate, spore production is often initiated.

Therefore, different parts of the colony are at different physiological ages, with the youngest actively extending hyphae at the edge of the fungal colony and the oldest, non-extending, sporulating mycelium at the centre. This has been demonstrated by estimating variation in growth rate within a mature colony of

Aspergillus niger by measuring the specific rate of ^{32}P uptake in different zones of the colony (Table 4.2).

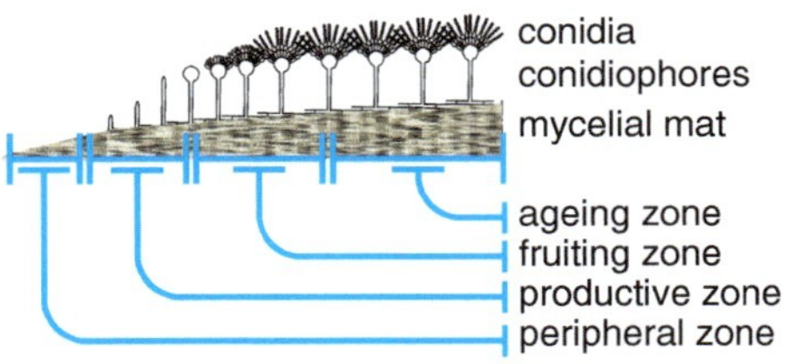

Figure 4.9. Morphological differentiation of colonies of *Aspergillus* shown in a diagrammatic radial section.

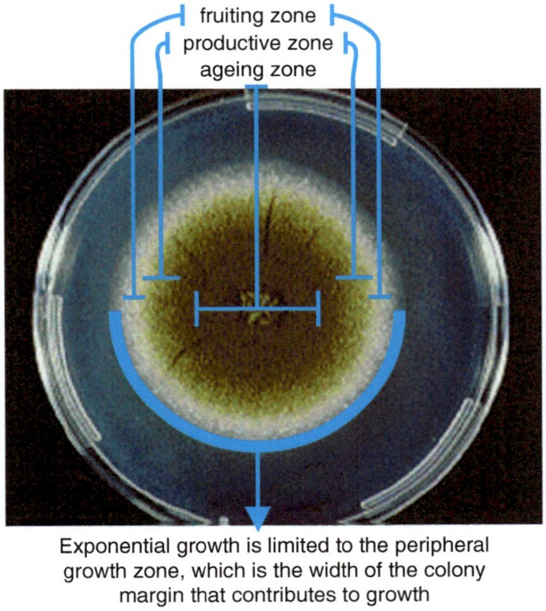

Figure 4.10. Morphological differentiation of colonies of *Aspergillus* shown in a photograph with the peripheral growth zone and other differentiated zones indicated.

Table 4.2. Physiological differentiation in *Aspergillus niger* revealed by variation in growth rate within a mature colony measured by its ability to take up radioactive phosphate

Uptake of ^{32}P by a mature colony of *Aspergillus niger*		
Zone of colony	**Width of region (mm)**	**Specific rate of ^{32}P uptake as a percentage of uptake in the peripheral zone**
Peripheral	1.5	100
Productive	2.0	31
Fruiting	4.5	6
Aged	3.5	3

4.7 DUPLICATION CYCLE IN MOULDS

Cell separation does not occur in filamentous fungi, yet it is possible to interpret events during fungal growth as analogous to those that comprise the duplication cycle of unicellular organisms. The main morphological events in the duplication cycle in *Aspergillus nidulans* are as follows (Figure 4.11) (Fiddy & Trinci, 1976):

- As the final step in the previous cycle, the apical compartment is reduced to about half its original length by the formation of two to six **septa** (**cross-walls**) at the rear of the compartment.
- The newly formed apical compartment continues to increase in length at a linear rate and the nuclei migrate towards the hyphal tip at a slightly slower rate.
- The volume of cytoplasm per nucleus increases until, at a critical ratio, the nuclei are induced to divide more or less synchronously (a single mitosis takes 5 minutes, and it takes about 12 minutes for the mitosis of all 50 nuclei found in the average apical compartment of *Aspergillus nidulans*).
- Mitosis is followed, during the final 7% of the duplication cycle, by the formation of two to six septa in the rear of the apical compartment, reducing its length by half.

The duration of a **duplication cycle** in apical compartments of leading hyphae of *Aspergillus nidulans* (2.1 hours) is identical to the **doubling time** of the organism in liquid culture; another indication that extension of apical compartments is unrestricted. In nature, *Aspergillus nidulans* has about 50 nuclei in each apical compartment, but duplication cycles of this sort have been described in apical compartments of leading hyphae of species that were monokaryotic (one nucleus per compartment),

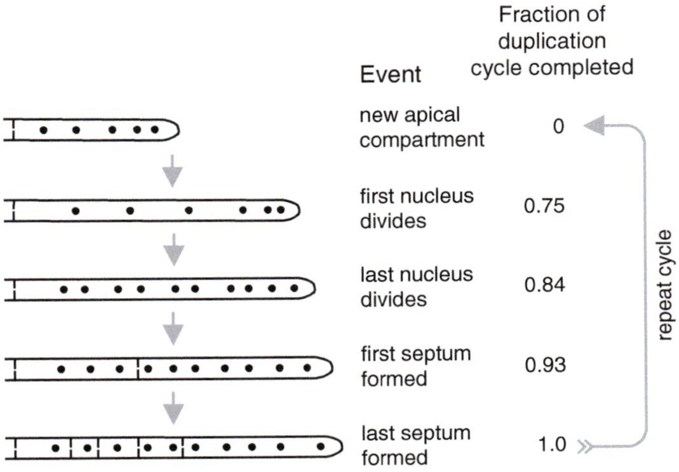

Figure 4.11. Diagrammatic representation of the duplication cycle in a leading hypha of *Aspergillus nidulans* extending at a linear rate on solid medium. On average, completion of a duplication cycle takes 2.1 hours; therefore a 0.1 fraction of the duplication cycle is equivalent to about 12.5 min. Redrawn after Trinci (1979).

dikaryotic (two nuclei per compartment) and multinucleate (up to about 75 nuclei per compartment).

The indications are that **synchronous mitosis** in apical hyphal compartments of leading hyphae is regulated by a **size-detecting mechanism** similar to that observed in the fission yeast *Schizosaccharomyces pombe*, where mitosis is triggered when the ratio **[cytoplasmic volume]:[number of nuclei]** exceeds a critical value.

An important **point of contrast** between the duplication cycle of fungi and those of animals and plants is that in fungi there is no necessary quantitative or spatial relationship between division of the nuclei and division of the cytoplasm by cross-wall formation (**septation**). Contrast the diagram in Figure 4.11 with the situation in an animal cell in which ***one*** mother cell is divided into ***two*** mitotic daughters by a cleavage furrow ***across the equator*** of the division spindle; and with that in a plant cell in which ***one*** mother cell is divided into ***two*** mitotic daughters by a new cell wall (cell plate) appearing ***across the equator*** of the division spindle. In contrast, mitoses of the 50 nuclei in the average apical compartment of *Aspergillus nidulans* are **not** accompanied by formation of 50 septa. There is, of course, a temporal relationship between **karyokinesis** and **cytokinesis** in fungi (septation occurring towards the end of the duplication cycle) but there is **no** strict arithmetical relationship of the sort 'one division spindle = one septum'.

4.8 REGULATION OF NUCLEAR MIGRATION

In fungi, nuclei are sufficiently motile to migrate through the hypha, and the spatial distribution of nuclei in a mycelium is determined by their migratory behaviour. The **migratory cycle of nuclei** (of the basidiomycete *Polystictus versicolor*) is divided into four phases (Figure 4.12):

- In the prolonged first phase, most of the cell contents (nuclei, mitochondria, vacuoles, vesicles and various granules) move towards the hyphal apex at about the same rate as the hyphal tip extends, so the distance between the cell contents and the tip remains approximately constant.
- In the second phase, nuclei stop migrating although other cytoplasm components continue to move forward, so the distance between nuclei and the hyphal tip increases.
- Synchronous mitosis occurs of the nuclei (two nuclei in the dikaryotic hyphae of *Polystictus versicolor*) during the third phase.
- In the fourth and final phase, one of the pair of daughter nuclei moves rapidly towards the apex and at a faster rate than tip extension, so that the nucleus migrates through the hypha to assume the phase 1 migratory distance behind the hyphal apex; while the second of the pair of daughter nuclei moves slowly away from the tip.

The generalisation that seems to apply to fungal hyphae commonly is that at a particular point in the mitotic cycle, nuclei

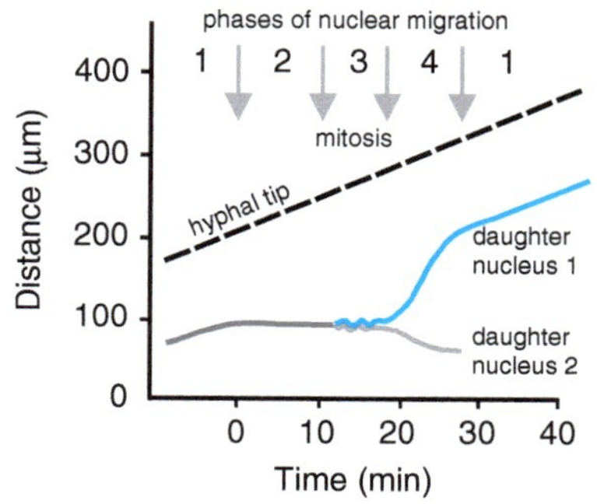

Figure 4.12. Hyphal extension and nuclear migration in dikaryotic apical compartments of *Polystictus* (*Trametes*) *versicolor* grown at 24°C. The curves are the means of 34 observations. The diagonal dashed line shows the location of the hyphal apex as it extends (upwards) away from the fixed point in space set as position zero. The solid line in the lower part of the figure records the position of one of the (haploid) dikaryotic nuclei located in the terminal compartment. In the prolonged phase 1 (prior to time zero) the nucleus moves towards the hyphal apex at about the same rate as the hyphal tip extends, so the distance between the two remains approximately constant. At the onset of mitosis (time zero), phase 2, the nucleus stops migrating, so the distance between the nucleus and the hyphal apex increases. Synchronous mitosis of the (two) dikaryotic nuclei takes place in that position (phase 3) and the hyphal apex continues to extend. We are only tracking one of the two dikaryotic nuclei but mitosis, of course, produces two daughter nuclei, so the line divides into two. In the fourth and final phase, one of the paired daughter nuclei moves rapidly towards the apex and at a faster rate than tip extension, so that the nucleus migrates through the hypha to assume the phase 1 migratory distance behind the hyphal apex. The second of the pair of daughter nuclei moves slowly away from the tip. Redrawn after Trinci (1979).

stop migrating and undergo mitosis, and by so doing get left behind by the advancing hyphal tip, but after mitosis the daughter nuclei reoccupy their normal position at a regulated distance behind the hyphal apex.

A functional microtubule system is necessary for normal nuclear migration and a considerable cellular apparatus is involved (described in Chapter 5).

4.9 GROWTH KINETICS

So far, we have described the basic kinetics of mycelial growth in words. Converting them to algebra results in the relationships we illustrated in Figures 4.4–4.6 being expressed in the equation:

$$\bar{E} = \mu_{max} G$$

where $\bar{E}$ is the mean extension rate of the colony margin;
μ_{max} is the maximum specific (biomass) growth rate;
and G is the hyphal growth unit length.

As can be seen from Figure 4.10 and Table 4.1, the factors which determine the radial growth rate of the colony (K_r) are the specific growth rate of the fungus (μ) and the width of the peripheral growth zone (w); that is, $K_r = w\mu$.

The fungal colony therefore grows outward radially at a linear rate (that is, an arithmetic plot of colony radius against time forms a straight line), continually growing into unexploited substratum. As it does, the production of new branches ensures the efficient colonisation and utilisation of the substratum. For the colony as a whole, the peripheral growth zone is a ring of active tissue at the colony margin which is responsible for expansion of the colony. At the level of the individual hypha, the peripheral growth zone corresponds to the volume of hypha contributing to extension growth of the apex of that hypha (the hyphal growth unit).

The rate of change in conditions below a colony will be related to the density (biomass per unit surface area) of the fungal biomass supported. It follows from this that a profusely branching mycelium (low value of G) will develop unfavourable conditions in the medium below the colony more rapidly than a sparsely branching mycelium (high value of G). Consequently, a relationship between G and w would be anticipated, and is observed. It means, for example, that K_r can be used to study the effect of temperature on fungal growth because w is not affected appreciably by temperature, however the concentration of glucose (for example) does affect w, so K_r cannot be used to investigate the effect of nutrient concentration. The biological consequence of this is that filamentous fungi can maintain maximal radial growth rate over nutrient-depleted substrates.

Unlike colonies formed by unicellular bacteria and yeasts, where colony expansion is the result of the production of daughter cells and occurs only slowly, the ability of filamentous fungi to direct all their growth capacity to the hyphal apex allows the colony to expand far more rapidly. Importantly, the fungal colony expands at a rate which exceeds the rate of diffusion of nutrients from the surrounding substratum.

Although nutrients under the colony are rapidly exhausted, the hyphae at the edge of the colony have only a minor effect on the substrate concentration and continue to grow outwards, exploring for more nutrients. In contrast, the rate of expansion of bacterial and yeast colonies is extremely slow and less than the rate of diffusion of nutrients (Table 4.3). Colonies of unicellular organisms quickly become diffusion limited and therefore, unlike fungal colonies, can only attain a finite size.

We have discussed how **hyphal extension growth** follows a few general relationships that are conveyed in relatively simple equations (above and section 4.4); it follows that hyphal growth kinetics are well suited to mathematical modelling, using these word-equations:

- $\bar{E}$, the mean tip extension rate, is given by μ_{max} (the maximum specific growth rate) multiplied by G, the hyphal growth unit.
- G is defined as the average length of a hypha supporting a growing tip.

Table 4.3. Colony radial growth rate (Kr) of bacterial and fungal colonies cultivated at their optimum temperature

Species	w (μm)	μ (h^{-1})	Kr (μm h^{-1})
Bacteria			
Escherichia coli	91	0.28	18
Streptococcus faecalis	Not done	0.65	18
Pseudomonas florescens	Not done	0.59	29
Myxococcus xanthus (non-motile)	Not done	Not done	20
Streptomyces coelicolor	Not done	0.32	22
Fungi			
Candida albicans (mycelial form)	119	0.39	46
Penicillium chrysogenum	496	0.16	76
Neurospora crassa	6,800	0.26	2,152

Data summarised from Oliver and Trinci (1985).

- G, consequently, is given by L_t, total mycelial length, divided by N_t, the total number of tips.

In a fungal colony, the hyphal growth unit is approximately equal to the width of the peripheral growth zone, which is a ring-shaped peripheral area of the mycelium that contributes to radial expansion of the colony.

- In a mycelium that is exploring the substrate, branching will be rare and so G will be large. G is therefore an indicator of branching density.
- A new branch is initiated when the capacity for a hypha to extend increases above $\bar{E}$, thereby regulating G to a uniform value indicative of the characteristic branching density of that fungus under those growing conditions.

All these features of normal filamentous hyphal growth can be expressed **algebraically** in a vector-based mathematical model in which the growth vector of each virtual hyphal tip is calculated at each iteration of the algorithm by reference to the surrounding virtual mycelium. For example, the **Neighbour-Sensing computer program** starts with a single hyphal tip, equivalent to a fungal spore. Each time the program runs through its algorithm the tip advances by a growth vector (initially set by the user) and may branch (with an initial probability set by the user).

In the Neighbour-Sensing program each **hyphal tip is an active agent**, described by its three-dimensional position in space, length and growth vector that can vector within three-dimensional data space using rules of exploration that are set (initially by the experimenter) within the program. The rules are biological characteristics such as:

- the basic kinetics of *in vivo* hyphal growth;
- branching characteristics (frequency, angle, position);
- tropic field settings that involve interaction with the environment.

The experimenter can alter parameters to investigate their effect on form; the final geometry is reached by the program (not the experimenter) adapting the biological characteristics of the active agents during their growth, as in life. The **Neighbour-Sensing model** brings together the essentials of hyphal growth kinetics into mathematical **cyberfungus** that can be used for experimentation on the theoretical rules governing hyphal patterning and tissue morphogenesis (Meškauskas *et al.*, 2004a, b).

The Neighbour-Sensing model 'grows' a simulated **cybermycelium** using realistic branching rules decided by the user. As the **cyberhyphal tips** grow out into the modelling space the model tracks where they have been, and those tracks become the hyphal threads of the cybermycelium. All positioning information is stored by the model as numerical data and so the data handling work becomes more and more extensive as branching produces more hyphal tips and the cybermycelium 'grows' in three dimensions on the computer monitor; it is this steady growth process that generates the very large amount of data.

The process of simulation is programmed as a closed loop. This loop is performed for each currently existing hyphal tip of the mycelium and the algorithm:

- Finds the number of neighbouring segments of mycelium (N). A segment is counted as neighbouring if it is closer than the given critical distance (R). In the simplest case we did not use the concept of the density field, preferring a more general formulation about the number of the neighbouring tips.
- If $N < N_{branch}$ (the given number of neighbours required to suppress branching), there is a certain given probability (P_{branch}) that the tip will branch. If the generated random number (0 … 1) is less than this probability, the new branch is created, and the branching angle takes a random value. The location of the new tip initially coincides with the current tip. This stochastic branch generation model is similar overall to earlier ones in which distance between branches and branching angles followed experimentally measured statistical distributions.

Initial versions of the model did not implement tropic reactions (to test the kind of morphogenesis that might arise without this component). Later versions of the model tested how autotropic reactions affected the simulation. This model is predictive and successfully describes the growth of hyphae, so confirming its credibility and indicating plausible links between the equations and real physiology; but it is just one of several mathematical models of fungal growth that have been published. For a wider view of this research we refer you (in alphabetical order) to Bartnicki-García *et al.* (1989), Boswell (2008), Boswell *et al.* (2003), Davidson (2007), Goriely and Tabor (2008), Moore and Meškauskas (2017), Moore *et al.* (2006), Prosser (1990, 1995a & b), Vidal-Diez de Ulzurrun *et al.* (2015) and Wang *et al.* (2019).

Most models published so far simulate growth of mycelia on a two-dimensional plane; the Neighbour-Sensing model, however, while being as simple as possible, is able to simulate formation of a spherical, uniformly dense fungal colony in a visualisation in **three-dimensional space**. A description of the mathematics on which the model is based can be found in Moore *et al.* (2006); we will not dwell on this aspect here. The complete application can be downloaded for **personal experimentation** at this URL: http://www.davidmoore.org.uk/CyberWEB/index.htm (Meškauskas *et al.*, 2004a, b; Moore & Meškauskas, 2017).

The Neighbour-Sensing model successfully imitates the three branching strategies of fungal mycelia illustrated by Nils Fries in 1943 (Figure 4.13 compares computer simulations with the original 1943 illustrations shown previously in Figure 4.2).

The Neighbour-Sensing model shows that **random growth and branching** (i.e. a model that does not include the local hyphal tip density field effect or any other tropism) is sufficient to form a spherical colony. The colony formed by such a model is more densely branched in the centre and sparser at the border; a feature observed in living mycelia. Models incorporating local hyphal tip density field to affect patterning produced the most

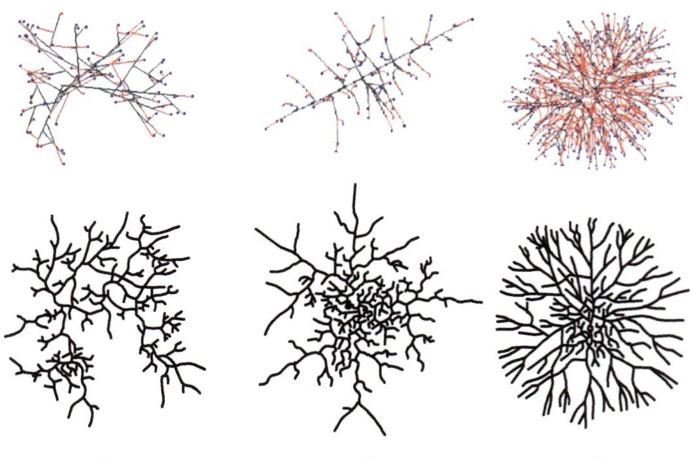

Figure 4.13. Simulation of the three different colony types described by Fries (1943) (compare with Figure 4.2). (A) shows the *Boletus* type, (B) the *Amanita* type and (C) the *Tricholoma* type. The modelling parameters used for each of these simulations are described in the text. The simulation is the upper figure in each case.

regular spherical colonies. As with the random growth models, making branching sensitive to the number of neighbouring tips forms a colony in which a near uniformly dense, essentially spherical core is surrounded by a thin layer of slightly less dense mycelia.

Using the branching types discussed by Fries (1943) as a comparison, the morphology of virtual colonies produced when branching (but not growth vector) was made sensitive to the number of **neighbouring tips** was closest to the so-called *Boletus* type (Figures 4.2 and 4.13). This suggests that the *Boletus* type branching strategy does not use tropic reactions to determine patterning, nor some pre-defined branching algorithm. Evidently, hyphal tropisms are not always required to explain 'circular' mycelia (that is, mycelia that are spherical in three dimensions).

When the Neighbour-Sensing model implements the **negative autotropism** of hyphae, a spherical, near uniformly dense colony is also formed, but the structure differs from the previously mentioned *Boletus* type, being more similar to the *Amanita rubescens* type, characterised by a certain degree of differentiation between hyphae (Figures 4.2 and 4.13):

- first, rank hyphae tending to grow away from the centre of the colony;
- second, rank hyphae growing less regularly and filling the remaining space.

In the early stages of development such a colony is more star-like than spherical. It is worth emphasising that this remarkable differentiation of hyphae ***emerges in the visualisation even though all virtual hyphae are driven by the same algorithm***. The program does not include routines implementing differences in hyphal behaviour.

Finally, when both **autotropic reaction** and **branching** are regulated by the **hyphal density field**, a spherical, uniformly dense colony is also formed. However, the structure is different again, such a colony being like the *Tricholoma* type illustrated by Fries (1943) (Figures 4.2 and 4.13). This type has the appearance of a dichotomous branching pattern, but it is not a true dichotomy. Rather the new branch, being very close, generates a strong density field that turns the growth vector of the older tip away from the new branch.

Hence the *Amanita rubescens* and *Tricholoma* branching strategies may be based on a negative autotropic reaction of the growing hyphae while the *Boletus* strategy may be based on the absence of such a reaction, relying only on **density-dependent branching**. Differences between *Amanita* and *Tricholoma* in the way that the growing tip senses its neighbours may be obscured in life. In *Amanita* and *Boletus* types, the tip may sense the number of other tips in its immediate surroundings. In the *Tricholoma* type, the tip may sense ***all*** other parts of the mycelium, but the local segments have the greatest impact.

This model shows that the broadly different types of branching observed in the fungal mycelium are likely to be based on differential expression of relatively simple control mechanisms. The 'rules' governing branch patterning (that is, the mechanisms causing the patterning) are likely to change in the life of a mycelium, as both intracellular and extracellular conditions alter. Some of these changes can be imitated by making alterations to specific model parameters during a simulation. By switching between parameter sets it is possible to produce more complex structures.

Experiments with the model simulated both colonial growth of the sort that occurs in Petri dish cultures (Figure 4.14) and development of a mushroom-shaped sporophore (Figure 4.15). These experiments make it evident that it is not necessary to impose complex spatial controls over development of the mycelium to achieve specific geometrical forms. Rather, geometrical form of the mycelium emerges because of the operation of specific locally effective hyphal tip interactions.

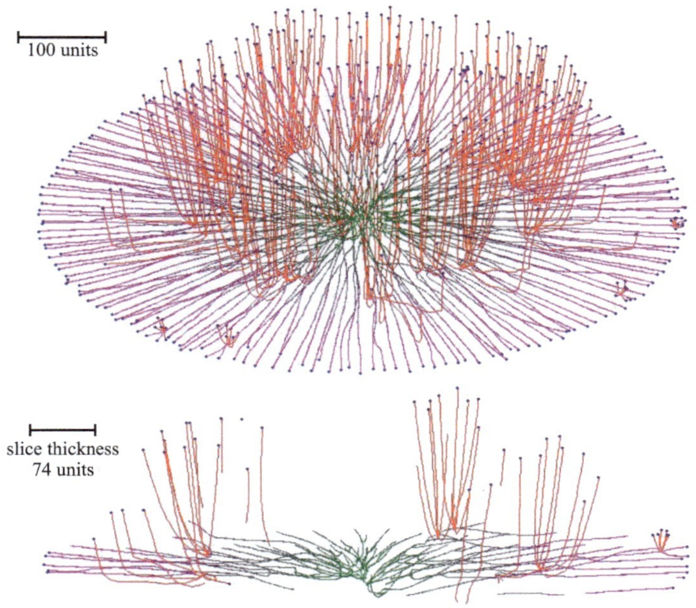

Figure 4.14. Simulation of colonial growth of the sort that occurs in Petri dish cultures. Oblique view (top) and slice of the colony (bottom), where secondary branching was activated at the 220-time unit. The secondary branches had negative gravitropism. For both primary and secondary branches, the growth was simulated assuming negative autotropic reaction and density-dependent branching. If the density allowed branching, the branching probability was 40% per iteration (per time unit). The final age of the colony was 294-time units. Secondary branches are colour-coded red, and hyphae of the primary mycelium are coloured green (oldest) to magenta (youngest), depending on the distance of the hyphal segment from the centre of the colony (modified from Meškauskas *et al.* (2004b); reproduced with permission from Elsevier).

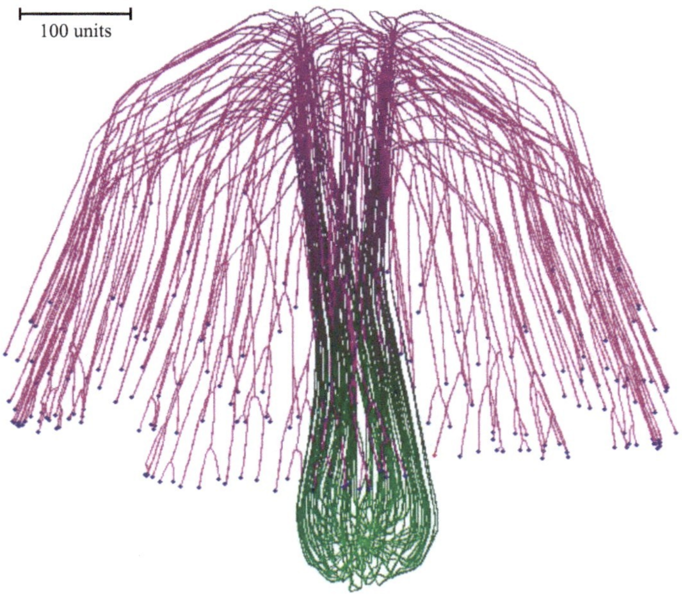

Figure 4.15. Simulation of a mushroom primordium. A spherical colony was first grown for 76-time units. This was converted into an organised structure, like the developing mushroom stem by applying the parallel galvanotropism for 250-time units. Subsequent application of a positive gravitropic reaction formed a cap-like structure (1,000-time units) (modified from Meškauskas *et al.* (2004a); reproduced with permission from Elsevier).

These computer simulations suggest that because of the kinetics of hyphal tip growth, very little regulation of cell-to-cell interaction is required to generate the overall architecture of fungal sporophore structures or the basic patterning of the mycelium. Specifically:

- Complex fungal sporophore shapes can be simulated by applying the same regulatory functions to all the growth points active in a structure at any specific time.
- The shape of the sporophore emerges as the entire population of hyphal tips respond together, in the same way, to the same signals.
- No global control of sporophore geometry is necessary (Meškauskas *et al.*, 2004a, b).

The experiments described above have exposed a fascinating feature of the 'crowd behaviour' of fungal hyphal tips, which is that the shapes of complex fungal sporophores can be simulated by applying the same regulatory functions to every one of the growth points active in a structure at any specific time. All parameter sets that generate shapes reminiscent of fungal sporophores feature a sequence of changes in parameter settings that are applied to all hyphal tips in the simulation. No localised regulation is necessary. Absence of global control of sporophore geometry does not necessarily imply an absence of localised control of details of sporophore structure. Indeed, by its very nature, the 'sensing of neighbouring hyphae' aspects of the model would support the interpretation of 'Reijnders' ***hyphal knots***' (Reijnders, 1963; and see Chapter 13, especially section 13.16) as a central 'inducer' hypha organising differentiation of a small group of surrounding hyphae to regulate detailed structures within sporophore tissues.

The remarkable reality of the simulations generated by the Neighbour-Sensing program encourages confidence in the accuracy and reliability of the Neighbour-Sensing mathematical model on which it is based. That confidence leads us to believe that the model is revealing unexpected capabilities of the hyphal lifestyle of fungi; but we feel that we have only just scratched the surface of what this mathematical model is able to reveal; the model is not yet perfect. A feature that remains to be implemented in the model is hyphal fusion (or hyphal anastomosis), which is such an important feature of living mycelia (section 5.15). Initial work on the mathematics of this suggests that hyphal fusion can be catered for in the algorithms underlying the Neighbour-Sensing model. Inclusion of anastomosis would enable the model to generate biologically inspired networks and so provide a tool to analyse these networks to yield information about connectivity, minimum path length, etc. In addition, it would be possible to address network robustness *in silico* by investigating the effect of removal of network links on connectivity.

Looking further into the future, it should be possible to add physiological data, such as substrate uptake and substrate transport kinetics, to growth and branching. Since the Neighbour-Sensing model 'grows' a realistic mycelium and tracks all the hyphal segments so generated, there is no mathematical impediment to assigning to those hyphal segments the algebraic characteristics defined to describe substrate uptake, utilisation and translocation kinetics, and their variation with age of the hyphal section.

The opportunity to tailor parameter sets (or 'strategies') to simulate specific species of fungi (the individual parameter sets being our ***cyberspecies***) became evident in our first experiments with the model in which distinctions could be made between cybermycelia with morphological similarities to the *Boletus, Amanita* and *Tricholoma* types of young mycelia discussed above. Comparable microscopic observations of young mycelia of any live fungus should enable the derivation of parameter sets that produce cybermycelia which are exact simulations of the living material.

This may also contribute to understanding hyphal and mycelial evolution because we might imagine that the evolutionary origins of specific aspects of the kinetics of hyphal growth and branching could be revealed by comparison of cyberspecies representing living taxa with known evolutionary relationships.

4.10 AUTOTROPIC REACTIONS

The hyphae of many fungi can alter their direction of growth to avoid growing into each other and accentuate exploration of uncolonised regions of the substratum. The avoidance

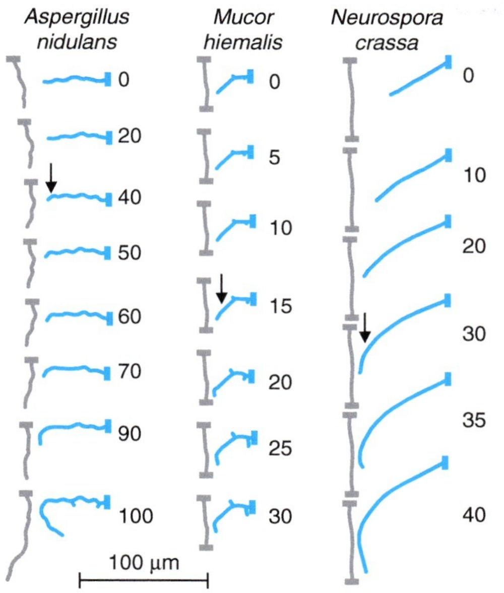

Figure 4.16. Autotropisms: diagrammatic illustrations of negative autotropism leading to the hyphal avoidance reaction, which is so crucial to the colonial growth form of filamentous fungi (as illustrated in Figure 4.3). The drawings show the responses of pairs of hyphae of *Aspergillus nidulans, Mucor hiemalis* and a wild-type strain of *Neurospora crassa*; the approaching hypha is shown in blue in each case. The point where the avoiding reaction was first observed is indicated by the arrow. The numerals give times in minutes. The medium was overlaid with cellophane membranes prior to inoculation to keep the hyphae on the surface and prevent hyphae diving beneath the surface. Modified and redrawn from Trinci *et al.* (1979).

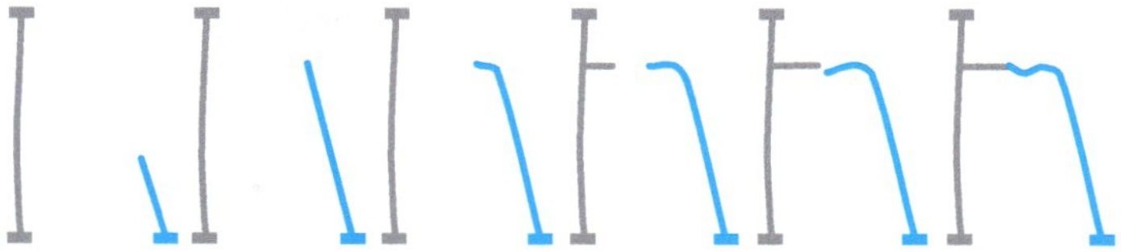

Figure 4.17. Autotropisms: sketch showing positive autotropism leading to a 'tip-to-peg' hyphal fusion reaction. This is one of the fusion reactions leading to the interconnected mycelial network of the mature colony as illustrated in Figure 4.3. Other reactions are tip to tip and peg to peg (see Figure 8.2). For more detail, see Figure 5.12, and refer to Hickey *et al.* (2002) and Glass *et al.* (2004). Particularly fine images and videos can be viewed on the University of Exeter's Fungal Cell Biology Group website at http://www.gerosteinberg.com/.

mechanism or **negative autotropism** (Figure 4.16) is particularly evident at low hyphal densities, in regions such as the margin of the growing colony. The ability of hyphae to sense the presence of another hypha is thought to be due either to a localised depletion of oxygen around the target hypha, a higher concentration of carbon dioxide, or the presence of a secreted metabolite.

The minimum distance of approach of two hyphae before negative autotropism caused one to grow away was 30, 27 and 24 µm respectively in *Neurospora crassa, Aspergillus nidulans* and *Mucor hiemalis* (Trinci *et al.*, 1979).

In a maturing mycelium, autotropism can be reversed so that young hyphae are attracted (possibly chemotropically?) to an older hypha (called **positive autotropism**). This can lead to hyphal fusions. In one mechanism the target hypha is induced to branch, and the consequential tip-to-tip contact is followed by breakdown of the two apices and fusion between the two hyphae (Figure 4.17). For more detail, see Figure 5.13 and its supporting text.

This process converts the central regions of a maturing colony into a fully interconnected network through which materials and signals can be communicated efficiently. This enables the vegetative mycelium to make best use of its resources. In the formation of sporophores and similar structures, positive autotropism enables many hyphal tips to congregate together to initiate the developing tissue. In these cases, hyphal fusions and adhesions may be rare (though they can be used to bind the structure together); instead, developmental regulation organises the concerted contribution of many independent hyphal tips to formation of the tissues and structures of the sporophore (discussed in Chapters 9 and 12).

4.11 HYPHAL BRANCHING

Growth of the mycelium depends on formation of hyphal branches. To proliferate a hypha must branch. ***There is no other way to turn one hypha into two hyphae.***

Although apical (or near-apical) branching of fungal hyphae does occur, most branches are laterally placed. Hyphal branches may be described as primary (which subtend no branches and arise directly from the main hypha), secondary (which subtend a primary branch), tertiary (which subtend a secondary branch), etc. In many biological and non-biological branching systems, there is an inverse logarithmic relation between the number of branches belonging to a specific order and order number; a relationship that allows maximum surface area to be colonised by minimum total length of filament. The relation holds in most fungi and reflects the efficiency of the mycelium in colonising the substratum while minimising the amount of biomass required to do so.

In early phases of growth, branches usually subtend an angle of approximately 90° to the long axis of the parent hypha. As we have seen, hyphae tend to avoid their neighbours (negative autotropism) and to grow radially away from the centre of the colony. A circular colony is formed eventually, with radially directed hyphae, approximately equally spaced, and extending at the margin at a constant rate.

As the colony circumference increases the apices of some branches catch up with their parent hyphae to maintain hyphal spacing at the colony margin. This occurs either by relaxation of controls on extension rate of branches as they become further separated from main hyphae or as a result of simple variability in extension rate.

A specific example of change in hyphal behaviour during colony development is provided by an analysis of mycelial differentiation in *Neurospora crassa* (McLean & Prosser, 1987). Up to about 20 hours of growth, all hyphae in mycelia of this fungus have similar diameters, growth zone lengths and extension rates, and all branches are at an angle of 90° to the parent hypha. After about 22 hours growth branch angle decreases to 63°, hyphal extension rates and diameters increase, and a hierarchy is established in which main hyphae are wider and have greater extension rates than their branches; the ratios between diameters of leading hyphae, primary branches and secondary branches being 100:66:42 and between extension rates being 100:62:26.

For a **lateral branch** to emerge from a region of a hypha with a mature rigidified wall requires the (internal) assembly of a **new hyphal apex** at the site of emergence. What specifies the site of the branch initiation is not known yet. Although a branch can potentially form at any point on a hyphal wall, in fungi that form septa there is usually a close relationship between septation and branching; branches may form at a specific time after septation and be positioned immediately behind septa.

In most fungi, though, the septation/branching relationship is less obvious and branch positioning is more variable. Indeed, although zygomycetes follow the same growth and branching kinetics as *Ascomycota* and *Basidiomycota*, only young mycelia of *Mucor hiemalis* and *Mucor rammanianus* form septa, mature zygomycete mycelia only form septa at the base of sporangia so in lead hyphae there is usually no relationship between septation and branching.

At one time, localised fluxes of ions through the hyphal membrane were thought to be involved in determining branching. Application of electrical fields certainly affects the site of branch formation and the direction of hyphal growth in young mycelia of several fungi, but endogenous ion fluxes seem to be more related to nutrient uptake than tip growth.

Several compounds act as paramorphogens, inhibiting hyphal extension and increasing hyphal branching. The (long) list includes non-metabolised sugar analogues (like L-sorbose, which is a hexose analogue, and validamycin A, which is a pseudo-oligosaccharide), inhibitors of phosphoinositide turnover, cyclic-AMP and cyclosporine (Figure 4.18), but the wide range of cellular targets represented does not help in understanding how branches initiate.

One possibility is that heat-shock proteins may be involved in the mechanism for branch initiation. Heat-shock proteins are polypeptides which interact with other proteins. They are '**molecular chaperones**' which bind to and stabilise other proteins to prevent incorrect intermolecular associations, then aid their correct folding by releasing them in a controlled manner. It is feasible that branch initiation requires assisted conformational alterations of wall proteins or that heat-shock proteins assist in the delivery of a branch-initiating polypeptide to the appropriate position.

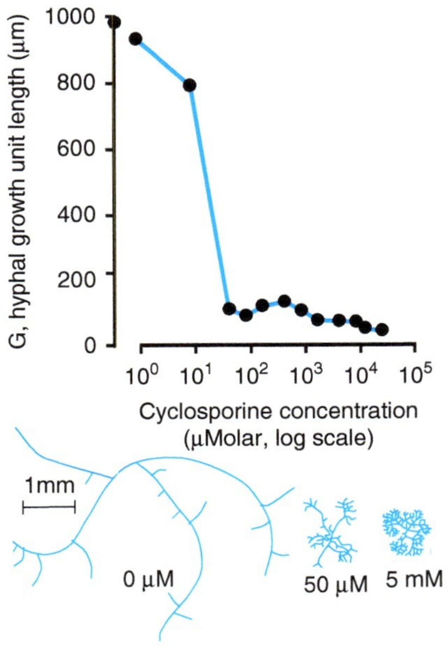

Figure 4.18. Effect of cyclosporine-A on the hyphal growth unit (shown graphically at top), and the morphology of a wild-type strain of *Neurospora crassa* grown at 25°C on a solid glucose-minimal salts medium (illustrated in the habit sketches beneath the graph, which also indicate the concentration of cyclosporine that causes the morphology shown). Modified and redrawn after Trinci *et al.* (1994).

4.12 SEPTATION

The hyphal growth form of filamentous fungi is an adaptation to the active **colonisation of solid substrata**. By hyphal extension and regular branching, the fungal mycelium can increase in size without disturbing the cell volume:surface area ratio so that metabolite and end-product exchange with the environment can involve translocation over very short distances. Fungal hyphae differ between species, but the hyphal filament, when separated into compartments by cross-walls, has an apical compartment which is perhaps up to 10 times the length of the intercalary compartments.

The septa which divide hyphae into cells may be:

- complete (imperforate);
- penetrated by cytoplasmic strands;
- perforated by a large central pore.

The pore may be open and offer little physical hindrance to the passage of cytoplasmic organelles and nuclei, or may be protected by a complex cap structure, called the **parenthesome**, derived from the endoplasmic reticulum (the **dolipore septum** of many *Basidiomycota*). In *Ascomycota*, which characteristically lack the parenthesome apparatus, the pore may be associated with cytoplasmic organelles known as **Woronin bodies**.

Over the years mycologists have been very sensitive to the question of whether fungi have cells, and how fungal cells and their interactions compare with those of plant and animal cells. Lower filamentous fungi (e.g. *Mucor*) have coenocytic hyphae; but they do not form multicellular structures. Hyphae of fungi which do exhibit complex developmental pathways form septa at regular intervals, though the septa usually have a pore. The pore is what worries people about the definition of fungal cells, because the implication carried with the word 'pore' is that all of the cytoplasm of a hypha is in continuity even though it might be subdivided by the septa into compartments.

Although movement of cytoplasm and organelles through septa has often been described and is frequently easy to demonstrate, it is also clearly the case that the movement or migration of cytoplasmic components between adjacent cells is under very effective control. There are instances in which nuclei move freely, but mitochondria do not, and others in which rapid migration of vacuoles is not accompanied by migration of any other organelle. Some biochemical experiments have even demonstrated that different sugars can be translocated in opposite directions in a hypha at the same time. There are also numerous examples available where grossly different pathways of differentiation have been followed on the two sides of what appear (to the electron microscope) to be open septal pores (illustrated in Figure 13.13).

Clearly, whatever the appearance, the hypha can be separated into compartments whose interactions are carefully regulated and which can exhibit contrasting patterns of differentiation. There may still be a semantic argument for preferring 'compartment' to 'cell', but from this point on we will take the pragmatic view that if it looks like a cell and if it behaves like a cell, then we will call it a cell. But please don't forget that every fungal cell is just a segment of a tubular hypha!

Cross-walls in fungal hyphae are pretty well always formed at right angles to the long axis of the hypha and this has a major impact on understanding the development of fungal tissues. Except in cases of injury or in hyphal tips already differentiated to form sporing structures, hyphal tip cells are not subdivided by oblique cross-walls, nor by longitudinally oriented ones. Even in fission yeast cells which are forced to produce irregular septation patterns under experimental manipulation, the plane of the septum is always perpendicular to the plane including the longest axis of the cell. In general, then, a fungus converts the one-dimensional hypha into a two-dimensional plate of tissue or three-dimensional block of tissue by controlling the formation of **branches**. The septum in any branch will be formed at right angles to the long axis of the branch, but its orientation relative to the parent hypha will depend entirely on the positioning of the apex of the new branch.

Primary septa in fungal hyphae are formed by a constriction process in which a belt of microfilaments around the hyphal periphery interacts with microvesicles and other membranous cell organelles (see Chapter 5). Except for the fact that there is no close linkage with mitosis (see above), there is a superficial similarity between fungal septation and animal cell cleavage (cytokinesis); but remember, fungi use organised microvesicles to divide blocks of cytoplasm in the free cell formation process (see section 3.4).

4.13 ECOLOGICAL ADVANTAGE OF MYCELIAL GROWTH IN COLONISING SOLID SUBSTRATES

The apical growth characteristic of the fungal hypha is the prime attribute of fungi and is, of course, an extreme cellular polarity. Because true extension growth is absolutely limited to the hyphal tip, the whole morphology of the hypha depends on events taking place at its apex. It follows from this that the pattern of hyphae in a mycelium, which is largely a consequence of the distribution of hyphal branches, depends on the pattern of formation of the hyphal tips which initiate those branches.

The dominance of filamentous fungi within the ecosystem is attributed to their lifestyle. By growing in a filamentous fashion, fungi can colonise substrates rapidly and grow away from nutrient-poor areas. Branching of the filament enables substrates to be efficiently captured for absorption from the environment. Maintaining a high extension rate even under poor nutrient conditions allows fungi to maximise their chances of finding new food sources. The success of this growth habit for exploiting the natural environment can be judged on a number of factors: the extraordinary diversity of fungal species (second only to the insects, but then every insect harbours a few parasitic fungi!), their distribution in virtually every habitat on the planet and the parallel evolution of a similar growth habit in other important soil microorganisms, the prokaryotic streptomycetes and the more fungus-like members of Kingdom *Straminipila*, like the *Oomycota*. Clearly the ability of a microbe to colonise new substrates rapidly by concentrating extension at the apex of a filament makes it ideally suited for life as a heterotroph in a heterogeneous environment.

Polarised growth of fungal hyphae is achieved by restricting extension to the hyphal apex. The cell wall at the hyphal tip has **viscoelastic properties**. This means it has some of the characteristics of both a liquid (being able to flow like a viscous fluid) and a solid (resisting and recovering from stretching, compression or distortion). These properties allow the wall at the hyphal apex to yield to the internal turgor pressure within the hypha by extending forward. Further behind the tip the wall is rigidified and resistant to the turgor forces. **Turgor** (the force within the cell that pushes the plasma membrane against the cell wall) powers the propulsion of hyphae through solid materials and therefore acts as the driving force for **hyphal extension** (Money, 2008).

Hyphal growth at the apex requires synthesis and insertion of new wall material and new membranes in a way that does not weaken the tip. This highly organised process is supported by the flow of vesicles generated within the cytoplasm behind the tip and is coordinated with the growth and replication of all the other cytoplasmic organelles and their migration towards the extending apex. It seems now to be generally accepted that the materials necessary for hyphal extension growth are produced at a constant rate (equal to the specific growth rate) throughout the mycelium and are transported towards the tip of the growing hyphae. Among the materials taking part in this polarised transport are numerous cytoplasmic vesicles, which are thought to contain wall precursors and the enzymes needed for their insertion into the existing wall to extend it. A considerable cytoplasmic architecture is involved in the **apical growth of the hypha** (Bartnicki-García *et al.*, 1989; Wessels, 1993; Takeshita, 2016). We will describe and discuss this in detail in Chapters 5 and 7.

4.14 REFERENCES

Bartnicki-García, S. (1968). Cell wall chemistry, morphogenesis and taxonomy of fungi. *Annual Reviews of Microbiology*, **22**: 87–108. WARNING: You have to be **CAUTIOUS** with this paper – it is a classic, and essential reading for every mycologist, but it predates the formal separation of the major eukaryotic kingdoms (which appeared in 1969) and Bartnicki-García discusses slime moulds (Acrasiales) and two other groups (*Oomycota* and *Hyphochytriomycota*) as 'lower fungi' whereas they are not *now* considered to be fungi at all and are placed in different kingdoms.

Bartnicki-García, S., Hergert, F. & Gierz, G. (1989). Computer simulation of fungal morphogenesis and the mathematical basis for hyphal (tip) growth. *Protoplasma*, **153**: 46–57. DOI: https://doi.org/10.1007/BF01322464.

Boswell, G.P. (2008). Modelling mycelial networks in structured environments. *Mycological Research*, **112**: 1015–1025. DOI: https://doi.org/10.1016/j.mycres.2008.02.006.

Boswell, G.P., Jacobs, H., Gadd, G.M., Ritz, K. & Davidson, F.A. (2003). A mathematical approach to studying fungal mycelia. *Mycologist*, **17**: 165–171. DOI: https://doi.org/10.1017/S0269-915X(04)00403-3.

Bull, A.T. & Trinci, A.P.J. (1977). The physiology and metabolic control of fungal growth. In *Advances in Microbial Physiology*, **Vol. 15**, ed. A.H. Rose & D.W. Tempest. London: Academic Press, pp. 1–84. DOI: https://doi.org/10.1016/S0065-2911(08)60314-8.

Buller, A.H.R. (1909). *Researches on Fungi*, **Vol. 1**. London: Longman, Green & Co. ISBN: 978-1314758955.

Buller, A.H.R. (1922). *Researches on Fungi*, **Vol. 2**. London: Longman, Green & Co. ISBN: 978-1330331767.

Buller, A.H.R. (1924). *Researches on Fungi*, **Vol. 3**. London: Longman, Green & Co. ASIN: B0008BT4QW.

Buller, A.H.R. (1931). *Researches on Fungi*, **Vol. 4**. London: Longman, Green & Co. ISBN: 978-1378204955.

Buller, A.H.R. (1934). *Researches on Fungi*, **Vol. 6**. London: Longman, Green & Co. ISBN: 978-1377064468.

Carlile, M.J. (1995). The success of the hypha and mycelium In *The Growing Fungus*, ed. N.A.R. Gow & G.M. Gadd. London: Chapman & Hall, pp. 3–19. ISBN-10: 0412466007, ISBN-13: 978-0412466007.

Davidson, F.A. (2007). Mathematical modelling of mycelia: a question of scale. *Fungal Biology Reviews*, **21**: 30–41. DOI: https://doi.org/10.1016/j.fbr.2007.02.005.

Davies, J.A. (2006). *Branching Morphogenesis*. Austin, TX: Landes Bioscience Publishing/Eurekah.com. ISBN-10: 0387256156, ISBN-13: 978-0387256153.

Feofilova, E.P.A., Ivashechkin, A., Alekhin, A.I. & Sergeeva, Y.E. (2012). Fungal spores: dormancy, germination, chemical composition, and role in biotechnology (review). *Applied Biochemistry and Microbiology*, **48**: 1–11. DOI: https://doi.org/10.1134/S0003683812010048.

Fiddy, C. & Trinci, A.P.J. (1976). Mitosis, septation and the duplication cycle in *Aspergillus nidulans*. *Journal of General Microbiology*, **97**: 169–184. DOI: https://doi.org/10.1099/00221287-97-2-169.

Fries, N. (1943). Untersuchungen über Sporenkeimung und Mycelentwicklung bodenbewohneneder Hymenomyceten. *Symbolae Botanicae Upsaliensis*, **6**(4): 633–664. URL: http://www.worldcat.org/title/untersuchungen-uber-sporenkeimung-und-mycelentwicklung-bodenbewohnender-hymenomyceten/oclc/11449121.

Glass, N.L., Rasmussen, C., Roca, M.G. & Read, N.D. (2004). Hyphal homing, fusion and mycelial interconnectedness. *Trends in Microbiology*, **12**: 135–141. DOI: https://doi.org/10.1016/j.tim.2004.01.007.

Goriely, A. & Tabor, M. (2008). Mathematical modeling of hyphal tip growth. *Fungal Biology Reviews*, **22**: 77–83. DOI: https://doi.org/ 10.1016/j.fbr.2008.05.001.

Hickey, P.C., Jacobson, D.J., Read, N.D. & Glass, N.L. (2002). Live-cell imaging of vegetative hyphal fusion in *Neurospora crassa*. *Fungal Genetics and Biology*, **37**: 109–119. DOI: https://doi.org/10.1016/S1087-1845(02)00035-X.

Lindahl, B.D. & Olsson, S. (2004). Fungal translocation: creating and responding to environmental heterogeneity. *Mycologist*, **18**: 79–88. DOI: https://doi.org/10.1017/S0269-915X(04)00204-6.

McLean, K.M. & Prosser, J.I. (1987). Development of vegetative mycelium during colony growth of *Neurospora crassa*. *Transactions of the British Mycological Society*, **88**: 489–495. DOI: https://doi.org/10.1016/S0007-1536(87)80032-3.

Meškauskas, A., Fricker, M.D. & Moore, D. (2004b). Simulating colonial growth of fungi with the Neighbour-Sensing model of hyphal growth. *Mycological Research*, **108**: 1241–1256. DOI: https://doi.org/10.1017/S0953756204001261.

Meškauskas, A., McNulty, L.J. & Moore, D. (2004a). Concerted regulation of all hyphal tips generates fungal fruit body structures: experiments with computer visualisations produced by a new mathematical model of hyphal growth. *Mycological Research*, **108**: 341–353. DOI: https://doi.org/10.1017/S0953756204009670.

Money, N.P. (2004). The fungal dining habit: a biomechanical perspective. *Mycologist*, **18**: 71–76. DOI: https://doi.org/10.1017/S0269-915X(04)00203-4.

Money, N.P. (2008). Insights on the mechanics of hyphal growth. *Fungal Biology Reviews*, **22**: 71–76. DOI: https://doi.org/10.1016/j.fbr.2008.05.002.

Moore, D. (2001). *Slayers, Saviors, Servants, and Sex: An Exposé of Kingdom Fungi*. New York: Springer-Verlag. ISBN-10: 0387951016, ISBN-13: 978-0387951010.

Moore, D., McNulty L.J. & Meškauskas, A. (2006). Branching in fungal hyphae and fungal tissues: growing mycelia in a desktop computer. In *Branching Morphogenesis*, ed. J. Davies. Austin, TX: Landes Bioscience Publishing/Eurekah.com, Chapter 4, pp. 75–90. ISBN-10: 0387256156, ISBN-13: 978-0387256153.

Moore, D. & Meškauskas, A. (2017). *The Algorithmic Fungus*. London: CreateSpace Independent Publishing Platform. ISBN-10: 1545439257, ISBN-13: 978-1545439258. Full text available: https://www.researchgate.net/publication/321361280_The_Algorithmic_Fungus.

Oliver, S.G. & Trinci, A.P.J. (1985). Modes of growth of bacteria and fungi. In *Comprehensive Biotechnology: The Principles, Applications and Regulations of Biotechnology in Industry, Agriculture and Medicine*, ed. M. Moo-Young. Oxford, UK: Pergamon Press. pp. 159–187. ISBN-10: 0080325106, ISBN-13: 978-0080325101.

Prosser, J. (1990). Growth of fungal branching systems. *Mycologist*, **4**: 60–65. DOI: https://doi.org/10.1016/S0269-915X(09)80533-8.

Prosser, J.I. (1995a). Kinetics of filamentous growth and branching. In *The Growing Fungus*, ed. N.A.R. Gow & G.M. Gadd. London: Chapman & Hall, pp. 301–318. ISBN-10: 0412466007, ISBN-13: 978-0412466007.

Prosser, J.I. (1995b). Mathematical modelling of fungal growth. In *The Growing Fungus*, ed. N.A.R. Gow & G.M. Gadd. London: Chapman & Hall, pp. 319–335. ISBN-10: 0412466007, ISBN-13: 978-0412466007.

Read, N.D. (2011). Exocytosis and growth do not occur only at hyphal tips. *Molecular Microbiology*, **81**: 4–7. DOI: https://doi.org/10.1111/j.1365-2958.2011.07702.x.

Reijnders, A.F.M. (1963). *Les problèmes du développement des carpophores des Agaricales et de quelques groupes voisins*. The Hague: Dr W. Junk. ISBN-10: 9061936284, ISBN-13: 978-9061936282.

Steele, G.C. & Trinci, A.P.J. (1975). Morphology and growth kinetics of hyphae of differentiated and undifferentiated mycelia of *Neurospora crassa*. *Journal of General Microbiology*, **91**: 362–368. DOI: https://doi.org/10.1099/00221287-91-2-362.

Takeshita, N. (2016). Coordinated process of polarized growth in filamentous fungi. *Bioscience, Biotechnology and Biochemistry*, **80**: 1693–1699. DOI: https://doi.org/10.1080/09168451.2016.1179092.

Trinci, A.P.J. (1969). A kinetic study of the growth of *Aspergillus nidulans* and other fungi. *Journal of General Micribiology*, **87**: 11–24. DOI: https://doi.org/10.1099/00221287-57-1-11.

Trinci, A.P.J. (1974). A study of the kinetics of hyphal extension and branch initiation of fungal mycelia. *Journal of General Microbiology*, **81**: 225–236. DOI: https://doi.org/10.1099/00221287-81-1-225.

Trinci, A.P.J. (1979). The duplication cycle. In *Fungal Walls and Hyphal Growth*, ed. J. Burnett & A.P.J. Trinci. Cambridge, UK: Cambridge University Press, pp. 319–358. ISBN-10: 0521224993, ISBN-13: 978-0521224994.

Trinci, A.P.J., Saunders, P.T., Gosrani, R. & Campbell, K.A.S. (1979). Spiral growth of mycelial and reproductive hyphae. *Transactions of the British Mycological Society*, **73**, 283–292. DOI: https://doi.org/10.1016/S0007-1536(79)80113-8.

Trinci, A.P.J., Wiebe, M.G. & Robson, G.D. (1994). The mycelium as an integrated entity. In *The Mycota*, **Vol. I**, ed. J.G.H. Wessels & F. Meinhardt. Berlin, Heidelberg: Springer-Verlag, pp. 175–193. ISBN-10: 3540577815, ISBN-13: 978-3540577812.

Trinci, A.P.J., Wiebe, M.G. & Robson, G.D. (2001). Hyphal growth. In *Encyclopaedia of Life Sciences*. Hoboken, NJ: John Wiley & Sons. DOI: https://doi.org/10.1038/npg.els.0000367.

Vidal-Diez de Ulzurrun, G., Baetens, J.M., Van den Bulcke, J. *et al.* (2015). Automated image-based analysis of spatio-temporal fungal dynamics. *Fungal Genetics and Biology*, **84**: 12–25. DOI: https://doi.org/10.1016/j.fgb.2015.09.004.

Wang, L., Romano, M.C. & Davidson, F. (2019). Translational control of gene expression via interacting feedback loops. Preprint in the arXiv® e-print service, *arXiv*:1904.03064 [physics.bio-ph]. URL: https://arxiv.org/abs/1904.03064 and https://arxiv.org/pdf/1904.03064.pdf.

Watkinson, S.C., Boddy, L., Burton, K. *et al.* (2005). New approaches to investigating the function of mycelial networks. *Mycologist*, **19**: 11–17. DOI: https://doi.org/10.1017/S0269-915X(05)00102-3.

Wessels, J.G.H. (1993). Wall growth, protein excretion and morphogenesis in fungi. *New Phytologist*, **123**: 397–413. DOI: https://doi.org/10.1111/j.1469-8137.1993.tb03751.x/.

CHAPTER 5

Fungal Cell Biology

Events at the hyphal tip are crucial to the extension of the hypha. It is vital that we describe the molecular processes taking place in the hyphal tip as far as we can, and this is the main purpose of Chapter 5.

In this chapter, we will give you a complete outline of eukaryotic cell biology with emphasis on how fungal cells work and how the cell biology contributes to mycelial growth. Because they are eukaryotes that are easy to cultivate in the laboratory, several fungi have been adopted as model organisms for experimentation and we will show how yeasts have been used in this way since the nineteenth century. We discuss the essentials of cell structure in some detail, emphasising the molecular biology of the nucleus, nucleolus, nuclear import and export, and mRNA translation and protein sorting. We also briefly cover mitotic and meiotic nuclear division; nuclear genetics will be dealt with in Chapter 6.

The plasma membrane and signalling pathways, and endomembrane systems, cytoskeletal systems and molecular motors form major topics because directed and rapid transport of materials needed for hyphal tip extension is a crucial and characteristic feature of highly polarised filamentous growth. Other features of cell biology that are specific to fungi include the fungal cell wall, the cell biology of the hyphal apex, the nature of hyphal fusions and mycelial interconnections, the meaning of cytokinesis in fungi, and septation and the yeast–mycelial dimorphism.

5.1 MECHANISMS OF MYCELIAL GROWTH

Polarised growth of fungal hyphae is achieved by restricting extension to the hyphal apex. The cell wall at the hyphal tip has viscoelastic (flowing like a liquid but resisting stretching, compression or distortion) properties, and yields to the internal turgor pressure within the hypha. Further behind the tip the wall is rigidified and resistant to the turgor forces resulting from the osmotic flow of water into the hypha. Turgor pressure generated within the hypha therefore acts as the driving force for hyphal extension.

Hyphal extension at the apex requires synthesis and insertion of **new wall material** and **new membranes** in a way that does not weaken the tip. This highly organised process is supported by the **continuous flow of vesicles** generated within the cytoplasm behind the tip and is coordinated with the growth and replication of all the other cytoplasmic organelles and their **migration towards the extending apex**. In this chapter, we will be describing all these individual processes and, in Chapter 7, this mosaic of processes will be assembled into what we hope will be a full picture of hyphal tip growth. What *is* important at this point is to recognise how much of the detailed cell biology of filamentous fungi is adapted, devoted and committed to forward thrusting of the hyphal apex; that is, **hyphal extension growth**. This is the supreme characteristic of filamentous fungi that sets them off from the other crown eukaryotes: animals and plants.

5.2 THE FUNGUS AS A MODEL EUKARYOTE

The cell we are describing is the generalised cell of a eukaryote. In most textbooks, when this is attempted it is usually the animal cell that takes centre stage (e.g. the classic cell biology text, Alberts *et al.*, 2014); plant cells might be described occasionally, when there is a need to deal with photosynthesis, and yeasts may get a mention as the source of some of the molecular detail. There is nothing wrong with that (although animal cells do not have the cell wall that is so important to the other eukaryotic kingdoms, this feature being lost in the distant past by the single-celled opisthokonts that gave rise to Kingdom *Animalia*), but it does downplay the enormous contribution that fungi have made to development of our knowledge of **eukaryotic cell biology**. As well, perhaps, as downplaying the enormous contribution that the fungal lifestyle has made to the evolution of eukaryotic cell biology.

Although several unicellular eukaryotes have been used as models in cell and molecular biology (Simon & Plattner, 2014), the fact is that most of what we know about the biology of the cell of higher organisms derives from work with yeasts. Most biologists would recognise the contribution made by yeast research to molecular biology in the 1990s. The first complete **sequence** analysis of any **eukaryote chromosome** was that of the entire DNA sequence of *Saccharomyces cerevisiae* chromosome III, published in 1992 by a large international team led by **Steve Oliver**. This was followed in 1996 by the sequencing of the whole of the genome of *Saccharomyces cerevisiae*, which was the **first eukaryote genome** to be sequenced (the 13-year Human Genome Project, which got all the headlines, was completed in 2003).

Some biologists will know that the **Nobel Prize in Physiology or Medicine in 2001** was awarded to three scientists 'for their discoveries of key regulators of the cell cycle', and two of them worked with yeasts (**Leland Hartwell** worked with *Saccharomyces cerevisiae* and **Paul Nurse** with *Schizosaccharomyces pombe*; the third Laureate was **Tim Hunt**, who worked with sea urchin eggs) (see https://www.nobelprize.org/nobel_prizes/medicine/laureates/2001/). The whole of genomics and cell cycle biology rests on foundations built with yeasts. But the crucial contribution goes much further back than the end of the twentieth century; it goes back to the mid-nineteenth century. The history of yeast in biology effectively starts with **Louis Pasteur** who connected yeast to **fermentation** in 1857 and demonstrated that the growth of microorganisms in nutrient broths is not due to spontaneous generation.

The word 'yeast' is a general term for any growth that appears in a fermenting liquid. In its origins, the word means frothy or foamy so it's descriptive of the fermentation process but has become associated with the agent of fermentation. When grape juice is collected it ferments quite naturally, and the growth that occurs and eventually forms sediment is 'yeast'. Making **alcoholic drinks** is such a simple process that all societies, even the most primitive, have one or more fermentations that they include in their rituals. There are some ancient Egyptian murals and tomb ornaments depicting both baking and wine making. From the biological point of view, it is remarkable that the one organism responsible for most fermentations is the yeast now called *Saccharomyces cerevisiae*, but known as **brewer's yeast** in one trade, and **baker's yeast** in the other (though we will call it '**budding yeast**'). Remember that yeasts (as well as filamentous fungi and other microbes) are present on the surfaces of grapes, fruits and seeds in nature so there was no need to add them and food preparation processes like these were carried out with no knowledge of the importance of microbes in the fermentation processes involved. *Saccharomyces cerevisiae* comes to the fore so often because its metabolic controls allow it to produce alcohol even in the presence of oxygen, a theme to which we will return in section 19.13.

At the end of the nineteenth century, industrialisation created the need to guarantee and improve production and product quality, and this prompted brewers and wine makers to sponsor research into the nature of fermentation. This naturally came to focus on the single-celled microorganisms we call yeast. Studies of yeast metabolism essentially founded the sciences of biochemistry and enzymology. Purification of cultures (necessary for a uniform product) and a drive to improve cultures (to increase the efficiency of fermentation or develop new products) was enhanced by the parallel development of the science of genetics at the beginning of the twentieth century. Pasteur was employed by the wine growers to improve the wine fermentation. From his experiments Pasteur concluded: 'I am of the opinion that alcoholic fermentation never occurs without simultaneous organisation, development and multiplication of globules [cells]' (Pasteur, 1860, 1879). And those cells were fungal (yeast) cells.

Pasteur died in 1895, before the first Nobel prizes were awarded (in 1901 for the first time; visit https://www.nobelprize.org/alfred_nobel/), but the winner of the **1907 Nobel Prize in Chemistry** was **Eduard Buchner** ('for his biochemical researches and his discovery of **cell-free fermentation**', https://www.nobelprize.org/nobel_prizes/chemistry/laureates/1907/). He determined that fermentation was caused by a yeast secretion that he termed **zymase**; we now call such things enzymes. Buchner's experiment, for which he won the Nobel prize, consisted of producing a cell-free extract of yeast cells and showing that this 'press juice' could ferment sugar. Here we have the beginnings of our understandings of cell biochemistry and metabolism, and we can chart the progress of basic biological knowledge through subsequent Nobel prizes (visit https://www.nobelprize.org/nobel_prizes/).

The Nobel Prize in Chemistry 1929 went to Arthur Harden (who worked on the involvement of phosphates in respiration of yeast) and Hans von Euler-Chelpin (who worked on enzymology and oxidative respiration of yeast) ' for their investigations on the fermentation of sugar and fermentative enzymes' (https://www.nobelprize.org/nobel_prizes/chemistry/laureates/1929/). Hans von Euler-Chelpin's Nobel Lecture (see https://www.nobelprize.org/uploads/2018/06/euler-chelpin-lecture.pdf) included the paragraph:

> Within the living organism, the majority of reactions are brought about by special substances already active in minimum quantities, such substances being known as enzymes or ferments. Every group of substances, and in fact practically every substance, requires its specific enzyme for its reaction. Only a few enzyme types were known in the early days, such as pepsin in the gastric juice, which splits proteins, or amylase in saliva and in malt, which converts starch into sugar, but in more recent times the number of enzymes whose existence has been proved or substantiated, has risen to over 100.

This lecture was delivered in May 1930.

At the end of t World War II, the **Nobel Prize in Physiology or Medicine 1945** was awarded jointly to **Sir Alexander Fleming, Ernst Boris Chain** and **Sir Howard Walter Florey** 'for the **discovery of penicillin** and its curative effect in various infectious diseases' (https://www.nobelprize.org/nobel_prizes/medicine/laureates/1945/) (see section 19.15).

Over the next quarter of a century, the great network of metabolism was established, and the genetic segregation side of the story was developed, too. Mendel's work on garden peas was rediscovered and republished in 1900 and inspired many experimenters, including those who worked with *Saccharomyces cerevisiae* to replicate and confirm his discovery of gene segregation (see Chapter 6). Subsequently, the first metabolic pathways were constructed using nutritionally deficient mutants of the filamentous ascomycete fungus *Neurospora* and the bacteria *Escherichia* and *Salmonella*.

In 1958, the Nobel Prize in Physiology or Medicine was awarded to George Beadle, Edward Tatum (both of whom worked with *Neurospora*) 'for their discovery that genes act by regulating definite chemical events' and Joshua Lederberg 'for his discoveries concerning genetic recombination and the organisation of the genetic material of bacteria ' (https://www.nobelprize.org/nobel_prizes/medicine/laureates/1958/).

During the first half of the twentieth century, then, yeasts and related filamentous fungi provided the foundation of knowledge of cell biochemistry, metabolism and its genetic control, and then, as we have mentioned, the same conceptual approach (isolating mutants defective in steps of a pathway to study that pathway) was applied to the cell cycle by Leland Hartwell and Paul Nurse (https://www.nobelprize.org/nobel_prizes/medicine/laureates/2001/) (Davis, 2000; Samson & Varga, 2008; Machida & Gomi, 2010).

The distribution of prizes continues; all of the following Nobel-prize-winning discoveries were supported by research on yeasts. The **Nobel Prize in Chemistry in 2006** went to **Roger D. Kornberg** 'for his studies of the molecular basis of **eukaryotic transcription**' (https://www.nobelprize.org/nobel_prizes/chemistry/laureates/2006/). The **Prize in Physiology or Medicine 2009** was awarded jointly to **Elizabeth H. Blackburn, Carol W. Greider** and **Jack W. Szostak** 'for the discovery of how **chromosomes are protected by telomeres and the enzyme telomerase**' (https://www.nobelprize.org/nobel_prizes/medicine/laureates/2009/) and the **Physiology or Medicine Nobel Prize 2013** was awarded jointly to **James E. Rothman, Randy W. Schekman** and **Thomas C. Südhof** 'for their discoveries of **machinery regulating vesicle traffic**, a major transport system in our cells' (https://www.nobelprize.org/nobel_prizes/medicine/laureates/2013/).

What we are planning to do next is introduce you to the working eukaryotic cell. There are several ways we could approach this, and we have chosen to focus on the typical fungal cell (and to headline, when needed, how this cell type differs from animal and plant cells), starting our description from the DNA level. Towards the end we will concentrate on the features that contribute most to apical extension of the filamentous hypha characterising the fungal lifestyle as already described.

5.3 THE ESSENTIALS OF CELL STRUCTURE

Cells were discovered in 1665 by **Robert Hooke** who saw them in slices of cork using his seventeenth century optical microscope. Hooke coined the term 'cell' because he compared the compartments in the cork he saw to the small rooms in which monks lived. The word cell comes from the Latin cellula, a small room. It was another 170 years before the cell theory, that all organisms are composed of one or more cells and all cells come from pre-existing cells, was first formulated (in 1839) by **Matthias Jakob Schleiden** and **Theodor Schwann**.

The essential life-sustaining activities of an organism occur within its cells, and almost all cells contain the genetic (hereditary) information necessary for performing and regulating those activities and for transmitting the genetic information to the next generation of cells. Prokaryotes are cells, too, and both prokaryotic and eukaryotic cells have a **membrane** that envelops the cell and performs the crucial task of separating the cell from its environment. This cell membrane is also selectively permeable; that is, it regulates the chemicals that enter and leave the cell. By regulating the flow of ions, the membrane adjusts the electric potential and pH of the cell. By regulating the flow of water (directly or indirectly), the membrane adjusts the volume and osmotic potential of the cell. The material within (and enclosed by) the cell membrane is a complex mix of molecules and ions called cytoplasm. The cytoplasm also contains a variety of regions specialised to carry out one or more vital functions. These are bounded by their own membranes in eukaryotes; these are generally called **organelles**.

There are two different kinds of **genetic material**: deoxyribonucleic acid (**DNA**) and ribonucleic acid (**RNA**). The DNA is used for long-term information storage (although this function is served by RNA in some viruses). An organism's genetic information is encoded in the sequence of its DNA or RNA. RNA is also characteristically used for information transport within the cell, e.g. as **messenger** and **transfer RNAs** (mRNA and tRNA), and **ribosomal RNAs** (rRNAs) serve essentially enzymatic functions during **protein synthesis**.

The genetic material in prokaryotes is organised as a simple **circular DNA molecule** which is packaged into the **nucleoid** region of the cytoplasm. This region is not separated by a membrane and this is a major distinguishing feature between prokaryotes and eukaryotes. In eukaryotes, the genetic material is distributed between different linear molecules called **chromosomes**, which are housed in a nucleus surrounded by an organised nuclear membrane.

Prokaryotes do not have a nuclear membrane, and they also lack the membrane-bound organelles that characterise eukaryotic cells. However, ribosomes, and many of the other components of the protein synthesis mechanism, are present in both prokaryotic and eukaryotic cells, though there are some differences in detail (and some of those differences are important in providing drug targets for therapeutic antibiotics). In prokaryotes, the functions performed by specialised organelles in eukaryotes are performed by the plasma membrane (Table 5.1). Prokaryotic cells do have structurally distinct regions:

- **cytoplasmic region** contains the cell genome (DNA), ribosomes and various inclusions;
- **appendages** attached to the cell surface – **pili** (singular pilus) are hair-like appendages responsible for attachment; motile bacteria have **flagella**, but these are largely composed of the self-assembling protein flagellin and are not related to eukaryotic flagella (pilus and flagellum proteins of animal pathogens are major antigens);
- **cell envelope** consisting of a capsule, a peptidoglycan cell wall (an important target for therapeutic antibiotics) and a plasma membrane.

Prokaryotes also frequently carry **extrachromosomal** (usually circular) **DNA** molecules called **plasmids**, which often carry genes conferring additional phenotypes, such as antibiotic resistance, pathogenicity, genome transfer and metabolism of exotic substrates. Plasmids are rare in most eukaryotes, but they do occur in fungi. Also, major eukaryotic organelles (e.g. mitochondria and chloroplasts) contain a genetic architecture separate from the nucleus comprising a (small) circular

Table 5.1. Comparison of the main features of prokaryotic and eukaryotic cells

	Prokaryotes	**Eukaryotes**
Typical organisms	bacteria, archaea	protists, fungi, animals, plants
Typical size	about 1–5 μm	about 10–100 μm
Type of nucleus	nucleoid region; no true nucleus, no membrane enclosure (pro + karyotos = before having nucleus (literally, nuts))	true nucleus surrounded by double membrane (eu + karyotos = having true nucleus (literally, nuts))
DNA	circular DNA (usually)	linear DNA molecules (chromosomes) organised around histone proteins
RNA/protein synthesis	coupled in cytoplasm	RNA synthesis inside the nucleus; protein synthesis in cytoplasm
Cytoplasmic ribosomes	Subunits = 50S + 30S	cytoplasmic 60S + 40S; organelle ribosomes more similar to prokaryotic ones
Cytoplasm structure	Cell is bounded by a membrane, very few structures internally	Cell is bounded by a membrane enclosing a cytoplasm containing numerous endomembranes, organelles and a cytoskeleton of microtubules and microfilaments
Cell movement	flagella largely composed of flagellin protein	flagella and cilia contain microtubules largely made of tubulin protein; lamellipodia and filopodia containing actin
Mitochondria	none (redox electron transport takes place in the plasma membrane)	one to several dozen per cell (probably originating from endosymbiotic prokaryotes). Some eukaryotes lack mitochondria; in anaerobes mitochondria are replaced by hydrogenosomes
Chloroplasts	none (photosynthetic electron transport takes place in folds of the plasma membrane stacked into thylakoids in cyanobacteria)	in algae and plants
Organisation	usually single cells	single cells, colonies, higher multicellular organisms with specialised cells
Cell division	binary fission (simple division)	mitosis (fission or budding) meiosis

Table modified and adapted from the Wikipedia article entitled *Cell (biology)*, https://en.wikipedia.org/wiki/Cell_(biology).

DNA molecule and an independent ribosome population and protein synthetic apparatus.

5.4 SUBCELLULAR COMPONENTS OF EUKARYOTIC CELLS: THE NUCLEUS

The subcellular components of eukaryotic cells include, most importantly, the **nucleus** with its associated **nucleolus** (Pollard *et al.*, 2017a) and all the extranuclear, or 'cytoplasmic', components: **ribosomes, endoplasmic reticulum, Golgi apparatus, cytoskeleton, mitochondria, vacuoles** and **vesicles** (Alberts *et al.*, 2014).

The **cell nucleus** is the most conspicuous organelle in many eukaryotes, but it can be small and inconspicuous in fungi. This organelle has two major functions: storage of hereditary material and coordination of all cellular activities (metabolism, growth, with all the synthetic processes on which it depends, and cell division). It houses the eukaryotic cell's chromosomes and is the location for all molecular processes involving DNA: replication, recombination and transcription (copying DNA gene sequences into messenger RNA). The nucleus is also where post-transcriptional steps in gene expression, such as RNA processing to remove introns, take place.

Genomic DNA molecules are extremely long. For example, yeast chromosome III, the first ever to be fully sequenced, comprises 3.15×10^5 bases and, since adjacent bases are separated by 0.34 nm, the chromosomal molecule can be estimated to be about 107 μm long. Yeast cells vary, but are about 5 to 7 μm in size; this one molecule is between 15 and 20 times longer than the cell that contains it. The molecule is only 2 nm in diameter, of course, but it is quite evident that the nuclear DNA must be **highly condensed** to fit in the nucleus.

The compaction progresses at several levels. Fourteen turns of the DNA helix, a length of 146 bp (base pairs), complex with an octamer of **histone proteins** (two copies of each of histones H2A, H2B, H3 and H4,) into the **nucleosome core particle** and the nucleosomes then wind into a hierarchy of 10 nm fibres, 30 nm fibres and chromosome loops, which end up as the chromatin of fully condensed chromosomes. In yeast, there is a length of about 45 bp of DNA between nucleosomes because the linker **histone H1**, which is involved in forming 30 nm fibres in animals and plants, **is missing**.

It is important to appreciate that chromatin is not a static structure. Chromatin participates in the minute-by-minute working activities of the nucleus and there are many proteins that are recruited to modify and adapt chromatin to enable its DNA to be expressed. Histones are acetylated and deacetylated to modify **chromatin structure**. There are several distinct histone acetylase and deacetylase complexes that have gene-specific effects on transcription. DNA-binding activators and repressors can recruit these histone acetylases/deacetylases to specific gene promoters to locally modify chromatin structure. Other chromatin-modifying complexes are responsible for creating larger regions of altered chromatin structure, associated with long-range effects on gene activity.

Most aspects of nuclear activity involve extremely large multiprotein complexes (often called **molecular machines**) and some of their components have chromatin-modifying activities. For example, the RNA transcription machinery and the molecular machine responsible for mRNA splicing are each comparable to ribosomes in terms of size and subunit complexity. As an aside: the **Nobel Prize in Chemistry 2014** was awarded jointly to **Jean-Pierre Sauvage, Sir J. Fraser Stoddart** and **Bernard L. Feringa** 'for the design and synthesis of molecular machines' (see https://www.nobelprize.org/nobel_prizes/chemistry/laureates/2016/popular-chemistryprize2016.pdf).

The central dogma of molecular biology has traditionally been that RNA is a messenger molecule that exports the information coded into DNA out of the nucleus in order to code the synthesis of proteins in the cytoplasm: DNA – RNA – protein. A **ribonucleic acid polymerase** (RNA polymerase or **RNAP**) is a multisubunit enzyme that catalyses the process of **transcription** during which an RNA polymer is synthesised from a DNA template. Other RNAs well known to be involved in protein synthesis are transfer RNA (tRNA) and ribosomal RNA (rRNA). However, it is now clear that RNA serves a range of other functions. Some RNA molecules regulate gene expression, others act as enzymes and many have functions that are still unknown. These types of RNA are called non-coding or ncRNA, a category that includes microRNA (miRNA), small RNA (sRNA), interfering RNA (iRNA), small interfering RNA (siRNA) and antisense RNA.

Prokaryotes use the same RNAP to catalyse the polymerisation of coding as well as non-coding RNAs, eukaryotes have **five distinct RNA polymerases**.

- RNA polymerase I synthesises the major RNA molecules of the ribosome (which can account for nearly half of the RNA transcribed in a eukaryotic cell).
- RNA polymerase II produces all primary transcripts, the mRNA precursors, as well as small nuclear RNAs and micro RNAs.
- RNA polymerase III transcribes transfer RNAs, small ribosomal RNA and other small RNAs found in the nucleus and cytoplasm and which are necessary for normal functioning of the cell.
- RNA polymerases IV and V are found exclusively in plants; their function is essential for the formation of small interfering RNA and heterochromatin in the plant nucleus.

Messenger RNA transcription in fungi, particularly in the budding yeast *Saccharomyces cerevisiae*, is one of the main model systems for research on transcription in eukaryotes, which requires a large set of proteins (called **general transcription factors**) to be assembled at the promoter before transcription can begin. These help the RNA polymerase to bind to the promoter, open up the double-stranded DNA and then switch RNA polymerase into elongation mode (Peñate & Chávez, 2014; Sesma & von der Haar, 2014).

Other proteins required for transcription initiation include activators binding to specific sequences to enhance attachment of polymerase; transcription mediators that interface the activators to the transcription factors; and other enzymes modifying chromatin structure to aid transcription by opening the chromatin structure. As some of these proteins are themselves made up of more than one polypeptide, approximately 100 protein subunits must assemble at the promoter site to initiate transcription (Kornberg, 2007) (Figure 5.1 shows a greatly simplified overview). Once transcription is underway, most of the transcription factors detach from the polymerase complex.

The 'gene specificity' aspects of the binding events that contribute to assembly of the **pre-initiation complex** (**PIC**) seem to reside in the TATA-binding protein (**TBP**, a subunit of $TF_{II}D$) and TATA-box binding protein-associated factor(s) (**TAFs**) and co-activators that make up $TF_{II}D$ and $TF_{II}B$. Among the TAFs, $TAF_{II}250$ seems to be particularly important as it regulates binding of TBP to DNA, binds core promoter initiator proteins, binds acetylated lysine residues in core histones, and possesses enzyme activities that modify histones and other transcription factors. These activities aid in positioning and stabilising $TF_{II}D$ at specific promoters, and alter chromatin structure at the promoter, creating a sharp bend in the promoter DNA, to allow assembly of transcription factors into the PIC. By so doing, $TAF_{II}250$ converts signals for gene activation into effective transcription. $TF_{II}E$ joins the growing complex and recruits $TF_{II}H$, which has protein kinase activity which phosphorylates RNA polymerase II within the **CTD** (the **C-terminal repeat domain**, CTD is an extension appended to the C terminus of the largest subunit of RNA polymerase II, which serves as a flexible binding scaffold for numerous nuclear factors; which factors bind being determined by the phosphorylation patterns on the CTD repeats). $TF_{II}H$ has DNA helicase activity to unwind the promoter DNA, and it recruits nucleotide-excision repair proteins. Subunits within $TF_{II}H$ that have ATPase and helicase activity create negative superhelical tension in the DNA that causes approximately one turn of DNA to unwind and form the transcription bubble. The template strand of the transcription bubble engages with the RNA polymerase II active site and RNA synthesis begins. After synthesis of about 10 nucleotides of RNA, RNA polymerase II escapes the promoter region to transcribe the remainder of the gene.

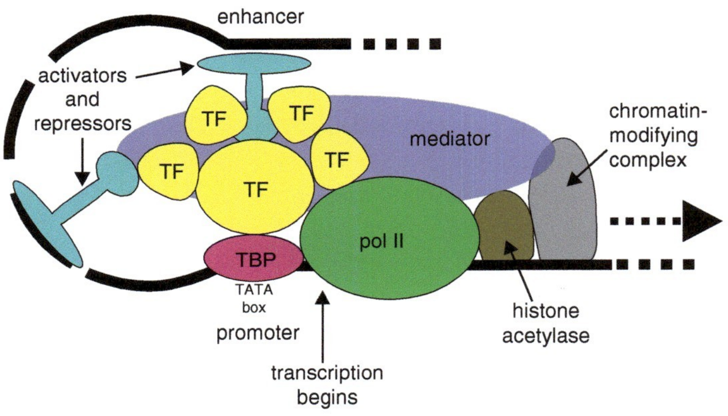

Figure 5.1. Activation of RNA polymerase II-dependent transcription involves assembly by gene-specific activator proteins of a pre-initiation complex (PIC) that will eventually synthesise a specific messenger RNA. This conversion requires structural changes in chromatin and assembly of general transcription factors (TFs) and RNA polymerase II (pol II) at the gene's core promoter sequence, which surrounds the transcription start site of the gene. A key event is the interaction of DNA-bound activators like the TATA-box binding protein (TBP) with co-activators (shown here labelled TF, but generally called TAF_{II}s [= TATA-box binding protein-associated factor(s)]. $TAF_{II}250$ is a scaffold for assembly of other TAF_{II}s with TBP into a complex called $TF_{II}D$. TBP first binds to the promoter and then recruits $TF_{II}B$ to join $TF_{II}D$ (and $TF_{II}A$ if present). Before joining the PIC, RNA polymerase II and $TF_{II}F$ are bound together, being recruited by $TF_{II}B$. Finally, RNA polymerase II recruits $TF_{II}E$, which further recruits $TF_{II}H$ to complete the PIC assembly. $TF_{II}D$ and $TF_{II}B$ are the only components of the pre-initiation complex that can bind specifically to core promoter DNA. View the Wikipedia entry: https://en.wikipedia.org/wiki/Eukaryotic_transcription, and/or the transcription animation at http://vcell.ndsu.nodak.edu/animations/transcription/index.htm for further explanation.

In view of the significance of **transcription factors** in directing transcription to specific genes, you should not be surprised that in later discussions of cellular events we will frequently refer to the involvement of transcription regulators in control of so many features. Those regulators may be modifying chromatin structure and/or affecting the specificity and/or activity of transcription and/or RNA processing machinery. Many of the transcription regulators themselves work through multiprotein complexes (often called co-activators) and individual proteins can be components of completely different, and functionally distinct, co-activator complexes.

Transcriptional regulation and chromatin structure are intimately meshed together with the result that events occurring during initiation of transcription can regulate mRNA processing and so affect gene expression. The critical association occurs when the mRNA is first synthesised. Before mRNA can be used by ribosomes as a template for protein synthesis, it must be processed by the addition of a methylated cap (at its 5′ end) and a polyadenylated (poly-A) tail (at its 3′ end). Provision for the cap is made at the very start of transcription; provision for the poly-A tail is made at the end of transcription. After the transcription initiation complex has been established, the carboxy-terminal end of RNA polymerase II is phosphorylated and this causes the enzyme to shift from the initiation mode to its elongation mode. Once transcription is underway, most of the transcription factors detach from the polymerase complex, their function to initiate transcription being complete.

RESOURCES BOX 5.1

We strongly recommend that you view the animations in the *Virtual Cell Animation Collection* produced by the Molecular and Cellular Biology Learning Center of the North Dakota State University. Visit this URL: http://vcell.ndsu.nodak.edu/animations/home.htm. The collection is also introduced on YouTube at https://www.youtube.com/channel/UCcSThfV7yiW9I5hXBnEk9Zg.

The phosphorylated 'tail' of the polymerase interacts directly with proteins that carry out the **RNA-capping, poly-A processing** and **splicing**; in other words the transcription machinery recruits the different **RNA processing machines** to the initial RNA transcript (which is usually called a pre-mRNA). These different machines share components; for example, the cleavage polyadenylation specificity factor (CPSF) contains particular TAFs as subunits. This close 'mechanical' relationship, and the fact that newly synthesised RNA, the RNA polymerase and many mRNA processing factors are all close together in the nucleus, strongly suggest that the different 'machines' are formed into an **'mRNA factory'** that integrates synthesis and processing of mRNA.

As soon as the **primary transcript** or **pre-mRNA** is completed, RNA polymerase releases the already 5'-capped RNA molecules, and cleavage factors bind to specific nucleotide sequences in the molecule. The 3' end of the pre-mRNA is then put into the correct configuration for cleavage and stabilising factors to be added to the complex. The poly-A polymerase now binds to the pre-mRNA and cleaves the 3' end, allows the complex to dissociate and synthesises the polyadenylated tail by adding adenine nucleotide residues to the 3' end. As the tail is synthesised, proteins bind to it, increasing the rate at which it is synthesised. When the polyadenylation process is completed the processed pre-mRNA (which still contains introns) is ready for the splicing process (Figure 5.2).

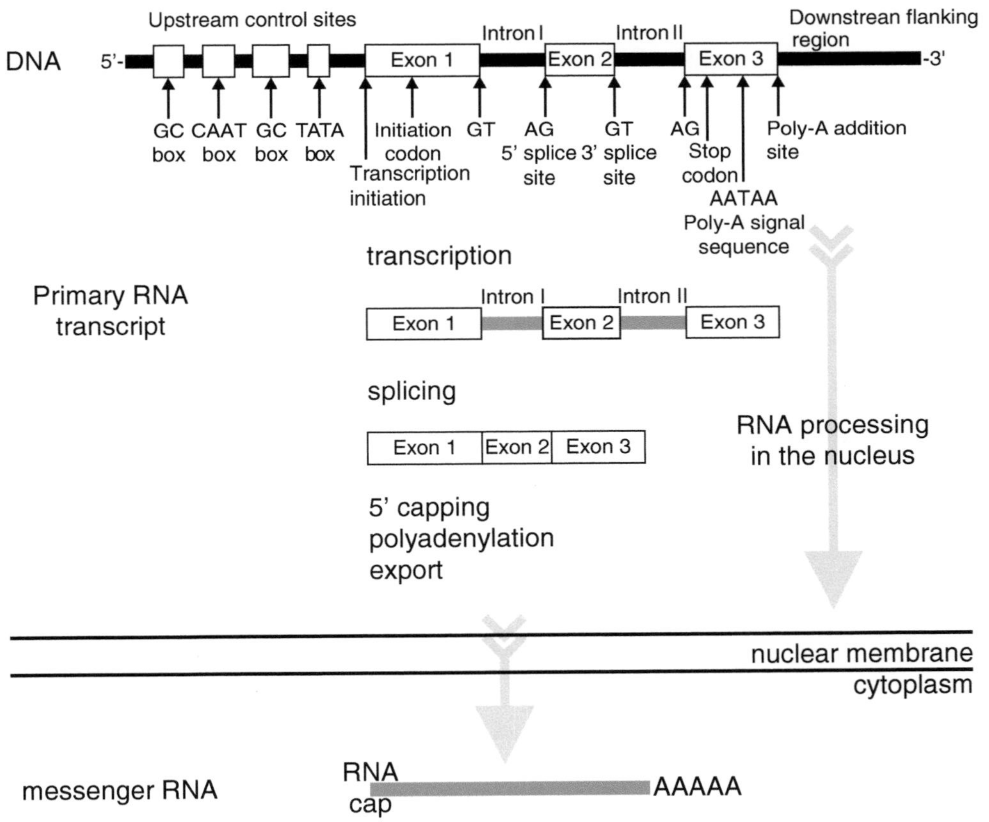

Figure 5.2. In eukaryotes, the transcript of a protein-coding gene is called a primary transcript or pre-mRNA transcript and before it leaves the nucleus it is heavily modified by enzymes and ribozyme complexes. The 5' end is capped with a modified guanine nucleotide; the 3' end is polyadenylated with a sequence of up to 200 adenine nucleotides; finally, the introns are removed by a process known as splicing. These processes are coordinated in time and space and occur as the pre-mRNA transcript is emerging from the RNA polymerase. Splicing involves linking the ends of two exons from a pre-mRNA transcript with high precision, discarding the intervening intron. The machinery involved uses five catalytic snRNA molecules (sn, small nuclear) and over 50 protein subunits. The assembly of snRNAs and proteins that perform the splicing is called the spliceosome. Splicing generates the fully processed mRNA which is then exported to the cytoplasm to be translated (see Figure 5.3).

Intron and exon boundaries are defined by specific sequences in pre-mRNA which are recognised by a large array of snRNPs (**small nuclear ribonucleoproteins**) and other proteins that come together to form the reaction centres, the **spliceosomes**, on each spliced intron (Papasaikas & Valcárcel, 2016). The proteins are called **SR proteins** and have a common structure, including one or more RNA-binding domains and a domain rich in arginine–serine dipeptides that function in the protein–protein interactions involved in splicing, transport and localisation of these proteins. Splicing involves linking the ends of two exons from a pre-mRNA transcript with high precision and discarding the intervening intron. The machinery involved uses an assembly of five catalytic **snRNA molecules** and over 50 protein subunits called the spliceosome. The process involves changes in protein conformation that effectively loop out the intron (into a branched loop structure called a lariat) so that the enzymes in the spliceosome can cleave out the intron and join the ends of adjacent exons (Figure 5.3). Splicing generates the fully processed mRNA, which is then exported to the cytoplasm for translation. A single primary transcript may be spliced in different ways to increase the complexity of gene expression during cellular differentiation and development. This phenomenon of alternative splicing is discussed in sections 6.4 and 13.14.

Throughout their time in the nucleus, RNA transcripts are associated with **heterogeneous nuclear ribonucleoproteins (hnRNPs)**, an abundant family of RNA-binding proteins. These proteins are involved in almost every aspect of pre-mRNA processing as well as in mRNA transport and in translation. Evidently, the fate and function of the transcription product are intimately dependent on RNP complexes composed of the transcribed RNA and a wide variety of proteins.

5.5 THE NUCLEOLUS AND NUCLEAR IMPORT AND EXPORT

The mRNA is not the only RNA family for which the nucleus is responsible. **Ribosomal RNA** transcription and processing, and **ribosome assembly** are crucial aspects of nuclear function. Indeed, the site of these activities is the most prominent morphological feature within the nucleus; it has been known for many years as a body called the nucleolus. Genes specifying ribosomal sequences are transcribed into pre-ribosomal-RNAs (pre-rRNAs). Virtually all cellular RNAs undergo post-transcriptional processing and modification. Pseudouridylation (which is the conversion of uridine to pseudouridine at specific sites in an RNA sequence) and 2′-O-methylation of the ribose sugar are the most common internal modifications. They are applied to three major types of stable RNA: the spliceosomal snRNAs, ribosomal RNAs (rRNAs) and transfer RNAs (tRNAs). As is the case for pre-mRNA processing, pre-rRNA processing is done by families of proteins working with numerous **small nucleolar RNAs (snoRNAs)** in the form of **RNP complexes (snoRNPs)** that base pair with specific parts of the pre-rRNA sequence to carry out their modification. SnoRNAs are a group of untranslated RNA molecules of variable length (80 to 1,000 nucleotides in yeast) required for rRNA maturation.

The snoRNAs are themselves produced from precursor transcripts and have to be targeted to the nucleolus so that they can

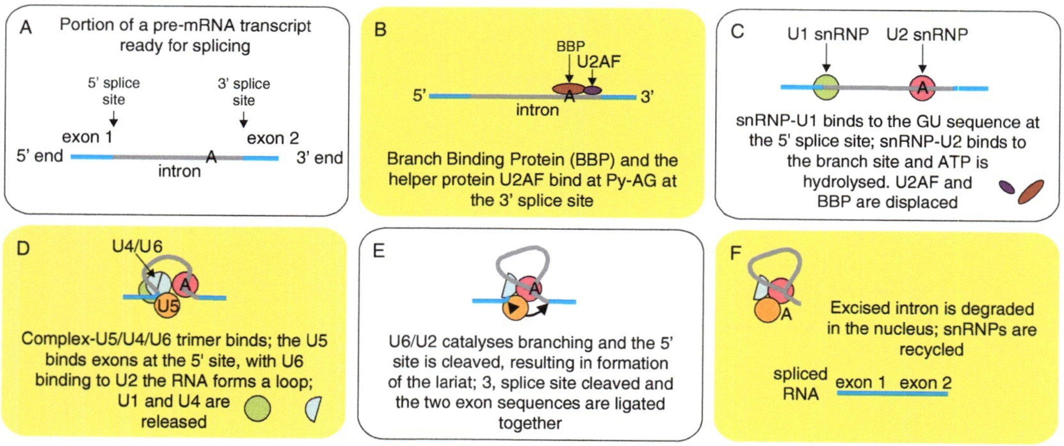

Figure 5.3. The splicing process conducted by the major spliceosome protein-RNA assembly, which splices introns containing GU at the 5′ splice site and AG at the 3′ splice site and accounts for more than 99% of splicing activity in eukaryotes. The first step involves two complexes that bind at the Py-AG at the 3′ splice site: Branch Binding Protein (BBP also called SF1 (splicing factor 1) in mammalian systems) and the helper protein U2AF. The RNA is looped, and three other protein-RNA complexes bind. This final complex undergoes a conformation change and the intron is cleaved at the 5′ GU sequence and forms a lariat at the A branch site. The 3′ end of the intron is next cleaved at the AG sequence, and the two exons are ligated together. As the spliced mRNA is released from the spliceosome, the intron debranches, and is then degraded. Visit the Wikipedia entry at https://en.wikipedia.org/wiki/Spliceosome, and/or the mRNA splicing animation at http://vcell.ndsu.nodak.edu/animations/mrnasplicing/index.htm for further explanation.

guide the critical modifications of the rRNA. As we indicated above, the snoRNPs can also modify spliceosomal snRNAs and as this could affect the way pre-mRNAs are processed it is evident that the **nucleolus** is responsible for more than just ribosome assembly. Nevertheless, the nucleolus is where ribosomal subunits are manufactured, and the proteins needed to assemble ribosomes are synthesised in the cytoplasm, like all proteins, so they must be imported into the nucleus and also targeted to the nucleolus. Subsequently, ribosomal subunits need to be re-exported to the cytoplasm. Ribosomal subunits are large macromolecular assemblies and the way they are translocated to the cytoplasm is not yet known. More is known about export of mRNA from the nucleus.

The nucleus is separated from the cytoplasm by a double membrane called the nuclear envelope, which isolates and protects the DNA from the turmoil of the cytoplasm (in prokaryotes, the processes that depend on DNA take place in the cytoplasm). In eukaryotes, many proteins and RNAs are transported across the nuclear envelope. This is known as **nuclear–cytoplasmic trafficking** and features both import into and export from the nucleus. It occurs through the **nuclear pore complexes (NPCs)**, which are complex assemblies that are embedded in the double membrane of the nuclear envelope. The NPCs provide channels (about 9 nm in diameter) that allow passage of ions and small molecules (less than about 50 kDa) by diffusion, but proteins, RNAs and ribonucleoprotein (RNP) particles larger than 9 nm are selectively transported through NPCs by an energy-dependent mechanism. The trafficking is selectively regulated by developmental and environmental signals.

The overall three-dimensional architecture of the NPC is conserved from yeast to higher eukaryotes though the yeast NPC is smaller than those found in vertebrates. The protein components of NPCs are called **nucleoporins**; the yeast NPC is composed of 30–50 different nucleoporins (up to 100 in vertebrates). We are just beginning to understand the three-dimensional molecular architecture of the NPC but, crudely, it looks rather like the ball valve of a swimmer's snorkel. There is a ring of nucleoporins on the cytoplasmic side of the pore that extends through the nuclear membrane, matched with another ring on the nuclear side of the pore. Between the rings there is a central plug (also called the **transporter** because some specimens seem to have cargo trapped within), and eight short fibrils extend from the cytoplasmic ring into the cytoplasm, whereas the nuclear ring anchors a basket made from eight long, thin filaments (Kosinski *et al.*, 2016; Upla *et al.*, 2017).

How this highly organised tunnel works remains a mystery, but the transport process itself depends on a family of soluble transporter proteins known as **importins** or **exportins** (both also called **karyopherins**) that carry proteins and RNA between the cytoplasm and nucleus. Proteins or RNAs ready for transport (the cargo) contain specific sequences that identify them. The **nuclear localisation signal (NLS)** is recognised by importins, and the **nuclear export signal (NES)** sequence is recognised by exportins. Transporter and cargo then dock at the mouth of the NPC and deliver the transport cargo. Importantly, for large RNA molecules such as mRNA, which is translocated as an RNP, the signals for export are not the mRNA itself but on the proteins, mostly hnRNP and splicing factors, that have guided the transcript through final processing. This emphasises again the integration and cooperation of the machineries that function in gene expression (Wente & Rout, 2010; Marfori *et al.*, 2011).

Interaction of cargo with transporter is modulated by the small protein Ran, which is a GTPase. Guanosine-5′-triphosphate (GTP) hydrolysis by Ran does not provide the energy for transport, but regulates the assembly and disassembly of **transport complexes**. Effectively, the nucleotide-bound state of Ran identifies the compartment: Ran-GTP in the nucleus, Ran-guanosine diphosphate (GDP) in the cytoplasm. For importins, the binding of cargo and Ran-GTP is antagonistic (so importins drop their cargo in the nucleus) and for exportins the binding of cargo and Ran-GTP is cooperative (so exportins are loaded in the nucleus). Of course, it's not as simple as that, there are several accessory factors involved in maintaining Ran in its proper GTP- or GDP-bound state, including a nuclear nucleotide exchange factor (RCC1), a cytoplasmic Ran-GTPase activating protein (RanGAP), a protein called RanBP1 and nuclear transport factor 2 (NTF2). Though the significance of this is uncertain, there are multiple signals and multiple pathways for both nuclear import and export; 14 members of the importin/exportin family are known in yeast and about 19 in humans (Chook & Süel, 2011).

By this stage in our description we've reached the point of exporting the gene transcript(s) to the cytoplasm. We will continue our story of what happens next to the ribosomal subunits and mRNAs in the cytoplasm later (section 5.9). Now, we want to consider how nuclei divide. The other prime function of the nucleus, namely storage and transmission of the cell's genetic information content will be discussed in Chapter 6.

5.6 MITOTIC NUCLEAR DIVISION

Prior to the nuclear division cycle, interphase (vegetative or S-phase), fungal nuclei look more or less spherical and have diffuse chromatin in the light microscope. Nuclei that are migrating through the hypha are elongated and have a highly visible nucleolus. As the nucleus enters its division cycle, nuclear volume decreases and chromatin condenses. The nucleoli remain evident in dividing fungal nuclei until late in the mitotic division (the stage called anaphase) and are then quickly restored and steadily enlarge until daughter nuclei are formed.

Fungal mitotic divisions are **intranuclear**: in this '**closed mitosis**' the division spindle forms inside the nucleus. This is quite different from the 'open mitosis' seen in most animals and plants where the nuclear envelope disassembles, and microtubules invade the nuclear space to form the division spindle. When the division spindle is formed within an **intact nuclear**

membrane progress of the division is more difficult to see and study, but it does not appear to affect the biological consequences of the mitotic division.

The typical mitosis proceeds through a series of morphologically distinct stages: **interphase, prophase, metaphase, anaphase, telophase**, ending with cytokinesis, when the cell divides to produce two identical daughter cells. In animals and plants, cytokinesis usually occurs in conjunction with mitosis. As we have mentioned, mitosis may occur independently of the branching and septation that are the equivalent of cytokinesis in filamentous fungi, but we will return to this topic later (see section 5.16).

During the process of mitosis the newly replicated **paired chromatids** condense and attach to fibres of the division spindle which then pull the sister chromatids to opposite ends of the spindle. The division spindle is an organelle in its own right. It consists of the two spindle poles, each containing a **centrosome** made up of two **centrioles** (in animals) or a **spindle pole body** (SPB) in fungi, that manage the spindle fibres. The spindle fibres are microtubules, connecting the poles with specialised regions of the chromosomes called **kinetochores**, protein structures assembled on the centromeres of the chromatids (***plants make do without a discrete organising centre***, but all eukaryotes use γ-tubulin to anchor and initiate microtubule assembly) (Biggins, 2013; Kilmartin, 2014; Rüthnick *et al.*, 2017).

All the preparation for mitosis is done during interphase, the **spindle apparatus** is assembled during prophase. In metaphase the condensed chromosomes are aligned near the spindle equator (forming the **metaphase plate**), then the chromosomes are moved towards the poles as the spindle elongates in anaphase. Finally, the daughter nuclei are separated in some way. In animals and plants, the cell divides near the spindle equator (cytokinesis). This may also apply in fungi (fission yeast *Schizosaccharomyces*), or one nucleus may migrate into a bud (budding yeast, *Saccharomyces*), into a newly formed branch, conidium or other spore (many filamentous fungi), or into a 'new' volume of cytoplasm in the hypha which may, or may not, be separated off by septation (many filamentous fungi).

In their daughter cytoplasm, the nuclei enter interphase. This whole series of events is called the **cell cycle**. There are networks of regulatory mechanisms, the crucial ones being known as **checkpoints**, which coordinate the timing and progress of events in mitosis (Barnum & O'Connell, 2014). For example, the spindle checkpoint inhibits the progress of anaphase until all of the kinetochores are attached to spindle microtubules. We will describe the cell cycle in section 5.16, but we first have to describe the basic machinery of mitosis.

Many features of nuclear and chromatid movement in mitosis are common to all eukaryotes and extensive study of mitosis in fungi, *Aspergillus nidulans* in particular, has contributed greatly to our understanding of eukaryote mitosis. **Nuclei are very mobile**, being pulled around by an attached organelle: in the fungi by the SPBs and in other eukaryotes by the centrosomes, and chromatids are pulled around by kinetochores assembled on their centromere regions. The centromere typically contains hundreds to thousands of kilobases of repeated DNA sequences. Analysis of these repeats in organisms from insects, plants and fungi to mammals and other vertebrates reveals no obvious conservation of sequence, suggesting that centromeres of different organisms differ in how they specify kinetochore assembly. However, there may be important centromere chromatin functions that are conserved throughout phylogeny (Bloom & Costanzo, 2017).

Centromere characteristics depend less on the sequence and more on what is done to these regions as DNA molecules and proteins are assembled on them and the kinetochore is constructed. Kinetochores are assemblies of about 45 different proteins, including a histone H3 variant (called CENP-A or CenH3) which is specialised for helping kinetochore binding to centromeric DNA. Other kinetochore proteins attach to spindle microtubules, and there are motor proteins, including both dynein and dynactin (a regulator of dynein) generally, and three kinesins in *Saccharomyces cerevisiae*, which generate forces that move chromosomes during mitosis. Other proteins monitor microtubule attachment and the tension between sister kinetochores and activate the spindle checkpoint to arrest the cell cycle when either of these is absent.

Kinetochore structure and function are not fully understood yet, but in essence the **dynein motor** (see section 5.11) uses energy from ATP to 'crawl' up the microtubule towards the originating SPB. This motor activity, along with polymerisation and depolymerisation of microtubules, is what provides the pulling force necessary to separate the chromosomes.

Knowledge of **nuclear migration** in *Saccharomyces cerevisiae* derives from study of **bud formation**. The SPB is a **microtubule-organising centre** embedded in the nuclear envelope. It produces astral microtubules emanating from the nucleus into the surrounding cytoplasm and spindle microtubules within the nucleus. Nuclear movements during mitosis mirror the growing and shrinking rate of **astral microtubules**, consistent with the idea that the microtubules are pulling the nuclei into position.

Nuclear migrations seen in yeast are very short range compared with the scale of **nuclear migration in filamentous fungi**. Nuclei migrate through the cytoplasm towards the advancing hyphal tip as the fungal colony grows. In *Gelasinospora tetrasperma*, a typical nuclear migration rate of 4 mm h^{-1} through a newly formed heterokaryon compares with a typical hyphal extension rate of only 0.7 mm h^{-1} (and in both cases we do mean millimetres). From early observations it was clear that nuclei are pulled from a point on the nuclear periphery, where the SPB is located. Observations on living fungi show that the nuclei move apart after mitosis, then migrate in the same direction, but at different rates, towards the hyphal tip, evening out their distribution along the hypha (see Figure 4.15). The motive force for the movement in mitosis is an interaction between the SPB and the microtubules. The pulling force that moves interphase nuclei through the hyphal cytoplasm is a continuation of this process and depends on cytoplasmic **dynein motor activity** (see section 5.11).

The first nuclear migration mutants were a by-product of a mitotic mutant search in *Aspergillus nidulans*; they were named *nud* (for nuclear distribution). Similar mutations of *Neurospora crassa* were named ropy (*ro*) because the hyphae resemble intertwined strands of rope. Many of the *nud* and *ro* genes are now known to encode structural subunits of, or components essential for the consistency, localisation or activity of, cytoplasmic **dynein** or **dynactin** (see section 5.11).

In the first stage of anaphase, chromatids move to the poles of the spindle, and in the second stage of anaphase the two poles of the division spindle move further apart. Mitotic anaphase movements in filamentous fungi are randomised in relation to the long axis of the hypha, so the mitotic division spindle does not have a preferred orientation (which it would be expected to show if septation were mechanically linked to division). The final stage of mitosis, telophase, follows one of three patterns:

- median constriction, separating the entire nucleoplasm into the two daughter nuclei;
- a double constriction which incorporates only a portion of the parental nucleoplasm into each daughter nucleus, the rest being discarded and degraded;
- formation of new daughter nuclear membranes, separate from the parental one, enclosing the chromosomes and a small portion of the nucleoplasm into the daughter nuclei while the bulk of the parental nucleoplasm and its membrane are discarded and degraded.

In multinucleate hyphae, mitotic divisions are not usually synchronised. The first three or four rounds of mitosis are synchronised in germinating spores of *Aspergillus nidulans*, but mitotic synchrony degenerates, perhaps because of increased difficulty of effective cytoplasmic signalling throughout the juvenile mycelium, though may still occur locally. In some fungal tissues, such as mushroom stems, nuclear division is rapid and there is some synchronisation although whether this is controlled or coincidental is not clear.

However, the higher fungi do seem to have a looser connection between cell differentiation and nuclear number and ploidy than is usual in plants and animals. In the cultivated mushroom, *Agaricus bisporus*, cells of vegetative mycelial hyphae have 6 to 20 nuclei, and those in the mushroom sporophore have an average of 6 nuclei, though cells in the mushroom stem have up to 32. In *Coprinopsis cinerea* (appearing in most publications under the name *Coprinus cinereus*), the vegetative dikaryon is regular with strictly two nuclei per compartment. However, cells of the mushroom stem can become multinucleate through a series of consecutive conjugate divisions, ending up with up to 156 nuclei, a peculiarity seen in other agarics. Species of *Armillaria* are unusual in having diploid tissues in the mushroom, so ploidy level and nuclear number are both variable.

5.7 MEIOTIC NUCLEAR DIVISION

Meiosis is the division of a diploid nucleus in which chromosomes reassort, producing four haploid daughter cells. This step in meiosis is what generates the genetic diversity of sexual reproduction.

Most fungi are **haploid** for most of their life cycles, diploidy being limited to a short period immediately prior to meiosis. This is a major difference with animals and plants. The main biological impact of this arrangement in fungi has been the evolution of processes to bring together two haploids so that genetically different nuclei can co-exist in the same cytoplasm. These processes are the **incompatibility mechanisms** regulating cytoplasmic and nuclear compatibility. Through their action, hyphae from different haploid parental mycelia can safely approach each other, undergo hyphal fusion to create a channel between themselves and then exchange cytoplasm and nuclei, so that a **heterokaryotic mycelium** is formed. We will describe all these processes in detail in Chapter 8. Once formed, the heterokaryon grows normally as a vegetative mycelium until conditions are right for sexual reproduction to take place. At this stage, cells of the heterokaryon go through karyogamy (nuclear fusion) to produce the diploid nucleus that can undergo meiosis.

Meiosis is often called 'reductional division' because in its first stage the number of chromosomes in the daughter nucleus is reduced by half. Meiosis I is the reductional division, achieving its reduction in chromosome number by sending paternal and maternal kinetochores to opposite poles. Meiosis II, the second meiotic division, is an equational division because it does not reduce chromosome numbers; it shares the same machinery with mitosis. Unlike most plants and animals, fungi carry out **meiosis with the nuclear membrane remaining intact through prophase I**.

Meiosis in heterothallic fungi goes through stages fairly typical for eukaryotes. In particular, the major round of **DNA replication precedes the start of meiosis I**. Indeed, it was research with the filamentous ascomycete *Neotiella* (in 1970) that ***first demonstrated*** this aspect of eukaryote meiosis. Because the haploid nuclei are in different (but adjacent) cells, it could be demonstrated that DNA replication was completed before karyogamy established the diploid nucleus. Other aspects of preparation for meiosis became evident as molecular events were established in yeast. In *Saccharomyces cerevisiae*, mating cells respond to each other by modifying their shape into pear-shaped cells termed **shmoos** (because of their similarity in shape to a character in Al Capp's 'Li'l Abner' cartoon strip). In the first stage of mating, and before the cells fuse, the SPB forms a cluster of microtubules (called **astral microtubules**) that move the haploid nucleus into the tip of the shmoo. As the nuclei move towards the shmoo tips of the two mating cells, the tips fuse and the shmoo tip microtubule clusters fuse to form a bundle, which progressively shortens as the two nuclei and their SPBs come together and fuse (**karyogamy**) (Merlini *et al.*, 2013).

The first (diploid) bud emerges adjacent to the fused SPB, and then reproduces by mitosis. In fission yeast (*Schizosaccharomyces pombe*) the nuclear locations of centromeres and the ends of the chromosomes (called telomeres) change during vegetative growth and in the early stages of meiosis telomeres cluster near the SPB. This telomeric cluster then leads the chromosomes in a rapid oscillating movement that appears to aid **synapsis**. When conditions are right the diploid progeny enter meiosis. In filamentous fungi, the meiocytes, the cells in which meiosis takes place, are the zygosporangia, ascus mother cells or basidia of the zygomycete fungi, *Ascomycota* and *Basidiomycota*, respectively.

During prophase I, homologous chromosomes pair and form synapses, a step unique to meiosis. The paired chromosomes are called **bivalents**, and the formation of **chiasmata** caused by genetic recombination becomes apparent as chromosomal condensation allows these to be viewed in the light microscope. Note that the bivalent has two chromosomes, one chromosome from each parent, each chromosome has one centromere, but each has two chromatids. One kinetochore forms on each chromosome (rather than one per chromatid) and the chromosomes attach to spindle fibres and begin to align at the metaphase plate. During anaphase I, the chromosomes move to separate poles of the spindle, completing this in telophase. Each of the daughter nuclei so formed is now haploid, although each chromosome they contain is composed of two chromatids. Following the completion of telophase I, the daughter nuclei continue to meiosis II.

5.8 TRANSLATION OF mRNA AND PROTEIN SORTING

Translation is the synthesis of a protein in the cytoplasm, using an mRNA template that has been exported from the nucleus (Clancy & Brown, 2008). For protein biosynthesis, the mRNA molecules bind to protein–RNA complexes called **ribosomes** in the cytoplasm, where they are translated into polypeptide sequences. The ribosome is another one of these **RNP machines** that mediates the formation of a polypeptide sequence based on the mRNA nucleotide sequence (Figure 5.4). Initiation of translation begins when the small subunit of the ribosome attaches to the methylated cap at the 5′-end of the mRNA and moves to the translation initiation site. Another key molecule is the family of **transfer RNA** (tRNA) molecules, which carry anticodons complementary to the mRNA codons. The first codon to be translated in the mRNA is typically AUG (the general translation start signal), and the tRNA that corresponds to this carries the amino acid methionine.

The large ribosome subunit now binds to the assembly, creating the peptidyl (or P) site and aminoacyl (or A) sites on the ribosome. The first tRNA occupies the P-site, and the second tRNA, complementary to the second codon, enters the A-site. A peptide bond is made between the methionine (at the P-site) and the second amino acid at the A-site; next, the first tRNA exits, and the ribosome moves along the mRNA to free up the A-site so that the next tRNA can enter. As **peptide elongation** continues, the growing peptide is always transferred to the tRNA at the A-site by being peptide-bonded to the amino acid carried by that tRNA. The ribosome then moves along the mRNA, so that the tRNA carrying the peptide synthesised to date remains at the same place on the mRNA, held by the hydrogen bonding between its anticodon and the mRNA codon, but is shifted to the ribosome's P-site by movement of the ribosome.

Note also that amino acids are linked to their tRNA via their carboxy group (i.e. **aminoacyl-tRNA**), so the first amino acid to be placed in the peptide will have a free amino group, and the last amino acid to be added will have a free carboxylic acid group (once the tRNA is removed). Synthesis is initiated at the amino-terminal end and ends at the carboxy-terminal end of

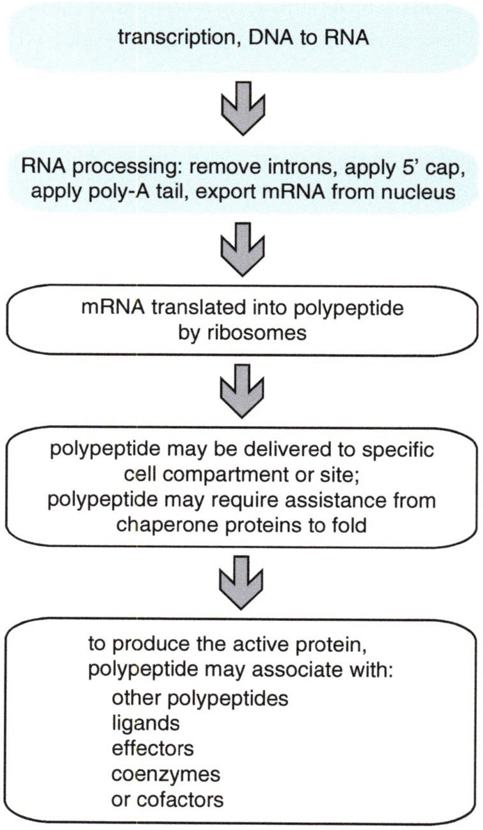

Figure 5.4. Flow chart giving an overall summary of protein synthesis. Transcription of DNA into RNA takes place within the nucleus (shaded panels; more detail in Figures 5.2 and 5.3), and the primary transcript RNA is then modified to add cap and poly-A tail and remove introns. The resulting mature mRNA is exported from the nucleus into the cytoplasm for translation into protein. The mRNA is translated by ribosomes that use tRNA charged with specific amino acids to create the polypeptide corresponding to the sequence of codons carried by the mRNA. Newly synthesised proteins often carry signal sequences that enable them to be sorted and delivered to specific sites and may be further modified, such as by enzymic activation chemically modifying them, by binding to effectors or coenzymes, and perhaps by binding to other polypeptides.

the polypeptide. When a stop codon is encountered in the A-site, a release factor rather than a tRNA enters the A-site and translation is terminated. When **termination** is reached, the ribosome dissociates, and the newly formed polypeptide is released and folds into a functional three-dimensional protein molecule.

RESOURCES BOX 5.2

Visit the following URL for an animation describing **translation**: http://vcell.ndsu.nodak.edu/animations/translation/index.htm or on YouTube at https://www.youtube.com/watch?v=5bLEDd-PSTQ.

The termination of synthesis is not the end of the story. Only a minority of the proteins synthesised in the cytoplasm function there. Most proteins must be delivered to one or other of the cellular organelles to carry out their function, and they may need to be delivered to the inner space of an organelle, to the organelle's interior membrane(s), to the cell's outer membrane, or be exported totally to its exterior. This delivery process is carried out using information contained in the protein itself; these proteins have a stretch of amino acid residues in the chain, called **signal peptides** or **targeting peptides**, which target the protein to its proper destination.

This addressing mechanism is called the protein targeting or **protein sorting** mechanism. During transit through the cytoplasm proteins called chaperonins become associated with a newly synthesised polypeptide to modify its folding and deliver it in a suitable form to the proteins of translocator complexes embedded in the organelle membrane that import it into the organelle. When it is delivered a complex called the signal peptidase removes the signal sequence (so although the coding for this sort of signal sequence must be contained within the mRNA sequence, it does not appear in the final working protein).

RESOURCES BOX 5.3

The **Nobel Prize in Physiology or Medicine 1999** was awarded to Günter Blobel 'for the discovery that proteins have intrinsic signals that govern their transport and localisation in the cell'.

Visit: https://nobelprize.org/nobel_prizes/medicine/laureates/1999/.

There are two types of targeting peptides: **presequences** and **internal targeting peptides**. Presequences are usually short amino acid sequences at the amino-terminal end of the peptide made up of basic and hydrophobic amino acids. The internal targeting signals are stretches of the primary sequence; to function as signal sequences these stretches must be brought together on the protein surface by its folding. They are called **signal patches**.

An amino-terminal signal sequence is the first to be translated and is recognised by a **signal recognition particle (SRP)** (Akopian *et al.*, 2013), while the polypeptide is still being synthesised. SRP binds to the ribosome and targets it to the membrane-bound SRP receptor of the **translocon** in the membrane of the membranous organelle called the endoplasmic reticulum (ER). When the translating ribosome arrives at the ER membrane the translocon channel is initially closed, but when the signal sequence of the polypeptide is recognised by the channel, it opens, and while it is still being synthesised the protein is inserted into the channel. A junction between ribosome and translocon blocks return of the protein to the cytoplasm. The signal sequence is immediately removed from the polypeptide by signal peptidase when it is translocated into the ER. Within the ER, the protein is associated with more chaperonin proteins to protect while it folds correctly. Once folded, the protein may be further modified (by glycosylation, for example), before transport to the Golgi apparatus for more processing and transport to other organelles or retention in the ER.

Some proteins are translated in the cytoplasm, independently of the ER, and later transported to their working destination. This happens for proteins intended for the nucleus, mitochondria, chloroplasts in plants or peroxisomes. Peroxisomes are highly significant to fungi for a variety of reasons (Sibirny, 2016):

- The glyoxylate cycle occurs within the peroxisome so they are necessary for fatty acid metabolism.
- Filamentous *Ascomycota* have specialised peroxisomes called Woronin bodies that plug the septal pore to separate individual cells in a hypha (see section 5.16).
- Increasing evidence indicates that this organelle can also function as an intracellular signalling compartment and as an organiser of developmental decisions inside the cell.
- Peroxisomes contain hydrogen peroxide-generating oxidases and the hydrogen peroxide-detoxifying enzyme catalase.

Most peroxisomal matrix proteins have their (type 1) peroxisome-targeting signal sequence at the **carboxy-terminal end** of the polypeptide. The type 2 (N-terminal) peroxisomal targeting signal occurs in only a few proteins (four are known in humans, one in yeast) (Pollard *et al.*, 2017b).

Some proteins are transmembrane proteins, often transmembrane receptors that pass through their membrane one or more times. These are inserted into the membrane by a translocation process like that already described (the first transmembrane domain acts as the first signal sequence) but the process is stopped by an amino acid signal sequence, which is also called the membrane anchor.

Most mitochondrial proteins are synthesised in the cytoplasm with appropriate presequences. In this case, the ribosome does not dock onto the mitochondrial translocator but synthesises the preprotein as a cytoplasmic soluble protein. Cytoplasmic chaperonins then deliver the preprotein to a translocator in the mitochondrial membrane.

RESOURCES BOX 5.4

Visit the following URL for an animation describing **mitochondrial protein transport**: http://vcell.ndsu.nodak.edu/animations/mito-pt/index.htm or on YouTube at https://www.youtube.com/watch?v=LfDYGanMi6Q.

The proteins involved include receptors and the 'general import pore' at the outer membrane, which together make up the **'translocase of outer membrane'** (**TOM**). The preprotein translocates through TOM in hairpin loops and is then transported through the space between membranes to **TIM** (**'translocase of inner membrane'**), through which it exits to the mitochondrial matrix, where the targeting presequence is removed. Mitochondria are complex organelles and proteins must be targeted to different parts of them, involving several signals and transport pathways. Targeting to the outer membrane, intermembrane space and/or inner membrane often requires separate signal sequence(s) in addition to the matrix targeting sequence.

In plants, proteins are also targeted to chloroplasts. We will not dwell on this apart from mentioning that most chloroplast proteins are synthesised in precursor form on cytoplasmic ribosomes and are translocated through the TOC (translocon at the outer envelope membrane of chloroplasts) and TIC (translocon at the inner envelope membrane of chloroplasts) complexes. A basic similarity in strategy of mechanism (and naming) is evident between the two major organelles, mitochondria and chloroplasts, which encourages the view that similar mechanisms may be used for other membrane-bound organelles.

For proteins that must get from the cytoplasm to the nucleus, there is a nuclear localisation signal. On the cytoplasmic side of the nuclear membrane, the **NPC** (see section 5.5) recognises proteins carrying a nuclear localisation signal and selectively transports them into the nucleus. Proteins cross the NPC after they have been synthesised and have folded into a three-dimensional shape.

We have to consider **protein destruction** as the final act in protein sorting. Synthesis might produce a defective protein, a protein that may be damaged as it functions or, more likely, protein function may no longer be required. Such proteins are recycled and are targeted for recycling by **ubiquitin** (Pickart & Eddins, 2004). Ubiquitin is so called because it is ubiquitous in virtually all types of cells and is one of the most highly conserved (i.e. least changed) proteins. Ubiquitin is a protein consisting of 76 amino acids and in the normal course of events, proteins are inactivated by having several molecules of ubiquitin attached to them; this process is called ubiquitination. Ubiquitin serves as the signal for transport machinery to ferry a protein to a site for proteolysis. There are two pathways of **intracellular proteolysis: lysosomal** and **proteasomal**.

RESOURCES BOX 5.5

In **2004, Aaron Ciechanover** and **Avram Hershko** of the Technion Israel Institute of Technology in Haifa, Israel and **Irwin Rose** of the University of California, Irvine, USA shared the **Nobel Prize in Chemistry** 'for the discovery of ubiquitin-mediated protein degradation'.

Visit: https://www.nobelprize.org/nobel_prizes/chemistry/laureates/2004/ for more details of this.

A minority of proteins modified with ubiquitin (in *Saccharomyces cerevisiae* it tends to be plasma membrane proteins) are internalised and degraded in a **lysosome vacuole**. Most ubiquitin-labelled proteins are transported to the proteasome for degradation. The **proteasome** is another molecular machine. It is a cylindrical particle with a central cavity comprising a multicatalytic protease complex together with, at each end, a complex containing several ATPases and a binding site for ubiquitin molecules. The ubiquitinated protein is unfolded by the ATPase complexes, which also recover the ubiquitin and inject the substrate protein into the central chamber where there are several catalytic sites. The substrate protein is cleaved into peptides of 6–9 amino acid residues; the length of the product corresponding to the distance between adjacent catalytic sites in the central chamber.

5.9 THE ENDOMEMBRANE SYSTEMS

We have already mentioned the membrane-bounded organelles, mitochondria and chloroplasts, which are the power generators for the cell, but eukaryotes depend on several other membrane systems that deserve reference.

Mitochondria contain their own genome separate from the nuclear genome, and their key role is to manage the flow of oxygen atoms, protons and electrons in the process of respiration. The purpose is to generate compounds (particularly ATP and nicotinamide adenine dinucleotide, NADH) containing chemical energy that can be used for useful work elsewhere.

Chloroplasts are the photosynthetic counterparts of mitochondria, using the photolysis of water to generate protons and electrons that are captured in chemical energy (mainly as ATP and nicotinamide adenine dinucleotide phosphate, NADPH). Organelles in plants that are modified chloroplasts are broadly called plastids, and are often involved in nutrient storage. Since both mitochondria and chloroplasts contain their own genomes, they are thought to have once been separate organisms, which formed symbiotic relationships at some stage in the evolution of eukaryotic cells. The most convincing version of this **'endosymbiosis theory'** (Margulis, 2004) suggests a series of symbiotic relationships being established between prokaryotic partners: the mitochondria of eukaryotes evolving from aerobic bacteria living within a host cell; chloroplasts of eukaryotes evolving from endosymbiotic cyanobacteria; eukaryotic cilia and flagella arising from endosymbiotic spirochetes (Martin *et al.*, 2015).

The other main components of the eukaryotic cell's collection of single-layer endomembranes are the endoplasmic reticulum and the Golgi apparatus, both of which are responsible for managing the macromolecules of the cell. Endoplasmic reticulum (ER) is a transport network within the cell that sorts molecules intended for specific modifications and/or destinations. There are two forms of ER: rough ER has ribosomes docked on its cytoplasmic surface, anchored there because the polypeptides they are synthesising carried a signal sequence that directed them to the ER translocon, and the smooth ER, which lacks docked ribosome. The ER is a system of membrane-bounded channels and sacs (called cisternae) connected to the double-layered nuclear envelope, which makes, processes and transports chemical compounds for use throughout the cell, and provides a link between nucleus and cytoplasm.

RESOURCES BOX 5.6

Visit the following URL for an animation describing the **mitochondrial electron transport chain**: http://vcell.ndsu.nodak.edu/animations/etc/index.htm or on YouTube at https://www.youtube.com/watch?v=xbJ0nbzt5Kw.

RESOURCES BOX 5.7

Visit the following URL for an animation describing **ATP synthase powered by a hydrogen (proton) gradient**: http://vcell.ndsu.nodak.edu/animations/atpgradient/index.htm or on YouTube at https://www.youtube.com/watch?v=3y1dO4nNaKY.

The **Golgi apparatus** was one of the first intracellular organelles to be visualised, being observed by several workers at the end of the nineteenth century, though **Camillo Golgi** was the first to ***publish*** the observation. Golgi apparatus is another system of membrane-enclosed compartments, which modifies molecules made in the endoplasmic reticulum and prepares them for export to the outside of the cell.

This description implies that there is constant traffic between the two membrane systems, and this is the case, transport or carrier vesicles bud from the endoplasmic reticulum membranes and selectively fuse with the Golgi apparatus (and there is a reverse, or retrograde, traffic too). These events require specific protein–protein interactions. First, the vesicle forms through assembly of a protein coat on the cytoplasmic surface of the ER membrane which establishes curvature in the membrane and determines which membrane proteins are accepted by the vesicle. Second, and after transport, pairing of vesicle proteins with proteins in the Golgi membrane allows the vesicle and target membranes to fuse so that the cargo is deposited into the Golgi apparatus. Vesicle traffic from the ER to the Golgi involves vesicles coated with **'coatomer protein' COPII**. A different vesicle **coat protein complex** (**COPI**) is responsible for transport from Golgi to ER, and for transport between Golgi compartments (Day *et al.*, 2013).

COPII proteins were first identified by genetic studies with yeast, but mammalian counterparts to the yeast genes were soon identified and shown to serve the same functions, so this is another highly conserved eukaryotic process. This molecular/genetic evidence is significant because there is a considerable difference in the morphology of the Golgi apparatus between fungi and mammals. The Golgi body in mammals (as originally observed in the cytoplasm of neurons by Golgi) is a stack of flattened large vesicles with swollen peripheral regions (the cisternae, often called dictyosomes by plant cell biologists), which are linked together by tubular connections and surrounded by numerous smaller spherical vesicles which have budded off from the cisternae.

There are usually 5–8 cisternae in the stack, but as many as 60 have been observed in some unicellular flagellates. Stacked Golgi bodies like this are not observed in plants or fungi. Most plant cells have hundreds of individual Golgi dictyosomes distributed throughout the cytoplasm. Fungi also have individual dictyosomes throughout the hypha, but these are organised to supply all the materials needed for rapid hyphal tip extension (Harris, 2013). Such a complicated structure is obviously functionally differentiated and it is easier to describe, and understand, the stacked Golgi apparatus of animals. Obviously, a stack of anything has a top and a bottom and it turns out that the stack of cisternae is functionally differentiated in that one end is the ***entry face***, called the *cis*-Golgi, at which vesicles are received, the other end is the *trans*-Golgi, or ***exit face*** from which export vesicles leave. The medial (or middle) saccules are between the two, and both *cis* and *trans* are associated with networks of tubular and small cisternal structures called the *cis*-Golgi network (CGN) and *trans*-Golgi network (TGN).

RESOURCES BOX 5.8

Visit the following URL for an animation describing **protein trafficking in the Golgi apparatus**: http://vcell.ndsu.nodak.edu/animations/proteintrafficking/index.htm or on YouTube at https://www.youtube.com/watch?v=rvfvRgk0MfA.

RESOURCES BOX 5.9

Visit the following URL for an animation describing **protein modification within the Golgi apparatus**: http://vcell.ndsu.nodak.edu/animations/proteinmodification/index.htm or on YouTube at https://www.youtube.com/watch?v=u38LjCOvDZU.

The main **functions of the Golgi apparatus** are to modify, sort, package and transport substances formed by the ER. These reach the Golgi stack at its *cis*-side via vesicles budded off the ER. Enzymes within the Golgi cisternae modify proteins by the addition of carbohydrates (glycosylation), sulfate (sulfation) and phosphate (phosphorylation). To achieve this, the Golgi also imports substrates like phosphate and sulfate donors and nucleotide sugars from the cytoplasm. Proteins are also labelled with signal sequences to specify their final destination; for example, adding a mannose-6-phosphate labels protein destined for lysosomes. The animal cell Golgi is also important in the synthesis of proteoglycans needed for the animal cell extracellular matrix. It is also a major site of carbohydrate synthesis.

The products of these modifications are packaged in membrane-bounded vesicles at the periphery of the *trans*-Golgi network that are targeted to the appropriate membrane somewhere else in the cell, such as lysosomes, peroxisomes and plasma membrane (for excretion from the cell). There is a flow in the other (retrograde) direction, too. Some of the product vesicles that emerge from the *trans*-Golgi network return ER-resident proteins to the ER and export Golgi-modified products for distribution by the ER.

In addition, we should appreciate that the **cell plasma membrane (plasmalemma)** is a membrane organelle in its own right and it is capable of both **endocytosis** (uptake) and **exocytosis** (secretion). The cell plasma membrane is made from a double layer of phospholipids. Because of its polar nature, the phosphatidyl 'head' of a phospholipid is hydrophilic (attracted to water) and the lipid tails are hydrophobic but lipophilic. Existing in an aqueous environment, the phospholipid bilayer is stable with the phosphatidyl heads on the two outer surfaces with the lipid tails together forming a lipophilic microenvironment between them. This forms a very fluid membrane that has a variety of protein molecules embedded within it acting as **channels, transporters** and **receptors** for regulation of the exchange and contact with the exterior of the cell.

Large molecules cannot pass through this membrane without active assistance. The process that imports these is called **endocytosis**. There are specialised regions called **coated pits** on the outside of the membrane. The 'coat' is a protein called **clathrin**, which can form a localised polyhedral lattice on the plasma membrane to cause an invagination of the membrane with the result that the **imported cargo** is drawn into the cytoplasm in a vesicle. This mechanism is used for uptake of essential metabolites, uptake of some regulators and growth factors, or uptake of previously exported molecules for recycling. The resultant **endocytic vesicles** are transported to the Golgi apparatus and the ER where they fuse with Golgi and ER endosomes. After vesicle fusion, the coat detaches and may be reused, and the membrane that formed the invagination is ultimately returned to the cell surface. A similar clathrin-coating process also buds membrane segments from the *trans*-Golgi network.

The small transport vesicles we have already mentioned are defined by their coat proteins: COPII-coated vesicles allow export from the ER, COPI vesicles shuttle proteins between Golgi and ER, and clathrin-coated vesicles mediate transport from the *trans*-Golgi network and endocytic transport from the plasma membrane. These organised pathways require pairs of membranes to specifically recognise one another and subsequently fuse. Different targeting reactions involve distinct protein complexes that act to mark the target organelle for incoming vesicles (Faini *et al.*, 2013). This is the responsibility of **SNARE proteins** (SNARE is an acronym derived from **'soluble N-ethylmaleimide sensitive factor attachment protein receptor'**), which are integral membrane proteins found predominantly on vesicles (**v-SNAREs** incorporated into the membranes of transport vesicles during budding) or target membranes (**t-SNAREs** located in the membranes of target compartments). SNAREs are **fusion proteins**. They comprise a large family of proteins, with more than 60 members, in yeast. Initial recognition between a vesicle and a target membrane is the responsibility of another group of proteins called **tethering factors**, which are large fibrous proteins that can span relatively long distances (>200 nm) between vesicle and target membrane, so they may form a molecular 'fishing net' to catch relevant vesicles. Tethering factors bind membranes together prior to the interaction of v-SNAREs and t-SNAREs across the membrane junction. The SNAREs form a complex that extends across both membranes and then 'zippers' the two membranes together (Wang *et al.*, 2017).

The previous paragraphs describe the ***animal*** Golgi comprised of ***stacked cisternae***. This is ***not found in fungi***, but it is not true to say, as some accounts do, that 'fungi lack Golgi dictyosomes'. Fungi ***do*** carry out all the Golgi functions, but the dictyosomes are scattered through the cytoplasm and are not stacked into the historical Golgi body (Day *et al.*, 2013; Pantazopoulou, 2016). Cisternal (or dictyosome) stacking is an animal characteristic, but the stack consists of a collection of cisternae at different stages of maturation, with the youngest at the *cis*-side and most mature at the *trans*-side. As the *trans*-cisternae mature they 'dissolve' into vesicles and are replaced by the next-most-mature from the stack, and a new cisterna is assembled on the *cis*-side. This dynamic production line is well illustrated in the animation *Protein trafficking in the Golgi apparatus* at http://vcell.ndsu.nodak.edu/animations/proteintrafficking/movie-flash.htm. In plants and fungi the dictyosomes go through essentially ***the same maturation processes*** without being collected together into a stack.

Fungi have a main **central vacuole** as their principal degradative compartment and there are several pathways to the vacuole that supply it with hydrolytic enzymes, two from the *trans*-Golgi network, and a cytoplasm-to-vacuole pathway. There is a further range of vacuolar import pathways for substrate delivery: including endocytosis from the cell surface delivering metabolites in bulk as well as specific components, and regular transport of proteins destined for recycling in the vacuole. For the most part, flow to the vacuole happens in response to specific environmental stimuli. The main vacuole may accumulate any of a wide range of metabolic end products varying from nutrient stores all the way to waste materials. Each of these accumulation, transport and delivery events depends on cellular machinery sensing the environmental signal(s), activating appropriate metabolic pathways and designating the appropriate transport mechanisms and targets (Veses *et al.*, 2008; Richards *et al.*, 2010, 2012; Tong *et al.*, 2016).

The network of vesicle traffic in the growing hypha was first visualised using fluorescent dyes that insert into the outer layer of the plasma membrane (Fischer-Parton *et al.*, 2000). Within 30 s of adding the dye to hyphae, a cloud of fluorescent endocytic vesicles appears within the cytoplasm. Although components of the endocytic pathway have been characterised by molecular and genetic means, uptake of the dye is a clear demonstration that full endocytosis process does occur in filamentous fungi.

The summarised model that results from the observations (Figure 5.5) shows that the next stained organelles that could be visualised were small and roughly spherical endosomes. Subsequently, in subapical compartments, the next obviously stained organelle was the large main vacuole. This sequence is similar to observations of budding yeast, where staining of vacuole membranes followed that of endosomes. In hyphae the vacuolar system consists of an extensive tubular network in addition to the large roughly spherical vacuoles. Both the Golgi and ER of fungal hyphae would become stained via pathways connecting the endosomal system, Golgi and ER, which are known to occur in budding yeast.

The model shown in Figure 5.5 also introduces a membranous organelle that is a ***characteristic feature in fungi with true hyphae***: an apical cluster of vesicles and cytoskeletal elements, which plays a crucial role in hyphal tip extension growth and is known by its German name, the **Spitzenkörper** ('apical body'). The Spitzenkörper is the organising centre for hyphal extension and morphogenesis; present in actively growing tips but lost when extension ceases. It can be detected with phase contrast light microscopy in growing hyphae of *Ascomycota* and *Basidiomycota*. Some zygomycetes lack a recognisable Spitzenkörper but do have a loose distribution of vesicles in the hyphal apex that may serve the same function. The structure of the Spitzenkörper, as seen by light microscopy, differs in detail between species, but is also dynamic and variable within a species. Even though it lacks a membrane boundary, it is clearly a complex organelle, which is always adjacent to the site of polarised hyphal extension. It is composed of vesicles of various kinds and sizes, microfilaments, microtubules and ribosomes.

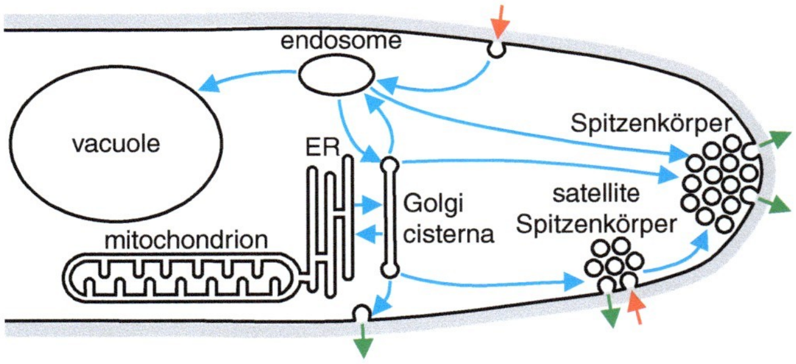

Figure 5.5. A hypothetical model of the organisation of the vesicle trafficking network in a growing hypha based on the pattern of fluorescent dye staining. Intracellular trafficking is shown with blue arrows, exocytosis with green arrows and endocytosis with red arrows. Plasma membrane and other membrane compartments are shown in black with the hyphal cell wall shown in grey. ER, endoplasmic reticulum. Redrawn and modified from Fischer-Parton *et al.* (2000).

Extension of the hyphal tip requires polarised incorporation of plasma membrane and cell wall constituents into the growing apex. Apical extension of the hyphal tip is considered to depend on the supply of wall-building secretory vesicles generated by Golgi cisternae and discharged from a 'vesicle supply centre' which is the cytologically visible Spitzenkörper. As Figure 5.5 suggests, **satellite Spitzenkörper(s)** arise immediately beneath the plasma membrane a few micrometres behind the apex, before migrating towards and merging with the main Spitzenkörper to supply additional wall-building vesicles to the growing hyphal tip. In experiments with fluorescent dye, the Spitzenkörper stained shortly after the endosomes suggesting that, as well as the Golgi, the endosomal system may supply materials to the Spitzenkörper.

It is plausible that **recycled plasma membrane**, captured by endocytosis and sorted by the endosome(s), is among the material supplied to the Spitzenkörper by the vesicle flow (Harris *et al.*, 2005; Virag & Harris, 2006; Steinberg, 2007b; Riquelme, 2013; Steinberg *et al.*, 2017). Fluorescent probes also label vacuoles that form a tubular reticulum in hyphal tip cells, and short tubules that undergo sequences of characteristic movements and transformations to produce ring-like membrane structures not yet fully identified. The cell biology of the apical dome of the hypha is complex and dynamic (Zhuang *et al.*, 2009), and the vacuolar system of the hypha exhibits multiple stress responses and environmental adaptations (Tong *et al.*, 2016).

5.10 CYTOSKELETAL SYSTEMS

In the descriptions above, we have often used words implying **delivery of a cargo** contained in a vesicle to a specific target site. The vesicles do not slosh around in a cytoplasmic soup; rather they are conveyed to their destinations by the **cytoskeletal system** (Lichius *et al.*, 2011; Riquelme *et al.*, 2016).

The cytoskeleton is a characteristic feature of eukaryotic cells. Although homologues to all the major proteins of the eukaryotic cytoskeleton can be found in prokaryotes, the sequence comparisons indicate very distant evolutionary relationships. As the name 'cytoskeleton' implies, it is generally presented as the structure that maintains cell shape and permits motion, but this is only fully applicable to animal cells. In plants and fungi, the cell wall largely determines cell shape and motion is limited (though in fungi the invasive hyphal tip could be viewed as being a motile apex). The cytoskeletal functions that apply throughout the eukaryotes are to provide for intracellular transport of vesicles and organelles, and segregation of chromosomes during nuclear division (see sections 5.7 and 5.8) (Wickstead & Gull, 2011; Erickson, 2017).

There are three main components making up the eukaryotic cytoskeletal system: actin filaments or microfilaments, intermediate filaments and microtubules:

- **Actin filaments** (also called **microfilaments**) are solid rods about 7 nm in diameter made of two chains of the globular protein called actin (one of the most abundant proteins in nature). These filaments are contractile and contribute to shape changes, cell-to-cell or cell-to-wall connections, signal transduction, cytokinesis and cytoplasmic streaming.
- **Intermediate filaments** are a very broad class of fibrous proteins forming structural fibres in the range 8–12 nm diameter. They also participate in cell-to-cell and cell-external connections, but mostly function as tension-bearing elements maintaining shape and rigidity of the animal cell, and the structure of membranous structures like the nuclear envelope. The first intermediate filament gene to be characterised in filamentous fungi, the *Aspergillus nidulans* mbmB gene product, co-localises with mitochondria, and deletion of mdmB affects mitochondrial morphology and distribution.
- **Microtubules** are straight, hollow cylinders about 25 nm in diameter, usually composed of 13 protofilaments, which are polymers of α- and β-tubulin. They have a very dynamic behaviour, and carry out a variety of functions, ranging from transport to structural support.

The importance of actin in budding yeast is evident from the phenotype changes resulting from mutation of the gene that encodes actin, which include:

- gross morphological defects;
- abnormal chitin deposition;
- defective bud site selection;
- abnormal nuclear segregation;
- abnormal cytokinesis;
- abnormal distribution of intracellular organelles;
- abnormal secretion and uptake;
- altered sensitivities to the environment (temperature, osmotic and ion concentrations).

The actin protein has a molecular weight of about 42 kDa (375 amino acids), which exists as a monomer (referred to as **globular** or **G-actin**) or as a linear polymer of the monomer, known as **filamentous** or **F-actin**. It is the filaments that are so important in morphogenesis, organelle movements and cytokinesis. The microfilaments are extremely dynamic structures that can be rapidly modified by interactions with several **actin-binding proteins (ABPs)** (Pollard, 2016; Svitkina, 2018). The actin cytoskeleton of filamentous fungi is required for polarity establishment and maintenance at hyphal tips, and for formation of a contractile ring at sites of septation. Actin-related proteins, known as **septins** and **formins**, have been identified as independent nucleators of actin polymerisation. Filamentous fungi contain a single formin that localises to both hyphal tips and the sites of septation (Xiang & Plamann, 2003; Lichius *et al.*, 2011; Breitsprecher & Goode, 2013; Riquelme *et al.*, 2016).

Microtubules and their associated proteins are involved with as wide a range of intracellular functions as are actin microfilaments (indeed, the two work together in most instances), particularly transport and positioning of organelles, vesicles and nuclei. Microtubules are made up of **dimers of α- and β-tubulin** that assemble into cylindrical tubes with a diameter of about 25 nm, but vary greatly in length. The polymerisation dynamics of microtubules are central to their biological functions. Even *in vitro*, purified tubulin dimers continuously self-assemble and disassemble; microtubules polymerise and elongate until they randomly switch to depolymerisation (the switch is termed **catastrophe**). Depolymerisation leads to rapid shortening of the microtubule, resulting either in their complete disappearance or in a switch back to elongation (which is termed **rescue**). This dynamic instability is an essential property of microtubules as it enables them to perform mechanical work. An elongating microtubule is polarised, having a rapidly **elongating plus end**, and a slowly or **non-elongating minus end** (in cells, the minus end is usually the anchor point).

In vivo, microtubules utilise the energy of **GTP hydrolysis** to drive their dynamic instability (polymerisation/depolymerisation). Microtubules are formed by assembly of α- and β-tubulin dimers. The first stage of their formation is called nucleation, which requires Mg^{2+} and GTP. During nucleation, α- and β-tubulins join to form a dimer. Each dimer carries two GTP molecules, but it's the one on the β-tubulin that is hydrolysed to GDP when a tubulin molecule adds to the microtubule. Dimers attach to other dimers to form oligomers, 13 at a time, forming rings 25 nm in diameter. The longitudinal rows of dimers are called protofilaments. Nucleation is relatively slow, but once the microtubule is fully formed the second phase, called elongation, proceeds more rapidly. The length of microtubules is determined by elongation rate, shortening rate, catastrophe frequency and rescue frequency; the cell controls its microtubule network by modifying these parameters, and this in turn affects the intracellular transport processes.

A wide range of **microtubule-binding proteins** is involved in these organising and maintenance processes. Some microtubule-binding proteins specifically associate with the plus ends of growing microtubules. These proteins are called **plus-end tracking proteins** (or **+TIPs**). They form clusters at the ends of growing microtubules and regulate microtubule dynamics. A specific animal example is CLIP-170 which is a '**C**ytoplasmic ***Li***nker ***P***rotein' that binds to the growing plus end of microtubules, enhances microtubule assembly, and is involved in interactions between endosomal membranes and microtubules. CLIP-170 homologues in *Saccharomyces cerevisiae* stabilise microtubules by reducing all four parameters of dynamic instability, while they suppress catastrophes in *Schizosaccharomyces pombe* (remember, these two yeasts are related, but belong to different phylogenetic lineages; see Figure 2.9 and section 3.7). CLIP-170 homologues of *Aspergillus nidulans* promote microtubule growth by doubling the rescue frequency. Evidently, organism, tissue and cell-type specific functions can be expected in these and other **microtubule-associated proteins (MAPs)**. MAPs include the **microtubule motors, kinesin** and **dynein**, which 'walk' along the microtubules to provide the characteristic motility. They work in opposite directions, **kinesins move cargo to the plus end, dynein towards the minus end**. Dynein motors are concentrated at microtubule plus ends in fungi, where they influence catastrophe and rescue rates in *Saccharomyces cerevisiae, Aspergillus nidulans* and *Ustilago maydis*, thus functioning like +TIPs as well as being minus-end-directed motors.

It is worth mentioning here that the antifungal drug griseofulvin inhibits mitosis in fungi by binding to MAPs and the fact that it is clinically useful suggests that there are major differences between the MAPs of humans and those of fungi (Shoham *et al.*, 2017).

5.11 MOLECULAR MOTORS

Microtubules and microfilaments are used by **molecular motors** for the manipulation and long-distance transport of membranous organelles, vesicles, and RNA and protein complexes (Steinberg, 2007a, b; Steinberg *et al.*, 2017). Movement is the responsibility of molecular motors, which are molecular machines that move their cargo along F-actin microfilaments or microtubules. Consequently, those motors and their associated transport apparatus deserve a little more detailed attention.

There are three major types of molecular motor: the **microtubule-associated kinesins and dyneins**, and the **actin-associated myosins**. As we have mentioned, microtubules are polarised (anchored at the minus end, polymerised at the plus end) and microtubular motors are classified as plus-end-directed or minus-end-directed according to the direction of their movement along the microtubule. Generally speaking (there are exceptions), kinesins and dyneins work in opposite directions, kinesins move their cargo towards the plus end (Figure 5.6), and dynein walks towards the minus end of the microtubule.

Fungal representatives of all of these support the numerous cellular transport processes in hyphae including **apical polar extension, septation** and **nuclear division**. The different motors share some structural similarities in that they have a head or motor domain, a linker region and a tail/stalk region, although there are many differences between them (Figure 5.7). In most cases, motors consist of a homodimer of heavy chains and a variable number (up to about 10) of associated lighter chains; the latter often have regulatory roles, and some of them anchor the motor to its cargo (specific details of the components are given below). The heavy chains form the globular motor region, containing the ATPase activity; this region binds either to microtubules or the F-actin of microfilaments, as appropriate. ATP cleavage leads to repeated cycles of '*change in shape – detach from the fibre – reattach to the fibre*' **alternately** in each of the heavy chains making up the motor domain. This results in the coordinated 'walking' of the motor along the cytoskeletal fibre, though it's more of a rapid shuffle than a stylish stroll. Nevertheless, the chemical energy provided by ATP hydrolysis is effectively transduced into kinetic energy in the form of the migration of the motors for long distances along their fibre. Yeast dynein can walk along microtubules without detaching, however in metazoans, cytoplasmic dynein must be activated by the binding of dynactin.

There is a close interrelationship between the actin and microtubule cytoskeletal systems and the same organelles move on both types of filament. For example, mutations in each of these motors can result in similar defects in septation, nuclear migration and organelle distributions in fungi. On the other hand, there is also evidence in fungi that movement of some organelles is specific to a particular motor, and which motor does what differs between fungi. For example, in *Saccharomyces cerevisiae*, transport of mitochondria, secretory vesicles and vacuoles is based mainly on F-actin and its associated myosin motors, whereas in the fission yeast *Schizosaccharomyces pombe* the microtubular system is involved in this sort of traffic.

Vacuoles are important in transport over long hyphal transport pathways in filamentous hyphae, since they contain many nutrients and essential metabolites, organic and inorganic. In the apical cells of hyphae of most fungi the vacuoles form a ***highly* motile tubular reticulum**. The vacuoles are most active in hyphal tips, but non-motile vacuoles at a distance from the tip can be induced to become motile by environmental changes. The vacuolar system in filamentous fungi has been

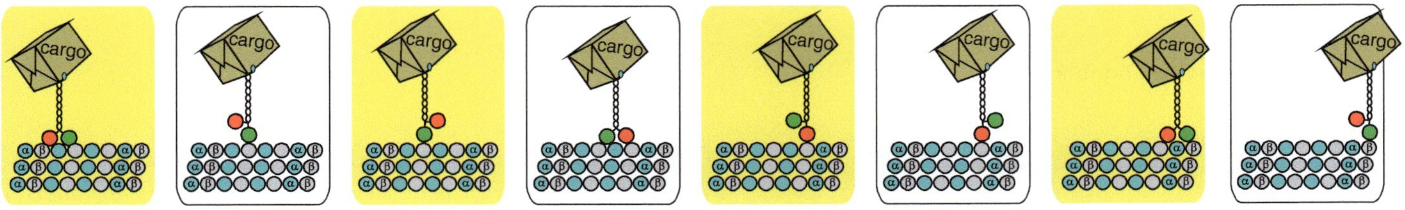

Figure 5.6. A cartoon strip depicting the mechanism of (conventional) kinesin-dependent movement ('walking') that transports vesicles along microtubules. Here the transport is from left to right, and is towards the plus end of the microtubule, which is the end with a high polymerisation rate. ATP cleavage leads to repeated cycles of 'change in shape – detach from the fibre – reattach to the fibre' alternately in each of the heavy chains making up the motor domain. This results in the coordinated 'walking' of the motor along the cytoskeletal fibre, though it's more of a rapid shuffle than a stylish stroll. You can see animations of this mechanism on YouTube at https://www.youtube.com/watch?v=y-uuk4Pr2i8 and https://www.youtube.com/watch?v=-7AQVbrmzFw. The animation at https://www.youtube.com/watch?v=wJyUtbn0O5Y is also highly recommended.

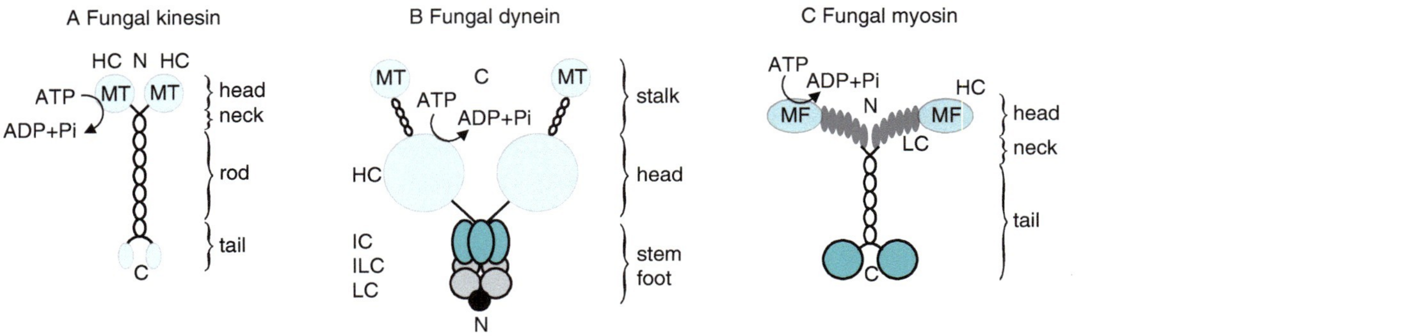

Figure 5.7. Diagrams of fungal kinesin (A), dynein (B) and myosin (C). Conventional kinesins from fungi comprise a homodimer of two heavy chains (105 kDa) that provides the enzymatic activity (no evidence for carboxy-terminal light chains that are typical for kinesins from animals has been found). The dynein complex contains two heavy chains of 400 kDa, which probably bind associated polypeptides. In fungal myosins, the heavy chain (180 kDa) of Myo2p, a class V myosin from *Saccharomyces cerevisiae*, provides six binding sites (grey ovals) for a light chain. Abbreviations key: C, carboxyl terminus; HC, heavy chain; IC, intermediate chain; ILC, intermediate light chain; LC, light chain; MF, F-actin-binding site; MT, microtubule-binding site; N, amino terminus. Redrawn and modified from Steinberg (2000).

described as 'a unique solution to internal solute translocation involving a complex, extended vacuole. In all filamentous fungi examined, this extended vacuole forms an interconnected network, dynamically linked by tubules, which has been hypothesised to act as an internal distribution system' (Darrah *et al.*, 2006). There is good evidence that **cytoplasmic microtubules** are important for the maintenance of vacuolar tubules (Figure 5.8), while F-actin microfilaments are not. Staining of hyphae with α-tubulin antibodies shows longitudinal arrays of microtubules. Tubular vacuoles lie parallel to bundles of longitudinal microtubules. Drugs which depolymerise filamentous actin do not affect the vacuole system in hyphae. Tubular vacuoles tend to cluster or accumulate in the apical region except in an apical cap of actin-rich cytoplasm about 5 μm long in the hyphal tip.

The *Saccharomyces cerevisiae* genome has genes encoding five myosins, six kinesins and one dynein. This is a modest number compared with some vertebrate systems that have 50 different motors per cell. This must mean that each motor participates in several cellular processes in fungi. Fungal motors are known to be involved in the following (and see Figure 5.8):

- secretion and endocytosis;
- cytokinesis;
- organelle positioning and inheritance;
- mitosis;
- genetic recombination;
- RNA transport.

Myosins and kinesins belong to large and diverse families of proteins, and a complex nomenclature has developed as they have been identified. **Myosin class II** is the predominant class of myosin found in muscles and is historically referred to as '**conventional myosin**'; there are another 19 classes of myosin which are currently referred to as the 'unconventional' myosins.

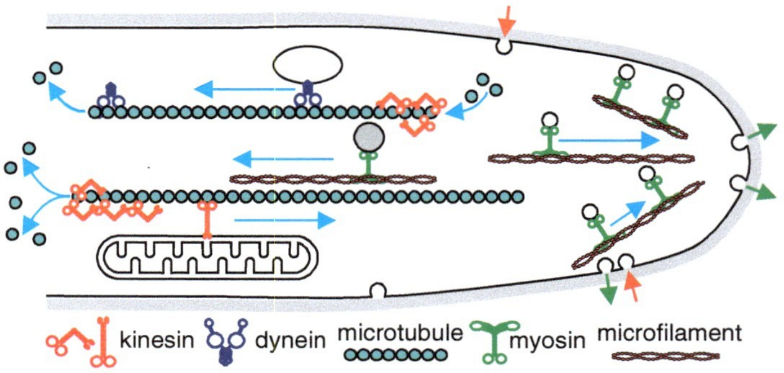

Figure 5.8. Hypothetical model of motor activity in a fungal hypha. Kinesins and dyneins move in opposite directions along polar microtubules (kinesins towards the plus end that has a high polymerisation rate, and dyneins towards the minus end, the end with low polymerisation rate). Myosins use F-actin (microfilaments), often concentrated in the apex of the hypha. In a classical model, motors translocate their 'cargo' (e.g. vesicles and other membranous organelles) along these filamentous tracks. However, some motors influence microtubule stability, probably by modifying the microtubule ends (indicated here by depolymerisation at one end of each microtubule shown). Note that all the indicated processes will be occurring at the same time in a growing hypha, and that there will be a very large number of microtubule and microfilament tracks throughout the hypha. Redrawn and modified from Steinberg (2000).

Numerous kinesins (more than 600 from a variety of species) have been described; they share a common motor domain of 340–350 amino acids. A prominent member of the kinesin family originally identified in neuronal cells is **kinesin-1**; this is the '**conventional kinesin**', which is a plus-end-directed motor. **Kinesin-1** is not present in *Saccharomyces cerevisiae* but is involved in **hyphal extension** of filamentous fungi. Conventional kinesins are two-headed molecular motors that move over micrometre-long distances on their microtubules. Movement over such long distances is called 'processive movement' and *in vitro* assays can be used to measure the processivity of purified motors. For example, *Neurospora* kinesin (NKin) moved 1.75 μm on average ($n = 182$) before detaching from the microtubule, while human kinesin motors moved only half the distance (0.83 μm, $n = 229$) under identical conditions. A yeast kinesin, Kar3p, translocates at rates of about 1–2 μm min^{-1}. In contrast, two kinesins of *Saccharomyces cerevisiae* that belong to the category kinesin-14 are 'unconventional' in that they do not motor to the microtubule plus end like kinesin-1, rather they transport cargo to the minus end of the microtubule. Yet, despite being 'unconventional', one is essential for meiosis and mating in yeast, and the other has an important role at the spindle pole body during yeast mitosis. Kinesin-7 and kinesin-1 motors participate in the transport of regulatory compounds to microtubule plus ends and in so doing affect microtubule dynamics and organisation. Kinesin-14 and kinesin-8 motors probably affect the stability of microtubules in mitosis and interphase in fungi; and the minus-end-directed kinesin-14 (and dynein, too) can transport assembled microtubules to specific regions of the cell, so polarising the microtubule array (Steinberg, 2007a & b; Steinberg *et al.*, 2017).

Myosins and, particularly, kinesins both seem to be numerous and functionally specialised, but in contrast there is only one major form of dynein to serve many different cellular roles. Functional diversity for **dynein** is achieved using an activator or cofactor, **dynactin**, in most if not all of its functions.

In the next few paragraphs, we will briefly mention a few specific examples of motor usage for a range of cellular functions in yeasts and filamentous fungi. Better understanding of mechanisms involved in vesicle migration in fungi might reveal sites for selective toxicity allowing development of new antifungal agents. The overall summary is brought together in Figure 5.9.

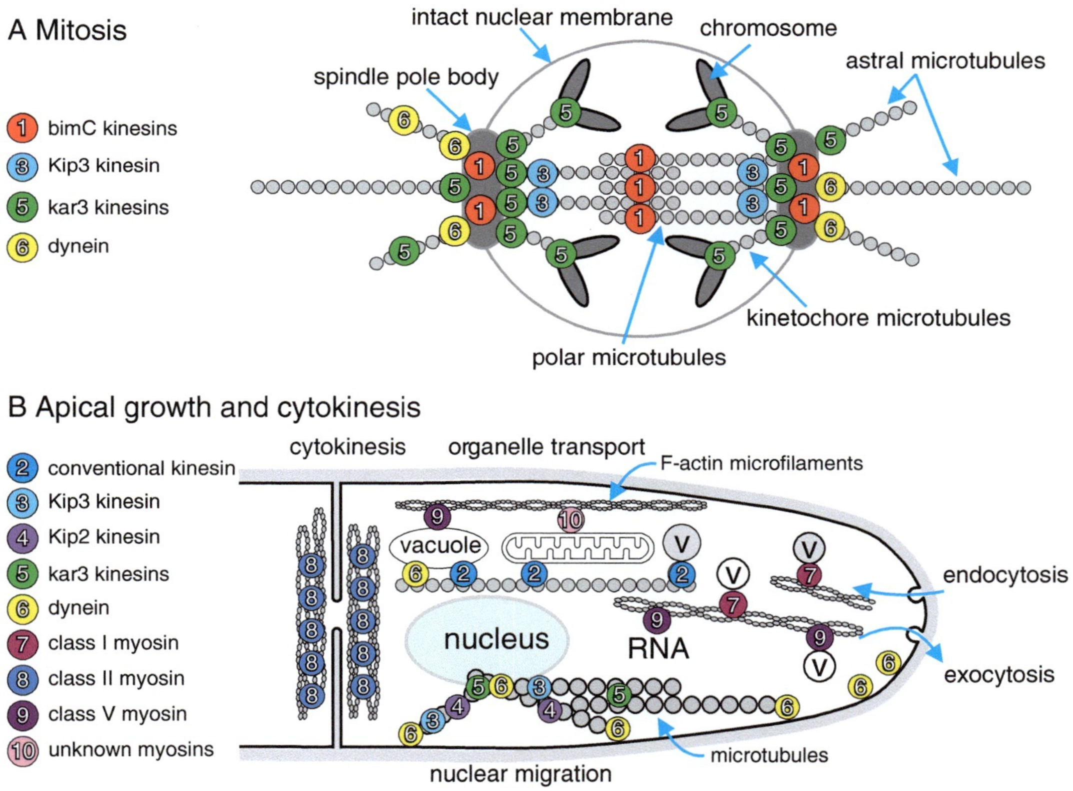

Figure 5.9. Overall summary diagrams, combining data from several fungi, indicating the localisation and/or assumed sites of action of specific fungal motors during mitosis and during polarised apical growth of filamentous hyphae. (A) During mitosis kar3-kinesins influence the dynamics of spindle microtubules and counteract bimC-like motors, which appear to cross-link polar microtubules. In addition, Kar3 protein might function within the chromosomal kinetochore. Anaphase is supported by cytoplasmic dynein that exerts pulling forces on astral microtubules (see Figures 5.10 and 5.11) and, in conjunction with Kip2 protein and Kip3 protein, probably modifies microtubule dynamics. Note that the localisation of Kip3 protein within the spindle is not known. (B) Molecular motors are involved in a wide range of processes during apical extension and cytokinesis of the hypha; various sorts of organelle transport and positioning are highlighted here (and see Figure 5.8), including nucleus, vacuole and mitochondrion, and there is a rapid traffic of a very diverse population of vesicles and microvesicles (all labelled 'v' here; compare Figure 5.5). Redrawn and modified from Steinberg (2000).

One general point that we need to make early on is that although we have given the impression so far that motors move along their microtubules to transport cargo, this is not the whole story. They certainly do move, and this is the **processivity** that can be demonstrated so readily with *in vitro* assays. But motors also modify the dynamics of the cytoskeleton; meaning that they affect the lengthening and shortening of microtubules and this delivers cargo, too. The motor domain of some motors might only be needed to reach the final site of delivery at the end of the microtubules. A particular example is the yeast kinesin Kar3p (the symbol 'Kar3p' means 'the protein corresponding to gene sequence Kar3') that functions in mitosis mainly by destabilising the microtubule at the SPB; that is, Kar3p **pulls** the microtubule into the SPB. Kar3p also occurs in the kinetochore of chromosomes, raising the possibility that in this case chromosome movement on the mitotic spindle results from the microtubule being depolymerised and pulled in at both ends.

Two other kinesin motors, Kip2p and Kip3p, are also involved in modifying microtubule stability during nuclear migration and mitosis in yeast and similar reports of destabilising activities of spindle motors in vertebrates suggests that modification of cytoskeletal dynamics might be a crucial feature of their cellular function. Similarly, mutants of *Aspergillus nidulans* indicate that a dynein motor destabilises and thereby exerts pulling forces on SPB astral microtubules during nuclear migration (Figures 5.10 and 5.11). As well as providing motility, kinesins and dynein also actively participate in organising their tracks in fungal cells. However, the molecular mechanisms by which motors regulate microtubule stability and turnover remain unknown (Urnavicius *et al.*, 2015; Xiang *et al.*, 2015; Steinberg *et al.*, 2017).

After mitosis, cell separation during cytokinesis requires conventional (class II) myosins in *Saccharomyces cerevisiae* and *Schizosaccharomyces pombe*. In *Schizosaccharomyces pombe*, the myosin assembles into a ring-like structure at the cell cleavage plane, where it interacts with F-actin and supports cytokinesis.

The **cell wall** determines the shape of the yeast cell and the hypha in filamentous fungi and its assembly shows extreme polarity (Takeshita *et al.*, 2014). Intracellular transport of vesicles enables the hyphal cell to construct and modify the wall and it has been calculated that transport to the hyphal apex of up to 38,000 vesicles per minute is necessary for each fast-growing hypha of *Neurospora crassa*. This is the scale of the process, which is supported by apically polarised transport of vesicles along both F-actin microfilaments and the microtubular cytoskeleton. Some so-called class V, or unconventional, myosins as well as conventional kinesins take part in this.

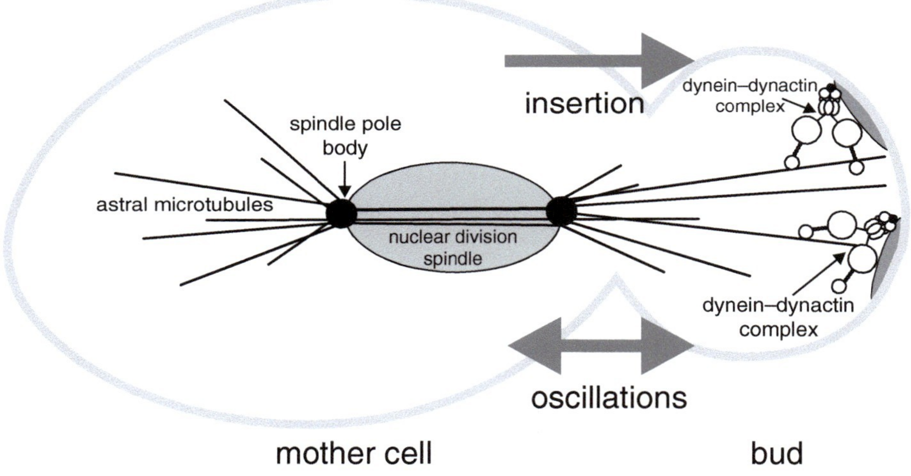

Figure 5.10. A model for the role of dynein–dynactin during anaphase nuclear migration in *Saccharomyces cerevisiae*. The dynein–dynactin complex has been shown to be involved in mitosis in a function that partially overlaps with the kinesin-related protein Kip3p which orients the nucleus near the bud neck (not shown here). Dynein is involved in inserting the nucleus through the bud neck during anaphase and is also responsible for the oscillatory movements of the nucleus observed during nuclear insertion. Redrawn and modified from Karki and Holzbaur (1999).

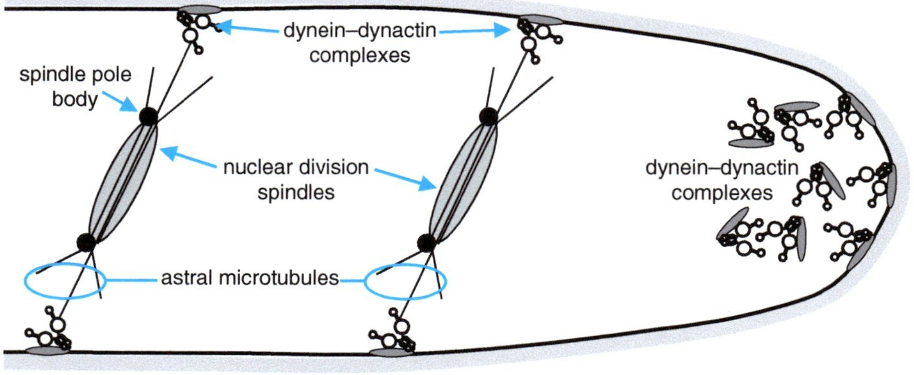

Figure 5.11. A model for the role of dynein and dynactin in nuclear migration in *Aspergillus nidulans*. Cytoplasmic dynein, which is anchored to the cell membrane through dynactin, is located at sites where the nuclei need to be positioned. Dynein is also known to localise to the apex of the hypha. Redrawn and modified from Karki and Holzbaur (1999).

Additionally, class I myosins appear to support polarised growth and endocytosis (e.g. Myo3p and Myo5p from *Saccharomyces cerevisiae* and MYOA from *Aspergillus nidulans*). However, MYOA is also required for secretion, so this myosin is involved in both endo- and exocytosis in *Aspergillus nidulans*. Exocytosis in *Saccharomyces cerevisiae* involves a class V myosin, called Myo2p, which also contributes to polarised growth of the bud. Myo2p binds secretory vesicles with its carboxy-terminal tail and moves its cargo along F-actin filaments to sites of growth. Myo2p has been identified with several functions in organelle trafficking and spindle orientation. At present, Myo2p appears to be the main membrane-bound motor in yeast and is therefore likely to participate in the transport of many components, including the Chs3p chitin synthase protein required for cell wall synthesis. A significant fraction of the total Myo2p in the cell is associated with a large, mRNA-containing particle that is distinct from the exosome and actively translating polysomes. Myo2p may promote the release of mRNAs from this particle for either degradation or translation (Chang *et al.*, 2008).

In some fungi the chitin synthase activity is fused to their molecular motor. Examples include CsmA of *Aspergillus nidulans* and Csm1 from the rice blast fungus *Pyricularia oryzae*. These are **myosin motor/chitin synthase** fusion proteins and both domains are required for correct cellular function. Another class V myosin, Myo4p, is responsible for specifically localising of ASH1 mRNA in the daughter bud of a budding *Saccharomyces cerevisiae* cell (Ash1p is a repressor of the mating-switch endonuclease that allows mother cells to switch their mating; see section 9.3). Showing that cytoskeletal motors transport mRNA, and by so doing generate RNA gradients in the fungal cell.

Microtubule motors of the conventional kinesin class I have been found in fungi belonging to the zygomycetes, *Ascomycota* and *Basidiomycota*. They participate in apical exocytosis generally, as well as having other functions (e.g. mitochondrial positioning in *Nectria haematococca* and vacuole organisation in *Ustilago maydis*). The mitotic spindle consists mainly of microtubules and we have already indicated how kinesins can be involved in microtubule dynamics and chromosome movement during fungal mitosis.

Assembly and organisation of the spindle requires counteracting motors to create tension within the spindle rather than moving a cargo; Kar3 and bimC motors have these counteracting functions. Kar3 kinesin motors (specifically, Kar3p of *Saccharomyces cerevisiae*; KlpA of *Aspergillus nidulans*; and Pkl1 of *Schizosaccharomyces pombe*) are unconventional in that they move towards the minus ends of microtubules; they also locate at the spindle poles where their major role is to destabilise (and 'reel-in') microtubules, as mentioned above. Conventional kinesins of the bimC family are required, in all organisms so far examined, for separation of spindle pole body/centrosome (depending on organism) at the onset of mitosis, as well as for the assembly and maintenance of the spindle structure. They may also alter microtubule dynamics, but bimC-like motors have two motor domains at each end and they are in the middle region of the spindle during anaphase, suggesting that bimC kinesins separate the spindle poles by cross-linking the polar microtubules they are destabilising.

Fungal dynein is located outside the mitotic nucleus in the growing apex and serves two general functions; the motor pulls on astral microtubules to drive nuclear migration and it effects transport of exocytotic vesicles. This is an extreme contrast with vertebrate animals in which cytoplasmic dynein is located at the spindle and functions in spindle assembly and chromosome segregation.

Cytoplasmic dynein is a multisubunit protein (overall molecular mass about 1.2 MDa) consisting of two heavy chains of about 500 kDa each, which fold to form the two heads of the motor, as well as multiple intermediate chains (about 70–74 kDa each), light intermediate chains (about 53–59 kDa) and light chains (from 8–22 kDa). Dynactin is also a large multisubunit complex of at least seven polypeptides ranging in size from 22 to 150 kDa. Studies with *Saccharomyces cerevisiae, Neurospora crassa* and *Aspergillus nidulans* demonstrate interaction between dynein and dynactin. Specifically, **dynactin acts as an anchor** that stabilises the dynein, so it can pull on astral microtubules coming out of the spindle pole body (Figures 5.10 and 5.11) (Xiang, 2018).

In budding yeast, the SPB is embedded in the nuclear envelope, where it serves as the origin for spindle microtubules from its inner 'spindle plaque' and cytoplasmic (called astral) microtubules from its outer plaque. Astral microtubules and motors, together with the actin cytoskeleton and additional cell-polarity determinants act to control nuclear movements into and through the neck of buds (Figure 5.10). Two major nuclear movements occur during yeast budding: alignment of the nucleus along the mother-to-bud axis and near the neck of the bud, and then the post-anaphase propulsion of the daughter nucleus through the neck and into the bud. The model shown in Figure 5.10 suggests that the Kip3 kinesin positions the dividing nucleus near the bud neck, while dynein is responsible for inserting the nucleus through the bud neck during anaphase, and causes the oscillatory movements of the nucleus observed during nuclear insertion.

Figure 5.11 shows a model for the role of dynein and dynactin in nuclear migration in *Aspergillus nidulans*. Dynein is known to localise to the apex of the hypha, in association with the cell membrane. Another notable feature of filamentous fungi is the even distribution of nuclei along the fungal hypha. Achieving this also requires dynein and dynactin. Cytoplasmic dynein, which is anchored to the cell cortex through dynactin, is located at sites where the nuclei need to be positioned and this ensures the **correct positioning of daughter nuclei** (Roberts & Gladfelter, 2016).

Other proteins are involved in specifying the position of the dynactin anchors. One seems to be Num1, a protein defective in one of the first mutants identified in the nuclear migration pathway of budding yeast, which is named for *Nu*clear *m*igration. Num1 is the cortical anchor for the motor protein dynein.

The *NUM1* gene encodes a complex, 313 kDa protein which has **pleckstrin** homology (PH) domains. PH domains were originally identified as an internal repeat in pleckstrin, a phosphoprotein from blood platelets. PH domains are found in animals and fungi but have not yet been detected in plants or bacteria. Proteins carrying PH domains are either involved in signal transduction or they are part of the cytoskeleton. The ligands for many PH domains are membrane-bound inositol phospholipids, which support the role of PH domains as membrane anchors. Molecular motors contribute to numerous processes that are of key importance to the organisation and polarisation of extension in fungal cells in yeasts and filamentous hyphae alike (Martin & Arkowitz, 2014; Roberts & Gladfelter, 2016).

5.12 PLASMA MEMBRANE AND SIGNALLING PATHWAYS

The prime function of the membranes of fungi, like those of other eukaryotic cells, is to provide a barrier between the cell and its environment. Plasma membranes are composed of a phospholipid bilayer, but this is not a static barrier. An enormous number and variety of proteins are anchored into the membrane, which carry out a tremendous range of functions. Sterols are also a critical component of the fungal membrane and serve to regulate membrane fluidity and the activity of membrane-associated enzymes and transport mechanisms.

For fungi to grow, external nutrients must be assimilated across the plasma membrane. To absorb nutrients successfully from the surrounding environment, the fungi possess a diverse range of specific **transport proteins** in the plasma membrane (van Dijck *et al.*, 2017). Three main classes of nutrient transport occur in fungi, **facilitated diffusion, active transport** and **ion channels**. Fungi usually contain two transport mechanisms for the assimilation of nutrients, such as sugars and amino acids. One is a constitutive low affinity transport system which allows the accumulation of a nutrient when it is present at a high concentration outside the hypha. This process of facilitated diffusion is not energy dependent and does not allow accumulation of solutes against a concentration gradient. However, the 'facilitator', more often called the permease, carrier or transporter, is a polypeptide that confers enzyme-like specificity to the uptake process.

When the external solute concentration is lower than it is within the cell (which is probably more often the case in the natural environment), a second class of carrier protein is induced that has a higher affinity for the nutrient and can take up the solute against the concentration gradient. This, though, is energy dependent, occurring at the expense of ATP, and is called an active transport process. Most active transport processes in the fungal cell are powered by an **electrochemical proton gradient** that fungi create by pumping hydrogen ions out of the hyphae, at the cost of ATP hydrolysis, using proton-pumping ATPases in the plasma membrane. The resultant proton gradient provides the electrochemical gradient which can drive nutrient uptake by a carrier that couples nutrient uptake to the flow of hydrogen ions back down the gradient. As active transport processes are usually induced by environmental conditions (often low nutrient concentration), fungi are evidently capable of adapting their transport mechanisms according to the external solute concentration; this assures continued nutrient supply over a variety of environmental circumstances.

Ion channels are highly regulated pores in the membrane which allow influx of specific ions into the cell when open. Several ion channels have been described in fungi by **patch-clamp electrophysiology** experiments analogous to studies conducted with mammalian cells. Patch clamping involves measuring the current flowing across a patch of the plasma membrane, which can be used to study the flow of specific ions across the membrane. In fungi, the cell wall must be removed first by incubating mycelium in an osmotic stabiliser and a mixture of lytic enzymes, which digest away the cell wall. This produces naked sphaeroplasts (which are mostly wall free) or protoplasts (which are entirely wall free) from which patches of membrane can be removed with micropipette electrodes. Several different channel types have been identified, including anion-selective channels (such as Cl^-) as well as various cation-selective channels, especially K^+ and Ca^{2+} permeable channels.

Channels that carry an inward flux of K^+ ions are thought to be involved in regulating the internal turgor pressure of the hypha (the bulk of the osmotic potential of the hypha is provided by inorganic ions most of the time). The presence of mechanosensitive or stretch-activated Ca^{2+} channels has attracted interest. These are opened when the membrane is under mechanical stress and they may play important roles in the generation of the Ca^{2+} gradients in response to physical changes to the membrane, which might be caused, say, by rapid water fluxes in the cell (rapid water influx stretching the membrane/rapid water loss releasing tension), or by physical pressure (encountering an obstacle, gravitational stress, etc.). This, of course, amounts to an environmental sensory system since the mechanosensitive ion flux can be linked to any of a range of response systems.

The hypha must also be able to detect the nutrient status of the substratum over which, or through which, it is growing. This is the responsibility of signal transduction pathways which recognise a specific chemical at the plasma membrane/outside world interface and then convey the information into the cell. Intermediary metabolism is obviously crucial to fungal growth and development (Moore, 1998, 2000; see section 11.12) and imposes numerous regulatory events on the cell. Primary metabolites like glucose are especially important, and because the amount of available glucose (for example) can fluctuate wildly in the heterogeneous substrata that fungi inhabit, individual fungal hyphae must be able to sense the amount available to them and respond appropriately.

When readily utilisable sugars, such as glucose or fructose, are added to fungi which are starved (derepressed) of carbohydrates, a wide range of metabolic responses follow rapidly. Some of these are mediated by a transient rise in the levels of **cyclic-AMP** (cAMP), this being the final product in the cycle of ATP dephosphorylation. These responses include:

- **inhibition of gluconeogenesis**, generation of glucose from non-sugar carbon substrates like pyruvate, and glucogenic amino acids, like alanine and glutamate;
- **activation of glycolysis**, to amplify energy generation from whatever sugar and non-sugar carbon substrates may be available;
- **activation of trehalase**, which can produce two molecules of glucose by hydrolysing one molecule of the disaccharide trehalose (the characteristic storage sugar of fungi).

The immediate response to external nutrient supply level will be to modify the activity of existing enzyme systems (as described above), the ultimate effect of (for example, external glucose) is control of gene expression. Glucose has two major effects on gene expression in *Saccharomyces cerevisiae*: it represses expression of many genes, including those encoding proteins in the respiratory pathway (including cytochromes) and enzymes for metabolism of alternative carbon sources (e.g. galactose, sucrose, maltose); it also induces expression of genes required for glucose utilisation, including genes encoding glycolytic enzymes and glucose transporters.

There are two signal transduction pathways responsible for these effects of glucose on yeast. The first uses the Mig1 transcriptional repressor, which is a zinc-finger transcription factor whose function is inhibited by the Snf1 protein kinase. The Snf1 protein kinase is a glucose sensor in the plasma membrane and is an AMP-activated serine/threonine protein kinase required for transcription of several glucose-repressed genes; the Snf1 complex phosphorylates Mig1 in response to glucose and thereby releases the transcriptional repression of those genes for which Mig1 is responsible (Papamichos-Chronakis *et al.*, 2004).

The signalling pathway responsible for glucose induction centres on the Rgt1 transcription factor which regulates expression of several glucose transporter genes in response to glucose. It binds to promoters and acts as a transcription activator (Kayikci & Nielsen, 2015). In the absence of glucose, Rgt1 function is inhibited by the SCF protein complex. The SCF complexes are a family of ubiquitin ligases that target specific proteins for destruction at the 26S-proteasome (Willems *et al.*, 2004).

Many of these responses are mediated by the activation of **cAMP-dependent protein kinases** modifying the activity of existing proteins by phosphorylation. The increase in cAMP is due to the activation of the membrane-bound enzyme adenylate cyclase because of the activation of a class of small GTP-binding proteins, the RAS proteins. RAS proteins are binary switches, cycling between ON and OFF states during signal transduction. Activation of RAS leads to the exchange of RAS-bound GDP for GTP, causing a conformational change that leads to stimulation of adenylate cyclase activity. The RAS protein complex includes an intrinsic GTPase, which converts GTP-bound RAS to GDP-bound RAS, thus returning RAS to its resting state in the absence of an activator. The evidence strongly suggests that the RAS pathway forms part of a global mechanism for general signalling that controls processes as diverse as cytoskeletal integrity and cell–cell fusion and exocytosis. RAS signalling at the plasma membrane of *Aspergillus* hyphal tips is involved in the polarisation of the actin cytoskeleton that is required for hyphal growth and, possibly, for asexual morphogenesis (Harispe *et al.*, 2008; Noble *et al.*, 2016). We should also add that defective RAS proteins are the cause of several human diseases (Simanshu *et al.*, 2017).

RAS is a **G-protein**, a regulatory GTP hydrolase that cycles between the two conformations we've described, activated (RAS-GTP) or inactivated (RAS-GDP). RAS is attached to the cell membrane by a chemical modification called **prenylation**, which is the addition of hydrophobic molecules to a protein. Protein prenylation involves the transfer of either a farnesyl or a geranylgeranyl group to carboxy-terminal cysteine(s) of the protein by farnesyltransferase or geranylgeranyltransferase. Prenyl groups enable proteins to attach to cell membranes, serving as a lipid anchor for the protein (Wang & Casey, 2016).

Throughout the eukaryotes many hormones, neurotransmitters, chemokines (proteins secreted by cells to control other cells), mediators (local mediators are membrane-modifying lipids, such as omega-3-fatty acids, but mediator is also a large protein complex that interacts with the RNA polymerase II machinery), and sensory stimuli exert their effects on cells by binding to **G-protein-coupled receptors**. More than a thousand such receptors are known, and more are being discovered all the time. Heterotrimeric **G-proteins** (made up of a, b and g subunits) transduce ligand binding to these receptors into intracellular responses. There are four main classes of G-proteins:

- Gs, which activates adenylyl cyclase;
- Gi, which inhibits adenylyl cyclase;
- Gq, which activates phospholipase C;
- G-proteins of unknown function.

G-proteins are inactive in the GDP-bound state and are activated by receptor-catalysed guanine nucleotide exchange resulting in GTP binding to the Ga subunit, which leads to dissociation of Gb and Gg subunits that activate downstream effectors. Since the first description of G-proteins in yeasts and filamentous fungi in the 1990s, they have been shown to be essential for growth, asexual and sexual development, and yeast–hyphal dimorphism and virulence of both animal and plant pathogenic species (Li *et al.*, 2007; Shi *et al.*, 2007).

G-protein and RAS phosphorylation also initiate the **MAP kinase signalling pathway**. An extracellular factor affixes to a receptor on the outer membrane surface, the resultant conformational change in the receptor dimer causes it to become phosphorylated and it is then able to activate a specific G-protein, which in turn catalyses the phosphorylation of RAS by GTP into its active form. In this form, the RAS is able to catalyse the phosphorylation of the beginnings of the MAP kinase cascade. This cascade begins with **MAP Kinase Kinase Kinase** (**MAPKKK**), a family of enzymes which in turn activates the second step in the cascade, the family of **MAP Kinase Kinases** (**MAPKK**), which then activates the **MAP kinases** (**MAPK**), which have the actual effects of stimulating the transcription of specific genes in the nucleus. MAP kinase pathways are remarkably conserved throughout evolution, functioning as key signal transduction components in fungi, plants and mammals. '**MAPK**' stands for '**Mitogen-Activated Protein Kinase**' so the genes that are particularly controlled are those required for nuclear division and cell differentiation. MAPKs produce many of the responses that are induced in cells by changes in environmental conditions and/or exposure to external stimuli. These pathways were first revealed in budding yeast in which pathways include activation by pheromones during formation of mating projections in shmoos (see section 5.7) (Chen & Thorner, 2007); and a range of other cellular events in filamentous fungi (Erental *et al.*, 2008; Read *et al.*, 2009). Of specific interest is that in plant pathogenic fungi the MAPK cascades in the two organisms mutually contribute to an interconnected molecular dialogue between plant and fungus. Fungal MAPKs promote penetration of host tissues, while plant MAPKs activate plant defences. However, some pathogenesis-related processes controlled by ***fungal*** MAPKs lead to the activation of ***plant*** MAPK cascade signalling. Conversely, ***plant*** MAPKs that promote defence mechanisms against fungal cells lead to a fungal protective stress response mediated by ***fungal*** MAPK cascades (Hamel *et al.*, 2012).

Although there are differences in molecular details, signalling pathways triggered by G-protein-linked receptors have several points in common and all have the essential similarity that their function is to make a massive response to a very small signal. The 'cascade' structure of the pathway enables this **signal amplification**; each step produces molecules that can modify even more molecules in the next step. Consequently, a single extracellular signal molecule can cause many thousands of intracellular protein molecules to be altered.

We have mentioned protein prenylation as a means of directing specific proteins to membranes. A related process is to attach **glycosylphosphatidylinositol** (**GPI**) as a lipid anchor for many cell-surface proteins. The GPI anchor is a post-translational modification of proteins with a glycolipid carried out in the ER; the carboxy-terminal signal sequence of the preprotein remains in the ER membrane with the rest of the protein in the ER lumen. Final removal of the signal peptide is combined with attachment of the finished protein's new carboxy-terminus to an amino group on a pre-assembled GPI precursor located in the inner leaflet of the ER. Depending on the organism, cell type and protein, the GPI backbone can be chemically modified in several ways. The anchored protein can then be delivered to the plasma membrane in a vesicle and located with the GPI anchor in the outer leaflet of the plasma membrane and the protein on the cell exterior.

GPI anchors are used ubiquitously in eukaryotes and most likely in some *Archaea*, but not in *Eubacteria*. GPI-anchored proteins are the major form of cell-surface proteins in protozoa. In fungi, many GPI-anchored proteins are ultimately incorporated into the cell wall; their GPI anchor is partially trimmed off at the plasma membrane just before their incorporation into the cell wall, though the glycan part of the GPI anchor remains and is linked to the cell wall glucans. In humans, there are at least 150 GPI-anchored proteins, with a variety of roles: receptors, adhesion molecules, enzymes, transcellular transport receptors and transporters, and protease inhibitors (Kinoshita, 2016).

5.13 FUNGAL CELL WALL

The **fungal wall** is a sophisticated cell organelle. It defines the volumetric shape of the cell, provides osmotic and physical protection and, together with the plasma membrane and periplasmic space, influences and regulates the influx of materials into the cell. However, it is also able to control the environment in the immediate external vicinity of the cell membrane, and it represents the interface between the organism and the outside world. This is an **active interface**, since the interaction of the organism and the outside world (and the latter will include other cells) is subject to modulation and modification. The fungal cell wall is metabolically active, interactions between its components occur to give rise to the mature cell wall structure. The wall must be understood to be a **dynamic structure** which is subject to modification at various times to suit various functions. Besides enclosing and supporting the cytoplasm, those functions include selective permeability, as a support for immobilised enzymes and cell–cell recognition and adhesion.

The wall is a multilayered complex of polysaccharides, glycoproteins and proteins. The polysaccharides are glucans and mannans and include some very complex polysaccharides (like gluco-galacto-mannans). In hyphae the major component of the wall, and certainly the most important for its structural integrity, is **chitin** (a polymer of *N*-acetylglucosamine) though this is frequently cross-linked to other wall constituents, particularly a **β(1–3)-glucan**, the terminal reducing residue of a chitin chain being attached to the non-reducing end of a β(1–3)-glucan chain by a (1–4) linkage. Removal of the outer wall layers with lytic enzymes has revealed the architecture of the inner chitin wall to be composed of microfibrils formed by the aggregation of the chitin polymers by hydrogen bonding. The chitin inner wall is cross-linked to the outer β-glucan components and forms a major structural component of the walls of most true fungi. Synthesis of the cell wall occurs at the outer surface of the plasma membrane of the growing hyphal tip. Chitin synthase is the enzyme that catalyses formation of chitin from the precursor Uridine diphosphate N-acetylglucosamine (UDP-*N*-acetylglucosamine). **Chitin synthase** adds two molecules of UDP-*N*-acetylglucosamine (UDPGlcNAc) to the existing chitin chain in the reaction:

$$(\mathrm{GlcNAc})_n + 2\mathrm{UDPGlcNAc}\text{-}(\mathrm{GlcNAc})_{n+2} + 2\mathrm{UDP}$$

It appears as an inactive **zymogen** requiring activation by cleavage of a peptide by an endogenous protease to generate the active enzyme (Robson, 1999).

Phylogenetic analysis has revealed seven classes of chitin synthases divided into two families. Class I, II and IV genes make up the first family and are present in all fungi, whereas classes III, V, VI and VII (the second family) are specific to filamentous fungi and certain dimorphic species. The number of putative chitin synthase ***genes*** (as opposed to classes of chitin synthase) within each species also varies, with three in *Saccharomyces cerevisiae*, four in *Candida albicans*, seven in *Wangiella dermatitidis* and *Neurospora crassa* and eight each in *Aspergillus nidulans, Aspergillus fumigatus* and *Cryptococcus neoformans* (Rogg *et al.*, 2012). For the most part, the precise functions of the individual genes remain to be determined; however, three chitin synthase genes, *CHSI, CHSII* and *CHSIII* that have been cloned from budding yeast, *Saccharomyces cerevisiae*, serve different functions in the cell:

- *CHSI* acts as a repair enzyme and is involved in synthesising chitin at the point where the daughter and mother cells separate.
- *CHSII* is involved in septum formation.
- *CHSIII* is involved in chitin synthesis of the main cell wall.

Like chitin, β-(1–3)-glucan is synthesised by a membrane-associated **β-(1–3)-glucan synthase** which utilises UDP-glucose as its monomeric substrate, inserting glucose into the β-glucan chains. The β-(1–3)-glucan synthase activity is found both in the membrane and cytoplasmic fractions of fungal mycelia and is stimulated by GTP. Although genes encoding a β-glucan synthase have been isolated from a number of fungi, it is not clear whether a gene family for β-(1–3)-glucan synthase exists in fungi as it does for chitin synthase (Teparić & Mrša, 2013). The fungal cell wall represents an attractive target for pharmacologic inhibition, as many of the components are fungal specific. Though targeted inhibition of β-glucan synthesis has been used to treat some fungal infections, for example *Pneumocystis* pneumonia (Schmatz *et al.*, 1990), the ability of the cell wall to dynamically compensate via the cell wall-integrity pathway limits their usefulness. To date, chitin synthesis inhibitors have not been successfully deployed in the clinical setting.

Although during normal apical extension of the hypha the incorporation of newly synthesised chitin is limited to the hyphal apex, there is evidence that inactivated chitin synthases are widely distributed in the plasma membrane, indeed inactivated chitin synthase activity appears to be an intrinsic property of the plasma membrane, the enzyme being activated in some way specifically at the hyphal apex and at sites where **branch formation** is initiated.

Chitin is important at particular sites in yeast walls, although the major structural component in these organisms is a fibrillar inner layer of β-glucan. The glucan has secreted **mannoproteins** attached to it. Mannoproteins play an essential role in cell wall organisation, and there is evidence for the formation of covalent bonds between these molecules and the structural polymers (glucans and chitin) outside the plasma membrane. This makes the point that the protein components of fungal walls are of considerable importance, too. Some of the proteins identified in walls have enzymic activities associated with them; these include α-glucosidases and β-glucosidases, enolase and alkaline phosphatase. Other obvious components are proteins involved in cell-to-cell recognition, like the products of mating type factors. Cell-to-cell adhesion during yeast mating depends on interaction of two glycoproteins inserted into the outer coat of the cell wall, which are the gene products of components of the mating type factors and are first located in the plasma membranes with GPI anchors.

The outer surface of many fungal walls is usually layered with proteins that modify the **biophysical properties** of the wall surface appropriately to the environment. Hydrophobic surfaces (like spores and aerial mycelium) have rodlets of the protein **hydrophobin** as the outermost rodlet layer (see section 7.8). Some GPI-anchored wall proteins span the wall with extended glycosylated polypeptides protruding into the surrounding medium and providing the wall with hydrophilic and even adhesive surfaces (Bayry *et al.*, 2012).

Fungal walls are remarkably variable in their detailed aspects of structure (Gow *et al.*, 2017). Different strains of the same species may exhibit differences in overall composition of the wall and in the polysaccharide and polypeptide structures the wall contains. Indeed, a single colony may have different wall structures in its different regions (aerial structures, surface mycelium, submerged mycelium, etc.). In the face of such differences, it is not feasible to give an exact description of a typical fungal wall (there's no such thing), but it is possible to describe the conceptual framework of the fungal wall. In brief, the fungal wall concept is this:

- The main structural substance of the wall is provided by polysaccharides, mostly glucan, and in filamentous forms the shape-determining component is chitin.
- Various polysaccharide components are linked together by hydrogen bonding and by covalent bonds.
- A variety of proteins/glycoproteins contribute to wall function, some of these are structural, some are enzymic, and some vary the biological and biophysical characteristics of the outer surface of the wall.
- Proteins may be anchored in the plasma membrane, covalently bonded to wall polysaccharides or more loosely associated with the wall.
- The wall is a dynamic structure which is modified (a) as it matures, and/or (b) as part of hyphal differentiation, and/or (c) on a short-term basis to react to changes in physical and physiological conditions.

Finally, by definition, **the wall is extracellular**, its entire structure lies outside the plasma membrane, so all additions to its structure must be externalised through the membrane before the wall can be restructured and some of the chemical reactions which link wall components together are extracellular reactions.

5.14 CELL BIOLOGY OF THE HYPHAL APEX

Extension of the cell wall at the hyphal apex is the most striking characteristic of the fungi, but the problem is to understand how wall extension can be achieved without jeopardising the integrity of the existing hypha. It is quite clear that new wall material must be delivered to the apex and inserted into pre-existing wall (note that the phrase 'new wall material' should be interpreted to include *everything* that is necessary to extend the tip: membrane, periplasmic materials and all wall layers). The models which have been suggested over the years to explain this process differ in how they account for this being achieved, but underlying them all is the recognition that the act of inserting new wall material could itself weaken the wall. The potential validity of the models must therefore be judged not only on how they provide for wall synthesis, but on how they safeguard the integrity of the hypha while wall synthesis is in progress.

The problem is that a fungal hypha is part of a **closed hydraulic system** which is under pressure. The osmotic influx of water, due to the difference in water activity between the inside and the outside of the semipermeable plasma membrane, attempts to increase the cytoplasmic volume but is counteracted by the wall pressure, due to the mechanical strength of the wall outside the plasma membrane. The difference between these two forces is the turgor pressure which is the resultant 'inflation pressure' which keeps the hypha inflated. In a closed vessel, the pressure is the same over the whole of the inside wall surface. This applies whatever the shape of the vessel; irrespective of shape, the wall pressure is uniform.

In the context of the hypha, *if* the mechanical force which the wall can exert is equal to or greater than the force exerted by turgor, then it will remain intact. However, if turgor exceeds the breaking strain of the wall ***at any point*** then the wall will rupture (possibly explosively) and the cell will die. The problem we face in understanding apical hyphal extension is that the structure of the wall needs to be weakened to allow insertion of new wall material to the continuously elongating tip. How can that be achieved without exploding the tip?

In the true fungi, turgor is regulated to contribute to tip extension by driving the tip forward and shaping it by plastic deformation of the newly synthesised wall. We emphasise 'in the true fungi' here because members of the *Oomycota* (specifically *Achlya bisexualis* and *Saprolegnia ferax*) can grow in the absence of turgor, so their hyphal systems are obviously very different from those of true fungi. *Achlya bisexualis* and *Saprolegnia ferax* do not regulate turgor, their response to the addition of nutrients that raise the osmotic pressure of the medium is to produce a more plastic wall and continue to grow, and near-normal hyphae are formed by *Saprolegnia ferax* in the absence of a measurable turgor pressure. In many respects, hyphal extension in *Oomycota* is like pseudopod extension in animal cells, in that polymerisation of the actin cytoskeletal network at the apex of the hypha plays an indispensable role in the absence of turgor, actin polymerisation becoming the main driving force for extension. The most effective model for these 'almost-fungi' envisages that hyphal tip expansion in *Oomycota* is regulated (restrained under normal turgor pressure and protruded under low turgor) by a peripheral network of F-actin-rich components of the cytoskeleton, which is attached to the plasmalemma and the cell wall by integrin-containing linkages (an integrin is an 'integral membrane protein' that is permanently attached to the plasma membrane).

This model is not obviously applicable to the true fungi if it is interpreted as postulating involvement of the actin cytoskeleton to drive tip extrusion. This is unnecessary because the true fungi can use the hydrostatic pressure of their regulated turgor to 'inflate' a plastic hyphal tip and drive tip extrusion that way. However, a model involving the actin cytoskeleton could solve the potential problem that the extending hyphal tip might not be strong enough to resist turgor during the process of wall synthesis.

The cytoskeleton, both microfilaments and microtubules, is involved in the control of polarity in hyphal extension of true fungi (Takeshita *et al.*, 2014), so an actin cytoskeleton with firm attachments through the membrane to the growing wall does appear to be involved at both the tip and the septum regions of wall synthesis. In the case of septum formation, mechanical resistance to turgor does not arise, so an obvious interpretation of the actin network is that it functions in transport, directing wall components to the growing sites for terminal exocytosis.

Actin microfilaments are tension elements and actin filaments anchored to the wall by integrin-like molecules through the membrane close to extension sites could certainly serve to direct microvesicles to their target *and* provide tension anchors to supplement the mechanical strength of the synthesis-weakened, plastic cell wall. By being involved in both the directional control of the precursors *and* the mechanical support of the weakened wall, such structures would be ideally placed to serve as sensors of the local mechanical strength of the wall and thereby act as regulators of the amount of new synthesis required to restore wall integrity. This idea is also attractive because it can be readily appreciated how it might have arisen as an evolutionary development of some ancestral version of the 'cytoskeletal extrusion' mechanism which is used in present-day *Oomycota*. It is likely that the F-actin cytoskeleton at the hyphal apex:

- is anchored through the membrane to the wall at its growing points;
- thereby reinforces the wall so that its components can be partially disassembled for new material to be inserted;
- directs precursors to those points;
- acts as a strain gauge, adjusting the traffic of precursors in step with local mechanical requirements.

Importantly, the contribution made by microtubular components of the cytoskeleton to apical extension does not seem to be as crucial as the contribution made by F-actin. Apical extension of *Schizophyllum commune* hyphae continues for several hours after drug-induced disruption of cytoplasmic microtubules.

Two major models have been proposed to explain the mechanism of hyphal tip extension in mycelial fungi, both envisage the tip wall being enlarged as the result of fusion with the plasma membrane of vesicles carrying precursors and enzymes so externalising their contents, particularly chitin synthase (so the vesicles are called chitosomes).

1. In the **hyphoid model** the rate of addition to any part of the wall depends on its distance from an autonomously moving **vesicle supply centre (VSC)** which is presumed to be a representation of the Spitzenkörper. The model originates from an interpretation of the particular shape of the hyphal tip as a 'hyphoid' curve (as opposed to being hyperbolic or hemispherical). The hyphoid equation was then elaborated into a mathematical model which assumes that wall-building vesicles are distributed from the VSC, this being an organiser from which vesicles move radially to the hyphal surface in all directions at random. Fusion of vesicles with the plasma membrane externalises their content of lytic enzyme (endo-glucanase, perhaps chitinase) and:
   - hydrolyses structural glucan molecules in the existing wall;
   - mechanical stretching then pulls the broken molecules apart;
   - resynthesis occurs either by insertion of oligoglucan or by synthetic extension of the divided molecule(s).

   The resynthesised molecules have the same mechanical strength as before but have been lengthened and the tip has grown. Forward migration of the VSC generates the hyphoid shape. Computer modelling suggests that the position and movement of the VSC determines the morphology of the fungal cell wall. The model both mimics observations made on living hyphae and predicts observations that were subsequently confirmed, making the hyphoid model and its VSC concept a very plausible hypothesis to explain hyphal morphogenesis (Bartnicki-García *et al.*, 1989).
2. The **steady state or 'soft spot' model** assumes that turgor pressure stretches the wall at the hyphal tip where it is still plastic and, in addition, that the synthesis vesicles fuse with the membrane only if they reach parts of the wall that are sufficiently new to be still plastic. If these conditions are met, new wall will be incorporated preferentially into the most recently synthesised wall and this cooperative insertion of newer wall into new wall is the steady state synthesis to which the name of the model refers.

   Turgor stretches the new plastic wall, which thins it but then it is re-thickened and restored by vesicular exocytosis of proteins and polysaccharides as wall precursors. The preferential targeting of these vesicles to the most recently synthesised wall means that, once established, the growing point will be maintained (steady state, again). Pre-softening to generate the plastic stretching/thinning in the first place occurs through endolytic cleavage (by chitinase?) though in this model such an enzyme activity is employed only briefly to initiate the growth mode but not to sustain it. Stretching of the wall and addition of new wall material from the cytoplasmic side occur maximally at the extreme tip. Newly added wall components are chitin and β(1–3)-glucan molecules. With time, these two polymers interact to form covalent linkages and to cross-link with proteins.

   At the extreme tip, the wall is minimally cross-linked and supposed to be most plastic. Subapically, wall added at the apex becomes stretched and partially cross-linked, while new wall material is added from the inside to maintain wall thickness. Wall material at the outside is always the oldest. Cross-linking increases progressively from the tip and as 'wall hardening' proceeds the wall hardly yields to turgor pressure and stretching, and synthetic activity declines. If the tip ceases to extend for any reason (e.g. change in turgor) the steady state breaks down, newer wall is not added to new wall and cross-linking between the wall polymers spreads into the apical dome and over the whole apex. From this stopped state, a fresh round of endolytic cleavage would be necessary to restart tip extension (Wessels, 1993).
3. We said there are two models, but we must join in with all this modelling and suggest our own **consensus model of tip extension** (see Chapter 2, Sections 2.2.4 and 2.2.5, in Moore, 1998, for a full discussion). In this, we associate chitin synthase activity with stretch receptors, involve membrane architectural proteins, and recognise the contribution of vesicle gradients, combined with other components of the two prime models. Our consensus model has the following features:
   - Noncovalent interactions occur between mannoproteins and other wall components.
   - The initial network is consolidated by formation of covalent cross-linkages among the wall polymers.
   - Cytoplasmic vesicles and vacuoles are assumed to be crucial to the extension of hyphal apices and to be responsible for delivering the enzymes and substrates needed for wall construction.
   - The actin cytoskeleton is assumed to be involved directly in hyphal tip extension in a number of ways (in animal cells **focal contacts** are specialised membrane domains where bundles of actin filaments extend from the intracellular cytoskeleton ***through*** the membrane and anchor the cell to its substratum).

This consensus model of tip extension assumes that similar structures to animal focal contacts are connected to the fungal wall. In the fungal cell, we anticipate that intracellular cytoskeletal elements could penetrate the membrane to anchor the wall, providing additional tensile strength while the membrane proteins modify the wall structure. These (integrin-mediated?) attachments to the extracellular wall matrix at the hyphal apex, which we will call '**wall contacts**', would permit the actin cytoskeleton to be directly linked, through the plasma membrane, to components of the wall. Such wall contacts might also be specific for different wall components or even fragments of wall components. Wall contacts are anticipated to function as:

- **Additional mechanical support to the wall** (and via spectrins to the membrane) to compensate for loss of wall strength due to enzymatic cleavage of wall components as new wall materials are incorporated.
- **Detectors of mechanical strain** which might then regulate enzyme activity or modulate the flow of vesicles to the local region.
- **Feedback detectors of the progress of wall synthesis**: in this respect, it should be noted that, again in animal systems, interaction between integrins and the extracellular matrix generates signals which cause phosphorylation of intracellular signal transduction pathways. This is known as 'outside-in signalling'. Such a mechanism at the fungal wall would enable the progress of wall synthesis to be reported to the synthesis machinery allowing it to modulate vesicle supply in quantity and/or kind.
- **Vesicle traffic directors and regulators**: close to the membrane the actin filaments will be directing vesicles with great specificity, in response to the outside-in signalling information; more distant filaments will be marshalling and collecting vesicles into the supply pathways which result in the required overall vesicle fusion gradient. In most, if not all, cases vesicle supply will be channelled through a vesicle supply centre/Spitzenkörper structure.

Evident in all these models is the proposition that hyphal extension is supported by the continuous forward flow of constructional materials generated within the cytoplasm behind the tip. Much of this flow is in the form of cargo within vesicles derived from the endoplasmic reticulum and processed through Golgi dictyosomes before being transported towards the extending apex.

The vesicles are readily seen in longitudinal sections of fixed hyphae by electron microscopy and consist of two main size classes, the smallest ranging from 20 to 80 nm and the largest from 80 to 200 nm in diameter. The smaller group is readily purified from hyphae and includes the **chitosomes** that contain chitin synthase. When incubated in UDP and magnesium ions, chitosomes extracted from true fungi produce chitin microfibrils *in vitro* that are identical to those produced *in vivo*; so they are clearly essential to creation of the chitinous part of the wall.

Fusion of the vesicles with the membrane at the hyphal apex releases the biosynthetic machinery for wall assembly and also adds new membrane to the growing hypha; synthesis of hyphal wall and plasma membrane are therefore coordinated. Extracellular enzyme secretion into the environment to catalyse the degradation of complex polymers in the substratum, e.g. lignocellulose and polypeptide, is also maximal at the hyphal tip. Potentially, the highly polarised vesicular pathway not only supports rapid hyphal extension, but also acts as a transport mechanism for extracellular enzymes, integrating forward exploration of the substratum with immediate exploitation.

We have already mentioned the astonishing fact that 38,000 vesicles have to fuse with the apical membrane ***each minute*** to support hyphal extension in *Neurospora crassa* when it is growing at its maximum rate. This is a simple calculation based on the observed size (volume) of vesicles in the hypha, and the observed increase in apex volume per minute as the hypha grows.

And that's the point: the hypha grows. So obviously, the motors and cytoskeletal systems we have described above have no difficulty in supplying materials to the hyphal apex at this sort of rate. Clearly, when we say that materials 'are transported' along microfilament and microtubule tracks, you should understand that we mean that materials are ***speedily*** transported along those tracks. In addition, motile vacuolar systems have been observed in a range of different fungi, as extensive 'trains' of vacuoles (very much larger than the apical vesicles) quickly moving along hyphae and tip cells, which could easily supply the vesicular contents used for apical extension growth. There is an extremely lively forward flow of materials towards the hyphal apex.

Possibly related to organelle movements and the supply of vesicles to the tip is the 'pulsed growth' of hyphal tips of several different species of fungi recorded using video analysis and image enhancement. In all fungi tested, the hyphal elongation rate fluctuated continuously with more or less regular intervals of fast and slow growth, which was interpreted as fluctuations in the overall rate of secretory vesicle delivery or discharge at the hyphal apex being reflected in fluctuations of hyphal elongation rate. If this is the case, then it will influence our understanding of the mechanism of vesicle supply.

However, a problem with digital video recording is that the pixel (picture element) structure of the electronically observed image can impose pulsations upon smooth movements. As the edge of the moving object moves from one pixel to the next there is a defined time interval during which no observation is possible, yet the eye, and computer-aided image enhancements, can either compensate for this or amplify it depending on circumstances which have nothing to do with the moving object itself. This effect can readily generate pulsations where they do not exist. A simple test of the validity of pulsations in video measurements (which is to compare recordings made at different optical magnifications) does not seem to have been applied to these observations of fungal growth, but as stepwise changes in elongation rate of hyphae of the bacterium *Streptomyces* have been recorded using photographic methods not prone to this particular artefact, the phenomenon *could* happen generally in filamentous systems, so cyclical variations in vesicle delivery *may* cause pulsations in apex extension rate. Our doubts concern the observational techniques used to detect pulsed growth; this is not to say that we are doubtful of ***all*** suggestions of pulsed growth. Evidence is beginning to mount that actin assembly and exocytosis in some filamentous fungi are coordinated by pulsed influxes of Ca^{2+} resulting in stepwise cell extension (Takeshita *et al.*, 2017, and references therein).

In most, but not all, filamentous fungi the electron-dense body we have described as the Spitzenkörper is present in the cytoplasm just distal to the growing apex. This body is composed of a complex of vesicles, and controls both the direction of growth of the hypha and its rate of extension. Satellite Spitzenkörpers separate from the main body and migrate to the lateral cell wall just before new hyphal branches appear. Observations of this sort led to the concept of the Spitzenkörper as a VSC, which acts as the focus from which apical vesicles migrate towards the apical membrane and wall. In this model, vesicles are first delivered to the Spitzenkörper and then they are 'sprayed' forward equally in all directions. A computer model based on this concept generates a tube with a tapering tip (a 'hyphoid') very similar to the apex of a growing hypha.

The computer model also shows that the direction of growth can be altered by moving the VSC, mimicking the movements observed of the Spitzenkörper during changes in the direction of hyphal growth. The Spitzenkörper is therefore thought to play a critical role in controlling the mechanism of polarised growth of true fungi; not only hyphal extension rate and direction of growth, but also the generation of lateral branches. It seems that the Spitzenkörper is anchored in position behind the growing apex by cytoskeletal elements, most likely F-actin. F-actin is usually found located at the growing tip in the form of a fibrillar network radiating into the cytoplasm from the extreme apex as slender cables, although the majority of actin detectable in the light microscope by immunofluorescence is in the form of '**actin plaques**' (localised patches of actin) within the first 10 to 12 μm of the hyphal tip. In serial sectioned electron microscope images, these F-actin plaques/patches correspond in position to bodies called **filasomes**, which are generally spherical, 100–300 nm in diameter and consist of a single microvesicle (35–70 nm diameter) surrounded by fine filaments that contain actin. Filasomes are found adjacent to newly formed glucan fibrils in protoplasts that are regenerating their walls, so the overall interpretation is that a filasome is one of the F-actin plaque/patch structures appearing in the cytoplasm at sites where the cell wall is formed.

The Spitzenkörper also seems to be dependent on the microtubules. Experimental interference with dynein, dynactin or microtubules causes eccentric wall deposition and a meandering and more copiously branched mycelium, suggesting that the dynein/dynactin/microtubule system also positions and maybe stabilises the Spitzenkörper. Remember that this is the system thought to be responsible for nuclear migration towards the growing tip. Elongation of the hyphal tip generally occurs before nuclear migration and the simplest explanation for their tip-ward migration is that nuclei are attached to membrane-anchored dynein and dynactin through astral microtubules from their SPBs (Figure 5.11). This assumes that the dynein motors exert a pulling force on the microtubules that position the nuclei. Dependence of Spitzenkörper behaviour on the same system suggests close integration of the transport of all organelles, vacuoles, vesicles and cytoplasmic components that must migrate forward as the hyphal apex extends (Riquelme, 2013; Riquelme & Sánchez-León, 2014; Steinberg *et al.*, 2017).

Despite the evidence that the Spitzenkörper is a crucial component of tip extension in true fungi it is not the only way in which polarised filamentous extension can occur. A typical Spitzenkörper has not been observed in the hyphae of most zygomycete fungi, which have instead an **apical vesicle crescent** (AVC). This is a crescent-shaped band of loosely organised secretory vesicles gathered together near the hyphal apex. AVC vesicles are more variable in size and are generally larger on average than the vesicles observed in the Spitzenkörper (Fisher & Roberson, 2016).

Growth of hyphae of the fungus-like *Oomycota*, for example *Saprolegnia ferax*, proceeds without a Spitzenkörper; as does growth of other non-fungal polarised systems, like pollen tubes. However, all these filamentous systems, including pollen tubes, seem to share one feature, namely the ***presence*** of an apical gradient of Ca^{2+} at the apex, the highest concentration being immediately distal to the tip. The maintenance of such a gradient is thought to be due to the presence of Ca^{2+} ion channels in the apical membrane that allow Ca^{2+} to flow down a concentration gradient from the outside to the inside of the cell. The concentration of Ca^{2+} in eukaryotic cells is highly regulated and maintained at low levels in the cytoplasm in two ways:

- Ca^{2+}-pumping ATPases located in the plasma membrane pump calcium out of the cell.
- Ca^{2+}-pumping ATPases located in the vacuolar membranes promote sequestration and storage of the ion within hyphal vacuoles.

Although all the roles and functions of a Ca^{2+} gradient in the hypha have yet to be established, Ca^{2+} ions are known to regulate the assembly of the cytoskeleton and to aid vesicle fusion with membranes, so the presence of a '**tip-high calcium gradient**' is likely to be an important factor in establishing and maintaining apical polarity in the hypha.

5.15 HYPHAL FUSIONS AND MYCELIAL INTERCONNECTIONS

The hyphal tip we have just described is capable of forward extension for an indefinite period, or at least for as long as nutrients are available and environmental conditions remain amenable. But the hyphal tip has another characteristic capability: it can fuse with another hypha. **Hyphal fusion** (also called **anastomosis**) is a ubiquitous phenomenon in filamentous fungi, serving many important functions at different stages in the life cycle.

At the very earliest stages of the vegetative phase, when the spore is germinating, fusions can occur between the germ tubes of clustered spores as they germinate. Subsequently, in the interior of the maturing vegetative colony, numerous hyphal fusions occur to convert the radially oriented main hyphae of the mycelium into an interconnected network. This is essential for intra-hyphal communication within the colony, particularly translocation of water and nutrients, and maintenance of general homeostasis (Trinci *et al.*, 1994, 2001). Also, entry into the sexual cycle usually involves the fusions between hyphae of the two parents, and then maintenance of the dikaryotic state, the prelude to karyogamy, also requires anastomoses, but this time in specialised structures called **croziers** (in the **ascogenous hyphae** of *Ascomycota*) or **clamp connections** (associated with the **vegetative dikaryotic hyphae of *Basidiomycota***). Hyphal fusion is also involved in the formation of the multicellular tissues found in mushrooms and other sporing structures.

Vegetative and sexual compatibility systems regulate hyphal anastomosis. Self-recognition between hyphae of the same mycelium is important to enable enhanced colonisation of solid substrates and building of tissues and sporophores. Recognition of genetically different individuals is important for mating and, equally, recognition of nonself with the potential to inhibit anastomosis is important to protect against invasive pathogenic hyphae and hostile cytoplasmic elements like plasmids and viruses. We will discuss later the mechanisms that exert control over hyphal fusion by defining vegetative and sexual compatibility (see Chapters 8 and 9); here we will outline the cell biology of hyphal fusion, which has been a major research focus for several years (Figure 5.12; see Hickey *et al.*, 2002; Glass *et al.*, 2004; Wright *et al.*, 2007; Read *et al.*, 2009, 2010; Read & Roca, 2013; Raudaskoski, 2015; Fleißner & Serrano, 2016; Daskalov *et al.*, 2017).

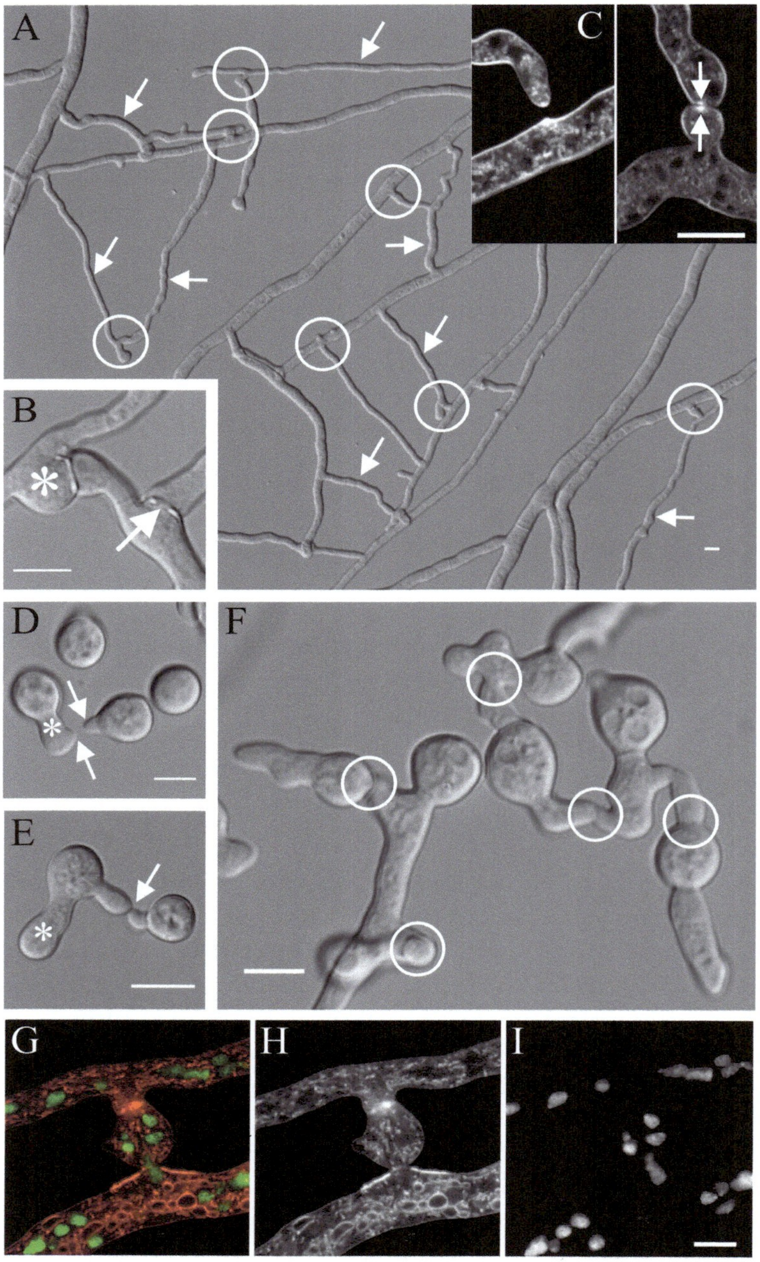

Figure 5.12. Hyphal fusions in *Neurospora crassa*. (A) Region about 1–2 cm behind the leading edge of a mature mycelium in which specialised fusion hyphae (some are arrowed) fuse with other hyphae to establish a colony network (some of the fusion connections are circled). (B) Detail of a hyphal fusion connection showing isotropic swelling upon contact (asterisk) and a fusion pore (arrow). (C) Confocal images of hyphal homing and fusion in *Neurospora crassa*: the image at left shows a hyphal tip growing towards a hyphal peg (pre-contact stage), with a brightly fluorescent Spitzenkörper evident in both the hyphal tip and emerging peg; the right-hand image shows the post-contact stage in which the adherent hyphal tips have swollen but their fluorescent Spitzenkörpers (arrowed) have persisted even though hyphal extension has ceased. Self-fusion in *Neurospora crassa* occurs during early as well as late stages of colony development. (D) after an initial phase of isotropic expansion, conidia of *Neurospora crassa* polarise leading to the outgrowth of germ tubes (asterisk) and conidial anastomosis tubes (called 'CATs') that can arise directly from conidia or from germ tubes (arrows); in this specimen the tip of the right-hand CAT seems to have induced the formation of the CAT from the germ tube. (E) CATs chemotropically attract (since these are conidia from the same mycelium this is an autotropism) and become attached to each other (arrow indicates site of contact); when they come into contact, tip growth is arrested, and a fusion pore is formed. Each germinating conidium can interact with several neighbours (F), thereby creating an interconnected network (CAT connections are circled in F) of newly germinated conidia (germlings). (G–I) Confocal images showing fused hyphal branches of *Neurospora crassa* through which nuclei are flowing stained with specific fluorescent probes: G is the combined image (colour-coded). (H) shows fluorescence of membrane-specific labelling alone. (I) shows only the fluorescence of nuclei specifically labelled with histone-specific staining. Scale bars, 10 μm. Modified from Hickey *et al.* (2002); Glass *et al.* (2004); Read *et al.* (2009), using graphic files kindly provided by Professor N. D. Read. Images reproduced with permission of Elsevier.

Three types of pre-contact behaviour have been described:

- a hyphal tip induces a branch and they subsequently fuse (shown in Figure 5.12B);
- two hyphal tips approach each other and subsequently fuse;
- a hyphal tip approaches the side of another existing hypha and there is a tip-to-side fusion (shown in Figure 5.12C).

The Spitzenkörper is closely involved in the process of hyphal fusion, which have been characterised into nine stages (Glass *et al.*, 2004) as follows:

- **Stage 1**: a hyphal tip competent to initiate fusion secretes an unknown diffusible, extracellular signal which induces Spitzenkörper formation in the hypha it is approaching. It is presumed that the second hypha secretes a corresponding signal.
- **Stages 2** and **3**: these events result in two fusion-competent hyphal tips each secreting diffusible, extracellular chemotropic signals that regulate Spitzenkörper behaviour so that the hyphal tips grow towards each other (Figure 5.13).
- **Stage 4**: cell walls of the approaching hyphal tips make contact, apical extension ceases, but both Spitzenkörpers persist.
- **Stage 5**: adhesive material is secreted at the hyphal tips.
- **Stage 6**: polarised apical extension is converted to 'all-over' isotropic growth, which results in swelling of the adherent hyphal tips.
- **Stage 7**: cell walls and adhesive material at the point of contact are dissolved, bringing the two plasma membranes into contact.
- **Stage 8**: plasma membranes of the two hyphal tips fuse, creating a pore with which the Spitzenkörper stays associated as the pore begins to widen and cytoplasm starts to flow between the now connected hyphae.
- **Stage 9**: the pore widens, Spitzenkörper disappears, organelles, including nuclei, vacuoles and mitochondria, can flow between the fused hyphae, though the flow may be regulated.

The principal lesson to take away from this brief description of hyphal fusion can be expressed as: what the Spitzenkörper has put together, the Spitzenkörper can take apart.

For anastomosis to occur between two hyphal tips, two separate cell walls must be disassembled, and two separate plasma membranes stitched together around a newly formed hypha-wide pore, and all of this must be done without breaching the integrity of walls and membranes that are under osmotic hydraulic stress. Add the need for signalling and targeting systems to ensure that fusions occur at the right place and right time, and you can see that the phrase 'hyphal fusion' covers an extensive range of cell biological processes that are under delicate control.

It will be some time before this entire process is understood, but a model of **self-signalling** between hyphal tips of the same mycelium is emerging (Read *et al.*, 2009, 2010). The model (summarised in Figure 5.15) imagines that approaching tips alternately signal each other and after receiving a signal a tip grows along the gradient of the signalling molecule (autotropism). In the first half of the signal exchange, one of two approaching tips releases the chemoattractant signal while the other responds to the signal by adjusting its growth direction towards the signaller. In the second half of the conversation, the roles are reversed and the second tip releases signal to give the first one a target to adjust its growth direction. The exchange will be repeated until the two tips come into contact.

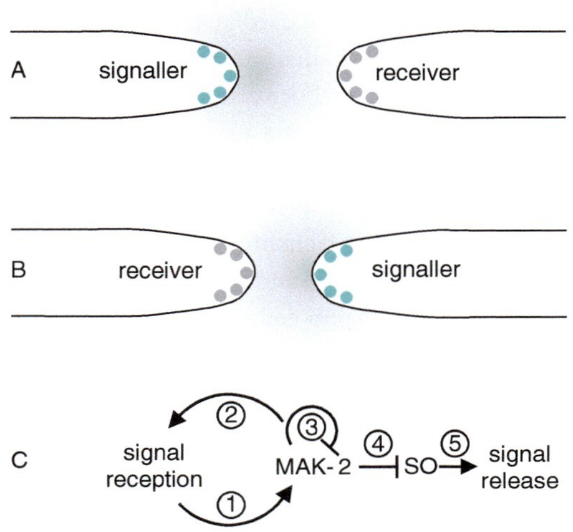

Figure 5.13. A model of self-signalling between hyphal tips of the same mycelium. The model imagines that approaching tips periodically and alternately signal each other (called a ping-pong mechanism) and after receiving a signal a tip grows along the gradient of the signalling molecule (autotropism). (A) shows the first half of the signal exchange; the tip on the left releases the chemoattractant signal while the one on the right responds to the signal by adjusting its growth direction towards the signaller. In (B), the second half of the conversation, the roles are reversed and the right-hand tip releases signal so that the tip on the left can adjust its growth direction. The exchange will be repeated until the two tips come into contact. (C) puts some detail on this proposed mechanism of signalling: (1) a chemoattractant-receptor complex induces local recruitment and activation of a mitogen-activated protein (MAP) kinase pathway (here designated MAK-2); (2) positive feedback inherent in MAP kinase amplifies the received signal; (3) as its local concentration increases, one of the MAK-2 proteins downregulates the MAP kinase protein complex; (4) decrease in MAK-2 derepresses the protein SO (specified by a *Neurospora crassa* gene locus called soft, which encodes a protein containing a domain involved in mediating protein-protein interactions via proline-rich regions). The soft gene is required for vegetative hyphal fusion. Accumulation of SO leads to the formation of SO-containing protein complexes at the tip; (5) a burst of SO-stimulated chemoattractant release occurs. Modified and redrawn from Read *et al.* (2009).

The proposed mechanism of signalling, called a **ping-pong mechanism** (Figure 5.13C), involves periodic and alternate signalling between two approaching hyphal tips. An elaborate signalling network has been identified that controls communication, attraction and merger of fusing hyphae. The network contains many well-conserved elements, including MAP kinases, systems that generate reactive oxygen, Ca^{2+}-binding regulators, cell-polarity factors and proteins homologous to the mammalian ***striatin-interacting phosphatase and kinase*** (**STRIPAK**) complex. The mammalian striatin family consists of three proteins that have no intrinsic catalytic activity but serve as scaffolding proteins, which organise many varied, and large, signalling complexes that contribute to a variety of cellular processes. In mammals they are involved in cell signalling, cell cycle control, apoptosis, vesicular trafficking, Golgi assembly, cell polarity, cell migration, neural and vascular development, and cardiac function. They have been connected to clinical conditions, including cardiac disease, diabetes, autism and cerebral malformations (Hwang & Pallas, 2014). It is no surprise that a system that plays so many crucial roles in a sister clade of organisms also has many essential roles in fungi. STRIPAK homologues are involved in cell cycle arrest and mating in budding yeast, signalling transition from mitosis to cytokinesis in *Schizosaccharomyces pombe*, and the cell cycle, septal dissolution, nuclear migration, clamp cell development and fusion, and hyphal fusions in filamentous ascomycetes and basidiomycetes (Raudaskoski, 2015; Fleißner & Serrano, 2016).

5.16 CYTOKINESIS AND SEPTATION

The hyphal growth form of filamentous fungi is an adaptation to the active colonisation of solid substrata. By hyphal extension and regular branching, the fungal mycelium can increase in size without disturbing the cell volume:surface area ratio so that metabolite and end-product exchange with the environment can involve translocation over very short distances. Growth of the mycelium is regulated, of course, and three mechanisms involved in regulating the growth pattern of undifferentiated mycelia are regulation of:

- hyphal polarity;
- branch initiation;
- the spatial distribution of hyphae.

Fungal hyphae are variable between species, but the hyphal filament, when separated into compartments by cross-walls, has an apical compartment which is perhaps up to 10 times the length of the intercalary compartments.

The septa which divide hyphae into cells may be complete (imperforate), penetrated by cytoplasmic strands, or perforated by a large central pore. The pore may be open (and offer little hindrance to the passage of cytoplasmic organelles and nuclei) or may be protected by a complex cap structure derived from the endoplasmic reticulum (the dolipore septum of *Basidiomycota*). In *Ascomycota*, the pore may be associated with, perhaps plugged by, Woronin bodies, which are modified peroxisomes. Septal form may be modified by the hyphal cells on either side of the septum, and may vary according to age, position in the mycelium or position in the tissues of a differentiated structure. These features make it clear that the movement or migration of cytoplasmic components between neighbouring compartments is under very effective control. The hypha is divided up by the septa, and the cellular structure of the hypha extends, at least, to its being separated into compartments whose interactions are carefully regulated and which can exhibit contrasting patterns of differentiation.

Septation in fungi involves chitin deposition in a ring defined by a preformed ring of actin microfilaments. Early research showed that septation in the main hypha is in some way defined by the position of the dividing nucleus, though neither septation nor branch formation are universally dependent on orientation of the nuclear division spindle. There is evidence for cytoskeleton-related functions being shared by karyokinesis and cytokinesis and these may form the basis of a structural memory which allows septa to be laid down in positions defined by a nuclear division which occurred sometime before. Nuclear migration features in most division processes in fungi, though the coupling between karyokinesis and cytokinesis varies from a rather loose association in hyphae, where multinucleate cells are formed to a strict coupling in yeasts and during spore formation when uninucleate spores are produced.

Observations from the 1970s to the present day echo the ancient phylogenetic relationships by emphasising the remarkable similarity between cell cycle events in animals and those in fungi. Primary septa in fungal hyphae are formed by a constriction process in which a belt of microfilaments around the hyphal periphery interacts with microvesicles and other membranous cell organelles. This implies some level of correspondence between **fungal septation** and **animal cell cleavage (cytokinesis)**. Indeed, genetic analyses of fungal (yeast), animal and plant cytokinesis have led to the realisation that fundamental mechanisms of cytokinesis may be highly conserved among all eukaryotes (Pollard, 2010). Evidently cytokinesis by septum formation (in hyphae or yeast cells) is quite distinct from the free cell formation we discussed before (see section 3.3).

The fundamental function of cytokinesis is to divide both the cell surface ***and*** the cytoplasm of one cell to form two cells. The machinery needed for this is assembled towards the end of mitosis at, or close to, the site of nuclear division. Despite the presence of a rigid cell wall, cytokinesis in *Aspergillus* and both budding yeast and fission yeast occurs by a constriction mechanism resembling that of animal cells. In fungi, as in animals, contraction of an actomyosin-based ring of fibres produces a 'cleavage furrow' in the outer membrane that squeezes into the cell (actomyosin is a combination of actin and myosin that, with other substances, makes up the contractile muscle fibres of animals). Microtubules also contribute to the cleavage process in all eukaryotes; in particular, microtubules signal the position of the division plane in animal cells, and elements of this, at least, are conserved in plant cell division even though a constriction process is not evident in plants. This theme of a combination of both similarities and differences extends to the assembly of the contractile ring in yeast and animal cells (Pollard, 2010).

In animal cells, actin and myosin accumulate to the furrow region (and it starts to constrict) during late anaphase. In the fission yeast, *Schizosaccharomyces pombe*, the actomyosin ring assembles during early mitosis, but cytokinesis does not occur until late anaphase (Gu & Oliferenko, 2015). In budding yeast, *Saccharomyces cerevisiae*, the actomyosin ring is assembled during two separate stages: myosin first forms a ring at the site of budding at the start of DNA synthesis and then F-actin is incorporated into the ring during late anaphase just prior to contraction.

The **cleavage furrow** penetrates the whole animal cell to divide it into two daughters. In fungi the contraction is coupled to mechanisms for the synthesis of a cell wall so that the membrane furrow surrounds a septum that grows out from the existing wall, penetrating the cytoplasm. In yeasts the completed septum separates the bud from its parent (in budding yeast), or two daughters each made up of half the parent cell (fission yeast); while in filamentous fungi, septal growth (and cleavage furrow contraction) cease with the septum incomplete, leaving a central pore that may be elaborated in different ways in different fungal groups.

In addition to actin and myosin, cytokinesis requires other proteins, many of which are evolutionarily conserved (Pollard, 2010). Of special importance are proteins called **septins**, a family of GTPases that form filaments in fungi and animals, and are required for assembly of the myosin ring in budding yeast. Septins that localise to the bud neck of *Saccharomyces cerevisiae* localise to the cleavage furrows in the fruit fly *Drosophila* and in mammalian cells, so they are obviously conserved in the two kingdoms. In addition to their role in cell division, septins coordinate nuclear division, membrane trafficking and cytoskeletal organisation (Lindsey & Momany, 2006); although the *Coprinopsis cinerea* septin Cc.Cdc3 localises to the apices of vegetative hyphae, it is essential for elongation of stipe cells where it assembles into abundant thin filaments that are thought to provide localised rigidity to the plasma membrane of these specialised cells of the sporophore (Shioya *et al.*, 2013). Mutations in budding yeast septin genes result in the loss of several proteins from the bud neck, suggesting that septins might have a structural function here, focusing other components on the cleavage site. Septins can also bind and hydrolyse GTP, so they may also have a role in signalling. **IQGAP proteins** are another conserved family involved in cytokinesis. IQGAP family members have a structure that includes a domain which binds actin filaments, 'IQ repeats' which bind calmodulin, and binding sites for small GTPases, making them candidates as proteins that link Ca^{2+} signalling to cytokinesis.

Another protein family that is required for eukaryote cytokinesis is the family of **formin-homology (FH) proteins**. In fission yeast, an FH protein is a component of the actomyosin ring that is required for recruitment of F-actin to the ring. This protein binds **profilin**, a protein involved in polymerising actin, so it may act as an initial focus for actin ring formation. A requirement for an FH protein in cytokinesis has been shown in *Aspergillus*; mutants of the gene (called *SepA*) lack septa and the protein is required for actin ring assembly in *Aspergillus* septa. The **SPB** is involved in signalling cytokinesis to the actomyosin ring. A change in SPB biochemistry occurs at the end of spindle elongation, releasing a signal that activates actomyosin ring maturation (F-actin recruitment) and/or contraction, providing a positive signal for cytokinesis initiation.

Microtubule arrays from the mitotic division spindle have functions in cytokinesis (Pollard, 2010). In animal cells, which have a close connection between the position of the division spindle and the plane of cytokinesis, signals must pass from the mitotic spindle to the outer membrane of the cell to induce furrowing of the membrane to initiate the cleavage furrow. First, at the end of mitosis, the plasma membrane region becomes able to furrow. Then, signals from the mitotic apparatus regulate the contractility of the actomyosin ring system and start assembly of a persistent furrow that eventually bisects the spindle. Specialised subsets of spindle microtubules are important for cleavage furrow positioning, including **astral microtubules**, which radiate from the spindle poles, and microtubule bundles that assemble in the mid-region of the anaphase spindle and are called **interzonal microtubules**. Two sets of astral microtubules are sufficient to induce furrow formation but there is a continuing requirement for interzonal microtubules for the furrow to deepen (this is called '**ingression**') and bisect the spindle.

A very similar role for microtubules during cytokinesis in plants is also well established, although plant cytokinesis is achieved by formation of a cell plate that divides the cell into two rather than a cleavage furrow. Nevertheless, vesicle fusion to assemble the cell plate is managed by an array of interzonal microtubules (and actin filaments) called the **phragmoplast**. Both animal and plant cells reorganise their interzonal microtubules to generate microtubular arrays (called stem bodies in older publications) that perform a specialised function during cytokinesis. It is possible that these microtubules are required for vesicle accumulation and membrane deposition during cytokinesis. Remember that the surface area of the cell needs to increase as it divides, and additional membrane must be recruited into the plasma membrane. At least some of this recruitment derives from insertion of vesicles into the cleavage furrow. **Syntaxins** are t-SNAREs (see section 5.9) that are required on the target membrane to promote fusion of the vesicle with the target membrane during exocytosis and some syntaxin-encoding genes have been implicated in cytokinesis. A careful balance in membrane dynamics is required to ensure proper cell-surface remodelling.

Correct positioning of the cleavage plane is important to ensure that each daughter cell receives a nucleus, so the most economical position is to bisect the axis of chromosome segregation.

This fundamental requirement is achieved either by the cleavage plane being specified to a pre-selected site prior to mitosis by a mechanism independent of the spindle, but to which the spindle or cleavage structure is adjusted, or by the division plane being specified by the location of the spindle (as in animal cells). In budding yeast and higher plant cells, the division site is predetermined by a landmark early in the cell cycle (the bud scar in yeast); in fission yeast, the site of cell division correlates with the position of the premitotic nucleus. Sensing cell wall stress is a strategy that enables septation and cell division to be maintained in fungi, even under adverse environmental conditions (Walker *et al.*, 2013).

For the cleavage plane to be specified prior to mitosis and independently of spindle location, we must introduce the idea of some sort of marker being established prior to mitosis that acts as a landmark to dictate the position of both the division and cleavage plane. Budding yeast, *Saccharomyces cerevisiae*, employs such a mechanism; the division site, which will be the bud neck between mother cell and bud, is selected early in the cell cycle prior to spindle assembly. During mitosis, the spindle is positioned in the bud neck by a cytoplasmic dynein-dependent mechanism involving a dynamic interaction of astral microtubules with the bud cytoplasm and membrane (Figures 5.9 and 5.10). The position of the bud site is selected according to several 'landmark' proteins associated with a bud scar left at the previous bud site.

In *Schizosaccharomyces pombe*, a medial ring containing actin, myosin II and other components assembles in the centre of the cell early in mitosis, its location correlating with the location of the premitotic nucleus, which may be positioned in the cell by the SPB. Plants also landmark the location of the division plane prior to mitosis, but here it is the phragmoplast that is directed to the landmark, rather than the metaphase spindle. The site where the new cell plate will fuse with the cell wall is marked by the pre-prophase band (PPB), a temporary array of microtubules whose position is dictated by the location of the pre-division nucleus (Davì & Minc, 2015). During cell division in animal cells, it is likely that spindles are positioned by the concerted action of the forces acting on microtubule arrays; then the cleavage furrow is positioned based on the position of the mitotic spindle.

Filamentous fungi form multinucleate hyphae that are eventually partitioned by septa into multicellular hyphae. In *Aspergillus nidulans*, septum formation follows the completion of mitosis and requires the assembly of a septal band (Figure 5.14). This band is a dynamic structure composed of actin, septin and formin. Assembly depends on a conserved protein kinase cascade that, in yeast, regulates mitotic exit and septation (Takeshita *et al.*, 2014).

However, the crucial feature of filamentous fungi is that their hyphae are multinucleate. Even after the formation of septa, and even in the most regularly septate fungi, many parts of the hypha are multinucleate. The fact that individual hyphae or hyphal compartments contain several nuclei means that different patterns of nuclear division have been observed in the multinucleated fungal cells. There are three commonly observed patterns: synchronous, parasynchronous and asynchronous, and these vary between organisms and vary with environmental conditions (Figure 5.15). In **synchronous division**, all nuclei divide simultaneously. In **parasynchronous division**, mitosis is initiated in one location and then a wave of mitotic divisions travels down the hypha (starting near the apex and proceeding distally) so that nuclei divide in sequence. In **asynchronous division**, nuclei divide independently of their neighbours, with the result that the spatial and temporal pattern of mitosis in the hypha is randomised.

Nuclei of *Neurospora crassa* and *Ashbya gossypii* (both members of the *Ascomycota*) behave independently of one another so these fungi exhibit an asynchronous division pattern. This type of nuclear division pattern maintains the control of mitosis within a restricted volume of a shared cytoplasm, which potentially means that the hypha is better able to make local responses to nutrients or other environmental stimuli. Asynchronous divisions also allow the hypha to 'spread the cost' over a period of time (in terms of the energy and resources required for nuclear replication and division). Additionally, asynchrony might also shield the hypha from the dramatic change in the nucleus:cytoplasm ratio that would result if all nuclei divided at the same time (Gladfelter, 2006).

In contrast, synchronous or parasynchronous patterns of nuclear division produce rapid doubling of the number of nuclei in the hypha. Synchronous divisions occur in the apical compartments of vegetative hyphae of the oak wilt parasite, *Ceratocystis fagacearum*. In *Aspergillus nidulans*, a parasynchronous wave of nuclear division spreads, in a period of about 20 minutes, from the apex of the leading hyphal compartment. These waves can involve 60 to 100 nuclei and extend over a length of up to 700 μm of hypha. This wave is unidirectional in *Aspergillus nidulans*, moving away from the hyphal tip, but in *Fusarium oxysporum* a wave of nuclear division starts at the midpoint of an intercalary (i.e. central) compartment rather than the apical cell, and then the wave progresses in both directions in the cell. Parasynchrony breaks down when it meets a septum. Parasynchrony is also lost when the organism encounters nutritionally poor media, in which circumstances the nuclei divide asynchronously (Gladfelter, 2006).

Coordinated nuclear division cycles of this sort obviously imply an additional long-distance level of control and regulation. They provide the hypha with a unified response to environmental stimuli and might also be important for development or morphogenesis. The different nuclear synchronisation patterns observed in multinucleated fungal hyphae leave open the question of how the basic cell cycle control mechanisms are adapted in filamentous fungi. There is evidence that cell cycle control proteins are shared among nuclei in a hypha. Equally, there is evidence that some key cell cycle factors are unequally distributed between nuclei in a hypha. In *Neurospora crassa*, there are some suppressor mutants that suppress mutations in their own nucleus but do not act on neighbouring nuclei sharing the cytoplasm of a heterokaryon. Also intriguing are experiments with *Basidiomycota* in which the two nuclei of a dikaryon

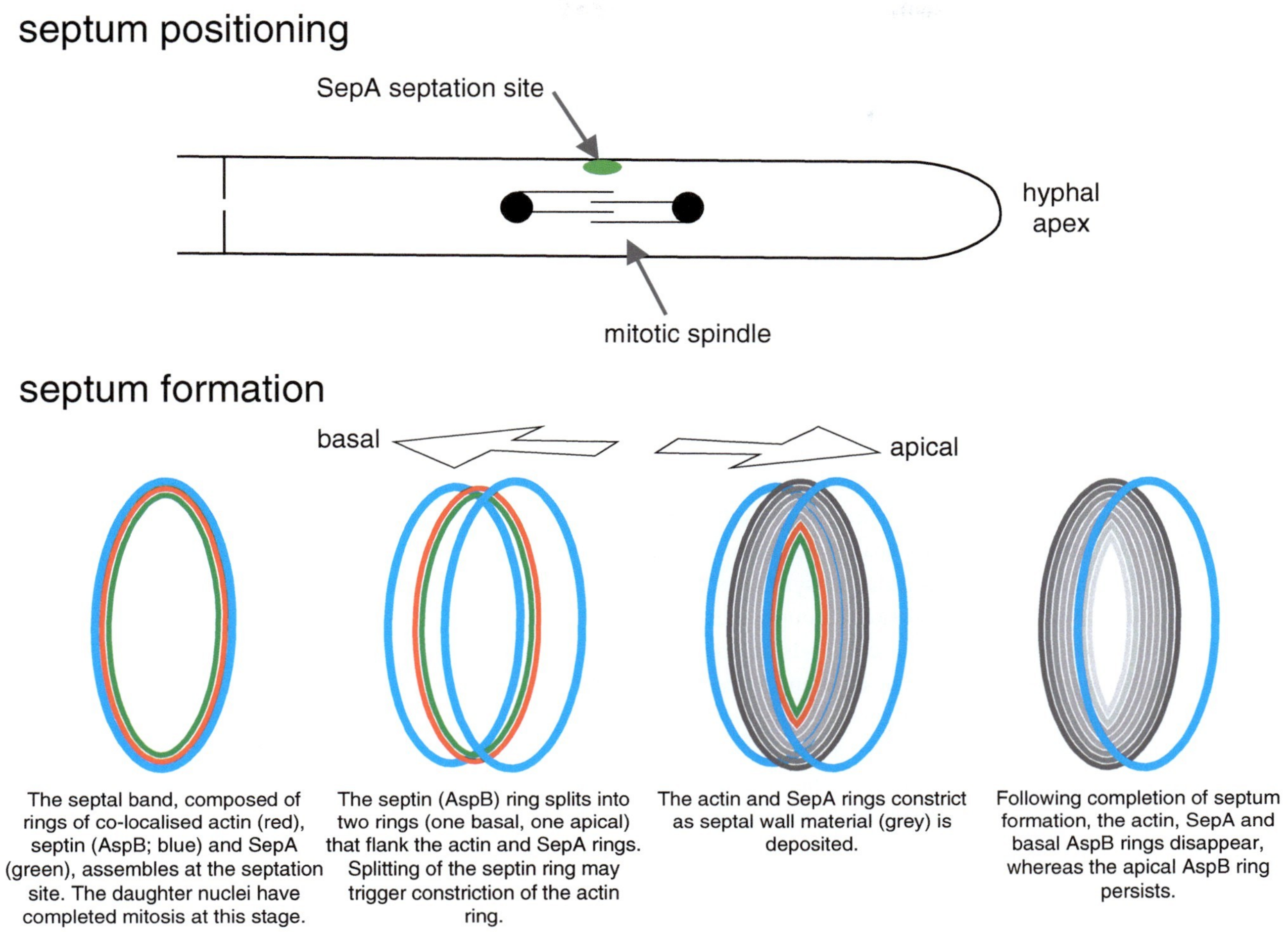

septum positioning
SepA septation site
hyphal apex
mitotic spindle
septum formation
basal
apical
The septal band, composed of rings of co-localised actin (red), septin (AspB; blue) and SepA (green), assembles at the septation site. The daughter nuclei have completed mitosis at this stage.
The septin (AspB) ring splits into two rings (one basal, one apical) that flank the actin and SepA rings. Splitting of the septin ring may trigger constriction of the actin ring.
The actin and SepA rings constrict as septal wall material (grey) is deposited.
Following completion of septum formation, the actin, SepA and basal AspB rings disappear, whereas the apical AspB ring persists.

Figure 5.14. Schematic model depicting the positioning and assembly of the septal band and the septum. A hyphal segment is drawn at the top, oriented with the tip to the right. In response to signals emanating from the mitotic spindle, SepA localises to the septation site as a cortical dot (green); patches of actin and/or the septin AspB may co-localise with SepA. This locates the site at which the circumferential septal band assembles, composed of co-localised rings of actin (red), SepA (green) and septin (AspB; blue). By this time the daughter nuclei have completed mitosis. Subsequently the septal band assembles the septum. The AspB ring splits into two rings that flank the actin and SepA rings (one on the basal side, one on the apical side). Splitting of the septin (AspB) ring may trigger constriction of the actin ring. As the actin and SepA rings constrict septal wall material (grey in the diagram) is deposited, vectorially synthesised through the plasma membrane (remember, the septum is extracellular; it's on the outside of the plasma membrane). When assembly of the septum is completed, the actin, SepA and basal AspB rings disappear, leaving the apical AspB ring to persist. Based on a diagram in Harris (2001).

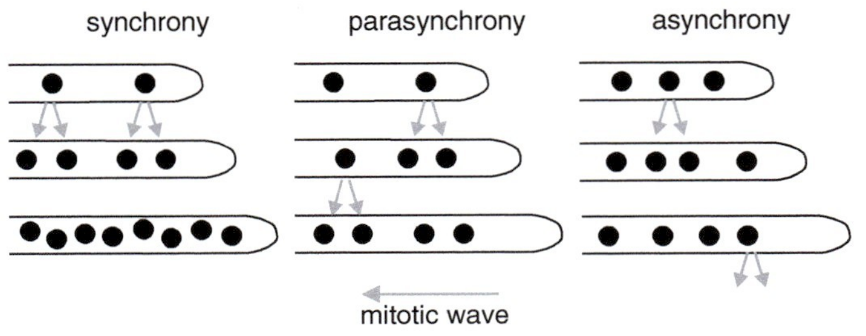

synchrony
parasynchrony
asynchrony
mitotic wave

Figure 5.15. Patterns of nuclear division in multinucleated fungi. In synchronous division, all nuclei divide simultaneously. In parasynchronous division, mitosis is initiated in one location and then a wave of mitoses travels backward down the hypha so that nuclei divide sequentially, almost as dominos fall in a line. In asynchronous division, nuclei divide independently of neighbours, so the spatial and temporal pattern of mitosis is randomised. From Gladfelter (2006).

are held in an F-actin cage and, apparently, are held a particular distance apart. For example, in *Schizophyllum commune*, cells in which the two nuclei are 2 μm apart have different patterns of gene expression than cells where the two nuclei are positioned 8 μm apart, though still residing in the same cytoplasm.

Essentially, it comes down to understanding how large cells spatially organise signalling pathways (Gladfelter, 2006). Another aspect is the control and maintenance of **cell polarity**. Even fission yeast cells have 'ends' and a 'middle' of the cell, and this sort of axial arrangement depends on the architecture of the cytoskeleton: during interphase in *Schizosaccharomyces pombe* microtubules position the nucleus at the middle of the cell and orientate microtubule 'plus' ends towards the ends of the cell (Davì & Minc, 2015). This architecture must be reproduced during cytokinesis. Filamentous fungi maintain several (indeed, numerous) axes of polarised growth simultaneously and take the regulation of genes involved in the cell cycle and in polarity maintenance to a different level that we still do not understand (Takeshita *et al.*, 2014; Davì & Minc, 2015; Osés-Ruiz *et al.*, 2016).

Successful cell division requires that **cell cycle events are coordinated in time.** In particular, cytokinesis must not occur before the mitotic spindle has done its job and the chromosomes have segregated. The synthesis of DNA and the act of cell division are discontinuous processes; a series of 'events' that have to occur in a specific sequence and the replication of DNA and segregation of chromosomes to the two daughter nuclei are the two key discontinuous processes for cell survival. To put it simply, if either process is performed inaccurately, the daughter cells will be different from each other, and will almost certainly have lost or gained essential genes or even whole chromosomes. Further, to maintain a constant size during cell proliferation, the growth rate must match the rate of division. Factors that govern proliferation must therefore regulate both the cellular biosynthesis that drives build-up of cell mass, and progression through the cell division cycle. Several mechanisms couple cell division to growth and different mechanisms may be used at different times during development to coordinate growth, cell division and patterning in multicellular organisms so that cells keep a fairly constant size, and developing tissues and organs achieve the proper size and cell density. The potential relationships between cell division and growth are:

- cell division drives growth;
- growth drives cell division;
- growth and cell division rates are controlled in parallel by a common regulator;
- independent regulation of cell division and growth.

They conclude that while cell cycle progression appears to be unable to drive growth, in some cases inhibition of cell division can lead indirectly to growth suppression (Boye & Nordström, 2003).

The **cell cycle**, as it occurs in rapidly growing and dividing eukaryotic cells, is usually described as a series of phases (Figure 5.16): interphase and mitosis (**M**), with interphase subdivided into the times when DNA synthesis is going on, known as **S-phase** (for synthesis phase) and the **gaps** that separate S-phase from mitosis. G1 is the gap after mitosis before DNA synthesis starts, and G2 is the gap after DNA synthesis is complete, but before mitosis and cell division (cytokinesis). Other descriptors/subdivisions may be used to describe cells that temporarily interrupt their cell cycle; for example, some cells pause after mitosis in a state that is often called G0; they only enter S-phase when they receive signals from outside (for animal cells the signals are usually peptide growth factors). Cells can also pause in G2 phase, in which case the continuation signal causes entry into mitosis (or meiosis in a meiocyte).

The fission yeast, *Schizosaccharomyces pombe*, normally passes through S-phase immediately after mitosis without a significant lag and spends most of its cell cycle between the end of S-phase and the next mitosis while the cells grow to the size required for division. The crucial steps in the cell cycle are those that ensure that one process is complete before the next process starts. These are known as **checkpoints**; safe places to pause the cell cycle for the time needed to check some condition. Different organisms can generate temporary arrest points at different places in the cell cycle while using similar signals to announce continuation. In general, it seems to be possible to construct 'plug-in modules' that can detect almost anything that might be relevant to a particular cell at a particular time, and couple this to the cell cycle control machinery. These modules must contain a **sensor** and an **effector**, and often lying between these is a **signal transduction pathway** that depends on protein kinases phosphorylating regulatory proteins. Checkpoints connect the cell division and chromosome replication cycles so that neither is allowed to run free, and the control circuitry is arranged to produce a strict alternation that prevents cells from returning to mitosis until they have undergone S-phase.

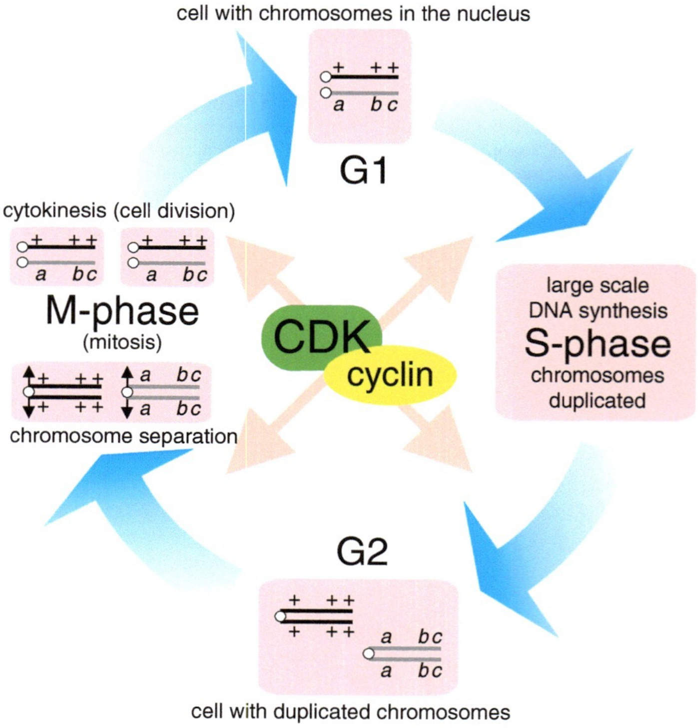

Figure 5.16. The different phases of the cell cycle. In the first phase (G1) the cell grows. When it has reached a certain size, it enters the phase of DNA synthesis (S) where the chromosomes are duplicated. During the next phase (G2) the cell prepares itself for division. During mitosis (M) the chromosomes are separated and segregated to the daughter nuclei, and karyokinesis is followed by cytokinesis. This completes the cell cycle and the daughter cells are then back in G1. This diagram depicts a (heterozygous) diploid but applies equally to the mitotic cycle of haploid cells. Redrawn and modified from a diagram on https://nobelprize.org/nobel_prizes/medicine/laureates/2001/press.html.

The most revealing contribution to understanding the control of the cell cycle came from genetic studies on yeasts, as we have already mentioned (see section 5.3). While isolating **temperature-sensitive mutants** for DNA, RNA or protein synthesis, it became apparent that some mutants were blocked at specific points in the cell cycle. If a population of such a mutant was transferred from 23°C (permissive temperature) to 36°C (nonpermissive or restrictive temperature), the cells stopped growing, but all showed the same morphology; all might fail to produce a bud, for example, or all might produce a bud, but remain attached because of inability to complete cytokinesis. A particularly important one of these temperature-sensitive mutants was called cdc28 (cell division cycle mutant number 28). This fails to form a bud at the nonpermissive temperature, but has no defect in DNA, RNA or protein synthesis; rather, cdc28 cells continue to grow when transferred to the nonpermissive temperature and can complete an already started S-phase. The important point is that they can never start a new one. A functional *CDC28* gene is required for the cell cycle transition that was called '**start**' (which is the same as what is called the '**restriction point**' in higher eukaryotes). Budding yeast halts the cell cycle at 'start' in response to several conditions; including nutrient starvation or detection of the peptide pheromone announcing proximity of a potential mating partner.

Many other cdc mutants have been isolated. In the fission yeast, *Schizosaccharomyces pombe*, a key observation was the recognition of *cdc2* as a particularly important regulator of mitosis and a gene essential for cell cycle progression. The first hints of an underlying unity and order to cell cycle control came when *Saccharomyces cerevisiae CDC28* and *Schizosaccharomyces pombe cdc2* were cloned and sequenced, and found to be homologous and, indeed, interchangeable between these two very distantly related yeasts. The proteins specified by these genes belong to the protein kinase family of enzymes, showing that protein phosphorylation, which is one of the most common ways of altering the activity of enzymes and other proteins generally in eukaryotes, is involved in regulating the cell cycle. Function of the cell cycle-specific protein kinase is activated by a partner protein called a **cyclin** (because their concentration varies cyclically during the cell cycle). Cyclins are produced or degraded as needed to drive the cell through the different stages of the cell cycle. The cyclin forms a complex with its partner cyclin-dependent kinase (cdk) (Figure 5.16), which is 'turned on' and phosphorylates the appropriate proteins needed for the cell cycle.

The budding yeast cell cycle can be summarised as follows, and the same general principles probably apply to the cell cycles of all eukaryotes (Pollard, 2010):

- At the end of mitosis, mitotic cyclins are degraded by a protease that stays active until 'start'. This prevents the cell from immediately restarting mitosis.
- At 'start', transcription and translation of the G1 cyclins CLN1 and CLN2 is activated, and these proteins combine with Cdc28 to form a protein kinase that initiates formation of the bud and prepares the cell for S-phase (G1 to S transition). This latter includes activation of the genes responsible for cyclins CLB5 and CLB6 that replace CLN1 and CLN2 as partners of Cdc28 to promote DNA synthesis during S-phase.
- Towards the end of S-phase, synthesis of the 'late' cyclins CLB1 and CLB2 starts up, which turns off transcription of the 'early' cyclin genes, and as these 'late' cyclins accumulate, they transform the Cdc28 kinase into its mitotic form to promote transition from G2 to M.
- The initiation of mitosis in eukaryotic cells is governed by a phosphorylation cascade which culminates in the activation of mitosis promoting factor (MPF).
- MPF consists minimally of the cyclin-dependent kinase Cdc28/Cdc2 in yeasts, Cdk1 in higher eukaryotes, and a B-type cyclin regulatory subunit. This takes the cell into telophase, after which the protease is activated, the surviving cyclins are degraded and the cycle finishes.

5.17 YEAST–MYCELIAL DIMORPHISM

Most fungi grow in one of two growth forms: either as rounded or spherical cells (yeast fungi) or as a polarised branching mycelium (filamentous fungi). However, some fungi can grow in both forms depending on the local environmental conditions. Such fungi are found throughout the fungal kingdom and are termed '**dimorphic**'. Dimorphism is common among pathogens because the different growth forms represent different advantageous strategies for the pathogen: the yeast form is ideal for distribution around the host through fluid circulation systems (water-conducting tissues of plants, blood or lymphatic circulation in animals), while the hyphal growth form can penetrate solid tissues. Examples include the human and animal pathogens *Candida albicans, Paracoccidioides brasiliensis, Histoplasma capsulatum* and *Blastomyces dermatitidis*, and plant pathogens like *Ustilago maydis, Ophiostoma ulmi* and *Rhodosporidium sphaerocarpum*. While most dimorphic fungi are members of the *Ascomycota*, dimorphism also occurs in other phyla (examples are the common mould *Mucor rouxii*, a member of the *Mucoromycota*, and *Ustilago* and *Rhodosporidium*, which are in the *Basidiomycota*).

Candida albicans is an important opportunistic pathogen of humans and most research on it has focused on understanding the mechanisms involved in dimorphism in this organism. Normally, infections are relatively superficial and restricted to mucosal membranes causing both oral and vaginal candidiasis (commonly known as thrush). However, for individuals who are immunocompromised, for example following immunosuppressive therapy after transplant surgery or because of HIV infection, *Candida* infections often become systemic and invasive and have a high rate of fatality. Invasive candidiasis has been associated with the presence of the hyphal form of the organism, whereas superficial infections are generally associated with the yeast phase, so implicating the transition from yeast to hypha as an important event in the pathology of the organism. However, a non-hyphal mutant strain of *Candida albicans* was still able to cause infections and death in mice, though the extent and spread of the organism was reduced; and the human pathogens *Histoplasma capsulatum* and *Paracoccidioides brasiliensis* are pathogenic only in the yeast phase.

A wide range of environmental parameters have been shown to induce a yeast-to-hypha transition in *Candida*, including serum, *N*-acetylglucosamine, proline and a shift from an acidic to a more alkaline medium. This suggests that several independent signal transduction systems exist within *Candida* for each factor, particularly as mutants that have lost the ability to undergo a yeast–mycelium transition by one stimulus are still capable of forming hyphae or germ tubes when exposed to other stimuli (Gauthier, 2015; Nigg & Bernier, 2016).

5.18 REFERENCES

Akopian, D., Shen, K., Zhang, X. & Shan, S. (2013). Signal recognition particle: an essential protein targeting machine. *Annual Review of Biochemistry*, **82**: 693–721. DOI: https://doi.org/10.1146/annurev-biochem-072711-164732.

Alberts, B., Johnson, A., Lewis, J. *et al.* (2014). *Molecular Biology of the Cell*, 6th editon. New York: Garland Science. ISBN-10: 0815344643, ISBN-13: 978-0815344643.

Barnum, K.J. & O'Connell, M.J. (2014). Cell cycle regulation by checkpoints. *Methods in Molecular Biology*, **1170**: 29–40. DOI: https://doi.org/10.1007/978-1-4939-0888-2_2.

Bartnicki-García, S., Hergert, F. & Gierz, G. (1989). Computer simulation of fungal morphogenesis and the mathematical basis for hyphal (tip) growth. *Protoplasma*, **153**: 46–57. DOI: https://doi.org/10.1007/BF01322464.

Bayry, J., Aimanianda, V., Guijarro, J.I., Sunde, M. & Latgé, J.-P. (2012). Hydrophobins: unique fungal proteins. *PLoS Pathogens*, **8**: article number e1002700. DOI: http://doi.org/10.1371/journal.ppat.1002700.

Biggins, S. (2013). The composition, functions, and regulation of the budding yeast kinetochore. *Genetics*, **194**: 817–846. DOI: https://doi.org/10.1534/genetics.112.145276.

Bloom, K. & Costanzo, V. (2017). Centromere structure and function. In *Centromeres and Kinetochores*, ed. B. Black. *Progress in Molecular and Subcellular Biology*, **Vol. 56**. Cham, Switzerland: Springer International Publishing, pp. 515–539. ISBN: 978-3-319-58591-8. DOI: https://doi.org/10.1007/978-3-319-58592-5_21.

Boye, E. & Nordström, K. (2003). Coupling the cell cycle to cell growth. *EMBO Reports*, **4**: 757–760. DOI: https://doi.org/10.1038/sj.embor.embor895.

Breitsprecher, D. & Goode, B.L. (2013). Formins at a glance. *Journal of Cell Science*, **126**: 1–7. DOI: http://doi.org/10.1242/jcs.107250.

Chang, W., Zaarour, R.F., Reck-Peterson, S. *et al.* (2008). Myo2p, a class V myosin in budding yeast, associates with a large ribonucleic acid-protein complex that contains mRNAs and subunits of the RNA-processing body. *RNA*, **14**: 491–502. DOI: https://doi.org/10.1261/rna.665008.

Chen, R.E. & Thorner, J. (2007). Function and regulation in MAPK signaling pathways: lessons learned from the yeast *Saccharomyces cerevisiae*. *Biochimica et Biophysica Acta, Molecular Cell Research*, **1773**: 1311–1340. DOI: https://doi.org/10.1016/j.bbamcr.2007.05.003.

Chook, Y.M. & Süel, K.E. (2011). Nuclear import by karyopherin-βs: recognition and inhibition. *Biochimica et Biophysica Acta, Molecular Cell Research*, **1813**: 1593–1606. DOI: https://doi.org/10.1016/j.bbamcr.2010.10.014.

Clancy, S. & Brown, W. (2008). Translation: DNA to mRNA to protein. *Nature Education*, **1**:101. URL: https://www.nature.com/scitable/topicpage/translation-dna-to-mrna-to-protein-393.

Darrah, P.R., Tlalka, M., Ashford, A., Watkinson, S.C. & Fricker, M.D. (2006). The vacuole system is a significant intracellular pathway for longitudinal solute transport in basidiomycete fungi. *Eukaryotic Cell*, **5**: 1111–1125. DOI: https://doi.org/10.1128/EC.00026-06.

Daskalov, A., Heller, J., Herzog, S., Fleißner, A. & Glass, N. (2017). Molecular mechanisms regulating cell fusion and heterokaryon formation in filamentous fungi. In *The Fungal Kingdom*, ed. J. Heitman, B. Howlett, P. Crous *et al.* Washington, DC: ASM Press, pp. 215–229. DOI: https://doi.org/10.1128/microbiolspec.FUNK-0015-2016.

Davì, V. & Minc, N. (2015). Mechanics and morphogenesis of fission yeast cells. *Current Opinion in Microbiology*, **28**: 36–45. DOI: https://doi.org/10.1016/j.mib.2015.07.010.

Davis, R.H. (2000). *Neurospora: Contributions of a Model Organism*. New York: Oxford University Press. ISBN-10: 0195122364, ISBN-13: 978-0195122367.

Day, K.J., Staehelin, L.A. & Glick, B.S. (2013). A three-stage model of Golgi structure and function. *Histochemistry and Cell Biology*, **140**: 239–249. DOI: https://doi.org/10.1007/s00418-013-1128-3.

Erental, A., Dickman, M.B. & Yarden, O. (2008). Sclerotial development in *Sclerotinia sclerotiorum*: awakening molecular analysis of a 'dormant' structure. *Fungal Biology Reviews*, **22**: 6–16. DOI: https://doi.org/10.1016/j.fbr.2007.10.001.

Erickson, H.P. (2017). The discovery of the prokaryotic cytoskeleton: 25th anniversary. *Molecular Biology of the Cell*, **28**: 357–358. DOI: http://doi.org/10.1091/mbc.E16-03-0183.

Faini, M., Beck, R., Wieland, F.T. & Briggs, J.A.G. (2013). Vesicle coats: structure, function, and general principles of assembly. *Trends in Cell Biology*, **23**: 279–288. DOI: https://doi.org/10.1016/j.tcb.2013.01.005.

Fischer-Parton, S., Parton, R.M., Hickey, P.C., Dijksterhuis, J., Atkinson, H.A. & Read, N.D. (2000). Confocal microscopy of FM4-64 as a tool for analysing endocytosis and vesicle trafficking in living fungal hyphae. *Journal of Microscopy*, **198**: 246–259. DOI: https://doi.org/10.1046/j.1365-2818.2000.00708.x.

Fisher, K.E. & Roberson, R.W. (2016). Fungal hyphal growth: Spitzenkörper versus apical vesicle crescent. *Fungal Genomics & Biology*, **6**: 1–2. DOI: https://doi.org/10.4172/2165-8056.1000136.

Fleißner, A. & Serrano, A. (2016). The art of networking: vegetative hyphal fusion in filamentous ascomycete fungi. In *The Mycota*, **Vol. I**, *Growth, Differentiation and Sexuality*, 3rd edition, ed. J. Wendland. Cham, Switzerland: Springer International Publishing, pp. 133–153. DOI: https://doi.org/10.1007/978-3-319-25844-7_7.

Gauthier, G.M. (2015). Dimorphism in fungal pathogens of mammals, plants, and insects. *PLoS Pathogens*, **11**: article number e1004608 (7 pp.). DOI: https://doi.org/10.1371/journal.ppat.1004608.

Gladfelter, A.S. (2006). Nuclear anarchy: asynchronous mitosis in multinucleated fungal hyphae. *Current Opinion in Microbiology*, **9**: 547–552. DOI: https://doi.org/10.1016/j.mib.2006.09.002.

Glass, N.L., Rasmussen, C., Roca, M.G. & Read, N.D. (2004). Hyphal homing, fusion and mycelial interconnectedness. *Trends in Microbiology*, **12**: 135–141. DOI: https://doi.org/10.1016/j.tim.2004.01.007.

Gow, N.A.R., Latgé, J.-P. & Munro, C.A. (2017). The fungal cell wall: structure, biosynthesis, and function. *Microbiology Spectrum*, **5**: FUNK-0035–2016. DOI: https://doi.org/10.1128/microbiolspec.FUNK-0035-2016.

Gu, Y. & Oliferenko, S. (2015). Comparative biology of cell division in the fission yeast clade. *Current Opinion in Microbiology*, **28**: 18–25. DOI: https://doi.org/10.1016/j.mib.2015.07.011.

Hamel, L.-P., Nicole, M.-C., Duplessis, S. & Ellis, B.E. (2012). Mitogen-activated protein kinase signaling in plant-interacting fungi: distinct messages from conserved messengers. *The Plant Cell*, **24**: 1327–1351. DOI: https://doi.org/10.1105/tpc.112.096156.

Harispe, L., Portela, C., Scazzocchio, C., Peñalva, M.A. & Gorfinkiel, L. (2008). Ras GTPase-activating protein regulation of actin cytoskeleton and hyphal polarity in *Aspergillus nidulans*. *Eukaryotic Cell*, **7**: 141–153. DOI: https://doi.org/10.1128/EC.00346-07.

Harris, S.D. (2001). Septum formation in *Aspergillus nidulans*. *Current Opinion in Microbiology*, **4**: 736–739. DOI: http://doi.org/10.1016/S1369-5274(01)00276-4.

Harris, S.D. (2013). Golgi organization and the apical extension of fungal hyphae: an essential relationship. *Molecular Microbiology*, **89**: 212–215. DOI: https://doi.org/10.1111/mmi.12291.

Harris, S.D., Read, N.D., Roberson, R. W. *et al.* (2005). Polarisome meets Spitzenkörper: microscopy, genetics, and genomics converge. *Eukaryotic Cell*, **4**: 225–229. DOI: https://doi.org/10.1128/EC.4.2.225-229.2005.

Hickey, P.C., Jacobson, D.J., Read, N.D. & Glass, N.L. (2002). Live-cell imaging of vegetative hyphal fusion in *Neurospora crassa*. *Fungal Genetics and Biology*, **37**: 109–119. DOI: https://doi.org/10.1016/S1087-1845(02)00035-X.

Hwang, J. & Pallas, D.C. (2014). STRIPAK complexes: structure, biological function, and involvement in human diseases. *International Journal of Biochemistry & Cell Biology*, **47**: 118–148. DOI: https://doi.org/10.1016/j.biocel.2013.11.021.

Karki, S. & Holzbaur, E.L.F. (1999). Cytoplasmic dynein and dynactin in cell division and intracellular transport. *Current Opinion in Cell Biology*, **11**: 45–53. DOI: http://doi.org/10.1016/S0955-0674(99)80006-4.

Kayikci, Ö. & Nielsen, J. (2015). Glucose repression in *Saccharomyces cerevisiae*. *FEMS Yeast Research*, **15**: fov068 (8 pp.). DOI: https://doi.org/10.1093/femsyr/fov068.

Kilmartin, J.V. (2014). Lessons from yeast: the spindle pole body and the centrosome. *Philosophical Transactions of the Royal Society B: Biological Sciences*, **369**: article number 20130456 (6 pp.). DOI: https://doi.org/10.1098/rstb.2013.0456.

Kinoshita, T. (2016). Glycosylphosphatidylinositol (GPI) anchors: biochemistry and cell biology: introduction to a thematic review series. *Journal of Lipid Research*, **57**: 4–5. DOI: https://doi.org/10.1194/jlr.E065417.

Kornberg, R.D. (2007). The molecular basis of eukaryotic transcription. *Proceedings of the National Academy of Sciences of the United States of America*, **104**: 12955–12961. DOI: https://doi.org/10.1073/pnas.0704138104.

Kosinski, J., Mosalaganti, S., von Appen, A. *et al.* (2016). Molecular architecture of the inner ring scaffold of the human nuclear pore complex. *Science*, **352**: 363–365. DOI: https://doi.org/10.1126/science.aaf0643.

Li, L., Wright, S., Krystofova, S., Park, G. & Borkovich, K.A. (2007). Heterotrimeric G protein signaling in filamentous fungi. *Annual Review of Microbiology*, **61**: 423–452. DOI: https://doi.org/10.1146/annurev.micro.61.080706.093432.

Lichius, A., Berepiki, A. & Read, N.D. (2011). Form follows function: the versatile fungal cytoskeleton. *Fungal Biology*, **115**: 518–540. https://doi.org/10.1016/j.funbio.2011.02.014.

Lindsey, R. & Momany, M. (2006). Septin localization across kingdoms: three themes with variations. *Current Opinion in Microbiology*, **9**: 559–565. DOI: https://doi.org/10.1016/j.mib.2006.10.009.

Machida, M. & Gomi, K. (2010). *Aspergillus: Molecular Biology and Genomics*. Norwich, UK: Caister Academic Press. ISBN: 978-1-904455-53-0.

Marfori, M., Mynott, A., Ellis, J.J. *et al.* (2011). Molecular basis for specificity of nuclear import and prediction of nuclear localization. *Biochimica et Biophysica Acta, Molecular Cell Research*, **1813**: 1562–1577. DOI: http://doi.org/10.1016/j.bbamcr.2010.10.013.

Margulis, L. (2004). Serial endosymbiotic theory (SET) and composite individuality. Transition from bacterial to eukaryotic genomes. *Microbiology Today*, **31**: 172–174. URL: http://www.davidmoore.org.uk/21st_Century_Guidebook_to_Fungi_PLATINUM/REPRINT_collection/Margulis_serial_endosymbiotic_theory.pdf.

Martin, S.G. & Arkowitz, R.A. (2014). Cell polarization in budding and fission yeasts. *FEMS Microbiology Reviews*, **38**: 228–253. DOI: http://doi.org/10.1111/1574-6976.12055.

Martin, W.F., Garg, S. & Zimorski, V. (2015). Endosymbiotic theories for eukaryote origin. *Philosophical Transactions of the Royal Society B: Biological Sciences*, **370**(1678): article number 20140330. DOI: https://doi.org/10.1098/rstb.2014.0330.

Merlini, L., Dudin, O. & Martin, S.G. (2013). Mate and fuse: how yeast cells do it. *Open Biology*, **3(3)**: 130008 (13 pp.). DOI: https://doi.org/10.1098/rsob.130008.

Moore, D. (1998). *Fungal Morphogenesis*. New York: Cambridge University Press, Chapter 3. ISBN-10: 0521552958, ISBN-13: 978-0521552950.

Moore, D. (2000). *Slayers, Saviors, Servants and Sex. An Exposé of Kingdom Fungi*. New York: Springer-Verlag, Chapter 3. ISBN-10: 0387951016, ISBN-13: 978-0387951010.

Nigg, M. & Bernier, L. (2016). From yeast to hypha: defining transcriptomic signatures of the morphological switch in the dimorphic fungal pathogen *Ophiostoma novo-ulmi*. *BMC Genomics*, **17**: 920 (16 pp.). DOI: https://doi.org/10.1186/s12864-016-3251-8.

Noble, L.M., Holland, L.M., McLauchlan, A.J. & Andrianopoulos, A. (2016). A plastic vegetative growth threshold governs reproductive capacity in *Aspergillus nidulans*. *Genetics*, **204**: 1161–1175. DOI: https://doi.org/10.1534/genetics.116.191122.

Osés-Ruiz, M., Sakulkoo, W. & Talbot, N.J. (2016). Septation and cytokinesis in pathogenic fungi. In *The Mycota*, **Vol. I**, *Growth, Differentiation and Sexuality*, 3rd edition, ed. J. Wendland. Cham, Switzerland: Springer International Publishing, pp. 67–79. DOI: https://doi.org/10.1007/978-3-319-25844-7_4.

Pantazopoulou, A. (2016). The Golgi apparatus: insights from filamentous fungi. *Mycologia*, **108**: 603–622. DOI: https://doi.org/10.3852/15-309.

Papamichos-Chronakis, M., Gligoris, T. & Tzamarias, D. (2004). The Snf1 kinase controls glucose repression in yeast by modulating interactions between the Mig1 repressor and the Cyc8-Tup1 co-repressor. *EMBO Reports*, **5**: 368–372. DOI: https://doi.org/10.1038/sj.embor.7400120.

Papasaikas, P. & Valcárcel, J. (2016). The spliceosome: the ultimate RNA chaperone and sculptor. *Trends in Biochemical Sciences*, **41**: 33–45. DOI: https://doi.org/10.1016/j.tibs.2015.11.003.

Pasteur, L. (1860). Mémoire sur la fermentation alcoolique. *Annales de chimie et de physique*, 3e série, **58**: 323–426.

Pasteur, L. (1879). *The Physiological Theory of Fermentation* (originally published in Paris by Gauthier-Villars). Translated by F. Faulkner & D.C. Robb and available online at these URLs: https://ebooks.adelaide.edu.au/p/pasteur/louis/ferment/complete.html and https://sourcebooks.fordham.edu/mod/1879pasteur-ferment.asp. See also Barnett, J.A. (2000). A history of research on yeasts 2: Louis Pasteur and his contemporaries, 1850–1880. *Yeast*, **16**: 755-771. DOI: https://doi.org/10.1002/1097-0061(20000615)16:8<755::AID-YEA587>3.0.CO;2-4.

Peñate, X. & Chávez, S. (2014). RNA polymerase II-dependent transcription in fungi and its interplay with mRNA decay. In *Fungal RNA Biology*, ed. A. Sesma & T. von der Haar. Cham, Switzerland: Springer International Publishing, pp. 1–26. ISBN: 978-3319056869.

Pickart, C.M. & Eddins, M.J. (2004). Ubiquitin: structures, functions, mechanisms. *Biochimica et Biophysica Acta, Molecular Cell Research*, **1695**: 55–72. DOI: https://doi.org/10.1016/j.bbamcr.2004.09.019.

Pollard, T.D. (2010). Mechanics of cytokinesis in eukaryotes. *Current Opinion in Cell Biology*, **22**: 50–56. DOI: https://doi.org/10.1016/j.ceb.2009.11.010.

Pollard, T.D. (2016). Actin and actin-binding proteins. *Cold Spring Harbor Perspectives in Biology*, **8**: a018226. DOI: http://doi.org/10.1101/cshperspect.a018226.

Pollard, T.D., Earnshaw, W.C., Lippincott-Schwartz, J. & Johnson, G. (2017a). Nuclear structure and dynamics. In *Cell Biology*, 3rd edition, ed. T.D. Pollard, W.C. Earnshaw, J. Lippincott-Schwartz & G. Johnson. Philadelphia, PA: Elsevier, pp. 143–160. ISBN: 978-0-323-34126-4. DOI: https://doi.org/10.1016/B978-0-323-34126-4.00014-1.

Pollard, T.D., Earnshaw, W.C., Lippincott-Schwartz, J. & Johnson, G. (2017b). Posttranslational targeting of proteins. In *Cell Biology*, 3rd edition, ed. T.D. Pollard, W.C. Earnshaw, J. Lippincott-Schwartz & G. Johnson. Philadelphia, PA: Elsevier, pp. 303–315. ISBN: 978-0-323-34126-4. DOI: https://doi.org/10.1016/B978-0-323-34126-4.00014-1.

Raudaskoski, M. (2015). Mating-type genes and hyphal fusions in filamentous basidiomycetes. *Fungal Biology Reviews*, **29**: 179–193. DOI: https://doi.org/10.1016/j.fbr.2015.04.001.

Read, N.D., Fleißner, A., Roca, M.G. & Glass, N.L. (2010). Hyphal fusion. In *Cellular and Molecular Biology of Filamentous Fungi*, ed. K.A. Borkovich & D.J. Ebbole. Washington, DC: American Society for Microbiology Press, pp. 260–273. ISBN-10: 1555814735, ISBN-13: 978-1555814731.

Read, N.D., Lichius, A., Shoji, J.-Y. & Goryachev, A.B. (2009). Self-signalling and self-fusion in filamentous fungi. *Current Opinion in Microbiology*, **12**: 608–615. DOI: https://doi.org/10.1016/j.mib.2009.09.008.

Read, N.D. & Roca, G.M. (2013). Vegetative hyphal fusion in filamentous fungi. In *NCBI Bookshelf, Madame Curie Bioscience Database [internet]*. Austin (TX): Landes Bioscience; 2000–2013. URL: https://www.ncbi.nlm.nih.gov/books/NBK5993/.

Richards, A., Gow, N.A.R. & Veses, V. (2012). Identification of vacuole defects in fungi. *Journal of Microbiological Methods*, **91**: 155–163. DOI: https://doi.org/10.1016/j.mimet.2012.08.002.

Richards, A., Veses, V. & Gow, N.A.R. (2010). Vacuole dynamics in fungi. *Fungal Biology Reviews*, **24**: 93–105. DOI: https://doi.org/10.1016/j.fbr.2010.04.002.

Riquelme, M. (2013). Tip growth in filamentous fungi: a road trip to the apex. *Annual Review of Microbiology*, **67**: 587–609. DOI: https://doi.org/10.1146/annurev-micro-092412-155652.

Riquelme, M., Roberson, R.W. & Sánchez-León, E. (2016). Hyphal tip growth in filamentous fungi. In *The Mycota*, **Vol. I**, *Growth, Differentiation and Sexuality*, 3rd edition, ed J. Wendland. Cham, Switzerland: Springer International Publishing, pp. 47–66. DOI: https://doi.org/10.1007/978-3-319-25844-7_3.

Riquelme, M. & Sánchez-León, E. (2014). The Spitzenkörper: a choreographer of fungal growth and morphogenesis. *Current Opinion in Microbiology*, **20**: 27–33. DOI: https://doi.org/10.1016/j.mib.2014.04.003.

Roberts, S.E. & Gladfelter, A.S. (2016). Nuclear dynamics and cell growth in fungi. In *The Mycota*, **Vol. I**. *Growth, Differentiation and Sexuality*, 3rd edition, ed. J. Wendland. Cham, Switzerland: Springer International Publishing, pp. 27–46. ISBN: 978-3-319-25842-3. DOI: https://doi.org/10.1007/978-3-319-25844-7_2.

Robson, G.D. (1999). Hyphal cell biology. In *Molecular Fungal Biology*, ed. R.P. Oliver & M. Schweizer. Cambridge, UK: Cambridge University Press, pp. 164–184. ISBN-10: 0521561167, ISBN-13: 978-0521561167.

Rogg, L.E., Fortwendel, J.R., Juvvadi, P.R. & Steinbach, W.J. (2012). Regulation of expression, activity and localization of fungal chitin synthases. *Medical Mycology*, **50**: 2–17. DOI: https://doi.org/10.3109/13693786.2011.577104.

Rüthnick, D., Neuner, A., Dietrich, F. *et al.* (2017). Characterization of spindle pole body duplication reveals a regulatory role for nuclear pore complexes. *Journal of Cell Biology*, **216**: 2425–2442. DOI: https://doi.org/10.1083/jcb.201612129.

Samson, R.A. & Varga, J. (2008). *Aspergillus in the Genomic Era*. Wageningen, the Netherlands: Wageningen Academic Publishers. ISBN-10: 9086860656, ISBN-13: 9789086860654.

Schmatz, D.M., Romancheck, M.A., Pittarelli, L.A. *et al.* (1990). Treatment of *Pneumocystis carinii* pneumonia with 1–3-beta-glucan synthesis inhibitors. *Proceedings of the National Academy of Sciences of the United States of America*, **87**: 5950–5954. DOI: https://doi.org/10.1073/pnas.87.15.5950.

Sesma, A. & von der Haar, T. (ed.) (2014). *Fungal RNA Biology*. Cham, Switzerland: Springer International Publishing. ISBN-10: 3319056867, ISBN-13: 978-3319056869.

Shi, C., Kaminskyj, S. Caldwell, S. & Loewen, M.C. (2007). A role for a complex between activated G protein-coupled receptors in yeast cellular mating. *Proceedings of the National Academy of Sciences of the United States of America*, **104**: 5395–5400. DOI: https://doi.org/10.1073/pnas.0608219104.

Shioya, T., Nakamura, H., Ishii, N. *et al.* (2013). The *Coprinopsis cinerea* septin Cc.Cdc3 is involved in stipe cell elongation. *Fungal Genetics and Biology*, **58–59**: 80–90. DOI: https://doi.org/10.1016/j.fgb.2013.08.007.

Shoham, S., Groll, A.H., Petraitis, V. & Walsh, T.J. (2017). Systemic antifungal agents. In *Infectious Diseases*, **Vol. 2**, 4th edition, ed. J. Cohen, W.G. Powderly & S.M. Opal. Amsterdam: Elsevier, pp. 1333–1344. ISBN: 9780702062858. DOI: https://doi.org/10.1016/B978-0-7020-6285-8.00156-8.

Sibirny, A.A. (2016). Yeast peroxisomes: structure, functions and biotechnological opportunities. *FEMS Yeast Research*, **16**(4): fow038 (14 pp.). DOI: https://doi.org/10.1093/femsyr/fow038.

Simanshu, D.K., Nissley, D.V. & McCormick, F. (2017). RAS proteins and their regulators in human disease. *Cell*, **170**: 17–33. DOI: http://doi.org/10.1016/j.cell.2017.06.009.

Simon, M. & Plattner, H. (2014). Unicellular eukaryotes as models in cell and molecular biology: critical appraisal of their past and future value. *International Review of Cell and Molecular Biology*, **309**: 141–198. DOI: https://doi.org/10.1016/B978-0-12-800255-1.00003-X.

Steinberg, G. (2000). The cellular roles of molecular motors in fungi. *Trends in Microbiology*, **8**: 162–168. DOI: https://doi.org/10.1016/S0966-842X(00)01720-0.

Steinberg, G. (2007a). Preparing the way: fungal motors in microtubule organization. *Trends in Microbiology*, **15**: 14–21. DOI: https://doi.org/10.1016/j.tim.2006.11.007.

Steinberg, G. (2007b). Hyphal growth: a tale of motors, lipids, and the Spitzenkörper. *Eukaryotic Cell*, **6**: 351–360. DOI: https://doi.org/10.1128/EC.00381-06.

Steinberg, G., Peñalva, M., Riquelme, M., Wösten, H. & Harris, S. (2017). Cell biology of hyphal growth. In *The Fungal Kingdom*, ed. J. Heitman, B. Howlett, P. Crous *et al.* Washington, DC: ASM Press, pp. 231–265. DOI: https://doi.org/10.1128/microbiolspec.FUNK-0034-2016.

Svitkina, T. (2018). The actin cytoskeleton and actin-based motility. *Cold Spring Harbor Perspectives in Biology*, **10**: a018267. DOI: https://doi.org/10.1101/cshperspect.a018267.

Takeshita, N., Evangelinos, M., Zhou, L. *et al.* (2017). Pulses of Ca^{2+} coordinate actin assembly and exocytosis for stepwise cell extension. *Proceedings of the National Academy of Sciences of the United States of America*, **114**: 5701–5706. DOI: https://doi.org/10.1073/pnas.1700204114.

Takeshita, N., Manck, R., Grün, N., de Vega, S.H. & Fischer, R. (2014). Interdependence of the actin and the microtubule cytoskeleton during fungal growth. *Current Opinion in Microbiology*, **20**: 34–41. DOI: https://doi.org/10.1016/j.mib.2014.04.005.

Teparić, R. & Mrša, V. (2013). Proteins involved in building, maintaining and remodeling of yeast cell walls. *Current Genetics*, **59**: 171–185. DOI: https://doi.org/10.1007/s00294-013-0403-0.

Tong, S.M. Chen, Y., Ying, S.-H. & Feng, M.-G. (2016). Three DUF1996 proteins localize in vacuoles and function in fungal responses to multiple stresses and metal ions. *Scientific Reports*, **6**: article number 20566. DOI: https://doi.org/10.1038/srep20566.

Trinci, A.P.J., Wiebe, M.G. & Robson, G.D. (1994). The mycelium as an integrated entity. In *The Mycota*, **Vol. I**, ed. J.G.H. Wessels & F. Meinhardt. Berlin, Heidelberg: Springer-Verlag, pp. 175–193. ISBN-10: 3540577815, ISBN-13: 978-3540577812.

Trinci, A.P.J., Wiebe, M.G. & Robson, G.D. (2001). Hyphal growth. In *Encyclopaedia of Life Sciences*. Chichester, UK: John Wiley & Sons. DOI: https://doi.org/10.1038/npg.els.0000367.

Upla, P., Kim, S.J., Sampathkumar, P. *et al.* (2017). Molecular architecture of the major membrane ring component of the nuclear pore complex. *Structure*, **25**: 434–445. DOI: https://doi.org/10.1016/j.str.2017.01.006.

Urnavicius, L., Zhang, K., Diamant, A.G. *et al.* (2015). The structure of the dynactin complex and its interaction with dynein. *Science*, **347**: 1441–1446. DOI: https://doi.org/10.1126/science.aaa4080.

van Dijck, P., Brown, N.A., Goldman, G.H. *et al.* (2017). Nutrient sensing at the plasma membrane of fungal cells. In *The Fungal Kingdom*, ed. J. Heitman, B. Howlett, P. Crous *et al.* Washington, DC: ASM Press, pp. 231–265. DOI: https://doi.org/10.1128/microbiolspec.FUNK-0031-2016.

Veses, V., Richards, A. & Gow, N.A.R. (2008). Vacuoles and fungal biology. *Current Opinion in Microbiology*, **11**: 503–510. DOI: https://doi.org/10.1016/j.mib.2008.09.017.

Virag, A. & Harris, S.D. (2006). The Spitzenkörper: a molecular perspective. *Mycological Research*, **110**: 4–13. DOI: https://doi.org/10.1016/j.mycres.2005.09.005.

Walker, L.A., Lenardon, M.D., Preechasuth, K., Munro, C.A. & Gow, N.A.R. (2013). Cell wall stress induces alternative fungal cytokinesis and septation strategies. *Journal of Cell Science*, **126**: 2668–2677. DOI: https://doi.org/10.1242/jcs.118885.

Wang, M. & Casey, P.J. (2016). Protein prenylation: unique fats make their mark on biology. *Nature Reviews Molecular Cell Biology*, **17**: 110–122. DOI: https://doi.org/10.1038/nrm.2015.11.

Wang, T., Li, L. & Hong, W. (2017). SNARE proteins in membrane trafficking. *Traffic*, **18**: 767–775. DOI: http://doi.org/10.1111/tra.12524.

Wente, S.R. & Rout, M.P. (2010). The nuclear pore complex and nuclear transport. *Cold Spring Harbor Perspectives in Biology*, **2**(10): a000562 (19 pp.). DOI: http://doi.org/10.1101/cshperspect.a000562.

Wessels, J.G.H. (1993). Wall growth, protein excretion and morphogenesis in fungi. *New Phytologist*, **123**: 397–413. DOI: https://doi.org/10.1111/j.1469-8137.1993.tb03751.x.

Wickstead, B. & Gull, K. (2011). The evolution of the cytoskeleton. *The Journal of Cell Biology*, **194**: 513–525. DOI: http://doi.org/10.1083/jcb.201102065.

Willems, A.R., Schwab, M. & Tyers, M. (2004). A hitchhiker's guide to the Cullin ubiquitin ligases: SCF and its kin. *Biochimica et Biophysica Acta, Molecular Cell Research*, **1695**: 133–170. DOI: https://doi.org/10.1016/j.bbamcr.2004.09.027.

Wright, G.D., Arlt, J., Poon, W.C.K. & Read, N.D. (2007). Optical tweezer micromanipulation of filamentous fungi. *Fungal Genetics and Biology*, **44**: 1–13. DOI: https://doi.org/10.1016/j.fgb.2006.07.002.

Xiang, X. (2018). Insights into cytoplasmic dynein function and regulation from fungal genetics. In *Dyneins: The Biology of Dynein Motors*, 2nd edition, ed. S.M. King. Amsterdam: Academic Press, pp. 470–501. ISBN: 978-0-12-809471-6. DOI: https://doi.org/10.1016/B978-0-12-809471-6.00016-4.

Xiang, X. & Plamann, M. (2003). Cytoskeleton and motor proteins in filamentous fungi. *Current Opinion in Microbiology*, **6**: 628–633. DOI: https://doi.org/10.1016/j.mib.2003.10.009.

Xiang, X., Qiu, R., Yao, X. *et al.* (2015). Cytoplasmic dynein and early endosome transport. *Cellular and Molecular Life Sciences*, **72**: 3267–3280. DOI: https://doi.org/10.1007/s00018-015-1926-y.

Zhuang, X., Tlalka, M., Davies, D.S. *et al.* (2009). Spitzenkörper, vacuoles, ring-like structures, and mitochondria of *Phanerochaete velutina* hyphal tips visualised with carboxy-DFFDA, CMAC and $DiOC_6(3)$. *Mycological Research*, **113**: 417–431. DOI: https://doi.org/10.1016/j.mycres.2008.11.014.

CHAPTER 6

Fungal Genetics: From Gene Segregation to Gene Editing

We start this chapter with the basic genetics of fungi. Within a few decades of the rediscovery in 1900 of Mendel's research with peas, his experiments had been repeated with yeast and had demonstrated that yeast genes operated to the same set of rules, as did the genes of plants and animals. During the next 50 years it was clearly established that the basic genetic architecture of fungi is typical of the eukaryotes. We limit our discussion of the basic (segregational) genetics of fungi because it is dealt with adequately elsewhere, so we concentrate here on highlighting some of the differences between most fungi and most other eukaryotes.

In the bulk of this chapter, we venture into territory that is more generally recognised as representing 'molecular biology'. We discuss fungal genome structure and the sequencing, annotation and comparison of fungal genomes. We then describe methods of manipulating genomes, editing genes, indeed editing the genome, to enable filamentous fungi to be used in the commercial production of recombinant protein and other metabolites.

6.1 BASIC FUNGAL GENETICS

The research on garden peas that **Gregor Mendel** published in 1865 was rediscovered and republished in 1900, when it inspired many experimenters to demonstrate similar gene segregation in plants and animals. At about the same time it was becoming evident that vegetative cells of *Saccharomyces cerevisiae* are usually diploid, produced by 'copulation' of two haploid spores. In the early 1930s, the basic facts of the yeast life cycle were established, in particular that diploid nuclei underwent **meiosis** (the reduction division) during spore formation to produce four haploid ascospores. With the life cycle so clearly established, the way was open for breeding experiments. By the 1940s it was possible for **Carl Lindegren** to write:

> Thirteen asci were analysed from a heterozygous hybrid made by mating a galactose fermenter by a nonfermenter; two spores in each of these asci carried the dominant gene controlling fermentation of galactose, and two carried the recessive allele. A backcross of fermenter [offspring] to the fermenter parent produced 13 asci; all four spores in each of these asci carried the fermenting gene. A backcross of the nonfermenter to the nonfermenting parent produced seven asci, each of which contained four nonfermenting spores. A heterozygous zygote was produced by backcrossing a nonfermenter [offspring] to the fermenting parent; six asci were analysed and each contained two nonfermenting spores. This analysis shows quite convincingly that the genes controlling fermentation of galactose behave in a regular Mendelian manner. (Lindegren, 1949)

Unlinked genes undergo Mendelian segregations, just like the genes of Mendel's garden peas, and linked genes show recombination in frequencies characteristic of their distance apart on the chromosome, just like linked genes in fruit flies. Basic chromosomal structure and the nuclear division process are defining characteristics of eukaryotes, and all the major principles of genetics apply in fungi, namely gene structure and organisation, Mendelian segregations, recombination, etc. What differences exist arise from peculiarities in the biology of the fungal lifestyle; such as:

- Most fungi are haploid, which makes analysis of gene segregations in crosses more direct and provides a tool for the experimenter to use selection methods to find very rare mutations, or rare recombinants.
- Sibling meiotic products tend to be grouped together (ascospores in asci, basidiospores on basidia), and even in yeast the products of each meiosis tend to clump together, which gives the geneticist an opportunity to bring centromere segregation into the analysis, a tool that is only very rarely available in other eukaryotes.

We do not intend to discuss the segregational genetics of fungi any further here and urge you to read the book *Essential Fungal Genetics* by Moore & Novak Frazer (2002), especially their Chapter 5; and Dyer *et al.* (2017). Here, we will highlight some of the differences between most fungi and most other eukaryotes, and essentially this means concentrating on molecular aspects of fungal genetics.

Table 6.1. Approximate genome sizes of representative eukaryotes

Organism	Genome size (Mb*)
Rhizopus oryzae	35
Saccharomyces cerevisiae	12
Aspergillus nidulans	31
Neurospora crassa	39
Coprinopsis cinerea	36
Ustilago maydis	20
Drosophila melanogaster	122
Sea urchin	814
Human	3, 300

*Mb = megabases (10^6 base pairs). Genomic data are held on open databases on the internet which are freely available. For the most up-to-date details visit the following URLs:

- https://genome.jgi.doe.gov/programs/fungi/index.jsf
- https://www.ncbi.nlm.nih.gov/genome
- https://www.ebi.ac.uk/genomes/eukaryota.html
- http://fungi.ensembl.org/index.html
- http://fungalgenomes.org/
- http://fungidb.org/fungidb/
- https://en.wikipedia.org/wiki/List_of_sequenced_fungi_genomes
- https://www.broadinstitute.org/fungal-genome-initiative

Fungi have a generally **smaller genome** size than other eukaryotes (Table 6.1). Remember, though, that the higher organisms have much more ***non-coding DNA***; for example, in *Homo sapiens*, it is estimated that only 3% of the genome codes for protein. The Japanese pufferfish (*Fugu rubripes*) has the smallest genome known for any vertebrate species, being only one-tenth the size of the human genome, but the size difference between these two genomes can be explained by differences in intron sizes. In contrast, analysis of genomes of grasses reveals that differences in sizes (up to 40 times) can be explained by their having extensive regions between genes filled with repetitive DNA, suggesting that in plants most repeats integrate into intergenic DNA, but in animals most repeats integrate into intron DNA (see section 6.3 for more information about introns).

The genomes of different eukaryotes are more similar when we compare the numbers of protein-coding genes that different genomes specify. For example, the Puffer fish and human genomes contain about 30,000 genes, *Drosophila melanogaster* contains about 14,000; *Neurospora crassa* about 10,000; and yeast about 6,000.

The **karyotype** of most fungi, that is, the profile of the chromosome set, can be resolved by electrophoresis. The technique reveals that ***chromosome length polymorphisms are widespread in fungi*** in both the sexual and asexual-reproducing species, revealing general ***genome plasticity***. Tandem repeats, for example repeats of rRNA genes, frequently vary in length, and dispensable ***supernumerary chromosomes***, which are usually less than 1 million base pairs in size, as well as dispensable chromosome regions also occur in fungal karyotypes. Many karyotype changes are genetically neutral; others may be advantageous in allowing strains to adapt to new environments (Mehrabi *et al.*, 2017).

6.2 ESTABLISHING FUNGAL GENETIC STRUCTURE

We have a long heritage of using fungi and fungal products. Some of our current fungal biotechnology, such as baking, brewing and the numerous fermented food products described in Chapter 19, originated hundreds or even thousands of years ago and largely by the chance association between natural fungi and one or more of the constituents of the food material. Although the original discovery of penicillin was also a matter of chance (section 19.15) its industrial production in the middle of the twentieth century was a much more directed process, as was the development of other products, such as citric acid (section 19.14). Yet all improvements were at the organismal level. Techniques were found that enabled cultivation of specific organisms, and ***strains were selected*** that had advantageous biological characteristics.

In the second half of the twentieth century the rapidly accumulating knowledge of fungal genetics was brought to bear, and from what we have discussed so far it must be clear that a thorough understanding of the molecular genetics of fungi is essential to future exploitation. All the major principles of eukaryote genetics apply in fungi; these are aspects of what might be called ***Mendelian*** or ***segregational*** genetics, which apply because of the chromosomal architecture and mechanisms of meiosis (Moore & Novak Frazer, 2002; Dyer *et al.*, 2017).

Chromosome maps constructed solely from recombination frequencies have a limited resolution, although large numbers of progeny can be scored in most microorganisms, so reducing this problem to some extent. When the *Saccharomyces cerevisiae* genome-sequencing project began in 1989, the conventional genetic map consisted of more than 1,400 markers, an average of one every 3.3 kb, and this was detailed enough for the sequencing programme without the need for much more **physical mapping**. However, *Saccharomyces cerevisiae* was one of the two most intensively mapped eukaryotes at the time (the fruit fly, *Drosophila melanogaster*, being the other), so physical mapping is necessary to improve the marker density in other fungi as they are included in genome-sequencing programmes.

Physical mapping procedures include **restriction mapping**, which establishes the positions of restriction endonuclease recognition sites in a DNA molecule, locating markers on chromosomes by hybridising marker probes to intact chromosomes, and mapping known sequences in genome fragments using PCR and hybridisation. The ideal is to establish the locations of unique sequences, which are not duplicated at any other site, as markers spaced about 100 kb apart (that's just less than 1% genetic recombination) throughout the genome. The collection of such markers is known as a **mapping panel**.

Use of mapping panels is essentially a management technique for sharing the effort between participating laboratories, something which has been common for many years in several physical sciences, such as particle physics and astronomy, though genome sequencing represented the entry of biology into the 'big-science' league. The first example of ***big fungal science*** was the programme to sequence the yeast genome which was initiated in 1989 by the European Commission. The project involved 35 European laboratories at the outset, and the first sequence of a complete chromosome was published in 1992. Eventually, over 600 scientists were involved, at locations in Europe, North America and Japan, and progress involved distribution of DNA fragments to the contributing laboratories by the DNA coordinator. The complete sequence of the yeast genome was published in 1997.

The genome is made up of the entire DNA content of a cell. Eukaryotes and prokaryotes have quite different types of genome, but it is generally assumed that something like the prokaryotic grade of organisation is the primitive form from which the eukaryote organisation evolved. Modern prokaryotes and eukaryotes have a great deal in common (see Chapter 5); the DNA of a gene is transcribed into RNA, which is called a messenger RNA (mRNA) if it is a transcript of a protein-coding gene, and the mRNA is translated into protein by the ribosomes and other translation machinery. The part of a protein-coding gene sequence that is translated into protein is called the **open reading frame**, usually abbreviated to **ORF**.

As a genome sequence is assembled the functional genes in the sequence are recognised as ORFs. The process is called genome **annotation**, and is discussed in more detail below. Not all the ORFs that are identified can be associated with a gene of identified function; an ORF specifying a product that does not resemble a known protein is called an unidentified reading frame, or URF. But comparative genomics does more than identify the genes. It can show the evolutionary relationships between different organisms, and aids understanding of how the genotype relates to lifestyle and environment.

Characteristically, the ORF is read in the 5′ to 3′ direction along the mRNA, and it starts with an initiation codon and ends with a termination codon (Figure 6.1). Nucleotide sequences that occur in the mRNA before the ORF make up the **leader** sequence, and sequences following the ORF make up the **trailer**

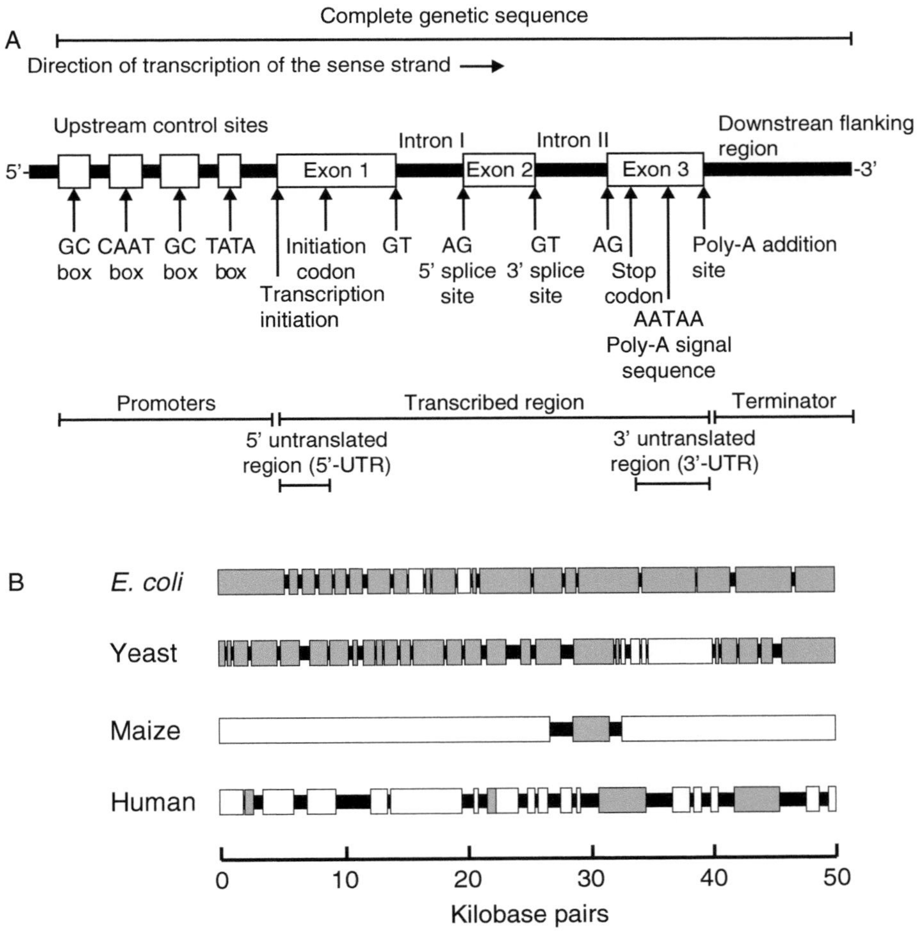

Figure 6.1. (A) The basic structure of a typical eukaryotic gene. The schematic diagram indicates the structure of a type II gene; that is, a protein-encoding gene transcribed by polymerase II. The diagram is not drawn to scale and the relative sizes of the different sections differ between genes and between the eukaryotic kingdoms. (B) Comparison of 50 kbp segments of the genomes of the prokaryote *Escherichia coli* and three eukaryotes to show how the 'density' of genetic information varies. In each case the grey boxes correspond to gene sequences, and the white boxes correspond to stretches of repeated sequences. Adapted from Moore & Novak Frazer (2002).

segment. Many eukaryotic genes are split into **exons** (meaningful segments) and **introns**, which are sequence segments that do not contribute to the protein-coding sequence represented in the functional mRNA.

An **intron** is any nucleotide sequence within a gene that is removed by RNA splicing during maturation of the final functional mRNA (Figure 6.1; and see section 5.4), and several different mRNAs may be produced from any one primary RNA transcript by a process known as **alternative splicing** (section 6.4). Then there are the **transposable elements**, often called **transposons** (section 6.5), which are seemingly ubiquitous genetic elements that have been discovered in all prokaryotes and eukaryotes so far investigated. In remarkable contrast to all other genes, transposable elements are able to move to new locations within their host genomes in a process called **transposition**. Transposable elements contribute enormously to eukaryotic genome diversity. Their ubiquitous presence affects the genomes of all species; mediating genome evolution by causing mutations, repetitions and chromosomal rearrangements and by modifying gene expression.

Indeed, introns and transposons seem to be extremely ancient genetic structures, which certainly existed long before the eukaryote grade of organisation emerged. Over evolutionary time ***they*** have ***created*** new patterns of gene expression by 'shuffling' functional motifs together and then combining them with new control elements to produce differentiated cellular structures with new morphologies and/or new developmental possibilities. And not just within the one genome. Transposition is also responsible for **horizontal gene transfer events between species**, of the sort we describe in section 19.15 in relation to penicillin biosynthesis. Remember that horizontal gene transfer is the transmission of genetic material between organisms and across major taxonomic boundaries and that there is evidence for ***many horizontal gene transfer events in fungi*** (Richards *et al.*, 2011; Slot *et al.*, 2017; Steenkamp *et al.*, 2018). All of this is discussed further below.

The genomes of yeasts are some of the smallest eukaryotic genomes known (about 10 Mb), with the largest ones being found in vertebrates and plants (over 100,000 Mb), so we can observe some surprising structural differences when we compare different eukaryotes (Table 6.1). Generally speaking (but remember there are exceptions to all generalisations), it appears that space is saved in the genomes of less complex organisms by having the genes more closely packed together and by having much less repetition (Figure 6.1). The genome of *Saccharomyces cerevisiae* contains more genes per unit length of DNA than occur in human or maize DNA. On the other hand, up to 40% of the gene sequences of *Saccharomyces cerevisiae* are duplicated. In most cases, the duplicated sequences are so similar that their protein products are identical and, presumably, either functionally redundant, or (more likely) under very different regulatory control. Furthermore, as we detail in section 6.7, the whole of *Saccharomyces cerevisiae* chromosome XIV is made up of regions duplicated on other chromosomes.

Nevertheless, the small size of the yeast genome is one reason why yeast geneticists and molecular biologists ***pioneered*** eukaryote genome analysis. Although some of the more unusual aspects of genome structure observed in higher animals and plants might not be represented in fungi, the genomes of yeast and other fungi remain good models of eukaryotic genetic architecture and their smaller size means that the information they contain is technically more accessible. In terms of genetic information content, the organisation of the fungal genome is, generally, much more economical than that of animals and plants. Although fungal genes may be generally more compact with fewer introns and shorter spaces between genes, a major difference is that fungi contain much less of the so-called redundant DNA, which is devoted to ***repetitive non-coding sequences*** in so many animals and plants. But then, fungi have made specialist use of **heterokaryosis** (see Chapter 8), which is uncommon in the other kingdoms in nature. What better way is there to increase genome size than by being a syncytium?

In section 6.7 we will describe the methods developed for sequencing and studying whole fungal genomes, but first we want to discuss a few more fungus-specific genetic details; namely introns, alternative splicing, transposons, genomic variation, including gene clusters and horizontal transfer, and ploidy variation (and see Sánchez *et al.*, 2020).

6.3 INTRONS

The frequency of introns in the genome varies widely between different organisms. For example, introns are extremely common within the nuclear genome of higher vertebrates (e.g. humans and mice), where protein-coding genes almost always contain multiple introns; while introns are generally rare in the nuclear genes of yeasts. In contrast, the mitochondrial genomes of vertebrates are entirely devoid of introns, while those of yeasts may contain many introns (Freel *et al.*, 2015). We might reasonably ask what are introns doing in the genome? No-one can answer that question yet, though introns seem to have been a major evolutionary feature throughout the history of eukaryotes (Rogozin *et al.*, 2012). The consensus seems to be forming that the exon–intron structure of protein-coding genes may have evolved as ***a protection against genetic instability*** that can result from transcription errors that damage the template DNA (Bonnet *et al.*, 2017). Resources Box 6.1 suggests a few articles that will give you a little more information about introns and topics for discussion.

The term **intron** refers to both the DNA sequence within a gene and the corresponding sequence in the primary RNA transcript. Sequences that are joined together in the final mature RNA after RNA splicing are **exons. Introns** are found in the genes of most organisms and many viruses and can be located in a wide range of genes, including those that generate proteins, ribosomal RNA (rRNA) and transfer RNA (tRNA). When proteins are generated from intron-containing genes, RNA splicing takes place as

RESOURCES BOX 6.1 What Are Introns All About?

The fact that genes are split or interrupted by what came to be called introns was discovered independently in 1977 by P. A. Sharp and R. J. Roberts, who were working on protein-coding genes of adenovirus. Introns are now known to occur in a wide variety of genes and in organisms and viruses in all of the biological kingdoms.

Shepelev, V. & Fedorov, A. (2006). Advances in the Exon–Intron Database (EID). *Briefings in Bioinformatics*, **7**: 178–185. DOI: https://doi.org/10.1093/bib/bbl003. And see this URL: http://bpg.utoledo.edu/~afedorov/lab/eid.html (unfortunately, there are no fungi in the database).

The Nobel Prize in Physiology or Medicine 1993 was awarded jointly to **Sir Richard J. Roberts** and **Phillip A. Sharp** for 'their discovery that genes in eukaryotes are not contiguous strings but contain introns, and that the splicing of messenger RNA to delete those introns can occur in different ways, yielding different proteins from the same DNA sequence' (https://www.nobelprize.org/nobel_prizes/medicine/laureates/1993/).

Sharp, P.A. (2005). The discovery of split genes and RNA splicing. *Trends in Biochemical Sciences*, **30**: 279–281. DOI: https://doi.org/10.1016/j.tibs.2005.04.002.

The term intron was introduced by American biochemist Walter Gilbert: *'The notion of the cistron ... must be replaced by that of a transcription unit containing regions which will be lost from the mature messenger – which I suggest we call introns (for intragenic regions) – alternating with regions which will be expressed – exons.'*

Gilbert, W. (1978). Why genes in pieces? *Nature*, **271**: 501. DOI: https://doi.org/10.1038/271501a0.

Analyses of gene structures in different organisms have identified numerous intron gain and loss events that have occurred both recently and throughout the distant evolutionary past.

Yenerall, P. & Zhou, L. (2012). Identifying the mechanisms of intron gain: progress and trends. *Biology Direct*, **7**: 29 (10 pp.). DOI:https://doi.org/10.1186/1745-6150-7-29.

Little is known about these processes but genome-wide analyses of distant species differing in their intron content have shown that intron-containing genes, and the intron-richest genomes, are best protected against genetic instability that can result from transcription errors that damage the template DNA. This provides a possible rationale for the conservation of introns in eukaryotes.

Bonnet, A., Grosso, A.R., Elkaoutari, A. *et al.* (2017). Introns protect eukaryotic genomes from transcription-associated genetic instability. *Molecular Cell*, **67**: 608–621.e6. DOI:https://doi.org/10.1016/j.molcel.2017.07.002.

The exon–intron structure of protein-coding genes appears to have evolved along with the eukaryotic cell; introns being a major factor of evolution throughout the history of eukaryotes.

Rogozin, I.B., Carmel, L., Csuros, M. & Koonin, E.V. (2012). Origin and evolution of spliceosomal introns. *Biology Direct*, **7**: 11. DOI:http://doi.org/10.1186/1745-6150-7-11.

Stajich, J.E., Dietrich, F.S. & Roy, S.W. (2007). Comparative genomic analysis of fungal genomes reveals intron-rich ancestors. *Genome Biology*, **8**: R223. DOI:https://doi.org/10.1186/gb-2007-8-10-r223.

part of the RNA processing pathway that follows transcription and precedes translation. The word intron is derived from the term intragenic region, meaning a region inside a gene; they are sometimes called intervening sequences (https://en.wikipedia.org/wiki/Intron). The term 'intervening sequence', though, can refer to any of several families of internal nucleic acid sequences that are not present in the final gene product, including inteins ('protein introns' which are segments of a protein able to excise themselves and join the remaining portions [the exteins] with a peptide bond in a process termed protein splicing), untranslated sequences (UTR), and nucleotides removed by RNA editing, in addition to introns.

At least four distinct classes of introns have been identified:

- introns in nuclear protein-coding genes that are removed by spliceosomes (see section 5.4) (called **spliceosomal introns**);
- introns in nuclear and transfer RNA genes that are removed by proteins (tRNA introns);
- self-splicing group I introns that are removed by RNA catalysis;
- self-splicing group II introns that are removed by RNA catalysis;
- there is a fifth type, called group III introns, which are possibly related to spliceosomal introns but little is known about how their splicing takes place.

Eukaryotic protein-coding genes are interrupted by **spliceosomal introns**, which are removed from transcripts before protein translation.

The first fungal genomes to be characterised had low intron densities: the yeasts *Schizosaccharomyces pombe* (average 0.9 introns per gene) and *Saccharomyces cerevisiae* (even fewer at an average of 0.05 introns per gene). However, among filamentous ascomycete fungi, *Neurospora crassa* and *Aspergillus nidulans* have much higher intron densities (2–3 per gene), and **average** intron densities in basidiomycete and zygomycete fungi have proved to be among the highest known among eukaryotes (4–6 per gene).

Some aspects of intron structure are taxon specific; for example, introns of *Fusarium circinatum*, as well as *Fusarium verticillioides, Fusarium oxysporum* and *Fusarium graminearum*, are characterised by some unique species-specific features. Several fungal species share many of their intron positions with distantly related species; but although there are many examples of plant and animal genes having introns located in the same places, there has been a general loss of introns in fungi. Both the fungal ancestor and fungus–animal ancestor (of the opisthokont lineage, section 2.6) were very intron rich, with intron densities matching or exceeding the highest known average densities in modern species of fungi and approaching the highest known across eukaryotes. Fungal evolution has been dominated by intron loss with nearly complete intron loss in some fungal lineages. Avoiding extremes, the average picture is of moderate intron densities in the common ancestors followed by a tripling of intron number in vertebrates and plants, massive intron loss in yeasts like *Schizosaccharomyces pombe* and *Saccharomyces cerevisiae*, and variable intron loss in other fungi (Stajich *et al.*, 2007; Irimia & Roy, 2014; Phasha *et al.*, 2017).

6.4 ALTERNATIVE SPLICING

As we have seen, introns must be removed from the primary transcript by the spliceosome to generate the messenger RNA, which is translated into protein. If there are several introns in the gene sequence, there may be several ways of removing them, and several alternative mRNAs that could be spliced together as a result. For example, one gene in the ascomycete *Verticillium dahliae* (cause of wilt diseases in many plants) has been shown to use six different splice sites to produce up to five mature mRNAs. This is called **alternative splicing** of the primary transcripts of protein-coding genes. It is a major post-transcriptional regulatory mechanism which, in addition to regulation of transcription itself, provides the complex diversity of the transcriptome and proteome that characterises eukaryotes.

Alternative splicing is common throughout eukaryotes. **Transcriptome sequencing** has shown that almost 94% of human genes are alternatively spliced. In plants, estimates of alternative splicing vary from about 60% of intron-containing genes in *Arabidopsis*, 52% in soy bean, 40% in cotton, 40% in maize to 33% in rice. In fungi, it has been estimated that on average, about 6–7% of the genes are affected by alternative splicing, but the extent of alternative splicing varies across the fungal kingdom (and varies between species, too). In general, the number of splice variants found in fungi is lowest in the yeasts (3 in *Schizosaccharomyces pombe*; 9 in *Saccharomyces cerevisiae*), somewhat higher in filamentous ascomycetes (20 in *Neurospora crassa*; 100 in *Aspergillus nidulans*; 231 in *Fusarium graminearum*; 861 in *Coccidioides immitis*), and higher still in basidiomycetes (4,819 in *Schizophyllum commune*). The higher rates of alternative splicing are associated with developmental complexity and with a pathogenic lifestyle, particularly in genes involved in functions of stress response and dimorphic switching. It has been shown that alternatively spliced transcripts are regulated differentially in development.

Alternative splicing is an important regulatory mechanism, which in many eukaryotes increases the coding capacity from a limited set of genes to provide the additional complexity to the proteome that may be required for more elaborate cell functions. However, even in mammalian cells, including humans, not normally thought of as having 'a limited set of genes', most genes are alternatively spliced, and mutations in alternative transcripts can give rise to diseases such as cancer. In fungi, genes involved in virulence in fungal pathogens, genes specifying transcription factors, and genes involved in cell growth and morphogenesis have all been reported to be regulated by alternative splicing (Grützmann *et al.*, 2014; Irimia & Roy, 2014; Gehrmann *et al.*, 2016; Jin *et al.*, 2017).

6.5 TRANSPOSONS

Transposable elements (TEs) are ubiquitous and vital components of almost all prokaryotic and eukaryotic genomes. Eukaryotic transposons are classified into two main classes:

- Class I elements, also known as **retrotransposons**, use an RNA intermediate during transposition, which is transcribed from its DNA template; the **reverse transcriptase** which does this is often encoded by the TE itself.
- Class II TEs form a large and diverse group of mobile elements, but the most important of these in fungi are those with a 'cut-and-paste' transposition mechanism that does not involve an RNA intermediate. These transpositions are catalysed by an endonuclease (**transposase**), which is encoded within the TE; the basic architecture of which comprises a transposase and **terminal inverted repeats** (TIRs), which are the excision sites for the transposase. The transposase makes a staggered cut at the excision sites producing sticky ends, cuts out the DNA transposon and ligates it into target sites elsewhere in the genome. Some transposases bind non-specifically to any target site in the DNA; others bind to specific target sequences. DNA polymerase fills in the single-strand gaps resulting from the sticky ends and DNA ligase closes the sugar–phosphate backbone. This results in excision site duplication and the insertion sites of DNA transposons can be identified by short direct repeats (resulting from the DNA polymerase repair of the staggered cut in the target DNA) followed by inverted repeats of the excision sites (required for any future TE excision by transposase).

Cut-and-paste TEs may be duplicated if their transposition takes place during S-phase of the cell cycle while the DNA is being replicated and this can result in gene duplication, which plays an important role in genomic evolution.

Fungal genomes are exceptionally variable in their TE content, with 0.02–29.8% of their genome consisting of transposable elements. Like other eukaryotes, each fungal transposable element is either of class I or of class II. Here again, though, there is tremendous variability, with the genomes of two strains of *Pleurotus ostreatus* populated mainly by class I elements. A survey of 1,730 fungal genomes for transposable elements found DNA TEs across the whole data set, but no correlation was observed between TE content and fungal classification, beyond the fact that TE content generally correlated with genome size and fungal lifestyle, being elevated in mycorrhizas and diminished in pathogens of animals.

TEs are opposed by several genome defence mechanisms including **R**epeat-**I**nduced **P**oint mutation (RIP) and RNA interference (where RNA molecules inhibit gene expression or translation, by neutralising targeted mRNA molecules; now called RNAi, but also known as co-suppression, post-transcriptional gene silencing and quelling). Fungi that possessed RIP and RNAi systems had more total TE sequences but fewer elements retaining a functional transposase. This indicates stringent control over transposition and an expression of epigenetic defence intended to suppress TE expression and limit their proliferation (Gladyshev, 2017).

There are very few DNA transposons in genomes belonging to the oldest fungal lineages; the *Cryptomycota, Microsporidia, Chytridiomycota* and *Blastocladiomycota*. Lower ***terrestrial*** fungi vary in their TE composition: *Glomeromycotina* have large genomes with more than 80,000 copies of DNA TEs, but only 59 have been found in *Mortierella alpina* (though this genome had about 4,000 remnant copies that lacked transposase) and 165 in *Mortierella elongata*. Genome architectures of *Ascomycota* also varied significantly. Most members of *Saccharomycetes* had fewer than 20 TE copies with a transposase domain, whereas species of *Erysiphe, Tuber* and *Pseudogymnoascus* could have thousands of DNA TEs. In the *Basidiomycota*, two contrasting genome architectures have been distinguished: those with compact genomes with only a handful of transposons (*Ustilaginomycotina*, Microbotryomycetes) and those with large genomes with a very large number of transposons, e.g. *Agaricomycetes* (with up to a thousand TEs) and *Pucciniomycetes* (with several thousand TEs) (Kempken & Kück, 1998; Castanera *et al.*, 2016; Muszewska *et al.*, 2017).

The occurrence of 'cut-and-paste' transposons in many eukaryotic lineages and their similarity to the prokaryotic **insertion sequences** suggest that eukaryotic TEs may be older than the last common eukaryotic ancestor. TEs shape genomes by recombination and transposition. They lead to chromosomal rearrangements. They create new gene neighbourhoods. They alter gene expression by introducing new regulatory sequences for established host genes and they play key roles in adaptation to new lifestyles like mutualism/symbiosis and pathogenicity by duplicating host genes, so they can take on new roles without endangering their original functions. Pritham (2009) described eukaryotic genomes as containing 'a menagerie of populations of transposable elements' and stated that 'it is evident that these elements have played an important role in genome evolution'.

TEs are thought to have been responsible for assembling the **m**etabolic **g**ene **c**lusters (MGCs), which are common features of most fungal genomes but rarely found in other eukaryotes, though they are common in prokaryotes. MGCs are defined as 'tightly linked sets of mostly nonhomologous genes involved in a common, discrete metabolic pathway'. They encode various functions in fungi, nutrient acquisition, synthesis and/or degradation of metabolites, etc., and are reminiscent of the ***developmental subroutines*** we will discuss in section 13.13. As well as encoding the enzymes that perform these anabolic or catabolic processes, MGCs often contain appropriate regulatory sequences, and those that code for production of toxins also include the mechanisms needed to protect their fungal resident from the toxins.

This modular nature of MGCs contributes to the metabolic and ecological adaptability of fungi. MGCs enable easy pathway amplification by gene duplication, and the duplication event can also be engineered by the TEs that assembled the cluster. Indeed, as well as assembly, TEs are capable of transposing MGCs to new sites in the same genome, to other nuclei in the same heterokaryon, or to another species entirely. This last possibility is what causes the **horizontal gene transfer**, which is discussed in section 19.15 in relation to the genes coding for enzymes involved in penicillin biosynthesis. There is ***evidence for many horizontal gene transfer events in fungi***; events that have greatly enhanced the basic lifestyle of the fungi concerned (Richards *et al.*, 2011; Slot *et al.*, 2017; Steenkamp *et al.*, 2018).

6.6 PLOIDY AND GENOMIC VARIATION

A characteristic feature of fungi is the presence of large number of nuclei in a common cytoplasm. Even in fungi with septate hyphae the septa are usually perforated to some degree, so the mycelium is essentially coenocytic (although the fact that neighbouring hyphal cells can show very different differentiation states on the two sides of what appear to be open septal pores, as illustrated in Figure 13.13 (section 13.6), suggests that hyphal compartments can be physiologically distinct). Nuclei in filamentous fungi are highly dynamic in terms of both physical movement and pattern of distribution in the mycelium. The nuclei in a mycelium may cooperate or compete and may even combat each other. It has even been proposed that at least some nuclei in filamentous fungi are redundant in a genetic sense and serve instead as a stockpile of carbohydrate, nitrogen and phosphorus (in the form of their DNA) which is accessed by regulated autophagy.

As far as the primary genetic function of nuclei is concerned, the fungal mycelium is commonly **heterokaryotic**. Heterokaryosis refers to the presence of two or more genetically distinct nuclei within the same hypha. It is uncommon in all other organisms, but heterokaryosis, along with most of the other features mentioned in the previous paragraph, is a hallmark of Kingdom *Fungi* (Roberts & Gladfelter, 2016; Strom & Bushley, 2016). We deal with most aspects of the biology of heterokaryons in Chapters 5 and 8 and will not repeat them here, but we do have a few additional points to make.

Hyphal fusion between different fungal individuals is limited by vegetative compatibility barriers (section 8.5). However, these compatibility barriers are not absolute, and exchange of nuclei between hyphae of different species is now believed to enhance fungal diversification. Such an event produces a fungal **chimera**, which is an organism that contains cells or tissues from two or more different species, and this can enhance diversification at the species level by allowing horizontal gene transfer between mycelia that are too distantly related to hybridise sexually (Roper *et al.*, 2013).

Polyploidy, featuring past and recent whole-genome duplications, is a major evolutionary process in eukaryotes, particularly in plants and, but to a less extent, in animals; and it also occurs in fungi. Many fungi undergo ploidy changes during adaptation to adverse or new environments. Some fungi exist as stable haploid, diploid or polyploid (triploid, tetraploid) hyphae, while others change ploidy under some conditions and revert to the original ploidy level under other conditions. Aneuploidy (an abnormal chromosome number) is sometimes observed in fungi exposed to new or stressful environments and because of previous ploidy change. The ability of an organism to replicate and segregate its genome with high fidelity is vital to its survival and to produce future generations. Errors in replication or segregation can lead to a change in ploidy or chromosome number.

Ploidy can increase through mating, endoreduplication (which is replication of the nuclear genome in the absence of mitosis) or failure of cytokinesis after replication. Evidently, some fungi have evolved the ability to tolerate large changes of genome size and generate vast genomic heterogeneity without using the meiotic reduction division; indeed, the evolutionary history of *Saccharomyces* species has been shaped by past and recent whole-genome duplication events (Albertin & Marullo, 2012; Todd *et al.*, 2017). We have already mentioned that species of *Armillaria* are unusual in having ***diploid*** tissues in the (mushroom) sporophore though this is produced by a dikaryotic mycelium (section 5.6), and we will discuss the **parasexual cycle** in section 8.8. The parasexual cycle is the sequence: diploidisation, mitotic recombination and haploidisation through nondisjunction (improper transport of chromosomes to the poles of the division spindle during mitosis) resulting in random chromosome loss over several divisions, so the diploid is reduced to a haploid state through a series of ***aneuploid intermediates*** (Stukenbrock & Croll, 2014).

The phrase **'the Tree of Life'** is often used to describe the evolutionary history of living things. Purposefully, the phrase creates the mental image of evolution represented as a structure like a tree: the main stem is 'rooted' at the (one and only) origin of life on Earth, the stem's branches depict the major clades of evolving organisms, the secondary branches depict the orders and families that make up those clades, and the tips of the final, finest branches bear the leaves of the tree, which represent the elemental ***species of organisms***. The older we get, and the more molecular genetics we read, the more we distrust that mental image of a ***tree*** of life; and the more we feel that a better mental image is of the ***currents in an ocean***. Of flows, which are sufficiently distinct to have a direction of their own, but with occasional maelstroms of mixing vortices where different currents interact. All this disriven in organismal evolution by the insertional and transpositional genetic elements that make up the genomes of the drifting leaves. Not so much **Richard Dawkins'** ***selfish gene*** (view Richard Dawkins at the Royal Institution on YouTube: https://www.youtube.com/watch?v=j9p2F2oa0_k), but more a case of the ***ambitious gene***.

6.7 SEQUENCING FUNGAL GENOMES

Very little of what we have described in the above sections could have been written before the complete sequences of fungal genomes became available, so the topic of genome sequencing deserves some discussion.

The first whole DNA genome to be sequenced was that of the bacteriophage ΦX174 (phi-X-174). This bacteriophage has a circular single-stranded DNA genome consisting of 5,386 nucleotides that encode 11 proteins. The genome was sequenced at the University of Cambridge by a team led by Frederick Sanger; they developed and used a DNA sequencing technology that became the backbone of the first part of the genome era. The technique used the chain termination method and is now referred to as the 'first-generation technology' of genome sequencing.

As of 2019, Fred Sanger is the only person to have been awarded the Nobel Prize in Chemistry twice. The **Nobel Prize in Chemistry 1958** was awarded to **Frederick Sanger** 'for his work on the structure of proteins, especially that of insulin' (https://www.nobelprize.org/nobel_prizes/chemistry/laureates/1958/). The **Nobel Prize in Chemistry 1980** was divided, one-half awarded to **Paul Berg** 'for his fundamental studies of the biochemistry of nucleic acids, with particular regard to recombinant DNA', the other half jointly to **Walter Gilbert** and **Frederick Sanger** 'for their contributions concerning the determination of base sequences in nucleic acids' (https://www.nobelprize.org/nobel_prizes/chemistry/laureates/1980/).

The priority of genome sequencing is to establish the number, disposition and function of genes in an organism. Genomics is the systematic study of an organism's genome. Consideration of the many uses of a genome sequence started by focusing on the human genome (Sharman, 2001) and came up with these activities:

- studying the proteins and RNA of the proteome and transcriptome (and perhaps deciding how to change them to serve our own purposes);

- establishing the genetic basis of interactions between organisms, especially pathogenesis and the mechanisms of disease, but including more benign relationships such as mutualisms and mycorrhizas;
- comparing genome sequences from related organisms to examine genome evolution and relationships between organisms at the genomic level: for example, how/if genes are conserved in different species; how relationships between genomes compare with conventional taxonomic classifications, which are of course based upon the outcome of information encoded in the genome; and studying mechanisms of speciation.

The Human Genome Project began with Sanger sequencing technology. This method relies on dideoxynucleotides (ddNTPs), a type of deoxynucleotide triphosphate (dNTP) that lacks a hydroxyl group on carbon atom-3 and has a hydrogen atom instead. It is the sugar molecule which is 'dideoxy'; in the normal deoxynucleoside the deoxyribose has a hydroxyl group replaced with a hydrogen atom on carbon atom-2 (symbolised: 2′) and the sugar's 3′ and 5′ hydroxyls are used to covalently link adjacent nucleotides in the growing DNA sequence. The 3′-OH is normally phosphorylated and the phosphate is linked to a 5′-sugar carbon atom on the adjacent nucleotide to form the phosphate-sugar backbone of the polynucleotide. Because dideoxynucleotides also lack that 3′ hydroxyl group, when they bind to the growing DNA sequence, they terminate replication as they cannot covalently link to other nucleotides.

To perform Sanger sequencing, you add your primers to a solution containing the genetic information to be sequenced, then divide up the solution into four PCR reactions. Each reaction contains a nucleotide mix with one of the four nucleotides substituted with a ddNTP (A, T, G and C reactions). The DNA polymerase incorporates the **dideoxynucleotide** efficiently but is then prevented from elongating the growing chain any further. That is the '**chain termination**' part of chain termination DNA sequencing.

As an example, consider the result of carrying out DNA synthesis in the presence of dideoxy-ATP. Including ddATP in the reaction means that DNA daughter chains will be terminated at random at all those points where the template has a **thymine** (the complementary base to the dideoxy-ATP). This produces a family of adenine-nucleotide-terminated chains having lengths equivalent to all the stretches in the template that extend from the 3′-end of the primer to each thymine in the sequence. Consequently, these fragments together report the **position of every thymine** in the template. When this family of sequences is electrophoresed in polyacrylamide gel, the fragments in the population will migrate at a rate dependent on their exact length, and a series of bands are obtained, each band corresponding to one of those '3′-end to thymine' stretches in diminishing size down the gel.

Of course, there are four nucleotide bases, so to get a complete picture you need four such reactions. Then, alongside the '3′-end to thymine' family, there will be three other families of terminated chains corresponding to the '3′-end to adenine', '3′-end to cytosine' and '3′-end to guanine' stretches.

When first developed, the banding pattern was visualised by autoradiography, achieved by including a radioactively labelled nucleotide in each reaction. Products of the four reactions, that is, the ddATP, ddCTP, ddGTP and ddTTP-terminated reactions, were loaded into adjacent lanes of the gel. The smallest molecules ran fastest during electrophoresis, so the sequence could then be read from the bottom of the autoradiograph by noting the position of the band in any one of the four lanes: bands in the lane loaded with the ddATP reaction products reported the positions of **thymine** in the original template (remember, you are reading the radio-label of the complementary copy), bands in the lane loaded with ddCTP products report the locations of **guanine**, ddGTP reports **cytosine** and ddTTP-terminated reactions report **adenine** locations in the template.

Consequently, the sequence could be read from its 3′-end by reading up the four lanes of the gel. The smallest, fastest running molecule represented an oligonucleotide terminated at the first base position after the primer site in the template, the second band corresponded to the second base position after the primer, and so on. The sequence of the DNA template could be deduced by continuing to **read the banding pattern** upwards on a gel until the point near the top of the gel at which the bands could no longer be resolved. In a good separation, this corresponded to about 1,500 nucleotides.

This procedure is technically undemanding, but time consuming and labour intensive. Over the years various aspects were automated and the development of dideoxynucleotides labelled with four different **fluorescent labels**, 'fluorochrome labelling' enabled all four dideoxynucleotides to be used in one reaction tube, but this also permitted the development of sequencing machines that have a fluorescence detector that can discriminate between the different labels.

Automatic sequencing machines rely on capillary electrophoresis rather than slab gels. In capillary electrophoresis the capillary is filled with buffer solution at a certain pH value. Fluorescently labelled PCR products of various lengths are separated in the capillary according to their size, but the separating force is the difference in charge to size ratio (not their ability to filter-flow through a gel). In other words, size is measured by the overall negative charge, and the longer the fragment, the more negative the charge it bears. As the fragments are driven towards the positive electrode of a capillary by the electric field, they pass a laser beam that triggers a flash of light from the fluorochrome attached to the ddNTP that is characteristic of the base type (for example, green for A, yellow for T, blue for G, red for C). In this way, the genome is carefully read by the machine in one pass; and, of course, the machine can examine many capillary gels in each run. A single machine can sequence half a million bases per day and then continue into the night without complaint.

This is the start of the improvement in speed and accuracy, and reduction in labour and cost, of the genome-sequencing technology that was used to complete the Human Genome Project in 2003. Subsequent replacement of the electrophoretic capillary with a flow cell, miniaturisation, and use of high-throughput and massively parallel processing brought us to present-day **next-generation sequencing** (**NGS**), also called **second-generation sequencing**. Next-generation sequencing is the general term used to describe several different modern sequencing technologies, which differ in engineering configurations and sequencing chemistry. Some of these platforms can sequence 1 million to 43 billion 'short reads', of sequence fragments of 50–400 bases each, in each instrument run (see https://en.wikipedia.org/wiki/Massive_parallel_sequencing). For more details, we suggest you check out the European Bioinformatics Institute (EMBL-EBI) online video lectures at https://www.ebi.ac.uk/training/online/course/ebi-next-generation-sequencing-practical-course.

Genomics is the systematic study of the genome of an organism. 'Systematic study' may well involve comparison with the genomic sequences of other organisms; and phylogenetic study may involve comparisons with many other genomes. Genomics characteristically involves **large data sets** because it deals with DNA sequences by the megabase. Overall, the word genomics has come to embrace a considerable range of activities that can be 'structural' (these have defined endpoints that are reached when the structural determination is complete) or 'functional' (which are more open ended because additional aspects of function can be added continually). Genomics requires the use of a combination of different methods, including:

- DNA mapping and sequencing;
- collecting genome variation;
- transcriptional control of genes;
- transcriptional networks that integrate functions of, potentially, many genes;
- protein interaction networks, which are similarly potentially very extensive;
- signalling networks.

Genomics has enabled the expansionist approach to be taken to biology. Rather than being restricted by the techniques to concentrate on how **individual parts** of the organism work in isolation, the biologist can now expect to investigate how many (ultimately, perhaps all) parts of the organism work together. The newly coined expression '**omics**', although originally informal, is being increasingly used to refer to fields of study of genome biology by adding the ending '-omics'. The related suffix '-ome' is used to describe the objects of study of such fields. Some examples we have already used in this book are:

- genomics/genome, the complete gene complement of an organism;
- transcriptomics/transcriptome, all mRNA expressed transcripts;
- proteomics/proteome, all translated proteins;
- metabolomics/metabolome, the set of metabolites, the small-molecule intermediates and products, of primary and secondary metabolism.

All of these fields of study contribute to **systems biology**, which is a **holistic** (rather than **reductionist**) scientific approach focusing, often with mathematical and computational modelling, on a wide range of complex interactions in biological systems (see the Wikipedia definition at https://en.wikipedia.org/wiki/Systems_biology). Horgan and Kenny (2011) explain the rationale this way:

> The basic aspect of these approaches is that a complex system can be understood more thoroughly if considered as a whole ... . Systems biology and omics experiments differ from traditional studies, which are largely hypothesis-driven or reductionist. By contrast, systems biology experiments are hypothesis-generating, using holistic approaches where no hypothesis is known or prescribed but all data are acquired and analysed to define a hypothesis that can be further tested.

Apart from the four 'omic' fields of study outlined above, there are several others you may come across:

- taxonomics/taxome, the sum of all the described species and higher groups (genera, families, phyla) of all life, or the sum of all valid taxa of a particular life form (often specified, for example, beetle taxome, rust taxome, etc.);
- phylogenomics (at the time of writing 'phylogenome' is not defined), the reconstruction of evolutionary relationships by comparing sequences of whole genomes or portions of genomes;
- interactome, the whole set of molecular interactions in a specific biological cell;
- functome, the complete set of functional molecular units in biological cells.

The omics wiki site (http://omics.org/) describes many more. Check out the *History of Omics: As a Generic Name for Various Omics and a Standalone Biology Discipline* by Jong Bhak at http://omics.org/index.php/History_of_Omics. He describes using a computer program to generate tens of thousands of omic terms. Well, we suppose it will keep textbook authors in honest employment. There's a word for that, too, actually: ***textome***, which is 'the complete set of biological literature that contain useful information when combined to generate new information through bioinformatics'. Where will it all end? Endome?

The ability to **sequence** and **compare complete genomes** is improving our understanding of many areas of biology. Such data reveal evolutionary relationships and indicate how pathogens spread and cause disease. They enable us to approach a comprehensive understanding of the activities of living cells and how they are controlled at the molecular level. The information has practical value, too. This is why so many pharmaceutical companies are involved in genome projects: the hope is that it will be possible to identify genes responsible for, or which have influence on, diseases, and then design therapies to combat disease directly (Sharma, 2015; Taylor *et al.*, 2017).

Physical and molecular analyses originally moved the genetical focus away from the functional gene and towards the DNA sequence; now, **functional genomics** dominates. Sequencing an entire genome is only the beginning of functional studies of the **transcriptome** (all the transcripts made from the genome), the **proteome** (all the polypeptides made from the transcriptome) and the **metabolome** (all the metabolic reactions governed by the proteome). Now, in the twenty-first century our understanding is that the genome is not context sensitive because it is the full set of genetic information. Instead, the transcriptome, proteome and metabolome are all context sensitive because what they comprise depends upon the instantaneous regulatory status of the cell.

Or, as the old-time segregational (Mendelian) geneticists at the beginning of the twentieth century put it:

phenotype = genotype + environment.

6.8 ANNOTATING THE GENOME

The process of '**annotating the genome**' starts once the genome sequence has been established and its **assembly** completed. Annotation is the association of its component sequences with specific functions, and, if the *Saccharomyces cerevisiae* example is a guide, this process can continue for a long time. Annotation requires sophisticated computation, that is, it is an *in silico* analysis. Gene identification is probably the most difficult problem and relies on computer programs that align sequences and use 'gene finder' programs. Gene finding is easier with bacterial genomes, in which computer programs can find 97–99% of all genes automatically. In eukaryotes, both gene finding and gene function assignment remain challenging tasks.

The problem can be likened to identifying the beginning and end of every word in a book when the text has lost all punctuation and you have no clear idea of the language and vocabulary used in the book. Sense is made of genome sequences by annotation *in silico* to:

- identify ORFs by their start and finish codons, and allowing for the minimum length of functional proteins (Figure 6.2);
- detect the presence of recognisable functional motifs in segments of the deduced gene or protein;
- compare against known protein or DNA sequences using homologous genes from the same or other genomes (Figure 6.3).

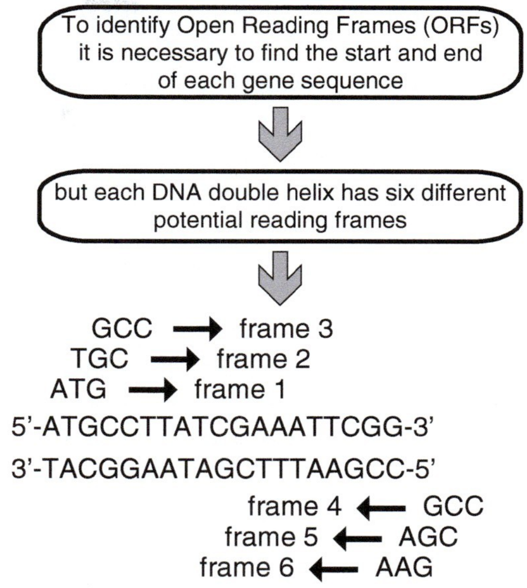

Figure 6.2. Searching for ORFs in DNA sequences, every one of which has six reading frames.

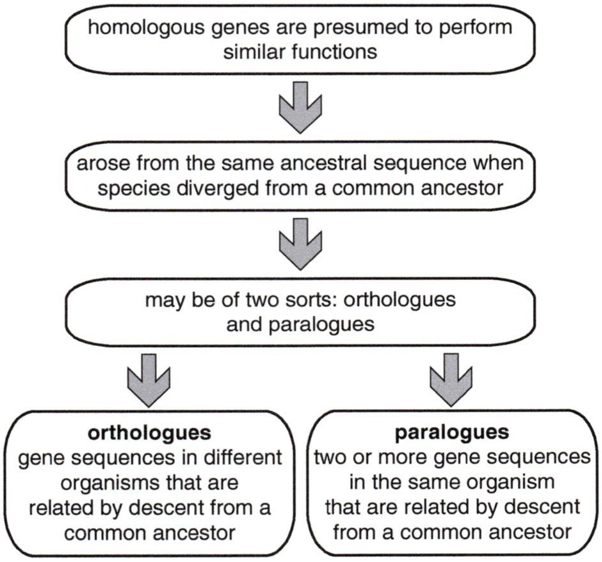

Figure 6.3. Sequence annotation with homologous genes from the same or other genomes.

Further annotation is done experimentally by:

- classical gene cloning and functional analysis;
- analysis of cDNA clones or EST sequences (an **expressed sequence tag** or EST is a short component sequence of a transcribed cDNA, so it is a portion of an expressed gene), and gene expression data.

No single method of genome annotation is comprehensive; all have their limitations, so they must be used in concert.

Many of the genes identified in sequencing projects will be 'new' in the sense that when the sequence is identified the gene function is unknown. Establishing the cellular role of such new ORFs requires a different set of bioinformatics tools that integrate sequence information with the accumulated knowledge of metabolism so that conjectures can be made about **likely functions**. Those predictions are then tested experimentally by using heterologous expression, gene knockouts and characterisation of purified proteins. Parallel analysis of phylogenetically diverse genomes can also help in understanding the physiology of the organism whose genome is being sequenced.

When the sequence of the whole genome has been established and annotated, the **genome can be compared** with others on the databases. Prokaryotic genomes are generally much smaller than those of eukaryotes. The *Escherichia coli* genome, for example, is composed of 4.64 Mb (megabase pairs) of DNA; that of *Streptomyces coelicolor* is 8 Mb, while the yeast genome, at 12.1 Mb, is about three times the size of the *Escherichia coli* genome, and the human genome is 3,300 Mb (see Table 6.1). The physical organisation is also different, because in prokaryotes the genome is contained in a single, circular, DNA molecule. Eukaryotic nuclear genomes are divided into linear DNA molecules, each contained in a different **chromosome**. In addition, all eukaryotes have mitochondria, and these possess small, usually circular, mitochondrial genomes. Photosynthetic eukaryotes (plants, algae, some protists) have a third small genome in their chloroplasts.

The size range of the genome corresponds to some extent with the degree of complexity of the organism, but the fit is not exact by any means because this correlation depends on the structure and organisation of the genes. For example, the *Escherichia coli* genome has 4,397 genes and the **yeast genome comprises about 5,000–6,000 genes**, so you might feel confident about believing that yeast has more genes because it is a eukaryote, and you can understand why it doesn't have *many* more, because it's a simple eukaryote. However, the genome of the streptomycete bacterium *Streptomyces coelicolor* contains more than 7,000 genes. This organism is a prokaryote, but it has nearly 30% more genes than the model eukaryote, yeast. Admittedly, *Streptomyces* is a complex bacterium and highly advanced in an evolutionary sense; but it *is* a bacterium. The arithmetic difference lies in the fact that the average yeast gene is 2,200 base pairs long, while the average *Streptomyces coelicolor* gene is only 1,200 base pairs long. But we can't explain *why* such a difference in gene size exists.

The yeast *Saccharomyces cerevisiae* is a well-established model organism with a long history in physiology, biochemistry and molecular biology (see section 5.2). Its genome continues to be a useful model for eukaryotes, comprising a grand total of 12.1 Mb distributed over 16 chromosomes, which range in size between 250 kb and more than 2.5 Mb. The yeast genome-sequencing project was started in 1989. The sequence of chromosome III was published in 1992, followed by the sequences of chromosomes II and XI in 1994, and the entire genome in 1996. Quality control measures ensured a 99.97% level of accuracy of the sequence.

Today, the place to learn about this genome is the ***Saccharomyces* Genome Database (SGB)** website at https://www.yeastgenome.org/ and the Yeast Genome Snapshot at https://www.yeastgenome.org/genomesnapshot. As of January 2017, there were 6,572 open reading frames (ORFs) which possibly encode metabolically active proteins, of which 5,138 were verified, 754 were uncharacterised and 680 were considered dubious. On average, a protein-encoding gene is found every 2 kb in the yeast genome. The ORFs vary from 100 to more than 4,000 codons, although two-thirds are less than 500 codons, and they are evenly distributed on the two strands of the DNA. In addition to these, the yeast genome contains 27 rRNA genes in a large tandem array on chromosome XII, 77 genes for small nuclear RNAs, 277 tRNA genes (belonging to 42 codon families) scattered across the chromosomes, and 51 copies of the yeast retrotransposons (Ty elements). There are also non-chromosomal elements, most notably the yeast mitochondrial genome (80 kb) and the 6 kb 2μ plasmid DNA, but there may be other plasmids, too.

Twenty-one years after the genome was sequenced, only about 80% of the ORFs had been ***verified***; a rate of progress that makes it even more amazing that on 17 April 2018, SGB announced a single publication in the journal *Nature* by a team of researchers jointly led by Joseph Schacherer and Gianni Liti, that had reported the ***whole-genome sequences*** and phenotypes of no fewer than 1,011 different *Saccharomyces cerevisiae* yeast strains (Peter *et al.*, 2018). Isolates of *Saccharomyces cerevisiae* were gathered from many diverse geographical locations and ecological niches; from wine, beer and bread, but also from rotting bananas, sea water, human blood, sewage, termite mounds and more. The authors then surveyed the evolutionary relationships among the strains to describe the worldwide population distribution of this species and deduce its historical spread. This unusually large-scale population genomic survey demonstrates that the likely geographic origin of *Saccharomyces cerevisiae* lies somewhere in East Asia. Budding yeast began spreading around the globe about 15,000 years ago, undergoing several independent domestication events during its worldwide journey. For example, whereas genomic markers of domestication appeared about 4,000 years ago in sake yeast, such markers appeared in wine yeast only 1,500 years ago. While domesticated isolates exhibit high variation in ploidy, aneuploidy and genome content, genome evolution in wild isolates was mainly driven by the accumulation of single nucleotide polymorphisms, most of which are present at very low frequencies.

The alleged purpose of study of a model organism like yeast is the expectation that its analysis will enable the identification of genes relevant to disease in humans; and this expectation seems to be fulfilled. Comparing the sequences of human genes available in the sequence databases with yeast ORFs shows that over **30% of yeast genes have homologues among the human sequences**, most of these representing basic cell functions. Finding this sort of homology can contribute to the understanding of human disease. The first example of this seems to be

Friedreich's ataxia, which is the most common type of inherited ataxia (loss of control of bodily movements) in humans, the biochemistry of which was uncovered by demonstrating homology to a yeast ORF of known function. Friedreich's ataxia is caused by enlargement of a GAA repeat in an intron that results in decreased expression of the frataxin gene. Frataxin is a highly conserved iron-binding protein present in most organisms, and Friedreich's ataxia pathology is associated with disruption of iron–sulfur cluster biosynthesis, mitochondrial iron overload and oxidative stress.

Frataxin is a human mitochondrial protein that has homologues in yeast. In yeast, mutants defective in the frataxin homologue accumulate iron in mitochondria and show increased sensitivity to oxidative stress. Biosynthesis of iron–sulfur clusters in yeast is a vital process involving the delivery of elemental iron and sulfur to scaffold proteins and the architecture of the protein complex to which frataxin contributes is essential to ensure concerted and protected transfer of potentially toxic iron and sulfur atoms to the mitochondrion. This comparison suggests that Friedreich's ataxia is caused by mitochondrial dysfunction and may point towards novel methods of treatment (Pastore & Puccio, 2013; Ranatunga *et al.*, 2016). In many ways, this kind of comparison alone can justify all the effort devoted to sequencing the yeast genome.

The aim of **functional genomics** is to determine the biological function of all the genes and their products, how they are regulated and how they interact with other genes and gene products. The outcome of studies that define gene/protein function is known as the **Gene Ontology**. Originally, ontology was a branch of metaphysics; a philosophical inquiry into the nature of being. For the computer scientist, ontology is the rigorous collection and organisation of knowledge about a specific feature. The aims of Gene Ontology (GO) are to:

- **develop and standardise the vocabulary** about the attributes of genes and gene products that is species neutral, and equally applicable to prokaryotes and eukaryotes, and uni- and multicellular organisms;
- **annotate genes and gene products** within sequences, and coordinate understanding and distribution of annotation data;
- **provide bioinformatics tools** to aid access to all these data.

To achieve all this, there are three organising principles of GO to describe the function of any gene/protein sequence as follows:

- **Biological process**: effectively the answer to the question **why** does the sequence exist? This can be cast in very broad terms describing the biological goals accomplished by function of the sequence, for example mitosis, meiosis, mating, purine metabolism, etc.
- **Molecular function**: effectively **what** does the sequence do? The tasks performed by individual gene products, for example transcription factor, DNA helicase, kinase, phosphatase, phosphodiesterase, dehydrogenase, etc.
- **Cellular component: where** is that function exercised? The location in subcellular structures and macromolecular complexes. For example, nucleus, telomere, cell wall, plasma membrane, endoplasmic reticulum lumen, etc.

The ontology data are freely available from the **Gene Ontology Consortium**'s website at http://www.geneontology.org/. General information about genomics is accessible through the **Broad Institute**'s listings at https://www.broadinstitute.org/.

Annotation has been automated by annotation programs (available online) that quickly identify ORFs for **hypothetical genes** in a genome. Many sequences are conserved across large evolutionary distances, so many functional assignments can be inferred using information already available from other organisms; importantly, the process of sequence search and comparison can be automated. Annotating the genes of filamentous fungi, even those of close relatives of *Saccharomyces cerevisiae*, is demanding work because their genomes are much larger and their gene structure more complex than those of yeast. **Genes of filamentous fungi often contain multiple introns** (section 6.3), with some within the open reading frame of the gene (very few yeast genes contain introns, those that do have a single intron at the start of the coding sequence, often interrupting the initiation codon). Also, the intron-boundary sequences may not become evident until the transcriptome is analysed, and alternative splicing events catalogued (section 6.4). The greater complexity of gene structure in filamentous fungi demands independent data on gene expression to make confident functional assignments. Methods have been described that use cDNA or EST sequence alignments, and gene expression data to predict reliably the function of *Aspergillus nidulans* genes. We recommend you read the discussion and explanation of the approach by Sims *et al.* (2004).

Yandell and Ence (2012) have published *'A Beginner's Guide to Eukaryotic Genome Annotation'* and further information and advice is freely available online at:

- BioInformatics Platform for Agroecosystem Arthropods at https://bipaa.genouest.org/is/how-to-annotate-a-genome/;
- Annotation for Amateurs tutorial on the PlantGDB website at this URL: http://www.plantgdb.org/tutorial/annotatemodule/;
- DNA annotation pages on the Wikipedia page at https://en.wikipedia.org/wiki/DNA_annotation.

The most up-to-date information on the genes of any organism in which you are interested can be obtained from the website devoted to that organism (use your preferred search engine to find it). For example, entering 'coprinopsis cinerea genome' into the search engine finds the *Coprinopsis cinerea* home page, which gives you general information about the organism and its genome, on the **Joint Genome Institute (JGI) Genome Portal** (https://genome.jgi.doe.gov/Copci1/Copci1.home.html). This page also has a menu of hyperlinks across the top that give access to the deepest detail about the genome of this species. The main internet sites for fungal genomic data are discussed in the next section (section 6.9).

Bioinformatics is essentially the use of computers to process biological information when computation is necessary to manage, process and understand very large amounts of data. Although there are many bioinformatics tools and databases, using them effectively often requires specialised knowledge; where this is lacking, the BioStar platform can help. ***Biostar*** is an online forum where experts and those seeking solutions to problems of computational biology exchange ideas. BioStar can be accessed at https://www.biostars.org/ (Parnell *et al.*, 2011).

Bioinformatics is particularly important as an adjunct to genomics research, because of the large amount of complex data this type of research generates, so to a great extent the word, and the approaches it encompasses, have become synonymous with the use of computers to store, search and characterise the genetic code of genes (**genomics**), the transcription products of those genes (**transcriptomics**), the proteins related to each gene (**proteomics**) and their associated functions (**metabolomics**) (see section 6.7). But there are other large data sets in need of analysis that rightly fall within range of the fundamental definition of the word 'bioinformatics'. These are large data sets arising from:

- Survey data and censuses, particularly, but not only, those involving automatic data capture and 'surveys of surveys' (***metadata***) (for example, see section 14.17).
- Data generated by mathematical models that seek to simulate a biological system and its behaviour in time (for example, see section 4.9).

Add interactions with the environment to functional genomics and you have a fully integrated biology; what has come to be known as **systems biology** (Klipp *et al.*, 2009; Nagasaki *et al.*, 2009; Horgan & Kenny, 2011).

Comprehensive studies of such large collections of molecules as occur in the transcriptome, proteome and metabolome require what are described as ***high-throughput*** methods of analysis at each stage from the generation of mutants through to the determination of which proteins are associated with which functions. Each stage generates massive amounts of data that are qualitatively and quantitatively different, which must be integrated to allow construction of realistic models of the living system (Delneri *et al.*, 2001).

Functional genomic analysis of the yeast *Saccharomyces cerevisiae* established the key concepts, approaches and techniques, although research on filamentous fungi is expanding. Considerable progress was made in analysis of yeast gene function using mutants with deletions of ORFs. However, **genetic redundancy** in the genome, resulting perhaps from gene duplication(s) during evolution, can be a problem in this type of analysis. In retrospect, analysis of yeast shows that much of the redundancy in the yeast genome is made up of identical, or almost identical, gene products fulfilling distinct physiological roles due to **differential expression** of the genes under different physiological conditions, and/or targeting the similar proteins to different cellular compartments. Nevertheless, more extensive studies require more extensive collections of mutants; those in which entire gene families are deleted and, ultimately, a collection in which all genes are represented by appropriate mutants.

There is scope for large-scale international collaboration in this sort of exercise and 1999 saw the establishment of a collection of **mutant yeast strains**, each bearing a defined deletion in one of 6,000+ potential protein-encoding genes in *Saccharomyces cerevisiae* (Winzeler *et al.*, 1999). This is the EUROSCARF collection (**EURO**pean **S**accharomyces **C**erevisiae **AR**chive for **F**unctional analysis; see http://www.euroscarf.de/). Using a PCR-based gene disruption strategy, mutant strains with a deletion of most of the ORFs in the genome were prepared in this systematic deletion project. In addition, each deleted ORF was flanked by two 20 base pair sequences unique for each deletion. These allow the sequences to be detected easily; effectively, they act as molecular barcodes that allow large numbers of deletion strains, potentially the whole library, to be analysed in parallel at the same time.

Another approach used a **transposon** that created gene fusions in a yeast clone library so that the protein products of the mutated yeast genes could be identified and analysed by immunofluorescence using antibodies to the peptide introduced by the transposon. In the original work, a *Saccharomyces cerevisiae* genomic DNA library was mutagenised in *Escherichia coli* with a multipurpose minitransposon derived from the bacterial transposable element known as Tn3. The minitransposon contained cloning sites and a 274-base pair sequence encoding 93 amino acids, called a HAT tag, which was inserted into the yeast target proteins. The HAT tag allows immunodetection of the mutated yeast protein.

Transposon mutagenesis generated 10^6 independent transformants. Subsequently, individual transformant colonies were selected and stored in 96-well plates. Plasmids were prepared from these strains and transformed into a diploid yeast strain in which homologous recombination integrated each fragment at its corresponding genomic locus, thereby replacing its genomic copy. Then, 92,544 plasmid preparations and yeast transformations were carried out, identifying a collection of over 11,000 strains, each carrying a transposon inserted within a region of the genome expressed during vegetative growth and/or sporulation. These insertions affected nearly 2,000 annotated genes, distributed over all 16 chromosomes of *Saccharomyces cerevisiae* and representing about **one-third of the yeast genome**. The study demonstrated the value of a particular strategy for mutant generation and detection, but it also indicated the scale of what has been called '**new yeast genetics**'. Finding methods that generate large numbers of gene mutants and simultaneously identify the mutants and/or their products in ways amenable to automation was the start of the ***high-throughput*** approach (Ross-Macdonald *et al.*, 1999; Cho *et al.*, 2006; Caracuel-Rios & Talbot, 2008; Honda & Selker, 2009).

Messenger RNA molecules are the subject of **transcriptome analyses** and can be studied in a fully comprehensive manner using **hybridisation-array analysis**, which is described as **a massively parallel technique** because it allows so many sequences to be examined at one time. Remember, though, that mRNA molecules transmit instructions for synthesising proteins; they do not function otherwise in the workings of the cell, so transcriptome analyses are considered to be an indirect approach to functional genomics.

The transcriptome comprises the complete set of mRNAs synthesised in the cell under any given well-defined set of physiological conditions. Unlike the genome, which has a fixed collection of sequences, the transcriptome is **context dependent**, which means that its content of sequences depends on the cell response to the current set of physiological circumstances, and the make-up of that set will change when the physiological circumstances change. Those physiological circumstances will be adapted in response to changes in both the intracellular and extracellular environment of the cell; its nutritional status, state of differentiation, age, etc. The mRNA of genes that are newly expressed (**upregulated**) will appear in the sequence collection, and the mRNA of genes that are not expressed (**downregulated**) in the new circumstance will disappear from, or be greatly reduced in, the sequence collection. Determination of the nature and sequence content of the transcriptome in all these circumstances is precisely what transcriptome analysis is intended to achieve, because the pattern of mRNA content in the transcriptome reveals the pattern of **gene regulation**.

Hybridisation arrays are now used widely to study the transcriptome because of their ability to measure the expression of many genes with great efficiency. Microarrays permit assessment of the relative expression levels of hundreds, even thousands, of genes in a single experiment. Hybridisation arrays are also called DNA micro- or macroarrays, **DNA chips, gene chips** and bio chips (Nowrousian, 2007, 2014a). The web definition of **DNA microarray** is: 'a collection of microscopic DNA spots attached to a solid surface forming an array; used to measure the expression levels of many genes simultaneously' (https://en.wiktionary.org/wiki/DNA_microarray). The array of single-stranded DNA molecules is typically distributed on glass, a nylon membrane or silicon wafer (any of which might be called 'a chip'), each being immobilised at a specific location on the chip in a predetermined (and computer-recorded) grid formation. Microarrays and macroarrays differ in the size of the sample spots of DNA; in **macroarrays**, the size of the spot is over 300 μm, in microarrays it is less than 200 μm.

Macroarrays are normally spotted by high-speed robotics onto nylon membranes, **microarrays** are made on glass (usually called custom arrays) or quartz surfaces (GeneChip®, from Affymetrix Inc.; see https://www.affymetrix.com/site/mainPage.affx) (Lipshutz *et al.*, 1999). The **immobilisation** onto the solid matrix is the most crucial aspect of the technique as it must preserve the biological activity of the molecules. The spotted material can be genomic DNA, cDNA, PCR products (any of these sized between 500–5,000 base pairs) or oligonucleotides (20–80-mer oligos).

The identities and locations of the single-stranded DNAs are known, so when the chip is treated with a suspension of experimental cDNA molecules prepared from a set of mRNAs, the cDNAs complementary to those on the chip will bind to those specific spots. The **complementary binding** pattern can be detected and since the DNAs at each position on each grid are known, the complementary binding pattern indicates the pattern of **gene expression** in the sample. Macroarrays are hybridised using a radioactive probe; normally ^{33}P, an isotope of phosphorus which decays by β-emission so that the decay, and therefore the position of the complementary binding can be imaged with a **phosphorimager**, a device in which β-particle emissions excite the phosphor molecules on the plate in a way that can be detected by scanning the plate with a laser and the attached computer converts the energy it detects to an image in which different colours represent different levels of radioactivity.

Microarrays are exposed to a set of targets either separately (single dye experiment) or in a mixture (two dye experiment) to determine the identity/abundance of complementary sequences. Laser excitation of the spots yields an emission with a spectrum characteristic of the dye(s), which is measured using a **scanning confocal laser microscope**. Monochrome images from the scanner are imported into software in which the images are pseudo-coloured and merged and combined with information about the DNAs immobilised on the chip. The software outputs an image which shows whether expression of each gene represented on the chip is unchanged, increased (upregulated) or decreased (downregulated) relative to a reference sample. In addition, data are accumulated from multiple experiments and can be examined using any number of data-mining software tools.

There are many uses for DNA microarrays. Apart from the **expression profiling** to examine the effect of physiological circumstance on gene expression on which we have so far concentrated, hybridisation arrays can be used to:

- dissect metabolic pathways and signalling networks;
- establish transcription factor regulatory patterns, target genes and binding sites;
- compare gene expression in normal tissue with that of diseased tissue, initially to establish which genes are involved in response to disease, and subsequently to diagnose disease;
- identify gene expression of different tissues and different states of cell differentiation to establish tissue-specific and/or differentiation-specific genes;
- study responses to antifungals or toxins and thereby identify drug targets, side effects and resistance mechanisms.

The proteome is the complete set of proteins synthesised in the cell under a given set of conditions. The traditional method for quantitative proteome analysis combines protein separation by high-resolution two-dimensional isoelectric focusing (IEF)/SDS-PAGE (2DE) with mass spectrometric (MS) or tandem mass spectrometric (MS/MS) identification of selected protein spots detected in the 2DE gels by use of specific protein stains. Continued improvement in technology is steadily increasing the throughput of protein identifications from complex mixtures and permitting quantification of protein expression levels and how they change in different circumstances (Aebersold, 2003; Bhadauria *et al.*, 2007; Rokas, 2009). An important feature that arises from analysis of the proteome is the enormous extent and complexity of the **network of interactions** among proteins and between proteins and other components of the cells. These **networks** can be visualised as maps of cellular function, depicting potential interactive complexes and signalling pathways.

'Metabolomics consists of strategies to quantitatively identify cellular metabolites and to understand how trafficking of these biochemical messengers through the metabolic network influences phenotype' (Jewett *et al.*, 2006). Metabolomics is particularly important in fungi because these organisms are widely used **to produce chemicals**. The main difficulty in metabolome analysis is not technical as there are sufficient analytical tools and mathematical strategies available for extensive metabolite analyses. However, the indirect relationship between the metabolome and the genome raises conceptual difficulties. The biosynthesis or degradation of a single metabolite may involve many genes, and the metabolite itself may impact on many more. Consequently, the bioinformatics tools and software required must be exceptionally powerful.

Ultimately, you may think in terms of applying all this knowledge to the creation of something entirely new. That is, to developing a biological system of some form that does not already exist in the biosphere. In the past this was achieved by the evolutionary process of artificial selection (***selective breeding***), producing crop species (like maize) or domesticated animals (like high-milk yield cattle) that simply could not exist in the wild. The 'modern' version of this is called **synthetic biology**, and with the current passion for applying management definitions to long-standing activities it has been defined as the area of science that applies engineering principles to biological systems to design and build novel biological functions and systems. Wikipedia defines synthetic biology as 'the design and construction of novel artificial biological pathways, organisms or devices, or the redesign of existing natural biological systems' (visit https://en.wikipedia.org/wiki/Synthetic_biology). Synthetic biology is emerging from molecular biology as a distinct discipline based on **quantification**. And that is the real defining feature, this is a branch of biology that depends on large-scale computer processing of large amounts of ***numerical*** data. In fact, this is a branch of biology that verges on ***engineering*** (Silver *et al.*, 2014).

6.9 FUNGAL GENOMES AND THEIR COMPARISON

Saccharomyces cerevisiae is the best-studied fungus (and has a genome comprised of about 12 Mb, with 5,000–6,000 genes that code for proteins), and the fission yeast *Schizosaccharomyces pombe* is also an important model organism for which a complete genome is available (of 13.8 Mb with 4,824 protein-coding genes). However, neither of these yeasts is an adequate model for filamentous fungi, which have more genes (approximately 8,400) and bigger genomes (30–40 Mb); both features are presumably related to the wider morphogenetic, metabolic and ecological capabilities of filamentous fungi. Certainly, it is already clear that several genes present in filamentous fungi are not present in yeasts, so **comparative genomics** is a growing business. There are several genome projects underway and planned; they include many filamentous fungi. The principal one being the **Earth BioGenome Project**, which has been described as 'a moonshot for biology'. This proposed 10-year project aims to sequence, catalogue and characterise the genomes of all of Earth's eukaryotic biodiversity (Lewin *et al.*, 2018).

Comparative genomics is a science in its own right (Gibson & Muse, 2009) and we can do no more than simply introduce it to you here. We have already used several examples from the wealth of data available to illustrate a range of topics in this textbook and these examples, together with what we will show you in this section of the range of detailed information that is online, may well inspire you to see how studies of this sort could contribute to your own interests. Significantly, with more than 2,000 already fully sequenced or in progress, the range of fungal genomes available is the **widest sampling of genomes from any eukaryotic kingdom**. This is true now and was even true when we wrote the first edition of this textbook (Galagan *et al.*, 2005), and as you can see on the websites detailed below, many of the fungal genomes fall into groups of related species that are ideal for comparative studies (see, for example, Jones, 2007). Nevertheless, the fungi chosen for sequencing initially were mostly pathogens or model organisms and dealing with this bias was one aim of the **1,000 Fungal Genomes Project** (http://1000.fungalgenomes.org/home/), the motto for which is 'Sequencing unsampled fungal diversity'. Another approach is to design genome-sequencing programmes with some specific objective in mind such as development of alternative bioenergy sources, bioremediation and fungus–environment interactions (Baker *et al.*, 2008).

The first global initiative to sequence and annotate fungal genomes was managed and coordinated by the Broad Institute of MIT and Harvard under what was called the ***Fungal Genome Initiative*** (FGI), which is still described at this legacy URL: https://www.broadinstitute.org/fungal-genome-initiative. The FGI was supported by the National Human Genome Research Institute, the National Science Foundation, the National Institute of Allergy and Infectious Disease and the US

Department of Agriculture. FGI prioritised sequence data from fungi that are important to medicine, agriculture and industry and established a sequence database for that purpose. Over 100 fungi were sequenced in this programme, including human and plant pathogens as well as fungi that serve as basic models for molecular and cellular biology. Fungal genome websites at the Broad Institute have been changed as the sequencing projects have been completed. Formerly interactive websites have been replaced with static pages providing information on fungal projects, along with links to sites where data sets can still be downloaded, and the primary repositories for all fungal genomic data now are **MycoCosm, FungiDB** and **Ensembl Fungi**.

- **MycoCosm** (https://genome.jgi.doe.gov/programs/fungi/index.jsf) is hosted by the **Joint Genome Institute (JGI)**, a Department of Energy Office of Science User Facility managed by Lawrence Berkeley National Laboratory at the University of California. MycoCosm is 'JGI's web-based fungal genomics resource, which integrates fungal genomics data and analytical tools for fungal biologists. It provides navigation through sequenced genomes, genome analysis in context of comparative genomics and genome-centric view' and offers the largest available collection of fungal genomes, for comparative genomics across phylo- and eco-groups, along with interactive web-based tools for genome downloading, searching and browsing, and a form for nominating new species for sequencing to fill gaps in the Fungal Tree of Life. This ***portal*** also hosts the **1,000 Fungal Genomes Project**, an international collaboration set up to sequence 1,000 fungal genomes (though this number has now been greatly exceeded) (https://jgi.doe.gov/our-science/science-programs/fungal-genomics/1000-fungal-genomes/); and the **Genomic Encyclopedia of Fungi**, which focuses on genomes of fungi that contribute to plant health (including symbiosis, pathogenicity and biocontrol), biorefinery mechanisms (conversion of biopolymers to sugars for fuel production) and fungal diversity (https://jgi.doe.gov/our-science/science-programs/fungal-genomics/genomic-encyclopedia-of-fungi/#feedstock) (Grigoriev *et al.*, 2014).
- **FungiDB** (http://fungidb.org/fungidb/; Stajich *et al.*, 2012) is now one of the EuPathDB family of databases (this being the eukaryotic pathogen genomics database resource) that supports a wide range of microbial eukaryotes; FungiDB (Aurrecoechea *et al.*, 2017) includes many fungal (and oomycete) species, including non-pathogens. This resource provides automated analysis of multiple genomes, curated information, with comments and supporting evidence from the user community. In addition, FungiDB offers sophisticated tools for integrating and mining diverse omics data sets that fungal biologists will find useful. The FungiDB website also gives access to a ***YouTube tutorials channel, web tutorials*** (videos and PDF downloads) and ***teaching exercises***.
- **Ensembl Fungi** is a browser for fungal genomes (http://fungi.ensembl.org/index.html). The genomes are taken from the databases of the *International Nucleotide Sequence Database Collaboration* (the European Nucleotide Archive at the European Bioinformatics Institute, https://www.ebi.ac.uk/; GenBank at the US National Center for Biotechnology Information, https://www.ncbi.nlm.nih.gov/; and the DNA Data Base of Japan, https://www.ddbj.nig.ac.jp/index-e.html). The portal offers an extensive range of tools, downloads and documentation.

As of 2019, well over ***1,000 fungal genomes have already been sequenced*** and annotated or are in the process of being sequenced and annotated (and that total does ***not*** include the 1,011 *Saccharomyces cerevisiae* genomes published by Peter *et al.*, 2018).

We strongly recommend that you visit the websites listed above because the genomic data are updated regularly as improvements and amendments are made to the sequences; but also because the index pages provide hyperlinks that allow you to access, and even download the genome sequences and information about many aspects that we cannot deal with here, including:

- basic statistics about genome size, gene density, etc.;
- search facilities allowing you to find similarities to other sequences;
- feature searches to explore and view annotated features on the sequence;
- gene indexes to find specific genes by a variety of methods;
- ability to browse the DNA sequence, find clones and graphically view sequence regions;
- opportunity to download sequence, genes, markers and other genome data.

You could start at these addresses:

- List of fungal genomes in the **MycoCosm** system at https://genome.jgi.doe.gov/fungi/fungi.info.html;
- Progress at the **1,000 Fungal Genomes Project** is regularly reported at this project website: http://1000.fungalgenomes.org/home/;
- List of all fungi on the website of **Ensembl Fungi** at http://fungi.ensembl.org/species.html.

6.10 MANIPULATING GENOMES: GENE EDITING

Analysing natural genomes rapidly leads to ideas about ***modifying*** genomes. Of course, since the dawn of agriculture, practical people have been involved in modifying the genomes of their cultivated plants and animals by a combination of **artificial selection** and **selective breeding**. Indeed, although we were unaware of it at the time, by selecting brews or ferments that produced the most satisfactory end products in brewing, baking and other food fermentations (cheese, salami, soy, miso) we have also been unconsciously applying selection pressure to the fungi and bacteria involved in those processes for a very long time (in the yeast world the process is called '***domestication***').

During the twentieth century, increasing knowledge of genetics enabled **applied genetics** to be much more formalised and very considerable advances were made in breeding improved varieties. Classical genetics of this sort puts the emphasis on the phenotype. What matters is the phenotypic characteristic of the new strain; those features that make it more useful or advantageous. In time, deeper analysis might establish the genetic basis of a phenotype or trait and enable genetic manipulation (mutation, controlled breeding) to further enhance the trait and/or combine it with others. Today, it is only a slight exaggeration to say that if you can dream of a genetic manipulation, then there's a genome editing technology that can now make it happen (McCluskey & Baker, 2017). Hibbett *et al.* (2013) call the science 'genome-enabled mycology' describing it as being characterised 'by the pervasive use of genome-scale data and associated computational tools in all aspects of fungal biology. Genome-enabled mycology is integrative and often requires teams of researchers with diverse skills in organismal mycology, bioinformatics and molecular biology.' Their paper discusses the technical and social changes that need to be made to enable all fungal biologists to make use of the new data, and is the first in a special issue of *Mycologia* devoted to genome-enabled mycology. Resources Box 6.2 brings your attention to several journal special issues and monographic reviews that are worth examining because they show you the incredibly wide range of the interests (and the technologies) of existing mycologists in a way that might inspire you to join the ranks of the practising professionals.

Automated DNA sequencing generates large volumes of genomic sequence data quite quickly. The consequence is that many genetic sequences are discovered well in advance of information about their function in the life of the organism. Molecular analysis enables us to start from the other end of this line of activity; we can seek to find the possible phenotypes generated

RESOURCES BOX 6.2 Learning More About Genome-Enabled Mycology

Several journals in the recent past have published special issues that are worth examining because they show you the incredibly wide range of the interests (and the technologies) of existing mycologists. The following is just a sample (arranged in date order):

Fungal Ecology 2012, volume 5, an issue entitled *Fungi and Global Change*, contents: https://www.sciencedirect.com/journal/fungal-ecology/vol/5/issue/1.

Mycologia 2013, issue 6 of volume 105 contained papers on *Genome-Enabled Mycology* contents: https://www.tandfonline.com/toc/umyc20/105/6.

Fungal Ecology 2014, volume 10, an issue entitled *Fungi in a Changing World: The Role of Fungi in ecosystem Response to global Change* contents: https://www.sciencedirect.com/journal/fungal-ecology/vol/10/suppl/C.

Fungal Biology 2016, issue 2 of volume 120, an issue entitled *Barcoding – Species Concepts and Species Recognition in Medical Mycology* contents: https://www.sciencedirect.com/journal/fungal-biology/vol/120/issue/2.

Fungal Biology 2016, issue 11 of volume 120, an issue entitled *Integrative Taxonomy – Uncovering Fungal Diversity* contents: https://www.sciencedirect.com/journal/fungal-biology/vol/120/issue/11.

Fungal Genetics and Biology 2016, special supplement entitled *The Era of Synthetic Biology in Yeast and Filamentous Fungi* contents: https://www.sciencedirect.com/journal/fungal-genetics-and-biology/vol/89/suppl/C.

The Royal Society of London published in 2016 the proceedings of a 2015 discussion meeting entitled 'Tackling emerging fungal threats to animal health, food security and ecosystem resilience' contents: http://rstb.royalsocietypublishing.org/content/371/1709.

In addition to special issues of journals, there are several books and monographs you should note:

Appasani, K. (2018). *Genome Editing and Engineering: From TALENs, ZFNs and CRISPRs to Molecular Surgery*. Cambridge, UK: Cambridge University Press. ISBN: 9781107170377.

Heitman, J., Howlett, B.J., Crous, P.W. *et al.* (2017). *The Fungal Kingdom*. Washington, DC: ASM Press. ISBN: 9781555819576. DOI:https://doi.org/ doi:10.1128/9781555819583.

Nowrousian, M. (ed.) (2014b). *The Mycota*, **Vol. XIII**, 2nd edition, *Fungal Genomics*. Berlin, Heidelberg: Springer-Verlag. ISBN: 978-3-642-45217-8. DOI:https://doi.org/10.1007/978-3-642-45218-5.

Petre, M. (ed.) (2015). *Mushroom Biotechnology: Developments and Applications*. London: Academic Press. ISBN: 9780128027943.

Satyanarayana, T., Deshmukh, S. & Johri, B.N. (2017). *Developments in Fungal Biology and Applied Mycology*. Singapore: Springer Nature Singapore Pte Ltd. ISBN: 978-981-10-4767-1. DOI:https://doi.org/10.1007/978-981-10-4768-8.

Thangadurai, D., Sangeetha, J. & David, M. (2016). *Fundamentals of Molecular Mycology*. Waretown, NJ: Apple Academic Press. 194 pp. ISBN: 978-1771882538.

And you may be interested in these two papers:

Harries, D. (2017a). DNA and the field mycologist: part 1. *Field Mycology*, 18: 20–23. DOI:https://doi.org/10.1016/j.fldmyc.2017.01.006.

Harries, D. (2017b). DNA and the field mycologist: part 2. *Field Mycology*, 18: 92–96. DOI:https://doi.org/10.1016/j.fldmyc.2017.07.008.

by specific genetic sequences obtained by DNA sequencing. If classical, twentieth century, genetics is considered to be forward genetics, proceeding from phenotype to genetic sequence, then the molecular genetics of the twenty-first century has come to be called **reverse genetics**. Reverse genetics attempts to link a specific genetic sequence with precise effects on the organism. But more than reversing the direction from which you view the objects the 'basic aspect of these approaches is that a complex system can be understood more thoroughly if considered as a whole' (Horgan & Kenny, 2011).

In practice, the process proceeds from functional analysis by experimental design and can eventually lead to functional design. The essential flow of activity is: gene sequence – change or disrupt the DNA (deletion, inactivation by insertion, point mutation) – mutant phenotype – function – alter function – change sequence – new (improved?) phenotype.

This is called **functional genomics**; being the study of gene function on the genomic scale. In filamentous fungi, it is a field of research that has made great advances in very recent years and which continues to advance at rapid pace. Transformation and gene manipulation systems have been developed and applied to many economically important filamentous fungi and oomycetes. Overall, the integration of information from the various processes that occur within a cell provides a more complete picture of how genes give rise to biological functions and will ultimately help us to understand the biology of organisms, in both health and disease (Weld *et al*., 2006; Bunnik & Le Roch, 2013).

We have been using fungi to produce materials of commercial value for a long time; we mentioned 'domestication' in the first paragraph and there has always been a drive to improve the fungal strains involved, and this certainly applies to yeasts used in alcohol fermentation and baking, and fungi used for cheese finishing, and a range of food fermentations around the world.

We are concentrating on twenty-first century mycology in this book, but first we wish to highlight the recombinant DNA technologies first developed in filamentous fungi more than a generation ago. A few noteworthy historical examples are:

- 1973: The **first DNA-mediated transformation** of a fungal species using genomic DNA without the use of vectors was carried out by Mishra and Tatum (1973), who transformed an inositol-requiring mutant strain of *Neurospora crassa* to inositol independence using DNA extracted from an inositol-independent strain.
- 1979: Case *et al*. (1979) developed an **efficient transformation system** for *Neurospora crassa* that used sphaeroplasts and a recombinant *Escherichia coli* plasmid carrying the *Neurospora crassa qa*-2^+ gene (which encodes the enzyme dehydroquinase).
- 1983: Ballance *et al*. (1983) performed the **first auxotrophic marker transformation in *Aspergillus nidulans*** when they relieved an auxotrophic requirement for uridine in a mutant strain of *Aspergillus nidulans* by transformation with a cloned segment of *Neurospora crassa* DNA containing the corresponding (i.e. homologous) gene coding for orotidine-5′-phosphate decarboxylase.
- 1985: The first successful **transformations of a filamentous industrial fungus** when Buxton *et al*. (1985) transformed sphaeroplasts of a mutant of *Aspergillus niger* defective in ornithine transcarbamylase function with plasmids carrying a functional copy of the *argB* gene of *Aspergillus nidulans*, and Kelly and Hynes (1985) transformed *Aspergillus niger*, which cannot use acetamide as a nitrogen or carbon source, with the *amdS* (acetamidase) gene of *Aspergillus nidulans*.

Restriction enzymes were discovered in 1970 and the 'recombinant DNA technology' toolkit emerged in the 1970s and 1980s. Several strategies have been used for these historical improvement projects, including:

- **Mutagenesis**; meaning the use of mutagens to generate random deletions, insertions and point mutations, usually by creating large populations of mutagen-treated organisms (forming a large library of mutants) using chemical mutagens (point mutations), gamma radiation (deletions) or DNA insertions (insertional knockouts). The hope is that in one step the treatment will produce strains with improved expression and secretion of the product of interest. This has been used successfully to improve productivity of:
  - α-amylase by *Aspergillus oryzae* (section 19.16);
  - 'cellulase' by *Trichoderma reesei*, some mutants of which produce up to 40 g l^{-1} total 'cellulase' activity of which half is the cellobiohydrolase known as CBH-l (see section 19.12);
  - penicillin by *Penicillium chrysogenum*, strain development by mutagenesis and strain selection of which is shown in Table 19.9 (section 19.15).
- **Site-directed mutagenesis** is a more refined technique that can modify chosen parts of the sequence of interest, such as regulatory regions in the promoter of a gene or codon changes in the ORF to identify/modify specific amino acids to affect directly the protein function. The technique can also be used to create **'gene knockouts'** by deleting a gene function (forming what is known as a **null allele**). Directed deletions have been created in every non-essential gene in the yeast genome (Winzeler *et al*., 1999) and methods are available for efficient gene targeting in filamentous fungi (Krappmann, 2007). A significant advantage of site-directed (or insertional) mutagenesis over random chemical or radiation mutagenesis is that the genes mutated by insertion are tagged (i.e. physically identified) by the transforming DNA (T-DNA), which is used to disrupt the genes. This means that the molecules are readily identifiable *in vitro*, and, if the inserted sequence carries an expressed phenotype distinct from the recipient (such as an antibiotic resistance, ability to use an exotic substrate or render a toxin harmless) then the successfully transformed cells can be identified *in vivo*.

- **Knockouts** are gene deletions; an alternative approach is to ***substitute*** genes at specific times and in specific cells with experimental sequences and this is called '**gene knockin**'. The method involves insertion of a protein-coding cDNA 'signal' or 'reporter' sequence at a particular site and is particularly applicable to study the function of the regulatory sites (promoters, for example) controlling expression of the gene being replaced. This is accomplished by observing how the easily observed reporter phenotype responds to regulation.

Gene knockouts and knockins are permanent sequence alterations. Several gene silencing techniques target the expression machinery and are generally temporary. This approach is often called **gene knockdown** since the effect is usually to grossly reduce ***expression*** of the gene. Gene silencing may use double-stranded RNA, also known as RNA interference (RNAi) or Morpholino oligos.

- RNA interference relies on a specific cellular pathway (called the RNAi pathway) interacting with the introduced double-stranded RNAs (dsRNAs, typically over 200 nucleotides long), which are made to be complementary to some target messenger RNA (mRNA). An RNase-like enzyme called Dicer in this pathway generates small interfering RNAs (siRNAs) about 20–25 nucleotides long. The siRNAs assemble into complexes containing ribonuclease (known as RISCs, or RNA-induced silencing complexes). The siRNA strands guide the RISCs to their complementary target RNA molecules, which they cut and destroy; thereby systematically interfering with expression of the target gene, so that the effect of the absence of that gene activity can be catalogued.
- Morpholino antisense oligos block access to the target mRNA without the need for mRNA degradation. Morpholinos contain standard nucleic acid bases, but instead of the bases being linked to ribose rings connected by phosphate groups, those bases are bound to morpholine rings linked through phosphorodiamidate groups. The latter are uncharged and therefore not ionised in the usual physiological pH range; this and the other structural differences mean that Morpholinos are not sensitive to the same enzymes or chemical reactions as natural polynucleotides, but they still bind to complementary sequences of RNA by standard nucleic acid base-pairing. Morpholinos (usually 25 bases in length) base pair with regions of the natural RNA; binding that blocks splicing and translation, and therefore prevents expression of the target gene.
- **Natural genetic recombination**; meaning classical 'applied genetics', involving cross-breeding to generate segregation and recombination of 'desirable' genes using the sexual cycle (although only a few of the fungi used in commercial industries reproduce sexually); the parasexual cycle (section 8.8); heterokaryosis or protoplast fusion; combined with artificial selection of the required combination of useful traits. This approach has been used successfully to improve productivity of glucoamylase by *Aspergillus niger* and exoglucanase by *Trichoderma reesei*. Generally speaking, mutagenesis and recombination strategies increase productivity by less than two-fold in a single step.
- **Genetic manipulation**; meaning the use of recombinant DNA technology to create a potentially unnatural fungal genotype that has commercially desirable characteristics, which is the main topic of the rest of this section.

All these approaches require that a recipient cell is **transformed** by uptake of the constructed DNA so that the latter can at least form a partial heterozygote that ideally undergoes homologous recombination and integrates the constructed DNA into the resident chromosome. The first barrier to successful transformation is the fungal cell wall, and most transformation techniques depend on three main ways of breaching the wall (which can be combined to improve efficiency) (Weld *et al.*, 2006):

- enzymic removal of the cell wall to create **protoplasts** (which lack all wall material) or **sphaeroplasts** (which retain a residual amount of the original wall);
- use of **electroporation** by applying electric shocks; a brief electric pulse (lasting in the region of 1 to 20 ms) at a potential gradient of about 0.5 to 10kV cm^{-1} is applied to temporarily permeabilise cell membranes to enable entry of large, charged molecules across the hydrophilic membrane; or
- by 'shooting' micrometre-sized particles (usually of denser, relatively inert, metals like tungsten or gold) coated with DNA or RNA ***into*** the cells; a process called **biolistic transformation**. The microparticles coated with DNA or RNA are introduced into cells by being accelerated to velocities of approximately 500 m s^{-1} by the forces generated by explosion of gunpowder or by explosive expansion of cold helium gas.

Success was achieved with all these approaches, but by far the most important development was ***Agrobacterium tumefaciens*-mediated transformation (AMT)**.

Agrobacterium tumefaciens is a gram-negative bacterium which is a common plant pathogen that causes *crown gall tumours on plants*. This tumorous growth of the plant tissue is induced when the bacterium transfers some bacterial DNA (called T-DNA) to the host plant. T-DNA is located on a 200 kbp plasmid (the tumour inducing or Ti plasmid). The T-DNA integrates into the plant genome, then T-DNA genes that encode enzymes to produce plant growth regulators are expressed, and their expression results in uncontrolled growth of the plant cells. However, for use as a cloning vector, the T-region of the Ti plasmid can be deleted and replaced by other DNA sequences because plasmid virulence, transfer and integration are controlled by genes elsewhere on the plasmid.

What is significant for our present discussion is that *Agrobacterium tumefaciens can transfer its T-DNA to a very wide range of fungi* and produces a significantly higher frequency of more stable transformants than alternative transformation methods (Michielse *et al.*, 2005). AMT is a relatively simple system to work with, primarily because it does not require the production of protoplasts or sphaeroplasts. Indeed, a major attraction of AMT is the variety of starting materials that can be used: protoplasts, spores, mycelium and pieces of sporophore tissues have all been successfully transformed. Even fungi that have not been transformed by other systems have been successfully transformed by co-cultivation with *Agrobacterium*. The approach seems to be applicable to the full range of fungi (zygomycetes, *Ascomycota* and *Basidiomycota*) and shows great potential for fungal biotechnology and medicine (Michielse *et al.*, 2005; Sugui *et al.*, 2005). *Agrobacterium tumefaciens*-mediated transformation has been described as:

> one of the most transformative technologies for research on fungi developed in the last 20 years, a development arguably only surpassed by the impact of genomics … [AMT] has been widely applied in forward genetics, whereby generation of strain libraries using random T-DNA insertional mutagenesis, combined with phenotypic screening, has enabled the genetic basis of many processes to be elucidated. Alternatively, AMT has been fundamental for reverse genetics, where mutant isolates are generated with targeted gene deletions or disruptions, enabling gene functional roles to be determined. (Idnurm *et al.*, 2017)

Despite the confident descriptions given above and the use of phrases like 'relatively simple system', applying a transformation system to an organism for the first time is often not as 'simple' as might be suggested. There are many variables that must be optimised and even after reliable transformation systems have been developed, there may still be difficulties to overcome before it is possible to analyse gene function. A major potential problem for genetic analysis of any filamentous fungus is the multinucleate nature of the hyphae. Multiple nuclei can confuse results because gene replacement and insertional mutagenesis rely on the isolation of *homokaryotic transformants derived from a single transformation event* to study loss of function mutants (Weld *et al.*, 2006). The consequence is that methods must be carefully refreshed and optimised every time they are applied to a new organism.

Gene cloning involves inserting DNA molecules of interest into specialised carriers called **vectors** that enable replication within a host cell, producing many copies of the inserted piece of DNA carried by the vector. Cloning vectors are 'engineered' to contain one or several recognition sites for restriction enzymes. Digesting both the vector and the DNA to be cloned with the same restriction enzyme produces complementary 'sticky ends' in both molecules, allowing the foreign (or **heterologous**) DNA fragment to be inserted into the vector. A vector carrying an inserted fragment of DNA is known as a recombinant plasmid. The replicated molecules are called clones because all the copies made in the host cell are identical.

After harvesting from the host cell, the cloned DNA can be purified for further analysis. There are several types of cloning vector, which differ in origin, nature of host cell and in their capacity for the size of inserted DNA they can carry. The simplest vectors are bacterial plasmids, which are circular, double-stranded DNA molecules that replicate in the host independently of the main bacterial chromosome. Commonly used plasmids can carry up to 15 kb of foreign DNA, or up to 25 kb can be accommodated in vectors derived from the **bacteriophage ('phage') lambda (λ)**. This is a double-stranded DNA virus that infects the bacterium *Escherichia coli*. The λ phage DNA molecule circularises after infection because it has complementary single-stranded overlaps at each end known as *cos* (for cohesive end) sites. A completely artificial, larger capacity vector has been engineered by inserting *cos* sites into a plasmid. These are called **cosmids**. They can carry up to 45 kb of inserted DNA and have the additional advantages that they use a virus coat to infect host bacteria (a very efficient way of entering the host) but replicate like a plasmid and can be constructed to use plasmid-derived markers for recombinant selection.

Yeast artificial chromosomes (YACs) can carry DNA inserts of up to 1 million base pairs (1 megabase = 1 Mb) in length. YAC vectors are plasmids that contain yeast centromere DNA, two yeast telomeres separated by a restriction site, and yeast replication origins (autonomous replication sequences, or ARS) as well as two selectable markers. Restriction enzyme digestion produces two fragments, one a telomere + selectable marker + cloning site, the other a telomere + selectable marker + replication origin + centromere + cloning site, which are mixed with the DNA to be cloned. Among the constructs which result will be some which behave like yeast chromosomes during mitosis. Any that are constructed with two centromeres, without a centromere, or lacking a telomere will fail to segregate. Consequently, the presence of both selectable markers coupled with proper mitotic segregation is sufficient to identify the desired constructs.

All yeast vectors are **shuttle vectors**, meaning that they can be propagated (that is, grown) in cell cultures of both yeast and the bacterium *Escherichia coli*. These vectors contain a bacterial plasmid backbone that contains all the functions required for maintenance and selection in *Escherichia coli*. They also contain yeast chromosomal elements that determine their characteristics and behaviour within yeast cells. The main types of yeast vectors are:

- **integrative plasmids (YIp)**, which are maintained as a single copy providing they have integrated successfully into the genome;
- **replicative plasmids (YRp)**, which contain a chromosomal origin of replication (ARS), and because of this origin of replication are maintained autonomously at high copy number (which means 20 to 200 copies of the plasmid per yeast cell);

- **centromeric plasmids (YCp)**, which contain both ARS and centromere sequences and are consequently maintained in the cell as a single copy autonomously replicating supplementary chromosome;
- **episomal plasmids (YEp)**, which is also an autonomously replicating plasmid (contains ARS) but contains the origin of replication from yeast's own 2μ plasmid so it is maintained at a copy number of about 20 to 50 copies per cell. This type of vector is used for gene over-expression purposes (as are YRps). Gene over-expression creates a gain of function mutation and requires the use of multicopy vectors and strong promoters.

The ideal vector carries easily selectable markers, which enable transfer and incorporation of the vector and its cargo-DNA to be detected, and easily controlled regulation of the cargo-DNA (which usually means a strong and readily controlled promoter) so that expression of the genes in which you are interested can be controlled; for examples, see Meyer *et al.* (2011, 2016) and Gressler *et al.* (2015). The best way of finding out about useful vectors is to view the genomics website for the organism you want to study. At the very least, this will give you references to research on the organism, which will direct you towards vectors and techniques that have already been used successfully.

By a very considerable distance, the most crucial development in recent years has been **gene editing**. The process depends on **engineered nucleases**, which can be designed to cut at any location in the genome of any species and introduce modified DNA sequences into the endogenous (host organism) sequence. There are three major classes of engineered nuclease enzymes (we will describe the fourth gene editing system, the ***CRISPR-Cas system***, separately below):

- zinc-finger nucleases (ZFNs);
- transcription activator-like effector nucleases (**TALENs**); and
- engineered meganucleases.

Engineered nucleases create site-specific double-strand breaks at desired locations in the genome. These fusion proteins serve as readily targetable 'DNA scissors' for gene editing applications that enable targeted genome modifications to be accomplished such as sequence insertion, deletion, repair and replacement in living cells. The induced double-strand breaks are repaired through nonhomologous end-joining or homologous recombination, and the whole process results in precisely targeted mutations ('edits') being incorporated into the experimental genome. This type of gene editing was selected by the journal *Nature Methods* as the ***2011 Method of the Year*** (Anonymous, 2011). Fundamental to the use of engineered nucleases in genome editing is that the engineered enzymes produce double-stranded breaks (DSBs) in the DNA of the target organism. Double-strand breaks are cytotoxic lesions that threaten genome integrity and most organisms have mechanisms to repair these breaks (Ceccaldi *et al.*, 2016).

The concept underlying ZFNs and TALENs technologies is that of a nonspecific DNA cutting catalytic domain (obtained from an endonuclease with discrete and separate DNA recognition and cleaving sites) being linked to peptides that recognise specific DNA sequences such as **zinc fingers** (ZFNs) and **transcription activator-like effectors** (TALEs). Zinc-finger motifs occur in several transcription factors. The C-terminal part of each finger is responsible for the specific recognition of a short region (about 3 base pairs) of the DNA sequence. Combining 6–8 zinc fingers whose recognition sites have been characterised produces a protein that can target around 20 base pairs of a specific gene.

Although the nuclease portions of both ZFNs and TALENs constructs have similar properties, the difference between these engineered nucleases is in their DNA recognition peptide. ZFN '**zinc fingers**' rely on a combination of cysteine and histidine residues to react with their metal ions so codons for those amino acids identify the nuclease target sequence.

Transcription Activator-Like Effectors (TALEs) are proteins secreted by *Xanthomonas* plant pathogenic bacteria that bind promoter sequences in the host and activate expression of plant genes that aid bacterial infection. They recognise plant DNA sequences through a central repeat domain consisting of a variable number of about 34 amino acid repeats. **TALEs** can be engineered to bind to practically any desired DNA sequence, so when combined with a **n**uclease, the TALENs (which are artificial, engineered, restriction enzymes) can cut DNA at the specific location(s) desired by the experimenter. TALEN constructs are used in a similar way to ZFNs but have three advantages in targeted mutagenesis: (i) DNA-binding specificity is higher; (ii) off-target effects are lower; and (iii) construction of DNA-binding domains is easier.

Meganucleases, discovered in the late 1980s, are endonucleases characterised by a large recognition site (DNA sequences of 12–40 base pairs). Sites of this length generally occur only once in any given genome, so meganucleases are the most specific of the naturally occurring restriction enzymes. Such meganucleases are quite common, but the most valuable tools for gene engineering have been derived from the LAGLIDADG family of endonucleases, so called for the conservation of a specific amino acid sequence motif which is defined by each letter as a code that identifies a specific residue (the motif is: Leucine–Alanine–Glycine–Leucine–Isoleucine–Aspartic acid–Alanine–Aspartic acid–Glycine). This motif binds to a specific DNA sequence; change the amino acid sequence and it will bind to a different DNA sequence. The 'engineering' aspect of this is that mutagenesis and high-throughput screening methods have been used to create meganuclease variants that recognise ***a defined catalogue of unique DNA sequences***. Others have been fused to various meganucleases to create hybrid enzymes that recognise a new sequence and yet others have had the DNA interacting amino acids of the meganuclease altered to design sequence-specific meganucleases; all contributing to what is called rationally designed meganuclease.

Meganucleases have the benefit of causing less toxicity in cells than ZFNs because of more stringent DNA sequence recognition; however, the construction of sequence-specific enzymes for all possible sequences is costly and time consuming. Nevertheless, it can be done. View https://en.wikipedia.org/wiki/Genome_editing to learn more about engineered nucleases.

The **CRISPR-Cas9**-based system has become a common platform for genome editing in a variety of organisms. CRISPRs (**C**lustered **R**egularly **I**nterspaced **S**hort **P**alindromic **R**epeats) are genetic elements, which provide bacteria with adaptive immunity to viruses and plasmids. They consist of short sequences that originate as remnants of genes from past infections, sandwiched between unusual, repeated bacterial DNA sequences, the 'clustered regularly interspaced short palindromic repeats' that give CRISPR its name. The CRISPR-associated protein Cas9 is an endonuclease that uses a guide sequence within an RNA duplex, tracrRNA:crRNA, to form base pairs with DNA target sequences, enabling Cas9 to introduce a site-specific double-strand break in the DNA. The dual tracrRNA:crRNA was ***engineered*** as a single guide RNA (sgRNA) that retains two critical features: a sequence at the 5′ side that determines the DNA target site by Watson-Crick base-pairing with the target DNA and a duplex RNA structure at the 3′ side that binds to Cas9. From this, Doudna and Charpentier (2014) created a simple two-component system in which experimenter-determined changes in the guide sequence of the sgRNA direct Cas9 to target the specific DNA sequence of interest to the experimenter. Cas9-sgRNA-mediated DNA cleavage produces a blunt double-stranded break in the target DNA that triggers repair enzymes to disrupt or replace DNA sequences at or near the cleavage site. Catalytically inactive forms of Cas9 can also be used for programmable regulation of transcription and visualisation of genomic loci.

The simplicity of the CRISPR-Cas9 system has made this a cost-effective and easy-to-use technology to precisely and efficiently target, edit, modify, regulate and mark genomic loci of a wide array of cells and organisms. By introducing plasmids containing Cas genes and specifically constructed CRISPRs into living eukaryotic cells, the eukaryotic genome, ***any genome***, can be cut at any desired position. This is the quickest and cheapest method for gene editing and requires the least amount of expertise in molecular biology because it is RNA rather than protein that is engineered to guide the nuclease to the target. This is a major advantage that CRISPR has over the ZFN and TALEN methods; it can target different DNA sequences using its about 80-nucleotide sgRNAs, while both ZFN and TALEN methods require construction and testing of the proteins created for targeting each DNA sequence. The CRISPR-Cas system was selected by the journal *Science* as its *2015 Breakthrough of the Year* (McNutt, 2015); you can read about the latest developments in the '*CRISPR revolution*' topic page written by Jon Cohen (a staff writer for *Science*) at this URL: http://www.sciencemag.org/topic/crispr.

Gene editing technologies have been developed for application to animals (Dunn & Pinkert, 2014), plants (Mohanta *et al.*, 2017) and fungi (Nødvig *et al.*, 2015; Chen *et al.*, 2017; Pudake *et al.*, 2017; Zheng *et al.*, 2017), and we think they all make fascinating reading (because even those that deal with organisms in which you have little or no interest can show tactics and strategies that might help you with your organism). We make a few suggestions for future research targets at the end of section 19.26.

Anzalone *et al.* (2019) describe a technique they call **prime editing** (or ***search-and-replace genome editing***) as being: '...a versatile and precise genome editing method that directly writes new genetic information into a specified DNA site using a catalytically impaired Cas9 fused to an engineered reverse transcriptase, programmed with a prime editing guide RNA (pegRNA) that both specifies the target site and encodes the desired edit.'

The authors claim that prime editing greatly expands the scope and capabilities of genome editing, and in principle could correct about 89% of known pathogenic human genetic variants. So, what could it do for fungi?

6.11 REFERENCES

Aebersold, R. (2003). Quantitative proteome analysis: methods and applications. *Journal of Infectious Diseases*, **187**: S315–S320. DOI: https://doi.org/10.1086/374756.

Albertin, W. & Marullo, P. (2012). Polyploidy in fungi: evolution after whole-genome duplication. *Proceedings of the Royal Society B: Biological Sciences*, **279**: 2497–2509. DOI: https://doi.org/10.1098/rspb.2012.0434.

Anonymous (2011). Method of the Year 2011. *Nature Methods*, **9**: 1. DOI: https://doi.org/10.1038/nmeth.1852.

Anzalone, A.V., Randolph, P.B., Davis, J.R., Sousa, A.A., Koblan, L.W., Levy, J.M., Chen, P.J., Wilson, C., Newby, G.A., Raguram, A. & Liu, D.R. (2019). Search-and-replace genome editing without double-strand breaks or donor DNA. *Nature*, **576**: 149–157. DOI: https://doi.org/10.1038/s41586-019-1711-4.

Aurrecoechea, C., Barreto, A., Basenko, E.Y. *et al.* (2017). EuPathDB: the eukaryotic pathogen genomics database resource. *Nucleic Acids Research*, **45**: D581–D591. DOI: https://doi.org/10.1093/nar/gkw1105.

Baker, S.E., Thykaer, J., Adney, W.S. *et al.* (2008). Fungal genome sequencing and bioenergy. *Fungal Biology Reviews*, **22**: 1–5. DOI: https://doi.org/10.1016/j.fbr.2008.03.001.

Ballance, D.J., Buxton F.P. & Turner, G. (1983). Transformation of Aspergillus nidulans by the orotidine-5′-phosphate decarboxylase gene of *Neurospora crassa*. *Biochemical and Biophysical Research Communications*, **112**: 284–289. DOI: https://doi.org/10.1016/0006-291X(83)91828-4.

Bhadauria, V., Zhao, W.-S., Wang, L.-X. *et al.* (2007). Advances in fungal proteomics. *Microbiological Research*, **162**: 193–200. DOI: https://doi.org/10.1016/j.micres.2007.03.001.

Bonnet, A., Grosso, A.R., Elkaoutari, A. *et al.* (2017). Introns protect eukaryotic genomes from transcription-associated genetic instability. *Molecular Cell*, **67**: 608–621.e6. DOI: https://doi.org/10.1016/j.molcel.2017.07.002.

Bunnik, E.M. & Le Roch, K.G. (2013). An introduction to functional genomics and systems biology. *Advances in Wound Care*, **2**: 490–498. DOI: https://doi.org/10.1089/wound.2012.0379.

Buxton, F.P., Gwynne, D.I. & Davies, R.W. (1985). Transformation of *Aspergillus niger* using the argB gene of *Aspergillus nidulans*. *Gene*, **37**: 207–214. DOI: https://doi.org/10.1016/0378-1119(85)90274-4.

Caracuel-Rios, Z. & Talbot, N.J. (2008). Silencing the crowd: high-throughput functional genomics in *Magnaporthe oryzae*. *Molecular Microbiology*, **68**: 1341–1344. DOI: https://doi.org/10.1111/j.1365-2958.2008.06257.x.

Case, M.E., Schweizer, M., Kushner, S.R. & Giles, N.H. (1979). Efficient transformation of *Neurospora crassa* by utilizing hybrid plasmid DNA. *Proceedings of the National Academy of Sciences of the United States of America*, **76**: 5259–5263. URL: https://www.jstor.org/stable/70428.

Castanera, R., López-Varas, L., Borgognone, A. *et al.* (2016). Transposable elements versus the fungal genome: impact on whole-genome architecture and transcriptional profiles. *PLoS Genetics*, **12**: article number e1006108. DOI: https://doi.org/10.1371/journal.pgen.1006108.

Ceccaldi, R., Rondinelli, B. & D'Andrea, A.D. (2016). Repair pathway choices and consequences at the double-strand break. *Trends in Cell Biology*, **26**: 52–64. DOI: https://doi.org/10.1016/j.tcb.2015.07.009.

Chen, J., Lai, Y., Wang, L. *et al.* (2017). CRISPR/Cas9-mediated efficient genome editing via blastospore-based transformation in entomopathogenic fungus *Beauveria bassiana*. *Scientific Reports*, **7**: article number 45763. DOI: https://doi.org/10.1038/srep45763.

Cho, Y., Davis, J.W., Kim, K.-H. *et al.* (2006). A high-throughput targeted gene disruption method for *Alternaria brassicicola* functional genomics using linear minimal element (LME) constructs. *Molecular Plant-Microbe Interactions*, **19**: 7–15. DOI: https://doi.org/10.1094/MPMI-19-0007.

Delneri, D., Brancia, F.L. & Oliver, S.G. (2001). Towards a truly integrative biology through the functional genomics of yeast. *Current Opinion in Biotechnology*, **12**: 87–91. DOI: https://doi.org/10.1016/S0958-1669(00)00179-8.

Doudna, J.A. & Charpentier, E. (2014). The new frontier of genome engineering with CRISPR-Cas9. *Science*, **346**: article 1258096. DOI: https://doi.org/10.1126/science.1258096.

Dunn, D.A. & Pinkert, C.A. (2014). Gene editing. In *Transgenic Animal Technology. A Laboratory Handbook*, 3rd edition, ed. C.A. Pinkert. London: Elsevier, pp. 229–248. ISBN: 9780124104907. DOI: https://doi.org/10.1016/B978-0-12-410490-7.00008-6.

Dyer, P.S., Munro, C.A. & Bradshaw, R.E. (2017). Fungal genetics. In *Oxford Textbook of Medical Mycology*, ed. C.C. Kibbler, R. Barton, N.A.R. Gow *et al.* Oxford, UK: Oxford University Press, pp. 35–42. ISBN-10: 0198755384, ISBN-13: 978-0198755388.

Freel, K.C., Friedrich, A. & Schacherer, J. (2015). Mitochondrial genome evolution in yeasts: an all-encompassing view. *FEMS Yeast Research*, **15**: fov023. DOI: https://doi.org/10.1093/femsyr/fov023.

Galagan, J.E., Henn, M.R., Ma, L.-J., Cuomo, C.A. & Birren, B. (2005). Genomics of the fungal kingdom: insights into eukaryotic biology. *Genome Research*, **15**: 1620–1631. DOI: https://doi.org/10.1101/gr.3767105.

Gehrmann, T., Pelkmans, J.F., Lugones, L.G. *et al.* (2016). *Schizophyllum commune* has an extensive and functional alternative splicing repertoire. *Scientific Reports*, **6**: article number 33640. DOI: https://doi.org/10.1038/srep33640.

Gibson, G. & Muse, S. (2009). *A Primer of Genome Science*, 3rd edition. Basingstoke, UK: Sinauer Associates/Macmillan Publishers. ISBN-10: 0878932364, ISBN-13: 978-0878932368.

Gilbert, W. (1978). Why genes in pieces? *Nature*, **271**: 501. DOI: https://doi.org/10.1038/271501a0.

Gladyshev, E. (2017). Repeat-Induced Point mutation (RIP) and other genome defense mechanisms in fungi. *Microbiology Spectrum*, **5**: FUNK-0042–2017. DOI: https://doi.org/10.1128/microbiolspec.FUNK-0042-2017.

Gressler, M., Hortschansky, P., Geib, E. & Brock, M. (2015). A new high-performance heterologous fungal expression system based on regulatory elements from the *Aspergillus terreus* terrein gene cluster. *Frontiers in Microbiology*, **6**: 184. DOI: https://doi.org/10.3389/fmicb.2015.00184.

Grigoriev, I.V., Nikitin, R., Haridas, S. *et al.* (2014). MycoCosm portal: gearing up for 1000 fungal genomes. *Nucleic Acids Research*, **42**: D699–D704. DOI: https://doi.org/10.1093/nar/gkt1183.

Grützmann, K., Szafranski, K., Pohl, M. *et al.* (2014). Fungal alternative splicing is associated with multicellular complexity and virulence: a genome-wide multi-species study. *DNA Research*, **21**: 27–39. DOI: https://doi.org/10.1093/dnares/dst038.

Harries, D. (2017a). DNA and the field mycologist: part 1. *Field Mycology*, **18**: 20–23. DOI: https://doi.org/10.1016/j.fldmyc.2017.01.006.

Harries, D. (2017b). DNA and the field mycologist: part 2. *Field Mycology*, **18**: 92–96. DOI: https://doi.org/10.1016/j.fldmyc.2017.07.008.

Heitman, J., Howlett, B.J., Crous, P.W. *et al.* (2017). *The Fungal Kingdom*. Washington, DC: ASM Press. ISBN: 9781555819576. DOI: https://doi.org/10.1128/9781555819583.

Hibbett, D.S., Stajich, J.E. & Spatafora, J.W. (2013). Toward genome-enabled mycology. *Mycologia*, **105**: 1339–1349. DOI: https://doi.org/10.3852/13-196.

Honda, S. & Selker, E.U. (2009). Tools for fungal proteomics: multifunctional *Neurospora* vectors for gene replacement, protein expression and protein purification. *Genetics*, **182**: 11–23. DOI: https://doi.org/10.1534/genetics.108.098707.

Horgan, R.P. & Kenny, L.C. (2011). 'Omic' technologies: genomics, transcriptomics, proteomics and metabolomics. *The Obstetrician & Gynaecologist*, **13**: 189–195. DOI: https://doi.org/10.1576/toag.13.3.189.27672.

Idnurm, A., Bailey, A.M., Cairns, T.C. *et al.* (2017). A silver bullet in a golden age of functional genomics: the impact of *Agrobacterium*-mediated transformation of fungi. *Fungal Biology and Biotechnology*, **4**: 6. DOI: https://doi.org/10.1186/s40694-017-0035-0.

Irimia, M. & Roy, S.W. (2014). Origin of spliceosomal introns and alternative splicing. *Cold Spring Harbor Perspectives in Biology*, **6**: article a016071. DOI: https://doi.org/10.1101/cshperspect.a016071.

Jewett, M.C., Hofmann, G. & Nielsen, J. (2006). Fungal metabolite analysis in genomics and phenomics. *Current Opinion in Biotechnology*, **17**: 191–197. DOI: https://doi.org/10.1016/j.copbio.2006.02.001.

Jin, L., Li, G., Yu, D. *et al.* (2017). Transcriptome analysis reveals the complexity of alternative splicing regulation in the fungus *Verticillium dahlia*. *BMC Genomics*, **18**: 130. DOI: https://doi.org/10.1186/s12864-017-3507-y.

Jones, M.G. (2007). The first filamentous fungal genome sequences: *Aspergillus* leads the way for essential everyday resources or dusty museum specimens? *Microbiology*, **153**: 1–6. DOI: https://doi.org/10.1099/mic.0.2006/001479-0.

Kelly, M.K. & Hynes, M.J. (1985). Transformation of *Aspergillus niger* by the amdS gene of *Aspergillus nidulans*. *EMBO Journal*, **4**: 475–479. URL: https://www.ncbi.nlm.nih.gov/pmc/articles/PMC554210.

Kempken, F. & Kück, U. (1998). Transposons in filamentous fungi: facts and perspectives. *BioEssays*, **20**: 652–659. DOI: https://doi.org/10.1002/(SICI)1521-1878(199808)20:8<652::aid-bies8>3.0.CO;2-K.

Klipp, E., Liebermeister, W., Wierling, C. *et al.* (2009). *Systems Biology: A Textbook*. Weinheim, Germany: Wiley-VCH Verlag GmbH & Co. KgaA. ISBN-10: 3527318747, ISBN-13: 978-3527318742.

Krappmann, S. (2007). Gene targeting in filamentous fungi: the benefits of impaired repair. *Fungal Biology Reviews*, **21**: 25–29. DOI: https://doi.org/10.1016/j.fbr.2007.02.004.

Lewin, H.A., Robinson, G.E., Kress, W.J. *et al.* (2018). Earth BioGenome Project: sequencing life for the future of life. *Proceedings of the National Academy of Sciences of the United States of America*, **115**: 4325–4333. DOI: https://doi.org/10.1073/pnas.1720115115.

Lindegren, C.C. (1949). *The Yeast Cell, Its Genetics and Cytology*. St Louis, MI: Educational Publishers, Inc. ASIN: B001P8I0BW.

Lipshutz, R.J., Fodor, S.P., Gingeras, T.R. & Lockhart, D.J. (1999). High density synthetic oligonucleotide arrays. *Nature Genetics*, **21** (January supplement): 20–24. DOI: https://doi.org/10.1038/4447.

McCluskey, K. & Baker, S.E. (2017). Diverse data supports the transition of filamentous fungal model organisms into the post-genomics era. *Mycology*, **8**: 67–83. DOI: 10.1080/21501203.2017.1281849.

McNutt, M. (2015). Editorial: breakthrough to genome editing. *Science*, **350**: 1445. DOI: https://doi.org/10.1126/science.aae0479.

Mehrabi, R., Mirzadi Gohari, A. & Kema, G.H.J. (2017). Karyotype variability in plant-pathogenic fungi. *Annual Review of Phytopathology*, **55**: 483–503. DOI: https://doi.org/10.1146/annurev-phyto-080615-095928.

Meyer, V., Nevoigt, E. & Wiemann, P. (2016). The art of design. *Fungal Genetics and Biology*, **89**: 1–2. DOI: https://doi.org/10.1016/j.fgb.2016.02.006.

Meyer, V., Wanka, F., van Gent, J. *et al.* (2011). Fungal gene expression on demand: an inducible, tunable, and metabolism-independent expression system for *Aspergillus niger*. *Applied and Environmental Microbiology*, **77**: 2975–2983. DOI: https://doi.org/10.1128/AEM.02740-10.

Michielse, C.B., Hooykaas, P.J.J., van den Hondel, C.A.M.J.J. & Ram, A.F.J. (2005). *Agrobacterium*-mediated transformation as a tool for functional genomics in fungi. *Current Genetics*, **48**: 1–17. DOI: https://doi.org/10.1007/s00294-005-0578-0.

Mishra, N.C. & Tatum, E.L. (1973). Non-Mendelian inheritance of DNA-induced inositol independence in *Neurospora*. *Proceedings of the National Academy of Sciences of the United States of America*, **70**: 3875–3879. URL: https://www.jstor.org/stable/62669.

Mohanta, T.K., Bashir, T., Hashem, A., Abd-Allah, E.F. & Bae, H. (2017). Genome editing tools in plants. *Genes*, **8**: 399. DOI: https://doi.org/10.3390/genes8120399.

Moore, D. & Novak Frazer, L. (2002). *Essential Fungal Genetics*. New York: Springer-Verlag. ISBN-10: 0387953671, ISBN-13: 978-0387953670. See Chapter 2 *Genome interactions* and Chapter 5 *Recombination analysis*.

Muszewska, A., Steczkiewicz, K., Stepniewska-Dziubinska, M. & Ginalski, K. (2017). Cut-and-paste transposons in fungi with diverse lifestyles. *Genome Biology and Evolution*, **9**: 3463–3477. DOI: https://doi.org/10.1093/gbe/evx261.

Nagasaki, M., Saito, A., Doi, A., Matsuno, H. & Miyano, S. (2009). *Foundations of Systems Biology*. London: Springer-Verlag. ISBN-10: 1848820224, ISBN-13: 978-1848820227.

Nødvig, C.S., Nielsen, J.B., Kogle, M.E. & Mortensen, U.H. (2015). A CRISPR-Cas9 system for genetic engineering of filamentous fungi. *PLoS ONE*, **10**: article number e0133085. DOI: https://doi.org/10.1371/journal.pone.0133085.

Nowrousian, M. (2007). Of patterns and pathways: microarray technologies for the analysis of filamentous fungi. *Fungal Biology Reviews*, **21**: 171–178. DOI: https://doi.org/10.1016/j.fbr.2007.09.002.

Nowrousian, M. (2014a). Genomics and transcriptomics to analyze fruiting body development. In *The Mycota, Fungal Genomics*, **Vol. XIII** 2nd edition, ed. M. Nowrousian. Berlin, Heidelberg: Springer-Verlag, pp. 149–172. ISBN: 978-3-642-45217-8. DOI: https://doi.org/10.1007/978-3-642-45218-5_7.

Nowrousian, M. (ed.) (2014b). *The Mycota, Fungal Genomics*, **Vol. 2**, 2nd edition. Berlin, Heidelberg: Springer-Verlag. ISBN: 978-3-642-45217-8. DOI: https://doi.org/10.1007/978-3-642-45218-5.

Parnell, L.D., Lindenbaum, P., Shameer, K. *et al.* (2011). BioStar: an online question & answer resource for the bioinformatics community. *PLoS Computational Biology*, **7**: article number e1002216. DOI: https://doi.org/10.1371/journal.pcbi.1002216.

Pastore, A. & Puccio, H. (2013). Frataxin: a protein in search for a function. *Journal of Neurochemistry*, **126**: 43–52. DOI: https://doi.org/10.1111/jnc.12220.

Peter, J., De Chiara, M., Friedrich, A. *et al.* (2018). Genome evolution across 1,011 *Saccharomyces cerevisiae* isolates. *Nature*, **556**: 339–344. DOI: https://doi.org/10.1038/s41586-018-0030-5.

Petre, M. (ed.) (2015). *Mushroom Biotechnology: Developments and Applications*. London: Academic Press, an imprint of Elsevier Inc. 242 pp. ISBN: 9780128027943.

Phasha, M.M., Wingfield, B.D., Coetzee, M.P.A. *et al.* (2017). Architecture and distribution of introns in core genes of four *Fusarium* species. *G3: Genes, Genomes, Genetics*, **7**: 3809–3820. DOI: https://doi.org/10.1534/g3.117.300344.

Pritham, E.J. (2009). Transposable elements and factors influencing their success in eukaryotes. *Journal of Heredity*, **100**: 648–655. DOI: https://doi.org/10.1093/jhered/esp065.

Pudake, R.N., Kumari, M., Sahu, B.B. & Sultan, E. (2017). Targeted gene disruption tools for fungal genomics. In *Modern Tools and Techniques to Understand Microbes*, ed. A. Varma & A. Sharma. Cham, Switzerland: Springer, pp. 81–102. ISBN: 978-3-319-49195-0. DOI: https://doi.org/10.1007/978-3-319-49197-4_5.

Ranatunga, W., Gakh, O., Galeano, B.K. *et al.* (2016). Architecture of the yeast mitochondrial iron-sulfur cluster assembly machinery: the sub-complex formed by the iron donor, Yfh1 protein, and the scaffold, Isu1 protein. *Journal of Biological Chemistry*, **291**: 10378–10398. DOI: https://doi.org/10.1074/jbc.M115.712414.

Richards, T.A., Leonard, G., Soanes, D.M. & Talbot, N.J. (2011). Gene transfer into the fungi. *Fungal Biology Reviews*, **25**: 98–110. DOI: https://doi.org/10.1016/j.fbr.2011.04.003.

Roberts, S.E. & Gladfelter, A.S. (2016). Nuclear dynamics and cell growth in fungi. In *The Mycota*, **Vol. I**, *Growth, Differentiation and Sexuality*, 3rd edition, ed. J. Wendland. Cham, Switzerland: Springer International Publishing, pp. 27–46. ISBN: 978-3-319-25842-3. DOI: https://doi.org/10.1007/978-3-319-25844-7_2.

Rogozin, I.B., Carmel, L., Csuros, M. & Koonin, E.V. (2012). Origin and evolution of spliceosomal introns. *Biology Direct*, **7**: 11. DOI: https://doi.org/10.1186/1745-6150-7-11.

Rokas, A. (2009). The effect of domestication on the fungal proteome. *Trends in Genetics*, **25**: 60–63. DOI: https://doi.org/10.1016/j.tig.2008.11.003.

Roper, M., Simonin, A., Hickey, P.C., Leeder, A. & Glass, N.L. (2013). Nuclear dynamics in a fungal chimera. *Proceedings of the National Academy of Sciences of the United States of America*, **110**: 12875–12880. DOI: https://doi.org/10.1073/pnas.1220842110.

Ross-Macdonald, P., Coelho, P.S., Roemer, T. *et al.* (1999). Large-scale analysis of the yeast genome by transposon tagging and gene disruption. *Nature*, **402**: 413–418. DOI: https://doi.org/10.1038/46558.

Sánchez, C., Moore, D., Robson, G.D. & Trinci A.P.J. (2020). A 21st century miniguide to fungal biotechnology/Una miniguía del siglo XXI para la biotecnología de hongos. *Mexican Journal of Biotechnology*, **5**: 11–42. DOI: https://doi.org/10.29267/mxjb.2020.5.1.11.

Satyanarayana, T., Deshmukh, S. & Johri, B.N. (2017). *Developments in Fungal Biology and Applied Mycology*. Singapore: Springer Nature Singapore Pte Ltd. ISBN: 978-981-10-4767-1. DOI: https://doi.org/10.1007/978-981-10-4768-8.

Sharma, K.K. (2015). Fungal genome sequencing: basic biology to biotechnology. *Critical Reviews in Biotechnology*, **36**: 743–759. DOI: https://doi.org/10.3109/07388551.2015.1015959.

Sharman, A. (2001). The many uses of a genome sequence. *Genome Biology*, **2**: reports 4013.1–4013.4. DOI: https://doi.org/10.1186/gb-2001-2-6-reports4013.

Sharp, P.A. (2005). The discovery of split genes and RNA splicing. *Trends in Biochemical Sciences*, **30**: 279–281. DOI: https://doi.org/10.1016/j.tibs.2005.04.002.

Shepelev, V. & Fedorov, A. (2006). Advances in the Exon-Intron Database (EID). *Briefings in Bioinformatics*, **7**: 178–185. DOI: https://doi.org/10.1093/bib/bbl003. And see this URL:

http://bpg.utoledo.edu/~afedorov/lab/eid.html (unfortunately, there are no fungi in the database).

Silver, P.A., Way, J.C., Arnold, F.H. & Meyerowitz, J.T. (2014). Engineering explored. *Nature*, **509**: 166–167. DOI: https://doi.org/10.1038/509166a.

Sims, A.H., Gent, M.E., Robson, G.D., Dunn-Coleman, N.S. & Oliver, S.G. (2004). Combining transcriptome data with genomic and cDNA sequence alignments to make confident functional assignments for *Aspergillus nidulans* genes. *Mycological Research*, **108**: 853–857. DOI: https://doi.org/10.1017/S095375620400067X.

Slot, J.C., Townsend, J.P. & Wang, Z. (2017). Fungal gene cluster diversity and evolution. *Advances in Genetics*, **100**: 141–178. DOI: https://doi.org/10.1016/bs.adgen.2017.09.005.

Stajich, J.E., Dietrich, F.S. & Roy, S.W. (2007). Comparative genomic analysis of fungal genomes reveals intron-rich ancestors. *Genome Biology*, **8**: R223. DOI: https://doi.org/10.1186/gb-2007-8-10-r223.

Stajich, J.E., Harris, T., Brunk, B.P. *et al.* (2012). FungiDB: an integrated functional genomics database for fungi. *Nucleic Acids Research*, **40**: D675–D681. DOI: https://doi.org/10.1093/nar/gkr918.

Steenkamp, E.T., Wingfield, M.J., McTaggart, A.R. & Wingfield, B.D. (2018). Fungal species and their boundaries matter: definitions, mechanisms and practical implications. *Fungal Biology Reviews*, **32**: 104–116. DOI: https://doi.org/10.1016/j.fbr.2017.11.002.

Strom, N.B. & Bushley, K.E. (2016). Two genomes are better than one: history, genetics, and biotechnological applications of fungal heterokaryons. *Fungal Biology and Biotechnology*, **3**: 4. DOI: https://doi.org/10.1186/s40694-016-0022-x.

Stukenbrock, E.H. & Croll, D. (2014). The evolving fungal genome. *Fungal Biology Reviews*, **28**: 1–12. DOI: https://doi.org/10.1016/j.fbr.2014.02.001.

Sugui, J.A., Chang, Y.C. & Kwon-Chung, K.J. (2005). *Agrobacterium tumefaciens*-mediated transformation of *Aspergillus fumigatus*: an efficient tool for insertional mutagenesis and targeted gene disruption. *Applied and Environmental Microbiology*, **71**: 1798–1802. DOI: https://doi.org/10.1128/AEM.71.4.1798–1802.2005.

Taylor, J.W., Branco, S., Gao, C. *et al.* (2017). Sources of fungal genetic variation and associating it with phenotypic diversity. *Microbiology Spectrum*, **5**: FUNK-0057–2016. DOI: https://doi.org/10.1128/microbiolspec.FUNK-0057-2016.

Thangadurai, D., Sangeetha, J. & David, M. (2016). *Fundamentals of Molecular Mycology*. Waretown, NJ: Apple Academic Press. 194 pp. ISBN: 978-1771882538.

Todd, R., Forche, A. & Selmecki, A. (2017). Ploidy variation in fungi: polyploidy, aneuploidy, and genome evolution. *Microbiology Spectrum*, **5**: FUNK-0051–2016. DOI: https://doi.org/10.1128/microbiolspec.FUNK-0051-2016.

Weld, R.J., Plummer, K.M., Carpenter, M.A. & Ridgway, H.J. (2006). Approaches to functional genomics in filamentous fungi. *Cell Research*, **16**: 31–44. DOI: https://doi.org/10.1038/sj.cr.7310006.

Winzeler, E.A., Shoemaker, D.D., Astromoff, A. *et al.* (1999). Functional characterization of the *S. cerevisiae* genome by gene deletion and parallel analysis. *Science*, **285**: 901–906. DOI: https://doi.org/10.1126/science.285.5429.901.

Yandell, M. & Ence, D. (2012). A beginner's guide to eukaryotic genome annotation. *Nature Reviews Genetics*, **13**: 329–342. DOI: https://doi.org/10.1038/nrg3174.

Yenerall, P. & Zhou, L. (2012). Identifying the mechanisms of intron gain: progress and trends. *Biology Direct*, **7**: 29 (10 pp.). DOI: https://doi.org/10.1186/1745-6150-7-29.

Zheng, Y.-M., Lin, F.-L., Gao, H. *et al.* (2017). Development of a versatile and conventional technique for gene disruption in filamentous fungi based on CRISPR-Cas9 technology. *Scientific Reports*, **7**: article number 9250. DOI: https://doi.org/10.1038/s41598-017-10052-3.

CHAPTER 7

Structure and Synthesis of Fungal Cell Walls

The fungal wall can justifiably be described as a sophisticated cell organelle because of the range of functions for which it is responsible and for its importance as a feature which is characteristic of the fungi.

In this chapter, we will discuss the fungal wall as a working organelle, and then consider the fundamental aspects of wall structure, function and architecture. We will describe each of the main components in detail; the chitin component, the glucan and the glycoprotein. Wall synthesis and remodelling is also described, although you should already be aware of some of the mechanisms that may be involved (discussed in section 5.14) and that the dynamic nature of the fungal cell wall is also mentioned during discussions of hyphal and spore differentiation (section 10.3), hyphal branching (section 4.11), septation (sections 4.12 and 5.16) and hyphal anastomosis (section 5.15).

In the final two sections of this chapter, we consider two aspects that are often overlooked: what happens on the ***outside*** of the wall is considered in the section entitled 'On the far side'; and, finally, we look briefly at the fungal wall as a clinical target (although antifungal agents that target the wall are dealt with in detail in Chapter 18).

7.1 THE FUNGAL WALL AS A WORKING ORGANELLE

The fungal cell wall is an important phylogenetic and taxonomic character, but it is a dynamic organelle, which, generally speaking, has four major functions:

- **Maintenance of cell shape**. By so doing, the wall determines the morphology of the hypha or other fungal cell, like yeast cells and spores. The more spectacular multicellular sporophores of higher fungi are a striking example of how the cell wall contributes to morphogenesis.
- **Stabilisation of the internal osmotic conditions** is an important outcome of maintaining cell shape. The wall is sturdy and elastic and creates a counteracting pressure, which stops excessive water influx.
- **Protection against physical stress**, and functions as a protective coat. The combination of mechanical strength and high elasticity allows the wall to transmit and redistribute physical stresses, offering effective protection against mechanical damage *and* allowing the hypha to penetrate aggressively the substrata into which it grows (Money, 2004, 2008).
- **A scaffold for proteins**. Most of the mechanical strength of the wall is provided by its stress-bearing polysaccharides, but these also function as a scaffold to which an external layer of glycoproteins is attached. These glycoproteins function to limit the permeability of the wall (in both directions), create a microenvironment in the inner region of the wall adjacent to the plasma membrane which is under the control of the fungus. Negatively charged phosphate groups in their carbohydrate side chains probably contribute to water retention, and proteins of this external layer allow recognition of mating partners, substrata, substrates and hosts, and subsequent adhesion and capture of any or all of these.

Cell wall construction is tightly controlled. The wall is the first part of the fungal cell to challenge the environment or potential host, so it is also the first to experience whatever demands the environment imposes, or defensive inhibitors the host can produce. The wall is the target for environmental stress, natural plant and animal defences, and clinical pharmaceutical agents. As we will show below, because of the unique components in its structure, and the vital roles that the cell wall plays in the life of the fungus, the fungal wall is an excellent **target for antifungal agents**.

Polysaccharide composition, structure and thickness of the wall all vary with environmental conditions and are coordinated with the cell cycle and developmental progress of the organism (Latgé & Beauvais, 2014).

Formation and remodelling of the cell wall involves several biosynthetic pathways and the combined and coordinated actions of hundreds of gene products within the fungal cell. An indication of the significance of the cell wall in the cell biology of the fungal cell is that about 20% of the genome of *Saccharomyces cerevisiae* (that is, approximately 1,200 genes) is devoted to the construction of the cell wall. Some provide substrates for the components of the wall; others are concerned with the delivery and assembly of wall materials; and many regulate the assembly and maintenance processes, as well as wall modifications in response to environmental challenges (Latgé & Beauvais, 2014). The gene products include enzymes that create, modify or split glycosidic bonds, including multigene families of chitin and glucan synthases as well as enzymes such as glucanases and chitinases (glycohydrolases) and transglycosidases, which are concerned with remodelling the structure of the wall (Gow *et al.*, 2017). Any disruption to the structure of the cell wall will have a profound effect on the growth and morphology of the fungal cell; ultimately making the cell vulnerable to lysis and death.

Although many of the building blocks of the cell wall are conserved in different fungal species, other aspects of wall composition vary between species (Coronado *et al.*, 2007; Xie & Lipke, 2010). Many of the enzymes involved in wall construction and for cross-linking walls were strongly conserved; including glycosyl hydrolases and transferases, proteases, lipases, enzymes of the GPI anchor synthesis pathway, and the chaperone proteins that assist polypeptide folding or unfolding and the assembly or disassembly of macromolecular structures. Indeed, many of these proteins are also conserved in other eukaryotes and are also associated with wall synthesis in plants. On the other hand, the wall-resident proteins have diversified rapidly; these are the structural glycoproteins, adhesins and sexual agglutinins (in *Saccharomyces*). Coronado *et al.* (2007) suggest that fungal cell walls are assembled by the products of a conserved set of genes that is ancestral, like the basic architecture of the wall itself, to the evolutionary divergence of the ascomycete and basidiomycete clades. Xie and Lipke (2010) point out that the relatively fast sequence divergence of wall-resident proteins, together with their unusually high content of tandem and nontandem repeats, may have served as a driver for yeast speciation; enabling the yeast clade to adapt and exploit diverse environments.

Most of what we can describe here about the fungal wall has come from detailed molecular, genomic and proteomic analyses of model fungal systems, especially *Saccharomyces cerevisiae, Schizosaccharomyces pombe, Candida albicans, Aspergillus fumigatus* and *Neurospora crassa*, and, in recent years, a growing number of pathogenic fungi, including human pathogens (Free, 2013). We will summarise information dealing with as wide a range of species as we can, but you should be aware that the story we can tell is still incomplete.

Structurally, the fungal wall is a **three-dimensional network of polysaccharides, glycoproteins and proteins**, many of which are unique to fungi. Typically, the walls contain **fibrillar polysaccharides** of high tensile strength embedded in a more gel-like matrix comprising a variety of polysaccharides, glycoproteins and proteins with a range of minor components, including lipids, pigments, inorganic ions and salts.

The structural fibrillar material of the wall is largely inert, forming the main structural component, but the composition of the other materials changes with time and position as these components can serve:

- as nutrient reserves;
- in transport and translocation;
- for the metabolism of non-permeable substrates;
- for communication and interaction with the exterior; and
- and for protection from outside attack.

7.2 FUNDAMENTALS OF WALL STRUCTURE AND FUNCTION

Most cell walls are layered. The innermost layer (that is, the layer immediately surrounding the plasma membrane) is a relatively conserved structural skeletal layer and the outer layers are more varied between species and are dynamically tailored to needs of the organism as it develops and matures, and in response to interactions with the environment. Proteins rarely make up more than 20% of the wall material, and most are glycoproteins. Some proteins have a structural role, but most contribute to the many other functions mentioned above. Proteins at, or close to, the outer surface determine the surface properties of the wall; that is, whether it is **hydrophobic** or **hydrophilic** (i.e. non-wettable or wettable), or **adhesive**, or **antigenic**. The low concentrations of lipids and waxes found in fungal walls usually serve to control water movement, especially to prevent desiccation.

Approximately 80% of the wall consists of polysaccharides. In hyphae, a major component of the wall, and certainly the most important to structural integrity, is **chitin**, which is a β-1–4-linked long **polymer of N-acetylglucosamine** (Figure 7.1), the chemistry of which has been described in Chapter 5 (see section 5.13). Chitin degradation, which is relevant to wall remodelling, is explored in Chapter 11 (see section 11.5).

Cell walls of filamentous fungi, such as *Neurospora* and *Aspergillus*, contain 10–20% **chitin**, but chitin accounts for only 1–2% of the yeast cell wall by dry weight (yeast chitin is mainly found in bud scars and only a very small fraction, around 2%, of the chitin content is found in the lateral walls). In both types of fungal cell, though, chitin molecules form into microfibrils, about 10 nm in diameter, by head-to-tail 'crystallisation' between the polymers. These crystalline polymers are held together by spontaneous hydrogen bonding and the microfibrils have enough tensile strength to provide the wall with its main **structural integrity**.

Chitin is a linear homopolymer chain but is frequently cross-linked to other wall constituents, especially **glucan (polymer of glucose)** and **mannan (polymer of mannose)** polysaccharides. In almost all fungi, the main bulk component of the cell wall is a **branched β-1–3-glucan** that is linked to chitin by a β-1–4 linkage. Interchain β-1–6 glucosidic linkages (the branch points in the branched glucan) account for 3% and 4% of the total

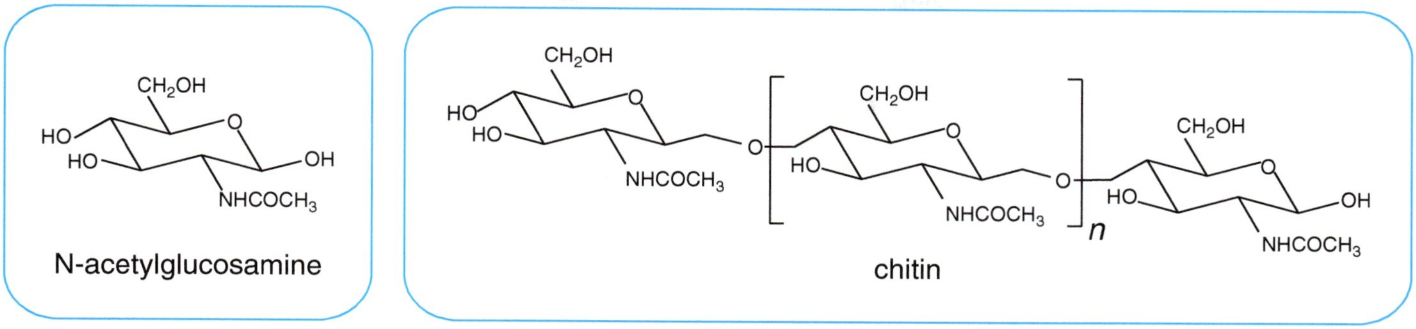

Figure 7.1. Covalent structures of N-acetylglucosamine and its linear homopolymer, chitin, which is synthesised by the enzyme chitin synthase (and see Figure 11.2). Natural sources of chitin have molecular masses of a few million; extraction processes fragment the polymer and the molecular mass of commercial preparations of chitin can vary between 350,000 and 650,000.

glucan linkages, respectively, in *Saccharomyces cerevisiae* and *Aspergillus fumigatus* (Latgé, 2007). The β-1–3 glucan is synthesised by a membrane-bound, GTP-stimulated, **β-1–3-glucan synthase**. The wall may also contain some complex polysaccharides, like glucogalactomannans.

The yeast cell wall is mainly composed of **mannans**, rather than glucans so we begin to see how the wall can contribute to taxonomy, because although there is a common structural theme, there are consistent differences in gross chemical composition of the wall that distinguish specific taxa. For example (see Gow *et al.*, 2017, for references), the cell walls in the group traditionally known as zygomycetes contain chitin and chitosan (a polymer of glucosamine formed from chitin by removing the N-acetyl groups) among the more fibrous polymers, and polyglucoronic acid, and glucuronomannoproteins among the water-soluble gel-like polymers.

Ascomycota and *Basidiomycota* both contain chitin together with the β-1–3, β-1–6-linked glucans in fibrous form, but:

- *Ascomycota* feature a gel-like **α-1–3-glucan** and **galactomannoproteins**. The yeasts *Candida, Saccharomyces* and the human pathogen *Pneumocystis jirovecii* have an outer cell wall containing highly mannosylated glycoproteins that covers the inner wall layers. In the human pathogen *Histoplasma capsulatum*, an outer wall layer of α-1–3-glucan prevents the immune system recognising the invading cells.
- In *Basidiomycota*, xylomannoproteins are found alongside the α-1–3-glucan (Bowman & Free, 2006). The cell wall of the basidiomycetous yeast *Cryptococcus* is enveloped by a gelatinous capsule composed of glucuronoxylomannan (this makes up about 90% of the mass of the capsule) and galactoxylomannan that is anchored to the main wall via α-1–3-glucan.

The taxonomic significance of the wall extends to the kingdom level. Not only are the walled cells of plants and fungi distinguished from one another by the use of cellulose as the main structural component in the former and chitin in the latter (compare Figure 7.1 with Figure 11.1), but the walls of hyphae of fungus-like members of Kingdom *Straminipila* (*Oomycota* like *Phytophthora* and *Pythium*) contain cellulose microfibrils, which are about 12 nm in diameter, rather than chitin alongside β-1–3, β-1–6-linked glucans. At the other end of the taxonomic spectrum, Calonje *et al.* (1995) were able to detect significant differences in the gross chemical composition of the walls of four commercial strains of the cultivated mushroom *Agaricus bisporus*. Differences detected included overall composition of the wall and in the polysaccharide structures. In the face of such varietal differences, we must be careful in assigning taxonomic significance to differences in wall structure.

Another feature that has been pointed out (Lesage & Bussey, 2006) is that the term 'cell wall' is applied to the **extracellular matrices** of fungi and plants, but an extracellular matrix is present in all three metazoan, fungal and plant kingdoms. Although the compositions of extracellular matrices do vary between taxonomic groups, all eukaryotic extracellular matrices result from the activity of a common cytoskeleton, which positions highly conserved secretory and construction machinery at, in and through the plasma membrane.

Consequently, the evolutionary diversity that is observed seems to be only in the composition and structure of the extracellular matrices themselves, although this in itself must mean there will be some specialisation in the underlying cytoskeletal and secretory mechanisms involved in their construction (e.g. plant cells specialised for cellulose synthesis and assembly, fungal cells specialised for chitin synthesis and assembly, animal cells specialised for glycoprotein modification, etc.). On this view, then, cell walls are extracellular matrices specialised for physical containment that allow 'fungi and plants to build structures based on the use of cells as **hydrostatic bricks**' (Lesage & Bussey, 2006).

The shape of the fungus is determined by its cell wall. The cell wall encloses the cell, providing osmotic and physical protection. The traditional view of the cell wall as a rigid structure, functioning primarily to withstand turgor pressure, has changed through time to a more dynamic view of the wall being a structure that is able to adapt to various conditions of growth, development and environmental stress, which, together with the plasma membrane and **periplasmic space** (the extracellular space between the plasmalemma and the wall), influences and regulates the flow of materials into and out of the cell. 'Hydrostatic bricks' might be an adequate description of plant cells, but fungal septa are incomplete (see section 5.16). Each fungal hypha is part of a closed hydraulic system which is under pressure. Osmotic influx of water occurs because of differences in water activity between the inside and the outside of the semi-permeable plasma membrane. The resultant tendency to increase cytoplasmic volume is counteracted by wall pressure due to the mechanical strength of the wall outside the plasma membrane. The difference between these two forces is the turgor pressure, which is the resultant 'inflation pressure' keeping the hypha inflated. In a closed vessel, the pressure is the same over the whole of the inside wall surface; this applies whatever the shape of the vessel. A fungal mycelium is a closed vessel divided up by septa which are perforated and, whatever the shape, the wall pressure is uniform throughout the mycelium.

The only way that pressure can be varied locally is to enclose part of the pre-existing mycelium by creating complete (meaning unperforated) septa or by plugging septal pores. This creates a second 'pressure vessel' (that might be a spore, haustorium, appressorium or other specialised structure), which could then be provided with a different internal pressure to that of its originating mycelium. The pressurised mycelium or section of mycelium will remain intact for as long as the mechanical strength of its wall, everywhere over its entire surface, is sufficient to resist the internal pressure.

The resulting turgor pressure has been estimated to be between 0.2 and 10 MPa (Money, 2004). 'Typical' atmospheric pressure at sea level is about 101 kPa (0.1 MPa), so this turgor pressure is equivalent to 2–20 times atmospheric pressure. Turgor pressure generates the force that enables hyphae to exert mechanical force on the substrates they are penetrating (Money, 2004, 2008) and these turgor pressures are sufficient to enable most vegetative hyphal tips to exert forces at least equivalent to atmospheric pressure at the ***micro***metre scale of the hyphal apex.

Of course, if the hyphal apex could be modified to a point, then even greater penetrating pressure could be exerted at the extreme tip. This is exactly what happens in the appressoria (i.e. infection structures) of some plant pathogens, such as *Magnaporthe oryzae* which have hyphal walls reinforced with melanin, making them the most robust of all walls found in nature and able to withstand an internal turgor pressure of up to 20 MPa and exert forces many times atmospheric pressure at the ***nano***metre scale of the appressorial apex (Howard *et al.*, 1991). This indicates that hyphae can penetrate some materials by pressure alone, although *in vivo* penetration is usually accompanied by enzymatic softening of the substratum being attacked.

In the context of the vegetative hypha, if the mechanical force that the wall can exert is equal to or greater than the force exerted by turgor, it will remain intact. However, if turgor exceeds the breaking strain of the wall *at any point* then the wall will rupture (possibly explosively locally), and the cell will die. This raises the problem we face in understanding apical hyphal growth; because the structure of the wall needs to be weakened to allow insertion of new wall material to permit the tip to elongate continuously and this must be achieved without exploding the tip (see our earlier discussion about growth at the hyphal apex in section 5.14).

As long ago as 1976, Rosenberger compared the **two-phase system of fungal walls** (crystalline **microfibrils** embedded in an amorphous, gel-like **matrix**) with the two-phase systems of **building composites**: 'Fungal walls contain a network of fibrils with the spaces in the net filled by matrix polymers and in this they resemble such man-made composites as glass-fibre reinforced plastic (GRP) and reinforced concrete' (Rosenberger, 1976).

This is a useful, if not perfect, analogy. Both systems are very strong for their weight, a property important for the ability of fungal walls to resist the turgor pressure of the cytoplasm. But we can best see how this analogy helps us understand the function of components of the fungal wall if we ask *why* civil engineers use this sort of building composite. In these building materials, the **matrix** (the plastic in GRP or the cement plus aggregate in concrete) provides strength in **compression** and the reinforcing **fibres** (glass fibres in GRP or steel bars in concrete) improve strength in **tension** (stretching) and **shear** (force acting parallel to the surface) caused by bending. Applying this analogy to fungal walls shows that its 'engineering' produces a composite of highly cross-linked glycan and polypeptides combined with chitin microfibrils (concentrated in wall layers nearest to the plasma membrane) that provides the strength overall to resist damage inflicted by compression, bending and inflationary forces.

The civil engineering analogy can also be extended to tissues (see Chapter 13), but for the hyphal wall the shape and mechanical properties can presumably be altered, as they can in composites, by changing fibre thickness, fibre length and fibre orientation. These changes bring about what we observe as hyphal wall rigidification and hyphal differentiation. We cannot leave this engineering analogy without bringing to your attention the growing interest attached to the use of the filamentous hyphae of mycelium cultivated in low-cost organic wastes to convert the wastes into useful and environmentally friendly materials. Although presently used primarily for a limited range of packaging and construction applications, a range of potential uses have been proposed for these **mycelium composites**, including acoustic dampers, super absorbents, paper, textiles, wound dressings and even electronic parts (Jones *et al.*, 2017).

In the true fungi, hyphal turgor is controlled to contribute to tip extension by driving the tip forward and shaping it by plastic deformation of newly synthesised wall (Wessels, 1993;

Bartnicki-García *et al.*, 2000; and see section 5.14). Chitin immediately provides the wall with most of its mechanical strength. If chitin synthesis is disrupted, or chitin is removed by experimental treatment with enzymes, the wall is sufficiently disrupted for the fungal cell to become osmotically unstable. We emphasise 'in the true fungi' here because members of the *Oomycota* like *Achlya bisexualis* and *Saprolegnia ferax* have little ability to regulate their turgor in response to external changes, probably because they grow only or predominantly in freshwater habitats, and they grow in the absence of turgor (Money & Harold, 1992, 1993; Lew *et al.*, 2004); yet another difference between Kingdoms *Fungi* and *Straminipila*, this one demonstrating that their superficially similar filamentous growth forms are based on very different mechanisms.

The above is a highly generalised and abbreviated description of wall structure. Its purpose is to present a ***concept*** of the fungal wall at this point, rather than a specifically detailed constructional diagram. We will add detail to our description below, for the moment the concept is this:

- The main structural substance of the wall is provided by polysaccharides, mostly glucan, and in filamentous fungi the shape-determining component is chitin.
- The various polysaccharide components are linked together by hydrogen bonding and by covalent bonds.
- A variety of proteins and glycoproteins contribute to wall function, some of these are structural, some are enzymic, and some vary the biological and biophysical characteristics of the outer surface of the wall.
- Proteins may be anchored in the plasma membrane, covalently bonded to wall polysaccharides or more loosely associated with the wall.
- The wall is a dynamic structure which is modified as it matures, and/or:
  - as part of hyphal differentiation; and/or
  - on a short-term basis to react to changes in physical and physiological conditions.

Finally, remember that, by definition, the wall is extracellular, it is a functionally differentiated ***extracellular*** matrix and its entire structure lies outside the plasma membrane, so all additions to its structure must be externalised through the membrane before the wall can be restructured and the enzymic reactions that link wall components together are extracellular reactions carried out by secreted enzymes. In fact, fungal wall construction involves biosynthetic reactions that take place inside the cell in the endomembrane system, at the cell membrane by plasma membrane anchored enzymes, and in the cell wall itself. The major wall polysaccharides, chitin and glucan, are synthesised ***at*** the plasma membrane by transmembrane enzyme complexes that are delivered to the plasma membrane in an inactive form via secretory vesicles and then activated after insertion into the membrane (see below). In contrast, glycoconjugates, like the mannans, are synthesised ***intracellularly***, in the endoplasmic reticulum and Golgi dictyosomes, where they may be conjugated to cell wall proteins, and then brought to the cell wall by secretory vesicles. Within the cell wall, polysaccharides and glycoconjugates can be cross-linked or branched by ***extracellular*** enzymes that reside within the cell wall.

7.3 FUNDAMENTALS OF WALL ARCHITECTURE

Electron microscope observations of fungal cell wall **architecture** in both filamentous fungi and yeast reveal a **layered structure**. The inner, most electron-dense layer composed mainly of chitin and glucan, both of which form intrachain hydrogen bonds and assemble into microfibrils that form a basket-like framework just external to the plasma membrane. It makes eminent sense for the main skeletal element (microfibrils) to be located in the innermost layer of the wall, as this is where turgor pressure is first exerted. This is covered by an outer layer composed of glycoproteins together with glucans in filamentous fungi, or mannans in yeasts (Figure 7.2). The glycoprotein, glucan and chitin components are extensively cross-linked to create a complex **three-dimensional network**, which forms the structural basis of the cell wall. Some of the glycoproteins are bound to the cell through hydrogen bonds; some are attached covalently to the cell wall via disulfide bonds; and some are covalently linked to the glucans of the cell wall.

Two different classes of glucan-linked cell wall glycoproteins have been described:

- the 'protein with internal repeats' (PIR) class;
- **glycosylphosphatidylinositol** (**GPI**)-linked class, which are tethered to the plasma membrane through their (modified) GPI anchor.

GPI anchors and their association with the fungal wall are mentioned in the last two paragraphs of section 5.12.

7.4 THE CHITIN COMPONENT

Chitin synthesis is the responsibility of the enzyme **chitin synthase**, an integral membrane enzyme that catalyses the transfer of N-acetylglucosamine from UDP-N-acetylglucosamine to a growing chitin chain (described briefly in section 5.13). Most fungi have several chitin synthase (*CHS*) genes, which suggests some overlap and/or specialisation in function. However, the significance of each of the enzymes in the family is not known. The classic approach is to find or induce mutations to determine the effect of the defective enzyme(s), but mutations in common chitin synthase genes do not always result in similar phenotypes. Paradoxically, among the mutants identified are those with reduced chitin content in the walls, but normal chitin synthase activity in *in vitro* assays and a second type that has drastically reduced enzyme activity in *in vitro* assays but has the usual cell wall chitin content. In addition, other genes, also named ***CHS***, are not associated with chitin synthase **enzyme** activity but are involved in the regulation or localisation of that activity. Some

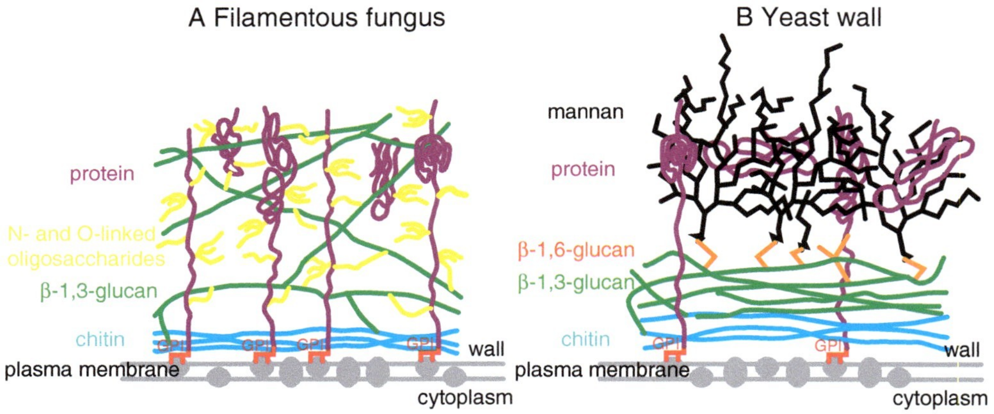

Figure 7.2. Diagrammatic depictions of the structure of fungal walls. (A) Interpretation of the filamentous hyphal cell wall (modified from Bowman & Free, 2006). Most of the chitin (blue) is thought to be near to the plasma membrane where it resists the turgor pressure of the protoplast, β-1–3-glucan (green) extends throughout the wall. Protein (shown in purple), glucan and chitin components are integrated into the wall by cross-linking the components together with N- and O-linked oligosaccharides (yellow). Many of the glycoproteins have GPI anchors (red), which tether them to the plasma membrane; while other glycoproteins are secreted into the wall matrix. (B) Generalised cell wall structure of *Candida albicans*; note that mannans (shown in black) dominate the outer regions of this yeast wall and β-1–6-glucan (orange) cross-links the components (modified from Odds *et al.*, 2003). Also note that in liquid cultures the wall 'blends into' the surrounding aqueous medium because the surface polymers (polypeptide and polysaccharide) are hydrophilic and able to dissolve into the aqueous medium. Walls of aerial hyphae can be chemically modified at the surface by deposition of polyphenols and/or assembly of layers of proteins like hydrophobin (see section 7.8). Similar diagrammatic representations of the walls of a range of fungal pathogens of humans are shown by Erwig and Gow (2016).

chitin synthases are zymogens that are activated by proteolysis, and others are regulated by phosphorylation. Some fungi have more than 20 *CHS* genes, and some have only 1 (see Gow *et al.*, 2017, for references).

The multiple ***CHS*** genes have been grouped into two families and seven classes, based on amino acid sequence (see section 5.13). Class I, II and IV genes are present in all fungi; classes III, V, VI and VII are specific to filamentous fungi and some dimorphic species. Chitin synthases of classes I, II and III all share a catalytic domain surrounded by a hydrophilic amino-terminal region and a hydrophobic carboxy-terminal region. The number of chitin synthase genes within each species varies and the exact functions of many chitin synthases are still unknown (Rogg *et al.*, 2012). In some fungi the chitin synthase activity is fused to their molecular motor (section 5.11); both domains being required for correct cellular function. Class V and class VII chitin synthases, which are exclusive to filamentous fungi, encode proteins of around 1,500 amino acids that contain a myosin motor-like domain in their amino-terminal region, which has been shown to interact directly with the actin cytoskeleton (Steinberg, 2011; Fernandes *et al.*, 2016; Morozov & Likhoshway, 2016).

Examples of these **myosin motor/chitin synthase** fusion proteins have been found in *Aspergillus nidulans*, fungi pathogenic to humans (*Coccidioides posadasii* and *Paracoccidioides brasiliensis*) and plant pathogens too (*Colletotrichum graminicola* and *Pyricularia oryzae*). These enzymes are often essential for growth, morphogenesis, virulence and stress tolerance. The motor domain functions in actin-mediated cytoplasmic transport to the hyphal apex. However, the motor domain is not required for vesicle motility in *Aspergillus nidulans* and *Ustilago maydis*, and the motor may instead serve to tether vesicles at the hyphal apex (Treitschke *et al.*, 2010), increasing the residence time at its location and favouring vesicle fusion with the plasma membrane (Schuster *et al.*, 2012). In *Ustilago maydis*, the class VII chitin synthase Mcs1 and the class V chitin synthase 6 (Chs6) can be co-transported on the same secretory vesicle along with the β-1–3-glucan synthase Gsc1. The complex of glucan and chitin synthases seems to be tethered to a localised site of exocytosis and wall synthesis (Schuster *et al.*, 2016).

The different **chitin synthase isozymes** can be associated with a variety of specific functions that are time or position dependent, such as formation of septa and spores in *Aspergillus* (Ichinomiya *et al.*, 2005), appressoria in *Colletotrichum* (Werner *et al.*, 2007), complexing with chitin deacetylases to produce chitosan in *Cryptococcus* (Banks *et al.*, 2005), or for wall repair or response to stress (Bowman & Free, 2006). Some are essential for pathogenicity in several species, in the sense that deletion mutants are non-pathogenic a feature that confirms the critical importance of cell wall integrity in fungal infection processes (Werner *et al.*, 2007; Martin-Urdiroz *et al.*, 2008).

But the most essential function to which the chitin synthases contribute is hyphal extension growth and we are beginning to understand how these proteins are localised in the cell and how they are transported to their sites of action in regions of active wall growth. As described in Chapter 5, hyphae of septate filamentous fungi have a unique organelle, the Spitzenkörper, which is an accumulation of vesicles, ribosomes and cytoskeletal components at their growing apex (see Figure 5.6 in section 5.9). It is also clear that fungal cells have two well-defined types of secretory vesicle: **macrovesicles** are conventional eukaryotic secretory vesicles, which carry the components of the cell wall matrix, glucans, extracellular enzymes and glycoproteins; and these are accompanied by a large population of **microvesicles** carrying chitin synthase. These microvesicles have been called **chitosomes**; they are the smallest vesicles that can form chitin microfibrils *in vitro*, and it has been suggested that *in vivo* they transport chitin synthase to the plasma membrane at the hyphal apex and at the positions of developing septa, which are the places where chitin synthesis is highly localised in actively growing vegetative hyphae (Bartnicki-García, 2006).

Using chitin synthases from *Neurospora crassa* labelled with green fluorescent protein and high-resolution confocal laser scanning microscopy of living hyphae, Riquelme *et al.* (2007) and Riquelme and Bartnicki-García (2008) studied the trafficking of chitin synthase during hyphal growth. They showed that at the hyphal apex, fluorescence label was localised to a single conspicuous body (the Spitzenkörper), while at distances more than 40 μm from the apex the chitin synthase fluorescence occurred in a network of **endomembrane compartments**, mostly irregularly tubular, but some spherical. In between (20–40 μm from the apex), fluorescence occurred as globules that disintegrated into vesicles as they moved forward and ultimately contributed to the Spitzenkörper. These *in vivo* observations suggested that Spitzenkörper fluorescence specifically: 'originated from the advancing population of microvesicles (chitosomes) in the subapex' (Riquelme *et al.*, 2007). It was also demonstrated that brefeldin A (known to be a specific inhibitor of protein transport in Golgi dictyosomes) was unable to interfere with the traffic of chitosomes, and that fluorescence associated with the labelled chitin synthase was not associated with the endoplasmic reticulum. These observations support the idea that chitin synthase proteins are delivered to the cell surface by a secretory pathway distinct from the classic endoplasmic reticulum-to-Golgi. The finding of a direct interaction between chitin synthase proteins equipped with their own myosin motors and the actin cytoskeleton also supports this interpretation.

Once delivered to its site of action, chitin synthase is an **integral membrane-bound enzyme** (a **transmembrane protein**). The enzyme protein spans the membrane, accepting substrate monomers from the cytoplasm on the inner face of the membrane and extruding the lengthening chains of chitin through the plasma membrane as they are made. Hydrogen bonding between the newly formed polymers of chitin results in microfibril formation and subsequent crystallisation of chitin in the extracellular space immediately adjacent to the plasma membrane. This process of chitin synthesis primarily occurs at sites of active growth and cell wall remodelling. Although during normal apical growth of the hypha the incorporation of newly synthesised chitin is limited to the hyphal apex, there is evidence that inactivated chitin synthases are widely distributed in the plasma membrane. One piece of evidence for this is that inhibition of protein synthesis in *Aspergillus nidulans* resulted in chitin synthesis occurring uniformly over the hypha. This observation implies that protein synthesis is required to maintain the inactivity of chitin synthases already in place in the membrane. **Inactivated chitin synthase** proteins are an intrinsic component of the mature hyphal plasma membrane, the enzyme(s) being activated somehow specifically at the hyphal apex and at sites where branch or septum formation is initiated.

Chitin is important at specific sites in yeast walls, although the major structural component in these organisms is a fibrillar inner layer of β-glucan. In *Saccharomyces cerevisiae*, chitin is concentrated mostly in septa. This yeast has three chitin synthases (Chs1, Chs2 and Chs3). Chs3 is a class IV chitin synthase and is required for formation of chitin rings at the base of emerging buds and for chitin synthesis in the lateral cell wall during vegetative growth. Chs3 is found at the plasma membrane and in chitosomes, and its correct localisation requires the functioning of several other proteins. The Chs3 chitin synthase is responsible for generating approximately 80–90% of the total chitin of the yeast cell. Chs2, a class II enzyme, synthesises chitin in primary septa. Appearance of Chs2 is cell cycle dependent; it appears at a late stage of mitosis and is localised to the septation site and degraded immediately after septum formation. Chs1, a class I enzyme, repairs the weakened cell walls of daughter cells after their separation from mother cells. This separation is executed by the activity of a chitinase that digests the primary septa. Chs1 chitin synthase repairs the wall, replenishing chitin polymers lost during cytokinesis. Chs1 exists at a constant level throughout the cell cycle and is present in the plasma membrane and chitosomes. The simultaneous deletion of all genes coding for these three chitin synthases is lethal for the cell, demonstrating that chitin is an indispensable component of the cell wall of *Saccharomyces cerevisiae* even though it is a minor component (Bowman & Free, 2006; Lesage & Bussey, 2006).

7.5 THE GLUCAN COMPONENT

Although chitin provides the crucial mechanical strength of the wall, glucans make up 50–60% of the dry weight of the fungal cell wall, and between 65% and 90% of the cell wall glucan is the β-1–3-glucan. Glucans with other chemical linkages between their repeating glucose residues have been found in various fungal cell walls, including β-1–6, mixed β-1–3 with β-1–4, α-1–3, and α-1–4-linked glucans. However, the β-1–3-glucan serves as the **main structural element** to which other cell wall components are covalently attached. As a result, the synthesis of β-1–3-glucan is required for cell wall formation and normal development of fungi (Latgé *et al.*, 2005; Bowman & Free, 2006; Lesage & Bussey, 2006).

As with chitin, glucan polymers are formed by multisubunit **enzyme complexes in the plasma membrane**, the polymer being extruded into the extracellular space through a vectorial synthesis. The linear polymer is synthesised by the enzyme complex known as **β-1–3-glucan synthase** (Douglas, 2001; Latgé *et al.*, 2005). The polysaccharide polymer produced ranges up to 1,500 glucose residues long; it becomes **insoluble** as the degree of polymerisation increases, and enzyme remains attached to the nascent glucan. Synthesis is vectorial, the new glucan chains being extruded into the extracellular (**periplasmic**) space immediately adjacent to the plasma membrane. This promotes their integration into the cell wall at points of active cell wall synthesis. Glucan synthase complexes, like the chitin synthases, are mainly localised to regions of active extension growth, budding, branching or septation. The carbon-6 positions of approximately 40–50 of the grand total of 1,500 glucose residues in a glucan polymer have additional β-1–3-glucans linked to them to generate the β-1–6-linked branched structure of the mature wall glucan.

A good range of β-1–3-glucan synthase enzymes has been studied, including those from the yeasts *Saccharomyces cerevisiae, Schizosaccharomyces pombe, Candida albicans*, filamentous *Ascomycota Neurospora crassa, Aspergillus nidulans* and *Aspergillus fumigatus*, the pathogenic basidiomycete *Cryptococcus neoformans* and even the oomycete *Phytophthora*. All these enzymes are membrane-bound complexes that use UDP-glucose as a substrate and produce a linear polysaccharide. The reaction is processive, meaning that for every molecule of UDP-glucose hydrolysed, one molecule of glucose is added to the length of the polymer chain.

Genes encoding β-1–3-glucan synthase proteins were first identified in *Saccharomyces cerevisiae*, which contains two catalytic subunits and one regulatory protein. The *Saccharomyces cerevisiae* catalytic proteins are encoded by genes *FKS1* and *FKS2*, both of which are essential for normal wall formation. Disruption of either the *FKS1* or *FKS2* gene results in mutants with slow growth rate and cell wall defects; simultaneous deletion of both genes is lethal. Clearly, the catalytic subunits have overlapping functions but activity of both is required and β-1–3-glucan is essential for yeast survival. The regulatory protein is a GTPase subunit called Rho1; this is also essential for survival. The *FKS* and *RHO1* genes are highly conserved in other fungi. *Aspergillus fumigatus* and *Neurospora crassa* genomes each contain one catalytic subunit gene and one gene encoding the GTPase regulatory subunit; both genes are required for cell viability (Latgé *et al.*, 2005). There are fewer ***glucan*** synthase genes than there are ***chitin*** synthase genes in pathogenic fungi. The single *FKS1* of *Aspergillus fumigatus* and *Cryptococcus neoformans* is unique and essential, but in *Candida albicans* three different *FKS* sequences have been identified.

7.6 THE GLYCOPROTEIN COMPONENT

Within the wall, the branched glucans are cross-linked together and to chitin. In addition, the glucan has **mannoproteins** attached to it, and this introduces the idea that the protein components of fungal walls are of considerable importance, too (de Groot *et al.*, 2005; Klis *et al.*, 2006). In the yeast cell wall (*Saccharomyces cerevisiae* and *Candida albicans*), proteins account for 30–50% of the dry weight; the *Neurospora crassa* hyphal cell wall comprises approximately 15% protein by dry weight and, generally, filamentous fungi cell walls contain 20–30% protein.

Most cell wall proteins are glycoproteins that are modified with N-linked and O-linked oligosaccharides. These are proteins that have an amino-terminal signal peptide in their primary sequence which attaches the translating ribosome to the ER membrane. As the polypeptide is translated, the protein is extruded into the ER lumen; and this is the start of their journey through the secretory pathway (Figure 7.3; discussed in section 5.8). During translocation of the protein through the secretory pathway glycosylation occurs by covalent linkage of large, branched oligosaccharide structures to what then becomes a **glycoprotein**.

N-linked oligosaccharides are added to asparagines located in peptides carrying the sequence motifs: *asparagine–any amino acid–serine* and *asparagine–any amino acid–threonine*. The branched oligosaccharide contains N-acetylglucosamine, mannose and glucose, and is transferred to the protein ready made from a dolichol lipid donor (dolichol is a general name for a polymer consisting of various numbers of isoprene units; see section 11.13). This oligosaccharide structure is synthesised on the ER membrane by sequential addition of sugar residues by glycosyltransferase enzymes to the dolichol lipid group which tethers the growing oligosaccharide chain to the membrane. When completed, the branched oligosaccharide is transferred to the nascent protein (during translation) by the N-oligosaccharyltransferase complex catalysing formation of a glycosidic bond between the first N-acetylglucosamine in the oligosaccharide chain and the amino group of the target asparagine.

Once attached to the protein, the oligosaccharide is extensively modified during transit of the glycoprotein through the ER and Golgi (see section 5.9). Modifications include both removal and addition of sugars, and account for the great diversity in N-linked oligosaccharides of glycoproteins. N-glycans are the major form of mannoprotein modification in the yeasts *Saccharomyces cerevisiae* and *Candida albicans* and consist of a core structure, which is similar in all eukaryotes and is further elaborated in the Golgi to form an outer chain comprising a linear α-1–6-mannan backbone that is highly branched with α-1,2- and α-1–3-containing side chains to the N-linked oligosaccharides generating the mannoproteins characteristic of yeast cell walls (Figure 7.2B, above). In *Neurospora crassa* and *Aspergillus fumigatus*, the N-linked oligosaccharides are modified to become galactomannans: being processed by Golgi

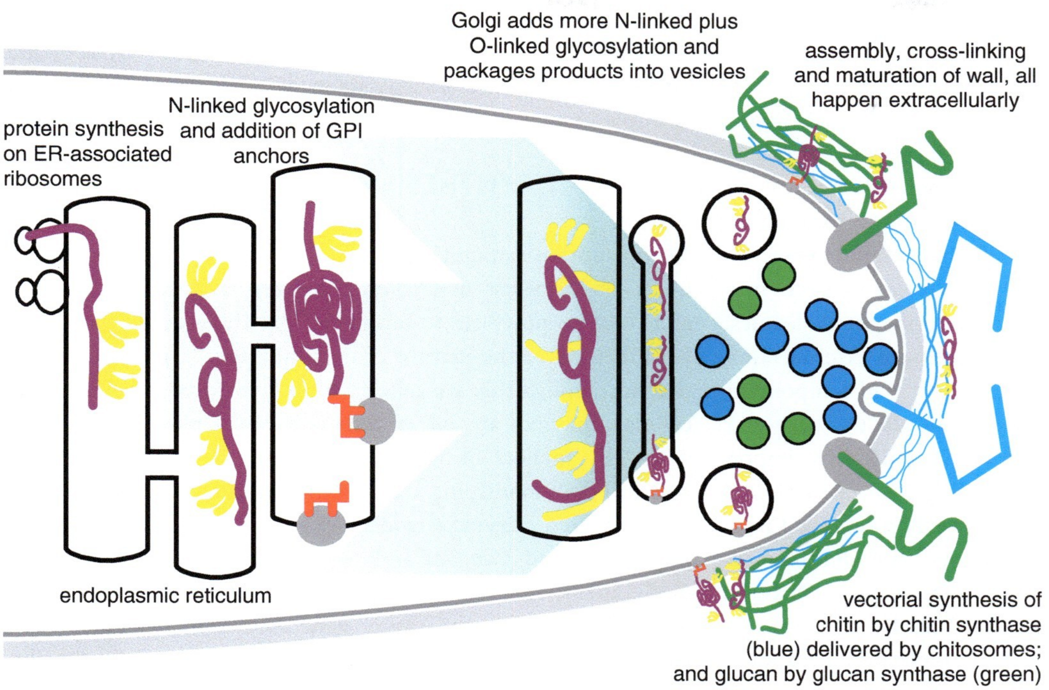

Figure 7.3. Biosynthesis of cell wall components. Glycoprotein synthesis begins in the endoplasmic reticulum (ER; protein shown in purple) with the addition of N-linked oligosaccharides (yellow) during translation. GPI anchors (red) are also added to some proteins in the ER. In the Golgi apparatus, the glycosyltransferases modify the proteins further by addition of sugars to generate O-linked oligosaccharides and to extend N-linked oligosaccharides. Glycoproteins are secreted into the cell wall space where they are integrated into the cell wall structure. The chitin (blue) and glucan (green) components of the cell wall are vectorially synthesised on the plasma membrane and extruded into the cell wall space during their synthesis. The various components of the wall are cross-linked together in the cell wall space by cell wall-associated glycosylhydrolases and glycosyltransferases (based on a figure in Bowman & Free, 2006).

resident enzymes into an α-1–6-linked mannose core with α-1,2-linked mannose side chains terminated with a variable number of β-1,5-linked galactose residues (Bowman & Free, 2006; Lesage & Bussey, 2006).

O-linked mannans of fungi tend to be short, linear chains composed of α-linked mannose sugars. The synthesis of other **decorating polymers** is less well understood, but they include α-1–3-glucans, β-1–6-glucans and a lengthening list of heteropolymeric polysaccharides (that is, polysaccharides composed of one, two or several different monosaccharides, in straight or branched chains). Most of what we know about O-linked glycosylation comes from work on *Saccharomyces cerevisiae*, where synthesis begins in the ER with the addition of a single mannose to the 'hydroxy' oxygen of selected serine and/or threonine residues in the polypeptide which is being translated. This sugar is transferred from dolichol-mannose by a family of protein-O-mannosyltransferase enzymes. The remainder of the O-linked oligosaccharide elongation is done by mannosyltransferase enzymes in the Golgi apparatus. In *Neurospora crassa*, the O-linked structure contains galactose as well. O-linked oligosaccharides of fungi are not branched, and they are not as long as N-linked glycans.

A special form of glycosylation is the GPI anchor; in this case glycosylation serves to attach a hydrophobic lipid anchor to proteins that carry a particular carboxy-terminal signal sequence. The GPI anchor is a lipid and oligosaccharide-containing structure that directs and localises the protein to the plasma membrane. The attachment is done by a protein complex located in the ER membrane, known as GPI transamidase (though the complex consists of about 20 different proteins), which recognises the carboxy-terminal signal sequence as the translated polypeptide migrates through the ER membrane, and transfers the pre-assembled GPI anchor onto the newly generated carboxy-terminus of the protein.

Genetic analyses have shown that many of the genes in the GPI anchor biosynthetic pathway are essential for viability in both yeasts and filamentous fungi, suggesting that the GPI-anchored proteins are required for integrity of the cell wall. Yeast genomes can contain from 28 to 169 GPI-anchored protein sequences (most in the pathogenic *Candida albicans*), with 74 and 87 genes in the genomes of *Aspergillus nidulans* and *Neurospora crassa*, respectively. Known GPI-anchored proteins include glycosylhydrolases, glycosyltransferases and peptidases, all of which are likely to contribute to cell wall synthesis and remodelling. Clearly, GPI-anchored proteins are critical for the formation and maintenance of the cell wall (Bowman & Free, 2006; Lesage & Bussey, 2006). GPI anchors and their association with the fungal wall is mentioned in the last paragraph of section 5.12.

Most cell wall proteins are structurally integrated into the wall, being covalently linked to chitin and/or glucan through sugars in their N- and O-linked oligosaccharides and/or in their GPI anchor. The proteins recognised as '**proteins with internal repeats**', or (PIR)-proteins cross-link β-1–3-glucan chains through ester linkages, forming the protein-polysaccharide complex β-1–3-glucan–PIR-β-1–3-glucan and are found throughout the inner layer of the wall (Yin *et al.*, 2007). Proteins in the cell wall contribute to maintenance of cell shape, adhesion, protection against environmental stress, absorption of molecules, transmission of intracellular signals and receipt of external stimuli, as well as the synthesis and remodelling of wall components (Adams, 2004; de Groot *et al.*, 2005; Klis *et al.*, 2006).

A family of aspartic peptidases (called **yapsins**) are important for cell wall assembly and/or remodelling. They are tethered to the membrane and the wall by a GPI anchor and are found only in fungi. No natural substrates have yet been identified for any of the yapsin enzymes, but they are thought to function in two ways: activation of periplasmic enzymes by endoproteolysis to involve them in glucan synthesis (perhaps in response to changing environmental conditions), and/or to release GPI-anchored enzymes from the wall when active cell wall assembly is no longer required by shedding the proteins by endoproteolysis. Some yapsins of *Candida* spp. have been implicated in virulence and infection by influencing adherence to epithelial cells (Gagnon-Arsenault *et al.*, 2006).

Another significant group of cell wall-associated proteins are cytoplasmic, even mitochondrial, enzymes, which do not have a signal sequence for export across the cell membrane. In most natural situations, the outer hyphal wall has a layer of such proteins, which have been called **moonlighting proteins** (Jeffery, 1999). These have been assumed to be contaminants of fungal cell wall preparations in the past, because finding enzymes of glycolysis or other metabolic pathways hanging around in the wall is, to say the least, unexpected. Another possibility is that because fungal cell walls have a significant capacity to absorb soluble proteins from the environment, fungal surfaces may be contaminated with cytoplasmic proteins picked up from the environment. However, there is increasing evidence that they really are resident proteins cross-linked into the cell wall. Obviously, such proteins could provide the 'wall organelle' with a range of functions, but not enough is known yet to form a coherent picture.

It is abundantly clear, though, that moonlighting proteins occur very widely in many different organisms. The first examples of moonlighting proteins were described in the late 1980s, when it was found that the *crystallin* structural proteins in the lens of the vertebrate eye were enzymes of intermediary metabolism. For example, duck ε-crystallin is lactate dehydrogenase, and turtle τ-crystallin is the glycolytic enzyme α-enolase. Any metabolic function for such enzymes in the lens of the eye is highly unlikely. Instead they probably serve a structural function in the lens where they accumulate to very high concentrations (Huberts & van der Klei, 2010). Moonlighting proteins are multifunctional proteins that perform several, often unrelated, and completely independent functions, which is why they are given that name, by analogy to *moonlighting people* who have multiple jobs.

7.7 WALL SYNTHESIS AND REMODELLING

Newly synthesised chitin and glucan polysaccharides are either linear or shapeless molecules. They are cross-linked together and to other polymers to form the rigid three-dimensional network typical of the mature cell wall (Figure 7.3). The enzymes that accomplish this are called **transglycosidases**. These extracellular enzymes are either tethered to the plasma membrane with GPI anchors or located within the cell wall itself. The first cell wall-modifying transglycosidase to be recognised is a GPI-anchored enzyme encoded by the genes *GEL* and *PHR* in *Aspergillus* and *Candida* and *GAS* in *Saccharomyces* (Mouyna *et al.*, 2013). This enzyme splits a β-1–3-glucan molecule internally and transfers the newly generated reducing end to the non-reducing end of a different molecule; this elongates the glucan chain. Transglycosidases able to cross-link β-1–6-glucan and chitin are also known (Arroyo *et al.*, 2016). Indeed, six families of conserved GPI proteins of this sort have so far been identified as being common across fungal species.

The transglycosidases are generally considered to be responsible for strengthening, or rigidifying, the fungal cell wall through the cross-links they generate. Cell wall **hydrolases** (like endo-β-1–3-glucanase and chitinase) and **deacetylases**, on the other hand, can plasticise the rigid cell wall. Chitinase is involved in bud separation in *Saccharomyces*, and endoglucanase is essential for cytokinesis in *Schizosaccharomyces*. In filamentous fungi, some models of apical wall synthesis suggest a delicate balance between hydrolysis of established polymers and synthesis of new wall material at the hyphal apex (see section 5.14), although there is no clear evidence that cell wall hydrolases are required for hyphal tip extension. Deacetylases do appear to be widespread, and in zygomycetes, ascomycetes and basidiomycetes substantial **deacetylation of chitin** (which produces **chitosan**) occurs. Chitosan is a more flexible molecule than chitin, and is resistant to chitinases. Also, deacetylation of the chitin in the walls of plant pathogens can prevent the host plant's receptors from recognising the pathogen and so delay the onset of plant defences.

During active hyphal growth wall construction activity is concentrated at the hyphal tip. Autoradiographic studies indicate that all synthesis of chitin and glucans takes place within 10 μm of the apex of the hypha of *Neurospora crassa*. The tip is highly plastic as the wall is laid down, but as walls mature they become more rigid. The rigidity is provided by the cross-linking of polymers to which reference is made above, thickening of fibrils and the deposition of materials in the interfibrillar matrix. The process is highly polarised, and reliant on maintenance of a positive turgor pressure within the cytoplasm (Bartnicki-García *et al.*, 2000). Wall rigidification in vegetative hyphae is a remarkably rapid process, taking only a couple of minutes. An estimate of the time elapsing between the deposition of plastic wall material at the hyphal tip and its subsequent rigidification at the base of the elongation zone (indicated by the point at which the hypha first attains a constant diameter) can be obtained by dividing the length of the tapered extension zone by the rate of hyphal elongation (Trinci, 1978).

Apical growth of the hypha requires long-distance transport between the subapical part and the apex of the tip cell. We have discussed some of the mechanisms that may be involved in section 5.14). These mechanisms include cytoskeleton-based motors delivering vesicles containing enzymes and substrates over long distances to the hyphal tip (Bartnicki-García, 2006; Riquelme *et al.*, 2007; Steinberg, 2007, 2011). In addition, hyphal growth is accompanied by the secretion of exoenzymes that participate in both lysis and synthesis of fungal cell wall components which make the wall in apical regions flexible. Extension of the hypha can then occur as both turgor and cytoskeleton-based cytoplasmic expansion push the cytoplasm against the flexible apical wall. A summarised diagrammatic compilation of these events is shown in Figure 7.4. This, together with the discussion in section 5.14 should provide a sufficiently detailed view of fungal extension growth.

We do not wish to overemphasise apical wall growth here, because it is not the end of the story for the fungal wall. The picture that has emerged for the cell walls of yeasts and filamentous fungi alike is that the wall is an extremely dynamic construction with its components maintained in continual balance as wall enzymes repair and remodel the original wall. Cell wall damage in budding yeast, *Saccharomyces cerevisiae*, triggers a salvage mechanism called the cell wall-integrity pathway consisting of

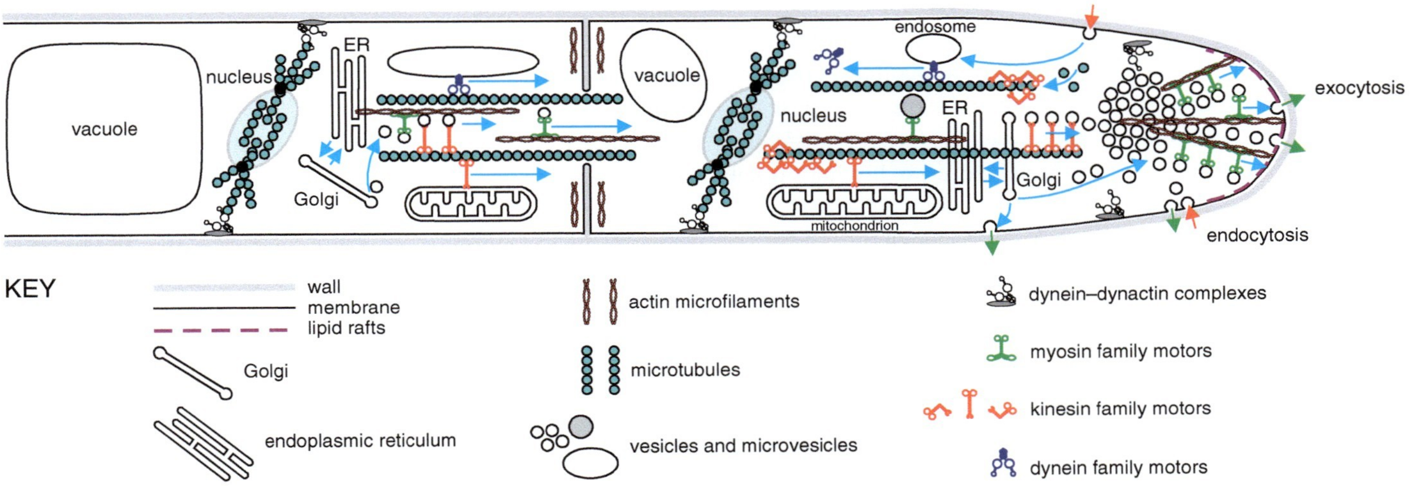

Figure 7.4. Cartoon representation of an overall molecular model of hyphal growth. The key feature of hyphal apical growth is rapid movement towards the apex of all the materials needed to create new wall, new membranes and new cytoplasmic components. Most of these materials are exported in vesicles by the endoplasmic reticulum (ER) and Golgi organelles, the vesicles being delivered to the apical vesicle cluster (called the Spitzenkörper; see section 5.14 and Figure 5.5) along microtubules powered by motor proteins of the kinesin and dynein families (see section 5.12 and Figure 5.8). The Spitzenkörper organises the final distribution of microvesicles along actin microfilaments to the plasma membrane at the extending tip. Vesicle fusion with the membrane is enabled by t-SNARE and v-SNARE proteins. Sterol-rich 'lipid rafts' at the hyphal tip could provide domains for apical proteins like signalling and binding complexes and might facilitate endocytosis. Endocytosis at the hyphal tip is dependent upon actin patches where myosin-1 polymerises actin into filaments that take the endocytotic vesicles away from the membrane. The extreme apex of hyphal tips undergoes extensive exocytosis, which is mainly devoted to synthesis of wall polymers outside the membrane and wall construction and maturation (sections 7.3 and 7.4 and Figures 7.2 and 7.3). Endocytosis features in the flanking regions of the hyphal tip, and this both recycles membrane components (originally delivered as exocytotic vesicles) and imports nutrients; both of which are transported to the endomembrane system for sorting and appropriate use (see sections 5.9 and 5.11 and Figure 5.5). This figure also shows that (potentially many) subterminal hyphal cells contribute to the apical migration of resources; (streams of) vesicles, (trains of rapidly moving) vacuoles (section 5.11) and mitochondria are all transported towards the apex and this transport extends through hyphal septa. Also note that the position of nuclear division spindles is probably specified by interaction between astral microtubules and membrane-bound dynein–dynactin complexes (Figures 5.9 and 5.10), and septal positioning is associated with rings of actin microfilaments (section 5.16). Remember: this *is* a cartoon, no attempt is made to portray relative scale or relative timing (some structures, like division spindles) are more transient than others (like the Spitzenkörper). Also, everything happens, quickly; in the text (sections 5.11 and 5.14) we show that 38,000 vesicles must fuse with the apical membrane each minute (that's over 600 every second) to support extension of each hyphal tip of *Neurospora crassa* when it is growing at its maximum rate. See text of Chapters 5 and 7 for complete explanation, and refer to Steinberg (2007, 2011).

at least 18 cell wall-maintenance genes controlled by a single transcription factor and a specific signal transduction pathway. Sequences belonging to this integrity pathway are conserved in several yeasts and filamentous fungi (Bowman & Free, 2006; Klis *et al.*, 2006; Gow *et al.*, 2017).

We discuss the dynamic nature of the fungal cell wall elsewhere in this text in discussions of:

- hyphal and spore differentiation (section 10.3);
- hyphal branching (section 4.11);
- septation (sections 4.12 and 5.16);
- hyphal anastomosis (section 5.15).

All these processes require that wall synthesis is restarted within a mature wall at a very closely controlled place and at a specific time. Another relevant circumstance to be aware of is that two hyphal branches that must be joined together will synthesise a joint wall, and the resultant join can be stronger than the original hyphal walls (section 13.7). All such remodelling depends on the coordinated activity of several glycoproteins already present within the wall structure. The 'wall-associated enzymes' involved include chitinases, glucanases and peptidases (Adams, 2004; Seidl, 2008; Gow *et al.*, 2017); enzymes that hydrolyse and break down cell wall components, as well as glycosyltransferases which are involved in the synthesis and cross-linking of wall polymers.

In addition to these instances of remodelling, there are many observations of **secondary hyphal walls** being synthesised as **internal thickenings** mostly made up of thick fibrils. These are probably glucans accumulated as intermediate to long-term nutritional reserves. The first detailed observations of this were made by **Jos Wessels** in the 1960s (reviewed by Bartnicki-García, 1999) who showed that during the final stages of maturation of the *Schizophyllum commune* sporophore when its nutritional support is completely endogenous, requiring no external sources of nitrogen or carbon, an alkali-insoluble cell wall component, which was called **R-glucan**, was the main fraction of the wall to be broken down. R-glucan contained both β-1–6 and β-1–3 linkages and was distinct from S-glucan which was alkali-soluble and constituted the bulk of the cell wall material left after mobilisation of the R-glucan. Studies of the mobilisation process indicated that cell wall degradation correlated with cap development and was controlled by changes in the level of a specific R-glucanase enzyme. It seems that while glucose remains available in the medium carbohydrate is temporarily stored in the form of R-glucan in the walls of mycelial and sporophore hyphae. During this phase of net R-glucan synthesis the R-glucanase is repressed by glucose in the medium, but when this is exhausted the repression is lifted, R-glucanase is synthesised and by breaking down the R-glucan it provides substrate(s) specifically required for sporophore development.

The story is slightly different in *Coprinopsis cinerea*. In this organism **glycogen** seems to serve a similar sort of function to the R-glucan of *Schizophyllum commune* during sporophore development (glycogen is an α-1–4/α-1–6-linked glucan; see section 11.6); the reason may be that mushrooms of *Coprinopsis cinerea* develop much more rapidly than those of *Schizophyllum commune* and the glycogen represents a much more efficient transient reserve that enables large quantities of sugar to be rapidly translocated through the sporophore with no disturbance to solute balance. Glycogen is involved in various aspects of vegetative morphogenesis in *Coprinopsis cinerea* (Waters *et al.*, 1975b; Jirjis & Moore, 1976), but so are wall glucans.

Mycelium of *Coprinopsis cinerea* forms multicellular sclerotia, about 250 μm in diameter, as resistant survival structures which pass through a period of dormancy before utilising their accumulated reserves to 'germinate' by producing a fresh mycelium. Glycogen is synthesised and accumulated in young sclerotia but is not the long-term storage product. For long-term storage, much of the carbohydrate is converted into a form of secondary wall material, probably glucan. Cells in the central bulk of the sclerotium may become **extremely thick walled**, the primary walls being thickened on their inner surfaces by loosely woven and very large fibrils, the development of which coincides with the gradual disappearance of glycogen from the cells (Waters *et al.*, 1972, 1975a, b).

What we mean by 'thick walled' is illustrated in the electronmicrographs of Figures 7.5 to 7.9. Secondary wall is ***external*** to the plasma membrane but ***internal*** to primary wall. The illustrations in Figures 7.7 to 7.9 make it clear that the secondary walls can come to make up a very considerable proportion of the cell volume. This inevitably constricts the protoplasm to a correspondingly smaller central lumen, but the indications are that these cells remain alive; being effectively dormant until the sclerotium is presented with amenable growth conditions (Erental *et al.*, 2008).

Dormant sclerotia may survive for several years, being protected by a rind composed of tightly packed hyphal tips which develop another form of secondary thickening which is a heavily melanised thickened wall that forms an impervious surface layer (Figure 7.6). Melanin is a dark-coloured pigment (almost black at high concentration); it is a high molecular weight polymer of phenolic and/or indolic compounds. Phenols have hydroxyl group(s) bonded directly to aromatic hydrocarbons. Indoles are heterocyclic organic compounds containing nitrogen; indole itself consists of a six-membered benzene ring fused to a five-membered nitrogen-containing pyrrole and has the formula C_8H_7N. These are negatively charged hydrophobic pigments that protect the hyphae and spores when they are cross-linked into the cell wall structure. Normally, they are so completely interlinked, in fact, that it is possible in the laboratory to digest away all the other wall components to leave '***melanin ghosts***' of the entire original cell wall (Dadachova *et al.*, 2008). Natural melanins are biocompatible conductors. Nanocomposites using melanin nanoparticles extracted from squid inks could potentially be used in bioelectronic devices such as biosensors and implantable devices (Eom *et al.*, 2017). Fungal melanin ghosts,

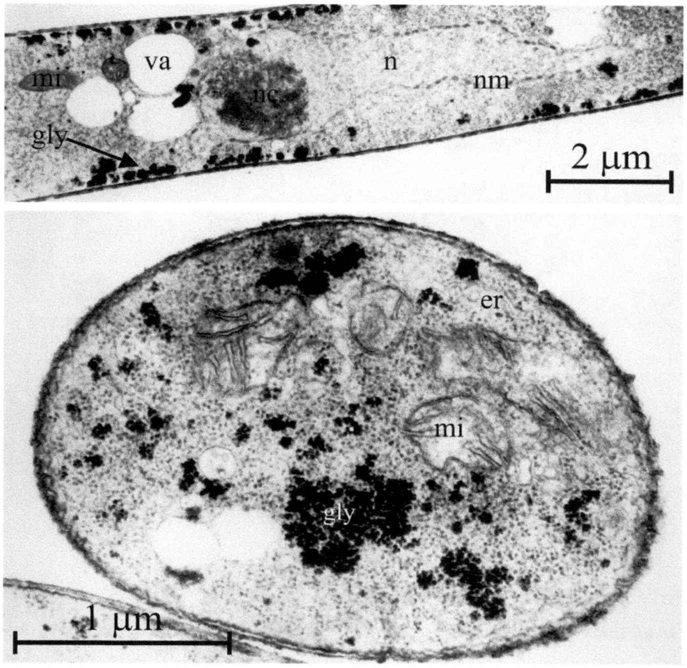

Figures 7.5 to **7.9**. A selection of transmission electronmicrographs (TEMs) of submerged mycelium and sclerotia of *Coprinopsis cinerea* showing the two main types of thick-walled hyphal cells; rind cells with dense, pigmented secondary walls of heavily melanised glucan on outer and side walls which make up a plate-like layer of protective rind which resists environmental extremes, and the secondary walls of medullary cells which are uniformly thickened around the cell with large, branched fibres of glucan. This image, Figure 7.5 (at left) shows a TEM of typical young vegetative hyphae from the submerged mycelium of *Coprinopsis cinerea*. Hyphae shown in longitudinal (top) and transverse sections. Note the thin (primary) hyphal walls that characterise this undifferentiated tissue. Electronmicrograph by Dr Henry Waters. Key: n, nucleus, nc, nucleolus, nm, nuclear membrane, mi, mitochondrion, va, vacuole, er, endoplasmic reticulum, gly, glycogen granules.

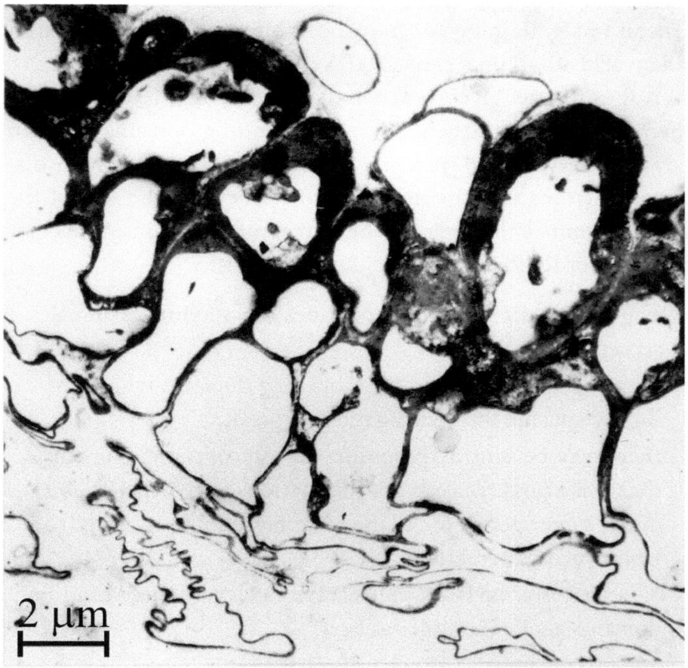

Figure 7.6. Secondarily-thickened walls of rind cells on the outside of sclerotia of *Coprinopsis cinerea*. These secondary walls provide a protective layer and are heavily melanised. Electronmicrograph by Dr Henry Waters.

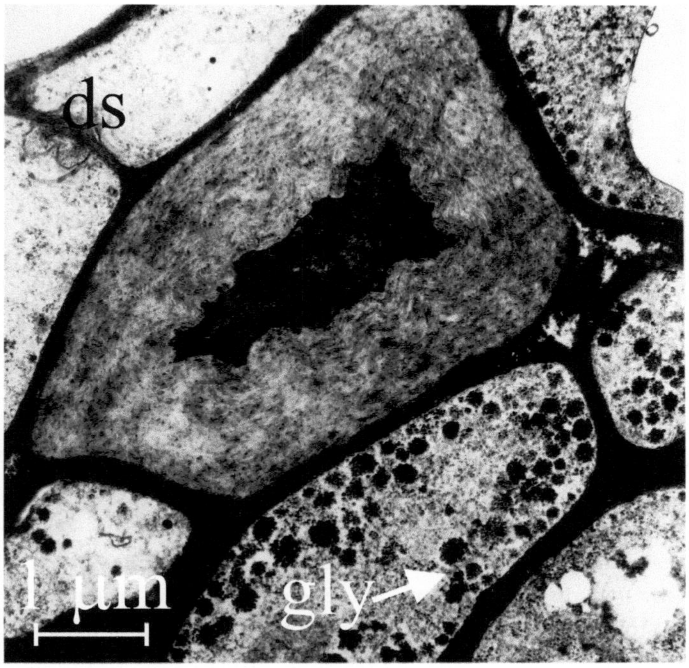

Figure 7.7. Secondarily-thickened walls of the central (medulla) region of a young sclerotium showing a cell with a secondarily-thickened wall in transverse section alongside many normal cells. This secondary wall is not melanised; it is constructed of thick fibres of glucan and is eventually recycled to provide carbohydrate resources when the sclerotium germinates. The granules (labelled gly) are accumulations of glycogen, which forms a short-term carbohydrate resource. Accumulation of glycogen occurs before formation of the glucan fibres and as the fibres are formed the glycogen content declines. Note the dolipore septum (ds) at top left. Electronmicrograph by Dr Henry Waters.

with their greater structural integrity, could have more interesting applications.

Melanin is extremely resistant to chemical and enzymic attack and contributes to virulence in many pathogenic fungi (for example *Paracoccidioides brasiliensis, Sporothrix schenckii, Histoplasma capsulatum, Blastomyces dermatitidis* and *Coccidioides posadasii*) by reducing the susceptibility of melanised fungi to host defences and drugs (Taborda *et al.*, 2008). Melanin increases cell wall rigidity, enabling hyphae of black fungi to penetrate host tissues and pigmented conidia or yeast cells to remain turgid when desiccated. Fungi with melanised walls are also resistant to electromagnetic and ionising radiations; the pigment seems to

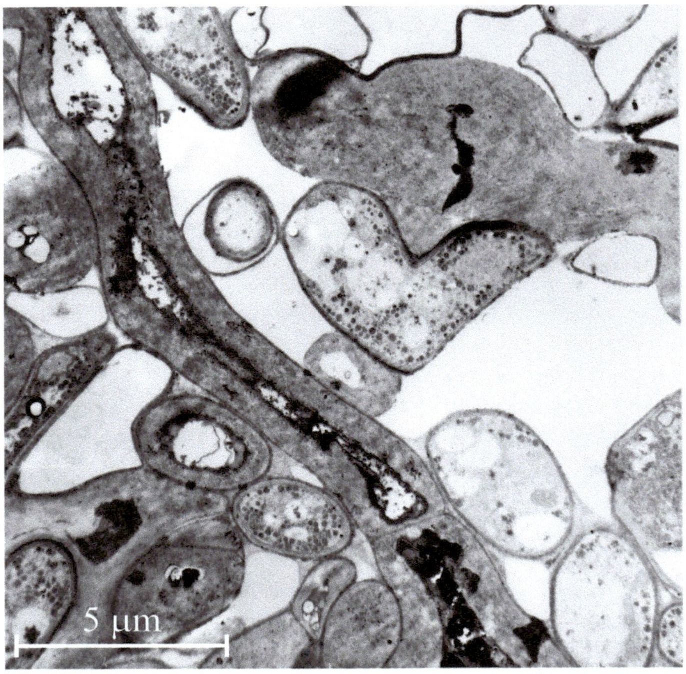

Figure 7.8. Medullary cells with secondarily-thickened wall shown in longitudinal section, illustrating that the thickening (and consequent constriction of the cell lumen) is uniform over the length of the hypha, and that the thickening can cross and involve the dolipore septa. Electronmicrograph by Dr Henry Waters.

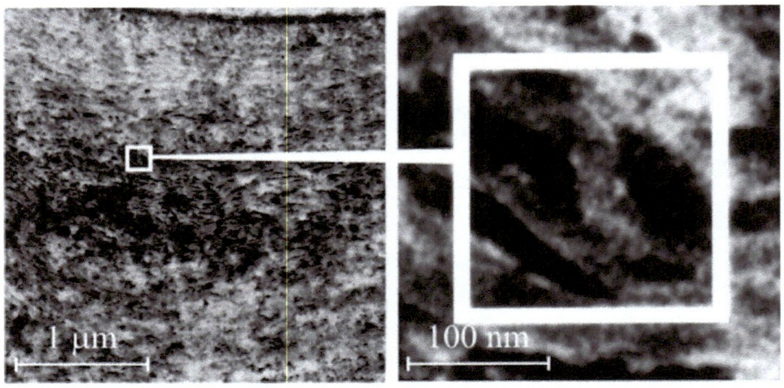

Figure 7.9. Magnified images of a longitudinal section of a secondarily-thickened wall showing its fibrillar structure. The region of secondarily-thickened (glucan) fibres highlighted in the left-hand image is shown at extreme magnification on the right. Electronmicrographs by Dr Henry Waters.

provide both physical shielding (from UV light) and quenching of cytotoxic free radicals (caused by ionising radiations). Although the detailed structure of melanin is unknown, two main types of melanin are found in the fungal cell wall, which are named after precursors in their biosynthetic pathways. DHN-melanin is synthesised from 1,8-dihydroxynaphthalene; and DOPA-melanin, which is synthesised from 3,4-dihydroxyphenylalanine (L-DOPA). In various fungi, there is evidence that melanin can be cross-linked to mannans, mannosylated proteins, galactoxylomannan, chitin or chitosan (Gow *et al.*, 2017).

Other cell wall pigments, the carotenoids (section 11.13), also protect against UV radiation. In general, mutants of the entomopathogenic fungus *Metarhizium anisopliae* with white conidia were more sensitive to UV radiation than mutants with purple conidia, which were more sensitive than mutants with yellow conidia, which in turn were more sensitive than the green wild strain (Braga *et al.*, 2006). They're not just pretty colours.

7.8 ON THE FAR SIDE

A wall has an outer side; an interface between that which is contained and that which is excluded. Consequently, there are molecules which can be described as being on the outside of the fungal wall.

Examples are the mating proteins, called **agglutinins**, which are responsible for 'agglutination' (cell-to-cell adhesion) between cells of different mating types in *Saccharomyces cerevisiae*, as an early process in mating (discussed in section 9.2). The interaction takes place between two GPI-anchored cell-surface glycoproteins anchored in the outer face of the plasma membranes of the two mating cells. These are representative of a growing class of cell adhesion proteins, generally called **adhesins**, which are critical to a wide range of fungal cell interactions with the outside world (Dranginis *et al.*, 2007).

Adhesins are glycoproteins that are secreted to the wall, then anchored with their interactive binding domains elevated beyond the wall surface. They have several domains with discrete functions and the need to project beyond the rest of the wall determines the order of their structural domains. The characteristic order (from the outside inwards) is:

- as glycoproteins their polypeptides carry amino-terminal secretion signals;
- then there are the interactive binding domains which are highly specific for their ligands (targets);
- there may be central threonine-rich glycosylated domains that make possible cell–cell interactions with other cells of the same species (or of the same mycelium), so-called **homotypic** interactions;
- N- and O-glycosylated stalks serve to elevate these binding domains above the wall surface;
- carboxy-terminal regions mediate covalent cross-linking to the wall matrix and possibly tethering to the plasma membrane through modified GPI anchors (Dranginis *et al.*, 2007).

These structural motifs are common, but not universal, and it is probable that we currently have an incomplete knowledge of a very large and diverse range of glycoproteins. Adhesins are known to mediate homotypic interactions in mating (as in budding yeast), as well as in multicellular differentiation, an example being glycoproteins known as **galectins** which are specific to the mushroom of *Coprinopsis*.

Adhesins also enable adhesion to surfaces, and for pathogens that includes the surfaces of host organisms; adhesion to host cells and tissues is a vital property of the fungal cell wall that promotes virulence. Fibrils extending from the wall into the suspending fluid are commonly found where fungi attach to surfaces. The fibrils are highly specific to a complementary molecular structure on the host surface. This system of recognition and communication between fungus and partner is widespread in pathogenic and mutualistic interactions, enabling adherence to the surfaces of host organisms (animal and plant). But they also enable the formation of biofilms (see below), both on natural surfaces and on catheters and other medical devices.

The best characterised of several cell wall proteins that have adhesin-like properties are the *Candida albicans* Als family of eight proteins and the *Candida glabrata* Epa family. Both are families of GPI proteins that have a characteristic domain organisation with N-terminal adhesin domains that impart specificity in host protein/glycan binding; different adhesin family members having different host protein specificities.

A hint of other capabilities of these molecules is provided by an adhesin secreted by *Blastomyces dermatiditis*, which grows as a filamentous mould at room temperature in soils contaminated with animal excreta (like many pastures); spores inhaled by humans and other animals convert into large invasive yeasts in the lungs, and cause a potentially fatal blastomycosis. The adhesin produced by the fungus has a domain structured like an animal epidermal growth factor that binds to host cells and downregulates the host's immune response, consequently potentiating the infection (Dranginis *et al.*, 2007). It is becoming increasingly evident that, in fungal pathogens of humans and other animals, the fungal cell wall incorporates immune decoys and shields. In plant pathogenic fungi, too, the cell wall is detected by receptors of the plant cell that induce local and systemic defence mechanisms; and the fungal cell wall can deploy its own mechanisms to frustrate those host defences (Nishimura, 2016; Geoghegan *et al.*, 2017; Gow *et al.*, 2017).

The surface interactions described so far take place at least in aqueous films and in some cases in a fluid medium, either natural or artificial. In the latter case the boundary between 'wall' and 'medium' may not always be clearly demarcated (as is indicated in Figure 7.2A), because the zone beyond the wall contains successively diminishing concentrations of polysaccharides and glycoproteins; but these polysaccharides and glycoproteins are ideal contributions to biofilms.

Present-day biofilms are microbial communities that thrive attached to surfaces in a matrix composed of mixed microbial cells, polysaccharides and other excreted and secreted cellular products, including enzyme activities. They have been the subject of an enormous amount of research (Hobley *et al.*, 2015; Flemming, 2016). Biofilms form on all sorts of surfaces in the natural environment; for example, rocks and minerals, and structures like stonework, brickwork, steelwork and concrete, and even the horn, bones and exoskeletons of dead organisms. They also form on inside and outside surfaces of plants, other microbes and animals. At these interfaces, whole communities of microbes gather as films, mats, flocs or sludge and these are the different kinds of 'biofilm' which is the general name given to a thin coating comprised of living material. Biofilms are clinically important because formation of a surface biofilm (often internally in the body) is a frequent first step in the attack of a pathogen; but biofilms are also often involved in nosocomial infections (originating in a hospital) because they can form on hospital surfaces and medical equipment. Biofilm formation is also important in biotechnology. On the one hand, biofilms may be essential to a specific process (biofilms recycle your sewage), on the other hand, chemical engineers can spend sleepless nights trying to prevent biofilm formation in processes where uniform growth in suspension is most important for efficient formation of the desired product.

Morphologically, biofilms can be smooth and flat, rough, fluffy or filamentous, and the biofilm matrix is composed of a range of extracellular polymeric substances (**EPS**); '[t]he EPS components are extremely complex and dynamic and fulfil many functional roles, turning biofilms into the most ubiquitous and successful form of life on Earth' (Flemming, 2016), but the principal component of the matrix is water. Many EPS are hygroscopic and actively retain water, so the biofilm matrix provides a highly hydrated environment that dries more slowly than its surroundings and consequently protects cells in the biofilm against fluctuations in water availability. Fungi are active participants in biofilms in the present day, their extracellular matrix being composed of glucans, chitin and nucleic acids. *Saccharomyces cerevisiae* can initiate biofilm formation, growth in low-glucose medium causing the yeast cells to adhere to plastic surfaces, and to form 'mats' on dilute agar medium where yeast-form cells adhere together. Both formation of mats and attachment to plastic require specific families of fungal cell-surface glycoproteins for adherence (Reynolds & Fink, 2001).

Filamentous fungi also form biofilms in the present day (Harding *et al.*, 2009) and the more primitive zygomycetes, including *Rhizopus* and *Rhizomucor*, produce an extensive extracellular matrix to aid adherence to surfaces (Singh *et al.*, 2011). In *Aspergillus fumigatus*, the extracellular matrix is composed of 25% polysaccharides and 70% monosaccharides with some hydrophobic proteins and melanin. The matrix plays an essential role in the organisation of the colony by sticking together mycelial threads. In human pathogens, this biofilm material has been implicated in blocking recognition and immune capture by phagocytic cells (Gow *et al.*, 2017). An important ability of hyphal growth is that apically growing filamentous hyphae can explore and exploit the biofilm, digesting the adhesives and other polymers that make up the biofilm matrix, even escaping from the biofilm under their own volition.

It has been argued that, in the most ancient times, biofilms would have fostered the evolution of the hyphal growth form by providing the appropriate selection pressure for the filamentous hyphae to escape from, and destructively exploit and dominate, the biofilm matrix. Even more importantly, filamentous hyphae could parasitise the photosynthetic microbes of the biofilm community to recruit photobionts into primitive lichen-like arrangements, which then had the terrestrial surface of the Earth at their disposal (Moore, 2013).

In contrast to the above fluid biofilms, many hyphal surfaces are dry and even hydrophobic (that is, non-wettable). Examples are the surfaces of the innumerable air-dispersed spores found in fungi, their sporophores, as well as all sorts of aerial mycelium and aerial structures like sporophores and stromata. Less obvious examples are cavities within organised tissues that are kept fluid-free by hydrophobic coatings to enable gas exchange (for example see Lugones *et al.*, 1999; Pareek *et al.*, 2006).

The outer surface of the hyphal or spore walls in these cases is usually found to be composed of a layer of 10 nm wide rodlets composed of proteins which modify the biophysical properties of the wall surface. The most commonly encountered family of such proteins are called **hydrophobins** (Wessels, 1996; Wösten & de Vocht, 2000; Sunde *et al.*, 2008; Cox & Hooley, 2009). Hydrophobins belong to a large, diverse group of related proteins found widely in the fungi; when expressed to the maximum they may constitute up to 10% of total wall protein. Each molecule consists of a hydrophobic domain and a hydrophilic domain; that is, they are **amphipathic** (a term you may have met in relation to the phospholipids that make up biological membranes, which also have a hydrophilic group at one end and a hydrophobic group at the other). Their amino-terminal part determines the hydrophilic side of the assemblage. The amphipathic structure of hydrophobins enables them to self-assemble to form a monolayer around aerial structures such as hyphae and sporophores, coating the outer hydrophilic layers of the fungal wall and generating a hydrophobic interface between the fungal cells and their environments. This provides the molecules with an extraordinary potential array of functions, they:

- enable hyphae to break through the air–water interface of fluid habitats;
- provide the hydrophobicity required by hyphae and spores in contact with air;
- participate in morphogenetic signalling, initiating conidiation and sporophore formation;
- have important roles in tissue formation, particularly in controlling fluid and air spaces;
- promote adhesion between the hydrophilic cell wall of the fungus and the hydrophobic surfaces of plants and insects, and so potentiate infection and aid penetration of the host surface;
- mediate symbiotic interactions with plant roots (mycorrhizas) and algae (lichens);
- avoid host immune systems: aerial conidia of *Aspergillus, Penicillium* and *Cladosporium* that have surface layers of hydrophobin do not activate the immune system.

> Hydrophobin rodlets are said to 'immunologically silence airborne moulds', which means that although fungal spores are ubiquitous in the air we breathe they neither continuously activate host immunity nor induce inflammatory responses after inhalation. For example, the hydrophobin protein *RodA* of *Aspergillus fumigatus* prevents immune recognition by alveolar macrophages; the immune response to the invader occurs only when the hydrophobin layer cracks open as the spore swells and germinates and so exposes the underlying galactosaminoglycan and glucan layers of the wall. (Aimanianda *et al.*, 2009)

The molecules that do all this are relatively small proteins, usually around 100 amino acids, that have extensive homologies and characteristically contain signal sequences for secretion, and eight cysteine residues conserved in the same positions. These eight conserved cysteine residues form four disulfide bridges and they prevent self-assembly of the hydrophobin in the absence of a hydrophilic–hydrophobic interface. Hydrophobins are unique to mycelial fungi but are expressed by both *Ascomycota* and *Basidiomycota*. Each fungus has genes for more than one, often more than 10 different hydrophobins, and the genes are usually expressed at different times. In *Schizophyllum commune*, the hydrophobin found in the vegetative hyphal wall differs from that expressed in the hyphal walls of the sporophore (Wessels, 1996).

Hydrophobins are secreted from the hyphal tip; if the hypha is in an aqueous environment, the hydrophobins pass into solution. But the protein molecules can self-assemble into covering films at the water–air (i.e. hydrophilic–hydrophobic) interface and when a hypha emerges from the solution, the polypeptide polymerises on the surface of the hyphal wall, forming an array of parallel rodlets. Each hydrophobin molecule is bound to the fungal wall by its hydrophilic end; the hydrophobic domain being exposed to the outside world (Linder *et al.*, 2005; Cox & Hooley, 2009). A difference in solubility of these assemblages divides hydrophobins into two groups: class I hydrophobins form highly insoluble membranes dissolved only by trifluoroacetic acid and formic acid, while assemblies of class II hydrophobins dissolve readily in ethanol or sodium dodecyl sulfate (SDS).

The hydrophobin assembly on the hypha reduces water movement through the wall, giving protection from desiccation, but the exposed hydrophobic surface enables bonding to other hydrophobic surfaces. This happens because hydrophobes are not electrically polarised and the lowest energy state for two hydrophobes is for them to bond together to exclude electrically polarised water molecules. A fungal wall coated with hydrophobins will be able to use this hydrophobic interaction to bond to other aerial hyphae, leading to the formation of multicellular hyphal structures. A hydrophobin-coated spore could also attach immediately and firmly to the hydrophobic (for example, waxy) surface of its plant or insect host; giving time for formation of appressoria or other penetration structures.

Within fungal tissues, and that includes the tissues of lichen thalli, hydrophobins provide control over the movement of water and ***gases*** within the tissue because the exposed hydrophobic domains prevent water logging in airspaces. The distribution of hydrophobins and hydrophilic wall coatings allow the fungus to control which channels through the tissue are used for movement of water and aqueous nutrients, and which are kept free of fluid and used for movement of gases (Wösten & de Vocht, 2000; Sunde *et al.*, 2008). The remarkable ability of hydrophobins to change the nature of a surface (they turn hydrophobic surfaces hydrophilic and hydrophilic surfaces hydrophobic) make hydrophobins candidates for use in commercial and medical applications (Scholtmeijer *et al.*, 2005; Cox & Hooley, 2009; Cox *et al.*, 2009).

Another fungal wall protein that deserves specific mention is the glycoprotein known as **glomalin**, which is produced ***abundantly*** in the walls of hyphae and spores of arbuscular mycorrhizal fungi in soil and in roots. What makes this protein deserving of mention is the amount of it that is produced. It permeates soil organic matter and can account for around **30% of the carbon in soil**. In the soil, glomalin forms clumps of soil granules called aggregates that add structure to soil. Glomalin is only produced by members of Subphylum *Glomeromycotina*, fungi that form arbuscular mycorrhizas. When first discovered it was clearly present in such quantity that it must make a massive contribution to the aggregation of soil particles. It was first thought that it must be secreted or otherwise released into the soil by arbuscular mycorrhizal fungi specifically to control soil structure; a mechanism that has been called habitat engineering. The view being that increased soil aggregation (i.e. improved tilth) would benefit the host plant, and thereby the associated mycorrhizal fungus, and so justify the energetic 'cost' of producing the glomalin. There is some experimental support for this idea though there is also evidence that glomalin is not secreted but is covalently bound into the hyphal wall matrix where it protects the hypha. It may be that the characteristics that enable glomalin to protect the hyphal wall also allow the protein to promote soil aggregation (Driver *et al.*, 2005; Singh *et al.*, 2013; Yang *et al.*, 2017; Zhang *et al.*, 2017).

7.9 REFERENCES

Adams, D.J. (2004). Fungal cell wall chitinases and glucanases. *Microbiology*, **150**: 2029–2035. DOI: https://doi.org/10.1099/mic.0.26980-0.

Aimanianda, V., Bayry, J., Bozza, S. *et al.* (2009). Surface hydrophobin prevents immune recognition of airborne fungal spores. *Nature*, **460**: 1117–1121. DOI: https://doi.org/10.1038/nature08264.

Arroyo, J., Farkaš, V., Sanz, A.B. & Cabib, E. (2016). Strengthening the fungal cell wall through chitin-glucan cross-links: effects on morphogenesis and cell integrity. *Cellular Microbiology*, **18**: 1239–1250. DOI: https://doi.org/10.1111/cmi.12615.

Banks, I.R., Specht, C.A., Donlin, M.J. *et al.* (2005). A chitin synthase and its regulator protein are critical for chitosan production and growth of the fungal pathogen *Cryptococcus neoformans*. *Eukaryotic Cell*, **4**: 1902–1912. DOI: https://doi.org/10.1128/EC.4.11.1902–1912.2005.

Bartnicki-García, S. (1999). Glucans, walls, and morphogenesis: on the contributions of J.G.H. Wessels to the golden decades of fungal physiology and beyond. *Fungal Genetics and Biology*, **27**: 119–127. DOI: https://doi.org/10.1006/fgbi.1999.1144.

Bartnicki-García, S. (2006). Chitosomes: past, present and future. *FEMS Yeast Research*, **6**: 957–965. DOI: https://doi.org/10.1111/j.1567-1364.2006.00158.x.

Bartnicki-García, S., Bracker, C.E., Gierz, G., López-Franco, R. & Lu, H. (2000). Mapping the growth of fungal hyphae: orthogonal cell wall expansion during tip growth and the role of turgor. *Biophysical Journal*, **79**: 2382–2390. DOI: https://doi.org/10.1016/S0006-3495(00)76483-6.

Bowman, S.M. & Free, S.J. (2006). The structure and synthesis of the fungal cell wall. *BioEssays*, **28**: 799–808. DOI: https://doi.org/10.1002/bies.20441.

Braga, G.U.L., Rangel, D.E.N., Flint, S.D., Anderson, A.J. & Roberts, D.W. (2006). Conidial pigmentation is important to tolerance against solar-simulated radiation in the entomopathogenic fungus *Metarhizium anisopliae*. *Photochemistry and Photobiology*, **82**: 418–422. DOI: https://doi.org/10.1562/2005-05-08-RA-52.

Calonje, M., Mendoza, C.G., Cabo, A.P. & Novaes-Ledieu, M. (1995). Some significant differences in wall chemistry among four commercial *Agaricus bisporus* strains. *Current Microbiology*, **30**: 111–115. DOI: https://doi.org/10.1007/BF00294192.

Coronado, J.E., Mneimneh, S., Epstein, S.L., Qiu, W.G. & Lipke, P.N. (2007). Conserved processes and lineage-specific proteins in fungal cell wall evolution. *Eukaryotic Cell*, **6**: 2269–2277. DOI: https://doi.org/10.1128/EC.00044-07.

Cox, A.R., Aldred, D.L. & Russell, A.B. (2009). Exceptional stability of food foams using class II hydrophobin HFBII. *Food Hydrocolloids*, **23**: 366–376. DOI: https://doi.org/10.1016/j.foodhyd.2008.03.001.

Cox, P.W. & Hooley, P. (2009). Hydrophobins: new prospects for biotechnology. *Fungal Biology Reviews*, **23**: 40–47. DOI: https://doi.org/10.1016/j.fbr.2009.09.001.

Dadachova, E., Bryan, R.A., Howell, R.C. *et al.* (2008). The radioprotective properties of fungal melanin are a function of its chemical composition, stable radical presence and spatial arrangement. *Pigment Cell Melanoma Research*, **21**: 192–199. DOI: https://doi.org/10.1111/j.1755-148X.2007.00430.x.

De Groot, P.W.J., Ram, A.F. & Klis, F.M. (2005). Features and functions of covalently linked proteins in fungal cell walls. *Fungal Genetics and Biology*, **42**: 657–675. DOI: https://doi.org/10.1016/j.fgb.2005.04.002.

Douglas, C.M. (2001). Fungal β-(1–3)-D-glucan synthesis. *Medical Mycology*, **39 Supplement 1**: 55–66. DOI: https://doi.org/10.1080/744118880.

Dranginis, A.M., Rauceo, J.M., Coronado, J.E. & Lipke, P.N. (2007). A biochemical guide to yeast adhesins: glycoproteins for social and antisocial occasions. *Microbiology and Molecular Biology Reviews*, **71**: 282–294. DOI: https://doi.org/10.1128/MMBR.00037-06.

Driver, J.D., Holben, W.E. & Rillig, M.C. (2005). Characterization of glomalin as a hyphal wall component of arbuscular mycorrhizal fungi. *Soil Biology & Biochemistry*, **37**: 101–106. DOI: https://doi.org/10.1016/j.soilbio.2004.06.011.

Eom, T., Woo, K., Cho, W. *et al.* (2017). Nanoarchitecturing of natural melanin nanospheres by layer-by-layer assembly: macroscale anti-inflammatory conductive coatings with optoelectronic tunability. *Biomacromolecules*, **18**: 1908–1917. DOI: https://doi.org/10.1021/acs.biomac.7b00336.

Erental, A., Dickman, M.B. & Yarden, O. (2008). Sclerotial development in *Sclerotinia sclerotiorum*: awakening molecular analysis of a 'dormant' structure. *Fungal Biology Reviews*, **22**: 6–16. DOI: https://doi.org/10.1016/j.fbr.2007.10.001.

Erwig, L.P. & Gow, N.A.R. (2016). Interactions of fungal pathogens with phagocytes. *Nature Reviews Microbiology*, **14**: 163–176. DOI: https://doi.org/10.1038/nrmicro.2015.21.

Fernandes, C., Gow, N.A.R. & Gonçalves, T. (2016). The importance of subclasses of chitin synthase enzymes with myosin-like domains for the fitness of fungi. *Fungal Biology Reviews*, **30**: 1–14. DOI: https://doi.org/10.1016/j.fbr.2016.03.002.

Flemming, H.-C. (2016). EPS-then and now. *Microorganisms*, **4**: article 41 (18 pp.). DOI: https://doi.org/10.3390/microorganisms4040041.

Free, S.J. (2013). Fungal cell wall organization and biosynthesis. *Advances in Genetics*, **81**: 33–82. DOI: https://doi.org/10.1016/B978-0-12-407677-8.00002-6.

Gagnon-Arsenault, I., Tremblay, J. & Bourbonnais, Y. (2006). Fungal yapsins and cell wall: a unique family of aspartic peptidases for a distinctive cellular function. *FEMS Yeast Research*, **6**: 966–978. DOI: https://doi.org/10.1111/j.1567-1364.2006.00129.x.

Geoghegan, I., Steinberg, G. & Gurr, S. (2017). The role of the fungal cell wall in the infection of plants. *Trends in Microbiology*, **25**: 957–967. DOI: https://doi.org/10.1016/j.tim.2017.05.015.

Gow, N.A.R., Latgé, J.-P. & Munro, C.A. (2017). The fungal cell wall: structure, biosynthesis, and function. *Microbiology Spectrum*, **5**: FUNK-0035–2016. DOI: https://doi.org/10.1128/microbiolspec.FUNK-0035-2016.

Harding, M.W., Marques, L.L.R., Howard, R.J. & Olson, M.E. (2009). Can filamentous fungi form biofilms? *Trends in Microbiology*, **17**: 475–480. DOI: https://doi.org/10.1016/j.tim.2009.08.007.

Hobley, L., Harkins, C., MacPhee, C.E. & Stanley-Wall, N.R. (2015). Giving structure to the biofilm matrix: an overview of individual strategies and emerging common themes. *FEMS Microbiology Reviews*, **39**: 649–669. DOI: https://doi.org/10.1093/femsre/fuv015.

Howard, R.J., Ferrari, M.A., Roach, D.H. & Money, N.P. (1991). Penetration of hard substrates by a fungus employing enormous turgor pressures. *Proceedings of the National Academy of Sciences of the United States of America*, **88**: 11281–11284. DOI: https://doi.org/10.1073/pnas.88.24.11281.

Huberts, D.H.E.W. & van der Klei, I.J. (2010). Moonlighting proteins: an intriguing mode of multitasking. *Biochimica et Biophysica Acta, Molecular Cell Research*, **1803**: 520–525. DOI: https://doi.org/10.1016/j.bbamcr.2010.01.022.

Ichinomiya, M., Yamada, E., Yamashita, S., Ohta, A. & Horiuchi, H. (2005). Class I and class II chitin synthases are involved in septum formation in the filamentous fungus *Aspergillus nidulans*. *Eukaryotic Cell*, **4**: 1125–1136. DOI: https://doi.org/10.1128/EC.4.6.1125–1136.2005.

Jeffery, C.J. (1999). Moonlighting proteins. *Trends in Biochemical Sciences*, **24**: 8–11. DOI: https://doi.org/10.1016/S0968-0004(98)01335-8.

Jirjis, R.I. & Moore, D. (1976). Involvement of glycogen in morphogenesis of *Coprinus cinereus*. *Journal of General Microbiology*, **95**: 348–352. DOI: https://doi.org/10.1099/00221287-95-2-348.

Jones, M., Tien, H., Chaitali, D., Fugen, D. & Sabu, J. (2017). Mycelium composites: a review of engineering characteristics and growth kinetics. *Journal of Bionanoscience*, **11**: 241–257. DOI: https://doi.org/10.1166/jbns.2017.1440.

Klis, F.M., Boorsma, A. & De Groot, P.W.J. (2006). Cell wall construction in *Saccharomyces cerevisiae*. *Yeast*, **23**: 185–202. DOI: https://doi.org/10.1002/yea.1349.

Latgé, J.-P. (2007). The cell wall: a carbohydrate armour for the fungal cell. *Molecular Microbiology*, **66**: 279–290. DOI: https://doi.org/10.1111/j.1365-2958.2007.05872.x.

Latgé, J.-P. & Beauvais, A. (2014). Functional duality of the cell wall. *Current Opinion in Microbiology*, **20**: 111–117. DOI: https://doi.org/10.1016/j.mib.2014.05.009.

Latgé, J.-P., Mouyna, I., Tekaia, F. *et al.* (2005). Specific molecular features in the organization and biosynthesis of the cell wall of *Aspergillus fumigatus*. *Medical Mycology*, **43**: 15–22. DOI: https://doi.org/10.1080/13693780400029155.

Lesage, G. & Bussey, H. (2006). Cell wall assembly in *Saccharomyces cerevisiae*. *Microbiology and Molecular Biology Reviews*, **70**: 317–343. DOI: https://doi.org/10.1128/MMBR.00038-05.

Lew, R.R., Levina, N.N., Walker, S.K. & Garrill, A. (2004). Turgor regulation in hyphal organisms. *Fungal Genetics and Biology*, **41**: 1007–1015. DOI: https://doi.org/10.1016/j.fgb.2004.07.007.

Linder, M.B., Szilvay, G.R., Nakari-Setälä, T. & Penttilä, M.E. (2005). Hydrophobins: the protein-amphiphiles of filamentous fungi. *FEMS Microbiology Reviews*, **29**: 877–896. DOI: https://doi.org/10.1016/j.femsre.2005.01.004.

Lugones, L.G. Wösten, H.A.B., Birkenkamp, K.U. *et al.* (1999). Hydrophobins line air channels in fruiting bodies of *Schizophyllum commune* and *Agaricus bisporus*. *Mycological Research*, **103**: 635–640. DOI: https://doi.org/10.1017/S0953756298007552.

Martin-Urdiroz, M., Roncero, I.G., González-Reyes, J.A. & Ruiz-Roldán, C. (2008). ChsVb, a class VII chitin synthase involved in septation, is critical for pathogenicity in *Fusarium oxysporum* f. sp. *lycopersici*. *Eukaryotic Cell*, **7**: 112–121. DOI: https://doi.org/10.1128/EC.00347-07.

Money, N.P. (2004). The fungal dining habit: a biomechanical perspective. *Mycologist*, **18**: 71–76. DOI: https://doi.org/10.1017/S0269-915X(04)00203-4.

Money, N.P. (2008). Insights on the mechanics of hyphal growth. *Fungal Biology Reviews*, **22**: 71–76. DOI: https://doi.org/10.1016/j.fbr.2008.05.002.

Money, N.P. & Harold, F.M. (1992). Extension growth of the water mold *Achlya*: interplay of turgor and wall strength. *Proceedings of the National Academy of Sciences of the United States of America*, **89**: 4245–4249.

Money, N.P. & Harold, F.M. (1993). Two water molds can grow without measurable turgor pressure. *Planta*, **190**: 426–430. DOI: https://doi.org/10.1007/BF00196972.

Moore, D. (2013). *Fungal Biology in the Origin and Emergence of Life*. Cambridge, UK: Cambridge University Press. ISBN-10: 1107652774, ISBN-13: 978-1107652774.

Morozov, A.A. & Likhoshway, Y.V. (2016). Evolutionary history of the chitin synthases of eukaryotes. *Glycobiology*, **26**: 635–639. DOI: https://doi.org/10.1093/glycob/cww018.

Mouyna, I., Hartl, L. & Latgé, J.P. (2013). β-1–3-Glucan modifying enzymes in *Aspergillus fumigatus*. *Frontiers in Microbiology*, **4**: article 81 (9 pp.). DOI: https://doi.org/10.3389/fmicb.2013.00081.

Nishimura, M. (2016). Cell wall reorganization during infection in fungal plant pathogens. *Physiological and Molecular Plant Pathology*, **95**: 14–19. DOI: https://doi.org/10.1016/j.pmpp.2016.03.005.

Odds, F.C., Brown, A.J.P. & Gow, N.A.R. (2003). Antifungal agents: mechanisms of action. *Trends in Microbiology*, **11**: 272–279. DOI: https://doi.org/10.1016/S0966-842X(03)00117-3.

Pareek, M., Allaway, W.G. & Ashford, A.E. (2006). *Armillaria luteobubalina* mycelium develops air pores that conduct oxygen to rhizomorph clusters. *Mycological Research*, **110**: 38–50. DOI: https://doi.org/10.1016/j.mycres.2005.09.006.

Reynolds, T.B. & Fink, G.R. (2001). Bakers' yeast, a model for fungal biofilm formation. *Science*, **291**: 878–881. DOI: https://doi.org/10.1126/science.291.5505.878.

Riquelme, M. & Bartnicki-García, S. (2008). Advances in understanding hyphal morphogenesis: ontogeny, phylogeny and cellular localization of chitin synthases. *Fungal Biology Reviews*, **22**: 56–70. DOI: https://doi.org/10.1016/j.fbr.2008.05.003.

Riquelme, M., Bartnicki-García, S., González-Prieto, J.M. *et al.* (2007). Spitzenkörper localization and intracellular traffic of green fluorescent protein-labeled CHS-3 and CHS-6 chitin synthases in living hyphae of *Neurospora crassa*. *Eukaryotic Cell*, **6**: 1853–1864. DOI: https://doi.org/10.1128/EC.00088-07.

Rogg, L.E., Fortwendel, J.R., Juvvadi, P.R. & Steinbach, W.J. (2012). Regulation of expression, activity and localization of fungal chitin synthases. *Medical Mycology*, **50**: 2–17. DOI: https://doi.org/10.3109/13693786.2011.577104.

Rosenberger, R.F. (1976). The cell wall. In *The Filamentous Fungi*, **Vol. 2**, *Biosynthesis and Metabolism*, ed. J.E. Smith & D.R. Berry. London: Edward Arnold Ltd, pp. 328–344. ISBN-10: 0713125373, ISBN-13: 978-0713125375.

Scholtmeijer, K., Rink, R., Hektor, H.J. & Wösten, H.A.B. (2005). Expression and engineering of fungal hydrophobins. *Applied Mycology and Biotechnology*, **5**: 239–255. DOI: https://doi.org/10.1016/S1874-5334(05)80012-7.

Schuster, M., Martin-Urdiroz, M., Higuchi, Y. *et al.* (2016). Co-delivery of cell-wall-forming enzymes in the same vesicle for coordinated fungal cell wall formation. *Nature Microbiology*, **1**: 16149. DOI: https://doi.org/10.1038/nmicrobiol.2016.149.

Schuster, M., Treitschke, S., Kilaru, S. *et al.* (2012). Myosin-5, kinesin-1 and myosin-17 cooperate in secretion of fungal chitin synthase. *EMBO Journal*, **31**: 214–227. DOI: https://doi.org/10.1038/emboj.2011.361.

Seidl, V. (2008). Chitinases of filamentous fungi: a large group of diverse proteins with multiple physiological functions. *Fungal Biology Reviews*, **22**: 36–42. DOI: https://doi.org/10.1016/j.fbr.2008.03.002.

Singh, P.K., Singh, M. & Tripathi, B.N. (2013). Glomalin: an arbuscular mycorrhizal fungal soil protein. *Protoplasma*, **250**: 663–669. DOI: https://doi.org/10.1007/s00709-012-0453-z.

Singh, R., Shivaprakash, M.R. & Chakrabarti, A. (2011). Biofilm formation by zygomycetes: quantification, structure and matrix composition. *Microbiology*, **157**: 2611–2618. DOI: https://doi.org/10.1099/mic.0.048504-0.

Steinberg, G. (2007). Hyphal growth: a tale of motors, lipids, and the Spitzenkörper. *Eukaryotic Cell*, **6**: 351–360. DOI: https://doi.org/10.1128/EC.00381-06.

Steinberg, G. (2011). Motors in fungal morphogenesis: cooperation versus competition. *Current Opinion in Microbiology*, **14**: 660–667. DOI: https://doi.org/10.1016/j.mib.2011.09.013.

Sunde, M., Kwan, A.H.Y., Templeton, M.D., Beever, R.E. & Mackay, J.P. (2008). Structural analysis of hydrophobins. *Micron*, **39**: 773–784. DOI: https://doi.org/10.1016/j.micron.2007.08.003.

Taborda, C.P., da Silva, M.B., Nosanchuk, J.D. & Travassos, L.R. (2008). Melanin as a virulence factor of *Paracoccidioides brasiliensis* and other dimorphic pathogenic fungi: a minireview. *Mycopathologia*, **165**: 331–339. DOI: https://doi.org/10.1007/s11046-007-9061-4.

Treitschke, S., Doehlemann, G., Schuster, M. & Steinberg, G. (2010). The myosin motor domain of fungal chitin synthase V is dispensable for vesicle motility but required for virulence of the maize pathogen *Ustilago maydis*. *Plant Cell*, **22**: 2476–2494. DOI: https://doi.org/10.1105/tpc.110.075028.

Trinci, A.P.J. (1978). Wall and hyphal growth. *Science Progress*, **65**: 75–99. URL: http://www.jstor.org/stable/43420445.

Waters, H., Butler, R.D. & Moore, D. (1972). Thick-walled sclerotial medullary cells in *Coprinus lagopus*. *Transactions of the British Mycological Society*, **59**: 167–169. DOI: https://doi.org/10.1016/S0007-1536(72)80059-7.

Waters, H., Butler, R.D. & Moore, D. (1975a). Structure of aerial and submerged sclerotia of *Coprinus lagopus*. *New Phytologist*, **74**: 199–205. URL: http://www.jstor.org/stable/2431389.

Waters, H., Moore, D. & Butler, R.D. (1975b). Morphogenesis of aerial sclerotia of *Coprinus lagopus*. *New Phytologist*, **74**: 207–213. URL: http://www.jstor.org/stable/2431390.

Werner, S., Sugui, J.A., Steinberg, G. & Deising, H.B. (2007). A chitin synthase with a myosin-like motor domain is essential for hyphal growth, appressorium differentiation, and pathogenicity of the maize anthracnose fungus *Colletotrichum graminicola*. *Molecular Plant-Microbe Interactions*, **20**: 1555–1567. DOI: https://doi.org/10.1094/MPMI-20-12-1555.

Wessels, J.G.H. (1993). Wall growth, protein excretion and morphogenesis in fungi. *New Phytologist*, **123**: 397–413. Stable URL: http://www.jstor.org/stable/2557792.

Wessels, J.G.H. (1996). Fungal hydrophobins: proteins that function at an interface. *Trends in Plant Science*, **1**: 9–15. DOI: https://doi.org/10.1016/S1360-1385(96)80017-3.

Wösten, H.A.B. & de Vocht, M.L. (2000). Hydrophobins, the fungal coat unravelled. *Biochimica et Biophysica Acta (BBA): Reviews on Biomembranes*, **1469**: 79–86. DOI: https://doi.org/10.1016/S0304-4157(00)00002-2.

Xie, X. & Lipke, P.N. (2010). On the evolution of fungal and yeast cell walls. *Yeast*, **27**: 479–488. DOI: https://doi.org/10.1002/yea.1787.

Yang, Y., He, C., Huang, L., Ban, Y. & Tang, M. (2017). The effects of arbuscular mycorrhizal fungi on glomalin-related soil protein distribution, aggregate stability and their relationships with soil properties at different soil depths in lead-zinc contaminated area. *PLoS ONE*, **12**: article number e0182264. DOI: https://doi.org/10.1371/journal.pone.0182264.

Yin, Q.Y., de Groot, P.W.J., de Koster C.G. & Klis, F.M. (2007). Mass spectrometry-based proteomics of fungal wall glycoproteins. *Trends in Microbiology*, **16**: 20–26. DOI: https://doi.org/10.1016/j.tim.2007.10.011.

Zhang, J., Tang, X., Zhong, S., Yin, G., Gao, Y. & He, X. (2017). Recalcitrant carbon components in glomalin-related soil protein facilitate soil organic carbon preservation in tropical forests. *Scientific Reports*, **7**: article number 2391. DOI: https://doi.org/10.1038/s41598-017-02486-6.

CHAPTER 8

From the Haploid to the Functional Diploid: Homokaryons, Heterokaryons, Dikaryons and Compatibility

Most fungal mycelia contain haploid nuclei. This is a characteristic of Kingdom *Fungi*; unlike the other major eukaryotic groups, most true fungi are haploid. Even in fungus-like organisms in the *Oomycota* (Kingdom *Straminipila*), such as *Phytophthora infestans*, the cause of potato blight, the nuclei are diploid. This difference in ploidy is an important contrast between 'true' and 'non-true' fungi. Of course, there are exceptions to every rule and some true fungi are diploid, like *Candida albicans*, a yeast which causes disease in humans; and rhizomorphs and mushrooms of *Armillaria mellea* (a pathogen of trees that belongs to the *Basidiomycota*).

In this chapter, compatibility and the individualistic mycelium will be our main concerns. Formation and breakdown of heterokaryons, and the nature and maintenance of the dikaryon are major topics, as are the mechanisms that regulate these processes: vegetative compatibility and the incompatibility systems. We also discuss gene segregation during the mitotic division cycle, which culminates conceptually in what is known as the parasexual cycle. Finally, we consider the segregations of the cytoplasmic genetic entities, mitochondria, plasmids, viruses and prions.

8.1 COMPATIBILITY AND THE INDIVIDUALISTIC MYCELIUM

Because of the difference in ploidy, the life cycles of true fungi and those of other major groups of eukaryotes differ significantly. For example, for most true fungi diploid nuclei are only produced transiently during sexual reproduction, whereas the haploid state is limited to the gametes in most animals and plants. For true fungi, there is great variation in how the process of sexual reproduction is achieved, that is, in the mechanics of the process; and there is also a great variation in the duration of each phase of sexual reproduction. Indeed, diploids even occur in nature. A study that collected 154 isolates of *Aspergillus nidulans* from nature found that 140 were haploid strains and 14 were diploid. The diploid strains were not stable in cultivation, though, and eventually became haploid. When they are formed, whether naturally or by some experimental manipulation, diploid nuclei are larger than haploids and the nuclear:cytoplasmic ratio is correspondingly higher (Figure 8.1).

One consequence of being haploid is that all mutations can be expressed as there is only one copy of each chromosome in the nucleus. However, as there are many nuclei in a common cytoplasm, mutations occurring in individual nuclei may not be expressed because any lost functions will be provided by the many normal nuclei in the same hyphal system. The fungal kingdom is very large and diverse, and because of this the part played by sexual reproduction in the life cycle of fungi is also very diverse. At one extreme of the spectrum of behaviour, fungi display many forms of sexuality that govern the bringing together of genetic information from different parents into some arrangement that eventually produces a (potentially heterozygous) diploid nucleus. This nucleus undergoes meiotic division during which chromosomal segregation and genetic recombination take place as in every other eukaryote.

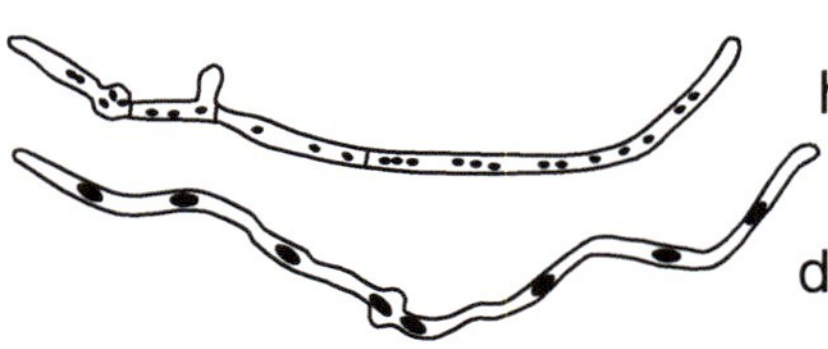

Figure 8.1. Diagrammatic comparison of germinating haploid and diploid spores of *Aspergillus nidulans*. Redrawn from camera lucida drawings in Fiddy and Trinci (1976).

At the other extreme, there are fungi that appear to be asexual organisms. In the past, most examples of these have been separately classified and they were given formal-sounding taxonomic names such as *Fungi Imperfecti* or *Deuteromycota* (Seifert & Gams, 2001). Taxonomic molecular sequence analysis may allow asexual (**imperfect**) fungi to be classified with their closest sexual (that is, **perfect**) relatives, and such informal assignments are no longer warranted. However, some of the most frequently encountered fungi are asexual, and many of them are of enormous economic significance, either as industrial fungi or as pathogens. As fungal taxonomy is still based on the morphology of the sexual reproductive structures, it is best to use informal descriptions such as deuteromycetes, imperfect fungi, **anamorphic fungi** (which is used generally to indicate the asexual or imperfect state) or **mitosporic fungi** (which refers to their spores being formed by mitosis). Anamorphic fungi can be yeasts or filamentous forms, and the latter fall into three morphological groups known most commonly as:

- **hyphomycetes**, which are mycelial forms producing conidia on separate hyphae or groups of hyphae but not in sporophores (Bärlocher, 2007);
- **coelomycetes** form conidia in sporophores that may be called pycnidia, acervulae or stromata;
- **agonomycetes**, which do not form conidia but may produce chlamydospores, sclerotia or other vegetative structures.

As knowledge accumulates, an anamorphic fungus may be classified into the *Ascomycota* or *Basidiomycota* following phylogenetic analyses or discovery of a sexual phase (known as the **teleomorph**). For example, a particularly important anamorphic fungus, the human pathogenic yeast *Candida albicans*, has been shown by genomic comparisons to possess homologues of genes known to be involved in meiosis in *Saccharomyces cerevisiae* and some other eukaryotes, which implies that *Candida albicans* may have a complete sexual cycle.

Many fungi long thought to be anamorphic have been found eventually to have a sexual stage. *Candida glabrata* remains as a potentially asexual fungus. Previously known as *Torulopsis glabrata, Candida glabrata* is a highly opportunistic pathogen of the human urogenital tract and the bloodstream and is especially prevalent in elderly and immunocompromised patients. Despite major efforts, no sexual cycle has yet been found, but all known isolates are haploids of one or the other of two opposite mating types and much of the machinery associated with mating and meiosis are present in the genome; indeed, there is evidence of genetic recombination in the population. It is believed that anamorphs have arisen by the **loss** of sexuality, which has occurred in many different groups of fungi at different times as an adaptation to the lifestyle represented by asexual vegetative growth. It seems that mutational loss of sexuality results in a competitive advantage for the asexual derivatives; because they no longer need to express the numerous mRNAs and proteins required for sexual reproduction (Heitman, 2015).

Fungal genomes vary considerably in size and organisation. They have an average genome size of about 37 Mb, but size ranges from 2.3 Mb for the microsporidian *Encephalitozoon intestinalis*, while the genomes of several basidiomycetes and ascomycetes greatly exceed 100 Mb, with the largest genome on record at the time of writing being the approximate 8,000 Mb for *Entomophaga aulicae* (Gregory *et al.*, 2007; and see Chapter 6). Similarly, chromosome numbers and ploidy levels can differ even between closely related species. Differences in the genomes between fungi are a reflection of the interaction of mutational processes with the lifestyle and population biology of the different species. Even the asexual organisms to which we have just referred are not static in an evolutionary sense because they can generate variation by modifying genetic expression or by adapting processes that occur during the mitotic division to produce asexual propagules in which chromosomes have segregated in new combinations or which contain recombinant chromosomes (Taylor *et al.*, 1999; Pringle & Taylor, 2002; Stukenbrock & Croll, 2014; see section 8.8 below).

Before we can take the discussion very much further, we must define some of our vocabulary. A **homokaryon** is a mycelium in which all nuclei are of the same genotype; in a **heterokaryon**, the nuclei in the mycelium have two or more genotypes. A species is **homothallic** if an individual can complete the sexual cycle on its own, but as we will explain below, there are different ways in which homothallism can be achieved. We should also emphasise that a homothallic species is not ***limited*** to self-fertilisation. Two homothallic strains may well interbreed, either in nature or with 'assistance' in the laboratory. The point is that a homothallic **can** self-fertilise, it is not a necessity.

In contrast, for a **heterothallic** species, sexual reproduction ***requires*** interaction of two different individuals. Individual isolates of heterothallic fungi are self-sterile or self-incompatible but can be cross compatible. Heterokaryosis results from the fusion of hyphae of different isolates, followed by migration of nuclei from one hypha into the other, so that the hyphae come to have two kinds of nuclei, and such a hypha is a **heterokaryon**.

In the most highly adapted version of this behaviour, in model basidiomycetes like *Coprinopsis cinerea* and *Schizophyllum commune*, a basidiospore germinates to produce homokaryotic mycelium with uninucleate cells, called a **monokaryon**. When two monokaryons meet, hyphal anastomoses occur, and, if they are vegetatively compatible, nuclei of one migrate into the mycelium of the other, the dolipore septa between the cells being broken down to allow nuclei to migrate. If, *in addition*, the nuclei have compatible mating types, the new growth and the cells of much of the pre-existing mycelium will form **binucleate cells**, containing one nucleus of each monokaryotic parent; this mycelium is called a **dikaryon**.

However, there are some important differences in lifestyle between higher fungi and other eukaryotes that have great genetic significance. First, many fungal mycelia can tolerate (in fact, more than tolerate, can benefit from) the presence of several genetically distinct nuclei within their hyphae. This probably arises from the second important difference that we have already mentioned, namely the fact that hyphal anastomoses occur very readily within the *Ascomycota* and *Basidiomycota* but are not observed in the zygomycetes.

An important benefit of heterokaryosis is that recessive mutations will not be expressed in a heterokaryon if sufficient nuclei containing the dominant alleles are present (and unless manipulated by an experimenter, recessive mutations are always in the minority). The presence of unexpressed recessive mutations in a population of nuclei in a heterokaryon means that such mycelia have a larger gene pool than homokaryons and this provides the potential for evolutionary selection and hence a more immediate response to environmental stress. Thus, despite being haploid, true fungi are a successful group and this success can be attributed in large part to their use of heterokaryosis to overcome the genetic and physiological disadvantages associated with having just a single nuclear genotype (whether haploid or diploid) within the vegetative growing body. The possibility has also been pointed out that the DNA of supernumerary nuclei in filamentous fungi could serve as a store of nitrogen, phosphorus and carbohydrate, which could be degraded by autophagy with the breakdown products recycled, and providing nutrition for hyphal tips to continue foraging into new areas (Maheshwari, 2005). We are not sure how likely this is, but supernumerary nuclei must have ***some*** selective advantage.

8.2 FORMATION OF HETEROKARYONS

Hyphal anastomosis, of course, is the fusion between hyphae or hyphal branches, and we have described the process of breakdown of two hyphal walls and union between two separate plasma membranes to bring the cytoplasm of the fusing hyphae into continuity with each other. It is the consequence of this that's important here: once they are in continuity, they can exchange nuclei and other organelles. Of course, this happens in the gametic fertilisation which is a prelude to sexual reproduction in animals and plants.

Fungi differ because anastomosis is not limited to sexual reproduction; rather, hyphal fusions are essential to the efficient functioning of the mycelium of filamentous *Ascomycota* and *Basidiomycota*. Anastomoses convert the initially radiating system of hyphae into a fully interconnected (and three-dimensional) network. Hyphal fusions are common within the individual mycelium as it matures. The interconnections they establish enable transport of nutrient and signalling molecules anywhere in the colony.

Heterokaryons do occur in nature (Table 8.1). Evidently, higher fungi are fully equipped with the machinery necessary for hyphal tips to target other hyphae and this is part of normal mycelial development; it is not a specialisation of the sexual cycle, nor a peculiarity of *in vitro* cultivation. It is complex machinery, though, because during normal mycelial growth vegetative hyphae usually avoid each other (known as **negative autotropism**), behaviour that promotes exploration and exploitation of the available substrate. Anastomosis requires that hyphae grow towards each other (called **positive autotropism**). How and why the usual avoidance reactions between hyphae become reversed is unknown but, evidently, it is a change in the behaviour of hyphal tips that occurs as the mycelium matures (and may depend on local population density).

Higher fungi, therefore, have a mechanism that promotes cell fusions to an extent that is never encountered in animals, plants, or even lower fungi and their more primitive relatives. The process brings two different cytoplasms into one (joint) mycelium, and brings nuclei originating from two different mycelia into the same (joint) cytoplasm (Figure 8.2). The genetic consequences of this for the heterokaryon are important. Alleles in

Table 8.1. Frequency of the occurrence of heterokaryons among fungi isolated from nature

Species	No. of isolates	% of isolates found to be heterokaryons
Aspergillus glaucus	15	13
Penicillium cyclopium	16	25–50
Sclerotinia trifoliorum	10	60

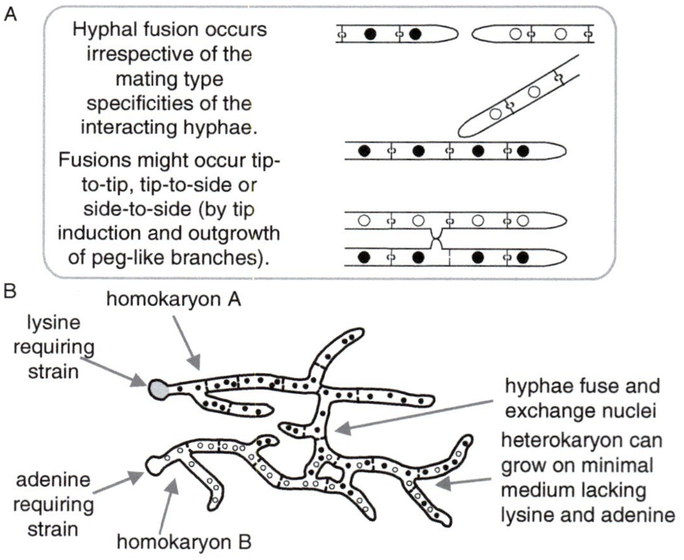

Figure 8.2. Homokaryosis and heterokaryosis in *Aspergillus nidulans* (*Ascomycota*). (A) Diagrams of the different ways in which hyphal fusion may occur between homokaryons. (B) Diagram of a heterokaryon being formed by two homokaryons. When the nuclei in the heterokaryotic mycelium are of two or more genotypes, alleles in separate nuclei in a heterokaryon may complement each other. In this diagrammatic example a heterokaryon formed from a lysine auxotroph and an adenine auxotroph will be able to grow on minimal medium, although the two constituent homokaryons will not.

separate nuclei in a heterokaryon will complement each other; in the example shown in Figure 8.2 a heterokaryon formed from a lysine-requiring strain (a nutritionally deficient 'auxotroph') and an adenine auxotroph will be able to grow on a minimal medium (one that contains neither of these nutrients), although the two constituent homokaryons will not.

The ratio of different nuclei in a heterokaryotic mycelium may influence both its morphology (branching frequency, for example) and growth rate (by affecting hyphal extension). Thus, the phenotype of a heterokaryon is determined by the balance of nuclei in the heterokaryon. Equally, the ratio of different nuclei will itself be influenced by the environmental and growth conditions.

Table 8.2 shows the effect of environmental conditions (in terms of the nature of the medium) on the nuclear ratio in a heterokaryotic mycelium of *Penicillium cyclopium* (a fungus that commonly causes fruit spoilage with production of toxins), where the B-type nuclei predominated on a minimal medium, but both nuclei were equally represented in heterokaryons grown on an apple pulp medium. Thus, the ratio of nuclear types in a heterokaryon can vary within wide limits. In this case, the interpretation is that the complex medium provides a selection pressure favouring maintenance of the heterokaryon, the minimal medium does not, and the B-type nucleus has some selective advantage. Note the contrast with the example shown in Figure 8.2, where the minimal medium favours the heterokaryon comprising ade^- and lys^- nuclei. In another example of experiments with heterokaryons of *Neurospora crassa* formed between colonial mutants (with a highly branched mycelium that expands slowly) and non-colonial strains (with a sparsely branched mycelium that expands rapidly), the heterokaryons exhibited the non-colonial phenotype when nuclei of the two mutants are present in a ratio of 1:1, but a colonial phenotype when the heterokaryon contained more colonial than non-colonial nuclei. Thus, the ratio of nuclei present in a heterokaryon also influences its morphological phenotype, that is, the branching frequency and rate of hyphal extension.

Evidently, there are several consequences of heterokaryosis (Strom & Bushley, 2016; Daskalov *et al.*, 2017):

- different nuclei provide genetic variation in the mycelium;
- phenotype of the mycelium depends on interaction of all nuclear types;
- phenotype of the mycelium can vary in different areas of the same mycelium, depending on the localised nuclear ratio;
- the heterokaryon has the physiological flexibility to respond to different nutritional conditions;
- heterokaryons increase genetic diversity independently of sexual reproduction;
- provides a route for cytoplasmic inheritance (mitochondria, viruses, plasmids, prions; see section 8.9, below).

Table 8.2. Effect of environmental conditions on the nuclear ratio in a heterokaryotic mycelium of *Penicillium cyclopium* (A- and B-type nuclei)

Growth medium	% of the two nuclei in the heterokaryon	
	A type	B type
Minimal	9	91
Complex (apple pulp)	52	48

8.3 BREAKDOWN OF A HETEROKARYON

Unless there is a continual selection pressure favouring their continued existence, there is a tendency for heterokaryons to break down by forming sectors of homokaryotic growth. The nuclei of hyphae at the margins of heterokaryons are predominantly from one parent or the other, and if conditions change to favour one nuclear type, hyphae containing that nucleus will be selected (Figure 8.3).

Heterokaryons also break down because of asexual sporulation if uninucleate spores are formed, as in *Aspergillus nidulans* (Figure 8.4); the uninucleate spores then germinate to form homokaryons.

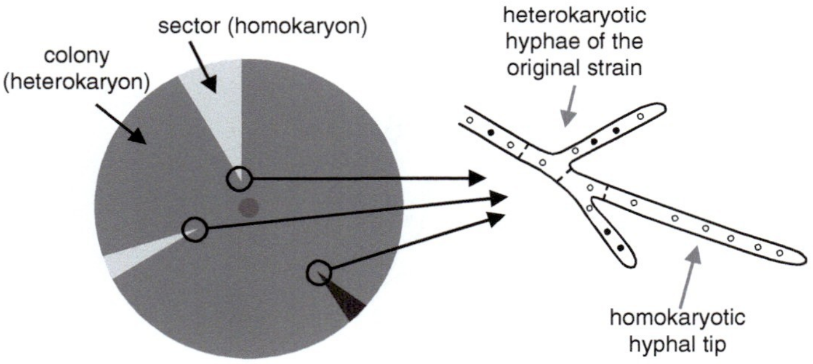

Figure 8.3. Sectoring within a heterokaryotic colony. 'Sectored' refers to the pattern of hyphae in a fungal colony on a Petri dish; sectors of homokaryotic hyphae are formed like the sectors of a pie-diagram. If the homokaryon and heterokaryon differ morphologically (for example by the colour of the spores they make), then these sectors are visible to the naked eye (diagrammed at left). The sectors arise when a hyphal tip receives nuclei of only one type and can continue to grow as a homokaryon (diagrammed at right). This may happen at any time in the growth of the colony; the closer it happens to the inoculum, the larger and more extensive is the final sector. Sectoring may be purely a matter of chance, depending only on the random distribution of nuclei between newly formed hyphal tips; but may be made more frequent by chemicals that disturb nuclear distribution or by selection pressure that favours the homokaryon. When there is no positive selection pressure to maintain a heterokaryon it is more likely to break down into its component homokaryons. For example, if the heterokaryon shown in Figure 8.2 is grown on a 'complete' medium containing lysine and adenine, the absence of selection pressure favouring heterokaryon maintenance results in its breakdown into the component auxotrophic homokaryons.

In contrast, many fungi produce multinucleate spores, and if the mycelium producing these is a heterokaryon, the spores will also be heterokaryotic, and the state will be propagated through asexual sporulation (an example of this is *Botrytis cinerea*, a pathogen of many plants and cause of the 'noble rot' of wine grapes, Figure 8.5).

8.4 THE DIKARYON

In the *Basidiomycota* the heterokaryon is highly specialised, being made up of cells that each contains nuclei from two compatible mating types within a common cytoplasm; this is called a dikaryon. A similar arrangement exists in the **ascogenous hyphae** of *Ascomycota*, but in many *Basidiomycota* the dikaryotic condition is purposefully perpetuated by a special kind of hyphal growth (Figure 8.6) so that the dikaryotic mycelium can grow ***indeterminately*** like any other mycelium. Under normal circumstances only dikaryotic mycelium forms sporophores.

The dikaryotic state is maintained by a specialised cell biology: the nuclei divide together (**conjugate mitosis**) and at the same time a '**clamp connection**' grows out as a backward-projecting branch which loops backward and then fuses with the parent hypha. One of the nuclei completes its mitosis in the clamp connection, the other nucleus stays in the main body of the hypha. This separation means that two normal dolipore septa can be formed between daughters of the two mitotically dividing nuclei and both terminal and subterminal cells (that is, the two daughter compartments) each contain two nuclei of opposite mating type; one nucleus of the subterminal cell being delivered to it through the clamp connection (Figure 8.6). The regularity of this process results in every cell of the dikaryon containing two sets of homologous chromosomes: that is, the dikaryon is a functional diploid.

This is an important point. The so-called 'higher organisms' (plants and animals) are diploid but most true fungi are haploid, and this has genetic and evolutionary consequences. However, the haploid *Basidiomycota* have developed ***functionally diploid cells*** by extending the vegetative life of dikaryotic heterokaryons and in so doing they have gained the genetic (and evolutionary) advantages of diploid organisms, while remaining haploid. Like the diploid, a dikaryotic cell also has two sets of homologous chromosomes, even though each set is in a separate nucleus. Together, the two nuclei control and regulate the activities of a specific volume of cytoplasm. Complex cellular events have evolved in *Basidiomycota* to establish and maintain the dikaryotic condition. These include different mating types

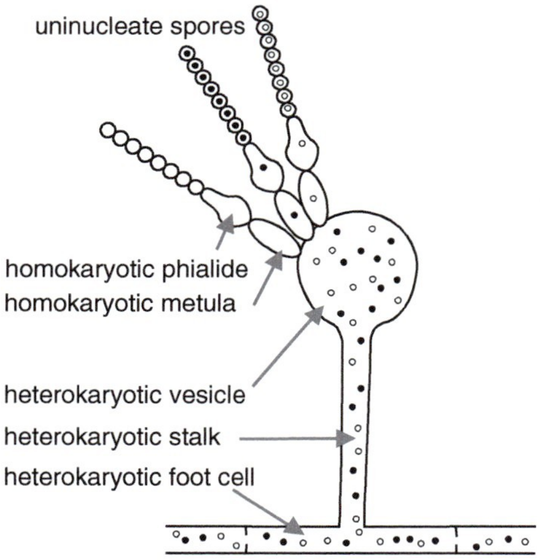

Figure 8.4. Heterokaryon breakdown occurs because of asexual sporulation when uninucleate spores are formed; the uninucleate spores then germinate to form homokaryons. In *Aspergillus nidulans*, for example, the conidia are produced at the apex of the phialides (spore mother cells) on the sporangiophore vesicle. Each phialide contains only one nucleus which has migrated from the vesicle. This undergoes successive mitotic divisions; after each mitosis one daughter nucleus migrates into the developing spore and the other remains in the phialide for the next round of mitosis. Thus, all the nuclei in the chain of uninucleate conidia formed from a single phialide are genetically identical. See Figure 10.9 for more detail about the structure of the conidiophore of *Aspergillus nidulans*.

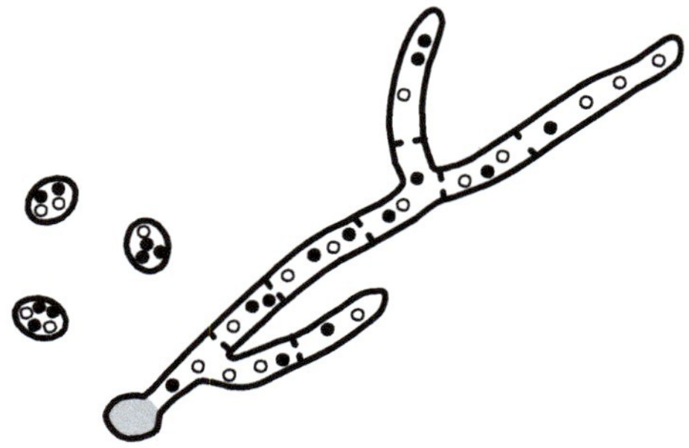

Figure 8.5. In some fungi, that produce multinucleate spores, heterokaryons can be propagated through germination of a heterokaryotic asexual spore (for example, *Botrytis cinerea*).

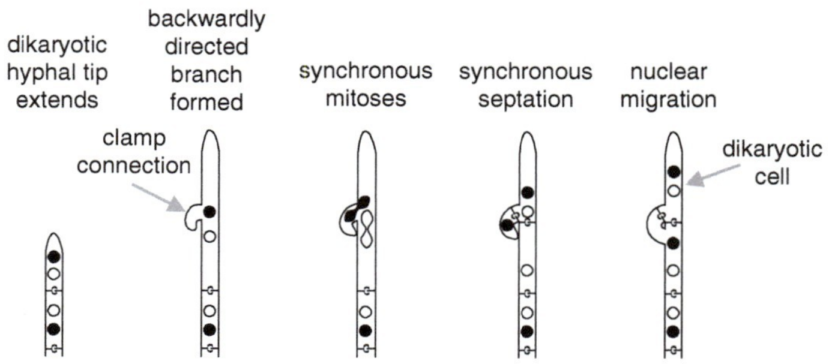

Figure 8.6. The dikaryon: a specialised binucleate heterokaryon. The dikaryotic state is maintained by the clamp connection: a backward-projecting hyphal branch which fuses with its parent hypha to deliver one of the nuclei and ensure that each compartment contains two nuclei of opposite mating type separated by a dolipore septum. Each hyphal compartment therefore contains two sets of homologous chromosomes and is a functional diploid.

and vegetative compatibility systems, elaborate dolipore septa, and an elaborate cytokinesis mechanism in which a backwardly directed hyphal branch enables a daughter nucleus to leapfrog a septum and maintain the dikaryotic condition in each daughter cell of the mitosis (Figure 8.6). No wonder *Basidiomycota* are considered the most advanced group of fungi in terms of their evolution.

Clamp connections are required to maintain the dikaryon because the ***dolipore septum controls organelle migration*** and normally hinders proper nuclear distribution. Clamp connections may not be necessary if the hypha is so wide that the conjugate mitoses can take place without hindrance, so some vegetative dikaryons may not form clamp connections. For most of the *Basidiomycota*, only the dikaryotic mycelium can differentiate into a sporophore, but many sporophores contain specialised tissues in which the dikaryotic condition breaks down. For example, many cells in the stem of the ink cap mushroom of *Coprinopsis* are enormously enlarged and are multinucleate, whereas cells elsewhere in the same fruit are regularly dikaryotic. Similarly, not all basidiomycete mycelia are dikaryotic; some are multinucleate, and it is curious that more fungi have not become true diploids. Perhaps the more flexible heterokaryotic condition offers greater advantages than we suspect.

Dikaryon formation occurs via the activities of cell-type-specific homeodomain transcription factors (the **mating type factors**), which form regulatory complexes to establish the dikaryotic state. Many years of classical genetic studies with *Coprinopsis* and *Schizophyllum* mushrooms created the foundation of our understanding of how the mating type factors work. *Basidiomycota* includes many pathogenic fungi too, including the corn (smut) pathogen *Ustilago maydis* and the globally distributed human pathogenic *Cryptococcus neoformans*, a leading cause of fungal meningitis. These two have featured in more recent molecular studies that have revealed novel mechanisms of regulation that function downstream of classic homeodomain complexes to ensure that dikaryons are established and propagated (Kruzel & Hull, 2010). For example, in *Ustilago maydis*, dikaryon formation is controlled by a ***DNA damage response cascade*** that includes two conserved DNA damage checkpoint kinases, called *Chk1* and *Atr1* (Pérez-Martín & de Sena-Tomás, 2011).

8.5 VEGETATIVE COMPATIBILITY

For genetically different hyphae to interact, to take advantage of heterokaryosis or the sexual cycle, for example, the process of hyphal anastomosis must be closely regulated so that the physiological and genetic advantages of heterokaryosis can be realised without hazard. And there ***are*** hazards: hyphal anastomosis carries the risk of exposure to contamination with alien genetic information from defective or harmful cell organelles, viruses or plasmids. But nuclear and cytoplasmic control strategies have very different requirements. To maximise the advantage of sexual reproduction the controls must ensure that the nuclei are genetically as ***different*** as possible. In contrast, safe operation of the cell requires that cytoplasms that are to mingle must be as ***similar*** as possible.

These features are under the control of:

- Genetic systems tht regulate the ability of hyphae to fuse, generally called **vegetative compatibility**. The phenotype of vegetative compatibility (also called vegetative, somatic or heterokaryon incompatibility) is formation of a joint heterokaryotic mycelium with vegetative compatibility as a **self/nonself recognition** process occurring when hyphae of the same species fuse.
- Genes called mating type factors that regulate the ability of nuclei that have been brought together by hyphal fusion to undergo karyogamy and meiosis, and the phenotype of compatible interaction between mating type factors is occurrence of sexual reproduction (details in Chapter 9).
- Vegetative compatibility, which is different from, but has a controlling influence over, mating type function in terms of both population structure and genetic diversity.

Vegetative compatibility is controlled by one to several nuclear genes that limit completion of hyphal anastomosis between colonies to those that belong to the same **vegetative compatibility group** (usually abbreviated to **v-c group**). Members of a v-c group possess the same vegetative compatibility alleles. Hyphal anastomosis is promiscuous in fungi, but compatibility of the cytoplasms determines whether cytoplasmic exchange will progress beyond the first few hyphal compartments involved in the initial interaction. Since the intracellular test for self/nonself recognition (that is, vegetative compatibility) occurs after anastomosis, this is called **post-fusion incompatibility**. If the colonies involved are not compatible the cells immediately involved in anastomosis are killed (Figure 8.7) by a programmed cell death response (Paoletti & Clavé, 2007; Paoletti, 2016). This strategy prevents transfer of nuclei and other organelles between incompatible strains, but if the incompatibility reaction is slow, a virus or cytoplasmic plasmid may be communicated to adjacent undamaged cells before the incompatibility reaction kills the hyphal compartments where anastomosis occurred (Bidard *et al.*, 2013).

The basis of a compatibility test carried out in the laboratory is that small pieces of the strains that are to be tested are placed side by side on the surface of an agar medium, and hyphal interactions in these '**confrontations**' usually show phenotypes that imply self/nonself recognition. When the confrontations are incubated, leading hyphae may mingle, and hyphal anastomoses occur between their branches. If the confronting strains are compatible the heterokaryon may proliferate so that the whole mycelium becomes heterokaryotic; this is what happens in *Neurospora crassa* and *Podospora anserina*. Alternatively, in species such as *Verticillium dahliae* and *Gibberella fujikuroi*, nuclei do not migrate between cells and heterokaryosis is limited to the branches that grow out of the fusion cells.

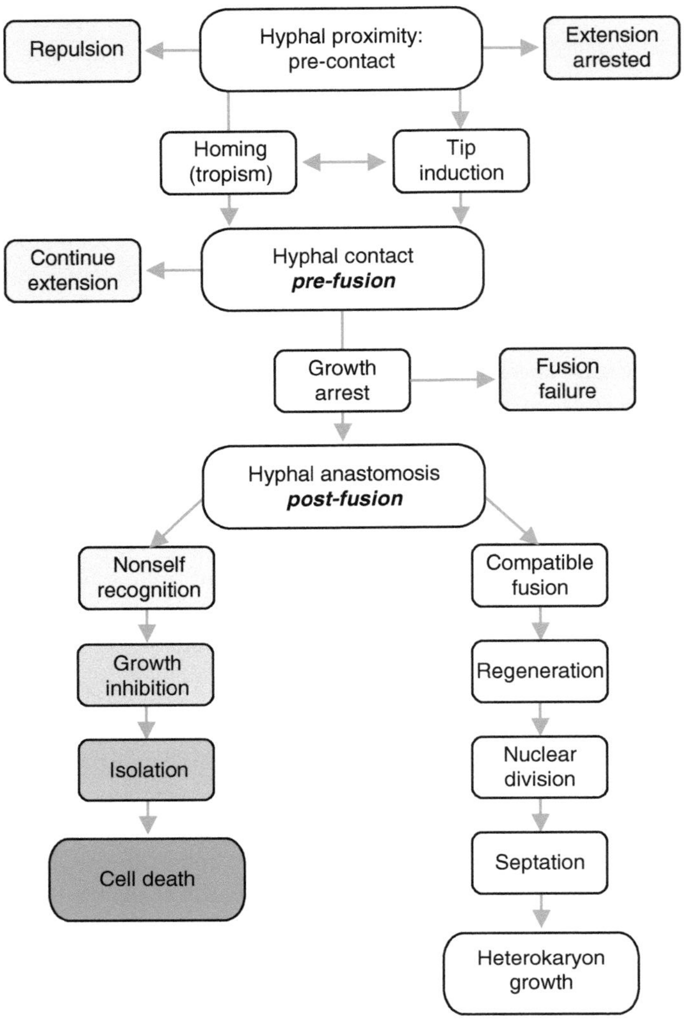

Figure 8.7. Flow diagram illustrating the progress of hyphal interaction leading to operation of the vegetative compatibility systems. Recognition processes between hyphae take place at all three major steps: pre-contact hyphal proximity, pre-fusion hyphal contact and post-fusion self-nonself-recognition. Modified from Chapter 2 in Moore & Novak Frazer (2002).

If the colonies involved are not compatible, the fusion cells are killed (Figure 8.7). Cell death resulting from vegetative incompatibility involves plugging of the septal pores, to compartmentalise dying hyphal segments, vacuolisation of the cytoplasm, DNA fragmentation, organelle degradation and shrinkage of the plasma membrane from the cell wall. It is an **internalised cell death**, different from necrotic cell death, with many features in common with **programmed cell death** (**PCD**, or apoptosis) in other multicellular eukaryotes (Paoletti & Clavé, 2007; Paoletti, 2016).

Vegetative compatibility (also called vegetative or somatic incompatibility, is now increasingly being called **heterokaryon incompatibility** or **HI**) will prevent formation of a heterokaryon unless the strains belong to the same v-c group. You must cope with a variety of terms that were applied to this research through the twentieth century; half of which was done by 'glass half-empty people' who knew they were researching ***incompatibility***, and the other half by 'glass half-full people' who were researching **compatibility**.

Incompatibility between strains in a confrontation is caused by genetic differences between the two individuals at specific gene loci, which are called *het* (for *het*erokaryon) or *vic* (for *ve*getative *i*n*c*ompatibility) loci, although once the major genes were identified several others that affected or otherwise modified their expression were also identified and given other descriptive names (Table 8.3).

There are usually about 10 *het* loci, but the number varies from one species to another: there are at least 11 *het* loci in *Neurospora crassa*, 9 in *Podospora anserina*, 8 in *Aspergillus nidulans* and 7 in *Cryphonectria parasitica*. At the time of writing, only genes from the main model organisms for the study of incompatibility, *Neurospora* (Hall *et al.*, 2010) and *Podospora* (Bidard *et al.*, 2013) have been cloned.

These genes encode **STAND proteins** (STAND proteins are **S**ignal **T**ransducing **A**TPases with **N**umerous **D**omains), their domains comprising:

- a carboxy-terminal **WD-repeat** domain (which features a sequence with a high frequency of tryptophan (symbol **W**) and aspartic acid (symbol **D**) pairs and named for the single-letter symbols commonly used in sequence data);
- a central **NACHT nucleotide-binding domain** (the name NACHT is derived from the four animal and fungal proteins which initially defined the unique features of this domain (specifically, the **N**euronal **A**poptosis inhibitory protein, the major histocompatibility **C**omplex transcription activator, the *Podospora anserina* incompatibility locus protein **H**ET-E, and **T**elomerase-associated protein); the NACHT domain has NTPase activity and preferentially binds GTP or ATP;
- and an N-terminal fungus-specific **HET domain**, which is a cell death execution domain found in many proteins encoded by fungal incompatibility genes (Paoletti & Clavé, 2007; Paoletti, 2016).

Not surprisingly, given the range of controlling elements represented by these HI proteins, the incompatibility response involves massive changes to the transcriptome (the spectrum of messenger RNA molecules expressed from the genes of the organism); 2,231 genes were upregulated by a factor 2 or more, and 2,441 genes were downregulated during the incompatibility reaction in *Podospora* (Bidard *et al.*, 2013). There was a significant overlap between regulated genes during incompatibility in *Podospora anserina* and *Neurospora crassa*, indicating similarities in the incompatibility responses in these two species. Some of the transcriptome changes observed during the incompatibility reaction mimic the impact of the host plant on plant pathogenic fungi, so it may not be surprising that

Table 8.3. Genes involved in vegetative incompatibility that have been cloned and characterised

Neurospora crassa	
Mat A-1	Mating type transcription regulator, contains region similar to *Mat α1* of *Saccharomyces cerevisiae*
Mat a-1	Mating type transcription regulator with a High Mobility Group (HMG) box (characteristic of HMG proteins, a class of proteins distinct from histones which are found in chromatin and represent a subclass of the non-histone proteins; the HMG proteins function in gene regulation and maintenance of chromosome structure)
het-c	Signal peptide (involved in endoplasmic reticulum targeting of secreted proteins) with glycine-rich repeats
het-6	Region of similarity to *tol* and *het-e* (of *Podospora anserina*)
un-24	Large subunit of type I ribonucleotide reductase
tol	Features a coiled-coil, leucine-rich repeat (a protein conformation found in extracellular matrix molecules), has regions similar to sequences in *het-e* (of *Podospora anserina*) and *het-6*
Podospora anserina	
het-c	Glycolipid transfer protein (glycolipids are involved in cell-to-cell interactions)
het-e	GTP-binding domain, region with similarity to *tol* and *het-6* of *Neurospora crassa*
het-s	Prion-like protein (abnormally folded variant can infectively communicate its abnormal conformation to normal proteins which then form aggregates)
idi-2	Signal peptide, induced by *het-R/V* incompatibility
idi-1, idi-3	Signal peptide, induced by nonallelic incompatibility
mod-A	SH3-binding domain (**s**rc **h**omology domain **3**; a protein domain of about 50 amino acid residues present in proteins involved in signal transduction, and also in a number of cytoskeletal proteins, generally involved in protein-protein interactions)
mod-D	α-subunit of G-protein with GTP binding (such proteins are involved in signal transduction in eukaryotic cells), modifier of *het-C/E* incompatibility
mod-E	Heat-shock protein (belongs to the Hsp90 family of 90 kDa polypeptides with ATPase activity which are essential for the viability of yeast cells and found in association with many regulatory proteins in eukaryotes, like steroid receptors and protein kinases), modifier of *het-R/V* incompatibility
pspA	Vacuolar serine protease, induced by nonallelic incompatibility

Adapted from Moore & Novak Frazer (2002).

the vegetative incompatibility groups into which strains of *Rhizoctonia solani* can be divided differ in their pathogenicities to the plant host.

Neurospora crassa is similar to *Aspergillus* in that heterokaryon formation requires genetic identity at all *het* genes. One of these genes is the mating type locus of *Neurospora crassa*, and although this is unusual, association between mating type and vegetative incompatibility is not restricted to *Neurospora crassa*, but has been reported in *Ascobolus stercorarius, Aspergillus heterothallicus* and *Sordaria brevicollis*. Usually, two different alleles of a *het* gene are found in wild-type isolates, although *het* loci with more than two alleles have been found (for further details see Chapter 2 in Moore & Novak Frazer, 2002).

In *Neurospora crassa*, heterokaryons made between strains of the opposite mating type grow slowly and have an irregular colony outline compared with the rapid, uniform growth of heterokaryons between strains with the same mating type. Evidently, the mating type gene of *Neurospora crassa* controls both sexual compatibility and heterokaryon compatibility, although the former requires that the mating types are different, and the latter requires that the mating types are identical. It seems that nuclei of opposite mating types do not readily co-exist in vegetative hyphae of *Neurospora crassa*. Aggressive maintenance of individuality between mates is neither unusual nor difficult to understand; in our own species, allegedly, men are from Mars, women from Venus. In *Neurospora crassa*, the molecular basis of this mating aggression is that the MATA-1 and MATa-1 mating polypeptides (detailed in Chapter 9, section 9.3) encode transcription regulators that specify different cell types in the sexual phase, but they are lethal when expressed together in a vegetative cell. The mating function of MATa-1 depends on its DNA-binding ability, but this is not needed for the vegetative incompatibility function. Different functional domains of the polypeptide serve these two different activities of the mating type idiomorphs (Wang *et al.*, 2012).

Heterokaryons made between *Neurospora crassa* strains of the same mating type (and the same *het* genotype) have nuclear ratios close to 1:1, full cytoplasmic continuity, and they also produce up to 30% heterokaryotic conidia. In the incompatible heterokaryon confrontations, pores in the septa of any cells that do fuse become blocked, and the cytoplasm becomes granular, then vacuolated and finally dies. When such cytoplasm, or even a phosphate-buffer extract of it, was injected into a different strain, the same degenerative changes resulted. The activity of the extract was associated with its proteins, demonstrating that heterokaryon compatibility self/nonself recognition depends on the **protein** products of the *het* genes.

When two incompatible colonies of *Podospora anserina* meet, hyphal fusion is followed by death of the fused cells and consequent absence of pigment, so a clear zone forms between the colonies called a barrage. The **barrage** is due to vegetative incompatibility, but the colonies might still be sexually compatible, and if they are compatible, a line of perithecia can be formed on each side of the barrage because although fused vegetative cells are killed, lethality does not extend to fused sexual cells. The vegetative incompatibility genes are probably regulators of enzymes that trigger the cell senescence and death of the incompatible fusions, and the mating type factors probably protect sexually compatible cells from the adverse effects of vegetative incompatibility genes.

Several *het* loci of *Podospora anserina* have been characterised, but the symbols have been assigned independently in the different fungi; that is, *het-c* in *Podospora anserina* has no relationship to the *het-c* of *Neurospora crassa*. Just as in *Neurospora crassa*, the *Podospora anserina het* loci encode varied gene products (Table 8.3); the *het-s* gene product behaves like a **prion protein**. A prion is a 'proteinaceous infectious particle', a cellular protein that can assume an abnormal conformation that is infectious in the sense that it can convert the normal form of the protein into the abnormal (see below). Hyphal anastomosis between *het-s* and the neutral *het-s** strain results in the cytoplasmic transmission and infectious propagation of the *het-s* phenotype.

Although the *het* loci encode very different gene products, three regions of similarity can be detected between predicted products of the *het-6* locus and the *tol* locus of *Neurospora crassa*, and the predicted product of the *het-e* locus of *Podospora anserina*. These regions might represent domains necessary for some aspect of vegetative incompatibility in which all three of these *het* loci are involved. Alleles of *het-c* that are found in *Neurospora crassa* are present in other *Neurospora* species and related genera, indicating there was a common ancestor and conservation during evolution of this sequence. However, despite this indication that there may be some underlying similarity in function, a *het* locus from one species does not necessarily confer vegetative incompatibility in a different species.

8.6 BIOLOGY OF INCOMPATIBILITY SYSTEMS

It is the compatibility reaction (including, for the moment, both vegetative compatibility discussed here, and the mating type systems dealt with in Chapter 9) that define what constitutes the **fungal individual** in real life. In yeasts, each cell is clearly an individual, but a mycelial individual is not so obvious. Spores are individuals and colonies developed from single spores must also be individuals. But are 10 mitotically produced spores from the same colony 10 different individuals, or just 10 bits separated from one individual? And then there are heterokaryons; mycelia that contain more than one nuclear type. Is a heterokaryon an individual, rather than a chimera or mosaic? These are important questions because in genetical terms a population consists of individuals that can interbreed. Individuals are important in evolution because ***selection operates on individuals***. To understand fungal populations we have to know where the individual begins and ends.

Populations are important because the fundamental unit of biological classification, the species, is conventionally defined in terms of mating success and production of viable offspring. This is the **biological species** (or genospecies) discussed earlier in section 3.9. Compatibility systems maintain the individuality of a mycelium and enable it to recognise unrelated mycelia of the same species with which it competes for territory and resources. In other words, they provide an individual mycelium with a way of establishing whether hyphae it encounters belong to itself or not; fungi have many ways of expressing their individuality because individuality ***is*** of prime importance to fungi by reducing the spread within a species of harmful cytoplasmic mutations and/or viral or plasmid infections, and possibly being the first step towards speciation.

Fungal individuals are most clearly seen in dead tree stumps that are colonised by several mycelia of a wood-decay species. Each mycelium explores the timber and attempts to capture a volume of the woody substrate for its own use. The mycelia interact where they meet and vegetative compatibility systems come into operation. If the strains are incompatible, demarcation zones of dead hyphae are formed between adjacent strains resulting from the death of fusion cells, killed by vegetative incompatibility. The living hyphae near these zones often form thickened and pigmented walls, which emphasise that the two mycelia are distinct. If the timber is sawn into segments these pigmented zones appear as pigmented lines ('demarcation' or 'zone' lines) on the cut surface. If the distinct zones extend to a

Figure 8.8. Felled logs colonised by mycelia of *Trametes versicolor* (*Basidiomycota*; commonly called turkey tail in the United States) (A, B, C) and *Hypholoma fasciculare* (D), commonly known as the sulphur tuft. Early in the season the mycelia reach the end of the log and the differentiating sporophores outline the separate decay columns in the timber (A), which are formed by mycelia belonging to different compatibility groups. Sporophores are formed on these surfaces later in the season (B, C and D). (D) (*Hypholoma fasciculare*) shows that the formation of sporophores from distinct decay columns is not limited to *Trametes*, though in this specimen the log is completely covered in moss. The diagram shown in (E) is a tree stump colonised by eight incompatible mycelia of *Trametes versicolor* in which the decay columns within the timber have been mapped in successive sections (of about 1 cm thickness). The decay columns remain distinct because operation of the vegetative compatibility system produces a demarcation zone of dead and pigmented hyphae between each pair of adjacent strains. In sections of the timber these appear as demarcation (or **zone**) lines showing the boundary of each mycelium (a, b, c, d, e, f, g & h) (and see Figure 14.4). Where a mycelium reaches the surface and fruits, it produces sporophores with the genotypes of that mycelium. Photographs by David Moore of specimens in the Lovell Tree Collection Arboretum at Jodrell Bank Discovery Centre, Cheshire (http://www.jodrellbank.net/); (E) based on Figure 3 in Rayner & Todd (1982).

surface at which they can form sporophores, a collection of sporophores expressing the different genotypes represented in that local population will appear (Figure 8.8).

8.7 GENE SEGREGATION DURING THE MITOTIC DIVISION CYCLE

In the middle of the twentieth century it became evident that meiotic segregations were not the only way of making maps of chromosomes. **Mitotic segregations** can also be analysed and are a convenient way of mapping chromosomes. The approach is applicable to any fungus that is normally haploid, although the first step is the selection of diploids that arise spontaneously through nuclear fusion at a rate of about one in every 10^6 or 10^7 mitoses.

The pioneering work was done with the ordinarily haploid filamentous fungus *Aspergillus nidulans*. Selection of diploid strains is a little easier in *Aspergillus* because its conidia are always uninucleate. Consequently, rare diploid conidia can be selected from among a large spore population obtained from a heterokaryon by selecting for a heterozygous phenotype. Uninucleate spores cannot be heterokaryotic, so conidia expressing a heterozygous phenotype must contain both homologues of at least one pair of chromosomes and may be completely diploid. If you make a heterokaryon between ***two*** (haploid) recessive auxotrophs you would expect that diploid spores would be the only conidia able to grow on minimal medium (Figure 8.2). Diploid conidia are larger than haploid conidia, being about twice the volume; they also, of course, contain twice the haploid amount of DNA (Figure 8.1) (Stukenbrock & Croll, 2014).

This sort of nutritional selection is an automatic method that certainly works efficiently, but it limits the number of nutritional markers that can be used in any experimental crosses. However, an especially useful feature of *Aspergillus nidulans* (not true for all fungi) is that ***the colour of the conidium depends on its***

own genotype. Consequently, a heterokaryon made between two non-allelic, recessive, spore colour mutants, say white-spore and yellow-spore strains, will produce large numbers of haploid white and haploid yellow conidia together with very occasional sectors of diploid conidia with the wild-type dark-green colour. Using colour selection leaves open the possibility of having several (unselected) nutritional markers in the cross, but it requires close scrutiny of the cultures.

Nutritional selective methods have been used to isolate diploids from many normally haploid fungi. This includes basidiomycetes such as *Schizophyllum commune* and the agaric *Coprinopsis cinerea*, but it is especially important in those fungi in which the known life cycle lacks sexual reproduction, and this group includes several commercially important species, including *Aspergillus niger, Aspergillus oryzae, Aspergillus flavus, Penicillium chrysogenum* and plant pathogens like *Penicillium expansum* and *Penicillium digitatum*.

Diploids are generally sufficiently stable to grow into diploid vegetative colonies, but these do produce rare sectors showing segregation of the originally heterozygous genes. This type of segregation, also based on mitotic recombination, can be used for genetic mapping. The key to understanding is to remember the crucial differences in chromosome behaviour during meiosis and mitosis. ***In meiosis***, homologous chromosomes undergo synaptic pairing, and they take their place on the first division spindle as a bivalent comprising two chromosomes, each divided into two chromatids (the four-strand stage as illustrated in Figure 8.9). At the first division of meiosis, the undivided (that is, maternal and paternal) centromeres are taken to opposite poles of the division spindle. ***None of this happens in mitosis***.

In mitosis, homologous chromosomes do not line up with one another, so there is no synapsis. The fact that recombinant diploid segregants can be obtained from mitotic crossing over indicates that occasional exchanges occur between homologous chromosomes during mitosis, and the recombination results in two homologous chromosomes with reciprocally recombinant arms (Figure 8.9), but it is extremely rare. In fact, mitotic crossovers are too rare for double exchanges to be a problem in genetic analysis. Mitotic crossing over can be visualised as very

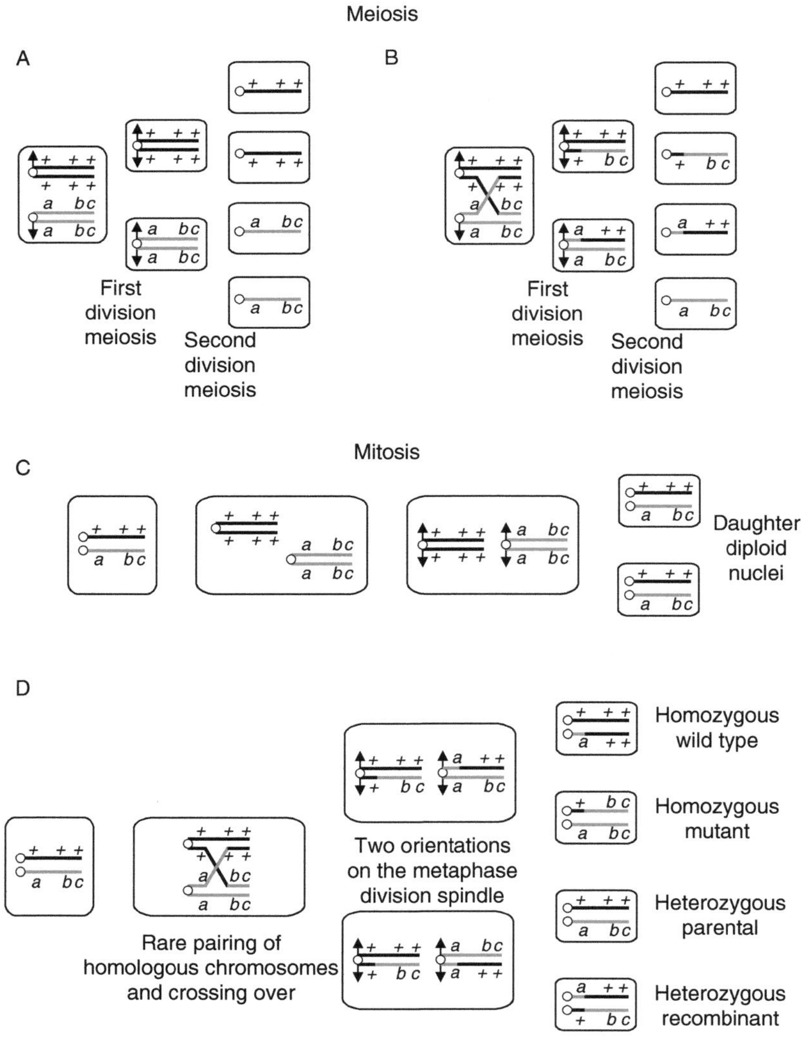

Figure 8.9. Comparison of the segregation mechanisms of meiosis and mitosis. (A) Meiosis, without recombination. (B) Meiosis incorporating a crossover event. These are the perfectly normal meiotic segregation processes applicable to all eukaryotes. Mitosis, with (C) and without (D) a crossover. Mitosis starts with replication of the parental chromosomes, but the two homologues do not normally associate with one another. Subsequently, the replicated chromosomes align on the mitotic division spindle independently, and the rule is that one daughter chromatid of each replicated chromosome segregates into each daughter nucleus, so that the two daughter diploid nuclei have the same genotype as the parent. Very rarely, the replicated chromosomes do associate with one another sufficiently closely for a crossover to occur. This results in two of the daughter chromatids being recombinant. However, the recombinant replicated chromosomes still independently align on the mitotic division spindle, and still follow the rule that one daughter chromatid of each replicated chromosome segregates into each daughter nucleus. Because of the crossover, there are two possible orientations on the mitotic division spindle. One produces two diploid daughter nuclei that are homozygous from the point of the crossover to the end of the chromosome arm (one homozygous mutant, one homozygous wild type). The alternative orientation produces two heterozygous diploid daughter nuclei, but one contains the two parental chromatids, and is the same as the parental nucleus, whereas the other receives the two recombinant chromatids. Modified from Chapter 5 in Moore & Novak Frazer (2002).

similar to meiotic crossing over, but the consequences in terms of the genotypes of progeny nuclei differs ***because chromosome segregation differs between meiosis and mitosis***. If a crossover takes place in mitosis the two chromosomes involved do not stay together (as they do in meiosis), but they separate and reach the equator of the division spindle independently. The two (homologous) chromosomes, which are, of course, divided into two daughter chromatids, then behave independently.

In meiosis, the two reciprocally recombinant chromatids must end up in different haploid daughter nuclei (Figure 8.9). But mitosis produces ***diploid*** daughter nuclei by sending one daughter chromatid of ***each*** homologous chromosome to each pole of the division spindle. Providing the rule that daughter chromatids must go to opposite poles of the division spindle is upheld, there is no other constraint. The daughter chromatids of a pair of homologous chromosomes segregate independently of each other. Consequently, following a mitotic crossover:

- the two reciprocally recombinant chromatids may pass to opposite poles at the subsequent anaphase stage of mitosis;
- alternatively, and with equal probability, they may go to the same pole.

The importance of this is that, in the former case (where a recombinant chromatid is accompanied by a parental chromatid), all markers between the crossover and the end of the chromosome will become homozygous (Figure 8.9).

In most organisms, the analysis of meiotic products is usually the easiest way of mapping the genome. However, use of mitotic segregations for genetic analysis does offer some worthwhile advantages over meiotic analysis to the experimenter. Relatively few segregants can provide a considerable amount of information about relative positions of genes on the chromosome. Even one diploid segregant in which linked genes *a* and *b* become homozygous simultaneously provides evidence that they are in the same chromosome arm, and a second segregant in which *a* becomes homozygous alone shows almost certainly that *a* is distal and *b* proximal with respect to the centromere.

Meiotic and mitotic linkage maps show the same gene ***orders*** but the ***spacing*** between the genes differs, implying ***different distributions of crossovers in mitosis and meiosis***. The overwhelming advantage of mitotic analysis, though, is in the formation of haploids. Because mitotic recombination is so rare, genes on the same chromosome show complete linkage during haploidisation. Genes reassort freely if they are not on the same chromosome. Thus, linkage group assignments can be made far more efficiently than is possible in meiotic analysis.

The method was developed first with *Aspergillus nidulans*, in which segregant sectors of the mycelium could be recognised by the colour of their spores. In *Aspergillus nidulans*, mitotic crossing over has a frequency of about two per thousand mitotic divisions and haploidisation about one per thousand mitotic divisions. In a reversal of the procedure used for identifying diploid sectors, segregants can be identified by the appearance of yellow or white-spored sectors against the background of dark-green spores of a parental diploid colony heterozygous for recessive colour mutations.

We will examine some results from a typical experiment with *Aspergillus nidulans* in which the original diploid was heterozygous for both white (*w*) and yellow (*y*) conidia. These two genes are on different chromosomes, and the *y* chromosome also carried auxotrophic mutations (that is, with a nutritional requirement because of their mutation) called *ade* (adenine requirement), *pro* (proline requirement), *paba* (requirement for the vitamin *para*-aminobenzoic acid), and *bio* (requirement for the vitamin biotin). The parental diploid had the chromosomal constitution shown in Figure 8.10: phenotypically, it had wild-type green (diploid) spores and could grow (as a diploid mycelium) on minimal medium with no nutritional supplements.

Segregants from this diploid were identified on the basis of spore colour: the parental diploid produces dark-green conidia, but sectors with yellow spores and sectors with white spores are

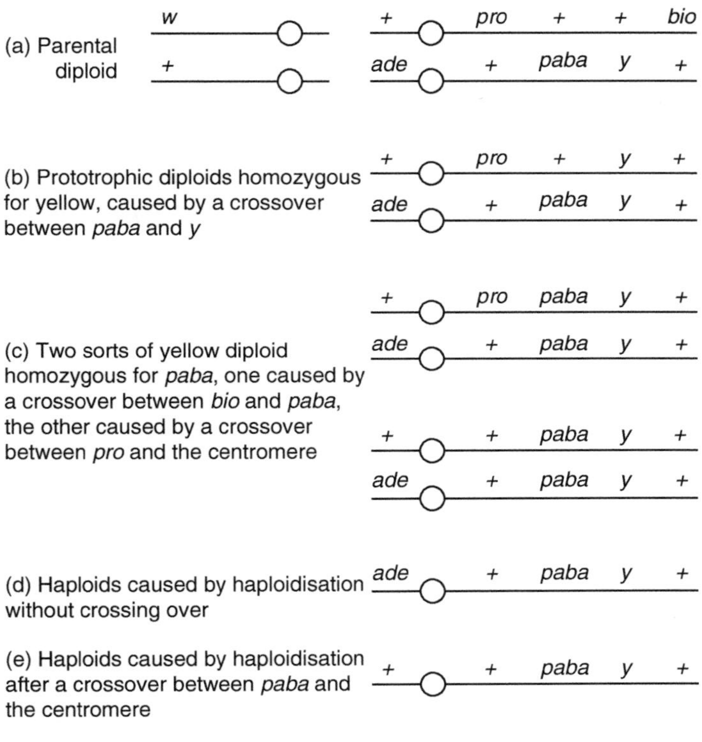

Figure 8.10. Yellow-spored mitotic segregants obtained from an experiment with *Aspergillus nidulans* in which the original diploid was heterozygous for both white (*w*) and yellow (*y*) conidia. The parental diploid had the chromosomal constitution shown in Figure 8.10a, and the crossovers referred to in the other sections of the figure occurred in one chromosome of this genotype before mitotic segregation. The open circle represents the centromeres; + represents the wild-type alleles; *ade*, mutant with a nutritional requirement for the purine adenine; *pro*, nutritional requirement for the amino acid proline; *paba*, nutritional requirement for the vitamin para-aminobenzoic acid; *bio*, nutritional requirement for the vitamin biotin. Modified from Chapter 5 in Moore & Novak Frazer (2002).

occasionally found. Among the yellow-spored segregants (Figure 8.10) were strains which were:

- prototrophic diploids homozygous for yellow, caused by a crossover between *paba* and *y* (Figure 8.10b);
- two sorts of yellow diploid homozygous for *paba*, and therefore auxotrophic for *p*-aminobenzoic acid, one caused by a crossover between *bio* and *paba* in the parental diploid, while the other was caused by a crossover between *pro* and the centromere (Figure 8.10c);
- haploids caused by haploidisation without crossing over (Figure 8.10d);
- haploids caused by haploidisation after a crossover; in this case a crossover somewhere between *paba* and the centromere (Figure 8.10e).

White-spored segregants resulted from homozygosity or haploidisation of *w* (Figure 8.11). The white-spored genotypes observed were:

- prototrophic white-spored diploids caused by a crossover between *w* and the centromere (Figure 8.11b);
- two sorts of white-spored haploids caused by haploidisation without crossing over (Figure 8.11c).

White-spored haploid segregants requiring proline and biotin were observed in about the same frequency as those requiring adenine and *p*-aminobenzoic acid, showing that the chromosome carrying these auxotrophic markers segregated independently of the white chromosome during the haploidisation process.

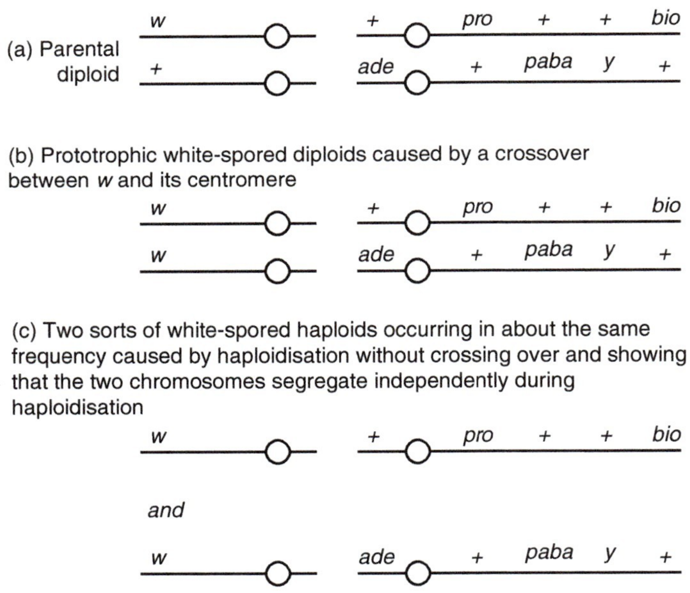
(a) Parental diploid
w
+
+ pro + + bio
ade + paba y +
(b) Prototrophic white-spored diploids caused by a crossover between w and its centromere
w
w
+ pro + + bio
ade + paba y +
(c) Two sorts of white-spored haploids occurring in about the same frequency caused by haploidisation without crossing over and showing that the two chromosomes segregate independently during haploidisation
w
+ pro + + bio
and
w
ade + paba y +

Figure 8.11. White-spored mitotic segregants obtained from an experiment with *Aspergillus nidulans* in which the original diploid was heterozygous for both white (*w*) and yellow (*y*) conidia. The parental diploid had the chromosomal constitution shown in Figure 8.11a, and the crossovers referred to in the other sections of the figure occurred in one chromosome of this genotype before mitotic segregation. Modified from Chapter 5 in Moore & Novak Frazer (2002).

These segregants, the yellow diploid segregants in particular, show that **mitotic crossing over** is a reality, so let's see what lessons can be learned from this example.

For practical analysis, the rule is that in diploid segregants homozygosis for one gene will always be accompanied by homozygosity of any genes distal to it (that is, on the side closest to the chromosome end) in the same chromosome arm, but not necessarily by homozygosity of genes more proximal to it (that is, on the side towards the centromere). This general characteristic of mitotic segregation is what allows us to deduce the genetic map of the segment of chromosome involved and is illustrated in the yellow diploid segregants in Figure 8.10. Homozygosity for *y* was invariably accompanied by homozygosity for the wild-type allele of *bio*. This fact places the *bio* gene distal (that is, closer to the end of the chromosome) to *y* in the same arm of the chromosome. Similarly, homozygosity for *paba* was frequent but not invariable in diploids homozygous for *y*, while homozygosity of *pro* was even less frequent in the yellow homozygotes. This pattern indicates that *paba* and *pro* were proximal to *y* in the same arm; that is, on the centromere side of *y*, with *pro* the nearer to the centromere. Noting that co-segregation of *ade* and *paba* in yellow haploids suggests that the two genes are linked, the absence of yellow diploid segregants homozygous for *ade* allows us to infer that this gene may be in the other arm of the chromosome (other data confirm that homozygosis of one arm of a chromosome occurs independently of homozygosity of the other arm).

The frequencies of the different sorts of homozygous diploid segregants are a measure of relative map distance between the genes; but it must be stressed that mitotic recombination is rare, so it is not easy to assemble a sample of independent segregants large enough to make frequency measurements reliable. Nevertheless, even a few segregants can give absolute guidance about gene order relative to the end of the chromosome arm.

8.8 PARASEXUAL CYCLE

We have described several separate events occurring in mitosis during vegetative fungal growth that might be arranged into a sequence. Mitotic segregants from the diploid prove to be haploid (produced by a process of regular chromosome loss during successive aberrant mitoses called **haploidisation**), partial diploids (aneuploids stabilised during the chromosome loss sequence) or diploids showing segregation for a few linked genetic markers and remaining heterozygous for the others. Haploidisation is caused by nondisjunction (improper transport of chromosomes to the poles of the division spindle during mitosis) resulting in random chromosome loss over several divisions, so the diploid is reduced to a haploid state through a series of aneuploid intermediates (Stukenbrock & Croll, 2014). Overall, the fusion of genetically different haploid nuclei in a heterokaryon followed by mitotic crossing over, then completed by haploidisation, is a sequence termed the **parasexual cycle**.

On the face of it, the parasexual cycle has much the same effect as the sexual cycle by reassorting and recombining genes, thereby increasing genetic variation within the species. A plausible argument can be made that the parasexual cycle could be an alternative to sex in fungi lacking a sexual cycle (so-called imperfect fungi), but there is not much clear evidence for this. Not a great deal of practical use has been made of the parasexual cycle in fungi, even though several commercial processes depend on imperfect fungi like *Penicillium chrysogenum*. Prior to the advent of molecular biology industrial fungal geneticists at Glaxo's Laboratories in Ulverston used the parasexual cycle to improve strains of *Penicillium chrysogenum* for penicillin production. At the time this was the only way that advantageous mutations could be recombined.

Ironically, however, the technique found its most extensive application in ***human genetics***. A very large proportion of the gene assignments to human chromosomes were made, ***before*** the genomics era, using this analogous cycle: mouse + human cell, forming a hybrid ***fusion cell heterokaryon*** that suffers successive loss of chromosomes during subsequent mitoses. Eventually, aneuploid cell lines, sufficiently stable for genetic and cytogenetic characterisation are formed and co-segregation of genes reveals linkage. This is another example of a phenomenon discovered in fungi being exploited to enhance some aspect of animal cell biology. Indeed, study of ***heterokaryons*** involving human cells is an expanding area of interest in medical research. Contributing to production of monoclonal antibodies, the generation of cell hybrids for cancer immunotherapy and organ repair therapies for several genetic and degenerative diseases. Experimental cell fusion heterokaryons of this sort have informed our understanding of malignancy and tumour-suppressor activity, and they are now fundamental to the study of nuclear reprogramming of differentiated cells into the pluripotent ('stem cell') state (Serov *et al.*, 2011; Narbonne *et al.*, 2012; Jang *et al.*, 2016).

Sexual reproduction is ubiquitous among eukaryotes, and fully asexual lineages are extremely rare; probably the most prominent of ancient, presumed asexual, lineages are the arbuscular mycorrhizal (AM) fungi (*Glomeromycotina*, see section 3.6). This group of highly successful plant mycorrhizal symbionts has a multinucleate cytoplasm, but no known sexual stage and it is thought that they might have been asexual for approximately 400 million years. It has been presumed that, in these organisms, evolution might depend on genetic variation resulting from accumulation of mutations occurring in a population of genetically different nuclei within individual arbuscular mycorrhizal fungi. While there is some support for this concept (Kuhn *et al.*, 2001), there is also evidence that genetic diversity in isolates of *Rhizophagus irregularis*, a model AM fungus, varies between isolates in a way that is like a fungus with a homokaryon-dikaryon ***sexual*** life cycle. A multiallelic mating type locus, containing two genes with structural and evolutionary similarities to the mating type locus of some *Dikarya* has also been identified (Ropars *et al.*, 2016). There is an ongoing controversy about the definition, dispersal and distribution of species in the *Glomeromycotina* clade which is described by Bruns *et al.* (2017) (and see section 3.6).

8.9 CYTOPLASMIC SEGREGATIONS: MITOCHONDRIA, PLASMIDS, VIRUSES AND PRIONS

There are several cytoplasmic factors that affect the fungal phenotype, and which depend for their transmission upon some of the features of heterokaryosis we have described so far. Chief among these are the mitochondria. Mitochondrial genomes are independent of and quite distinct from the nuclear genome. Loss of function mutations in mitochondrial genes result in characteristic phenotypes in both *Neurospora crassa* and *Saccharomyces cerevisiae*. **Mitochondrial DNA (mtDNA)** of yeast usually makes up 18% of the total DNA, but it has a distinctive very high AT content (82%) so is relatively easy to separate from chromosomal DNA. The mtDNA is circular, of 25 nm circumference, and comprises about 7.5×10^5 base pairs. In yeast, mtDNA codes for three of seven polypeptides of the cytochrome *c* complex (the rest derive from nuclear genes), four polypeptide components of a mitochondrial ATPase, and one component of cytochrome b. Mutations in these genes can produce recognisable respiratory deficiency phenotypes (for example, mutants named *petite* in *Saccharomyces cerevisiae, poky* in *Neurospora crassa*) and thereby provide mitochondrial mutants. Also, although chromosomal genes code mitochondrial ribosomal proteins, mtDNA determines mitochondrial ribosomal RNA (rRNA) and transfer RNAs (tRNAs).

Mitochondrial ribosomes are similar in size to prokaryotic ribosomes and share some other prokaryotic properties; in particular, protein synthesis on mitochondrial ribosomes is inhibited by chloramphenicol, erythromycin and several other antibacterials that have no effect on cytoplasmic (eukaryotic) ribosomes. Consequently, another kind of mutant phenotype due to mitochondrial mutation is resistance to inhibition by mitochondrion-specific drugs. Mitochondrial gene sequences are also like equivalent genes in prokaryotes. These features encouraged the endosymbiont theory of mitochondrial origin, which envisages mitochondria to be relics of ancient bacteria-like organisms that formed a symbiotic association that resulted in the ancestral oxygen-respiring 'eukaryotic' cell.

Saccharomyces cerevisiae can have over 100 genomes per mitochondrion, corresponding to about 6,500 in each yeast cell; some of the mitochondrial genomes are circular, others are linear. Oddly enough, when segregation of organelle genes is followed in genetic crosses, the segregation patterns are consistent with there being only one copy of the mitochondrial genome in the cell. The basic procedure for making mitochondrial crosses involves making heterokaryons (diploids in yeast) between haploids carrying mitochondrial markers. After some vegetative growth of the heterokaryon or diploid, diploid daughter cells (of yeast) or spores or hyphal fragments (of filamentous fungi) are plated out and the resulting colonies are scored for the mitochondrial markers present in the original haploids.

In *Neurospora crassa*, phenotypes controlled by mitochondrial genes are generally transmitted through the female, that is, the protoperithecial parent. This is **vertical mitochondrial transmission**, from one generation to the next. **Horizontal transmission**, which is between individuals of the same generation, occurs because of hyphal fusion. Although complementation and recombination can be detected between mitochondrial genomes, mycelia containing genetically different mitochondria (called **heteroplasmons** or heteroplasmic mixtures) tend to segregate the different mitochondria into different cells (Wilson & Xu, 2012).

Uniparental inheritance of mitochondrial phenotypes has been observed in yeast, which is isogamous (does not show a male/female differentiation). Mitochondrial genomes can segregate in association with mitosis and the consequential bud formation The propagation of mitochondria during cell division depends on replication and partitioning of mitochondrial DNA, cytoskeleton-dependent mitochondrial transport, intracellular positioning of the organelle, and a variety of activities that coordinate these processes (Westermann, 2014). In filamentous fungi, mitochondria are not closely associated with the mitotic spindle, so vegetative segregation may simply be a matter of random physical sorting. However, mitochondria interact with elements of the cytoskeleton in both *Neurospora* (Fuchs *et al.*, 2002) and *Aspergillus* (Suelmann & Fischer, 2000), although the cytoskeleton is involved in determining mitochondrial morphology as well as mitochondrial movement. Nuclear and mitochondrial genes influence mitochondrial genome transmission and it is also affected by membrane chemistry. In some *Ascomycota* sub-cultured for a long time in the laboratory, altered mitochondrial DNAs due to molecular rearrangements have been associated with modifications in mycelial growth in *Neurospora crassa* and *Neurospora intermedia*, as well as in growth of yeast.

Among the *Basidiomycota*, mitochondrial inheritance has been studied in matings of *Schizophyllum commune, Agaricus bitorquis, Agaricus brunnescens, Coprinopsis cinerea, Lentinula edodes, Pleurotus ostreatus*, plant pathogenic *Armillaria* species and *Ustilago violacea*, and in all these cases, hyphal anastomoses result in the production of mycelial colonies composed of sectors containing different mitochondrial DNAs (**mitochondrial mosaics**). In *Agaricus bitorquis, Agaricus bisporus, Armillaria bulbosa, Pleurotus ostreatus* and *Ustilago violacea*, dikaryons had mixed populations of mitochondria, which sometimes resulted in recombination between mitochondrial genomes. Recombinant mitochondrial DNAs have also been recovered from dikaryons and dikaryotic protoplasts of *Coprinopsis cinerea* (Xu & Wang, 2015).

As well as the mitochondrial genome, mitochondria can also contain autonomously replicating DNA molecules, that are either derived from the mitochondrial DNA or represent true plasmids that show no homology with the mitochondrial chromosome. An increasingly important aspect of mitochondrial transmission is their content of plasmids. Among true plasmids at least three different categories can be recognised: (a) **circular plasmids** encoding a DNA polymerase; (b) **linear plasmids** with terminal inverted repeats encoding a DNA and an RNA polymerase; and (c) **retroplasmids**, which are linear or circular plasmids that encode a reverse transcriptase. DNA and RNA polymerases, or reverse transcriptase, encoded by plasmid DNA sequences are used to maintain and propagate the plasmid. True plasmids are mostly **cryptic** (that is, neutral) passengers in nature, but some linear plasmids (notably of *Podospora anserina*) insert into mitochondrial DNA and cause **mycelial senescence**. Most linear plasmids exhibit typical virus characteristics as far as structure, replication and function are concerned even to the extent that plasmid-free strains may contain plasmid remnants integrated into their mitochondrial DNA. These different groups of true plasmids probably arose independently of one another and may have a different evolutionary origin from that of the mitochondrial host genome. They were either vertically transmitted from the original endosymbiont that gave rise to the mitochondrion or invaded the mitochondrion at various times during fungal evolution (Hausner, 2003).

Although most ***mitochondrial*** plasmids are cryptic and symptomless, ***cytoplasmic*** **plasmid DNA** is responsible for the **killer phenomenon** in the yeast *Kluyveromyces lactis* by coding for a toxin, which kills cells lacking the plasmid (cells hosting the killer plasmid are immune to the **toxin**). These plasmids reside in the cytoplasm and have an expression system independent of both nucleus and mitochondrion. Plasmids of *Kluyveromyces lactis* can be transferred to other yeasts (including *Saccharomyces cerevisiae*), conferring the killer/immunity phenotypes (Fichtner *et al.*, 2003). This shows that the plasmids are **autonomous replicons**, which can be expressed in a wide range of host yeasts.

The *Kluyveromyces lactis* killer plasmid toxin is chemically and functionally different from killer toxins produced in *Saccharomyces cerevisiae* (especially wine yeasts), which are encoded by **double-stranded RNA (dsRNA)** virus. Different killer toxins, K1, K2, K28 and Klus, have been described; each being encoded by a 1.5- to 2.3 kb double-stranded **M satellite** RNA located in the cytoplasm. These M satellite dsRNAs require larger **helper virus** (generally called L-A virus) for maintenance; L-A belongs to the Totiviridae family, and its dsRNA genome of 4.6 kb codes for proteins that form the virions that encapsidate separately the L-A or M satellites (Rodríguez-Cousiño *et al.*, 2011; Rodríguez-Cousiño & Esteban, 2017).

Both budding yeast (*Saccharomyces cerevisiae*) and fission yeast (*Schizosaccharomyces pombe*) have been used for studies of plant, animal and human viruses. Many RNA viruses and some DNA viruses replicate in yeasts. As many of the fundamental eukaryotic cell functions are highly conserved from yeasts to higher eukaryotes, these easily cultivated fungi offer many unique advantages in virus research over 'higher' eukaryotes and are particularly suited to study the impact of viral activities on cell function during virus-host interactions (Zhao, 2017). **Mitoviruses** are simple RNA **mycoviruses** that replicate in host mitochondria and are frequently found in fungi; genomics approaches are now being used to study them (Kotta-Loizou, 2019; Nibert *et al.*, 2019).

Virus-like particles (VLPs) have been observed in electronmicrographs of many fungi. They are very similar in appearance to small spherical RNA viruses, but there is little evidence that these particles are effective in hypha-to-hypha infection. Many of the observed VLPs are presumably degenerate or defective viruses that can only be transmitted by hyphal fusions. No vectors are known for fungal viruses; transmission seems to depend on hyphal fusions. Unexpectedly, virus infections of fungi usually cause no recognisable phenotype. An important exception is a mycovirus of *Agaricus bisporus*, which causes **La France disease** and ruins the crop. Diseased crops contain three double-stranded RNA virus particles and require up to 10 different RNA molecules to produce the isometric infective virus particles; although some are defective viruses and others are helper viruses, or perhaps different viruses perform complementary, but essential, functions (Frost & Passmore, 1980). In a study of an infection on commercial mushroom farms in Poland, the virus particles were found in 120 of 200 samples tested. This level of La France disease could be a threat to the mushroom industry (Borodynko *et al.*, 2010).

Saccharomyces cerevisiae also harbours retrovirus-like elements, as retrotransposons (now called **transposable elements** or **TEs**) able to integrate into the nuclear genome by targeting particular chromatin structures. The transposable elements of *Saccharomyces cerevisiae* consist of LTR (**L**ong **T**erminal **R**epeat) retrotransposons called Ty elements belonging to five families, Ty1 to Ty5. They take the form of either full-length coding sequences or non-coding solo LTRs corresponding to remnants of former transposition events. The first cytoplasmic plasmid to be observed was the so-called **two-micron DNA** of *Saccharomyces* (Ty1). The name refers to the contour length of the circular DNA molecules in electronmicrographs. Two-micron DNA has a base composition like nuclear DNA and quite different from mtDNA. There can be 50–100 two-micron DNA molecules per diploid cell, amounting to something like 3% of the nuclear DNA. The two-micron DNA molecules are transmitted to buds independently of both nuclei and mitochondria. The two-micron circular DNA carries inverted repeat sequences at either end of two different unique sequence segments; this structure implies that it inserts itself into the yeast chromosome (Bleykasten-Grosshans & Neuvéglise, 2011; Bleykasten-Grosshans *et al.*, 2013; Stukenbrock & Croll, 2014).

So far, we have described nucleic acid molecules that encode features segregating in the cytoplasm. In the final decade of the twentieth century, however, great attention was given (and continues to be given) to a **proteinaceous hereditary element**, called a **prion protein**. The attention devoted to prions derives from their ability to cause diseases in mammals: scrapie in sheep, bovine spongiform encephalopathy (BSE) in cattle and, in humans, kuru and new variant Creutzfeldt–Jakob disease (nvCJD). In these cases, the pathogenic agent is a variant of a normal membrane protein (the prion protein) that is encoded in the mammalian genome. Prions are infectious proteins, which means that they are altered forms of a normal cellular protein that may have lost their normal function but have acquired the ability to modify the normal form of the protein into the same abnormal configuration as themselves. The variant prion protein folds abnormally and in addition causes normal prion proteins to fold abnormally so that the proteins aggregate in the central nervous system and cause the encephalopathy; the aggregated proteins are called **amyloids**.

Several prion-forming proteins have been identified in fungi, mostly in the yeast *Saccharomyces cerevisiae*. We have already referred briefly (see section 8.5 and Table 8.3) to the infectious *het-s* (heterokaryon incompatibility) phenotype of *Podospora anserina* as a protein that adopts a prion-like form to function properly as part of the self/nonself recognition system that ensures that only related hyphae share resources. The prion form of *het-s* can convert the non-prion form of the protein in a compatible mate after hyphal anastomosis. However, when an incompatible mycelium mates with a prion-containing mycelium, the prion causes the incompatible hyphal compartments to die when a programmed cell death response is triggered by interaction between specific *het* alleles (Paoletti & Clavé, 2007; Bidard *et al.*, 2013; Paoletti, 2016).

First among several prions identified in *Saccharomyces cerevisiae* is the *PSI+* form of the Sup35p protein. Sup35p is an essential yeast protein involved in the termination of translation. In the [PSI$^+$] state, Sup35p adopts the structural conformation that causes it to direct the refolding of native molecules into a form that can aggregate into filaments of discarded nonfunctional protein. This depletes the cytoplasm of functional translation terminator and results in translation errors. This is the [PSI$^+$] phenotype, which is inherited by daughter cells following budding, and is infectious following cell fusion, in which case it propagates by autocatalytic conversion of the normal form of the protein. The prion phenotype is officially described as 'increased levels of nonsense suppression' because the prion suppresses nonsense mutations by allowing the mutant genes to produce functional proteins; that is, the translation error is to translate the mutant code as working protein. Another unusual trait identified in yeast in the 1990s was called [URE3], which results from the prion form of the normal cellular protein Ure2p, which is a nitrogen catabolite repressor. Note that the names for yeast prions are normally shown in square brackets to indicate that they segregate in a non-Mendelian manner. Two more recent discoveries are [MOT3$^+$], the prion form of a nuclear transcription factor, which as a prion causes transcriptional derepression of anaerobic genes; and [GAR$^+$], the normal function of the proteins being as components of plasma membrane proton pumps but as a prion it causes resistance to glucose-associated repression.

For a list of prions, refer to the Wikipedia page *Fungal prion* (at https://en.wikipedia.org/wiki/Fungal_prion), which seems to be regularly updated.

The part of the Sup35p protein that makes it a prion (the prion determining domain) is a glutamine/asparagine-rich amino-terminal region that contains several oligopeptide repeats. Removal of these repeats eliminates the ability of Sup35p to propagate *PSI+* and expanding the repeat region increases the spontaneous occurrence of *PSI+*. Although deleting the analogous repeats from BSE prion protein does not prevent prion propagation and transmission in experimental mice, expansion of the repeat region does increase the spontaneous appearance of spongiform encephalopathies by several orders of magnitude in humans. It is likely that the oligopeptide repeats give the prion protein the intrinsic tendency to acquire a conformation that enables the protein to refold and self-propagate its conformation, so effectively polymerising with sister molecules to form the fibrous proteinaceous deposits called amyloids. Database searches for regions with amino acid content comparable to the yeast prions have revealed numerous such domains in eukaryotes, but these are lacking from prokaryotes (Alberti *et al.*, 2009; King *et al.*, 2012).

Fungal prions are generally benign; indeed, some confer a potential advantage to the fungus. Although it has been claimed that [URE3] and [PSI$^+$] are 'diseases' of yeast caused by laboratory cultivation, some [PSI$^+$] cells actually fare better than their prion-free siblings when subjected to adverse conditions. When 700 wild strains of *Saccharomyces* were genetically screened for unknown prion elements, one-third of the strains were found to harbour them (Halfmann *et al.*, 2012). These 'natural' prions created diverse and often beneficial phenotypes. Evidently, fungal prions govern heritable traits in nature, but in a manner that confers the capacity on the fungus to adapt quickly and reversibly to variable environments. It has been suggested that the amyloid-folding promoted by **fungal prions** is a widespread and evolutionarily conserved mode of **signal transduction**, which is based on the transmission of an amyloid-fold from a STAND receptor protein to an effector protein (Daskalov *et al.*, 2012). Fungal prions potentially provide a model for understanding disease-causing prions of humans and animals, particularly by identifying the sequence features and mechanisms that enable prion domains to switch between functional and amyloid-forming states (for information about disease-causing prions of humans and animals refer to the US Department of Health & Human Services Centers for Disease Control and Prevention website at https://www.cdc.gov/prions/index.html).

8.10 REFERENCES

Alberti, S., Halfmann, R., King, O., Kapila, A. & Lindquist, S. (2009). A systematic survey identifies prions and illuminates sequence features of prionogenic proteins. *Cell*, **137**: 146–158. DOI: https://doi.org/10.1016/j.cell.2009.02.044.

Bärlocher, F. (2007). Molecular approaches applied to aquatic hyphomycetes. *Fungal Biology Reviews*, **22**: 19–24. DOI: https://doi.org/10.1016/j.fbr.2007.02.003.

Bidard, F., Clavé, C. & Saupe, S.J. (2013). The transcriptional response to nonself in the fungus *Podospora anserina*. *G3, Genes, Genomes, Genetics*, **3**: 1015–1030. DOI: https://doi.org/10.1534/g3.113.006262.

Bleykasten-Grosshans, C., Friedrich, A. & Schacherer, J. (2013). Genome-wide analysis of intraspecific transposon diversity in yeast. *BMC Genomics*, **14**: 399. DOI: https://doi.org/10.1186/1471-2164-14-399.

Bleykasten-Grosshans, C. & Neuvéglise, C. (2011). Transposable elements in yeasts. *Comptes Rendus Biologies*, **334**: 679–686. DOI: https://doi.org/10.1016/j.crvi.2011.05.017.

Borodynko, N., Hasiów-Jaroszewska, B., Rymelska, N. & Pospieszny, H. (2010). La France disease of the cultivated mushroom *Agaricus bisporus* in Poland. *Acta Virologica*, **54**: 217–219. DOI: https://doi.org/10.4149/av_2010_03_217.

Bruns, T.D., Corradi, N., Redecker, D., Taylor, J.W. & Öpik, M. (2017). Glomeromycotina: what is a species and why should we care? *New Phytologist*, **2017**: Viewpoint Early View (online version of record before inclusion in an issue), article 0028-646X. DOI: https://doi.org/10.1111/nph.14913.

Daskalov, A., Heller, J., Herzog, S., Fleißner, A. & Glass, N. (2017). Molecular mechanisms regulating cell fusion and heterokaryon formation in filamentous fungi. In *The Fungal Kingdom*, ed. J. Heitman, B. Howlett, P. Crous *et al.* Washington, DC: ASM Press, pp. 215–229. DOI: https://doi.org/10.1128/microbiolspec.FUNK-0015-2016.

Daskalov, A., Paoletti, M., Ness, F. & Saupe, S.J. (2012). Genomic clustering and homology between HET-S and the NWD2 STAND protein in various fungal genomes. *PLoS ONE*, **7**: article number e34854. DOI: https://doi.org/10.1371/journal.pone.0034854.

Fichtner, L., Jablonowski, D., Frohloff, F. & Schaffrath, R. (2003). Phenotypic analysis of the *Kluyveromyces lactis* killer phenomenon. In *Non-Conventional Yeasts in Genetics, Biochemistry and Biotechnology, One of the Springer Lab Manuals*, ed. K. Wolf, K. Breunig & G. Barth. Berlin, Heidelberg: Springer International Publishing, pp. 179–183. DOI: https://doi.org/10.1007/978-3-642-55758-3_28.

Fiddy, C. & Trinci, A.P.J. (1976). Mitosis, septation and the duplication cycle in *Aspergillus nidulans*. *Journal of General Microbiology*, **97**: 169–184. DOI: https://doi.org/10.1099/00221287-97-2-169.

Frost, R.R. & Passmore, E.L. (1980). Mushroom viruses: a re-appraisal. *Journal of Phytopathology*, **98**: 272–284. DOI: https://doi.org/10.1111/j.1439-0434.1980.tb03742.x.

Fuchs, F., Prokisch, H., Neupert, W. & Westermann, B. (2002). Interaction of mitochondria with microtubules in the filamentous fungus *Neurospora crassa*. *Journal of Cell Science*, **115**: 1931–1937. URL: http://jcs.biologists.org/content/115/9/1931.abstract.

Gregory, T.R., Nicol, J.A., Tamm, H. *et al.* (2007). Eukaryotic genome size databases. *Nucleic Acids Research*, **35**: D332–D338. DOI: https://doi.org/10.1093/nar/gkl828.

Halfmann, R., Jarosz, D.F., Jones, S.K. *et al.* (2012). Prions are a common mechanism for phenotypic inheritance in wild yeasts. *Nature*, **482**: 363–368. DOI: https://doi.org/10.1038/nature10875.

Hall, C., Welch, J., Kowbel, D.J. & Glass, N.L. (2010). Evolution and diversity of a fungal self/nonself recognition locus. *PLoS ONE*, **5**: article number e14055. DOI: https://doi.org/10.1371/journal.pone.0014055.

Hausner, G. (2003). Fungal mitochondrial genomes, plasmids and introns. *Applied Mycology and Biotechnology*, **3**: 101–131. DOI: https://doi.org/10.1016/S1874-5334(03)80009-6.

Heitman, J. (2015). Evolution of sexual reproduction: a view from the fungal kingdom supports an evolutionary epoch with sex before sexes. *Fungal Biology Reviews*, **29**: 108–117. DOI: https://doi.org/10.1016/j.fbr.2015.08.002.

Jang, H.S., Hong, Y.J., Choi, H.W. *et al.* (2016). Changes in parthenogenetic imprinting patterns during reprogramming by cell fusion. *PLoS ONE*, **11**: article number e0156491. DOI: https://doi.org/10.1371/journal.pone.0156491.

King, A.M.Q., Adams, M.J., Carstens, E.B. & Lefkowitz, E.J. (ed.) (2012). Fungal prions. In *Virus Taxonomy, Ninth Report of the International Committee on Taxonomy of Viruses*. San Diego, CA: Elsevier, pp. 1235–1245. DOI: https://doi.org/10.1016/B978-0-12-384684-6.00108-7.

Kotta-Loizou, I. (2019). Mycoviruses: past, present, and future. *Viruses*, **11**(4): article 361. DOI: https://doi.org/10.3390/v11040361 (an open access special issue of the journal on Mycoviruses).

Kruzel, E.K. & Hull, C.M. (2010). Establishing an unusual cell type: how to make a dikaryon. *Current Opinion in Microbiology*, **13**: 706–711. DOI: https://doi.org/10.1016/j.mib.2010.09.016.

Kuhn, G., Hijri, M. & Sanders, I.R. (2001). Evidence for the evolution of multiple genomes in arbuscular mycorrhizal fungi. *Nature*, **414**: 745–748. DOI: https://doi.org/10.1038/414745a.

Maheshwari, R. (2005). Nuclear behavior in fungal hyphae. *FEMS Microbiology Letters*, **249**: 7–14. DOI: https://doi.org/10.1016/j.femsle.2005.06.031.

Moore, D. & Novak Frazer, L. (2002). *Essential Fungal Genetics*. New York: Springer-Verlag. ISBN-10: 0387953671, ISBN-13: 978-0387953670. See Chapter 2 *Genome interactions* and Chapter 5 *Recombination analysis*.

Narbonne, P., Miyamoto, K. & Gurdon, J.B. (2012). Reprogramming and development in nuclear transfer embryos and in interspecific systems. *Current Opinion in Genetics & Development*, **22**: 450–458. DOI: https://doi.org/10.1016/j.gde.2012.09.002.

Nibert, M.L., Debat, H.J., Manny, A.R., Grigoriev, I.V. & De Fine Licht, H.H. (2019). Mitovirus and mitochondrial coding sequences from basal fungus *Entomophthora muscae*. *Viruses*, **11**(4): article 351. DOI: https://doi.org/10.3390/v11040351.

Paoletti, M. (2016). Vegetative incompatibility in fungi: from recognition to cell death, whatever does the trick. *Fungal Biology Reviews*, **30**: 152–162. DOI: https://doi.org/10.1016/j.fbr.2016.08.002.

Paoletti, M. & Clavé, C. (2007). The fungus-specific HET domain mediates programmed cell death in *Podospora anserina*. *Eukaryotic Cell*, **6**: 2001–2008. DOI: https://doi.org/10.1128/EC.00129-07.

Pérez-Martín, J. & de Sena-Tomás, C. (2011). Dikaryotic cell cycle in the phytopathogenic fungus *Ustilago maydis* is controlled by the DNA damage response cascade. *Plant Signaling & Behavior*, **6**: 1574–1577. DOI: https://doi.org/10.4161/psb.6.10.17055.

Pringle, A. & Taylor, J.W. (2002). The fitness of filamentous fungi. *Trends in Microbiology*, **10**: 474–481. DOI: https://doi.org/10.1016/S0966-842X(02)02447-2.

Rayner, A.D.M. & Todd, N.K. (1982). Population structure in wood-decomposing basidiomycetes. In *Decomposer Basidiomycetes: Their Biology and Ecology*, ed. J.C. Frankland, J.N. Hedger & M.J. Swift. Cambridge, UK: Cambridge University Press, pp. 109–128. ISBN: 9780521106801.

Rodríguez-Cousiño, N. & Esteban, R. (2017). Relationships and evolution of double-stranded RNA Totiviruses of yeasts inferred from analysis of L-A-2 and L-BC variants in wine yeast strain populations. *Applied and Environmental Microbiology*, **83**: e02991–16. DOI: https://doi.org/10.1128/AEM.02991-16.

Rodríguez-Cousiño, N., Maqueda, M., Ambrona, J. *et al.* (2011). A new wine *Saccharomyces cerevisiae* killer toxin (*Klus*), encoded by a double-stranded RNA virus, with broad antifungal activity is evolutionarily related to a chromosomal host gene. *Applied and Environmental Microbiology*; **77**:1822–1832. DOI: https://doi.org/10.1128/AEM.02501-10.

Ropars, J., Toro, K.S., Noel, J. *et al.*(2016). Evidence for the sexual origin of heterokaryosis in arbuscular mycorrhizal fungi. *Nature Microbiology*, **1**: article 16033. DOI: https://doi.org/10.1038/nmicrobiol.2016.33.

Seifert, K.A. & Gams, W. (2001). The taxonomy of anamorphic fungi. In *The Mycota Systematics and Evolution*, **VII part A**, 2nd edition, ed. D.J. McLaughlin & J.W. Spatafora. Berlin: Springer-Verlag, pp. 307–347. DOI: https://doi.org/10.1007/978-3-662-10376-0_14.

Serov, O.L., Matveeva, N.M. & Khabarova, A.A. (2011). Reprogramming mediated by cell fusion technology. *International Review of Cell and Molecular Biology*, **291**: 155–190. DOI: https://doi.org/10.1016/B978-0-12-386035-4.00005-7.

Strom, N.B. & Bushley, K.E. (2016). Two genomes are better than one: history, genetics, and biotechnological applications of fungal heterokaryons. *Fungal Biology and Biotechnology*, **3**: 4. DOI: https://doi.org/10.1186/s40694-016-0022-x.

Stukenbrock, E.H. & Croll, D. (2014). The evolving fungal genome. *Fungal Biology Reviews*, **28**: 1–12. DOI: https://doi.org/10.1016/j.fbr.2014.02.001.

Suelmann, R. & Fischer, R. (2000). Mitochondrial movement and morphology are dependent on an intact actin cytoskeleton in *Aspergillus nidulans*. *Cell Motility and the Cytoskeleton*, **45**: 42–50. DOI: https://doi.org/10.1002/(SICI)1097-0169(200001)45:1<42::aid-cm4>3.0.CO;2-C.

Taylor, J.W., Jacobson, D.J. & Fisher, M.C. (1999). The evolution of asexual fungi: reproduction, speciation and classification. *Annual Review of Phytopathology*, **37**: 197–246. DOI: https://doi.org/10.1146/annurev.phyto.37.1.197.

Wang, Z., Kin, K., López-Giráldez, F., Johannesson, H. & Townsend, J.P. (2012). Sex-specific gene expression during asexual development of *Neurospora crassa*. *Fungal Genetics and Biology*, **49**: 533–543. DOI: https://doi.org/10.1016/j.fgb.2012.05.004.

Westermann, B. (2014). Mitochondrial inheritance in yeast. *Biochimica et Biophysica Acta, Bioenergetics*, **1837**: 1039–1046. DOI: https://doi.org/10.1016/j.bbabio.2013.10.005.

Wilson, A.J. & Xu, J. (2012). Mitochondrial inheritance: diverse patterns and mechanisms with an emphasis on fungi. *Mycology*, **3**: 158–166. DOI: https://doi.org/10.1080/21501203.2012.684361.

Xu, J. & Wang, P. (2015). Mitochondrial inheritance in basidiomycete fungi. *Fungal Biology Reviews*, **29**: 209–219. DOI: https://doi.org/10.1016/j.fbr.2015.02.001.

Zhao, R.Y. (2017). Yeast for virus research. *Microbial Cell*, **4**: 311–330. DOI: https://doi.org/10.15698/mic2017.10.592.

CHAPTER 9

Sexual Reproduction: The Basis of Diversity and Taxonomy

Sexual reproduction is a nearly universal feature of eukaryotes and its core features are conserved throughout each group within the eukaryotic tree of life. This is taken to imply that sexual reproduction evolved once only and was present in the **Eukaryote Last Common Ancestor** (ELCA). Studies of the fungal kingdom have revealed novel and unusual patterns of sexual reproduction, which we will discuss in this chapter.

Fundamentally, sexual reproduction is the fusion of gametes (the differentiated sex cells) or their nuclei to form a diploid that can undergo meiosis. The overall summary 'equation' is:

$$\text{plasmogamy} \rightarrow \text{karyogamy} \rightarrow \text{meiosis}$$

For most fungi, plasmogamy occurs when hyphal fusion (anastomosis) occurs, and is controlled by the incompatibility systems. Growth of the resultant heterokaryon as an independent mycelium prolongs plasmogamy, in some cases indefinitely.

In this chapter, we describe the process of sexual reproduction in fungi. This includes description of mating and mating type switching in budding yeast and mating type factors of *Neurospora* and *Basidiomycota*. We finish with some thoughts about the biology of mating type factors.

9.1 THE PROCESS OF SEXUAL REPRODUCTION

The core features of sexual reproduction are conserved throughout the eukaryotic tree of life and are therefore thought to have evolved once and to have been a character of the **ELCA** (see Figure 2.11 and Moore, 2013, pp. 174 *et seq.*). It follows that sexual reproduction in present-day organisms displays a mixture of features that are ancient and **ancestral**, together with others that have arisen during the subsequent evolution of that organism. For example, sexual reproduction in the great majority of eukaryotes alive today involves two contrasting sexes or mating types, so this ***can*** be considered an ancestral feature. Yet, among the fungi, there are species that indulge in unisexual reproduction, where a single mating type can undergo self-fertile (or homothallic) reproduction on its own, either with itself or with other members of the population of the same mating type. Unisexual reproduction occurs in several different lineages and ***may*** therefore be interpreted as a **derived feature** that has arisen independently in those different lineages. On the other hand, the incredible variety of different types of sex (or mating type) determining mechanisms that can be observed in animals, plants, protists and fungi of the present day ***may*** suggest that specification of sex (or mating type) is not the ancestral feature but is a derived trait; and if this is the case, then the original form of sexual reproduction ***may*** have been unisexual, onto which sexes were imposed independently in the different lineages as they evolved.

We do not know what ELCA was like but the current belief is that our last common ancestor was a unicellular, aquatic, motile creature with one or two flagella (and really rather like the sort of thing that eventually became a chytrid fungus; see Moore, 2013). Of course, the ELCA certainly had a membrane-bound nucleus, mitochondria, secretory apparatus, the ability to regulate gene expression with RNAi, and the ability to reproduce both asexually and sexually (mitosis and meiosis are conserved processes throughout eukaryotes). When we settle down with a glass of wine and think about sex ... first evolving, we think it must have occurred first in an aqueous environment (probably in some primitive biofilm) and that it involved swimming cells, and changes in ploidy that needed a reduction division (meiosis) to correct. Although cell–cell and nucleus–nucleus fusion are prominent in sexual reproduction today, there may have been a time in the distant past when internal replication cycles (endoreplication) caused the change in ploidy that needed to be corrected by a reduction division (meiosis) during ancestral attempts at sexual reproduction. This view predicts that cell–cell fusion may be ancient, but perhaps not as ancient as other features of sexual reproduction (Heitman, 2015).

It is the potential benefits of sexual reproduction that have given it the competitive edge and caused sex to be so pervasive in the eukaryotic tree of life. These benefits include that it provides a means to purge the genome of (vegetatively accumulated) deleterious mutations; and a means to shuffle the genome by means of chromosome reassortment and recombination to give rise to different gene arrangements among the meiotic progeny. Sex may also enable organisms to compete with pathogens, some of which may be internal, like transposons (see discussion and references in Heitman, 2015). The potential benefits of sex must be balanced against the costs of sexual reproduction: that only 50% of a parental genome is transmitted to any given progeny, the time and energy required to locate mates, and the reassortment of already adapted gene arrangements.

The core features of sexual reproduction are conserved in organisms as diverse as the model budding yeast *Saccharomyces cerevisiae* and humans, despite at least a billion years of evolution separating us from our last shared ancestor. These conserved features include:

- regular changes in ploidy, from haploid to diploid to haploid, or from diploid to haploid to diploid;
- the process of meiosis that enables meiotic recombination and halves the ploidy of the genome;
- cell–cell fusion (**syngamy**) between mating partners or their gametes.

The conservation of these core features of sexual reproduction across this enormous evolutionary time is what indicates the antiquity of the process (Billiard *et al.*, 2012; Heitman, 2015). Most fungi can undergo both asexual reproduction and sexual reproduction. The evolutionary persistence of eukaryotes that rely on asexual reproduction alone is exceptional (see section 8.1). Examples include rotifers, glomeromycotan fungi, some arthropods and some plants; but even these exceptional examples of asexuality are uncertain because molecular analyses show that genes required for the sexual cycle are maintained in their genomes, so it may simply be that their cryptic sexual stages have not yet been observed (Billiard *et al.*, 2012).

Eventually, for the majority of fungi of the present day, karyogamy and meiosis take place and the nuclear products of meiosis are packaged into sexual spores. In many fungi sexual spores have thickened walls; that is, they are **resistant spores** that are often dormant, and formed in relatively small numbers. In some cases, the whole gametangium (the zygospores of zygomycetes would be a typical example; see Figure 3.8 in Chapter 3) develops into a resistant structure, in other cases the sexual spores (particularly ascospores) are resistant and have a period of obligate dormancy. However, in *Basidiomycota*, basidiospores are produced in large numbers and are **dispersal spores**, not dormant spores (section 3.8).

As befits its use in traditional taxonomy, there are numerous variations in sexual reproduction in fungi. The first of these variables is the presence or absence of incompatibility systems. For example, in the zygomycetes, *Mucor mucedo* is heterothallic (self-sterile), but its relative *Rhizopus sexualis* is homothallic (self-fertile).

There is then the matter of the morphology of the hyphal structures involved in the various stages and the manner in which the processes are carried out. For example, gametangia are morphologically alike in the true fungus (zygomycete) *Mucor mucedo*, but morphologically different in some of the *Oomycota* (Kingdom *Straminipila*; see section 3.10), like *Pythium*, which is an important pathogen causing damping-off of seedlings.

Similarly, the duration of the various stages of sexual reproduction may vary and some may be prolonged, for example **prolonged karyogamy** in diploid yeasts and, as indicated above, **prolonged plasmogamy** in the dikaryotic heterokaryon of *Basidiomycota*.

Hormones are probably involved in regulating sexual reproduction in most organisms, and fungi are no exception. Unfortunately, only a few of the active chemicals have been isolated from fungi; however, all major chemical classes of hormones identified in animals and plants are also known among fungi and *Oomycota* (see sections 3.4 and 5.12):

- **sterols** in *Achlya bisexualis*, female mycelium produces **antheridiol**, male produces **oogoniol** (section 3.10);
- the sesquiterpene hormone **sirenin** produced by female zoogametes of *Allomyces macrogynus* to attract male zoogametes (section 3.4);
- chemotropism to volatile precursors in the **trisporic acid** pathway that attracts heterothallic (self-sterile + and -) zygophores of *Mucor mucedo* to one another. On their own, neither strain can produce trisporic acid, but they 'converse', by exchanging a volatile precursor and collaborate in its biosynthesis (Lee & Heitman, 2014);
- **peptide pheromones** involved in yeast mating (see section 9.2 below);
- **mating type pheromones** of filamentous *Ascomycota* and *Basidiomycota* that are part of a G-protein signalling pathway (see section 9.5 below).

As in other eukaryotes, fungi have tightly regulated mechanisms that determine which haploid cells can fuse at syngamy; but fungi display a variety of life cycles and have additional possibilities for syngamy as compared to plants and animals. **Heterothallic** fungi require two compatible partners to produce sexual spores, whereas in **homothallic** fungi a single organism is capable of sexual reproduction. In fungi considered to be heterothallic, haploid selfing is prevented because they have a **mating system** that ensures syngamy can only occur between haploid cells carrying different mating type genes (Billiard *et al.*, 2011).

Mating systems (also called breeding systems) rely on nuclear genes that control progress towards meiosis in the heterokaryon established between vegetatively compatible mycelia. Basic analysis of such systems depends on making experimental confrontations between mycelia and scoring whether or not the sexual stage is completed. Such experiments test for the *phenotype* of sexual reproduction, and the pattern of its occurrence and its inheritance allow deductions about the control of sexual reproduction. A mycelium that possesses genes that prevent mating between mycelia that are genetically identical will be self-sterile; since it ensures that different mycelia must come together for a successful mating to occur and so such a system is called heterothallism.

The genes that determine mating in fungi reside at one or two specialised chromosomal regions known as the **mating type loci** (***MAT***). These genes determine haploid cell identity (conferring the cell's '*mating type*'); they enable compatible mating partners to attract each other, and they prepare cells for sexual reproduction ***after*** the fertilisation event (which is usually hyphal cell fusion). Some of these genes (and probably the ancestral ones) encode proteins belonging to a class of **transcription factor** known as the **High Mobility Group (HMG)** proteins. HMG proteins (named according to their mobility in polyacrylamide electrophoresis gels) are the largest class of non-histone proteins found in the nucleus and are often found in association with regions of active transcription in chromatin (Casselton, 2008). The transcription factors are responsible for coordinating haploid cell-type specificity ('*mating type*') when working alone, and the fate of the diploid (yeast zygote) or dikaryon (filamentous fungi) when two compatible *MAT* loci work together as a partnership after syngamy.

They can work together because they each produce **homeodomain proteins** that must form a dimer to function. Homeodomain proteins are evolutionarily conserved proteins which are present in the entire eukaryote kingdom, and most act as transcription factors and bind to DNA to control the activity of specific sets of genes involved in a very diverse range of functions. They bind to DNA in the promoter region of their target genes as complexes with other transcription factors. The specificity (in terms of ***which genes to regulate***) derives from the specific collection of transcription factors that assemble at the promoter to make up the complex (Bobola & Merabet, 2017; Vonk & Ohm, 2018). Compatible *MAT* genes produce different homeodomain proteins (generally called HD1 and HD2) that form a **heterodimer**, that is, a dimer of HD1+HD2, which is ***necessary*** for the mating to be compatible. Neither HD1+HD1 nor HD2+HD2 form functional dimers, though both are perfectly good haploid cells.

Many **heterothallic** fungi, indeed all known heterothallic *Ascomycota*, have only two mating types specified by a single locus with different 'alleles': *Neurospora crassa*, budding yeast *Saccharomyces cerevisiae* and the (basidiomycete) grass rust *Puccinia graminis* are examples. In such cases, the mating type of a culture depends on which 'allele' it has at the single mating type locus (involvement of one mating type locus gives rise to the alternative name of **unifactorial incompatibility**): successful mating only taking place between cells or mycelia that have different 'alleles' at the mating type locus. Of course, the diploid nucleus that results is heterozygous for the mating type factor, and meiosis produces equal numbers of progeny of each of the two mating types (hence yet another alternative name, **bipolar heterothallism**).

We put the word allele into quotes in the last few sentences because, although it is not evident from classical genetic analysis, one of the first things that molecular analysis revealed about the mating type factors is that the different forms of the mating type locus do not share the amount of DNA sequence homology you would expect of alleles. Their 'alleles' can be very different indeed, in some cases differing in length by thousands of base pairs. For this reason, they have been called **idiomorphs** rather than alleles. Idiomorphic structure (not allelism) is common to all fungal mating type genes that are known.

In **homothallic** (self-fertile) fungi, sexual reproduction can occur between genetically identical hyphae, but mating type factors may still be involved. **Primary homothallism** occurs in species completely lacking heterothallism, but **secondary homothallism** occurs in species that have an underlying heterothallism that is bypassed by the inclusion of two haploid nuclei of opposite mating types and from a single meiosis when spores are made; it is also called 'pseudo-homothallism' or automixis (Billiard *et al.*, 2012).

Neurospora tetrasperma, Coprinopsis bisporus and *Agaricus bisporus* are good examples. In these cases, there are more post-meiotic nuclei than spores, so the spores become binucleate and heterozygous for mating type factors. Spore germination gives rise to heterokaryotic mycelia that are, consequently, able to complete the sexual cycle alone, that is, they act like homothallic mycelia because they are heterokaryons right from the start. The presence of nuclei of the two mating types within dispersal spores ensures that the progeny spore will be able to mate with any haploid it encounters, and pseudo-homothallism will also allow mating between sibling spores (that is, spores from the same meiosis) when no other mating is available.

The terms 'homothallism' and 'heterothallism' are also used to describe reproductive phenomena in oomycetes, though the genetic basis and the evolutionary consequences are strikingly different from true fungi. In oomycetes, which are phylogenetically closer to brown algae and diatoms than to fungi, 'heterothallism' and 'homothallism' are used to describe how sexual reproduction ***is started***: heterothallic oomycetes cannot undergo gamete production and sexual reproduction unless an individual of the opposite mating type is present, while homothallic oomycetes can (Judelson, 2007). Once gametes are produced, diploid selfing and outcrossing are possible in both homothallic and heterothallic oomycetes. The terms 'homothallism' and 'heterothallism', which are often used interchangeably with the terms 'selfing' and 'outcrossing' in both oomycetes and fungi, correspond to different phenomena in these different organisms. Confusion over different usages of these terms, especially 'selfing', stems from the fact that ascomycetes have a long-lasting haploid mycelial stage, so that the 'individual' which may be indulging in 'self-fertilisation (selfing)' is the haploid mycelium which may be heterokaryotic (see discussion in Billiard *et al.*, 2012).

The yeasts *Saccharomyces cerevisiae* and *Schizosaccharomyces pombe* exemplify a different mechanism. Most strains are heterothallic with two mating types (see below), but in some strains mating occurs between progeny of a single haploid ancestor;

that is, the culture appears to be 'homothallic'. This apparent homothallism results from a recombinational process that allows a **switch**, in a few cells in the population, from one mating type to the other (see section 9.3 below for details) so that the (still heterothallic) haploid clone comes to contain cells of a different mating type.

There is one last general point to make before we describe some of the details. This is that the mating type factors were discovered phenotypically and genetically long before their molecular basis was revealed. Consequently, they were named, in some cases up to a hundred years ago, in ways that were appropriate then, but may not be appropriate now. Bipolar mating types may be +/- (plus and minus, now more often called P and M), or A and a, or a and α ('a' and 'alpha'). Where mating type is determined by two mating type factors (**tetrapolar heterothallism**) they are called A and B, one encodes the homeodomain proteins, and the other encodes pheromones and pheromone receptors; ***but*** which locus encodes which gene products varies with the species, ***because*** the genes were named historically as they were identified without knowledge of their molecular function.

9.2 MATING IN BUDDING YEAST

The life cycle of the yeast *Saccharomyces cerevisiae* features an alternation of a haploid phase with a true diploid phase (Figure 9.1), and in this respect differs from filamentous *Ascomycota* in which the growth phase after **anastomosis** (syngamy) is a heterokaryon.

There are two mating types: haploid yeast cells may be of mating type 'a' or 'α' (alpha). Karyogamy (nuclear fusion) follows the fusion of cells of opposite mating type and then the next daughter cell that is budded off contains a diploid nucleus. Most natural yeast populations are diploid because the haploid meiotic products mate while they are still close together immediately after the meiosis is completed. Diploid cells reproduce vegetatively by mitosis and budding until specific environmental conditions (deficiency in nitrogen and carbohydrate, but well aerated and with acetate or other carbon sources which favour the glyoxylate shunt) induce sporulation. When that happens, the entire cell becomes an ascus mother cell; meiosis occurs and haploid ascospores are produced. Ascospore germination re-establishes the haploid phase, which is itself maintained by mitosis and budding if the spores are separated from one another (by laboratory experiment, or by some disturbance in nature) to prevent immediate mating.

Mating type factors of yeast specify peptide hormones; these are called pheromones (the term originally applied to mate-attracting hormones of insects and mammals) and there are both pheromone α- and a-factors (Figure 9.2), and corresponding receptors specific for each pheromone. Pheromones organise the mating process; they have no effect on cells of the same mating type or on diploids but their binding to pheromone receptors on the surface of cells of opposite mating type (Figure 9.1) act through GTP-binding proteins to alter metabolism and:

- cause recipient cells to produce an agglutinin that enables cells of opposite mating type to adhere;
- stop growth in the G1 stage of the cell cycle;

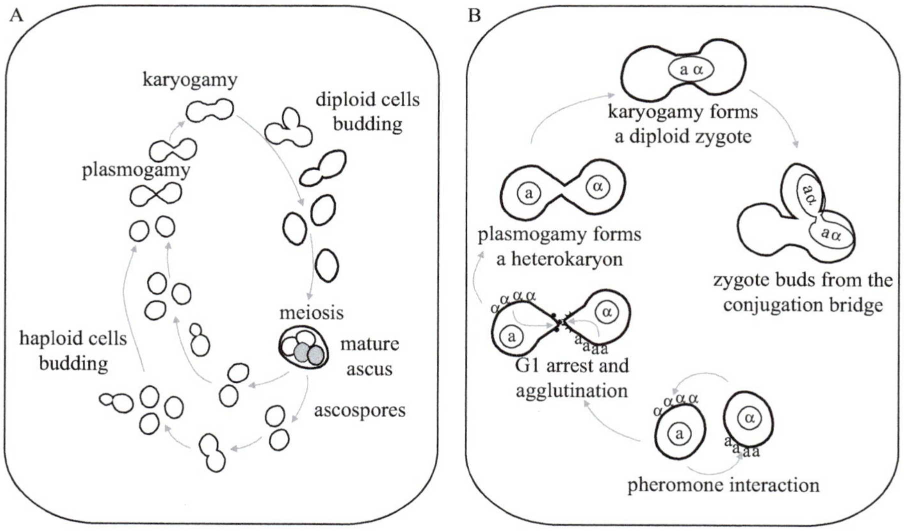

Figure 9.1. (A) Life cycle (left-hand panel) and mating process of the yeast *Saccharomyces cerevisiae*. Yeast can reproduce asexually by budding. Haploid cells of different mating types fuse to form dumbbell-shaped zygotes, which can themselves bud to establish a diploid clone. Well-nourished diploid cells, which are exposed to starvation conditions, enter meiosis, forming a four-spored ascus. Ascospores germinate by budding. In the laboratory, ascospores can be separated to form haploid clones but in nature ascospores usually mate immediately, so the haploid phase is greatly reduced. (B) depicts pheromone interaction, agglutination and the mating process in a little more detail. Modified from Chapter 2 in Moore & Novak Frazer (2002).

- change wall structure to alter the shape of the cell into elongated projections.

Fusion eventually occurs between the projections.

In *Saccharomyces cerevisiae*, a relatively small region of the genome, less than a thousand base pairs, codes for a complex genetic locus called *MAT*. The *MAT* locus lies in the middle of the right arm of chromosome III, about 100 kb from both the centromere and the telomere. The two mating type alleles, *MATα* and *MATa*, differ by about 700 bp of sequences (designated Yα and Ya respectively) that contain the promoters and most of the open reading frames for proteins that regulate many aspects of the cell's sexual activity (for review, see Haber, 2007). The *MAT* locus is divided into five regions (called W, X, Y, Z1 and Z2) on the basis of sequences that are shared between *MAT* and the two unexpressed (cryptic) copies of mating type sequences that serve as donors during the recombinational process that allows a *MATa* cell to **switch** to *MATα* or vice versa (see section 9.3).

The *MATa* locus encodes a1 and a2 polypeptides, the messengers for which are transcribed in opposite directions (Figure 9.3), and *MATα* encodes polypeptides α1 and α2, all of which are transcription factors responsible for orchestrating both haploid cell-type specificity (a or α) and the fate of the diploid zygote (a or α). Two are homeodomain proteins of the HD1 and HD2 class that form a heterodimer, a1/α2, so heterozygosity at *MAT* is a sign of diploidy and eligibility to sporulate; even partial diploids carrying *MATa*/*MATα* will attempt to sporulate. Analysis of the *MATa1*, *MATα1* and *MATα2* genes provided one of the earliest models of cell-type specification by transcriptional activators and repressors. *MATα* encodes two proteins, *MATα1* and *MATα2*. *MATα1*, in conjunction with a constitutively expressed protein called Mcm1, activates a set of α-specific genes that includes genes encoding the mating pheromone (α-factor), and Ste2, a transmembrane receptor of the opposite mating pheromone (a-factor). As mentioned above, these mating pheromones stop the cell cycle in the G1 stage, thereby promoting conjugation and ensuring that the zygote contains two unreplicated nuclei.

MATα2 encodes a homeodomain protein that acts with the constitutive Mcm1 to form a repressor of a-specific genes including those that produce a-factor pheromone (*MFa1* and *MFa2*) and the Ste3 transmembrane receptor protein that detects the presence of α-factor in the medium. Repression also requires the action of two other proteins, Tup1 and Ssn6. The *MATa* locus has two open reading frames *MATa1* and *MATa2* that are divergently transcribed (that is, they are transcribed in opposite directions). Although *MATa2* is well conserved evolutionarily, it has yet to be assigned a biological function. *MATa2* and *MATα2* share part of the same protein sequence and the evolutionary preservation of *MATα2* may account for conservation of *MATa2*, despite its lack of apparent function. If *MAT* is entirely deleted, haploid cells mate identically to MATa cells (that is, the MATa phenotype is the default phenotype), because a-specific genes are ***constitutively expressed in the absence of* Matα2 *and α-specific genes are not transcribed in the absence of* Matα-1**. Although MATa is not required for a-mating, the *MATa1* gene ***is required***, to work with *MATα2* to repress haploid-specific gene expression, and this includes a gene called *RME1*, which is itself a repressor of meiosis and sporulation (RME1 means repressor of meiosis-1). Repression of this repressor permits expression of diploid-specific attributes (Haber, 2012).

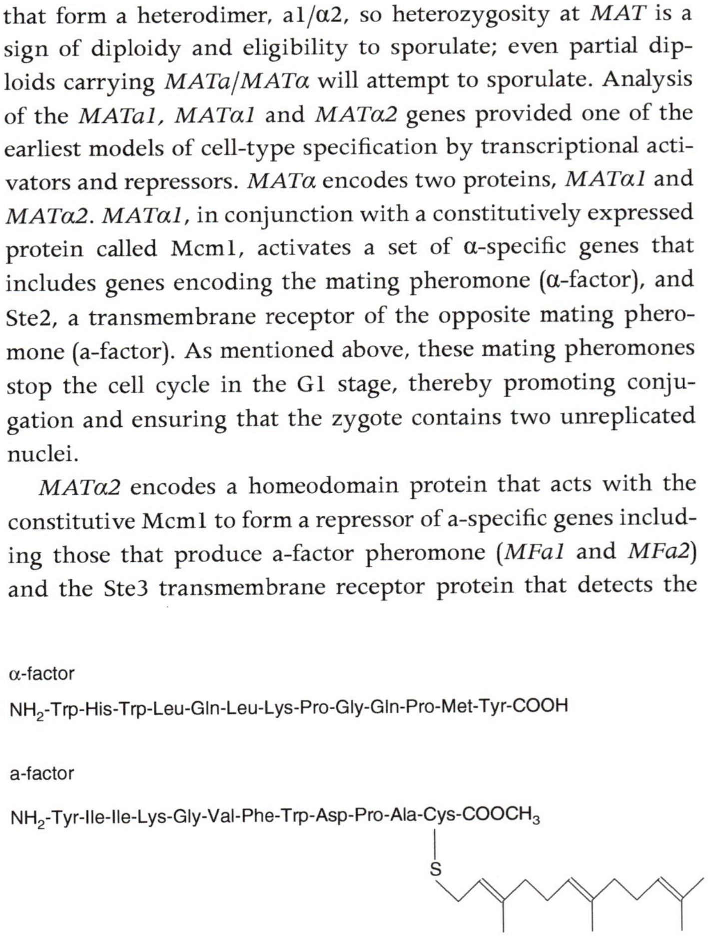

Figure 9.2. Simplified chemical structures of yeast pheromones.

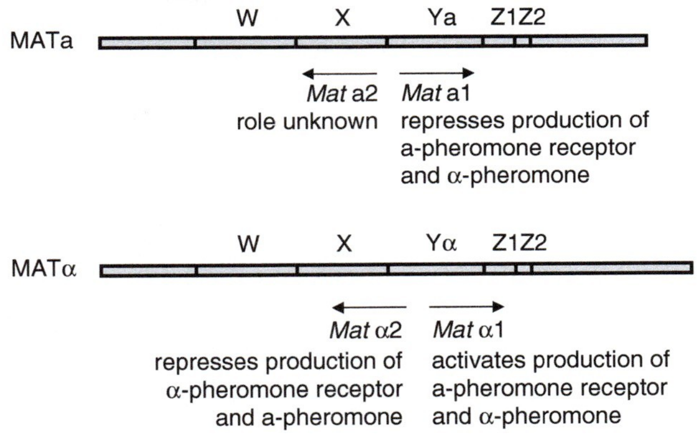

Figure 9.3. Functional domains in mating type factors of *Saccharomyces cerevisiae*. Region Y is the location of the mating type idiomorphs, which have very little homology with each other. Ya is 642 bp long, Yα is 747 bp long. Regions W, X, and Z1 and Z2 are homologous terminal regions. The arrows indicate direction of transcription and the legends beneath the arrows indicate functions of the gene products. In *Saccharomyces cerevisiae* of mating type a, a general transcription activator is responsible for production of a-pheromone and the membrane-bound α-pheromone receptor. In a/α diploids, the MATa1/MATα2 heterodimer protein activates meiotic and sporulation functions and represses haploid functions (turning off α-specific functions by repressing MATα1, a-specific functions being repressed by MATα2 alone). Modified from Chapter 2 in Moore & Novak Frazer (2002).

9.3 MATING TYPE SWITCHING IN BUDDING YEAST

Homothallic yeast cells can, remarkably, switch their mating type as often as every generation by a highly regulated, site-specific homologous recombination event that replaces one *MAT* allele with a different DNA sequence that encodes the opposite *MAT* allele. This replacement process involves the participation of two intact but unexpressed copies of mating type information at the heterochromatic loci, *HMLα* and *HMRa*, which are located at opposite ends (left and right, respectively) of the same chromosome that encodes MAT.

Saccharomyces cerevisiae is heterothallic but a clone of haploid cells of the same mating type frequently sporulates, and there will be equal numbers of a and α cells among the progeny. This results from mating type switching controlled by the gene *HO* (HOmothallic) that exists in two allelic forms (dominant *HO* and recessive *ho*) and encodes an endonuclease. On either side of the *MAT* locus, and on the same chromosome, there are ***silent*** storage loci for each mating type, called *HML* and *HMR*. The *HO/ho* endonuclease creates a double-strand break at the *MAT* locus that initiates switching of information, by a homologous recombination event between the two parts of the same chromosome, at the *MAT* locus with that at either *HML* or *HMR* (Figure 9.4). The phenotypic switch from MATα to MATa is quite rapid; within a single-cell division. Again, *MAT* heterozygosity plays a key role in controlling the switching of mating type genes. Switching occurs in homothallic strains expressing the HO endonuclease gene, *HO*, which can switch from MATa to MATα and vice versa, but once cells of opposite mating type conjugate to form a diploid, HO expression is repressed by the a1-α2 repressor. The Mata1, Matα-1 and Matα2 transcription regulators that organise haploid cell specificity are quite rapidly turned over, being degraded by ubiquitin-mediated proteolysis by the proteasome; in contrast, the a1-α2 co-repressor is much more stable.

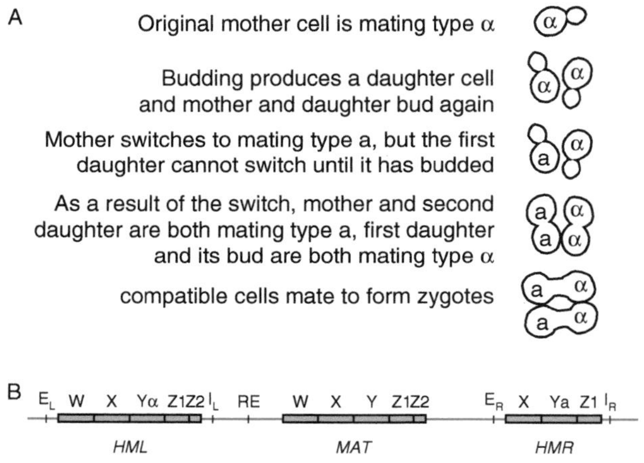

Figure 9.4. (A) Pattern of mating type switching in *Saccharomyces cerevisiae* showing the consequences of a mating type switch in one mother cell. (B) The three loci involved in mating type switching, *HML*, *MAT* and *HMR*, are located on the same chromosome (not drawn to scale). *HML* is about 180 kb from *MAT*, and *HMR* about 120 kb from *MAT*; the centromere is located between RE and the *MAT* locus. A double-strand break at the *MAT* locus, caused by the HO endonuclease, initiates a gene conversion event that replaces the Y region of the *MAT* locus with Y sequences from one of the storage loci. HML and HMR contain complete copies of the mating type genes but are not expressed because they have a repressed heterochromatin structure imposed by the E and I silencer sequences. *HML* shares more of the *MAT* sequences (W, X, Z1 and Z2) than does *HMR*. RE is a recombination enhancer that controls preferential recombination between *MATa* and *HML*, or between *MATα* and *HMR*. Modified from Chapter 2 in Moore & Novak Frazer (2002).

Since yeasts can live in very small habitats, like flower nectaries and surfaces of individual fruits, yeast populations can be very isolated from one another in nature, so the rare mating type switching will give isolated populations the opportunity to undergo sexual reproduction; this is presumably its selective advantage. Mating type switching occurs about once in 10^5 divisions in cultures carrying allele *ho*, whereas in strains carrying *HO* the switch occurs at every cell division. However, there is an asymmetry in the cell division in that a new daughter bud is not able to switch mating types until it has itself budded. In *Saccharomyces cerevisiae*, this is achieved by actively transporting the mRNA of a gene called *Ash1* into the budding daughter cell. This mRNA encodes an inhibitor of the HO endonuclease. Consequently, immediately after each division switching by the daughter is inhibited and only the mother cell is switchable. This means that even if there is only one cell to start with, a single division cycle will produce two cells of opposite mating type.

Saccharomyces cerevisiae has evolved an elaborate set of mechanisms to enable cells to switch their mating types. MAT switching depends on four phenomena:

1. Presence of two unexpressed (silenced or cryptic) copies of mating type sequences that act as donors during MAT switching. This implies that there is a mechanism which influences chromatin structure to maintain it in a silent configuration. The mechanism involves silencer sequences surrounding *HML* and *HMR* that interact, directly or indirectly, with several protein factors to repress the transcription of these genes. Among these are four **s**ilent **i**nformation **r**egulator (**Sir**) proteins, a set of silencer-binding proteins, histone proteins, the Rap1 protein, as well as several chromatin modifiers. Together, these create short regions (about 3 kb) of heterochromatin, in which the DNA sequences of *HML* and *HMR* are in a highly ordered nucleosome structure, known as **heterochromatin**, which is not transcribed by either RNA polymerase II or RNA polymerase III, and is resistant to cleavage by several endonucleases, including the HO endonuclease.
2. Programmed creation of a site-specific double-strand break at *MAT* that results in the replacement of Ya or Yα sequences. HO is a site-specific endonuclease that recognises a 24 bp sequence that spans the *MAT-Y/Z* border. Haploid yeast has three possible targets for HO: the *MAT* locus, *HMLα* and *HMRa*, but only the *MAT* locus is accessible under normal conditions because of gene silencing of *HML* and *HMR*. Normally the *HO* gene is tightly regulated and is expressed only in haploid mother cells and only at the G1

stage of the cell cycle. The result in mother cells is a single programmed double-strand break at the *MAT* locus prior to the initiation of DNA replication.

3. A cell lineage rule ensuring that any cell that has previously divided once can switch *MAT*, while new daughter cells cannot (so only half of the cells in a population switch at any one time). The way this works is that a germinating haploid spore grows, produces a bud, and divides without changing mating type. Then, in the next cell division cycle, the older mother cell and its ***second*** daughter change mating type while the first daughter buds and divides without changing. This ensures that there will be cells of both mating types in close proximity; they readily conjugate, forming *MATa*/*MATα* diploids in which the *HO* endonuclease gene is turned off so that further mating type switching is repressed. Control of this lineage pattern depends on expression of the *HO* gene being restricted to mother cells that have divided by the Swi5 transcription factor, which is localised to the mother cell nucleus and not that of her daughter. Lack of Swi5 expression in daughters is caused by the Ash1 repressor protein that is found only in the daughter cell. Ash1 may directly repress SWI5 gene transcription, thus restricting HO expression to the mother cell in the next G1 stage of the cell cycle. *Ash1* mRNA is localised to the daughter prior to cell division by a myosin-like protein, actin, cargo-binding proteins, translation repressors and a nuclear localising protein. Since *Ash1* mRNA localisation was first discovered, many other mRNAs have been found to show similar localisation in *Saccharomyces*.
4. A mechanism regulates the ***selective*** use of the two donors (called **donor preference**); study of which has yielded much of what we know about double-strand break-induced **recombination** during eukaryote mitosis. Switching one mating type to the other involves the replacement at the *MAT* locus of Ya or Yα by a **gene conversion** induced by the site-specific double-strand break at *MAT* caused by HO endonuclease; it is a DNA damage repair process. The process is highly directional, in that the sequences at *MAT* are replaced by ***copying*** new sequences from either *HMLα* **or** *HMRa* (whichever is the alternate to the resident sequence), while the two donor gene loci remain unchanged. HO endonuclease cannot cleave its recognition sequence at either HML or HMR, as these sites are occluded by nucleosomes in silenced DNA and this prevents crossing over. Any DNA single-strand exchanges (Holliday junctions) that might otherwise become crossovers are removed by two helicase enzymes, Sgs1 and Mph1, working with their partner proteins. Thus the (resident) *MAT* locus is cleaved and becomes the ***recipient*** in this gene conversion process. Overall, following HO cleavage of *MATa*, the end of the broken DNA molecules is excised in a 5′ to 3′ direction, creating a 3′-ended single-strand DNA tail on which assembles a filament of the **Rad51** recombinase protein. Rad51 is essential for repairing damaged DNA and is highly conserved in most eukaryotes; this family of proteins interact with several other single-strand DNA-binding proteins to form a helical nucleoprotein filament on the DNA. This protein–DNA complex engages in a search for a ***homologous*** DNA sequence (since we started with *MATa*, in this case it would search for *HMLα*) to effect the repair. This search culminates in strand exchange in which the single-stranded DNA base pairs with the complementary sequence of the donor, creating a displacement loop (D-loop). The 3′ end of the invading strand is then used as a primer to initiate copying of one strand of the donor locus, and the newly copied strand is displaced until it can anneal with homologous sequences on the opposite end of the double-strand break. The 3′-ended nonhomologous tail is clipped off and the new 3′ end is used to prime a second strand of DNA synthesis, completing the replacement of *MATa* with *MATα*.

These important mechanisms (reviewed by Haber, 2012) are more examples of how research on fungi has informed our fundamental knowledge of the molecular genetics of eukaryotes.

The switching system of the related yeast, *Kluyveromyces lactis*, is similar to that of *Saccharomyces cerevisiae* to some extent, and includes sequences recognisably similar to *MATa1*, *MATα-1* and *MATα2*, but the shared flanking sequences are not closely related to the W, X, or Z1/Z2 sequences of *Saccharomyces cerevisiae* and there is **no functional *HO* gene**. While the *Kluyveromyces lactis HMR* has a1 and a2; *HML* includes a novel gene, *α3*, in addition to *α1* and *α2*. Both *HML* and *HMR* are silenced by a mechanism dependent on Sir-proteins as in *Saccharomyces cerevisiae*. Switching is dependent on a protein (Mts1) that is the homologue of the *Saccharomyces cerevisiae* RME1 repressor; but in *Kluyveromyces lactis* it is required to ***activate*** switching and is turned off in *MATa*/*MATα* cells by an a1-α2 repressor. The major difference between the two yeasts, at least for *MATα* to *MATa* switching, is that it is the *α3* gene, which is a **transposable element** that can excise from the DNA as a circle and somehow catalyse switching (Barsoum *et al.*, 2010). *MATa* to *MATα* switching seems to be under the control of a different transposable element (Haber, 2012).

Mating type switching also occurs in the more distantly related fission yeast *Schizosaccharomyces pombe*, but the switching system differs in almost every detail from *Saccharomyces cerevisiae*. Mating type switching in *Schizosaccharomyces pombe* involves replacing genetic information at the expressed *mat1* locus with sequences copied from ***one*** of ***two*** silent donor loci, *mat2-P* or *mat3-M*, which are close together and located only a short distance away from *mat1* on the same chromosome arm in a 20 kb length of heterochromatin. There is no *HO*-like enzyme; instead, a persistent single-strand break is created in the DNA at mat1, which is converted to a double-strand break when cells enter S-phase. Only one of the two daughter cells can switch. Donor selection is dictated by cell type: *mat2* is the preferred donor in M cells, and *mat3* is the preferred donor in P cells. Donor preference involves major changes in chromatin modification and structure and the silencing system is different from that in *Saccharomyces cerevisiae* (Jia *et al.*, 2004).

Yeasts in a third clade of the *Saccharomycotina*, the **methylotrophs**, have a simpler two-locus switching system based on reversible inversion of a section of chromosome with *MATa* genes at one end and *MATα* genes at the other end. In *Hansenula polymorpha*, the invertible region, which is 19 kb long, lies beside a centromere so that, depending on the orientation, either *MATa* or *MATα* is silenced by centromeric heterochromatin. In *Pichia pastoris*, the orientation of a 138 kb invertible region puts either *MATa* or *MATα* beside a telomere where heterochromatin silences *MATa2* or *MATα2*. Both species are homothallic, and inversion of their *MAT* regions can be induced by crossing two strains of the same mating type. The three-locus *Saccharomyces cerevisiae* system may have been derived from mating type switching by chromosomal inversion as seen in methylotrophic yeasts; the increased complexity of the *Saccharomyces cerevisiae* switching apparatus, with three loci, donor bias and cell lineage tracking, resulting from continuous evolutionary selection to increase sporulation ability in young colonies (Hanson *et al.*, 2014).

Mating type switching, which is often referred to as stochastic (meaning randomised) mating type determination, has evolved independently in a number of organisms other than the yeasts; for example, in the ciliates, *Tetrahymena thermophila*, *Paramecium* spp. and *Euplotes crassus*. There is also evidence for some degree of randomised mating type identity during vegetative growth in the green algae *Chlamydomonas monoica* and *Closterium ehrenbergii*, and the dinoflagellate *Gymnodinium catenatum*, although the switching mechanism in these species is not known. Mating types in filamentous fungi tend to be far more stable. Oddly enough, switching does not occur in any of the best-studied organisms like *Neurospora*, *Aspergillus* or *Podospora*, but has been claimed in *Chromocrea spinulosa*, *Sclerotinia trifoliorum*, *Glomerella cingulata* and *Ophiostoma ulmi* (Tsui *et al.*, 2013), and the basidiomycete *Agrocybe aegerita* (Hadjivasiliou *et al.* 2016).

9.4 MATING TYPES OF *NEUROSPORA*

Species of *Neurospora* exhibit four different mating strategies:

- **bipolar heterothallism** with mating types *A* and *a* (in *Neurospora crassa*, *Neurospora sitophila*, *Neurospora intermedia* and *Neurospora discreta*) but mating type genes are present in a single copy per genome (unlike *Saccharomyces cerevisiae*);
- **secondary homothallism** (in *Neurospora tetrasperma*) through the production of asci containing four ascospores each containing compatible nuclei;
- **primary homothallism** in which each haploid genome carries genetic information of both mating types (*Neurospora terricola*, *Neurospora pannonica*);
- **primary homothallism**, but in which genetic information for only one mating type can be detected (for example, *Neurospora africana* possesses only an *A* idiomorph which shows 88% homology with the *A* idiomorph of *Neurospora crassa*).

Species that show primary homothallism form linear eight-spored asci (**octads**) in which all progeny are self-fertile. In the bipolar heterothallic species, strains of both mating types develop 'female' structures (protoperithecia and their receptive hyphae, the trichogynes) under nitrogen starvation, as well as asexual spores (macroconidia or microconidia). A trichogyne of one mating type is attracted to and fuses with a cell of opposite mating type (a macroconidium, microconidium or hyphal fragment), which serves as the 'male' in a sexual cross. After fusion between the trichogyne and conidium, the 'male' nucleus is transported to the ascogonium at the base of the protoperithecium where several **mitotic** divisions occur. Migration of (the still haploid) nuclei into the *Neurospora crassa* ascogonium (see Figure 9.5) depends on mating type gene function and after their arrival there must be some mechanism to sort nuclei, to ensure that the subsequent meiosis only involves one *a* and one *A* nucleus, because mating type *always* segregates 1:1 in the progeny. It is likely that transient dikaryosis in the crozier involves a nuclear recognition mechanism. However, crozier abortion occurs regularly even in normal ascosporogenesis, so an alternative process might be that croziers that do not contain one *a* and one *A* nucleus are aborted.

Mating is followed by formation of perithecia, within which as many as 200 asci are formed. Each ascus contains the products of a single meiosis. In many filamentous ascomycetes, a post-meiotic mitotic event, before ascospore formation, results in each ascus containing an **octad** comprised of four pairs of sister ascospores.

The first mating type genes to be cloned and sequenced in filamentous fungi were the *A* and *a* 'alleles' of *Neurospora crassa* (Figure 9.6); these are idiomorphs of the *mat* locus. In all heterothallic filamentous ascomycetes examined to date, the *mat* locus is the sole determinant of mating type, and sexual reproduction is regulated by alternative *mat* A or *mat a* idiomorphs at this locus, which lack sequence similarity and encode different transcriptional regulators.

In all the known *Neurospora A* and *a* idiomorphs the flanking regions are conserved, the region on the centromere side contains species-specific and/or mating type-specific DNA sequences. Immediately adjacent to these segments are regions that are very different between species. Next to these, species-variable regions are the idiomorphs themselves. These are highly conserved between species but are completely dissimilar between the two mating types within the species. These are

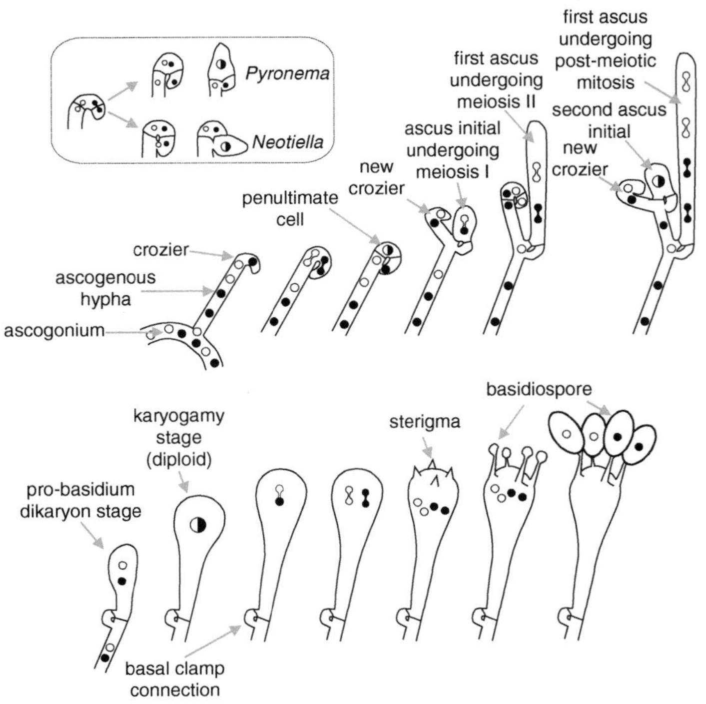

Figure 9.5. Meiosis and sporulation in *Ascomycota* and *Basidiomycota*. The major panel of diagrams at the top shows ascus formation. Hyphal fusion or similar mating between male and female structures results in nuclei moving from the male into the female to form an ascogonium in which male and female nuclei may pair but do not fuse (transient dikaryon stage). Ascogenous hyphae grow from the ascogonium. Most cells in these hyphae are dikaryotic, containing one maternal and one paternal nucleus, the pairs of nuclei undergoing conjugate divisions as the hypha extends. In typical development, the ascogenous hypha bends over to form a crozier. The two nuclei in the hooked cell undergo conjugate mitosis and then two septa are formed, creating three cells. The cell at the bend of the crozier is binucleate but the other two cells are uninucleate. Generally, the binucleate cell becomes the ascus mother cell, in which karyogamy takes place. In the young ascus meiosis results in four haploid daughter nuclei, each of which divides by mitosis to form the eight ascospore nuclei (the octad). The boxed inset shows that karyogamy may occur in the penultimate cell of the crozier (*Pyronema*-type) or the terminal and stalk cell nuclei might fuse (*Neotiella*-type). The panel of diagrams at the bottom shows basidium formation in a 'classic' mushroom fungus. The basidium arises as the terminal cell of a dikaryotic hyphal branch that inflates and undergoes karyogamy and meiosis. At the end of the meiotic division four outgrowths (sterigmata) emerge from the basidial apex and inflation of each sterigma tip produces the basidiospore (which is an exospore, produced outside the meiocyte in contrast to the endospores of ascomycetes). Nuclei then migrate from the basidium into the newly formed basidiospores. Mitosis may take place within the basidiospores before they are discharged. Comparison of these diagrams indicates how tempting it is to suggest some evolutionary relationship between crozier formation and the early stages of basidium and clamp connection formation. Modified from Chapter 2 in Moore & Novak Fraze (2002); for photographic illustrations see Raju (2008).

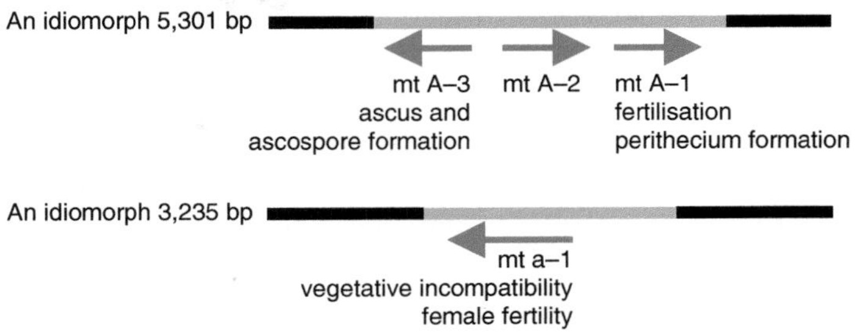

Figure 9.6. Functional regions of mating type factors of *Neurospora crassa*. The arrows indicate direction of transcription and the legends beneath the arrows indicate functions of the gene products. The black bars represent the conserved DNA sequences either side of the idiomorphs, which are shown as lines. These diagrams are oriented so that the centromere is on the left; consequently, the centromere-distal sequence is on the right. Modified from Chapter 2 in Moore & Novak Frazer (2002).

then followed by a 'mating type common region' of 57–69 bp, which separates an idiomorph from its nearby variable region and is very similar between species and between the two mating types.

The A idiomorph is 5,301 bp in length and gives rise to at least three transcripts (mat A1, A2 and A3), the first two being transcribed in the same direction (Figure 9.6). The 85 amino acids at the N-terminal region of the mating type A product are the minimum required for expression of female fertility. The region from position 1 to 111 determines the vegetative incompatibility activity of the mating type locus, and amino acids from position 1 to 227 are required for male-mating activity. Mating type-specific mRNA is expressed constitutively in vegetative cultures, and continues to be expressed after mating, both before and after fertilisation. Transcript mat A1 is very similar to MATα1 of *Saccharomyces cerevisiae* and is the main regulator of sexual development in mat A strains being essential for fertilisation and sporophore formation.

The other two transcripts, mat A2 and mat A3, increase fertility and are essential in events after fertilisation, including ascus and ascospore formation, but are not essential for sexual development, which is controlled by *mat A1*. The mat A3 transcript has DNA-binding ability and might function as a transcription-regulating factor. The function of the mat A2 polypeptide is not known. Its sequence contains motifs that occur in transcription activator proteins, but there are no obvious relatives in sequence databases other than the *Podospora* mating type homologue, SMR1.

The a idiomorph of *Neurospora crassa* has a single ORF as the major mating regulator in mat a strains, which is 3,235 base pairs long and gives rise to a single transcript (called mat a1) which encodes one polypeptide of 288 amino acids with DNA-binding activity. The only gene essential to mating in the a idiomorph is *mat a1*, and it has a role in the dikaryon stage. Amino acids 216–220 of the mat a1 polypeptide act in vegetative incompatibility while the region with DNA-binding activity is responsible for the mating function, implying that vegetative incompatibility and mating work through different mechanisms. The DNA sequences bound by *mat a1* centre on CAAAG sequences, like 'high mobility group' (HMG) proteins that bind in the minor groove of the DNA helix and introduce a bend in the DNA molecule. The DNA-binding targets differ in different developmental stages, and the specificity may result from the *mat* a1 protein interacting with other unidentified protein factors. Mutations in either *mat a1* or *mat A1* cause mating defects.

Neurospora crassa trichogynes respond to pheromones by bending towards the pheromone source in a mating type-specific way. Mating type mutants do not orient their growth towards pheromones.

The pheromone receptor gene expressed in mat A cells, pre-1, is required for mat A trichogynes to recognise and fuse with mat a cells. The ccg-4 and mfa-1 pheromone genes are necessary for chemotropic attraction and 'male' fertility in *mat A and mat a strains, respectively, so it is thou*ght that pre-1 binds to mfa-1, while pre-2 recognises ccg-4. DNA sequences of fungal pheromone receptors predict a product with seven transmembrane segments with the ability to interact with a heterotrimeric G-protein linked to a protein kinase cascade. Transcription of pheromone receptor genes is regulated by the mating type factors (in *Basidiomycota*, pheromone receptors are products of a mating type locus, see below). A G-protein encoding gene with mating function, called *gna-1*, has been characterised in *Neurospora crassa*. Female fertility is lost if *gna-1* is disrupted. Female infertility also results if the *gna-1* homologue in *Cryphonectria parasitica*, which is called *cpg-1*, is disrupted. Thi suggests that this G-protein is a component of a female pheromone response pathway that has been conserved in these filamentous ascomycetes. Pheromones and receptors are important for initial recognition of mates, but additional determinants, such as the presence of *mat A* and *mat a* genes in two different nuclei, are required for nuclear fusion, meiosis, and ascospore production (Kim *et al.* 2012).

Mating type genes evolve rapidly in *Neurospora*; the rapid divergence being due to either adaptive evolution or lack of selective constraints, depending on the reproductive mode of the taxa considered (Wik *et al.*, 2008). Various mating type genes may also have alternative or additional functions during vegetative development of the mycelium. Mating type (*mat*) genes were increasingly expressed as the mycelium matured. This applies to both mating types, though *mat A* more frequently exhibited a higher expression level than *mat a*, and pheromone and receptor genes were expressed in strains of both mating types in all development stages (Wang *et al.*, 2012). Genes that respond to illumination showed similar expression timing during vegetative development in this study, so the *mat* genes may be part of the extremely complex light-controlled gene regulatory network in *Neurospora crassa* (Wu *et al.*, 2014).

9.5 MATING TYPES IN *BASIDIOMYCOTA*

The breeding system in *Basidiomycota* relies on two *MAT* loci. One locus encodes tightly linked pheromones and pheromone receptors (now called the ***P/R*** locus), and the other encodes homeodomain-type transcription factors (called the ***HD*** locus), which determine events following syngamy. In basidiomycetes, as in other filamentous fungi, **syngamy** means '**hyphal fusion** or ***anastomosis***' as described in section 5.15, above). For a successful mating that completes the sexual cycle and produces progeny spores, the two haploid hyphae that fuse must differ at both *MAT* loci.

When the two *MAT* loci are on different chromosomes, meiosis can generate four mating types among the haploid progeny because the meiotic nucleus must be heterozygous at both *MAT* loci. The occurrence of four progeny mating types defines this as a **tetrapolar breeding system**. Some basidiomycetes have a **bipolar breeding system** instead, which is controlled by a single *MAT* locus because either the *P/R* and *HD* loci are closely linked on the same chromosome and segregate together, or because one has lost its function in determining mating type. About 65% of the species in the *Agaricomycotina* (see section 3.8) are tetrapolar, whereas *Ustilaginomycotina* and the *Pucciniomycotina* are predominantly bipolar. *Agaricomycetes*, which includes most fungi that produce conspicuous sporophores, have evolved a system to increase the diversity of 'alleles' at both *MAT* loci, in some cases yielding species with hundreds or thousands of possible mating types. Tetrapolar breeding systems, and especially this great diversity in *MAT* 'alleles', are believed to be adaptations to promote outbreeding in organisms that, for the most part, rely on air currents to distribute their spores (section 9.6).

The *P/R* system of *Basidiomycota* that governs hyphal fusion depends on small (10- to 15-amino acid) lipopeptide pheromones, which are made from precursor polypeptides containing 35 to 40 amino acids by post-translation modifications at both the N- and C-termini (Raudaskoski & Kothe, 2010; Raudaskoski, 2015). All active basidiomycete mating pheromones isolated so far are hydrophobic diffusible lipopeptides that undergo farnesylation at the C-terminal cysteine residue of a CAAX motif ('C' is cysteine, 'A' an aliphatic amino acid and 'X' is any residue) and amino-terminal processing. These diffusible pheromones are received by pheromone receptors, located in the plasma membrane with seven transmembrane domains, which are coupled to G-protein signal transduction cascades. This has been investigated in greatest detail in *Ustilago maydis*, where two interconnected signalling pathways, one involving cAMP-dependent protein kinase and the other a MAP kinase (section 5.12, above). The two pathways converge to a HMG-transcription factor (section 9.1, above) called **pheromone response factor** (**Prf1**), which recognises and binds to the specific pheromone response motifs located in the regulatory regions of the genes that the pheromone induces. Although the protein kinase cascade pathways are conserved over broad evolutionary distances, the transcription factors that ultimately activate or repress the specific target genes are often species specific (Coelho *et al.*, 2017).

This way of controlling mating fusion is similar to the a- and α-factor P/R system of ascomycetes and is thought to predate the separation of Asco- and *Basidiomycota* clades. It is feasible that what is common in this shared *P/R* system contributes to the information interchange that takes place between the two fusing hyphal cells involved in fusion we described in section 5.15, above. However, there ***are*** important differences between ascomycetes and basidiomycetes, in that only homologues of the a-factor pheromone and Ste3 pheromone receptor of the ascomycete system exist in basidiomycetes. Identity-sensing specificity in basidiomycetes is done by allelic variants of the same type of genes, rather than depending on the 'a-factor/Ste3' ***plus*** 'α-factor/Ste2' coupled sensing system characteristic of the ascomycetes (Casselton & Olesnicky, 1998; Kües, 2015).

Mating (syngamy or fusion) is initiated by a reciprocal exchange of pheromones recognised by matching pheromone receptor variants in both mating types, and thus the two mycelia involved need to carry different alleles of the pheromone and receptor genes at the *P/R* locus. Many additional pheromone receptor-like genes have been identified in *Agaricomycetes* that are non-mating type-specific receptors and are not sufficient to induce mating-specific development. Many of these non-mating type-specific receptors are located close to the *P/R* locus, while others are unlinked. Three distinctive features of these ***receptors*** are that: (i) they have longer C-terminal cytoplasmic regions, (ii) they lack pheromone genes in close association, and (iii) they show lower levels of intraspecific polymorphism. Non-mating-type-specific receptors have been found in *Coprinopsis cinerea, Phanerochaete chrysosporium, Laccaria bicolor, Postia placenta* and several polypores (Coelho *et al.*, 2017). Possible functions include a role in monokaryotic fruiting or in communication within the vegetative mycelium.

The ***HD*** **locus** encodes a pair of homeodomain proteins, HD1 and HD2. Their open reading frames are normally adjacent and divergently transcribed. Together these provide what has been called the **second compatibility checkpoint** (Coelho *et al.*, 2017). Compatibility of the genetic components at the *HD* locus is required to produce viable progeny from basidiomycete matings. What happens after the P/R system has arranged successful hyphal fusion (syngamy) depends upon the formation of functional heterodimeric transcription factors in which dimerisation is limited to HD1 and HD2 proteins that originate from two haploid nuclei that have different alleles at the HD locus. This requirement ensures that active heterodimers are only formed in **dikaryotic hyphal cells** and do not arise in the haploid (**monokaryotic**) hyphal cells of either parent.

Dimerisation between HD1 and HD2 proteins involve polar-to-hydrophobic interactions with allele-variable cohesive contact interfaces contributing to binding affinity. HD1 and HD2 proteins from the same allele are unable to form heterodimers because mismatches in their amino acid sequences prevent cohesion in the monokaryon. The interfaces involved in these HD1–HD2 interactions that determine allele specificity are located on the N-terminal side of the DNA-binding motifs. This is the region that is highly variable between different alleles of both HD1 and HD2 open reading frames. Once formed, the functional heterodimers will bind to promoter sequences of their target genes leading to the induction of the specific sets of genes involved in subsequent differentiation (Bobola & Merabet, 2017; Vonk & Ohm, 2018). This differentiation varies according to the lifestyles of different fungi and includes:

- a morphological switch from yeast-like cells to filamentous growth and development of pathogenicity, as in *Ustilago maydis* and other smuts (Pérez-Martín & de Sena-Tomás, 2011);
- vigorous vegetative growth of the dikaryon followed by the ability to form the sporophore in the mushrooms and related fungi (Kruzel & Hull, 2010; Kües, 2015; Coelho *et al.*, 2017).

The sequences around the *HD* locus are generally conserved in *Agaricomycetes*. Two genes usually border this locus: one encodes a **m**itochondrial **i**ntermediate **p**eptidase (***MIP***), and the other is known as a β-flanking gene (***β-fg***) for an unknown protein. Although this configuration is found in most species analysed, exceptions have been reported. In *Schizophyllum commune*, for example, two large rearrangements at the *HD* locus have taken place causing the separation of the *HD* locus into two clearly distinct subloci that now lie more than 500 kb apart and frequently recombine during meiosis.

A similar situation is observed in *Flammulina velutipes*, but the distance between the two *HD* subloci is about 70 kb. Other exceptions to the typical *MIP–HD1/HD2–β-fg* gene arrangement were observed in the cacao pathogen *Moniliophthora roreri*, where both *MIP* and *β-fg* are located 40 and 60 kb upstream of *HD1/HD2*, respectively, and in *Lentinula edodes*, where *MIP* is distant from the *HD* locus. Gene order conservation at the *P/R* locus is also observed among *Agaricomycetes*, although to a lesser extent compared to the *HD* locus. The *P/R* locus in *Laccaria bicolor* is in a chromosomal region prone to gene duplication, translocations and accumulation of transposable elements, and in the cantharelloid wood-decay basidiomycete *Botryobasidium botryosum*, there are transposons in the immediate vicinity of pheromone and receptor genes.

Ustilago maydis causes the smut disease of maize. It has a **tetrapolar mating system** comprising one 'multiallelic' mating type factor and one with only two 'alleles'. *Ustilago* produces unicellular, haploid sporidia that grow vegetatively by budding like a yeast phase; these can be cultured on synthetic media and are non-pathogenic for the host plant (see Figure 3.14).

Conjugation tubes are formed when sporidia of opposite mating type are mixed, and fusion of these produces the dikaryon, which then grows as a filamentous fungus. The dikaryon is the pathogenic stage. Fusion of sporidia is controlled by the biallelic '*a*' mating type locus, the heterozygous *a1*/*a2* genotype being required for conjugation and the transition between the yeast and filamentous forms. The multiallelic '*b*' locus stops diploid cells fusing, regulates the true hyphal growth form, and determines pathogenicity (Figure 9.7).

The *a1* idiomorph consists of 4.5 kb of DNA and the *a2* idiomorph is 8 kb. Two genes have been identified in these regions: *mfa1* (in *a1*) and *mfa2* (in *a2*) code for pheromones, and *pra1*/*pra2* encode pheromone receptors. The pheromones diffuse away from their producer cells and induce conjugation tubes after binding to pheromone receptors on cells of opposite mating type.

Pheromone signalling is also necessary for the maintenance of the filamentous dikaryon after cell fusion. Pheromone induces expression of all mating type genes to levels 10–50 times higher than their basal level. The upstream control element responsible for this pheromone stimulation, which is called the **pheromone response element** (***PRE***), has the sequence ACAAAGGG. It is on the same DNA molecule as the pheromone genes, so it is called a '*cis*-acting element', by analogy with chemical terminology which describes two substituents on the same side of an axis of symmetry in a molecule as the *cis*-configuration, and the alternative of two substituents on opposite sides of an axis as the *trans*-configuration.

The *PRE*-sequence is similar to the consensus sequence recognised by **HMG polypeptides**, including *mat a1* of *Neurospora crassa* (see above). A gene called *pfr1* (pheromone response factor), the product of which binds to the *PRE*-sequences found in both *a1* and *a2* idiomorphs, encodes the controlling transcription factor. The downstream pheromone response pathway includes at least one MAP kinase encoded by a gene called *fuz7*, which is homologous to the 'archetypal' MAP kinase gene of *Saccharomyces cerevisiae, STE7* (a **MAP kinase** is a '**mitogen-activated protein kinase**', where a mitogen is any agent that induces mitosis). Disruption of *fuz7* results in phenotypes that show that *fuz7* is involved in *a*-dependent mating events like conjugation tube formation, conjugation and establishment and maintenance of filamentous growth. Other components of the pheromone response pathway that have been found are four genes encoding G-proteins (*gpa1* to *4*). There are indications that *fuz7* and *gpa3* do not belong to the same pathway, so there may be several pheromone responses in *Ustilago maydis*, either in parallel or in series.

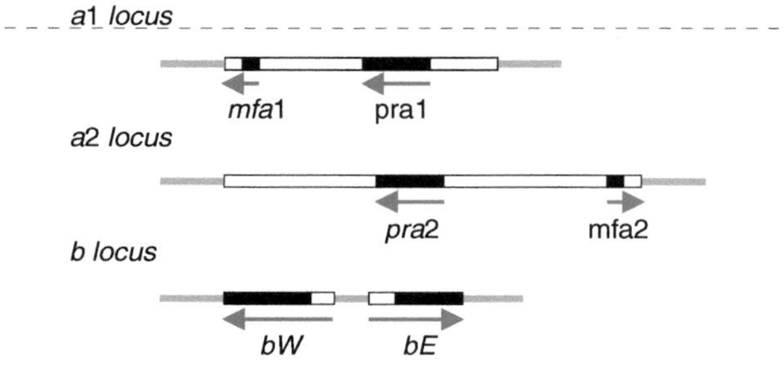

Figure 9.7. Schematic representations of the structures of the 'a' and 'b' mating type loci of *Ustilago maydis*. Idiomorphs of the 'a' locus consist of mating type-specific (i.e. variable) DNA sequences (4,500 base pairs in *a1*, 8,000 base pairs in *a2*), here shown as open boxes, within which are the genes for mating (*mfa* and *pra*). The b locus has two reading frames, *bW* and *bE*, which produce polypeptides containing domains of more than 90% sequence identity (shown as black boxes) and variable domains (open boxes) which show 60 to 90% identity. Arrows indicate the direction of transcription. The *PRE*-sequence mentioned in the text is a very short control site upstream of the pheromone genes, *mfa1* and *mfa2*. Modified from Chapter 2 in Moore & Novak Frazer (2002).

The *b* mating type factor contains two genes which are transcribed in opposite directions (Figure 9.7): *bE* and *bW* (= east and west) with coding sequences equivalent to polypeptides of 473 and 629 amino acids, respectively. The amino-terminal end of the coding sequence is highly variable, whereas the carboxy-terminal end is conserved in different *b* idiomorphs. The bE and bW proteins are, respectively, the HD1 and HD2 homeodomain proteins that form a dimer that is a transcription activator. This activates transcription of genes required for the sexual cycle and/or repression of haploid-specific genes (interaction of the equivalent *Coprinopsis* homeodomain proteins is illustrated in Figure 9.10).

Dimers comprised of bE amd bW from the same idiomorph are inactive; the heterodimer functions properly only when the proteins come from different idiomorphs. The proteins encoded by these genes contain sequences homologous with DNA-binding homeodomain regions of known transcription-regulating factors, which is why they are called HD1 and HD2. The homeodomain is an extended helix-turn-helix DNA-binding motif, which is encoded by a conserved DNA sequence of about 180 bp called the **homeobox**. This sequence is particularly associated with the transcriptional regulators of **homeotic**, or *Hox*, genes that were found originally in the fruit fly *Drosophila* and are involved in orchestrating development in higher eukaryotes. Mutations in animal *Hox* genes convert one body part into another. *Drosophila* has two *Hox* clusters, but vertebrates have four clusters of 9–11 genes each on different chromosomes. Vertebrate *Hox* genes are expressed in different patterns and at specific embryological stages. There is a compelling comparability between these developmental regulators and the *HD1*/*HD2* genes of the fungal mating type factors in *Schizophyllum* and *Coprinopsis*, as well as *Ustilago*.

Coprinopsis cinerea and *Schizophyllum commune* exhibit **tetrapolar heterothallism**, determined by two mating type factors, called *A* and *B*. These are both considered to be *MAT* loci and the natural population contains many different mating types. In crosses, these behave like multiple 'alleles' of the two mating type loci. Molecular analysis has revealed that each mating type locus is a very complex region containing several or even many genes (which is why we refer to mating type **factors**). The genes at *A* encode the transcription factor homeodomain proteins and constitute the *HD* locus; genes at *B* make up the *P*/*R* locus and encode the lipopeptide pheromones and pheromone receptors. Mating type factors are located on different chromosomes, and

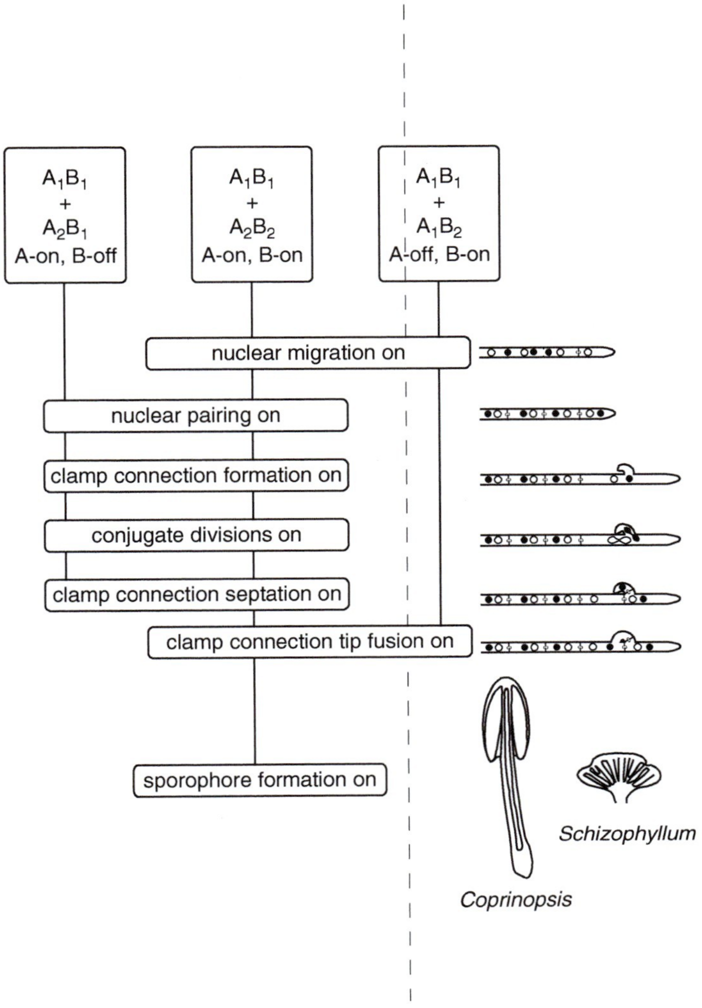

Figure 9.8. Flow chart diagram of 'A' (the *HD* locus) and 'B' (the *P/R* locus) mating type factor activity in the basidiomycetes *Coprinopsis cinerea* and *Schizophyllum commune*. From top to bottom the flow chart depicts the events that take place when two haploid mycelia confront each other. Dikaryon formation requires that both 'A' and 'B' factors are different (which is taken to mean that both 'A' and 'B' functions are turned on (A-on, B-on)). This is depicted in the central vertical line for a confrontation of A1B1 + A2B2, the vertical line connecting 'function boxes', the phenotypes of which are indicated in the cartoons on the right-hand side. When only 'A' functions are turned on (A-on, B-off), in a common-B heterokaryon (A1B1 + A2B1) conjugate divisions and clamp formation occur, but the clamp connections remain incomplete and without clamp cell fusion nuclear migration cannot take place. In a common-A heterokaryon (A1B1 + A1B2) the 'A' factors are the same, but 'B' factors are different (A-off, B-on) and nuclear migration can occur, but without conjugate division or clamp connection formation. Modified from Chapter 2 in Moore & Novak Frazer (2002).

even conventional genetic analysis has demonstrated internal structure, identifying subloci that are called *Aα*, *Aβ*, *Bα* and *Bβ*.

In *Schizophyllum commune* these subloci are relatively far apart, in terms of linkage distance, but they are much closer in *Coprinopsis cinerea*. The α and β subloci are functionally redundant in the sense that a difference need exist at only one of them for compatibility. Nine versions of *Aα*, and 32 of *Aβ* result in 288 different *A* mating type specificities in *Schizophyllum commune*. In *Coprinopsis cinerea*, there are estimated to be 160 *A* mating type specificities in the natural population. Because there are so many different mating types, most encounters between isolates in nature will be compatible and fertile, and this drives the frequency of outbreeding to greater than 99%. On the other hand, from any **one meiosis**, say the heterozygote A1B1 × A2B2, because the two *MAT* loci segregate independently, four different types of progeny are produced (A1B1, A2B2, A1B2 and A2B1) and so these systems are called tetrapolar mating systems.

In saprotrophic basidiomycetes, hyphal anastomoses occur readily as part of the maturation process of the mycelium, and anastomosis is independent of the mating type factors. Promiscuous cell–cell fusion can result when two monokaryotic hyphae of *Coprinopsis cinerea* and other Agaricomycetes encounter each other, irrespective of whether they are compatible at *MAT* loci (and see section 4.10). However, for the development and maintenance of the dikaryon, nuclei of interacting partners must carry different allelic versions of genes in at least one sublocus of both *P/R* and *HD* loci. A compatible mating in these fungi, which is characterised by clamp connections and conjugate nuclear divisions in the mated hyphae, ***requires*** that both *A* and *B* are different. In the belief that in this state the mating type factors are fully active, it is called *A*-on, *B*-on (Figure 9.8). Other than dikaryosis itself, there are some characteristic morphological differences between the monokaryotic and dikaryotic mycelia of *Coprinopsis cinerea*: branches emerge from monokaryotic hyphae at a wide angle (40–90°), but at an acute angle (10–45°) from dikaryotic hyphae; monokaryons, but not the dikaryons, produce asexual **arthrospores**, called **oidia**, in droplets of fluid; and the aerial mycelium of monokaryons is generally less dense and fluffy than that of dikaryons.

Cytological observations and genetic analyses of partially compatible interactions in *Coprinopsis cinerea* and *Schizophyllum commune* have shown that the *P/R* system (mating type factor *B*) is involved in events that follow hyphal fusion:

- the reciprocal exchange and migration of nuclei; and
- clamp cell fusion.

On the other hand, genes encoded at the *HD* locus (mating type factor *A*):

- repress asexual sporulation;
- regulate pairing of nuclei within the dikaryotic tip cell; and
- coordinate nuclear division, clamp cell formation and septation from the subapical cell.

Dikaryons arise when both *A* and *B* are different but heterokaryons can also be formed in matings in which one of the mating type factors is homozygous. When *A*-factors are the same (called common-*A*, or *A*-off, *B*-on, see Figure 9.8), ***nuclear migration takes place but no clamp connections form***. Mating between strains carrying the same *B*-factor forms a heterokaryon (called common-*B* or *A*-on, *B*-off) only where the mated monokaryons meet because ***nuclear migration is blocked***. In this case, conjugate divisions occur, and apical cells of heterokaryotic hyphae start to make clamp connections and the nuclei divide but the clamp (hook) cell fails to fuse with the adjacent cell and the nucleus in the clamp remains trapped (Figure 9.8). If the two 'mates' share identical alleles at *P*/*R* and *HD* loci, no nuclear exchange occurs, and the mycelial individuals will resume monokaryotic growth.

This 'division of labour' between *A*-and *B*-factors is not universal; nuclear migration is regulated by the *B*-factor alone in *Coprinopsis cinerea* and *Schizophyllum commune*, but in *Coprinus patouillardii* migration is regulated by both factors but the *A* alone determines fertility. In *Ustilago* and *Tremella*, cell fusion is controlled by a locus with two idiomorphs, but the ability of fused cells to grow as dikaryons depends on a second locus with multiple idiomorphs (see the *Ustilago* section above).

Clearly, the mating type factors determine the initial self/nonself recognition that follows the first anastomoses of the encounter; they also regulate mycelial morphogenesis (upregulating features that characterise the dikaryon and downregulating some monokaryotic features, like making oidia) but as it is only dikaryons that fruit in normal circumstances, ***the* A *and* B *mating type factors also regulate fertility.***

Nuclear migration through an established mycelium has also been studied in some basidiomycetes. In the species that have been examined in most detail, *Coprinopsis cinerea* and *Schizophyllum commune*, basidiospores germinate to form homokaryotic mycelia with uninucleate cells, which is usually called a **monokaryon** or homokaryon. Two monokaryons will form hyphal anastomoses, and if they are compatible, nuclei will migrate from one mycelium into the other. This establishes a new mycelium, called a **dikaryon**, which has regularly binucleate cells containing one nucleus of each parental type.

The growth of dikaryotic hyphal tips requires that the two nuclei complete mitosis together (conjugate division) and a mechanism of nuclear migration and sorting that depends on a small backward-growing branch (called a clamp connection or hook cell) at each hyphal septum (side-panel in Figure 9.8; and see Figure 8.6). Nuclei will migrate from a compatible (dikaryotic) mycelium and convert another monokaryon into a dikaryon. In experiments with *Coprinopsis radiatus* (a very close relative of *Coprinopsis cinerea*) nuclei invaded mycelia in such circumstances at a rate of 1.5 mm h^{-1}, which is at least four times higher than the hyphal growth rate. In *Coprinellus congregatus*, a **nuclear migration** rate of 4 cm h^{-1} (yes, we do mean *centimetres*!) has been reported, and migration rates in *Schizophyllum commune* range up to 2.7 mm h^{-1} with a hyphal growth rate of only 0.22 mm h^{-1}.

During nuclear migration the invading nucleus undergoes regular division and one of the daughter nuclei moves into the next cell, the intervening dolipore septa between adjacent cells of the monokaryon being broken down into simple pores through which the nuclei can be squeezed. Although some corresponding cytoplasmic movement has been observed in *Schizophyllum commune*, nuclear migration occurs more commonly without visible cytoplasmic flow. Nuclear migration is a highly active transport process involving microtubular motors in a manner analogous to the involvement of spindle fibres in the movement of chromosomes during division, and nuclear identification by mating type factors presumably accounts for its specificity (Debuchy, 1999; Shiu & Glass, 2000) (and see section 5.11). Clearly, in most cases a specific nuclear type is being transported in a specific direction, and in all this discussion of nuclear migration it is important to emphasise that ***only*** nuclei migrate; mitochondria are not exchanged between compatible mycelia. During migratory dikaryotisation, cells lacking nuclei and multinucleate cells are observed, so the dikaryotic state is not set up as soon as compatible cells fuse. Rather, ordered dikaryotic growth emerges after an interval of disorganised and irregular growth.

Detailed molecular analysis of the *A* mating type factors show that they contain many more genes than classical genetic analysis could reveal (Figure 9.9). Each *A* (= *HD*) locus of *Coprinopsis cinerea* contains a variable number of genes, which are arranged in pairs like the *bE–bW* pair in the *Ustilago maydis b* locus. The *Coprinopsis cinerea* gene pairs are designated groups 1, 2 and 3. Each group within the *A* (= *HD*) locus encodes two dissimilar homeodomain proteins (HD1 and HD2) which are homologous to the *Saccharomyces cerevisiae* MAT α2 and MAT a1 mating proteins, respectively (Casselton & Olesnicky, 1998; Brown & Casselton, 2001; Kües, 2015; Coelho *et al.*, 2017).

Several features combine to ensure that there is no intragenic recombination within the *A* locus, which is likely to disturb its regular 'two-by-two' structure. The groups are organised into cassettes so that they act as a single unit and the DNA sequences are sufficiently different between groups 1, 2 and 3 to avoid homologous recombination. In addition, the paired genes are transcribed in opposite directions (Figure 9.9). The *A* locus is bounded by DNA sequences that **are** homologous in all *A* mating type specificities, called *α-fg* and *β-fg*; also, the group 1 gene pair is separated from groups 2 and 3 by a 7 kbp DNA sequence that is homologous in all *A* loci (known as the 'homologous hole'). The group 1 gene pair corresponds to the *Aα* sublocus defined by conventional linkage analysis, while the group 2 and group 3 gene pairs comprise the *Aβ* sublocus; 7 kbp is approximately equivalent to the 0.1% recombination observed between these subloci. These short homologous sequences limit recombination to the regions between the homeodomain loci.

Basidiomycete sexual development is triggered by a dimerisation between HD1 and HD2 proteins from the different *A* mating type factors of compatible individuals (Figure 9.10). The HD2 homeodomain is crucial to DNA binding, but the homeodomains of HD1 proteins are relatively dispensable. The amino-terminal regions of these proteins, though, are essential for choosing a compatible partner and for ensuring that only the ***hetero***dimers made between the products of the two compatible mating type factors present in the cell can form the DNA-binding transcription regulators. Sequences of different idiomorphs of the mating type genes are dissimilar and have been interpreted

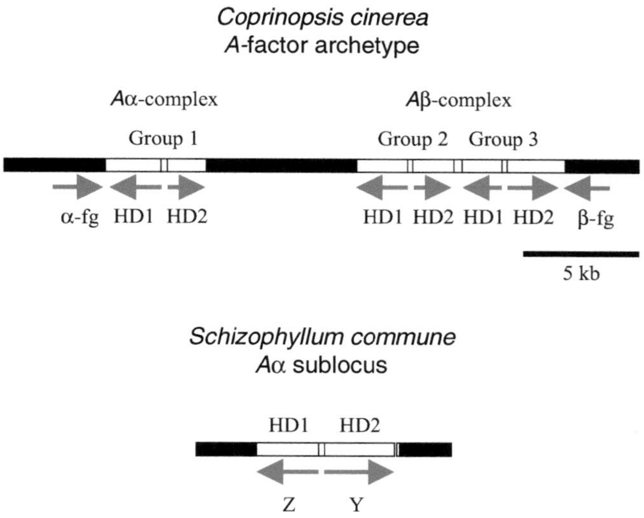

Figure 9.9. Diagrams of parts of the 'A' (*HD* locus) mating type factors in *Coprinopsis cinerea* and *Schizophyllum commune*. Arrows show the direction of transcription. The (predicted) archetypal 'A' factor from *Coprinopsis cinerea* has three pairs of functionally redundant genes (group 1, group 2 & group 3) which encode the homeodomain proteins (HD1 & HD2). The α-fg and b-fg sequences are homologous in all 'A' mating type specificities in *Coprinopsis cinerea*. Interaction between HD1 and HD2 proteins is the basis of the compatible reaction (see Figures 9.8 and 9.10). 'A' factors examined in different strains of *Coprinopsis cinerea* isolated from nature contain different combinations, and different numbers, of these genes. In *Schizophyllum commune* the mating type genes are called *Z* and *Y* and encode HD1 and HD2 respectively. Again, different idiomorphs are found in different natural mating types; indeed, the *Z* gene is absent in the Aα1 mating type. This simplified diagram is modified from Chapter 2 in Moore & Novak Frazer (2002). For much more detailed information about a wider range of fungi refer to Kües (2015) and Coelho *et al.* (2017).

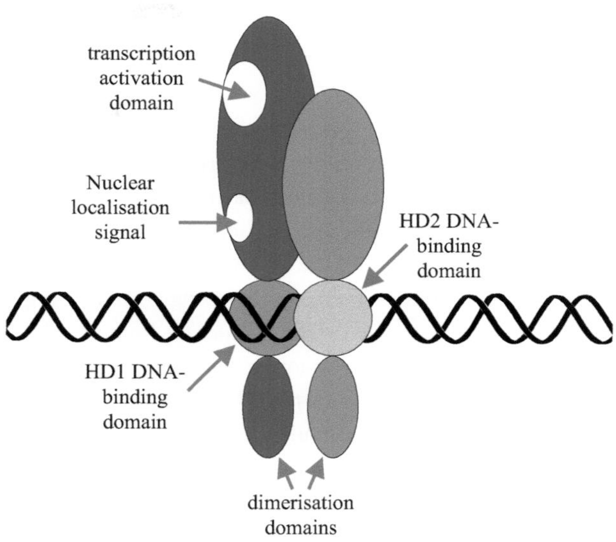

Figure 9.10. Schematic diagram illustrating a simplified model of homeodomain protein interactions involved in 'A' (*HD* locus) mating type factor activity in *Coprinopsis*. Modified from Chapter 2 in Moore & Novak Frazer (2002).

as being equivalent to the highly variable region in major histocompatibility loci in mammals that forms a self/nonself recognition system.

The *A* locus of *Schizophyllum commune* also controls nuclear pairing, clamp connection formation, conjugate nuclear division and clamp septation. The *Aα* locus contains the *Y* gene (which has alleles *Y1, Y3* and *Y4*) and the *Z* gene (with alleles *Z3* and *Z4*), which encode the homeodomain proteins HD2 and HD1, respectively. The *Schizophyllum commune Aα* locus corresponds to a single gene pair from the *Coprinopsis cinerea* complex. The *Aβ* locus also encodes a polypeptide with a homeodomain. The genes *Y* and *Z* are the only determinants of *Aα* activity and *Aα* and *Aβ* function independently of each other. Interactions of the Y and Z proteins have been demonstrated experimentally for nonself combinations (for example, Y4 with Z5) proteins, but no interaction occurs between Y and Z proteins encoded by the same *A*-factor (for example Y4 and Z4).

Sequencing the *Schizophyllum commune Bα1* region revealed a pheromone receptor gene (called *bar1*, standing for B-alpha-receptor-1) and three pheromone genes, *bap1, bap2* and *bap3* (*bap* stands for B-alpha-pheromone). The *Bβ1* locus also contains a receptor gene (*bbr1*) and genes for pheromones, *bbp1*(*1*) and *bbp1*(*2*). The *B*-factor of *Coprinopsis cinerea* contains three groups (called groups 1, 2, 3) of genes which each code for a pheromone receptor and two pheromones. The *B* pheromone genes are all predicted to encode for lipopeptides like the *Saccharomyces cerevisiae a*-factor while the *B* pheromone receptors are homologous to the *Saccharomyces cerevisiae a*-factor receptor, which is a typical G-protein-coupled receptor. So far, only the *a*-factor-type pheromones have been found in basidiomycetes whereas both *a*- and *α*-type **receptors** are present (Kothe, 1999).

The model of pheromone function that has been developed for these filamentous basidiomycetes is that after anastomosis, the pheromones produced by the invading nucleus diffuse ahead and act as advance signals of nuclear migration (see section 5.11). This activates receptors encoded by resident nuclei in nearby cells and the interactions prepare the cells for nuclear migration by initiating septum dissolution to allow nuclei to pass through. *A*-factor functions then establish the dikaryotic state and clamp connection formation.

Pheromone signalling is then further involved in clamp cell fusion. There are nine mating type specificities at the *Bα* factor of *Schizophyllum commune*, each encoding a receptor and one or more pheromones. Hence, each receptor must distinguish at least eight nonself pheromones. Individual pheromones may also activate more than one receptor. Work with mutants in *B*-regulated functions is beginning to identify the genes subject to pheromone signalling. Some of these mutations map to the *B* loci and affect mating specificity and could be modifiers of pheromone or pheromone receptor gene specificity. Others include nine genes that influence nuclear migration. Many of these genes are linked to *B*, suggesting that related functions are clustered.

9.6 BIOLOGY OF MATING TYPE FACTORS

Mating type factors are transcription regulators, together with pheromones and their receptors that create the circumstances promoting meiosis. Surprisingly, they are not universal. Many fungi get by perfectly well without mating types. In *Podospora*, the progress of meiosis and sporulation does not require heterozygous mating type factors, the organism is homothallic, and homothallism occurs even in higher mushroom fungi; the paddy straw mushroom *Volvariella volvacea* is homothallic. Even in *Coprinopsis cinerea* and *Schizophyllum commune*, apparently normal sporophores can be formed by haploid cultures, and sporophore formation can usually be separated, by mutation, from other parts of the sexual pathway in all species that have a well-developed mating type system. Therefore, the significance of mating type factors in regulating events beyond the initial mating reaction is difficult to judge; indeed, given the success of anamorphic fungi, understanding the selective advantage of sexual reproduction itself needs careful argument (Taylor *et al.*, 1999; Moore, 2001, see his Chapter 10; Moore & Novak Frazer, 2002, see their Chapter 2; Pringle & Taylor, 2002; Heitman, 2015).

Mating type factors are usually interpreted as promoting outbreeding by preventing breeding between closely related progeny. With only two idiomorphs, the likelihood that two individuals will be able to mate (which is the outbreeding potential) is 50%. But if there were n mating type idiomorphs the outbreeding potential would be $[1/n \times (n-1)] \times 100\%$; the greater the number of mating type idiomorphs, the greater the outbreeding potential.

In the case of the *Basidiomycota* with a tetrapolar mating system with two unlinked mating type factors (designated *A* and *B*), a compatible interaction is one between two mycelia with different idiomorphs, but now both *A* and *B* mating type factors must differ. As a result, the diploid nucleus that is formed will be heterozygous at the two mating type loci and meiosis will generate progeny spores of **four** different mating types (which is why ***tetra***polar heterothallism is the alternative name of this system).

Coprinopsis cinerea and *Schizophyllum commune* are the classic examples of this mating type system. In both species the wild population contains many different *A* and *B* idiomorphs, and the outbreeding potential approaches 100%. The inbreeding potential of a tetrapolar mating system (the likelihood of being able to mate with a sibling) is 25% (because there are four different mating types among the progeny of a single meiosis) whereas inbreeding and outbreeding potentials are both 50% in unifactorial incompatibility where there are only two mating types among the progeny. A bifactorial system tends to favour outbreeding. About 90% of higher fungi are heterothallic, and about 40% of these are bipolar and 60% tetrapolar. Implicit in the calculations of these paragraphs is the fact that fungi are eukaryotes and their basic genetics follows the pattern seen in other higher organisms (see Chapter 6).

Tetrapolar mating systems not only promote outbreeding but also lead to inbreeding depression. These differences in the frequencies of outbreeding and inbreeding are thought to provide the evolutionary pressure for transitions between bipolar and tetrapolar mating systems. Phylogenetic reconstructions across the fungal kingdom support the conclusion that bipolar mating is an ancestral state and the tetrapolar configuration is a derived state. All species with known mating type systems in the zygomycetes and ascomycetes are bipolar. This is in marked contrast to the basidiomycetes in which most species have a tetrapolar mating system. Though there are examples of species with bipolar mating type within the *Basidiomycota*, so far, no species with the tetrapolar mating system have been found outside of the *Basidiomycota* and this leads to the conclusion that the tetrapolar system is a derived state, possibly with a single origin at the base of the *Basidiomycota*.

Presumably, an ancestral system with just one *MAT* locus encoding homeodomain factors evolved into a system with a second sex determinant on another chromosome encoding pheromones and pheromone receptors. We must then make the further deduction that the bipolar mating systems of pathogenic basidiomycete species have been derived from the tetrapolar system and that this transition from tetrapolar to bipolar has occurred several times independently during evolution of the *Basidiomycota* (Heitman, 2015).

9.7 REFERENCES

Barsoum, E., Martinez, P. & Aström, S.U. (2010). Alpha3, a transposable element that promotes host sexual reproduction. *Genes and Development*, **24**: 33–44. DOI: https://doi.org/10.1101/gad.557310.

Billiard, S., López-Villavicencio, M., Devier, B. *et al.* (2011). Having sex, yes, but with whom? Inferences from fungi on the evolution of anisogamy and mating types. *Biological Reviews*, **86**: 421–442. DOI: https://doi.org/10.1111/j.1469-185X.2010.00153.x.

Billiard, S., López-Villavicencio, M., Hood, M.E. & Giraud, T. (2012). Sex, outcrossing and mating types: unsolved questions in fungi and beyond. *Journal of Evolutionary Biology*, **25**: 1020–1038. DOI: https://doi.org/10.1111/j.1420-9101.2012.02495.x.

Bobola, N. & Merabet, S. (2017). Homeodomain proteins in action: similar DNA binding preferences, highly variable connectivity. *Current Opinion in Genetics & Development*, **43**: 1–8. DOI: https://doi.org/10.1016/j.gde.2016.09.008.

Brown, A.J. & Casselton, L.A. (2001). Mating in mushrooms: increasing the chances but prolonging the affair. *Trends in Genetics*, **17**: 393–400. DOI: https://doi.org/10.1016/S0168-9525(01)02343-5.

Casselton, L.A. (2008). Fungal sex genes – searching for the ancestors. *Bioessays*, **30**: 711–714. DOI: https://doi.org/10.1002/bies.20782.

Casselton, L.A. & Olesnicky, N.S. (1998). Molecular genetics of mating recognition in basidiomycete fungi. *Microbiology and Molecular Biology Reviews*, **62**: 55–70. URL: http://mmbr.asm.org/content/62/1/55.long.

Coelho, M.A., Bakkeren, G., Sun, S., Hood, M.E. & Giraud, T. (2017). Fungal sex: the Basidiomycota. In *The Fungal Kingdom*, ed. J. Heitman, B. Howlett, P. Crous *et al.* Washington, DC: ASM Press, pp. 147–175. DOI: https://doi.org/10.1128/microbiolspec.FUNK-0046-2016.

Debuchy, R. (1999). Internuclear recognition: a possible connection between Euascomycetes and Homobasidiomycetes. *Fungal Genetics and Biology*, **27**: 218–223. DOI: https://doi.org/10.1006/fgbi.1999.1142.

Haber, J. (2007). Decisions, decisions: donor preference during budding yeast mating-type switching. In *Sex in Fungi: Molecular Determination and Evolutionary Implications*, ed. J. Heitman, J.W. Kronstad, J.W. Taylor & L.A. Casselton. Washington, DC: ASM Press, pp. 159–170. DOI: https://doi.org/10.1128/9781555815837.ch9.

Haber, J.E. (2012). Mating-type genes and MAT switching in *Saccharomyces cerevisiae*. *Genetics*, **191**: 33–64. DOI: https://doi.org/10.1534/genetics.111.134577.

Hadjivasiliou, Z., Pomiankowski, A. & Kuijper, B. (2016). The evolution of mating type switching. *Evolution; International Journal of Organic Evolution*, **70**: 1569–1581. DOI: https://doi.org/10.1111/evo.12959.

Hanson, S.J., Byrne, K.P. & Wolfe, K.H. (2014). Mating-type switching by chromosomal inversion in methylotrophic yeasts suggests an origin for the three-locus *Saccharomyces cerevisiae* system. *Proceedings of the National Academy of Sciences of the United States of America*, **111**: E4851–E4858. DOI: https://doi.org/10.1073/pnas.1416014111.

Heitman, J. (2015). Evolution of sexual reproduction: a view from the fungal kingdom supports an evolutionary epoch with sex before sexes. *Fungal Biology Reviews*, **29**: 108–117. DOI: https://doi.org/10.1016/j.fbr.2015.08.002.

Jia, S., Yamada, T. & Grewal, S.I.S. (2004). Heterochromatin regulates cell type-specific long-range chromatin interactions essential for directed recombination. *Cell*, **119**: 469–480. DOI: https://doi.org/10.1016/j.cell.2004.10.020.

Judelson, H.S. (2007). Sexual reproduction in plant pathogenic oomycetes: biology and impact on disease. In *Sex in Fungi: Molecular Determination and Evolutionary Implications*, ed. J. Heitman, J.W. Kronstad, J.W. Taylor & L.A. Casselton. Washington, DC: ASM Press, pp. 445–458. DOI: https://doi.org/10.1128/9781555815837.ch27.

Kim, H., Wright, S.J., Park, G. *et al.* (2012). Roles for receptors, pheromones, G proteins, and mating type genes during sexual reproduction in *Neurospora crassa*. *Genetics*, **190**: 1389–1404. DOI: https://doi.org/10.1534/genetics.111.136358.

Kothe, E. (1999). Mating types and pheromone recognition in the Homobasidiomycete *Schizophyllum commune*. *Fungal Genetics and Biology*, **27**: 146–152. DOI: https://doi.org/10.1006/fgbi.1999.1129.

Kruzel, E.K. & Hull, C.M. (2010). Establishing an unusual cell type: how to make a dikaryon. *Current Opinion in Microbiology*, **13**: 706–711. DOI: https://doi.org/10.1016/j.mib.2010.09.016.

Kües, U. (2015). From two to many: multiple mating types in Basidiomycetes. *Fungal Biology Reviews*, **29**: 126–166. DOI: https://doi.org/10.1016/j.fbr.2015.11.001.

Lee, S.C. & Heitman, J. (2014). Sex in the mucoralean fungi. *Mycoses*, **57**: 18–24. DOI: https://doi.org/10.1111/myc.12244.

Moore, D. (2001). *Slayers, Saviors, Servants, and Sex: An Exposé of Kingdom Fungi*. New York: Springer-Verlag. ISBN-10: 0387951016, ISBN-13: 978-0387951010. See Chapter 9, pp. 127–137, Birds do it. Bees do it. Even educated fleas do it. But why?

Moore, D. (2013). *Fungal Biology in the Origin and Emergence of Life*. Cambridge, UK: Cambridge University Press. 230 pp. ISBN-10: 1107652774, ISBN-13: 978-1107652774.

Moore, D. & Novak Frazer, L. (2002). *Essential Fungal Genetics*. New York: Springer-Verlag. ISBN-10: 0387953671, ISBN-13: 978-0387953670. See Chapter 2 *Genome interactions* (especially sections 2.6 to 2.10) and Chapter 5 *Recombination analysis*.

Pérez-Martín, J. & de Sena-Tomás, C. (2011). Dikaryotic cell cycle in the phytopathogenic fungus *Ustilago maydis* is controlled by the DNA damage response cascade. *Plant Signaling & Behavior*, **6**: 1574–1577. DOI: https://doi.org/10.4161/psb.6.10.17055.

Pringle, A. & Taylor, J.W. (2002). The fitness of filamentous fungi. *Trends in Microbiology*, **10**: 474–481. DOI: https://doi.org/10.1016/S0966-842X(02)02447-2.

Raju, N.B. (2008). Six decades of *Neurospora* ascus biology at Stanford. *Fungal Biology Reviews*, **22**: 26–35. DOI: https://doi.org/10.1016/j.fbr.2008.03.003.

Raudaskoski, M. (2015). Mating-type genes and hyphal fusions in filamentous basidiomycetes. *Fungal Biology Reviews*, **29**: 179–193. DOI: https://doi.org/10.1016/j.fbr.2015.04.001.

Raudaskoski, M. & Kothe, E. (2010). Basidiomycete mating type genes and pheromone signaling. *Eukaryotic Cell*, **9**: 847–859. DOI: https://doi.org/10.1128/EC.00319-09.

Shiu, P.K.T. & Glass, N.L. (2000). Cell and nuclear recognition mechanisms mediated by mating type in filamentous ascomycetes. *Current Opinion in Microbiology*, **3**: 183–188. DOI: https://doi.org/10.1016/S1369-5274(00)00073-4.

Taylor, J.W., Jacobson, D.J. & Fisher, M.C. (1999). The evolution of asexual fungi: reproduction, speciation and classification. *Annual Review of Phytopathology*, **37**: 197–246. DOI: https://doi.org/10.1146/annurev.phyto.37.1.197.

Tsui, C.K-M., DiGuistini, S., Wang, Y. *et al.* (2013). Unequal recombination and evolution of the mating-type (MAT) loci in the pathogenic fungus *Grosmannia clavigera* and relatives. *G3: Genes, Genomes, Genetics*, **3**: 465–480. DOI: https://doi.org/10.1534/g3.112.004986.

Vonk, P.J. & Ohm, R.A. (2018). The role of homeodomain transcription factors in fungal development. *Fungal Biology Reviews*, **32**: 219–230. DOI: https://doi.org/10.1016/j.fbr.2018.04.002.

Wang, Z., Kin, K., López-Giráldez, F., Johannesson, H. & Townsend, J.P. (2012). Sex-specific gene expression during asexual development of *Neurospora crassa*. *Fungal Genetics and Biology*, **49**: 533–543. DOI: https://doi.org/10.1016/j.fgb.2012.05.004.

Wik, L., Karlsson, M. & Johannesson, H. (2008). The evolutionary trajectory of the mating-type (*mat*) genes in *Neurospora* relates to reproductive behavior of taxa. *BMC Evolutionary Biology*, **8**: 109. DOI: https://doi.org/10.1186/1471-2148-8-109.

Wu, C., Yang, F., Smith, K.M. *et al.* (2014). Genome-wide characterization of light-regulated genes in *Neurospora crassa*. *G3: Genes, Genomes, Genetics*, **4**: 1731–1745. DOI: http://doi.org/10.1534/g3.114.012617.

CHAPTER 10

Continuing the Diversity Theme: Cell and Tissue Differentiation

Wikipedia points out that 'biodiversity' is a new portmanteau word made up from 'biology' and 'diversity' that probably arrived in the English language in 1985. Wikipedia defines 'biodiversity' as 'the variation of taxonomic life forms within a given ecosystem, biome or for the entire Earth. Biodiversity is often used as a measure of the health of biological systems' (see https://en.wikipedia.org/wiki/Biodiversity). As an aside to this Wikipedia definition of biodiversity, given the fact that in the fossil record fungal biodiversity greatly increases during major extinction events (see section 2.8), we are not convinced that fungal biodiversity is a good measure of the health of ***non-fungal*** biological systems!

In this Chapter 10, we will explain the meaning of the word 'diversity' in the context of the fungi, and then deal with its different aspects as they bear upon cells and tissues: mycelial differentiation and the different ways that fungi use for making spores. We describe *Aspergillus* conidiophores in some detail because something is known about the molecular regulation of *Aspergillus* sporulation. This is then compared with conidiation in *Neurospora* and other fungi. Finally, we make some general points about the nature and construction of fungal tissues and organs by describing conidiomata, and then linear structures like strands, cords, rhizomorphs and stipes; finishing off with the globose structures called sclerotia, stromata, ascomata and basidiomata.

10.1 WHAT IS DIVERSITY?

Much of the success of fungi is related to their ability to produce vast numbers of asexual, dispersal spores that spread the organism in space. A by-product of this productivity is that the air we breathe contains very large numbers of fungal spores (which can be potential allergens in the air of our urban environments). In contrast to the sexual spores of zygomycetes and *Ascomycota* (but like the basidiospores of *Basidiomycota*), asexual fungal spores are usually:

- produced in very large numbers (much of the biomass of the fungus being converted into asexual spores);
- relatively small;
- thin walled;
- non-dormant;
- able to germinate rapidly on suitable substrates, and
- short lived.

Though there are, of course, exceptions to these generalisations: for example, chlamydospores, which are large, thick-walled, resting spores of organisms like *Candida* and *Fusarium*.

These spores need to be efficiently and effectively dispersed and, as we describe more examples, we will encounter a range of mechanisms and processes that have been developed by fungi to contribute in some way to dispersal. This is part of fungal diversity.

The biodiversity of fungi starts with the numerous ways that fungal **mycelia differentiate morphologically**. This diversity in form is related to the ecological function(s) of the mycelium. The accepted interpretation is that fungal mycelia can express a range of alternative phenotypes that enable them to:

- explore (rapid extension of sparsely branched hyphal systems);
- assimilate (locally enhanced branching);
- conserve (cells differentiated to store metabolites and cope with adverse conditions);
- redistribute the resources within ecological niches that are heterogeneous in both space and time (via strands and similar hyphal systems adapted to bulk translocation).

Most of this mycelial biodiversity results from variation in the **kinetics of hyphal growth** parameters like rate of extension growth, branching frequency and branching angle (Trinci *et al.*, 1994). Together these can have enormous impact on the macroscopic morphology of the mycelium and the functional efficiency of the mycelium in the various roles it fulfils (see section 4.9). In the 1990s, **Alan Rayner** and his colleagues (Rayner, 1997; and references in Appendix 2) argued that different local morphologies in the mycelium result from changes in the hyphal membranes and walls affecting the flow and passage of molecules. Discussions of this sort raise the important point that differentiation of hyphal cells is, at least in part, a reaction to the local environment. Response to the environment (and the word 'environment' must be taken to cover other cells, the substrate, the substratum, the gaseous atmosphere and physical environmental conditions) has a large measure of control over differentiation of the hyphae. The genetic constitution of the hyphal cell will determine the nature of the differentiation of which the cell is capable, but the local environment determines where, whether and to what degree that differentiation will occur. The ways in which mycelia differentiate morphologically have been recorded in a very extensive literature, some of which can be found in the publications referenced in Appendix 2.

10.2 MYCELIAL DIFFERENTIATION

Rhythmic, or cyclical, growth of colonies on solid medium is an excellent example of this interaction between the environment and the genetic capability of the mycelium. Regular concentric banding of colonies grown *in vitro* is seen quite often in many different fungi. It results from regular changes in hyphal extension rate and branch formation as hyphae react to local conditions as part of an endogenous or externally regulated **circadian clock** (a roughly 24-hour cycle). These circadian rhythms are generated endogenously; under constant environmental conditions they are self-sustaining, and their period is determined genetically. However, they can be modulated by changes in external cues such as light, temperature and nutrients. Temporal rhythms are ubiquitous in eukaryotes and common regulatory patterns in circadian systems extend from fungi through to mammals. The formal study of biological rhythms (that may be daily, weekly, seasonal, annual or with an even longer period) is called chronobiology and covers an enormous range of animal (including human) and plant physiology. But cyclical changes are clearly evident in fungi and their causes are frequently more accessible than they are in other eukaryotes. We present a brief overview of the molecular aspects of **circadian rhythms** in the next few paragraphs.

Circadian rhythms are biological rhythms with periods of about 24 hours. **Circadian clocks** are molecular circuits that allow organisms to coordinate many processes, including gene expression, with a rhythm that is close to the daily 24-hour cycle. Rhythmic processes described in fungi include growth rate, stress responses, developmental capacity and sporulation, as well as many metabolic processes. Generally, fungi use clocks to anticipate daily environmental changes. Rhythmicity is endogenous and self-sustaining when environmental conditions are constant; the length of the rhythmic cycle being genetically determined. Rhythmically changing environmental signals, particularly of light and temperature, set the phase of the endogenous rhythm and adjust it to exactly 24 hours.

Circadian clocks are self-sustaining timekeepers found in almost all organisms on Earth. They have arisen at least three times through evolution, in prokaryotic cyanobacteria, in cells that evolved into higher plants, and in the opisthokont clade, the group of organisms that eventually became the fungi and the animals (Dunlap & Loros, 2017). They do not require complex tissue organisation, and even single cells can express rhythmicity. A negative feedback loop comprises the core of the circadian system in fungi and animals, centred on the transcription of clock genes and translation of clock proteins. The positive element in the loop is the transcriptional activation of one or more clock gene(s) through binding of **paired transcriptional activators** on the clock gene promoter; they are paired by interaction through PAS domains (see below). Translation of the transcribed message of the clock gene (which is subject to additional regulation) generates a **clock protein** that is the negative element of the feedback loop. This **blocks activation** of the clock gene so the amount of clock gene mRNA declines and eventually the levels of clock protein also decline. This generates a daily cycle of clock gene mRNA and clock proteins and forms an oscillator that generates what is known as an 'output' that is the basis of the timing signal that controls rhythmical cellular functions (which might be organism, organ or even cell specific) (Figure 10.1). A circadian system can be made up of one or more interconnected feedback loops forming quite a complex network. In addition, the system receives inputs of ambient light and temperature to adjust its phase so that the internal day matches the external day, and then uses the time information it generates to regulate the life of the cell and of the whole organism.

Many clock gene proteins have a common structural motif known as the **PAS domain**. The name PAS is an acronym created from 'PER-ARNT-Sim'. PAS domains were first identified in the *Drosophila* proteins PER and ARNT and they were later found in a wide range of organisms. PAS domains are involved in many signalling proteins where they are used as a signal sensor domain. In circadian rhythmicity, the PAS-domain proteins act as heterodimeric transcriptional activation complexes to drive expression of clock genes. PAS domains mediate protein–protein interactions in response to stimuli when cofactors bind within their hydrophobic cores, so they have important roles as sensory

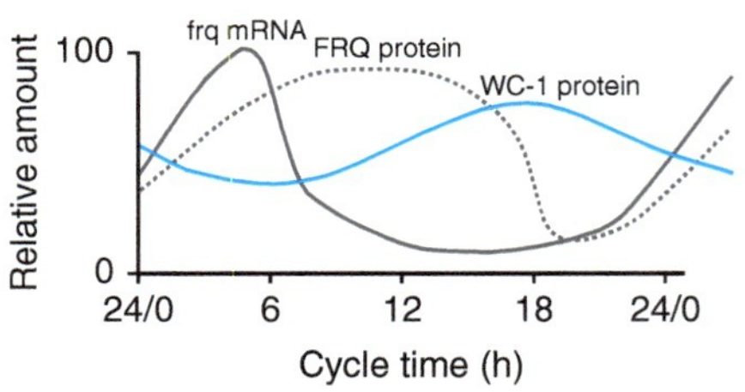

Figure 10.1. Behaviour of the FRQ–WC-1-based oscillator of *Neurospora crassa* over the course of a circadian cycle. The graph shows the daily cycles in levels of FRQ protein and WC-1 protein (product of the *white collar-1* gene) in constant darkness. Light activates production of frq mRNA, but ultimately the presence of excess FRQ protein represses this. Modified from Dunlap & Loros (2006; and see Dunlap & Loros, 2017).

modules for a wide range of environmental conditions including oxygen tension, redox potential, carbon monoxide and nitric oxide, as well as light intensity and temperature; all of which are the main inputs to the *Neurospora* circadian oscillators.

The circadian clock of *Neurospora*, still the best-studied fungal model, involves proteins (which are transcription factors) of mutant genes called *white collar*-1 (the protein is called WC-1), *white collar*-2 (WC-2) and *frequency* (gene symbol *frq*) (Koritala & Lee, 2017). FRQ (the protein encoded by the *frq* gene) is the core clock component (Cha *et al.*, 2015) and complexes with other proteins, physically interacting with the WC transcription factors reducing their activity; the kinetics being strongly influenced by progressive phosphorylation of FRQ. When FRQ becomes sufficiently phosphorylated that it loses the ability to influence WC activities, the circadian cycle starts again. Environmental cycles of light and temperature influence frq and FRQ expression and thereby reset the internal circadian clocks (Dunlap & Loros, 2017). Light acts in *Neurospora* to induce transcription of the negative elements that reset the clock and synchronise the cell to the daily light/dark cycle. Temperature-influenced translational regulation of FRQ synthesis in *Neurospora* sets the physiological temperature limits over which the clock operates.

The circadian system of *Neurospora* is an important model system (Montenegro-Montero *et al.*, 2015) used to understand circadian rhythms in other organisms. There is evidence for conservation of rhythmicity mechanisms in filamentous fungi in general; when tested for homology with *Neurospora* FRQ, WC-1 and WC-2 sequences, scores for similarity were high in genomes of *Basidiomycota*, and zygomycetes as well as other *Ascomycota* (Dunlap & Loros, 2006, 2017). The *Neurospora* circadian system contains at least three oscillators:

- the FRQ/WC-dependent circadian oscillator, the core components of which are FRQ, WC-1, WC-2 and two other proteins, FRH and FWD-1;
- the WC-dependent circadian oscillator;
- and one or more FRQ/WC-independent oscillators.

A survey of 64 fungal proteomes for homologues of *Neurospora* clock proteins found that the FRH and FWD-1 proteins were probably present in the last common ancestor of all the fungi surveyed. Homologues of WC-1 and WC-2 were absent from chytrids and *Microsporidia* but were present in all other major clades. In contrast, FRQ homologues were restricted taxonomically within the *Ascomycota* to *Sordariomycetes*, *Leotiomycetes* and *Dothideomycetes* (Salichos & Rokas, 2010). Our interpretation of these findings is that the components of the *Neurospora* circadian clock are widely conserved in fungal evolution, but the way they are assembled into a working oscillator in *Neurospora* is only one of several possibilities. The regulators that make up the clocks are involved in other cellular control events; for example, one of the FRQ homologues in the fungus *Botrytis cinerea* regulates virulence when the fungus is infecting its plant host *Arabidopsis thaliana* (Hevia *et al.*, 2015, 2016). The development and evolution of a clock circuit will depend on the balance between the different selection pressures exerted on the different functions of its component parts.

How rhythmic growth can influence mycelial morphology is illustrated by a clock mutant of *Podospora anserina* which forms concentric bands of aerial growth within the colony grown on agar media. The banding arises from a difference in growth pattern of aerial and submerged hyphae. Enhanced growth and increased branching of hyphae on the agar surface eventually cause growth of aerial hyphae to stop, perhaps because of the accumulation of excretory products (known as 'stalling substances'), so further extension on the surface is limited by this 'induction event'. On the other hand, submerged hyphae do not show this pattern of increased branching; they escape the restriction to growth and continue to extend and reach the surface some distance beyond the stopped surface mycelial front (Figure 10.2). Emergence of the submerged hyphae prevents further growth of the old surface mycelium but produces a new generation of surface hyphal tips which go through the same process of branching, stalling and growth limitation. Repetition of the cycle gives rise to zones of alternately dense and sparse surface mycelium which are visible as a regular series of bands on the surface. Reduced extension rates and increased branching in this mycelium of *Podospora anserina* are accompanied by increased oxygen uptake and exposure to light.

This example makes it clear that rhythmic growth is a **differentiation** process, which separates hyphae with different

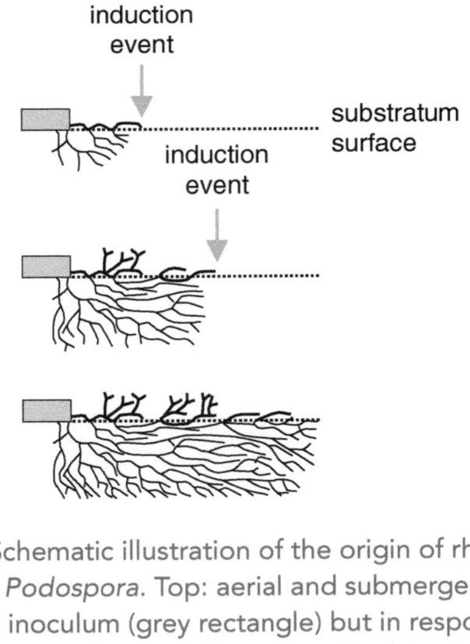

Figure 10.2. Schematic illustration of the origin of rhythmic growth in a mutant of *Podospora*. Top: aerial and submerged hyphae grow away from the inoculum (grey rectangle) but in response to an induction event (may be an endogenous physiological cycle or external signal, like light or temperature fluctuation) surface growth is stopped. The check to surface extension prompts aerial branching and allows submerged hyphae to grow beyond the surface mycelium, branches from submerged hyphae re-establishing surface growth. A second induction event initiates another cycle of aerial branching after the check to extension on the surface (bottom diagram). Redrawn after Lysek (1984).

functions and properties in space and time. In this case, extending hyphae, exploring for new substrates, are separated spatially from stationary surface hyphae, which may differentiate into spore-forming and/or resting structures after their extension growth is stopped.

Concentric rings and radial zonations of the mycelium are common expressions of mycelial growth rhythms. Evidently, an induction event must be detected, and this suggests that membrane sensors play an important role in these sorts of reactions. The evidence indicates that many different sensory modules reacting to a wide range of environmental conditions (gases, redox, light, temperature) can input signals to a common oscillator as indicated above. The important point is that the rhythmic output is sent to the **filamentous cell cycle** discussed in Chapters 4 and 5, so that the essentially homogeneous growth and branching pattern of the vegetative hyphae is disturbed by the inducing event and differentiation of the mycelium results. This may be an intrahyphal differentiation, as in spore differentiation, or a concerted co-differentiation of several or many hyphae to produce the equivalent of a differentiated tissue, as in the concentric rings and zonations we have just been discussing.

Even in more complex fungal structures, like sporophores, similar mechanisms may exist to change the hyphal growth and branching patterns to form functionally distinct fungal tissues. Certainly, experiments based on transferring such tissues to artificial environments *in vitro* indicate that hyphal cells differentiate in response to signals they receive from their immediate environment within the 'home' tissue and that maintenance of their state of differentiation requires continual reinforcement from those signals (Chiu & Moore, 1988a & b). It is what those signals do that is important, and the general concept is that signals that control fungal differentiation do so by modifying the *pattern* (that is, the temporal and/or spatial distribution) of any or all of the numerous components that contribute to the fungal cell cycle.

10.3 MAKING SPORES

In fungal spore formation, the wall-building mechanisms that are normally strictly apical are highly adapted. Clearly, wall synthesis at the hyphal apex is far from being the complete story. Further synthesis of new wall as well as modification of existing wall is a frequent occurrence. When and where it occurs is under exquisite control. There are so many instances in which fungal wall synthesis is positionally and temporally regulated to produce regular change in morphology that the activities located at the apex of the vegetative hypha, described in detail in Chapters 4 and 7, are much more mundane by comparison. There are several ways of creating spores and several ways of organising the walls of spores and, as you might imagine, several terms have been coined to describe these.

The term 'wall-building' has been coined to describe the process and three types have been distinguished (Figure 10.3). These are:

- **apical wall-building**, in which the ultrastructural secretory vesicles responsible for producing cell wall material are concentrated at the hyphal tip and form a cylindrical hypha by distal growth, in which the youngest wall material is at the extreme apex (Figure 10.3A);
- **diffuse wall-building**, in which the synthetic secretory vesicles are distributed all over the apical region at a low concentration, resulting in swelling of the cylindrical hypha through alteration of the pre-existing wall (Figure 10.3B);
- **ring wall-building**, in which wall synthesis is concentrated in a ring below the tip and produces new wall by proximal growth, so a cylindrical hypha is formed in which the youngest wall material is always at the base (Figure 10.3B).

You will appreciate that an infinite range of hyphal cell shapes can be produced by sequential operation of these processes. Add hyphal branching and control of autotropism (either positive to cause the branch to grow towards its parent and thereby form a multihyphal structure, or negative to cause the branch to grow away from the parent to generate an open network) and it is possible to account for the formation of a wide range of spore-bearing structures (sporophores, specifically, conidiophores). In turn, this diversity of spore shaping accounts for both the enormous biodiversity in asexual sporulation throughout the fungi, and the central role in classical taxonomy played by those spores and spore-bearing structures.

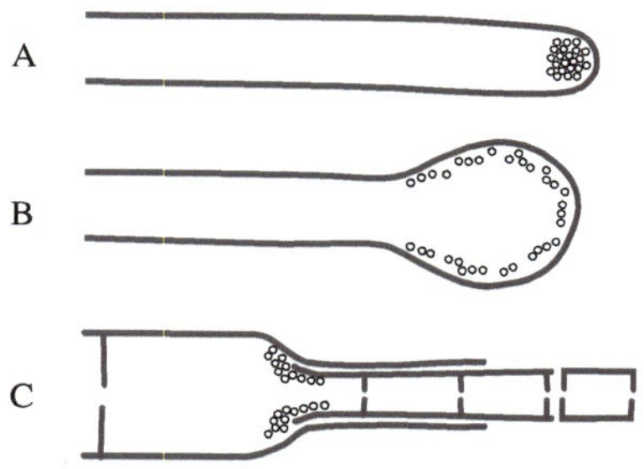

Figure 10.3. Different 'wall-building' processes involved in hyphal differentiation. (A) Apical wall-building, in which the ultrastructural secretory vesicles responsible for producing cell wall material are concentrated at the hyphal tip and form a cylindrical hypha by distal growth, in which the youngest wall material is at the extreme apex. (B) Diffuse wall-building, in which the synthetic secretory vesicles are distributed all over the apical region at a low concentration, resulting in swelling of the cylindrical hypha through alteration of the pre-existing wall. (C) Ring wall-building, in which wall synthesis is concentrated in a ring below the tip and produces new wall by proximal growth, so a cylindrical hypha is formed in which the youngest wall material is always at the base (Minter *et al.*, 1983a & b). Redrawn from the *Dictionary of the Fungi* (Kirk *et al.*, 2008).

The almost infinite variety of hyphal and cell shapes have:

- attracted the attention of taxonomists over the years, who have used this cellular diversity to define taxa of various ranks;
- been given an equally wide variety of descriptive names.

The prime source for explanation of names and terms in mycology is the *Dictionary of the Fungi* (Kirk *et al.*, 2008), and a good alternative is the *Illustrated Dictionary of Mycology* (Ulloa & Hanlin, 2012). In Appendix 2, we use extracts from the *Dictionary of the Fungi* to show something of **the range of diversity that has been observed in fungal cells**. The main purpose of Appendix 2 is to introduce you to the diversity of spore and hyphal morphology, and to the diverse terminology developed by mycologists to describe this. However, you should take the opportunity to examine the range of morphologies shown, so that you can understand ***how*** shape of a spore or hypha can be determined by precise control over the timing and position of renewed wall growth varying the formation of wall and associated machinery.

Conidiophores are usually morphologically distinct from the vegetative hyphae on which they arise. They may be branched or unbranched and often have inflated apices supporting groups of conidiogenous cells. Conidiogenous cells differ in the way they produce conidia, which may be formed singly, in clusters or in succession. When conidia are produced in succession, the chains that are formed may have the oldest conidium located at the apex of the chain, being pushed upward as younger conidia form at the base, or conversely the chain may be formed by successive production of the youngest conidium at the apex.

Mature conidia may be smooth or bear ornate **sculpturing**, and their surface layers may be modified to make them hydrophobic or hydrophilic. Conidia are often round or ellipsoid but may be curved, coiled or of more elaborate shape. Conidia are usually single cells but may be made up of two to several cells. Some are colourless, but many have pigmented walls, accumulations appearing white, blue, green, yellow, brown or black.

Recognising the ways in which the hyphal wall can be formed turns an observational study of the different shapes of the final structures into an understanding of the ways that conidia develop. Some fungi produce spores by transforming an intercalary cell (that's a cell 'in among others' or within the body of a hypha) into a chlamydospore by rounding up of the cells and deposition of wall thickening (i.e. diffuse wall-building and secondary wall formation). Chlamydospores are released when the parent hypha disintegrates. This is one of two basic sorts of conidiogenesis, called **thallic** development (the other is called **blastic**). In thallic conidiogenesis (Figure 10.4A), the spore initial differentiates ***after*** it has been delimited by completion of one or more septa, and the spore is differentiated from a whole cell.

The **spore initials** are delimited by septa formed after apical growth of the fertile hypha has ceased. Chlamydospores are enlarged and enlargement occurs only after the septa are formed. Spores that undergo this type of development but without enlargement are often given special names such as oidia, found

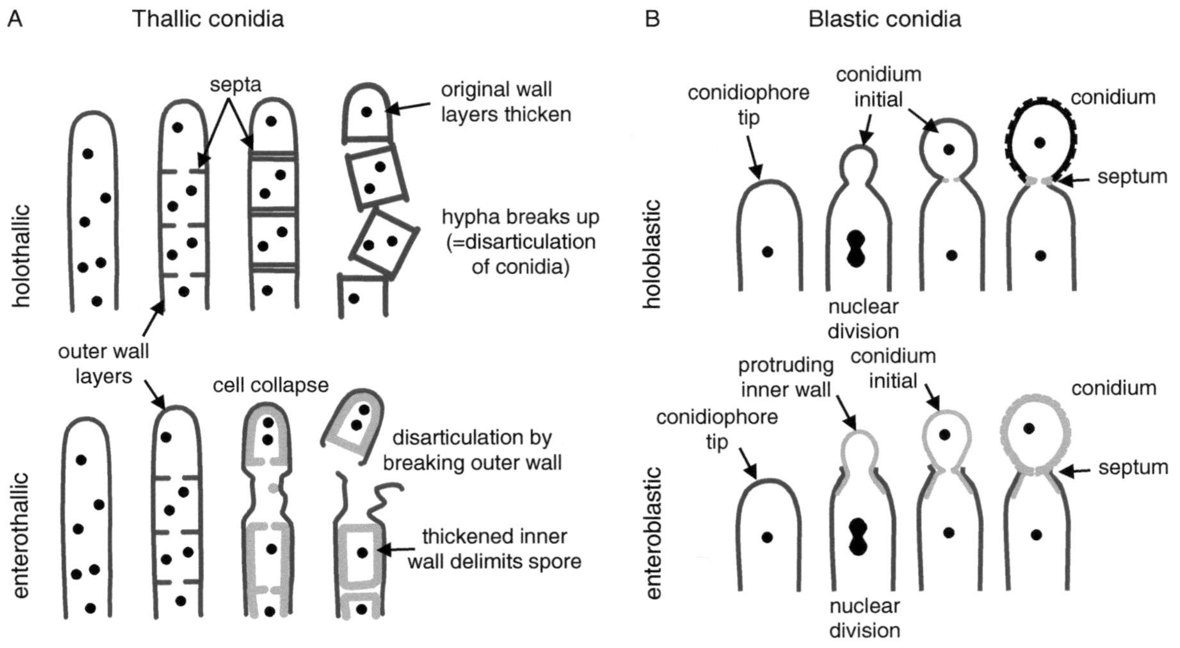

Figure 10.4. Comparison of thallic and blastic conidial development: (A) compares holothallic (top) and enterothallic (bottom) conidiogenesis. (B) illustrates holoblastic (top) and enteroblastic conidiogenesis (bottom).

on haploid monokaryotic mycelia of many of the smaller mushrooms, like *Coprinopsis*, or arthrospores, which are formed by fragmentation of hyphal branches of members of the *Ascomycota*, like *Geotrichum*, but they are all **thallic conidia**. However, they may be distinguished as **holothallic** or **enterothallic** on the basis that the former type uses the original wall of the parent conidiogenous cell, while in enterothallic conidiogenesis the conidial wall is newly formed and does not use the original wall of the conidiogenous cell (Figure 10.4A). Enterothallic conidia are relatively uncommon.

Blastic development is characterised by differentiation of a spore initial from part of a cell rather than the whole cell, and occurs ***before*** it is separated off by a septum (Figure 10.4B). These spores are called blastic conidia, and, again, may be distinguished as holoblastic (using the original wall) or enteroblastic (using only newly formed wall). **Blastic conidia** may be produced from a specialised (conidiogenous) cell that is sometimes another conidium (that is, they may form in chains). One to several conidiogenous cells may be produced from and supported by a conidiophore, a specialised hypha or hyphal branch. When mature, conidia separate readily from the conidiogenous cell, so the spore is described as deciduous.

Thallic conidial development is remarkably similar to division of *Schizosaccharomyces pombe* (fission yeast) in which septation involves chitin deposition in a ring defined by a preformed ring of actin microfilaments (Figure 10.5). Blastic development has similarities to budding in *Saccharomyces cerevisiae*, because the conidial initial emerges from a specific localised region of the conidiogenous cell and may enlarge considerably before being cut off by a septum (compare Figure 5.15). Ring wall-building creates the emergent conidial initial, and then diffuse wall-building within the initial causes it to balloon from the parent cell. As the conidial initial increases in size, a nucleus migrates from the parent cell into the young conidium, which is finally separated from the parent cell by centripetal growth of a septum. The septum sometimes includes a special abscission layer that enables conidial release.

Enteroblastic conidia are very commonly encountered. Conidiogenous cells differ morphologically and contribute to the morphological biodiversity of fungi. In some fungi, conidiogenous cells are not specialised to produce large numbers of conidia and are very similar to vegetative hyphae. Other common conidiogenous cells are specialised, non-elongating bottle-shaped cells with a narrow neck from which the conidia develop; these are called phialides (Figure 10.6).

The first conidium on a phialide may form by either the holoblastic or enteroblastic method. If formed enteroblastically, it ruptures the phialide wall as it emerges, and the ruptured wall may persist as a minute collar at the base of the conidial chain. Either way, development of subsequent conidia is enteroblastic, each one originating as the innermost layer(s) of the phialide wall are extruded through the 'mouth' of the phialide. Although they originate on the inner side of the phialide wall, these layers eventually become the outermost layers of the conidial wall. The septum is formed within the phialide to separate the exogenous maturing conidium from the next (endogenous) conidium initial. The spores of *Aspergillus* are the most often observed conidia that are formed in this way.

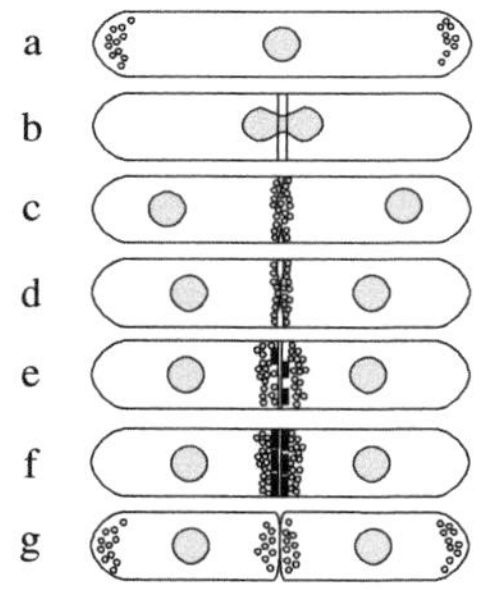

Figure 10.5. Septum formation and cytokinesis in *Schizosaccharomyces pombe* as an example of thallic development. (a) Fission yeast cells grow mainly by extension at both tips, actin is located at the growing ends of the cell (shown here as small circles) and a sheaf of microtubules runs from end to end. (b) When mitosis is initiated, actin is relocated to form an equatorial ring of filaments around the nucleus, the position of which anticipates the position of the septum. (c and d) As mitosis is completed the primary septum originates in the periplasmic space and grows into the cell. (c–e) As synthesis of the septum proceeds, actin distribution changes from a filamentous ring to clusters of vesicles located where septum material is deposited. (e and f) Secondary septa are formed either side of the primary, and the latter is removed to bring about cell separation (g). Redrawn from Moore (1998).

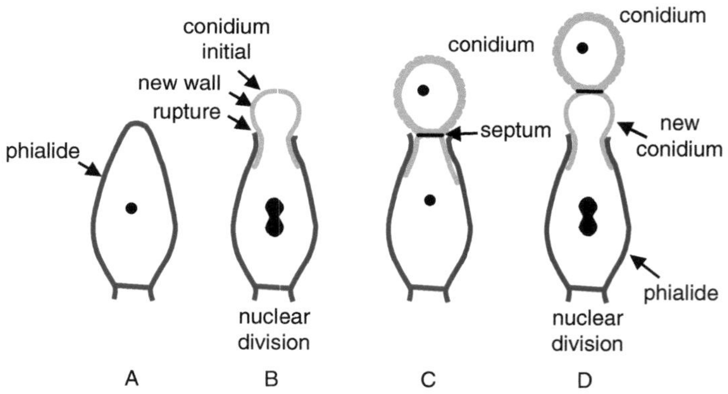

Figure 10.6. Development of conidia from a phialide. (A) Slightly elongated phialide with a single wall. (B) The apex of the phialide is ruptured by emergence of the enteroblastic conidial initial; the wall that protrudes becomes the external wall of the newly developing conidium. (C) The new conidium has been completed and separated by a septum (which may include an abscission layer). (D) The process is repeated to form a second conidium beneath the original and as subsequent conidia extend the chain, the oldest is always at the end of the chain and the youngest is emerging from the phialide. Redrawn after Moore-Landecker (1996).

10.4 *ASPERGILLUS* CONIDIOPHORES

A great deal of progress has been made in establishing the details of the genetic control of conidiospore formation in *Aspergillus nidulans*, the conidia of which arise on specialised aerial hyphae, the **conidiophores**. Formation of conidia by surface cultures of *Aspergillus nidulans* occurs after about 16 h of hyphal growth. This period of vegetative growth is required to make the cells competent to respond to the induction process, which requires exposure to air and is probably a reaction to cell-surface changes at air–water interfaces. After induction, some mycelial hyphae produce aerial branches which become conidiophore stalks (Figure 10.7).

The cell from which the branch emerges is the conidiophore **foot cell**, which is distinguishable from other vegetative cells by having a brown pigmented secondary wall thickening on the inside of its original wall. The **stalk** grows apically until it reaches a length of about 100 μm when the apex swells to form the **conidiophore vesicle** which has a diameter of about 10 μm. A single tier of numerous **primary sterigmata**, called **metulae**, then bud from the vesicle and **secondary sterigmata**, the **phialides**, bud from the exposed apices of the metulae. The phialides are the stem cells which subsequently undergo repeated asymmetric divisions to form the long chains of **enteroblastic conidia** which are approximately 3 μm in diameter (Figure 10.7).

Classical **genetic analysis**, by isolation and analysis of mutants, established the basic genetic outline of *Aspergillus* conidiation. Between 300 and 1,000 gene loci were estimated to be concerned with conidiation by comparing mutation frequencies at loci affecting conidiation with those for other functions (Martinelli & Clutterbuck, 1971). Analysis of mRNA species indicated that approximately 6,000 were expressed in vegetative mycelium and an additional 1,200 were found in cultures which included conidiophores and conidia; 200 of these additional mRNAs being found in the conidia themselves. Only about 2% of mutants of *Aspergillus nidulans* that lacked conidia had defects in stages concerned with conidiophore growth and development, and 85% of conidiation mutants were also defective in vegetative hyphal growth and in attaining competence.

Two genes, in particular, play a key role in conidiophore morphogenesis: these are the **'bristle'** (*brlA*) gene which has defects in vesicle and metula formation, and **'abacus'** (*abaA*) in which conidia are replaced by beaded lengths of hypha, so it is presumably defective in conidial budding from the phialide and final septation. A third gene, **wetA**, is defective in an early stage of **spore maturation**. Conidia of *wetA* mutants lack pigment and hydrophobicity; they autolyse after a few hours and fail to express a range of spore-specific mRNAs. The *wetA* gene transcript is lacking in *brlA* and *abaA* mutants (i.e. *brlA* and *abaA* are **epistatic** to *wetA*), and studies of double mutants show that these three genes act in the order: *brlA*–*abaA*–*wetA*.

A striking feature of the mutational analysis of conidiophore development in *Aspergillus nidulans* is that mutants of only these three genes cause defects in conidiophore and spore morphogenesis, whereas up to 1,000 other gene loci cause absence of conidiation when mutated. This implies that *brlA, abaA* and

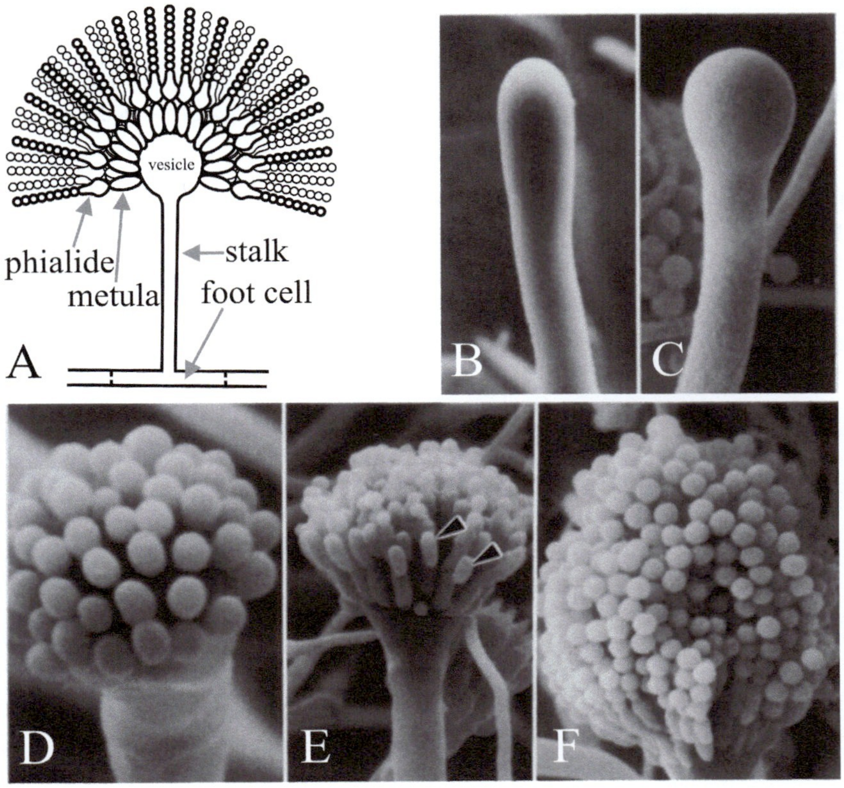

Figure 10.7. Structure and development of the conidiophore of *Aspergillus nidulans.* (A) is a diagrammatic illustration of the structure of the mature conidiophore. (B to F) are scanning electron micrographs of various stages in conidiation. (B) Young conidiophore stalk just prior to vesicle formation. (C) Developing vesicle at the tip of the conidiophore. (D) Developing metulae. (E) Developing phialides (arrowheads). (F) Tip of mature conidiophore bearing numerous chains of conidia. Electron micrographs B to F taken from Mims *et al.* (1988) with kind permission from Springer Science+Business Media: © Springer-Verlag 1988.

wetA are regulators that integrate the expression of other genes that are needed for conidiation but are not themselves dedicated to it. Many of the *Aspergillus* conidiation mutants were also defective in sexual reproduction, so another conclusion to be drawn is that there is some economy of usage of morphogenetic genes in different developmental processes. Presumably, different developmental modes employ structural genes that are not uniquely developmental, but function in numerous pathways, having their developmental specificity bestowed upon them by the regulators to which they respond. This is epitomised in the idea that the key to eukaryote development is in the ability to use relatively few regulatory genes to integrate the activities of many others.

Molecular analysis supports the interpretation that *brlA*, *abaA* and *wetA* are **regulators**; the *brlA* sequence encodes a **zinc-finger protein**, which is a sequence-specific DNA-binding transcription activator of developmentally regulated target genes (Figure 10.8). However, that's not the whole story, because the *brlA* product has different affinities for different target genes. Indeed, the *brlA* locus consists of **overlapping transcription units** (Figure 10.9), the downstream unit being designated *brlAα* and the upstream unit *brlAβ*; between them, their products solve two classic developmental problems, how to respond to a signal and how to maintain that response when the signal has dissipated.

The two *brlA* transcription units share the same reading frame for most of their length but *brlAβ* has an additional 23 amino acid residues at the amino-terminal end of that reading frame, and its transcript also possesses a separate reading frame of 41 amino acid residues (called μORF) near its 5′ terminus. The μORF product represses translation from the downstream (*BrlA*) reading frame and even though *brlAβ* transcript can be detected in vegetative hyphae, the BrlA peptide is not produced. The repression caused by μORF is only overcome when the mycelium is competent, which is probably signalled as a nitrogen limitation (a common environmental signal for initiation of sporulation in *Ascomycota*), which reduces aminoacyl-tRNA pools and disturbs translational regulation by μORF. When the effect of μORF is stalled, the BrlA peptide can be translated from existing transcript.

The discoverer of this process described the activation of the conidiation pathway as **'translational triggering'** (Timberlake, 1993), suggesting that the translational trigger is a way of making differentiation sensitive to the nutritional status of the hypha. The competent hypha is primed to undertake conidiophore development but irreversible activation of the conidiation pathway is prevented by a translational repression that maintains vegetative growth until sporulation conditions are ideal.

Activation of *brlA* is, therefore, seen as the first step in conidiophore development, and its product in turn activates a panel of conidiation-specific genes among which is the next regulator, *abaA*. The *abaA* product is also a DNA-binding transcriptional regulator protein which enhances expression of *brlA*-induced structural genes. The *brlA* and *abaA* genes are reciprocal

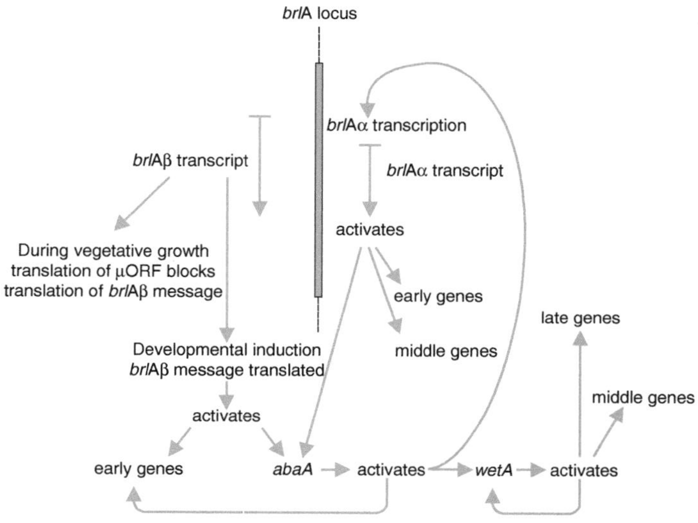

Figure 10.8. Summary of the genetic regulatory circuit for conidiophore development in *Aspergillus nidulans*.

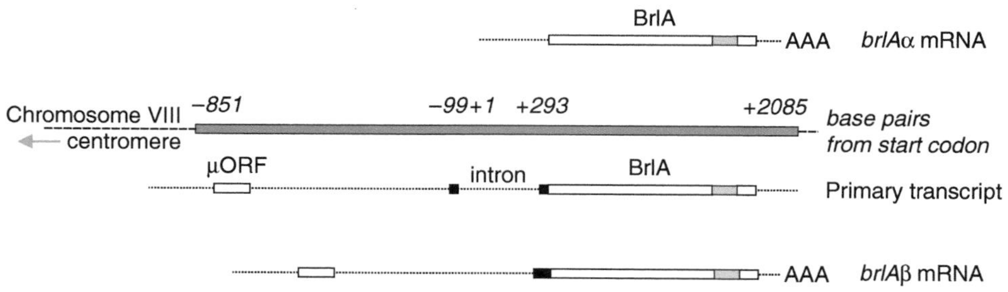

Figure 10.9. Structure of the *brlA* locus of *Aspergillus nidulans*. The brlAα mRNA is shown at the top, above a shaded box which represents the BrlA segment of chromosome VIII. The brlAβ primary transcript and mRNA are shown in the lower half of the figure. The brlAα sequence is a single exon encoding a Cys2-His2 zinc-finger polypeptide (location of the zinc fingers shown as a shaded box within BrlA). The brlAβ sequence contains one intron. The polypeptide encoded by brlAβ contains an additional 23 amino-terminal residues (corresponding sequence shown as a black box) and the transcript has a short upstream open reading frame (μORF) which regulates translation of brlAβ. Redrawn after Timberlake (1993; and see Lee *et al.*, 2016).

activators, because *abaA* also activates *brlA*. Of course, *brlA* expression must occur before *abaA* can be expressed, but the consequential *abaA* activation of *brlA* reinforces the latter's expression and effectively makes progress of the pathway independent of outside events.

The *abaA* product also activates additional structural genes and the final regulatory gene, *wetA*, which activates spore-specific structural genes. Since *brlA* and *abaA* are not expressed in differentiating conidia, *wetA* is probably involved in inactivating their expression in the spores (and perhaps in the phialide, too; as the current spore nucleus reaches the maturation stage that needs *wetA*-regulated genes, the phialide nuclei must be 'turned back' to the conidium-initiation state to start formation of the next spore). Expression of *wetA* is initially activated in the phialide by sequential action of *brlA* and *abaA*, and it is then autoregulated (the *wetA* product activates *wetA* transcription). Positive autoregulation of *wetA* subsequently maintains its expression after the conidium has been separated (physically and cytologically) from the phialide.

Timberlake (1993) called this mechanism **feedback fixation**: reciprocal activation, feedback activation and autoregulation of the core regulatory sequences reinforce expression of the whole pathway, making it independent of the external environmental cues that initiated it and enabling the spore to continue maturation even after separation from the phialide. Conidiophore development, like many other morphogenetic processes, is naturally divided into sequential steps. This regulatory network shows how translational triggering can relate a morphogenetic pathway to the development of competence on the one hand, and to initiation in response to environmental cue(s) on the other. After initiation, feedback fixation results in developmental determination in the classic embryological sense of continuing morphogenesis even when removed from the initiating environment.

We chose deliberately in this section to base our description on research done 50 years or so ago that started with the isolation of a large number of mutations with altered conidiation phenotypes and continued with 'classic' studies of gene segregations and gene complementation (Martinelli & Clutterbuck, 1971). Besides showing how our understanding of sporulation has been accomplished, it demonstrates the value of using these 'old-fashioned' techniques to investigate such phenomena. However, we can't ignore the extensive research which has been completed into the molecular mechanisms underlying growth and development of *Aspergillus*.

These studies have confirmed that the key event in sporulation is activation of the zinc-finger transcription factor encoded by *brlA*, and that *abaA* and *wetA* genes are necessary regulators of conidiation. The *abaA* encoded transcription factor is activated by *brlA* after differentiation of metulae and during the middle stages of conidiophore development; and the *wetA* gene, activated by *abaA*, functions in late phase of conidiation directing the synthesis of crucial cell wall components and transforming the metabolism of the maturing conidium. In *Aspergillus nidulans*, these three genes create the central regulatory pathway, which, with other genes, control conidiation-specific gene expression and the sequence of gene activation involved in the acquisition of developmental competence, conidiophore development and spore maturation (Lee *et al.*, 2016).

Many aspects of these developmental pathways have been conserved in other aspergilli. Members of the genus *Aspergillus* are among the most commonly encountered fungi, and all reproduce asexually by forming long chains of conidia. Several species, including *Aspergillus oryzae* and *Aspergillus niger*, are used in industry for enzyme production and food processing; while *Aspergillus flavus* is responsible for food spoilage by producing the most potent known naturally occurring carcinogens, the aflatoxins. Another species of concern is the opportunistic human pathogen *Aspergillus fumigatus*, which produces a massive number of small hydrophobic conidia as its primary means of dispersal and has become a widespread airborne fungal pathogen in developed countries. In immunocompromised patients, *Aspergillus fumigatus* causes an invasive aspergillosis that has a high mortality rate. The *BrlA–AbaA–WetA* developmental signalling pathway has been conserved in conidiation of these *Aspergillus* species (Yu, 2010; Tao & Yu, 2011; Krijgsheld *et al.*, 2013). Further, proteomic, transcriptomic and metabolomics studies have provided a detailed picture of the dynamic changes that occur in many thousands of genes, transcripts, enzymes and metabolic reactions during formation, maturation, dormancy and germination of *Aspergillus* conidia (van Leeuwen *et al.*, 2013; Novodvorska *et al.*, 2016; Teertstra *et al.*, 2017).

10.5 CONIDIATION IN *NEUROSPORA CRASSA*

Neurospora crassa forms two types of conidium, microconidia and macroconidia. **Microconidia** are small uninucleate spores which are essentially fragmented hyphae. They are not well adapted to dispersal and are thought to serve primarily as 'male gametes' in sexual reproduction. **Macroconidia** are more common and more abundant, they are large multinucleate, multicellular spores produced from aerial conidiophores. Conidiation (and sexual reproduction, too) in *Neurospora crassa* seems to respond more to environmental signals than to complex genetic controls like those operating in *Aspergillus*. Macroconidia are formed in response to nutritional limitation, desiccation, change in atmospheric CO_2 and light exposure (blue light is most effective, and though light exposure is not essential, conidia develop faster and in greater numbers in illuminated cultures). In addition, a circadian rhythm provides a burst of sporulation each morning. When induced to form conidia, the *Neurospora* mycelium forms aerial branches which grow away from the substratum, form many lateral branches which become conidiophores and undergo apical budding to produce conidial chains.

The genetics of conidiation, studied by means of mutation and molecular analysis, reveal some parallels in terms of types of mutants obtained with *Neurospora crassa* and *Aspergillus nidulans*, and a particular example would be the hydrophobic outer rodlet layer which is missing in the *Neurospora crassa* '*easily-wettable*' (*eas*) and *Aspergillus nidulans rodA* mutants. Despite such functional analogies, there is no underlying similarity between the genetic architectures used by these two organisms to control conidiation. In *Neurospora*, the fluffy gene (*fl*) is necessary to induce conidiophore development (Bailey-Shrode & Ebbole, 2004), and the *acon-2*, *acon-3* and *fld* genes are essential for conidium development. While there may be some similarity in regulatory strategy, it is important to note that the genome sequence of *Neurospora* possesses no homologues of the *brlA–abaA–wetA* regulators of *Aspergillus nidulans* (Galagan *et al.*, 2003).

Nevertheless, a great many mutants have been isolated which have defects in specific stages of conidiation. Several conidiation (*con*) genes are known which encode transcripts which become more abundant at specific stages during conidiation. At least four of these genes are expressed in all three sporulation pathways in *Neurospora* (macroconidia, microconidia and ascospores) but others have specific localisation to macroconidia. However, many of the *con* genes can be disrupted without affecting sporulation; despite being highly expressed during sporulation, they presumably encode non-essential or replaceable functions.

Transcriptional profiling has been used to study relative mRNA expression during colony development (Kasuga & Glass, 2008). The relative expression of genes involved in protein synthesis and energy production was enriched in the middle section of the colony, while sections of the colony undergoing asexual development (conidiogenesis) were enriched in expression of genes involved in protein/peptide degradation and other proteins of unclassified function. When colony development in *Neurospora crassa* and *Aspergillus niger* was compared, shared temporal and spatial patterns were observed in the regulation of gene orthologues. The expression of phylogenetically conserved groups of genes was enriched in the middle section of a *Neurospora crassa* colony whereas expression of genes unique to *Pezizomycotina* species and of *Neurospora crassa* orphan genes was enriched at the colony periphery and in the older, conidiating sections of a fungal colony.

Orphan genes (also called 'ORFans') are taxonomically restricted genes or genes without detectable homologues in other lineages. We take this observation to indicate that activities of genes specific to *Neurospora crassa* are enhanced in the mature central regions of the colony which is engaged in making spores, and at the colony margin in support of the *Neurospora*-specific hyphal extension growth. In other words, observations from these surveys emphasise the diversity theme of this chapter by ***demonstrating*** that at the growing edge of the mycelium where the species-specific morphology of the mycelium is first established, ***species-specific*** panels of genes and transcripts are put into action. Similarly, in the central, maturing, regions of the mycelium where the species-specific sporulation takes place, the macromolecules used for the creative modelling required by ***spore formation*** are more species specific (more diverse) than those deployed during spore germination.

The genomic transcriptional profile of conidial germination in *Neurospora crassa* has also been investigated (Kasuga *et al.*, 2005), and suggests that there are rather more common regulatory features affecting spore germination between widely different species than are conserved in other stages of conidiogenesis. Cross-species comparisons of expression profiles during spore germination of the ascomycete *Neurospora crassa*, the basidiomycete *Ustilago maydis* and the cellular slime mould *Dictyostelium discoideum* revealed that some of the same sets of orthologous genes were activated or deactivated under comparable stages of **spore germination** among these very different organisms (and see Breakspear & Momany, 2007, and McCluskey & Baker, 2017, for other comparisons of expression profiles between organisms).

10.6 CONIDIOMATA

Individual conidiophores are obviously complicated, but in some fungi several-to-many conidiophores can orchestrate their activities to produce coordinated tissues that form asexual sporophores. Of course, these (all called **conidiomata**, i.e. sporophores that produce asexual conidia; singular **conidioma**) have all been described by specific terms and given specific names. One type of arrangement is a **synnema**; this is a cluster of conidiophores that adhere to each other along some of their length, forming an elongated stem or bristle-like structure (called a fascicle, this being Latin for 'bundle'). **Synnemata** (that's the plural form) may have many hyphal strands in their structure and be fleshy, hard or brittle in structure. The apical portions of the conidiophores that make up the synnema are separate and radiate outward, often producing a mucus slime in which the conidia are trapped (perhaps awaiting insect dispersal) (Figure 10.10).

Several other sorts of conidiomata are produced, in particular, by pathogens and parasites of both animals and plants. A tightly clustered collection of conidiophores, shorter than those composing a synnema, and arising from a central cushion-shaped (pulvinate) stroma made up of tightly woven hyphae (a pseudoparenchymatous tissue, see below) is called a **sporodochium**.

An **acervulus** (Figure 10.11) is a distinct mass of conidiophores which develops just beneath the surface of a host plant and bursts through as the spores mature, pushing aside flaps of host tissue as they emerge (the name is derived from the Latin *acervus*, meaning 'heap'). It is rather less structured than a **stroma** and is produced by species of the plant pathogen *Colletotrichum*, cause of anthracnose diseases of citrus, banana, cucumber and other crops.

A **pycnidium** is a highly organised structure, basically a chamber, lined with conidiophores, constructed from a distinct wall layer of pseudoparenchymatous tissue (Figure 10.12) which may be dark and tough or brightly coloured and fleshy. Some pycnidia are enclosed in **stromata**, others in host tissues, still

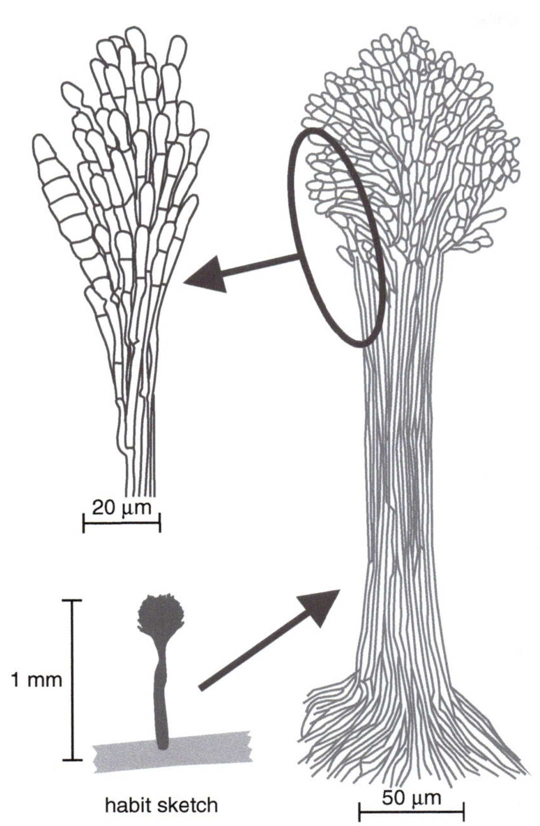

Figure 10.10. Conidiomata. Drawings of one of the synnemata (bunched conidiophores) of *Podosporium elongatum*. The species grows on decaying bamboo stems. The synnemata are about 1 mm tall. Here we show a habit sketch at bottom left and two successive magnifications, with a view of some separated-out conidiophores at top left. Redrawn after Chen & Tzean (1993).

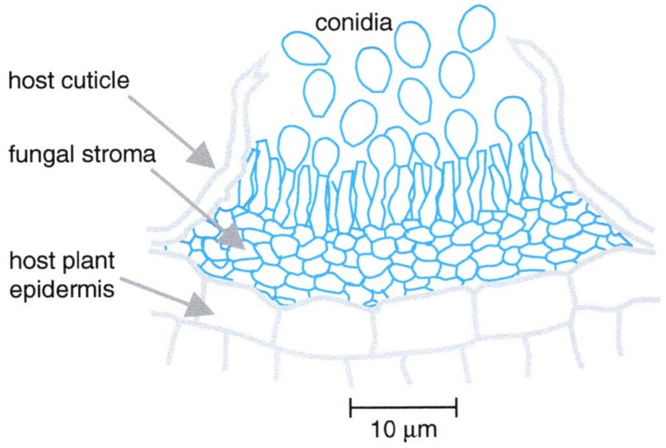

Figure 10.11. Conidiomata. Diagram illustrating the structure of an acervulus. The fungal pathogen forms a stroma of interwoven hyphae (shown in blue) in the epidermis of the host plant (shown in grey) and then bursts through the plant cuticle to release conidia. Redrawn after Moore-Landecker (1996).

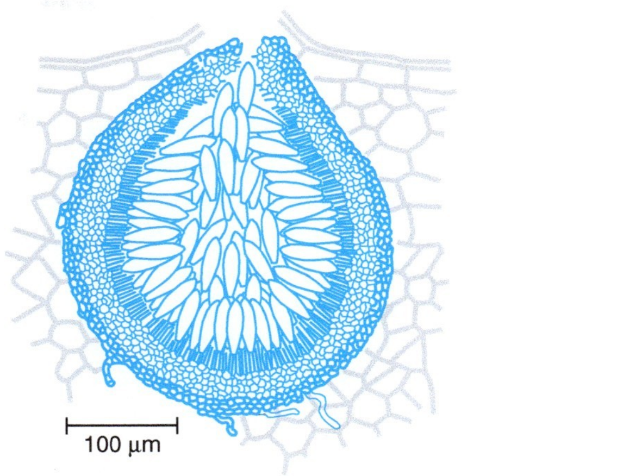

Figure 10.12. Conidiomata. Diagram illustrating the structure of a pycnidium. The fungus forms a chamber with walls of hyphae woven into a pseudoparenchymatous tissue (shown here in blue) that may be dark and tough or brightly coloured and fleshy. Conidiophores produce conidia within the pycnidium. Some pycnidia are enclosed in stromata, others in host tissues (shown here in grey), still others are surface structures. They may be disc-shaped, globose, flask-like or cup-shaped; and may be entirely closed or open to the air through a hole (ostiole), slit or tear to release conidia. Redrawn after Moore-Landecker (1996).

others are surface structures. They may be disc-shaped, globose, flask-like or cup-shaped; can be entirely closed or open to the air through a hole, or ostiole, slit or tears. Fungi that produce pycnidia often occur as saprotrophs on plant litter, or as plant pathogens, such as the leaf spot diseases caused by species of *Septoria* on wheat, celery, azaleas and gladioli.

A pycnidium superficially resembles the ascoma known as a **perithecium**, except that one produces conidia and the other produces sexual ascospores. We will describe and illustrate ascomata below, but the basis of the similarity between pycnidia and perithecia is that both are constructed from multihyphal tissues. The majority of the macroscopic fungal structures are formed by hyphal aggregation into either linear organs (strands, rhizomorphs and mushroom stipes) or globose masses (sclerotia and the familiar sporophores, as well as less-familiar sporulating structures, of the larger *Ascomycota* and *Basidiomycota*).

The general term **plectenchyma** (Greek *plekein*, to weave, with *enchyma* = infusion, meaning an intimately woven tissue) is used to describe organised fungal tissues. There are two types of plectenchyma: **prosenchyma** (Greek *pros*, towards + *enchyma*; i.e. approaching or almost a tissue), which is a rather loosely organised tissue in which the components can be seen to be hyphae; and **pseudoparenchyma** (Greek *pseudo* = false + *parenchyma* = a type of plant tissue), which, when seen in microscope sections, seems to be comprised of tightly packed cells resembling plant tissue. In pseudoparenchyma, the hyphae are not immediately obvious as such, though the hyphal nature of the components can be demonstrated by reconstruction from serial sections or by scanning electron microscopy.

When viewing sections of these (and other) fungal tissues (such as in Figure 10.12) it is essential that you remember that the structure is made up of **tubular hyphae**; what appear to be plant-like cells are just sectioned profiles of those hyphae. Most of the lower fungi have coenocytic hyphae, but lower fungi do not form multicellular (multihyphal) structures. Some authors have argued that the term **cellular element** should be used in preference to 'cell' in fungal tissues, because fungal cells are always hyphal compartments and consequently different from the concept of the cell which emerges from elementary biological education. However, we think this is an unnecessary complication and will use the word 'cell' here.

10.7 LINEAR STRUCTURES: STRANDS, CORDS, RHIZOMORPHS AND STIPES

Formation of parallel aggregates of morphologically similar hyphae is common among *Ascomycota* and *Basidiomycota*. The most loosely organised of them, mycelial strands and cords, can provide the main **translocation routes** of the mycelium, and so develop in circumstances that require large-scale movement of nutrients (including water) to and from particular sites. Mycelial strands originate when young branches adhere to older leading hyphae and weave them together (Figure 10.13). Further localised growth and incorporation of other hyphae it may meet leads to increase in size of the strand.

Anastomosis between the hyphae of the strands consolidates them and narrow hyphal branches (**tendril hyphae**) from the older regions of the main hyphae intertwine around the other hyphae (Figure 10.13). From the beginning, some of the central hyphae may be wide-diameter, thin-walled, so-called **vessel hyphae** and in older strands narrow, but thick-walled, **fibre hyphae** appear, running longitudinally through the mature strands. Strand formation occurs in ageing mycelium on an exhausted substrate when the hyphae are likely to be the main repositories of nutrients (especially nitrogen) and it has been argued that stranding results from the limitation of new growth to the immediate vicinity of the remaining nutrient. If the strand is the main supplier of nutrient then the integrity of the strand will be reinforced, but when the strand encounters an external source greater than its own endogenous supply the stimulus to cohesive growth is lost and spreading, invasive, hyphal growth envelops the new substrate.

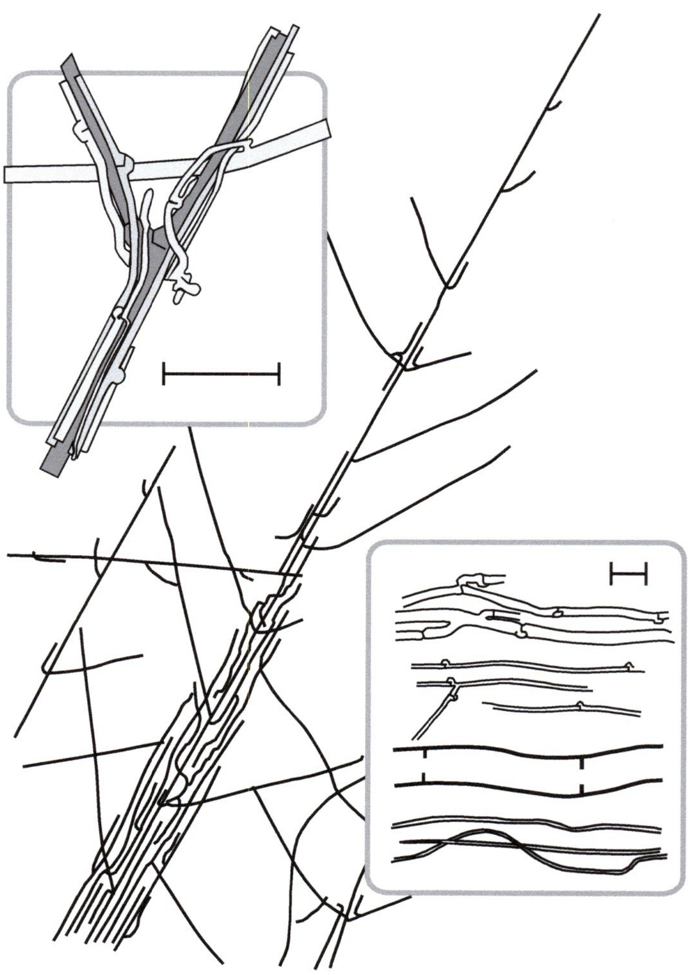

Figure 10.13. Drawing of hyphal strands of *Serpula lacrymans*. Strands originate when branches of a leading hypha form at an acute angle to grow parallel to the parent hypha which also tends to grow alongside other hyphae it may encounter. Anastomoses between the hyphae of the strands consolidates them and narrow hyphal branches ('tendril' hyphae) emerging from the older regions grow around the main hyphae and weave them together. The strand is shown in a general habit sketch running diagonally across the figure and the top panel shows tendril hyphae intertwined around main hyphae. The bottom panel shows some of the cell types encountered in strands, with undifferentiated hyphae at the top then tendril hyphae, a vessel hypha and fibre hyphae, scale bars, 20 μm. Modified from Moore (1995).

Although mycelial strands contain morphologically differentiated hyphae (Figure 10.13), their constituent hyphae are relatively loosely aggregated. Certain fungi produce highly differentiated aggregations of hyphae with well-developed tissues (Figure 10.14). These structures are very root-like in appearance and are called **rhizomorphs.** Rhizomorphs differ from strands by having extreme apical dominance with a highly organised apical growing point. The apical region of the rhizomorph contains a compact growing point of tightly packed cells, protected by a cap of intertwined hyphae in (and producing) a mucilaginous matrix. Behind is a medullary zone containing vessel hyphae composed of swollen, vacuolated and often multinucleate cells surrounded by copious air- or mucilage-filled spaces. The medullary region forms a central channel through the rhizomorph and, in mature tissues, is traversed by narrow-diameter, thick-walled fibre hyphae (Figure 10.14). Towards the periphery of the rhizomorph, the cells are smaller, darker and thicker-walled, and there is a fringing mycelium extending outwards between the outer layers of the rhizomorph, resembling the root-hair zone of a plant root.

The similarity, at least in microscope sections, with the plant root has prompted the suggestion in older literature

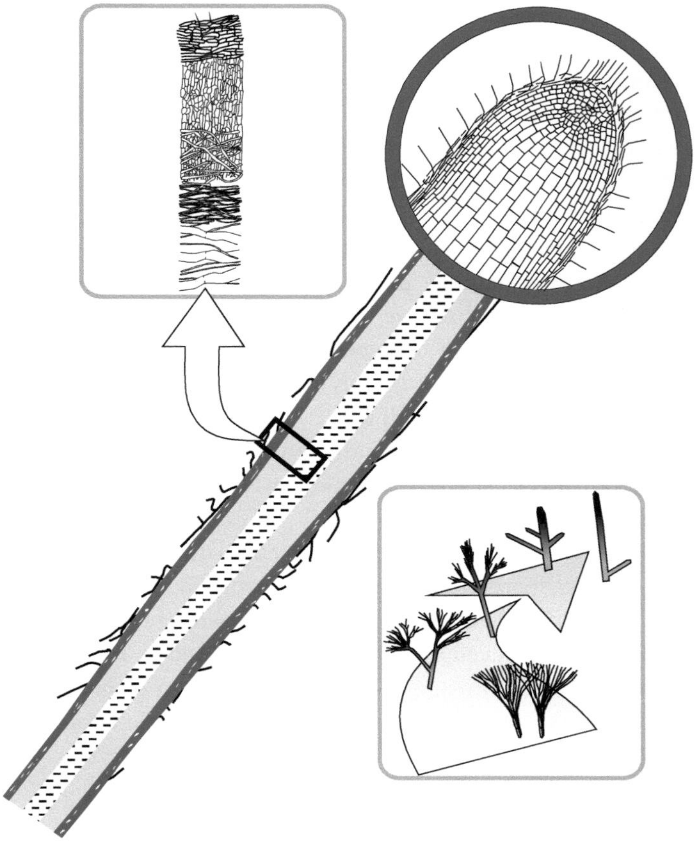

Figure 10.14. Rhizomorph structure. The diagonal diagram is a sectional drawing showing general structure, with the apical region magnified to show the appearance of a growing point of tightly packed cells. Behind the tip is a medullary zone containing swollen, vacuolated and often multinucleate cells surrounded by copious air- or mucilage-filled spaces. The medullary region forms a central channel through the rhizomorph and, in mature tissues, is traversed by narrow fibre hyphae and wide-diameter vessel hyphae, the microscopic appearance being indicated in the drawing in the top left panel. The panel at bottom right depicts mycelial fans, strands, cords and rhizomorphs as a series showing increasing apical dominance (Rayner *et al*., 1985). From Moore (1995).

that rhizomorph extension results from meristematic activity. However, a meristem-like structure would be totally alien to the growth strategy of the fungal hypha; meristems do not occur in fungi. The impression of central apical initials giving rise to axially arranged tissues is undoubtedly an artefact caused by sectioning compact aggregations of parallel hyphae. Ultrastructural, especially scanning electron microscope, observations clearly reveal the hyphal structure of the rhizomorph tip, but the appearance of the rhizomorph body being made up of parallel bundles of hyphae must mean that increase in rhizomorph diameter is associated with highly regulated hyphal branching.

Usually, rhizomorphs are initiated as compact masses of aggregated cells, the ultimate origin being ascribed to locally enhanced acute-angled branching of some marginal hyphae in a mycelium; a phenomenon described as 'point-growth'; the linear organs originating from originally unpolarised hyphal aggregations which become apically polarised. Mycelial strands and rhizomorphs are extremes in a range of hyphal linear aggregations that are related together in a hierarchy depending on increasing apical dominance (Figure 10.14). Their essential function is the **translocation** of nutrients, but they also penetrate the substratum, explore and migrate.

For example, strands form when there is a need to channel nutrients towards developing sporophores, and they are also formed by mycorrhizal fungi to radiate into the soil, where they greatly supplement the host plant's root system and gather nutrients for the host. In saprotrophic phases, strands are also **migratory organs**, extending from an existing food base to explore nutrient-poor surroundings for new nutrient sources. Strands of *Serpula lacrymans*, the dry rot fungus, are able to penetrate several metres of brickwork from a food base in decaying wood and to overgrow plastic and many inert building materials. The strands hasten capture of new substrate by increasing the inoculum potential of the fungus at the point of contact with it, aiding capture of resources and providing translocation routes in both directions. The distribution of strands around a food base changes with time as hyphae are resorbed; that is, digested from the inside to regain nutrient components. Redistribution of the nutrients recovered from old strands enable migration of the colony from place to place.

The prime example of rhizomorphs is usually *Armillaria mellea*, a pathogen of trees and shrubs, which spreads from one root system to another by means of its 'bootlace' rhizomorphs. Here, again, the structure serves translocatory and migratory functions and, as with strands, translocation is bidirectional, glucose being translocated towards and away from the apex simultaneously. In moist tropical forests, aerial rhizomorphs, mainly of *Marasmius* spp., form a network which intercepts and traps freshly fallen leaves, forming a **suspended litter layer**. In describing the structure of litter-trapping rhizomorph networks in moist tropical forests, Hedger *et al*. (1993) showed that the rhizomorphs have a reduced mushroom cap at their tips, so the stems (= stipes) of sporophores should also be included in this discussion.

These two linear organs are functionally very similar, the mushroom stipe translocates nutrients to the cap, and as many sporophores are served by radiating strands which convey nutrients towards them, the junction between strand and stipe can be obscure. The term 'radicating' is used to describe mushrooms whose stipes are elongated into root-like **pseudorhizas** which extend to the soil surface from some buried substrate. Even in species which do not normally produce pseudorhizas, they can be induced by keeping sporophore cultures in darkness (Buller, 1924), whereupon the stipe base can extend for many centimetres, driving the mushroom primordium on its tip towards any source of light. Rhizomorphs? Pseudorhizas? Extending stipes? What they are called is less important than the implication that a close morphogenetic relationship underlies all fungal linear hyphal aggregations.

10.8 GLOBOSE STRUCTURES: SCLEROTIA, STROMATA, ASCOMATA AND BASIDIOMATA

Sclerotia are pseudoparenchymatous hyphal aggregations in which concentric zones of tissue form an outer rind and inner medulla, with a cortex sometimes distinguishable between them. Sclerotia are tuber-like and detach from their parental mycelium at maturity. Some sclerotia consist of very few cells and are therefore of microscopic dimensions. At the other extreme, the sclerotium of *Polyporus mylittae*, found in the deserts of Australia, can reach 20–35 cm in diameter.

Sclerotia are stress-resistant, mycelial survival structures which pass through a period of dormancy before utilising their own accumulated reserves to 'germinate' (Erental *et al.*, 2008). Dormant sclerotia may survive for several years, thanks to a rind composed of tightly packed hyphal tips which become thick walled and pigmented (melanised) to form an impervious surface layer (see Figure 7.6). Such a layer of impervious tissue, which is especially protective against desiccation and irradiation, may be developed in other circumstances to form what has been called a **pseudosclerotial plate**, for example, over the surface of hyphal structures of the fungus *Hymenochaete corrugata* binding hazel branches together and in similar adhesions between rhizomorphs and leaf litter in tropical forests. When such plates completely enclose the mass on which they are formed, the structure that results is called a sclerotium. Many fungi enclose portions of the substrate and/or substratum (which may include host cells if the fungus is a pathogen) within a layer of pigmented, thick-walled cells; the whole structure may be regarded as a kind of sclerotium.

The term sclerotium is, therefore, a functional one that encompasses a variety of structures. The different forms arose by **convergent evolution**, perhaps more than 14 times during fungal evolution (Smith *et al.*, 2015), and most probably evolving, in *Ascomycota*, from aborted spore-forming organs like perithecia, cleistothecia and conidiomata. In *Basidiomycota* there is evidence for the same genes being involved in both sclerotium and mushroom initiation (in *Coprinopsis cinerea*; Moore, 1981), indicating that a common initiation pathway gives rise to hyphal aggregations which, under one set of environmental conditions (22–26°C plus illumination), develop axial symmetry and become mushroom initials, but under other conditions (30–37°C plus continuous darkness) develop radial symmetry and become sclerotia.

The medulla forms the bulk of the sclerotium, and its cells (and those of the cortex where present) may accumulate reserves of glycogen and other polysaccharides, polyphosphate, protein and lipid. Sclerotium development comprises **initiation**, when the hyphae begin to aggregate to form small, distinct **initials; development**, when the initials expand and grow to full size, accumulating nutritional reserves from the parent mycelium; and **maturation**, which is most obviously characterised by clear demarcation of the surface and pigmentation of its constituent cell walls, but which also involves conversion of the reserve nutrients to forms suitable for long-term storage.

Sclerotia may germinate to form mycelium, conidia, ascomata or basidiomata, and the mode of germination seems in some cases to be a matter of size; small sclerotia germinate to produce a mycelium, large ones produce spores and sporophores. The giant sclerotium of *Polyporus mylittae* can form a basidioma without being supplied with water as the flesh is honeycombed with blocks of translucent tissue where the hyphae form copious amounts of an extrahyphal gel which serves as both nutrient and water store.

In many *Ascomycota* and *Basidiomycota*, hyphae may aggregate to form sporophores which are responsible for producing and, just as important, distributing sexual spores. In *Ascomycota*, the sexually produced **ascospores** are contained in **asci** (singular: **ascus**) (see Figure 8.5) enclosed in an aggregation of hyphae termed an **ascoma**. Ascomata are formed from non-dikaryotic sterile hyphae surrounding the ascogonial hyphae of the centrum; several distinct types can be recognised (Figure 10.15;

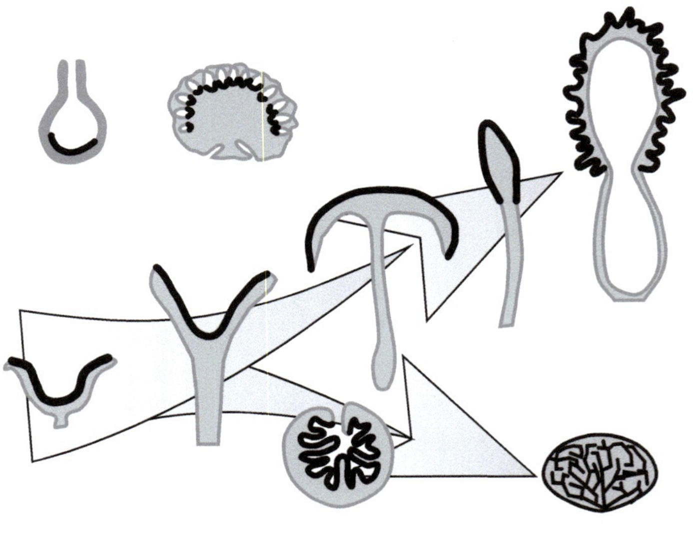

Figure 10.15. The variety of multicellular sporophores of ascomycete fungi in the form of simplified diagrammatic sectional drawings (redrawn after Burnett, 1968); in each case the hymenial tissue is represented by the black line. The panel shows a range of ascomata with, at top left a basic flask-like perithecium (e.g. *Sordaria*) alongside the perithecial stroma of *Daldinia*. The rest of the drawings are arranged to show how the simple cup-like ascoma of *Peziza* might, hypothetically, have given rise, on the one hand, to the morel (*Morchella*) via the sporophore forms of *Sarcoscypha*, *Helvella* and *Mitrula*, and, on the other hand. to the subterranean sporophore of *Tuber* via a form like *Genea*. Modified from Moore (1995). Photographs of some of these sporophores shown in Figure 3.13.

photographs shown in Figure 3.13), including the open cup-like apothecia of *Aleuria*, the flask-like perithecium (found, for example, in *Neurospora* and *Sordaria*) and the completely closed cleistothecium formed by, for example, *Aspergillus*. A similarly extensive range of diversity is found in ascospore shape and wall architecture; and ascus morphology, which may be globose or linear, and operculate (with a lid) or inoperculate (with an unlidded pore).

The sporophores of *Basidiomycota*, the mushrooms, toadstools, bracket fungi, puffballs, stinkhorns, bird's nest fungi, etc., are all examples of **basidiomata** (singular: **basidioma**) which bear the sexually produced **basidiospores** on **basidia** (see Figure 9.5). Simplified diagrammatic drawings of some of the different types of basidiomata are shown in Figure 10.16 (for photographs see Figures 3.15–3.22). Again, an extensive range of diversity is found in basidiospore shape and wall architecture, and basidium morphology, and the shape and form of sterile accessory cells like cystidia and paraphyses.

The basidium is the **meiocyte** within which meiosis occurs and from which basidiospores develop. Basidiospores are **exospores** (formed beyond the body of the meiocytes), ascospores are **endospores** (formed within the body of the meiocytes) (Figure 9.5). In boletes and polypores, basidia and basidiospores are formed by a **hymenium** lining the inner surface of pores on the lower surface of the sporophore. In the agaric or mushroom

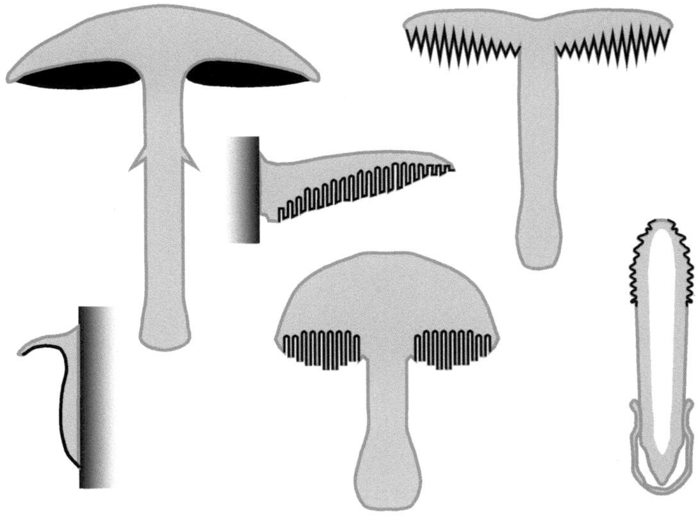

Fig 10.16. The variety of multicellular sporophores of basidiomycete fungi in the form of simplified diagrammatic sectional drawings (redrawn after Burnett, 1968); in each case the hymenial tissue is represented by the black line. This panel shows several basidiomata, including the agaric (mushroom) form of sporophore, the poroid bracket and toadstool polypore, and toothed (hydnoid) form. At bottom left is an encrusting sporophore (as in *Stereum* and *Phellinus*) and at bottom right the mature stinkhorn (*Phallus*). From Moore (1995).

fungi, the lower surface of the mushroom cap is developed into a set of plates (the **gills or lamellae**) on which the hymenium is located. Both structures are strategies to increase the surface area available for the spore-bearing hymenial tissue. Relative to spore production over a flat surface, gills achieve a maximum 20-fold increase in surface area. Poroid configurations of the hymenium produce larger increases in surface area, relative to gills; depending on the size and spacing of the pores, a poroid cap has an 18-fold to a 45-fold larger hymenium area relative to a flat surface (Fischer & Money, 2010).

However, **pores** have less efficient packing densities of basidia than gills because the inward curvature limits how close neighbouring basidia can be located before their spores run the risk of interfering with each other's dispersal. Hence, the packing density of basidia 'per unit mass of sporophore tissue' is inherently less for pores than for gills. On the other hand, the protection (particularly against desiccation) offered by pores is greater, and this strategy suits **perennial sporophores.** A single sporophore of a perennial bracket fungus can produce 10 million spores per minute continuously for periods of 5–6 months. These brackets tend to be large sporophores; for comparison, an agaric mushroom (something like *Agaricus campestris*) can release about 2 million spores per minute, however such a mushroom is unlikely to last more than two weeks.

The spores are released from the undersurface of the cap and are distributed by air currents in the majority of species, so the cap must be supported above the substratum; this is the function of the stem or stipe. Basidiospores are **violently discharged**, and the four spores of a basidium are liberated within a few minutes of each other; many reproductive structures of terrestrial fungi are actively liberated using a wide range of mechanical processes for propulsion (Ingold, 1999). The basidiospore discharge mechanism is a violent (ballistic) discharge that clearly depends on the secretion of a droplet of fluid (water + solutes), which is called **Buller's drop**, close to where the spore is joined to its supporting sterigma. A hydrophobic interaction with the spore surface is also probably important, generating what is known as a **surface tension catapult** (Turner & Webster, 1991; Money, 1998). This propels the spores directly away from the gill surface over distances ranging from 0.1–0.6 mm; spores with the largest Buller's drop are propelled the greatest distance (Stolze-Rybczynski *et al.*, 2009).

During discharge these spores can be subjected to forces of acceleration several thousand times greater than that experienced by astronauts during the launch of rockets servicing the International Space Station! Even more impressive is the fact that while those rockets consume 50% of their mass in fuel during the first 2 minutes of flight, ballistospore discharge is fuelled by metabolites that represent only 1% of the mass of the spore (Money, 1998). High-speed video analysis suggests that the size and shape of basidiospores affects discharge distance (Stolze-Rybczynski *et al.*, 2009), because the asymmetric coalescence of Buller's drop is what produces the launching momentum ***and launch direction*** (Liu *et al.*, 2017). There is consequently a close linkage between spore size and morphology and the tube radii and distances between gills; and this close linkage is reflected in the evolution and adaptation of these aspects of the sporophore. Liu *et al.* (2017) describe ballistospore discharge as a generic catapulting mechanism for colloidal particles and point out that understanding it has implications for many aspects of both biology and engineering.

The horizontal portion of the basidiospore flight path is halted by air viscosity within 1 ms of the launch and the spores then fall vertically until they clear the gills or pore. Basidiospores are thus 'shot' into the space between adjacent gills, or towards the centre of the pore. They cannot be shot further than about the middle of that space to avoid being impacted onto the other gill (or the other side of the pore). Basidiospores rely on **gravity** to clear the gills or pores and emerge into the turbulent air beneath the mushroom cap, and this raises another point because the sporophores (and gills in many mushrooms) are exquisitely sensitive to gravity and can adjust growth differentials to bring a disturbed stem and gills back to the **vertical**. The shape changes which occur in agaric sporophores in response to change in the direction of gravity, usually referred to as **gravitropism**, are morphogenetic changes. When a mushroom growing vertically, as is normal, is disoriented (say, laid flat on the substratum by a foraging animal or person in a white coat) gravitropic bending results from differential growth of cells in the stem; those on the 'bottom' extend more than those on the 'top' so that the stem bends upwards. Calcium signalling is involved in regulating these growth differentials and proteomic analysis has revealed that at least 51 proteins are differentially regulated during the response of *Coprinopsis cinerea* to gravity disorientation (Kim *et al.*, 2017; and see Häder, 2018).

Gravity ***perception*** seems to be dependent on the actin cytoskeleton and nuclear motility. Nuclei act as statoliths and interact with the actin cytoskeleton and trigger specific vesicle/microvacuole release from the endomembrane system to effect localised cell growth (Figure 10.17; Moore *et al.*, 1996). The current interpretation being that this statolith–cytoskeleton interaction is communicated to the transmembrane domains of a sugar transporter located in the plasma membrane to change its substrate specificity and cause an otherwise uniformly distributed growth regulator (probably a laevorotatory 6-deoxy hexose sugar) to be asymmetrically transported into the disoriented cells (Moore & Novak Frazer, 2017; and see further discussion in section 13.9). The function of all this is to get the spores into the turbulent air beneath the cap so they can be

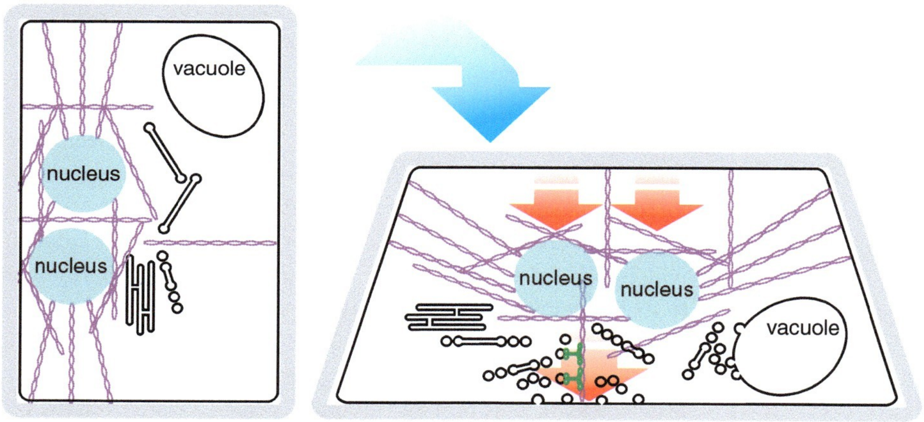

Fig 10.17. Diagrammatic interpretation of gravity perception in agarics. In *Basidiomycota* the two nuclei of a dikaryon are enclosed in an F-actin cage in which they are held a particular distance apart (the distance affects patterns of gene expression). Evidence indicates that the nuclei act as statoliths to detect the gravitational field and how this might work is illustrated here. In the vertical orientation (normal for a mushroom stem, for example) the nuclei within their actin cage are positioned within the cell by stressed cables of actin microfilaments (on the left). If the orientation of the tissue is disturbed (for example placed horizontal as on the right), movement of the statoliths (nuclei) within the cage of actin changes the stress on the actin cable connections to the endomembrane system and generates useful directional signals. The current interpretation is that this statolith–cytoskeleton interaction is communicated to the transmembrane domains of a sugar transporter located in the plasma membrane to change its substrate specificity and cause an otherwise uniformly distributed growth regulator (probably a laevorotatory 6-deoxy hexose sugar) to be asymmetrically transported into the disoriented cells (Moore & Novak Frazer, 2017; and see further discussion in section 13.11). Subsequently, components required for wall modification and resynthesis will be exported when the growth factor induces it. Because the wall resynthesis is localised to the bottom wall, the cell will be curled upwards. This effect, coordinated over many cells along the length of tissues like mushroom stems, can reorient them to the vertical if they are subjected to physical disturbance. Modified and redrawn after Moore *et al.* (1996).

distributed by the breeze. But it is another aspect of species diversity because the rate of perception and rate of reaction to changes in the physical environment of this sort are themselves species specific.

Basidiomycota exhibit **long-term plasmogamy** because the 'sexual' (usually dikaryotic) mycelium has the ability to grow indefinitely. This is not so different from the indeterminate growth of heterokaryons in *Ascomycota* and it is clear that in both groups the sexually compatible mycelium can have a number of alternative developmental pathways open to it: continuation of hyphal growth, production of asexual spores, and progress into the sexual cycle (Moore, 1998). Choice between these is often a matter of the impact of very localised environmental conditions on a mycelium which has become competent to embark on a developmental pathway as a result of its capture and accumulation of nutritional resources. The flow chart in Figure 10.18 shows an overview of the physiological processes involved in development of sporophores and other multicellular structures in fungi and gives some idea of the number of places where species-specific modifications can intervene in a general process to widen diversity.

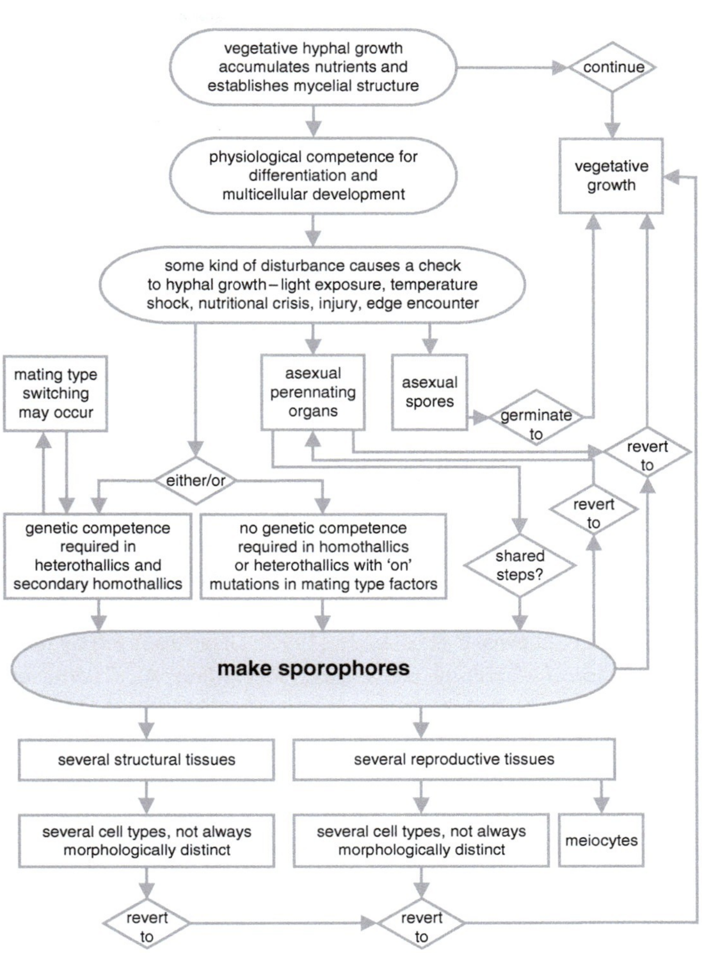

Fig 10.18. Flow chart showing a summarised view of the processes involved in development of sporophores and other multicellular structures in fungi. From Moore & Novak Frazer (2002).

10.9 REFERENCES

Bailey-Shrode, L. & Ebbole, D.J. (2004). The fluffy gene of *Neurospora crassa* is necessary and sufficient to induce conidiophore development. *Genetics*, **166**: 1741–1749. DOI: https://doi.org/10.1534/genetics.166.4.1741.

Breakspear, A. & Momany, M. (2007). The first 50 microarray studies in filamentous fungi. *Microbiology*, **153**: 7–15. DOI: https://doi.org/10.1099/mic.0.2006/002592-0.

Buller, A.H.R. (1924). *Researches on Fungi*, **Vol. 3**. London: Longman Green & Co. ASIN: B0008BT-4QW.

Burnett, J.H. (1968). *Fundamentals of Mycology*. London: Edward Arnold. ISBN-10: 071312203X, ISBN-13: 978-0713122039.

Cha, J., Zhou, M. & Liu, Y. (2015). Mechanism of the *Neurospora* circadian clock, a FREQUENCY-centric view. *Biochemistry*, **54**:150–156. DOI: https://doi.org/10.1021/bi5005624.

Chen, J.L. & Tzean, S.S. (1993). *Podosporium elongatum*, a new synnematous hyphomycete from Taiwan. *Mycological Research*, **97**: 637–640. DOI: https://doi.org/10.1016/S0953-7562(09)81190-8.

Chiu, S.W. & Moore, D. (1988a). Evidence for developmental commitment in the differentiating fruit body of *Coprinus cinereus*. *Transactions of the British Mycological Society*, **90**: 247–253. DOI: https://doi.org/10.1016/S0007-1536(88)80096-2.

Chiu, S.W. & Moore, D. (1988b) Ammonium ions and glutamine inhibit sporulation of *Coprinus cinereus* basidia assayed *in vitro*. *Cell Biology International Reports*, **12**: 519–526. DOI: https://doi.org/10.1016/0309-1651(88)90038-0.

Dunlap, J.C. & Loros, J.J. (2006). How fungi keep time: circadian system in *Neurospora* and other fungi. *Current Opinion in Microbiology*, **9**: 579–587. DOI: https://doi.org/10.1016/j.mib.2006.10.008.

Dunlap, J.C. & Loros, J.J. (2017). Making time: conservation of biological clocks from fungi to animals. In *The Fungal Kingdom*, ed. J. Heitman, B. Howlett, P. Crous, E. Stukenbrock, T. James & N.A.R. Gow. Washington, DC: ASM Press, pp. 515–534. DOI: https://doi.org/10.1128/microbiolspec.FUNK-0039-2016.

Erental, A., Dickman, M.B. & Yarden, O. (2008). Sclerotial development in *Sclerotinia sclerotiorum*: awakening molecular analysis of a `dormant' structure. *Fungal Biology Reviews*, **22**: 6–16. DOI: https://doi.org/10.1016/j.fbr.2007.10.001.

Fischer, M.W.F. & Money, N.P. (2010). Why mushrooms form gills: efficiency of the lamellate morphology. *Mycological Research*, **114**: 57–63. DOI: https://doi.org/10.1016/j.mycres.2009.10.006.

Galagan, J.E., Calvo, S.E., Borkovich, K.A. *et al.* (2003). The genome sequence of the filamentous fungus *Neurospora crassa*. *Nature*, **422**: 859–868. DOI: https://doi.org/10.1038/nature01554.

Häder, D.-P. (2018). Gravitropism in fungi, mosses and ferns: gravity sensing and graviorientation in microorganisms and plants. In *Gravitational Biology I. SpringerBriefs in Space Life Sciences*. Cham, Switzerland: Springer Nature Switzerland AG, pp. 67–74. ISBN: 9783319938936. DOI: https://doi.org/10.1007/978-3-319-93894-3_5.

Hedger, J.N., Lewis, P. & Gitay, H. (1993). Litter-trapping by fungi in moist tropical forest. In *Aspects of Tropical Mycology*, ed. S. Isaac, R. Watling, A.J.S. Whalley & J.C. Frankland. Cambridge, UK: Cambridge University Press, pp. 15–35. ISBN-10: 0521450500, ISBN-13: 978-0521450508.

Hevia, M.A., Canessa, P. & Larrondo, L.F. (2016). Circadian clocks and the regulation of virulence in fungi: getting up to speed. *Seminars in Cell & Developmental Biology*, **57**: 147–155. DOI: https://doi.org/10.1016/j.semcdb.2016.03.021.

Hevia, M.A., Canessa, P., Müller-Esparza, H. & Larrondo, L.F. (2015). A circadian oscillator in the fungus *Botrytis cinerea* regulates virulence when infecting *Arabidopsis thaliana*. *Proceedings of the National Academy of Sciences of the United States of America*, **112**: 8744–8749. DOI: https://doi.org/10.1073/pnas.1508432112.

Ingold, C.T. (1999). Active liberation of reproductive units in terrestrial fungi. *Mycologist*, **13**: 113–116. DOI: https://doi.org/10.1016/S0269-915X(99)80040-8.

Kasuga, T. & Glass, N.L. (2008). Dissecting colony development of *Neurospora crassa* using mRNA profiling and comparative genomics approaches. *Eukaryotic Cell*, **7**: 1549–1564. DOI: https://doi.org/10.1128/EC.00195-08.

Kasuga, T., Townsend, J.P., Tian, C. *et al.* (2005). Long-oligomer microarray profiling in *Neurospora crassa* reveals the transcriptional program underlying biochemical and physiological events of conidial germination. *Nucleic Acids Research*, **33**: 6469–6485. DOI: https://doi.org/10.1093/nar/gki953.

Kim, J.S., Kwon, Y.S., Bae, D.W., Kwak, Y.S. & Kwack, Y.B. (2017). Proteomic analysis of *Coprinopsis cinerea* under conditions of horizontal and perpendicular gravity. *Mycobiology*, **45**: 226–231. DOI: https://doi.org/10.5941/MYCO.2017.45.3.226.

Kirk, P.M., Cannon, P.F., Minter, D.W. & Stalpers, J.A. (2008). *Dictionary of the Fungi*, 10th edition. Wallingford: CABI Publishing. ISBN-10: 0851998267, ISBN-13: 978-0851998268. Kindle edition: ASIN: B00K7ANXJI.

Koritala, B.S.C. & Lee, K. (2017). Natural variation of the circadian clock in *Neurospora*. *Advances in Genetics*, **99**: 1–37. DOI: https://doi.org/10.1016/bs.adgen.2017.09.001.

Krijgsheld, P., Bleichrodt, R., van Veluw, G.J. *et al.* (2013). Development in *Aspergillus*. *Studies in Mycology*, **74**: 1–29. DOI: https://doi.org/10.3114/sim0006.

Lee, M.K., Kwon, N.J., Lee, I.S. *et al.* (2016). Negative regulation and developmental competence in *Aspergillus*. *Scientific Reports*, **6**: article number 28874. DOI: https://doi.org/10.1038/srep28874.

Liu, F., Chavez, R.L., Patek, S.N. *et al.* (2017). Asymmetric drop coalescence launches fungal ballistospores with directionality. *Journal of The Royal Society Interface*, **14**: 132. DOI: https://doi.org/10.1098/rsif.2017.0083.

Lysek, G. (1984). Physiology and ecology of rhythmic growth and sporulation in fungi. In *The Ecology and Physiology of the Fungal Mycelium*, ed. D.H. Jennings & A.D.M. Rayner. Cambridge, UK: Cambridge University Press, pp. 323–342. ISBN-10: 0521106265, ISBN-13: 978-0521106269.

Martinelli, S.D. & Clutterbuck, A.J. (1971). A quantitative survey of conidiation mutants in *Aspergillus nidulans*. *Journal of General Microbiology*, **69**: 261–268. DOI: https://doi.org/10.1099/00221287-69-2-261.

McCluskey, K. & Baker, S.E. (2017). Diverse data supports the transition of filamentous fungal model organisms into the post-genomics era. *Mycology*, **8**: 67–83. DOI: https://doi.org/10.1080/21501203.2017.1281849.

Mims, C.W., Richardson, E.A. & Timberlake, W.E. (1988). Ultrastructural analysis of conidiophore development in the fungus *Aspergillus nidulans* using freeze-substitution. *Protoplasma*, **144**: 132–141. DOI: https://doi.org/10.1007/BF01637246.

Minter, D.W., Kirk, P.M. & Sutton, B.C. (1983a). Thallic phialides. *Transactions of the British Mycological Society*, **80**: 39–66. DOI: https://doi.org/10.1016/S0007-1536(83)80163-6.

Minter, D.W., Sutton, B.C. & Brady, B.L. (1983b). What are phialides anyway? *Transactions of the British Mycological Society*, **81**: 109–120. DOI: https://doi.org/10.1016/S0007-1536(83)80210-1.

Money, N.P. (1998). More *g*'s than the space shuttle: the mechanism of ballistospore discharge. *Mycologia*, **90**: 547–558. Stable URL: http://www.jstor.org/stable/3761212.

Montenegro-Montero, A., Canessa, P. & Larrondo, L.F. (2015). Around the fungal clock: recent advances in the molecular study of circadian clocks in *Neurospora* and other fungi. *Advances in Genetics*, **92**: 107–184. DOI: https://doi.org/10.1016/bs.adgen.2015.09.003.

Moore, D. (1981). Developmental genetics of *Coprinus cinereus*: genetic evidence that carpophores and sclerotia share a common pathway of initiation. *Current Genetics*, **3**: 145–150. DOI: https://doi.org/10.1007/BF00365718.

Moore, D. (1995). Tissue formation. In *The Growing Fungus*, ed. N.A.R. Gow & G.M. Gadd. London: Chapman & Hall, pp. 423–465. ISBN-10: 0412466007, ISBN-13: 978-0412466007.

Moore, D. (1998). *Fungal Morphogenesis*. New York: Cambridge University Press (see Chapters 6 *Development of form* and 7 *The keys to form and structure*). ISBN-10: 0521552958, ISBN-13: 978-0521552950.

Moore, D., Hock, B., Greening, J.P. *et al.* (1996). Gravimorphogenesis in agarics. *Mycological Research*, **100**: 257–273. DOI: https://doi.org/10.1016/S0953-7562(96)80152-3.

Moore, D. & Novak Frazer, L. (2002). *Essential Fungal Genetics*. New York: Springer-Verlag (see Chapter 10, *The genetics of fungal differentiation and morphogenesis*). ISBN-10: 0387953671, ISBN-13: 978-0387953670.

Moore, D. & Novak Frazer, L. (2017). *Fungiflex: The Untold Story*. London: CreateSpace Independent Publishing Platform. ISBN-10: 1547168560, ISBN-13: 978-1547168569.

Moore-Landecker, E. (1996). *Fundamentals of the Fungi*, 4th edition. Upper Saddle River, NJ: Prentice Hall. ISBN-10: 0133768643, ISBN-13: 978-0133768640.

Novodvorska, M., Stratford, M., Blythe, M.J. *et al.* (2016). Metabolic activity in dormant conidia of *Aspergillus niger* and developmental changes during conidial outgrowth. *Fungal Genetics and Biology*, **94**: 23–31. DOI: https://doi.org/10.1016/j.fgb.2016.07.002.

Rayner, A.D.M. (1997). *Degrees of Freedom. Living in Dynamic Boundaries*. London: Imperial College Press. ISBN-10: 1860940374, ISBN-13: 978-1860940378.

Rayner, A.D.M., Powell, K.A., Thompson, W. & Jennings, D.H. (1985). Morphogenesis of vegetative organs. In *Developmental Biology of Higher Fungi*, ed. D. Moore, L.A. Casselton, D.A. Wood & J.C. Frankland. Cambridge, UK: Cambridge University Press, pp. 249–279. ISBN-10: 0521301610, ISBN-13: 978-0521301619.

Salichos, L. & Rokas, A. (2010). The diversity and evolution of circadian clock proteins in fungi. *Mycologia*, **102**: 269–278. DOI: https://doi.org/10.3852/09-073.

Smith, M.E., Henkel, T.W. & Rollins, J.A. (2015). How many fungi make sclerotia? *Fungal Ecology*, **13**: 211–220. DOI: https://doi.org/10.1016/j.funeco.2014.08.010.

Stolze-Rybczynski, J.L., Cui, Y., Stevens, M.H.H. *et al.* (2009). Adaptation of the spore discharge mechanism in the Basidiomycota. *PLoS ONE*, **4**(1): article number e4163. DOI: https://doi.org/10.1371/journal.pone.0004163.

Tao, L. & Yu, J.H. (2011). *AbaA* and *WetA* govern distinct stages of *Aspergillus fumigatus* development. *Microbiology*, **157**: 313–326. DOI: https://doi.org/10.1099/mic.0.044271-0.

Teertstra, W.R., Tegelaar, M., Dijksterhuis, J. *et al.* (2017). Maturation of conidia on conidiophores of *Aspergillus niger*. *Fungal Genetics and Biology*, **98**: 61–70. DOI: https://doi.org/10.1016/j.fgb.2016.12.005.

Timberlake, W.E. (1993). Translational triggering and feedback fixation in the control of fungal development. *Plant Cell*, **5**: 1453–1460. DOI: https://doi.org/10.1105/tpc.5.10.145.

Trinci, A.P.J., Wiebe, M.G. & Robson, G.D. (1994). The mycelium as an integrated entity. In *The Mycota*, **Vol. I**, *Growth, Differentiation and Sexuality*, ed. J.G.H. Wessels & F. Meinhardt. Berlin, Heidelberg, New York: Springer-Verlag, pp. 175–193. ISBN-10: 3540577815, ISBN-13: 978-3540577812.

Turner, J.C.R. & Webster, J. (1991). Mass and momentum transfer on the small scale: how do mushrooms shed their spores? *Chemical Engineering Science*, **46**: 1145–1149. DOI: https://doi.org/10.1016/0009-2509(91)85107-9.

Ulloa, M. & Hanlin, R.T. (2012). *Illustrated Dictionary of Mycology*, 2nd edition. St. Paul, MN: APS Press. ISBN: 978-0-89054-502-7. DOI: https://doi.org/10.1094/9780890545027.

van Leeuwen, M.R., Krijgsheld, P., Bleichrodt, R. *et al.* (2013). Germination of conidia of *Aspergillus niger* is accompanied by major changes in RNA profiles. *Studies in Mycology*, **74**: 59–70. DOI: https://doi.org/10.3114/sim0009.

Yu, J.H. (2010). Regulation of development in *Aspergillus nidulans* and *Aspergillus fumigatus*. *Mycobiology*, **38**: 229–237. DOI: https://www.ncbi.nlm.nih.gov/pmc/articles/PMC3741515/.

CHAPTER 11

Fungi in Ecosystems

Fungi make crucial contributions to all ecosystems because of their abilities as decomposers. One of the most important kingdom-specific characteristics of fungi is that they obtain their nutrients by external digestion of substrates. In the real world though there is some digestion of inorganic substrates (see section 1.7), the bulk of the substrates that fungi recycle are the remains of animals and, most particularly, plants. In this chapter we give an account of the ways in which fungal hyphae obtain, absorb, metabolise, reprocess and redistribute nutrients.

In doing this, fungi obviously contribute to recycling and mineralisation of nutrients and, in what follows, we will describe the enzyme systems that enable this activity. Our description is relatively brief, and more details can be found in the premier text on fungal physiology (Jennings, 2008). For clarity we must describe separately the enzyme systems involved, and in an order that we have chosen for descriptive purposes. This introduction is intended to convey the impression that for most fungi in most circumstances the initial nutritional step is the secretion of enzymes able to convert polymers to the simple sugars, amino acids, carboxylic acids, purines, pyrimidines, etc. that the cell can absorb. Also, in most circumstances most fungal mycelia will be carrying out all of these biochemical changes simultaneously.

In this chapter, we will see how fungi contribute to ecosystems; how they break down the polysaccharides cellulose, hemicellulose, pectin, chitin, starch and glycogen. We will discuss their (unique) ability to degrade lignin, and the ways they digest protein, lipid, esters, phosphates and sulfates. The flow of nutrients is dealt with in terms of transport and translocation; and in the final two sections we deal briefly, but comprehensively, with the main pathways of primary (intermediary) metabolism, as well as secondary metabolites, including commercial products, like statins and strobilurins.

11.1 CONTRIBUTIONS OF FUNGI TO ECOSYSTEMS

Wood, that is, plant secondary cell wall, is the most widespread substrate on the planet. Except for lignin (see below), the bulk of plant cell biomass consists of the polysaccharides cellulose, hemicellulose and pectins in varying proportions, depending on the type of cell and its age. Even though wall components predominate, the cytoplasm of dead cells contributes lipids, proteins and organic phosphates to the remains; however, wood itself is relatively poor in nitrogen and phosphorus. Plant biomass does not consist of neatly isolated packets of polysaccharide, protein and lignin. These three (and other materials) are intimately mixed together, so that it is better to think of the degradation of lignocellulosic and/or lignoprotein complexes.

A typical **agricultural residue**, like cereal straw or sugar cane bagasse, contains 30–40% cellulose, 20–30% hemicellulose and 15–35% lignin. Organisms that contribute to recycling may differ widely in their ability to degrade components of this mixture. On this sort of basis, wood-decay fungi have been separated into white-rot, brown-rot and soft-rot species. The white-rot fungi (about 2,000 species, mostly *Basidiomycota*) can metabolise lignin, on the other hand, brown-rot fungi (about 200 species of *Basidiomycota*) degrade the cellulose and hemicellulose components without much effect on the lignin. Soft-rot species (mostly soil-inhabiting *Ascomycota*) have rather intermediate capabilities, being able to degrade cellulose and hemicellulose rapidly, but lignin only slowly. These differences in behaviour reflect the different enzymes produced by these organisms and serve to emphasise that the organisms must digest complexes of potential nutrient sources and assemble panels of different enzymes to do so.

Bacteria and fungi are responsible for cellulose digestion in nature, but the ability to degrade lignin is restricted to fungi; specifically, *Basidiomycota* and a few *Ascomycota*. Lignin is a highly branched phenylpropanoid polymer (see section 11.7 below), and the ability to degrade a polyphenolic means that as well as contributing to nutrient recycling fungi can also help to compensate for damage done to the environment by industrial operations; a process called **bioremediation** with fungi. Many wastes are hazardous because they contain tannins and phenolics, which are toxic to plants and animals, and this applies to **agricultural wastes**, for example residues from extraction of oils such as cotton, rape, olive and palm oils, and fruit processing residues, like citrus wastes, as well as wastes contaminated with pesticides.

Indeed, the agricultural industry produces vast quantities of wastes: on average, world agriculture currently loses 40% of its primary production to pests and diseases, and then throws away more than 70% of what's left because the crop always represents so little of what is grown. Remember, the 'crop' may only be the seeds of the plant that is grown, or even only a portion of the seed, like its oil content. It is typical for 80–90% of the total biomass of agricultural production to be discarded as waste. Whether this waste is ploughed in, composted or otherwise left to rot, it is up to the fungi in the environment to recycle the bulk of these wastes.

There are three major fungal nutritional modes: probably the majority are ***saprotrophs*** for which the substrates are dead organic materials not killed by the fungus itself. ***Necrotrophs*** invade living tissues which they kill and then utilise, whereas ***biotrophs*** exploit host cells which remain alive. In the latter case, one might expect that, though local digestion of host tissue may be necessary for penetration or establishment of the pathogen, only simple nutrients would be removed from the host because of the damage that would be inflicted on the host by large-scale digestion of polymeric cell constituents. Biotrophs may be host specific, but saprotrophs and necrotrophs generally have a very large range of habitats open to them and in most of these, polymeric sources of nutrients predominate.

This predominance of polymers as sources of nutrients is obviously true for such materials as herbaceous plant litter, wood and herbivore dung. As we have just mentioned, immediate **plant litter**, though rich in carbon, is poor in nitrogen and phosphorus. Digested litter (a euphemism for herbivore dung) is relatively enriched in nitrogen, nucleic acids, vitamins, growth factors and minerals, since in passage through the intestine it accumulates the remains of bacteria, protozoa and other microorganisms. The composition of animal tissue varies enormously according to the specific organ system considered, but, in nature, most animal remains will be eaten by animal scavengers too rapidly for any microbes to be able to compete, so the microorganisms will be left with the parts, 'skin, gristle and bone', that other organisms cannot reach.

Nitrogen in the soil is mostly in the form of organic compounds; the proportion of nitrogen occurring as ammonium and nitrate (the nitrogen sources commonly added to synthetic agar cultivation media) or nitrite rarely exceed 2% of soil nitrogen, although some clay soils do trap more ammonium. Inorganic nitrogen compounds only predominate in agricultural soils which are repeatedly treated with chemical fertilisers. Nitrite is not usually detectable and nitrate content is usually very low in natural soils because these salts are so readily leached out by rain, so in most cases exchangeable ammonium on clay particles and the organic nitrogen provide saprotrophs with their nitrogen. In most surface soils, 20–50% of total nitrogen occurs in proteinaceous form and 5–10% as combined and complexed amino sugars. Amino sugars also contribute to the carbohydrate component of the soil, which represents 5–16% of total organic matter. Here, again, though, most soil carbohydrate is polymeric; simple sugars make up less than 1% of soil carbohydrate but cellulose can account for up to 14% and chitin is also well represented in view of the amino sugar content. About 50–70% of total phosphorus in soil is organic, mostly as phosphate esters related to or derived from compounds like nucleic acids, inositol phosphates and phospholipids.

Inorganic forms of **sulfur** may accumulate in some soils (e.g. as calcium and magnesium sulfates in arid regions, or calcium sulfate co-crystallised with calcium carbonate in calcareous soils). There is little inorganic sulfur in the surface horizons of soils in humid regions. Organic sulfur occurs in the form of methionine and cystine (and derivatives), and sulfate esters, including sulfated polysaccharides and lipids.

Extracellular enzymes are produced within the cell but act outside it. Consequently, they must be secreted across the plasmalemma. The processes involved in protein translocation across membranes are very similar in all eukaryotes. Polypeptides destined for secretion are identified by short amino-terminal transient 'signal' sequences which consist of uninterrupted stretches of at least six hydrophobic amino acid residues. The signal sequence is in the first part of the polypeptide to be synthesised on the ribosome and, as its hydrophobicity confers an affinity for the lipid environment of a membrane bilayer, the signal sequence 'targets' the ribosome translating it onto the endoplasmic reticulum membrane (see section 5.8).

Some of the extracellular enzymes that are produced are soluble and are freely dispersed in fluid films surrounding the hyphae, but others are fixed in space by being bound to the hyphal wall, extracellular matrix or to the substrate itself. This natural **immobilisation of enzymes** is advantageous to the producing fungus because it localises substrate degradation to the immediate vicinity of the hypha, so ensuring that the organism producing the enzyme has advantage in the competition with surrounding organisms for the soluble nutrients produced by the enzyme activity. In this way, the fungus can exert a degree of control over its immediate environment. For more details about the contributions made by fungi to ecosystems see Chapters 14–17 of this text, Gadd (2006) and Dighton and White (2017).

11.2 BREAKDOWN OF POLYSACCHARIDE: CELLULOSE

Polysaccharides are polymers of monosaccharides in which the constituent sugars are connected with glycosidic bonds. Because of the number and variety of available sugars and the diversity of bonding possibilities between different carbon atoms of the adjacent sugar residues, there is a considerable variety of polysaccharides. There is a matching variety of enzymes, hydrolases or glucosidases, capable of hydrolysing this range of glycosidic links. Enzymes responsible for polymer degradation (any polymer, not just polysaccharide) may employ one of two strategies of attack. They may attack randomly, effectively fragmenting the polymer molecule into a number of oligomers, these are the ***endo*-enzymes**, or they may approach terminally, digesting away monomers or dimers, the ***exo*-enzymes**.

Cellulose is the most abundant organic compound on Earth and accounts for over 50% of organic carbon; about 10^{11} tons are synthesised each year. It is an unbranched polymer of glucose in which adjacent sugar molecules are joined by β-1–4 linkages (Figure 11.1). There may be from a few hundred to a few thousand sugar residues in the polymer molecule, corresponding to molecular masses from about 50,000 to approaching 1 million. Breakdown of cellulose is chemically straightforward but is complicated by its physical form. Mild acid hydrolysis of cellulose releases soluble sugars but does not go to completion; oligomers of 100–300 glucose residues remain. The fraction which is readily hydrolysed is called amorphous cellulose while that which is resistant to acid is called crystalline cellulose. Since it influences chemical breakdown, the conformation and three-dimensional structure of cellulose must influence cellulolytic enzyme activity.

The cellulolytic enzyme (**cellulase**) complex of white-rot *Basidiomycota* like *Phanerochaete chrysosporium* (also named *Sporotrichum pulverulentum*, which is the anamorph of *Phanerochaete chrysosporium*) and *Ascomycota* like *Trichoderma reesei* consists of a number of hydrolytic enzymes: endoglucanase, exoglucanase and cellobiase (which is a β-glucosidase) which work synergistically and, in both bacteria and fungi, are organised into an extracellular multienzyme complex called a cellulosome (see next paragraph). **Endoglucanase** attacks cellulose at random, producing glucose, cellobiose (a disaccharide made up of two glucose molecules) and some cellotriose (a trisaccharide). **Exoglucanase** attacks from the non-reducing end of the cellulose molecule, removing glucose units; it may also include a cellobiohydrolase activity which produces cellobiose by attacking the non-reducing end of the polymer. **Cellobiase** is responsible for hydrolysing **cellobiose** to glucose. Glucose is, therefore, the readily metabolised end product of cellulose breakdown by enzymatic hydrolysis.

Cellulosomes were first described in anaerobic cellulolytic bacteria, and are also highly developed in fungi. These enzyme complexes are extracellular molecular machines ('nanomachines'). In addition to catalytic regions, cellulolytic enzymes contain domains not involved in catalysis, but taking part in substrate binding, multienzyme complex formation (so-called 'docking domains'), or attachment to the cell surface. Cellulosomes comprise a complex of **scaffoldin**, which is the structural subunit, and various enzymatic subunits. The intersubunit interactions in these multienzyme complexes are enabled by **cohesin** and **dockerin** modules. Cellulosome-producing bacteria have been isolated from many different environments, suggesting that this microbial enzymatic strategy was important and ecologically successful early in evolution. However, the detailed structure of cellulosomes in a given species is variable, and between species there is considerable diversity in the composition and configuration of cellulosomes; species-specific dockerin domains assist enzyme assembly onto cohesin motifs interspersed within variable protein scaffolds (Artzi *et al.*, 2017). Analogous structures in anaerobic fungi also assemble using non-catalytic dockerin domains and scaffoldin proteins, but these have no similarity to their bacterial counterparts. Cellulosomes in anaerobic fungi contain several enzyme activities not present in their bacterial equivalents. However, some of their catalytic domains are thought to have originated by horizontal gene transfer from bacteria present in the common ancestral environment (like some aspects of secondary metabolism, see section 19.15). Genomic and proteomic analysis of anaerobic fungi (rumen chytrids) suggest that the fungal cellulosome is an evolutionarily chimeric structure, consisting of an independently evolved fungal complex that co-opted useful activities from bacterial neighbours within the gut microbiome (Haitjema *et al.*, 2017).

Cellulosomes efficiently degrade crystalline cellulose and associated plant cell wall polysaccharides, provide for attachment to the cell surface and their adhesion to the insoluble substrate provides the individual microbial cell that produced them with a competitive advantage in the utilisation of soluble products. Enzymatic degradation of plant biomass has attracted the interest of researchers involved in biofuel production. In particular, because of their efficient organisation and hydrolytic activity, cellulosomes are considered to have commercial potential in this

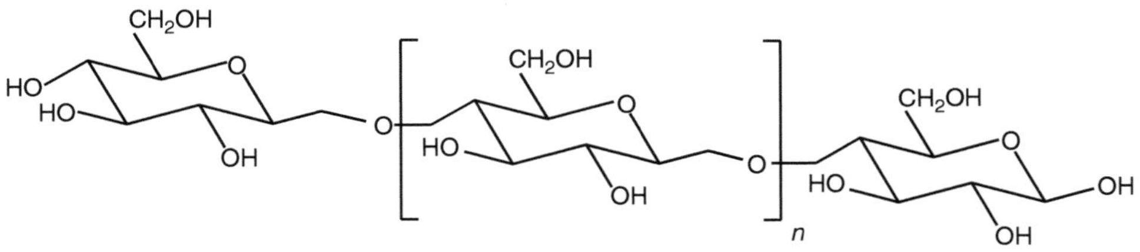

Figure 11.1. Structural formula of cellulose. There may be from a few hundred to a few thousand sugar residues in the polymer molecule, corresponding to molecular masses from about 50,000 to approaching 1 million.

biotechnology. They are the focus of much effort to use the versatile cohesin–dockerin protein assembly mechanism as a standard design principle to engineer ideal 'designer cellulosomes' for managing domestic and industrial lignocellulosic wastes (Bayer *et al.*, 2007; Rana & Rana, 2017).

When grown on cellulose, the **white-rot fungi**, like *Phanerochaete chrysosporium*, produce two **cellobiose oxidoreductases**: a cellobiose:quinone oxidoreductase (CBQ) and cellobiose oxidase (CBO). Cellobiose oxidase can oxidise cellobiose to the δ-lactone, which can then be converted to cellobionic acid and then glucose + gluconic acid; cellobiose δ-lactone can also be formed by the enzyme cellobiose:quinone oxidoreductase. Similar cellobiose-oxidising enzymes, capable of utilising a wide variety of electron acceptors, have been detected in many other fungi, though their role is uncertain. These enzymes are probably of most significance in regulating the level of cellobiose and glucose, the accumulation of which can inhibit endoglucanase activity. The role originally ascribed to CBQ was as a link between cellulose and lignin degradation. Cellobiose oxidase also reduces Fe(III) and together with **hydrogen peroxide**, generates hydroxyl radicals. These radicals can degrade both lignin and cellulose, indicating that cellobiose oxidase has a central role in degradation of wood by wood-decay fungi.

Brown-rot fungi use a rather different initial cellulolytic system to the hydrolytically based one employed by the white rots. Brown-rot fungi can depolymerise cellulose rapidly and virtually completely. Even cellulose deep within the walls and protected by lignin polymers is prone to attack. The process seems to depend on H_2O_2 (secreted by the fungus) and ferrous ions in the wood oxidising sugar molecules in the polymer, thereby fragmenting it and leaving it open to further attack by hydrolytic enzymes. **Oxalate** crystals that coat so many fungal hyphae may be responsible for reducing the ferric ions normally found in wood to ferrous ions, so aiding oxidative cleavage of the cellulose. This 'oxidoreductive cellulose-degrading system' exists in parallel with the hydrolytic cellulase system already described. The two systems complement one another: copper oxidases attack the highly crystalline region of cellulose, while endoglucanases attack amorphous cellulose with cellobiohydrolases that cleave the ends of chains of crystalline cellulose. Oxidative cleavage of cellulose uses polysaccharide monooxygenases (**PMOs**), cellobiose dehydrogenases (**CDHs**) and members of what is called ***'Auxiliary Activity Family 10'*** which is a family of polysaccharide monooxygenases known to cleave chitin and cellulose. Sequences specifying PMOs are widely distributed in the genomes of most ascomycetes and basidiomycetes (white-rot and brown-rot fungi alike) and they boost the efficiency of cellulose degradation by their action on the crystalline part of cellulose which disrupts the tightly packed cellulose chains, making them more accessible for attack by hydrolytic cellulases (Dimarogona *et al.*, 2012).

11.3 BREAKDOWN OF POLYSACCHARIDE: HEMICELLULOSE

Hemicellulose is a name which covers a variety of branched-chain polymers containing a mixture of various hexose and pentose sugars, which might also be substituted with uronic and acetic acids. The main hemicelluloses found in plants are **xylans** (1–4-linked polymers of the pentose sugar xylose), but **arabans** (polyarabinose), **galactans** (polygalactose), **mannans** and copolymers (e.g. glucomannans and galactoglucomannans) are also encountered. The major angiosperm hemicellulose is a xylan with up to 35% of the xylose residues acetylated, and it is also substituted with 4-O-methylglucuronic acid in dicotyledonous plants. Enzymes responsible for hemicellulose degradation are named according to their substrate specificity; for example, mannanases degrade mannans, xylanases degrade xylans, etc. As xylans predominate in plant walls, more is known about xylanases.

Xylanases can be induced by their substrate, the response being for the fungus to produce a complex of enzymes rather than one. The complex consists of at least two endoxylanases and a β-xylosidase. The **endoxylanases** degrade xylan to xylobiose and other oligosaccharides while the **xylosidase** degrades these smaller sugars to xylose. Some arabinose is also formed, showing that the xylanase complex can hydrolyse the branch points in xylan (Patel & Savanth, 2015).

11.4 BREAKDOWN OF POLYSACCHARIDE: PECTINS

Pectins consist of chains of β-1–4 linked **galacturonic acids**, in which about 20–60% of the carboxyl groups are esterified with methanol. They occur primarily in the middle lamella between plant cells. As this represents only a small proportion of the plant wall they are correspondingly of little importance as a component of the bulk of plant litter. However, extensive breakdown of the middle lamella of living plants is brought about by necrotrophic parasites. **Pectinases**, therefore, are of great importance during fungal invasion of plant tissue and different pectinase activities are involved in pathogenicity of fungi responsible for a wide array of diseases (Reignault *et al.*, 2008).

Polygalacturonases and **pectin lyases** attack the true pectins, while arabanases and galactanases degrade the neutral sugar polymers associated with them. These activities have drastic effects on the structural integrity of the tissues which may extend to death of the cell due to osmotic stresses imposed by damage to the wall. It seems likely that the products of pectinase activity will be absorbed as nutrient by the fungus, but these enzymes are better considered to be part of the machinery by which plant defences are breached than as being concerned primarily with nutrient supply. In addition to their importance in pathogenicity, pectinases are growing in importance in biotechnology; for example, in the processing of fruit and vegetable juices, wine, tea, coffee and animal feeds, in extraction of vegetable oil, and in recycling of pulp and waste paper (see section 19.16; Benoit *et al.*, 2012; Garg *et al.*, 2016).

11.5 BREAKDOWN OF POLYSACCHARIDE: CHITIN

Chitin, in which the repeating unit is the same as that in cellulose except that the hydroxyl group at C-2 is replaced by an acetamido group (Figure 11.2), is the second most abundant polymer on Earth as it occurs in the exoskeletons of arthropods and, of course, in fungal cell walls. Polysaccharides that contain amino sugars or their derivatives are called **mucopolysaccharides**. Chitin is degraded by **chitinase**, a glucan hydrolase which attacks the β-1–4 glycosidic bonds, eventually producing the disaccharide chitobiose which is then converted to the monosaccharide *N*-acetylglucosamine by chitobiase. Chitinase may also be involved in fungal wall synthesis (see Chapter 7). Several chitinases, which are glycoside hydrolase family enzymes, are produced by fungi and can have different substrate-binding site structures and are obviously specialised for different functions. Derivatives of chitin and chitosan have found many commercial uses from medicine to cosmetics and dietary supplements, so fungal chitinases have biotechnological applications, also (Seidl, 2008; Hartl *et al.*, 2012).

11.6 BREAKDOWN OF POLYSACCHARIDE: STARCH AND GLYCOGEN

Solid semi-crystalline starch and water-soluble glycogen are two distinct physical states of the same type of storage polysaccharide. Starch, the major reserve polysaccharide of plants, contains glucose polymers with α-1–4 glycosidic bonds. **Amylose** is composed of long unbranched chains, whereas **amylopectins** (comprising 75–85% of most starches) have branch points formed from α-1–6 glycosidic linkages (Figure 11.3); these 1–6 bonds form approximately every 20 to 30 glucose units along the chain and the branches are about 30 glucose units long. Starch-degrading enzymes include:

- α-amylases, which are **endoamylases** acting on 1–4 bonds and bypassing the 1–6 bonds;
- β-amylases, which are **exoamylases** producing the disaccharide maltose by splitting alternate 1–4 bonds until they reach a 1–6 branch point (which they cannot bypass);
- **amyloglucosidases** (or **glucoamylases**), which can act on both 1–4 and 1–6 bonds, and seem to occur almost exclusively in fungi;
- debranching enzymes (e.g. **pullulanase**) which sever 1–6 bonds; α-glucosidases which hydrolyse 1–4 glycosidic linkages in disaccharides and oligosaccharides, producing glucose as the end product of starch breakdown.

Glycogen is very similar to starch, being a branched polymer composed of glucose residues linked by α-1–4 glycosidic bonds, but branches are shorter and more frequent than they are in starch; about every tenth residue is involved in a branch formed by α-1–6 glycosidic bonds in glycogen, and branches are about 13 glucose units long. This is the polysaccharide reserve that is found in animal tissues, and in the fungi themselves. Most fungi are likely to encounter glycogen in their surroundings, as they are likely to be surrounded by dead and dying fungal cells. ***Intracellularly***, glycogen is degraded by a phosphorylase which releases glucose 1-phosphate for metabolic use (Figure 11.4), with the aid of a transferase and α-1–6-glucosidase (activities of a single polypeptide) to deal with the branches (plants use a similar phosphorolytic mechanism to mobilise their starch

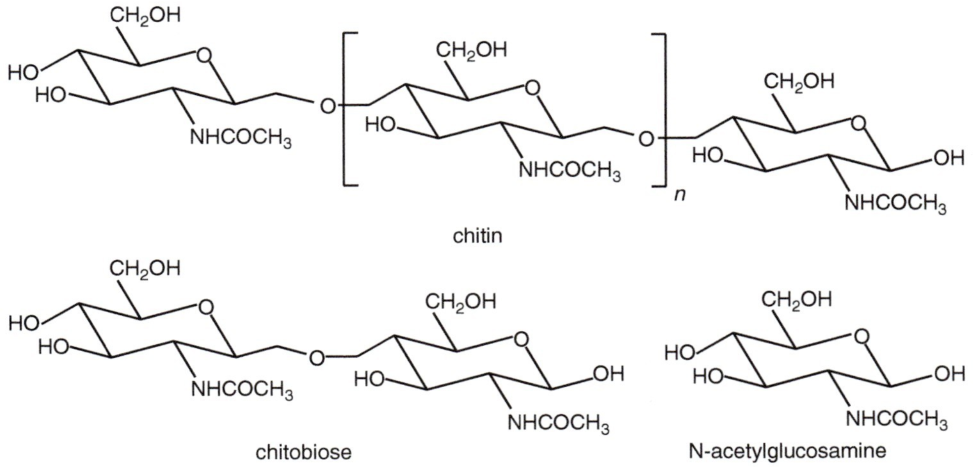

Figure 11.2. Structural formulae of chitin and its constituents. Chitin structure is similar to that of cellulose (Figure 11.1), except that acetylamino groups replace the hydroxyl groups at carbon position 2 of the glucose molecule. Approximately 16% of naturally occurring chitin units are deacetylated. Modified from Moore (1998).

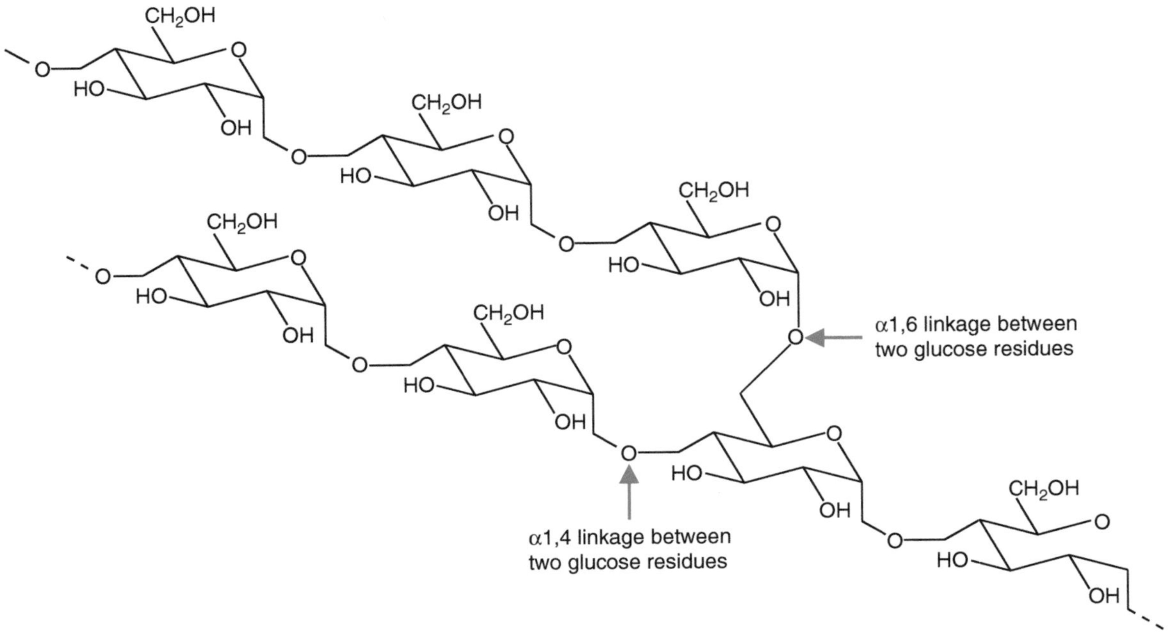

Figure 11.3. Structural formula of amylopectin. The 1–6 bonds form approximately every 20 to 30 glucose units along the chain and the branches are about 30 glucose units long. Modified from Moore (1998).

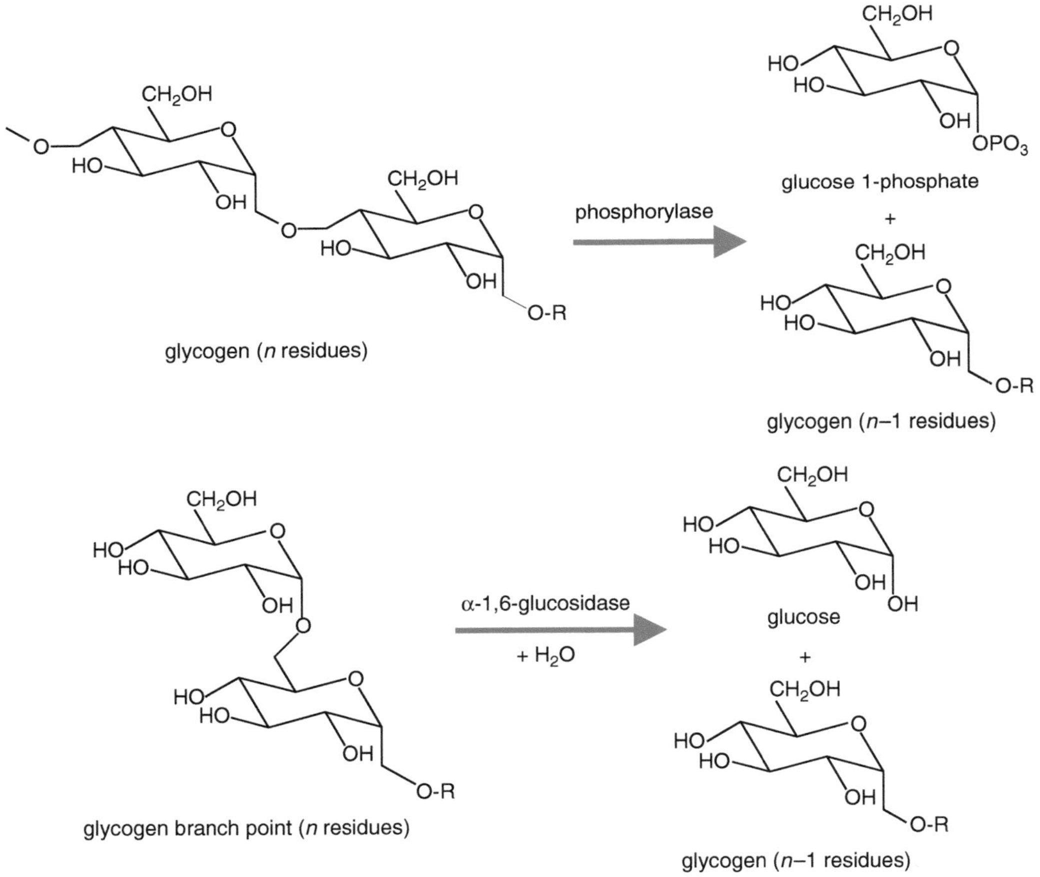

Figure 11.4. Intracellular degradation of glycogen. The top panel shows glycogen phosphorylase activity severing one of the 1–4 linkages. The bottom panel shows the 'debranching activity' of the glucosidase hydrolysing a 1–6 linkage. In glycogen, about every tenth residue is involved in a branch formed by α1–6 glycosidic bonds, and branches are about 13 glucose units long. Modified from Moore (1998).

reserves intracellularly using starch phosphorylase to release glucose 1-phosphate). ***Extracellularly***, glycogen, like starch, is degraded by components of the amylase enzyme complex.

The processes of glycogen synthesis and degradation share similarities among different microorganisms. However, the ***regulation*** of that metabolism is different. Glycogen accumulates in *Neurospora crassa* as a carbon and energy reserve to protect against adverse environmental conditions during growth and development (Bertolini *et al.*, 2012). In other organisms, glycogen metabolism seems often to be related to morphogenesis. For example, glycogen is a key storage compound in the rice blast fungus, *Magnaporthe oryzae*, and the two major enzymes that degrade cellular stores of glycogen regulate plant infection through the NADPH-dependent enzyme trehalose-6-phosphate synthase (Badaruddin *et al.*, 2013). Another example is that in the ectomycorrhizal Périgord truffle fungus, *Tuber melanosporum*, transcripts coding for trehalose metabolic enzymes were upregulated in sporophores, whereas genes involved in mannitol and glycogen metabolism were preferentially expressed in mycelia and ectomycorrhizas respectively (Ceccaroli *et al.*, 2011). The several roles served by glycogen in morphogenesis of sclerotia and sporophores of the ink cap mushroom *Coprinopsis* are discussed in Chapter 13.

11.7 LIGNIN DEGRADATION

Lignins are high molecular weight, insoluble aromatic polymers present predominantly in the walls of secondarily thickened plant cells, making them rigid and impervious. They are composed essentially of many methoxylated derivatives of benzene (**phenylpropanoid alcohols**, also called **monolignols**), especially **coniferyl, sinapyl** and **coumaryl** alcohols (Figure 11.5), the proportions of these three differ between angiosperms and gymnosperms, and between different plants. Lignins are highly heterogeneous polymers, the monolignols being assembled to suit the lignin to different functions in the plant cell wall (Humphreys & Chapple, 2002; Davin & Lewis, 2005; Vanholme *et al.*, 2010; Shafrin *et al.*, 2017).

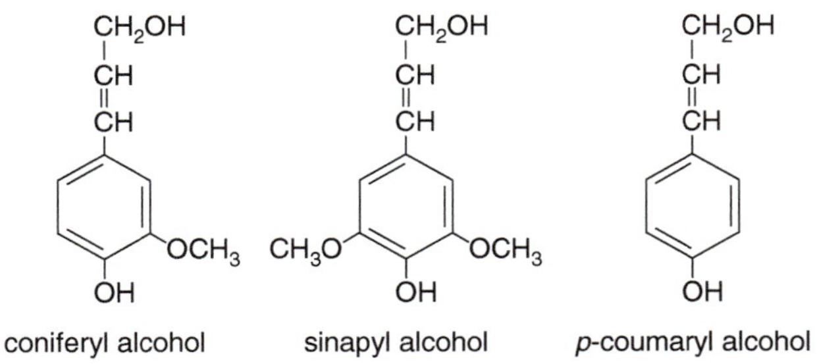

Figure 11.5. Chemical structures of the phenylpropanoid alcohols used to construct the lignin polymer. These are also called monolignols. The proportions of these three phenylpropanoid alcohols differ between the lignins of angiosperms and gymnosperms, and between lignins of different plant species.

You do not need to 'know' the primary structure of angiosperm lignin (Figure 11.6) in the sense of being able to draw it, but it is helpful to understand what Figure 11.6 is telling you about the ***nature of lignin*** as a nutritional substrate and the enzymological challenge it poses to any organism setting out to degrade it. Note the way the three phenylpropanoid alcohols are used in the structure; note the predominance of **ether** linkages and **carbon–carbon bonding**, and the presence of a few hydroxyls that can take part in cross-links to other polymers (polysaccharides and proteins). Above all, note the predominance of **benzene rings**. How many foods do you know that are full of benzene rings and how would you metabolise a benzene ring? In your kitchen, benzene rings, especially phenols, are more likely to be found in disinfectants than foods, and that's the key to the function of lignin. Microorganisms that attempt to degrade lignin produce antimicrobial disinfectants, which makes lignin very resistant to attack.

This extreme **resistance to microbial degradation** is part of the principal function of lignin as it can protect other polymers from attack. We are used to hydrolases cleaving polymers (polysaccharides, proteins, lipids), but the carbon–carbon and ether bonds that join subunits together in lignin must be cleaved by an ***oxidative*** process and a range of enzymes are needed for lignin to be degraded. Although ability to digest simple synthetic lignins has been reported for a few bacteria, ability to degrade natural lignins is generally considered to be limited to a very few fungi, including a range of *Basidiomycota* and *Ascomycota*. The fungi involved are generally known as 'white rots' because the lignin they digest away provides the main pigmentation of wood.

Oxidative lignin breakdown depends on two major groups of enzymes called **heme peroxidases** and **laccases**. Lignin-degrading peroxidases include:

- **lignin peroxidase** (LiP; a heme [Fe]-containing protein) which catalyses H_2O_2-dependent oxidation of lignin). LiPs are strong oxidants that oxidise the major non-phenolic structures of lignin;
- **manganese-dependent peroxidase** (MnP, which also catalyses H_2O_2-dependent oxidation of lignin). MnP catalyses the oxidation of various phenolic substrates but is not capable of oxidising the more recalcitrant non-phenolic lignin;
- **versatile peroxidase** (VP; also called manganese-independent peroxidase, another heme-containing peroxidase with the catalytic activities of both MnP and LiP). It can oxidise Mn^{2+}, like MnP, but also non-phenolic compounds, like LiP (VP has only been described in species of the genera *Pleurotus* and *Bjerkandera*);
- **dye-decolourising peroxidase** (DyP; which uses H_2O_2 to decolourise anthraquinone dyes). DyPs occur in both fungi and bacteria.

LiP, MnP and VP are class II extracellular fungal peroxidases that belong to the plant and microbial peroxidases superfamily. DyPs

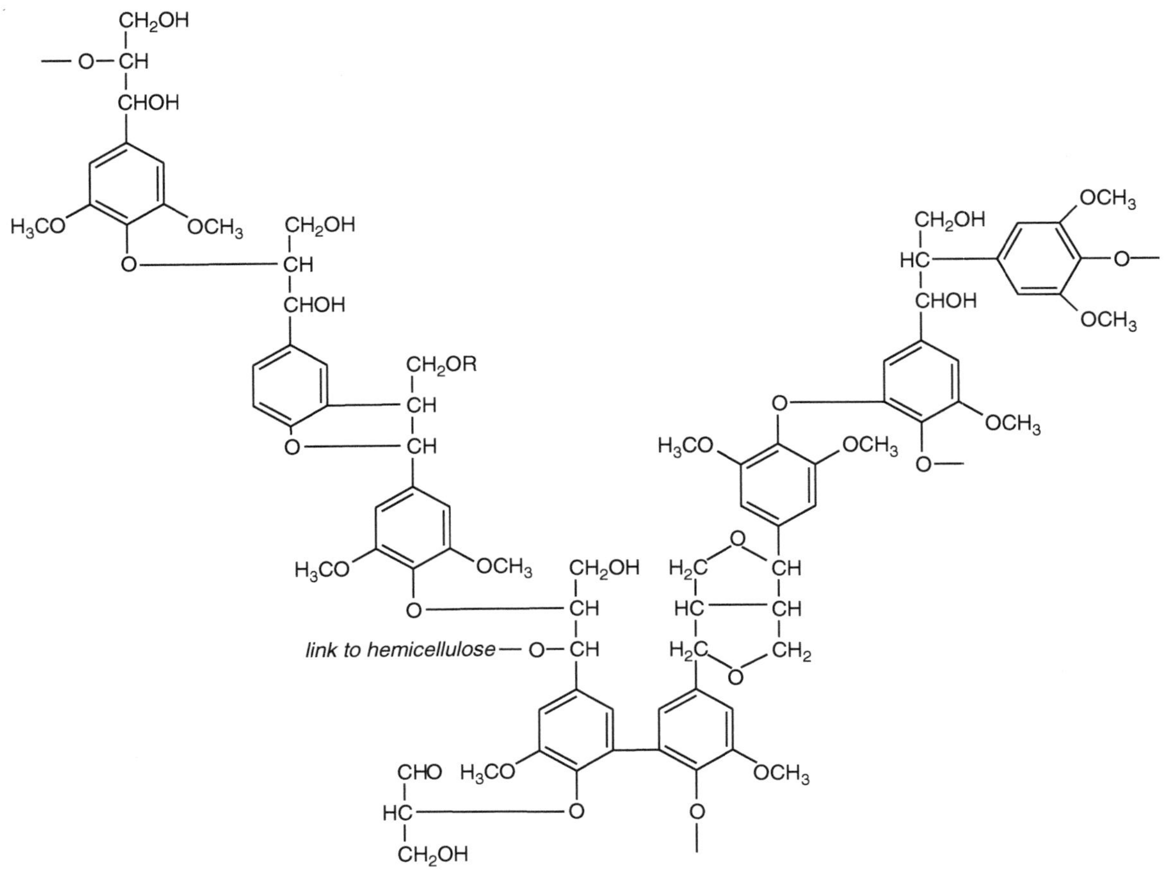

Figure 11.6. Schematic formula of the polymer structure of angiosperm lignin. Note the way the three phenylpropanoid alcohols are used in the structure; note the predominance of ether linkages and carbon–carbon bonding, and the presence of a few hydroxyls that can take part in cross-links to other polymers (polysaccharides and proteins). Above all, note the predominance of benzene rings. Modified from Moore (1998; and see Vanholme *et al.*, 2010, for more structural details).

are members of a new superfamily of heme peroxidases reported to oxidise lignin model compounds.

The second major group of lignin-degrading enzymes, **laccases**, are copper-containing proteins found in plants, fungi and bacteria. They belong to the **multicopper oxidase superfamily** and are described in a little more detail below. Fungal laccases catalyse demethylation of lignin components and can oxidise phenolic lignin model compounds; they have higher redox potential than bacterial laccases. In the presence of **redox mediators**, fungal laccases can also oxidise non-phenolic lignin model compounds (a redox mediator is a molecule that is stable in two interconvertible redox states, and is sufficiently small to diffuse easily into and out of protein channels and efficiently shuttle electrons between the enzyme active site and the substrate).

In addition to the peroxidases and laccases, fungi produce other accessory oxidases such as **aryl-alcohol oxidase**, also called veratryl alcohol oxidase (veratryl alcohol is a degradation product of lignin) and **glyoxal oxidase** that generate the hydrogen peroxide required by the peroxidases (Sigoillot *et al.*, 2012; Abdel-Hamid *et al.*, 2013).

The process of catabolic lignin degradation involves:

- cleavage of ether bonds between monomers;
- oxidative cleavage of the propane side chain;
- demethylation;
- benzene ring cleavage to ketoadipic acid which is fed into the tricarboxylic acid cycle as a fatty acid.

Most research has been concentrated on white-rot basidiomycete fungi, such as *Phanerochaete chrysosporium* (*Sporotrichum pulverulentum*), which is able to **mineralise lignin** completely to CO_2 and water. The lignin-degradative system of *Phanerochaete chrysosporium* appears after cessation of primary growth (that is, it is an aspect of the **secondary metabolism** of the organism) and can be induced by nitrogen starvation. Genomes of white-rot fungi feature families of genes encoding laccases, lignin peroxidases and manganese peroxidases. The multiplicity of these extracellular enzymes may be a response to the diversity of the lignin substrate but could also be a result of their having multiple roles in fungal physiology (laccases, for example, contribute to plant pathogenesis, sporulation and pigment formation).

Lignin peroxidase (ligninase) is the key lignin-degrading enzyme of white-rot fungi. The *Phanerochaete chrysosporium* family of lignin peroxidases comprises extracellular glycosylated heme proteins, which are secreted in response to nitrogen limitation. When the fungus is grown in low-nitrogen medium there is an increase in H_2O_2 production by cell extracts which correlates with the appearance of ligninolytic activity. Experimental destruction of H_2O_2 by adding the enzyme catalase strongly inhibits lignin breakdown. Activated oxygen derived from H_2O_2 is involved in degrading lignin but is held in the active site of a specific extracellular enzyme, the lignin peroxidase. Lignin peroxidases are strongly oxidative; the enzyme is activated by itself being oxidised by H_2O_2, the initial step involving oxidation by one electron to produce an unstable intermediate which is then able to catalyse oxidation of phenols, aromatic amines, aromatic ethers and polycyclic aromatic hydrocarbons. **Veratryl alcohol** is a secondary metabolite that stimulates lignin degradation by recycling the lignin peroxidase and protecting the enzyme against inactivation by H_2O_2.

There can be as many as 15 lignin peroxidase **isozymes**, ranging in molecular mass from 38,000 to 43,000, the spectrum of isozymes produced depending on culture conditions and strains employed. However, 10 lignin peroxidase genes, with a conserved sequence, have been identified in *Phanerochaete chrysosporium* and mapped into three different linkage groups. Lignin peroxidases are not found in all white-rot fungi, though. They seem to be absent from *Ceriporiopsis subvermispora*, a white-rot basidiomycete that is widely studied for its potential use in the pulp and paper industry ('biopulping') and in which manganese peroxidases are responsible for ligninolysis.

Manganese-dependent peroxidases are another family of extracellular glycosylated heme proteins, which are produced by most white-rot fungi. Like lignin peroxidase, the manganese peroxidases also require H_2O_2 to function, but the mechanism is very different. The manganese peroxidase system creates low molecular weight oxidising agents that diffuse into the lignin substrate and can oxidise phenolics residues in the lignin some distance from the enzyme. Some of these oxidising agents may be peroxidated lipids, but the chief one, from which the enzyme gets its name, is the metal ion manganese; the enzyme uses H_2O_2 to oxidise extracellular Mn(II) to Mn(III) and this becomes the diffusible oxidant that can degrade lignin at a distance.

The H_2O_2 which is required by peroxidases is probably generated by **glyoxal oxidase**, an extracellular enzyme that transfers electrons from low molecular weight aldehydes (e.g. glyoxal and glycoaldehyde) to O_2 and so forms H_2O_2. **Aryl-alcohol oxidase** is another H_2O_2 generating enzyme; in the process of converting benzyl alcohols to the aldehydes it transfers electrons to O_2, generating H_2O_2. *Ceriporiopsis subvermispora* secretes an **oxalate oxidase**, which catalyses the degradation of oxalate to carbon dioxide and H_2O_2 and could be the main provider of H_2O_2 for manganese peroxidase in this organism.

Although few fungi produce ligninolytic enzymes, a much wider range secrete **laccases** as extracellular enzymes. These are copper-containing oxygenases which can oxidise *o*- and *p*-phenols and are required for the metabolism of lignin-degradation products. They are particularly interesting as their appearance or disappearance in fungal cultures has been correlated with sexual and asexual reproduction in several cases. Thus, during mycelial growth of the cultivated mushroom, *Agaricus bisporus*, a large proportion of the compost lignin is degraded, and correspondingly high activities of laccase are recorded (this one enzyme can amount to 2% of the total fungal protein). Yet, as the culture forms sporophores, laccase activity is rapidly lost, initially by inactivation and subsequently by proteolysis; this pattern of behaviour reflecting the changing nutritional demands of fungal mycelia as they process through successive developmental phases and the ability of the mycelium to act on its environment to satisfy those demands.

Laccases have been found in many fungi, including non-ligninolytic members of the *Ascomycota*, such as *Aspergillus* and *Neurospora*, as well as wood-rotting *Basidiomycota*. Laccases also occur in plants where they contribute to lignin biosynthesis. To react with a broader range of substrates laccases can interact with a low molecular weight co-substrate secreted by the fungus that functions as a diffusible lignin-oxidising agent (redox mediator).

In their study of laccase, Mn-dependent peroxidase and versatile peroxidase activities in seven wood-rotting fungi digesting wheat straw lignin, Knežević *et al.* (2013) found that the profiles of the three enzymes varied between species and during the cultivation period. *Dichomitus* (*Bjerkandera*) *squalens* (western red rot, a poroid crust fungus of conifers) was the best lignin degrader (34.1% of lignin digested), and *Pleurotus ostreatus* and *Pleurotus eryngii* were the weakest (7.1% and 14.5% of lignin digested, respectively). These two *Pleurotus* species produced the most laccase, whereas *Dichomitus squalens* produced the most versatile peroxidase (with the least in *Pleurotus ostreatus*). *Lenzites betulinus* (birch mazegill) and *Fomitopsis pinicola* (red belted polypore, a bracket or shelf fungus on fir and hemlock) were the best Mn-dependent peroxidase producers, with *Pleurotus ostreatus* the weakest; but the laccase isoforms (isoforms being proteins that have similar but not identical amino acid sequences) produced by *Pleurotus ostreatus* depend on the conditions in which it is grown. The presence of inducers, the pH of the culture medium as well as the type and concentration of carbon and nitrogen sources all differentially affect the production of these enzymes, and the growth temperature also modifies the patterns of activity and production of laccase isoforms (Montalvo *et al.*, 2020).

Lignin-degrading enzymes have considerable promise for several biotechnological applications, such as the paper industry, textile industry, waste water treatment and the degradation of herbicides. **Biopulping** has already been mentioned as an industry where ligninolytic enzymes can improve the quality of pulp by releasing and purifying the cellulose. Laccases are extremely important biocatalysts, which can be exploited for:

- paper pulp bleaching,
- detoxification (particularly bioremediation of pulp mill effluents),

- fibre modification,
- dye decolourisation,
- removal of phenolics from wines,
- as well as organic synthesis and chemical transformation of pharmaceuticals (Sharma *et al.*, 2018).

Further, many pesticidal treatments depend on benzene rings for their effectiveness; examples are chemicals like **pentachlorophenol (PCP)** and **polychlorinated biphenyls (PCBs)**. These persist in the environment because there are so few organisms able to catabolise them. But organisms that can catabolise lignin have all the tools necessary to destroy such **persistent pesticides** (Peralta *et al.*, 2017).

DyP-type peroxidases and laccases are the major lignin-modifying enzymes produced by bacteria (de Gonzalo *et al.*, 2016). Lignocellulose-consuming animals (termites, shipworms) secrete some glycoside hydrolases, but most harbour a diverse gut microbiota that includes fungi (chytrids) oomycetes and bacteria able to secrete lignocellulolytic enzymes in a mutualism that is particularly complex in termites (Cragg *et al.*, 2015).

There are two intriguing side issues to fungal degradation of timber. One, which may become more crucial in the future, is that some of the small organic molecules the fungi produce to enhance activity of the lignin and manganese peroxidases incorporate ***chlorine*** from the wood into **chloroaromatics**, and even the synthesis of veratryl alcohol requires **chloromethane**. These volatile compounds are released into the atmosphere (thereby flushing the chlorine out of the substrate) but may themselves act as environmental pollutants. Some white-rot genera, particularly *Phellinus* and *Inonotus*, release enormous quantities of chloromethane into the atmosphere as they digest wood. The annual global release to the atmosphere from this source has been estimated at 160,000 tons, 75% of which is released from tropical and subtropical forests, with 86% being attributable to *Phellinus* spp. alone. Chloromethane is a powerful **greenhouse gas** and atmospheric pollutant which can have adverse effects on stratospheric ozone, yet in this case it is the product of a natural ecosystem (Watling & Harper, 1998; Anke & Weber, 2006).

The second intriguing aspect is that there are at least 75 species of fungi that are **bioluminescent**; they emit light from their mycelium and/or sporophores, and all species found so far have proved to be wood-decay fungi. All known truly bioluminescent fungi are white-rot fungi belonging to Order *Agaricales* (*Basidiomycota*) with one exceptional ascomycete belonging to Order *Xylariales*. They are mostly mushroom-forming, saprotrophic or, rarely, plant pathogenic species belonging to three distinct evolutionary lineages, which are named for their characteristic genera: 12 species belong to the *Omphalotus* lineage, 5 species to the *Armillaria* lineage (*Armillaria mellea* is the most widely distributed bioluminescent fungus) and 47 species belong to the mycenoid lineage. Different luminous fungi emit light from different tissues, for example, from:

- mycelium and rhizomorphs in *Armillaria*;
- mycelium only in many *Mycena* species;
- spores only in *Mycena rorida*;
- sclerotia only of *Collybia tuberosa*;
- mycelium and the whole (mushroom) sporophore in *Panellus stipticus* and *Omphalotus olearius* (*Clitocybe illudens*); in the latter, light is emitted from the mushroom cap, stem, gills and spores.

Light emission *in vitro* can be obtained enzymatically by mixing cold extracts (which extract the enzymes) and hot extracts (which extract the substrates) from different species of fungi, which indicates a common mechanism for all these fungi. Kinetic data suggest a consecutive two-step enzymatic mechanism: first, a light emitting substrate (arbitrarily called **'luciferin'**) is reduced by a soluble reductase at the expense of NAD(P)H; second, reduced luciferin is oxidised by an insoluble (membrane-bound?) **luciferase** that releases the energy in the form of bluish-green visible light with an emission maximum wavelength close to 530 nm.

Although fungal 'luciferin' is clearly one of the numerous secondary metabolites fungi can synthesise, its exact identity and structure has not yet been established but it is thought that all bioluminescent fungi share the same enzymatic mechanism, suggesting that the bioluminescent pathway arose early in the evolution of the *Agaricales* (Oliveira et *al.*, 2012). Nor is it entirely clear what the physiological and ecological function of **fungal bioluminescence** might be (which makes it difficult to appreciate its evolutionary selective advantage). It has been suggested that in the dark beneath closed tropical forest canopies bioluminescent sporophores may be at an advantage by ***attracting*** grazing animals (and that would include insects and other arthropods) that could help disperse their spores. Conversely, where mycelia (and vegetative structures like rhizomorphs and sclerotia) are the bioluminescent tissues, the argument has been made that light emission could ***deter*** grazing. Neither of these suggestions is entirely satisfactory. However, as far as is known at present, all luminescent basidiomycetes are white-rot fungi capable of lignin degradation. Bioluminescence is an oxygen-dependent metabolic process and may be used for **detoxifcation of peroxides** formed during **ligninolysis**; with the energy being released as light rather than heat. The favoured hypothesis now, therefore, is that fungal bioluminescence is an advantageous process because it provides antioxidant protection against the potential deleterious effects of reactive oxygen species produced during wood decay (Desjardin *et al.*, 2008; Audrey *et al.*, 2015).

11.8 DIGESTION OF PROTEIN

Proteases (also called **proteinases** or **peptidases**) are peptide hydrolases; a group of enzymes which hydrolyse the peptide bonds of proteins and peptides, cleaving the substrate molecule into smaller fragments and, eventually, into amino acids. This is a complex group of enzymes, varying greatly in physicochemical and catalytic properties. Proteolytic enzymes are produced intra- and extracellularly, playing important roles in regulatory processes of the cell as well as contributing to nutrition through degradation of protein food sources. Intracellular proteolysis seems to be the responsibility of large multicatalytic complexes of proteases which are called **proteasomes** (see section 5.8).

Extracellular proteases are involved mainly in the hydrolysis of large polypeptide substrates into the smaller molecules which can be absorbed by the cell. Extracellular proteases are produced by many species of fungi, but most is known about protein utilisation and protease production by *Aspergillus* species and *Neurospora crassa*. The *Basidiomycota Agaricus, Coprinopsis* and *Volvariella* have been shown to be able to use protein as a sole source of carbon about as efficiently as they can use the sugar glucose and can also use protein as a source of nitrogen and sulfur (Kalisz *et al.*, 1986). Mycorrhizal fungi have been shown to use protein as a source of both nitrogen and carbon and some ectomycorrhizas supply nitrogen derived from proteins in soil to their plant hosts.

In *Aspergillus* species, protease production is controlled by derepression; the *Neurospora crassa* protease is controlled by induction and repression. In neither case is protease produced in the presence of ammonia; which appears, therefore, to be the primary source of nitrogen for these fungi. In sharp contrast, production of extracellular proteases in *Basidiomycota* is regulated mainly by induction; if substrate protein is available the proteases are produced, even in the presence of adequate alternative supplies of ammonia, glucose and sulfate. In this case the protein might be presumed to be the 'first choice' substrate (Kalisz *et al.*, 1987, 1989).

Protein is probably the most abundant nitrogen source available to plant-litter-degrading organisms in the form of plant protein, lignoprotein and microbial protein. Many pathogenic microorganisms secrete proteases which are involved in the infection process and some, including the apple pathogenic fungus *Monilinia fructigena*, are known to utilise host proteins for nutrition. The virulence of a few pathogenic fungi is correlated with their extracellular protease activity. Several species release specific proteases which can hydrolyse structural and other proteins resistant to attack by most other proteases, such as insect cuticles. The animal dermatophytes *Microsporum* and *Trichophyton* produce collagenases, elastases and keratinases.

Enzymes that degrade proteins form two major groups: **peptidases** and **proteases** (Kalisz, 1988; Nirmal *et al.*, 2011; Monteiro de Souza *et al.*, 2015). **Exopeptidases** remove terminal amino acids or dipeptides and are subdivided according to whether they act at the carboxy-terminal end of the substrate protein (carboxypeptidases); the amino-terminal end (aminopeptidases), or on a dipeptide (dipeptidases). Proteases cleave internal peptide bonds; they are **endopeptidases**. The protease catalytic mechanism can be determined indirectly by study of response to inhibitors which react with specific residues in the active site of the enzyme. This leads to subclassification into four groups:

- **Serine proteases** are the most widely distributed group of proteolytic enzymes. They have a serine residue in the active site, are generally active at neutral and alkaline pH, and show broad substrate specificities. Most eukaryotic lineages that have been sequenced encode a set of 13–16 serine proteases, suggesting that all fungal serine protease families evolved in the eukaryote lineage before animals, plants and fungi diverged from one another (Muszewska *et al.*, 2017).
- **Cysteine proteases** occur in few fungi though extracellular cysteine proteases have been reported in *Microsporum* sp., *Aspergillus oryzae* and *Phanerochaete chrysogenum* (*Sporotrichum pulverulentum*). Cysteine proteases are involved in many functions in animals and replacement enzymes (which take control of the host function) are often produced by their parasites and pathogens so the enzymes from these sources may become good pharmacological targets in several major diseases of humans (Verma *et al.*, 2016).
- **Aspartic proteases** show maximum activity at low pH values (pH 3–4) and are widely distributed in fungi, performing important functions related to nutrition and pathogenesis. In addition, their high activity and stability at acid pH make them attractive for industrial application in the food industry. They are used as milk-coagulating agents in cheese production and are used to improve the flavour profile of some foods (Mandujano-González *et al.*, 2016).
- **Metalloproteases** have pH optima between 5 and 9 and are inhibited by metal-chelating reagents, such as ethylenediamine tetra-acetic acid (EDTA). In many cases, the EDTA-inhibited enzyme can be reactivated by zinc, calcium or cobalt ions. Metalloproteases are widespread, but only a few have been reported in fungi and most of these are zinc-containing enzymes, with one atom of zinc per molecule of enzyme. Some (known as fungalysins) are able to degrade the human extracellular matrix proteins elastin and collagen and are thought to act as virulence factors in diseases caused by fungi (Fernandez *et al.*, 2013).

11.9 LIPASES AND ESTERASES

Lipases and esterases catalyse the hydrolysis of esters made between alcohols and organic ('fatty') acids. They generally have low specificity and any lipase will hydrolyse virtually any **organic ester**, though different esters will be acted upon at different rates, and strictly speaking all these enzymes are **carboxylester hydrolases**. The main factors influencing what specificity is expressed are the lengths and shape of hydrocarbon chains either side of the ester link. The term esterase is generally applied to enzymes 'preferring' short carbon chains in the acyl group. Lipases are **lipolytic** enzymes which are a special class of carboxylester hydrolases capable of releasing long-chain fatty acids from natural water-insoluble carboxylic esters. Their substrates include fats, the lipid components of lipoprotein and the ester bonds in phospholipids. Rather than attempting to differentiate 'lipases' from 'esterases' using other criteria, it is probably best to remember that lipids, by definition, are insoluble in water and to distinguish **lipolytic esterases** (acting on lipids) from **nonlipolytic esterases** (not acting on lipids) (Ali *et al.*, 2012).

Extracellular lipase production has been detected in *Agaricus bisporus* during degradation of bacteria. In fermenter cultures, most of the lipase is produced in the stationary phase (i.e. it is a **secondary metabolic activity**) and regulation of lipase production in *Rhizopus* is very much affected by carbon and nitrogen sources in the medium and by the oxygen concentration.

Lipases are generally strongly activated by water–lipid interfaces (the active site of these enzymes is a hydrophobic cavity). These enzymes are able not only to catalyse the hydrolysis of triglycerides to free fatty acids, diglycerides and monoglycerides under aqueous conditions, but they are also able to carry out synthetic, and trans- and inter-esterification reactions in the presence of organic solvents. Add to these features stability and broad substrate specificity and you can begin to understand the biotechnological interest in these enzymes as commercial chemical catalysts (Barriuso *et al.*, 2013).

11.10 PHOSPHATASES AND SULFATASES

Phosphatases are also **esterases**, acting on esters of alcohols with phosphoric and sulfuric acids. They are enzymes of comparatively low specificity but fall into groups depending on their activity as phosphomonoesterases, phosphodiesterases and polyphosphatases. The phosphomonoesterases are further distinguished according to their pH optima as alkaline or acid phosphatases. For example, phosphatases are among the extracellular enzymes produced by *Agaricus bisporus* during growth on compost and must be important, therefore, in the nutrition of such litter-degrading fungi. But fungal phosphatases also serve as virulence factors in opportunistic fungal pathogens of humans because they do not respond to the control signals that regulate the host enzymes they mimic. However, the structural features that protect them from regulation by the host provide new opportunities for the development of inhibitors that specifically target the fungal enzymes without cross-reacting with host phosphatases. As several of the most common opportunistic fungal pathogens of humans are showing increasing resistance to conventional drugs, this is an area of therapeutics research with growing significance (Ariño *et al.*, 2011; Chen *et al.*, 2016).

Sulfatases act on **sulfate esters** in the same way that phosphatases act on **phosphate esters**. They may be important in recovery of sulfate from the sulfated polysaccharides which are found in soils, and they also serve several functions in intermediary metabolism. Fungal sulfatases, like fungal phosphatases, have been associated with virulence in pathogenic fungi (Toesch *et al.*, 2014; Chu *et al.*, 2016).

11.11 THE FLOW OF NUTRIENTS: TRANSPORT AND TRANSLOCATION

The plasma membrane is a lipoidal layer separating the aqueous 'bubble' of the cell from its aqueous surroundings. This separation is not complete or absolute as the cell must exchange chemicals with the environment; removing excretion products and absorbing nutrients. However, only molecules which dissolve readily in lipid can penetrate the membrane without assistance. Since the clear majority of the molecules that the cell needs to transfer across the membrane are hydrophilic rather than lipophilic, plasma membranes have evolved a range of associated **transport systems** that permit selective communication between the two sides of the membrane. This selectivity permits the cell to exercise considerable control over its interaction with the environment.

In an infinite solution, molecules of solute can move within the solution in two ways. Whole volumes of solution may be transported from place to place, taking solute molecules with them. This is **bulk flow** or mass flow and results from such things as convection flows and other large-scale disturbances within the solution. As far as living organisms are concerned, bulk flow may be achieved through cytoplasmic streaming, transpiration streams and similar processes. Although it is becoming clear that multiple motor proteins may work together to drive intracellular transport of organelles (Rai *et al.*, 2013; Pathak & Mallik, 2017), we must emphasise that what we are describing here is different from, and additional to, the vesicle and vacuolar traffic described in Chapter 5, above (sections 5.9, 5.10 and 5.11). Strongly bound kinesins fail to work collectively, whereas detachment-prone dyneins team up. It seems that leading dyneins in a team take short steps, while trailing dyneins take larger steps; the dyneins consequently bunch together, which shares the load more effectively and bonds the motor proteins more tenaciously to the microtubule. Even though such behaviour allows vesicle and vacuolar traffic to sustain the most rapid of extension growth rates in filamentous fungi, the bulk flow we describe here is more likely to be associated with distribution of materials (whether in solution or not) over ***multi***cellular or intercellular dimensions than with transfers via the endomembrane and cytoskeletal systems.

The second mode of solute movement is **diffusion**, where random thermal motion at the molecular level causes all solute molecules to move continuously. If the solution is completely homogeneous then any molecules that move out of a unit volume will be replaced by an identical number moving into that unit volume and the exchange of solute molecules will not be detectable. On the other hand, if there is a concentration gradient within the solution there will be a net flow of solute molecules from the high concentration end of the gradient, towards the low concentration end. Note that this gradient can be a chemical gradient of uncharged molecules (e.g. a sugar), an electrical gradient of a charged ion (e.g. K^+) or a combination of the two. This diffusion process is extremely relevant to the behaviour of cells, since there is likely to be a **concentration gradient** across the plasma membrane for just about every solute of importance to the cell.

To traverse the biological membranes a solute must leave the aqueous phase for the lipoidal environment of the membrane, traverse that, and then re-enter the aqueous phase on the other side of the membrane. Unaided simple diffusion of molecules across biological membranes depends considerably on their solubility in lipids. There are exceptions to this generalisation, though, as some small polar molecules (such as water) enter cells more readily than would be expected from their solubility in lipid. They behave as though they are traversing the membrane by simple diffusion through gaps or **pores** which are transiently generated by random movements of the acyl chains of the membrane phospholipids. Transfer of these materials (like that of molecules which are soluble in the lipid bilayer of the membrane, such as O_2 and CO_2) depends on **simple diffusion**. Their rate of movement is then proportional to the concentration differential on the two sides of the membrane and the direction of movement is from the high to the low concentration side. No metabolic energy is expended, and no specific membrane structures are involved in this mode of transfer, but net transfer ceases when the transmembrane concentrations equalise.

Only a minority of compounds pass through biological membranes *in vivo* by simple diffusion; the vast majority of metabolites that the cell needs to absorb or secrete are too polar to dissolve readily in lipid and too large in molecular size to make use of transient pores. To cope with these circumstances the membrane is equipped with **solute transport systems**. This applies to intracellular membranes bounding compartments within the cell as well as to the plasma membrane. The essential component of any transport system is a **transporter molecule**, a protein which spans the membrane and assists transfer of the metabolite across the lipid environment of the membrane.

With both passive and active transporters, substrate translocation depends on a conformational change in the transporter such that the substrate-binding site is alternately presented to the two faces of the membrane. These transporters are **transmembrane glycoproteins** of around 500 amino acids arranged into three major domains: **12** α-helices spanning the membrane, a highly charged cytoplasmic domain between helices 6 and 7, and a smaller external domain, between helices 1 and 2, which bears the carbohydrate moiety. Sequence homology between the amino- and carboxy-terminal halves of the protein suggests that the '12-α-helix structure' has arisen by the duplication of a gene encoding a 6-helix structure. Ion channels are different as their polypeptide subunits form a β-barrel containing a pore. One loop of the polypeptide is folded into the barrel and amino acids of this loop determine the size and ion selectivity of the channel. This transporter alternates between open and closed conformations.

If the transfer is **passive** with no requirement for metabolic energy then the transport process is described as **facilitated diffusion**. Such a process still depends upon a concentration differential existing between the two sides of the membrane, transfer occurring 'down the concentration gradient' (towards the compartment which has the lower concentration). However, transfer is much faster than would be predicted from the solubility of the metabolite in lipid, the high rate of transfer depending on the fact that the transporter and the transporter/metabolite complex are highly mobile in the lipid environment of the membrane. The major differences from simple diffusion are that facilitated diffusion exhibits:

- high substrate specificity;
- saturation kinetics.

Showing saturation kinetics means that as the concentration of the metabolite being transported is increased, the rate of transport increases asymptotically towards a theoretical maximum value at which all the transporters are complexed with the metabolite being transported (i.e. transporters are saturated).

Facilitated diffusion can transport a specific substrate very rapidly; but can only equalise the concentrations of the transported metabolite on the two sides of the membrane. Yet, in many cases, the cell needs to transfer a metabolite **against its concentration gradient**. The prime example will be where the cell is absorbing a nutrient available at only a low concentration; if growth of the cell is not to be limited by the external concentration of the nutrient, the cell must be able to accumulate the nutrient to concentrations greater than that existing outside. In this case, an adverse gradient of concentration will have to be established and maintained. Neither simple diffusion nor facilitated diffusion can do this; to achieve it the cell must expend energy to drive the transport mechanism. Such a process is called **active transport**.

Active transport is a transporter-mediated process in which movement of the transporter/substrate complex across the membrane is energy dependent. The transporter exhibits the same properties as a facilitated transport transporter (saturation kinetics, substrate specificity, sensitivity to metabolic inhibitors). In addition to these properties, active transport processes characteristically transfer substrate across the membrane against a chemical and/or electrochemical gradient and are subject to inhibition by conditions or chemicals that inhibit metabolic energy generation.

The mechanism is often a **co-transport** in which the movement of an ion down its electrochemical gradient is coupled to transport of another molecule against its concentration gradient. When the ion and the transported substrate move in the same direction the co-transporter is called a **symport**, whereas transporters which transport the two in opposite directions are termed **antiporters**. The electrochemical gradients, most usually of **protons** or K^+ in fungi, are created by **ion pumps** in which hydrolysis of ATP phosphorylates a cytoplasmic domain of the ion channel. Consequential conformational rearrangement of the protein then translocates the ion across the membrane and reduces the affinity of the binding site to release the ion at the opposite membrane face. Dephosphorylation restores the pump to its active conformation (and may translocate another ion or molecule in the opposite direction).

Complex interactions occur in transport of anions, cations and non-electrolytes; such interactions may depend on metabolic, chemical, biophysical and/or electrochemical relationships between a number of different molecular species and with the rest of metabolism. There are indications of what might be called transport strategy in operation in most cells. Single uptake systems are rarely encountered; dual or multiple systems are the norm, the different components being suited to different environmental conditions the organism may encounter. Multiple uptake systems inevitably result in complex uptake kinetics which might be indicative of physically separate transport transporters, each showing Michaelis–Menten kinetics (like the glucose transporters in *Neurospora*), or of single molecules exhibiting kinetics modulated by their environment (like the glucose transporter in *Coprinopsis*; Moore & Devadatham, 1979; Taj Aldeen & Moore, 1982).

Whatever the physical basis, the regulatory properties of the components of such 'families' of transport processes appear to be interlinked to ensure that nutrient uptake is maintained at a reliable level whatever the variation in substrate availability in the environment. Probably the most important generalisation that can be made about transport processes, though, is that for almost all of them the active extrusion of protons from the fungal cell seems to be essential. The proton gradient so established provides for uptake of sugars, amino acids and other nutrients by proton co-transport down the gradient and is directly involved in cation transport like the K^+/H^+ exchange or antiport. Each fungus possesses **multiple uptake systems** for most nutrients but the same basic process (active H^+ extrusion) energising most, if not all.

A crucial point, which has not yet been taken into account, is that the transport systems so far described will inevitably alter the solute concentrations of the cell and thereby influence the movement of that all-pervading nutrient, water. Water is a significant (even if often overlooked) component of innumerable biochemical processes. For example, every hydrolytic enzyme reaction ***uses*** a molecule of water, every condensation reaction ***produces*** a molecule of water and respiration of 1 g of glucose produces 0.6 g of water. The water relationships of the fungal cell are an important aspect of its overall economy. Water availability is determined by its potential energy; referred to as the **water potential**, symbolised by the Greek letter **psi** (Ψ). Zero water potential is the potential energy of a reference volume of free, pure water. The water in and around living fungal cells will have positive or negative potential energy relative to that reference state, depending on the effect(s) of osmotic, turgor, matrix and gravitational forces. Water will flow spontaneously along a water potential gradient, from high to low potentials, though in the normal state for most fungi this will mean from a negative to a more negative potential. The lower the water potential the less available is the water for physiological purposes and the greater is the amount of energy that must be expended to make the water available.

On the face of it, two things need to be considered. One is some sort of compensation for change in the solute relationships of the cell resulting from uptake of some substrate; such a process would further reduce the potential of the cell water and increase the tendency of external water to influx. The other is to provide the cell with a means to regulate its water uptake even though the external water potential is uncontrollable. In fact, of course, these are just two facets of the same problem. In either case, the fungus must cope with water potential stress and the evidence indicates that solute transport systems provide the mechanism which permits this. The internal maintenance of **turgor pressure** by movement of water across the membrane is related to transport of ions across the membrane and to the breakdown of macromolecules and biosynthesis of solutes. Inorganic ions usually make the greatest percentage contribution to the osmotic potential of the protoplasm. The main ions involved are K^+ and Na^+, with Cl^- being moved to balance the cation content. Some organic solutes also make major contributions, including glycerol, mannitol, inositol, sucrose, urea and proline.

The most immediate response to water potential stress is change in cell volume by the rapid flow of water into or out of the cell. The consequent change in turgor affects the cell membrane permeability and electrical properties so that the cell can restore the volume by transporting ions or other solutes across the membrane and/or by synthesising solutes or by obtaining them by degrading macromolecules. Response to water potential stress can be extremely rapid. Experimentally this is particularly evident in fungal protoplasts, the size of which alters soon after change in the solute concentration of the suspending medium. Such behaviour attests to the ready permeability of the cell membrane to water. Polar water molecules can move across cell membranes despite their lipid (hydrophobic) environment forming a natural barrier to their transport; such transport is called osmosis, which is just a special type of simple diffusion. Diffusion of water through lipid sounds like a very unlikely event, and it is, but what drives it is what drives all diffusion events, which is the relative concentrations of the diffusing molecule at the 'source' and at the 'sink'. In the case of water molecules penetrating a lipid membrane, the concentration of water in the lipid phase (the sink) will be ***extremely*** low (the solubility of water in lipid is about one molecule of water per million molecules of lipid), but the concentration in the aqueous phase (the source) will be extremely high (we'll leave you to calculate the concentration of water; remember a molar solution contains the gram-molecular mass of a solute in 1 litre of solution, and that the mass of a litre of water, by definition, is 1,000 g and the molecular mass of H_2O is 18. If you think all that's too easy, Google Avogadro's number and work out how many water molecules there are in a 200 ml glass of water). Such extreme diffusion gradients, with the additional facts that the water molecule is very small and the surface area to volume ratio of the cell is large, offset the very low permeability of the membrane and allow water to diffuse through the lipid bilayer. That's not the whole story, of course, because in some membranes the water flux is too high to be accounted for by simple diffusion alone. In such cases, water migrates by ***facilitated diffusion*** (see section 5.12) through pores or channels provided by proteins called **aquaporins** that form membrane-spanning complexes. Water moves through these channels passively in response to osmotic gradients (Nehls & Dietz, 2014).

The managed flow of water, coordinated with control of the wall synthetic apparatus, must be a prime factor in controlling the inflation of fungal cells, which is responsible for many of the changes in cell shape that characterise fungal cell differentiation. Turgor also contributes to flow along the fungal hypha. As this is a filamentous structure, flow of water and solutes within the hypha (i.e. translocation) is of enormous importance. Although our current view of apical growth requires that fungi can organise rapid translocation and specific delivery of various vacuoles and microvesicles, more general water flow along the hypha is driven by a turgor gradient and solutes are translocated by this turgor-driven bulk flow. Translocation of nutrients of all sorts in this manner is of crucial importance to morphogenesis because it must be the main way in which developing multicellular structures, such as a sporophore developing on a vegetative colony, are supplied with nutrients and water. Translocation is ably discussed by Jennings (2008; see his Chapter 14) and the mechanism can best be illustrated by quoting his description of the way in which *Serpula lacrymans* (the major timber decay organism in buildings in northern Europe) translocates carbohydrate:

> Mycelium attacks the cellulose in the wood, producing glucose, which is taken into the hyphae by active transport. Inside the hypha, glucose is converted to trehalose, which is the major carbohydrate translocated. The accumulation of trehalose leads to the hypha having a water potential lower than outside. There is a flux of water into the hyphae and the hydrostatic pressure so generated drives the solution through the mycelium. The sink for translocated material is the new protoplasm and wall material produced at the extending mycelial front. The mechanism of translocation in *Serpula lacrymans* is thus the same as that now accepted for translocation in the phloem of higher plants, namely osmotically driven mass flow. (Jennings, 2008, p. 459)

By measuring the increase in volume of developing cord networks of *Phanerochaete velutina*, Heaton *et al*. (2010) established that hyphal (and cord) growth induces mass flows across the whole mycelial network. This compensates for the fact that osmotically driven water uptake is often distal from the tips, and results in a rapid global response to local fluid movements. Coupling of growth and mass flow enables the development of efficient and highly adaptive mycelial transport networks. The velocity of fluid flow in each cord becomes a local signal that conveys information about the role of each cord within the mycelium; cords carrying fast-moving or large fluid flows are significantly more likely to increase in size than cords with slow-moving or small currents.

Jennings' (2008) description quoted above could be paraphrased to apply to other circumstances by, for example, featuring nutrients other than carbohydrates and/or alternative sinks, such as sporophores or particular tissues in sporophores. Importantly, this bulk flow does not have to be unidirectional *within a tissue*. Because the tissue is comprised of a community of hyphae, different hyphae in that community may be translocating in different directions simultaneously. Nutrients labelled with different radioisotopes have been used to demonstrate that carbon is translocated simultaneously in ***both*** directions along rhizomorphs of *Armillaria mellea*. In mycorrhizas, carbon sources from the host plant and phosphorus absorbed by the hyphae from the soil must move simultaneously in opposite directions. Indeed, the flow of carbon in mycorrhizas must be complex as carbon can be transferred between two different plants which are connected to the same mycorrhizal system. Although much remains to be learned, there is clear evidence that nutrients (including water) can be delivered over long distances through mycorrhizal hyphal systems, and that the flows can be managed and targeted to specific destinations in this 'mycorrhizal transportome' as circumstances demand (Courty *et al*., 2016).

11.12 PRIMARY (INTERMEDIARY) METABOLISM

Metabolic pathways which characteristically operate when a fungus is growing at or near its maximum rate are described as primary pathways. Secondary pathways become operational (or amplified) when growth rate is limited in some way to a level below the maximum (Bu'Lock, 1967). The fundamental function of primary metabolism is the utilisation of nutrients to form ATP and reduced nucleotide coenzymes (NADH and NADPH), which together equal chemical energy and reducing power, and the compounds which serve as precursors of cellular components, especially macromolecular constituents. All living organisms adapt continuously to changes in their environment and particularly to variation in the availability of energy-yielding substrates. Extensive studies have identified protein acetylation as an important mechanism to regulate metabolic enzymes by activation, inhibition, by influencing protein stability and by regulation of transcription. Proteomic analyses reveal many acetylated proteins in the cytoplasm and mitochondria, including most enzymes involved in intermediate metabolism

(Guan & Xiong, 2011). Numerous links between the products of intermediary metabolism and the cell's transcriptional machinery form a complex network that integrates environmental inputs to produce the most appropriate responses from the genome (Gut & Verdin, 2013).

This section is concerned with aspects of metabolism responsible for formation of substrates which serve biosynthetic pathways and compounds providing energy and reducing power; the so-called primary or intermediary metabolism.

The major source of energy and reducing power is the **catabolism of carbohydrate**. Although other carbon-containing compounds can be utilised for these purposes by most cells, the full sequence of enzymic processes are conventionally represented as involving the controlled release of energy using atmospheric oxygen to convert glucose to CO_2 and water. This overall process is described as respiration and its chemically balanced (or stoichiometric) summary equation is:

$$C_6H_{12}O_6 + 6O_2 \_ 6CO_2 + 6H_2O + \text{energy}$$

Note that this equation does not even begin to describe the biochemical mechanisms which achieve the indicated chemical transformation, but it does emphasise that for the conversion of each mole of glucose, six moles of oxygen must be absorbed from the atmosphere, and six moles of CO_2 and six moles of water appear within the cell, and 2,900 kJ of free energy (i.e. energy capable of doing some work) is released. To put this into more readily grasped units, respiration of 1 g of glucose uses 1.07 g (747 cm^3) oxygen and produces 1.47 g (747 cm^3) CO_2 and 0.6 g water, releasing 16.1 kJ of energy.

To achieve this basic chemistry the living cell uses a sequence of enzymically controlled reactions. These are conveniently divided into three phases or subpathways:

- **glycolysis** including the pentose phosphate pathway (PPP),
- the tricarboxylic acid (TCA, or Krebs) cycle,
- oxidative phosphorylation.

The word **glycolysis** describes the conversion of glucose to pyruvate without implying a particular pathway. In fact, there are three enzymic pathways that might be used, though one does tend to predominate. The **Embden–Meyerhof–Parnass**

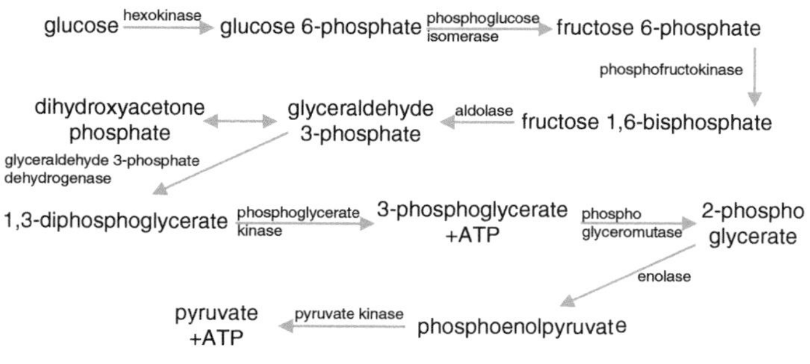

Figure 11.7. The Embden–Meyerhof–Parnass (EMP) glycolytic pathway.

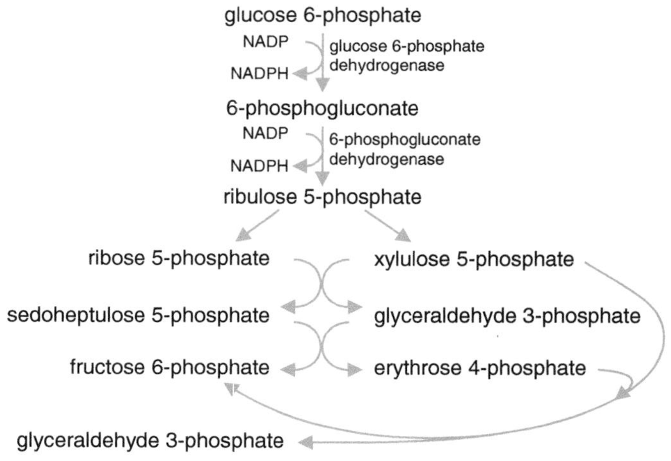

Figure 11.8. The pentose phosphate pathway (PPP) of glycolysis, also called the hexose monophosphate pathway (HMP). Modified from Moore (1998).

(EMP) pathway is the major pathway in most species; it comprises nine enzymic steps, all of which occur in the cytoplasm (Figure 11.7). The net outcome of the reactions summarised in Figure 11.7 is that one molecule of glucose is converted to two molecules of pyruvic acid plus two molecules of ATP and two molecules of $NADH_2$. Thus, the energy yield (2ATP + 2NADH$_2$, since the latter do represent potential chemical work) is rather small; the main function of the EMP pathway being conversion of glucose to pyruvate for processing in the TCA cycle.

A commonly encountered alternative glycolytic pathway is the **pentose phosphate pathway (PPP)** (Figure 11.8). This is also called the hexose monophosphate pathway (HMP) and in strictly chemical terms such a name is accurate: glucose 6-phosphate is diverted out, undergoes a range of chemical conversions, and then fructose 6-phosphate and glyceraldehyde 3-phosphate feed back into the EMP pathway. But the name pentose phosphate pathway does acknowledge that the PPP provides pentose sugars for nucleotide synthesis (which includes coenzymes and energy carriers as well as RNA and DNA), erythrose phosphate for the synthesis of aromatic amino acids through the shikimic acid pathway, and $NADPH_2$, which is the coenzyme most often used in biosynthetic reactions that require reducing power, especially fat and oil synthesis. Although the PPP can theoretically achieve complete glycolysis (six cycles through the reaction sequence would completely oxidise a molecule of glucose to CO_2), it is more likely to be involved in furnishing biosynthetic intermediates. The PPP also, of course, provides a route for utilisation of pentose sugars, which become available as carbon sources, and for interconverting hexose and pentose phosphates.

There is a third glycolytic pathway, the **Entner–Doudoroff (ED) pathway** (not illustrated here) which proceeds via 6-phosphogluconate to 2-keto-3-deoxy-6-phosphogluconate, and gives rise to pyruvate and glyceraldehyde 3-phosphate. It is a common glycolytic pathway in bacteria but has been demonstrated in only a few fungi.

The use of different glycolytic pathways in any cell will reflect the relative contribution their intermediates are required to make to the functions of the cell. They will change with age, activity and nutrition. In general, since the PPP provides intermediates for biosynthesis, use of this pathway increases in rapidly growing and differentiating cells, and is minimised in those which are resting or quiescent.

Whichever glycolytic pathway contributes the pyruvate, this latter molecule is formed in the cytoplasm and must then be transported into the mitochondrion where it is converted to **acetyl coenzyme A (acetyl-CoA)**. This step is achieved by the **pyruvate dehydrogenase complex**; a combination of enzymes which first decarboxylate pyruvate and then transfer the resulting acetyl group to coenzyme A. The pyruvate dehydrogenase complex is emerging as a major control point of oxidative phosphorylation in mitochondria (Acin-Perez *et al.*, 2010).

The TCA cycle is cyclic because its 'end product', oxaloacetate, reacts with acetyl-CoA to introduce what remain of the pyruvate carbon atoms into a reaction sequence (Figure11.9), the primary function of which is to convert pyruvate formed in glycolysis entirely to CO_2, the released energy being captured primarily in $NADH_2$. The overall stoichiometry is that one molecule of pyruvate with three molecules of water forms three molecules of CO_2 and releases 10 protons, the latter appearing in the form of three molecules of $NADH_2$, and one each of $FADH_2$ (flavin adenine dinucleotide) and the 'high-energy' compound GTP. The succinate dehydrogenase enzyme is bound to the inner mitochondrial membrane (and, because of this, is often used as a marker for the presence of mitochondria in fractionated cell extracts), the other enzymes of the TCA cycle occur in the mitochondrial matrix.

A common variant of the TCA cycle is the **glutamate decarboxylation loop** in which 2-oxoglutarate is aminated to glutamate rather than being oxidatively decarboxylated to succinate. The glutamate is decarboxylated to 4-aminobutyrate; transamination between the latter and 2-oxoglutarate yielding succinate semialdehyde which, on oxidation, feeds back into the TCA cycle as succinate (central panel in Figure 11.9). The enzymes of this loop have been found to be at high activity in mushrooms, especially caps, of the ink cap mushroom *Coprinopsis cinerea*, where it is the normal route of TCA metabolism in this organism. The glutamate decarboxylation loop also operates in *Agaricus bisporus*.

Through glycolysis and the TCA cycle, all of the carbon contained in the substrate glucose is released as CO_2. However, very little energy is released, most of it being captured in $NADH_2$ (with a small amount in GTP and $FADH_2$). The energy represented in the form of the reduced coenzymes is recovered as

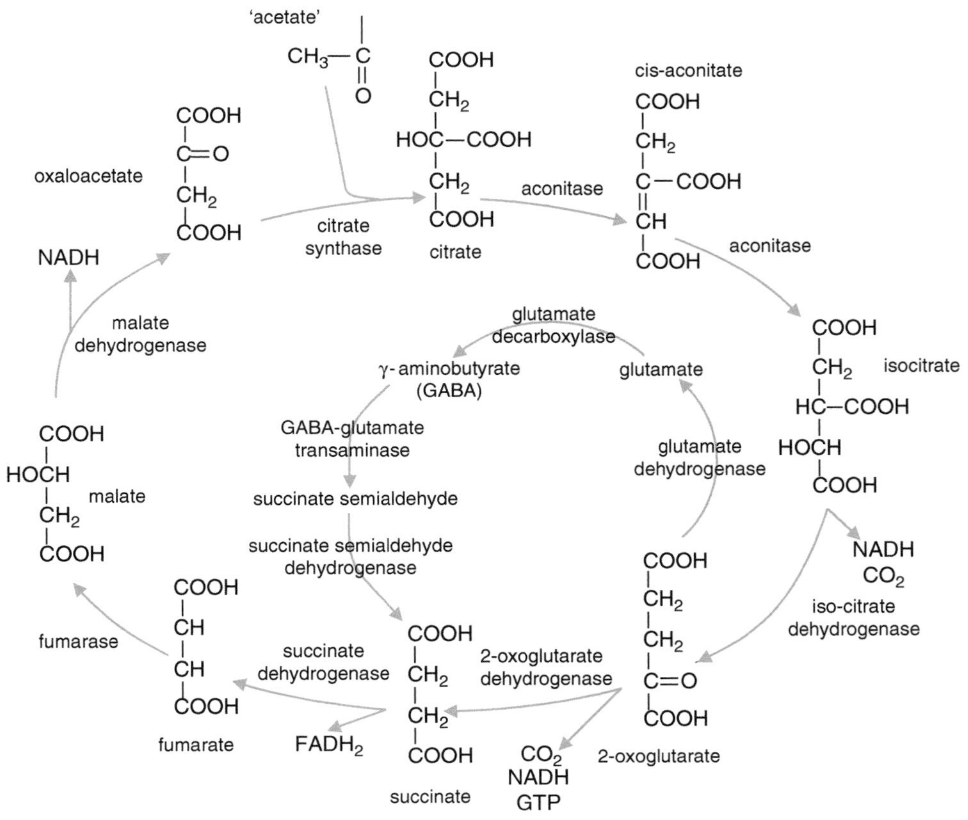

Figure 11.9. Oxidation of pyruvate: the tricarboxylic acid (TCA) cycle. The glutamate decarboxylation loop is shown in the centre of the main cycle. Modified from Moore (1998).

ATP through the electron transport chain, located on the inner mitochondrial membrane, in the process known as **oxidative phosphorylation**. The electron transport chain transfers electrons from the reduced coenzymes through a series of reactions until the electrons are finally passed to oxygen, reducing it to water. Stepwise transfer of electrons between components of the electron transport chain leads to the pumping of protons from the mitochondrial matrix into the intermembrane space.

The resulting **proton gradient** (the pH in the intermembrane space is about 1.4 pH units lower than that of the matrix) is used to generate ATP. Transfer of a pair of electrons from one molecule of $NADH_2$ to oxygen leads to proton pumping at three sites in the chain, at each of which the consequent proton gradient can be used to synthesise one molecule of ATP. The ATP is synthesised by an enzyme complex located on the matrix side of the inner mitochondrial membrane. As protons move down a channel in this complex (the channel, known as the F_0 sector, is composed of at least four hydrophobic subunits forming the proton channel located in the membrane) the associated F_1 sector (containing five different subunits) projects into the matrix and is responsible for ATP synthesis (see section 5.9; and visit http://vcell.ndsu.nodak.edu/animations/atpgradient/index.htm for an animation describing ATP synthase powered by a proton gradient or on YouTube at https://www.youtube.com/watch?v=3y1dO4nNaKY).

Although we are primarily concerned here with catabolism, the pathways described permit sugar synthesis with just a few modifications. We stressed above that glycolysis and the TCA cycle provide opportunities for the fungus to make use of a very wide range of potential carbon and energy sources; but one which is successfully growing on acetate, for example, is clearly required to synthesise all compounds which have more than two carbon atoms chained together. In such circumstances, glycolysis cannot simply be reversed because the steps governed by kinases (hexokinase, phosphofructokinase and pyruvate kinase) are irreversible, so for these steps, additional enzymes are required for **gluconeogenesis** (Figure 11.10).

In these circumstances, the first steps in conversion of pyruvate to carbohydrate are carried out by **pyruvate carboxylase**, which synthesises oxaloacetate which is then decarboxylated and phosphorylated to **phosphoenolpyruvate** by phosphoenolpyruvate carboxykinase. The phosphoenolpyruvate can then be converted to **fructose 1–6-bisphosphate** by reversal of the EMP pathway, but an additional enzyme, fructose bisphosphatase, is required to generate fructose 6-phosphate. As the sugar phosphates are readily interconvertible, once this compound is formed oligosaccharide and polysaccharide synthesis can proceed. The structures of many of the polysaccharides formed were shown earlier in this chapter.

Glycolysis and gluconeogenesis are obviously alternatives which demand close control to assure metabolic balance. Phosphofructokinase is the key glycolytic control point, and fructose bisphosphatase responds inversely to the same molecules, being allosterically activated by citrate but inhibited by AMP.

Before leaving carbohydrate metabolism, it is worth mentioning here that the sugar alcohol **mannitol** and the disaccharide **trehalose** are almost always found among the water-soluble cytoplasmic carbohydrates in fungi; trehalose being the most widely distributed sugar in fungi. Mannitol and trehalose seem to serve as transient storage compounds (i.e. molecules capable of immediate mobilisation when required) and both have been identified as substrates used for the metabolism associated with spore germination.

Trehalose synthesis/accumulation/degradation cycles occur at a number of stages in development of a fungus so it is quite clear that this sugar is the 'common currency' of the fungal carbohydrate economy. It is synthesised, as trehalose 6-phosphate, by the enzyme trehalose phosphate synthase from glucose 6-phosphate and the sugar nucleotide UDP-glucose. There may be other functions for mannitol. It can certainly serve an osmoregulatory function in the marine fungus *Dendryphiella* and may serve the same purpose in sporophores of the cultivated mushroom *Agaricus bisporus*, in which it can be accumulated to concentrations of up to 50% of the total dry weight (yes, that's right, half of the cultivated mushroom on your plate is mannitol). Similar concentrations have been encountered in mushrooms of *Lentinula edodes* (shiitake). In *Agaricus bisporus*, mannitol is synthesised by reduction of fructose by an NADP-linked mannitol dehydrogenase.

Fats are molecules of glycerol in which the three hydroxyl groups are replaced with three fatty acid molecules. In degradation, the first step is carried out by lipase which removes the fatty acids from the glycerol. The latter can be converted to glyceraldehyde 3-phosphate and thereby enter glycolysis (see Figure 11.8), but it represents only about 10% by weight of a fat molecule, the bulk being represented by the fatty acids which consist of long carbon chains (e.g. palmitic acid, C_{16}; stearic acid, C_{18}).

These chains are degraded by sequential removal of the two terminal carbon atoms in the form of an acetyl group attached

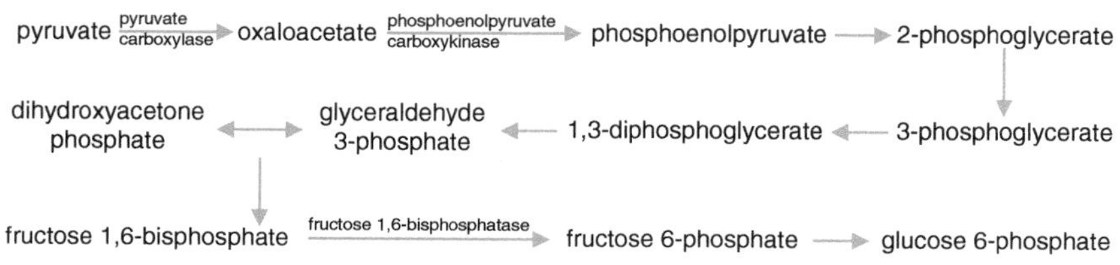

Figure 11.10. Gluconeogenesis: the making of carbohydrates. Modified from Moore (1998).

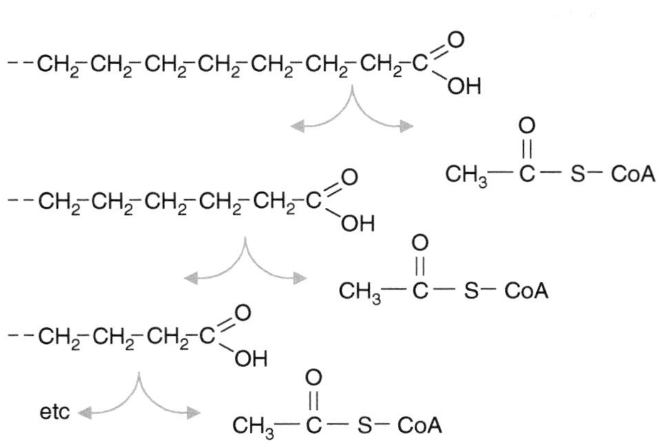

Figure 11.11. The process of β-oxidation of fatty acids. Modified from Moore (1998).

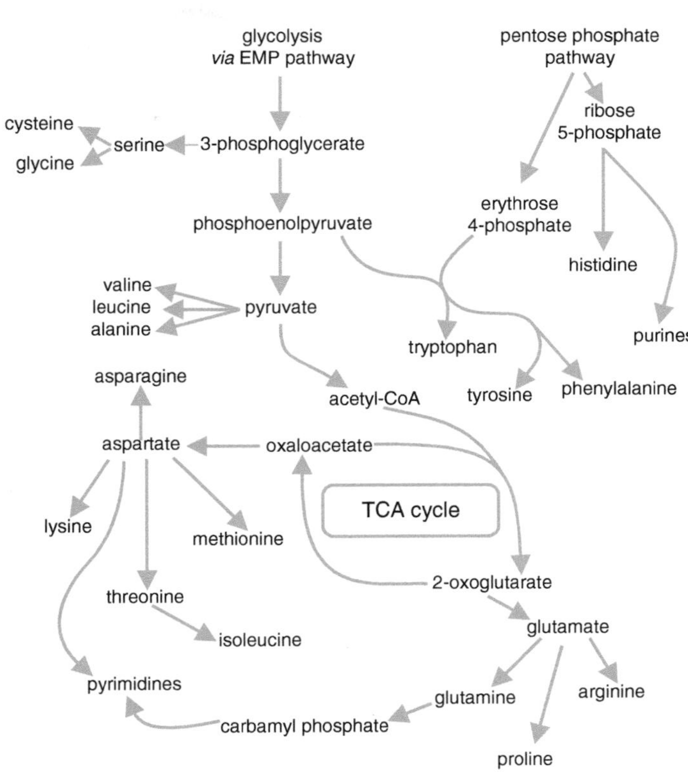

Figure 11.12. A flow chart illustrating pathways of nitrogen redistribution by showing the metabolic origins of amino acids, purines and pyrimidines. Modified from Moore (1998).

to CoA (Figure 11.11). Because the cleavage occurs at the second (β) carbon atom, this process is called **β-oxidation** and it takes place in the mitochondrial matrix. Each such cleavage is oxidative, enzymes passing the H-atoms to the coenzymes NAD and FAD. Thus, degradation of palmitic acid requires seven cleavages and yields eight molecules of acetyl-CoA (which enter the TCA cycle), seven $NADH_2$ and seven $FADH_2$ (both of which enter the electron transport chain for oxidative phosphorylation). Oxidation of fatty acids releases considerable amounts of energy; for example, one molecule of palmitic acid will give rise to about 100 molecules of ATP. This is why fats are such effective energy storage compounds.

Ultimately, all **nitrogen** in living organisms is derived from the native element in the atmosphere. Each year an amount between 100 and 200 million tons of atmospheric nitrogen is reduced to ammonium by the nitrogenase enzyme system of *nitrogen-fixing* bacteria and blue-green algae. Based on present knowledge, it is unlikely that any fungi are able to fix elemental nitrogen.

If a fungus is unable to access amino groups by direct absorption of amino acids, they must be formed, and the most immediate source is by the assimilation of ammonium. The only route of ammonium assimilation which can be considered universal in fungi is the synthesis of glutamate from ammonium and 2-oxoglutarate by the enzyme **glutamate dehydrogenase**. Many filamentous fungi and yeasts have been shown to produce two glutamate dehydrogenase enzymes; one linked to the coenzyme NAD and the other linked to NADP.

As can be appreciated from Figure 11.12, the interconversion of 2-oxoglutarate and glutamate is a reaction which occupies a central position in metabolism and is one at which important pathways in both carbon metabolism and nitrogen metabolism come together. The reaction is readily reversible, and it is often considered that NAD-linked glutamate dehydrogenase (NAD-GDH) has a deaminating or catabolic role (glutamate → 2-oxoglutarate + ammonium), while the NADP-linked enzyme provides the aminating or anabolic function (2-oxoglutarate + ammonium → glutamate). However, some organisms have evolved different patterns of ***endogenous*** regulation, especially in relation to their morphogenetic processes. For example, in the ink cap mushroom *Coprinopsis cinerea* the NADP-GDH normally appears at high activity only in the cap tissue of the mushroom where it is in the basidia apparently protecting meiosis and sporulation from inhibition by ammonium, i.e. acting as an ammonium detoxifier rather than ammonium assimilator.

This example aside, NADP-GDH is generally the most important enzyme involved in ammonium assimilation in mycelia and its activity is often increased when ammonium is provided as a growth-limiting, sole nitrogen source. However, in some fungi an alternative enzyme system appears to scavenge for ammonium when it becomes limiting, this is the glutamine synthetase/glutamate synthase system. **Glutamine synthetase** is widely, perhaps universally, distributed and is responsible for synthesis of glutamine. However, glutamine synthetase can have a high affinity for ammonium and, in combination with **glutamate synthase** (which converts glutamine + 2-oxoglutarate to two molecules of glutamate), forms an ammonium assimilation system which can recover ammonium even when this molecule is present at extremely low concentrations. The net result (2-oxoglutarate + NH_4^+ → glutamate) is the same as the reaction promoted by NADP-GDH, but the cost is higher as glutamine synthetase uses ATP to make glutamine. The glutamate synthase mechanism is common in bacteria and is encountered in fewer fungi, but has been demonstrated in *Neurospora crassa, Aspergillus nidulans,* and several yeasts and mycorrhizal fungi.

Some yeasts and a larger number of filamentous fungi can utilise nitrate as sole source of nitrogen. Chemically, nitrate is first converted to nitrite which is then converted to ammonium, but the enzymic steps are quite complicated. The complexity of the reaction is reflected in the large number of mutant genes, in both *Neurospora* and *Aspergillus,* which have been found to affect nitrate assimilation. The first stage is performed by **nitrate reductase** which, generally in fungi, has a cofactor containing molybdenum and requires NADPH (NADH in at least some yeasts). Nitrate is thought to bind to the molybdenum-cofactor of the enzyme prior to being reduced by removal of an oxygen atom. Removal of this allows the nitrate formed to be bound to nitrite reductase, through its nitrogen atom, for the reduction to ammonium. The ammonium formed from nitrate is immediately used for the reductive amination of 2-oxoglutarate to glutamate.

Conversion of nitrate to ammonia is a chemical ***reduction*** requiring considerable energy expenditure. In fact, the equivalent of four $NADPH_2$ molecules (880 kJ of energy) are used to reduce one NO_3^- ion to NH_3, which is additional to the energy demand for assimilation of the ammonium (one $NADPH_2$ is used for assimilation via NADP-GDH, 1 $NADPH_2$ + 1 ATP for assimilation through glutamine synthetase and glutamate synthase). Given these additional energy demands, it is not surprising that the nitrate reduction machinery is produced only when nitrate is the sole available source of nitrogen, being induced by nitrate and rapidly repressed by the presence in the medium of ammonium or any alternative sources of reduced nitrogen.

The constituents of living cells are in a continual state of flux; all components being subjected to turnover as old materials are catabolised and new ones synthesised. When proteins and other nitrogen-containing compounds are broken down, either as part of this turnover process or as externally supplied nutrients, the carbon can be disposed of as CO_2, hydrogen as water and nitrogen either as ammonium or as urea. The use of protein as a **carbon** source has been discussed above. In these circumstances, the organism (animal, plant or fungus) suffers an excess of nitrogen and must excrete it. Experiments with the basidiomycetes *Agaricus bisporus, Coprinopsis cinerea* and *Volvariella volvacea* have shown that one-third to one-half of the nitrogen contained in the protein given as substrate is excreted as ammonium into the medium (Kalisz *et al.*, 1986).

In terrestrial mammals metabolising protein, the toxicity of ammonium is avoided by excretion of urea formed through the urea cycle (Figure 11.13). However, the enzyme **urease** seems generally to be constitutive in fungal mycelia, so any urea formed is likely to be dissimilated to NH_3 and CO_2. Nevertheless, there are circumstances in which fungi accumulate urea (at which time they repress the urease). Especially large accumulations have been found in sporophores of *Basidiomycota*, where it seems likely that it acts as an **'osmotic metabolite'** controlling water entry into

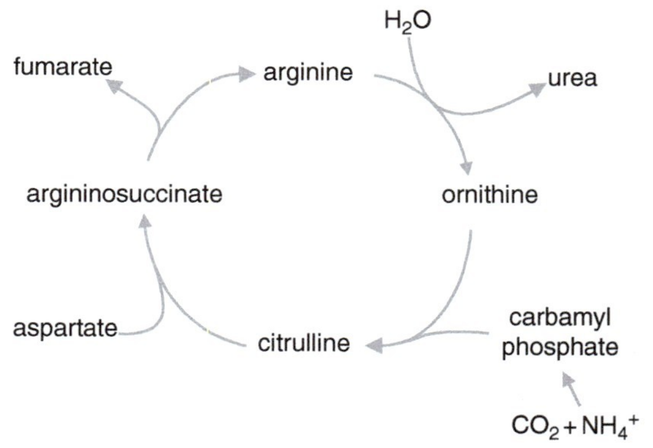

Figure 11.13. The urea cycle for disposal of excess nitrogen. The ammonium molecule, which contributes to carbamyl phosphate, and the amino group of aspartate are both 'discarded' into the urea molecule. Modified from Moore (1998).

cells during expansion. Thus, the capacity to synthesise urea, and even accumulate it, is well developed in fungi but it seems that it is ammonia which is excreted to dispose of excess nitrogen.

11.13 SECONDARY METABOLITES, INCLUDING COMMERCIAL PRODUCTS LIKE STATINS AND STROBILURINS

Primary and secondary metabolisms are coextensive; they can occur at the same time in the same cell and draw carbon-containing intermediates from the same sources. Generally, though, as nutrients are depleted the rate of growth slows and eventually stops. The progress of metabolism is correspondingly altered, and a number of special biochemical mechanisms appear (or become amplified) to establish what is called secondary metabolism. The resultant secondary metabolites seem to be unnecessary for normal growth of the organism, but rather appear to be aimed at functions such as intercellular communication and defence and competition with other organisms.

Secondary metabolism is a common feature of fungi. It consists of a relatively small number of enzymological processes (often of relatively low substrate specificity) which convert a few important intermediates of primary metabolism into a wide range of products (Bu'Lock, 1967). These later stages of secondary metabolism are so varied that individual secondary metabolites can have a narrow species distribution. The subject is huge, and this section is not intended to rival the comprehensive treatment which you can find in texts like Mérillon and Ramawat (2017) and Bills and Gloer (2016), rather we intend to illustrate the chemical versatility of fungi with a few examples that have some sort of importance to our daily lives.

Most secondary metabolites have their origins in a small number of the intermediates in pathways dealt with above:

- acetyl-CoA is, perhaps, the most important, being a precursor for terpenes, steroids, fatty acids and polyketides;

- phosphoenolpyruvate and erythrose 4-phosphate initiate synthesis of aromatic secondary metabolites through the shikimic acid pathway by which aromatic amino acids are synthesised;
- further secondary metabolites are derived from other (non-aromatic) amino acids (Figure 11.14).

Terpenes, carotenoids and **steroids** are widely distributed in nature and though the fungal products are not unusual in that sense, many of the end products have chemical structures which are unique to fungal metabolism. All these compounds are related by having five carbon atoms arranged as in the hydrocarbon isoprene. The terpenes are the simplest of the naturally occurring **isoprenoid** compounds and are derived by condensation of a precursor isoprene unit to form an isoprenoid chain (Figure 11.15).

Strictly, terpene is the name of the hydrocarbon with the formula C_9H_{16}, but the term is used generally to refer to the 10-carbon isoprenoids which may be open-chain (that is, **acyclic**) or **cyclic** compounds. Synthesis of acyclic terpenes starts when dimethylallyl pyrophosphate condenses with a molecule of isopentenyl pyrophosphate to form the 9-carbon monoterpene **geranyl pyrophosphate** (Figure 11.16); addition of another isopentenyl pyrophosphate gives the 15-carbon sesqui- ('one-and-a-half') terpene **farnesyl pyrophosphate**; and of another, the 20-carbon diterpene **geranylgeranyl pyrophosphate**, and so on.

As many as 24 isoprene units may condense in this way, though secondary metabolites usually have between two and five. Not all such molecules are secondary metabolites;

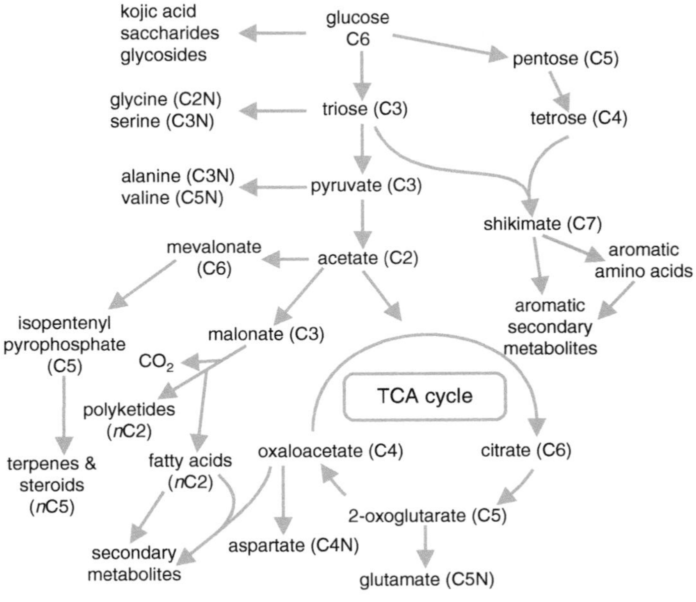

Figure 11.14. Metabolic route map summarising the relationships between primary metabolism and the major pathways for synthesis of secondary metabolites. Compare with Figures 11.9 and 11.12. Modified from Moore (1998).

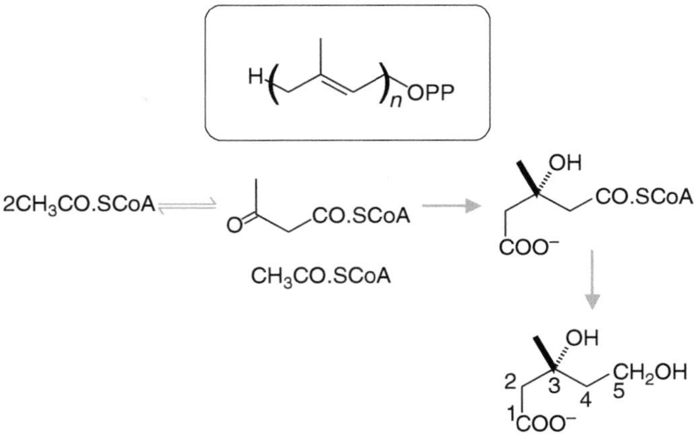

Figure 11.15. Derivation of terpenes from isoprene units. The basic isoprenoid repeat structure is shown at the top; the lower panel shows the synthesis of mevalonate from two molecules of acetyl-CoA. Growth of the terpene chain is continued in Figure 11.16, using the carbon atom numbering scheme indicated at bottom right of this figure. Modified from Moore (1998).

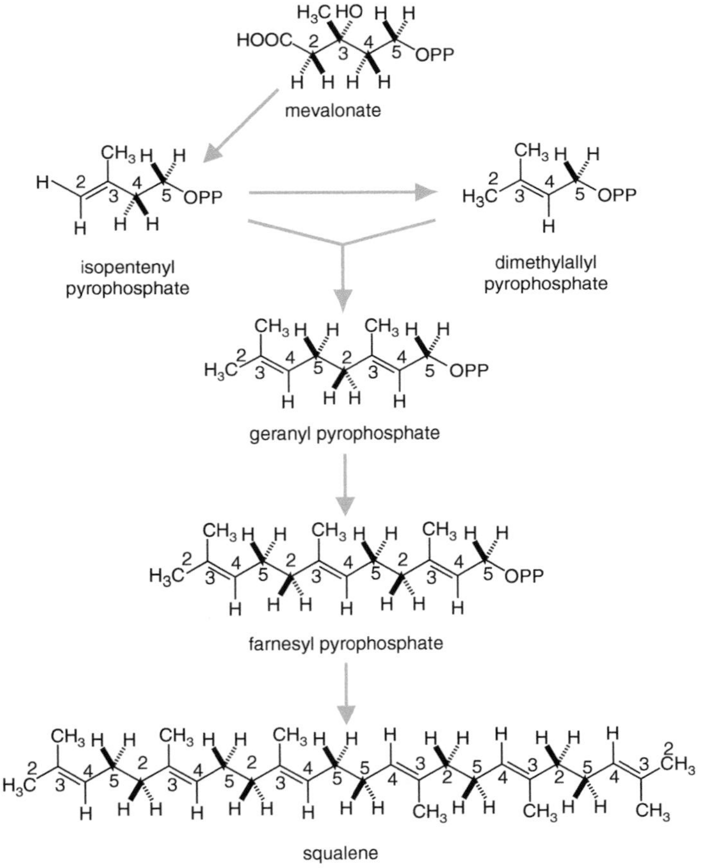

Figure 11.16. Synthesis of acyclic (that is, straight-chain) terpenes by successive condensations. Dimethylallyl pyrophosphate condenses with isopentenyl pyrophosphate to form geranyl pyrophosphate. Squalene is the precursor of sterols (see Figure 11.19). Squalene is formed by head-to-tail condensation of two sesquiterpene (farnesyl pyrophosphate) units. Check the numbering on the carbon atoms to understand how these molecules are assembled. Modified from Moore (1998).

the ubiquinones of the respiratory chain are **polyprenoid quinones**. Sesquiterpenes are the largest group of terpenes isolated from fungi. Most of the fungal **sesquiterpenes** are based on carbon skeletons which can be derived by cyclisation of farnesyl pyrophosphate as shown in the schemes in Figure 11.17 (most of which are plausible possibilities rather than defined metabolic pathways).

Diterpenes are derived by cyclisation of geranylgeranyl pyrophosphate (Figure 11.18) and include the **gibberellins**, a mixture of plant growth promoters which were first isolated from culture filtrates of the plant pathogen *Gibberella fujikuroi* (the sexual form, or teleomorph, of *Fusarium moniliforme*) and only later shown to be endogenous plant hormones.

The major significance of the **triterpenes** (C_{30}) is that the acyclic triterpene squalene is the precursor of sterols. Squalene is formed by a head-to-tail condensation of two sesquiterpene (farnesyl pyrophosphate) units. It is so called because it is found in high concentration in shark liver oil (*Squalus* is a genus of dogfish sharks), but in fungi as well as in other organisms sterols (and cyclic triterpenes) are derived by cyclisation of squalene oxide (Figure 11.19).

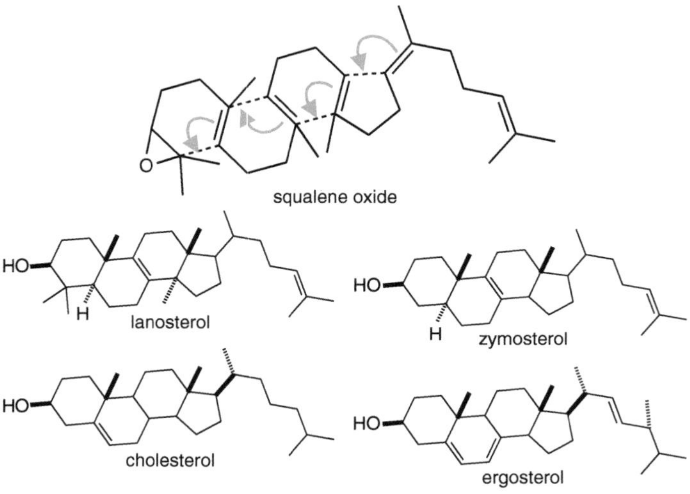

Figure 11.19. Sterols (and cyclic triterpenes) are derived by cyclisation of squalene oxide. This is illustrated in the upper panel and structural formulae of some sterols are shown below. See Figure 18.1 for more details about the clinical importance of differences between cholesterol and ergosterol. Modified from Moore (1998).

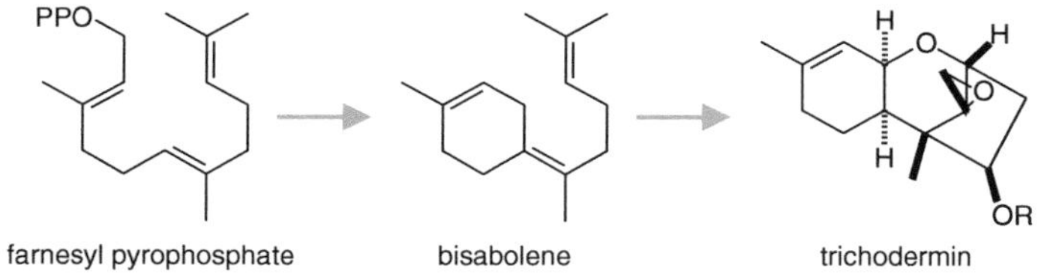

Figure 11.17. Sesquiterpenes: possible cyclisations of farnesyl pyrophosphate to produce the trichothecane nucleus. Modified from Moore (1998).

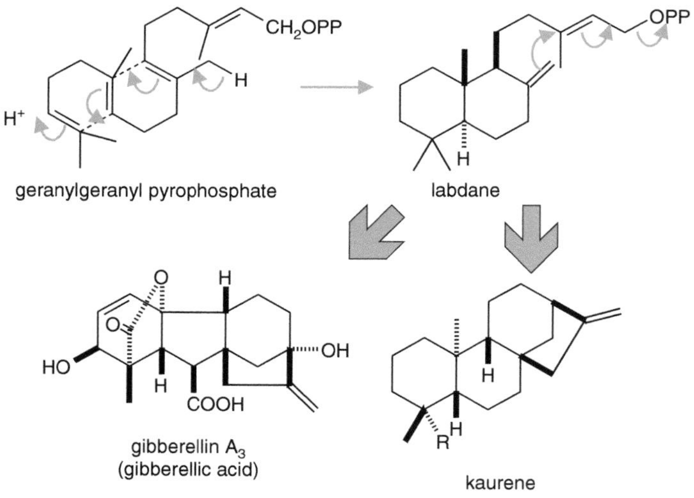

Figure 11.18. Cyclisation of the diterpene geranylgeranyl pyrophosphate producing gibbane and kaurane structures. Gibberellic acid is a plant growth hormone that was discovered only after it had been isolated as a secondary metabolite of a pathogenic fungus that caused growth abnormalities in its plant host. Modified from Moore (1998).

Cholesterol is the quantitatively predominant sterol in animals where it serves to control membrane fluidity. **Ergosterol**, so named because it was first isolated from the ergot fungus (*Claviceps purpurea*), probably fulfils a similar role in fungi, influencing permeability characteristics of the membrane and as it is unique to fungi; ergosterol synthesis is a potential target for antifungal agents. However, other sterols, even cholesterol, occur commonly and a very wide range of sterols has been detected in fungi which have scope for characterising strains. Lipid, sterol and phospholipid contents differ between yeast and mycelial forms of *Candida albicans* and *Mucor lusitanicus*, so there is a morphogenetic connection, too. The sequential order of steps in sterol synthesis is well established. A crucial step in sterol synthesis is controlled by the enzyme b-hydroxy-b-methylglutaryl-CoA reductase (**HMG Co-A reductase**), which converts HMG-CoA to mevalonic acid. Inhibit this enzyme and sterol synthesis is stopped.

In much the same manner as squalene is formed by head-to-tail condensation of sesquiterpenes, the C_{40} **carotenoid pigments** are formed by head-to-tail condensation of two diterpenes. These include acyclic compounds like lycopene, the pigment responsible for the red colour of tomatoes and monocyclic and dicyclic compounds like α-, β- and γ-carotene (Figure 11.20). The carotenes and some keto-derivatives are widely synthesised, and carotenoids are good taxonomic markers for some fungi, though not all fungi produce them; the biosynthetic pathway is well established.

Continued condensation of isoprene units can produce terpenes with very long carbon chains. Natural rubber is a highly polymerised isoprene compound, containing of the order of

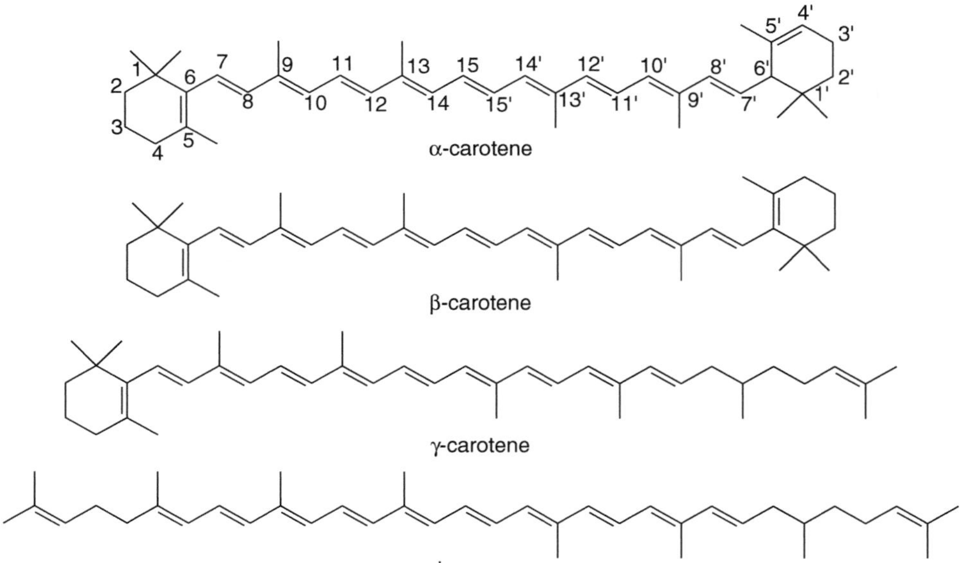

Figure 11.20. Fungal carotenoid pigments. Modified from Moore (1998).

5,000 isoprene units. The latex of the major commercial source, the rubber palm *Hevea brasiliensis*, is an aqueous colloidal suspension of particles of this hydrocarbon. In strong contrast, the milk-like juice exuded from broken mushrooms of *Lactarius*, stems of *Mycena* and gill edges of *Lacrymaria* is sometimes also called '**latex**' but is chemically very different from rubber palm latex, being only superficially similar in appearance. The fungal product probably differs in structure between genera, but the 'latex' or 'milk' of *Lactarius rufus* has been shown to contain mannitol, glucose and lactarinic acid ($CH_3[CH_2]_{11}CO[CH_2]_4COOH$; structural formula shown in Figure 11.25, below). The latter is a modified fatty acid (6-oxo-octadecanoic acid, also known as 6-ketostearic acid).

More secondary metabolites are synthesised through the polyketide pathway in fungi than by any other pathway. **Polyketides** are characteristically found as secondary metabolites among *Ascomycota*; they are rarely encountered in *Basidiomycota* and are produced by only a few organisms other than fungi. Polyketides are more correctly described as poly-β-ketomethylenes, the fundamental acyclic chain from which they are derived being comprised of $-CH_2CO-$ units. Just as linear poly-isoprenoid chains can 'fold' and cyclise, so too the polyketides can cyclise to produce a wide range of different molecules (see Figure 11.21). As might be expected from the chemical nature of the repeating unit, synthesis of polyketides involves transfer of acetyl groups which are ultimately derived from acetyl-CoA. In fact the synthesis of polyketides has a lot in common with fatty acid biosynthesis.

Fatty acids are catabolised by stepwise removal of 'acetyl-units' (β-oxidation, see Figure 11.11) but their synthesis requires malonyl-CoA which is formed by carboxylation of acetyl-CoA by the enzyme acetyl-CoA carboxylase: $CH_3COSCoA + HCO_3^- + ATP \rightarrow COOH.CH_2COSCoA + ADP + Pi$. Synthesis of fatty acids is carried out by a complex of enzymes called the fatty acid synthetase system. In the first reaction 'acetate' from acetyl-CoA is condensed with a malonyl group (from malonyl-CoA) to form acetoacetate which is still bound to the enzyme and then reduced, dehydrated and reduced again to form a butyryl-enzyme complex [$CH_3CH_2CH_2CO$-S-enzyme]. For chain lengthening, the butyryl residue is condensed with another malonyl group followed by repetition of the reduction, dehydration, reduction cycle. Thus, malonyl-CoA provides all the carbon of long-chain fatty acids except for the two terminal atoms, which derive from the 'acetate' initially introduced from acetyl-CoA.

This brief description of fatty acid synthesis can be echoed by a description of polyketide synthesis which is initiated by condensation of an acetyl unit with malonyl units, requires the respective CoA derivatives, and seems to occur on the enzymes involved. In both pathways, free precursors are not found in the cytoplasm. The chain-building mechanism is similar between fatty acids and polyketides, but for polyketides the reduction process does not occur, and successive rounds of condensation generate polymers with the $-CH_2CO-$ ('ketide') repeating unit. The number of these can be used to designate the chain formed as, for example:

- **tetraketides** (examples are orsellinic acid, found in many lichens, and fumigatin, characteristically produced by *Aspergillus fumigatus*);
- **pentaketides** (for example citrinin, an antifungal agent that inhibits HMG Co-A reductase, originally isolated from *Penicillium citrinum*, but commonly found in *Ascomycota*; and mellein, which may have value as a bioherbicide);

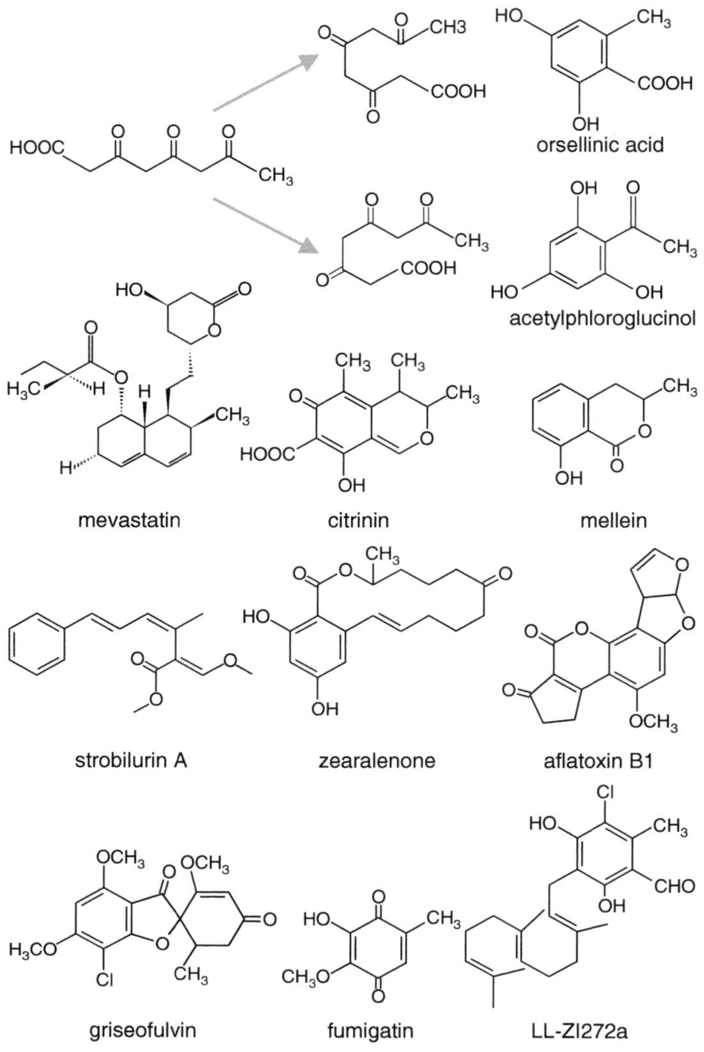

Figure 11.21. Polyketide chains can fold in a variety of ways and internal aldol condensations form closed aromatic rings. To illustrate this, the alternative cyclisations of a tetraketide are shown at the top, together with the potential products orsellinic acid and acetylphloroglucinol. The bottom section shows the structures of a small selection of polyketides discussed in the text. Modified from Moore (1998).

- **heptaketides** (such as the antifungal agent griseofulvin, originally isolated from *Penicillium griseofulvum*, which finds commercial use in treating fungal diseases of animals and plants).

The methylene (-CH_2-) and carbonyl (=C:O) groups of these chains can interact in internal aldol condensations to form closed aromatic rings. A great variety of cyclisations thus become possible and this, in part, accounts for the wide range of polyketides which are encountered; the number is further increased by subsequent chemical modifications. The result is that polyketide secondary metabolites are far too numerous to document here (see Agarwal & Moore, 2014) but they include many toxins, some of which have medical importance (Figure 11.21). Notable examples in Figure 11.21 are:

- **Zearalenone**, a toxin produced in stored grain contaminated by *Fusarium*, and the aflatoxins, which are produced in mouldy groundnut meal by *Aspergillus* species and are among the most toxic, mutagenic and carcinogenic natural products known.
- Compounds now known as **statins** (such as mevastatin, Figure 11.21), from *Penicillium citrinum*, and a similar molecule, later named lovastatin, from *Aspergillus terreus*, originally isolated for their ability to inhibit lipid synthesis. These are **HMG-CoA reductase inhibitors**, and as this is the rate-limiting enzyme of the mevalonate pathway of cholesterol synthesis they are used to lower cholesterol levels in people with, or at risk of, cardiovascular disease. Inhibition of this enzyme in the liver stimulates low-density lipoprotein (LDL) receptors, resulting in an increased clearance of LDL from the bloodstream and a decrease in blood cholesterol levels. The first positive results of treatment can be seen after just one week of use. These are the wonder drugs of the twenty-first century (antibiotics being the wonder drugs of the twentieth century) and are currently ***the most widely used pharmaceutical agents in the world***.
- **Strobilurin A** (Figure 11.21) represents the strobilurin fungicides. It is the original, natural strobilurin, produced by the pine cone fungus, *Strobilurus tenacellus*, a mushroom fungus that colonises pine cones and produces strobilurins to keep the pine cone to itself. Natural strobilurins are unstable in the presence of light, but there are many synthetic analogues that are more stable. Synthetic strobilurins are now the most widely used agricultural fungicides (Balba, 2007; see Chapter 18).
- Other, more complex compounds also arise and there are some compounds in which polyketide-derived aromatic rings are attached to sesquiterpenes. Examples are a group of compounds called ascochlorins that inhibit mitochondrial electron transport (Berry *et al.*, 2010).

The primary role of the **shikimate–chorismate pathway** is the synthesis of the aromatic amino acids phenylalanine, tyrosine and tryptophan (Figure 11.14), but the pathway provides intermediates for the synthesis of other aromatic compounds as secondary metabolites. Plants synthesise a particularly large variety of compounds by this route, though it is not so widely used in fungi, where the polyketide pathway is used more to make aromatic ring compounds. Nevertheless, numerous shikimate–chorismate derivatives that commonly occur in plants have been isolated from fungi, such as gallic acid, pyrogallol and methyl *p*-methoxycinnamate (Figure 11.22).

Among compounds more particularly associated with fungi are the **ergot alkaloids** (from *Claviceps purpurea*) which include ergocristine and lysergic acid amide and psilocybin, which is

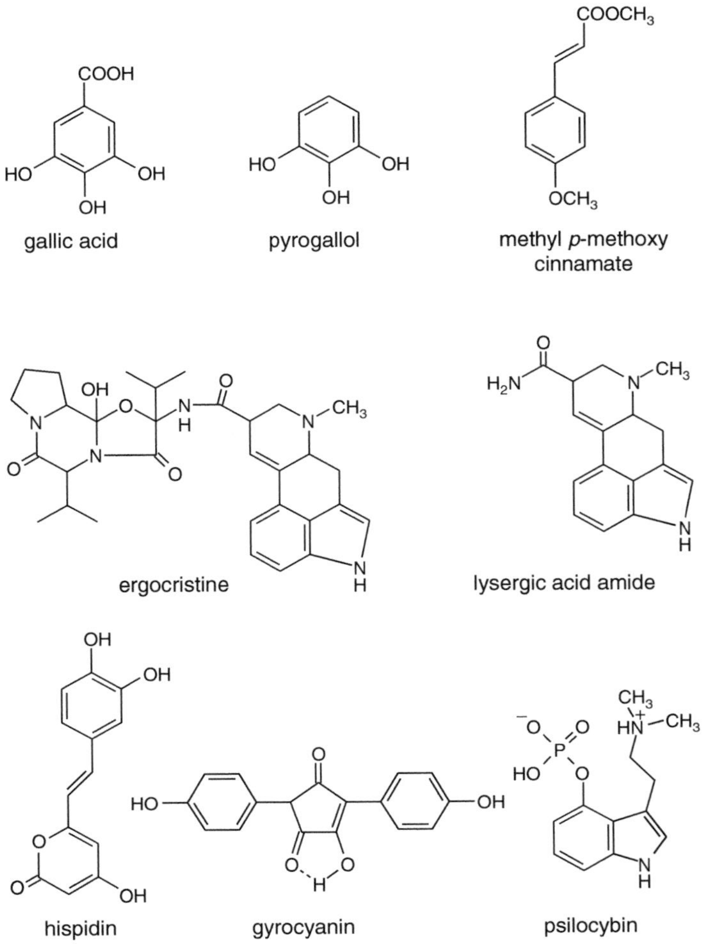

Figure 11.22. The shikimate–chorismate pathway produces a variety of aromatic compounds as well as the amino acids phenylalanine, tyrosine and tryptophan. Gallic acid, pyrogallol and methyl *p*-methoxycinnamate are relatively simple compounds produced by many plants as well as fungi. Ergocristine, which is one of the ergot alkaloids (from *Claviceps purpurea*), lysergic acid amide and psilocybin, the halucinogenic principle of the original 'magic' mushroom (*Psilocybe*), are all essentially tryptophan derivatives. Gyrocyanin is a product of *Gyroporus cyanescens* which oxidises to a blue pigment in injured sporophores, and hispidin is the precursor of a polymer which seems to be responsible for toughening the sporophores of *Polyporus hispidus*. Modified from Moore (1998).

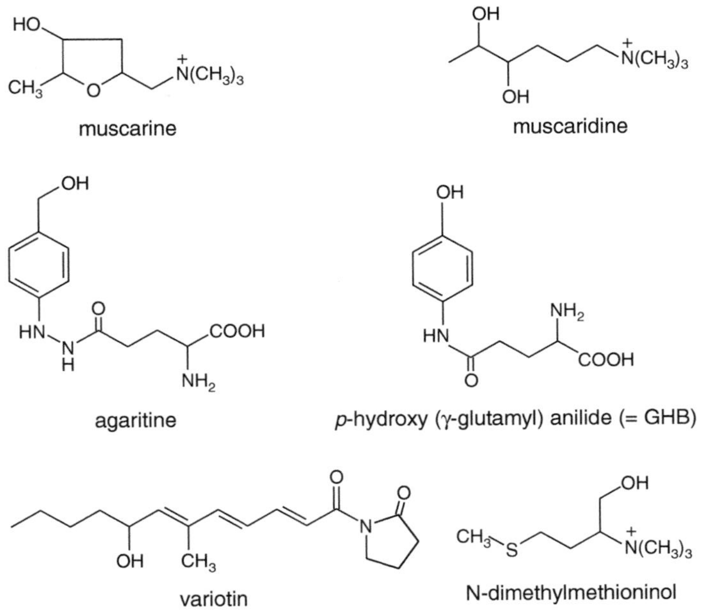

Figure 11.23. Non-aromatic amino acids may also be modified and accumulated as secondary metabolites. Muscarine and muscaridine, the main toxins of *Amanita muscaria*, are synthesised from glutamate, and agaritine and *p*-hydroxy (γ-glutamyl) anilide (also known as glutaminyl hydroxybenzene, or GHB) from *Agaricus* spp., are N-acylated glutamate molecules. Agaritine may account for up to 0.3% of the dry weight in *Agaricus bisporus* mushrooms and GHB may be involved in controlling basidiospore dormancy and is the most likely precursor for *Agaricus* melanin in the spore walls. Variotin is an antifungal agent isolated from *Paecilomyces varioti* and is γ-aminobutyric acid N-acylated with a hexaketide. N-dimethylmethioninol is a volatile amine formed by decarboxylation of methionine by *Penicillium camemberti*. It is responsible for the aroma of Camembert cheese. Modified from Moore (1998).

the hallucinogenic principle of the original 'magic' mushroom (*Psilocybe*). All of these are essentially tryptophan derivatives. Basidiomycete pigments like gyrocyanin, which oxidises to a blue pigment in injured sporophores of *Gyroporus cyanescens*, and hispidin, a polymer of which may be responsible for toughening of the sporophores of *Polyporus hispidus* (Bu'Lock, 1967) are derivatives of the shikimate–chorismate pathway.

Non-aromatic amino acids may also be modified: the muscarines and muscaridines (Figure 11.23), which are the main toxic constituents of *Amanita muscaria* (but are also found in *Inocybe* and *Clitocybe* species), are synthesised from glutamate. Agaritine and related molecules like glutaminyl hydroxybenzene (or GHB) from *Agaricus bisporus*, the cultivated mushroom, are substituted (strictly, *N*-acylated) glutamate molecules, while the antifungal agent variotin (from *Paecilomyces varioti*) is γ-aminobutyric acid (GABA, shown in Figure 11.9) *N*-acylated with a hexaketide (Figure 11.23). Volatile amines formed by decarboxylation of neutral amino acids form part of the distinctive odours of some fungi; the fungus causing stinking smut of wheat (*Tilletia tritici*) produces large quantities of trimethylamine, and the aroma of Camembert cheese depends on the formation of *N*-dimethyl methioninol by *Penicillium camemberti*.

A variety of fungal secondary metabolites are derived from **peptides**; two or more amino acids linked through a peptide bond. Among those derived from dipeptides are the **penicillins** and cephalosporins. These could rate as the most important compounds ever to be isolated from fungi as they gave rise to a new era in medicine and a new branch of biotechnology. The basic structure of both antibiotics is derived from cysteine and valine (Figure 11.24) though the variable acyl group may come from

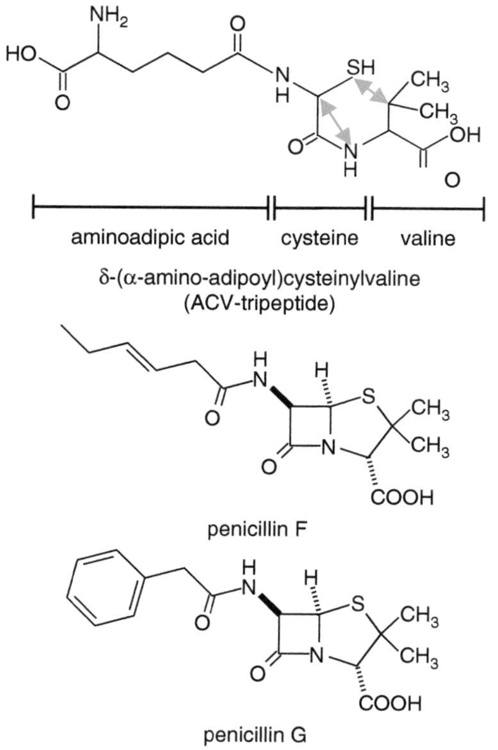

Figure 11.24. The penicillin antibiotics are representative of secondary metabolites which are derivatives of peptides. This figure shows the structural formulae of two penicillins. At the top is the precursor δ-(α-amino-adipoyl) cysteinylvaline. The aminoadipoyl, cysteine and valine residues of this compound are identified, and the double-headed arrows show the bonds which have to be made to create the penicillin nucleus. See also Figures 19.20 and 19.21 for more details.

another amino acid as in penicillin G where the acyl group is phenylalanine (illustrated in Figure 11.24). Biosynthesis involves the tripeptide δ(L-α-aminoadipyl)-L-cysteinyl-D-valine, which is also illustrated in Figure 11.24, but does not involve ribosomes. Instead, amino acid activating domains of peptide synthetases determine the number and order of the amino acid constituents of the peptide secondary metabolites. *In vitro* reconstruction of the gene sequences of these multifunctional enzymes produces hybrid genes that encode peptide synthetases with altered amino acid specificities able to synthesise peptides with modified amino acid sequences.

Also derived from peptides are the *Amanita* toxins of which there are many, but which can be represented by **α-amanitin** and **phalloidin**. Because of the resemblance between *Amanita phalloides* and edible field mushrooms, these toxins are involved in most cases of mushroom poisoning. Many fungi produce **siderophores** to acquire iron, which, though an essential nutrient, is not readily available in aquatic or terrestrial environments or in animal hosts. These iron-binding compounds also originate as peptides formed from modified amino acids.

Chemical modification of fatty acids produces a variety of secondary metabolites, most particularly the polyacetylenes, many of which have been obtained from basidiomycetes. These compounds have straight carbon chains varying between C_6 and C_{18}, though C_9 and C_{10} are most common in fungi. They are conjugated **acetylenes** (that is, they have triple bonds between adjacent carbons, e.g. hexatriyene from *Fomes annosus*, Figure 11.25), or systems containing both ethylenic (with double bonds between carbon atoms) and acetylenic structures (e.g. nemotinic acid, Figure 11.25). Polyacetylenes are derived from fatty acids by a series of dehydrogenation reactions (Negri, 2015).

Also formed from fatty acids are **cyclopentanes** like brefeldin, which has been extracted from *Penicillium, Nectria* and *Curvularia* species (Figure 11.25). Reference was made to lactarinic acid above but, if you want more specifics, view: http://www.chemspider.com/Chemical-Structure.321288.html; and if you ever need authoritative chemical information go to the ***Royal Society of Chemistry website*** at http://www.chemspider.com/Default.aspx.

As must be clear from the above discussion, fungi are prolific producers of secondary metabolites and these show a variety of biological activities. Advances in genome sequencing have shown that fungal genomes harbour far more gene clusters able to synthesise secondary metabolites than are expressed under normal laboratory conditions. Filamentous fungi have already played such an important role in the history of drug discovery and development; examples being antibiotics such as penicillin, immunosuppressants such as cyclosporine, antifungals such as griseofulvin, and the echinocandins and antihypercholesterolemic drugs such as lovastatin. Because of these past (mostly accidental) successes, activation of these 'silent' gene clusters is an ongoing challenge in the hope of discovering new natural products in the future (Brakhage & Schroeckh, 2011; Brakhage, 2013; Yaegashi *et al.*, 2014; Macheleidt *et al.*, 2016; Rodrigues, 2016).

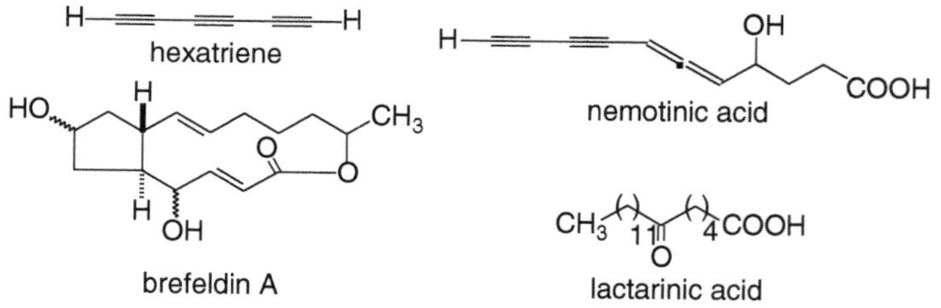

Figure 11.25. Chemical modification of fatty acids produces a variety of secondary metabolites including polyacetylenes and cyclopentanes. Lactarinic acid is a major component of the 'milk' or 'latex' which exudes from injured mushrooms of *Lactarius* spp. Lactarinic acid is 6-ketostearic acid and is very different from true latex from the rubber tree which is an isoprene compound. Modified from Moore (1998).

11.14 REFERENCES

Abdel-Hamid, A.M., Solbiati, J.O., Cann, I.K.O., Sariaslani, S. & Gadd, G.M. (2013). Insights into lignin degradation and its potential industrial applications. *Advances in Applied Microbiology*, **82**: 1–28. DOI: https://doi.org/10.1016/B978-0-12-407679-2.00001-6.

Acin-Perez, R., Hoyos, B., Gong, J. *et al.* (2010), Regulation of intermediary metabolism by the PKCδ signalosome in mitochondria. *The FASEB Journal*, **24**: 5033–5042. DOI: https://doi.org/10.1096/fj.10-166934.

Agarwal, V. & Moore, B.S. (2014). Fungal polyketide engineering comes of age. *Proceedings of the National Academy of Sciences of the United States of America*, **111**: 12278–12279. DOI: https://doi.org/10.1073/pnas.1412946111.

Ali, Y.B., Verger, R. & Abousalham, A. (2012). Lipases or esterases: does it really matter? Toward a new bio-physico-chemical classification. In *Lipases and Phospholipases. Methods in Molecular Biology (Methods and Protocols)*, **Vol. 861**, ed. G. Sandoval. New York: Humana Press/Springer International Publishing AG, pp. 31–51. DOI: https://doi.org/10.1007/978-1-61779-600-5_2.

Anke, H. & Weber, R.W.S. (2006). White-rots, chlorine and the environment: a tale of many twists. *Mycologist*, **20**: 83–89. DOI: https://doi.org/10.1016/j.mycol.2006.03.011.

Ariño, J., Casamayor, A. & González, A. (2011). Type 2C protein phosphatases in fungi. *Eukaryotic Cell*, **10**: 21–33. DOI: https://doi.org/10.1128/EC.00249-10.

Artzi, L., Bayer, E.A. & Moraïs, S. (2017). Cellulosomes: bacterial nanomachines for dismantling plant polysaccharides. *Nature Reviews Microbiology*, **15**: 83–95. DOI: https://doi.org/10.1038/nrmicro.2016.164.

Audrey, L.C.C., Desjardin, D.E., Tan, Y.S., Musa, M.Y. & Sabaratnam, V. (2015). Bioluminescent fungi from Peninsular Malaysia: a taxonomic and phylogenetic overview. *Fungal Diversity*, **70**: 149–187. DOI: https://doi.org/10.1007/s13225-014-0302-9.

Badaruddin, M., Holcombe, L.J., Wilson, R.A. *et al.* (2013). Glycogen metabolic genes are involved in trehalose-6-phosphate synthase-mediated regulation of pathogenicity by the rice blast fungus *Magnaporthe oryzae*. *PLoS Pathogens*, **9**: article number e1003604. DOI: https://doi.org/10.1371/journal.ppat.1003604.

Balba, H. (2007). Review of strobilurin fungicide chemicals. *Journal of Environmental Science and Health, Part B*, **42**: 441–451. DOI: https://doi.org/10.1080/03601230701316465.

Barriuso, J., Prieto, A. & Martínez, M.J. (2013). Fungal genomes mining to discover novel sterol esterases and lipases as catalysts. *BMC Genomics*, **14**: 712 (8 pp.). DOI: https://doi.org/10.1186/1471-2164-14-712.

Bayer, E.A., Lamed, R. & Himmel, M.E. (2007). The potential of cellulases and cellulosomes for cellulosic waste management. *Current Opinion in Biotechnology*, **18**: 237–245. DOI: https://doi.org/10.1016/j.copbio.2007.04.004.

Benoit, I., Coutinho, P.M., Schols, H.A. *et al.* (2012). Degradation of different pectins by fungi: correlations and contrasts between the pectinolytic enzyme sets identified in genomes and the growth on pectins of different origin. *BMC Genomics*, **13**: 321 (11 pp.). DOI: https://doi.org/10.1186/1471-2164-13-321.

Berry, E.A., Huang, L., Lee, D.W. *et al.* (2010). Ascochlorin is a novel, specific inhibitor of the mitochondrial cytochrome bc1 complex. *Biochimica et Biophysica Acta, Bioenergetics*, **1797**: 360–370. DOI: https://doi.org/10.1016/j.bbabio.2009.12.003.

Bertolini, M.C., Freitas, F.Z., de Paula, R.M., Cupertino, F.B. & Goncalves, R.D. (2012). Glycogen metabolism regulation in *Neurospora crassa*. In *Biocommunication of Fungi*, ed. G. Witzany. Dordrecht: Springer International Publishing AG, pp. 39–55. DOI: https://doi.org/10.1007/978-94-007-4264-2_3.

Bills, G. & Gloer, J. (2016). Biologically active secondary metabolites from the fungi. In *The Fungal Kingdom*, ed. J. Heitman, B. Howlett, P. Crous, E. Stukenbrock, T. James & N.A.R. Gow. Washington, DC: ASM Press, pp. 1087–1119. DOI: https://doi.org/10.1128/microbiolspec.FUNK-0009-2016.

Brakhage, A.A. (2013). Regulation of fungal secondary metabolism. *Nature Reviews Microbiology*, **11**: 21–32. DOI: https://doi.org/10.1038/nrmicro2916.

Brakhage, A.A. & Schroeckh, V. (2011). Fungal secondary metabolites – strategies to activate silent gene clusters. *Fungal Genetics and Biology*, **48**: 15–22. DOI: https://doi.org/10.1016/j.fgb.2010.04.004.

Bu'Lock, J.D. (1967). *Essays in Biosynthesis and Microbial Development*. New York: John Wiley & Sons. ISBN-10: 0471121002, ISBN-13: 978-0471121008. DOI: 10.1002/jobm.19680080527.

Ceccaroli, P., Buffalini, M., Saltarelli, R. *et al.* (2011). Genomic profiling of carbohydrate metabolism in the ectomycorrhizal fungus *Tuber melanosporum*. *New Phytologist*, **189**: 751–764. DOI: https://doi.org/10.1111/j.1469-8137.2010.03520.x.

Chen, E., Choy, M.S., Petrényi, K. *et al.* (2016). Molecular insights into the fungus-specific serine/threonine protein phosphatase Z1 in *Candida albicans*. *mBio* **7**: article number e00872-16. DOI: https://doi.org/10.1128/mBio.00872-16.

Chu, Z.J., Wang, Y.J., Ying, S.H., Wang, X.W. & Feng, M.G. (2016). Genome-wide host-pathogen interaction unveiled by transcriptomic response of Diamondback Moth to fungal infection. *PLoS ONE*, **11**: article number e0152908. DOI: https://doi.org/10.1371/journal.pone.0152908.

Courty, P.-E., Doidy, J., Garcia, K., Wipf, D. & Zimmermann, S.D. (2016). The transportome of mycorrhizal systems. In *Molecular Mycorrhizal Symbiosis*, ed. F. Martin. Hoboken, NJ: John Wiley & Sons, pp. 239–256. DOI: https://doi.org/10.1002/9781118951446.ch14.

Cragg, S.M., Beckham, G.T., Bruce, N.C. *et al.* (2015). Lignocellulose degradation mechanisms across the Tree of Life. *Current Opinion in Chemical Biology*, **29**: 108–119. DOI: https://doi.org/10.1016/j.cbpa.2015.10.018.

Davin, L.B. & Lewis, N.G. (2005). Lignin primary structures and dirigent sites. *Current Opinion in Biotechnology*, **16**: 407–415. DOI: https://doi.org/10.1016/j.copbio.2005.06.011.

de Gonzalo, G., Colpa, D.I., Habib, M.H.M. & Fraaije, M.W. (2016). Bacterial enzymes involved in lignin degradation. *Journal of Biotechnology*, **236**: 110–119. DOI: https://doi.org/10.1016/j.jbiotec.2016.08.011.

Desjardin, D.E., Oliveira, A.G. & Stevani, C.V. (2008). Fungi bioluminescence revisited. *Photochemical & Photobiological Sciences*, **7**: 170–182. DOI: https://doi.org/10.1039/b713328f.

Dighton, J. & White J.F. (ed.) (2017). *The Fungal Community: Its Organization and Role in the Ecosystem*, 4th edition. Boca Raton, FL: CRC Press. ISBN 10: 1498706657, ISBN 13: 9781498706650.

Dimarogona, M., Topakas, E. & Christakopoulos, P. (2012). Cellulose degradation by oxidative enzymes. *Computational and Structural Biotechnology Journal*, **2**: article number e201209015. DOI: https://doi.org/10.5936/csbj.201209015.

Fernandez, D., Russi, S., Vendrell, J., Monod, M. & Pallarès, I. (2013). A functional and structural study of the major metalloprotease secreted by the pathogenic fungus *Aspergillus fumigatus*. *Acta Crystallographica, Section D Structural Biology*, **69**: 1946–1957. DOI: https://doi.org/10.1107/S0907444913017642.

Gadd, G.M. (ed.) (2006). *Fungi in Biogeochemical Cycles* (British Mycological Society Symposium volume). Cambridge, UK: Cambridge University Press. ISBN-10: 0521845793, ISBN-13: 978-0521845793. DOI: https://doi.org/10.1017/CBO9780511550522.

Garg, G., Singh, A., Kaur, A. *et al.* (2016). Microbial pectinases: an ecofriendly tool of nature for industries. *3 Biotech*, **6**: 47 (13 pp.). DOI: https://doi.org/10.1007/s13205-016-0371-4.

Guan, K.L. & Xiong, Y. (2011). Regulation of intermediary metabolism by protein acetylation. *Trends in Biochemical Sciences*, **36**: 108–116. DOI: http://doi.org/10.1016/j.tibs.2010.09.003.

Gut, P. & Verdin, E. (2013). The nexus of chromatin regulation and intermediary metabolism. *Nature*, **502**: 489–498. DOI: http://doi.org/10.1038/nature12752.

Haitjema, C.H., Gilmore, S.P., Henske, J.K. *et al.* (2017). A parts list for fungal cellulosomes revealed by comparative genomics. *Nature Microbiology*, **2**: article number 17087. DOI: https://doi.org/10.1038/nmicrobiol.2017.87.

Hartl, L., Zach, S. & Seidl-Seiboth, V. (2012). Fungal chitinases: diversity, mechanistic properties and biotechnological potential. *Applied Microbiology and Biotechnology*, **93**: 533–543. DOI: https://doi.org/10.1007/s00253-011-3723-3.

Heaton, L.L.M., López, E., Maini, P.K., Fricker, M.D. & Jones, N.S. (2010). Growth-induced mass flows in fungal networks. *Proceedings of the Royal Society, Series B, Biological Sciences*, **277**: 3265–3274. DOI: https://doi.org/10.1098/rspb.2010.0735.

Humphreys, J.M. & Chapple, C. (2002). Rewriting the lignin roadmap. *Current Opinion in Plant Biology*, **5**: 224–229. DOI: https://doi.org/10.1016/S1369-5266(02)00257-1.

Jennings, D.H. (2008). *The Physiology of Fungal Nutrition*. Cambridge, UK: Cambridge University Press. ISBN-10: 0521038162, ISBN-13: 978-0521038164. DOI: https://doi.org/10.1017/CBO9780511525421.

Kalisz, H.M. (1988). Microbial proteinases. *Advances in Biochemical Engineering/Biotechnology*, **36**: 1–65. DOI: https://doi.org/10.1007/BFb0047943.

Kalisz, H.M., Moore, D. & Wood, D.A. (1986). Protein utilization by basidiomycete fungi. *Transactions of the British Mycological Society*, **86**: 519–525. DOI: https://doi.org/10.1016/S0007-1536(86)80052-3.

Kalisz, H.M., Wood, D.A. & Moore, D. (1987). Production, regulation and release of extracellular proteinase activity in basidiomycete fungi. *Transactions of the British Mycological Society*, **88**: 221–227. DOI: https://doi.org/10.1016/S0007-1536(87)80218-8.

Kalisz, H.M., Wood, D.A. & Moore, D. (1989). Some characteristics of extracellular proteinases from *Coprinus cinereus*. *Mycological Research*, **92**: 278–285. DOI: https://doi.org/10.1016/S0953-7562(89)80066-8.

Knežević, A., Milovanović, I., Stajić, M. *et al.* (2013). Lignin degradation by selected fungal species. *Bioresource Technology*, **138**: 117–123. DOI: https://doi.org/10.1016/j.biortech.2013.03.182.

Macheleidt, J., Mattern, D.J., Fischer, J. *et al.* (2016). Regulation and role of fungal secondary metabolites. *Annual Review of Genetics*, **50**: 371–392. DOI: https://doi.org/10.1146/annurev-genet-120215-035203.

Mandujano-González, V., Villa-Tanaca, L., Anducho-Reyes, M.A. & Mercado-Flores, Y. (2016). Secreted fungal aspartic proteases: a review. *Revista Iberoamericana de Micología*, **33**: 76–82. DOI: https://doi.org/10.1016/j.riam.2015.10.003.

Mérillon, J.M. & Ramawat, K.G. (ed.) (2017). *Fungal Metabolites*. Switzerland: Springer International Publishing. ISBN-10: 3319250000, ISBN-13: 978-3319250007.

Montalvo, G., Téllez-Téllez, M., Díaz, R., Sánchez, C. & Díaz-Godínez, G. (2020). Isoenzymes and activity of laccases produced by *Pleurotus ostreatus* grown at different temperatures. *Revista Mexicana de Ingeniería Química*, **19**: 345–354. DOI: https://doi.org/10.24275/rmiq/Bio570.

Monteiro de Souza, P., de Assis Bittencourt, M.L., Canielles Caprara, C. *et al.* (2015). A biotechnology perspective of fungal proteases. *Brazilian Journal of Microbiology*, **46**: 337–346. DOI: https://doi.org/10.1590/S1517-838246220140359.

Moore, D. (1998). *Fungal Morphogenesis*. New York: Cambridge University Press. See Chapter 3 *Metabolism and biochemistry of hyphal systems*. ISBN-10: 0521552958, ISBN-13: 978-0521552950. DOI: https://doi.org/10.1017/CBO9780511529887.

Moore, D. & Devadatham, M.S. (1979). Sugar transport in *Coprinus cinereus*. *Biochimica et Biophysica Acta (Biomembranes)*, **550**: 515–526. DOI: https://doi.org/10.1016/0005-2736(79)90153-6.

Muszewska, A., Stepniewska-Dziubinska, M.M., Steczkiewicz, K. *et al.* (2017). Fungal lifestyle reflected in serine protease repertoire. *Scientific Reports*, **7**: article number 9147. DOI: https://doi.org/10.1038/s41598-017-09644-w.

Negri, R. (2015). Polyacetylenes from terrestrial plants and fungi: recent phytochemical and biological advances. *Fitoterapia*, **106**: 92–109. DOI: https://doi.org/10.1016/j.fitote.2015.08.011.

Nehls, U. & Dietz, S. (2014). Fungal aquaporins: cellular functions and ecophysiological perspectives. *Applied Microbiology and Biotechnology*, **98**: 8835–8851. DOI: https://doi.org/10.1007/s00253-014-6049-0.

Nirmal, N.P., Shankar, S. & Laxman, R. (2011). Fungal proteases: an overview. *International Journal of Biotechnology & Biosciences*, **1**: 1–40. URL: https://www.researchgate.net/publication/256619613_Fungal_Proteases_An_Overview.

Oliveira, A.G., Desjardin, D.E., Perry, B.A. & Stevani, C.V. (2012). Evidence that a single bioluminescent system is shared by all known bioluminescent fungal lineages. *Photochemical & Photobiological Sciences*, **11**: 848–852. DOI: https://doi.org/10.1039/c2pp25032b.

Patel, S.J. & Savanth, V.D. (2015). Review on fungal xylanases and their applications. *International Journal of Advanced Research*, **3**: 311–315. URL: http://www.journalijar.com/article/3957/.

Pathak, D. & Mallik, R. (2017). Lipid–motor interactions: soap opera or symphony? *Current Opinion in Cell Biology*, **44**: 79–85. DOI: https://doi.org/10.1016/j.ceb.2016.09.005.

Peralta, R.M., da Silva, B.P., Gomes Côrrea, R.C., Kato, C.G., Vicente Seixas, F.A. & Bracht, A. (2017). Enzymes from basidiomycetes: peculiar and efficient tools for biotechnology. In *Biotechnology of Microbial Enzymes: Production, Biocatalysis and Industrial Applications*, ed. G. Brahmachari. Amsterdam: Academic Press, pp. 119–149. DOI: https://doi.org/10.1016/B978-0-12-803725-6.00005-4.

Rai, A.K., Rai, A., Ramaiya, A.J., Jha, R. & Mallik, R. (2013). Molecular adaptations allow dynein to generate large collective forces inside cells. *Cell*, **152**: 172–182. DOI: https://doi.org/10.1016/j.cell.2012.11.044.

Rana, V. & Rana, D. (2017). Role of microorganisms in lignocellulosic biodegradation. In *Renewable Biofuels: Bioconversion of Lignocellulosic Biomass by Microbial Community*, ed. V. Rana & D. Rana. SpringerBriefs in Applied Sciences and Technology. Switzerland: Springer International Publishing AG, pp.19–67. DOI: https://doi.org/10.1007/978-3-319-47379-6_2.

Reignault, P., Valette-Collet, O. & Boccara, M. (2008). The importance of fungal pectinolytic enzymes in plant invasion, host adaptability and symptom type. *European Journal of Plant Pathology*, **120**: 1–11. DOI: https://doi.org/10.1007/s10658-007-9184-y.

Rodrigues, A.G. (2016). Secondary metabolism and antimicrobial metabolites of Aspergillus. In *New and Future Developments in Microbial Biotechnology and Bioengineering: Aspergillus System Properties and Applications*, ed. V.K. Gupta. Amsterdam: Elsevier B.V., pp. 81–93. DOI: https://doi.org/10.1016/B978-0-444-63505-1.00006-3.

Seidl, V. (2008). Chitinases of filamentous fungi: a large group of diverse proteins with multiple physiological functions. *Fungal Biology Reviews*, **22**: 36–42. DOI: https://doi.org/10.1016/j.fbr.2008.03.002.

Shafrin, F., Ferdous, A.S., Sarkar, S.K. *et al.* (2017). Modification of monolignol biosynthetic pathway in jute: different gene, different consequence. *Scientific Reports*, **7**: article number 39984. DOI: https://doi.org/10.1038/srep39984.

Sharma, A., Jain, K.K., Jain, A., Kidwai, M. & Kuhad, R.C. (2018). Bifunctional *in vivo* role of laccase exploited in multiple biotechnological applications. *Applied Microbiology and Biotechnology*, **102**: 10327–10343. DOI: https://doi.org/10.1007/s00253-018-9404-8.

Sigoillot, J.C., Berrin, J.G., Bey, M. *et al.* (2012). Fungal strategies for lignin degradation. *Advances in Botanical Research*, **61**: 263–308. DOI: https://doi.org/10.1016/B978-0-12-416023-1.00008-2.

Taj Aldeen, S.J. & Moore, D. (1982). The *ftr* cistron of *Coprinus cinereus* is the structural gene for a multifunctional transport molecule. *Current Genetics*, **5**: 209–213. DOI: https://doi.org/10.1007/BF00391808.

Toesch, M., Schober, M. & Faber, K. (2014). Microbial alkyl- and aryl-sulfatases: mechanism, occurrence, screening and stereoselectivities. *Applied Microbiology and Biotechnology*, **98**: 1485–1496. DOI: https://doi.org/10.1007/s00253-013-5438-0.

Vanholme, R., Demedts, B., Morreel, K., Ralph, J. & Boerjan, W. (2010). Lignin biosynthesis and structure. *Plant Physiology*, **153**: 895–905. DOI: https://doi.org/10.1104/pp.110.155119.

Verma, S., Dixit, R. & Pandey, K.C. (2016). Cysteine proteases: modes of activation and future prospects as pharmacological targets. *Frontiers in Pharmacology*, **7**: article number 107. DOI: https://doi.org/10.3389/fphar.2016.00107.

Watling, R. & Harper, D.B. (1998). Chloromethane production by wood-rotting fungi and an estimate of the global flux to the atmosphere. *Mycological Research*, **102**: 769–787. DOI: https://doi.org/10.1017/S0953756298006157.

Yaegashi, J., Oakley, B.R. & Wang, C.C.C. (2014). Recent advances in genome mining of secondary metabolite biosynthetic gene clusters and the development of heterologous expression systems in *Aspergillus nidulans*. *Journal of Industrial Microbiology & Biotechnology*, **41**: 433–442. DOI: https://doi.org/10.1007/s10295-013-1386-z.

CHAPTER 12

Exploiting Fungi for Food

Fungal biomass is a high-quality food source because it contains a good content of protein (typically 20–30% crude protein as a percentage of dry matter), which contains all the amino acids that are essential to human and animal nutrition. Add characteristically low-fat content to the protein, a chitinous wall as a source of dietary fibre, useful vitamin content, especially of B vitamins, and carbohydrate in the form of glycogen and a good food source can be considered an ideal food. Judging from archaeological and similar finds, mushrooms, toadstools and bracket fungi have been used by humans since before recorded history for both food and medicinal purposes.

We currently depend on fungi and fungal products every hour of every day. This chapter will concentrate on the human fungal foodstuffs in current use (Moore, 2001; Zied & Pardo-Giménez, 2017). However, before turning to this aspect, we want to deal with the ways that other animals exploit fungi for food. Fungi feature prominently in food webs, and wild fungi are picked commercially, too. But it is not solely a matter of 'picking mushrooms'; fungal cells and mycelium are used as human food, and fungi are used to prepare many commonly used fermented foods, so there is a need to consider several different industrial cultivation methods. Finally, to make the point that we are not the only animals that cultivate fungi, we will discuss the relationship between gardening insects and their fungi, though we will describe these **mutualisms** in detail in Chapter 15.

12.1 FUNGI AS FOOD

Fungus sporophores are used as food by many invertebrates and are large enough to be used by many mammals, including large mammals like deer and primates (Hanson *et al.*, 2003). In most soils, though, there is such a large amount of fungal mycelium that hyphae make a major contribution to **food webs** by being eaten by invertebrates including insects, mites, nematodes and molluscs. Microarthropods are responsible for shredding organic matter in soil (and so prepare it for the final mineralisation processes carried out by microbes), but about 80% of the tens of thousands of microarthropod species in forest soils are **fungivores** (or **mycophagous**); meaning they depend on the fungal mycelium for food. We have introduced many of these small animals before, in Chapter 1 (section 1.5; see Haynes, 2014), and if you have not already done so, we suggest you check out the YouTube videos to which we refer in Resources Box 12.1.

RESOURCES BOX 12.1 Life in the Soil

Deep Down & Dirty: The Science of Soil. A close-up of creatures living beneath the soil, made by the British Broadcasting Corporation (BBC): https://www.youtube.com/watch?v=gYXoXiQ3vC0.

The Living Soil Beneath Our Feet. Made by the California Academy of Sciences: https://www.youtube.com/watch?v=MIREaT9hFCw.

In this chapter, we will illustrate some of the most important features with a short foray into the outside world. Nothing too strenuous; join us on a short walk to the bottom of our suburban garden in Stockport, Cheshire (Figures 12.1–12.8, discussed in the next section).

12.2 FUNGI IN FOOD WEBS

Many of the saprotrophic *Basidiomycota* that are involved in plant-litter recycling, particularly in woodland, form extensive **mycelial networks at the soil-litter interface**. At the growing fronts, individual hyphae are evident in these mycelia, but elsewhere the hyphae often aggregate into strands and cords specialised for translocation of water and nutrients between locations. We have described this before (section 10.7), but here we want to emphasise the size of this mycelial network both in terms of the quantity of mycelial biomass and its physical extent over the ground. This is where we venture into the

Figure 12.1. The suburban location of the mycelium of *Clitocybe nebularis* which is illustrated in Figures 12.2 to 12.10. This biological survey was completed in my back garden and the aerial map image shows the site of the observations (arrow) and makes the point that biology and natural history extends to the urban environment. Image © GeoPerspectives; kindly supplied by James Burn of the Sales Team at http://www.emapsite.com.

Figure 12.2. A troop of *Clitocybe nebularis* mushrooms in a suburban garden in Stockport, Cheshire in autumn 2006. Observations began on 21 October and continued for 29 days to 19 November. Mushrooms of *Coprinellus micaceus* emerged, matured and decayed around 26 October and 1 November (one such instance arrowed on the left above). At bottom right note the corner of a paving slab which was raised to reveal the mycelium shown in Figure 12.3. Photograph by David Moore.

garden. In the autumn of 2006, a troop of *Clitocybe nebularis* (common name: clouded agaric or clouded funnel) appeared in a secluded suburban garden in Stockport, Cheshire, UK (Figures 12.1 and 12.2).

Observations began on 21 October and continued for 29 days to 19 November. At the start, the mushrooms were young but close to maturity (5 cm diameter), Figure 12.2 shows the troop on 29 October. The first overnight frosts of the season occurred on 31 October and 1 November; temperatures were otherwise generally above average throughout October in that year. Some of the mushrooms were close to a paved area and when one of the paving slabs was lifted the mycelium that had penetrated beneath it was evident (Figure 12.3, and photographed, as were other mycelium images, on 1 November).

Figure 12.3. Removal of the paving slab revealed the extensive mycelium of *Clitocybe nebularis* growing beneath it. Numerals show the locations of Figures 12.4 to 12.8. Photograph by David Moore.

Figure 12.4. View of the exposed soil beneath the fruit bodies of *Clitocybe nebularis* showing mycelium extending through the soil and litter layer. Positional key for this image shown in Figure 12.3. Photograph by David Moore.

Mycelium was evident throughout both soil and litter layer (Figure 12.4), and mycelium beneath the 60 cm square paving slab covered soil particles, leaves and dead and dying roots with a lush growth of fine **hyphae** (Figures 12.5 and 12.6) that aggregated into **strands** and **cords** where it colonised the more discrete, and patchily distributed, debris such as root fragments and larger leaves (Figures 12.5, 12.7 and 12.8). As this mycelium is supporting growth of the mushrooms shown in Figure 12.2, it is quite clear that it must persist for several days to a few weeks. Yet it is equally clear that the biomass making up such a mycelium must be a tempting food resource for a range of organisms. Conventional ideas about nutrient cycling in the soil food web assume that bacteria consume easily broken down substrates faster than fungi, and that mycorrhizal fungi do not decompose organic matter; these views are now being challenged. However, considerable evidence has emerged showing that saprotrophic fungi consume significant quantities of the soil's readily available nutrients, and that mycorrhizal fungi decompose organic matter to obtain carbon, nitrogen, phosphorus and sulfur for their own and their host's needs (see Chapter 13). Regarding available carbon, De Vries and Caruso (2016) suggested a new conceptual model for the lowest trophic level of the soil food web, in which organic carbon consists of a continuous pool (rather than different pools labelled 'labile' and 'recalcitrant'), with saprotrophic fungi using substantial amounts of labile nutrients in a balanced co-existence with bacterial saprobes, together with decomposition of organic matter by mycorrhizal fungi.

Figures 12.5 (top left) to 12.8 (bottom right). Lush and extensive hyphal growth of *Clitocybe nebularis* beneath the paving slab, with mycelial strands emerging from litter fragments. Positional key for these images shown in Figure 12.3. Photographs by David Moore.

In natural woodland, large and persistent mycelial networks, like those illustrated in Figures 12.5 to 12.8, are common and their ability to recycle soil nutrients results in them being a food resource that is the foundation of several important food webs. There have been numerous studies on interactions between fungi and soil invertebrates but most of the scientific studies done so far have concentrated on grazing of fungal mycelia by microarthropods known as **collembola** (Tordoff *et al.*, 2006, 2008; Wood *et al.*, 2006). Collembola are integral components of the soil environment, feeding on many organisms including bacteria, lichens and decomposing material but they do prefer fungal hyphae over all other food sources.

Collembola (also known as springtails) are primitive wingless insects that range in size from 0.2 mm to 9 mm. They are plentiful in all soils and population densities commonly reach 10^4 m^{-2}, so hyphal grazing by these invertebrates can result in dramatic changes to mycelial morphology. They express intriguing **food preferences** in laboratory experiments (Rotheray *et al.*, 2009). Collembola, such as *Folsomia candida*, graze the older parts of the mycelium but appear to prefer hyphal tips and fine mycelium; other collembola choose to feed on the hyphae of conidial fungi rather than hyphae of arbuscular mycorrhizal fungi. Grazing on this scale can impede the spread of mycelial networks through woodland and their ability to decompose dead organic material (Tordoff *et al.*, 2006, 2008; Wood *et al.*, 2006). In response to **grazing damage**, the mycelium attempts to evade the attacking fungivore by diverting energy into accelerated growth away from the grazing animal, and in a mycorrhizal mycelium this can result in enhanced exploratory growth of young hyphae and improved nutrient supply to the host plant (Ngosong *et al.*, 2014). Aside from contributing to food webs through predation, the main environmental impact of the collembola is to mineralise nitrogenous and phosphorus compounds by digestion and excretion into the soil, and this leads to increased plant biomass. This positive effect will, of course, be offset by any damage done to mycorrhizal mycelium, although collembola tend to occupy the upper litter layers where saprotrophic fungi predominate.

Mites are at least as numerous in soil as collembola and are also very small, being in the region of 250 μm long. Most are predatory, feeding on nematodes and other arthropods, but there are others that favour bacteria and fungi. **Red pepper mites** (*Pygmephorus* spp.) feed only on *Ascomycota*, particularly *Trichoderma* spp., and are often associated with composts (including mushroom farm composts) in which these moulds can be abundant. They are not primary pests of the mushroom farm, but they indicate the presence of the *Trichoderma* weed mould, which can itself cause mushroom crop losses. The most important mite **pest of mushroom farms** is *Tarsonemus myceliophagus*; the clue is in the species name! This is a very small animal (about 180 μm long) that feeds preferentially on fungal mycelium. Because of its small size it can go unnoticed until damage symptoms appear in the mushroom crop (Rinker, 2017).

Many species of **nematode** occur in habitats rich in decaying organic matter (Glavatska *et al.*, 2017). Some are saprophagous (feeding on decaying materials), others are mycophagous (specifically fungal feeders). These are small colourless worms, up to 1 mm long, that swim in the surface films of water on soil and litter fragments. Two species of **mycophagous nematodes** that feed exclusively on fungal mycelium are *Dictylenchus myceliophagus* and *Aphelenchoides composticola*. Either of these can increase in number by 50 to 100 times in a week so they can rapidly reach population levels sufficient to destroy mycelium.

The larvae (or maggots) of many small **dipteran flies** are probably the most frequently seen mycophagous animals; it is rare to find a mushroom in the field that is completely free of maggots. Adult females lay their eggs on the fungal sporophore and when they hatch the larva eats its way into the tissue, and subsequently eats and tunnels through the sporophore tissue, causing enormous damage (Figures 12.9 and 12.10). These are '**mushroom flies**'; they are about the size (one to a few millimetres long) of fruit flies (Family Drosophilidae), to which they are related. Indeed, there is one fruit fly, *Drosophila funebris*, which lays its eggs in mushroom cap tissues. However, the specialised mushroom flies belong to the related families Sciaridae (commonly known as dark-winged **fungus gnats**), Phoridae (**humpbacked flies**, also known as **scuttle flies** because of their habit of escaping by running rather than flying) and Cecidomyiidae (called cecids, are **gall midges** or gall gnats, the larvae of most of which feed in plant tissues, creating abnormal growths called galls). Larvae of sciarids tend to tunnel from the bottom of the mushroom stem upwards, rarely burrowing into the cap; damage caused by phorids tends to be from the top of the mushroom downwards. Both also eat mycelium. Cecid larvae feed on mycelium and the base of primordial mushroom stems; some also have a taste for mushroom gills.

There are a few species of sciarids that are common, and one, *Lycoriella solani*, has become a very damaging **pest of mushroom farms** because it has developed resistance to organophosphorus insecticides. Adults can affect the crop by spreading mites, bacteria and spores of weed fungi around the mushroom house. Serious crop damage is caused by the larvae tunnelling through the stems of maturing mushrooms, but the most critical damage results from infestation early in crop growth because the larvae can destroy developing mushroom primordia and completely devastate the crop.

Phorid flies are common, their larvae are generally omnivorous, eating anything from decaying vegetation to fresh meat, and these flies are very diverse; Genus *Megaselia* alone has over 100 species in the UK. But one species, *Megaselia halterata*, has become an important pest of mushroom farms because the larvae eat mushroom mycelium (the female is attracted to freshly growing mycelium to lay her eggs) and the adults spread the disease-causing fungus *Verticillium*. Larvae of other species of *Megaselia* feed on and in developing mushrooms, causing tunnelling damage to mushroom cap and stem. *Megaselia nigra* is the species most often found causing tunnelling in wild mushrooms and can become a pest in mushroom farms.

Cecidomyiidae are so small that the adults are seldom seen, so although over 3,000 species have been described there are probably more awaiting discovery. Many species are economically important as pests of crop plants. Others produce predaceous larvae that prey on other arthropods. Six species have been found on mushrooms, the most frequently encountered being *Heteropeza pygmaea*. An unusual feature of the **cecids** is that the larvae reproduce by paedogenesis, an asexual process in which a cecid larva becomes a 'mother larva' from which 10 to 20 daughters emerge within a week of the mother larva hatching from its egg. This method of reproduction leads to rapid multiplication which produces larval swarms capable of doing enormous damage to mycelium and mushrooms (Rinker, 2017).

When the mycelium makes sporophores (like mushrooms, brackets and truffles), these are also vital **food sources for many animals**, from large arthropods and molluscs, to small mammals and humans. Slugs and snails can eat large cavities in the caps and stems of mushrooms, and a large slug can easily consume an entire mushroom (Figure 12.11). Many small mammals depend on fungal sporophores, especially hypogeous ones like truffles, for a significant part, even the bulk, of their diets. This extends the influence of fungi on the food web right up to the top predators, like birds of prey. Those that eat the small mammals that

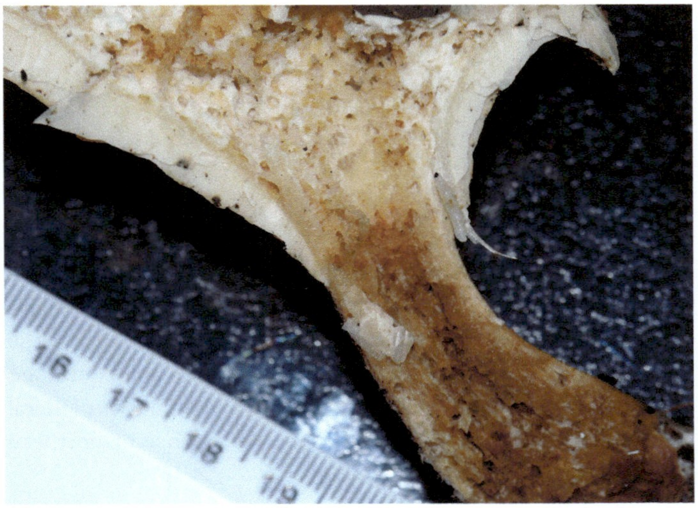

Figure 12.9. A sectioned mushroom of *Clitocybe nebularis* showing extensive tunnelling by larvae (maggots) of mushroom flies. Photograph by David Moore.

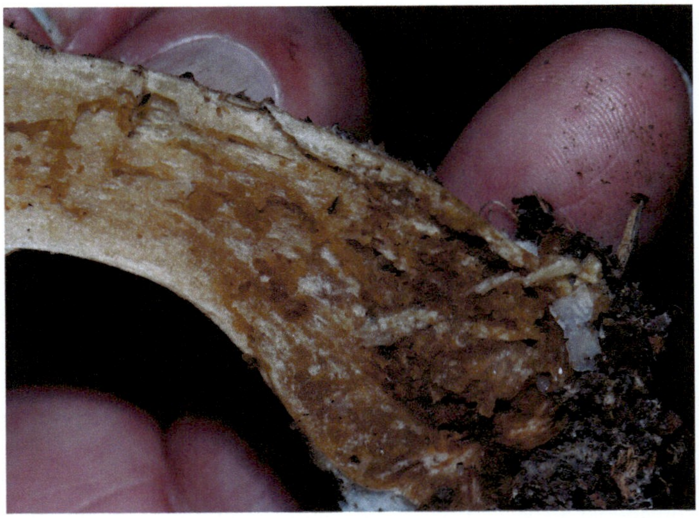

Figure 12.10. Base of the stem of a *Clitocybe nebularis* mushroom showing extensive tunnelling caused by mushroom fly larvae. Photograph by David Moore.

Figure 12.11. An orange form of the large black slug, *Arion ater*, completely demolishing a *Coprinellus micaceus* mushroom fruit body. Photographed by David Moore in the same garden in Stockport.

subsist almost entirely on fungal sporophores, are themselves indirectly dependent on the fungi of the soil.

A celebrated example of this is the **northern spotted owl** (*Strix occidentalis caurina*), in the Pacific Northwest of the United States. This bird has a wingspan of about 1.2 m and ranks as one of the largest birds in North America. Pacific Northwest communities rely heavily on timber harvested from federal forests in the area, and for many years when old-growth forest was harvested, the cleared areas were replanted with fast-growing commercial tree species, rather than the original, slow-growing species. Unfortunately, despite this reforestation the population of the northern spotted owl declined drastically.

The owls are very territorial and intolerant of habitat disturbance. They prefer old-growth forests with high and open tree canopies, so they can fly between and beneath the trees. Their staple diet consists of **small mammals**: squirrels, wood rats, mice and voles, and most of these small mammals themselves depend on **hypogeous fungal sporophores** (truffles) for a significant part of their diets. Unfortunately, the fast-growing tree species that were good for commercial reforestation had different mycorrhizas to those of the original forest; they did not produce sporophores small mammals would eat. Consequently, populations of small mammals declined, and with the loss of prey species, owl populations also declined. These circumstances contributed to the main lesson about conservation: that to conserve a species it is necessary to conserve its habitat. Harvest of the old-growth forest trees could only be made sustainable if cleared areas were replaced with similar communities; using a different range of tree species for commercial purposes produces an entirely different habitat. The Endangered Species Act of 1973 required the US federal government to identify species threatened with extinction, identify habitats they need to survive, and help protect both. The Northwest Forest Plan of 1994 established these principles as legally binding requirements that helped conserve the owl and its prey species but went further by listing 234 old-growth-dependent (and rare) fungal species for additional protection.

A more recent study suggests that the northern spotted owl will require more protection if it is to be conserved (Dugger *et al.*, 2016). During 1985–2013 the survival rates and reproductive outcomes of owls were analysed in 11 areas that represented 9% of the owl's range across Washington State, Oregon and northern California. They noted that spotted owl populations declined by 55–77% in the state of Washington, 31–68% in Oregon and 32–55% in California during this time; an average rangewide decline of nearly 4% per year between 1985 and 2013. It appears that the **barred owl** (*Strix varia*), an invasive species, is out-competing spotted owls for space, food and habitat. Barred owl population densities are high enough across the range of the northern spotted owl that, despite the continued management and conservation of suitable owl habitat on federal lands, the long-term prognosis for the persistence of northern spotted owls may be in doubt (Dugger *et al.*, 2016).

12.3 WILD HARVESTS: COMMERCIAL MUSHROOM PICKING

Collecting mushrooms for food is an age-old tradition which is on a par with collection of berries and other forest fruits. When the collection becomes an industry, which is pursued in order to supply a commodity, it becomes a commercial exploitation of the habitat. Several wild mushrooms have reached the 'exploitation' level: particularly **chanterelles** (*Cantharellus cibarius*), **morels** (*Morchella esculenta*, *Morchella deliciosa* and *Morchella elata*), **truffles** (*Tuber melanosporum*, the French black truffle, and *Tuber magnatum*, the Italian white truffle of Alba; and see Figure 3.13) and **matsutake** (*Tricholoma matsutake*). Many have a tradition and mystique associated with them, including festivals and markets that heighten interest in and appreciation of the qualities of the products themselves. In Europe, most truffles are collected in France and Italy, where truffle hunters use pigs and dogs to sniff out truffles. In Japanese tradition, a gift of matsutake is considered a special prize to be cherished by the recipient, and this tradition persists today. To many people the morel is the supreme edible fungus (Resources Box 12.2). *Morchella* belongs to the *Ascomycota* and fruits in spring (usually May), producing large, club-shaped sporophores with distinctive ridged and pitted heads (see Figure 3.13). There are numerous morel fairs around the world.

RESOURCES BOX 12.2

Culinary and gourmet fungi have high retail values. Truffles in particular can be extremely costly. At the time of writing prices quoted vary from about £1,600 (fresh black winter truffle), £1,350 (white spring truffles from Italy), to about £3,350 (fresh périgord black winter truffle) per kilogramme. You can get more information about these and other commercial fungi from the following URLs:

https://www.fineandwild.com/
http://www.confitdirect.co.uk/
https://trufflehunter.co.uk/
https://www.thegreatmorel.com/

Speciality mushrooms have always been important to humans. This is perhaps most clearly indicated by material carried by the long-dead Alpine traveller who has become known as **'Ötzi the Tyrolean Iceman'** (Maderspacher, 2008). About 3200 BC, a Neolithic traveller set out across the Alps. He didn't make it. Somehow, he was caught in the ice and snow, and died, most likely murdered, to be entombed and preserved in the glacier. Eventually, in 1991, the glacier's slow descent of the mountains exposed his corpse at the edge of the ice sheet close to the present Austrian–Italian border. A well-preserved 5,000-year-old corpse with all of its clothes and equipment is a remarkable find by any measure. Possibly most remarkable is that there were three separate fungal products among the Iceman's equipment (Peintner *et al.*, 1998). One of these was a mass of fibrous material in a leather pouch together with some flints. This fungus has been identified as one with a long history of use as a **tinder** (the tinder bracket, or hoof fungus, *Fomes fomentarius*), so clearly it was part of the Iceman's fire-making kit. The other two are more problematic. Both are pieces of a bracket fungus (*Piptoporus betulinus*) and both are threaded onto leather thongs. These objects were clearly carefully made and must have been important to the owner to be included as part of the kit he chose to take with him in his trek across the mountains. *Piptoporus* is known to accumulate antiseptics and **pharmacologically active** substances claimed to reduce fatigue and sooth the mind. With due ceremony and additional magic, these objects may well have been essential to a traveller in the mountains. The conical one might be a sort of styptic pencil to be applied to scratches and grazes. Perhaps the other was chewed or sucked when the going got tough and the undeniably tough needed a little help from psychotropic substances to keep going.

Coming a little closer to the present day, there is a story that, in 1872, New Zealand had a fungus-based industry earning, eventually, several £100,000 (present-day value several tens of £ millions/US$ millions) annually based on collecting the **wood ear fungus** (*Auricularia polytricha*) for sale in Hong Kong. The then Colonial Secretary of Hong Kong reported in 1871 that the fungus was 'much prized by the Chinese community' being used as a medicine and food (it is known in Asian markets as 'wood ear' or 'mu-ehr', but in Europe as 'cloud ear'; the 'wood ear' common in Europe is *Auricularia auricula-judae*). New Zealand colonial farmers were paid four pence per pound weight of sun-dried fungus and retail prices in the Hong Kong market were four to ten times that. The merchant is said to have purchased an average of £65 worth on **each market day** (present value £8,000/US$11,000) in New Plymouth. In those pre-decimalisation days, there were 240 pence in the pound, so £1 would have bought 60 pounds weight of sun-dried fungus, and £65 each market day would buy $60 \times 65 = 3{,}900$ pounds weight of sun-dried fungus (equivalent to 1.8 metric tons **per market day**). This trade in New Zealand wood ear was maintained until the 1950s, when commercial cultivation of *Auricularia* made collection from natural sources uncompetitive.

Clearly, mushrooms have been important sources of food and medicine for a great many years (Pegler, 2003a, b, c), but demand for wild mushrooms grew sharply during the early 1980s and harvesting wild mushrooms is now big business (Peintner *et al.*, 2013). In 1997, the world market for **chanterelles** (collected, not cultivated) was estimated to be worth more than US$$1.5 \times 10^9$ (Watling, 1997; present monetary value about US$$2.3 \times 10^9$). **Matsutake** is another expensive wild mushroom, which can sell for as much as US$200 a piece in Tokyo markets. Overall value of the harvest is estimated at US$4.6 to 7.7×10^9 annually. Add the value of tourism and peripheral matters like cookery programmes on TV, recipe books and magazine articles to these crop values, and the collection and appreciation of these fungi becomes a very large and diverse industry indeed.

The **commercial picking** industry has now expanded to a system of harvesters, buyers, processors and brokers. Harvesters locate and pick the mushrooms. Buyers, typically associated with a specific processor, set up buying stations near wooded areas known to produce mushrooms and advertise their willingness to buy. Processors grade, clean, pack and ship the product, while providing the cash directly to the field workers. Brokers market the mushrooms around the world. This is a model that has become common in Europe and the United States. One of the things that makes it viable is the easy access to rapid trans- and intercontinental transport. As transport and communications continue to improve, the commercial picking industry is bound to continue to expand.

A successful commercial picking job can see a region of woodland completely denuded of marketable mushrooms in just a few hours. Local residents may see this as destruction of a natural resource and there are frequent calls for the activity to be banned in the interests of 'conservation'. However, picking mushrooms does not do any damage that a conservationist need be concerned about. Mushrooms are not individuals, but simply the sporophores arising on underground mycelia. Removing one generation of sporophores probably encourages a new generation to emerge. Certainly, continued productivity of ***cultivated*** mushroom beds beyond the first flush of cultivated mushrooms is enhanced by regular harvesting. What must be safe-guarded, of course, is the health of the mycelium and so there is a need to avoid unnecessary trampling and disturbance, and activities (trampling, raking, vehicle movements) that physically damage the forest floor and the mycelium that is producing the crop. A long-term study in Switzerland showed that mushroom picking does not impair future harvests (Egli *et al.*, 2006); but on the other hand, high-traffic recreational sites in woodlands (such as camping and picnic sites) do show localised damage. Over a woodland as a whole, disturbance and compaction had no significant effect, but almost no sporophores occurred in the most disturbed areas with bare and trampled soil around campfire sites and picnic tables (Trappe *et al.*, 2009). Consequently, the research done so far in the United States and Europe emphasises the need for effective, and sympathetic, management of wild harvesting rather than its prevention (Górriz-Mifsud *et al.*, 2017).

12.4 CELLS AND MYCELIUM AS HUMAN FOOD

The two most successful applications for fungal biomass in the food industry, which are **yeast extract** and **Quorn®** mycoprotein, illustrate very different ways of using fungal vegetative growth as food. The first example is the conversion of brewery wastes, using low-technology processes into flavourings, diet supplements and products like **Marmite, Meridian Yeast Extract** and **Vegemite**. These yeast extracts illustrate one successful commercial model: start with a cheap (ideally, waste) feedstock and use conventional production processes to make a product that sells at a relatively high retail price. The alternative model with a fungal product currently on the market is the **mycoprotein** Quorn®. This is the mycelium of the filamentous soil fungus *Fusarium venenatum*, grown on food-grade glucose as the carbon source, usually derived from wheat or maize starch, in a 45 m tall airlift fermenter in continuous culture mode (i.e. medium is added continuously and mycelium + spent medium is harvested at a rate equal to the production of new hyphae; for details, see section 19.18). It is consequently a high-technology product (Trinci, 1991, 1992; Wiebe, 2004). The market virtues of the material centre on its filamentous structure which enables it to simulate the fibrous nature of meat (Figure 12.12).

Figure 12.12. Supermarket packages of Quorn® mycoprotein products. Top shows a chiller pack of Quorn® Pieces, which are described as 'perfect for a cracking curry or a sizzling stir fry'; they contain 95% mycoprotein (rehydrated free range egg white, natural flavouring, and the firming agents, calcium chloride and calcium acetate are the only other ingredients) and are entirely meat-free yet high in protein and low in saturated fat. Bottom shows two chiller packs of Quorn® mycoprotein prepared as a healthy meat-free alternative to conventional sliced meat products. Mycoprotein might be the way we can satisfy our taste for meat without costing the Earth. View https://www.quorn.co.uk/ for other product choices and recipes.

Coupled with the inherent nutritional value of fungal biomass, this permits the product to be sold as a low-fat, low-calorie, cholesterol-free health food to consumers who can afford to choose Quorn® as an alternative to meat. At point of sale, Quorn® can be more expensive than many meat products (and most mushrooms), but it is sold as a healthy 'meat alternative'. Evidently, it is positioned in the market to be sold to those who can choose to pay a premium price for an environmentally friendly vegetarian health food (visit https://www.quorn.com/ and https://www.quorn.co.uk/).

12.5 FERMENTED FOODS

As well as being used directly as food, fungi are also used in the processing of various food products. In these applications, the fungus primarily produces some characteristic odour, flavour or texture, and may or may not become part of the final edible product; production is dealt with in Chapter 19. Indonesian **tempeh** is produced by fermentation of partially cooked soy bean cotyledons with *Rhizopus oligosporus*. The fungus binds the soy bean mass into a protein-rich cake that can be used as a meat substitute, which is being increasingly widely sold into the vegetarian market. There are a variety of other fermented products of this sort. **Ang-kak** is a rice product popular in China and the Philippines which is fermented using *Monascus* species. *Monascus purpureus* produces the characteristic pigments and ethanol which are used for red rice wine and food colouring. The pigments are a mixture of red, yellow and purple polyketides and about 10 times more pigment is obtained from solid-state fermentation than from submerged liquid fermentation.

Soy sauce is, in its traditional form, a fermentation product. Soy beans are soaked, cooked, mashed and fermented with *Aspergillus oryzae* and *Aspergillus sojae*. When the substrate has become overgrown with the fungus, the material is transferred to brine and inoculated with the bacterium *Pediococcus halophilus* and 30 days later with *Saccharomyces rouxii*. The brine fermentation takes 6–9 months to complete, after which the soy sauce is filtered and pasteurised.

Cheese could be considered the occidental equivalent of the fermented soya products that are popular in Asia. Cheese is a solid or semi-solid protein food product manufactured from milk. Before the advent of modern methods of food processing, particularly refrigeration, cheese manufacture was the only method of preserving milk. Although basic cheese making is a bacterial fermentation, there are two important processes to which filamentous fungi contribute; these are the provision of enzymes for initial coagulation on the one hand, and mould ripening on the other.

Coagulating enzymes are an absolute necessity to produce all cheese varieties. The traditional coagulant was derived from the dried stomachs of unweaned calves; later, concentrated or dried enzyme extracts from that source, called **calf or animal rennet**, were developed for more controlled cheese making. But by the 1970s, demand for coagulating enzymes started to exceed the supply as world cheese production increased (by a factor of about 3.5 between 1961 and 2010 according to Food and Agriculture Organization (FAO) statistics) while reduced availability of calf stomachs decreased the rennet supply. The most important rennet substitutes include enzymes of microbial origin; recombinant (animal) proteases produced by genetically modified microbes, and plant proteases (proteolytic enzymes called cardosins, from *Cynara cardunculus* (cardoon thistle) are traditionally used to curdle the milk in some Spanish cheeses). With one exception, all commercial clotting enzymes are aspartic proteases that cleave the Phe105–Met106 bond of bovine casein; the exception is a protease from *Cryphonectria parasitica* (causal agent of chestnut blight) which cleaves the Ser104–Phe105 bond. Genetic engineering could provide unlimited amounts of appropriate coagulants, but these products meet consumer resistance and food regulations in various European countries prohibit the involvement of genetically modified organisms (GMOs) in food manufacture (Jacob *et al.*, 2011). Today, only 20–30% of the demand for coagulants can be satisfied by animal rennet, so up to 80% of cheese making now uses non-animal coagulants, mainly enzymes from filamentous fungi like *Aspergillus* spp. and *Mucor miehei*.

Mould ripening is another matter, being a traditional method of flavouring cheeses which has been in use for at least 2,000 years. **Blue cheeses**, like Roquefort, Gorgonzola, Stilton, Danish Blue and Blue Cheshire, use *Penicillium roqueforti* which is inoculated into the cheese prior to storage at controlled temperature and humidity. The fungus grows throughout the cheese (see Figure 19.41), producing methyl ketones, particularly 2-heptanone, as the major flavour and odour compounds. **Camembert** and **Brie** are ripened by *Penicillium camemberti*, which changes the texture of the cheese rather than its flavour (in some varieties, secondary bacterial growth changes the flavour). *Penicillium camemberti* grows on the surface of the cheese producing extracellular proteases which digest the cheese to a softer consistency, working from the outside towards the centre (see Figure 19.42).

12.6 INDUSTRIAL CULTIVATION METHODS

Having just included cheese production in this account, we could complete the ploughman's lunch with bread and beer, but we prefer to leave baking and brewing, not to mention Quorn® production, to later discussion of biotechnology in Chapter 19, and turn instead to mushroom cultivation (Moore & Chiu, 2001; Dupont *et al.*, 2017). Whether calculated in terms of value of product, mass of product, number of people involved in the industry, or geographical area over which the industry is practised, mushroom cultivation is by far the biggest non-yeast biotechnology industry in the world. The technicalities of other sorts of solid-state fermentation are discussed in Chapter 19;

here we will deal with the industrial aspects of mushroom farming. In the European tradition mushroom farming has come to mean cultivation of a mushroom crop on composted plant litter.

The mushroom industries of the world all depend on some form of **solid-state fermentation**; but there are two traditional cultivation methods. In the European tradition the mushroom crop is cultivated on composted plant litter. Similar approaches were developed for oyster and paddy straw mushroom cultivation in the Orient, largely by peasant farmers; while in the Japanese and Chinese traditions, the typical approach is to use wood logs to cultivate the crop of choice (*Lentinula*, called ***shiitake*** in Japanese or ***shiangu*** in Chinese).

The European mushroom industry is said to have originated in the caves beneath Paris at the end of the nineteenth century. It probably emerged from the food provisioning functions of the kitchen gardens on the estates of the European aristocracy, perhaps as early as the seventeenth century. Some of the surviving records of such estates refer to manured and composted plots set aside for mushroom production. The compost used, and its preparation, would very definitely be familiar to competent gardeners of those days.

The current industry depends on compost which is ***very selective*** for the crop species *Agaricus bisporus*. Although widely distributed in nature, this fungus is rarely noticed because established mycelia produce a few isolated mushrooms very infrequently. Today's industry is the result of a 'joint evolution' during which otherwise ordinary horticultural compost was developed which achieves high cropping densities by an otherwise unremarkable and not very abundant mushroom (Buth, 2017).

Composting proceeds in two phases: in **phase 1** the straw, manure and other components are mixed into large heaps. After the water is added the heaps are thoroughly mixed by mechanical compost turning machines. This 'pre-wetting' treatment continues for a few days and then the machines arrange the compost into long stacks about 2 m wide, 2 m high and many metres long. Within a few days bacterial activity heats the stack to around 70°C in the centre, though it is considerably cooler at the surface. Higher temperatures, which would kill the micro-organisms, are avoided by regular 'inside-out' turning of the compost heap. As well as heat, the bacterial degradation process releases large amounts of ammonia. An important aim in phase 1 of composting is to achieve uniformity by thorough mixing (so that all of the compost spends time within the hotter core of the stack). A week after the stack was first laid, it is mixed, or 'turned', by large, self-propelled 'turning' machines. It is left for a further week, then turned again. Three weeks after the process was started, the compost is ready for phase 2.

Phase 2, also known as **peak-heat, pasteurisation** or **sweat-out** is a continuation of the composting process but without further mixing and under more controlled conditions. The compost may be treated in bulk or loaded into the eventual growing containers. In either case the process is done in a building that allows air to be circulated around the growing containers or through the bulk of the compost. To begin with, air and compost temperatures are raised to about 60°C for several hours. This **pasteurisation** stage is usually completed in a day and then the amount of ventilation is increased, and compost temperature is kept at about 50°C for 4–6 days. The beds are then allowed to cool to around 25°C and are ready for use. A natural drop in temperature and absence of free ammonia are signs that the composting process has been completed.

Although the basics of **mushroom production** are the same however the crop is produced, growing containers differ and the process can be separated into specialised stages (Dhar, 2017). Mushrooms may be produced in large wooden trays, in beds on shelving or in plastic growing bags.

- Trays, made of wood in sizes varying from 0.9 × 1.2 m to 1.2 × 2.4 m and 15 to 23 cm deep, are arranged in tiers three or more high, separated by wooden legs. Fork lift trucks are needed to move the trays and some sort of mechanised tray handling line is necessary.
- Shelving is usually made of metal and arranged to give four to six layers of fixed shelves in a cropping room with centre and peripheral access gangways. Each shelf is about 1.4 m wide and extends almost the whole length of the room. Special machinery for compost filling, emptying, spawning, casing and other cultivation operations is necessary.
- Growing bags of about 25 kg are usually supplied to the farm with the compost completely colonised by the mycelia of the mushroom crop and may be arranged on the floor of the cropping house or on tiered shelving (this is phase 3 compost).

For a commercial mushroom farmer, the use of phase 1 compost gives the most flexibility to optimise farm conditions for cultivation of any mushroom strain. Purchase of phase 2 compost enables a farmer to choose which mushroom strain to spawn. The use of phase 3 compost, though more costly, guarantees the production of a crop in a short time and requires the least prior investment in facilities for mushroom production.

Spawning is the process that introduces the mushroom mycelium into the compost. This is generally done with some form of carrier that can be easily mixed into the compost, fungus-coated cereal grains (often barley) being the most usual (Moreaux, 2017). About 5 kg spawn per ton of compost (0.5% w/w) is used. From the inoculation centres that the spawn grains provide the mycelium grows out to invade the compost (this is called '**spawn running**'), filling the compost bed after 10–14 days at a compost temperature of 25°C.

Phase 3 compost is completely colonised by the mushroom mycelia. Nowadays, the mushroom production industry comprises spawn makers and phase 1 compost, phase 2 compost and phase 3 compost suppliers.

Casing is the process that encourages mushroom formation by *Agaricus*, the spawn run compost (which is compost completely permeated by mycelium) must be covered with a 'casing layer' which was originally simply a layer of garden soil but now is most usually a mixture of moist peat and chalk. The chalk is used to adjust the otherwise acid pH to a neutral one. Slow-release nutrients are sometimes also included (Pardo-Giménez *et al.*, 2017).

The optimum depth of the casing is 3–5 cm and it should be an even layer applied to a level compost surface. The mushroom mycelium grows into the casing layer but reaches the upper surface of this layer as strands and these are a necessary start to the sporulation process. To encourage completion of sporophores the growing room is ventilated to lower the concentration of carbon dioxide (usually to <0.1%) and to help reduce the temperature to 16–18°C.

Throughout these processes the casing layer is kept moist by misting with water, as required, because moisture, temperature and atmospheric gases all have to be closely controlled. After allowing 7–9 days for the *Agaricus* mycelium to grow into the casing layer, a machine with rotating tines is run across the mushroom bed to mix the casing layer thoroughly. This is called **'ruffling'** and it serves to break up the mycelial strands. This injury encourages the mushroom mycelia to colonise the surface of the casing layer, where it forms the mushroom initials. Casing is only needed for *Agaricus*, the procedure is not necessary when cultivating other species, such as *Volvariella* spp., *Pleurotus* spp., *Auricularia* spp. and *Lentinula edodes*, on composted straw.

A few days after ruffling, the injury and change in microclimate on the surface of the casing soil sensed by the mushroom mycelium together trigger the formation of *Agaricus* mushroom primordia. The first, called **'pins'** or **'pinheads'** are more or less spherical and have a smooth surface; they are seen about 7–10 days after casing, and 18–21 days after casing, marketable mushrooms can be harvested.

Successive crops of mushrooms (called **flushes**) then develop about 8 days apart, each taking about 5 days to clear from the beds. During the cropping period, the casing needs to be kept moist and the air temperature must be maintained in the 16–18°C range. Ventilation must also be maintained to keep CO_2 levels low. Accurate balance is required here: humidification is essential to minimise desiccation, but too high a level of humidity encourages disease.

Growers expect to harvest between three and five flushes from each spawning cycle, with a total yield of around 25 kg m^{-2} of growing tray. After the final pick (7–10 weeks after spawning) the compost is spent, and the cropping room is emptied, cleaned, sterilised and filled with the next crop. On most large commercial farms, a new crop is filled every one or two weeks throughout the year. A mushroom farmer is likely to see more crops in one year than a cereal farmer will see in a lifetime.

Commercial production of mushrooms produces a total crop of several million metric tons each year. In the mid-1970s, the button mushroom (*Agaricus*) accounted for over 70% of total global mushroom production. Today, it accounts for something closer to 45% even though production tonnage has increased at least 10-fold in the intervening years (Table 12.1; Royse *et al.*, 2017). At averaged-out prices this total crop currently has a retail value of about US$50 billion.

Table 12.1. FAO data for production quantities of mushrooms and truffles 1994–2016

Year	Global production (millions of tons)
1995	2.776
2000	4.190
2005	5.270
2010	7.443
2012	9.647
2014	10.409
2016	10.791

Geographical distribution (average 1994–2016) was approximately 69% farmed in Asia, 22% in Europe, 8% in the Americas, 0.8% in Oceania and 0.2% in Africa. Average production in the Asia region 1994–2016 was 4.231 million tons; mainland China alone produced an average of 3.972 million tons per year across this time period. Data from FAOSTAT website (http://www.fao.org/faostat/en/#data/QC/visualize).

The biggest change during the last quarter of the twentieth century was the increasing interest shown by consumers in a wider variety of mushrooms. Even in the most conservative of markets (like the UK) so-called **'exotic mushrooms'** have now penetrated the market and supplies of fresh shiitake (*Lentinula*) and oyster mushroom (*Pleurotus*) are routinely shelved alongside *Agaricus* in local supermarkets (Royse *et al.*, 2017). Many also offer enoki (*Flammulina velutipes*), buna shimeji (*Hypsizygus marmoreus*), shiroshimeji (*Pleurotus ostreatus*) and king oyster (*Pleurotus eryngii*) among others, most of which are cultivated locally (Stamets, 1994), but the industry is truly international and mushroom cultivation is the next-biggest biotechnology industry after alcohol production.

The production of *Pleurotus* (**oyster mushroom**) differs from that just described for *Agaricus* because the needs of the organism are much less stringent. *Pleurotus* will grow vigorously on both composted/pasteurised ***and*** sterilised/but uncomposted preparations of a wide range of substrates including sawdust, wood chips, cereal straw, and so on. Casing is not required. The crop can be adapted to different countries depending on their climates by growing different species of oyster mushrooms, e.g. *Pleurotus pulmonarius* (misnamed as *Pleurotus sajor-caju*) in India; *Pleurotus ostreatus* (commercially called *Pleurotus florida*, another inaccurate name) in Europe.

One reason for the remarkable increases seen in production of certain mushrooms at reasonable price has been the use of substrates that are **waste products** from other industries. For example, oyster mushroom species (*Pleurotus ostreatus, Pleurotus cystidiosus, Pleurotus pulmonarius*) are all easily grown on cotton wastes. Similarly, although the straw mushroom (*Volvariella volvacea*) is traditionally grown in South-East Asia on rice straw, it too can be grown on cotton waste. Cotton waste gives higher yields and is also more widely available than is rice straw, so it is a far cheaper substrate (the higher cost of rice straw does not derive from any intrinsic value but in the cost of transporting it to a non-rice-growing region). Cotton waste substrates are generated by the textile and garment industries and are produced in bulk by recycling schemes around the world.

Disposal of an abundant bulky solid waste coupled with currency earning by sale of a mushroom crop is a good example of an organic farming system integrated with a waste treatment system. The concept of using mushroom cultivation as a waste remediation has become a popular model in recent years, and we have already mentioned that agriculture generates enormous waste because so little of each crop is used (95% of the total biomass produced in palm and coconut oil plantations is discarded as waste, 98% of the sisal plant is waste, 83% of sugar cane biomass is waste, etc.).

Pleurotus spp. grow readily on so many lignocellulose agricultural wastes that it becomes an attractive notion to use the fungus to digest the waste and by so doing produce a cash crop of **oyster mushrooms**. Even more attractive is that after the mushrooms have been harvested the 'spent compost' can be useful:

- as animal feed (the mushroom mycelium boosts its protein content);
- as a soil conditioner, as it is a compost still rich in nutrients and with polymeric components that enhance soil structure and serve as a biofertiliser (Yu *et al.*, 2019); and
- to digest pollutants (like polychlorinated phenols) on landfill waste sites because it contains populations of microorganisms able to digest the natural phenolic components of lignin (see section 11.7).

Some care is needed because *Pleurotus* accumulates metal ions in the sporophore (Sakellari *et al.*, 2019). If waste used as substrate comes from an industrial source contaminated with **heavy metals** (cadmium is a problem in many industries, lead in others), then the mushroom crop may be unsuitable for consumption (Moore & Chiu, 2001; Kumhomkul & Panich-pat, 2013; Covaci *et al.*, 2017; Mohd Hanafi *et al.* 2018). Harvesting the mushrooms would still be an effective way of removing the heavy metal contamination, though, and activity of the mycelium will remediate the rest of the waste.

Although European farming methods are used around the world, the Asian tradition for commercial mushroom production tends to favour more natural substrates. *Lentinula edodes* (**shiitake**) is traditionally grown on hardwood logs (oak, chestnut, hornbeam) and is still very widely grown like this in the central highlands in China. To put this statement into perspective, the traditional log-pile approach is still the most frequently used method in China over a growing region which covers an area about equal to the entire land area of the European Union. Logs suitable for shiitake production are over 10 cm diameter and 1.5–2.0 m long and normally cut in spring or autumn to minimise pre-infestation by wild fungi or insects. Holes drilled in the logs (or saw or axe cuts) are packed with spawn, and the spawn-filled hole then sealed with wax or other sealant to protect the spawn from weather. In this case, the spawn may be mycelium grown on rice or other cereal grains, but is more likely to be mycelium grown on wooden dowels which can then be hammered into holes drilled in the production log (Royse *et al.*, 2017).

Inoculated logs are stacked in **laying yards** on the open hillside in arrangements which permit good air circulation and easy drainage and provide temperatures of 24–28°C. The logs remain here for the 5–8 months it takes for the fungus to grow completely through the log. Finally, the logs are transferred to the raising yard to promote mushroom formation. This is usually done in winter to ensure the temperature shock (12–20°C) and increased moisture which are required for sporophore initiation. The first crops of mushrooms appear in the first spring after being moved to the raising yard. Each log will produce 0.5–3.0 kg of mushrooms, each spring and autumn, for 5–7 years.

This traditional approach to shiitake production is expensive and demanding in both land and trees; for these and other reasons more industrial approaches are being applied. Hardwood chips and sawdust packed into polythene bags as '**artificial logs**' provide a highly productive alternative to the traditional technique, and the cultivation can be done in houses (which may only be plastic-covered enclosures) in which climate control allows year-round production.

Volvariella volvacea (**paddy straw mushroom**) is grown mainly on rice straw, although several other agricultural wastes make suitable substrates. Preparation of the substrate is limited to tying the straw into bundles which are soaked in water for 24–48 h. The soaked straw is piled into heaps about 1 m high which are inoculated with spent straw from a previous crop. In less than 1 month, a synchronised flush of egg-like sporophores appears. These immature sporophores (in which the universal veil is intact and completely encloses the immature mushroom are sold for consumption just like the young mushrooms ('**baby buttons**') of *Agaricus*, though this is not the case with oyster and shiitake mushrooms which are sold mature. Comparatively low yields of *Volvariella volvacea* are generated from the substrate, and it is difficult to maintain a good quality in post-harvest storage. Within 2–3 days the crop turns brown and autolyses even in cold storage. These factors restrict production of the crop.

Ganoderma lucidum is a cultivated mushroom which is unique in being consumed for its pharmaceutical value (real or imagined) rather than as a food. Under the names **lingzhi** or **reishi** in Asia, several *Ganoderma* species of the *Ganoderma lucidum* complex provide various commercial brands of **nutriceuticals**, in the form of health drinks, powders, tablets, capsules and diet supplements. *Ganoderma* is highly regarded as a traditional **herbal**

medicine, though the claims made for it are clinically unproven in most cases (see section 19.26).

It is cultivated by being inoculated into short segments of wooden logs which are then covered with soil in an enclosure (such as a plastic-covered 'tunnel') which can be kept moist and warm. Sporophores emerge in large number quite close together and the conditions encourage the fungus to form the desirable long-stemmed sporophore (Moore & Chiu, 2001).

The morphology of *Ganoderma* sporophores varies greatly. At least some of the reported variation is likely to be due to misidentifications as the taxonomy of the *Ganoderma lucidum* complex has been described as 'chaotic'. Analysis of 32 collections of the complex from Asia, Europe and North America using both morphology and molecular phylogenetics recovered a total of 13 taxonomically distinct species within the complex (Zhou *et al.*, 2015). In sharp contrast, a survey of the molecular phylogenetics of 20 specimens of the related clade, *Ganoderma sinense*, from China were found to exhibit varied sporophore morphology, even though they possessed identical nucleotide sequences (Hapuarachchi *et al.*, 2019). Evidently, the phenotypic plasticity (= varied fruit body morphology) of a specimen or strain of *Ganoderma* can be influenced greatly by extrinsic factors, such as climate, nutrition, vegetation and geographical environment, rather than being associated with genotypic variation.

Truffles are extremely valuable; they can be worth £1,000/US$1,400 to £3,000/US$4,200 per kg, with a single truffle potentially weighing over 200 g. Truffle cultivation is different from that of the other fungi so far described, because the truffle is the underground fruit body of one of the *Ascomycota* that is mycorrhizal on oak (*Quercus*), so it is dependent on its host tree. Traditionally, truffles are found using pigs or dogs trained to detect the volatile metabolites produced by the fruit body. Truffle 'cultivation' was first achieved in France early in the nineteenth century, when it was found that if seedlings adjacent to truffle-producing trees were transplanted, they too began producing truffles in their new location. Truffières or **truffle groves** have been established throughout France in the past hundred years and the value of the crop is such that the practice is now extending around the world. Truffières are started by planting oak seedlings in areas known to be rich in truffle fungi; the seedlings can be grown-on in greenhouses after infection with *Tuber melanosporum*. Seedlings can be colonised artificially with the related *Tuber magnatum* (a white truffle). Truffles begin to appear under such trees 7–15 years after planting out, and cropping continues for 20–30 years (Moore & Chiu, 2001; Dupont *et al.*, 2017).

12.7 GARDENING INSECTS AND FUNGI

Attine ants in Central and South America have a rather unusual and ancient relationship with a fungus generally called *Leucocoprinus gongylophorus*, in Family Lepiotaceae (*Basidiomycota*). The relationship is a mutualism which is thought to date back to about 65–45 million years ago. They actively inoculate their nest with the fungus and then cultivate it by providing it with pieces of leaves, pruning the hyphae and removing intruder fungi. As a reward, the fungus produces special structures, called gongylidia; hyphal branches that have evolved to be eaten by ants. In Africa, some species of termite also maintain a 'garden' of fungi belonging to Genus *Termitomyces* (Order *Agaricales*, *Basidiomycota*). Nourished by the excrement of the insects, the fungus digests dead plant material brought back to the nest by the workers and provides more digestible food for the termites. We will describe these mutualisms in detail later (Chapter 15), but we want to point out here that these are **gardening insects**, and just as we cultivate specific fungi for the food value of their sporophores, gardening insects cultivate fungi for the food value of the mycelium that results from its ability to digest plant remains.

12.8 DEVELOPMENT OF A FUNGAL SPOROPHORE

Development of a fungal sporophore requires that hyphal growth takes on a particular pattern. A pattern, which, time after time, produces the same species-specific morphology. This demands high levels of control and regulation. We have already indicated some of the regulatory circuits that are involved in cell and tissue differentiation (Chapter 10; sections 10.2 and 10.8) and in the next chapter we will examine the nature of the developmental pathways that give rise to fungal sporophores.

12.9 REFERENCES

Buth, J. (2017). Compost as a food base for *Agaricus bisporus*. In *Edible and Medicinal Mushrooms: Technology and Applications*, ed. D.C. Zied & A. Pardo-Giménez. Chichester, UK: John Wiley & Sons (Wiley-Blackwell), pp. 129–148. DOI: https://doi.org/10.1002/9781119149446.ch6.

Covaci, E. Darvasi, E. & Ponta, M. (2017). Simultaneous determination of Zn, Cd, Pb and Cu in mushrooms by differential pulse anodic stripping voltammetry. *Studia Universitatis Babes-Bolyai Chemia*, **62**: 133–144. DOI: https://doi.org/10.24193/subbchem.2017.3.10.

De Vries, F.T. & Caruso, T. (2016). Eating from the same plate? Revisiting the role of labile carbon inputs in the soil food web. *Soil Biology and Biochemistry*, **102**: 4–9. DOI: https://doi.org/10.1016/j.soilbio.2016.06.023.

Dhar, B.L. (2017). Mushroom farm design and technology of cultivation. In *Edible and Medicinal Mushrooms: Technology and Applications*, ed. D.C. Zied & A. Pardo-Giménez. Chichester, UK: John Wiley & Sons (Wiley-Blackwell), pp. 271–308. DOI: https://doi.org/10.1002/9781119149446.ch14.

Dugger, K.M., Forsman, E.D., Franklin, A.B. *et al.* (2016). The effects of habitat, climate, and barred owls on long-term demography of northern spotted owls. *The Condor*, **118**: 57–116. DOI: https://doi.org/10.1650/CONDOR-15-24.1.

Dupont, J., Dequin, S., Giraud, T. *et al.* (2017). Fungi as a source of food. In *The Fungal Kingdom*, ed. J. Heitman, B. Howlett, P. Crous *et al.* Washington, DC: ASM Press, pp. 1063–1085. DOI: https://doi.org/10.1128/microbiolspec.FUNK-0030-2016.

Egli, S., Peter, M., Buser, C., Stahel, W. & Ayer, F. (2006). Mushroom picking does not impair future harvests: results of a long-term study in Switzerland. *Biological Conservation*, **129**: 271–276. DOI: https://doi.org/10.1016/j.biocon.2005.10.042.

Glavatska, O., Müller, K., Butenschoen, O. *et al.* (2017). Disentangling the root- and detritus-based food chain in the micro-food web of an arable soil by plant removal. *PLoS ONE*, **12**: article number e0180264. DOI: https://doi.org/10.1371/journal.pone.0180264.

Górriz-Mifsud, E., Marini Govigli, V. & Bonet, J.A. (2017). What to do with mushroom pickers in my forest? Policy tools from the landowners' perspective. *Land Use Policy*, **63**: 450–460. DOI: https://doi.org/10.1016/j.landusepol.2017.02.003.

Hanson, A.M., Hodge, K.T. & Porter, L.M. (2003). Mycophagy among primates. *Mycologist*, **17**: 6–10. DOI: https://doi.org/10.1017/S0269-915X(03)00106-X.

Hapuarachchi, K.K., Karunarathna, S.C., McKenzie, E.H.C. *et al.* (2019). High phenotypic plasticity of *Ganoderma sinense* (Ganodermataceae, Polyporales) in China. *Asian Journal of Mycology*, **2**: 1–47. DOI: https://doi.org/10.5943/ajom/2/1/1.

Haynes, R.J. (2014). Nature of the belowground ecosystem and its development during pedogenesis. *Advances in Agronomy*, **127**: 43–109. DOI: https://doi.org/10.1016/B978-0-12-800131-8.00002-9.

Jacob, M., Jaros, D. & Rohm, H. (2011). Recent advances in milk clotting enzymes. *International Journal of Dairy Technology*, **64**: 14–33. DOI: https://doi.org/10.1111/j.1471-0307.2010.00633.x.

Kumhomkul, T. & Panich-pat, T. (2013). Lead accumulation in the straw mushroom, *Volvariella volvacea*, from lead contaminated rice straw and stubble. *Bulletin of Environmental Contamination and Toxicology*, **91**: 231–234. DOI: https://doi.org/10.1007/s00128-013-1025-4.

Maderspacher, F. (2008). Ötzi, a quick guide. *Current Biology*, **18**: R990–R991. DOI: https://doi.org/10.1016/j.cub.2008.09.009.

Mohd Hanafi, F.H., Rezania, S., Mat Taib, S. *et al.* (2018). Environmentally sustainable applications of agro-based spent mushroom substrate (SMS): an overview. *Journal of Material Cycles and Waste Management*, **20**: 1383–1396. DOI: https://doi.org/10.1007/s10163-018-0739-0.

Moore, D. (2001). *Slayers, Saviors, Servants and Sex. An Exposé of Kingdom Fungi*. New York: Springer-Verlag. ISBN-10: 0387951016, ISBN-13: 978-0387951010. See Chapter 7.

Moore, D. & Chiu, S.W. (2001). Fungal products as food. In *Bio-Exploitation of Filamentous Fungi*, ed. S.B. Pointing & K.D. Hyde. Hong Kong: Fungal Diversity Press, pp. 223–251. ISBN: 962-85677-2-1.

Moreaux, K. (2017). Spawn Production. In *Edible and Medicinal Mushrooms: Technology and Applications*, ed. D.C. Zied & A. Pardo-Giménez. Chichester, UK: John Wiley & Sons (Wiley-Blackwell), pp. 89–128. DOI: https://doi.org/10.1002/9781119149446.ch5.

Ngosong, C., Gabriel, E. & Ruess, L. (2014). Collembola grazing on arbuscular mycorrhiza fungi modulates nutrient allocation in plants. *Pedobiologia*, **57**: 171–179. DOI: https://doi.org/10.1016/j.pedobi.2014.03.002.

Pardo-Giménez, A., Pardo González, J.E. & Zied, D.C. (2017). Casing materials and techniques in *Agaricus bisporus* cultivation. In *Edible and Medicinal Mushrooms: Technology and Applications*, ed. D.C. Zied & A. Pardo-Giménez. Chichester, UK: John Wiley & Sons (Wiley-Blackwell), pp. 149–174. DOI: https://doi.org/10.1002/9781119149446.ch7.

Pegler, D.N. (2003a). Useful fungi of the world: the shii-take, shimeji, enoki-take, and nameko mushrooms. *Mycologist*, **17**: 3–5. DOI: https://doi.org/10.1017/S0269-915X(03)00107-1.

Pegler, D.N. (2003b). Useful fungi of the world: the monkey head fungus. *Mycologist*, **17**: 120–121. DOI: https://doi.org/10.1017/S0269-915X(03)00306-9.

Pegler, D.N. (2003c). Useful fungi of the world: morels and truffles. *Mycologist*, **17**: 174–175. DOI: https://doi.org/10.1017/S0269-915X(04)00402-1.

Peintner, U., Pöder, R. and Pümpel, T. (1998). The Iceman's fungi. *Mycological Research*, **102**: 1153–1162. DOI: https://doi.org/10.1017/S0953756298006546.

Peintner, U., Schwarz, S., Mešić, A. *et al.* (2013). Mycophilic or mycophobic? Legislation and guidelines on wild mushroom commerce reveal different consumption behaviour in European countries. *PLoS ONE*, **8**: article number e63926. DOI: https://doi.org/10.1371/journal.pone.0063926.

Rinker, D.L. (2017). Insect, mite, and nematode pests of commercial mushroom production. In *Edible and Medicinal Mushrooms: Technology and Applications*, ed. D.C. Zied & A. Pardo-Giménez. Chichester, UK: John Wiley & Sons (Wiley-Blackwell), pp. 221–238. DOI: https://doi.org/10.1002/9781119149446.ch11.

Rotheray, T.D., Boddy, L. & Jones, T.H. (2009). Collembola foraging responses to interacting fungi. *Ecological Entomology*, **34**: 125–132. DOI: https://doi.org/10.1111/j.1365-2311.2008.01050.x.

Royse, D.J., Baars, J. & Tan, Q. (2017). Current overview of mushroom production in the world. In *Edible and Medicinal Mushrooms: Technology and Applications*, ed. D.C. Zied & A. Pardo-Giménez. Chichester, UK: John Wiley & Sons (Wiley-Blackwell), pp. 5–13. DOI: https://doi.org/10.1002/9781119149446.ch2.

Sakellari, A., Karavoltsos, S., Tagkouli, D. *et al.* (2019). Trace elements in *Pleurotus ostreatus*, *P. eryngii* and *P. nebrodensis* mushrooms cultivated on various agricultural by-products. *Analytical Letters*, 0003–2719. DOI: https://doi.org/10.1080/00032719.2019.1594865.

Stamets, P. (1994). *Growing Gourmet & Medicinal Mushrooms*. Berkeley, CA: Ten Speed Press. ISBN-13: 978-1580081757.

Tordoff, G.M., Boddy, L. & Jones, T.H. (2006). Grazing by *Folsomia candida* (Collembola) differentially affects mycelial morphology of the cord-forming basidiomycetes *Hypholoma fasciculare, Phanerochaete velutina* and *Resinicium bicolor*. *Mycological Research*, **110**: 335–345. DOI: https://doi.org/10.1016/j.mycres.2005.11.012.

Tordoff, G.M., Boddy, L. & Jones, T.H. (2008). Species-specific impacts of collembola grazing on fungal foraging ecology. *Soil Biology and Biochemistry*, **40**: 434–442. DOI: https://doi.org/10.1016/j.soilbio.2007.09.006.

Trappe, M.J., Cromack, K. Jr, Trappe, J.M. *et al.* (2009). Relationships of current and past anthropogenic disturbance to mycorrhizal sporocarp fruiting patterns at Crater Lake National Park, Oregon. *Canadian Journal of Forest Research*, **39**: 1662–1676. DOI: https://doi.org/10.1139/X09-073.

Trinci, A.P.J. (1991). 'Quorn' mycoprotein. *Mycologist*, **5**: 106–109. DOI: https://doi.org/10.1016/S0269-915X(09)80296-6.

Trinci, A.P.J. (1992). Myco-protein: a twenty-year overnight success story. *Mycological Research*, **96**: 1–13. DOI: https://doi.org/10.1016/S0953-7562(09)80989-1.

Watling, R. (1997). The business of fructification. *Nature*, **385**: 299–300. DOI: https://doi.org/10.1038/385299a0.

Wiebe, M.G. (2004). Quorn mycoprotein: overview of a successful fungal product. *Mycologist*, **18**: 17–20. DOI: https://doi.org/10.1017/S0269-915X(04)00108-9.

Wood, J., Tordoff, G.M., Jones, T.H. & Boddy, L. (2006). Reorganization of mycelial networks of *Phanerochaete velutina* in response to new woody resources and collembola (*Folsomia candida*) grazing. *Mycological Research*, **110**: 985–993. DOI: https://doi.org/10.1016/j.mycres.2006.05.013.

Yu, Y.-Y., Li, S.-M., Qiu, J.-P. *et al.* (2019). Combination of agricultural waste compost and biofertilizer improves yield and enhances the sustainability of a pepper field. *Journal of Plant Nutrition and Soil Science*, online version. DOI: https://doi.org/10.1002/jpln.201800223.

Zhou, L.W., Cao, Y., Wu, S.H. *et al.* (2015). Global diversity of the *Ganoderma lucidum* complex (Ganodermataceae, Polyporales) inferred from morphology and multilocus phylogeny. *Phytochemistry*, **114**: 7–15. DOI: https://doi.org/10.1016/j.phytochem.2014.09.023.

Zied, D.C. & Pardo-Giménez, A. (ed.) (2017). *Edible and Medicinal Mushrooms: Technology and Applications*. Chichester, UK: John Wiley & Sons (Wiley-Blackwell). ISBN-13: 978-1119149415.

CHAPTER 13

Development and Morphogenesis

In Chapter 10, we showed how fungal hyphae are capable of a wide range of cell differentiation, and in other chapters (particularly Chapters 3 and 9), we have mentioned and illustrated some of the variety of multicellular sporophore structures (sexual and asexual) that fungi can produce. It must be evident, therefore, that a fungal mycelium has several alternative developmental pathways open to it:

- continuation of hyphal growth;
- production of asexual structures;
- progress into the sexual cycle.

These are not strict alternatives, because a mycelium of even modest size may express all possibilities at the same time, so we can deduce that whatever control of genetic expression is involved must be local.

In this chapter, we explain the nature of development and morphogenesis in fungi and the formal terminology of developmental biology. Observations and experiments on fungal developmental biology lead to the conclusion that there are ten ways to make a mushroom, depending on different arrangements of competence and regional patterning. As specific examples, we show how the *Coprinopsis* mushroom makes hymenia, and how *Coprinopsis* and *Volvariella* make gills (not forgetting how polypores make tubes). The *Coprinopsis* mushroom also provides examples of the construction of mushroom stems and the co-ordination of cell inflation throughout the maturing sporophore, which leads us to consider the mechanics of the mushroom.

The mechanics depend on biochemistry, and so we have to examine metabolic regulation in relation to morphogenesis, as well as ideas about developmental commitment, and comparisons with other tissues and other organisms. We show how classic genetic approaches allowed some progress to be made in the study of fungal development and how this has been accelerated by genomics and other aspects of global analysis of macromolecules. Finally, we turn to the end-game in development: degeneration, senescence and death, and finish with a summary of the basic principles of fungal developmental biology.

13.1 DEVELOPMENT AND MORPHOGENESIS

In a great many fungi, hyphae differentiate from the vegetative form that ordinarily composes a mycelium and **aggregate** to form tissues of multihyphal structures. These may be linear organs (that emphasise parallel arrangements of hyphae), such as strands, rhizomorphs and sporophore stems (see section 10.7); or globose masses (that emphasise interweaving of hyphae), such as sclerotia, sporophores and other sporulating structures of the larger *Ascomycota* and *Basidiomycota* (see section 10.8).

Development of any of these fungal multicellular structures requires that hyphal growth takes on a particular 'pattern'. A pattern that, time after time, produces the same species-specific structure and morphology of that structure, a process that demands precise control and regulation. Formation of a multicellular structure begins with a localised association of aerial hyphae into a **hyphal tuft** (also called a **hyphal knot**), which gradually enlarges and differentiates into a **primordium** of the sporophore (or other structure, according to circumstances) from which the mushroom (etc.) finally emerges.

The **differential growth** represented in this morphogenesis, and which gives rise to the development of the variety of tissues that make up a fungal multicellular structure involves detailed control and regulation of wall synthesis. Most of the descriptions of wall formation we have given so far have concentrated on hyphal tip (apical) growth, but in development of multicellular structures, 'mature' hyphal wall distant from its hyphal apex can restart wall formation to **remodel** and reshape the cell. In addition, two adjacent hyphal branches can be joined together by synthesis of a joint wall, which can be stronger than the original (see section 13.7). There are many observations of hyphal walls being thickened internally by synthesis of a secondary wall, mostly made up of thick fibrils, which are probably constructed of glucans accumulated as a nutritional reserve (see Figures 7.5–7.9). Fungal wall synthesis, **resynthesis** and **secondary wall formation** are topics worthy of separate treatment and we have dealt with them in detail in Chapter 7.

Differentiation events can be limited to particular hyphae within the structure (indeed, to particular cells in individual hyphae), and differential growth of tissues can generate mechanical forces that change the macroscopic shape of the whole structure. In developmental terminology, this is **pattern formation** (creation of a specific spatial arrangement of tissues that will generate the final morphology of the structure or organ) caused by **regional patterning (regional specification)** of the differentiation pathways followed by the cells within those patterns. That these processes take place can be deduced from relatively simple experiments and observations, even though it is not yet known ***how*** the hyphae that will differentiate are specified.

Here, we will illustrate how the formal principles of fungal developmental biology have been established by experiment and observation of the patterns of hyphal growth, branching and interactions that achieve the tissue patterns represented in the diverse morphologies of fungal sporophores and similar multicellular structures; and then theorise about how these patterns are achieved. First, we have a few words about the **terminology** that is employed.

We have already used some development-specific terminology; namely pattern formation and regional specification. We have also mentioned some important, indeed crucial, features that uniquely characterise fungal development. This is the dependence of fungal multicellular development on control and adaptation of the normal growth and branching of vegetative hyphae and, in particular, the fact that formation of any multicellular structure in fungi requires reversal of the outward, exploratory growth habit that characterises vegetative hyphae, specifically, altered autotropisms. To contribute to the aggregations that become multicellular structures, the hyphae concerned must convert the negative autotropism that ensures outward growth of hyphal tips in mycelia into a positive autotropism (see the discussion in section 4.10) that permits the tips of branches to approach each other and other hyphae and create the hyphal tuft that is the initiation point for the multicellular structure.

13.2 THE FORMAL TERMINOLOGY OF DEVELOPMENTAL BIOLOGY

Development is formally defined as the process of change and growth within an organism during the transition from embryo to adult. This definition immediately creates a major challenge for any mycologist interested in development, because it specifies 'embryo', an inevitable result of the fact that most developmental biologists deal with animal systems. Fungi do not have embryos. In fact, fungi are **modular organisms** in which the multicellular structures (especially fungal sporophores like mushrooms, etc.) are appendages of their mycelium, they are *not* individual organisms. This is a feature that fungi share with clonal corals and vegetatively propagated plants and algae (Harper *et al.*, 1986; Andrews, 1995; Monro & Poore, 2004). The most succinct description of the modular growth form as it applies to fungi (Andrews, 1995) notes that modular organisms are:

> characterised by an iterative, indeterminate mode of growth; internal age structure; localised rather than generalised senescence; cellular totipotency; and passive rather than active mobility.

If you think about it, that quotation describes the fungal mycelium, which features numerous instances of the same hyphal forms extending into new territory for an unstipulated time. The mycelium being comprised of hyphae of all ages; youngest (newly formed) hyphal apices at the growing margin with maturing hyphae towards the 'older' more central parts of the colony, and perhaps senescing and dying hyphae at the very centre. Yet even from that senescent centre, inocula can be transferred to fresh medium with the expectation that 'cellular totipotency' will ensure that new hyphal tips will be formed from even the senescent hyphae. Of course, the fungi have an absorptive, sessile lifestyle, so they are not as mobile as any animal, but the exploratory hyphae do invade new territory. Andrews (1995) goes on to point out that:

> differentiation of a few basic cell types gives rise to multicellular complexes associated with foraging, reproductive, survival, or dispersal activities.

And finally:

> morphological plasticity of fungi is analogous to but surpasses that of other modular creatures such as the benthic invertebrates and plants. For instance, in response to environmental signals, fungi can vary the timing, extent, and mode of differentiation; interconvert among different growth forms; and decouple the sexual and asexual phases of the life cycle ...

If the lifestyle of fungi is so different from that of the metazoan animals used to establish the basic principles of developmental biology, where do we go from here? Well, we suggest a 'pick-and-mix' approach. Let's compare fungal development with animal development and borrow those aspects of the concepts that apply to animals that suit our purposes.

Early development of something like a mushroom looks much like an embryonic process (Figure 13.1), and it is clear from images like this that we gain a lot by being able to use the vocabulary of animal embryology, For example, the series of images in Figure 13.1 clearly shows development of the basic **body plan** of a mushroom. The **pattern formation** that demarcates the main **tissues** and **organs** obviously takes place at a very early stage because the veil, cap, gills, stem and stem base are all clearly established in a tiny structure barely 1% the size of the final mature mushroom (just like an embryo!). We say that we can use the established language of developmental biology, but we have to apply it with care, understanding just how different

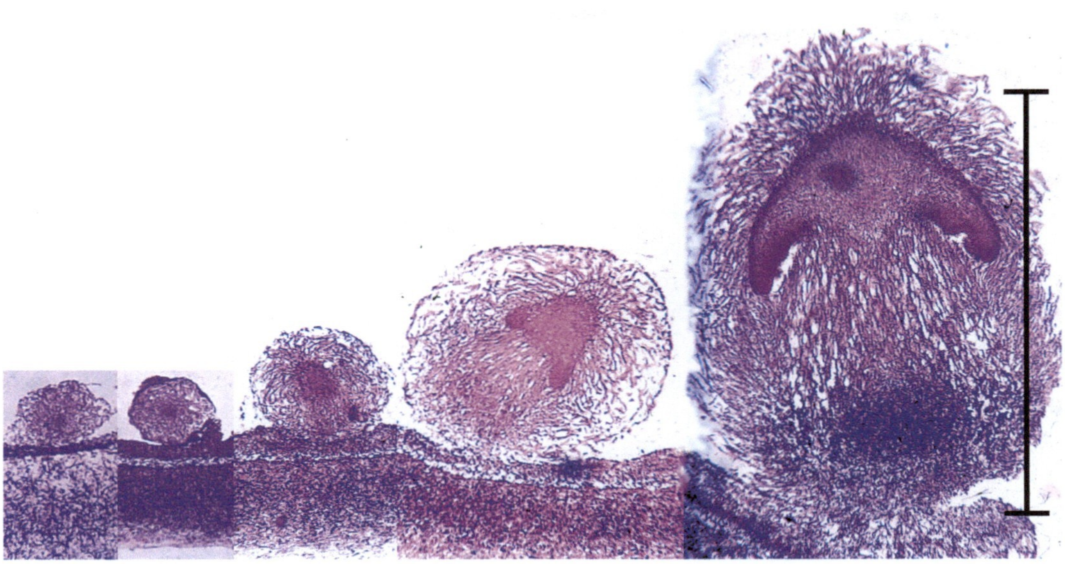

Figure 13.1. Early development of something like a mushroom looks very much like an embryonic process (scale bar, 1 mm) but remember that fungi are clonal organisms and these objects are sporophores, and many of them may be produced by an individual mycelium over an extended period of time. These images are photomicrographs of light microscope sections of successive stages in the very early development of the mushroom of the ink cap mushroom, *Coprinopsis cinerea*. The mature mushroom is approximately 100 mm tall, so this sequence covers just the first 1% of its developmental programme. The sections have been stained with the Periodic Acid-Schiff reagent, which stains polysaccharide accumulations blue-purple; in this case the polysaccharide, identified by other analyses, is glycogen. The object at extreme left is a large hyphal tuft, at second left is an initial (it shows some internal compaction and differentiation and can become either a sclerotium or a mushroom depending on environmental conditions). Note that the third section is obviously differentiated into cap-like and stem-like structures, even though it is only 300 μm tall, and this is even more evident in the fourth section (700 μm tall), which has young gills but no gill cavity. At extreme right is a 1.2 mm tall mushroom primordium, in which the basic 'body plan' of the mushroom (see Figure 13.2) is complete with clear demarcation into veil, pileipellis (cap epidermis), cap, gills (with the beginnings of an annular gill cavity) and stem (with a distinct stem basal bulb which features heavy accumulations of glycogen).

from each other are plants, animals and fungi in matters that influence multicellular development.

Remember that the main eukaryotic kingdoms are thought to have separated in evolution at some single-cell level (see Chapter 2), so the three eukaryote groups that make multicellular structures became distinct from one another long before the multicellular grade of organisation was established in any of them and must have evolved the mechanisms and mechanics of multicellularity quite independently. There is no reason to expect that the mechanisms that have developed will have anything in common. They will, of course, share all those features that clearly categorise them as eukaryotes, but there is no logical reason to expect that these three kingdoms will share any aspect of their multicellular developmental biology. If evolutionary separation between the major kingdoms occurred at a stage prior to the multicellular grade of organisation, then these kingdoms must have 'learned' how to organise populations of cells independently.

The fungal hypha differs in so many important respects from animal and plant cells that significant differences in the way cells interact in the construction of organised tissues must be expected. Inevitably, in many cases these very different organisms needed to solve the same sorts of morphogenetic control problems and evolution may have converged on some common strategies, but the outcomes will be **analogous** (in the evolutionary sense, this means similarity in function and/or structure between organs of different evolutionary origin), rather than **homologous** (a similarity of form or structure due to common descent).

For the most part, though, the many **differences in cell biology** between animals, plants and fungi result in very different emphases in the way these groups manage their developmental biology. Even in lower metazoans, a key feature of embryo development is the ***movement*** of cells and cell populations to remodel an initially unstructured group of cells (Leptin, 2005; Williams & Solnica-Krezel, 2017); evidently, cell migration (and everything that controls it) must play a central role in animal morphogenesis. In contrast, plant cells are encased in walls and have little scope for movement, so their changes in shape and form depend on control of the orientation and position of the mitotic division spindle and, consequently, the orientation and position of the daughter cell wall which forms at the spindle equator (ten Hove *et al.*, 2015).

Fungi are also encased in walls, of course; but two peculiarities, which we have mentioned in earlier chapters, of their basic structural unit, the hypha, result in fungal morphogenesis being totally different from plant morphogenesis. These are that:

- **a hypha grows only at its apex** (see Chapter 4, section 4.4, and Chapter 5, section 5.14);
- **cross-walls form only at right angles to the long axis of the hypha** (Chapter 4, section 4.12, and Chapter 5, section 5.16; Harris, 2001).

The consequence of these features is that no amount of cross-wall formation ('cell division') in fungi will turn one hypha into two hyphae (Field *et al.*, 1999; Momany, 2001; Steinberg *et al.*, 2017). The crucial understanding that makes fungal developmental morphogenesis distinct from that of both animals and plants is that fungal morphogenesis depends on the placement of **hyphal branches**. To proliferate a hypha must branch, and to form the organised structure of a tissue the new hyphal apices must be formed to initiate branches and the position at which the branch emerges from its parent hypha and its direction of growth must be precisely controlled.

Study of multicellular development in fungi (the true analogue of animal embryology) has mostly been done as part of taxonomic studies and has made no contribution to theoretical developmental biology. This is surprising given that so much of what we know about eukaryotic biochemistry, molecular cell biology, cell structure and the cell cycle derives from work with yeast. Still, we can use the concepts and vocabularies arising from developmental study of animals and plants to show how fungi differ.

The word **morphogenesis** is generally used to encompass the development of the body form of an organism. It has been used to describe the yeast–mycelial transition in some fungi, but this is an aspect of ***cell differentiation*** rather than morphogenesis. In animals and plants, the key problems in morphogenesis have always related to the interactions between cells from which the patternings of cell populations arise and from which the morphology of the embryo emerges, and the challenge is to apply this sort of approach to fungi.

Comparison of the way similar functions are controlled in different organisms can reveal whether and how different cellular mechanisms have been used to solve common developmental demands (Meyerowitz, 1999). Although, of course, fungi do not feature in such discussions. We intend to feature them, and we will do this by showing some of the experiments and observations that have led to the understanding we currently have of fungal multicellular development. This will involve delving into some of the older literature and will also show how simple experiments and quantitative observations can contribute to important ***inferences*** about the fundamental processes involved in fungal developmental biology. But don't worry; when we have a sufficiently sound foundation, we'll also deal with the molecular biology.

13.3 THE OBSERVATIONAL AND EXPERIMENTAL BASIS OF FUNGAL DEVELOPMENTAL BIOLOGY

A great many of the experiments we will mention describe development of the mushroom (Figure 13.1, for example) so we should first warn that this is where you suffer the prejudices and preferences of your authors; we focus on *Basidiomycota* because that's our interest!

The mushroom is a spore-dispersal organ that is umbrella-shaped to protect the spore-release process from rain (hydrophobic interactions with a droplet of watery fluid liberate basidiospores (section 10.8; this can't happen in the rain). Figure 13.2 shows a sketch diagram of the typical mushroom structure as seen in a vertical section. The spores form on cells in a hymenium that lines the undersurface of the mushroom cap. The surface area covered by hymenium is increased by the undersurface of the cap being divided into vertical plates (the gills), closely packed tubes (pores) or tooth-like spikes. Increasing the surface area covered by hymenium has selective advantage because it increases the reproductive capacity of the structure (see section 10.8), and it has evolved independently several times in different phylogenies.

The **geometrical proportions** of the stem (short and stocky, or long and slender) vary with the demands of this support function: large caps require more substantial stems, etc. But they also vary with the demands of the ecology and habitat of the species:

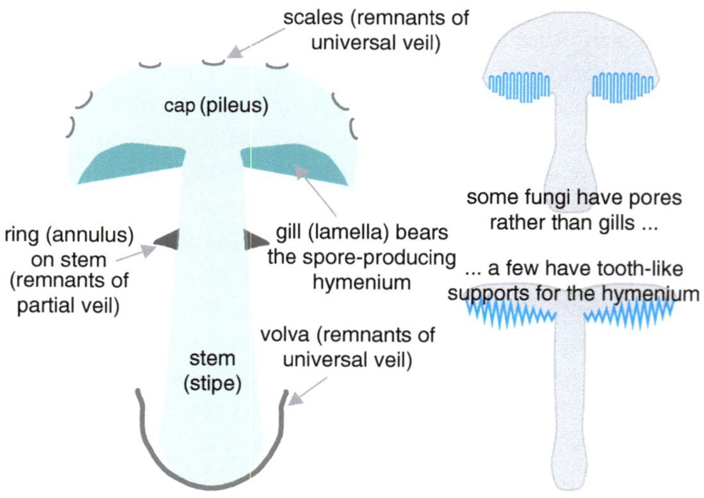

Figure 13.2. Main tissues of a mushroom. The mushroom is our main example, it is a spore-dispersal organ that is basically umbrella-shaped, with the spore-bearing hymenium lining the undersurface of a cap (also called pileus) and supported above the substratum by the stem (stipe), shown in this diagrammatic vertical section. The mature mushroom may also bear remnants of thin tissues (veils) that protect the juvenile stages. The text discusses how this mushroom structure is produced.

short stems may be appropriate in species that form sporophores on bare soil, but those that characteristically come up into copious plant litter or among heavy plant cover may depend on tall stems to lift the cap into an air flow. As well as the geometry, these structural considerations also determine the internal anatomy of the stem: tall and slim stems tend to be anatomically hollow cylinders at maturity; very short and stocky stems tend to be constructed of solid tissue when mature. The stem, however, has more to do than simply support the cap physically; it must also support the cap physiologically. All the nutrients and all the water the cap needs to shape itself and produce its crop of spores are supplied by the mycelium and must be **translocated** through the stem.

The anatomical pattern that we recognise as a 'mushroom' (Figure 13.2) results from differential growth and morphogenesis of the component tissues of the developing mushroom 'embryo' (Figure 13.1). The rest of this chapter will be devoted to showing how these mushroom structures are produced, but we will have to deal with just a few specific cases to do this and we first want to make the point that there are a wide variety of differential growth patterns that generate the tremendous morphological diversity evident among the sporophores of these higher fungi.

RESOURCES BOX 13.1 Latin and Greek Derivations of Technical Terms

Examples of prefixes and suffixes derived from classical languages that are currently used in science can be found on Dr Don Emmeluth's web pages at Armstrong Atlantic State University, Savannah, Georgia, USA, at http://www.angelfire.com/de/nestsite/modbiogreek.html.

You may also be interested in Douglas Harper's *Online Etymology Dictionary*, which provides, in his words: 'explanations of what our words meant and how they sounded 600 or 2,000 years ago' at https://www.etymonline.com/.

13.4 TEN WAYS TO MAKE A MUSHROOM

There seem to be ten ways to make this mushroom shape (Watling & Moore, 1994) and these are illustrated in the diagrammatic vertical sections shown in Figures 13.3–13.9. Sometimes the cap simply grows out of the top of the stem and as the stem increases in height, the cap increases in radius. The hymenium (spore-bearing tissue) forms on the lower surface of the developing cap but is exposed to the elements from the earliest stages. This is called **gymnocarpic** development; *gymnos* is the Greek word for 'naked' and *carpos* is Greek for fruit (Figure 13.3).

Some protection for the young hymenium results when the cap tissues arise shaped like a lid closely applied to the apex of the stem. In this case, hymenial tissue arises in an annular cavity formed between the cap and the stem. This cavity eventually expands to separate cap and stem but in the early stages the tissue (**veil**, from the Greek *velus*, a 'skin or parchment') formed between the stem and the closely applied cap protects the hymenium (Figure 13.4). This is called **gymnovelangiocarpic** development, the word combining *gymnos* for 'naked' (because it becomes naked as development proceeds) and *angio* which is derived from a Greek word meaning 'vessel' or 'receptacle' (particularly one containing seed), because the hymenium is enclosed by the veil when first formed.

In other cases, the margin or rim of the cap is curled over to protect the hymenium, this is known as **pilangiocarpic**

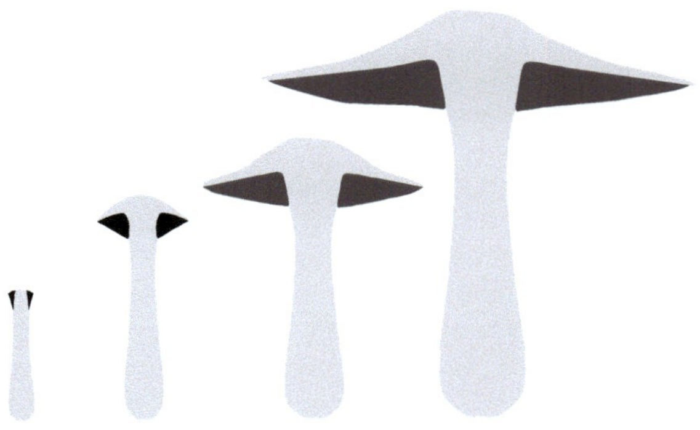

Figure 13.3. Schematic diagrams showing gymnocarpic development, where the hymenium (dark shading in these vertical section diagrams) simply grows out of the top of the stem, being naked at first appearance and developing to maturity exposed to the elements on the sporophore surface.

Figure 13.4. During gymnovelangiocarpic development the hymenium is protected by a much reduced veil, seen only during adolescence, which is formed between the stem and the closely applied cap.

development ('hat-receptacle-fruit' in Greek, see Figure 13.5; the word **pileus** being derived from the Greek *pilos* which was a close-fitting, brimless hat worn by the ancient Greek sailors). Another variation is that the base of the stem (which is called a stipe, the word deriving from the Latin *stipes*, meaning a tree trunk or post) may be elaborated into a protective flap (called a **volva**) that covers the developing hymenium in the early stages (called **stipitangiocarpic** development, or 'stipe-receptacle-fruit' in Greek) (Figure 13.6). The most extreme expression of this latter form of protection is **bulbangiocarpic** development, where the tissue protecting the hymenium is largely derived from the basal bulb of the stem and initially completely encloses the primordium (Figure 13.7).

The key feature in these descriptions so far has been protection of the hymenium, and the protective tissues may have different anatomical origins but are most often called veils. This has led to the recognition of four types of **velangiocarpic** development where one or more special covering tissues (the **veils**) specifically cover the hymenium (Figure 13.8). There may be a

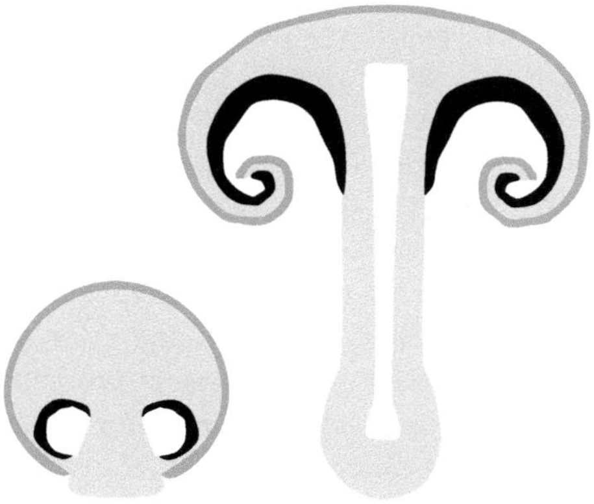

Figure 13.5. In pilangiocarpic development, the hymenium (black in the diagram) is protected by tissue extending downwards from the margin of the cap to enclose the spore-bearing tissue in the earliest stages and then curling in around the cap margin to protect the youngest hymenium as the mushroom matures.

Figure 13.6. Stipitangiocarpic development describes an arrangement in which the hymenium is protected by tissue extending upwards from the stem base. This does not enclose the whole primordium and the cap bursts through as it matures, often leaving a cup-shaped remnant of tissue at the base of the stem which is called a volva (which comes from a Latin word meaning 'a covering').

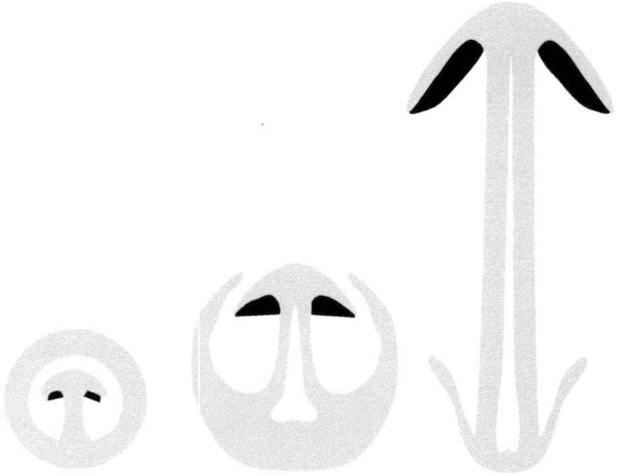

Figure 13.7. At its most extreme, the volva may extend right over the top of the sporophore and completely enclose it. This is called bulbangiocarpic development and is the basic structure of the paddy straw mushroom, *Volvariella volvacea*, and its relatives (see Figure 13.27).

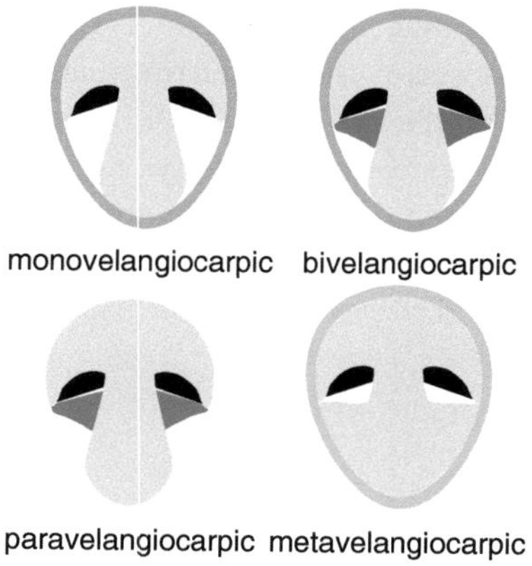

Figure 13.8. Four types of velangiocarpic development shown as diagrammatic vertical sections through immature sporophores. In velangiocarpic sporophores, the hymenium (shown black in the diagrams) is protected by veil tissue (shaded grey) until the spores mature and it then becomes exposed. Monovelangiocarpic types have a single (universal) veil enveloping the primordium. In the bivelangiocarpic type there is an inner (partial) veil (dark grey) in addition to the enveloping universal veil. In paravelangiocarpic types, the veil (dark grey) is reduced and often lost at maturity (the Greek prefix para- meaning 'closely resembling' or 'almost'). In the metavelangiocarpic type a union of secondary tissues emerging from the cap and/or stem forms an analogue of the universal veil (the prefix meta- in this context comes from the Greek and means 'with' or 'among').

single (**universal**) veil enveloping the primordium in **monovelangiocarpic** types; or an inner (**partial**) veil specifically covering the hymenium in addition to the enveloping universal veil in the **bivelangiocarpic** type. There is also a **paravelangiocarpic** type in which the veil is reduced and often lost at maturity; and a **metavelangiocarpic** type, where a union of secondary tissues emerging from the cap and/or stem forms an analogue of the universal veil.

Finally, there are several groups in which the sporophore structure is so modified that it does not develop as a mushroom at all, but may emerge as an even more peculiar sporophore, like a stinkhorn, or a puffball, or a truffle. This is called **endocarpic** development (from the Greek *endon* 'within' or 'inside') and it results in a sporophore in which the ***mature*** hymenium is enclosed or covered over. Three patterns of this sort are shown as diagrammatic vertical sections through mature sporophores in Figure 13.9. These sporophores, most of which are buried in the soil (**hypogeous**) or at least at the soil surface, were for many years assigned to a specific taxonomic group, the gasteromycetes ('stomach fungi', from the Greek *gaster*, meaning 'stomach' or 'belly').

However, it became evident that far from defining a natural taxon, the hypogeous sporophore represented a specialisation along many distinct phylogenetic lines. There are some species that produce hypogeous sporophores as a departure from their normal **epigeous** form to cope with adverse environmental conditions (particularly drought and desiccation).

The above discussion is intended to introduce you to the notion that **differential growth** of different tissues can and does generate a tremendous range of morphological diversity in fungal sporophores. This diversity is amplified when you add features like size, pigmentation, symmetry, regular or irregular relationships between the sizes of the component organs, and whether the hymenium is borne on gills, pores or teeth (and whether these are thick or large, thin or small, and/or numerous or few). If you count up the names assigned to our descriptions above you will find that there are, indeed, ten ways to make a mushroom; but in numerical terms, that's just the start of the equation and we hope our descriptions will enable you to think just a little more deeply about the mushrooms in your garden, or your mushroom risotto or stir fry, or alongside your steak.

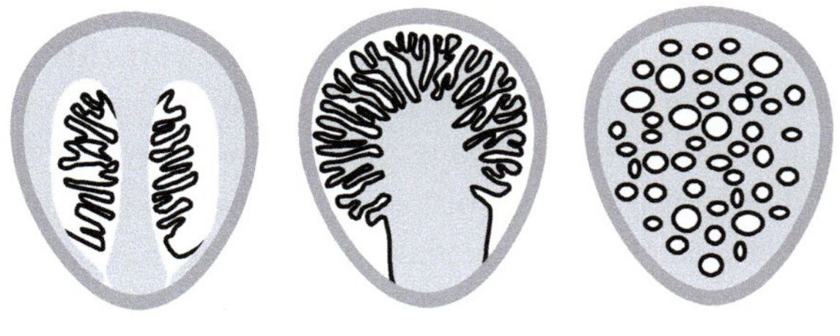

Figure 13.9. Endocarpic development produces a sporophore in which the mature hymenium is enclosed or covered over. Three patterns of this gasteromycetous form of sporophore are shown here as diagrammatic vertical sections through mature fruit bodies.

In and around the British Isles, there are 101 known species of mammal, 603 species of birds, 2,951 species of vascular plant, but more than **15,000 species of fungi**, a number that rivals the insects (22,500 British species). That's biodiversity.

From this point, we are going to concentrate on cellular details of development and most of the account will deal with an ink cap mushroom now called *Coprinopsis cinerea*. It is an unfortunate fact of life for those who work with this fungus that it has been misidentified on several occasions, so several specific names appear in the literature to which we will refer, including *Coprinus cinereus, Coprinus lagopus* and *Coprinus macrorhizus*. They all refer to the same fungus, the different names being the result of misidentifications. Some of the problems of identification are discussed by Moore *et al.* (1979) who also give a general description of the morphological aspects of sporophore development in the organism. Subsequently, Redhead *et al.* (2001) used molecular sequence analyses to revise the taxonomy of the whole group that had been called by the generic name '*Coprinus*' and subdivided it into *Coprinus* (in Family Agaricaceae), and *Coprinellus, Coprinopsis* and *Parasola* in the new Family Psathyrellaceae. This is how *Coprinus cinereus* became *Coprinopsis cinerea*.

13.5 COMPETENCE AND REGIONAL PATTERNING

Some of the processes involved in development of sporophores and other multicellular structures by fungal mycelia have been mentioned in earlier chapters. Production of a multicellular structure is an aspect of mycelial differentiation and the mycelium must have access to the necessary resources. The formal description is that the mycelium must be **competent** to undertake production of a multicellular structure.

In practical terms, this essentially means that the exploratory mycelium has found and captured sufficient of its substratum to have internalised, in the form of **accumulated nutritional stores**, adequate supplies for all the synthetic processes involved in further development and morphogenesis. The essential point is that nutrients are no longer outside the hyphal system and *potentially* available, but they are inside the hyphal system and *immediately* available. This change in balance results in shifts in some of the regulatory circuits that then lead to cell and tissue differentiation.

In Chapter 10, we mentioned differentiation of mycelium in response to environmental and other influences (section 10.2); and in Figure 10.18 we showed a flow chart summarising the processes involved in development of sporophores and other multicellular structures in fungi. The first two paragraphs, above, emphasised physiological aspects, but this flow chart reminds us that there is also a **genetic component to competence**. Section 10.4 describes the genetic circuit leading specifically to competence in *Aspergillus* conidiophores (Noble & Andrianopoulos, 2013; Lee *et al.*, 2016). Vegetative compatibility systems can also contribute to mycelial interaction (section 8.5), but it is usually governed by the mating type factors (Chapter 9) and results in formation of a sexually compatible mycelium, which is ***competent*** to make sporophores. However, in the higher fungi, even a haploid mycelium can have several alternative developmental pathways open to it: continuation of hyphal growth, production of asexual spores, and/or formation of asexual multihyphal structures (stromata, sclerotia, etc.), so a compatible mating opens up the additional pathways associated with progress into the sexual cycle. Choice between these is often a matter of the impact of sometimes very localised environmental conditions on a mycelium which has become competent to embark on any of several developmental pathways.

As is indicated in the Figure 10.18 flow chart, the onset of multicellular development is usually signalled by some sort of disturbance that causes a **check or restraint** to normal (exploratory, invasive) vegetative growth. This restraint might be imposed by a nutritional crisis: perhaps the substrate becomes exhausted of nutrients (of many or even only one or two crucial ones; the preferred carbon source and the preferred nitrogen source, for example). Now, this does not mean that the mycelium is starving, because the mycelium has absorbed and stored many nutrients, but it does mean that the *balance* of nutrient supply is changed, and this will result in a change of regulatory pattern within the hyphae.

Other major signals that have been the subject of numerous laboratory experiments are **temperature shocks, light exposure, edge encounters and physical injury**, all of which can also be related to natural events in the normal habitat. Many of the *Basidiomycota* require a decrease in temperature to fruit, which may reflect their adaptation to seasonal temperature fluctuations. Response to illumination patterns may reflect response to day light or day length, or even a growth habit in which the mycelium grows through a dark substratum (like a litter layer, pile of plant debris or herbivore dung, etc.) to eventually emerge at a light-exposed surface. Exposure to ***blue light induces hyphal knot formation*** in *Coprinopsis cinerea* mycelia grown on low-glucose media but not in mycelia grown on high-glucose media; on low-glucose media, many hyphal knots are visible near the edge of the colony one day after a 15-minute exposure to blue light. Transcriptome analysis revealed a two-stage response to illumination. Several genes are upregulated by 1 h of blue light exposure in the mycelial region where the hyphal knot will be developed; this upregulation is not influenced by nutrients. These genes are thought to be essential for induction, but not sufficient for development of the hyphal knots. In the second expression stage, genes involved in the architecture of hyphal knots are upregulated after a 10–16 h blue light exposure if the mycelia are cultivated on low-glucose media (Sakamoto, 2018; Sakamoto *et al.*, 2018).

In the laboratory 'edge encounter' usually means that the mycelium has reached the edge of the Petri dish, but in nature it may equate to encountering rocks in the soil or growth through a piece of timber reaching the surface. Physical injury in the wild may be inflicted by adverse weather conditions or disturbance by animals and may act by resulting in exposure of previously protected mycelium to any of the other environmental fluctuations (that is, a burrowing animal, say, may separate a mycelium from its food base, or leave it exposed to light or temperature stresses).

Sporophores do not arise from single cells, although this was believed to be the case for many years. Rather, **initiation of multicellular structures**, like sporophores and sclerotia, involves aggregation of cells from different sources as a result of hyphal congregation that is the direct result of the most fundamental change in hyphal growth pattern: namely that in vegetative mycelia hyphae show negative autotropism (hyphal tips grow away from each other), whereas to form a condensed multicellular tissue component hyphal tips must show positive autotropism and grow towards each other (Matthews & Niederpruem, 1972; Waters *et al.*, 1975; Van der Valk & Marchant, 1978).

Unlike animal embryos, fungal multicellular structures evidently do not normally consist of a population of cells that are the progeny of a single progenitor but are assembled from contributions made by a number of **cooperating hyphal systems**. Indeed, in a few cases sporophore chimeras have been observed (the word 'chimera' comes from the name of a Greek mythological monster, which was made of the parts of several animals; it now means 'composed of genetically distinct cells'). Kemp (1977) described how mushrooms of '*Coprinus*' formed on horse dung collected in the field but incubated in the laboratory could consist of two species (then identified as *Coprinus miser* and *Coprinus pellucidus* but now named *Parasola misera* and *Coprinellus pellucidus*). The two have distinctively different basidiospore morphologies and the hymenium comprised a mixed population of basidia bearing the two different sorts of spores, so the chimera extended throughout the sporophore.

The most highly differentiated cells occur on the outside of the tissue blocks that make up fungal multicellular structures (Williams *et al.*, 1985). Thus, major morphogenetic events in fungi, like those in animals and plants, are associated with **tissue surfaces** and their 'epidermal' layers of cells that separate adjacent tissues. As the adult tissues are demarcated in even the earliest sporophore initials (as shown in Figure 13.1), fungi are like animals and plants in having a basic 'body plan' which is established very early in development. The pattern of tissue distribution that makes up the 'body plan' is established sequentially by the processes of **regional specification, cell differentiation** and **cell coordination**. It is likely that all these processes are orchestrated by morphogens and/or growth factors; there is some indirect indication of this, although there is no direct evidence for the existence of morphogens in the differentiating sporophore primordium similar to the growth hormones and growth factors that are so important in animal and plant development. The very young hyphal tufts or initials of vegetative or sporophore structures are simply composed of a mass of tissue which is made up of apparently loosely interwoven hyphae. Very soon, however, tissue layers involved in rapid cell formation (that is, organised rapid **branch formation**) become recognisable, which demarcate the major tissue layers of the adult organ. To create such histologically distinct regions (recognisable as the future tissues of the adult organ) some organisation is being imposed on the initially homogeneous interwoven hyphae.

13.6 THE *COPRINOPSIS* SPOROPHORE: MAKING HYMENIA

In *Coprinopsis cinerea*, differentiation of the gills (on which the hymenium will be formed) becomes evident in a region corresponding to the boundary between stem and cap tissues before any annular cavity (Figure 13.1). The gills arise as vertical ridges of small, closely packed hyphal branches, and as this wave of differentiation moves through the tissue towards the outer surface of the cap it leaves behind two organised plates of columnar cells which constitute the primordial hymenia of adjacent gills separated by a developing gill cavity. An annular cavity is seen only after gills are well formed and it seems to appear because of inflation of hyphae in the cap forcing the cap away from the stem.

The central tissues of the first-formed (primary) gills remain intimately connected by interwoven hyphae to the periphery of the stem for a considerable time, becoming freed only in well-developed primordia. From this early stage right through to final maturation of the mushroom, developmental changes proceed along two major vectors: from inner edge of the gill (that is, the edge closest to the stem) towards the outer edge (that closest to the exposed surface of the cap) and from the cap margin towards the cap apex (Rosin & Moore, 1985a). These **morphogenetic polarities**, established during the earliest steps in morphogenesis, are maintained throughout sporophore development.

A mature, fully differentiated mushroom cap of *Coprinopsis cinerea* consists of a thin layer of cap flesh (the 'epidermis' of which is called the **pileipellis**) bounded exteriorly by the veil cells and with the gills suspended on the inner side. The gills are described as developing parallel to one another and retain this arrangement throughout development; strictly speaking, they are arranged along the radii of the cap, but they are so thin and numerous as to appear parallel when small segments are examined (Figure 13.10). The gill surface is the **hymenium** composed of three highly differentiated cell types: **basidia** (the cells on which spores are formed, name derived from the Greek *basis* for 'step or pedestal' (of a column) or 'base', because they come to support the spores); **cystidia** (large inflated cells, name derived from the Greek *kystis*, meaning 'bladder'); and **paraphyses** (name derived from the Greek preposition *para* for 'beside' and *physis* 'nature'), which are sterile secondary cells, though this description plays down their importance as they eventually form a pavement that provides most of the mechanical structure of the mature gill.

All these cells arise as ***hyphal branches*** from the central layer of the gill, the **trama**, the constituent cells of which

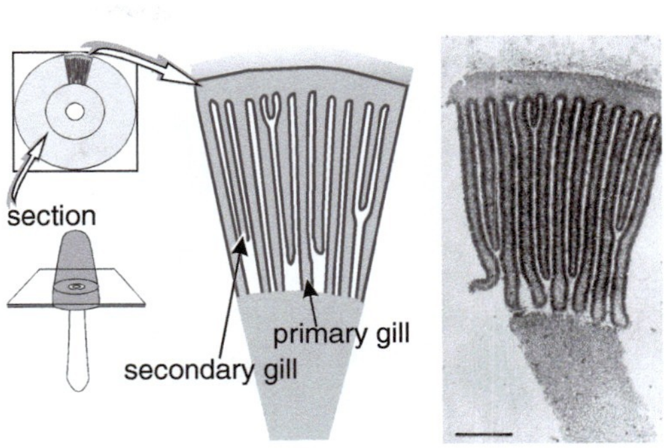

Figure 13.10. Segment of a transverse section of a mushroom cap of *Coprinopsis cinerea* at a late stage of development. Preparation of the section is diagrammed at left. In the section itself the stem is the segment of tissue at the bottom, and the more diffuse tissue at the top is the remnant of the universal veil. The gills are the seemingly 'parallel' (actually, radial) tissue blocks in the centre of the image. The hole in the centre of the section (at top left) represents the stem lumen. At this stage in development the cap encloses the top of the stem and the gills are held along the long axis of the stem, so transverse sectioning cuts across the gill plates and displays them as a stack of gill profiles. The hymenium contains closely packed and more heavily stained cells so appears as a darker line in the photographic image. Hymenium is forming over the margins of primary gills where they contact the stem, though there is still evidence for hyphae extending between the primary gill margins and the stem. Note the presence of both primary gills (extending from the stem outwards), and secondary gills (which do not extend over the full radius). How do you think secondary gills are formed? Scale bar on the photomicrograph, 100 μm.

maintain an open and basically hyphal structure. The 'ridges of small, closely packed hyphal branches' that form the first hymenium of *Coprinopsis cinerea*, and effectively carve out the main gill plate, are mostly young basidia (often called **protobasidia**), which will eventually undergo meiosis (see Figure 9.5) and produce basidiospores, with cystidia scattered among them. The latter arise as the terminal compartments of unbranched tramal hyphae (Figure 13.11); from the very first, cystidia are much larger cells. They span the gill cavity, connecting the hymenia of two different gill plates and serve a mechanical function, spreading the tension loads caused by cap inflation through all the gills (discussed below). Only about 8% of the tramal hyphal branches become cystidia; the rest become protobasidia (Horner & Moore, 1987).

Paraphyses originate as **branches** from the hyphal cells immediately beneath the basidia, and the branches force their way into the basidial layer, inflating as they do so (Figure 13.11). About 75% of the paraphyses have inserted by the time meiosis is completed, the rest insert at later stages of development. In contrast, basidia do not increase in number once paraphysis insertion had started (Rosin & Moore, 1985b).

Evidence for local control of morphogenesis has been obtained from a comparison of the distributions of **cystidia** on adjacent hymenia in the *Coprinopsis* mushroom (Horner & Moore, 1987).

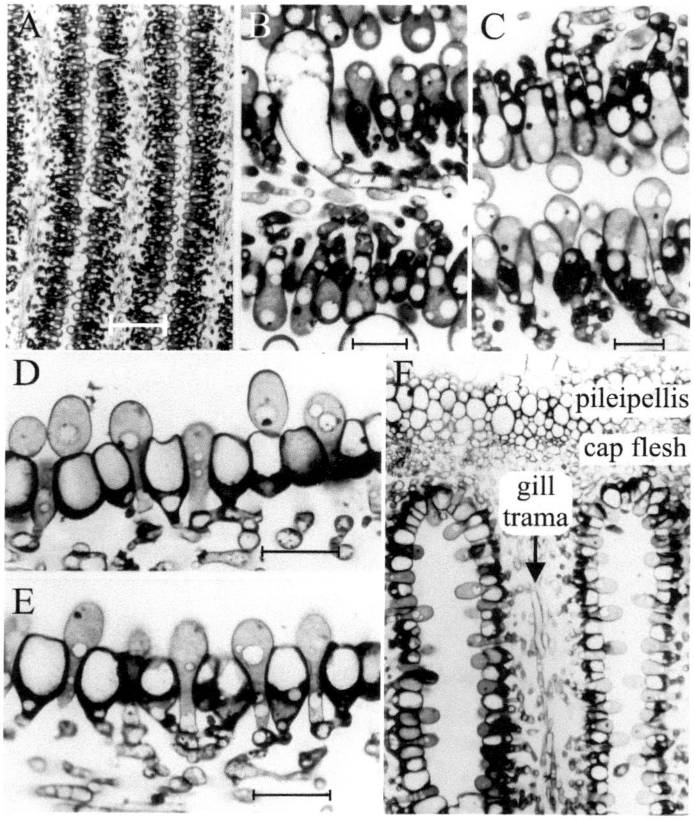

Figure 13.11. The differentiating hymenium of *Coprinopsis cinerea* in light micrographs of sectioned tissue. (A) (scale bar: 20 μm), (B and C) (scale bars: 5 μm) show the hymenium of a primordium 1 to 2 mm tall (like that shown in the extreme left photograph of Figure 13.1). They reveal that at this stage the hymenium consists of a closely packed layer of young basidia, with occasional large, but still young, cystidia. The branches from sub-basidial cells that become paraphyses are evident as more densely stained hyphal tips, beginning to force their way into the basidial layer. (D, E) (scale bars: 10 μm) and (F) show views of sections of a later stage in which paraphysis insertion into the hymenium has been accomplished and paraphyses have begun to inflate. Their connection to sub-basidial cells is still very clear. (F) shows that the pileipellis ('epidermis' of the cap) is a layer made up of closely packed cells, like the hymenium. However, the cap flesh is a very open tissue with large intercellular spaces, and the same is true of the gill trama (the gill flesh) in which the hyphal nature of the tissue is clear. Photomicrographs by Isabelle V. Rosin.

The large, inflated, cystidial cells span the gill cavity, impinging upon the opposing hymenium of the neighbouring gill. There are, as a result, certain classes of cystidial relationship (as seen in microscope sections) that are open to numerical analysis. As you can see in the photomicrographs in Figure 13.11, cystidia spanning the gill cavity may be 'distant', that is having other cells separating them, or 'adjacent', with no intervening cells; and, in either case, both cystidia may emerge from the same hymenium (described as *cis*) or from opposite hymenia (*trans*) (Figure 13.12).

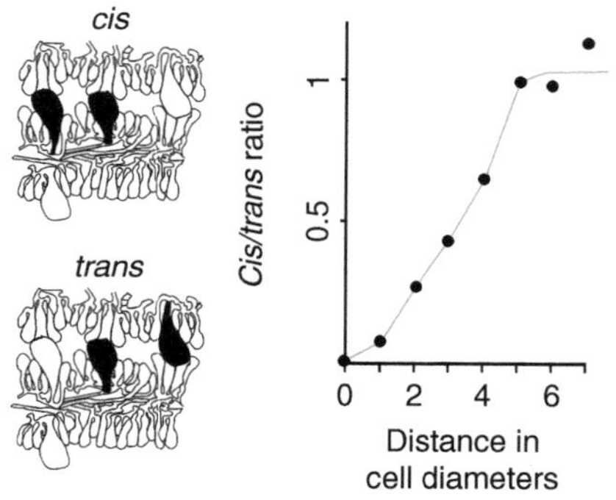

Figure 13.12. Cystidium distribution in the *Coprinopsis cinerea* hymenium. Drawings at left show the categorisation of neighbouring pairs of cystidia in micrographs as either *cis* (both emerge from the same hymenium) or *trans* (emerging from opposite hymenia). The graph compares the frequencies of these two types over various distances of separation and shows that closely spaced *cis* neighbours are less frequent than closely spaced *trans* neighbours, implying some inhibitory influence over the patterning of cystidia emerging from the same hymenium.

If the distribution of cystidia is entirely randomised, then the frequency of adjacent pairs will depend on the population density (which can be measured as a function of the distance between neighbouring cystidia) and there will be an equal number of *cis* and *trans* in both the distant and adjacent categories. However, quantitative data from serial sections of a primordium showed a positive **inhibition** of formation of neighbouring cystidia in the same hymenium (Figure 13.12). Evidently, formation of a cystidium actively lowers the probability of another emerging from the same hymenium in the immediate vicinity (Horner & Moore, 1987). The range of the inhibitory influence extended over a radius of about 30 μm and was strictly limited to the hymenium of origin.

It may be that cystidium formation is activated by the concentration of a component, possibly water vapour, of the atmosphere (assuming this to be gaseous rather than fluid) in the gill cavity immediately above the developing hymenium. The distribution pattern of cystidia would consequently be dependent on interplay between activating and inhibiting factors. Such a patterning process is open to analysis and simulation using the **activator–inhibitor model** (also called the morphogenetic field model) developed by Meinhardt and Gierer (1974) and Meinhardt (1976, 1984). This is an elegantly simple model which suggests that morphogenetic pattern results from interaction with just two compounds: an activator, which autocatalyses its own synthesis, and an inhibitor, which inhibits synthesis of the activator. Both diffuse from the region where they are synthesised; the inhibitor diffusing more rapidly and consequently preventing activator production in the surrounding cells. The model readily accounts for stomatal, cilial, hair and bristle distributions in plants and animals, and a wide variety of patterns can be generated in computer simulations by varying diffusion coefficients, decay rates and other parameters (Meinhardt, 1984). Successful application of this to fungi as well as to plants and animals concentrates attention on the fact that the distribution of stomata on a leaf, bristles on an insect and cystidia on a fungal hymenium have a great deal in common at a fundamental mechanistic level. In other words, these examples are expressions of general rules of pattern formation that will apply similarly to all multicellular systems even though the molecular mechanisms involved are likely to be different.

The descriptions we have just given illustrate several developmental control events that indicate exquisite regulation of morphogenesis.

- First, these differentiations are occurring in the gill tissues of the cap while an alternative set of differentiations are happening in the stem tissues, just a few hundred micrometres distant. The organs of the sporophore are strongly specified regions. We wonder what it is about the control mechanisms that keeps them localised.
- Second, these are ***fungal*** tissues, so the cellular elements involved are **hyphal branches** and **hyphal compartments** (the cells bounded by septa). The branches are formed in such frequency that images such as those in Figure 13.11 highlight dramatically the changed growth pattern exhibited by a mycelium during sporophore formation. In vegetative growth on the surface of an agar medium, leading hyphae of this dikaryon produce branches at about 73 μm intervals, but as you can see from Figure 13.11 the branching frequency at the hymenium is 10- or 20 times greater than that. We wonder what makes them branch so much.
- Third, there are at least four cellular differentiation pathways on display here: undifferentiated hyphae of the trama, basidia, cystidia and paraphyses (we'll refer to a fifth cell type, **cystesia**, below); all the cell types mentioned arise as branches from the tramal hyphae, yet despite their relationship as sister branches of the same hyphal system, cells separated only by a dolipore septum clearly follow totally different pathways of differentiation (Figure 13.13). We wonder how alternate pathways of differentiation are regulated on the two sides of such a septum.
- Fourth, all of the cells that make up the hymenium are derived from hyphal branches, but their tips do not grow continuously, instead their extension is brought under control and they stop extending and differentiate. Further, cessation of extension seems to be coordinated across the entire population of branches so that the surface of the hymenium ends up as a uniform layer. We wonder what mechanism(s) cause all the probasidial branches to stop extending at the same time, and yet allow the (few) cystidial branches to extend across the gill cavity and collide with the opposing hymenium.
- Fifth, the ***direction*** of extension of the hymenial branches is also uniform; look at Figure 13.11A and you'll see hundreds (in life, hundreds of thousands) of branches from the tramal hyphae all growing upwards to make one hymenium layer, and just a few micrometres distant from these an equal number of branches from the ***same*** tramal hyphae all growing downwards to make the other hymenium layer of the ***same*** gill. We wonder what makes the branches in the same layer ***all*** turn in the same direction.
- Finally, there is a matter of timing; the first population of branch tips become protobasidia and cystidia, and ***after*** they have been specified branches emerge from the first cell (hyphal compartment) beneath each protobasidium and specifically grow into and between the established layers of protobasidia to create the paraphyseal pavement. This process is also coordinated in time and in position across a large population of independent hyphal branches.

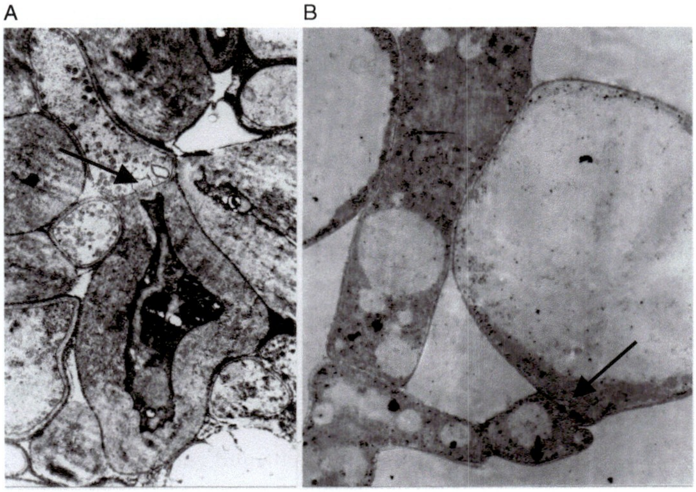

Figure 13.13. Transmission electron micrographs of differentiation across the dolipore septum in *Coprinopsis cinerea*. (A) is a section of the medulla of a sclerotium in which thick walls and thin walls occur in hyphal compartments either side of a septum (arrow). (B) is a section of an immature hymenium in which the enormously inflated paraphysis shares a dolipore septum (arrow) with a cell below which is of more normal hyphal appearance.

The hymenium of *Agaricus bisporus* adds another example to our catalogue of diversity **by using basidia in two different ways**: for their primary purpose of sporulation and to serve as structural members. Although the *Agaricus* hymenium is said to contain sterile cells, Allen *et al.* (1992) found that most of the constituent cells of the hymenium were protobasidia. Protobasidia with single fusion nuclei were in the majority throughout the life of the mushroom, right through to senescence. Progress through meiosis was slow and at any one time, even in old mushrooms, only a very small minority of basidia had four daughter nuclei, suggesting that in most protobasidia **meiosis was arrested** in meiotic prophase I and the arrested protobasidia were kept in that state so they could serve as structural elements in the hymenial layer.

13.7 *COPRINOPSIS* AND *VOLVARIELLA* MAKING GILLS (NOT FORGETTING HOW POLYPORES MAKE TUBES)

There is a fifth cell type in the hymenium of *Coprinopsis cinerea* and its origin illustrates yet another phenomenon found in other organisms. At early stages in growth of the cystidium across the gill cavity the cell(s) with which the cystidium will come into contact in the opposing hymenium are indistinguishable from their fellow protobasidia. However, when the cystidium comes firmly into contact with the opposing hymenium, the hymenial cells with which it collides develop a distinct granular and vacuolated cytoplasm. This suggests a contact stimulus aborting continuation of basidial differentiation in those cells and setting in train an alternative pathway of differentiation leading to an adhesive cell type (called the **cystesium**, a name derived from 'cystidium' plus the Latin root of 'adhere'; Horner & Moore, 1987) (Figure 13.14). The cystidium–cystesium adhesion is very strong and, indeed, is based on joint wall synthesis (shown in Figure 13.14C).

An interesting description of the consequence of the strength of this joint wall synthesis in '*Coprinus atramentarius*' is recorded by Buller (1910):

> When one succeeds in forcibly separating parts of two neighbouring gills, one finds that although the cystidia have mostly separated from one of the gills at their apical ends and have thus remained attached to the other gill at their basal ends …., not infrequently the reverse happens; the cystidia … have broken away at their basal ends and have remained attached at their apical ends …

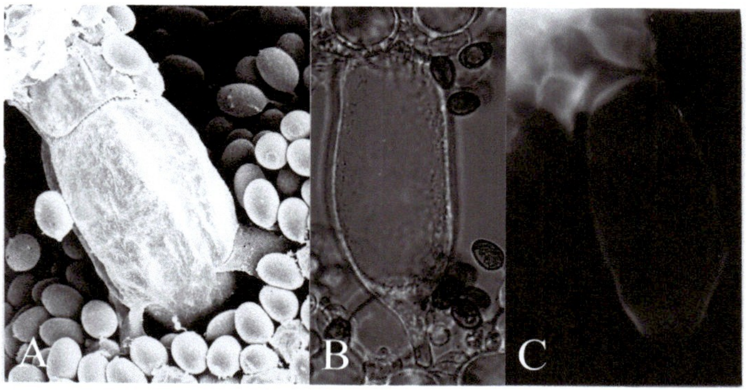

Figure 13.14. Cystidia and the adhesive cystesia of *Coprinopsis cinerea*. A panel of three images all showing a cystidium spanning the gill space and fused with cystesia in the opposite hymenium. (A) is a critical-point-dried scanning electron microscope specimen showing the junction line between cystidium (below) and cystesium (above). (B) is a bright field light micrograph of a cystidium and its cystesia (above), and (C) shows the same cell under UV-excitation of the fluorescence of calcofluor white, revealing newly synthesised chitin in the walls by which the cells adhere. Thus, a cystidium spanning between two hymenia has its connections to both hymenia reinforced. Scale bar, 20 μm. Photomicrographs by Professor S.-W. Chiu, Chinese University of Hong Kong.

Such graphic descriptions indicate that it has been known for a very long time that adhesion between the cystidium and the cystesium can be so strong as to exceed the breaking strain of the subhymenial hyphal branching system causing cells to be torn out of their hymenium of origin. What could be the function of such a singularly strong adhesion? The answer to this question starts with the observation that when first formed the primary gills of *Coprinopsis cinerea* appear as convoluted, folded plates (Figure 13.15).

For such **convoluted gills** of *Coprinopsis cinerea* to end up 'parallel', it is envisaged that as the cap expands the gills are tensioned into the regular parallel arrangement (as shown in Figure 13.10). The process is accomplished through the anchorages provided by the connection of primary gills to the stem and by cystidium–cystesium pairs acting as tension elements, interconnecting gill plates into a fully tensioned structure around the stem as cap expansion pulls the gills into shape. The tension is a geometrical consequence of the way the mushroom matures. The primordial cap in *Coprinopsis* encloses the top of the stem (diagrammed in Figure 13.10) with the gills arranged as vertical plates radially around the stem. Simple measurements show that as typical mushroom primordia of *Coprinopsis cinerea* grow from 1–34 mm in height, the circumference of the stem increases by a factor of 9, while the outer circumference of the cap increases by a factor of 15. As the primary gills are connected to both surfaces they will inevitably be stretched by the outer surface being increased more than the inner, and the regular distribution of strongly adhering cystidium–cystesium pairs will act as **tension ties** communicating and equalising these tension forces through all of the gills.

Figure 13.15B also shows that embryonic gills of *Volvariella* are convoluted in very young primordia but this species is quite different from the coprinoid type (compare Figure 13.7 with the diagram in Figure 13.10). The overall developmental programme for all agaric gills (using the word 'agaric' in the widest sense to mean any 'mushroom' with gills) seems to be that primary gills arise as ridges

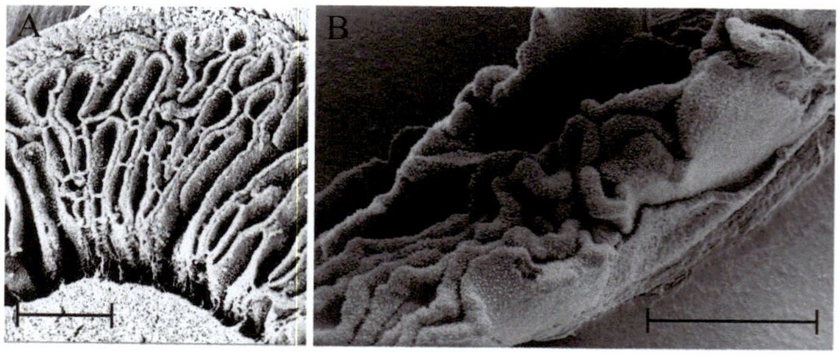

Figure 13.15. 'Embryonic gills' in both *Coprinopsis cinerea* (A) and *Volvariella bombycina* (B) are clearly seen to be convoluted in these scanning electron microscope preparations. The scale bars, 0.5 mm so these are pieces of tissue from very small mushroom primordia, just a few mm in overall size from organisms that produce mature mushrooms several cm in size. Photomicrographs by Professor S.-W. Chiu, Chinese University of Hong Kong.

on the lower surface of the cap, projecting towards the stem, and secondary and tertiary gills are added ***whenever and wherever space becomes available***. Generation of additional gills occurs by bifurcation of an existing gill either from one side or at the free edge, or by folding of the hymenium layer near or at the roots of existing gills. Application of these 'developmental rules' (particularly the 'whenever and wherever space becomes available' rule) inevitably gives rise to the sinuous (labyrinthiform) gill pattern illustrated in Figure 13.15B as a normal transitional stage to the regularly radial pattern so often seen in mature agaric mushrooms.

Transformation of the convoluted gills into regularly radial ones is probably accomplished by cell inflation in the gill trama of *Volvariella*. The mature hymenium of *Volvariella* (Figure 13.16) is a layer of tightly appressed cells, and the trama of the broken gill shown in Figure 13.16 is full and crowded with greatly inflated cells. Reijnders (1963) and Reijnders & Moore (1985) emphasised the role played by cell inflation in driving developmental events in fungal structures, so it doesn't take much to suggest that inflation and growth of tramal cells in gills enclosed by the hymenial 'epidermis' will generate compression forces which will effectively inflate, and so stretch, the embryonic gills to form the regularly radial pattern of the mature cap. Thus, in this case the gill is inflated into shape and, in engineering terms, the mature gill is a **stretched-skin construction** which owes its structural strength to the combination of compression in the trama and tension in the hymenium (Chiu & Moore, 1990b).

The direction of gill development has been a matter of controversy since the start of the twentieth century. The crucial question is whether gills generally extend by growth at their edge or at their root (or both, or neither). This is not an esoteric point but an important question about which region is growing and therefore the location at which developmental control is exercised. The question can be easily answered for conventional agaric fungi by placing ink marks at measured intervals on immature mushrooms and observing what happens to them as the mushrooms mature (Figure 13.17). Experiments done by painting water-based ink or water-colour paint onto the live tissue with minimum disturbance to the young mushroom of *Volvariella* (Chiu & Moore, 1990b) showed ink marks placed near the cap apex remained close together, whereas marks originally close together in the mid-regions of the outer surface of the cap became widely separated (and were made more diffuse) by subsequent growth. Also, ink marks placed on the cap margin and those placed on the free edges of the gills remained at the margin or the free edge. These show that:

(a) the greatest contribution to cap expansion occurred in an annulus (that is, a ring) some way in from the margin and not extending to the apex;
(b) the free edges of the gills remained intact and were not replaced;
(c) the margin of the cap was similarly not replaced though its circumference increased.

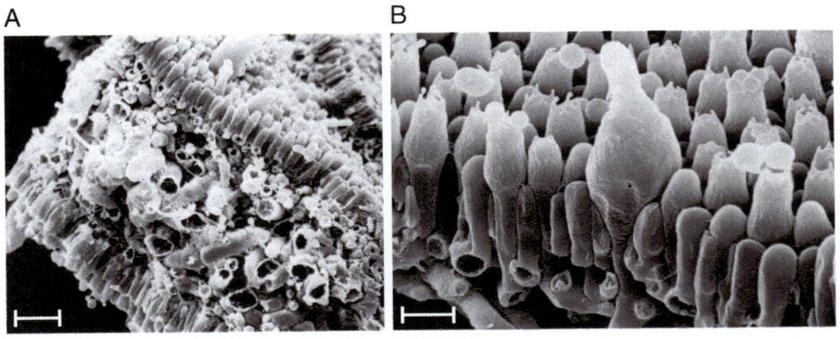

Figure 13.16. The mature gill of *Volvariella bombycina* in scanning electron microscope images. (A) is a view of the exposed trama of a gill broken while frozen (scale bar, 20 μm), and (B) is a magnified view of the hymenium (scale bar, 5 μm). Note that the hymenial cells are closely packed and interlocked to form a coherent layer, like the tightened skin of a balloon, stretched over a gill trama which contains a mass of greatly inflated cells. The gill is inflated into shape and, in engineering terms, is a stretched-skin construction which owes its structural strength to the combination of compression in the trama and tension in the hymenium. Photomicrographs by Professor S.-W. Chiu, Chinese University of Hong Kong.

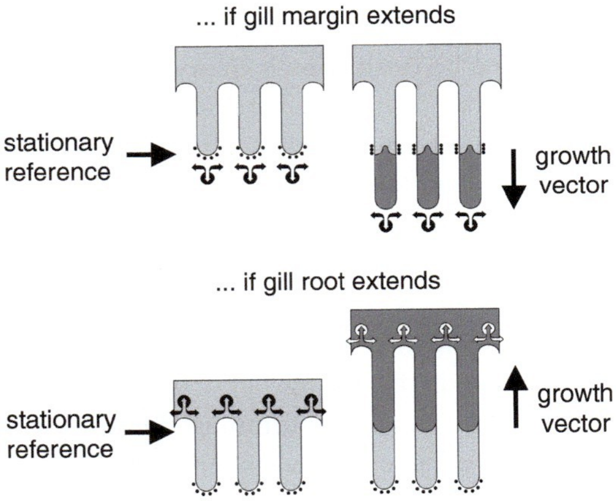

Figure 13.17. Gill formation in *Volvariella* (Chiu & Moore, 1990b). Diagrams of marking experiments done by placing ink marks on gill margins of a sporophore of *Volvariella* when it had a 10 mm diameter cap and observing the marks at maturity, when the cap diameter had increased to about 50 mm. The diagrams show the profiles of three gill plates; ink marks are shown as arcs of small black dots on the margin of the 'immature' gills. New tissue formed during maturation is shown shaded dark grey. The two diagrams illustrate the expected effects on the disposition of ink marks of different strategies for gill growth. If the gills grow at their margin (top panel), hyphal tips will grow beyond the ink marks and the ink will become buried (and hidden from view) within the gill spaces. If the gills extend by growth of their roots into the cap tissue (bottom panel), the original margins remain intact as the gill organisers 'excavate' the gill spaces. In the real experiment the ink marks remained *in situ* through a 5-fold increase in size, it is concluded that *Volvariella* gills extend at their roots; the same hyphal tips that were located at the gill margin in the immature sporophore remained at the gill margin in the mature mushroom.

It's a simple experiment, but it has decisive implications. The crucial observation is that ink marks placed on the free edges of the gills remained at the free edges even though the gills can increase in size by several millimetres as it matures. It means that the gill edge is fully differentiated; the gill edge is not a growth site, so ink marks placed upon it are not overgrown as the gill matures because the same hyphal tips that were located at the gill margin in the immature sporophore remained at the gill margin throughout maturation. Thus, it follows from observation (b) that individual gills of *Volvariella* do not grow by extension at their free edge, and from (c) that the cap margin is not a growth centre for radial extension. So now you should be able to answer the question about secondary gills posed in the caption to Figure 13.10.

The inevitable conclusion is that agaric gills extend in depth by growth at their roots and by insertion of hymenial elements into their central and root regions. The cap extends radially, presumably by insertion of new hyphal branches alongside the older hyphae in a broad region behind the margin. Thus, the hyphal tips which form the cap margin when it is established at the very earliest stage of development remain ***at*** the margin. They do not continue to grow apically to extend the margin radially, nor are they overtaken by other hyphae; instead they are 'pushed' radially outwards by the press of fresh growth behind, and they are joined by fresh branches appearing alongside as the circumference of the margin is increased. Similar experiments have been done with sporophores of *Pleurotus pulmonarius*, marked at the gill margins with the vital stain Janus Green: the originally marked gill margins remained intact although further growth of the cap periphery involved production of entirely new (unstained) tissue which extended the cap periphery radially outwards. Evidently, these gills also grow at their root.

As we have noted, the young *Coprinopsis* mushroom has a different geometry from a conventional 'mushroom' because the primordial cap completely encloses the top of the stem, and **primary gills**, from their formation, have their inner (tramal) tissue in continuity with the outer layers of the stem. We have also stated that both stem and cap circumferences increase by a factor of around ten (more at the cap margin, slightly less at the stem). Now, if the (stem) surface to which the gills are attached is expanding, then the gills would be expected to be prone to widening during maturation (Figure 13.18).

However, mature gills are narrow so this tendency to widen as the circumference increases must be compensated by increase in number of gills (Figure 13.19). How this is accomplished is indicated in images like Figure 13.20.

This shows formation of a new gill cavity and its bounding pair of hymenia ***within*** the trama of a pre-existing primary gill. The Y-shaped gill structure is the crucial observation. There seems to be a formative element at the fork in the Y where the change in structure occurs from the randomly intertwined tramal hyphae with large intercellular spaces to the highly compacted hymenial plates separated by the (new) gill cavity. This

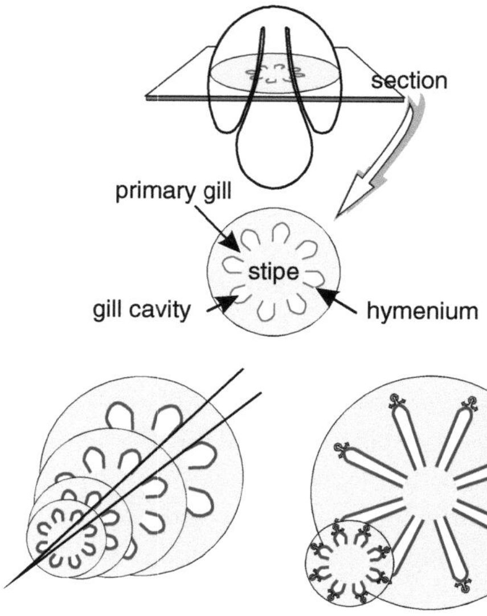

Figure 13.18. Primary gills of *Coprinopsis* are connected to the stem. The top panel is intended to orient the reader and to illustrate the basic layout of a transverse section of the primordial pileus. If these geometrical relationships remained unchanged during expansion, gill thickness would increase greatly for two reasons: (a) because the primary gills are connected to the stem circumference, increase in stem circumference would be accompanied by increase in gill thickness (bottom left panel); (b) because gill organisers migrate radially outwards (generating the branching pattern which forms the protohymenia as palisades bounding an incipient fracture plane as illustrated in Figure 13.21) the gills would inevitably become thicker as their radial paths diverge (bottom right panel). The solutions to these problems are illustrated in Figure 13.19.

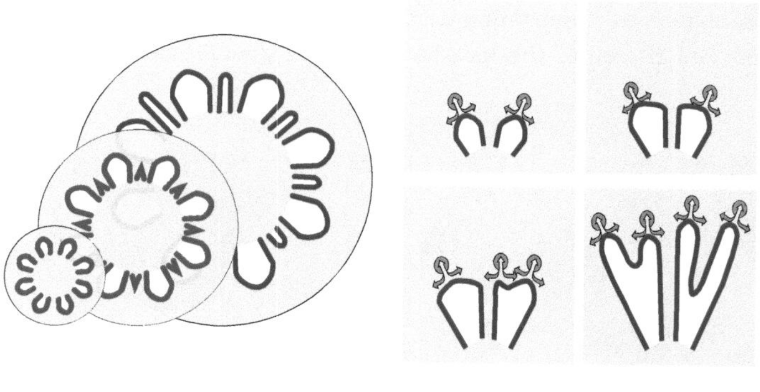

Figure 13.19. Generation of new primary and secondary gills in *Coprinopsis*. At left: increase in the thickness of the primary gill at its junction with the expanding stem is compensated by the appearance of new gill cavities within the trama of the original primary gills. At right: the formative element in outward extension of the gill is the gill organiser located in the tissue which borders the outermost part of the gill cavity. As their radial paths diverge, this part of the gill cavity expands tangentially until sufficient space exists between neighbouring gill organisers for a new organiser to appear between them. Continued outward migration of parent and daughter gill organisers creates a secondary gill between them. These two mechanisms are not alternatives; they occur together to generate the community of radial, narrow gills which characterise the mature *Coprinopsis* mushroom (see Figure 13.20).

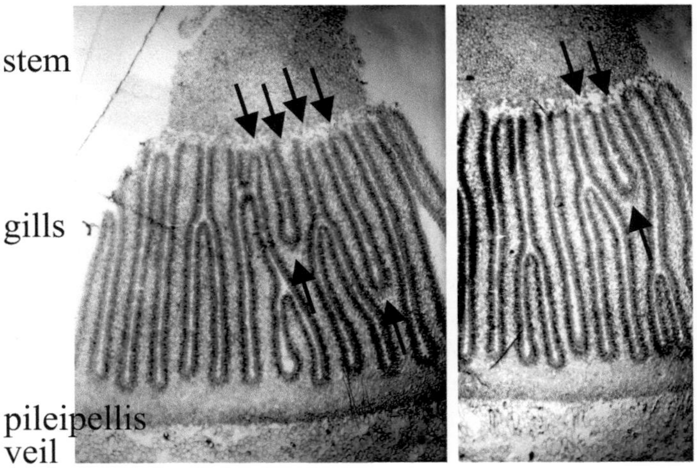

Figure 13.20. Two transverse sections of a mushroom cap of *Coprinopsis cinerea*. Stem tissue is at the top. All the gills that have their tramal tissue continuous with the stem are primary gills. Note that two of these in one section and one in the other show a Y-shaped profile, in which both arms of the Y are still connected to the stem (arrows). This indicates that primary gills proliferate by formation of a new gill space within the trama of a pre-existing primary gill. Light micrographs by Isabelle V. Rosin.

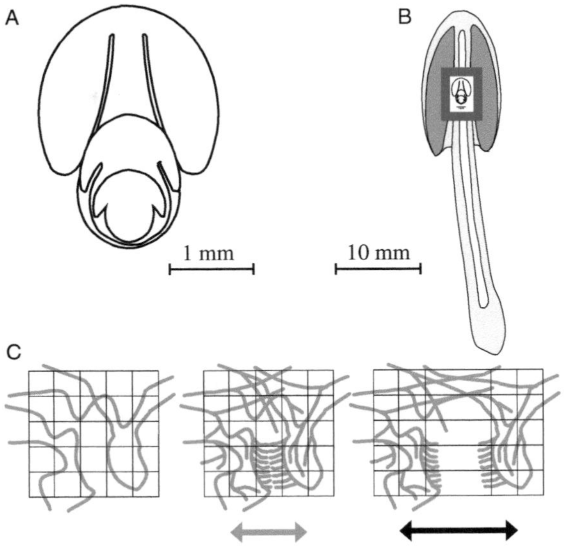

Figure 13.21. Dynamics of sporophore expansion in *Coprinopsis cinerea*. In (A) outline diagrams traced from sections of sporophore initials and primordia like those shown in Figure 13.1 are nested together to illustrate the steady outward expansion of the tissue layers. These diagrams are superimposed (to scale) onto a median diagrammatic section of a mature mushroom in (B) to demonstrate the extent of the outward movement of tissue boundaries. (C) shows how this expansion can generate cavities by putting tension stress across an incipient fracture plane. In these diagrams, a small region of undifferentiated tissue is shown first against a reference grid. In the central diagram a cycle of branching is assumed to have taken place forming branches of determinate growth arranged in two opposing palisades (two young hymenia). This constitutes the incipient fracture plane and when tension is applied the palisades will be pulled apart (right-hand figure) producing a gill cavity bounded by two hymenia.

has been called a ***gill organiser*** and it is presumed that a new gill organiser appears between existing hymenia as stem expansion forces them apart (and perhaps diminishes an inhibitory field to the point where a new gill organiser can form). Observation of many sporophore sections (Rosin & Moore, 1985a) shows that these Y-shaped gill structures are oriented exclusively as though the new gill organiser originates at the level of the stem surface and moves outwards towards the cap. There is a clear indication that gills in *Coprinopsis* broaden radially outwards, their roots extending into the undifferentiated tissue of the cap tissue.

In the cap tissue, it is believed that there is a gill organiser in the vicinity of each gill arch (that is, at the **apex of each gill space**) which is responsible for the progression of the gill morphogenetic field outward, away from the stem into the steadily replenished undifferentiated cap tissue (Figure 13.21).

Formation of secondary gills is presumed to occur when the distance increases between existing gills because of cap expansion, again creating the space within which a new organiser can arise. This set of observations gives presumptive evidence for two classic components of theoretical morphogenesis. On the one hand, we can hypothesise radial diffusion of an **activating signal** which assures progression of the gill sort of morphogenesis radially. On the other hand we may have tangential diffusion of an inhibitor which prevents formation of new organisers until the extent of the tissue domain exceeds the effective range of the inhibitor; at which stage a new organiser can arise in response to the radial activating signal (Rosin & Moore, 1985a). Interaction between the presumed gill organiser ***activator*** and the presumed gill organiser ***inhibitor*** is all that is necessary to control gill spacing, gill number, gill thickness, and the perfectly radial orientation of the gill field.

The characteristic general arrangement of the stem, gills and cap which is the essential mushroom, constitutes the basic body plan of gilled fungi. The basic mammalian body plan can provide for animals as diverse as humans, giraffes and the blue whale, yet it is arrived at by essentially identical morphogenetic processes in those animals, such that the early embryonic stages are virtually indistinguishable. In the face of this commonality, it is unlikely that gilled fungi would use a multitude of morphogenetic processes to arrive at the same basic mushroom body plan, even though there are several distinct phylogenetic routes. It is most significant that the growth vector across the width of the gill is directed **away from the stem** in all species in which this point has been examined implying that **the gill grows at its root** and not at its margin. Thus, the generalised interpretation is that all gills grow by extension at the root and not by growth at the free edge (Moore, 2013).

For *Coprinopsis* and *Volvariella*, which are characteristic agarics, the developmental vector of the gills is directed **away from**

the stem. Surprisingly, the same also seems to be true for '**polypores with gills**' like *Pleurotus* and *Lentinula*. When gill edges of *Pleurotus pulmonarius* and *Lentinula edodes* were marked with the vital stain Janus Green, the edge remained intact even when further growth and development was quite considerable. Evidently, these gills also grow at their root. However, outward growth of the cap margin was not the same as in agarics. In agarics, the cap margin advances outwards because of the press of growth behind and marks made on the margin remain at the margin as the cap develops. In *Pleurotus pulmonarius* and *Lentinula edodes* **the margin proved to be a region of active growth**, reference marks made on the margin were overtaken by growth of the tissue and the new cap margin was free of the staining added at the primordial stage. Presumably these different growth patterns are consequences of the suite of developmental strategies which, like cell inflation versus cell proliferation, create the **morphogenetic distinction between agarics and polypores**.

Certainly, the radially outward growth of cap tissue evident in *Pleurotus pulmonarius* is functionally similar to the lateral extension of the sporophore in the resupinate polypore *Phellinus contiguus* (Butler & Wood, 1988). It seems, therefore, that 'gilled polypores' maintain the **ancestral direction of development** for the structural tissue of the sporophore but abandon the ancestral direction for development of the hymenophore. So, what of the pores of polypores?

Long ago, Corner (1932) described how pores and dissepiments arose in an area on the underside of the young part of the cap of *Polystictus xanthopus*, a region he called the '**pore field**'. Pores, of course, are the tubular holes which are lined with the hymenium. The tissue that separates the pores is called the **dissepiment**. This word literally means a partition and is specifically assigned to the tissue between the pores of a polypore. Described in this way, it is implied that the pore is an entity in its own right. However, like the space between the gills of gilled fungi, the pore is defined by the tissue around it; that is, by the pattern-forming growth and branching processes which take place in the hyphae of the dissepiments.

The pore is the place where the dissepiment does not grow. Corner (1932) defined the 'pore field' as an annulus near the margin on the **underside of the cap**, in which localised development of differentiating hyphal growth resulted in initiation of dissepiments as **protruding ridges**, which branched and united around non-extending areas. The latter became the pore bases. Corner also described a **spine field** in the hydnoid sporophore of *Asterodon ferruginosus*, corresponding to the pore field of polypores, in which spines arose as localised areas of **downwards growth** 200–300 μm in diameter in which the hyphal tips were **positively gravitropic** (growing towards the centre of the Earth).

Constituent cells of the pore wall differentiated from mycelial hyphae. Hymenial elements were the first to differentiate, basidia and setae (a seta is a stiff hair-like sterile hyphal end with thick walls, projecting from the hymenium) arose directly from 3-day-old divergently growing mycelium, setae differentiating first. In the pore bases, this sparse discontinuous hymenium became more continuous as more basidia, and setae, differentiated in parallel with dissepiment growth (Butler, 1992a). Setae, basidia and dissepiment-building hyphae all differentiated on explants (Butler, 1992b).

Phellinus contiguus in nature produces a poroid sporophore which is flat on the substratum (that is, a **resupinate** sporophore), the usual substratum being the lower side of branches and twigs. Butler & Wood (1988) described production of pores on colonies cultured in Petri dishes, 3–4 mm behind the colony margin. Sporophores formed in agar cultures of *Phellinus contiguus* were generally similar to naturally occurring specimens; even the pore size and density were within the range found in nature. Butler (1995) found that fruiting could be promoted by one or more metabolic products found in extracts of *Phellinus contiguus* cultures. The fruiting inducing substance has not been characterised.

Pores were formed only in **inverted cultures** (agar upwards). Disorganised growth of the tissue occurred in dishes with the agar lowermost, showing that some hyphae that contribute to sporophore construction are not positively gravitropic, though extension of the dissepiments clearly is. Butler (1988) distinguished two developmental processes in pore morphogenesis which she called **island initiation** and **aerial fascicle growth** (a fascicle is a little group or bundle of hyphae).

Sporophore tissue arose as islands in the narrow initiation zone immediately behind the mycelial margin. New pores were formed by lateral extension from the island initials and the final pore density was not defined in the original pore field; rather, secondary pore formation occurred both by differential growth within thick dissepiment areas and by subdivision of primary pores. Continued growth of the dissepiment away from the supporting cap tissue delimited the pore. The edge of the dissepiment grows ***away from the underside*** of the cap, so the edge of the dissepiment grows vertically downwards. Clearly, then, ***pores extend at their free edge, but gills always extend at their root***.

13.8 THE *COPRINOPSIS* MUSHROOM: MAKING STEMS

Mushroom stems are not as clearly divided into tissues as are caps, but they do show a level of differentiation and a distinct morphogenesis. Corner (1932) introduced **hyphal analysis** as a procedure to describe tissues of polypore sporophores and this is used in routine classification (see Appendix 2). Over the years several different types of hyphae and a range of tissue types have been described (Corner, 1966). Corner coined the terms monomitic, dimitic and trimitic to describe tissues consisting of one, two or three kinds of hyphae, and hyphae in these different categories have been referred to as generative (because they ultimately give rise to the basidia and directly or indirectly to all other structures), skeletal (with thick walls and narrow lumen, but lacking branching and septation) or binding hyphae (which have limited growth and irregular often-repeated branching). Corner also introduced the terms sarcodimitic and sarcotrimitic to describe sporophores that have two or three types of hyphae (respectively) of which one is inflated and with thickened walls.

All these, remember, were first defined in polypore sporophores, but Redhead (1987) recognised such structures in a group of gilled fungi that includes the general *Agaricus, Collybia, Marasmius* and *Flammulina*, so cellular differentiation in sporophore tissues is widespread. Indeed Fayod (1889) had already described the presence of narrow hyphae among the more easily seen cells of the sporophore tissues he examined. These were called '**fundamental hyphae**' but they have not been used in identification in the way that the mitic system has in polypores; today we might call them something like '**embryonic stem cells**', being undifferentiated vegetative cells able to differentiate into a diverse range of specialised cell types. Although there are many reports of the occurrence of differentiated hyphal cells in various parts of fungal sporophores it is the relative sizes of differentiated hyphal populations, their distributions and the way the populations change during development that are crucial to understanding cell and tissue functions and there is only one clear account of a **quantitative hyphal analysis**, and this deals with the stem of *Coprinopsis cinerea* (Hammad *et al.*, 1993a).

In this study, 5 μm thick transverse sections were cut from mushroom stems of a range of developmental ages. Hyphal profiles were counted, and cross-sectional areas measured using image analysis software. For each section of each piece of stem examined, the area of every cell within two randomly chosen radial transects was measured; transects were 12 μm wide. Individual cells were measured in strict order through the transect, starting from the exterior of the stem and ending at its centre, or at the lumen in older stems (Figure 13.22). **Two distinct populations of hyphae** were identified and categorised as narrow hyphae, with cross-sectional area <20 μm^2, and inflated hyphae, with cross-sectional area ≥20 μm^2. These populations of hyphae may be randomly distributed, evenly distributed or clumped together. In a random distribution the presence of one individual does not affect the probability of another occurring nearby; in an even distribution that probability is lowered while in a clumped distribution it is raised. Statistical analysis of spatial distributions of the two populations of hyphae showed that inflated hyphae were very strongly ***evenly*** distributed, regardless of the age of the mushroom (from 27 mm tall to 70 mm tall) and regardless of position within the stem. The even (that is, non-random) distribution of inflated hyphae implies some form of organisational control. On the other hand, the spatial distribution of narrow hyphae was essentially ***random***.

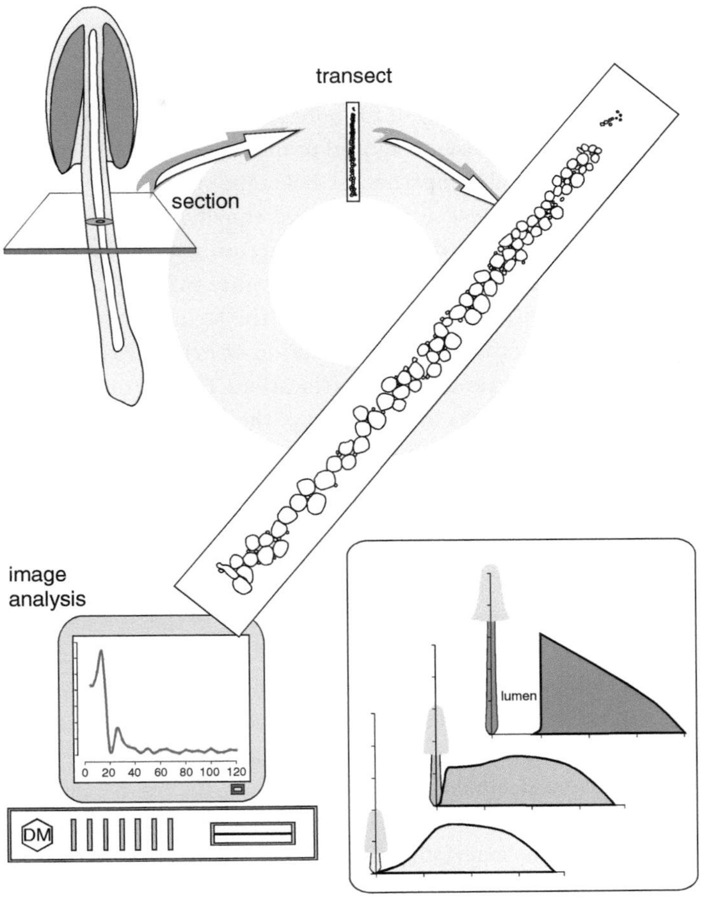

Figure 13.22. Stem structure and development in *Coprinopsis cinerea*. The main diagram shows how cell size distribution data were obtained from microscope sections of basidiome stems. The cell size distribution plot on the computer monitor illustrates the very distinct population of narrow hyphae and the very disperse population of inflated hyphae. The panel at lower right shows average cell area along a radius extending outwards from the centre of the stem for basidiomes which were 27, 45 and 70 mm tall. The changing distribution of cell size shows that stem growth is accompanied by inflation of cells in an annulus deep within the original stem tissue. The outcome of this is that the inner region is torn apart (forming the central lumen) and the outer region is stretched.

Both narrow and inflated hyphae can be seen in primordia only 3 mm tall. However, during stem elongation, the numerical proportion of narrow hyphae decreased implying that at least some (approximately 25%) of them are recruited to the inflated category. Low magnification images of transverse sections of any stem more than a few mm tall were dominated by highly **inflated cells** but narrow hyphae always constituted a significant numerical proportion (23–54%) of the cells in transverse sections, although, being 'narrow', they only contributed 1–4% of the overall cross-sectional area. **Narrow hyphae** stained densely with several conventional histological stains (especially Mayer's haemalum, toluidine blue and aniline blue/safranin; all of which can stain nucleic acid/protein and protein/polysaccharide complexes) and revealed especially strong, particulate, staining with periodic acid–Schiff reagent for polysaccharide.

However, not all narrow hyphal profiles in a transverse section and not all hyphal compartments belonging to any one narrow hypha in longitudinal sections stained equally. This differential staining might reflect **differential function** among the narrow hyphal population or, since narrow hyphae may be important in translocation of nutrients through the stem, it may simply reflect non-uniform vertical distribution of cytoplasmic materials caught in the process of translocation. The narrow hyphae seem to form networks independent of the inflated hyphae; they were seen to be branched and to be fused laterally with other narrow hyphae, but inflated hyphae were neither branched nor associated in networks.

Generally, narrow hyphae were interspersed with inflated hyphae across the full radius of all stems irrespective of position along the length of the stem and irrespective of the developmental age. However, there was a progressive change in the distribution of inflated hyphae during maturation. In 6 mm and 27 mm tall mushrooms, the inflated hyphae increased in cross-sectional area from the exterior surface to about halfway across the radius but then size declined towards the lumen. In the 45 mm tall mushrooms, the cross-sectional area of inflated hyphae increased steadily from the exterior to the lumen and this pattern was even more pronounced in 70 mm tall mushrooms. These measurements imply that expansion of the stem is mainly due to increase in cross-sectional area of inflated hyphae in an annular region just over halfway across the radius; we wonder what makes those ***specific*** stem hyphae expand. The **geometrical consequences of cell inflation** being emphasised in a distinct annulus ***within*** the stem tissue would be (Figure 13.23):

- first, that the **central core would be torn apart** (it is not expanding, but the tissue exterior to it is expanding so the centre must tear apart), leaving its constituent cells as a remnant around the inner wall of the lumen so created;
- second, that the tissues in the zones outside the expanding annulus would be stretched and reorganised (they are not expanding, but the tissue interior to them is expanding, so **outer tissue must be stretched**).

What makes some hyphae become inflated while others remain morphologically similar to the vegetative mycelial hyphae is not known. However the **organised pattern of inflation** generates the final morphology of the mature stem. Cell inflation in a distinct annulus within the stem tissue turns a solid cylinder of hyphal tissue into a hollow tube, which itself is an engineering advantage as it produces a supporting element with greater stiffness per unit mass, and additionally produces another **stretched-skin construction** which gains even more structural strength from the combination of compression within the stem tissue and tension in its stretched outer tissues.

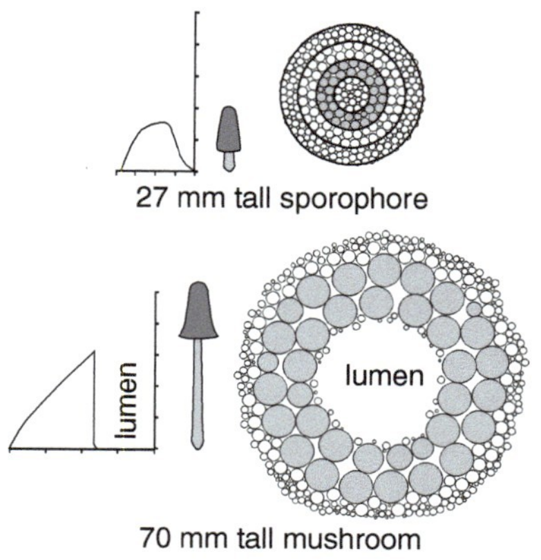

Figure 13.23. Interpretation of the geometrical consequences of cell size changes during development of the stem of *Coprinopsis cinerea*. The two graphs show cell size distributions across the radius of the stem for 27 and 70 mm stems alongside scale drawings of the mushrooms and diagrams of transverse sections of the stems, also drawn to scale. The diagrammatic transverse section of the stem of a 27 mm tall primordium (on left) is composed of solid tissue which is divided into four zones corresponding to the zones in the radius in the graph alongside. The central (zone 4) and outermost (zone 1) zones are comprised of rather smaller cells than the two cortical zones. During further growth the most dramatic cell inflation occurs in the cells of zone 3 which are here shown shaded. Growth from 27 mm to 70 mm in height is accompanied by a 3.6-fold increase in cell area in zone 3 but only 1.6-fold increase in zone 2. The cells in the other zones must be rearranged to accommodate the inflation of zone 3 and a major consequence is that a lumen appears in the centre of the stem. Changes in cell profile area are shown to scale in these diagrams, though only a total of 327 cell profiles are illustrated. As this is only a tiny proportion of the cells involved *in vivo*, the diagrams inevitably distort the apparent relationship between cell sizes and stem size. Narrow hyphae have been ignored in these diagrams, though they are distributed randomly throughout the tissue and their conversion to inflated hyphae contributes to stem expansion.

13.9 COORDINATION OF CELL INFLATION THROUGHOUT THE MATURING MUSHROOM

As you can see from the descriptions above, most of the changes in shape during sporophore development in basidiomycetes depend on **cell inflation**; this is typically a slow process in young primordial stages and characteristically more rapid during late maturation. Local cell inflation during sporophore development has often been described and Reijnders (1963) showed that the different zones of sporophores enlarge proportionally, so that different tissues mature without being impeded, compressed or distorted by the growth of other parts. Such coordination of differentiation of cells in relation to their location is one of the most important general principles of animal and plant morphogenesis. It is thought to be based upon the migration through the developing tissues of pattern-forming morphogenetic factor(s) or signals.

If this is true for fungi, the nature of the signals is generally obscure, though in the larger sporophore primordia of *Ascomycota* and *Basidiomycota* the mechanism would need to operate over distances of many millimetres, a similar scale to animal and plant hormone fields. However, apparent 'coordination' could also result if developmental events are arranged in a consequential series such that one (secondary) event is only instigated by the initiation or completion of an earlier (primary) event. For any conclusion to be reached on this issue what is needed is a holistic study of inflation over the whole sporophore and an assessment of the correlation between cell behaviour in widely separated locations against an accurate time frame. This was done by Hammad *et al.* (1993b) for the mushroom of *Coprinopsis cinerea*.

This study examined a sufficiently large sample of mushrooms to establish the exact timing of major meiotic and sporulation events, and this provided the basic **timeline** to which other processes could be referenced simply by microscopic examination of a small piece of cap tissue. Removing small tissue fragments ('biopsies') for rapid microscopic examination does not interfere with sporophore development. This timeline is objective in the sense that it depends upon processes which are endogenously controlled. It is reliable because these processes are central to sporophore function and it is versatile since by examining slivers at known (real) time intervals the effects of any change in cultivation conditions or culture genotype become apparent.

In the cultivation conditions described by Hammad *et al.* (1993b), mushroom primordia were formed 5 days after inoculation of the culture medium and developed into mature mushrooms within 2–3 days, during which the basidia underwent a sequence of morphologically and physiologically distinctive stages:

- the dikaryotic protobasidia underwent nuclear fusion (karyogamy) and then entered meiosis I, followed immediately by meiosis II;
- after completion of meiosis, four sterigmata emerged
- followed by formation of four basidiospore initials;
- nucleus migration;
- maturation and discharge of the spores.

All these events were completed within 18 h, and the overlap between different stages varied through development:

- in a 'meiosis I specimen' about 60% of the basidia had two nuclei;
- at meiosis II about 70% had four nuclei;
- when sterigmata first appeared about 90% of the basidia were at the same stage.

Defining karyogamy as time zero, basidia took 5 h to reach meiosis I and meiosis II was completed after a further hour. From meiosis II to the emergence of sterigmata required 1.5 h, spores emerging 1.5 h after that. Spore formation continued for 1 h and then nuclear migration started. Spore pigmentation commenced 1 h after spore formation and spores matured and were discharged 7 h later.

Coordination of cell inflation was studied by measuring cell sizes in microscope sections of mushrooms of *Coprinopsis cinerea* whose chronological development was defined by the stage they had reached in meiosis and sporulation.

As far as cell inflation was concerned (measured in terms of cell area in tissue sections), there was only a small increase before meiosis. Presumably, any sporophore expansion occurring in these early stages is due primarily to cell proliferation rather than cell inflation. In contrast, there was a large increase in cell profile area in sections of mushrooms undergoing meiosis. The average length-to-width ratio of the cells in stems before meiosis (mushrooms up to about 8 mm tall) was about 2, but this increased greatly after meiosis, to 10 in 48 mm tall mushrooms, 20 in 55 mm tall mushrooms and approximately 35 in mushrooms that were 83 mm tall. The phase of most rapid stem elongation occupied the 5 h prior to spore discharge and started 8 h after karyogamy. Cap expansion started as spores matured, 14 h after karyogamy.

The most remarkable feature of these data was that rapid **stem elongation** was correlated with the **ending of meiosis**. Indeed, ***inflation of all the different cell types in the cap as well as elongation of cells of the stem began immediately after meiosis***.

Elongation of the stem is necessary to raise the cap into the air for more effective spore dispersal; inflation of the different cell types in the gill tissue is also necessary for effective spore dispersal so coordination of cell inflation across the whole mushroom is clearly advantageous. The coordination that is observed may be achieved by some sort of **signalling system** that 'reports' the end of meiosis to spatially distant parts of the sporophore and initiates cell expansion. Indeed, stem elongation was significantly greater when the cap was left attached than when it was removed and leaving half of the cap in place was sufficient to ensure normal elongation.

This effect of surgical removal of cap segments on stem elongation has been observed with several species of mushroom over the years. The elongating stem curves away from the side with the cap segment left intact (providing it contains gill tissues) and this has been interpreted as indicating that the gills are the source of a 'growth hormone'. Unfortunately, the detailed information we have about hormonal effects in mushrooms is still sparse and unsatisfactory (Novak Frazer, 1996), with no definitive fungal growth hormone of any sort being chemically isolated even now; which is a remarkable statement for a kingdom of organisms which have been known for many years to include pathogenic/parasitic representatives that produce both animal and plant hormones to control their hosts. However, some recently reported research has suggested that chemically modified sugars, formed in the gill tissues and secreted into the mucilage that surrounds the hyphae making up the *Coprinopsis* mushroom, might be candidates for the signal that coordinates hyphal cell inflation across the whole mushroom (Moore & Novak Frazer, 2017).

This family of molecules (called **Fungiflexes**) was isolated during a study of gravitropism in *Coprinopsis cinerea* and found to be most probably laevorotatory 6-deoxy hexoses, which could be substituted with amino and/or amido groups, then be N-acetylated, and possibly phosphorylated or sulfated to serve different signalling purposes. When applied asymmetrically to isolated mushroom stems, hyphal cells in the immediate region of the application changed their growth pattern in response to at least two components, one promoting immediate hyphal lengthwise contraction (called Fungiflex 1) and the second (Fungiflex 2), after some hours, promoting hyphal lengthwise extension. It is thought that the Fungiflex molecules are produced near the stipe apex, perhaps in what constitutes cap tissue near or at the junction of stipe and cap. They are then released into the extracellular matrix of the stipe to diffuse away from their source (which effectively means ***down*** the stipe) to regulate hyphal extension progressively. The assumption is that Fungiflex 1 inhibits stipe extension while the cap is being formed and, following on several hours later, Fungiflex 2 enhances the extension of stipe hyphae to raise the cap above surrounding vegetation to facilitate the eventual spore release (Moore & Novak Frazer, 2017).

13.10 MUSHROOM MECHANICS

It is obvious that coordinated inflation plays a central role in differentiation of the cells and tissues which make up the mushroom, but it seems likely that cell inflation has a profound bearing on the morphogenesis of the sporophore as a whole. Again, our examples are concentrated on the ink cap mushroom *Coprinopsis* and its relatives, but they exemplify general points of structure that apply equally to other fungi.

Except for the descriptions given by Buller (1924, 1931), discussion of cap morphogenesis in ink caps is generally restricted to the role of **autolysis** (the self-digestion of the gill tissue that produces the 'ink') in the removal of spent gill tissue which might otherwise interfere with spore-fall from the mushroom. However, during this autodigestive period, the cap passes through a profound change in shape before autolysis: the initially vertical orientation of the gills is transformed to a horizontal one (Buller, 1924, 1931). In *Coprinopsis cinerea*, this process is completed in about 6 h and is achieved by a gradual rolling-back of the edge of the cap, which, together with radial splitting of cap and gill tissues, allows the cap to open like an umbrella. The mechanical process that achieves this emerges from measurement of the profile contour length of the cap at successive stages during expansion, which shows that it remains constant in the final autodigestive phase (Figure 13.24). This is true even though:

- as paraphyses differentiate their inflation must ***increase the area*** of the gill, and
- although spent gill tissue is removed by autodigestion a considerable portion remains intact until very late in cap expansion.

A gill that is expanding in area is bounded by a flexible but evidently inextensible band of cap tissue, a combination which will inevitably lead to an outward curvature (Figure 13.24). Cap tissues become the most important structural members after the autodigestive process has removed the bulk of the gill, but initially paraphyseal inflation provides the major expansionary force and structural integrity (Ewaze *et al.*, 1978; Moore *et al.*, 1979).

We have now brought attention to several structural characteristics of the *Coprinopsis cinerea* mushroom that are the result of mechanical processes:

- the regular 'parallel' arrangement of gills (section 13.7);
- conversion of the stem to a hollow cylinder (see Figure 13.22);
- and now the final morphogenesis occurring during maturation of the mushroom cap.

It is an important general point that several distinctive aspects of mature multicellular structures in fungi are the result of **mechanical interactions between tissues** that are inflating to different extents. They result in a very exact, and recognisable, final morphology; but that final morphology is not explicitly specified in the genome. Rather, the genome specifies a programme of metabolic and structural activities which, if played out correctly, will result in that characteristic morphology.

13.11 METABOLIC REGULATION IN RELATION TO MORPHOGENESIS

We have emphasised the importance of cell inflation and tissue expansion in sporophore development, but there are several resources that must be provided to support cell inflation. Obviously, these demands include extensive synthesis of new cell wall (see Chapter 7). However, wall synthesis alone will not ***inflate*** the cell:

- The volume enclosed by the wall needs to be filled with 'cytoplasm' and that usually means a large increase in the size, and perhaps number, of vacuoles.
- Vacuoles are filled with aqueous solutions, so this, in turn, requires a considerable increase in uptake of water by the cells.
- Since water is driven across cytoplasmic and vacuolar membranes from a solution of high water potential (low solute concentration) to one of low water potential (high solute concentration) there is a consequential need to adjust metabolism so that osmotically active solutes are accumulated in the cells and cytoplasmic compartments that must be inflated.

Ultimately, therefore, cell inflation is driven by modifications to the pattern of primary (intermediary) metabolism (meaning specific increases or decreases in activity levels of existing enzymes

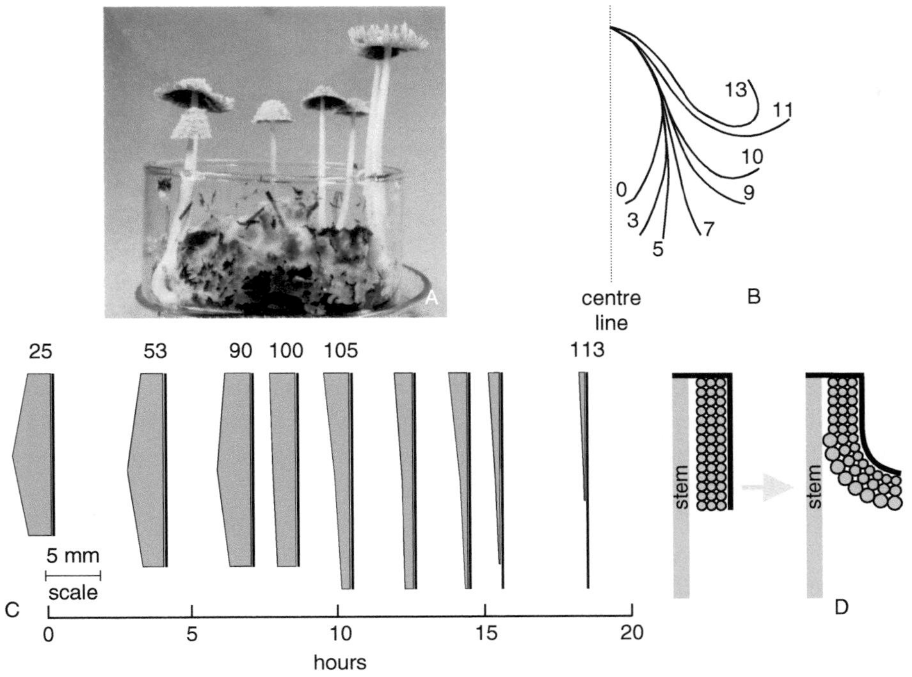

Figure 13.24. Final maturation of the mushroom cap of *Coprinopsis cinerea.* (A) Photograph of the organism growing on compost in a 9 cm diameter crystallising dish. (B) Diagrams recording the way in which the profile of the cap changes during its final phase of expansion. Consecutive photographs were made of the same mushroom as it matured, and tracings were then made of the upper edge of the cap profile. Number show hours elapsed since observation started at time zero. During the course of the sequence shown the stem extended from 25–93 mm, spore discharge commenced about 5 h from the start of observation. (C) Scale drawings of gill growth and autolysis. At regular intervals in the final stages of mushroom development a small segment was surgically removed from an otherwise undisturbed sporophore and preserved in formalin. Subsequently, the outer contour length (length of cap flesh or pileipellis) was measured and the depth of gill tissue measured at the top, centre and bottom of a suitable primary gill in each segment. In these diagrams the shape of the cap is ignored (for which see B); the vertical bars represent the contour length and the shaded area represents the gill lamella. The numbers above the diagrams record the length (in mm) of the stem at the time the sample was taken. In this specimen, spore discharge began at about the 7th hour. These observations show that the cap flesh remains intact and extended throughout the entire period of gill autolysis, and that for the final period of cap development the length of the cap flesh is unchanged. (D) Development of the mushroom cap in *Coprinopsis cinerea.* Diagram to show how cell expansion in the hymenium can account for the outward bending of the gills (as indicated in the profiles shown in B) if it is assumed that the cap flesh (pileipellis) acts as a flexible but inextensible outer boundary (shown as a thick black line). Diagrams redrawn after Moore *et al.* (1979).

and/or introduction or removal of particular enzymes) which are done as a crucial contribution to a ***differentiation process*** rather than as a response to nutritional conditions (see section 11.12).

Quite a catalogue of changes in enzyme activities has been noted during the development of ink cap mushrooms:

- In *Coprinopsis cinerea* and *Coprinus comatus*, development-related increases in the activities of glucosidase and glucanases, protease and chitinase have been noted during cap expansion (Iten, 1970; Iten & Matile, 1970; Bush, 1974) and have been identified in the 'ink' fluid (the autolysate); so degradation of the substance of the spent gill tissue is provided for by specific enzyme derepression.
- In earlier stages, prior to the onset of autodigestion, several enzymes are specifically derepressed as tissues of the *Coprinopsis cinerea* mushroom cap begin to form (Moore, 2013). Among these, NADP-linked glutamate dehydrogenase (NADP-GDH) and glutamine synthetase (GS) have been found to derepress specifically in the *Coprinopsis* mushroom cap during development.

A vegetative mycelium also shows derepression of NADP-GDH under specific conditions *in vitro*, and this has enabled experimental study of regulatory factors complementing observation of endogenously controlled events in the developing mushroom to establish the function(s) of the enzymes that are so distinctively regulated (for review of both the enzymology and the extensive histochemistry and cytochemistry, see Moore, 2013). The crucial observations for our present purposes are as follows:

- Derepression of NADP-GDH in the mycelium involved *de novo* synthesis of the enzyme protein (Jabor & Moore, 1984) and was caused by transfer into a medium lacking any nitrogen source but rich in carbon (usually 100 mM pyruvate was used as carbon source, but acetate, glucose and fructose were also effective).

- Inclusion of as little as 2 mM NH_4^+ in the transfer medium prevented derepression of NADP-GDH; among many other nitrogenous compounds tested only those able to generate ammonium (including glutamine) also inhibited derepression.
- A mutant unable to synthesise acetyl-CoA failed to derepress NADP-GDH (Moore, 2013).
- In the sporophore derepression of NADP-GDH occurred in the cap, but not in the stem (Stewart & Moore, 1974; Al-Gharawi & Moore, 1977), the accompanying NAD-GDH also increased in activity (in both tissues) as the mushroom developed. In the cap the ratio of the two enzymes altered from 10:1 (NAD-GDH: NADP-GDH) to 1.4:1, a considerable swing in favour of the enzyme with a 10-fold greater affinity for ammonium.
- There was a high positive correlation between derepression of NADP-GDH and derepression of glutamine synthetase in both mushroom cap and mycelium and coordinate regulation of four enzymes involved in arginine and urea synthesis metabolism in the cap (Ewaze *et al.*, 1978).
- Both light and electron microscope observations demonstrated cytochemical localisation of NADP-GDH to the developing basidia of the young hymenium in the mushroom cap (Elhiti *et al.*, 1979).
- Large quantities of glycogen (up to 2 mg dry weight per sporophore) were accumulated in sporophores and translocated specifically to the mushroom cap (Jirjis & Moore, 1976; Moore *et al.*, 1979).
- Derepression of NADP-GDH occurred initially at karyogamy and was then re-amplified post-meiotically. In *Coprinopsis*, unlike other agarics, the meiotic division is highly synchronised across the whole population of basidia. Observation of stained nuclei coupled with determination of activity of NADP-GDH and assay of glycogen in the same tissue showed that increase in enzyme activity was initiated as karyogamy became evident in normally developing sporophores. Glycogen utilisation was initiated towards the end of meiosis and was almost completely utilised within a few hours immediately following the meiotic division during which time basidiospore formation occurred (Moore *et al.*, 1987b).
- Activity of the enzyme urease was absent from the cap though present in both stem and mycelium (Moore & Ewaze, 1976; Ewaze *et al.*, 1978).

In interpreting these observations, the first important point we want to make is that such modifications of metabolism do not represent a novelty in terms of the reaction sequences involved; much of what occurs in the sporophore can be shown to occur in mycelia. However, the mushroom and the cap in particular, does present a very special regulatory picture, because these metabolic changes, which are a normal part of the development of the sporophore can only be induced in vegetative cultures by exposing them to particular (and in some cases most peculiar) synthetic media. What is specific to the sporophore then, is that at the outset of its development there is set in train such a fundamental change in the emphasis of the metabolism of its constituent cells that cap cells become metabolically quite distinct from those of either the stem or the parent mycelium. To understand better the following discussion, you might find it useful to refer to that part of Chapter 11 (see section 11.12 dealing with *Primary (intermediary) metabolism*, especially Figure 11.9. showing oxidation of pyruvate, the tricarboxylic acid cycle and glutamate decarboxylation loop; the flow chart illustrating pathways of nitrogen redistribution in Figure 11.12; and the urea cycle shown in Figure 11.13).

The observations listed above show that arginine synthesis and urea formation via the ornithine cycle occurred in the *Coprinopsis cinerea* mushroom cap and stem (Figure 11.13). Although urease occurred at high activity in mycelium and stem, it was not detectable in extracts of cap tissue, but arginine biosynthesis was specifically amplified during development of the cap, as judged from metabolism of isotopically labelled substrates and increased enzyme activities. Four enzymes, **NADP-linked glutamate dehydrogenase, glutamine synthetase, ornithine acetyltransferase** and **ornithine carbamoyltransferase**, were **upregulated (derepressed) in developing caps** while remaining low (or declining) in activity in the stems supporting those caps. Arginine, alanine and glutamate accumulated in the cap because of amplified arginine biosynthesis. More significantly, the ***quantity*** of urea in the cap doubled during development, although the ***concentration*** of urea remained unchanged, which suggests a causal relationship between urea accumulation and water influx into developing tissues. We conclude that **urea is the osmotically active metabolite** in *Coprinopsis cinerea* and metabolism is adapted to enhance the flow of nitrogen towards the urea cycle. All of this is devoted to driving water into the cap tissue to enable its expansion, accounting for the changes in form through which the cap progresses as maturation proceeds; but that's neither the end, nor is it the whole, of the story.

Urea is an ideal osmotic metabolite, being a low molecular weight, non-toxic, nitrogenous excretion product, so it is easy to understand how tissues that are about to enter an autodigestive phase, in which the whole emphasis is shifting from anabolism to catabolism, can have their metabolism altered to yield large quantities of urea rapidly (Figure 11.12), and the decline in activity of urease exclusively in the cap ensures that the urea will accumulate there.

In this system, however, co-regulation of NADP-glutamate dehydrogenase and glutamine synthetase (GS) depends on a circuit involving build-up of acetyl-CoA in tissues where ammonium is limiting; and this does not fit well into the 'urea accumulation through catabolism story'. In other circumstances (and many other organisms), NADP-GDH and GS both have very high affinities for ammonium and together contribute to an ammonium-scavenging system. This is an ***anabolic*** process, alleviating nitrogen deprivation and creating amino acids needed for protein synthesis. In *Coprinopsis cinerea* cap tissue, amino-nitrogen must be obtainable readily from autodigested proteins, so it is unlikely that the ammonium-scavenging GDH and GS activities are needed to assimilate nitrogen at this late stage in sporophore maturation.

However, a consequence of enzymatic scavenging is that the scavenged substrate is very effectively removed from the cytoplasm, so an alternative interpretation is that these two enzymes are derepressed to maintain an **ammonium-free environment** in cells committed to processes which are inhibited by ammonium. Now, NADP-GDH activity has been specifically localised to basidia in *Coprinopsis cinerea* cap tissue, so the cells we are talking about are basidia, which are specialised for meiosis. And, indeed, it seems that proper progress of meiosis in eukaryotes generally may depend on ammonium-sensitive processes. Glycogen breakdown, DNA synthesis, and extensive RNA and protein breakdown occur together uniquely in fungal meiocytes; and all these processes seem to be inhibited by ammonium ions (Moore *et al.*, 1987a, and references therein). This is true even in yeast:

- Degradation of accumulated glycogen is not observed in yeast cells incubated in medium supplemented with ammonium.
- Treatment with ammonium delays degradation of proteins at the onset of meiosis and inhibits protein and DNA synthesis.
- DNA replication is arrested by ammonium after initiation and continued incubation in the presence of ammonium leads to massive DNA degradation.

The interpretation is, therefore, that NADP-GDH and GS activities are specifically enhanced in basidia of *Coprinopsis cinerea* to protect those cells from the inhibitory effects of ammonium ions, and in the next section we will describe how meiosis and sporulation were shown to be sensitive to inhibition by ammonium ions by direct experiments *in vitro* and *in vivo*.

13.12 DEVELOPMENTAL COMMITMENT

There is, then, a wealth of evidence for highly specific differentiation of individual cells in fungi. There is even some direct evidence for that developmentally important concept: commitment. This is the process whereby a cell becomes firmly committed to just one of the several developmental pathways that are open to it ***before expressing the phenotype of the differentiated cell type***. This has always been an important concept in metazoan development, and with the recent growth of interest in the potential of embryonic stem cells in medical treatments, it is still a major line of research. The research is intended to establish what factors endow an embryonic stem cell with the plasticity to give rise to several differentiated types of cell, what signals drive cells to commit to a specific cell fate and whether the decision to differentiate along a particular developmental path is irreversible; all, of course, in the hope of using stem cells in disease therapies (Lassar & Orkin, 2001; Trounson & McDonald, 2015).

The classic demonstration of developmental commitment involves transplantation of the cell into a new environment; if the transplanted cell continues along the developmental pathway characteristic of its **origin** then it is said to have been committed prior to transplant. On the other hand, if the transplanted cell embarks upon the pathway appropriate to its **new environment** then it was clearly not committed at the time of transplant. Most fungal structures produce vegetative hyphae very readily when disturbed and 'transplanted' to a new 'environment' (which is usually an *in vitro* culture medium). Strictly, of course, this is a regenerative phenomenon, important in its own right and a very significant, unusual and experimentally attractive attribute of fungi. Yet it does create the impression that fungal cells express little commitment to their state of differentiation.

Very little formal transplantation experimentation has been reported with fungal multicellular structures. The clearest examples of commitment to a developmental pathway in fungi has been provided by Bastouill-Descollonges & Manachère (1984) and Chiu & Moore (1988a), who demonstrated that basidia of isolated gills of *Coprinellus congregatus* and *Coprinopsis cinerea*, respectively, continued development to spore production if removed to agar medium at early meiotic stages. Similar results were obtained whether water agar, buffered agar or nutrient agar was used as explantation medium. Other hymenial cells, cystidia, paraphyses and tramal cells, immediately reverted to hyphal growth but this did not often happen to immature basidia. Evidently, basidia are specified irreversibly as meiocytes and they become committed to complete the sporulation programme during meiotic prophase I.

Clearly, then, these experiments demonstrate **commitment to the basidium** differentiation (sporulation) pathway some time before the differentiated phenotype arises in these fungi. It is also important to stress that other cells of the hymenium do not show commitment. Rather, they immediately revert to hyphal growth on explantation as though they have an extremely tenuous grasp on their state of differentiation. That these cells do not default to hyphal growth *in situ* implies that some aspect of the environment of the tissue that they normally comprise somehow continually reinforces their state of differentiation. These non-committed cells must be considered to be totipotent stem cells. It is this uncommitted state of differentiation of most of the cells in mushroom sporophores that accounts for the readiness of field-collected mushrooms to revert to vegetative growth when fragments are inoculated to culture medium. Mycologists expect as a matter of common routine to be able to make cultures in this way, using the simplest media, from sporophores they collect. No animal or plant ecologist can expect to be able to do this. The difference denotes a significant quality of fungal cell differentiation.

Protobasidia of *Coprinopsis cinerea* in the prekaryogamy (dikaryotic) stage at the time of excision of a transplant were arrested at that stage. Explants made at later physiological ages did complete meiosis and/or sporulation, though at a slower rate than *in vivo*. Once initiated, the maturation of basidia is an autonomous, endotrophic process which can proceed *in vitro*, but it can be inhibited. Post-prophase I protobasidia were committed to the meiocyte programme of differentiation through meiosis and culminating in basidiospore maturation even when explanted to water agar. This *in vitro* explantation process has been developed as a rapid, small-scale bioassay for chemicals which interfere with sporulation processes. Simply including various compounds in the explantation medium inhibited basidium differentiation. Growth is not inhibited in the bioassay; rather, **differentiation inhibitors** cause vegetative hyphal tips to grow out from regions of the basidial apparatus expected to be in active growth during sporulation (Figure 13.25) (Chiu & Moore, 1988b, 1990c).

The differentiation inhibitors tested fell into two groups:

- ammonium salts and immediate structural analogues, including alkali metal salts, and glutamine and some of its structural analogues were effective in tissues exposed at any time after meiotic division I (even into post-meiotic sporulation stages);
- ionophores, cAMP and wall synthesis inhibitors were effective only if applied during meiosis.

Differential sensitivity of basidia between meiotic and sporulation stages implies that during the nuclear division the cell is prepared in advance for sporulation so that by the end of the cytologically recognisable nuclear division sporulation can proceed despite the metabolic disturbance caused by ionophores and wall synthesis inhibitors. On the other hand, the sporulation process must involve essential components that are sensitive to excess ammonium (and glutamine, etc.) but which remain crucial after completion of the nuclear division. Depending on the stage reached at the time of exposure to the inhibitors, vegetative hyphal tips emerged from the four apical sites for sterigmata, from the tips of sterigmata, from partially formed or abnormal spores, and from the basal regions of the basidium from which paraphyses would be expected to arise. These and other data allow basidium differentiation in *Coprinopsis cinerea* to be categorised into a sequence of steps of commitment comprising:

- commitment to recombination;
- commitment to meiosis;
- commitment to sporulation.

The ability of exogenous ammonium and glutamine to inhibit sporulation and promote vegetative growth in these explantation experiments is complementary to the observations described in the previous section, that ammonium-scavenging NADP-GDH and GS activities are naturally increased in sporophore caps as the mushroom matures (see section 13.11) and allow completion of the cycle of interpretations.

We can infer from the observations and experiments that within the normally developing *Coprinopsis cinerea* mushroom ammonium-scavenging enzymes are specifically derepressed to prevent ammonium ions halting meiocyte differentiation and

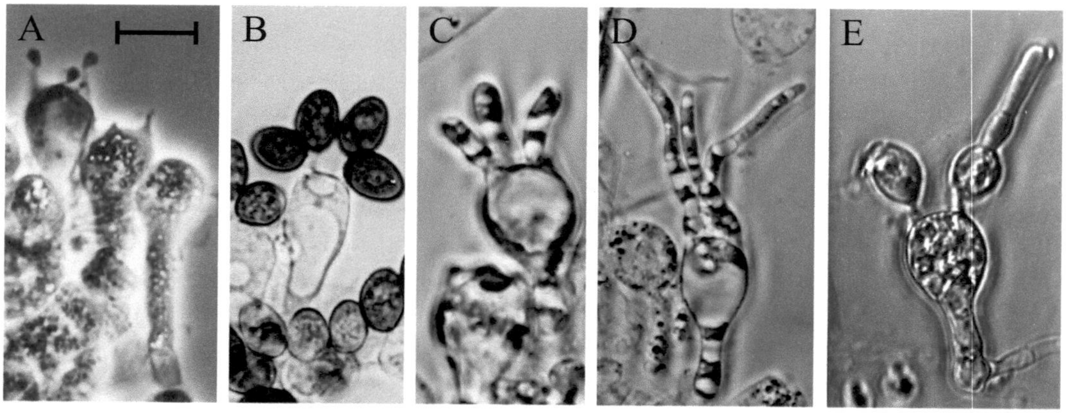

Figure 13.25. Basidium development of *Coprinopsis cinerea* is disrupted by application of ammonium salts. This can be demonstrated by *in vitro* experiments in which slivers of gill tissue cut from a developing mushroom are transferred to agar medium. Providing they are transplanted after prophase I of meiosis (A), *Coprinopsis* basidia are committed to continue development and they complete meiosis and spore formation after excision from the mushroom and transfer to water agar (B). If ammonium salts are included in the medium normal development is arrested; instead of forming sterigmata and spores, ammonium-treated basidia immediately dedifferentiate to produce vegetative hyphal tips (C, D and E) from regions of active wall growth. In (C and D), the sterigmata have grown out as vegetative hyphal tips. In (E) the hyphal tip has grown from the top of an incomplete spore. Note that this is not equivalent to premature germination because the germ pore (from which a mature spore will germinate) is located close to the sterigma attachment point. In specimen E the ammonium treatment has caused the emergence of a hyphal tip from a region of primary wall growth in the incomplete spore. Scale bar, 10 μm. Photomicrographs by Professor S.-W. Chiu, Chinese University of Hong Kong.

terminating sporulation. *Coprinopsis cinerea* grows naturally in compost where ammonia is abundant and volatile, so the ammonium assimilating enzymes are produced in the cap to serve as an efficient ammonium-scavenging system that maintains the microenvironment surrounding the hymenium free from the inhibitor. The proof of this seems to be that when the system was overloaded by injecting solutions of ammonium salts into the cap, the meiotic pathway was halted, leaving white patches of non-sporing hymenium on the mature cap (Chiu & Moore, 1990c). Thus, meiosis and sporulation can be shown to be sensitive to inhibition by ammonium ions *in vivo*.

13.13 COMPARISONS WITH OTHER TISSUES AND OTHER ORGANISMS

So far, we have described an example in which intermediary metabolism is a major contributor to cell differentiation and fungal morphogenesis. Given that cell inflation and tissue expansion are crucial aspects of sporophore maturation in most fungi, it seems reasonable to ask whether this particular example can be generalised. The key observation in *Coprinopsis cinerea* was that urea increased by a factor of 2.5 on a dry weight basis, but the urea concentration on a fresh weight basis was essentially unchanged during cap development, leading to the inference that urea behaves as an osmoregulator, driving water into hymenial cells and thereby driving expansion of the mushroom cap.

Agaricus bisporus also accumulates substantial amounts of urea in sporophores, and content varied with stage of development (Wagemaker *et al.*, 2005). However, in *Agaricus bisporus* and *Lentinula edodes*, **mannitol** (a sugar alcohol, or polyol) is synthesised to serve an analogous osmoregulatory function (Hammond & Nichols, 1976; Tan & Moore, 1994). To support accumulation of large amounts of mannitol, activity of the PPP (see Figure 11.8) is specifically elevated by a factor of about 15 times in the sporophore, though it is always at low activity in mycelium, (Tan & Moore, 1995). Patyshakuliyeva *et al.* (2013) studied the expression of genes encoding various carbon metabolism enzymes from different growth stages of *Agaricus bisporus*, correlating gene expression, enzyme activities and soluble carbohydrates in the hyphae of compost, casing layer and mushrooms. They found that compost-grown vegetative mycelium consumes a wide variety of monosaccharides, however only hexoses or their conversion products are transported from the vegetative mycelium to the sporophore where only hexoses accumulate and only hexose catabolism occurs. Genes encoding enzymes that degrade ***plant cell wall***-polysaccharides were mainly expressed in compost-grown mycelium but were absent from mushrooms. Whereas, genes encoding ***fungal cell wall***-modifying enzymes were expressed in both mushrooms and vegetative mycelium, but with ***different gene sets*** being expressed. A similar transcriptomic analysis of *Pholiota microspora* (commonly known as nameko, a small mushroom used as an ingredient in miso soup) revealed significant differences in gene expression between mycelium, primordia and sporophores, and suggests that mannitol, trehalose, glycogen and β-glucan play key roles in regulating carbohydrate metabolism and storage in the development of sporophores (Zhu, 2018).

Elevated PPP enzymes have also been noted in *Pleurotus pulmonarius*, but in this organism it seems that sporophores are characterised by elevated levels of **both** urea and polyols. The genome of the related sclerotium-forming mushroom *Pleurotus tuber-regium* has been sequenced and used to study the carbohydrate metabolising enzymes involved in the bioconversion of the lignocellulose of the substrate to β-glucan reserves within the hyphae of this white-rot fungus (Lam *et al.*, 2018).

The expansion and elongation of cells in the stem are essential aspects of stem elongation in *Coprinopsis cinerea*, but stem metabolism contrasts dramatically with cap metabolism. Differences between the two (interconnected) tissues are so great as to indicate that the common result is achieved in very different ways. Reported differences between cap and stem include: different patterns of metabolite accumulation, in terms of carbohydrates and amino acids; different patterns of metabolism of isotopically labelled substrates; profound differences in the activities of particular enzymes, especially polyol dehydrogenases, chitin synthase, NADP-GDH, alkaline phosphatase, enzymes involved in ornithine metabolism, glucanase and phenylalanine ammonia-lyase (Moore *et al.*, 1979, and references therein). Nevertheless, elongation of the stem depends on an enormous increase in the volumes of its constituent cells and as the walls remain unthickened and most of the cell interiors are occupied by vacuoles, the stem must be supported by a hydrostatic skeleton.

Uptake of water by the stem during development is dramatic: in developing from about 20 mm to about 100 mm long, the stem of an 'average' mushroom doubles its fresh weight with hardly any change in dry weight and absorbs nearly 200 mg of water.

The turgor pressure of stem cells remains constant throughout the period of stem elongation so appropriate osmoregulatory solute(s) must be formed and accumulated in the stem cells in parallel with the absorption of water and synthesis of wall. Low molecular weight carbohydrates are the best candidates as the stem osmoregulator: trehalose, for example, can account for 18% of the final dry weight. Trehalose is a non-reducing sugar, but alcohol-soluble reducing sugars can account for 12% of the dry weight of mature stems. Together, therefore, these sugars represent about 30% of the dry weight of the mature stem. Mannitol does not have any role in *Coprinopsis cinerea* (total polyols never exceed 6% of the dry weight and decline in quantity as the mushroom matures) though it was mentioned above as an osmoregulator in *Agaricus bisporus* and *Lentinula edodes*. For comparison, mannitol alone can amount to as much as 50% of the total mushroom dry weight in *Agaricus bisporus*, and 20–30% in *Lentinula edodes*.

The bulk of the osmotic potential of stem cells must be contributed by compounds of low molecular weight and it is probable that inorganic ions (especially potassium ions) make a very important contribution. Although the sugar content may account for a lesser fraction of the overall osmotic potential of the cell, it is a fraction that is readily adjusted by metabolism of materials already within the cell; something which is not possible with the inorganic components of the cell.

All these data point to fundamental differences in the ways by which cell inflation and sporophore expansion are achieved in different fungi, and even in different tissues in the same fungus. Cap expansion is due to hyphal inflation in *Agaricus bisporus* but depends on hyphal proliferation in *Lentinula edodes*.

Coprinopsis cinerea (which expands by hyphal inflation) and *Schizophyllum commune* (which expands by hyphal proliferation) show many metabolic similarities despite the fundamental difference between their cellular strategies of expansion. Evidently, totally different tactics can be used to achieve the same strategic end. Presumably, during evolution the 'choices' have been made between different ***metabolic*** mechanisms which enable a morphogenetic process to be put into effect, and these have been made independently of 'choices' between particular ***cellular*** processes which might contribute to that morphogenesis.

13.14 GENETIC APPROACHES TO STUDY DEVELOPMENT: THROUGH THE CLASSIC TO GENOMIC SYSTEMS ANALYSIS

The suggestion that there are several fundamentally different ways in which cell inflation and tissue expansion can be achieved is reinforced by work done on the induction of **developmental variants**. The classic genetic approach to the study of any pathway depends on:

- identification of variant strains;
- complementation tests to establish functional cistrons;
- comparison of heterokaryon phenotypes to determine dominance;
- determination of epistatic relationships in heterokaryons, to indicate the sequence of gene expression (Moore & Novak Frazer, 2002).

This, of course, now looks like a rather old-fashioned approach, though it has lost none of its power to dissect an individual pathway. The modern molecular approach is to use genomic analysis tools to analyse and compare the transcriptome (transcribed RNA population) and/or proteome (translated polypeptide population) to give expression profiles of different stages as the organism progresses through its developmental sequence. The classic approach ('first, find a variant') immediately focuses attention on a decisive event, though it can then be difficult to identify the network of interactions to which this belongs (the network only begins to become evident as more and more individual decisive events are dissected out, which can take a long time and a lot of work).

In contrast, genome-wide analysis gives you an immediate view of the overall grand network, which comprises many thousands of interactions, and it can then be difficult to bring the focus down to identify the few decisive events that enable neighbouring cells in a hypha to embark on different developmental pathways. The abstract of one paper, which describes gene expression studies during mushroom development in 'our favourite little mushroom' *Coprinopsis cinerea*, concludes with the sentence: 'We implicated a wealth of new candidate genes important to early stages of mushroom fruiting development, though their precise molecular functions and biological roles are not yet fully known' (Cheng *et al.*, 2013). Finding out what all those genes do can take a long time and a lot of work. We will describe some of this work in this section, as it applies to the organisms we have already mentioned. We have detailed the methods used in Chapter 6. If you need a reminder about the methods employed, we recommend you refer to the following volumes in the ***Methods in Enzymology*** book series: Weissman *et al.* (2010; **volume 470**); Jameson *et al.* (2011; **volume 500**); DeLong (2013; **volume 531**); Doudna & Sontheimer (2014; **volume 546**).

It is easier for us to begin our summary of genetic aspects of fungal developmental biology with a trip down Memory Lane to the pregenomics era, while we continue our story about *Coprinopsis cinerea*. The classic genetic approach, when applied to *Coprinopsis cinerea*, resulted in the isolation of, among others: ***stipe elongationless*** mutants, which were unable to elongate the stem but with normal cap expansion, or ***cap expansionless*** mutants unable to expand the cap but with normal stipe elongation, and ***sporeless*** mutants in which the mushrooms were morphologically completely normal except for the total absence of spores (Takemaru & Kamada, 1971, 1972) (Figure 13.26).

The three mutants were all dominant and segregated in crosses as single genes. Evidently, the pathways of cap and stem differentiation can be totally separated, and the sporulation pathway is separate from normal cap development. This leads to the conclusion that assembly of different parts of the same mushroom uses genetically distinct pathways. That conclusion can be generalised to: differentiation of fungal multicellular structures is **genetically compartmentalised** and normal morphogenesis is made up of several-to-many **developmental subroutines**.

For example, there may be subroutines for stem, cap, hymenium, hymenophore, etc. The subroutines can be put into operation independently of one another and are under separate genetic control and under separate physiological control. Normal morphogenesis of a specific fungus involves integration of these subroutines to produce the sporophore morphology that characterises that species. The same subroutines integrated in a different way will generate a different morphology characteristic for some other species. In their analysis of complex multicellularity in fungi, Nagy *et al.* (2018) use a slightly different description, suggesting that ***gene regulatory circuits*** that orchestrate

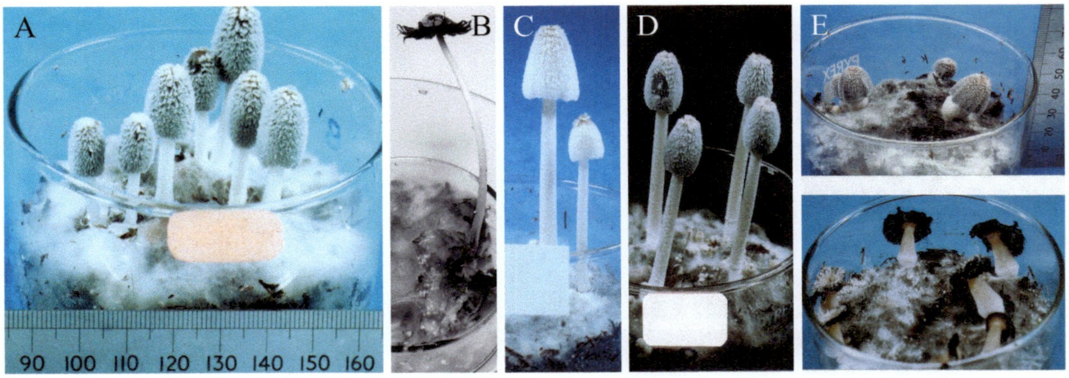

Figure 13.26. Developmental mutants of the ink cap mushroom, *Coprinopsis cinerea*. Shown from left to right are (A) and (B) the wild type (A, mid-development; B, fully mature and autolysing), and then (C) sporeless, (D) cap expansionless, and (E) stipe elongationless (immature at top, fully mature and autolysing below). The three mutants were all dominant and segregated in crosses as single genes. All cultures shown are contained in 9 cm diameter crystallising dishes. Photographs by Dr Junxia Ji.

sporophore development in fungi evolved in a stepwise manner, building on ancient ***regulatory modules***, but also on co-option of conserved genes coupled with the evolution of new ones.

Another important point is that fungi are very tolerant of **developmental imprecision**. Providing only that the sporulation pathway is intact, even a grossly abnormal sporophore can produce and distribute spores. Variation in the morphology of sporophores of higher fungi has been reported in many species, often appearing to be a strategy for adaptation to environmental stress. Detailed analysis of such developmental plasticity in spontaneous abnormal mushrooms of *Volvariella bombycina* has given more evidence for normal morphogenesis being an assemblage of distinct developmental segments (Chiu *et al.*, 1989). Although the sporophore structures observed in this study varied from the normal agaric form to completely abnormal enclosed, puffball-like structures, they were all able to act as production/dispersal structures for basidiospores (Figure 13.27). Thus, a sporophore may, quite ***abnormally***, bear a functional hymenium on the upper surface of the cap, rather than (or even in addition to) its normal hymenium on the lower surface (Figure 13.27). The important point is that although the development is abnormal because a morphogenetic subroutine has been invoked in the wrong place, the result is a functional and recognisable tissue, not a disorganised tumorous growth.

This tolerance of imprecision is an important biological attribute of fungal morphogenesis, as it provides the flexibility in expression of developmental subroutines that allows the sporophore to react to adverse conditions and still produce a crop of spores. Even when conditions are adverse to normal development, there is still sufficient flexibility in the developmental programme for the sporophore to fulfil its function (Moore, 2013).

The possibility has been discussed that fungal differentiation pathways exhibit what would be described as 'fuzzy logic' in cybernetic programming terms (Moore, 1998). Instead of viewing fungal cell differentiation as involving individual major (yes/no) 'decisions' which switch progress between alternative developmental pathways that lead inevitably to specific combinations of features, this idea suggests that the endpoint in fungal differentiation depends on the balance of a network of minor 'approximations'. Fuzzy logic is an extension of conventional (Boolean) logic that can handle the concept of partial truth, that is, truth-values between 'completely true' and 'completely false'. It is the logic underlying modes of decision-making that are approximate rather than exact, being able to handle uncertainty and vagueness and has been applied to a wide variety of problems.

The importance of fuzzy logic derives from the fact that the theory provides a mathematical basis for understanding how decision-making seems to operate generally in biological systems, which are not always 'all-or-none' decisions. In the case of sporophore structure, the outcome is that fungal cells are allowed the flexibility to assume a differentiation state even when all conditions of that state have not yet been met. In other words, developmental decisions between pathways of differentiation are able to cope with a degree of uncertainty. The conclusion is that fungal differentiation pathways must be based on application of rules that allow considerable latitude in expression (fuzzy constraints), which in the ultimate can lead to highly polymorphic, ***yet functional***, sporophores.

One consequence of this line of argument is that ***patterning genes may not be necessary*** to develop the patterns of a mushroom. Rather, as we have described, the patterns may arise from the regular application of simple, and fuzzy, developmental rules and basic metabolic regulation combined with physical forces (stretching, inflation, expansion). A specific gill pattern in an agaric, say, may not require a panel of regulatory genes that specify 'x number of gills per unit volume of cap' because that pattern 'invariably' results from the application of general rules about the local circumstances within the tissue that allow gills to be made and the subsequent set of circumstances that drive mushroom maturation.

Figure 13.27. Polymorphic sporophores of *Volvariella bombycina*. (A) is the normal morphology, of a typical agaric mushroom, which is characterised by a well-developed enclosing volva through which the mushroom cap emerges during development (bulbangiocarpic development, see Figure 13.7) (scale bar, 20 mm). (B–G) illustrate a few of the polymorphisms which have arisen on otherwise normal cultures, though often seeming to be associated with some environmental stress, such as abnormal temperature and/or desiccation. (B) is apparently normal, but completely lacks the volva despite this being characteristic of the genus (scale bar, 5 mm). (C) is a gymnocarpous sporophore (see Figure 13.3) with a grossly enlarged basal volva (scale bar, 2 mm). (D) is a sporophore with a sinuous extra hymenium formed by the proliferating margin of the cap curling onto the upper surface of the cap (scale bar, 10 mm). (E) is a scanning electron microscope (SEM) view of specimen D and shows the well-developed gills on the upper surface of the cap (scale bar, 1 mm). (F) is a sporophore in which hymenium has formed all over the outer surface of the club-shaped sporophore (scale bar, 5 mm). (G) is an SEM view of the bisected sporophore and shows that the labyrinthine hymenophore covers the entire surface of the club-shaped 'pileus' (scale bar, 0.5 mm). This polymorph is called 'morchelloid' because of its resemblance to the normal sporophore of the morel (*Morchella*), which, of course, is a member of the *Ascomycota*! Photomicrographs by Professor S.-W. Chiu, Chinese University of Hong Kong. Illustration modified from Chiu *et al*. (1988).

The first complete genome sequence of any eukaryote to be released, in 1996, was that of the budding yeast *Saccharomyces cerevisiae*. It was the work of a worldwide team of hundreds of researchers (Goffeau *et al*., 1996). In the time since then, the yeast genome has been intensively studied (Cuomo & Birren, 2010) exemplifying numerous approaches that have been followed with other organisms. An updated reference genome sequence prepared with more recent (so-called second-generation) sequencing technologies using a single yeast colony of *Saccharomyces cerevisiae* strain S288C has been published by Engel *et al*. (2014). The genome sequence of *Saccharomyces cerevisiae* enabled systematic genome-wide experimental approaches. The approach has been called a genome-scale systems analysis (Weissman *et al*., 2010). Systems analysis is a problem-solving technique that involves study of the entire set of parts or components that are connected to form the more complex whole process (the 'system').

Molecular analysis provides an overview of the complex network of gene regulatory events that are involved in a developmental (or any other) process, but it can be difficult to focus on the unique decisive events that characterise (and/or initiate) the process. New sequencing methods have hastened genome-scale comparative studies of fungi by reducing the cost and difficulty of sequencing. This has enabled wider use of sequencing applications, such as to study genome-wide variation in populations (for example, Wyss *et al*., 2016) or to profile RNA transcripts totally (Conesa *et al*., 2016). The rapid impact of this level of analysis on research with yeast prompted efforts to expand genomic resources for other fungi, and the Fungal Genome Initiative was launched as an organised genome sequencing effort to assemble genomes of over 100 fungi in public databases to promote comparative and evolutionary studies across the fungal kingdom (Cuomo & Birren, 2010; Stajich, 2017). If you've not already done so, check out volumes 470, 500, 531 and 546 in the ***Methods in Enzymology*** book series; the citations are given above.

Following on from the development of new sequencing technologies, the pace of progress accelerated considerably, and in 2011 (http://mycor.nancy.inra.fr/blogGenomes/?p=2659), the ***1,000 Fungal Genomes Project*** (http://1000.fungalgenomes.org/home/) aimed to 'sequence 1,000+ fungal genomes' and at the time of writing (2019), there are over 1,000 fungal genomes listed as complete in at least draft form in the ***MycoCosm*** portal of the **JGI** (the US Department of Energy) **Joint Genome Institute**; https://jgi.doe.gov/) (Markowitz *et al.*, 2012, 2015; Grigoriev *et al.*, 2014; Nordberg *et al.*, 2014). We have described these general topic areas in Chapter 6; here, as an illustration of some of the extensive genomic systems analysis work that has been done, we will return to the ink cap mushroom, *Coprinopsis cinerea*, and particularly its sporophore development.

Coprinopsis cinerea was included in the list of fungi chosen for the Fungal Genome Initiative in 2002 (Cuomo & Birren, 2010) and the genome of *Coprinopsis cinerea*, which comprises 36 megabase pairs (a megabase pair (Mbp) is a unit of length of nucleic acids, equal to 1 million base pairs) was sequenced and assembled into 13 chromosomes by an international team of 47 researchers (Stajich *et al.*, 2010). This was used to investigate how rates of meiotic recombination varied along the chromosomes, and the distribution, expression and conservation of single-copy genes and gene families. It was evident that regions of the genome that have been conserved over evolutionary time show low rates of meiotic recombination and lack sequences such as tandemly repeated members of large gene families and transposable elements that could promote recombination between nonhomologous chromosomes (called ectopic recombination), which can cause chromosomal rearrangements. Single-copy genes with identifiable **orthologous genes** in other basidiomycetes (orthologous means the same function in different species and descended from a common ancestor) were also mostly located in low-recombination regions of their chromosome. On the other hand, **paralogous genes** (meaning genes related by duplication that may evolve new functions) were found in the highly recombining regions and near chromosome ends. These latter included families of hydrophobin genes (section 7.8) and cytochrome P_{450} gene families that code for the terminal oxidases of the electron transfer chains.

The genome project also revealed two genes of the DNA (cytosine-5)-methyltransferase-1 family that catalyses transfer of methyl groups to specific CpG structures in DNA to form 5-methylcytosine, a process called **DNA methylation**. This confirms prior studies which showed methylation of repeated genes in *Coprinopsis cinerea*. DNA methylation is used to control gene expression by changing the activity of DNA without changing the sequence; methylation of a gene's promoter typically represses transcription. The *Coprinopsis cinerea* genome contained 38 copies of a gene predicted to catalyse the formation of the additional modified base, 5-hydroxymethylcytosine, which has been detected in *Coprinopsis cinerea* DNA. This is especially relevant to our current discussion of fungal development because 5-hydroxymethylcytosine has a role in the '**poised (or bivalent) chromatin**' domains found in mammalian embryonic stem cells that play an important role in cell differentiation.

Poised chromatin is defined by the simultaneous presence of histone modifications associated with both gene activation and gene repression and is usually found near promoters of transcriptionally silent developmental regulatory genes. In general, pluripotent cells (capable of giving rise to several different cell types) have high numbers of **poised domains** compared with more differentiated cells, and these domains tend to resolve towards a purely active or purely repressed state during differentiation (Voigt *et al.*, 2013; Lesch & Page, 2014).

5-Methylcytosine in the DNA of eukaryotes is an important mark for **epigenetic gene regulation**; epigenetic marks silence certain gene sequences and activate others so that cells can differentiate. The 'ten eleven translocation' enzymes (**TET**; the enzymes are named for a common translocation in human cancers) oxidise 5-methylcytosines and promote locus-specific reversal of DNA methylation. 'J binding proteins' (**JBP**; originally named for their effect on glycosylated thymine in trypanosome DNA, which is known as base J) catalyse hydroxylation of bases in nucleic acids. TET/JBP enzymes are a family of nucleic acid base-modifying 2-oxoglutarate and iron-dependent dioxygenases that demethylate by converting 5-methylcytosine to oxidised methylcytosines, such as hydroxy-, formyl- and carboxymethylcytosine, which serve as further epigenetic marks or intermediates for demethylation. Iyer *et al.* (2014) showed that fungi, especially all major clades of basidiomycetes, have numerous TET/JBP genes, which are often associated with a unique class of transposable elements. This pattern differs starkly from the situation in most other organisms that possess just a single or a few copies of the TET/JBP family. This unique transposon-TET/JBP association in fungi is likely to play important roles in speciation during evolution and epigenetic regulation. 5-methylcytosine (5mC) and oxidised methylcytosines (oxi-mCs) have been mapped in *Coprinopsis cinerea* at close to base pair resolution (Chavez *et al.*, 2014). Centromeres and transposable elements had distinctive patterns of 5mC and oxi-mC, and 5mC and oxi-mC within genes silenced duplicated genes.

The *Coprinopsis cinerea* genome includes several highly expanded unusual protein kinase families including some orthologous to kinases of *Saccharomyces cerevisiae*, several kinase classes found in animals and/or *Dictyostelium*, but lost from yeast, and a hugely expanded, to 133 members, family of **FunK1 kinases**. Protein kinases are enzymes that regulate the activity of other enzymes by phosphorylation, which may turn an inactive enzyme into an active form (or vice versa), or otherwise change enzyme activity, cellular location, or its interaction with other proteins. FunK1 kinases are only found in multicellular ascomycete and basidiomycete fungi. The unusually large gene family in *Coprinopsis cinerea* shows considerable coding diversity; also 59 of the 133 *FunK1* family genes are found within a short region near the end of chromosome 9, overlapping a recombination hotspot in a region where meiotic synapsis initiates. The FunK1 kinases are differentially expressed during dikaryon formation (mating), so they appear to have different functions during sexual morphogenesis and important roles in development of *Coprinopsis cinerea*.

Finally, when the genomes of *Coprinopsis cinerea* and *Laccaria bicolor*, an ectomycorrhizal basidiomycete that grows in mixed birch and pine woods, were compared they were found to share extensive regions of **synteny** (meaning blocks of genes in a similar order within two sets of chromosomes that are being compared with each other). The largest syntenic blocks were found to occur in regions with low meiotic recombination rates, no transposable elements and tight gene spacing (Stajich *et al.*, 2010).

More evidence of tissue and development-specific gene control was found by studies that focused on the transcriptional programme of *Coprinopsis cinerea*. Burns *et al.* (2010) examined the transcripts produced during meiosis, including time points extending over the 15 h from before haploid nuclear fusion through to formation of basidiospore tetrads (see section 13.9). They found that about 20% of the genes in the genome are differentially regulated during this developmental process, with successive waves of transcription apparent in nine transcriptional clusters, one of which was enriched for meiotic functions. Transcriptional waves of this sort in other organisms have been found to represent differential expression of genes essential for recombination, chromosome cohesion, chromosome segregation, etc. Comparison with meiotic expression data from *Saccharomyces cerevisiae* and *Schizosaccharomyces pombe* showed that meiotic regulatory machinery was not conserved (not too surprising as, physiologically, both yeasts are stimulated to enter meiosis by nutritional limitation, while induction of mushroom development and meiosis in *Coprinopsis cinerea* is largely controlled by illumination and temperature).

However, a 'core meiotic transcriptome' of 75 genes was conserved between the three fungi. Ascomycetes and basidiomycetes are thought to have diverged in evolution about 500–900 million years ago, with *Saccharomyces cerevisiae* and *Schizosaccharomyces pombe* diverging from one another shortly after. Conservation of the core meiotic transcriptome between these three fungi implies that the essential transcriptional programme of meiosis has persisted through at least half a billion years of evolution, and potentially for up to a billion years (Burns *et al.*, 2010).

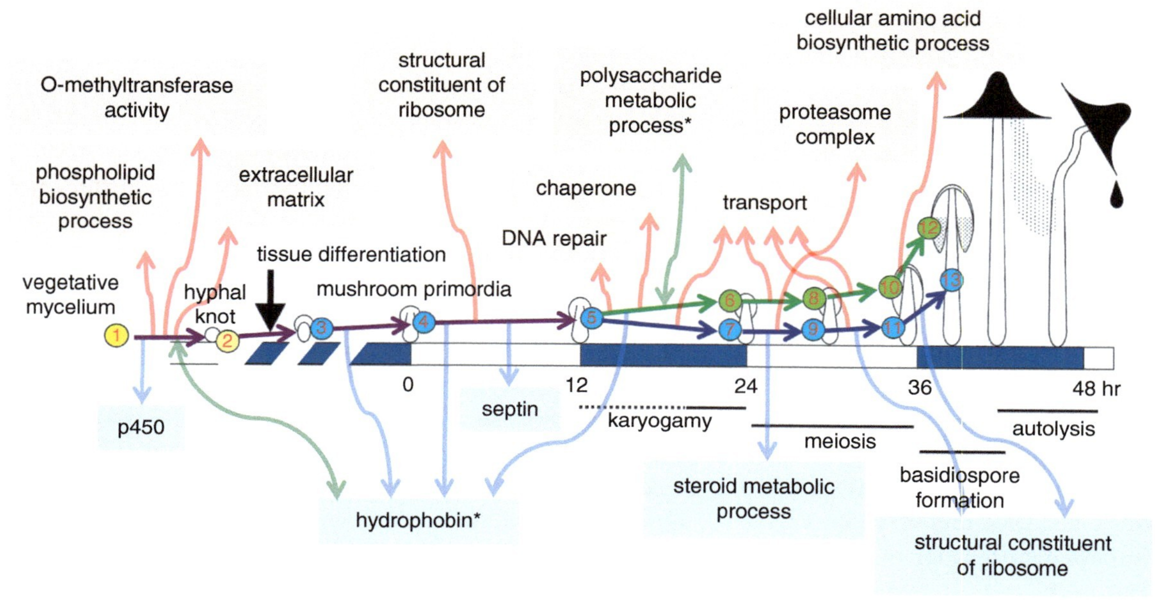

Figure 13.28. Summary of major transcriptional events during morphogenetic transitions during mushroom development in *Coprinopsis cinerea*. The cartoon diagrams of fruiting across the centre run from vegetative mycelium at left to autolysing mature mushroom at right and are placed on an average time scale with time zero being set by the illumination cycle that triggers primordium maturation. The nine developmental stages selected to investigate the transcriptome are shown with numbers in circles (cap and stipe tissue samples were taken separately from 24 h onwards). Developmental and cellular events in each stage and tissue are shown in Table 13.1. The study identified differentially expressed gene sets using GO enrichment analysis (http://geneontology.org/page/go-enrichment-analysis); that is, sets of gene sequences of related function that are upregulated or downregulated in pairwise comparisons between two succeeding developmental stages, and between vegetative mycelium and each individual developmental stage. The programme does this by statistically determining which sequence(s) are over-represented (or under-represented) in each member of the pair relative to the other. Notable transcriptional programme events are depicted in the text labels. Events detected by upregulated differentially expressed genes are indicated in the upper part with upward red arrows; downregulated differentially expressed gene sets are shown in the lower part with downward blue arrows. Sets which featured a mix of both upregulated and downregulated gene sequences of related function are indicated with asterisks and bidirectional green arrows. Taken from Muraguchi *et al.* (2015) under Creative Commons licence https://creativecommons.org/licenses/by/4.0/.

Table 13.1. Major morphogenetic events and transcriptional transitions during mushroom development in *Coprinopsis cinerea*

Tissue designation		Cellular events	Differential expression in each transition between developmental stages			
			Transition	Regulation	Transcripts	Top three enriched sequence annotations (functions)
1	Vegetative mycelium	Apical growth, branching, ageing	1–2	UP	408	Phospholipid biosynthesis; O-methyltransferase activity; extracellular region↕
				DOWN	134	Cytochrome P_{450}; hydrophobin↕; adenyl nucleotide binding
2	Mycelium and hyphal knots	Central and surface cells differentiate	2–3	UP	2	-
				DOWN	3	-
3	Small primordia	Stipe and cap clearly demarcated	3–4	UP	85	adenyl nucleotide binding↕
				DOWN	38	adenyl nucleotide binding↕; hydrophobin
4	Primordia begin maturation 0 h	Light stimulation; set as time zero	4–5	UP	148	Structural constituent of ribosome; zinc ion binding; FAD binding
				DOWN	590	Hydrophobin; fungal phospholipid biosynthesis; septin complex
5	Whole primordium at 12 h	Pre-meiotic DNA replication starts	5–6	UP	949	DNA repair; chaperonin Cpn60/TCP-1; glucose catabolism↕
				DOWN	1,660	Hydrophobin; cytochrome P_{450}; polysaccharide metabolism
			5–7	UP	3,301*	Transport; FAD binding; ATP coupled proton transport
				DOWN	3,569*	Structural constituent of ribosome; RNA processing; RNA recognition motif, RNP-1
6	Cap, 24 h after light trigger	Karyogamy in basidia	6–8	UP	2,471	Transport; FAD binding; cytochrome P_{450}
				DOWN	2,034	RNA recognition motif, RNP-1; RNA processing; structural constituent of ribosome

table continues

Tissue designation		Cellular events	Differential expression in each transition between developmental stages			
			Transition	Regulation	Transcripts	Top three enriched sequence annotations (functions)
7	Stipe, 24 h after light trigger	Stipe nuclei divide before elongation	7–9	UP	457	Proteasome complex; AMP-dependent synthetase and ligase; EF-hand calcium-binding proteins
				DOWN	326	Steroid metabolism; lipase activity; lipid biosynthesis
8	Cap, 30 h after light trigger	Karyogamy + 6 h	8–10	UP	3,465*	Transmembrane; FAD binding; cytochrome P_{450}
				DOWN	3,197*	Structural constituent of ribosome; WD40 repeat histone-binding proteins; vesicle-mediated transport
9	Stipe, 30 h after light trigger	Stipe enlarges	9–10	UP	2,088	Transport; vitamin B6 binding; cytochrome P_{450}
				DOWN	2,356	Structural constituent of ribosome; WD40 repeat histone-binding proteins; ribosome biogenesis
10	cap, 36 h after light trigger	Karyogamy + 12 h	10–12	UP	1,743	Nitrogen compound biosynthesis; cellular amino acid biosynthesis; alcohol dehydrogenase GroES-like fold (catalytic domain of alcohol dehydrogenases) proteins
12	cap, 39 h after light trigger	Karyogamy + 15 h spores form		DOWN	1,701	ATP binding; N-terminal FAD-linked oxidase; galactose oxidase, beta propeller (ligand binding)
11	stipe, 36 h after light trigger	Stipe starts to elongate	11–13	UP	136	Cytochrome P_{450}↕; C2H2-type zinc-finger proteins (transcription factors); metal ion binding
13	stipe, 39 h after light trigger	Stipe elongates Cap expands		DOWN	192	Cytochrome P_{450}↕; FAD binding; structural constituent of ribosome

↕, indicates that similar functions are found in both upregulated and downregulated differentially expressed genes.
*, indicates that only the first 3,000 genes were analysed. Data from Muraguchi *et al.* (2015).

The transcription pattern during the transition from the vegetative mycelium to the primordium during mushroom development of *Coprinopsis cinerea* was examined by Cheng *et al.* (2013). They evaluated the expression of more than 3,000 genes in both growth stages and discovered that almost one-third of these genes were preferentially expressed in one or other stage. The data indicated that various structural and functional protein families were uniquely employed in one or other stage and that during primordial growth, cellular metabolism is highly upregulated. Various signalling pathways were also upregulated, particularly:

- the 'protein kinase A' family of enzymes whose activity is dependent on cellular levels of cyclic-AMP (**cAMP-PKA**);
- the 'mitogen-activated protein kinase' family (**MAPK**); and
- the 'target of rapamycin' (**TOR**) signalling pathway, which integrates both intracellular and extracellular signals and serves as a central regulator of cell metabolism.

All of which is consistent with the idea that sensing of nutrient levels and environmental conditions are important in this developmental transition. Over 100 of the upregulated genes were found to be ***unique to mushroom-forming basidiomycetes***, highlighting the novelty of sporophore development in these fungi (Cheng *et al.*, 2013).

The transcriptomes of stage 1 primordia and vegetative mycelium of *Coprinopsis cinerea* were also studied by Plaza *et al.* (2014), but these authors paid attention to regulation of the expression of defence genes as well as the conserved transcriptional patterns of sexual reproduction. Comparing their results with analogous data from *Laccaria bicolor* and *Schizophyllum commune* revealed a ***conserved transcriptional circuitry in basidiomycete sporophores*** consisting of nearly 70 genes involved in mushroom development and function. They also found evidence for a role of the '**velvet protein regulon**' in sexual development of basidiomycetes, as a velvet domain-containing protein was found to be upregulated in all three fungi. The four velvet proteins VeA, VelB, VelC and VosA were originally identified in *Aspergillus nidulans* and shown to form a family of fungal regulators that couple together light-regulated processes, control of secondary metabolism, growth and differentiation. Velvet proteins share a region of about 150 amino acids, the velvet domain, which lacks sequence homology to any other proteins (Bayram *et al.*, 2008; Ahmed *et al.*, 2014). These velvet regulators are present in most clades of the fungal kingdom from chytrids to basidiomycetes.

Muraguchi *et al.* (2015) extended the investigation of the transcriptome during mushroom development in *Coprinopsis cinerea* by using high-throughput sequencing of cDNA libraries from 13 developmental points (nine developmental stages, but from stage 6 (onset of meiosis) cap and stipe tissues were analysed separately; see Table 13.1). The analysis revealed changes in expression of many differentially expressed genes between two succeeding developmental stages and between vegetative mycelium and each developmental stage. GO analysis of ***differentially expressed genes*** (**DEGs**) in each transition revealed an overall developmental framework and many notable cellular events; these results are summarised in Figure 13.28 and Table 13.1.

Looking at the discussion so far you can see that the conclusions from the ***pregenomics era*** have been broadly confirmed by genome systems analysis, though expanded and given their quantitative context (Table 13.1). Several hundred to a thousand genes are preferentially expressed at different stages in development of the mushroom. Various ***signalling pathways*** have been implicated including ***cAMP-PKA, MAPK*** and ***TOR***, the fungus-specific ***velvet protein regulon*** and the ***FunK1 kinases***, which are only found in multicellular ascomycete and basidiomycete fungi. Considering the accounts given above (section 13.11) of the cycles of glycogen accumulation and translocation in different mushroom tissues it is interesting to see in Figure 13.28 that the category 'polysaccharide metabolic process' is a set that featured a mix of both upregulated and downregulated transcripts. Similarly, the importance attached to enzymes involved in ammonium scavenging and the urea cycle (section 13.11) fits well with the upregulation of transcripts in the category 'cellular amino acid biosynthetic process'. Indeed, transcriptional profiling is beginning to unravel this 'glycogen regulatory circuit' by showing that glycogen synthase kinase 3 (GSK-3) activity is essential for sporophore formation in *Coprinopsis cinerea* by directly affecting expression of sporophore induction genes together with genes in downstream cellular processes (Chan *et al.*, 2018); and experimental inhibition of GSK-3 expression can control mushroom development (Chang *et al.*, 2018b).

There are several other features of Figure 13.28 that confirm deductions from the pregenomics era, particularly the upregulation of 'transport' transcripts in both cap and stipe between stages 5 and 6/7, and then again between 6/7 and 8/9, and yet again between 8/9 and 10/11 all confirm the presumed rapid translocation of metabolites and water during mushroom maturation. Perhaps not quite so obvious is that the upregulation of transcripts in the 'extracellular matrix' category fits well with what was known about the importance of the extracellular matrix to hyphal knot formation; and the upregulation of 'proteasome complex' in the stipe between stages 6/7 and 8/9, with the implication that protein degradation is important to the stipe at these stages for some reason is compatible with what is known about the biochemistry of the *Coprinopsis* stipe which, unusually, does not accumulate sugars or polyols and is likely, therefore to use the carbon skeletons of amino acids derived from protein degradation as energy source. Furthermore, the conclusion that differentiation of fungal multicellular structures is genetically compartmentalised with normal morphogenesis being made up of several-to-many ***developmental subroutines*** (see above) are equivalent to the ***transcriptional clusters, transcriptional waves*** of differential gene expression and ***differentially expressed gene sets*** detected in these different studies of the transcriptional programme of the *Coprinopsis cinerea* genome. Differential transcriptional patterns of a family of glycoside hydrolases of *Coprinopsis cinerea* matched well with the requirement for the wall-softening that these enzymes effect in the germination of basidiospores, hyphal growth and branching, primordium formation, stipe elongation, pileus expansion and autolysis (Kang *et al.*, 2019). Some members of the family were predominantly expressed in dormant basidiospores, primordia and during basidiosporogenesis. Other members were dominantly expressed in the germinating basidiospores, growing mycelia, and elongating stipes. The dynamics of the proteome during fruit body development in *C. cinerea* has also been studied (Majcherczyk *et al.*, 2019).

Muraguchi *et al.* (2015) also observed dramatic changes in gene expression of some transcription factor networks, and several examples of **natural antisense transcripts** (**NATs**) that are likely to be involved in developmental regulation of sporulation. NATs are widespread in eukaryotes; they are RNAs that are complementary to other ('sense') RNA transcripts. Evidence indicates that sense and antisense transcripts interact, suggesting a role for NATs in the regulation of gene expression (Wight & Werner, 2013). MicroRNAs (**miRNAs**) are about 22 nucleotides in length and have essential roles in post-transcriptional regulation of gene expression in animals and plants. Although little studied in fungi, microRNAs have been associated with early mushroom development in *Coprinopsis cinerea*, where Lau *et al.* (2018) established their potential targets and demonstrated ***differential miRNA expression*** during the early developmental stages of the mushroom, suggesting their involvement in regulating development in this fungus.

Krizsán *et al.* (2019a, b) constructed a reference atlas of mushroom formation based on comparisons of developmental transcriptome data of more than 200 whole genomes belonging to six species of *Agaricomycetes* (*Armillaria ostoyae, Coprinopsis cinerea, Lentinus tigrinus, Phanerochaete chrysosporium, Rickenella mellea* and *Schizophyllum commune*); 12,003 to 17,822 expressed genes were detected in the six species of which 938 to 7,605 were developmentally regulated. Importantly, nearly 300 **conserved gene families** in more than 70 functional groups contained developmentally regulated genes. These data outline the major multicellularity related and developmental processes of mushrooms, including the role of transcriptional reprogramming, gene coexpression networks (corresponding to the developmental subroutines discussed above) and alternative splicing. Many of the developmentally regulated sequences were expanded in multicellular plants and/or animals too, implying convergent evolution towards solutions to the challenges imposed by complex multicellularity within independently evolving lineages. Data relating to the evolutionary age distribution of developmentally regulated genes suggested that most developmental gene families are older than sporophore formation; the sequences falling into three groups: (a) genes shared by all living species; (b) genes shared by the *Dikarya* (*Ascomycota* and *Basidiomycota*); and (c) species-specific genes.

Another promising technique for detecting transcriptional regulatory relationships on a genome-wide scale that is beginning to be applied to fungal development is the analysis of expression Quantitative Trait Loci (**eQTL**). A ***quantitative trait locus*** (**QTL**) is a DNA segment associated with a particular phenotype, which varies across a population in a manner attributed to polygenic control (presumed to be the product of two or more genes together with their environment). The approach is essentially a metagenomic statistical analysis of variation in the phenotype of interest that can reveal the genetic architecture of that trait; that is, whether phenotype is controlled by many genes of small effect, or by a few genes of large effect. Expression quantitative trait loci (**eQTLs**) are genomic loci that describe the variation in expression levels of mRNA molecules. Analysis of eQTLs can recognise potential regulatory factors of gene expression and could identify master regulators in the genome (Michaelson *et al.*, 2009; Liseron-Monfils & Ware, 2015; Cheng *et al.*, 2016; Chang *et al.*, 2018a).

Because the current interpretation of the mechanism of gravity perception in the *Coprinopsis cinerea* mushroom is that cytoskeletal interactions communicated to the transmembrane domains of a sugar transporter change its substrate specificity to cause a 6-deoxy hexose sugar growth regulator to be transported asymmetrically (see sections 10.8 and 13.9), it is significant that the eQTL network analysis of Chang *et al.* (2018a) identified a ***sugar transporter as an 'eQTL hotspot'***. Such hotspots are believed to regulate expression of several genes, and the sugar transporter identified as a hotspot in this study regulated the expression of at least 12 associated genes. One of those associated genes specifies a Zn-finger protein (which can bind DNA, RNA, proteins or small molecules) and another is a pyranose dehydrogenase enzyme with broad substrate tolerance for oxidation of different sugars. The transporter is thought to interact with the pyranose dehydrogenase via a transcription factor.

We already knew about the extensive use of transcription factors to direct transcription to specific genes in fungi (see sections 2.4, 5.4, 5.12, 7.7, 9.5, 10.4), particularly the homeodomain proteins of the mating type factors (Bobola & Merabet, 2017; Vonk & Ohm, 2018), so add those to the list of **fungal regulators**. We should also add to that list of regulators ***translational triggering*** and ***feedback fixation***, discovered in the 1990s in research on *Aspergillus nidulans* conidiophores (section 10.4), which solve the problems of how to respond to a transient signal, and how to maintain that response when the signal has dissipated. The mechanism involves a genetic sequence, in this case the *brlA* locus, that codes for an RNA transcript containing overlapping reading frames (Figure 10.9). ***Alternative splicing of precursor mRNA*** is now established as an essential mechanism to increase the complexity of gene expression that plays an important role in cellular differentiation and development in eukaryotes (Wang *et al.*, 2015; section 6.4). Alternative splicing is common in higher plants and animals and increases the coding capacity from a limited set of genes. Grützmann *et al.* (2014) claim that alternative splicing is also a common phenomenon in fungi, and that, on average, about 6.0% of genes in ascomycetes, and 8.6% of basidiomycete genes show alternative splicing (*Cryptococcus neoformans* had the extraordinary rate of 18%), and that this process is associated with multicellular complexity, dimorphic switching, stress response and pathogenic virulence.

Fungal mycelia, sporophore primordia and mature sporophores are attractive sources of food to small animals including molluscs, arthropods and nematodes (section 12.2; Figure 12.11), and it is well known that mushrooms produce defence proteins and secondary metabolites to deter such predators and other competitors. Plaza *et al.* (2014) revealed the existence of two different sets of **fungal defence proteins** in the proteomes of mushroom primordia of *Coprinopsis cinerea*. One of the primordium-specific proteins is toxic to insects and nematodes and the expression pattern of these proteins correlates with the type of antagonists with which these tissues are confronted. For instance, cytoplasmic lectins, also referred to as sporophore lectins due to their specific expression pattern, have been shown to be toxic to nematodes, insect larvae and amoebae (Bleuler-Martínez *et al.*, 2011; see section 15.6). In contrast, vegetative mycelia secrete antibacterial peptides which deter competing bacteria. Known as **defensins**, these are found in vertebrates, higher plants and invertebrates as well as saprotrophic fungi (Wu *et al.*, 2014). Defensins have therapeutic promise; plectasin, isolated from the ascomycete *Pseudoplectania nigrella*, is especially active against *Streptococcus pneumoniae*, including strains resistant to conventional antibiotics. Genome editing techniques are now available to manipulate production of potentially useful metabolites like these (Doudna & Sontheimer, 2014; and see section 6.10), and a novel method of high-throughput CRISPR/Cas9-based genome editing of *Coprinopsis cinerea* using cryopreserved protoplasts has been reported (Sugano *et al.*, 2017).

Surprisingly, there is no mention in any of these studies of homologues of sequences of the animal signalling mechanisms known as Wnt, Hedgehog, Notch and TGF-α, all of which are essential and highly conserved components of ***normal development in animals***. Systematic mining of fungal genomes for these sequences established that all were absent from the fungal genomes, just as they are absent from plants (Moore *et al.*, 2005). Subsequently, a fully comprehensive data-mining exercise searched for homologies to all developmental sequences in all sequenced genomes, animal, fungal and plant that were then available (Moore & Meškauskas, 2006, 2009). The outcome was that homology between the genomes of fungi and other eukaryotes was limited to 78 sequences involved in the architecture of the eukaryotic cell. This survey demonstrated that there was no evidence in fungal genomes of *Wnt, Hedgehog, Notch, TGF, p53*, which are all crucial animal regulators, nor of the plant control sequences *SINA*, or *NAM*. All of which leads to the conclusion that the unique cell biology of filamentous fungi has caused them to evolve control of their multicellular development in a radically different fashion from that in animals and plants.

In this section, we focused on the developmental biology of *Coprinopsis cinerea* but as we indicate above 'there are over 1,000 fungal genomes listed as complete in at least draft form in the ***MycoCosm*** portal'. We can't prolong this discussion any further, so to ease you into the world of fungal genomics we've put together a list of publications (open access text favoured, so you should be able to obtain the full texts) in Resources Box 13.2.

RESOURCES BOX 13.2 Genomic Systems Analysis of Fungi

MycoCosm is a fungal genomics portal (http://jgi.doe.gov/fungi), developed by the US Department of Energy Joint Genome Institute to support integration, analysis and dissemination of fungal genome sequences and other 'omics' data by providing interactive web-based tools. MycoCosm also promotes and facilitates user community participation through the nomination of new species of fungi for sequencing, and the annotation and analysis of resulting data. View the current list on the following page, and links to publications about the sequences: https://genome.jgi.doe.gov/fungi/fungi.info.html.

Grigoriev, I.V., Nikitin, R., Haridas, S. *et al.* (2014). MycoCosm portal: gearing up for 1000 fungal genomes. *Nucleic Acids Research*, **42**: D699–D704. DOI: https://doi.org/10.1093/nar/gkt1183.

The following is a list of published papers (additional to those cited in the text) that will ease you into the world of fungal genomics. The list has been put together favouring these criteria: (a) open access publishing; and (b) illustrating different approaches. The list is in alphabetical order of the first author's name.

Banks, A.M., Barker, G.L.A., Bailey, A.M. & Foster, G.D. (2017). Draft genome sequence of the coprinoid mushroom *Coprinopsis strossmayeri. Genome Announcements*, **5**: e00044-17. DOI: https://doi.org/10.1128/genomeA.00044-17.

Bao, D., Gong, M., Zheng, H. *et al.* (2013). Sequencing and comparative analysis of the Straw Mushroom (*Volvariella volvacea*) genome. *PLoS ONE*, **8**: article number e58294. DOI: https://doi.org/10.1371/journal.pone.0058294.

Carter, G.W., Galas, D.J. & Galitski, T. (2009). Maximal extraction of biological information from genetic interaction data. *PLoS Computational Biology*, **5**: article number e1000347. DOI: https://doi.org/10.1371/journal.pcbi.1000347.

de Freitas Pereira, M., Narvaes da Rocha Campos, A., Anastacio, T.C. *et al.* (2017). The transcriptional landscape of basidiosporogenesis in mature *Pisolithus microcarpus* basidiocarp. *BMC Genomics*, **18**: 157. DOI: https://doi.org/10.1186/s12864-017-3545-5.

Jin, L., Li, G., Yu, D. *et al.* (2017). Transcriptome analysis reveals the complexity of alternative splicing regulation in the fungus *Verticillium dahlia. BMC Genomics*, **18**: 130. DOI: https://doi.org/10.1186/s12864-017-3507-y.

Li, H., Wu, S., Ma, X. *et al.* (2018). The genome sequences of 90 mushrooms. *Scientific Reports*, **8**: article number 9982. DOI: https://doi.org/10.1038/s41598-018-28303-2.

Liu, J.-Y., Chang, M.-C., Meng, J.-L. *et al.* (2017). Comparative proteome reveals metabolic changes during the fruiting process in *Flammulina velutipes. Journal of Agricultural and Food Chemistry*, **65**: 5091–5100. DOI: https://doi.org/10.1021/acs.jafc.7b01120.

Martin, F., Aerts, A., Ahrén, D. *et al.* (2008). The genome of *Laccaria bicolor* provides insights into mycorrhizal symbiosis. *Nature*, **452**: 88–92. DOI: https://doi.org/10.1038/nature06556.

Martinez, D., Challacombe, J., Morgenstern, I. *et al.* (2009). Genome, transcriptome, and secretome analysis of wood decay fungus *Postia placenta* supports unique mechanisms of

box continues

lignocellulose conversion. *Proceedings of the National Academy of Sciences of the United States of America*, **106**: 1954–1959. DOI: https://doi.org/10.1073/pnas.0809575106.

McCluskey, K. & Baker, S.E. (2017). Diverse data supports the transition of filamentous fungal model organisms into the post-genomics era. *Mycology*, **8**: 67–83. DOI: 10.1080/21501203.2017.1281849.

Morin, E., Kohler, A., Baker, A.R. *et al.* (2012). Genome sequence of the button mushroom *Agaricus bisporus* reveals mechanisms governing adaptation to a humic-rich ecological niche. *Proceedings of the National Academy of Sciences of the United States of America*, **109**: 17501–17506. DOI: https://doi.org/10.1073/pnas.1206847109.

Murat, C., Payen, T., Noel, B. *et al.* (2018). Pezizomycetes genomes reveal the molecular basis of ectomycorrhizal truffle lifestyle. *Nature Ecology & Evolution*, **2**: 1956–1965. DOI: https://doi.org/10.1038/s41559-018-0710-4.

Nowrousian, M. (2018). Genomics and transcriptomics to study fruiting body development: an update. *Fungal Biology Reviews*, **32**: 231–235. DOI: https://doi.org/10.1016/j.fbr.2018.02.004.

Ohm, R.A., de Jong, J.F., Lugones, L.G. *et al.* (2010). Genome sequence of the model mushroom *Schizophyllum commune*. *Nature Biotechnology*, **28**: 957–963. DOI: https://doi.org/10.1038/nbt.1643.

Toyotome, T., Hamada, S., Yamaguchi, S. *et al.* (2018). Comparative genome analysis of *Aspergillus flavus* clinically isolated in Japan. *DNA Research*, article dsy041. DOI: https://doi.org/10.1093/dnares/dsy041.

Wang, L., Wu, X., Gao, W. *et al.* (2017). Differential expression patterns of *Pleurotus ostreatus* catalase genes during developmental stages and under heat stress. *Genes*, **8**: 335. DOI: https://doi.org/10.3390/genes8110335.

Wang, M., Gu, B., Huang, J. *et al.* (2013). Transcriptome and proteome exploration to provide a resource for the study of *Agrocybe aegerita*. *PLoS ONE*, **8**: article number e56686. DOI: https://doi.org/10.1371/journal.pone.0056686.

Wang, Z., Gudibanda, A., Ugwuowo, U., Trail, F. & Townsend, J.P. (2018). Using evolutionary genomics, transcriptomics, and systems biology to reveal gene networks underlying fungal development. *Fungal Biology Reviews*, **32**: 249–264. DOI: https://doi.org/10.1016/j.fbr.2018.02.001.

Wu, B., Xu, Z., Knudson, A. *et al.* (2018). Genomics and development of *Lentinus tigrinus*: a white-rot wood-decaying mushroom with dimorphic fruiting bodies. *Genome Biology and Evolution*, **10**: 3250–3261. DOI: https://doi.org/10.1093/gbe/evy246.

Zhang, J., Ren, A., Chen, H. *et al.* (2015). Transcriptome analysis and its application in identifying genes associated with fruiting body development in basidiomycete *Hypsizygus marmoreus*. *PLoS ONE*, **10**: article number e0123025. DOI: https://doi.org/10.1371/journal.pone.0123025.

Zheng, Y.-M., Lin, F.-L., Gao, H. *et al.* (2017). Development of a versatile and conventional technique for gene disruption in filamentous fungi based on CRISPR-Cas9 technology. *Scientific Reports*, **7**: article number 9250. DOI: https://doi.org/10.1038/s41598-017-10052-3.

And check out this special issue of the journal *Fungal Genetics and Biology* (Volume **89**, pages 1–156, April 2016) entitled *The Era of Synthetic Biology in Yeast and Filamentous Fungi* at https://www.sciencedirect.com/journal/fungal-genetics-and-biology/vol/89/suppl/C.

The **Earth BioGenome Project** aims to sequence, catalogue and characterise ***the genomes of all of Earth's eukaryotic biodiversity***, as described in the following:

Lewin, H.A., Robinson, G.E., Kress, W.J. *et al.* (2018). Earth BioGenome Project: sequencing life for the future of life. *Proceedings of the National Academy of Sciences of the United States of America*, **115**: 4325–4333. DOI: https://doi.org/10.1073/pnas.1720115115.

See also the *Earth BioGenome Project* website at https://www.earthbiogenome.org/ and the *Darwin Tree of Life Project* at https://www.sanger.ac.uk/news/view/genetic-code-66000-uk-species-be-sequenced.

13.15 SENESCENCE AND DEATH

Senescence and death are important aspects of biology in the other two major eukaryotic kingdoms and in fungi, too (Shefferson *et al.*, 2017). This is another cellular process that contributes to morphogenesis, as well as contributing to the evolutionary biology of the organism. Removal of old individuals makes way for the young and allows populations to evolve, and programmed cell death (PCD), which is the removal of tissue in a manner controlled in time and position, has been recognised as a crucial contributor to morphogenesis in both animals and plants. There are two types of cell death: traumatic or necrotic death and apoptosis or PCD.

In higher animals, PCD involves a sequence of well-regulated processes, including synthetic ones, which lead to internal cell degeneration and eventual removal of the dying cell by phagocytosis. It is important that apoptotic elimination of cells is ***intracellular*** in higher animals to avoid escape of antigens and the consequent danger of an immune response to components of the animal's own cells (autoimmunity). This is not a consideration in plants and fungi. The most obvious example of fungal PCD is the autolysis that occurs in the later stages of development of ink cap mushrooms which was long ago interpreted as an integral part of sporophore development (autolysis removes gill tissue from the bottom of the cap to avoid interference with spore discharge from regions above). Autolysis involves production and organised release of a range of lytic enzymes (Iten, 1970; Iten & Matile, 1970), so autolytic destruction of these tissues is clearly a programmed cell death.

There is only one experimental study of the longevity of fungal sporophores. Umar & Van Griensven (1997a) grew the cultivated mushroom in artificial environments which protected the culture from pests and diseases. They found that the lifespan of mushrooms of *Agaricus bisporus* was 36 days. Ageing was first evident in mushrooms about 18 days old, when localised nuclear and cytoplasmic lysis was seen, and after 36 days most of the cells in the mushroom were severely degenerated and malformed. Nevertheless, a number of basidia and subhymenial cells were alive and cytologically intact even on day 36. So even in severely senescent mushrooms healthy, living cells were found and these are presumably the origin of an unusual phenomenon known as **renewed fruiting**.

Field-collected sporophore tissues of a mushroom usually generate abundant vegetative hyphae when inoculated onto nutrient agar plates. Such reversion from the sporophore stage into vegetative stage is not an abrupt process, rather there appears to be some sort of 'memory' of the differentiated state. Initial hyphal outgrowth from gill lamellae usually mimics the densely packed branching and intertwined hyphal pattern of the gill tissues at first, being quite unlike the pattern of normal vegetative hyphae in culture. The 'memory' could be no more than the residual expression of differentiation-specific genes before their products are diluted out by continued vegetative growth, but as we have shown above (section 13.14), there is considerable scope for ***epigenetic marks*** in fungi.

Renewed fruiting (the formation of sporophores directly on old sporophore tissue) is not uncommon, and it can occur at various locations (cap, stem and/or gills) in improperly stored excised mushrooms. Experiments *in vitro* show that numerous primordia can arise on excised sporophore tissues and can mature into normal, though miniature, fruit bodies. In comparison to vegetative cultures, the excised sporophore tissues form sporophores very rapidly. For example, in *Coprinopsis cinerea*, renewed fruiting occurred within four days, compared with cultures inoculated with vegetative dikaryon which, under the same conditions, formed mushrooms in 10–14 days (Chiu & Moore, 1988a; Brunt & Moore, 1989; Bourne *et al.*, 1996). Renewed fruiting may have an important role in survival, consuming and immediately recycling the resource in the dying sporophore tissue to disperse further crops of spores.

Umar & Van Griensven (1997b, 1998) found that cell death is a common occurrence in various structures starting to differentiate, for example the formation of gill cavities in *Agaricus bisporus*. The authors point out that specific timing and positioning imply that cell death is part of the differentiation process. **Fungal PCD** could play a role at many stages in development of many species (Moore, 2013). Individual hyphal compartments can be sacrificed to trim hyphae to create tissue shaping. Fungal PCD is used, therefore, to sculpture the shape of the sporophore from the raw medium provided by the hyphal mass of the sporophore initial and primordium. In several examples detailed by Umar & Van Griensven (1998), the programme leading to cell death involves the sacrificed cells overproducing mucilaginous materials which are released by cell lysis. Remember that in autolysing *Coprinopsis* gills the cell contents released on death contain heightened activities of lytic enzymes (section 13.11). Evidently, in fungal PCD the cell contents released when the sacrificed cells die are specialised to specific functions too.

Fungal cultures suffer spontaneous degeneration through successive subculture on artificial media; the culture may stop growing or suffer loss of (or severe reduction of) asexual sporulation, sexuality, ability to fruit, or reduced production of secondary metabolites. Thus, fungal culture degeneration can be a significant economic problem for industrial production processes. It's a problem that applies as much to yeast cultures as to cultures of filamentous fungi. A budding yeast mother cell can produce a finite number of daughter cells before it stops dividing and dies, whereas filamentous fungal cultures frequently and spontaneously degenerate during ongoing culture maintenance, resulting in formation of sterile and/or weakly-growing sectors in the colony. Senescence seems to result from a progressive decline in physiological function, including mitochondrial dysfunction. This physiological decline is linked to impairments of cellular machines and to the generation of reactive oxygen species (chemically reactive molecules containing oxygen) which arise during normal metabolism (e.g. in the respiratory chain), and serve essential functions, but can cause molecular damage if in excess.

In *Neurospora*, senescing strains usually contain mitochondrial plasmids, which cause insertional mutagenesis when they integrate into the mitochondrial DNA. The functionally defective mitochondria replicate faster than the wild-type mitochondria and spread between hyphal cells (Maheshwari & Navaraj, 2008). Senescence can also be due to spontaneous lethal nuclear or mitochondrial gene mutations. Ultimately the growth of a fungal colony ceases due to dysfunctional oxidative phosphorylation (Li *et al.*, 2014; Wiemer *et al.*, 2016). Although the detailed underlying processes may differ from species to species, this situation appears to be basically conserved between organisms and to be a major cause of degenerative diseases in humans (Morales-González, 2013); yet another area of eukaryote biology in which research on fungi can contribute to human health.

13.16 BASIC PRINCIPLES OF FUNGAL DEVELOPMENTAL BIOLOGY

The main features of fungal developmental biology have been summarised as a set of principles by Moore (2005). The list (below) combines most of the observations made in the pregenomics era discussed above. It starts with the reminder that in many fungi hyphae differentiate from the vegetative form that ordinarily composes a mycelium and aggregate to form tissues of multi-hyphal structures; which may be linear organs that emphasise parallel arrangements of hyphae, or globose masses that emphasise interweaving of hyphae, such as sclerotia and fruit of the larger *Ascomycota* and *Basidiomycota*. The series of principles on which fungal morphogenesis is suggested to depend, most of which differ from both animals and plants, are as follows:

- **Principle 1**: The fundamental cell biology of fungi on which development depends is that hyphae extend only at their apex, and cross-walls form only at right angles to the long axis of the hypha.
- **Principle 2**: Fungal morphogenesis depends on the placement of hyphal branches. Increasing the number of growing tips by hyphal branching is the equivalent of cell proliferation in animals and plants. To proliferate, the hypha must branch, and to form an organised tissue the position of branch emergence and its direction of growth must be controlled.
- **Principle 3**: The molecular biology of the management of cell-to-cell interactions in fungi is completely different from that found in animals and plants; comprehensive genome surveys found no evidence in fungal genomes sequences which are crucial animal regulators, nor of plant control sequences.
- **Principle 4**: Fungal morphogenetic programmes are organised into developmental subroutines, which are integrated collections of genetic information that contribute to individual isolated features of the programme. Execution of all the developmental subroutines at the right time and in the right place results in a normal structure. Developmental subroutines are equivalent to the transcriptional clusters, transcriptional waves of differential gene expression and differentially expressed gene sets detected in the transcriptional programme of the genome.
- **Principle 5**: Because hyphae grow only at their apex, global change to tropic reactions of all the hyphal tips in a structure is sufficient to generate basic sporophore shapes.
- **Principle 6**: Over localised spatial scales, coordination is achieved by an inducer hypha regulating the behaviour of a surrounding knot of hyphae and/or branches. These are called hyphal knots and they have two fates, becoming either sclerotia or sporophore initials depending on temperature and illumination (high temperature and darkness favour sclerotia; light and lowered temperature favour the sporophore pathway).
- **Principle 7**: The response of tissues to tropic signals and the response of hyphal knots to their inducer hyphae, coupled with the absence of lateral contacts between fungal hyphae analogous to the plasmodesmata, gap junctions and cell processes that interconnect neighbouring cells in plant and animal tissues suggest that development in fungi is regulated by morphogens communicated mainly through the extracellular environment. Upregulation of genomic transcripts in the 'extracellular matrix' category corresponds with what is known about the importance of the extracellular matrix to formation of hyphal knot which exude fluid droplets as they mature and become more condensed.
- **Principle 8**: Fungi can show extremes of cell differentiation in adjacent hyphal compartments even when pores in the cross-wall appear to be open (as judged by transmission electron microscopy).
- **Principle 9**: Meiocytes appear to be the only hyphal cells that become committed to their developmental fate. Other highly differentiated cells retain totipotency – the ability to generate vegetative hyphal tips that grow out of the differentiated cell to re-establish a vegetative mycelium. Genomic analysis demonstrates that fungi, in general, have numerous genes involved in creating epigenetic marks by DNA methylation.
- **Principle 10**: In arriving at a morphogenetic structure and/or a state of differentiation, fungi are tolerant of considerable imprecision (i.e. expression of fuzzy logic), which results in even the most abnormal sporophores (caused by errors in execution of the developmental subroutines) being still able to distribute viable spores, and poorly (or wrongly) differentiated cells still serving a useful function.
- **Principle 11**: Mechanical interactions influence the form and shape of the whole sporophore as it inflates and matures, and often generate the shape with which we are most familiar.

These principles form the warp and weft of the canvas upon which fungal developmental biology has been built by the cell and molecular biologists of the pregenomics era. It is now up to the genomic systems analysts to paint the rest of the individual details of the stories on that canvas (Kües & Navarro-González, 2015; Halbwachs *et al.*, 2016; Pelkmans *et al.*, 2016; Hibbett *et al.*, 2017; Stajich, 2017; Kües *et al.*, 2018).

13.17 REFERENCES

Ahmed, Y.L., Gerke, J., Park, H.-S. *et al.* (2014). The velvet family of fungal regulators contains a DNA-binding domain structurally similar to NF-κB. *PLoS Biology*, **11**: article number e1001750. DOI: https://doi.org/10.1371/journal.pbio.1001750.

Al-Gharawi, A. & Moore, D. (1977). Factors affecting the amount and the activity of the glutamate dehydrogenases of *Coprinus cinereus*. *Biochimica et Biophysica Acta*, **496**: 95–102. DOI: https://doi.org/10.1016/0304-4165(77)90118-0.

Allen, J.J., Moore, D. & Elliott, T.J. (1992). Persistent meiotic arrest in basidia of *Agaricus bisporus*. *Mycological Research*, **96**: 125–127. DOI: https://doi.org/10.1016/S0953-7562(09)80926-X.

Andrews, J.H. (1995). Fungi and the evolution of growth form. *Canadian Journal of Botany*, **73**: S1206–S1212. DOI: https://doi.org/10.1139/b95-380.

Bastouill-Descollonges, Y. & Manachère, G. (1984). Photosporogenesis of *Coprinus congregatus*: correlations between the physiological age of lamellae and the development of their potential for renewed fruiting. *Physiologia Plantarum*, **61**: 607–610. DOI: https://doi.org/10.1111/j.1399-3054.1984.tb05177.x.

Bayram, Ö., Krappmann, S., Ni, M., Bok, J.W., Helmstaedt, K., and six others. (2008). VelB/VeA/LaeA complex coordinates light signal with fungal development and secondary metabolism. *Science*, **320**: 1504–1506. DOI: https://doi.org/10.1126/science.1155888.

Bleuler-Martínez, S., Butschi, A., Garbani, M. *et al.* (2011). A lectin-mediated resistance of higher fungi against predators and parasites. *Molecular Ecology*, **20**: 3056–3070. DOI: https://doi.org/10.1111/j.1365-294X.2011.05093.x.

Bobola, N. & Merabet, S. (2017). Homeodomain proteins in action: similar DNA binding preferences, highly variable connectivity. *Current Opinion in Genetics & Development*, **43**: 1–8. DOI: https://doi.org/10.1016/j.gde.2016.09.008.

Bourne, A.N., Chiu, S.W. & Moore, D. (1996). Experimental approaches to the study of pattern formation in *Coprinus cinereus*. In *Patterns in Fungal Development*, ed. S.W. Chiu & D. Moore. Cambridge, UK: Cambridge University Press, pp. 126–155. ISBN-10: 0521560470, ISBN-13: 978-0521560474.

Brunt, I.C. & Moore, D. (1989). Intracellular glycogen stimulates fruiting in *Coprinus cinereus*. *Mycological Research*, **93**: 543–546. DOI: https://doi.org/10.1016/S0953-7562(89)80050-4.

Buller, A.H.R. (1910). The function and fate of the cystidia of *Coprinus atramentarius*, together with some general remarks on *Coprinus* fruit bodies. *Annals of Botany*, **24** (old series): 613–629.

Buller, A.H.R. (1924). *Researches on Fungi*, **Vol. 3**. London: Longmans Green & Co. ASIN: B0008BT4QW.

Buller, A.H.R. (1931). *Researches on Fungi*, **Vol. 4**. London: Longmans Green & Co. ASIN: B0008BT4R6.

Burns, C., Stajich, J.E., Rechtsteiner, A. *et al.* (2010). Analysis of the basidiomycete *Coprinopsis cinerea* reveals conservation of the core meiotic expression program over half a billion years of evolution. *PLoS Genetics*, **6**: article number e1001135. DOI: https://doi.org/10.1371/journal.pgen.1001135.

Bush, D.A. (1974). Autolysis of *Coprinus comatus* sporophores. *Experientia*, **30**: 984–985. DOI: https://doi.org/10.1007/BF01938959.

Butler, G.M. (1988). Pattern of pore morphogenesis in the resupinate basidiome of *Phellinus contiguus*. *Transactions of the British Mycological Society*, **91**: 677–686. DOI: https://doi.org/10.1016/S0007-1536(88)80044-5.

Butler, G.M. (1992a). Location of hyphal differentiation in the agar pore field of the basidiome of *Phellinus contiguus*. *Mycological Research*, **96**: 313–317. DOI: https://doi.org/10.1016/S0953-7562(09)80944-1.

Butler, G.M. (1992b). Capacity for differentiation of setae and other hyphal types of the basidiome in explants from cultures of the polypore *Phellinus contiguus*. *Mycological Research*, **96**: 949–955. DOI: https://doi.org/10.1016/S0953-7562(09)80596-0.

Butler, G.M. (1995). Induction of precocious fruiting by a diffusible sex factor in the polypore *Phellinus contiguus*. *Mycological Research*, **99**: 325–329. DOI: https://doi.org/10.1016/S0953-7562(09)80907-6.

Butler, G.M. & Wood, A.E. (1988). Effects of environmental factors on basidiome development in the resupinate polypore *Phellinus contiguus*. *Transactions of the British Mycological Society*, **90**: 75–83. DOI: https://doi.org/10.1016/S0007-1536(88)80182-7.

Chan, K.P., Chang, J., Xie, Y. *et al.* (2018). Transcriptional profiling elucidates the essential role of glycogen synthase kinase 3 to fruiting body formation in *Coprinopsis cinerea*. *Cold Spring Harbor Laboratory preprint service bioRxiv*: article 492397. DOI: https://doi.org/10.1101/492397.

Chang, J., Au, C.H., Cheng, C.K. & Kwan, H.S. (2018a). eQTL network analysis reveals that regulatory genes are evolutionarily older and bearing more types of PTM sites in *Coprinopsis cinerea*. *Cold Spring Harbor Laboratory preprint service bioRxiv*: article 413062. DOI: https://doi.org/10.1101/413062.

Chang, J., Chan, K.P., Xie, Y., Ma, K.L. & Kwan, H.S. (2018b). Modified recipe to inhibit GSK-3 for the living fungal biomaterial manufacture. *Cold Spring Harbor Laboratory preprint service bioRxiv*: article 496265. DOI: https://doi.org/10.1101/496265.

Chavez, L., Huang, Y., Luong, K. *et al.* (2014). Simultaneous sequencing of oxidized methylcytosines produced by TET/JBP dioxygenases in *Coprinopsis cinerea. Proceedings of the National Academy of Sciences of the United States of America*, **111**: E5149–E5158. DOI: https://doi.org/10.1073/pnas.1419513111.

Cheng, C.K., Au, C.H., Wilke, S.K. *et al.* (2013). 5′-serial analysis of gene expression studies reveal a transcriptomic switch during fruiting body development in *Coprinopsis cinerea. BMC Genomics*, **14**: 195 (17 pp.). DOI: https://doi.org/10.1186/1471-2164-14-195.

Cheng, W., Zhang, X. & Wang, W. (2016). Robust methods for expression quantitative trait loci mapping. In *Big Data Analytics in Genomics*, ed. K.C. Wong. Cham, Switzerland: Springer International Publishing, pp. 25–88. ISBN: 9783319412788. DOI: https://doi.org/10.1007/978-3-319-41279-5_2.

Chiu, S.W. & Moore, D. (1988a). Evidence for developmental commitment in the differentiating fruit body of *Coprinus cinereus. Transactions of the British Mycological Society*, **90**: 247–253. DOI: https://doi.org/10.1016/S0007-1536(88)80096-2.

Chiu, S.W. & Moore, D. (1988b). Ammonium ions and glutamine inhibit sporulation of *Coprinus cinereus* basidia assayed *in vitro. Cell Biology International Reports*, **12**: 519–526. DOI: https://doi.org/10.1016/0309-1651(88)90038-0.

Chiu, S.W. & Moore, D. (1990a). A mechanism for gill pattern formation in *Coprinus cinereus. Mycological Research*, **94**: 320–326. DOI: https://doi.org/10.1016/S0953-7562(09)80357-2.

Chiu, S.W. & Moore, D. (1990b). Development of the basidiome of *Volvariella bombycina. Mycological Research*, **94**: 327–337. DOI: https://doi.org/10.1016/S0953-7562(09)80358-4.

Chiu, S.W. & Moore, D. (1990c). Sporulation in *Coprinus cinereus*: use of an *in vitro* assay to establish the major landmarks in differentiation. *Mycological Research*, **94**: 249–253. DOI: https://doi.org/10.1016/S0953-7562(09)80623-0.

Chiu, S.W., Moore, D. & Chang, S.T. (1989). Basidiome polymorphism in *Volvariella bombycina. Mycological Research*, **92**: 69–77. DOI: https://doi.org/10.1016/S0953-7562(89)80098-X.

Conesa, A., Madrigal, P., Tarazona, S. *et al.* (2016). A survey of best practices for RNA-seq data analysis. *Genome Biology*, **17**: 13. DOI: https://doi.org/10.1186/s13059-016-0881-8.

Corner, E.J.H. (1932). A *Fomes* with two systems of hyphae. *Transactions of the British Mycological Society*, **17**: 51–81. DOI: https://doi.org/10.1016/S0007-1536(32)80026-4.

Corner, E.J.H. (1966). A monograph of cantharelloid fungi. *Annals of Botany Memoirs* no. **2**. Oxford, UK: Oxford University Press. ASIN: B000WUINOS.

Cuomo, C.A. & Birren, B.W. (2010). The Fungal Genome Initiative and lessons learned from genome sequencing. In *Guide to Yeast Genetics: Functional Genomics, Proteomics, and Other Systems Analysis, Methods in Enzymology*, **Vol. 470**, ed. J. Weissman, C. Guthrie & G.R. Fink. San Diego, CA: Academic Press, an imprint of Elsevier Inc., pp. 833–855. DOI: https://doi.org/10.1016/S0076-6879(10)70034-3.

DeLong, E.F. (ed.) (2013). *Microbial Metagenomics, Metatranscriptomics, and Metaproteomics, Methods in Enzymology*, **Vol. 531**. San Diego, CA: Academic Press. DOI: https://doi.org/10.1016/B978-0-12-407863-5.09983-4.

Doudna, J.A. & Sontheimer, E.J. (ed.) (2014). *The Use of CRISPR/Cas9, ZFNs, and TALENs in Generating Site-Specific Genome Alterations, Methods in Enzymology*, **Vol. 546**. San Diego, CA: Academic Press. DOI: https://doi.org/10.1016/B978-0-12-801185-0.09983-9.

Elhiti, M.M.Y., Butler, R.D. & Moore, D. (1979). Cytochemical localisation of glutamate dehydrogenase during carpophore development in *Coprinus cinereus. New Phytologist*, **82**: 153–157. DOI: https://doi.org/10.1111/j.1469-8137.1979.tb07570.x.

Engel, S.R., Dietrich, F.S., Fisk, D.G. *et al.* (2014). The reference genome sequence of *Saccharomyces cerevisiae*: then and now. *G3: Genes, Genomes, Genetics*, **4**: 389–398. DOI: https://doi.org/10.1534/g3.113.008995.

Ewaze, J.O., Moore, D. & Stewart, G.R. (1978). Co-ordinate regulation of enzymes involved in ornithine metabolism and its relation to sporophore morphogenesis in *Coprinus cinereus. Journal of General Microbiology*, **107**: 343–357. DOI: https://doi.org/10.1099/00221287-107-2-343.

Fayod, V. (1889). Prodrome d'une histoire naturelle des Agaricinés. *Annales des Sciences Naturelles, Botanique Série*, **7–9**, 179–411.

Field, C., Li, R. & Oegema, K. (1999). Cytokinesis in eukaryotes: a mechanistic comparison. *Current Opinion in Cell Biology*, **11**: 68–80. DOI: https://doi.org/10.1016/S0955-0674(99)80009-X.

Goffeau, A., Barrell, B.G., Bussey, H. *et al.* (1996). Life with 6000 genes. *Science*, **274**: 546–567. DOI: https://doi.org/10.1126/science.274.5287.546.

Grigoriev, I.V., Nikitin, R., Haridas, S. *et al.* (2014). MycoCosm portal: gearing up for 1000 fungal genomes. *Nucleic Acids Research*, **42**: D699–D704. DOI: https://doi.org/10.1093/nar/gkt1183.

Grützmann, K., Szafranski, K., Pohl, M. *et al.* (2014). Fungal alternative splicing is associated with multicellular complexity and virulence: a genome-wide multi-species study. *DNA Research*, 21: 27–39. DOI: https://doi.org/10.1093/dnares/dst038.

Halbwachs, H., Simmel, J. & Bässler, C. (2016). Tales and mysteries of fungal fruiting: how morphological and physiological traits affect a pileate lifestyle. *Fungal Biology Reviews*, **30**: 36–61. DOI: https://doi.org/10.1016/j.fbr.2016.04.002.

Hammad, F., Ji, J., Watling, R. & Moore, D. (1993b). Cell population dynamics in *Coprinus cinereus*: co-ordination of cell inflation throughout the maturing fruit body. *Mycological Research*, **97**: 269–274. DOI: https://doi.org/10.1016/S0953-7562(09)81119-2.

Hammad, F., Watling, R. & Moore, D. (1993a). Cell population dynamics in *Coprinus cinereus*: narrow and inflated hyphae in the basidiome stipe. *Mycological Research*, **97**: 275–282. DOI: https://doi.org/10.1016/S0953-7562(09)81120-9.

Hammond, J.B.W. & Nichols, R. (1976). Carbohydrate metabolism in *Agaricus bisporus* (Lange) Sing.: changes in soluble carbohydrates during growth of mycelium and sporophore. *Journal of General Microbiology*, **93**: 309–320. DOI: https://doi.org/10.1099/00221287-93-2-309.

Harper, J.L., Rosen, B.R. & White, J. (1986). *The Growth and Form of Modular Organisms*. London: The Royal Society. ISBN-10: 0521350743, ISBN-13: 978-0521350747.

Harris, S.D. (2001). Septum formation in *Aspergillus nidulans*. *Current Opinion in Microbiology*, **4**: 736–739. DOI: https://doi.org/10.1016/S1369-5274(01)00276-4.

Hibbett, D.S., Stajich, J.E. & Spatafora, J.W. (2017). Towards genome-enabled mycology. *Mycologia*, 105: 1339–1349. DOI: https://doi.org/10.3852/13-196.

Horner, J. & Moore, D. (1987). Cystidial morphogenetic field in the hymenium of *Coprinus cinereus*. *Transactions of the British Mycological Society*, **88**: 479–488. DOI: https://doi.org/10.1016/S0007-1536(87)80031-1.

Iten, W. (1970). Zur funktion hydrolytischer enzyme bei der autolysate von *Coprinus*. *Berichte Schweitzerische Botanische Gesellschaft*, **79**: 175–198.

Iten, W. & Matile, P. (1970). Role of chitinase and other lysosomal enzymes of *Coprinus lagopus* in the autolysis of fruiting bodies. *Journal of General Microbiology*, **61**: 301–309. DOI: https://doi.org/10.1099/00221287-61-3-301.

Iyer, L.M., Zhang, D., de Souza, R.F. *et al.* (2014). Lineage-specific expansions of TET/JBP genes and a new class of DNA transposons shape fungal genomic and epigenetic landscapes. *Proceedings of the National Academy of Sciences of the United States of America*, **111**: 1676–1683. DOI: https://doi.org/10.1073/pnas.1321818111.

Jabor, F.N. & Moore, D. (1984). Evidence for synthesis *de novo* of NADP-linked glutamate dehydrogenase in *Coprinus* mycelia grown in nitrogen-free medium. *FEMS Microbiology Letters*, **23**: 249–252. DOI: https://doi.org/10.1111/j.1574-6968.1984.tb01072.x.

Jameson, D., Verma, M. & Westerhoff, H.V. (ed.) (2011). *Methods in Systems Biology, Methods in Enzymology*, **Vol. 500**. San Diego, CA: Academic Press. DOI: https://doi.org/10.1016/B978-0-12-385118-5.00031-1.

Jirjis, R.I. & Moore, D. (1976). Involvement of glycogen in morphogenesis of *Coprinus cinereus*. *Journal of General Microbiology*, **95**: 348–352. DOI: https://doi.org/10.1099/00221287-95-2-348.

Kang, L., Zhu, Y., Bai, Y. & Yuan, S. (2019). Characteristics, transcriptional patterns and possible physiological significance of glycoside hydrolase family 16 members in *Coprinopsis cinerea*. *FEMS Microbiology Letters*, **366**: article number fnz083. DOI: https://doi.org/10.1093/femsle/fnz083.

Kemp, R.F.O. (1977). Oidial homing and the taxonomy and speciation of basidiomycetes with special reference to the genus *Coprinus*. In *The Species Concept in Hymenomycetes*, ed. H. Clémençon. Vaduz: J. Cramer, pp. 259–273. ISBN-10: 3768211738, ISBN-13: 978-3768211734.

Krizsán, K., Almási, É., Merényi, Z. *et al.* (2019a). Transcriptomic atlas of mushroom development highlights an independent origin of complex multicellularity. *Cold Spring Harbor Laboratory preprint service bioRxiv*: article 349894. DOI: https://doi.org/10.1101/349894.

Krizsán, K., Almási, É., Merényi, Z. *et al.* (2019b). Transcriptomic atlas of mushroom development reveals conserved genes behind complex multicellularity in fungi. *Proceedings of the National Academy of Sciences of the United States of America*, **116**: 7409–7418. DOI: https://doi .org/10.1073/pnas.1817822116.

Kües, U., Khonsuntia, W. & Subba, S. (2018). Complex fungi. *Fungal Biology Reviews*, **32**: 205–218. DOI: https://doi.org/10.1016/j.fbr.2018.08.001.

Kües, U. & Navarro-González, M. (2015). How do Agaricomycetes shape their fruiting bodies? 1. Morphological aspects of development. *Fungal Biology Reviews*, **29**: 63–97. DOI: https://doi .org/10.1016/j.fbr.2015.05.001.

Lam, K.-L., Si, K., Wu, X. *et al.* (2018). The diploid genome of the only sclerotia-forming wild-type species in the genus *Pleurotus* – *Pleurotus tuber-regium* – provides insights into the mechanism of its biomass conversion from lignocellulose substrates. *Journal of Biotechnology*, **283**: 22–27. DOI: https://doi.org/10.1016/j.jbiotec.2018.07.009.

Lassar, A.B. & Orkin, S. (2001). Cell differentiation: plasticity and commitment – developmental decisions in the life of a cell (editorial overview of a special topic issue). *Current Opinion in Cell Biology*, **13**: 659–661. DOI: https://doi.org/10.1016/S0955-0674(00)00267-2.

Lau, A.Y-T., Cheng, X., Cheng, C.K. *et al.* (2018). Discovery of microRNA-like RNAs during early fruiting body development in the model mushroom *Coprinopsis cinerea*. *PLoS ONE*, **13**: article number e0198234. DOI: https://doi.org/10.1371/journal.pone.0198234.

Lee, M.K., Kwon, N.J., Lee, I.S. *et al.* (2016). Negative regulation and developmental competence in *Aspergillus*. *Scientific Reports*, **6**: article number 28874. DOI: https://doi.org/10.1038/srep28874.

Leptin, M. (2005). Gastrulation movements: the logic and the nuts and bolts. *Developmental Cell*, **8**: 305–320. DOI: https://doi.org/10.1016/j.devcel.2005.02.007.

Lesch, B.J. & Page, D.C. (2014). Poised chromatin in the mammalian germ line. *Development*, **141**: 3619–3626. DOI: https://doi.org/10.1242/dev.113027.

Li, L., Hu, X., Xia, Y. *et al.* (2014). Linkage of oxidative stress and mitochondrial dysfunctions to spontaneous culture degeneration in *Aspergillus nidulans*. *Molecular & Cellular Proteomics*, **13**: 449–461. DOI: https://doi.org/10.1074/mcp.M113.028480.

Liseron-Monfils, C. & Ware, D. (2015). Revealing gene regulation and associations through biological networks. *Current Plant Biology*, **3–4**: 30–39. DOI: https://doi.org/10.1016/j .cpb.2015.11.001.

Maheshwari, R. & Navaraj, A. (2008). Senescence in fungi: the view from *Neurospora*. *FEMS Microbiology Letters*, **280**: 135–143. DOI: https://doi.org/10.1111/j.1574-6968.2007.01027.x.

Majcherczyk, A., Durnte, B., Subba, S., Zomorrodi, M. & Kües, U. (2019). Proteomes in primordia development of *Coprinopsis cinerea*. *Acta Edulis Fungi*, **26**: 37-50. DOI: https://doi. org/10.16488/j.cnki.1005-9873.2019.03.005.

Markowitz, V.M., Chen, I.-M.A., Chu, K. *et al.* (2015). Ten years of maintaining and expanding a microbial genome and metagenome analysis system. *Trends in Microbiology*, **23**: 730–741. DOI: https://doi.org/10.1016/j.tim.2015.07.012.

Markowitz, V.M., Chen, I.-M.A., Palaniappan, K. *et al.* (2012). IMG: the integrated microbial genomes database and comparative analysis system. *Nucleic Acids Research*, **40**: D115–D122. DOI: https://doi.org/10.1093/nar/gkr1044.

Matthews, T.R. & Niederpruem, D.J. (1972). Differentiation in *Coprinus lagopus*. I. Control of fruiting and cytology of initial events. *Archiv für Mikrobiologie*, **87**: 257–268. DOI: https://doi .org/10.1007/BF00424886.

Meinhardt, H. (1976). Morphogenesis of lines and nets. *Differentiation*, **6**: 117–123. DOI: https:// doi.org/10.1111/j.1432-0436.1976.tb01478.x.

Meinhardt, H. (1984). Models of pattern formation and their application to plant development. In *Positional Controls in Plant Development*, ed. P.W. Barlow & D.J. Carr. Cambridge, UK: Cambridge University Press, pp. 1–32. ISBN-10: 052125406X, ISBN-13: 978-0521254069.

Meinhardt, H. & Gierer, A. (1974). Applications of a theory of biological pattern formation based on lateral inhibition. *Journal of Cell Science*, **15**: 321–346.

Meyerowitz, E.M. (1999). Plants, animals and the logic of development. *Trends in Cell Biology*, **9**: M65–M68. DOI: https://doi.org/10.1016/S0962-8924(99)01649-9.

Michaelson, J.J., Loguercio, S. & Beyer, A. (2009). Detection and interpretation of expression quantitative trait loci (eQTL). *Methods*, **48**: 265–276 (special issue on *Global approaches to study gene regulation*). DOI: https://doi.org/10.1016/j.ymeth.2009.03.004.

Momany, M. (2001). Cell biology of the duplication cycle in fungi. In *Molecular and Cellular Biology of Filamentous Fungi*, ed. N.J. Talbot. Oxford, UK: Oxford University Press, pp. 119–125. ISBN-10: 0199638373, ISBN-13: 978-0199638376.

Moore, D. (1998). Tolerance of imprecision in fungal morphogenesis. In *Proceedings of the Fourth Conference on the Genetics and Cellular Biology of Basidiomycetes*, ed. L.J.L.D. Van Griensven & J. Visser. The Mushroom Experimental Station: Horst, the Netherlands, pp. 13–19. URL: https://www.researchgate.net/publication/326406024_Tolerance_of_imprecision_in_fungal_morphogenesis.

Moore, D. (2013). *Coprinopsis: An Autobiography*. London: CreateSpace Independent Publishing Platform. 216 pp. ISBN-10: 1482618974; ISBN-13: 978-1482618976. Download full text: https://www.researchgate.net/publication/321361317_Coprinopsis_an_autobiography.

Moore, D., Elhiti, M.M.Y. & Butler, R.D. (1979). Morphogenesis of the carpophore of *Coprinus cinereus*. *New Phytologist*, **83**: 695–722. DOI: https://doi.org/10.1111/j.1469-8137.1979.tb02301.x.

Moore, D. & Ewaze, J.O. (1976). Activities of some enzymes involved in metabolism of carbohydrate during sporophore development in *Coprinus cinereus*. *Journal of General Microbiology*, **97**: 313–322. DOI: https://doi.org/10.1099/00221287-97-2-313.

Moore, D., Horner, J. & Liu, M. (1987a). Co-ordinate control of ammonium-scavenging enzymes in the fruit body cap of *Coprinus cinereus* avoids inhibition of sporulation by ammonium. *FEMS Microbiology Letters*, **44**: 239–242. DOI: https://doi.org/10.1111/j.1574-6968.1987.tb02275.x.

Moore, D., Liu, M. & Kuhad, R.C. (1987b). Karyogamy-dependent enzyme derepression in the basidiomycete *Coprinus*. *Cell Biology International Reports*, **11**: 335–341. DOI: https://doi.org/10.1016/0309-1651(87)90094-4.

Moore, D. & Meškauskas, A. (2006). A comprehensive comparative analysis of the occurrence of developmental sequences in fungal, plant and animal genomes. *Mycological Research*, **110**: 251–256. DOI: https://doi.org/10.1016/j.mycres.2006.01.003.

Moore, D. & Meškauskas, A. (2009). Where are the sequences that control multicellular development in filamentous fungi? In *Current Advances in Molecular Mycology*, ed. Y. Gherbawy, R.L. Mach & M. Rai. Hauppauge, New York: Nova Science Publishers, pp. 1–37. ISBN-10: 1604569093, ISBN -13: 9781604569094.

Moore, D. & Novak Frazer, L. (2002). *Essential Fungal Genetics*. New York: Springer-Verlag. ISBN-10: 0387953671, ISBN-13: 978-0387953670.

Moore, D. & Novak Frazer, L. (2017). *Fungiflex: The Untold Story*. London: CreateSpace Independent Publishing Platform. ISBN-10: 1547168560, ISBN-13: 978-1547168569. Download full text: https://www.researchgate.net/publication/321361153_Fungiflex_the_untold_story.

Moore, D., Walsh, C. & Robson, G.D. (2005). A search for developmental gene sequences in the genomes of filamentous fungi. In *Applied Mycology and Biotechnology*, **Vol. 6**, *Genes, Genomics and Bioinformatics*, ed. D.K. Arora & R. Berka. Amsterdam: Elsevier Science, pp. 169–188.

Morales-González, J.A. (ed.) (2013). *Oxidative Stress and Chronic Degenerative Diseases: A Role for Antioxidants*. London: InTech. An open access text under Creative Commons BY 3.0 license. DOI: https://doi.org/10.5772/45722.

Monro, K. & Poore, A.G.B. (2004). Selection in modular organisms: is intraclonal variation in macroalgae evolutionarily important? *The American Naturalist*, **163**: 564–578.

Muraguchi, H., Umezawa, K., Niikura, M. *et al.* (2015). Strand-specific RNA-Seq analyses of fruiting body development in *Coprinopsis cinerea*. *PLoS ONE*, **10**: article number e0141586. DOI: https://doi.org/10.1371/journal.pone.0141586.

Nagy, L.G., Kovács, G.M. & Krizsán, K. (2018). Complex multicellularity in fungi: evolutionary convergence, single origin, or both? *Biological Review of Cambridge Philosophical Society*, **93**(4):1778–1794. DOI: 10.1111/brv.12418.

Noble, L.M. & Andrianopoulos, A. (2013). Reproductive competence: a recurrent logic module in eukaryotic development. *Proceedings of the Royal Society, Series B*, **280**: article number 20130819. DOI: https://doi.org/10.1098/rspb.2013.0819.

Nordberg, H., Cantor, M., Dusheyko, S. *et al.* (2014). The genome portal of the Department of Energy Joint Genome Institute: 2014 updates. *Nucleic Acids Research*, **42**: D26–D31. DOI: https://doi.org/10.1093/nar/gkt1069.

Novak Frazer, L. (1996). Control of growth and patterning in the fungal fruiting structure. A case for the involvement of hormones. In *Patterns in Fungal Development*, ed. S.W. Chiu & D. Moore. Cambridge, UK: Cambridge University Press, pp. 156–181. ISBN-10: 0521560470, ISBN-13: 978-0521560474.

Patyshakuliyeva, A., Jurak, E., Kohler, A. *et al.* (2013). Carbohydrate utilization and metabolism is highly differentiated in *Agaricus bisporus*. *BMC Genomics*, **14**: 663. DOI: https://doi.org/10.1186/1471-2164-14-663.

Pelkmans, J.F., Lugones, L.G. & Wösten, H.A.B. (2016). Fruiting body formation in basidiomycetes. In *The Mycota*, **Vol. I**. *Growth, Differentiation and Sexuality*, 3rd edition, ed. J. Wendland. Cham, Switzerland: Springer International Publishing, pp. 387–405. DOI: https://doi.org/10.1007/978-3-319-25844-7_15.

Plaza, D.F., Lin, C.-W., van der Velden, N.S.J., Aebi, M. & Künzler, M. (2014). Comparative transcriptomics of the model mushroom *Coprinopsis cinerea* reveals tissue-specific armories and a conserved circuitry for sexual development. *BMC Genomics*, **15**: 492. DOI: https://doi.org/10.1186/1471-2164-15-492.

Redhead, S.A. (1987). The Xerulaceae (Basidiomycetes), a family with sarcodimitic tissues. *Canadian Journal of Botany*, **65**: 1551–1562. DOI: https://doi.org/10.1139/b87-214.

Redhead, S.A. Vilgalys, R., Moncalvo, J.-M., Johnson, J. & Hopple, J.S. Jr (2001). *Coprinus* Pers. and the disposition of *Coprinus* species *sensu lato*. *Taxon*, **50**: 203–241. DOI: https://doi.org/10.2307/1224525.

Reijnders, A.F.M. (1963). *Les problèmes du développement des carpophores des Agaricales et de quelques groupes voisins*. The Hague: Dr W. Junk. ISBN-10: 9061936284, ISBN-13: 978-9061936282.

Reijnders, A.F.M. & Moore, D. (1985). Developmental biology of agarics – an overview. In *Developmental Biology of Higher Fungi*, ed. D. Moore, L.A. Casselton, D.A. Wood & J.C. Frankland. Cambridge, UK: Cambridge University Press, pp. 581–595. ISBN-10: 0521301610, ISBN-13: 978-0521301619.

Rosin, I.V. & Moore, D. (1985a). Origin of the hymenophore and establishment of major tissue domains during fruit body development in *Coprinus cinereus*. *Transactions of the British Mycological Society*, **84**: 609–619. DOI: https://doi.org/10.1016/S0007-1536(85)80115-7.

Rosin, I.V. & Moore, D. (1985b). Differentiation of the hymenium in *Coprinus cinereus*. *Transactions of the British Mycological Society*, **84**: 621–628. DOI: https://doi.org/10.1016/S0007-1536(85)80116-9.

Sakamoto, Y. (2018). Influences of environmental factors on fruiting body induction, development and maturation in mushroom-forming fungi. *Fungal Biology Reviews*, **32**: 236–248. DOI: https://doi.org/10.1016/j.fbr.2018.02.003.

Sakamoto, Y., Sato, S., Ito, M. *et al.* (2018). Blue light exposure and nutrient conditions influence the expression of genes involved in simultaneous hyphal knot formation in *Coprinopsis cinerea*. *Microbiological Research*, **217**: 81–90. DOI: https://doi.org/10.1016/j.micres.2018.09.003.

Shefferson, R.P., Jones, O.R. & Salguero-Gómez, R. (ed.) (2017). *The Evolution of Senescence in the Tree of Life*. Cambridge, UK: Cambridge University Press. ISBN-10: 1107078504, ISBN-13: 978-1107078505.

Stajich, J.E. (2017). Fungal genomes and insights into the evolution of the kingdom. In *The Fungal Kingdom*, ed. J. Heitman, B. Howlett, P. Crous, E. Stukenbrock, T. James & N.A.R. Gow. Washington, DC: ASM Press, pp. 619–633. DOI: https://doi.org/10.1128/microbiolspec.FUNK-0055-2016.

Stajich, J.E., Wilke, S.K., Ahrén, D. *et al.* (2010). Insights into evolution of multicellular fungi from the assembled chromosomes of the mushroom *Coprinopsis cinerea* (*Coprinus cinereus*). *Proceedings of the National Academy of Sciences of the United States of America*, **107**: 11889–11894. DOI: https://doi.org/10.1073/pnas.1003391107.

Steinberg, G., Peñalva, M., Riquelme, M., Wösten, H. & Harris, S. (2017). Cell biology of hyphal growth. In *The Fungal Kingdom*, ed. J. Heitman, B. Howlett, P. Crous, E. Stukenbrock, T. James & N. Gow. Washington, DC: ASM Press, pp. 231–265. DOI: https://doi.org/10.1128/microbiolspec.FUNK-0034-2016.

Stewart, G.R. & Moore, D. (1974). The activities of glutamate dehydrogenases during mycelial growth and sporophore development in *Coprinus lagopus* (*sensu* Lewis). *Journal of General Microbiology*, **83**: 73–81. DOI: https://doi.org/10.1099/00221287-83-1-73.

Sugano, S.S., Suzuki, H., Shimokita, E. *et al.* (2017). Genome editing in the mushroom forming basidiomycete *Coprinopsis cinerea*, optimized by a high-throughput transformation system. *Scientific Reports*, **7**: article number 1260. DOI: https://doi.org/10.1038/s41598-017-00883-5.

Takemaru, T. & Kamada, T. (1971). Gene control of basidiocarp development in *Coprinus macrorhizus*. *Reports of the Tottori Mycological Institute (Japan)*, **9**: 21–35.

Takemaru, T. & Kamada, T. (1972). Basidiocarp development in *Coprinus macrorhizus*. I. Induction of developmental variations. *Botanical Magazine (Tokyo)*, **85**: 51–57. DOI: https://doi.org/10.1007/BF02489200.

Tan, Y.H. & Moore, D. (1994). High concentrations of mannitol in the shiitake mushroom *Lentinula edodes*. *Microbios*, **79**: 31–35.

Tan, Y.H. & Moore, D. (1995). Glucose catabolic pathways in *Lentinula edodes* determined with radiorespirometry and enzymic analysis. *Mycological Research*, **99**: 859–866. DOI: https://doi.org/10.1016/S0953-7562(09)80742–9.

ten Hove, C.A., Lu, K.-J. & Weijers, D. (2015). Building a plant: cell fate specification in the early *Arabidopsis* embryo. *Development*, **142**: 420–430. DOI: https://doi.org/10.1242/dev.111500.

Trounson, A. & McDonald, C. (2015). Stem cell therapies in clinical trials: progress and challenges. *Cell Stem Cell*, **17**: 11–22. DOI: https://doi.org/10.1016/j.stem.2015.06.007.

Umar, M.H. & Van Griensven, L.J.L.D. (1997a). Morphological studies on the life span, developmental stages, senescence and death of *Agaricus bisporus*. *Mycological Research*, **101**: 1409–1422. DOI: https://doi.org/10.1017/S0953756297005212.

Umar, M.H. & Van Griensven, L.J.L.D. (1997b). Hyphal regeneration and histogenesis in *Agaricus bisporus*. *Mycological Research*, **101**: 1025–1032. DOI: https://doi.org/10.1017/S0953756297003869.

Umar, M.H. & Van Griensven, L.J.L.D. (1998). The role of morphogenetic cell death in the histogenesis of the mycelial cord of *Agaricus bisporus* and in the development of macrofungi. *Mycological Research*, **102**: 719–735. DOI: https://doi.org/10.1017/S0953756297005893.

Van der Valk, P. & Marchant, R. (1978). Hyphal ultrastructure in fruit body primordia of the basidiomycetes *Schizophyllum commune* and *Coprinus cinereus*. *Protoplasma*, **95**: 57–72. DOI: https://doi.org/10.1007/BF01279695.

Voigt, P., Tee, W.-W. & Reinberg, D. (2013). A double take on bivalent promoters. *Genes & Development*, **27**: 1318–1338. DOI: https://doi.org/10.1101/gad.219626.113.

Vonk, P.J. & Ohm, R.A. (2018). The role of homeodomain transcription factors in fungal development. *Fungal Biology Reviews*, **32**: 219–230. DOI: https://doi.org/10.1016/j.fbr.2018.04.002.

Wagemaker, M.J.M., Welboren, W., van der Drift, C. *et al.* (2005). The ornithine cycle enzyme arginase from *Agaricus bisporus* and its role in urea accumulation in fruit bodies. *Biochimica et Biophysica Acta (BBA) – Gene Structure and Expression*, **1681**: 107–115. DOI: https://doi.org/10.1016/j.bbaexp.2004.10.007.

Wang, Y., Liu, J., Huang, B. *et al.* (2015). Mechanism of alternative splicing and its regulation. *Biomedical Reports*, **3**: 152–158. DOI: https://doi.org/10.3892/br.2014.407.

Waters, H., Moore, D. & Butler, R.D. (1975). Morphogenesis of aerial sclerotia of *Coprinus lagopus*. *New Phytologist*, **74**: 207–213. DOI: https://doi.org/10.1111/j.1469-8137.1975.tb02607.x.

Watling, R. & Moore, D. (1994). Moulding moulds into mushrooms: shape and form in the higher fungi. In *Shape and Form in Plants and Fungi*, ed. D.S. Ingram & A. Hudson. London: Academic Press, pp. 270–290. ISBN-10: 0123710359, ISBN-13: 978-0123710352.

Weissman, J., Guthrie, C. & Fink, G.R. (ed.) (2010). *Guide to Yeast Genetics: Functional Genomics, Proteomics, and Other Systems Analysis, Methods in Enzymology*, **Vol. 470**. San Diego, CA: Academic Press. DOI: https://doi.org/10.1016/S0076-6879(10)70041-0.

Wiemer, M., Grimm, C. & Osiewacz, H.D. (2016). Molecular control of fungal senescence and longevity. In *The Mycota*, **Vol. I**. *Growth, Differentiation and Sexuality*, 3rd edition, ed. J. Wendland. Cham, Switzerland: Springer International Publishing, pp. 155–181. DOI: https://doi.org/10.1007/978-3-319-25844-7_8.

Wight, M. & Werner, A. (2013). The functions of natural antisense transcripts. *Essays in Biochemistry*, **54**: 91–101. DOI: https://doi.org/10.1042/bse0540091.

Williams, M.A.J., Beckett, A. & Read, N.D. (1985). Ultrastructural aspects of fruit body differentiation in *Flammulina velutipes*. In *Developmental Biology of Higher Fungi*, British Mycological Society Symposium **Vol. 10**, ed. D. Moore, L.A. Casselton, D.A. Wood & J.C. Frankland. Cambridge, UK: Cambridge University Press, pp. 429–450. ISBN-10: 0521301610, ISBN-13: 978-0521301619.

Williams, M.L.K. & Solnica-Krezel, L. (2017). Regulation of gastrulation movements by emergent cell and tissue interactions. *Current Opinion in Cell Biology*, **48**: 33–39. DOI: https://doi.org/10.1016/j.ceb.2017.04.006.

Wu, J., Gao, B. & Zhu, S. (2014). The fungal defensin family enlarged. *Pharmaceuticals*, **7**: 866–880. DOI: https://doi.org/10.3390/ph7080866.

Wyss, T., Masclaux, F.G., Rosikiewicz, P., Pagni, M. & Sanders, I.R. (2016). Population genomics reveals that within-fungus polymorphism is common and maintained in populations of the mycorrhizal fungus *Rhizophagus irregularis*. *The ISME Journal*, **10**: 2514–2526. DOI: https://doi.org/10.1038/ismej.2016.29.

Zhu, G. (2018). Major genes expression of storage carbohydrate metabolism in fruiting body formation of *Pholiota microspora*. *Plant Gene*, **14**: 83–89. DOI: https://doi.org/10.1016/j.plgene.2018.05.003.

CHAPTER 14

Ecosystem Mycology: Saprotrophs, and Mutualisms Between Plants and Fungi

In this chapter on ecosystem mycology, we cover fungi as saprotrophs, and the mutualisms between plants and fungi, concentrating on fungi as recyclers that can make the earth move. Fungi also cause food contamination and deterioration through their formation of toxins, although some of these, like statins and strobilurins, are exploited commercially for our own practical purposes.

The ability of fungi to degrade wood makes them responsible for the decay of structural timber in dwellings, but on the other hand enables them to be used to remediate toxic and recalcitrant wastes. A downside, though, is that wood-decay fungi release chlorohydrocarbons, potent greenhouse gases, to the atmosphere and thereby potentially contribute to global warming.

Interactions with plants dominate the rest of the chapter. We describe all types of mycorrhiza: arbuscular (AM) endomycorrhizas, ericoid endomycorrhizas, arbutoid endomycorrhizas, monotropoid endomycorrhizas, orchidaceous endomycorrhizas, ectomycorrhizas and ectendomycorrhizas. The effects of mycorrhizas and their commercial applications, and the impact of environmental and climate changes are also discussed. Finally, we introduce lichens, endophytes and epiphytes.

Evidently, fungi contribute to a broad and vibrant network of interactions with all members of the plant, animal and bacterial kingdoms (Prosser, 2002). Because of their unique attributes, fungi play vital roles in most ecosystems.

- They act as mutualists (forming symbiotic associations in which all partners benefit) of virtually all plant species, and a great many animals; but arguably the most important of these are the lichens, which are usually associations between a fungus and a green alga, and the mycorrhizal association between plants and fungi. Most plants depend on mycorrhizal fungi for their survival and growth, the fungus taking photosynthetic sugars from the host in return. The benefits of mycorrhizas to plants include efficient nutrient uptake, especially phosphorus, because the fungus extends the nutrient-absorbing surface area of the roots, but water uptake through the mycorrhiza enhances plant resistance to drought stress, and the mycorrhiza provides direct or indirect protection against some pathogens (see below).
- Parasitic and pathogenic fungi infect the living tissues of plants, animals and other fungi, causing injury and disease that adds to the selection pressure on their hosts (see Chapters 15, 16 and 17). Even fungal plant pathogens have a positive effect on the natural environment by enriching its species structure. Plants killed by disease provide organic matter for nutrient cycling organisms; dead branches or heart rot in live trees create habitats for cavity nesting animals, while gaps in stands of dominant plants caused by disease allow development of other plants, contributing both to species diversity and diversity of food for animals, from insects to elk.

14.1 ECOSYSTEM MYCOLOGY

Fungi are the saprotrophs that perform the decomposition processes that contribute to organic and inorganic nutrient cycling. Clearly, fungal decomposition of dead organic matter, be it wood or other plant litter, animal dung, or cadavers and bones, is an essential ecosystem function because it maintains soil nutrient availability (see below). But there is another significant contribution along the way: fungi that decay wood soften the timber sufficiently to allow small animals (birds, reptiles, amphibians, insects and mammals) to make burrows and nests.

- Fungal products aggregate soil particles and organic matter, improving drainage and aeration; and by so doing they create habitat diversity for many other organisms (see our discussion of glomalin in section 7.8).
- Fungi serve as both prey and predators of many soil organisms, including bacteria, other fungi, nematodes, microarthropods and insects (Wall *et al.*, 2010; Crowther *et al.*, 2012; Menta, 2012; Ngosong *et al.*, 2014) (see sections 1.5, 12.2 and Chapter 15).
- Mushrooms and truffles are consumed by many animals including large mammals like primates, deer and bears, and many small mammals rely on mushrooms and truffles for nearly their entire food supply. The fungal mycelium is an equally important nutritional resource for many microarthropods (see section 12.2).
- The significance of fungi in nature means that changes in the composition and functioning of fungi in a community can have sweeping effects on the diversity, health and productivity of our natural environment. Fungal *diversity* is also called *richness* (Andrew *et al.*, 2019) and molecular methods (*'metagenomics'*) are now allowing this to be studied directly (Lindahl & Kuske, 2013). Although it is species of fungi, bacteria, nematodes and arthropods that typically dominate terrestrial ecosystems in terms of species richness, most conservation work is unfortunately concentrated on vertebrates and vascular plants. Yet, there is evidence that land management practices can affect fungal diversity.

In most environments, most of the larger, showy and fleshy mushrooms that are readily seen, as well as truffles beneath the surface, are mycorrhizal. Obviously, diversity of mycorrhizal species will be influenced, if not determined, by plantings of their potential hosts. However, management practices can also affect the diversity of saprotrophic fungi. Indeed, in northern Europe intensive management of forest is associated with decline in species diversity of wood-degrading saprotrophic fungi. This appears to result from management regimes that remove woody debris from managed forests. Diversity of such species is positively correlated with both the quality and quantity of woody debris left in a forest and coarse woody debris even promotes the abundance and diversity of truffles. It is counterproductive to allow this to occur as change in the diversity and abundance of wood-degrading fungi will adversely affect the recycling of key nutrients and the provision of ecological niches in the managed community. The result is that the influence extends beyond the plant communities to all those other organisms that interact with fungi, from insects and slugs that depend on fungi for food, to the vertebrates that eat the invertebrates. And, of course, it's not just the commercial forests to which this applies. Amenity land (public garden and park land) is an increasingly important aspect of the urban environment which so many of us inhabit, but here, too, excessive tidiness can adversely affect the biodiversity and diminish the recreational value of the resource (Czederpiltz *et al.*, 1999; Juutilainen *et al.*, 2014; Floren *et al.*, 2015; Heilmann-Clausen *et al.*, 2017).

From several aspects of the above, it is evident that fungi contribute to human welfare, both directly and indirectly, and therefore represent part of the total economic value of planet Earth. But these are just a few examples of the ways that fungi make this contribution; descriptions of many other examples occur throughout this book. This is the fungal part of the interface between *economics* and *biology*. Overview of the many benefits that the natural world offers to humans interprets those benefits as a range of *services* (first called *Nature's Services*; Daily, 1997) to which monetary value (the *natural capital*) can be attached. The hope being that understanding the value of the natural systems on which we are all vitally dependent will encourage greater efforts to protect the Earth's basic life-support systems before it is too late (on the principle 'money talks'). The description *'ecosystem services'* was adopted by Costanza *et al.* (1997), and this name is now the most widely used. Costanza *et al.* (1997) estimated the (minimum) economic value of the entire biosphere to be an average of US$33 trillion (that is, US$33 $\times$ 10^{12}) per year. This is equivalent to US$50 trillion at 2018 prices. To put this into perspective, the gross domestic product of the United States ran at a rate of $20 trillion a year during the second quarter of 2018. Ecosystem services is now the principal concept in ecology. By March 2017, the paper in the journal *Nature* by Robert Costanza *et al.* (1997) had been cited over 17,000 times and Gretchen Daily's book (Daily, 1997) had been cited over 6,000 times, making them among the most highly cited works in ecology to date (Costanza *et al.*, 2017). In Resources Box 14.1 (below) we provide a few more reference sources for *Ecosystem Services* and in Table 14.1 we list the fungal examples you can find in this book.

RESOURCES BOX 14.1 Ecosystem Services

Ecosystem services are the benefits people obtain from ecosystems; ultimately, these services support human life, so ecosystem services are the life-support systems of the planet. Historically, the nature and value of planet Earth's life-support systems have largely been ignored until their disruption or loss revealed their importance. The pro-active evaluation of ecosystems and biodiversity has become an important field of investigation for economists and ecologists alike, largely motivated by the search for arguments in favour of broader conservation policies. In 2000, the then United Nations Secretary-General Kofi

box continues

Annan requested the first major scientific appraisal of the world's ecosystems and the services they provide, known as the *Millennium Ecosystem Assessment*, which was done between 2001 and 2005 and involved the work of more than 1,360 experts worldwide (see https://www.millenniumassessment.org/en/About.html). The first overview of the project, describing its conceptual framework, was published in 2005 in the 155-page book *Ecosystems and Human Well-Being: Synthesis* (by C. Corvalan, S. Hales and T. McMichael, published by Island Press, Washington; ISBN: 9781597260404). Wikipedia (https://en.wikipedia.org/wiki/Ecosystem_services) provides a quick introduction, and check out the following websites for further details:

http://www.greenfacts.org/en/ecosystems/index.htm
https://www.millenniumassessment.org/en/Index-2.html
https://www.iucn.org/about
https://www.iucn.org/theme/ecosystem-management

and the following publications:

Cavanagh, R.D., Broszeit, S., Pilling, G.M. *et al.* (2016). Valuing biodiversity and ecosystem services: a useful way to manage and conserve marine resources? *Proceedings of the Royal Society B: Biological Sciences*, **283**: article number 20161635. DOI:https://doi.org/10.1098/rspb.2016.1635.

Costanza, R., d'Arge, R., de Groot, R. *et al.* (1997). The value of the world's ecosystem services and natural capital. *Nature*, **387**: 253. DOI:https://doi.org/10.1038/387253a0.

Costanza, R., de Groot, R., Braat, L. *et al.* (2017). Twenty years of ecosystem services: how far have we come and how far do we still need to go? *Ecosystem Services*, **28**: 1–16. DOI:https://doi.org/10.1016/j.ecoser.2017.09.008.

Daily, G.C. (ed.) (1997). *Nature's Services: Societal Dependence on Natural Ecosystems*. Washington, DC: Island Press. ISBN-10: 1559634766, ISBN-13: 978-1559634762.

Dighton, J. (2016). *Fungi in Ecosystem Processes*, 2nd edition. Boca Raton, FL: CRC Press. ISBN: 9781482249057.

Dighton, J. & White J.F. (ed.) (2017). *The Fungal Community: Its Organization and Role in the Ecosystem*, 4th edition. Boca Raton, FL: CRC Press. ISBN: 9781498706650.

Farber, S., Costanza, R., Childers, D.L *et al.* (2006). Linking ecology and economics for ecosystem management. *BioScience*, **56**: 121–133. DOI:https://doi.org/10.1641/0006-3568(2006)056[0121:LEAEFE]2.0.CO;2.

Gadd, G.M. (ed.) (2006). *Fungi in Biogeochemical Cycles* (British Mycological Society Symposium volume). Cambridge, UK: Cambridge University Press. ISBN-10: 0521845793, ISBN-13: 978-0521845793. DOI:https://doi.org/10.1017/CBO9780511550522.

Salles, J.-M. (2011). Valuing biodiversity and ecosystem services: why put economic values on Nature? *Comptes Rendus Biologies*, **334**: 469–482. DOI:https://doi.org/10.1016/j.crvi.2011.03.008.

Vo, Q.T., Kuenzer, C., Vo, Q.M., Moder, F. & Oppelt, N. (2012). Review of valuation methods for mangrove ecosystem services. *Ecological Indicators*, **23**: 431–446. DOI:https://doi.org/10.1016/j.ecolind.2012.04.022.

Zhang, W., Ricketts, T.H., Kremen, C., Carney, K. & Swinton, S.M. (2007). Ecosystem services and dis-services to agriculture. *Ecological Economics*, **64**: 253–260. DOI:https://doi.org/10.1016/j.ecolecon.2007.02.024.

There are tens of thousands of other publications relevant to this topic. We'll leave you to find the ones that interest you most!

Table 14.1. Fungal contributions to Ecosystem Services you can find in this book

An ecosystem is defined as a dynamic complex of plant, animal and microorganism communities, together with the non-living environment, interacting as a functional unit. Humans are an integral part of ecosystems. Ecosystem services are the benefits people obtain from ecosystems. Ecosystem services are usually grouped into four broad categories, to which we have added *ecosystem goods*, defined (in the left-hand column, below) as follows:

Ecosystem Service category	Fungal contributions	See section
Provisioning, such as the production of food, water and other materials	Food: This includes the vast range of food products derived from plants, animals and microbes. So, the fungal contribution ranges from mushrooms of several different sorts, morels and truffles, cultivated and collected, through yeast extracts, mycoprotein (a filamentous fungus isolated from field soil), and all those fermented foods (bread, cheese, salami, soy sauce and other soya products, tempeh and miso) but also including meats and dairy products because our farm animals cannot digest grass without their anaerobic fungi.	3.3 12.2 to 12.6 15.5 19.18 19.20 19.23 19.24 19.25

table continues

Ecosystem Service category	Fungal contributions	See section
	And drink? Not usually mentioned, but where would we be without beers, wines and spirits? The yeasts deriving, even if in prehistory, from natural ecosystems. Citric acid's fizz and the preparation of fruit juices using fungal enzymes are other contributions.	19.13 19.14 19.16
	Materials such as wood, jute, hemp and many other products are derived from ecosystems and depend on plants, which all depend on mycorrhizal fungi to make their roots work adequately.	14.9 to 14.17
	Similarly, water supply depends on erosion control, which needs vegetation for soil retention and the prevention of landslides. Vegetation depends on mycorrhizal fungi.	14.9 to 14.17
	Climate regulation. Ecosystems influence climate by either sequestering or emitting greenhouse gases. Wood-decay fungi have influence here by releasing chlorohydrocarbon volatiles from timber, and by recycling C and N from the soil.	14.2 14.7 14.17
Regulating, such as the control of climate and disease, can have impact on local or global climate over time scales relevant to human decision-making (decades or centuries)	Water supply. As mentioned above, changes in vegetation, such as conversion of wetlands, the replacement of forests with croplands, or croplands with urban areas, change the water storage potential of the ecosystem. All of which depend on parallel changes in mycorrhizal fungi.	14.9 to 14.17
	Waste treatment. Ecosystems can be a source of impurities in water, but fungal bioremediation can decompose organic wastes introduced by agriculture and/or industry into inland waters and coastal and marine ecosystems.	14.2 14.6
	Regulation of animal and human diseases. Changes in ecosystems can directly change the abundance of pathogens, leading to emerging fungal infectious diseases that seriously endanger clinical practices and threaten animal extinctions in nature.	17.1 to 17.10 17.11 17.12 18.5
	Regulation of plant diseases. Rarely specifically mentioned in these discussions, but fungal diseases dominate in plants, potentially causing sweeping damage to plants of the forest and urban environments (like Dutch elm disease), as well as crucial crop plants like rice and wheat.	16.3 to 16.8 16.11 16.12 16.17
	Biological control. Fungal pathogens can alter the abundance of crop and livestock pests and diseases.	14.19 15.6 17.14 17.15
Supporting services maintain the conditions for life on Earth over extremely long times	Prime examples of supporting services are photosynthetic primary production using CO_2 and water, and photosynthetic production of atmospheric oxygen. At least 90% of the terrestrial plants that do this depend on mycorrhizal fungi for their adequate nutrition.	14.8 to 14.16 14.17
	Soil formation is another long-term supporting service to which fungi contribute their abilities to degrade and manipulate minerals, accumulate metals, and the lichen terrestrial pioneer primary producers, which are mutualistic associations made by fungi with algae and bacteria. Polymers produced by the most widespread mycorrhizal fungi (*glomalin*) control the structure of mature soils.	1.2 to 1.8 7.8 14.3 14.18 19.22

table continues

Ecosystem Service category	Fungal contributions	See section
	Nutrient cycling is a continued supporting service provided by fungi by their ability to secrete enzymes into ecosystems that can digest even the most recalcitrant materials (like lignocellulose) to capture the various nutrients those materials contain. Mycorrhizal and saprotrophic fungi recycle major nutrients for later release by the former to their plant associates, while the many mycophagous animals (small and large) benefit by eating the mycelia and sporophores of both. Importantly, fungi also contribute to water cycling in ecosystems; mycorrhizas are major suppliers of water to their hosts and saprotrophs translocate water over considerable distances. Section 14.10 shows how the *Rhizobium*–legume symbiosis, which assimilates atmospheric nitrogen into organic compounds, makes use of molecular components derived from the arbuscular mycorrhizal fungal partner of the legume to create its symbiotic interface.	1.4 to 1.7 10.7 10.8 11.1 to 11.11 12.1 12.2 14.2 14.10
	Removal of natural wastes (which means dead and dying organisms) from ecosystems is an aspect of the nutrient cycling described in the previous section. However, *contaminating wastes*, which are mostly the result of human activities, can be recycled by fungi because they are so adept at producing degradative enzymes. The process is generally called bioremediation.	11.1 to 11.11 14.6 19.22 19.26
Cultural, such as spiritual and recreational benefits	Contributions of fungi to human cultural activities extend into prehistoric times. Without anaerobic fungi to digest the grass they eat there would be no large herbivores to provide skins for clothes, horn for tools or sinews for prehistoric bows and for binding flint arrowheads, handles to blades or spear points to spears. So not much hunting, though there would still be mushrooms to gather.	3.3 12.3 15.5 19.20
	The Neolithic traveller known as 'Ötzi the Tyrolean Iceman' had three separate fungal products among his equipment; one was clearly used as tinder, the others possibly medicinal. Psychotropic fungi have been used since ancient times for mystical purposes by witch doctors and shamans.	12.3 14.4 17.13
	'Flowers used as ornaments' are normally included in this category. One of the largest families of flowering plants, and arguably the most ornamental, are the orchids. Orchid seedlings are non-photosynthetic and depend on the endomycorrhizal fungus partner for carbon sources; seedling stage orchids can be interpreted as parasitising the fungus.	14.9 14.14
Ecosystem goods, this category depends on biodiversity as the source of many such goods	All plant materials that serve as sources of energy (wood, dung, other biomass fuels) depend on mycorrhizal fungi enabling the plant to grow in the first place.	14.9 to 14.17
	Increasingly, yeast fermentation of agricultural wastes is being used for fuel alcohol production.	19.13 19.22
	Many of today's most widely used pharmaceuticals were 'found by accident' as products of fungi in natural ecosystems and then developed into industrial chemicals for global use. Traditional medicines (including nutraceuticals) also derive from fungi collected from natural ecosystems and now cultivated commercially.	18.1 19.15 19.19 19.26
	Many industrial products, fine biochemicals, biocides, food additives, and enzymes for processing food and other goods, have been developed from metabolites originating from fungi isolated from natural ecosystems.	19.14 19.16 19.26

table continues

Ecosystem Service category	Fungal contributions	See section
	Production of all natural fibres depends on fungi providing nutrition to the organism that makes the fibre; industrial processing, laundering and conditioning of fabrics made from such fibres depends on fungal enzymes.	12.3 12.6 19.16
	Foods depending on fermentation (with a variety of fungi) resulted from traditional use of microbes from natural ecosystems. The foods include mycoprotein, soya foods, tempeh, miso, ang-kak.	19.18 19.25
	Genetic resources are included in this category. This includes all genetic information used in strain breeding and biotechnology. It's also worth remembering that we must thank research with yeast for most of what we know about eukaryote cell biology, genetics and molecular biology.	6.1 to 6.10
	The final topic in this category is usually 'Storm Protection'. It's not obvious how fungi can contribute to this, but the normal explanation is that coral reefs can dramatically reduce the damage caused by hurricanes or large waves, so a negative impact of fungi could be aspergillosis disease of coral.	17.8

World agriculture's vulnerability to climate change is increasingly being expressed as its resilience. Agricultural resilience, defined as the capacity of the system to absorb shocks and stresses (to agricultural production and farming livelihoods), has become a distinct policy objective for sustainable and equitable development (Bousquet *et al.*, 2016). The resilience concept and its relationships to biodiversity, ecosystem services and socioeconomics are explored fully by Gardner *et al.* (2019).

The *2019 Global Assessment Report on Biodiversity and Ecosystem Services* produced by the United Nations' Intergovernmental Science-Policy Platform has demonstrated that the natural environment across most of the globe has now been significantly altered by human activities, with most indicators of ecosystem health and biodiversity showing rapid decline. At the time of writing this text the full *Report* (which is expected to exceed 1,500 pages) has not been published, but you can access the 39-page *Summary for Policymakers* that was released in May 2019 at https://www.ipbes.net/news/ipbes-global-assessment-summary-policymakers-pdf.

In Chapters 15–17, we will add detail to the brief descriptions given above of the ways that fungi contribute to the Earth's ecosystems (and see Suz *et al.*, 2018). In this chapter we will concentrate attention on some saprotrophic activities to begin with, but our main topic will be the various associations between plants and fungi.

14.2 FUNGI AS RECYCLERS AND SAPROTROPHS

We have discussed many aspects of this topic in earlier chapters (particularly in Chapters 11 and 12); we will not repeat them here but will mention them again to bring the strands together and provide reference to their earlier discussion.

Fungi have been recycling biological remains for at least half a billion years. The earliest fibrous fossils (nematophytes; see section 2.7 and Figures 2.6 and 2.7) may well have been large multi-hyphal fungi that were widespread and dominant terrestrial organisms 440 million years ago. These fungi were presumably creating the ancient 'soil', which the first terrestrial animals and plants could invade. Several examples of mycoparasitism have been described in specimens from the Lower Devonian Rhynie chert, and the same material shows vesicular-arbuscular mycorrhizas occurring in the earliest terrestrial plants among these 400 million-year-old fossils (section 2.7) (Taylor *et al.*, 2015; Edwards *et al.*, 2018; Krings *et al.*, 2018).

A defining characteristic of fungi is that they obtain their nutrients by external digestion of substrates. Technically, all fungi are chemo-organoheterotrophs, meaning that they derive their carbon, energy and electrons (with which they do further chemical work) from a wide variety of organic sources. The three main subgroups of this category are shown immediately below, but the relationship between a fungus and its host is complex and variable (see section 14.19):

- Saprotrophs are the decomposers; the category probably covers the majority of fungi and is the main topic of this section.
- Necrotrophs invade and kill host (usually plant) tissue rapidly and then live saprotrophically on the dead remains. These tend to be relatively unspecialised pathogens, able to attack any plant tissue if conditions at the tissue surface are favourable for infection; they cause diseases like foot-rots, damping-off in seedlings, and leaf and stem blotch of mature plants because their infection causes tissue necrosis. Examples are diseases caused by *Rhizoctonia*, *Fusarium*, *Septoria* and, among the *Oomycota*, *Pythium* (see Chapter 16).

- Biotrophs are found on or in living plants and they do not kill their host plant rapidly. They may have very complex nutrient requirements, so they either cannot be grown in culture or grow only to a limited extent on specialised media. These are the specialised pathogens that are highly host specific, such as powdery mildew caused by *Erysiphe graminis*, rice blast caused by *Magnaporthe oryzae* (*Magnaporthe grisea*) and animal pathogens like *Cryptococcus neoformans* (see Chapters 16 and 17).

Saprotrophic fungi decompose many things, and because they can digest and extract nutrients from so many of the materials that exist on, within and under the soil, fungal mycelia act as sinks of organic carbon and nitrogen in the soil. In many ecosystems, a fair proportion of the carbon fixed by photosynthesis ends up in fungal mycelium because of the prevalence of the mycorrhizal symbiotic association between fungus and plant roots (see below).

Something that usually gets much less attention is the fungal involvement in inorganic transformations and element cycling in rocks and minerals (see section 1.8). Several fungi can dissolve and mobilise minerals and metal ions more efficiently than bacteria, and in all soils fungi are involved in scavenging and redistributing minerals and inorganic nutrients (for example, essential metal ions and phosphate). Though this activity is not limited to mycorrhizal fungi, it enables mycorrhizas to make a particularly important contribution to their host as the insoluble salts of minerals (especially phosphates) cannot be absorbed by plant roots but can be solubilised by the mycorrhizal hyphae.

Fungi can also accumulate large amounts of metal ions; Avila *et al.* (1999) showed that concentrations of radioactive caesium in roe deer in Sweden, a legacy of fallout from the Chernobyl incident, increased by up to a factor of five specifically during the mushroom season. Also, silver accumulates in *Amanita* sporophores to levels reaching 1 g kg^{-1} dry weight which was 800–2,500 times higher than in the underlying soils (Borovička *et al.*, 2007); Lepp *et al.* (1987) give similar information about cadmium and vanadium. Filamentous fungi are also able to recover gold from waste electric and electronic devices (e-waste) and could be worth harnessing for commercial biorecovery of gold (Bindschedler *et al.*, 2017). Perhaps the most important thing overall is that the mycelial sink in the soil keeps the mineral nutrients 'on site' and so prevents their loss by rainwater leaching.

A particularly important ability of fungi is that they are the only organisms that can digest wood; that is, plant secondary cell wall, which is the most widespread substrate on the planet, constituting about 95% of the terrestrial biomass. The lignin, which is complexed with hemicelluloses and cellulose in wood, is extremely difficult to degrade and has evolved in part to be a deterrent to microbial attack on long-lived plant parts. Lignin digestion is restricted to fungi, mostly *Basidiomycota* but including a few *Ascomycota*, which between them consume an estimated 4×10^{11} metric tons of plant biomass each year. About 70% of the mass of wood is made up of cellulose, hemicelluloses and pectins and the enzymology of polysaccharide digestion has been described in Chapter 11 (see also section 19.22).

Lignin is a highly branched phenylpropanoid polymer that comprises about 20–25% of wood. Lignin breakdown is an oxidative rather than a hydrolytic process and involves cleavage of ether bonds between monomers, oxidative cleavage of the propane side chain, phenol demethylation and benzene-ring cleavage to ketoadipic acid which is fed into the tricarboxylic acid cycle. It depends on a panel of enzymes including: manganese peroxidase, which catalyses H_2O_2-dependent oxidation of lignin; lignin peroxidase, a heme [Fe]-containing protein, which also catalyses H_2O_2-dependent oxidation of lignin; and laccase (a copper-containing protein) that catalyses demethylation of lignin components (see sections 11.7 and 19.22).

The most abundant nitrogen sources available to organisms degrading plant litter are polypeptides in the form of plant protein, lignoprotein and microbial protein. Extracellular protease enzymes are produced by many species of fungi to hydrolyse large polypeptide substrates into the smaller molecules that can be absorbed by the cell (see section 11.8). Many pathogenic fungi secrete proteases as part of their infection process and, of course, they also use host proteins for nutrition. Saprotrophic and mycorrhizal fungi can use protein as a source of both nitrogen and carbon (and sulfur, too) and some ectomycorrhizas (see below) supply nitrogen derived from soil proteins to their plant hosts.

Rather than continuing to expand on details about saprotrophic fungi that appear in other chapters, we will describe here some incidental matters arising out of this mode of growth. These are (a) they can make the earth move, (b) they produce fungal toxins, (c) they decay structural timber in dwellings, (d) they can remediate toxic and recalcitrant wastes, and (e) they release chlorohydrocarbons into the atmosphere.

14.3 MAKE THE EARTH MOVE

Saprotrophic fungi can be physically destructive even without digesting things. Mushrooms may be soft and squashy, but they have been known to lift stone slabs and force their way through tarmac. In the 1860s, a famous mycologist called Mordecai Cubitt Cooke wrote *A Plain and Easy Account of British Fungi* (Cooke, 1862) in which he told of:

> a large kitchen hearthstone which was forced up from its bed by an under-growing fungus and had to be relaid two or three times, until at last it reposed in peace, the old bed having been removed to a depth of six inches and a new foundation laid.

Cooke also tells of a comparable observation made by a Dr Carpenter:

> Some years ago the town of Basingstoke was paved; and not many months afterwards the pavement was observed

> to exhibit an unevenness which could not readily be accounted for. In a short time after, the mystery was explained, for some of the heaviest stones were completely lifted out of their beds by the growth of large toadstools beneath them. One of the stones measured twenty-two inches by twenty-one, and weighed eighty-three pounds …

It's not only nineteenth century structures that are prone to fungal attack. The BBC News website of 30 October 2008 carried a report about mushrooms lifting the tarmac driveway of a house in Reading, Berkshire (Figure 14.1).

Another relevant pictorial example is on the back-cover page of the April 1991 issue of the *Mycologist*, a magazine published by the British Mycological Society. This photograph shows sporophores of the puffball *Scleroderma bovista* coming through a tennis court. In this case the constructional history was recorded as follows: 'the original hard porous court made of fly ash was overlaid in 1989 with 75 mm of gravel and then a 20 mm layer of tarmacadam was rolled smooth over the top. The first fruit bodies appeared in 1990 …' (Taylor & Baldwin, 1991). Buller (1931) did some experiments in which he put weights on the top of developing mushrooms to see how much pressure they could exert (Figure 14.2). He worked out that a single mushroom could apply a pressure of at least two-thirds of an atmosphere; that is, about 10 pounds per square inch (roughly 70 kPa). It's all a matter of hydraulics, of course, as we indicated in section 7.2, the mushrooms can fill themselves with water and force their way through cracks and crevices. They are not doing it because of some perverse intention to break up paving, but because in nature they need to push through soil and plant litter to bring their sporophores to a position from which they can release their spores to the breeze; if there happens to be a tennis court in the way, then game, set and match to the fungus!

14.4 FUNGAL TOXINS: FOOD CONTAMINATION AND DETERIORATION (INCLUDING MENTION OF STATINS AND STROBILURINS)

We all know that a few species of mushrooms are poisonous, and many fungi produce secondary metabolites, some of which are known as fungal toxins (see section 11.13). The adaptive significance of toxins in fungi has been given rather little attention, though it is evident that toxin production is scattered across the entire kingdom and must have evolved independently several times in evolutionary history. An obvious possible function for toxins is to act as a deterrent to the many animals that might otherwise eat the mushrooms or other fungal structures (see section 12.2) (Wong, 2013; Singh *et al.*, 2014).

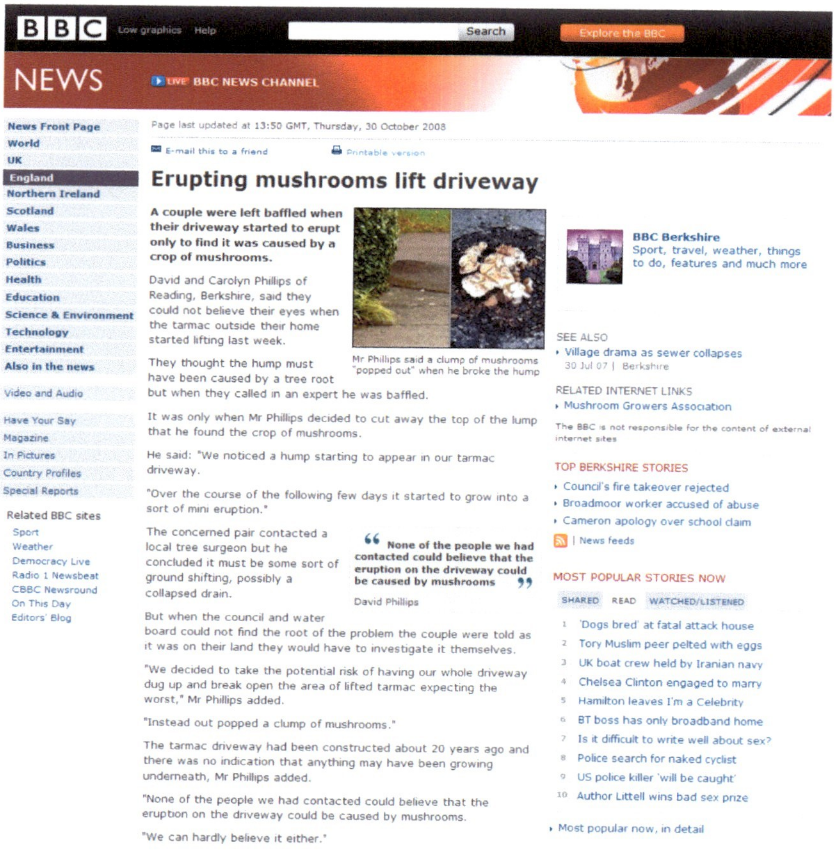

BBC Low graphics Help Search Explore the BBC

NEWS — LIVE BBC NEWS CHANNEL

News Front Page | World | UK | England | Northern Ireland | Scotland | Wales | Business | Politics | Health | Education | Science & Environment | Technology | Entertainment | Also in the news

Video and Audio

Have Your Say | Magazine | In Pictures | Country Profiles | Special Reports

Related BBC sites: Sport | Weather | Democracy Live | Radio 1 Newsbeat | CBBC Newsround | On This Day | Editors' Blog

Page last updated at 13:50 GMT, Thursday, 30 October 2008

E-mail this to a friend — Printable version

Erupting mushrooms lift driveway

A couple were left baffled when their driveway started to erupt only to find it was caused by a crop of mushrooms.

David and Carolyn Phillips of Reading, Berkshire, said they could not believe their eyes when the tarmac outside their home started lifting last week.

Mr Phillips said a clump of mushrooms "popped out" when he broke the hump

They thought the hump must have been caused by a tree root but when they called in an expert he was baffled.

It was only when Mr Phillips decided to cut away the top of the lump that he found the crop of mushrooms.

He said: "We noticed a hump starting to appear in our tarmac driveway.

"Over the course of the following few days it started to grow into a sort of mini eruption."

The concerned pair contacted a local tree surgeon but he concluded it must be some sort of ground shifting, possibly a collapsed drain.

"None of the people we had contacted could believe that the eruption on the driveway could be caused by mushrooms" David Phillips

But when the council and water board could not find the root of the problem the couple were told as it was on their land they would have to investigate it themselves.

"We decided to take the potential risk of having our whole driveway dug up and break open the area of lifted tarmac expecting the worst," Mr Phillips added.

"Instead out popped a clump of mushrooms."

The tarmac driveway had been constructed about 20 years ago and there was no indication that anything may have been growing underneath, Mr Phillips added.

"None of the people we had contacted could believe that the eruption on the driveway could be caused by mushrooms.

"We can hardly believe it either."

BBC Berkshire Sport, travel, weather, things to do, features and much more

SEE ALSO
- Village drama as sewer collapses 30 Jul 07 | Berkshire

RELATED INTERNET LINKS
- Mushroom Growers Association

The BBC is not responsible for the content of external internet sites

TOP BERKSHIRE STORIES
- Council's fire takeover rejected
- Broadmoor worker accused of abuse
- Cameron apology over school claim

| News feeds

MOST POPULAR STORIES NOW

SHARED READ WATCHED/LISTENED

1. 'Dogs bred' at fatal attack house
2. Tory Muslim peer pelted with eggs
3. UK boat crew held by Iranian navy
4. Chelsea Clinton engaged to marry
5. Hamilton leaves I'm a Celebrity
6. BT boss has only broadband home
7. Is it difficult to write well about sex?
8. Police search for naked cyclist
9. US police killer 'will be caught'
10. Author Littell wins bad sex prize

- Most popular now, in detail

Figure 14.1. A couple were left baffled when their driveway started to erupt only to find it was caused by a crop of mushrooms … visit http://news.bbc.co.uk/1/hi/england/berkshire/7699964.stm to see the original report.

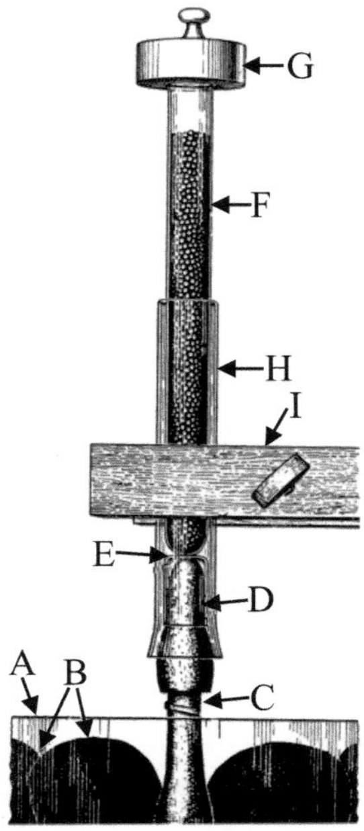

Figure 14.2. Diagram of an experimental rig (original appears as Figure 67 in Buller, 1931) to test the ability of mushrooms of *Coprinus sterquilinus* to lift quantities of lead shot. The *Coprinus* mushroom (C) had grown on some balls of horse dung (B) incubated in a glass dish (A) in the laboratory. A glass tube (H) was placed over the mushroom to stabilise it with clamp (I) and then the mushroom cap (D) was covered with a small glass beaker (E) and then loaded with a test tube containing lead shot (F) to a total weight of 150 g. The mushroom grew a further millimetre in 2 h after this and even the addition of a further 50 g weight (G) did not decrease the rate of growth. Only when the overall loading was increased to 300 g did the mushroom stem (C) bend and break.

Wild opossums (*Didelphis virginiana*) became ill after eating the toxic mushroom *Amanita muscaria*, (fly agaric) and subsequently developed an aversion to the fungus (Camazine, 1983). It has been argued that such a function leads to the expectation that poisonous mushrooms should signal their hazard in some way, and the 'red cap with white spots' of *Amanita muscaria could* be cited as an example of warning colouration. Analysis of ecological and morphological traits associated with edible and poisonous mushrooms showed that the poisonous ones were not more colourful than edible mushrooms but were more likely to have distinctive odours (and perhaps flavours). Raising the possibility that poisonous mushrooms have evolved warning odours/flavours as antifeedants to enhance avoidance learning by fungivores, in contrast to the warning colouration used by poisonous animals to signal 'avoid me' to potential predators (Sherratt *et al.*, 2005). There are dangers in this line of argument: it is close to being anthropomorphic (viewing animal behaviour in human terms), and it may be taken to imply that fungal toxins are produced only by mushroom fungi and/or that fungal toxins are aimed specifically at animals. The fly agaric is a remarkable mushroom in many respects. With its relative *Amanita pantherina*, it is responsible for poisoning in humans characterised by central nervous system dysfunction, with ibotenic acid and muscimol being the active components. However, *Amanita muscaria* contains other substances responsible for psychotropic effects in humans, and it has been used since ancient times for mystical purposes by witch doctors and shamans (Michelot & Melendez-Howell, 2003; Dugan, 2011; Yamin-Pasternak, 2011).

The most important toxins in terms of contamination of human food are the aflatoxins. These toxins are produced by the filamentous *Ascomycota Aspergillus flavus* and *Aspergillus parasiticus* (and less frequently by several other species of *Aspergillus*) as secondary metabolites when the fungi grow saprotrophically on stored food products at temperatures between 24° and 35°C, and moisture contents higher than 7% (10% with ventilation). Food products likely to be contaminated with aflatoxins include cereals, such as maize, sorghum, pearl millet, rice and wheat, oilseeds like groundnut (peanut), soy bean, sunflower and cottonseeds, spices including chillies, black pepper, coriander, turmeric and ginger, tree nuts such as almonds, pistachio, walnuts, coconut and pecans. Because the milk of animals fed on contaminated crops can also contain high concentrations of aflatoxin, dairy products from farm animals can also be contaminated.

The contamination occurs when the mould fungus grows on the crop and secretes the aflatoxin, either through its growth in the field or during post-harvest storage. According to FAO estimates, 25% of world food crops are affected by aflatoxins each year (Moretti *et al.*, 2017).

In well-developed countries, aflatoxin contamination rarely occurs in foods at levels that cause acute aflatoxicosis in humans, but Williams *et al.* (2004) conclude that about 4.5 billion people (that's about two-thirds of the human population) living in developing countries are chronically exposed to largely uncontrolled amounts of the toxin. Aflatoxin contamination of grain consequently poses a major threat to human and livestock health, and aflatoxin content of the diet is at least associated (and may cause) liver cancer. It is no coincidence that liver disease is a common health problem in areas where aflatoxicosis is rife (Kumar *et al.*, 2016; Umesha *et al.*, 2016; Vettorazzi & López de Cerain, 2016).

Why is *Aspergillus* attacking humans with aflatoxins? The truth is, of course, that there is no way that aflatoxin production could have evolved in *Aspergillus* in the time (maybe a few thousand years) that humans have been developing agriculture to the point where we store fair amounts of cereal grain and other crops. The biologically important animals in this story are not humans but rodents. About 40% of mammal species are rodents and they cause billions of dollars in lost crops every year because they collect seed grains into stores in their burrows and nests. Rodents first appear in the fossil record towards the end of the Palaeocene epoch, 55 to 65 million years ago, and that's plenty of time for *Aspergillus* to start competing with the rodents for 'ownership' of their grain stores by producing highly toxic feeding deterrents!

Humans and furry little mammals are not the only animals that eat fungi (see section 12.2 and Figure 12.11) (Singh *et al.*, 2014), and many toxic secondary metabolites are targeted at invertebrates. For example, Seephonkai *et al.* (2004) describe a glycoside from the insect pathogenic fungus *Verticillium* that is cytotoxic towards animal cells. And in a different type of investigation, Nakamori & Suzuki (2007) show that the cystidia of sporophores of *Russula bella* and *Strobilurus ohshimae* defend the sporophore against collembola, producing (unknown) compounds that kill arthropods on contact.

Similar to these compounds are the HMG-CoA reductase inhibitors that were isolated from *Pythium* and *Penicillium* cultures in the 1970s by researchers who 'hoped that certain microorganisms would produce such compounds as a weapon in the fight against other microbes that required sterols or other isoprenoids for growth' (Endo, 1992).

The compounds that were isolated became known as the statins: mevastatin, lovastatin, simvastatin and pravastatin, now marketed around the world as effective and safe cholesterol-lowering drugs that are probably the most widely used pharmaceuticals at the moment (and see sections 11.13 and 19.12). Note that quotation: 'as a weapon in the fight against other microbes'; these compounds are obviously toxic to their target competing species but clearly beneficial to humans. What constitutes a fungal toxin is a matter of definition.

And then there are the strobilurins. Fungi that produce strobilurins (and the related oudemansins) are found all over the world in all climate zones, and with only one exception (*Bolinea lutea*, an ascomycete) all belong to the *Basidiomycota*. Strobilurins and oudemansins inhibit the mitochondrial respiratory chain of fungi by binding to the ubiquinone (= coenzyme Q) carrier that carries electrons to the cytochrome b-c_1 complex (details in section 18.4).

One final facet of the fungal toxins story is their potential as weapons; either the biological component (the fungus) or the specific chemical component (the active toxin itself). Paterson (2006) argues that the low molecular weight toxins rather than the fungi themselves are the biggest threat as bioweapons. Although none are yet known, or even suspected, to have been 'weaponised', it is necessary to be aware of the potential threats so that treatment and decontamination regimes can be developed in advance. As with most threats, it's better to be prepared than paranoid (Zhang *et al.*, 2014).

14.5 DECAY OF STRUCTURAL TIMBER IN DWELLINGS

We have seen that a biological character that sets off the fungi from other organisms is their ability to digest wood. Fungi can cause decay and ultimate destruction of standing trees, felled timber, plant litter and all sorts of timber construction (Schwarze, 2007). In the 200 years or so that Britain 'ruled the waves' with its wooden ships, fungi were the bane of the Royal Navy, which always preferred oak for the hulls of its ships. If our native forests had been able to supply sufficient English oak, no other timber would have been used.

Conservation legislation in Britain probably started with various Acts of Parliament that were intended to control felling and use of timber, and encourage attempts at reforestation, specifically to safeguard the supply of timber for ship construction. But wooden ships have always suffered from rotting of their timbers. Cycles of alternate wetting and drying of parts of the woodwork, poor ventilation and even the use of unseasoned wood in constructions all favour the development of the fungi which have evolved alongside the trees to use wood as a nutrient. Use of well-seasoned timber and provision of adequate ventilation are key to keeping timber constructions in good condition.

Structural timber is no different from other wood; it will decay unless kept dry. Proper building design is a key. The fungi causing the rot are those common in woodlands, and for the most part they are indifferent to the carpenter's expertise. Wood which is always dry is immune from fungal attack. If used out of doors or in humid conditions internally, all wooden structures eventually rot unless treated with some preservative. All kinds of timber are liable to attack. Resistance to attack is relative, soft woods being generally more susceptible than hard woods like oak, yew and teak.

There are three fungi which may be responsible for dry rot damage, but one called *Serpula lacrymans* is the chief culprit. When spores of this fungus fall onto damp wood they germinate, and this is one reason why wet timber is prone to attack. The hyphae penetrate the wood, releasing enzymes that extract nutrients from it. The fungal hyphae may remain wholly within the wood, with no external sign of their presence until severe rotting has developed. Then bulging and cracking of the surface are early signs of attack, especially in painted wood, although mycelium and sporophores soon become evident (Figure 14.3).

The sporophore of this fungus is not a mushroom, but rather, a flat, orange-brown or cinnamon-coloured resupinate sporophore that ranges in size up to a metre or so across. The numerous spores can form rust-coloured deposits on furniture, floors and other surfaces and these are often the first sign of dry rot which are noticed by the occupant of an infected house. *Serpula* forms strands of hyphae which can be 5 mm thick. The strands are invasive and the cells of which they are made cooperate to grow away from the food source which is already infected to find other food sources (see section 10.7).

The strands can translocate food materials efficiently and this enables *Serpula* to spread over materials and structures from which it can derive no nutrition. When *Serpula* grows on wood it decomposes it and eventually reduces the wood to powder (that's why it's called 'dry' rot). But when the chemicals that make up wood are digested, water is formed equivalent to half the dry weight of the wood. During active growth, therefore, the fungus can provide itself with the water it needs; so although it

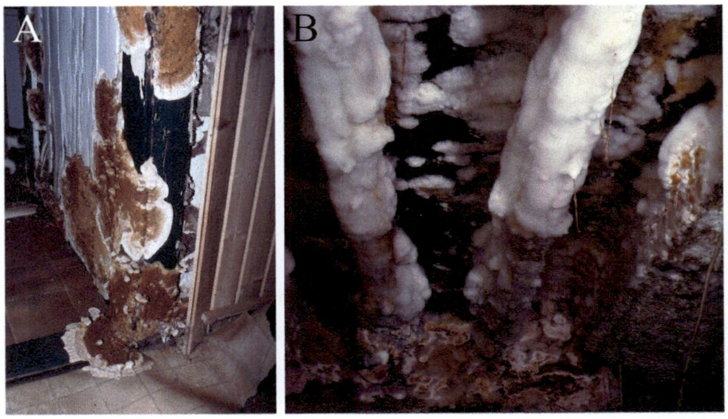

Figure 14.3. The dry rot fungus, *Serpula lacrymans*, growing in the wall panelling (A) and roof beams (B) of an infested basement. The orange-brown areas are the resupinate sporophores. Photographs kindly provided by Dr Ingo Nuss, of Mintraching-Sengkofen, Germany.

must have moist wood to begin its attack it can continue growth into dry timber. Indeed, when growth is luxuriant, there may be excess water produced from wood decay and this is exuded by the fungus in droplets. These are the 'tears' suggested by the 'lacrymans' part of the binomial.

In a real sense the strands are explorers and if wood is reached in a strand's wanderings it is immediately attacked and eventually destroyed. It is the strands that make wood-decay fungi so dangerous to constructions because strands penetrate through pores in brickwork, cement and stone, under tiles and other flooring, and over plaster and other ceilings; indeed, across anything that gives mechanical support. In the laboratory strands have been grown across a full metre of totally dry plaster board, and they can do this so long as the originally infected wood continues to provide nutrition to the exploratory strand. *Serpula* strands can translocate food materials in both directions (in the laboratory, it can be shown that nutrients can flow in both directions at the same time). So when the strand finds newly discovered timber to attack, maybe several yards away from the original, the whole infestation is effectively integrated into a single organism which can become the size of the whole building.

About 60 or so indoor wood-decay fungi that have been described and the strand morphology of the 20 most common species is sufficiently distinctive for the infesting fungi to be identified from the characteristics of the strands (Huckfeldt & Schmidt, 2006). The fungi most commonly encountered as causing damage to building structures are the dry rot fungus (*Serpula lacrymans*), cellar rot fungus (*Coniophora puteana*) and wet rot fungi (*Antrodia vaillantii, Antrodia xantha, Asterostroma* spp., *Donkioporia expansa, Paxillus panuoides, Phellinus contiguus, Tyromyces placentus*) (Singh, 1999). A fungus called *Phellinus megaloporus* (stringy oak rot) is found as frequently as *Serpula* in parts of Europe and was responsible for serious damage to the roof of the Palace of Versailles. It requires very moist conditions and a relatively high temperature. It probably causes more rapid decay in oak than any other fungus, but it does not spread rapidly because it does not form strands. *Serpula lacrymans* is by far the most serious cause of decay of building timbers in the UK and northern Europe.

The dry rot fungus occupies a specialised ecological niche in buildings because of its unique biology, but despite its prevalence in structural timber across Europe, it does not occur in European woodland. The species has only been found in the wild in the Himalayas (Singh, 1999), where *Serpula lacrymans* is typically found in spruce and other conifers in boreal forests. When these trees were harvested for constructing buildings, the dry rot fungus migrated indoors, adapting to thrive in anthropogenic environments, and the international trade in timber gave it the wide distribution in buildings in the temperate zones it enjoys today. Comparative genomic analyses combined with growth experiments using this species and its wild relatives have established that *Serpula lacrymans* evolved a very effective brown-rot decay compared to its wild relatives, enabling an extremely rapid decay of constructional timber. The species has adaptations in nutrient and water transport that promote hyphal growth and invasion of timber in houses, but poor combative ability with other brown-rot fungi *in vitro*. It appears to be an ecological specialist that is preadapted for success in the sheltered indoor conditions of domestic constructions (Balasundaram *et al.*, 2018).

Not all wood deterioration results in destruction of the timber. There are other fungi (sap-stain fungi) that discolour wood. Although the strength of the timber is not lessened, the discoloration (generally called spalting) renders the wood unfit for most purposes and so its value is downgraded because the unsightly timber cannot be used as a natural finish. Staining develops during storage of the sawn timber; sap-stain fungi develop most rapidly when the wood has high moisture content and the weather is warm. The oft-repeated preventatives apply: use well-seasoned (i.e. dried) timber and keep it well ventilated.

Some staining is actually sought after for special cabinet work. The fungus *Chlorociboria aeruginascens*, produces a characteristic bright blue-green colour in oak and other deciduous trees. Wood stained in this way has been used for ornament, usually marquetry, in products called Tunbridge ware. There is even a British patent issued in the early years of the twentieth century covering the artificial infection of trees with *Chlorociboria aeruginascens* to generate stained timber.

Wood turners appreciate the pleasing colour patterns that may result from discoloration in wood caused by the early stages of rotting. Dark winding lines and thin streaks of red, brown and black pigmentation, called zone lines, are often found in infested wood (Figures 8.8 and 14.4).

This type of spalting results from interactions between different fungi growing in the timber. The zone line shows where the different mycelia interact (Figure 8.8) by creating barriers formed from plates of mycelial cells that have darkly pigmented secondary walls, referred to as pseudosclerotial plates. Zone lines

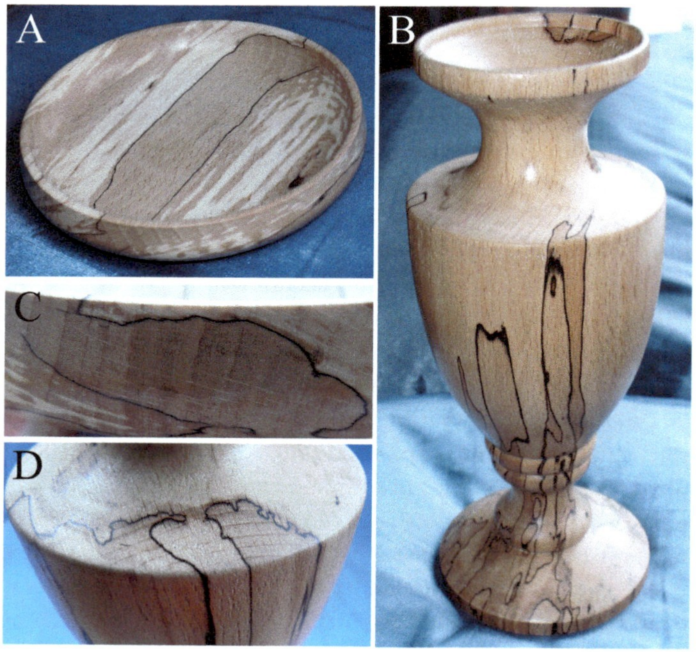

Figure 14.4. Spalted timber products. Turned wood specimen dish (A) and vase (B) made from spalted beech. Both show the attractive figuring of the timber that encourage wood turners to use such timber for its decorative value, but close-ups reveal the three-dimensional distribution of the zone lines between fungal colonies (C and D and see Figure 7.8) that produce the spalting. The vase was made by Matt Cammiss of the Birstall Woodturning Club, West Yorkshire. Photographs by David Moore.

themselves do not damage the wood, though the fungi responsible for creating them often do. Fungi also have a potential role in the biodeterioration/bioremediation of artificial polymers, which we will discuss in the next section.

14.6 USING FUNGI TO REMEDIATE TOXIC AND RECALCITRANT WASTES

Fungi are quite capable of growing on waxes, paints, leather goods and all forms of textiles, from the finest cotton to the heaviest canvas. Much of this degradative ability results from the activity of lignocellulose-degrading enzymes, particularly the panel of ligninolytic enzymes, described in Chapter 11:

- breakdown of polysaccharide: cellulose section 11.2;
- lignin degradation section 11.7;
- digestion of protein section 11.8;
- and processing agricultural, industrial and forest residues, described in 'Digestion of lignocellulosic residues' in section 19.22.

A pest is an organism that is doing what it normally does, but in a place that *we* consider inappropriate. A wood-degrading organism growing in the timbers of your house roof is a pest. However, the same organism could be considered a technological marvel if it were recruited to degrade some of our waste products. Overall, on biological and chemical grounds the more advanced fungi, especially the mushroom fungi, are the ideal candidates to degrade the waste vegetation that we produce through our agricultural activities.

On average, world agriculture currently loses 40% of its primary production to pests and diseases, and then throws away more than 70% of what's left because the crop always represents so little of what is grown. Remember, the 'crop' may only be the seeds of the plant that is grown, or even only a portion of the seed, like its oil content. Just imagine how much of the *coffee bean* ends up in a jar of *instant coffee*. Typically, 80 to 90% of the total biomass of agricultural production is discarded as waste. Some agricultural wastes are polluted with pesticides. Other agricultural wastes are hazardous because they contain tannins and phenolics (toxic to plants and animals) as residues from extraction of oils, such as cotton, rape, olive and palm oils, or fruit processing residues, like citrus wastes. These materials are hazardous because they contain compounds chemically similar to the complex phenolic compounds found in wood. Since the fungi can decompose the wood, they can also be used to degrade the environmental pollutants, both in soils and in liquid effluents. The latter including industrial waste water discharges such as those produced by the paper pulp industry, but also wastes contaminated with pesticides, such as chlorinated biphenyls, aromatic hydrocarbons, dieldrin and even the fungicide benomyl.

The advantage is that the *fungi do not partially degrade* these materials, leaving other possibly dangerous substances behind; rather *they completely mineralise the pollutant* so that its chemical constituents are returned to the atmosphere and soil as carbon dioxide, ammonia, chlorides and water. Laboratory tests have shown that the oyster mushroom (*Pleurotus* spp.) is particularly good at this sort of thing. The tests were conducted with pentachlorophenol (PCP), one of the chlorophenols that have been commonly used as disinfectants and preservatives around the world. They do a good job as pesticides but because most environmental microorganisms find them impossible to degrade they persist in the environment and remain toxic for many years. It is illegal to use PCP in most countries today, but it has been the most heavily used pesticide throughout the world. For example, in the United States during the 1980s, approximately 23 million kg y^{-1} was used, mainly as a wood preservative and, since the 1960s, approximately 5 million kg y^{-1} was sprayed over vast areas of central China as a molluscicide to kill the snails that carry the schistosomiasis parasite. The chemical is very persistent and most of what has been released into the environment is still there. It is highly toxic, uncoupling oxidative phosphorylation by making cell membranes permeable to protons, and thereby dissipating transmembrane proton gradients. It is also cancer inducing and has been declared a priority pollutant for remediation treatment.

The conventional remediation strategy for PCP-contaminated land is excavation and incineration or land filling. Such methods are expensive, obviously destructive to the environment and ineffective for anything other than highly localised 'point source' pollution. Bioremediation is a very promising alternative, using biological systems for the environmental clean-up. The ability of a range of known wood-decay or plant-litter-decay fungi to remove PCP from a batch culture was compared. Fungi tested as mycelia were: *Armillaria gallica, Armillaria mellea, Ganoderma lucidum, Lentinula edodes, Phanerochaete chrysosporium, Pleurotus pulmonarius*, a *Polyporus* sp., *Coprinopsis cinerea* and *Volvariella volvacea*, and the spent mushroom compost from farm beds growing the oyster mushroom *Pleurotus pulmonarius* was also tested. All these fungi showed active breakdown and absorption of PCP removal mechanisms (Figure 14.5), though the tolerance level of the fungus towards PCP did not correlate with its degradative capacity. In a 7-day incubation, *Armillaria mellea* mycelium showed the highest degradative capacity (13 mg PCP g^{-1} mycelium dry weight) and *Pleurotus pulmonarius* mycelium was second-highest with 10 mg PCP g^{-1} mycelium dry weight; the least effective was *Polyporus* with 1.5 mg PCP g^{-1} mycelium dry weight. On the other hand, the *Pleurotus* spent mushroom compost, harbouring both bacteria and fungi, had a degradative capacity of 19 mg PCP g^{-1} dry weight in only 3 days exposure to PCP (Chiu *et al.*, 1998).

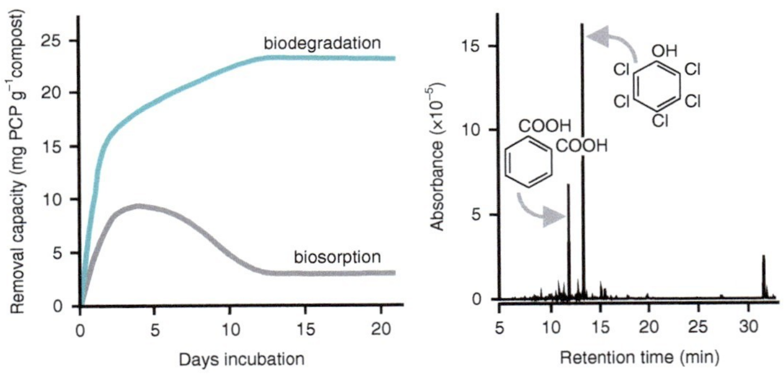

Figure 14.5. Data showing that incubation for a few weeks with spent oyster mushroom compost leads to destruction of pentachlorophenol (not just its adsorption) (left-hand panel), and that mycelium of *Pleurotus pulmonarius* dechlorinates pentachlorophenol (PCP) through a catabolic process that involves removal of the chlorine followed by opening of the benzene ring (right-hand panel). For the left-hand plot, absolute removal capacity of PCP by spent oyster mushroom substrate (that is, the substrate left after the last crop was harvested) was quantified by capillary electrophoresis. The right-hand panel shows a mass-selective gas chromatography (GC-MS) spectrum of the extract of the fungal biomass of *Pleurotus pulmonarius* mycelium after two days incubation in a medium containing 25 mg l^{-1} pentachlorophenol. The most prominent peak (retention time 13.53 min) is PCP, the next most prominent peak at 12.03 min is benzene-1,2-dicarboxylic acid (also called phthalic acid). Other peaks at longer retention times include fatty acids that result from opening the benzene ring and esterification of its straight-chain derivatives. For example, peaks at retention times of 15.73, 31.96 and 32.25 min have been identified as hexadecanoic acid (palmitic acid, $C_{16}H_{32}O_2$) and its derivatives. Modified and redrawn from Chiu *et al.* (1998).

Mass-selective gas chromatography (GC-MS) chromatograms revealed only residual PCP peaks in extracts of PCP-treated spent mushroom substrate extracts, a contrast with the mycelial incubations in which a variety of breakdown products were detectable. The spent compost left after oyster mushroom cultivation does two crucial things. It absorbs, immobilises and concentrates PCP so it can be transported away from the contaminated site, and it also digests PCP completely. Mushroom cultivation is a common practice all over the world and the idea that hazardous waste materials could have their pollutants removed and produce a mushroom crop at the same time is exceedingly attractive.

The decontamination of polluted soil and water using fungi is known as mycoremediation (Rhodes, 2014), and there are two approaches: one in which the contaminated material is treated on site, described as *in situ*; and *ex situ*, when the material is physically removed to be treated elsewhere. Removing contaminated soil is the costlier procedure; if the soil can be left and decontaminated *in situ*, the overall expense is far less. The goal of bioremediation is the full mineralisation of contaminants, that is, their transformation to CO_2, H_2O, N_2, HCl, etc. Heavy metal and radioactive ions can be immobilised by chemical change or accumulated by the fungus and physically removed by harvesting the fungal sporophore.

But the danger in this approach is that the harvested mushrooms must not be used as a food source. Oyster mushrooms can concentrate the metal cadmium (a common industrial contaminant) to such an extent that by eating less than an ounce (28 g) (dry weight) of the most contaminated samples you would exceed the weekly limit tolerated by humans. Cadmium is so toxic that this situation could pose a public health hazard. There are no worries about conventionally cultivated oyster mushrooms, of course. The point is that if the mushroom is grown on composts that *might* be mixed with industrial wastes (in remediation programmes, for example), then it would be advisable to monitor the heavy metal contents before mushrooms are marketed for food. By far the most promising technique is use of the spent mushroom substrates remaining *after* harvesting mushroom crops. Ironically, these are often discarded as wastes themselves, but they are clearly able to offer an integrated approach combining soil conditioning with degradation of pollutants as an effective strategy for bioremediation *in situ*. Spent mushroom farm substrates contain mixed populations of microbes and the synergism that results in such mixtures promises a range of bioremediation approaches to degrade at the molecular level some of the most recalcitrant and toxic organic materials, including oil spills, pesticide spills and industrial wastes. Cadmium tolerance of *Aspergillus fumigatus* has been shown to depend on a gene that encodes a cadmium efflux pump together with extracellular siderophore production. The natural function of the 'cadmium efflux pump' could be to protect the hyphae from *iron* overload by acting as a *ferrous iron pump*. This sort of genomics enables understanding of the processes in enough detail to raise the prospects of strain improvement strategies that use gene editing to construct fungal strains specifically for use in biosorption or biomining processes or to prevent accumulation of toxic metals in crops (Kurucz *et al.*, 2018).

Petroleum-degrading bacteria can be used for bioremediation of soil contaminated with petroleum sludge, which is the oily sludge generated in refineries when inlet and outlet tanks are cleaned. Disposal of the sludge causes environmental issues as well as human health concerns, but bacterial remediation of contaminated soil can restore its ability to support healthy growth of crop plants (Varma *et al.*, 2017). Successful, and rapid, degradation rates of total petroleum hydrocarbon (TPH) have also been demonstrated by mixed populations of petroleum-degrading bacteria with white-rot fungi grown in solid-state fermentation (see section 19.21). Enzymes secreted by the white-rot fungus (laccases; see section 11.7), acting together with the bacteria, degraded the petroleum hydrocarbons in the contaminated soil (Liu *et al.*, 2017).

Petroleum oil spill is no longer the most dramatically damaging form of pollution of the natural environment due to human activity; plastics have now reached the number one spot. Plastics have become an essential part of modern life; world production of plastic materials totalled around 335 million metric tons in 2016. The phrase 'plastic materials' covers a range of polymers, including polyvinylchloride (PVC), polyurethanes, polystyrene, polyamides and polyesters with a range of properties and susceptibility to degradation (Sabev *et al.*, 2006; Shah *et al.*, 2008).

Generally low production costs have led to plastics being used in a vast range of applications; as you can see from a casual glance around your surroundings, from flooring materials, heat insulation, shoe soles, cable sheaths, pipework, packaging, food containers, furnishings, electronic devices and a host of other products that have become essential to modern life.

An estimated total of 8,300 million metric tons of plastics have been produced in the world during the past 65 years. By 2015, approximately 6,300 million metric tons of *plastic waste* had been generated, around 9% of which had been recycled, 12% was incinerated, and *79% was accumulated* in landfills or just discarded into the natural environment. Between 5 and 13 million metric tons of plastic end up in the oceans every year (Geyer *et al.*, 2017).

Lack of degradability and growing land and marine pollution problems have led to mounting concern about plastics. With the excessive use of plastics and increasing pressure being placed on capacities available for plastic waste disposal, there is an obvious need for biodegradable plastics, and biodegradation of the plastics that have already been discarded. Despite recycling efforts, most of these materials are discarded into landfill sites or into the environment where they accumulate. Many plastics that contain a carbon–carbon backbone (homopolymers), such as PVC and polystyrene, are resistant to chemical or microbial degradation, whereas others that contain other elements in the backbone (for example nitrogen or oxygen) and are heteropolymers, such as polyurethanes, are highly susceptible to enzymatic microbial degradation. So much so that this can severely damage or limit the useful life of many products made from such polymers, causing major industrial and some medical problems. Consequently, broad-spectrum biocides are often incorporated into polymer blends to inhibit fungal and bacterial growth and so extend the lifetime of the final product; which, of course, only increases the adverse environmental impact when the plastic is discarded. For some recalcitrant plastics, such as PVC, although the polymer itself is highly resistant to degradation, it is the plasticisers (organic acids blended into the material to increase flexibility of the product) that are often themselves highly susceptible to enzymatic microbial attack.

We have already indicated (see Chapter 11) that the capacity of basidiomycetes to degrade the complex structure of lignocellulose is due to their ability to secrete all the extracellular enzyme systems necessary to degrade biomass. These enzymes have a high potential for biotechnological applications including mycoremediation. Recent genomic studies of basidiomycetes have provided valuable information about the variety of enzymes they make available (Peralta *et al.*, 2017); we should all hope that 'synthetic biology' (Osbourn *et al.*, 2012) can fulfil its promise and be used to redesign existing biological systems for this essential mycoremediation function. Thankfully, the importance of fungi and of mycology for continued world development and improvement is beginning to be realised (Lange *et al.*, 2012). In section 19.26 we will describe some microbial metabolism that has recently been found to digest plastics and plasticisers, and which may suggest methods that might be used to remediate plastic wastes.

14.7 RELEASE OF CHLOROHYDROCARBONS INTO THE ATMOSPHERE BY WOOD-DECAY FUNGI

The ability of fungi, particularly white-rot wood-decay fungi, to metabolise chlorinated hydrocarbons is not an exotic aspect of their biochemistry. Far from it. It is a normal everyday reaction during the degradation of timber. Some of the small organic molecules ('mediator' compounds or co-substrates) these fungi produce contribute to the H_2O_2-regenerating system essential for activity of the lignin and manganese peroxidases incorporate chlorine from the wood into chloroaromatics (already mentioned in the section on *Lignin degradation* in Chapter 11, section 11.7). The synthesis of veratryl alcohol requires chloromethane (CH_3Cl) which acts as a methyl donor.

When these volatile compounds are released into the atmosphere they flush the chlorine out of the substrate, but they may act as environmental pollutants. In most species, production of chloromethane is tightly coupled with its consumption so detectable quantities are not released by laboratory cultures. Excessive chloromethane production has been observed in several fungi. For example, *Phellinus pomaceus* (also known as *Phellinus tuberculosus*), the cushion bracket which is weakly parasitic on many common bushes and trees, can release over 90% of the chloride ions of its substrate as chloromethane; and the volatile hydrocarbons can even be used as an indicator of wood decay in timber stocks (Anke & Weber, 2006; Konuma *et al.*, 2015; Weigold *et al.*, 2016).

Unlike other greenhouse gases (such as chlorofluorocarbons or CFCs), chloromethane originates mainly from natural rather than human activities. About 5 million tons of chloromethane are released annually worldwide, making chloromethane the most abundant halocarbon gas in the outer atmosphere (stratosphere). Terrestrial sources of CH_3Cl predominate; the major ones being biomass burning, wood-rotting fungi, coastal salt marshes, tropical vegetation and decomposition of soil organic matter and reaction between plant pectin and chloride ion (which forms chloromethane abiotically at ambient temperatures in senescent leaves and leaf litter) (Keppler *et al.*, 2005; Tsai, 2017; Misztal *et al.*, 2018).

The fungal chloromethane contribution has been estimated at around 150,000 tons per annum (Watling & Harper, 1998), which is about the same as world salt marshes but 60% more than is released into the atmosphere by industrial coal burning furnaces worldwide. Chloromethane is a powerful greenhouse gas and atmospheric pollutant which contributes significantly to the destruction of stratospheric ozone. Its contribution to climatic change remains to be established.

However, on a global scale, decomposition of dead wood releases billions of tons of CO_2 to the atmosphere each year, a similar magnitude, in fact, to the annual CO_2 emissions from fossil fuel combustion. According to the study by Rinne-Garmston *et al.* (2019), temperature was the main overall control on the respiration rate of dead wood in Norway spruce-dominated forest in Finland. Consequently, increased decomposition of the estimated total of 36–72 billion metric tons of carbon represented in coarse woody debris in the world's forests might be expected as climate change increases environmental temperatures. The relationship is complex, however, as the decomposition rate of dead wood is also significantly dependent on local factors, such as the nature of the woody debris and its fungal community, nitrogen availability and soil moisture status (Rinne-Garmston *et al.*, 2019).

14.8 INTRODUCTION TO MYCORRHIZAS

The mycorrhizal mutualistic association between fungi and the roots of plants has contributed, from the earliest times, to the evolution of the Earth's terrestrial ecosystem. In this relationship the roots of the plant are infected by a fungus, but the rest of the fungal mycelium continues to grow through the soil, digesting and absorbing nutrients and water and sharing these with its plant host. This was discovered by a German botanist called Albert Bernhard Frank in 1885. He began a study of the possibility of cultivating truffle groves in Prussia, but his study developed into a revolutionary theory of tree nutrition via symbiosis between fungi and tree roots in a compound structure he called a *Wurzelpilze*, or root-fungi. In fact, Frank used the words *Wurzelsymbiose, Wurzelpilze* and *Mycorhiza* in the titles of three successive papers in 1885 (and *mykorhiza* and *mycorrhiza* in titles of two later papers). The spelling has now been standardised to 'mycorrhiza', but it still means fungus-root! Later studies have shown that just about all of Frank's interpretations were correct.

A mycorrhiza is a mutualistic (or symbiotic) interaction between a fungus and the roots of a plant. The plant benefits from increased nutrient uptake (particularly phosphate) via the fungal mycelium, while the fungus is supplied with photosynthetic sugars by its host. The arrangement seems to have evolved as soon as plants first colonised the land some 450 to 600 million years ago. Today about 6,000 species of fungi are known to form mycorrhizas with something like 240,000 plant species. Overall, 95% of vascular plants have mycorrhizas associated with their roots, in all habitats, including deserts, lowland tropical rain forests, through high latitudes and altitudes, and including aquatic ecosystems. The few non-mycorrhizal plant families, such as rushes (*Juncus*), sedges (*Carex, Kobesia*), campions (*Silene*) and crop plants, such as rapeseed (*Brassica napus*; one specific group of cultivars is known as canola), tend to colonise open habitats, where competition for nutrients is reduced, particularly in habitats where phosphorus availability is likely to be adequate (Brundrett, 2009; Varma *et al.*, 2017). For more information and illustration of mycorrhizas visit Mark Brundrett's website at http://mycorrhizas.info.

Mycorrhizal fungi also link plants together into communities that are more resilient to stress and disturbance than single plants. Nutrients can be transferred between two different plants which are connected to the same mycorrhizal system; for example, from a well-placed donor plant to a shaded recipient plant. The mycorrhizal interconnections form a network through which plant-to-plant, plant-to-fungus and fungus-to-plant transfers of nutrients and signalling molecules can take place. In nutrient-poor soils, mycorrhizal fungi can provide nitrogen to their host plant which their mycelia have obtained by saprotrophic digestion of nutrients in the soil. Van der Heijden and Horton (2009) completed a comprehensive review of the importance of mycorrhizal networks in natural ecosystems and state that 'mycorrhizal networks play a key role in plant communities by facilitating and influencing seedling establishment, by altering plant-plant interactions and by supplying recycled nutrients'.

For plants, the arrangement has become the rule rather than the exception. Conversely, although there are common and important examples of mycorrhizas from all the major types of fungi, most fungi are not mycorrhizal. The mycorrhizal fungi you are most likely to meet are the mushrooms common in wooded areas.

Names already mentioned, like *Amanita* and *Boletus*, are mycorrhizal partners with trees and other forest plants, as are chanterelles, and truffles too, although truffles are not mushrooms, of course, but a hypogeous sporophore of the *Ascomycota*.

14.9 TYPES OF MYCORRHIZA

Mycorrhizas were traditionally classified into the two types: ectotrophic and endotrophic, a classification based on the location of the fungal hyphae in relation to the root tissues of the plant; *ecto* means outside the root, *endo* means inside. This classification is now regarded as too simplistic, and there is now a nomenclature identifying seven mycorrhizal types; however, we will telescope this into four major types with three additional subclasses as follows.

Endomycorrhizas; in which the fungal structure is almost entirely within the host root, comprising three major and two minor groupings:

- Arbuscular (AM) endomycorrhizas, which are the commonest mycorrhizas, and first to evolve. The fungi are members of the *Glomeromycotina*, they are obligate biotrophs, and they are associated with roots of about 80% of plant species, including many crop plants. The AM association is endotrophic and has previously been referred to as vesicular-arbuscular mycorrhiza (VAM). This name has since been dropped in favour of AM, as some do not form vesicles (see Table 14.2) but you may still find that other textbooks refer to 'VAM' or 'VA' mycorrhizas.
- Ericoid endomycorrhizas are mycorrhizas of *Erica* (heather), *Calluna* (ling) and *Vaccinium* (bilberry), that is, plants that endure moorlands and similar challenging environments. Fungi are members of the *Ascomycota* (an example is *Hymenoscyphus ericae*). The plant's rootlets are covered with a sparse network of hyphae; the fungus digests polypeptides saprotrophically and passes absorbed nitrogen to the plant host. In extremely harsh conditions the mycorrhiza may even provide the host with carbon sources (by metabolising polysaccharides and proteins for their carbon content). Two specialised subgroups may be separated out of the ericoid endomycorrhizal group:
  - arbutoid endomycorrhizas, and
  - monotropoid endomycorrhizas (the mycorrhizal association formed by the achlorophyllous plants of the Montropaceae).
- Orchidaceous endomycorrhizas are similar to ericoid mycorrhizas but their carbon nutrition is even more dedicated to supporting the host plant as the young orchid

Table 14.2 Main characteristics of these seven types of mycorrhiza

	Mycorrhizal type						
	Endomycorrhizas				Ectomycorrhizas		
Feature	AM	Ericoid	Arbutoid	Mono-tropoid	Orchid	Ecto-	Ectendo-
Fungi septate	no	yes	yes	yes	yes	yes	yes
Fungi aseptate	yes	no	no	no	no	no	no
Intracellular colonisation	yes	yes	yes	yes	yes	no	yes
Fungal sheath	no	no	yes or no	yes	no	yes	yes or no
Hartig net	no	no	yes	yes	no	yes	yes
Vesicles	yes or no	no	no	no	no	no	no
Plant host chlorophyllous?	yes (? no)	yes	yes	no	no*	yes	yes
Fungal taxa	*Glomeromycotina*	*Ascomycota*	*Basidiomycota*	*Basidiomycota*	*Basidiomycota*	*Basidiomycota* *Ascomycota*	*Basidiomycota* *Ascomycota* (*Glomeromycotina*)
Plant taxa†	Bryo Pterido Gymno Angio	Ericales Bryo	Ericales	Monotropaceae	Orchidaceae	Gymno Angio	Gymno Angio

* All orchids are achlorophyllous in the early seedling stages, but usually chlorophyllous as adults. Table based on Table 1 in Smith and Read (1997) and Harley (1991). †Bryo = Bryophyta, Pterido = Pteridophyta, Gymno = Gymnospermae, Angio = Angiospermae.

seedling is non-photosynthetic and depends on the fungus partner utilising complex carbon sources in the soil and making carbohydrates available to the young orchid. All orchids are achlorophyllous in the early seedling stages, but usually chlorophyllous as adults, so in this case the seedling stage orchid can be interpreted as parasitising the fungus. A characteristic fungus example is the basidiomycete genus *Rhizoctonia* (although this is a complex genus which can be divided into several new genera).

Ectomycorrhizas are the most advanced symbiotic association between higher plants and fungi, involving about 3% of seed plants including most forest trees. In this association the plant root system is surrounded by a sheath of fungal tissue which can be more than 100 μm thick, though it is usually up to 50 μm thick. The hyphae penetrate between the outermost cell layers forming what is called the Hartig net. From this a network of hyphal elements (hyphae, strands and rhizomorphs) extends out to explore the soil domain and interface with the fungal tissue of the root. Ectomycorrhizal fungi are mainly *Basidiomycota* and include common woodland mushrooms, such as *Amanita* spp., *Boletus* spp. and *Tricholoma* spp. Ectomycorrhizas can be highly specific (for example *Boletus elegans* with larch) and nonspecific (for example *Amanita muscaria* with 20 or more tree species. In the other specificity direction, 40 fungal species can form mycorrhizas with pine.

Ectomycorrhizas can link together groups of trees, the submerged mycelium acting as what has been described as a 'wood-wide-web' (Wohlleben, 2017) or, more formally, 'common mycorrhizal network' (Gilbert & Johnson, 2017) (see section 14.15). Ectomycorrhizal fungi depend on the plant host for carbon sources, most being uncompetitive as saprotrophs. With few exceptions (*Tricholoma fumosum* being one), the fungi are unable to utilise cellulose and lignin; but the fungus provides greatly enhanced mineral ion uptake for the plant and the fungus is able to capture nutrients, particularly phosphate and ammonium ions, which the root cannot access. Host plants grow poorly when they lack ectomycorrhizas. This ectomycorrhizal group is reasonably homogeneous, but a subgroup, ectendomycorrhizas, has been appended.

- Ectendomycorrhiza is a purely descriptive name for mycorrhizal roots that exhibit characteristics of both ectomycorrhizas and endomycorrhizas. Ectendomycorrhizas are essentially restricted to the plant genera *Pinus* (pine), *Picea* (spruce) and, to a lesser extent, *Larix* (Larch). Ectendomycorrhizas have the same characteristics as ectomycorrhizas but also show extensive intracellular penetration of the fungal hyphae into living cells of the host root.

14.10 ARBUSCULAR (AM) ENDOMYCORRHIZAS

Arbuscular (AM) endomycorrhizas are the most common type of mycorrhizal association and were probably the first to evolve; the fungi are members of the *Glomeromycotina* (discussed in section 3.6). In other textbooks, you may find these fungi placed in Order Glomales and Phylum 'Zygomycota' but this is incorrect. The AM fungi are obligate biotrophs, and they are associated with roots of about 80% of plant species (that's equivalent to about two-thirds of all land plants, or around 90% of all vascular plants), including many crop plants. The AM association is endotrophic and has previously been referred to as a vesicular-arbuscular mycorrhiza (VAM). This name has since been dropped in favour of AM (see Table 14.2 and Figures 3.7 and 14.6), since members of Family Gigasporaceae do not form vesicles. Figure 14.6 illustrates the main features of this type of mycorrhiza.

There is a wide-ranging fungal mycelium within the host root, and AM fungi explore the soil or other substrata with an extensive extraradical mycelium. Externally to the host, the fungal hyphae produce the very large spores (often called chlamydospores). Formation of the mycorrhizal association is an infection process (Figure 14.7). Spores germinate near a plant root and the germinating hyphae penetrate the root in response to root exudates. Hyphae grow through the root tissues and in the root cortex hyphal branches form appressoria that penetrate the plant cells. The host plasmalemma invaginates and proliferates around the fungal intrusion. Repeated dichotomous branching of the fungal 'hypha' produces the arbuscule (see Figure 3.7B) inside the cortical cell. Arbuscules have a lifespan of 4–15 days, after which they break down, and the plant cell returns to normal.

Many AM fungi (except Family Gigasporaceae) also produce vesicles within the roots (Figure 3.7C), either between or within the cortical cells. These are swollen spherical or oval structures containing lipids, which are thought to be used for storage. Vesicles are usually over 100 μm in diameter and are swollen hyphal tips.

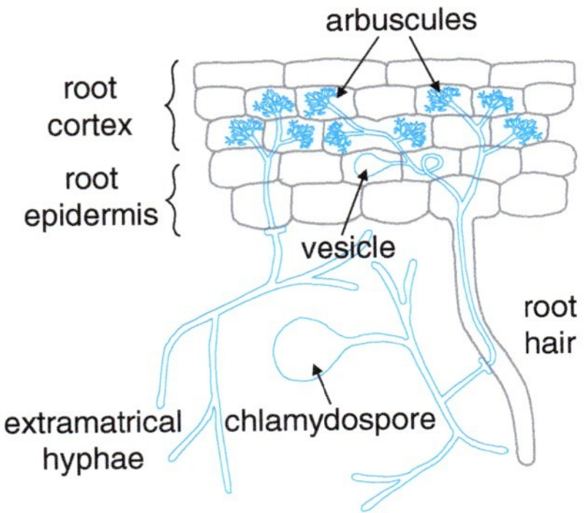

Figure 14.6. Diagrammatic representation of the main cellular features of the arbuscular endomycorrhiza. Hyphae from a germinating spore infect a root hair and can grow within the root between root cortical cells and penetrate individual cells, forming arbuscules. These are finely branched clusters of hyphae, which are thought to be the major site of nutrient exchange between fungus and plant (modified from Jackson & Mason, 1984). See Figure 3.7 for photographs illustrating many of these features and for more information and illustration of mycorrhizas visit Mark Brundrett's website at http://mycorrhizas.info.

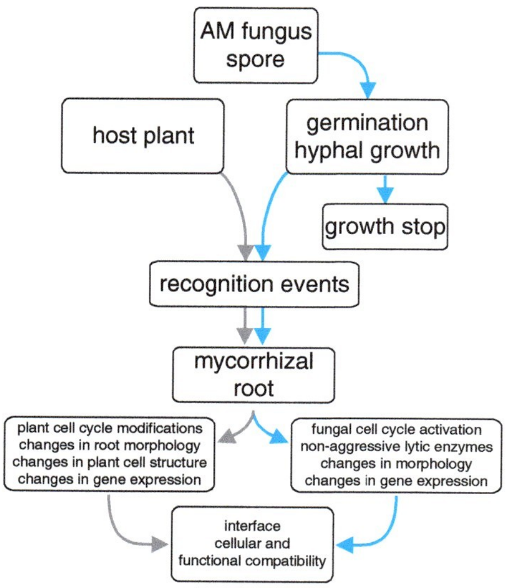

Figure 14.7. Flow chart for mycorrhizal establishment.

All AM fungi are obligately biotrophic; that is, they are completely dependent on plants for their survival. This does not present a problem for AM fungi, as they show little or no host specificity. Unlike ericoid and orchid mycorrhizas, AM fungi are not restricted to any particular taxonomic group of plants, and are found extensively in pteridophytes, gymnosperms and angiosperms of all habitats and even 'primitive' plants such as bryophytes and liverworts form AM-like symbioses.

Because AM fungi are obligately biotrophic, attempts to grow them in axenic culture have so far failed; though the spores will germinate in axenic culture, they subsequently die. However, since the majority of commercially grown plants form AM associations, interest in developing ways of inoculating crops with mycorrhizal fungus continues to increase (see section 14.17).

The highly branched arbuscule produces a large surface area which enhances exchange of nutrients between the partners. Bidirectional nutrient transfer is a key feature of the mycorrhizal symbiosis. In arbuscular mycorrhizas the major nutrients exchanged are:

- reduced carbon created through the photosynthetic activity of the plant partner; and
- phosphate, mobilised and taken up by the fungal hyphae through their exploration of soil microhabitats;
- however, although this is a neglected area of research, these mycorrhizas probably also contribute to nitrogen cycling in the soil (Veresoglou *et al.*, 2012).

This two-way exchange of nutrients takes place between the plant root cortical cell and the fungal arbuscule that has penetrated it. This is the symbiotic interface and on the one side it is bordered by the plant plasma membrane and its extracellular matrix, and on the other side by the fungal plasma membrane and its extracellular matrix.

Nutrient transfer between fungus and plant is *indirectly* linked; that is, there is no 'one-for-one' linkage between, say, glucose and phosphate exchange. This is indicated by the fact that the same host plant can be in symbiosis with different AM fungi and, when this happens, there can be very different carbohydrate:phosphate exchange ratios between the different plant–fungus pairwise combinations. In addition, carbohydrate transfer can be decoupled from phosphate transfer if the soil has high phosphate availability. The process of transfer of nutrients between the plant and the fungus is believed to involve passive efflux of solutes from each donor organism into the interfacial apoplast (the apoplast is the free diffusional space between the two, plant and fungal, plasma membranes), followed by active uptake by the receiver organism. All the active transporters that have been implicated so far in this mechanism are proton-pumping ATPase symporters (see section 11.11). Specifically, on the fungal side there is an ATPase that pumps protons into the arbuscule at the expense of ATP and simultaneously takes in glucose, and on the plant side there are mycorrhiza-inducible plant phosphate and ammonium transporters in mycorrhizal roots (Figure 14.8) (Ferrol *et al.*, 2002; Fellbaum *et al.*, 2012, 2014).

There is a tripartite symbiosis between leguminous plants, their arbuscular mycorrhizal fungi and their rhizobial bacteria that form nitrogen-fixing nodules on the legume roots. This is an association at the molecular level that is crucial for the nitrogen uptake of legumes. No eukaryotic enzymes can break the triple bond between nitrogen atoms in atmospheric gaseous nitrogen, N_2. The reduction of nitrogen to ammonia (nitrogen fixation) is limited to prokaryotes and is catalysed by the enzyme nitrogenase. Most of the nitrogen entering the biosphere (around 100 million metric tons of N_2 each year) does so through the activity of nitrogenase (the other significant source, lightning, contributes about 10%). Consequently, plants that can form a mutualism with nitrogen-fixing bacteria not only have a significant selective advantage under conditions of limiting nitrogen but

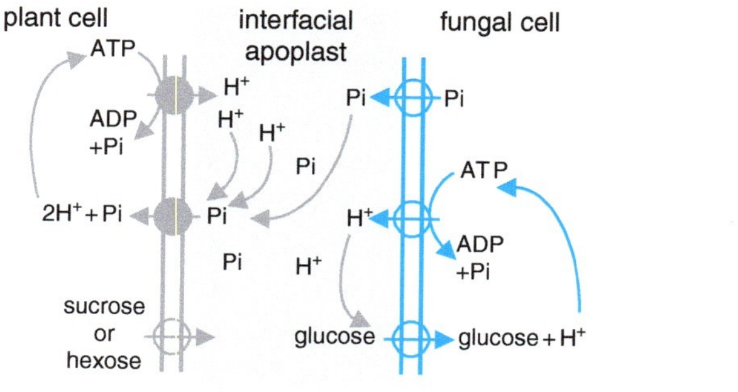

Figure 14.8. Schematic illustration of a model for the exchange-transfer of phosphate and carbon compounds across the arbuscular mycorrhizal interface. Plasma membrane proton-pump-ATPases and secondary transporters that have been experimentally localised in the membranes of the arbuscular interface are indicated by the circles, with the arrows indicating direction of transport. Redrawn from Ferrol *et al.* (2002).

also make a major contribution to the availability of reduced nitrogen in the biosphere. By horizontal gene transfer, the bacteria in the *Rhizobium*–legume symbiosis have inherited the ability to infect legume roots and form nitrogen-fixing nodules from the tissues of the plant root. These provide a habitat for the bacteria that is perfectly suited for N_2 reduction by providing to the bacteria a supply of carbon (as malate from the plant) and an environment with a low O_2 tension (required for nitrogenase activity). Bacteria are housed in specialised intracellular membrane compartments, formed by the host plant, which control the bidirectional exchange of nutrients and, therefore, form the essential core of the endosymbiosis.

Recent studies have demonstrated that the *Rhizobium*–legume symbiosis makes use of a signalling pathway, receptor and regulators of exocytotic vesicle trafficking *derived from the fungal component of the more ancient arbuscular mycorrhizal symbiosis* to form this bacterium–plant symbiotic interface. Furthermore, the beneficial effects of the nitrogen-fixing nodules can be *communicated by the mycorrhiza* to non-leguminous plants with which it also forms a mycorrhizal symbiosis. For example, inoculating with both arbuscular mycorrhizal fungus and *Rhizobium* in a soy bean/maize intercropping system improved the nitrogen fixation efficiency of soy bean, and promoted nitrogen transfer from soy bean to maize, resulting in yield improvement in this *legume/non-legume intercropping system* (Ivanov *et al.*, 2012; Downie, 2014; Meng *et al.*, 2015).

14.11 ERICOID ENDOMYCORRHIZAS

Most plants belonging to the order Ericales can associate symbiotically with soil fungi to form the distinctive ericoid mycorrhiza. This association was initially described in members of Family Ericaceae, which is more abundant in the northern hemisphere and is probably best known in the UK through heather moorland ('heathland') genera characterised by *Erica* (heather), *Calluna* (ling) and *Vaccinium* (bilberry), though the group also includes *Rhododendron*. These are plants that endure moorlands and similar challenging environments, as heathland habitats are typically found at high altitudes and colder latitudes, and have nutrient-poor, acidic soils. A morphologically similar mycorrhizal association occurs in the Family Epacridaceae, which is widely distributed in the southern hemisphere, particularly southern Australia; the family takes its name from the genus *Epacris*.

The fungi involved in ericoid associations are members of the *Ascomycota*. Ericoid fungal hyphae form a loose network over the hair root surface (Figure 14.9A). The hyphae also penetrate the epidermal cells, often at several points in each cell (Figure 14.9B, C), and those cells become filled with coils of hyphae. As in all endosymbioses, the intracellular fungal symbiont is separated from the plant cytoplasm by a plant-derived membrane, which invaginates to follow fungal growth and coil formation. Up to 80% of the hair root volume can be comprised of fungal tissue. It is through these coils that nutrient exchange occurs. This colonisation by the fungus is limited to mature and fully expanded epidermal cells. In the apical region of the hair root, just behind the growing meristem, the immature epidermis remains uncolonised.

Ericoid mycorrhizal fungi are facultative symbionts; that is, they can exist as free-living mycelia in the soil, and they can also be cultured in artificial media. When grown on nutrient agar the fungi produce dark-coloured, slow-growing and sterile mycelia. Absence of spores and reproductive structures has complicated the identification and classification of ericoid fungi and generated increasing use of molecular techniques for examining DNA and RNA profiles. These show that there is considerable genetic diversity between isolates that are superficially similar in appearance. Comparative genomics and transcriptomics show that the fungal genes most highly upregulated in the ericoid mycorrhizal symbiosis are those coding for fungal and plant cell wall-degrading enzymes, lipases, proteases and transporters. This gene transcription pattern suggests a versatile dual saprotrophic

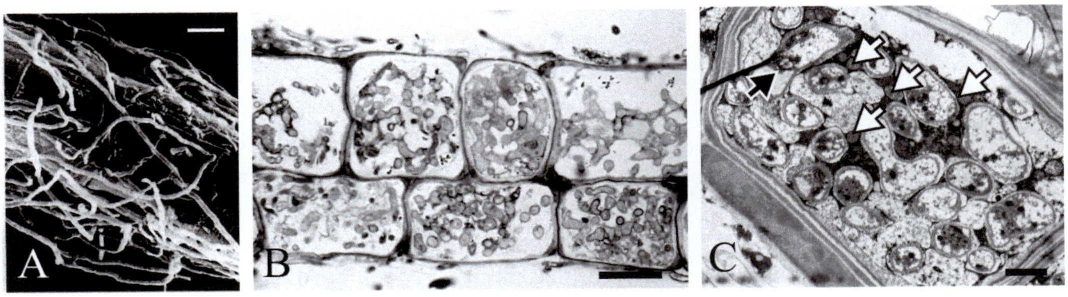

Figure 14.9. Cell biology of ericoid mycorrhizas. (A) Scanning electronmicrograph of *Rhododendron* root covered with a loose weft of fungal hyphae; modified from Smith & Read (1997). (B) Section of the ericoid mycorrhiza of *Gaultheria poeppiggi* (Ericaceae); the cortical cells are well colonised by fungal hyphae, scale bar, 10 μm. (C) Cortical cell of an *Erica cinerea* hair root colonised by hyphae (they make up most of the volume of the cell): black arrowhead indicates a dolipore septum of an intracellular hypha; white arrowheads indicate mitochondria of the plant cell, scale bar, 2 μm. B and C modified from Selosse *et al.* (2007) using graphics files kindly provided by Professor Marc-André Selosse, Université Montpellier II, France and Dr Sabrina Setaro, Wake Forest University, NC, USA. Reproduced with permission of John Wiley and Sons.

and biotrophic lifestyle quite like fungal endophytes (see section 14.19; Martino *et al.*, 2018).

Ericaceous shrubs can become dominant in many heathland communities, especially in environmental conditions where plant litter is only slowly decomposed, resulting in acidic soils rich in hard-to-digest organic matter but low in available mineral nutrients such as nitrogen and phosphorus. These mycorrhizas improve nitrogen and phosphorus uptake by the plant, enabling the host plants to access otherwise unavailable organic nutrients. The network of hyphae covering the root and the mycelium in the soil digest polypeptides saprotrophically and absorbed nitrogen is exchanged with the plant host. In extremely harsh conditions (for example winter in high altitude and northern latitudes) the mycorrhiza may even support the host plant with carbon nutrients (by metabolising polysaccharides and proteins for their carbon content). Normally, though, the fungus takes photosynthetically produced carbohydrates *from* the plant host.

These physiological attributes of the mycorrhizal fungus enable ericaceous plants to act as pioneer species, colonising unpromising habitats ranging from arid sandy soils in the southern hemisphere to moist 'mor' humus (a raw humus state of unincorporated organic material, poorly mineralised and with an acid pH; only slightly more degraded than peat) of the northern hemisphere; and the absence of ectomycorrhizal fungi seems to prevent colonisation of heathland by tree species like pines and birch (Collier & Bidartondo, 2009). Also, thanks to the mycorrhizal fungi, ericaceous plants are able to grow in highly polluted environments, where the soil contains otherwise toxic levels of metal ions. Taken together, the mycorrhizal symbiosis enables ericaceous plants to survive under nutrient stressed conditions and improves the ability of the host to compete successfully with other plants even on polluted sites contaminated by heavy metals (Perotto *et al.*, 2002; Mitchell & Gibson, 2006).

14.12 ARBUTOID ENDOMYCORRHIZAS

Arbutoid mycorrhizas are also formed by plants in Order Ericales, but specifically by the plant genera *Arbutus* (strawberry tree and madroña) and *Arctostaphylos* (bearberry and manzanita) in Family Ericaceae and in *Pyrola* (wintergreen) of Family Pyrolaceae (though some authors recognise this as a separate type of mycorrhiza). All of these are hardy, mainly evergreen shrubs, which occur in the wild in harsh, high altitude regions, and are cultivated widely for their decorative foliage, flowers and berries.

The fungi of arbutoid mycorrhizas are basidiomycetes, often the same fungal species that form ectomycorrhizal associations with forest trees (see below). Indeed, the structures of these two mycorrhizal types are very similar; arbutoid mycorrhizas have:

- a well-developed fungal sheath and a Hartig net restricted to the outer layers of cells of the root and the presence of the fungal sheath, as in ectomycorrhizas, insulates the host from the soil, so everything absorbed by mycorrhizal roots must pass through this sheath (all characteristics of the ectomycorrhiza);
- *but* there is extensive penetration of the outer cortical cells by fungal hyphae and hyphal coils fill those cells (characteristics of the endomycorrhiza).

The arbutoid type of mycorrhiza does not fit neatly into the classification scheme that distinguished ectotrophs from endotrophs because it shares characteristics of both. The intracellular coils, along with the mantle sheath and Hartig net, *and the fact that the host plant is a member of the Ericaceae* are the diagnostic features of arbutoid mycorrhizas. We will see below that there is another intermediate type of mycorrhiza, called the ectendomycorrhiza, which is an ectomycorrhiza in which the hyphae penetrate the outer cortical cells, and fill them with coils, which the hyphae of ectomycorrhizas do not normally do. In this case, and this distinguishes arbutoid from ectendomycorrhiza, *the plant host is a member of the Coniferophyta* (cone-bearing gymnosperm trees, including pines, yews, but also the 'living fossil' *Ginkgo*).

14.13 MONOTROPOID ENDOMYCORRHIZAS

This is the mycorrhizal association formed by the achlorophyllous plants that used to be placed in Family Montropaceae, though they are currently placed in Family Ericaceae; it's convenient to talk about them collectively so we will call them monotropes. The group includes the genera *Monotropa, Monotropsis, Allotropa, Hemitomes, Pityopus, Pleuricospora, Pterospora* and *Sarcodes*.

All these monotropes are highly unusual plants in that they completely lack chlorophyll, and so are unable to photosynthesise. Instead, they parasitise their mycorrhizas (they are called mycoheterotrophs), using the fungi not only to obtain minerals and nutrients from the soil, but also to tap into the carbohydrate supplies of nearby photosynthesising plants via their root connections to the mycorrhizal fungus. As they are native to temperate regions of the northern hemisphere, these other plants can include beech, oak and cedar but are more usually pine, spruce and fir, because *Monotropa* species are most commonly found in coniferous forests. Carbohydrates pass from conifer to *Monotropa* through their common mycorrhizal partner, and it is the fungus that controls the transfer of carbohydrate to *Monotropa*. Radiolabelled phosphorus (^{32}P) injected into *Monotropa* has been recovered in neighbouring trees, so the translocation is bidirectional.

The fungi involved in monotropoid associations belong to the *Basidiomycota*. Several *Boletus* species, such as *Boletus edulis*, have been identified as forming monotropoid endomycorrhizas. *Boletus* species also form ectomycorrhizas with a wide range of tree species, and this may explain how *Monotropa* becomes associated with so many different tree species. Though they are generally considered to be scarce or rare, *Monotropa* and other montropes such as *Pterospora* and *Sarcodes* are often the only higher plants growing beneath the trees in forests because the dense shading often excludes the chlorophyllous plants, which are so dependent on light to produce their carbohydrates.

The *Monotropa* roots are surrounded in a dense fungal sheath, from which hyphae spread into the soil. In *Monotropa* and *Pterospora*, the sheath encloses the root apex, while in *Sarcodes* the apex remains free. In all three genera, a Hartig net surrounds the outer epidermal layer, but does not penetrate the underlying cortex. However, individual hyphae do grow from the Hartig net into the outer cortical cells, the walls of which invaginate to accommodate the growing hypha.

These intrusions from the Hartig net into the cortical cell walls are known as fungal pegs. The peg proliferates, increasing the surface area within the cell. Eventually, the tip of the peg opens out into a membranous sac extending into the plant cell cytoplasm. Contents of the fungal peg fill the membranous sac but never directly enter the host cell cytoplasm. The number of fungal pegs produced by a monotropoid mycorrhiza is seasonal: maximum period of peg formation occurring in June, coinciding with flowering (and perhaps meeting a high demand for nutrients), while 'tip-bursting' occurs from July to August as seeds are released (and perhaps providing a late surge of nutrients to boost seed production just before the flower stalk senesces).

14.14 ORCHIDACEOUS ENDOMYCORRHIZAS

The Orchidaceae is the largest family of flowering plants, with about 880 genera and 22,000 accepted species, though there may be as many as 25,000. Overall, orchidaceous plants account for around 10% of all seed plants. Just to put the number of orchid species in context, there are about twice as many orchids as there are *bird* species and about four times the number of *mammal* species. Although there are representatives throughout the world, for example, there are about 90 European species, most species are naturally distributed in the tropics and subtropics. However, orchids are commonly cultivated and horticulturists now have more than 100,000 decorative hybrids and cultivars in cultivation around the world.

Orchidaceous mycorrhizas are similar to ericoid mycorrhizas, but their carbon nutrition is even more dedicated to supporting the host plant. All orchids have a stage where they are non-photosynthetic and therefore dependent on external sources of nutrients. Orchid seeds are very small and contain low nutrient reserves, so in most cases it is the seedling stage that is obligately mycorrhizal. In fact, most orchid seeds will not germinate unless they have been infected by an appropriate fungus, but the young orchid seedling itself is non-photosynthetic and depends on the fungus partner utilising complex carbon sources in the soil and making carbohydrates available to the young orchid. All orchids are achlorophyllous in the early seedling stages, but usually chlorophyllous as adults, so in this case the seedling stage orchid can be interpreted as parasitising the fungus. It is now known that many orchid species can be cultured artificially without mycorrhizal infection if they are supplied with exogenous supplies of carbohydrate. Mature orchids with mycorrhizas are more competitive on poor soils, and another measure of the value of mycorrhizas is that, *in vitro*, symbiotic seedlings reach a higher nitrogen concentration in their tissues than asymbiotic controls. These observations demonstrate that the mycorrhizal fungi are essential supports for nutrient uptake by the plants.

In the wild, the entry of mycorrhizal fungi into orchid roots is mostly through root hairs. There is no relation between root-hair characteristics and the extent of colonisation, and in some epiphytic species fungal entry is directly through the epidermis. The fungi form highly coiled hyphal structures (pelotons) in the root cortex; peloton size is related to cell size. Fungal invasion of cortical cells is by cell-to-cell penetration. The ratio of intact and lysed pelotons contained in the cortical cells varies between species and life forms. Chlamydospores and microsclerotial structure are frequently found within the plant's cortical and root-hair cells, being released when the root hairs split open (Sathiyadash *et al.*, 2012).

The mycorrhizal fungi in orchids are *Basidiomycota* and, in particular, many orchids are associated with species of *Rhizoctonia* (although this is a complex genus which can be divided into several new genera). Many orchid mycorrhizal fungi are serious pathogens of other plants. This is especially true of *Rhizoctonia* species and *Rhizoctonia* strains isolated as pathogens of other plants can support the germination of orchid seeds (for example, *Rhizoctonia solani*, which causes a stunt disease of all major cereal crops and probably most grasses, and *Rhizoctonia cerealis*, cause of sharp eyespot of wheat). Similarly, these and other species of *Rhizoctonia* can be isolated from mature, and healthy, mycorrhizal orchid roots. However, many of the *Rhizoctonia* orchid mycorrhizas are saprotrophic, producing a range of carbohydrate-degrading and other enzymes enabling the breakdown of plant debris, including lignin. Some *Mycena* species that associate with *Cymbidium* and *Gastrodia* orchids are known better as saprotrophs in leaf litter.

The orchid *Gastrodia elata* is particularly interesting because its seedlings develop with the saprotroph *Mycena osmundicola* (a small mushroom, originally found in Bermuda growing on plant litter), but as it matures comes to depend on the highly destructive tree pathogen *Armillaria mellea* (honey fungus). Even *Lentinula edodes*, which is the cultivated shiitake mushroom and is a white-rot wood degrader, can support the development of an achlorophyllous orchid called *Erythrorchis*, as can several other wood-rotting fungi. Thus, orchids use a great diversity of fungi with a range of different nutritional strategies as their mycorrhizal partners (Rasmussen, 2002; Bonnardeaux *et al.*, 2007). The dependency of the plant on its fungi also changes over the lifetime of the orchid; in most orchids decreased dependence is evident from the seedling stage onwards. Some orchids remain greatly dependent on their mycorrhizas despite the photosynthetic capacity of their leaves. Seedlings associate with a wider range of fungi than do older plants, suggesting that individual plants change their associations during development.

Infection of an orchid seed by fungi occurs after the embryo takes up water and swells, rupturing the seed coat. The small orchid seeds lack differentiated embryos or food reserves and their germination is limited to an intermediate stage called a protocorm. This may emit a few epidermal hairs before its growth stops. Further development of the protocorm will only occur if a mycorrhizal fungus, like *Rhizoctonia*, is able to colonise the epidermal hairs (Figure 14.10A–C). An orchid 'embryo' consists of only a few hundred cells and the fungi spread quickly from cell to cell. As hyphae penetrate cells of the embryo, the plasma membrane of the orchid cell invaginates, and the hypha becomes surrounded by a thin layer of host cytoplasm.

Within the orchid cells, the fungal hyphae form a dense mass of coiled hyphae called a peloton, which greatly increases the interfacial surface area between orchid and fungus (Figure 14.10D). Each intracellular peloton lasts only a few days before it degenerates; indeed, degeneration may begin within 24 h of formation. The degraded peloton is ultimately left as an amorphous clump of collapsed hyphae surrounded by a continuous sheath of host plasma membrane (called the perifungal membrane). During this process, the plant cell remains perfectly functional and can be recolonised by any surviving hyphae, or by fungi invading from adjacent cells. An orchid cell can go through this cycle several times, being repeatedly reinfected; indeed more than one fungus can form pelotons at the same time in the same orchid tissue.

Traditionally, two types of activity by orchid mycorrhiza have been recognised:

- tolypophagy, which is described above, and is based on *degradation of the pelotons*; it is found in the great majority of species;
- ptyophagy, described below, only found in several highly mycotrophic tropical orchids.

In ptyophagy, the *fungal hyphae are digested*, older hyphae develop large vacuoles and thickened cell walls, when the cytoplasm degenerates the hyphal cells collapse and are consumed by the orchid cell. In *Gastrodia elata*, for example, hyphae of *Armillaria mellea* extend in bundles along the roots in cortical

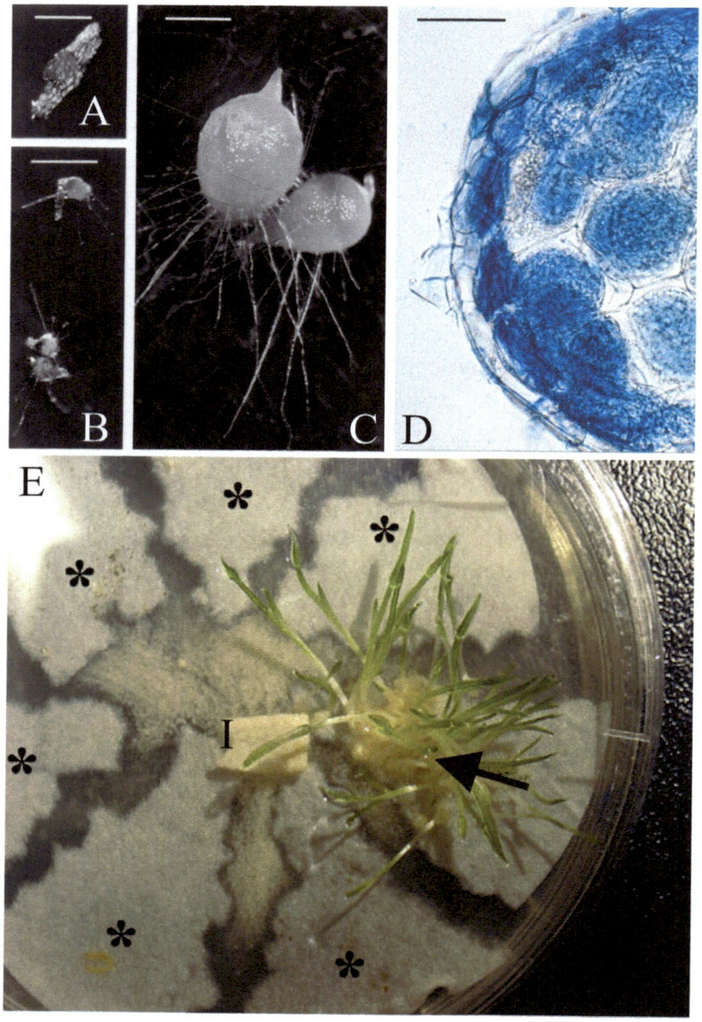

Figure 14.10. Infection of orchid seed by fungi. (A–D), *Rhizoctonia cerealis* forming mycorrhizas with seeds of the heath spotted orchid (*Dactylorhiza maculata* ssp. *ericetorum*) on mineral salts agar in the laboratory. (A) An ungerminated seed of *Dactylorhiza maculata* on agar enclosed in its testa (scale bar, 0.5 mm). (B) Germinated but uninfected protocorms on a control plate. The seeds have increased in size, rupturing their testae, and have produced a few epidermal hairs but no further growth was observed (bar, 1 mm). (C) Germinated protocorms infected with *Rhizoctonia cerealis* showing great increase in size and differentiation, polarised and showing evidence of an emerging shoot tip (bar, 1mm). (D) Section of an infected protocorm mounted in Cotton Blue/lactic acid to stain the fungus blue; note the hyphal coils (called pelotons) in cells of the outer cortex of the protocorm and their disintegration in cells of the inner cortex (scale bar, 100 µm). (E) Testing the specificity of orchid–fungus associations by sterile culture assays using orchid seed (on roughly torn segments of filter paper) and a potential mycorrhizal fungus pregrown on agar medium in a standard Petri dish (original fungal inoculum labelled I). Each piece of filter paper carries a different batch of seed. Only one orchid species is compatible with the fungus grown on this plate and can germinate and form seedlings with leaves (arrow). Seeds of other orchids failed to develop (asterisks). A–D modified from Weber & Webster (2001); E modified from Bonnardeaux *et al.* (2007); all reproduced with permission of Elsevier. For more information and illustration of mycorrhizas visit Mark Brundrett's website at http://mycorrhizas.info.

canals that develop from lines of 'passage cells' whose adjoining cell walls and original cell content deteriorate. In the host cells outside the passage canal the hyphal coils persist, but the inner cortex contains 'digestion cells'. When hyphae enter a digestion cell, an interface is formed between the plant plasma membrane and the hyphal wall. The plant releases lytic vesicles into this interface. The fungal wall breaks open in the interfacial space (which is now a digestion vacuole) and invaginations from the plant plasma membrane become pinocytotic vesicles containing fragments of fungal wall which are absorbed by the orchid cell for further digestion (Rasmussen, 2002).

Infection by mycorrhizal fungi does not necessarily result in the germination and growth of an orchid. Upon infection, three outcomes are possible:

- a mycorrhizal interaction, as described above;
- parasitic infection, in which the orchid cells are invaded, and the embryo dies;
- the orchid cells reject the fungal infection and fail to establish a mycorrhiza.

A successful fungal infection results in successful development of the orchid seedling. Although the mycorrhizal fungus can be the sole source of nutrition for the orchid during the first years of life, most orchid species develop chlorophyll as they mature and become less dependent on the mycorrhiza. However, most mature plants will still have mycorrhizal roots, using the fungus to enhance nitrogen and phosphorus uptake.

There are about 200 species of orchid that remain achlorophyllous when mature; examples are the genera *Galeola, Gastrodia, Corallorhiza* and *Rhizanthella*. These continue to depend on their mycorrhizal fungi for their nutrition (that is, they are obligate mycoheterotrophs). Some chlorophyllous orchids extend their dependency on their mycorrhizal fungi by spending several years underground before producing aerial (and photosynthetic) flowering stems (for example, the European wild orchids called helleborines); these are partial mycoheterotrophs.

14.15 ECTOMYCORRHIZAS

Ectomycorrhizas are generally considered to be the most advanced symbiotic association between higher plants and fungi, because although they involve only about 3% of seed plants, all of these are woody species: trees and shrubs, including most forest trees. The ectomycorrhizal association (ECM) is globally important because of the large area covered by the plants, and because of their economic value as the source of timber. In total, 43 plant families and 140 plant genera have been identified as forming ectomycorrhizas. In the northern temperate regions, plants such as pine (*Pinus*), spruce (*Picea*), fir (*Abies*), poplar (*Populus*), willow (*Salix*), beech (*Fagus*), birch (*Betula*) and oak (*Quercus*) typify the ectomycorrhizal association (or the related ectendomycorrhiza). In the southern hemisphere, *Eucalyptus* and *Nothofagus* (southern beech) are important genera as is the Family Dipterocarpaceae, which dominates the lowland rain forests around the globe from South America, through Africa to South-East Asia. This ectomycorrhizal group is reasonably homogeneous, but a subgroup, ectendomycorrhizas, has been separated out (see below). The characteristic features of the ECM association are:

- the presence of a substantial sheath of fungal tissue around the plant root;
- roots that are shorter and wider than uninfected ones; and
- an extensive network of intercellular hyphae penetrating between epidermal and cortical cells, called the Hartig net.

In this association, the plant root system is completely surrounded by a sheath of fungal tissue which can be more than 100 μm thick, though it is usually up to 50 μm thick. The hyphae penetrate between the outermost cell layers of the root forming what is called the Hartig net (Figure 14.11). From this a system of hyphal elements (hyphae, strands and rhizomorphs) extend out to explore the soil domain and interface with the fungal tissue of the root.

A wide range of fungi form ecto- or ectendomycorrhizas: at least 65 genera comprising between 5,000 and 6,000 species of fungi have been identified. About two-thirds of these belong to the *Basidiomycota*, the rest are *Ascomycota* except for the genus *Endogone*, which is a small zygomycetous truffle that forms arbuscular mycorrhizas as well. About three-quarters of the species produce epigeous (above-ground) sporophores, but up a quarter have hypogeous (underground) sporophores such as truffles; members of the *Ascomycota* are particularly well represented by hypogeous forms. Ectomycorrhizal fungi include common woodland mushrooms, such as *Amanita* spp., *Boletus* spp. and *Tricholoma* spp. Ectomycorrhizas can be highly specific (for example *Boletus elegans* with larch) but they generally do not show a high degree of host specificity (for example *Amanita muscaria* with 20 or more tree species as varied as birch, eucalyptus, spruce and Douglas fir). In the other specificity direction, 40 fungal species can form mycorrhizas with pine, and Norway spruce (*Picea abies*) can form ectomycorrhizal symbioses with over 100 different fungal species. It is common to find mycorrhizas belonging to several different fungi on the root system of a single tree.

Ectomycorrhizas start to develop when hyphae infect the secondary or tertiary roots of woody species, which they seem to prefer, especially on trees. Hyphae grow back up the root from just behind the root cap and meristem, forming a weft that may later become the bulky sheath (Figure 14.11). Hyphae from the sheath grow inwards between epidermal cells and cortical cells, by forcing their way mechanically and by excreting pectinases. This forms the Hartig net. Hyphae never penetrate cells (but see ectendomycorrhizas, below) or into the stele. The intercellular Hartig net may surround each cell completely (Figure 14.11), so they have little or no contact with other plant cells. The extensive surface area of the Hartig net is the main interface for exchange of substances between plant and fungus.

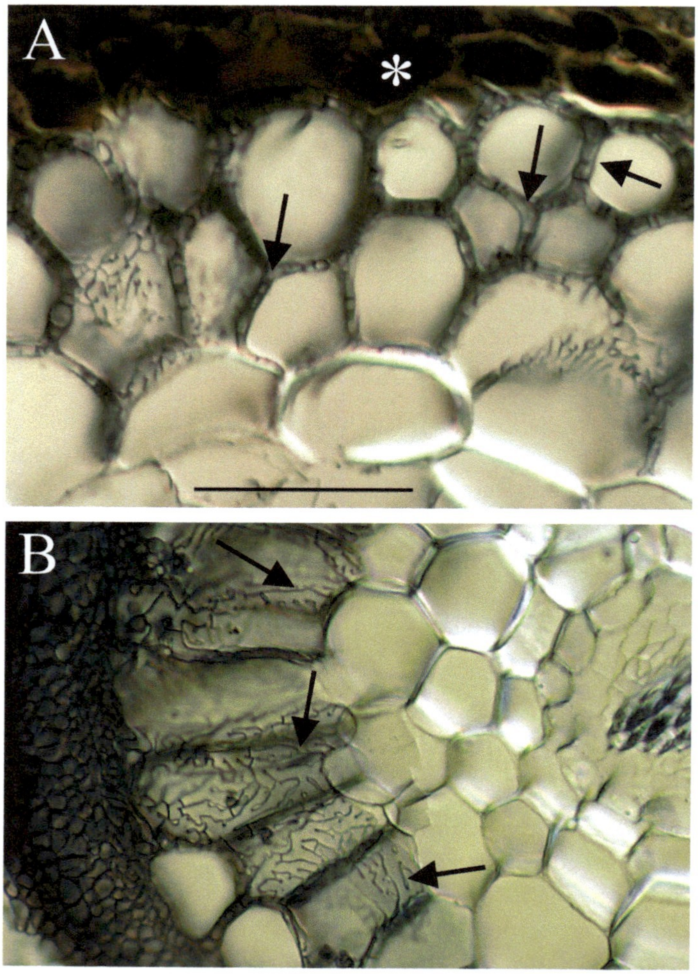

Figure 14.11. Light micrographs of hand sections of ectomycorrhizal forest tree roots cleared and stained with Chlorazol black E and viewed with (Nomarski) interference contrast microscopy. Fungal hyphae penetrate between host cells and branch to form a labyrinthine structure called the Hartig net. Host responses may include polyphenols (tannin) production in cells and the deposition of secondary metabolites in walls. Ectomycorrhizal gymnosperms, such as *Pinus* and *Picea*, have Hartig net hyphae extending deep into the cortex; this contrasts with the typical situation in angiosperms (such as *Populus*, *Betula*, *Fagus*, *Eucalyptus*, etc.), which usually have a one cell layer Hartig net confined to the epidermis. (A) Transverse section of ectomycorrhizal *Tsuga canadensis* (hemlock) with labyrinthine Hartig net hyphae (arrows) penetrating between the cortical cells of the root and surrounding many cells; note tannin-filled epidermal cells in the inner mantle (asterisk); scale bar, 100 μm). (B) Root cross section of ectomycorrhizal *Populus tremuloides* (quaking or trembling aspen) showing labyrinthine Hartig net hyphae (arrows) around elongated epidermal cells. This complex hyphal branching pattern is thought to increase the fungal surface area in contact with the root. The active mycorrhizal zone occurs several mm behind the root tip (because of the time required for mycorrhizal formation), but Hartig net hyphae senesce in older regions further from the root tip.
The images are modified from Brundrett *et al.* (1990) and were kindly supplied by Dr Mark Brundrett, School of Plant Biology, University of Western Australia. For more information and illustration of mycorrhizas visit Mark Brundrett's website at http://mycorrhizas.info.

Fungal infection changes the growth pattern of the root. The fungal sheath reduces the rate of cell division at the root tip, slowing elongation of the cells and therefore of the root growth. Cortical cells elongate radially resulting in the infected root appearing short and thick compared to uninfected ones (Figure 14.12; they are often called 'short roots').

The fungal sheath also suppresses root-hair development with the result that all nutrients entering the plant must be channelled through the sheath. A fungus will extend its sheath with a growing root, ensuring the root is always covered and preventing colonisation by other fungi. However, fungi can be replaced as roots resume growth after winter. If the fungal sheath does not resume growth straight away, the root can be exposed to other fungal symbionts. This sort of replacement (or supplementation) is different from the replacement mentioned above in relation to early and late stage colonisers.

The ectomycorrhizal association is a mutualistic symbiosis; the partners are reciprocally benefited. The biological function of this symbiosis is the exchange of fungus-derived mineral nutrients for plant-derived carbohydrates; photosynthetic carbohydrates are translocated from the plant to the fungus which uses this rich carbon source to develop extensive hyphal growth into the soil. The mycelium explores the soil and absorbs minerals and water which it shares with the plant roots, and also confers pathogen resistance to the plant partner and protection from water stress. This external mycelium is called the extraradical mycelium. It is an interconnected three-dimensional network of hyphae and, in many cases, specialised hyphal aggregates like strands and rhizomorphs, that not only massively increases the absorbing surface area of the host plant but also the overall potential reach of the host (Figure 14.12), finding and transporting nutrients from distant sites of capture to the sites of nutrient transfer within the mycorrhizal tissues of the host.

The exchange of metabolites is essential for the persistence of *both* plant and fungus, particularly in stressed environmental conditions. Most ectomycorrhizal fungi are so uncompetitive as saprotrophs that they depend on the plant host for carbon sources. With few exceptions (*Tricholoma fumosum* being one), the fungi are unable to utilise cellulose and lignin. On the other hand, the fungus is able to capture nutrients, particularly phosphate and ammonium ions (and water), which the root cannot access and this results in greatly enhanced mineral uptake for the plant. Success of this symbiosis is essential to the stability of forest ecosystems, trees grow poorly when they lack ectomycorrhizas. Ectomycorrhizal infection is necessary for the successful establishment of some trees, such as *Pinus*, and can allow seedlings to

Figure 14.12. Ectomycorrhizal roots; small parts of the root systems of various forest tree species to illustrate the morphological diversity of ectomycorrhizal roots. (A) to (D) are all roots of Douglas fir (*Pseudotsuga menziesii*), but mycorrhizal with different fungi: (A) with the basidiomycetous truffle *Hysterangium* (a truffle sporophore is shown; photograph by B. Zak); (B) mycorrhizal fungus, *Rhizopogon vinicolor* (*Boletales*, *Basidiomycota*); (C) mycorrhizal fungus, *Poria terrestris* (*Byssoporia terrestris*, *Polyporales*; photograph by B. Zak); (D) mycorrhizal fungus, *Lactarius sanguifluus* (*Russulales*; photograph by B. Zak). (E) and (F) show mycorrhizas involving the same fungus (*Amanita muscaria*) but different hosts (photographs by R. Molina): (E) sitka spruce (*Picea sitchensis*) and (F) Monterey pine (*Pinus radiata*). Images A to F prepared from graphics files kindly provided by Dr Randy Molina, Pacific Northwest Research Station, USDA Forest Service, Oregon, USA. Illustration (G) shows a pine seedling grown in symbiosis with the ectomycorrhizal fungus *Suillus bovinus* in an experimental microcosm (where the seedling is grown in soil placed in a container made from two sheets of glass separated by a 1 cm thick former). In this case two soils were used: 1 is from a podzol E horizon soil (explained below), and 2 is a loamy organic soil. The extraradical fungal mycelia (m; evident mainly as hyphal strands) extend from the colonised root tips (r) into both soil substrates and are far more extensive than the roots (scale bar, 1 cm). Podzols are the typical soils of coniferous, or boreal, forests in the northern hemisphere and eucalypt forests and heathlands in the southern hemisphere. In podzols organic material and soluble minerals are leached from the upper layers (horizons) to the lower; the E horizon is a 4–8 cm thick layer of heavily leached soil and is largely composed of insoluble minerals. The high abundance of mycelium in this mineral soil shows that it is an important growth substrate for ectomycorrhizal fungi. Ectomycorrhizal fungi modify their chemical environment through local acidification around the hyphae and by exuding metal-complexing weathering agents such as organic acids (see Figures 1.2, 1.4, 1.5) and play a central role in mineral weathering of boreal forest soils (Rosling *et al.*, 2009; this image reproduced with permission from Elsevier). For more information and illustration of mycorrhizas visit Mark Brundrett's website at http://mycorrhizas.info.

compete against mature trees in less favourable conditions, such as when the seedling is shaded by the mature canopy leaves.

Ectomycorrhizas can link together groups of trees, the submerged mycelium acting as what has been described as a 'wood-wide-web' (Wohlleben, 2017) or Common Mycorrhizal Network (Gilbert & Johnson, 2017) in which signals are exchanged between plants through their mycorrhizas. These networks can transport signals produced by plants in response to herbivore and pathogen infestation to neighbouring plants before they are themselves attacked. There does seem to be a finely tuned balance of signals between the partners at several levels. The speed of transfer to uninfected plants is such that the mechanism could have benefits for plant protection if the networks could be harnessed for pest management in agriculture (Gilbert & Johnson, 2017). Fungal hyphae and plant root cells have developed cell-anchor receptors and mobile signal ligands like adhesins (Dranginis *et al.*, 2007; see section 7.8) as mutual sensing molecules, and responses to these signals allow both to adapt rapidly to changes in their local cellular environment and to signal their nucleus through appropriate transduction pathways. As an aside, it's interesting to point out that this is just one aspect of what is turning out to be a rich sensory environment for plants. Recent research has shown that flowers (of *Oenothera*) increase nectar sugar concentration in response to the ***sound vibrations*** of approaching pollinator insects (Veits *et al.*, 2019).

On both sides, of course, there is a genetic component to the ability to form a symbiosis, and a panel of responses to environmental conditions which influence (enhance or repress) establishment of the relationship. Cell sensing of the environment and cell-to-cell communications within the ectomycorrhizal tissues must play a major role in this transaction. Research on the genetics of mycorrhizal networks has advanced greatly in recent years as molecular tools have increased their resolution and throughput, but a large gap remains between our understanding of the genes and transcripts on the one hand, and the physiology, ecology and evolution of mycorrhizal networks in our changing environment on the other (Tagu *et al.*, 2002; Simard *et al.*, 2012) (Figure 14.13).

Arguably the most important communication between plant and fungus in the ectomycorrhiza is the exchange of nutrients. As with the exchange of nutrients between endomycorrhizas and their hosts (see discussion and Figure 14.8, above) nutrient exchange between fungus and host depends on one partner releasing nutrient into the apoplastic interface and the uptake of that nutrient from the apoplastic interface by the other partner. A model showing the transporters either known or hypothesised to be involved in ectomycorrhizal tissues is summarised in Figure 14.14. Note that the system operating at the plant–fungus interface, which is the Hartig net, involves:

- Delivery of sucrose into the apoplast by the plant and its hydrolysis by a plant-derived acid invertase; the resulting hexoses then being taken up by the fungal cells.
- Fungal Pi (inorganic phosphate) uptake systems 'harvesting' Pi from the environment, probably assisted by several secreted fungal enzymes (like phosphatases) that could play crucial roles in improving P availability by digesting soil reserves of 'bound' phosphate. The hyphae translocate P by active processes (highly mobile vacuoles?) to deliver to the apoplast for take-up by their P-deficient host plant.
- Potentially nitrate, ammonium and peptide uptake into the fungus through specific fungal uptake systems feeding intermediary metabolism that supplies amino acids for release into the apoplast by the fungus through amino acid exporters, followed by subsequent amino acid uptake by the plant cell.
- High affinity active K^+ uptake by peripheral hyphae from the soil solution and its release through fungal outward K^+ channels into the apoplast for absorption by (currently hypothetical) host plant K^+ importers.

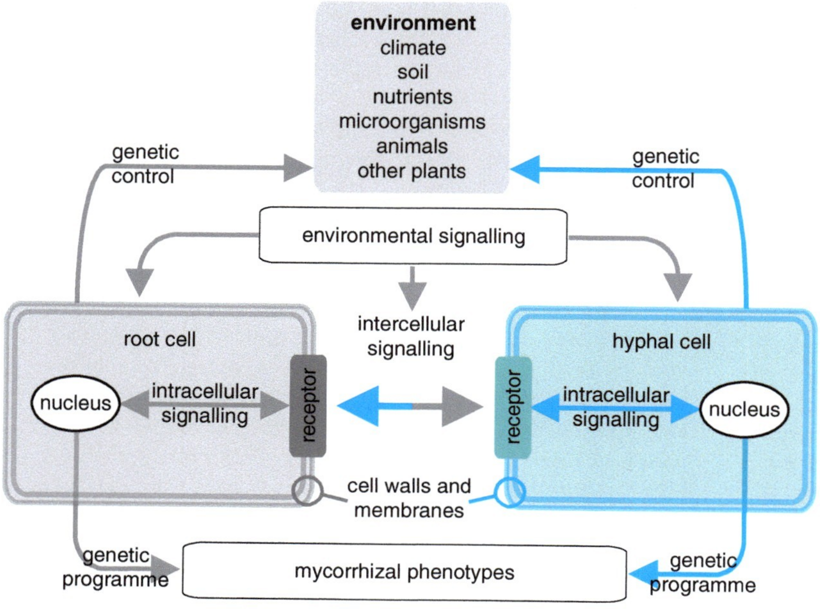

Figure 14.13. Model of the inter- and intracellular communications that might exist between fungal hyphae and root cells in the ectomycorrhizal symbiosis. Changes in environmental conditions may produce signals sensed by cells of both partners in the symbiosis, and both probably transduce this information to their nuclei to provoke modifications in gene expression and consequently in phenotypes. Redrawn and modified from Tagu *et al.* (2002).

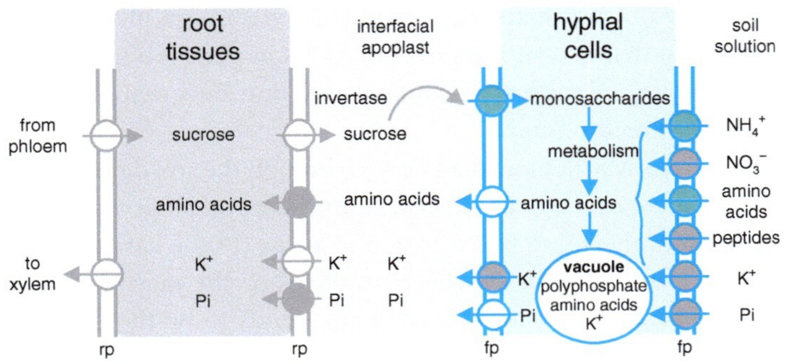

Figure 14.14. Nutrient exchange between fungus and host depends on one partner releasing nutrient into the apoplastic interface and the uptake of that nutrient from the interfacial apoplast by the other partner. This diagram summarises current ideas about the transporters acting in ectomycorrhizal tissues that achieve this nutrient exchange. Key: fp, fungal plasma membrane, rp, root plasma membrane. The circles represent transporters, with the arrows indicating direction of transport. Blue circles represent transporters where at least one member of the transporter family has been characterised by functional complementation of a yeast-deficient strain; grey circles are putative transporters for which candidate genes exist in the genome; white circles represent hypothetical transporters. Redrawn and modified from Chalot *et al.* (2002).

Ectomycorrhizal root systems, meaning plant and fungal partner together, can receive about half of the photosynthetically fixed carbon produced by the aerial parts of the plant. Monosaccharide uptake capacity and carbohydrate flux through glycolysis and intermediate carbohydrate storage pools (trehalose and/or mannitol) in the fungus is greatly increased at the plant–fungus interface to enable such a major carbohydrate flux. Depending on the identity of the fungus involved in the symbiosis, fine control of fungal carbohydrate uptake and metabolism seems to be determined either by developmental mechanisms or by the apoplastic sugar content. Trees increase their photosynthetic capacity to satisfy the increased carbohydrate demand of the symbiosis. In addition, host plants control and restrict carbohydrate flux towards their mycorrhizal partner to avoid fungal parasitism, though the mechanisms responsible for this are still largely unknown (Nehls *et al.*, 2010).

14.16 ECTENDOMYCORRHIZAS

Rather than being a definitive subgroup, this is a descriptive name for mycorrhizal roots that exhibit characteristics of both ectomycorrhizas and endomycorrhizas. Ectendomycorrhizas are essentially restricted to the plant genera *Pinus* (pine), *Picea* (spruce) and, to a lesser extent, *Larix* (Larch). Ectendomycorrhizas have the same characteristics as ectomycorrhizas *but also* show extensive intracellular penetration of the fungal hyphae into living cells of the host root. Formation of ectendomycorrhizas begins with formation of a Hartig net, which grows behind the apical meristem of the growing root.

The Hartig net hyphae grow between the epidermal and outer cortical cells and later extend to the inner cortex. Growing up the older parts of the root, intracellular penetration increases, once inside a cell, the hyphae branch repeatedly. The oldest plant cells become almost filled with coils of septate hyphae. The ectendomycorrhizal association can persist for a full year, and there is no evidence of hyphal degeneration or lysis. Ectendomycorrhizal formation induces the growth of short roots, like the ectomycorrhizal association. Emergent roots become covered in a matrix of highly branched hyphae; the coarse sheath develops between root hairs and eventually covers the entire root. Assessment of the influence of inorganic and organic phosphorus and the lack of phosphorus in the culture medium on the type of mycorrhizal colonisation formed (ecto-, ectendo- or endomycorrhiza) during *Helianthemum almeriense* × *Terfezia claveryi* symbiosis *in vitro* revealed that the mycorrhizal types formed depended on the phosphorus source in the medium, suggesting that it is the organic phosphorus that induces the formation of intracellular colonisation (Navarro-Ródenas *et al.*, 2012).

The fungi involved in ectendomycorrhizal symbioses are *Ascomycota*; most belong to the genus *Wilcoxina* and most isolates can be assigned to two species, *Wilcoxina mikolae* and *Wilcoxina rehmii*. These two taxa occupy distinctive habitats: *Wilcoxina mikolae* is a chlamydospore-producing fungus, found predominantly in disturbed mineral soils, while *Wilcoxina rehmii* thrives in peaty soils, but does not produce chlamydospores.

14.17 THE EFFECTS OF MYCORRHIZAS, THEIR COMMERCIAL APPLICATIONS AND THE IMPACT OF ENVIRONMENTAL AND CLIMATE CHANGES

So far, we have described some detail about each specific type of mycorrhiza. We now want to bring these strands together and discuss the mycorrhizal symbiosis in general.

The major advantage that a mycorrhizal association confers to both plant and fungus is a supply of nutrients enhanced beyond that which would normally be available. It is usually a bidirectional movement of nutrients: carbon source (as carbohydrate) from plant to fungus, and soil nutrients, especially phosphorus, nitrogen and water, from fungus to plant. This is the essential quality of a mycorrhiza and is presumably the foundation for the mutualistic association because it results in the synergy that the two organisms involved can perform better when together than they can perform when apart. There are exceptions.

From what we have already described about monotropoid mycorrhizas and all orchidaceous mycorrhizas at the juvenile stages of the plant, it is not at all clear that the fungus benefits from the association. That is, these associations are much closer to being better described as parasitism of the fungus by the plant. Remembering that the Orchidaceae is the most diverse family of flowering plants, there is a very large number of this particular exception. The monotropoid relationship is more complicated as monotropoid mycorrhizas are often ectomycorrhizal with surrounding trees, and carbohydrates from that alternate source sustain both the fungus and the associated *Monotropa*.

As a rule, mycorrhizal infection enhances plant growth by increasing nutrient uptake by:

- increasing the surface area of the absorbing organ within the soil (it becomes the mycelium rather than the root hairs);
- mobilising nutrients from sources that are not available to the plant because they need to be digested or made soluble by enzymes produced by the fungus;
- scavenging what nutrients are present by excreting chelating compounds and/or producing transporters with very high substrate affinities.

Nutrients are obtained by hyphae, which extend from the roots, sometimes organised into strands, exploring and exploiting a much larger volume of soil than the host roots themselves could possibly exploit. Following uptake by the hyphae, nutrients are translocated back to the host root with high efficiency (see sections 5.9, 5.15 and 11.11). The weft of hyphae that usually surrounds mycorrhizal roots, or the hyphal sheath of ectomycorrhizas, can both act as storage sites for nutrients acquired from the soil; such nutrients can be supplied to the plant as and when required.

In return for receiving extra nutrients, mycorrhizal host plants supply the fungus with at least 10% to 20% (and up to 50%) of the photosynthates they produce, which go to the formation, maintenance and functioning of the mycorrhizal fungi and their associated structures. This is an investment in the fungus, not a tax on the plant: the more carbohydrate supplied to the fungus, the greater its ability to win soil nutrients for the benefit of the plant.

Regarding the transfer of carbon from plant to fungus, carbohydrate leaves the plant as sucrose, is hydrolysed to glucose and fructose, most often by an invertase of plant origin, then carbohydrates entering the fungi appear in the hyphae as trehalose, and in some strains as mannitol. When carbohydrates are transferred from fungus to plant, as in orchid and monotropoid mycorrhizas, the reverse occurs, with trehalose and mannitol being converted to sucrose.

In arbuscular mycorrhizas nutrient exchange occurs through the arbuscules and in ericoid (and other endotrophic) mycorrhizas through the hyphal coils; both of these structures are intracellular interfaces as far as the plant cell is concerned. In contrast, the interface between plant and fungi in ectomycorrhizas takes place through the Hartig net, which is extracellular to the plant cell.

Direct uptake of nutrients by plant roots can often result in zones of nutrient deficiency around the root system. Subsequent absorption of nutrients is then limited entirely by the rate at which nutrients can move through the soil. Hyphae can grow past these depletion zones and by so doing enhance the absorption of nutrients by the root. Nourishing a mycorrhizal fungus so that it continues extending its hyphae is a more economical way of sustaining contact with nutrients than by continually extending roots.

The rate at which nutrients move through the soil depends on their chemistry and can be crucial to their availability to plant roots. Phosphorus, in the form of phosphate (PO_4^-), has very low mobility in soil, because the phosphates of the most common divalent metals are insoluble, and this tends to be the limiting macronutrient in most ecosystems (Karunanithi *et al.*, 2015). In contrast, ammonium (NH_4^+) is about 10 times more mobile than phosphate but is also required and absorbed in 10 times greater quantity. Ammonium is an acid (ammonia, NH_3, is a gas) and is in greatest abundance in acidic soils, such as those of forests and heathland; otherwise in most soils nitrate (NO_3^-) is the main immediately available source of nitrogen for the root.

What the fungal component of the mycorrhizal association adds to these considerations is a change in the meaning of the word 'available'. When horticulturalists talk about 'available nitrogen and phosphorus' they are talking about soluble salts in the soil water. If the salt is not soluble, it is not available; if the element is not in simple mineral salt form, it is not available. The fungus doesn't speak this restrictive language.

The fungi produce organic compounds such as citrate and oxalate (Plassard & Fransson, 2009), which help in the desorption of adsorbed phosphate or the dissolution of poorly soluble phosphates; they also secrete extracellular acid phosphatases which release phosphate from organic complexes (phosphoglucans, phospholipids and nucleic acids) in the soil. In addition, fungi store and translocate phosphorus in the form of polyphosphates in the hyphae, which maintains a low inorganic phosphate (low Pi) concentration in the hypha and enhances uptake of any external phosphate that can be found. The secretion of oxalic acid plays a major role in the dissolution (bioweathering) of the mineral *apatite*, which mainly consists of calcium phosphate, releasing phosphorus for uptake by the ectomycorrhizal fungus and immobilising the released calcium as calcium oxalate (Schmalenberger *et al.*, 2015).

Add to these considerations the large volume of soil explored per unit surface area of all those narrow hyphae, and the meaning of the phrase 'available phosphorus' to the mycorrhizal root is expanded enormously, and, in general, the large growth increases in the plant consequential on mycorrhizal infection are due to increases in phosphate absorption. In arbuscular mycorrhizal plants, up to 80% of the plant phosphorus comes to it through the hyphal network. The beneficial effects of mycorrhizas with regard to phosphorus uptake are lost if the concentration of phosphorus in the soil increases. Both arbuscular and ectomycorrhizal infections decrease in high phosphorus soils. Ultimately, the mycorrhizal association can be abandoned in such soils.

Similar considerations apply to the nutrient nitrogen. Plant roots may be limited to mineral sources (NH_4^+ and NO_3^-) and most plants will preferentially take up ammonium over nitrate, largely because nitrate-nitrogen must be *reduced* before it can be used in metabolism and this is an energy-demanding process. This preference is also expressed by fungi; indeed, many fungi are not able to utilise nitrate at all. Fungi are particularly efficient at assimilating ammonium through high affinity transporters, but their main contribution is their ability to use *organic nitrogen sources*, through the production of proteases and peptidases. This allows them to access nitrogen sources that are not available to non-mycorrhizal plant roots. As with phosphorus, if mineral nitrogen becomes available to the plant from other sources (by application of chemical fertilisers like NH_4NO_3, for example) the value of the mycorrhizal association to the plant is diminished and mycorrhizal infections tend to decrease.

Mycorrhizas offer other benefits to their plant partner, though. Metal ions such as potassium (see Figure 14.14), calcium, copper, zinc and iron are all assimilated more quickly and in greater amounts by mycorrhizal plants, again through the ability of the fungal hyphae to release, absorb and translocate minerals quickly and efficiently. The ectomycorrhizal fungal sheath also provides several advantages over naked (that is, non-mycorrhizal) plant roots. The sheath hyphae can accumulate and immobilise heavy metals, so when growing in soils with high concentrations of heavy metals, like zinc, cadmium and arsenic, the metals are unable to reach the plant tissues and the host plant remains undamaged (Hartnett & Wilson, 2002).

Mycorrhizal fungi can also help their plant partners by increasing tolerance to other adverse conditions. Probably most crucial, and controversial, is that all fungi can grow at water potentials lower than those that plants can tolerate. This means that fungi remain metabolically active and scavenging for water and nutrients in conditions where non-mycorrhizal plants would wilt, cease to grow and eventually die. Mycorrhizal plants can therefore continue to grow in conditions of high water stress because the extended mycelial zone of their mycorrhizal fungi enables them to tap reserves of water from deeper in the soil and from greater distance from the host plant. This operates in all ecosystems but is especially obvious in desert ecosystems. Many of the most common desert plants, including cacti, are heavily mycorrhizal, showing that mycorrhizas make a particularly important contribution to the water relations of their host plants even in those extreme ecosystems where soils have poor water retention.

What is controversial about this is that plant physiologists argue that the cross-sectional area of fungal hyphae connected to a root is typically too small for them to make a significant contribution to water flux in the plant, so water transport by the mycorrhiza may be significant only in delaying wilting at times of extreme water deficit. This line of argument seems to ignore the fact that fungal hyphae are highly adapted to translocating large quantities of all sorts of nutrients extremely rapidly over long distances; it's their way of life. Add to this consideration the sheer *quantity* of extraradical mycelium that might be supporting a plant (see Figure 14.12B) and we think there is little doubt about the potential value of the mycorrhizal fungal contribution to the water relations of their host plant.

Earlier we mentioned that mycorrhizas protect the plant against root pathogens. Ectomycorrhizas are particularly effective and have several strategies to combat pathogen attack:

- excretion of antifungal and antibiotic substances, 80% of *Tricholoma* species produce antibiotics and *Boletus* and *Clitocybe* produce antifungal substances;
- stimulation of the growth of other microorganisms, which themselves limit pathogens;
- stimulation of the plant to produce antibiotics under the control of the mycorrhizal fungus;
- structural protection of the root by the thick fungal sheath; the mechanical barrier gives effective protection because plant pathogens need to access plant tissue to infect it, they cannot usually infect fungal tissue.

Arbuscular mycorrhizas also increase plant resistance to pathogen attack, in particular to root-infecting fungi, such as *Fusarium oxysporum*, and *Oomycota*, like *Phytophthora parasitica*. Because most root pathogenic fungi infect roots more rapidly than do arbuscular mycorrhizal fungi, simultaneous infection often leads to the mycorrhiza being out-competed. However, if the mycorrhizal association is already established, pathogenic infection is much reduced.

For example, one study showed that arbuscular mycorrhizal tomato plants challenged with *Fusarium* showed only 9% necrotic roots compared to 32% in non-mycorrhizal plants. Similarly, arbuscular mycorrhizal infection has been shown to reduce the effects of pathogens and even pests, such as root pathogenic nematodes. Little is known about the mechanism(s) involved in this case, but the plant response to mycorrhizal infection, perhaps altered cell wall chemistry or the production of phytoalexins, may cause a general improvement to pathogen and pest resistance.

It is not just plants that benefit from all those mycorrhizal hyphae. We have discussed elsewhere that many small animals (microarthropods in particular) depend on fungal mycelium for food (see section 12.2). We have also stated above that mycorrhizal fungi consume about 20% of their host plant's photosynthetically fixed carbon.

Put the last two sentences together and you will see that mycorrhizal hyphae are an important route for the redistribution of plant-derived carbohydrate into the soil *animals*: along the pathway atmospheric carbon dioxide to photosynthetic carbohydrate in the plant, sucrose to hexose sugar in the apoplastic space, hexose to trehalose in the mycorrhiza synthesised into fungal tissue, which is finally eaten, digested and transformed into the bodies of the small animals that make up the soil biota.

By benefiting individual plants in the ways outlined above mycorrhizas inevitably affect the ecology of plant communities. In general, both ectomycorrhizas and arbuscular mycorrhizas show very low host specificity, which allows a single fungus to infect several plants in one area, and several different fungi to infect a single plant. This allows mycorrhizas to establish a network that may link many different plants of many different species within a single habitat. This is the 'wood-wide-web'; a network that allows carbohydrates, amino acids and mineral nutrients, and chemical signals, to flow between plants through their shared fungi.

This can allow seedlings to be supported by their parent plant while establishing themselves. Receiving nutrients through the mycelial network can also allow plants and seedlings to thrive in less favourable conditions, such as in local shade, suffering water stress, inadequate soil, etc., the disadvantaged plants having their nutrition supplemented by resources from neighbouring plants growing in more ideal conditions. A consequence of the contribution that mycorrhizas make to plant growth is that plant reproduction increases, and offspring survival is improved, which can increase population size and influence population density and species distributions within the community (Teste & Dickie, 2017).

Many of the fungal sporophores with which we are most familiar belong to mycorrhizal fungi and it is common knowledge that they are seasonal and follow successional changes alongside the associated changes in the vegetation. There is a considerable research literature about changes in species composition of the communities characterising a range of habitats (Frankland, 1998). Successions of ectomycorrhizal agarics have been described during the development of temperate forests, particularly those of *Betula*, *Pinus* and *Picea*.

The pattern and succession of mycorrhizal development varies during the life of a tree. Some fungi are successful primary colonisers (called early stage fungi), while other fungi dominate as the tree ages and reaches maturity (late stage fungi). However, some fungi will infect regardless of plant age, and seedlings germinating in a mature forest are usually colonised by the dominant late stage fungi. Early stage fungi have more of a role where their plant partner is a pioneer colonist and there are few other soil fungi present. These early stage fungi act like ruderal or *r*-selected species (populations that experience rapid growth and rapid reproduction), and those of the late stage are unable to survive in soil not already inhabited by other mycelia and are *K*-selected organisms (*K*-selected species tend to be very competitive, with stable populations having low reproductive rates).

As far as forest fungi are concerned, species diversity tends to increase until canopy closure (which takes about 27 years in northern temperate forests), but declines when tree litter accumulates, when a greater proportion of the nitrogen is in organic form. *Laccaria proxima* is an example of agarics that are not highly selective of their host trees, and that reproduce prolifically, producing relatively small sporophores lacking mycelial strands or cords, and occurring under young, first-generation trees. More host-selective species, for example *Amanita muscaria*, with larger, more persistent mushrooms, which are often served by mycelial strands and cords (implying greater dependence on nutrients supplied by the host tree) typified the later stage.

A word of warning is necessary here, because evidence based on the presence of sporophores above ground can be misleading unless it is accompanied by information about the mycelia associated with the tree roots. Some common species of mycorrhiza fruit rarely if at all. This is one reason why so much attention is currently given to analysis of molecular sequences from the environment to characterise fungal diversity, including metagenomics (Mitchell & Zuccaro, 2006; Anderson & Parkin, 2007; Lindahl & Kuske, 2013). The picture that emerges when ectomycorrhizas on pine roots are studied as mycelia using these molecular techniques is slightly different from that resulting from study of sporophores. Mycelia indicate that species diversity increases for up to 41 years before it stabilises (Frankland, 1998). The ecology of fungi is less well understood than that of animals and plants because most are microscopic organisms hidden in the substrates on which they grow. However, use of DNA barcodes to identify fungi is emerging as a promising approach for studying the *population dynamics of fungal mycelia* in natural ecosystems (Gao *et al.*, 2018).

The distinction between early and late types of fungi is important commercially as it aids the forester, for example, to choose between mycorrhizal inoculants for forest nurseries and for regeneration of trees following deforestation. The early stage fungi can be used as inoculants to stimulate new forest growth, whereas the late stage fungi are not suited to this.

Carefully recorded and regularly repeated observation of the occurrence of species in nature is essential. Individual records must be accurately identified, location of the find identified with clarity, and surrounding habitat recorded in detail. When that has been done often enough, for long enough, and over a sufficiently large geographical area, field records allow species to be evaluated for occurrence in space and time, vulnerability of habitat, threats to species decline, level of protection and taxonomic uniqueness. Species can then be ranked by number of occurrences (or rarity), number of individuals, population and habitat trends, and type and degree of threats. The data can be assembled into distribution maps (recording occurrence of one species over a geographical area) and checklists (species occurrence lists that provide an inventory of species in a specific geographical region); if these are prepared year-on-year, they contribute to conservation by detecting unusual changes. Ultimately, species can be assessed as *endangered, threatened, sensitive*, and even *extinct* in the region. Such records are essential to the preparation of Rarity, Endangerment and Distribution lists (RED lists; hence, Red Data lists). Red Data lists alert conservation biologists and policy makers to issues surrounding rare fungal species and provide direction for the management and protection of the species. Decrease in fungal species diversity in northern Europe was first reported in the mid-1970s and Red Data lists are essential to awareness of such conservation issues.

In the northern temperate zone, we expect to see the majority of *Basidiomycota* sporulating in autumn, following mycelial growth and decomposer activity in spring and summer. Temperature and rainfall are the two main factors affecting mushroom productivity. In a 21-year survey of a forest plot in Switzerland, appearance of sporophores was correlated with July and August temperatures, an increase of 1°C resulting in a delay of sporulating by saprotrophs of about 7 days. In contrast, mushroom productivity was correlated with rainfall in the period June to October (Straatsma *et al.*, 2001).

Climate change has been, and remains, a major concern for us all. Several studies have shown climate-associated changes in periodic life cycle events in plants, insects and birds, and this has also been demonstrated to be the case for fungi. Changes in the seasonal pattern of fungal sporophore formation in the United Kingdom have been detected from field records of sporophore finds made over 56 years from 1950 (Gange *et al.*, 2007). This study analysed a large data set of sporophore records of 200 species of decomposer *Basidiomycota* in Wiltshire, UK, recorded during 1950 to 2005.

Statistical analysis of this data set showed that the mushroom production season has extended since the 1970s. On average, the date at which the first sporophores appeared is now significantly earlier (the average advancement was 7.9 days per decade). Similarly, the final date on which sporophores were seen is significantly later in 2005 than it was in 1950 (average delay was 7.2 days per decade). In the 1950s, the average sporulating period of the 315 species in the data set was 33.2 days, but this more than doubled to 74.8 days by the first decade of the twenty-first century.

As well as changes to autumn sporulating patterns, significant numbers of species that previously only fruited in autumn now also fruit in spring; the response depending on habitat type. Since mycelia must be active in uptake of water, nutrients and energy sources before sporophores can be produced this suggests that these fungi may now be more active over winter and spring than they were in the past.

Climate changes in Wiltshire were also analysed thanks to well-maintained local weather records. There was a significant relationship between early sporophore formation and summer temperature and rainfall. Local July and August temperatures had significantly increased, while rainfall had decreased over the 56 years of the survey (Gange *et al.*, 2007). In summary, the date of *first* sporulation is now significantly earlier in the year, and the *last* date of sporulation is now significantly later in the year than they were 60 years ago, resulting in a greatly extended season.

Sporulation of species that are mycorrhizal with both deciduous and coniferous trees is delayed in deciduous, but not in coniferous forests, indicating important physiological differences in climate responses between these ecosystems. Significant numbers of species that previously only fruited in autumn now also fruit in spring, indicating increased mycelial activity and decay rates in ecosystems in response to changes in spring and summer temperature and rainfall. Such analyses show that relatively simple field observations of fungi can detect climate change, and that fungal responses are sufficiently sensitive to react to the climate change that has already occurred by adapting their pattern of development.

The sporulation trends seem to be responses to temperature and rainfall in July and August. In years when July and August temperatures were high and rainfall low in Wiltshire, sporulation was delayed. Trends among mycorrhizal species are different from non-mycorrhizal saprotrophic fungi in the same habitats, suggesting that mycorrhizal species respond to cues from their host trees more than environmental factors (Gange *et al.*, 2007).

Overall, the impact of climate change on UK fungi seems to be that the mycelium of many species has now become active in late winter and early spring as well as late summer and autumn (well into November, probably). This has major implications for ecosystem functioning, as it indicates increased decomposition rates and altered competition between fungi over this much longer interval of mycelial activity in the year. The fact that mycelial activity is changing and increasing rapidly means that decay processes and symbiotic associations will also change, leading to profound alterations in grassland and forest dynamics and food webs (Gange *et al.*, 2007; Moore *et al.*, 2008).

The same effect has been observed in Europe and North America. Across central to northern Europe, mean sporulation varied by approximately 25 days, primarily with latitude. Altitude affected sporophore production by up to 30 days, with spring delays and autumnal accelerations, most likely because of bioclimatic change. Temperature drove sporulation of autumnal ectomycorrhizal and saprotrophic, as well as spring saprotrophic groups, while primary production and precipitation were major drivers for spring-sporulating ectomycorrhizal fungi (Andrew *et al.*, 2018). These authors point out that there is a significant likelihood that further climatic change, especially in temperature, will impact the seasonal and cyclic behaviour of fungi over very large spatial scales. Diez *et al.* (2013) used historical records (herbarium data on 274 species of fungi found in Michigan, USA) to see if the times of sporulation depend on annual climate. They showed that fungal sporophores appeared generally later in warmer and drier years, leading to a shift towards later sporulation dates for autumn-sporulating species; which is consistent with the European observations. However, the Michigan data also revealed high variability between species, at least partly due to differences in lifestyle between the fungi. However, these authors still conclude that these 'differences in climate sensitivities are expected to affect community structure as climate changes' (Diez *et al.*, 2013; Karavani *et al.*, 2018).

Global environmental change, especially increased atmospheric carbon dioxide concentration and consequential increases in temperatures, will affect most ecosystems. Museum collections and curated citizen science data (amateur collections and surveys) have been used successfully to demonstrate that fungal richness is strongly correlated with land use and climate conditions, especially seasonality, and that ongoing global climate changes will affect fungal richness patterns over both large and small geographic scales, and influence fungal biogeography and future conservation policy decisions (Andrew *et al.*, 2018, 2019; Gange *et al.*, 2019a).

We need to know more about the effects of these changes on fungal, especially mycorrhizal, systems because, as we have outlined above, mycorrhizal fungi have a major impact on the structure of plant communities and already account for a substantial proportion of their host's photosynthetic capacity. But all the fungi, mycorrhizas and saprotrophs alike, contribute to so many food webs that the ecological impact will be much broader if climate change affects the timing of nutrient (and food) availability in ecosystems.

Given that mycorrhizas are nearly always beneficial to plant growth and health, their potential value in agriculture and horticulture could be immense; it is not surprising that there are several commercial applications of mycorrhizas. Studies have shown that several crops respond well to inoculation with arbuscular mycorrhizal fungi: growth rates of maize, wheat and barley increased by two, three and four times, respectively, when inoculated with arbuscular mycorrhizas. Onions inoculated with arbuscular mycorrhizas showed a six times greater growth than non-mycorrhizal controls. Despite this potential, however, deliberate inoculation of crops with mycorrhizas is rarely done and mycorrhizas are introduced intentionally in only a few industries.

Forestry is one of the industries to recognise and exploit the role of mycorrhizas in plant growth. The bulk of commercial timber comes from trees forming ectomycorrhizal associations and amounts to more than 3 billion cubic metres felled annually, worth something like US$640 billion. New plantations benefit considerably from ectomycorrhizal inoculation, especially if the land has not previously grown ectomycorrhizal trees. Similarly, non-indigenous species of tree (so-called *exotics*) nearly always grow unsatisfactorily unless suitable mycorrhizal fungi are introduced with them. However, once an exotic has been established, others of its species are easier to cultivate as the mycorrhizas are already established in the soil.

Several smaller industries regularly use mycorrhizal infection. Germination of orchid seeds requires mycorrhizal inoculation unless the seedlings are supplied with the necessary nutrients in artificial culture. However, non-mycorrhizal seedlings tend to be more susceptible to fungal disease when mature, and the industry generally accepts that the key to producing healthy, mature plants is to inoculate seedlings with mycorrhizas.

Undoubtedly, soil fungi, saprotrophs and mycorrhizas alike, have been, are today, and will remain into the future, major architects of the global environment. In forest ecosystems, ectomycorrhizal and saprotrophic fungi play a central role in the breakdown of soil organic matter. Ectomycorrhizal and ericoid mycorrhizal fungi produce proteases and associated enzymes, allowing them greater access to organic nitrogen sources than that achieved by arbuscular mycorrhizal fungi (section 11.8). However, arbuscular mycorrhizas establish a tripartite symbiosis with leguminous plants and their rhizobial bacteria that form nitrogen-fixing nodules on the legume roots (and then the mycorrhiza distributes the products of the nitrogen fixation to non-leguminous plants with which it also forms a mycorrhizal mutualism) (section 14.10) (Johnson *et al.*, 2017). Fungi and bacteria live together in most environments; their complex interactions are important for the health of plants and animals that share or provide those environments and are significant drivers of many ecosystem functions (Deveau *et al.*, 2018; Johnston *et al.*, 2019).

Soil contains more carbon than the atmosphere and above-ground vegetation combined, and these below-ground fungi play key roles in terrestrial ecosystems as they regulate carbon, nitrogen and mineral nutrient cycles, and influence soil structure and the biological and ecological diversity of habitats (it's called ecosystem multifunctionality, see Alsterberg *et al.*, 2017). Up to 80% of plant N and P is provided by mycorrhizal fungi and most plant species depend on these symbionts for growth and survival. It is thought that competition between these two fungal guilds (saprotrophs and mycorrhizas) can lead to suppression of the overall rate of organic matter decomposition, a phenomenon known as the 'Gadgil effect' (Gadgil & Gadgil, 1971; Fernandez & Kennedy, 2016).

Decomposition of soil organic matter is often limited by the availability of nitrogen to soil microbial saprotrophs, and plants, through their mycorrhizas, compete directly with free-living decomposers for nitrogen. Ectomycorrhizal fungi, though, are the major contributors to long-term oxidation of lignocellulose in soil organic matter. They may either hamper or stimulate decomposition, by different mechanisms, depending upon the stage of decomposition and the *location* of the organic matter in the soil profile (Averill *et al.*, 2014; Field *et al.*, 2015; Heijden *et al.*, 2015; Sterkenburg *et al.*, 2018). It has been suggested that the mycelia in soil provide 'logistics networks' for transport of materials in structurally and chemically heterogeneous soil ecosystems (where 'logistics' is used in its general business sense of the flow of things between point of origin and point of consumption). The idea being that the hyphae are not only responsible for internal translocation; but the external *micro-habitats surrounding hyphae* (fluid layers that often contain polypeptides and polysaccharides produced by the hyphae) form a mycosphere of structured biofilms within which materials and microbes (motile bacteria, for example) can travel in all directions through the soil (Worrich *et al.*, 2018).

Understanding the mechanisms controlling the accumulation and stability of soil carbon is critical to predicting the future climate of the Earth. Rising concentrations of atmospheric CO_2 stimulate plant growth; an effect (called CO_2 fertilisation) that could reduce the pace and impact of anthropogenic climate change. But plants also need nitrogen for growth and experiment shows that plant species that associate with ectomycorrhizal fungi experience a strong biomass increase in response to elevated CO_2 regardless of nitrogen availability, whereas low nitrogen availability limits CO_2 fertilisation in plants that associate with arbuscular mycorrhizal fungi (Terrer *et al.*, 2016). Waste materials and other pollutants of the environment can significantly alter the community structure of ectomycorrhizal fungi by applying selection pressure favouring fungi with the potential to cope with adverse conditions. This research suggests a way of screening for antiadversity fungi which can cope with, and even remediate organic and inorganic environmental pollutants (Sun *et al.*, 2016; and references therein).

We have already mentioned above, that mycorrhizal plants can tolerate higher levels of heavy metals in soils than those without mycorrhizas, so mycorrhizas can contribute to reclamation programmes for ex-industrial land. Large areas of wasteland created by mining and other heavy industrial operations are often polluted by metals such as aluminium, iron, nickel, lead, zinc or cadmium. The metallic pollutants may also affect soil pH and the mobility of essential nutrients such as nitrogen, phosphorus and potassium.

Restoration can involve major earth moving operations such as reshaping and regrading of slopes, and large-scale toxic waste removal, with the result that the topsoil is largely artificial and is often pH-corrected and dosed with chemical fertilisers. Despite these efforts, up to 90% of plants may fail to establish and if recolonisation is discouraged bare soil remains unstable and subject to erosion. A common reason for failure to recolonise is the absence of mycorrhizas, which can prevent heavy metal poisoning and help seeds to germinate successfully. By accumulating and immobilising heavy metals, the mycorrhizal fungus acts a natural clean-up mechanism for soils. As we have seen above, mycorrhizas initiate the cycling of nutrients in the new soil, taking up immobile nutrients that would otherwise be unavailable to plants and compensating for the deficiencies in nitrogen and phosphorus that often afflict mining soils and industrial wasteland. Once a mycelial network has been established, soil structure and stability improve, and the land is well on the way to restoration.

14.18 INTRODUCTION TO LICHENS

Lichens are mutualistic (symbiotic) associations between a fungus (the mycobiont) and a photosynthetic partner (the photobiont, also called the phycobiont as most are algal). It is usually an association between a fungus with a green alga, sometimes a cyanobacterium, as photobionts. The basis of the association is that the photobionts use sunlight to produce carbohydrate photosynthetically that is shared by both partners in the lichen. Some lichens contain both green algae and cyanobacteria as photobionts, in which case the cyanobacterium may focus on fixing atmospheric nitrogen for the joint metabolism of the tripartite association.

Using metatranscriptomics to examine the lichen symbiosis, Spribille *et al.* (2016) discovered that the classical binary view of lichens as a symbiosis between a single fungus, usually an ascomycete, and a photosynthesising partner, is too simple. They found that macrolichens are constituted of not two but three symbiotic partners: an ascomycete fungus, a photosynthetic alga and unexpectedly, a basidiomycete yeast. The yeast cells are embedded in the cortex of the lichen thallus and may contribute significantly to its morphology. The yeasts are ubiquitous and essential partners for most lichens and are not casual colonisers or parasites. And then there are the bacteria. In addition to nitrogen-fixing cyanobacteria, five bacterial orders are frequently found, providing functions to the overall symbiotic community of the lichen ranging from the provision of vitamins and cofactors to the degradation of phenolic compounds (Cernava *et al.*, 2017).

There are about 20,000 species of lichens of all types in the living world (the current global checklist of lichens and allied fungi lists 18,882 species; Feuerer & Hawksworth, 2007). The lichen association is an intimate symbiosis which greatly extends the ecological range and scope of all partners. The lichen tissue is called a thallus, and in most cases, it is very different from either a fungal mycelium or an alga growing separately. The fungal hyphae surround the algal cells, often incorporating them into multicellular fungal tissues unique to lichen associations. They vary in size, shape and colour. Some are flat and firmly attached to the surfaces on which they grow, like those yellow-brown discs that are often seen scattered over walls and roofs. But others are scaly, leafy or bushy, or hang in strands from their supports. Several lichen thallus morphologies are recognised, and some are very similar to simple plants in appearance and growth pattern (Sanders, 2001; Büdel & Scheidegger, 2008):

- crustose (a flat crust, commonly seen on walls and other stonework);
- foliose (leafy);
- fruticose (branched, shrubby);
- filamentous (hair-like);
- leprose (powdery);
- squamulose (made up of small scales);
- gelatinous (lichens in which the cyanobacteria produce a polysaccharide that absorbs and retains water).

For illustrations, see Brodo *et al.* (2001) and visit the 'Lichens of North America' website at URL: http://www.lichen.com/; Purvis (2007), and for more 'floristic' detail of biodiversity, Feuerer & Hawksworth (2007) and Smith *et al.* (2009). There are expectations that the increasing availability of molecular data will improve appreciation of lichen biodiversity. However, generating genome-scale data from multi-species symbioses is challenging because metagenomic DNA from inseparable symbiotic organisms yields sequenced loci that may belong to any number of the organisms involved in these intimate associations. Nevertheless, appropriate approaches and methods are being devised (Aschenbrenner *et al.*, 2016; Grewe *et al.*, 2017; Meiser *et al.*, 2017).

In microscope sections, a typical foliose lichen thallus (Figure 14.15) reveals four zones of interlaced fungal filaments. The uppermost zone is formed by densely interwoven hyphae forming an outer protective tissue layer called the cortex, which might be a few hundred micrometres thick (Büdel & Scheidegger, 2008). In some *Parmeliaceae* the cortex secretes an epicortex, up to 1 µm thick, which seems to be a sort of cuticular layer. The algal cells occur in a zone beneath the cortex embedded in a dense hyphal tissue. Each cell or group of cells of the photobiont is usually individually enveloped by hyphae. Under the algal layer is a third zone, called the medulla, of loosely interwoven fungal hyphae without algal cells. The lower surface of the thallus, beneath the medulla, is like the upper surface; it is called the lower cortex and may consist of densely packed fungal hyphae. Hyphal branches emerging from the lower cortex, as root-like structures known as rhizines, serve to attach the thallus to its substratum.

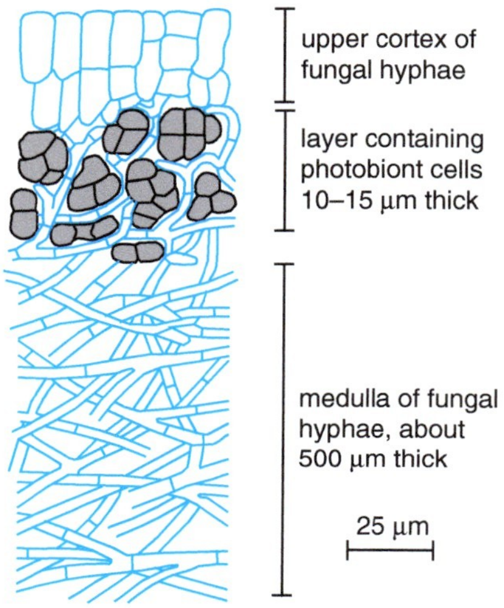

Figure 14.15. A sketch of a slice of lichen, showing the general structure of a stratified (heteromerous) thallus. Most lichens develop internally stratified thalli; the most usual layers being an upper cortex, photobiont layer, medulla and, sometimes, a lower cortex. Lower surfaces, though, more usually lack a cortex, and are mycelial (described as 'cottony'), often with fungal hyphae aggregated into strands so that the undersurface of the thallus appears to have a network of veins. The fungal tissues may be composed of elongated hyphae in a fairly loose mycelium (prosoplectenchymatous) or may be tightly packed so that individual hyphae are not distinguishable, and the tissue seems to be made up of isodiametric cells, this is called pseudoparenchymatous tissue because it resembles the genuine parenchyma of vascular plants. The sketch shown illustrates a lichen like *Peltigera*, which is a foliose lichen with broad lobed thalli occurring in many parts of the world growing on soil, rocks, trees, and similar surfaces. All species of *Peltigera* associate with the nitrogen-fixing cyanobacterium *Nostoc*, which is probably the most widely distributed 'blue-green alga' in the world. Some also have a eukaryotic algal photobiont. Because of their dual abilities to fix nitrogen from the atmosphere and photosynthesise, such lichens are crucial primary colonisers and actually start the process of soil formation. For more details see Nash (2008a) and Sanders (2001). This figure redrawn and modified from an illustration in Hudson (1986).

The most ancient associations, in evolutionary terms, are between fungi and blue-green algae. Blue-green algae are better called cyanobacteria because they are bacteria rather than algae, but they do have chlorophyll and can photosynthesise. *Cyanobacteria* were the first organisms to release oxygen into the Earth's atmosphere, so they started, probably 3 billion years ago, the revolution from which emerged the atmosphere on which we rely today. The most ancient lichen-like fossils so far reported were found in South China. These fossils involve filamentous hyphae in close association with coccoidal cyanobacteria or algae. They are between 551 and 635 million years old and indicate that fungi developed symbiotic partnerships with photosynthetic organisms *long before the evolution of vascular plants* (Yuan *et al.*, 2005).

Fossil lichens have also been described from the Rhynie chert which is 400 million years old (Taylor *et al.*, 2015; Krings *et al.*, 2018). As the more advanced (eukaryotic) algae evolved the fungi (mostly relatives of present-day *Ascomycota*) developed the partnerships that are now lichens. It has been suggested that lichenisation is ancestral and has been gained infrequently during the onward evolution of *Ascomycota*, and indeed, that the major *Ascomycota* lineages of non-lichen-forming species alive today might be *derived* from lichen-forming ancestors (Lutzoni *et al.*, 2001).

Another entertaining evolutionary speculation you might like to read about is the suggestion that *vascular plants originated from a genomic combination of fungal genome + algal genome* providing the genetic architecture for a 'higher plant body' which is visualised as a mosaic of generalised alga-type cells (the photosynthetic cells of modern plants) interspersed with highly specialised fungus-type cells, which make up the structural and translocatory tissues (Atsatt, 1988).

Stated like this, the notion may sound fanciful, but it has been regularly *restated* over the years (Jorgensen, 1993; Selosse & Strullu-Derrien, 2015; Aanen & Eggleton, 2017) and we recommend you read all these publications. Even more remarkable is the report that cells of the alga *Nannochloropsis oceanica* can, under certain conditions, become internalised ***within the hyphae*** of the fungus *Mortierella elongata* (*Mortierellomycotina*) (Du *et al.*, 2019). In her discussion of some of the questions raised by this observation, Bonfante (2019) emphasises that prior to this observation in all known interactions between algae and fungi, notably all lichens, the algal cells remain ***outside*** the hyphae of the fungus. But in this study, which involved co-cultivation of alga and fungus, the algae first aggregated on the hyphal surface, and eventually algae entered the hyphae where they continued to photosynthesise, grow and divide. Nutrient exchange, including carbon and nitrogen transfer between fungal and algal cells was demonstrated by isotope tracer experiments. Both partners remained physiologically active over two months of co-cultivation. When the alga *Nannochloropsis* grows within the hyphae of *Mortierella*, those hyphae changed colour to green due to the photosynthetic activity of the algae within.

A comprehensive phylogenetic analysis of 259 transcriptomes and 10 green algal and basal land plant genomes concluded that the most recent common ancestor of present-day land plants and green algae was *preadapted for symbiotic associations*, specifically by having the genes required for fungus-plant nutrient exchange. The study further suggests that subsequent rounds of gene duplication led to the acquisition of additional pathways, required for the perception and transduction of diffusible fungal signals for root colonisation that enabled formation of the arbuscular mycorrhizal symbiosis (Delaux *et al.*, 2015). Could it be that the algal ancestor of land plants was 'preadapted for symbiosis' by being the photobiont in a lichen association?

For reproduction several lichens produce flakes comprising small groups of algal cells surrounded by fungal filaments (called soredia; Figure 14.16) that form a powdery mass in structures called *soralia* on the upper surface of the lichen. Other structures are fragile, upright, elongated outgrowths from the thallus (called isidia) that break off for mechanical dispersal. Many lichens break up into fragments when they dry, dispersing themselves by wind action, to resume growth when moisture returns; fruticose lichens in particular can easily fragment. In any of these cases the fragments include *both* fungus and alga and are easily dislodged to be blown about in the breeze like spores. They can begin a new colony if they land in a suitable place.

Lichens are the classic examples of symbiosis, in fact they prompted the term to be coined. Lichens dominate more of the land surface of Earth than do tropical rain forests, they involve one in five of all fungi, and can live in places that are inaccessible to other organisms. All this because of their ultrastructurally and physiologically unique partnership. The fungus protects the alga and supports it physically and physiologically by taking in water and using its externalised enzymes to extract nutrients from the soil, and even from the rocks themselves. These organisms directly or indirectly induce chemical weathering leading to the mobilisation of minerals like magnesium, manganese, iron, aluminium and silicon (Nash, 2008c). The algal or cyanobacterial cells in the partnership can photosynthesise; photolysing water and reducing atmospheric carbon dioxide to provide carbohydrates for use by both partners.

More recent studies raise questions about how truly mutually beneficial the relationship might be, however. Most of the body of the lichen is fungus; the algal partner constitutes only 5–10% of the total biomass. Indeed, microscopic examination shows that fungal cells might even penetrate the algal cells in a way like pathogenic fungi; although for the most part the readiness with which nutrients leak out of algal cells makes penetration by fungal haustoria completely unnecessary. Certainly, the photosynthetic partner can exist independently of the fungal partner in nature, but not vice versa.

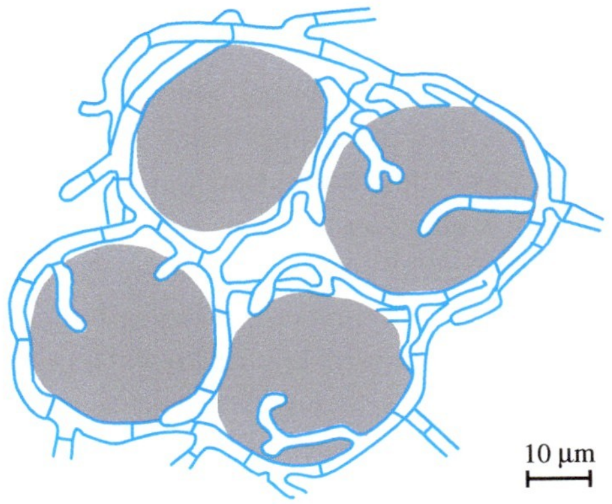

Figure 14.16. Soredium. Lichen reproductive structure composed of photobiont cells (shown in grey) enveloped by hyphae (shown in blue). Redrawn and modified from an illustration in Hudson (1986).

If cultivated in the laboratory without the photobiont, a lichen fungus grows as an undifferentiated mass of hyphae. Only when combined with its photobiont does the hyphal morphology characteristic of the lichen thallus arise. Cells of the photobiont are routinely destroyed in the course of nutrient exchange; stability of the association depending on photobiont cells reproducing more rapidly than they are destroyed. Because of these observations it has been suggested that the fungus in the lichen is really parasitising the alga and using the products of algal photosynthesis for fungal nutrition (Palmqvist *et al.*, 2008). However, other observations support the notion that 'mutualism' best describes the association, particularly the numerous metabolites that the two partners share, and the relationship has been likened to a farming operation in which the fungus is 'farming' its photobiont. The statement 'Lichens are a case of fungi that have discovered agriculture ...' has been attributed to the lichenologist Trevor Goward (Department of Botany Herbarium, University of British Columbia) (Sanders, 2001).

Lichens have a remarkable ability to thrive where no other organisms can exist. They are often the first to settle in places lacking soil, constituting the sole vegetation in some extreme environments. A characteristic feature is that they can tolerate severe desiccation. In dry conditions lichens go into metabolic stasis (known as cryptobiosis) where dehydration of the cells of the lichen halts most biochemical activity. In this state lichens can survive extremes of temperature, irradiation and drought in the harshest environments (Beckett *et al.*, 2008).

Lichens tolerate temperature extremes from the heat of deserts to the cold of Arctic and Antarctic wastes. Some grow on high mountain rocks, extracting moisture from mists and fog. They are also extremely common as epiphytes growing on the surfaces of plants, particularly on the trunks, branches and even twigs of trees. In these circumstances the epiphytic lichens are not parasites; they have no physiological impact on the plant; they are simply using its surface as a habitat. Understandably, lichen growth can be rather slow; maybe 2 inches per thousand years in the Arctic, but somewhat faster in less extreme environments.

The current global checklist of lichens lists 18,882 species; representing about 20% of all known fungi. They are mainly *Ascomycota* with a few (only about 15) *Basidiomycota*, which are called basidiolichens to differentiate them from the much more common ascolichens. Even though lichens are characteristically found in environments subject to temperature, desiccation and low nutrient stresses, fungal sexual structures are often formed in abundance, a feature which obviously contributes to their diversity and ecological success (Seymour *et al.*, 2005). This may be a common form of reproduction in basidiolichens, which form sporophores like their non-lichenised fungal relatives. In the ascolichens, spores are produced in apothecia, perithecia and pycnidia. After dispersal, these fungal spores must find a compatible algal partner before a functional lichen can be formed.

The taxonomic naming of lichens is based on the fungal component. As lichens are resistant to environmental extremes they are pioneer colonisers of terrestrial habitats, such as rock faces, tree bark, roofing tiles, etc., so they are very common in both urban and rural environments, where lichen species diversity can be used as a measure of atmospheric pollution (Nash, 2008a & b). Lichen dependence on the atmosphere and rainfall makes them highly sensitive to atmospheric pollution. Acidic rain and sulfur dioxide kill many lichens, so cities in industrialised countries may have few lichens because of the poor air quality. If you do have lichens growing on your stonework, it's probably good for your lungs!

The nutritional value of lichens is similar to cereal seeds, though they do not make major contributions to human food. In the harsh areas where lichens grow, native peoples use them as food supplements. One lichen, which occurs in the deserts of the Middle East, may have been the manna that fell from heaven to rescue the Children of Israel from starvation in the Old Testament Bible story:

> and in the morning the dew lay round about the host. And when the dew that lay was gone up, behold upon the face of the wilderness there lay a small round thing, as small as the hoar frost on the ground. And when the children of Israel saw it they said one to another, It is manna: for they wist not what it was. And Moses said unto them, This is the bread which the Lord hath given you to eat. (Exodus, 16, v. 13–15)

Lichens are much more important as animal food, especially in the Arctic, where lichens can form as much as 95% of the diet of reindeer. In less harsh conditions, most mammals of these wildernesses will supplement their normal winter diets with lichens. Many Lepidopteran larvae feed exclusively on lichens. In hotter climates, sheep in Libya graze a lichen that grows on the desert rocks, grinding their teeth away as they chew it off. The extreme lifestyle of lichens leads them to produce several exotic chemicals, some of which are very useful for lichen identification. These compounds may be useful to humans. They include antibiotics, essential oils for perfumery and dyes for textiles, and there are probably many others awaiting discovery and exploitation (Oksanen, 2006; Elix & Stocker-Wörgötter, 2008; Cernava *et al.*, 2015).

14.19 INTRODUCTION TO ENDOPHYTES

Many plants harbour fungi within their tissues that are at least harmless and may be beneficial. There are several definitions of endophytes, but the emphasis is usually on the fact that although they live inside a plant, they do not cause any disease symptoms. The presence of endophytic fungi within symptomless plants is usually recognised indirectly through culturing methods: small pieces of tissue (leaf, for example) are surface sterilised, placed on agar plates, and the fungi inside the leaf grow out on to the agar.

Endophytic fungi can also be observed directly within leaves using histological methods with the light microscope. A particularly successful technique is to label one of the constituents of fungal cell walls with a fluorescent dye as this enables them to be observed in thin sections using a microscope equipped with fluorescence optics (Figure 14.17).

These sections show that each plant host is usually colonised by several fungal species concurrently. They also show that the category 'symptomless' has different interpretations depending on the level of analysis. Microscopic examination may reveal that the plant is responding to an infection in an apparently healthy leaf that is symptomless to the naked eye (Johnston *et al.*, 2006; Currie *et al.*, 2014). Endophyte genome analysis has revealed the genes that are important for the endophytic lifestyle, and further application of these approaches will explore the role of endophytes in host plant ecology and their biotechnological promise (Sanjana *et al.*, 2016).

Fungal endophytes mainly belong to the *Ascomycota* and are a tremendously diverse range of fungi. They can be found in the above-ground tissues of all major lineages of land plants: liverworts, hornworts, mosses, lycophytes, equisetums, ferns and seed plants. They range from the arctic to the tropics and occur in agricultural fields and the wilderness. There is another continuum in that they have a wide range of impacts on their hosts, ranging from the strictly symptomless cohabitant through to potential pathogens that can become pathogenic to their host plant in response to some change in circumstances. Within that continuum lies the possibility that the endophyte is symbiotic with its host but not all endophytes are mutualists (Arnold, 2007; Lacava & Azevedo, 2014).

Endophytes became a significant topic of study when those living entirely within grasses (in particular, tall fescue) were shown to be responsible for the toxicity of grasses to livestock. Fungal alkaloids produced by the endophyte are toxic to livestock that graze on the plant, but they improve the health of the plant by protecting against attack by grazing insects. In this case, the endophyte is conferring an advantage to its host. Unfortunately, it's bad for farmed livestock so endophyte-infected tall fescue must be avoided in livestock feed, though it is actively encouraged for amenity sites, especially golf courses because the natural protection offered by the endophytic fungus means reduced use of chemical pesticides. Endophyte symbiosis in tall fescue can have a dramatic effect on many aspects of natural ecological systems; influencing plant–plant competition, diversity, productivity, succession, plant-herbivore interactions and energy flow through the food web (Rudgers & Clay, 2007; Omacini, 2014).

A functional relation to the host is not always obvious, many endophytes simply colonise inner spaces of plants like any other moist secluded habitat. The overall endophyte–plant interaction, though, depends on environment, and the age and genotype diversity of the host plants. Endophyte interactions of forest

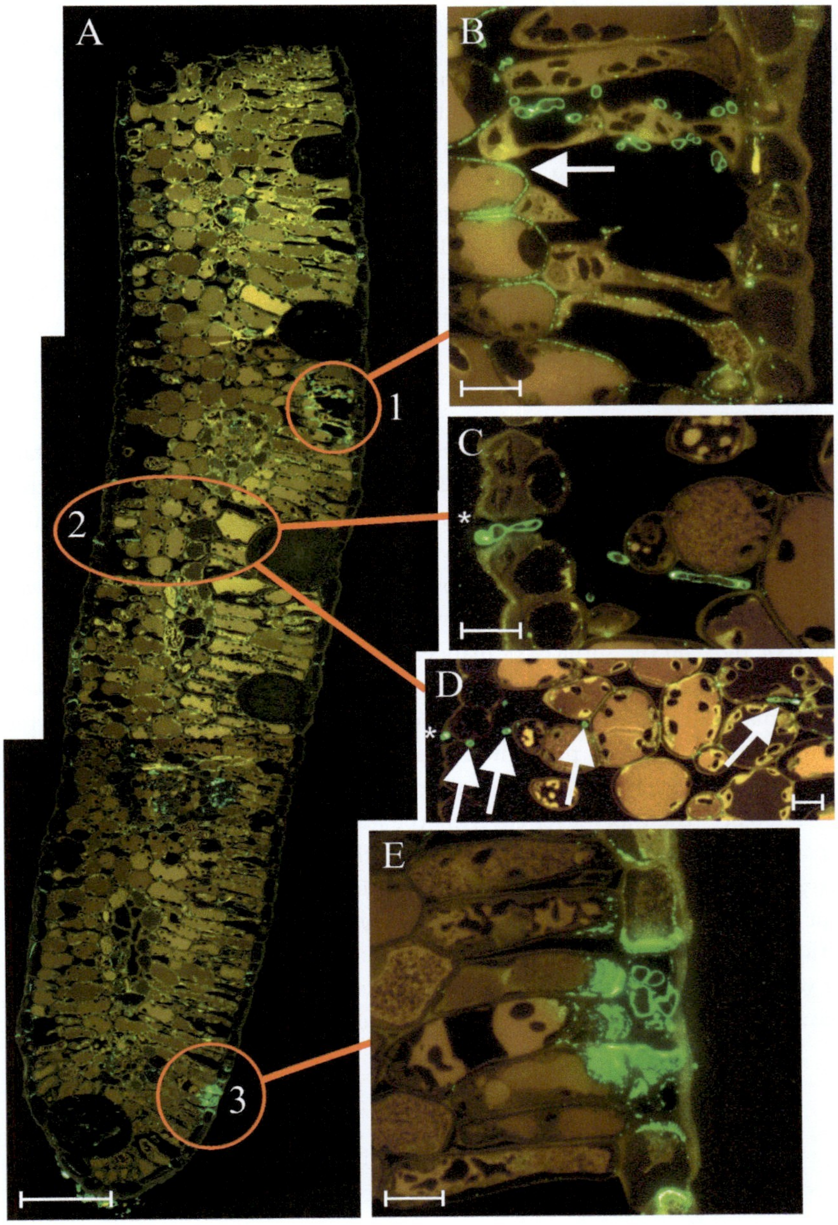

Figure 14.17. Section from a *Kunzea ericoides* leaf, 200 nm thick (A), with fungal cell walls fluorescently labelled to detect (1-3)-β-D-glucan (*Kunzea ericoides* is the 'white tea tree', a tree that occurs naturally in New Zealand). Walls of fungal hyphae label green, the plant cell walls autofluoresce to appear yellow or brown although plant cell walls with callose production also appear green. Three separate fungal infections occur in the photomontage that makes up the leaf section (A) each of which is illustrated in detail in (B–E). (B) shows infection type 1, in which entry occurs through a stoma, fungal hyphae are restricted to the stomatal cavity, and the plant cells around the stomatal cavity produce callose (arrowed) in reaction to the fungus. (C and D) show infection type 2, where entry again occurs through a stoma, but the plant cells surrounding the substomatal cavity show with no callose production, and fungal hyphae extend deeply into the leaf tissue (arrows in D), asterisk indicates the same stoma in (C and D). (E) shows the third type of infection, where the fungus directly penetrates the epidermal cell wall, and extensive callose production in surrounding plant cells restricts the fungus to a single epidermal cell. Scale bars: A, 100 μm, B–E, 10 μm. Modified from Johnston *et al.* (2006) using graphics files kindly supplied by Dr Peter R. Johnston, Landcare Research, New Zealand. Images reproduced with permission from Elsevier.

trees and other woody plants range from mutualism to pathogenicity (Saikkonen, 2007; Verma & Gange, 2014). Although mutualism is often assumed, the evidence is often circumstantial and the main reason for this is that endophytes are so common that endophyte-free control trees are simply lacking. Sieber (2007) summarised the situation like this:

> tree endophytes are mostly harmless colonisers of the internal [spaces] of healthy plant tissues. Some are potentially pathogenic, but disease is only caused in combination with other, mostly unknown, inciting factors. Proof of mutualism of endophyte-host symbioses has been inconclusive in most cases, but plant communities would probably not survive many environmental stresses without these symbioses. All we know for certain is that endophytes are present in any healthy plant tissue!

Endophytes have been evolving with their hosts for millions of years; and not just their hosts, they have been evolving with their host's community and ecosystem, and that includes the pests and pathogens that prey on the host. Gall-forming insects are among the pests that infest plants and cause abnormal outgrowths ('tumours') of plant tissues formed in response to the stimulus caused by the insect laying eggs, or its larvae or nymph feeding on the plant. The insect becomes enclosed by the gall and feeds only on gall tissue during its development. These outgrowths are called galls because they contain large amounts of tannin which has a very bitter taste.

The insects involved include aphids, phylloxerans, psyllids, midges (gall gnats) and cynipid wasps (gall wasps) but the last of these, the gall wasps (Order *Hymenoptera*, Family *Cynipidae*) are the most important insects that induce plant galls, and about 80% of gall wasps produce galls specifically on oak trees (in fact, 60% of all known insect galls occur on oaks).

The endophytic fungi have been evolving with the oak and its gall wasps for a long time and a particularly interesting interaction has emerged for the oak gall wasp of Oregon white oak, because the fungal endophyte grows from the leaf into the gall and infects all the gall tissue. As a result, the insect in the gall dies but it is not directly killed by the fungus; rather, it dies because its food source, the gall tissue, is killed by the fungal infection. Approximately 12.5% of the galls on the tree die as a result of invasion by the fungal endophyte, so it's a very considerable contribution to control of the pest. When the pest has been dealt with, the fungus returns to its life as a symptomless inhabitant of the leaf tissue (Lacava & Azevedo, 2014).

The status of most endophytes is still unclear but is evidently complex. For example, *Fusarium solani* strain K is a root endophyte of *Solanum lycopersicum* (tomato), which protects against root and foliar pathogens (Skiada *et al.*, 2019). Fungal hyphae of strain K proliferate within the vascular bundle of the plant, and plant cell death is involved in colonisation of the root by this fungus. Establishment of *Fusarium* strain K within the plant tissues requires both fungal growth adaptations and plant cell-autonomous responses of the sort that occur during both symbiotic and pathogenic plant–fungal interactions.

A variety of different factors can trigger the transformation of the fungus from innocent saprotroph, to mutualist, to pathogen. And they can be aggressive pathogens, often killing large parts of their host. But this usually follows physical damage or general stress on the host plant. There are even some observations that indicate structural integration of endosymbiotic fungi with the cells and tissues of the host, reminiscent of the lichen association (Barrow *et al.*, 2007; Lucero *et al.*, 2008). Higher plants are ubiquitously colonised by fungal endophytes, and at a time when emerging dramatic climate changes are likely to increase stress on plant communities, endophytes might become the source of new plant diseases (Slippers & Wingfield, 2007), and could also be the source of valuable new secondary metabolites (Suryanarayanan *et al.*, 2009; Higginbotham *et al.*, 2013; Verma & Gange, 2014; Hodkinson *et al.*, 2019). However, these ubiquitous hidden fungi may be important bodyguards of plants, protecting against grazing by insect herbivores (Gange *et al.*, 2019b).

14.20 EPIPHYTES

Plant surfaces, like any other surface, provide potential habitats to fungi. We have already mentioned the growth of lichens on tree trunks, branches and twigs; of yeasts on leaves and fruits and in the nectaries of flowers (which, though fluid, are external to the plant); and of rhizomorphs of such pathogens as *Armillaria*, 'scrambling' over the surface of potential hosts. Fungi that live on the surfaces of plants are called *epiphytes*. Some show special adaptations to the plant surface, which is a challenging environment, being dry, waxy and exposed to direct sunlight. Epiphytes are often pigmented (particularly melanised) to protect them from UV radiation, and some can digest lipids sufficiently to use the waxy layer covering the leaf epidermis. The yeast form usually has a short life cycle, which enables yeast epiphytes to multiply even if favourable conditions last only a short time. Fungal epiphytes are a varied and polyphyletic group with a worldwide distribution; the majority belong in the Phylum *Ascomycota* and foliar epiphytes probably evolved as early as the Permian period, about 300 million years ago (Hongsanan *et al.*, 2016).

14.21 REFERENCES

Aanen, D.K. & Eggleton, P. (2017). Symbiogenesis: beyond the endosymbiosis theory? *Journal of Theoretical Biology*, **434**: 99–103. DOI: https://doi.org/10.1016/j.jtbi.2017.08.001.

Alsterberg, C., Roger, F., Sundbäck, K. *et al.* (2017). Habitat diversity and ecosystem multifunctionality: the importance of direct and indirect effects. *Science Advances*, **3**: article e1601475. DOI: https://doi.org/10.1126/sciadv.1601475.

Anderson, I.C. & Parkin, P.I. (2007). Detection of active soil fungi by RT-PCR amplification of precursor rRNA molecules. *Journal of Microbiological Methods*, **68**: 248–253. DOI: https://doi.org/10.1016/j.mimet.2006.08.005.

Andrew, C., Büntgen, U., Egli, S. *et al.* (2019). Open-source data reveal how collections-based fungal diversity is sensitive to global change. *Applications in Plant Sciences*, **7**: article e1227 (online version of record before inclusion in an issue). DOI: https://doi.org/10.1002/aps3.1227.

Andrew, C., Heegaard, E., Høiland, K. *et al.* (2018). Explaining European fungal fruiting phenology with climate variability. *Ecology*, **99**: 1306–1315. DOI: https://doi.org/10.1002/ecy.2237.

Anke, H. & Weber, R.W.S. (2006). White rots, chlorine and the environment: a tale of many twists. *Mycologist*, **20**: 83–89. DOI: https://doi.org/10.1016/j.mycol.2006.03.011.

Arnold, A.E. (2007). Understanding the diversity of foliar endophytic fungi: progress, challenges and frontiers. *Fungal Biology Reviews*, **21**: 51–66. DOI: https://doi.org/10.1016/j.fbr.2007.05.003.

Aschenbrenner, I.A., Cernava, T., Berg, G. & Grube, M. (2016). Understanding microbial multi-species symbioses. *Frontiers in Microbiology*, **7**: 180. DOI: https://doi.org/10.3389/fmicb.2016.00180.

Atsatt, P.R. (1988). Are vascular plants 'inside-out' lichens? *Ecology*, **69**: 17–23. DOI: https://doi.org/10.2307/1943156.

Averill, C., Turner, B.L. & Finzi, A.C. (2014). Mycorrhiza-mediated competition between plants and decomposers drives soil carbon storage. *Nature*, **505**: 543–545. DOI: https://doi.org/10.1038/nature12901.

Avila, R., Johanson, K.J. & Bergstrom, R. (1999). Model of the seasonal variations of fungi ingestion and ^{137}Cs activity concentrations in roe deer. *Journal of Environmental Radioactivity*, **46**: 99–112. DOI: https://doi.org/10.1016/S0265-931X(98)00108-8.

Balasundaram, S.V., Hess, J., Durling, M.B. *et al.* (2018). The fungus that came in from the cold: dry rot's pre-adapted ability to invade buildings. *The ISME Journal*, **12**: 791–801. DOI: https://doi.org/10.1038/s41396-017-0006-8.

Barrow, J., Lucero, M., Reyes-Vera, I. & Havstad, K. (2007). Endosymbiotic fungi structurally integrated with leaves reveals a lichenous condition of C4 grasses. *In Vitro Cellular & Developmental Biology – Plant*, **43**: 65–70. DOI: https://doi.org/10.1007/s11627-006-9007-4.

Beckett, R., Kranner, I. & Minibayeva, F. (2008). Stress physiology and the symbiosis. In *Lichen Biology*, 2nd edition, ed. T.H. Nash. Cambridge, UK: Cambridge University Press, pp. 134–151. ISBN-13: 978-0521459747. DOI: https://doi.org/10.1017/CBO9780511790478.009.

Bindschedler, S., Vu Bouquet, T.Q.T., Job, D., Joseph, E. & Junier, P. (2017). Fungal biorecovery of gold from e-waste. *Advances in Applied Microbiology*, **99**: 53–81. DOI: https://doi.org/10.1016/bs.aambs.2017.02.002.

Bonfante, P. (2019). Symbiosis: algae and fungi move from the past to the future. *eLife*, **8**: article number e49448. DOI: https://doi.org/10.7554/eLife.49448.

Bonnardeaux, Y., Brundrett, M., Batty, A. *et al.* (2007). Diversity of mycorrhizal fungi of terrestrial orchids: compatibility webs, brief encounters, lasting relationships and alien invasion. *Mycological Research*, **111**: 51–61. DOI: https://doi.org/10.1016/j.mycres.2006.11.006.

Borovička, J., Řanda, Z., Jelínek, E., Kotrba, P. & Dunn, C.E. (2007). Hyperaccumulation of silver by *Amanita strobiliformis* and related species of the section *Lepidella*. *Mycological Research*, **111**: 1339–1344. DOI: https://doi.org/10.1016/j.mycres.2007.08.015.

Bousquet, F., Botta, A., Alinovi, L. *et al.* (2016). Resilience and development: mobilizing for transformation. *Ecology and Society*, **21**: article 40. DOI: https://doi.org/10.5751/ES-08754-210340.

Brodo, I.M., Sharnoff, S.D. & Sharnoff, S. (2001). *Lichens of North America*. New Haven, CT: Yale University Press. ISBN: 9780300082494.

Brundrett, M.C. (2009). Mycorrhizal associations and other means of nutrition of vascular plants: understanding the global diversity of host plants by resolving conflicting information and developing reliable means of diagnosis. *Plant and Soil*, **320**: 37–77. DOI: https://doi.org/10.1007/s11104-008-9877-9.

Brundrett, M.C., Murase, G. & Kendrick, B. (1990). Comparative anatomy of roots and mycorrhizae of common Ontario trees. *Canadian Journal of Botany*, **68**: 551–578. DOI: https://doi.org/10.1139/b90-076.

Büdel, B. & Scheidegger, C. (2008). Thallus morphology and anatomy. In *Lichen Biology*, 2nd edition, ed. T.H. Nash. Cambridge, UK: Cambridge University Press, pp. 40–68. ISBN-13: 978-0521459747. DOI: https://doi.org/10.1017/CBO9780511790478.005.

Buller, A.H.R. (1931). *Researches on Fungi*, **Vol. 4.** London: Longmans, Green and Co. ASIN: B0008BT4R6.

Camazine, S. (1983). Mushroom chemical defense: food aversion learning induced by hallucinogenic toxin, Muscimol. *Journal of Chemical Ecology*, **9**: 1473–1481. DOI: https://doi.org/10.1007/BF00990749.

Cernava, T., Erlacher, A., Aschenbrenner, I.A. *et al.* (2017). Deciphering functional diversification within the lichen microbiota by meta-omics. *Microbiome*, **5**: 82. DOI: https://doi.org/10.1186/s40168-017-0303-5.

Cernava, T., Müller, H., Aschenbrenner, I.A., Grube, M. & Berg, G. (2015). Analyzing the antagonistic potential of the lichen microbiome against pathogens by bridging metagenomic with culture studies. *Frontiers in Microbiology*, **6**: 620. DOI: https://doi.org/10.3389/fmicb.2015.00620.

Chalot, M., Javelle, A., Blaudez, D. *et al.* (2002). An update on nutrient transport processes in ectomycorrhizas. *Plant and Soil*, **244**: 165–175. DOI: https://doi.org/10.1023/A:1020240709543.

Chiu, S.W., Ching, M.L., Fong, K.L. & Moore, D. (1998). Spent oyster mushroom substrate performs better than many mushroom mycelia in removing the biocide pentachlorophenol. *Mycological Research*, **102**: 1553–1562. DOI: https://doi.org/10.1017/S0953756298007588.

Collier, F.A. & Bidartondo, M.I. (2009). Waiting for fungi: the ectomycorrhizal invasion of lowland heathlands. *Journal of Ecology*, **97**: 950–963. DOI: https://doi.org/10.1111/j.1365-2745.2009.01544.x.

Cooke, M.C. (1862). *A Plain and Easy Account of the British Fungi, with Descriptions of the Esculent and Poisonous Species, Details of the Principles of Scientific Classification, and a Tabular Arrangement of Orders and Genera*. London: Robert Hardwicke. View this vintage book (free) at this URL: http://www.archive.org/details/aplainandeasyacc00cookiala/.

Costanza, R., d'Arge, R., de Groot, R. *et al.* (1997). The value of the world's ecosystem services and natural capital. *Nature*, **387**: 253. DOI: https://doi.org/10.1038/387253a0.

Costanza, R., de Groot, R., Braat, L. *et al.* (2017). Twenty years of ecosystem services: how far have we come and how far do we still need to go? *Ecosystem Services*, **28**: 1–16. DOI: https://doi.org/10.1016/j.ecoser.2017.09.008.

Crowther, T.W., Boddy, L. & Jones, T.H. (2012). Functional and ecological consequences of saprotrophic fungus-grazer interactions. *The ISME Journal*, **6**: 1992–2001. DOI: https://doi.org/10.1038/ismej.2012.53.

Currie, A.F., Wearn, J., Hodgson, S. *et al.* (2014). Foliar fungal endophytes in herbaceous plants: a marriage of convenience? In *Advances in Endophytic Research*, ed. V. Verma & A. Gange. New Delhi: Springer International Publishing, pp. 61–81. ISBN: 978-81-322-1574-5. DOI: https://doi.org/10.1007/978-81-322-1575-2_3.

Czederpiltz, D.L.L., Stanosz, G.R. & Burdsall, H.H. Jr (1999). Forest management and the diversity of wood-inhabiting fungi. *McIlvainia*, **14**: 34–45. Forest Products Laboratory URL: https://www.fpl.fs.fed.us/documnts/pdf1999/czede99a.pdf.

Daily, G.C. (ed.) (1997). *Nature's Services: Societal Dependence on Natural Ecosystems*. Washington, DC: Island Press. ISBN-10: 1559634766, ISBN-13: 978-1559634762.

Delaux, P.-M., Radhakrishnan, G.V., Jayaraman, D. *et al.* (2015). Algal ancestor of land plants was preadapted for symbiosis. *Proceedings of the National Academy of Sciences of the United States of America*, **112**: 13390–13395. DOI: https://doi.org/10.1073/pnas.1515426112.

Deveau, A., Bonito, G., Uehling, J. *et al.* (2018). Bacterial-fungal interactions: ecology, mechanisms and challenges. *FEMS Microbiology Reviews*, **42**: 335–352. DOI: https://doi.org/10.1093/femsre/fuy008.

Diez, J.M., James, T.Y., McMunn, M. & Ibáñez, I. (2013). Predicting species-specific responses of fungi to climatic variation using historical records. *Global Change Biology*, **19**: 3145–3154. DOI: https://doi.org/10.1111/gcb.12278.

Downie, J.A. (2014). Legume nodulation. *Current Biology*, **24**: R184–R190. DOI: https://doi.org/10.1016/j.cub.2014.01.028.

Dranginis, A.M., Rauceo, J.M., Coronado, J.E. & Lipke, P.N. (2007). A biochemical guide to yeast adhesins: glycoproteins for social and antisocial occasions. *Microbiology and Molecular Biology Reviews*, **71**: 282–294. DOI: https://doi.org/10.1128/MMBR.00037-06.

Du, Z.-Y., Zienkiewicz, K., Vande Pol, N., Ostrom, N.E., Benning, C. & Bonito, G.M. (2019). Algal-fungal symbiosis leads to photosynthetic mycelium. *eLife*, **8**: article number e47815. DOI: https://doi.org/10.7554/eLife.47815.001.

Dugan, F.M. (2011). *Conspectus of World Ethnomycology: Fungi in Ceremonies, Crafts, Diets, Medicines, and Myths*. St Paul, MN: The American Phytopathological Society. ISBN-10: 089054395X, ISBN-13: 978-0890543955.

Edwards, D., Kenrick, P. & Dolan, L. (2018). History and contemporary significance of the Rhynie cherts – our earliest preserved terrestrial ecosystem. *Philosophical Transactions of the Royal Society of London, Series B*, **373**: article number 20160489. DOI: https://doi.org/10.1098/rstb.2016.0489.

Elix, J. & Stocker-Wörgötter, E. (2008). Biochemistry and secondary metabolites. In *Lichen Biology*, 2nd edition, ed. T.H. Nash. Cambridge, UK: Cambridge University Press, pp. 104–133. ISBN-13: 978-0521459747. DOI: https://doi.org/110.1017/CBO9780511790478.008.

Endo, A. (1992). The discovery and development of HMG-CoA reductase inhibitors. *Journal of Lipid Research*, **33**: 1569–1582. URL: http://www.jlr.org/content/33/11/1569.short.

Fellbaum, C.R., Gachomo, E.W., Beesetty, Y. *et al.* (2012). Carbon availability triggers fungal nitrogen uptake and transport in arbuscular mycorrhizal symbiosis. *Proceedings of the National Academy of Sciences of the United States of America*, **109**: 2666–2671. DOI: https://doi.org/10.1073/pnas.1118650109.

Fellbaum, C.R., Mensah, J.A., Cloos, A.J. *et al.* (2014). Fungal nutrient allocation in common mycorrhizal networks is regulated by the carbon source strength of individual host plants. *New Phytologist*, **203**: 646–656. DOI: https://doi.org/10.1111/nph.12827.

Fernandez, C.W. & Kennedy, P.G. (2016). Revisiting the 'Gadgil effect': do interguild fungal interactions control carbon cycling in forest soils? *New Phytologist*, **209**: 1382–1394. DOI: https://doi.org/10.1111/nph.13648.

Ferrol, N., Barea, J.M. & Azcón-Aguilar, C. (2002). Mechanisms of nutrient transport across interfaces in arbuscular mycorrhizas. *Plant and Soil*, **244**: 231–237. DOI: https://doi.org/10.1023/A:1020266518377.

Feuerer, T. & Hawksworth, D.L. (2007). Biodiversity of lichens, including a worldwide analysis of checklist data based on Takhtajan's floristic regions. *Biodiversity and Conservation*, **16**: 85–98. DOI: https://doi.org/10.1007/s10531-006-9142-6.

Field, K.J., Pressel, S., Duckett, J.G., Rimington, W.R. & Bidartondo, M.I. (2015). Symbiotic options for the conquest of land. *Trends in Ecology & Evolution*, **30**: 477–486. DOI: https://doi.org/10.1016/j.tree.2015.05.007.

Floren, A., Krüger, D., Müller, T. *et al.* (2015). Diversity and interactions of wood-inhabiting fungi and beetles after deadwood enrichment. *PLoS ONE*, **10**: article number e0143566. DOI: https://doi.org/10.1371/journal.pone.0143566.

Frankland, J.C. (1998). Fungal succession: unravelling the unpredictable. *Mycological Research*, **102**: 1–15. DOI: https://doi.org/10.1017/S0953756297005364.

Gadgil, R.L. & Gadgil, P.D. (1971). Mycorrhiza and litter decomposition. *Nature*, **233**: 133. DOI: https://doi.org/10.1038/233133a0.

Gange, A.C., Allen, L.P., Nussbaumer, A. *et al.* (2019a). Multiscale patterns of rarity in fungi, inferred from fruiting records. *Global Ecology and Biogeography*, online version. DOI: https://doi.org/10.1111/geb.12918.

Gange, A.C., Gange, E.G., Sparks, T.H. & Boddy, L. (2007). Rapid and recent changes in fungal fruiting patterns. *Science*, **316**: 71. DOI: https://doi.org/10.1126/science.1137489.

Gange, A.C., Koricheva, J., Currie, A.F., Jaber, L.R. & Vidal, S. (2019b). Meta-analysis of the role of entomopathogenic and unspecialised fungal endophytes as plant bodyguards. *New Phytologist*, accepted article online ahead of publication. DOI: https://doi.org/10.1111/nph.15859.

Gao, C., Montoya, L., Xu, L., Madera, M. *et al.* (2018). Strong succession in arbuscular mycorrhizal fungal communities. *The ISME Journal*, **2018**. DOI: https://doi.org/10.1038/s41396-018-0264-0.

Gardner, S.M., Ramsden, S.J. & Hails, R.S. (ed.) (2019). *Agricultural Resilience: Perspectives from Ecology and Economics*. Cambridge, UK: Cambridge University Press. ISBN: 9781107665873.

Geyer, R., Jambeck, J.R. & Law, K.L. (2017). Production, use, and fate of all plastics ever made. *Science Advances*, **3**: article number e1700782. DOI: https://doi.org/10.1126/sciadv.1700782.

Gilbert, L. & Johnson, D. (2017). Plant–plant communication through common mycorrhizal networks. *Advances in Botanical Research*, **82**: 83–97. DOI: https://doi.org/10.1016/bs.abr.2016.09.001.

Grewe, F., Huang, J.-P., Leavitt, S.D. & Lumbsch, H.T. (2017). Reference-based RADseq resolves robust relationships among closely related species of lichen-forming fungi using metagenomic DNA. *Scientific Reports*, **7**: article number 9884. DOI: https://doi.org/10.1038/s41598-017-09906-7.

Harley, J.L. (1991). The state of the art. In *Methods in Microbiology*, **Vol. 23**. *Techniques for the Study of Mycorrhiza*, ed. J.R. Norris, D.J. Read & A.K. Varma. London: Academic Press, pp. 1–23. ISBN: 9780125215237. DOI: https://doi.org/10.1016/S0580-9517(08)70171-5.

Hartnett, D.C. & Wilson, G.W.T. (2002). The role of mycorrhizas in plant community structure and dynamics: lessons from grasslands. *Plant and Soil*, **244**: 319–331. DOI: https://doi.org/10.1023/A:1020287726382.

Heijden, M.G.A., Martin, F.M., Selosse, M.-A. & Sanders, I.R. (2015). Mycorrhizal ecology and evolution: the past, the present, and the future. *New Phytologist*, **205**: 1406–1423. DOI: https://doi.org/10.1111/nph.13288.

Heilmann-Clausen, J., Adamčík, S., Bässler, C. *et al.* (2017). State of the art and future directions for mycological research in old-growth forests. *Fungal Ecology*, **27**: 141–144. DOI: https://doi.org/10.1016/j.funeco.2016.12.005.

Higginbotham, S.J., Arnold, A.E., Ibañez, A. *et al.* (2013). Bioactivity of fungal endophytes as a function of endophyte taxonomy and the taxonomy and distribution of their host plants. *PLoS ONE*, **8**: article number e73192. DOI: https://doi.org/10.1371/journal.pone.0073192.

Hodkinson, T.R., Doohan, F.M., Saunders, M.J. & Murphy, B.R. (ed.) (2019). *Endophytes for a Growing World*. Cambridge, UK: Cambridge University Press. ISBN: 9781108471763.

Hongsanan, S., Sánchez-Ramírez, S., Crous, P.W., Ariyawansa, H.A., Zhao, R.L. & Hyde, K.D. (2016). The evolution of fungal epiphytes. *Mycosphere*, **7**: 1690–1712. DOI: https://doi.org/10.5943/mycosphere/7/11/6.

Huckfeldt, T. & Schmidt, O. (2006). Identification key for European strand-forming house-rot fungi. *Mycologist*, **20**: 42–56. DOI: https://doi.org/10.1016/j.mycol.2006.03.012.

Hudson, H.J. (1986). *Fungal Biology*. London: Edward Arnold. 304 pp. ISBN-10: 071312895X, ISBN-13: 978-0713128956.

Ivanov, S., Fedorova, E.E., Limpens, E. *et al.* (2012). *Rhizobium*-legume symbiosis shares an exocytotic pathway required for arbuscule formation. *Proceedings of the National Academy of Sciences of the United States of America*, **109**: 8316–8321. DOI: https://doi.org/10.1073/pnas.1200407109.

Jackson, R.M. & Mason, P.A. (1984) *Mycorrhiza*. The Institute of Biology's *Studies in Biology* series. London: Edward Arnold. ISBN: 0-7131-2876-3.

Johnson, N.C., Gehring, C. & Jansa, J. (2017). *Mycorrhizal Mediation of Soil: Fertility, Structure, and Carbon Storage*. Amsterdam: Elsevier. ISBN: 9780128043127. DOI: https://doi.org/10.1016/B978-0-12-804312-7.01001-9.

Johnston, P.R., Sutherland, P.W. & Joshee, S. (2006). Visualising endophytic fungi within leaves by detection of (1–3)-β-D-glucans in fungal cell walls. *Mycologist*, **20**: 159–162. DOI: https://doi.org/10.1016/j.mycol.2006.10.003.

Johnston, S.R., Hiscox, J., Savoury, M., Boddy, L. & Weightman, A.J. (2019). Highly competitive fungi manipulate bacterial communities in decomposing beech wood (*Fagus sylvativa*). *FEMS Microbiology Ecology*, **95**: article fiy225. DOI: https://doi.org/10.1093/femsec/fiy225.

Jorgensen, R. (1993). The origin of land plants: a union of alga and fungus advanced by flavonoids? *Biosystems*, **31**: 193–207. DOI: https://doi.org/10.1016/0303-2647(93)90049-I.

Juutilainen, K., Mönkkönen, M., Kotiranta, H. & Halme, P. (2014). The effects of forest management on wood-inhabiting fungi occupying dead wood of different diameter fractions. *Forest Ecology and Management*, **313**: 283–291. DOI: https://doi.org/10.1016/j.foreco.2013.11.019.

Karavani, A., De Cáceres, M., de Aragón, J.M., Bonet, J.A. & de-Miguel, S. (2018). Effect of climatic and soil moisture conditions on mushroom productivity and related ecosystem services in Mediterranean pine stands facing climate change. *Agricultural and Forest Meteorology*, **248**, 432–440. DOI: https://doi.org/10.1016/j.agrformet.2017.10.024.

Karunanithi, R., Szogi, A.A., Bolan, N. *et al.* (2015). Phosphorus recovery and reuse from waste streams. *Advances in Agronomy*, **131**: 173–250. DOI: https://doi.org/10.1016/bs.agron.2014.12.005.

Keppler, F., Harper, D.B., Röckmann, T., Moore, R.M. & Hamilton, J.T.G. (2005). New insight into the atmospheric chloromethane budget gained using stable carbon isotope ratios. *Atmospheric Chemistry and Physics*, **5**: 2403–2411. DOI: https://doi.org/10.5194/acp-5-2403-2005.

Konuma, R., Umezawa, K., Mizukoshi, A., Kawarada, K. & Yoshida, M. (2015). Analysis of microbial volatile organic compounds produced by wood-decay fungi. *Biotechnology Letters*, **37**: 1845–1852. DOI: https://doi.org/10.1007/s10529-015-1870-9.

Krings, M., Harper, C.J. & Taylor, E.L. (2018). Fungi and fungal interactions in the Rhynie chert: a review of the evidence, with the description of *Perexiflasca tayloriana* gen. et sp. nov. *Philosophical Transactions of the Royal Society of London, Series B*, **373**: article number 20160500. DOI: https://doi.org/10.1098/rstb.2016.0500.

Kumar, P., Mahato, D.K., Kamle, M. *et al.* (2016). Aflatoxins: a global concern for food safety, human health and their management. *Frontiers in Microbiology*, **7**: article number 2170. DOI: https://doi.org/10.3389/fmicb.2016.02170.

Kurucz, V., Kiss, B., Szigeti, Z.M. *et al.* (2018). Physiological background of the remarkably high Cd^{2+} tolerance of the *Aspergillus fumigatus* Af293 strain. *Journal of Basic Microbiology*, **2018**: 1–11. DOI: https://doi.org/10.1002/jobm.201800200.

Lacava, P.T. & Azevedo, J.L. (2014). Biological control of insect-pest and diseases by endophytes. In *Advances in Endophytic Research*, ed. V. Verma & A. Gange. New Delhi: Springer International Publishing, pp. 231–256. ISBN: 978-81-322-1574-5. DOI: https://doi.org/10.1007/978-81-322-1575-2_13.

Lange, L., Bech, L., Busk, P.K. *et al.* (2012). The importance of fungi and of mycology for a global development of the bioeconomy. *IMA Fungus*, **3**: 87–92. DOI: https://doi.org/10.5598/imafungus.2012.03.01.09.

Lepp, N.W., Harrison, S.C.S. & Morrell, B.G. (1987). A role for *Amanita muscaria* L. in the circulation of cadmium and vanadium in a non-polluted woodland. *Environmental Geochemistry and Health*, **9**: 61–64. DOI: https://doi.org/10.1007/BF02057276.

Lindahl, B.D. & Kuske, C.R. (2013). Metagenomics for study of fungal ecology. In *The Ecological Genomics of Fungi*, ed. F. Martin. Hoboken, NJ: John Wiley & Sons, pp. 279–303. ISBN: 9781119946106. DOI: https://doi.org/10.1002/9781118735893.ch13.

Liu, B., Liu, J., Ju, M., Lic, X. & Wang, P. (2017). Bacteria-white-rot fungi joint remediation of petroleum-contaminated soil based on sustained-release of laccase. *RSC Advances*, **7**: 39075–39081. DOI: https://doi.org/10.1039/C7RA06962F.

Lucero, M., Barrow, J.R., Osuna, P. & Reyes, I. (2008). A cryptic microbial community persists within micropropagated *Bouteloua eriopoda* (Torr.) Torr. cultures. *Plant Science*, **174**: 570–575. DOI: https://doi.org/10.1016/j.plantsci.2008.02.012.

Lutzoni, F., Pagel, M. & Reeb, V. (2001). Major fungal lineages are derived from lichen symbiotic ancestors. *Nature*, **411**: 937–940. DOI: https://doi.org/10.1038/35082053.

Martino, E., Morin, E., Grelet, G.A. *et al.* (2018). Comparative genomics and transcriptomics depict ericoid mycorrhizal fungi as versatile saprotrophs and plant mutualists. *New Phytologist*, **217**: 1213–1229. DOI: https://doi.org/10.1111/nph.14974.

Meiser, A., Otte, J., Schmitt, I. & Grande, F.D. (2017). Sequencing genomes from mixed DNA samples: evaluating the metagenome skimming approach in lichenized fungi. *Scientific Reports*, **7**: article number 14881. DOI: https://doi.org/10.1038/s41598-017-14576-6.

Meng, L., Zhang, A., Wang, F. *et al.* (2015). Arbuscular mycorrhizal fungi and rhizobium facilitate nitrogen uptake and transfer in soybean/maize intercropping system. *Frontiers in Plant Science*, **6**: article 339. DOI: https://doi.org/10.3389/fpls.2015.00339.

Menta, C. (2012). Soil fauna diversity: function, soil degradation, biological indices, soil restoration. In *Biodiversity Conservation and Utilization in a Diverse World*, ed. G.A. Lameed. London: InTech. Creative Commons BY 3.0 license, ISBN: 978-953-51-0719-4. DOI: https://doi.org/10.5772/51091.

Michelot, D. & Melendez-Howell, L.M. (2003). *Amanita muscaria*: chemistry, biology, toxicology, and ethnomycology. *Mycological Research*, **107**: 131–146. DOI: https://doi.org/10.1017/S0953756203007305.

Misztal, P.K., Lymperopoulou, D.S., Adams, R.I. *et al.* (2018). Emission factors of microbial volatile organic compounds from environmental bacteria and fungi. *Environmental Science & Technology*, in press. DOI: https://doi.org/10.1021/acs.est.8b00806.

Mitchell, D.T. & Gibson, B.R. (2006). Ericoid mycorrhizal association: ability to adapt to a broad range of habitats. *Mycologist*, **20**: 2–9. DOI: https://doi.org/10.1016/j.mycol.2005.11.015.

Mitchell, J.I. & Zuccaro, A. (2006). Sequences, the environment and fungi. *Mycologist*, **20**: 62–74. DOI: https://doi.org/10.1016/j.mycol.2005.11.004.

Moore, D., Gange, A.C., Gange, E.G. & Boddy, L. (2008). Fruit bodies: their production and development in relation to environment. In *Ecology of Saprotrophic Basidiomycetes*, ed. L. Boddy, J.C. Frankland & P. van West. London: Academic Press, pp. 79–103. ISBN-10: 0123741858, ISBN-13: 978-0123741851.

Moretti, A., Logrieco, A.F. & Susca, A. (2017). Mycotoxins: an underhand food problem. In *Mycotoxigenic Fungi, Methods in Molecular Biology*, **Vol. 1542**, ed. A. Moretti & A. Susca. New York: Humana Press, pp. 3–12. DOI: https://doi.org/10.1007/978-1-4939-6707-0_1.

Nakamori, T. & Suzuki, A. (2007). Defensive role of cystidia against collembola in the basidiomycetes *Russula bella* and *Strobilurus ohshimae*. *Mycological Research*, **111**: 1345–1351. DOI: https://doi.org/10.1016/j.mycres.2007.08.013.

Nash, T.H. (2008a). *Lichen Biology*, 2nd edition. Cambridge, UK: Cambridge University Press. ISBN: 9780511790478. DOI: https://doi.org/10.1017/CBO9780511790478.

Nash, T.H. (2008b). Lichen sensitivity to air pollution. In *Lichen Biology*, 2nd edition, ed. T.H. Nash. Cambridge, UK: Cambridge University Press, pp. 299–314. ISBN-13: 978-0521459747. DOI: https://doi.org/10.1017/CBO9780511790478.016.

Nash, T.H. (2008c). Nutrients, elemental accumulation, and mineral cycling. In *Lichen Biology*, 2nd edition, ed. T.H. Nash. Cambridge, UK: Cambridge University Press, pp. 234–251. ISBN-13: 978-0521459747. DOI: https://doi.org/10.1017/CBO9780511790478.013.

Navarro-Ródenas, A., Pérez-Gilabert, M., Torrente, P. & Morte, A. (2012). The role of phosphorus in the ectendomycorrhiza continuum of desert truffle mycorrhizal plants. *Mycorrhiza*, **22**: 565–575. DOI: https://doi.org/10.1007/s00572-012-0434-2.

Nehls, U., Göhringer, F., Wittulsky, S. & Dietz, S. (2010). Fungal carbohydrate support in the ectomycorrhizal symbiosis: a review. *Plant Biology*, **12**: 292–301. DOI: https://doi.org/10.1111/j.1438-8677.2009.00312.x.

Ngosong, C., Gabriel, E. & Ruess, L. (2014). Collembola grazing on arbuscular mycorrhiza fungi modulates nutrient allocation in plants. *Pedobiologia*, **57**: 171–179. DOI: https://doi.org/10.1016/j.pedobi.2014.03.002.

Oksanen, I. (2006). Ecological and biotechnological aspects of lichens. *Applied Microbiology and Biotechnology*, **73**: 723–734. DOI: https://doi.org/10.1007/s00253-006-0611-3.

Omacini, M. (2014). Asexual endophytes of grasses: invisible symbionts, visible imprints in the host neighbourhood. In *Advances in Endophytic Research*, ed. V. Verma & A. Gange. New Delhi: Springer International Publishing, pp. 143–157. ISBN: 978-81-322-1574-5. DOI: https://doi.org/10.1007/978-81-322-1575-2_7.

Osbourn, A.E., O'Maille, P.E., Rosser, S.J. & Lindsey, K. (2012). Synthetic biology. *New Phytologist*, **196**: 671–677. DOI: https://doi.org/10.1111/j.1469-8137.2012.04374.x.

Palmqvist, K., Dahlman, L., Jonsson, A. & Nash, T. (2008). The carbon economy of lichens. In *Lichen Biology*, 2nd edition, ed. T.H. Nash. Cambridge, UK: Cambridge University Press, pp. 182–215. ISBN-13: 978-0521459747. DOI: https://doi.org/10.1017/CBO9780511790478.011.

Paterson, R.R.M. (2006). Fungi and fungal toxins as weapons. *Mycological Research*, **110**: 1003–1010. DOI: https://doi.org/10.1016/j.mycres.2006.04.004.

Peralta, R.M., da Silva, B.P., Gomes Côrrea, R.C. *et al.* (2017). Enzymes from basidiomycetes – peculiar and efficient tools for biotechnology. In *Biotechnology of Microbial Enzymes: Production, Biocatalysis and Industrial Applications*, ed. G. Brahmachari, A.L. Demain & J.L. Adrio. Amsterdam: Academic Press, an imprint of Elsevier Inc., pp. 119–149. ISBN: 978-0-12-803725-6. DOI: https://doi.org/10.1016/B978-0-12-803725-6.00005-4.

Perotto, S., Girlanda, M. & Martino, E. (2002). Ericoid mycorrhizal fungi: some new perspectives on old acquaintances. *Plant and Soil*, **244**: 41–53. DOI: https://doi.org/10.1023/A:1020289401610.

Plassard, C. & Fransson, P. (2009). Regulation of low molecular weight organic acid production in fungi. *Fungal Biology Reviews*, **23**: 30–39. DOI: https://doi.org/10.1016/j.fbr.2009.08.002.

Prosser, J.I. (2002). Molecular and functional diversity in soil micro-organisms. *Plant and Soil*, **244**: 9–17. DOI: https://doi.org/10.1023/A:1020208100281.

Purvis, W. (2007). *Lichens*. London: The Natural History Museum. ISBN-13: 978-0565091538.

Rasmussen, H.N. (2002). Recent developments in the study of orchid mycorrhiza. *Plant and Soil*, **244**: 149–163. DOI: https://doi.org/10.1023/A:1020246715436.

Rhodes, C.J. (2014). Mycoremediation (bioremediation with fungi) – growing mushrooms to clean the earth. *Chemical Speciation & Bioavailability*, **26**: 196–198. DOI: https://doi.org/10.3184/095422914X14047407349335.

Rinne-Garmston, K.T., Peltoniemi, K., Chen, J. *et al.* (2019). Carbon flux from decomposing wood and its dependency on temperature, wood N_2 fixation rate, moisture and fungal composition in a Norway spruce forest. *Global Change Biology*, in press. DOI: https://doi.org/10.1111/gcb.14594.

Rosling, A., Roose, T., Herrmann, A.M. *et al.* (2009). Approaches to modelling mineral weathering by fungi. *Fungal Biology Reviews*, **23**: 138–144. DOI: https://doi.org/10.1016/j.fbr.2009.09.003.

Rudgers, J.A. & Clay, K. (2007). Endophyte symbiosis with tall fescue: how strong are the impacts on communities and ecosystems? *Fungal Biology Reviews*, **21**: 107–123. DOI: https://doi.org/10.1016/j.fbr.2007.05.002.

Sabev, H.A., Barratt, S.R., Greenhalgh, M., Handley, P.S. & Robson, G.D. (2006). Biodegradation and biodeterioration of man-made polymeric materials. In *Fungi in Biogeochemical Cycles*, ed. G.M. Gadd. Cambridge, UK: Cambridge University Press, pp. 212–235. ISBN-10: 0521845793, ISBN-13: 978-0521845793.

Saikkonen, K. (2007). Forest structure and fungal endophytes. *Fungal Biology Reviews*, **21**: 67–74. DOI: https://doi.org/10.1016/j.fbr.2007.05.001.

Sanders, W.B. (2001). Lichens: interface between mycology and plant morphology. *BioScience*, **51**: 1025–1035. DOI: https://doi.org/10.1641/0006-3568(2001)051[1025:LTIBMA]2.0.CO;2.

Sanjana, K., Tanwi, S. & Manoj, K.D. (2016). 'Omics' tools for better understanding the plant–endophyte interactions. *Frontiers in Plant Science*, **7**: 955. DOI: https://doi.org/10.3389/fpls.2016.00955.

Sathiyadash, K., Muthukumar, T., Uma, E. & Pandey, R.R. (2012). Mycorrhizal association and morphology in orchids. *Journal of Plant Interactions*, **7**: 238–247. DOI: https://doi.org/10.1080/17429145.2012.699105.

Schmalenberger, A., Duran, A.L., Bray, A.W. *et al.* (2015). Oxalate secretion by ectomycorrhizal *Paxillus involutus* is mineral-specific and controls calcium weathering from minerals. *Scientific Reports*, **5**: article number 12187. DOI: https://doi.org/10.1038/srep12187.

Schwarze, F.W.M.R. (2007). Wood decay under the microscope. *Fungal Biology Reviews*, **21**: 133–170. DOI: https://doi.org/10.1016/j.fbr.2007.09.001.

Seephonkai, P., Isaka, M., Kittakoop, P., Boonudomlap, U. & Thebtaranonth, Y. (2004). A novel ascochlorin glycoside from the insect pathogenic fungus *Verticillium hemipterigenum* BCC 2370. *Journal of Antibiotics*, **57**: 10–16. DOI: https://doi.org/10.7164/antibiotics.57.10.

Selosse, M.-A., Setaro, S., Glatard, F. *et al.* (2007). Sebacinales are common mycorrhizal associates of Ericaceae. *New Phytologist*, **174**: 864–878. DOI: https://doi.org/10.1111/j.1469-8137.2007.02064.x.

Selosse, M.-A. & Strullu-Derrien, C. (2015). Origins of the terrestrial flora: a symbiosis with fungi? *BIO Web of Conferences*, **4**: 00009. DOI: https://doi.org/10.1051/bioconf/20150400009.

Seymour, F.A., Crittenden, P.D. & Dyer, P.S. (2005). Sex in the extremes: lichen-forming fungi. *Mycologist*, **19**: 51–58. DOI: https://doi.org/10.1017/S0269915XO5002016.

Shah, A.A., Hasan, F., Hameed, A. & Ahmed, S. (2008). Biological degradation of plastics: a comprehensive review. *Biotechnology Advances*, **26**: 246–265. DOI: https://doi.org/10.1016/j.biotechadv.2007.12.005.

Sherratt, T.N., Wilkinson, D.M. & Bain, R.S. (2005). Explaining Dioscorides' `double difference': why are some mushrooms poisonous, and do they signal their unprofitability? *American Naturalist*, **166**: 767–775. DOI: https://doi.org/10.1086/497399.

Sieber, T. (2007). Endophytic fungi of forest trees: are they mutualists? *Fungal Biology Reviews*, **21**: 75–89. DOI: https://doi.org/10.1016/j.fbr.2007.05.004.

Simard, S.W., Beiler, K.J., Bingham, M.A. *et al.* (2012). Mycorrhizal networks: mechanisms, ecology and modelling. *Fungal Biology Reviews*, **26**: 39–60. DOI: https://doi.org/10.1016/j.fbr.2012.01.001.

Singh, J. (1999). Dry rot and other wood-destroying fungi: their occurrence, biology, pathology and control. *Indoor and Built Environment*, **8**: 3–20. DOI: https://doi.org/10.1177/1420326X9900800102.

Singh, V.K., Meena, M., Zehra, A. *et al.* (2014). Fungal toxins and their impact on living systems. In *Microbial Diversity and Biotechnology in Food Security*, ed. R. Kharwar, R. Upadhyay, N. Dubey & R. Raghuwanshi. New Delhi: Springer, pp. 513–530. ISBN: 978-81-322-1800-5. DOI: https://doi.org/10.1007/978-81-322-1801-2_47.

Skiada, V., Faccio, A., Kavroulakis, N. *et al.* (2019). Colonization of legumes by an endophytic *Fusarium solani* strain FsK reveals common features to symbionts or pathogens. *Fungal Genetics and Biology*, in press. DOI: https://doi.org/10.1016/j.fgb.2019.03.003.

Slippers, B. & Wingfield, M.J. (2007). Botryosphaeriaceae as endophytes and latent pathogens of woody plants: diversity, ecology and impact. *Fungal Biology Reviews*, **21**: 90–106. DOI: https://doi.org/10.1016/j.fbr.2007.06.002.

Smith, C.W., Aptroot, A., Coppins, B.J. *et al.* (2009). *The Lichens of Great Britain and Ireland*, 2nd edition. London: The British Lichen Society and The Natural History Museum. ISBN: 9780954041885.

Smith, S.E. & Read, D.J. (1997). *Mycorrhizal Symbiosis*, 2nd editon. London: Academic Press. ISBN-10: 0126528403, ISBN-13: 978-0126528404.

Spribille, T., Tuovinen, V., Resl, P., Vanderpool, D. *et al.* (2016). Basidiomycete yeasts in the cortex of ascomycete macrolichens. *Science*, **353**: 488–492. DOI: https://doi.org/10.1126/science.aaf8287.

Sterkenburg, E., Clemmensen, K.E., Ekblad, A., Finlay, R.D. & Lindahl, B.D. (2018). Contrasting effects of ectomycorrhizal fungi on early and late stage decomposition in a boreal forest. *The ISME Journal*, **12**: 2187–2197. DOI: https://doi.org/10.1038/s41396-018-0181-2.

Straatsma, G., Ayer, F. & Egli, S. (2001). Species richness, abundance, and phenology of fungal fruit bodies over 21 years in a Swiss forest plot. *Mycological Research*, **105**: 515–523. DOI: https://doi.org/10.1017/S0953756201004154.

Sun, Q., Liu, Y., Yuan, H. & Lian, B. (2016). The effect of environmental contamination on the community structure and fructification of ectomycorrhizal fungi. *Microbiology Open*, **6**: article e00396. DOI: https://doi.org/10.1002/mbo3.396.

Suryanarayanan, T.S., Thirunavukkarasu, N., Govindarajulu, M.B. *et al.* (2009). Fungal endophytes and bioprospecting. *Fungal Biology Reviews*, **23**: 9–19. DOI: https://doi.org/10.1016/j.fbr.2009.07.001.

Suz, L.M., Sarasan, V., Bidartondo, M. & Hodkinson, T.R. (2018). Positive plant-fungal interactions. In *State of the World's Fungi 2018*, ed K.J. Willis. Report. Richmond, UK: Royal Botanic Gardens, Kew. ISBN: 978-1-84246-678-0. URL: https://stateoftheworldsfungi.org/2018/ (the PDF of the report is a free download).

Tagu, D., Lapeyrie, F. & Martin, F. (2002). The ectomycorrhizal symbiosis: genetics and development. *Plant and Soil*, **244**: 97–105. DOI: https://doi.org/10.1023/A:1020235916345.

Taylor, R.S. & Baldwin, N.A. (1991). Surface disruption of an artificial tennis court caused by *Scleroderma bovista*. *The Mycologist*, **5**: 79. DOI: https://doi.org/10.1016/S0269-915X(09)80099-2.

Taylor, T.N., Krings M. & Taylor, E.L. (2015). *Fossil Fungi*. San Diego, CA: Academic Press. 398 pp. ISBN-10: 0123877318, ISBN-13: 978-0123877314. DOI: https://doi.org/10.1016/B978-0-12-387731-4.00014-1.

Terrer, C., Vicca, S., Hungate, B.A., Phillips, R.P. & Prentice, I.C. (2016). Mycorrhizal association as a primary control of the CO_2 fertilization effect. *Science*, **353**: 72–74. DOI: https://doi.org/10.1126/science.aaf4610.

Teste, F.P. & Dickie I.A. (2016). Mycorrhizas across successional gradients. In *Mycorrhizal Mediation of Soil: Fertility, Structure, and Carbon Storage*, ed. N.C. Johnson, C. Gehring & J. Jansa. Amsterdam: Elsevier, pp. 67–69. DOI: https://doi.org/10.1016/B978-0-12-804312-7.00005-X.

Tsai, W.-T. (2017). Fate of chloromethanes in the atmospheric environment: implications for human health, ozone formation and depletion, and global warming impacts. *Toxics*, 5: 23 (13 pp.). DOI: https://doi.org/10.3390/toxics5040023.

Umesha, S., Manukumar, H.M., Chandrasekhar, B. *et al.* (2016). Aflatoxins and food pathogens: impact of biologically active aflatoxins and their control strategies. *Journal of the Science of Food and Agriculture*, **97**: 1698–1707. DOI: https://doi.org/10.1002/jsfa.8144.

van der Heijden, M.G.A. & Horton, T.R. (2009). Socialism in soil? The importance of mycorrhizal fungal networks for facilitation in natural ecosystems. *Journal of Ecology*, **97**: 1139–1150. DOI: https://doi.org/10.1111/j.1365-2745.2009.01570.x.

Varma, A., Prasad, R. & Tuteja, N. (ed.) (2017). *Mycorrhiza: Eco-Physiology, Secondary Metabolites, Nanomaterials*. Cham, Switzerland: Springer International Publishing. DOI: https://doi.org/10.1007/978-3-319-57849-1. ISBN: 978-3-319-57848-4.

Varma, S.S., Lakshmi, M.B., Rajagopal, P. & Velan, M. (2017). Degradation of total petroleum hydrocarbon (TPH) in contaminated soil using *Bacillus pumilus* MVSV3. *Biocontrol Science*, **22**: 17–23. DOI: https://doi.org/10.4265/bio.22.17.

Veits, M., Khait, I., Obolski, U., Zinger, E., Boonman, A., Goldshtein, A., Saban, K., Seltzer, R., Ben-Dor, U., Estlein, P., Kabat, A., Peretz, D., Ratzersdorfer, I., Krylov, S., Chamovitz, D., Sapir, Y., Yovel, Y. & Hadany, L. (2019). Flowers respond to pollinator sound within minutes by increasing nectar sugar concentration. *Ecology Letters*, **22**: 1483–1492. DOI: https://doi.org/10.1111/ele.13331.

Veresoglou, S.D., Chen, B. & Rillig, M.C. (2012). Arbuscular mycorrhiza and soil nitrogen cycling. *Soil Biology and Biochemistry*, **46**: 53–62. DOI: https://doi.org/10.1016/j.soilbio.2011.11.018.

Verma, V. & Gange, A. (ed.) (2014). *Advances in Endophytic Research*. New Delhi: Springer. ISBN: 978-81-322-1574-5. DOI: https://doi.org/10.1007/978-81-322-1575-2.

Vettorazzi, A. & López de Cerain, A. (2016). Mycotoxins as food carcinogens. In *Environmental Mycology in Public Health: Fungi and Mycotoxins Risk Assessment and Management*, ed. C. Viegas, C. Pinheiro, R. Sabino, S. Viegas, J. Brandão & C. Veríssimo. Amsterdam: Academic Press, an imprint of Elsevier Inc., pp. 261–298. ISBN: 978-0-12-411471-5. DOI: https://doi.org/10.1016/B978-0-12-411471-5.00017-X.

Wall, D.H., Bardgett, R.D. & Kelly, E. (2010). Biodiversity in the dark. *Nature Geoscience*, **3**: 297–298. DOI: https://doi.org/10.1038/ngeo860.

Watling, R. & Harper, D.B. (1998). Chloromethane production by wood-rotting fungi and an estimate of the global flux to the atmosphere. *Mycological Research*, **102**: 769–787. DOI: https://doi.org/10.1017/S0953756298006157.

Weber, R.W.S. & Webster, J. (2001). Teaching techniques for mycology: 14. Mycorrhizal infection of orchid seedlings in the laboratory. *Mycologist*, **15**: 55–59. DOI: https://doi.org/10.1016/S0269-915X(01)80077-X.

Weigold, P., El-Hadidi, M., Ruecker, A. *et al.* (2016). A metagenomic-based survey of microbial (de)halogenation potential in a German forest soil. *Scientific Reports*, **6**: article number 28958. DOI: http://doi.org/10.1038/srep28958.

Williams, J.H., Phillips, T.D., Jolly, P.E. *et al.* (2004). Human aflatoxicosis in developing countries: a review of toxicology, exposure, potential health consequences, and interventions. *American Journal of Clinical Nutrition*, **80**: 1106–1122. DOI: https://doi.org/10.1093/ajcn/80.5.1106.

Wohlleben, P. (2017). *The Hidden Life of Trees: What They Feel, How They Communicate – Discoveries from a Secret World*. Glasgow, UK: William Collins, an imprint of HarperCollins Publishers Limited. ISBN: 9780008218430.

Wong, J.H. (2013). Fungal toxins. In *Handbook of Biologically Active Peptides*, 2nd edition, ed. A. Kastin. Amsterdam: Academic Press, an imprint of Elsevier Inc., pp. 166–168. ISBN: 978-0-12-385095-9. DOI: https://doi.org/10.1016/B978-0-12-385095-9.00025-7.

Worrich, A., Wick, L.Y. & Banitz, T. (2018). Ecology of contaminant biotransformation in the mycosphere: role of transport processes. *Advances in Applied Microbiology*, **104**: 93–133. DOI: https://doi.org/10.1016/bs.aambs.2018.05.005.

Yamin-Pasternak, S. (2011). Ethnomycology: fungi and mushrooms in cultural entanglements. In *Ethnobiology*, ed. E.N. Anderson, D. Pearsall, E. Hunn & N. Turner. Hoboken, NJ: John Wiley & Sons, pp. 213–230. ISBN: 9780470547854. DOI: https://doi.org/10.1002/9781118015872.ch13.

Yuan, X., Xiao, S. & Taylor, T.N. (2005). Lichen-like symbiosis 600 million years ago. *Science*, **308**: 1017. DOI: https://doi.org/10.1126/science.1111347.

Zhang, X., Kuča, K., Dohnal, V. *et al.* (2014). Military potential of biological toxins. *Journal of Applied Biomedicine*, **12**: 63–77. DOI: https://doi.org/10.1016/j.jab.2014.02.005.

CHAPTER 15

Fungi As Symbionts and Predators of Animals

This chapter deals with fungal cooperative ventures, including ant and beetle agriculture, termite gardeners. An important coevolutionary story is that which associates anaerobic fungi, the evolution of grasses and the rise of the ruminants; a fascinating story that links with human evolution since humans use cereal grasses as staple foods and selected their main food animals from among the ruminants. Finally, we look at the predatory nematode-trapping fungi.

Fungi have co-existed with animals and plants throughout the whole of the evolutionary time since these three groups of higher organisms originally separated from one another. Living together closely for this length of time has given rise to many cooperative ventures. We have already seen how many fungi have combined with plants as partners in mutually beneficial relationships such as mycorrhizas and lichens. In these symbiotic, or mutualistic, associations the partners each gain something from the partnership so that the association is more successful than either organism alone. The organisms concerned (often two but sometimes more) live in such close proximity to each other that their cells may intermingle and may even contribute to the formation of joint tissues, as they do in the lichen thallus, which is one of the most ancient mutualistic associations of all and found in some of the most inhospitable environments.

Fungi have similarly close relationships with animals, which we will describe in this chapter. There are several examples and the two about which we know most are:

- the '**fungus gardens**' created by leaf-cutter ants and termites (the most notable case being leaf-cutter ants which cultivate and then graze upon hyphae, with neither fungus nor ant being able to live without the other);
- **chytrid** association with **ruminants**.

These examples show the key features of symbiotic relationships:

- They are mutually beneficial to the partners.
- The partners show behavioural and anatomical adaptations to enable the partnership.
- The partners can be shown to have evolved in step ('**coevolved**') over a considerable length of time, even to the point that the fungi, and perhaps the animal, cannot survive independently of the partnership.

15.1 FUNGAL COOPERATIVE VENTURES

We have already discussed the circumstance in which fungi are used as **food for grazing animals**, from cattle weighing hundreds of kilograms to the microarthropods weighing hundredths of a gram (see section 12.1). Lichens (*Cladonia* spp.) form an essential part of the winter diet of caribou or reindeer (*Rangifer tarandus*), and the reindeer, which are ruminants, have behavioural and anatomical adaptations enabling them to graze lichens under snow cover for their resident microbiota to digest the fragments (Kumpula, 2001; Turunen *et al.*, 2013). Fungus sporophores are eaten by many animals, but notably by mammals weighing less than 3 kg. The ***dependence*** of animals on fungi as a food resource varies between species, but even mammals as large as deer and primates benefit by supplementing their diet with fungal sporophores (Hanson *et al.*, 2003; Trierveiler-Pereira *et al.*, 2016).

Many invertebrates use fungal mycelium as a major part of their nutrition, with approximately 80% of the tens of thousands of microarthropod species in forest soils being **fungivores** (or mycophagous) that depend on the fungal mycelium as their main or only food source. The many small animals involved have been mentioned in section 12.1 (see Figure 12.11). At the microscopic level, some of the smallest **grazers on fungi** are the soil organisms discussed in section 12.2: collembola, mites, nematodes and larvae of mushroom flies. The integrated nature of mycelial networks ensures that, in many fungi, a morphological reaction occurs in response to grazing that serves to limit the damage caused. This implies that even at this level there is an **evolutionary link** between the grazing animal and the fungus. Indeed, collembola may even be perceived as mutualistic under some circumstances as they sometimes promote extra growth of the fungus as it overcompensates for grazing damage, producing a net increase in biomass, though this happens only when sufficient nutrients are available to the fungus (Bretherton *et al.*, 2006; Ngosong *et al.*, 2014; Põldmaa *et al.*, 2016).

Another factor known to influence the presence of collembola in soil is the presence of **endophytic fungi** in the leaves of grass growing on the soil. Leaf litter infected with fungal endophytes appears to decompose faster than uninfected plant matter, possibly due to the presence of toxins changing the composition of soil communities to produce a higher proportion of **detritivores** (Lemons *et al.*, 2005). Endophytes deter larger animals, such as cattle, from grazing, thereby performing a protective function for their host plants (see section 14.19), but in so doing they are also interacting with the animals by determining the palatability, digestibility or nutritional value of their food source.

Most of the above discussion concentrates on animals using fungi as a nutritional resource, but that last paragraph raises 'host protection' as another function for associations between microbes and animals which has been described as **defensive symbiosis**. All organisms are threatened by aggressive encounters with predators, pathogens and parasites, but it is becoming increasingly evident that microbial symbionts can make important contributions to the arsenal of chemicals the host can use to defend themselves. Many of these defensive compounds are produced by the host animals themselves, but they do take advantage of the immense chemical potential of microbial secondary metabolism and these microbial defensive symbioses are widespread. Unfortunately, most are bacterial symbionts and defensive symbiotic partnerships between animals and fungi are rarely studied. Yet they are a promising research target as fungi have a vast biosynthetic potential and are already a rich source of antibiotics (Flórez *et al.*, 2015).

Another promising research topic that has emerged recently concerns the fungal associates of sessile marine animals: corals, sponges and tunicates. Metagenomic and transcriptomic analyses have revealed residential, as well as transient and variable, fungal communities. The ecological function of fungi in association with sessile marine animals is complex and there is evidence for the full range of detrimental and beneficial interactions between fungi and their marine hosts. Based on evidence from terrestrial ecosystems a rich variety of relationships is likely to be found in sessile organisms of the marine environment. The evaluation of marine animal–fungal symbioses under changing environmental conditions may prove to be critical for predicting marine ecosystem responses to global pollution and climatic change (Yarden, 2014). Section 17.8 will describe a case where the relationship has tipped over into an emerging infectious disease. Our next topic in this chapter, fungus farming (or gardening) by animals, is a classic mutualism, but this time with the fungus encouraging grazing rather than attempting to deter the fungivore.

15.2 ANT AGRICULTURE

Attine ants are a group of more than 200 fungus-growing ant species living in the Neotropics (Central and South America) that have a rather unusual ability to cultivate fungi. Most use leaf-litter debris for fungal cultivation, but the leaf-cutter ants (*Atta* spp. and *Acromyrmex* spp.) cut and collect fresh leaves to grow two genera of fungi, *Leucocoprinus* and *Leucoagaricus* in the Family Agaricaceae (*Basidiomycota*: *Agaricales*) (Mueller & Rabeling, 2008). This ant–fungus mutualism shows how successful, and how complicated, a relationship like this can be (Mueller *et al.*, 2001; Sánchez-Peña, 2005).

Fungus cultivation evolved apparently only once in the attines, about 45–65 million years ago. The ants actively inoculate their nest with the fungus and then cultivate it by providing it with pieces of leaves, pruning the hyphae and removing intruder fungi. As a reward, the fungus provides bundles of specialised hyphae that the ants use as a food source; these are the gardening ants. The ants are engaged in an agricultural activity; they collect fresh leaf biomass to convert it to compost in order to cultivate a particular fungus that then provides the main food source for the nest (Vega & Blackwell, 2005). Schultz and Brady (2008) point out that agriculture is a specialised symbiosis that is known to have evolved in only four animal groups: ants, termites, bark beetles and humans.

Gardening ants, which collect and compost plant material, occur in Central and South American tropical rain forests where they are the dominant herbivore (only humans destroy more trees). They destroy the forests, of course, but also damage crops, devastating tracks up to 30 cm wide. An ant colony may contain several million ants, cover 8 m^2 and may be 1 m deep. Leaf cuttings (several millimetres square) are pulped by ants and the chewed-up plant material + saliva + faeces are turned into compost for a fungal garden which is spawned with mycelium carried from parts of an established garden. The **garden is a monoculture** of a member of Order *Agaricales* which does not normally produce sporophores. Under the care of the ants, the mycelium produces hyphae with swollen tips rich in lipids and carbohydrates (**bromatia**; these are also called **gongylidia**, both words mean 'swollen hyphal tips' although, strictly speaking, bromatia are produced in ant gardens and gongylidia in termite gardens) (Figure 15.1).

These swollen hyphal tips are cropped by ants and are the main food for the larvae; the larvae depend entirely on the fungus to digest the leaves (Bass & Cherrett, 1996). The first genomic studies of the attine leaf-cutting ant genera *Atta* and *Acromyrmex* demonstrated losses of two genes in the arginine synthesis pathway (argininosuccinate synthase and argininosuccinate lyase) although all other (non-leaf-cutting) ants, including species in the same subfamily (*Myrmicinae*) as the attine ants, have functional copies of those genes. When transcriptomes of four different attine ant species (three species of the genus *Sericomyrmex* and *Apterostigma megacephala*) were specifically searched for

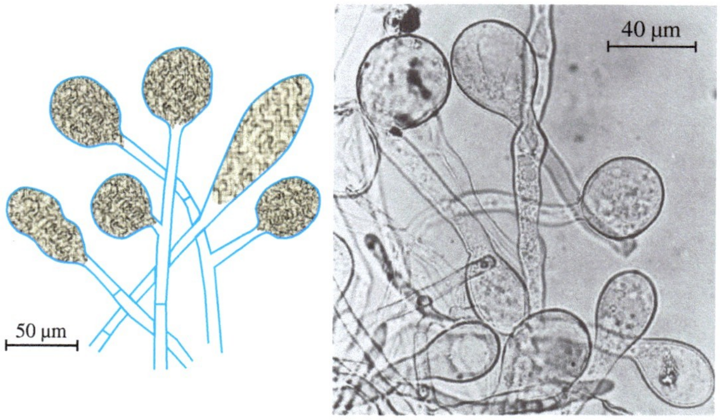

Figure 15.1. Diagram and micrograph of bromatia (inflated hyphal tips) of the ant-cultivated fungus, *Leucoagaricus gongylophorus* (*Agaricales*), which are the swollen tips of hyphae. They provide the main food of the ants because they contain accumulations of carbohydrates, particularly trehalose and glycogen (photograph by Jack Fisher, from Fisher *et al.*, 1994; reproduced with permission of Elsevier). These are also called gongylidia, both words mean 'swollen hyphal tips' although, strictly speaking, bromatia are produced in ant gardens and gongylidia in termite gardens.

genes coding for argininosuccinate synthase and argininosuccinate lyase, they were not found. There is strong genomic and transcriptomic evidence that loss of the ability to produce the amino acid arginine is the reason for the obligate dependence of attine ants on their cultivated fungi. The evidence also strongly suggests that the loss coincided with the origin of attine ants (Nygaard *et al.*, 2011; Suen *et al.*, 2011; Ješovnik *et al.*, 2016).

The fungus must be carried from nest to nest and this is done by newly mated females as part of the mating flight. The nest is started off by a single queen with fungus in the intrabuccal pocket (a filtering device located in the oral cavity of all ants). The new queen mixes the **fungus inoculum** she carries with some suitable plant material and lays eggs on it as soon as the fungus begins to grow. She lays about 50 eggs each day. The first to hatch become workers who eventually establish a nest with a thousand or so interconnected chambers which might be excavated deep into the forest soil and be able to house a colony of several million individuals. When the first workers emerge, foraging and collection of leaves begins. Foraging may be in nocturnal or diurnal cycles depending on species, habitat and environmental conditions.

The workers cut pieces from leaves on trees and carry them back to the nest (Figure 15.2A). Because they usually carry the leaves in their mandibles so that the leaf extends over the ant's head (Figure 15.2B), they are also called **parasol ants**. The caste which produces the largest animals among the several million ants in an average nest is the soldier; a 20 mm long ant which is responsible for protecting the colony and its trails against intruders. The most numerous caste in the colony is the worker caste (these ants are about 8 mm long) which forage in the forest in search of leaves. They can cut leaf pieces bigger than themselves and then **carry them back to the colony**. Leaves are collected along trails within a colony's territory. Trails make it easier to find resources and reduce aggression between neighbouring colonies. Physical trails are lines of cleared ground that increase efficiency of foraging. There are also chemical trails of pheromones that the ants leave to mark the routes to good foraging. Once delivered to the nest, smaller workers, about half the

Figure 15.2. Leaf-cutter ants. Workers (which are about 8 mm long) cut leaves (A), then transport the leaf fragments to their nest (B) (the large size of the leaf fragments they carry is the origin of the alternative common descriptive name 'parasol ants'). In the nest the leaves are used as compost to grow fungus for food as described in the text (shown as the mycelial background in C). Leaf-cutter ant queens (C) are among the most fertile and long-lived queens of all social insects. Photographs by Alex Wild (http://www.alexanderwild.com/). Figure and legend adapted from Mueller and Rabeling (2008); © (2007) National Academy of Sciences, USA.

size of the foragers, chew the leaves into smaller pieces and carry them into brood chambers.

Then the smallest ants, only 1.5 mm long, take over. These are the **cultivators** of the fungus garden; the mushroom farmers. They clean the leaf pieces and then inoculate them with fungus mycelium taken from the existing garden. The ants pluck hyphae from existing mycelia and transfer them to new areas of chewed-up leaf substrate to expand the garden. The cultivators continuously maintain the fungus garden, but they also tend the larvae and the queen (Figure 15.2C). Substrate is usually exhausted and dumped after three or four months. Exhausted substrate is deposited into dump chambers in the nest together with other refuse such as dead ants. Refuse can also be taken from garden chambers and emptied outside the nest.

The fungus cultivated by leaf-cutter ants does not produce spores, but it does have those special hyphal tips (Figure 15.1) that exude a sort of honeydew which the cultivator ants collect and feed to the larvae. The fungus concerned in these associations is always an agaric mushroom fungus but not all of them can be found living free in the forest; specifically, the leaf-cutter ant fungus is always associated with leaf-cutter ant nests. This is a **mutual absolute dependence**.

The demand for leaf material as the colony grows is enormous. In the tropical rain forests of Central and South America, leaf-cutter ants are the dominant herbivores. That 'dominant' label includes the humans of the forest. Around 50 agricultural and horticultural crops and about half that number of pasture plants are attacked. None of this is new, of course. In the last quarter of the nineteenth century leaf-cutting ants were described as **'one of the greatest scourges of tropical America'** and early Brazilian farmers were so frustrated in their battles against leaf-cutter ants that they concluded 'Brazil must kill the ants or the ants will kill Brazil' (Mueller & Rabeling, 2008).

It has been calculated that leaf-cutting ants harvest 17% of total leaf production of the tropical rain forest. Nests located in pastures can reduce the number of head of cattle the pasture can carry by 10–30%. Statistics like this reveal how leaf-cutting ants can become dominant exploiters of living vegetation and how they impact human agriculture. Leaf-cutting ants compete successfully with humans for plant material and are, therefore, counted as **important pests**. Because of their foraging activities, leaf-cutting ants can cause production losses in cacao and citrus fruit of 20–30%. Losses caused by leaf-cutting ants (assuming no control measures are used) could exceed US$1 billion per year and this justifies their being described as a **'dominant herbivore'** (Boulogne *et al.*, 2014).

The combination of a top-of-the-range social insect with a top-of-the-range fungal plant-litter degrader seems to be the key to this success. The social insect has the organisational ability to collect food material from a wide radius around its nest; but the extremely versatile biodegradation capability of the fungus enables the insect to collect just about anything that's available.

The total number of species of trees per hectare in most plant communities increases from the poles to the equator. For example, coniferous forest in northern Canada will have 1–5 species ha^{-1}, deciduous forest in North America, 10–30, but tropical rain forest in South America has 40–100 species ha^{-1}.

The tropical rain forest has enormous chemical and physical diversity in its plants and this presents a major problem to herbivores. Most plant eaters have a narrow diet tolerance because evolution has equipped them with only a limited range of digestive enzymes. Plant-eating insects usually only eat one plant. The leaf-cutting ants of the tropical rain forest have, on the other hand, a **very wide breadth of diet**. These ant colonies are able to harvest 50–80% of the plant species around their nests. This is almost entirely due to the broad range degradative abilities of the fungus they cultivate.

This is an **obligate symbiotic association** between ant and fungus. Ant faeces contain nitrogen sources (allantoin, allantoic acid, ammonia and over 20 amino acids), which supplement the compost and are used by the fungus. A proteolytic enzyme produced by the fungus is acquired, accumulated and transported by ants and deposited in faeces; it hydrolyses leaf proteins.

Cellulases produced by the fungus digest cellulose and the products are converted to the fungal carbohydrates that the ants harvest (like other animals, the ants can digest glycogen, but they cannot digest cellulose). The fungus enables the ants to use plant materials as nutrients, ***but*** the ants enable the fungus to use a far wider range of plants than any plant pathogenic fungus can attack. Some plants do protect themselves by producing deterrents which inhibit cutting, pick-up or feeding. These protectants include toughness, production of sticky latex, and a wide range of defensive chemicals.

There is a certain irony in the fact that the tropical rain forest is lush and green because all those mycorrhizal fungi in the roots of the trees give the plants that extra something that enables them to grow with tropical exuberance (see Chapter 14). And then along comes a six-legged army of harvesters to cut down all those lush green leaves. To do what? To feed another fungus, that's what!

Schultz & Brady (2008) analysed **molecular phylogenies** covering the entire fungus-growing ant tribe (Attini); their data are fossil calibrated (so they have a reliable time scale) and involve multiple genes (so they are a reliable sample of the genomes). They showed that ant agriculture originated with the cultivation of several species of fungi over **50 million years ago** in the early Eocene, the second epoch of the Paleogene period in the Cenozoic era (Schultz & Brady, 2008) (see Figure 2.6 for a geological time scale). Ješovnik *et al.* (2016) reported slightly older dates than these, so the best estimate is that about 55–65 million years ago ant agriculture, as practised by the fungus-farming 'attine' ants, arose in the wet rain forests of South America. In contrast, the transition from lower to higher agriculture (see bullet points below) is most likely to have occurred in a seasonally dry habitat, which would have been inhospitable to the independent free-living growth of the fungus cultivated by attine ants. In other words, dry habitats provided selection pressure favouring fungal cultivars specialised for growth in the ant gardens (Branstetter *et al*., 2017).

Most of the fungi (generally described as 'leucocoprineaceous fungi') belong to the tribe Leucocoprineae, which forms a large clade in the *Agaricales* (*Basidiomycota*) featuring the genera *Leucoagaricus* and *Leucocoprinus* (Mehdiabadi *et al*., 2012). Approximately 60 million years ago, the attine ants split into Paleoattine and Neoattine sister clades and during the past 30 million years, three major ant agricultural systems have diverged from the original Paleoattine system, each involving distinct leucocoprineaceous fungal cultivars. Schultz & Brady (2008) categorised ant agriculture into five different types as follows:

- **Lower agriculture** is practised by 76 species of ants in the majority of attine genera, including the most primitive, which cultivate a wide range of fungal species. These are the Paleoattine ants, and all the fungi involved in this category are still able to grow in nature independently of ant garden cultivation.
- **Coral fungus agriculture**, which evolved about 15 million years ago, is employed by a few ants in the ant genus *Apterostigma* that cultivate a clade of coral fungi (Family *Pterulaceae*; which is a small, mainly tropical, family of wood- and leaf-decomposing fungi) closely related to the fungal genera *Pterula* and *Deflexula* (Munkacsi *et al*., 2004).
- **Yeast agriculture** evolved about 20 million years ago and is used by a few species in the ant genus *Cyphomyrmex* which cultivate a distinct clade of dimorphic leucocoprineaceous fungi that grow wild as mycelia but as yeast morphs when associated with ants, the gardens being clusters of small, irregularly shaped nodules about 0.5 mm diameter composed of single-celled yeasts (even though the yeast mode of growth is unusual in the *Agaricales*).
- **Generalised higher agriculture** is utilised by species in the genera of non-leaf-cutting 'higher attine' ants (*Sericomyrmex* and *Trachymyrmex*), which cultivate another distinct clade of leucocoprineaceous fungi so adapted to garden life that they are not found outside the ant nest (though they can fruit from nests reared in the laboratory).
- **Leaf-cutter agriculture**, evolved from within the higher agriculture ants between 8 and 10 million years ago, and is practised by species of ecologically dominant ants in the genera *Atta* and *Acromyrmex* that cultivate a single highly derived species of fungus, which has been identified as *Leucoagaricus gongylophorus* (Pagnocca *et al*., 2001).

A feature of molecular studies is that the phylogenetic relationships of the attine ants are mirrored in the phylogenetic patterns among their fungal cultivars (they are said to be congruent) and the congruence might even extend to a fungal parasite of the gardens. Currie *et al*. (2003) illustrated the relationship, as shown in Figure 15.3. The evolution of mutualist partners can take two extremes (or something between):

- Coevolution may be tightly integrated, that is, an evolutionary change in one partner results in a simultaneous evolutionary change in the other.
- At the other extreme, frequent switching between partners can weaken the associations between any given pair of partners, causing a diffuse association between species.

Extensive phylogenetic data indicate that these ant–fungus partnerships have been stable over evolutionary time spans of several millions of years. This historical pattern of long-term ant–fungus fidelity provides the opportunity for coevolution, in which evolutionary modification in an ant host can prompt modification in its fungal symbiont, and vice versa. Overall, it seems that attine ants and their fungal partners represent the full spectrum from 'one-to-one' coevolution among the leaf-cutter agriculture group that exhibit coevolved modifications, to something more like diffuse ('one-to-many') coevolution in the (more ancient) lower agriculture groups; with a narrowly diffuse ('one-to-few') coevolution in the *Cyphomyrmex* ant clade (Mehdiabadi *et al*., 2012).

As Figure 15.3 indicates, the association may be even more complex than described so far. There is a common (fungal) parasite of the fungus garden, called *Escovopsis*, and it has been claimed that this too shows coevolution with the ants and their cultivars of fungi in a tripartite evolutionary model (Currie *et al*., 2003). However, analysis of a much greater range of specimens has found that gardens from different clades of ants can sometimes be infected by closely related strains of *Escovopsis*, suggesting that the alleged tripartite coevolution may be spurious. It seems that *Escovopsis* shows much greater diversity than previously thought and has a correspondingly more complex evolutionary history (Meirelles *et al*., 2015). *Escovopsis* (*Ascomycota*: anamorphic *Hypocreales*) is a known parasite of free-living fungi, the ancestors of which were probably associated accidentally with the ancient fungal cultivars. *Escovopsis* has evolved and has now emerged as a **specialised parasite of ant garden fungal cultivars**. *Escovopsis* is a parasite of filamentous hyphae; it does not infect yeast gardens. Avoidance of parasitism by *Escovopsis* may have been part of the selection pressure favouring yeast growth in the ant–yeast agriculture mutualisms.

The interactions and dependencies go even further than described so far, because **yet another organism** is involved. The ants have developed a symbiotic relationship with a filamentous **actinomycete bacterium** of the genus *Streptomyces*. Areas of the cuticle of fungus-growing ants are coated with what appears to the naked eye to be a powdery, whitish-grey crust which is actually formed from masses of *Streptomyces*. Actinomycetes are common organisms, mostly soil-dwelling, that produce many secondary metabolites with antibacterial or antifungal properties (most of our own clinical antibiotics are actinomycete metabolites, and many are from the genus *Streptomyces*).

The *Streptomyces* associated with fungus-growing ants produces **antibiotics** that suppress growth of the parasite

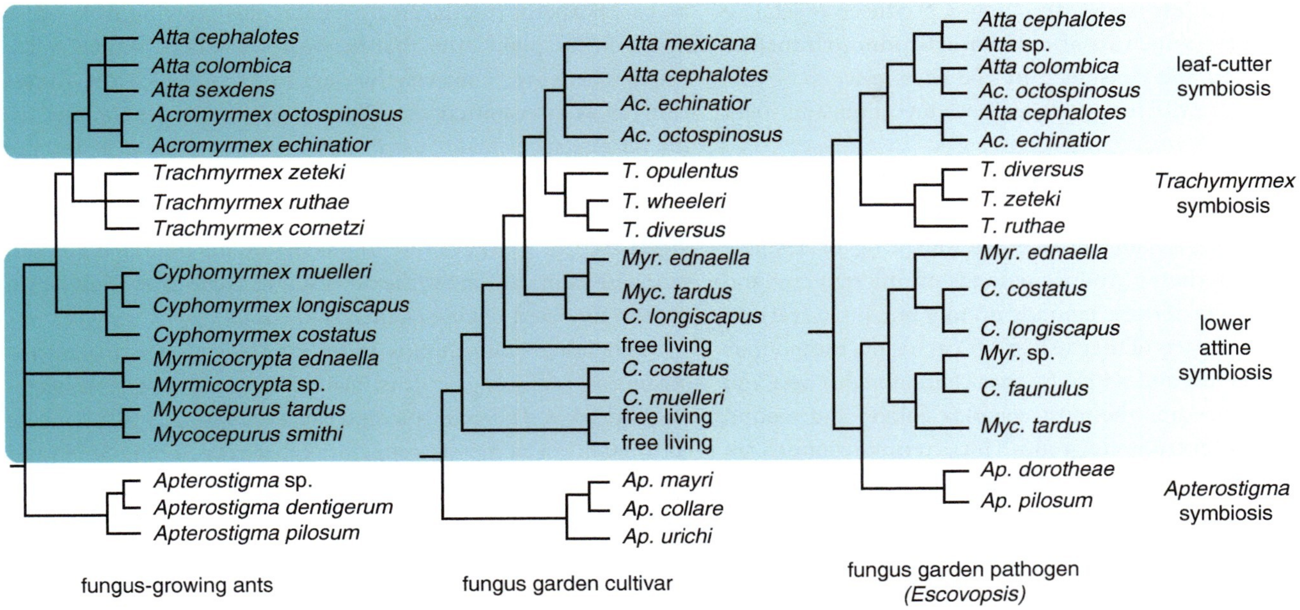

Figure 15.3. Possible phylogenetic trees of attine ants and their associated fungi. This figure shows a phylogenetic reconstruction of the coevolution of fungus-growing ants (based on the genetic sequences of the ants), their garden fungal cultivars (based on the genetic sequences of the fungi isolated from the gardens), and the fungus garden pathogen *Escovopsis* (based on the genetic sequences of *Escovopsis* isolated from the gardens). The different fungus garden cultivar and *Escovopsis* strains in the cladograms are indicated by the name of the host ant species maintaining the garden from which they were isolated. The similarities in the cladograms of this illustration show the apparent tripartite coevolution of the symbioses, but this may oversimplify a complex *Escovopsis* population (see text). The highly significant similarity* displayed here is that all three phylogenies separate into four major lineages, representing four major evolutionary innovations: (i) the initial lower attine symbiosis between Leucocoprinaceous fungi and the most primitive attine ant species; (ii) the *Apterostigma* symbiosis, involving Tricholomataceous fungi and a clade of ants derived within the genus *Apterostigma*; (iii) the symbiosis, incorporating derived Leucocoprinaceous fungi and ants in the genus *Trachymyrmex*; and (iv) the leaf-cutter symbiosis, involving highly derived Leucocoprinaceous fungi and the well-known leaf-cutting ant species. Illustration redrawn and modified from Currie *et al.* (2003).*Exact correspondence in branching patterns is not expected for several reasons. Overall, it seems that attine ants and their fungal partners are sufficiently diverse to represent the full spectrum from 'one-to-one' coevolution among the leaf-cutter agriculture group that exhibit coevolved modifications, to something more like diffuse ('one-to-many') coevolution in the (more ancient) lower agriculture groups; with a narrowly diffuse ('one-to-few') coevolution in the *Cyphomyrmex* ant clade (Mehdiabadi *et al.*, 2012).

Escovopsis. This actinomycete is associated with all species of fungus-growing ants so far studied, being carried in genus-specific crypts and exocrine glands on the surface of the ant. The bacteria are transmitted vertically (from parent to offspring colonies), and the association of *Streptomyces* with attine ants is both highly evolved and of very ancient origin (Currie, 2001; Mueller *et al.*, 2001; Currie *et al.*, 2003, 2006). Effectively, attine ants use multidrug therapy to maintain their fungal cultivars (Barke *et al.*, 2011).

15.3 TERMITE GARDENERS OF AFRICA

Leaf-cutting ants cultivate their fungus in the Americas, but in the Old World, Africa and Asia, the insect partner that engages in a similar fungus-gardening relationship is a termite. Termites are responsible for the bulk of the **wood degradation** in the tropics. Most of them carry populations of microbes in their guts to digest the plant material and release its nutrients, but termites in the insect Subfamily Macrotermitinae have evolved a different strategy. They eat the plant material to get what nutrition they can from it, and then use their faeces as a compost on which they cultivate a fungus. The fungus belongs to the **mushroom genus** *Termitomyces* (*Basidiomycota*: Family *Lyophyllaceae*) including *Termitomyces titanicus* of West Africa, which produces some of the largest mushrooms you can find, being up to about a metre across the cap. In the termite nest, fungal enzymes digest the more resistant woody plant materials and the fungus becomes a food for the termites.

Termites maintain their fungal cultivar on special structures in the nest, called **fungus combs**, within specially constructed chambers, either inside a nest mound or dispersed in the soil. Workers feeding on dry plant material produce faecal pellets (primary faeces) which are added continuously to the top of the comb, providing fresh substrata into which the fungal mycelium rapidly grows. In a few weeks, the fungus produces vegetative 'nodule' structures, which are **aborted mushroom primordia**. These are cropped and consumed by the termite workers who later consume the entire fungus comb, both mycelium and spent compost.

Nests of fungus-growing termites can have volumes of thousands of litres, may persist for decades and contain millions of sterile workers, which are normally the offspring of a single queen. Different termites produce mounds of different size and shape. Chimney-like termite mounds up to 9 m tall (30 feet) are common in several parts of the bush in Africa. Inside, the mounds have many chambers and air shafts that ventilate both nest and fungus culture; perhaps the most complex colony and mound structures of any invertebrate group. Thus, 'termite mounds are metre-sized structures built by millimetre-sized insects. These structures provide climate-controlled microhabitats that buffer the organisms from strong environmental fluctuations and allow them to exchange energy, information, and matter with the outside world … ' (King *et al.*, 2015).

As we stated above, phylogenomics identifies independent origins of insect agriculture in the three clades of fungus-farming insects: the termites, ants or ambrosia beetles, and dates all of them to the Paleogene period in the Cenozoic era (24–66 million years ago). Fossil fungus gardens, preserved within 25 million-year-old termite nests, have been found in the Rukwa Rift Basin of southwestern Tanzania, and confirm an African Paleogene origin for the termite–fungus symbiosis; perhaps coinciding with Rift initiation and consequential changes in the African landscape (Roberts *et al.*, 2016).

All termite larval stages and most adults **eat the fungus**. The termite queen, 'king' and soldiers are exceptions, being fed on salivary secretions by the workers (Aanen *et al.*, 2002, 2007). The two main symbioses of social insects with fungi, the agricultural symbioses of ants and termites are similar in many respects, but they differ in others. Mutualism with fungi has allowed both ants and termites to occupy otherwise inaccessible habitats that have abundant resources: the attine ants are dominant herbivores of the New World tropics; **fungus-growing termites are major decomposers** of the Old World tropics. However, the fungal cultivars of attine ants rarely fruit and are normally propagated clonally and vertically by being carried by dispersing queens (see above), whereas fungal symbionts of the *Macrotermitinae* often produce sporophores. In the rainy season the termites may take portions of the culture out of the nest mound to fruit on the ground nearby and abandoned nest mounds also produce mushrooms after the termites have left.

These wild sporophores are inferred to be the source of fungal inoculum for new nests; that is, it is assumed that the fungal cultivar is generally a 'horizontal acquisition' because the termite fungal cultivars have a freely recombining genetic population structure rather than being clonally related. This implies that new termite colonies will usually start up without a fungus and then acquire the fungal symbiont through the occurrence of its basidiospores (produced by mushrooms growing from other nests) on the plant litter that the workers collect (Aanen *et al.*, 2002). There are, however, two examples in which the fungal cultivar ***is*** transmitted clonally between nest generations. In the termite species *Macrotermes bellicosus* (via the male termite reproductives) and all species in the genus *Microtermes* (via the female reproductives) the reproductives of one or the other sex ingest asexual spores of the fungus before the nuptial flight and use these as inoculum for the new fungus comb after foundation of their new nest colony (Aanen *et al.*, 2007).

Symbiotic relationships with a wide range of intestinal microorganisms, including protists, methanogenic archaea and bacteria, have played a major role in termite evolution. Plant biomass conversion is a multistage cooperation between *Termitomyces* and gut bacteria, with termite farmers mainly providing the gut compartments in which this fermentation can occur. *Termitomyces* has ability to digest lignocellulose and gut microbes of worker termites primarily contribute enzymes for final digestion of oligosaccharides. Termite gut microbes are most important during the second passage of comb material through the termite gut, after a first gut passage where the crude plant substrate is inoculated with *Termitomyces* asexual spores so that initial fungal growth and lignocellulose decomposition can proceed with high efficiency. All termites rely on gut symbionts to decompose organic matter but the single Subfamily Macrotermitinae evolved a mutualistic **ectosymbiosis** with *Termitomyces* fungi to digest lignocellulose of woody substrates (Varma *et al.*, 1994; Bignell, 2000; Poulsen *et al.*, 2014).

The Macrotermitinae comprises 11 genera and 330 species; 10 of the 11 genera are found in Africa, 5 genera occur in Asia (one of these exclusively) and 2 genera occur in Madagascar. Approximately 40 species of the *Termitomyces* symbiont have been described. Molecular phylogenetic analyses of termites and their associated fungi show that the symbiosis had a single origin in Africa. These data are also consistent with horizontal transmission of fungal symbionts in both the ancestral state of the mutualism and most of the extant taxa. Clonal vertical transmission of fungi in *Microtermes* and *Macrotermes bellicosus* (mentioned above) had two independent origins. Despite these features there was a significant congruence between the termite and fungal phylogenies, probably because mutualistic interactions show high specificity; meaning that **different genera of termites tend to rear different clades of *Termitomyces***.

Fungus-growing termites are **pests because they attack wooden structures**; by eating through the wood they leave a maze of galleries that destroy the strength of the timber. Of the more than 2,300 species of termite in the world, 183 species are known to damage buildings. Termite damage and control costs are estimated at US$5 billion annually in the United States alone (Varma *et al.*, 1994; Su & Scheffrahn, 1998; and see https://www.americanpest.net/blog/post/the-real-facts-about-termite-damage). Insecticides and fungicides can help to control this pest. The chitin synthesis inhibitor, hexaflumuron

(1-[3,5-dichloro-4-(1,1,2,2-tetrafluoroethoxy) phenyl]-3-(2,6-difluorobenzoyl) urea), has proved particularly effective as a slow-acting bait. Termite colonies can be eliminated using less than 1 g of hexaflumuron, which is described in pesticide listings as a systemic (stomach-acting) insecticide, but it will also target chitin synthesis in the fungus and poison the nest that way as well (Su & Scheffrahn, 1998). Hexaflumuron is registered with the US Environmental Protection Agency as a reduced-risk pesticide (one believed to pose less risk to human health and the environment than existing alternatives).

15.4 AGRICULTURE IN BEETLES

The final example of an intimate interdependent association between an insect and a fungus concerns wood-boring beetles in the weevil Subfamilies *Scolytinae* and *Platypodinae*, which are unusual in that the adults burrow into the trunks of living trees that have been under some sort of stress (drought, air pollution, etc.) for feeding and egg-laying. They may also be found in trees that have been recently cut or blown down. Rather than create fungal gardens underground like ants and termites, these female beetles dig tunnels and galleries into wood (Figure 15.4), laying eggs on the tunnel wall and inoculating the wood with fungal material carried from a previous nest. They bring the fungi to their new host tree in one of several small glandular pits on the surface of their body, called **mycangia** or **mycetangia**.

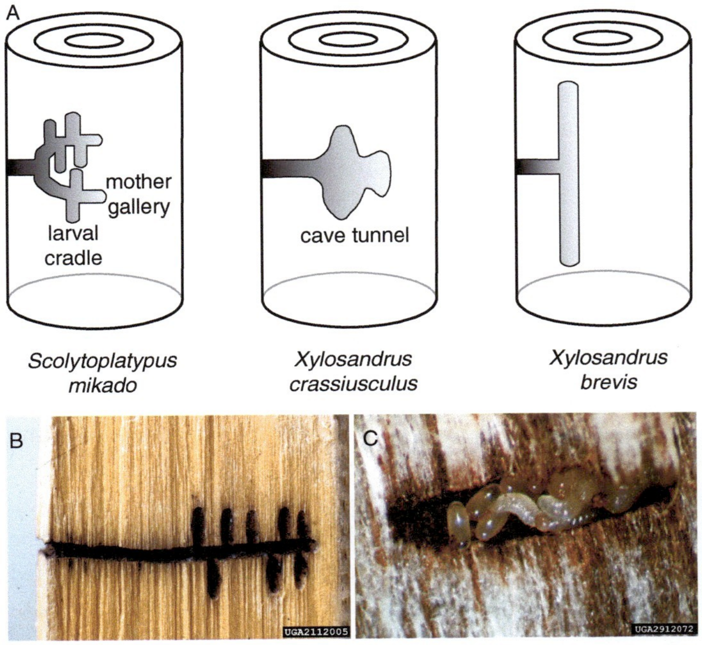

Figure 15.4. Ambrosia beetles are pests which damage timber. (A) Diagrams of gallery structures of typical ambrosia beetles, with the names of beetles forming that type of gallery indicated below the diagrams. (B) Galleries of the striped ambrosia beetle (*Trypodendron lineatum*) in timber of Norway spruce (*Picea abies*) (photograph is image number 2112005 by Petr Kapitola, State Phytosanitary Administration, Bugwood.org). (C) Eggs and larvae of the granulate ambrosia beetle (*Xylosandrus crassiusculus*) (image number 2912072 by Will Hudson, University of Georgia, Bugwood.org). Photographs B and C from Forestry Images (https://www.forestryimages.org/), a joint project of The Bugwood Network and USDA Forest Service; reproduced with permission.

Ambrosia beetles are completely dependent on their fungal symbionts and have deep and complicated pouch-like mycangia. These specialised structures on the body of the animal are adapted for the transport of symbiotic fungal spores and mycelia in pure, often yeast-like, cultures, secreting substances to support fungal spores and perhaps to nourish mycelium during transport. Once introduced, the fungi grow in the wood as mycelia and these help the beetle to evade tree defences, as well as serving directly as food. They are found in many xylophagous ('wood-eating') insects, although, despite this name, the beetles derive most of their nutrition from digestion of the fungi growing on the wood (Filipiak, 2018). In some cases, as in ambrosia beetles, the fungi are the sole food, and the excavations in the wood are only to create suitable microenvironments for the fungus to grow. In other cases (for example, the pine beetle, *Dendroctonus frontalis*), the fungus is carried by mites that ride on the beetles.

By the time the eggs hatch, the fungus will have grown over the tunnel walls, using its enzymes to digest constituents of the wood. This fungus 'lawn' (called, rather fancifully, '**ambrosia**') provides the developing young larvae with a **readily digested food**. Ambrosia beetle larvae do little tunnelling; instead they feed together in chambers on asexually produced fungal conidia induced in cultures kept pure by the parents. Eventually, the larvae pupate and subsequently emerge as adults with a supply of fungus in their mycangia. Because they are 'gratuitous' food for the larvae, the fungi have become known as ambrosia fungi and the insects as ambrosia beetles.

Ambrosia fungi are **derived from plant pathogens** in the group known as ophiostomatoid fungi (that is, they are related to, or belong to, Genus *Ophiostoma*, Order *Ophiostomatales*, Subclass *Sordariomycetidae* and Phylum *Ascomycota*); the fungal cultivars are now classified to the genera *Ambrosiella* and *Raffaelea*. Other beetles in this group are known as **bark beetles** and are associated with free-living, pathogenic ophiostomatoid fungi that aid beetle attack of phloem of trees with resinous defences. In these cases, the beetle is acting as a vector for the disease fungus; we will describe one such association in our discussion of Dutch elm disease in section 16.7.

Ambrosia beetles total about 3,400 species, and many have been found in 30 million-year-old amber, suggesting that the origins of this association occurred in the Tertiary, up to 60 million years ago, roughly contemporaneous with the origin of the attine ant symbiosis. However, in contrast to the single instance of origin of the attine ant symbiosis, there are at least 10 independent instances of the evolution of the ambrosia fungus-gardening association. Adoption of the fungus-gardening habit allows ambrosia beetles to adopt a 'generalist strategy' because the digestive abilities of the fungus widen the range of tree types they can use, wood boring by adults transports the fungi deep within trees, and fungal gardening by the beetle enhances their joint ability to use the biomass-rich resource that the forest trees represent.

Ophiostomatoid fungi emerged about 200 million years ago, soon after the appearance of conifers, which are the principal hosts of pathogenic *Ophiostoma*. This considerably predates the earliest possible origin of the ambrosia beetle–fungus association, 50 million years ago. *Raffaelea* and *Ambrosiella*, the two genera of ambrosia fungi, are both polyphyletic and each arose at least five times from the ophiostomatoid clade that includes the obligate pathogens *Ophiostoma* and *Ceratocystis*. Phylogenetic analysis estimated the oldest origin of fungus farming near to 50 million years ago, long after the origin of the *Scolytinae* subfamily (100–120 million years ago). Younger origins were dated to 21 million years ago. Origins of fungus farming corresponded mainly with two periods of global warming in the Cenozoic era, which were characterised by broadly distributed tropical forests. Hence, it seems likely that warm climates and expanding tropical angiosperm forests played critical roles in the successful radiation of different fungus-farming groups (Jordal & Cognato, 2012).

Today, invasions of non-native bark and ambrosia beetles are a threat to forests worldwide, and the climatic impact implied by the evolutionary story told in the previous paragraph seems likely to be played out again. Study of invasion patterns in the United States reveal differences between bark beetles and ambrosia beetles depending on their differing ecology. Bark beetles are less dependent on climate, which allows them to colonise more areas within the United States, while non-native ambrosia beetles are dependent on higher rainfall and warmer temperatures (Rassati *et al.*, 2016). It is the sudden appearance of pathogenicity in insect–fungus symbioses that is the new and currently uncontrollable threat to forest ecosystems, as well as fruit and timber industries, around the globe. Increasingly, perhaps triggered by climate change, some invasive bark and ambrosia beetle/fungus symbioses are shifting from non-pathogenic saprotrophy in their native ranges to a prolific tree-killing in invaded ranges: new and significant diseases are emerging in the world's forests (Hulcr & Dunn, 2011; Keskitalo *et al.*, 2018; Pasanen *et al.*, 2018).

15.5 ANAEROBIC FUNGI AND THE RISE OF THE RUMINANTS

We point out above Schultz and Brady's (2008) statement that agriculture is a specialised symbiosis that has evolved in the four animal groups: ants, termites, bark beetles and humans. So far, we have dealt with ants, termites and bark beetles and, though we do not intend to deal at any length with human agriculture, most of the rest of our topics do have some relevance to farms and agriculture. In this section, we will expand on the symbiotic **association between chytrid fungi and ruminant mammals.** This is crucial to human agriculture because the diet of grazing farm animals, consisting predominantly of plant structural carbohydrates, such as lignocellulose, cellulose and hemicellulose, can only be digested because the animals have evolved to rely on symbiotic microorganisms in their alimentary tracts to hydrolyse these compounds under the anaerobic conditions that prevail in the gastrointestinal tract.

Ruminants are well adapted to achieve maximum digestion of the otherwise indigestible fibrous components of plant foods within the rumen. The most characteristic behaviour pattern of all ruminants is the regurgitation, rechewing and reswallowing of partially fermented food from the foregut, which is termed **rumination**. Foregut fermenting mammals also produce two enzymes, stomach lysozyme and pancreatic ribonuclease, which accompany and are adaptations to this mode of digestion. The microbial community of the ruminant gastrointestinal tract contains the full range of microbes: bacteria and bacteriophage, archaea, ciliate protozoa and anaerobic fungi; and all these at characteristically high species diversity and population densities, and exhibiting complex interactions (Mackie, 2002; Kumar *et al.*, 2015).

We will concentrate on just one component of this complex community: the anaerobic chytrid fungi. These fungi are not exclusively found in the rumen of the animals, but throughout the entire digestive tract. Further, anaerobic fungi have also been recovered from the faeces of ruminants, suggesting that the fungi can enter a resistant stage enabling them to survive desiccation and the oxygenated atmosphere.

Through to the middle of the twentieth century, it was commonly assumed that all fungi required oxygen to survive, which led to the view that the microbial population of the rumen consisted primarily of anaerobic bacteria and flagellate protozoa. This view held until 1975, when the rumen 'flagellate', *Neocallimastix frontalis* was properly described as a ***Chytridiomycete* fungus** even though it was a **strict anaerobe** (Orpin, 1975; Gruninger *et al.*, 2014).

Chytrids are an ancient group of true fungi (see section 3.2). They usually produce uniflagellate zoospores, but some of the obligately anaerobic chytrids produce bi- or multiflagellated zoospores. All chytrid flagella are of the whiplash type, lacking hairs or scales, and are located posteriorly on the zoospore. Chytrids are usually described as aquatic organisms, but in fact they are equally abundant in terrestrial environments and have even been isolated from sand from arid canyons in Arizona, and from the permafrost in the Arctic. Chytrids are saprotrophs or parasites and their real importance lies in their ecological role as decomposers able to digest complex polymers such as cellulose, hemicellulose, chitin and keratin, as well as some of the most recalcitrant materials in the biological world such as lignin and sporopollenin (a complex, highly cross-linked polymer found in the outer wall layers of pollen grains and some fungal spores). As parasites, they exist on/in a wide range of hosts including algae, other fungi, plants, mosses, insects and invertebrates; the first chytrid parasite of a vertebrate, *Batrachochytrium dendrobatidis*, has been found parasitising and killing amphibians (section 17.7). Chytrids are found **throughout the world**, the majority (80%) in temperate regions, although this relative abundance is most likely due to biased collection from these regions; that is, chytrids have been under-collected in tropical and polar regions (Shearer *et al.*, 2007; Fliegerova *et al.*, 2015).

Although they are morphologically like other chytrids, differences are sufficient for the anaerobic chytrids to be placed in their own phylum, which is called *Neocallimastigomycota* (see section 3.3). They appear to be among the most primitive of all the fungal phyla; the *Neocallimastigomycota* being the **earliest diverging lineage** (so the lineage originated long, long before its current hosts, ruminant herbivores, appeared on the geological scene) as a sister group to the rest of the *Chytridiomycota*.

Six genera have been described among rumen fungi, based on number of flagella and sporangial characters, and all are placed in Order *Neocallimastigales*. No sexual stage is known. Anaerobic chytrids may be monocentric (having one centre of growth, producing either a single or several sporangia) or polycentric (having several centres of growth), and the sporangia may have filamentous or bulbous rhizoids and produce multiflagellate or uniflagellate zoospores. The six genera described are: *Neocallimastix, Piromyces, Caecomyces, Anaeromyces, Orpinomyces* and *Cyllamyces*. *Neocallimastix frontalis* was the original isolate from the domestic cow and *Piromyces* species have been isolated from horses and elephants.

Rumen chytrids are the **primary invaders** of freshly ingested plant material in the rumen and, overall, the rumen chytrid biomass can amount to about 20% of the total rumen microbial biomass. Zoospores alight on plant fragments and encyst, forming a thallus with a well-developed rhizoidal system that penetrates the plant material to extract energy by fermentation of its carbohydrate and other polymers within the animal's rumen and intestine. The nucleus of the zoospore is retained in the cyst as it develops into a sporangium, which is cut off from the anucleate rhizomycelium by a septum. Protoplasm in the sporangium is cleaved into uninucleate zoospore initials; eventually, zoospores (with up to 16 flagella per zoospore in *Neocallimastix frontalis*) are formed in the sporangium and eventually released from the apex of the sporangium into the surrounding fluid. *Neocallimastix frontalis* is obligately anaerobic and grows on fragments of grass in the rumen of cattle, sheep and other herbivorous animals including water buffalo, goat and deer (see illustrations in Rezaeian *et al.*, 2004a & b).

When grown in culture *Neocallimastix frontalis* forms an extensive **rhizomycelium**. The fungal mode of growth is what makes the role of the chytrids so crucial. The filamentous rhizoids extend into the plant material secreting the array of enzymes needed to degrade cellulose (animals do not produce their own cellulose-degrading enzymes) and other polymers of the fragments of plant material that make up the herbivore's food.

The rumen is a dynamic habitat, nutrient rich and oxygen poor. The pH is continuously modified by the host's diet, the metabolic activity of the resident microorganisms and by the tissues of the host. Anaerobic fungi are deficient in mitochondria, and so unable to produce energy by aerobic respiration. Instead, they possess hydrogenosomes that allow a mixed acid fermentation of carbohydrate to be carried out. Mixed acid fermentation converts hexose and pentose sugars to formate, acetate, lactate and ethanol, which the organelle converts to **energy** in the form of ATP, CO_2 **and hydrogen** by producing pyruvate oxidoreductase and hydrogenase.

Subsequently, **methanogenic bacteria** convert the excess H_2 into methane, which is expelled from both ends of the animal. The fungi also produce a wide range of digestive enzymes, giving them broad substrate specificity, and enabling the fungi to transform the core structural polymers of plant cell walls into a variety of simple oligosaccharides, disaccharides, monosaccharides, amino acids, fatty acids, etc., which enter mixed acid fermentation to create energy resources, and other aspects of anabolism that contribute to cell growth, reproduction and population growth. Microbial growth is eventually passed onto the host's stomach where digestion by the animal makes it available as a source of nutrients and energy (Trinci *et al.*, 1994; van der Giezen, 2002; Puniya *et al.*, 2015).

The microorganisms within the rumen form both cooperative and competitive interactions, producing a **complex ecosystem**. Some of the interactions are purely competitive. Ciliate protozoa are a major component of the rumen microflora. Ciliates ingest fungal zoospores as well as bacteria, and their predation of the fungal population can reduce overall cellulolytic activity. Bacteria like *Ruminococcus albus* reduce the ability of *Neocallimastix frontalis* to digest the xylan of barley straw, maize stem and wheat straw, compared to fungal monocultures. The bacteria secrete extracellular factors that inhibit fungal xylanases and cellulases. Some species of fungi produce inhibitors effective against bacteria. These characteristics seem to be simply an expression of the competition between the organisms. However, the overall degradation of plant material is greater when the fungi and bacteria are working together than when they are working individually.

Some of the interactions extend the mutualisms; for example, the methanogenic bacteria are the primary hydrogenotrophs in the rumen ecosystem and their activity enables the chytrids to work more efficiently. Even low levels of free hydrogen inhibit the action of hydrogenase, yet this enzyme is crucial to fungal metabolism, so accumulation of hydrogen results in a decrease in carbon flow and an increase in inhibitory products such as ethanol and lactate. The methanogenic bacteria use the hydrogen in the rumen, releasing the hydrogenase enzyme from inhibition. The result of methanogenic bacterial activity is an increase in carbon flow through the chytrids, and, incidentally, increased production of H_2.

Consequently, the **methanogenic bacteria and fungi are synergistic**. Together they carry out a more efficient fermentation process, consequently releasing a higher biomass yield from the food, generating a larger microbial community and greater benefit to the host. We now have a **tripartite mutualism**: mammal–chytrid–methanogen. If we add the farmer who manages the pasture land for his cattle, the butcher who prepares the meat, the restaurateur who turns the meat into a meal, and the diner who eats that meal, the range of mutual dependencies increases even further.

A newly born ruminant does not possess this microbiota; instead it must acquire the anaerobic fungi, bacteria and protozoa that would normally inhabit a mature animal. This colonisation is achieved rapidly, before the rumen becomes functional. It is brought about by accidental exposure to faeces in its pasture that contain resistant stages of the chytrids. Also, fungi are present throughout the alimentary canal of ruminants, including the mouth and throat, suggesting that saliva is a likely vehicle for inoculation through licking and grooming of the infants by their mother, and interactions with other juveniles. Air samples have also been found to contain several species of anaerobic fungi, suggesting the possibility that aerosols act as another route for transmission (especially likely in large herds or flocks of individuals such as occur in intensive farming).

High-efficiency fermentation is achieved by larger mammals in two different ways.

- The first, termed **hindgut fermentation**, occurs in non-ruminant herbivores, which possess an enlarged area of the hindgut, usually the caecum, where fermentation takes place long after the initial gastric digestion in the stomach.
- The second, referred to as **foregut fermentation**, applies in ruminants. These large animals provide accommodation for their microbial partners in a stomach which is modified into four chambers. The three forestomachs, sometimes considered to be elaborations of the oesophagus though other authorities consider them derivatives of the stomach, comprise: the rumen (by far the largest), reticulum and omasum. The true stomach or abomasum then follows. This is the only site in the digestive tract that produces acid and digestive enzymes (pepsin and rennin).

In the new-born calf, the abomasum makes up about 80% of the total stomach volume, while in the mature cow it amounts to only 10%. During the first weeks of life, when the animals are still suckling milk, the rumen is not functional: the suckled milk does not pass through it due to closure of the oesophageal groove by reflex action. Its relative proportions are considerably smaller than in the adult and some of its rumen wall villi, which serve for absorption of nutritional components, have not yet developed. Changes in the structural and physiological properties of the rumen with age are associated with development of the rumen microorganisms, as their fermentation products are important for the development of the wall villi. The rumen of new-born animals is rapidly colonised by aerobic and facultatively anaerobic microbial taxa close to birth, which are gradually replaced by exclusively anaerobic microbes (Jami *et al.*, 2013). In the mature dairy cow, total volume of the stomach is about 130 litres (human monogastric stomachs generally have a volume of about one litre) and in the cow these organs, collectively, occupy almost 75% of the abdominal cavity.

Taxonomically, a ruminant is a mammal of the Order ***Artiodactyla*** (even-toed ungulates); the anatomical features just described are exemplified by cattle, sheep, goats, giraffes, bison, yaks, water buffalo, deer, wildebeest and various antelopes. All of these are placed in Suborder ***Ruminantia***. Other animals, also generally called ruminants, have slightly different forestomach anatomy; camels, llamas, alpacas, vicunas have a reduced omasum and are occasionally referred to as pseudoruminants or as having 'three stomachs' rather than four. These are placed in Suborder ***Tylopoda***. This drastic adaptation of the alimentary canal is the 'evolutionary investment' that the animals have made in this mutualism. It provides the microbes with a steady supply of freshly cropped plant material and a warm safe habitat in which they can digest the supplied food matter; in return the animal gets a high-efficiency plant cell wall digester.

Piromyces and *Caecomyces* have been isolated from the horse and donkey (both of which are in Genus *Equus* in Order ***Perissodactyla*** or odd-toed ungulates), and Indian elephant (Order ***Perissodactyla***). These animals are examples of the non-ruminant herbivores in which **hindgut fermentation** occurs. Hindgut fermentation provides the host with sources of energy and a range of nutrients that the microbes extract and make available from the digested plant materials, but because it takes place downstream of the stomach the lack of subsequent digestive processes means that the benefit is limited.

Hindgut fermentation is effectively a 'downstream recovery' process; it offers an evolutionary advantage to the animal by scavenging some of the nutritional value from the food that would otherwise be lost. But it is a relatively low-efficiency system: for example, elephants spend 16 hours per day collecting plants for food (about half is grasses, and they browse for other leaves, shoots, roots, fruits, etc.), but 60% of that food leaves the elephant's body undigested. In comparison, the ruminant strategy offers a high-efficiency nutrient extraction process since fermentation occurs for an extended time because of rumination, and because the products of fermentation enter the stomach where the digestive fluids of the host animal can digest both fermented plant materials ***and*** the extremely large populations of microbes.

The symbiotic relationship between *Artiodactyla* and chytrids enabled the animals to incorporate difficult-to-digest grasses into their diet, and the efficiency of ruminant digestion together with expansion of the grasslands gave the artiodactyls the opportunity to become the **dominant terrestrial herbivores** throughout the world in the most recent epochs. However, the story starts long before the evolution of grasses. In fungal phylogenies the *Neocallimastigomycota* emerge as the earliest diverging lineage of the chytrid fungi (James *et al.*, 2006). We take this to mean that these fungi have existed on Earth, presumably as saprotrophs in anaerobic niches like muds and stagnant pools, since before herbivorous animals of any sort emerged.

Fossilised flagellated fungi have been reported from the Precambrian, but the identification of the material is disputed. On the other hand, chytrids are 'probably the most common microbial element' (Taylor *et al.*, 2004) in the Devonian Rhynie chert, which is 400 million years old. Several arthropod groups, including mites and collembolans are also well represented in the Rhynie chert even though it is best known for its plant communities. Consequently, this excellently preserved fossil record demonstrates that the chytrids and other fungal classes, **and** their associations with plants and microarthropods of the day were well established by about the middle of the Paleozoic era (see Figure 2.6) (Taylor *et al.*, 2015; Edwards *et al.*, 2018; Krings *et al.*, 2018).

From that time onwards the fungi were abundant, so any browsing animal that feasted on the community of plants represented in the Rhynie chert would have got a mouthful of saprotrophic microfungi along with their salad. The first mammalian herbivores were most probably fruit and seed eaters (**frugivores**) because the starch, protein and fats stored in fruits and seeds can be more easily digested than the plant fibres in foliage. It is argued that the evolution of large size was a prerequisite for the exploitation of leaves because of the need for a long residence time in the gut for fermentation to extract sufficient nutrients from foliage and herbage (Mackie, 2002). That includes the dinosaur megaherbivores (sauropods), the dominant herbivores throughout the Jurassic, which could only browse on pre-angiosperm plants such as gymnosperms, ferns and fern allies for food (Hummel *et al.*, 2008). It is also argued that the evolution of herbivores (and the microbiota of their guts that digested the plant food) drove the evolution of plant defences against herbivores through the animal's feeding choices (Poelman & Kessler, 2016).

True ***grazing*** animals appeared much later in the Miocene (around 20 million years ago; Figure 2.6) with the radiation of grassland-forming ***grasses*** of the plant family Poaceae (but see below). Plant-eating mammals during the late Cretaceous and early Palaeocene (say, 80 million years ago) were physically small frugivores; mammals did not become herbivores until the Middle Palaeocene (60 million years ago). Herbivore browsers first appeared in the Middle Palaeocene, but they did not become major components of the fauna until the late Eocene (40 million years ago; Figure 2.6). It is envisaged that the earliest herbivores were large, ground-dwelling mammals, reaching their dietary specialisation by evolution from large, ground-dwelling frugivores or by a major size increase from small insectivorous ancestors (Mackie, 2002). It is also argued that hindgut fermentation must have developed first, with foregut fermentation emerging after this initial adaptation of the hindgut, and this seems to reflect the evolutionary appearance of perissodactyls first, followed by artiodactyls.

Grass-dominated ecosystems, including steppes, temperate grasslands and tropical-subtropical savannas, play a central role in the modern world, occupying about a quarter of the Earth's land surface; these ecosystems evolved during the Cenozoic. Grasses are thought to have initially evolved 60 million years ago. The first to appear used the C_3 photosynthetic pathway, but the grasses that dominate the semi-arid savanna are the C_4 grasses (Strömberg, 2011; Oliveras & Malhi, 2016).

The C_3 photosynthetic pathway is the typical one used by most plants. C_4 plants can photosynthesise more efficiently in the higher temperatures and sunlight encountered by savanna grasses because they use water more effectively and have biochemical and anatomical adaptations to reduce photorespiration. There is a good argument for the evolution of the artiodactyls being driven by the development and expansion of savanna and steppe grasslands in Africa and Eurasia (Cerling, 1992; Bobe & Behrensmeyer, 2004).

The appearance of grasslands in Africa and Eurasia during the Eocene epoch, and subsequent spread during the Miocene saw the artiodactyls begin to dominate over the perissodactyls. A credible hypothesis for the **evolution of rumination** in artiodactyls is that it represented a joint adaptation to increasing aridity of the local environment due to climatic cooling and drying.

The Eocene climate was humid and tropical and is likely to have favoured browsers and frugivores (and hindgut fermentation). With the onset of the Oligocene, the climate became generally cooler and drier, and this trend persisted throughout the Tertiary. This increasing aridity, coupled with high sunlight exposure in the equatorial zone, would have favoured the C_4 grasses, and as they became the dominant vegetation the emphasis in herbivore evolution would be to increase the efficiency of the fermentation of the more fibrous plant material. Selection pressure, in other words, against hindgut fermentation and in favour of foregut fermentation and rumination (Mackie, 2002; Bobe & Behrensmeyer, 2004).

The expansion of grassland at the expense of Miocene forests created conditions that were favourable for the evolution of Artiodactyla that could survive aridity and exploit grassland vegetation; changes in the environment drive major evolutionary events, and in this case major changes in bovid abundance and diversity were caused by dramatic **climatic changes** affecting the entire ecosystem (Bobe & Eck, 2001; Franz-Odendaal *et al.*, 2002; Bobe, 2006).

An added twist to the story is that the emergence of the genus *Homo* in the Pliocene of East Africa also appears to be broadly correlated in time with the advent of these same climatic changes and the introduction of the ecosystems they brought about (Bobe & Behrensmeyer, 2004). Since grasslands do currently represent 25% of the vegetation cover of planet Earth; this family of plants (*Poaceae*) is the most important of all plant families to human economy as it includes our staple food grains. The artiodactyls became the most abundant and successful order of current and fossil herbivores, with about 190 species living today.

- The grasses owe their success to the environmental pressures plants faced during the evolution of the savanna grasslands of East Africa.
- The artiodactyls owe their success wholly to their symbiotic relationship with the rumen chytrids.
- Humans found their staple cereal foods among the Poaceae and their main food animals among the ruminants.
- And is it too much to add the claim that it was all made possible by the fungi?

15.6 NEMATODE-TRAPPING FUNGI

Fungi are pathogenic, parasitic or symbiotic with a range of different animals, but their relationship with soil nematodes goes a step beyond parasitism and into predation. There are about 700 species of taxonomically diverse fungi that are able to attack living nematodes (eelworms), which are active animals about 0.1–1.0 mm long. Among these nematophagous fungi, only a few species are obligate parasites of nematodes; the majority are facultative saprotrophs. Nematophagous fungi fall into four general groups: (i) fungi that use specialised structures that trap and then invade eelworms; (ii) fungi that invade eelworms after first immobilising them with toxins; (iii) obligate endoparasitic fungi that invade eelworms following spore germination; (iv) opportunistic saprotrophic fungi that colonise nematode eggs, females or cysts. Nematophagous fungi are natural enemies of nematodes in soil ecosystems and have potential as biocontrol agents against plant- and animal-parasitic nematodes (Jiang *et al.*, 2017).

Over 200 species of fungi (zygomycetes, *Basidiomycota* and *Ascomycota*) catch free-living nematodes in the soil using **traps produced by the fungal mycelium** that adhere to the worm, then penetrate, kill and digest the tissue of the nematode. The most widespread predatory fungi are in the Family Orbiliaceae (*Ascomycota*). Five kinds of trapping device have been recognised (Figure 15.5); the first four of those listed below capture nematodes using an adhesive layer covering part or the entire surface of the hyphal structure (Yang *et al.*, 2007; Su *et al.*, 2017):

- **Adhesive network** (an), the most widely distributed trap, is formed by lateral branches from a vegetative hypha, looping around to fuse with the parent hypha, developing a network of loops with an aperture diameter of about 20 μm (Figure 15.5A). This is a three-dimensional network (see Figure 15.6A) which entangles the prey (Figure 15.7B) and can be formed by germinating conidia (Figure 15.5C).
- **Adhesive knob** (ak) is a morphologically distinct inflated cell that is either a short ('sessile') or long ('stalked') hyphal branch, which are usually closely spaced along the hypha (Figure 15.5B).
- **Nonconstricting rings** (ncr) always occur alongside ak and are produced when lateral branches from a vegetative hypha loop and inflate, forming a three-celled ring on a supporting stalk (Figure 15.5B).
- **Adhesive column** (ac) is a short erect hyphal branch consisting of a few swollen cells (Figure 15.5C).
- **Constricting ring** (cr) is also a looped hyphal branch of (usually) three cells, but it is the most sophisticated trapping device (Figure 15.5D) and captures prey actively. When a nematode enters a constricting ring the three ring cells are triggered to swell inwards within 1–2 seconds and firmly lasso the victim. The cells inflate to maximum size, which is an approximate three-fold increase in cell volume, within 0.1 second, with the swelling of the ring cells being strictly inward (Figure 15.7).

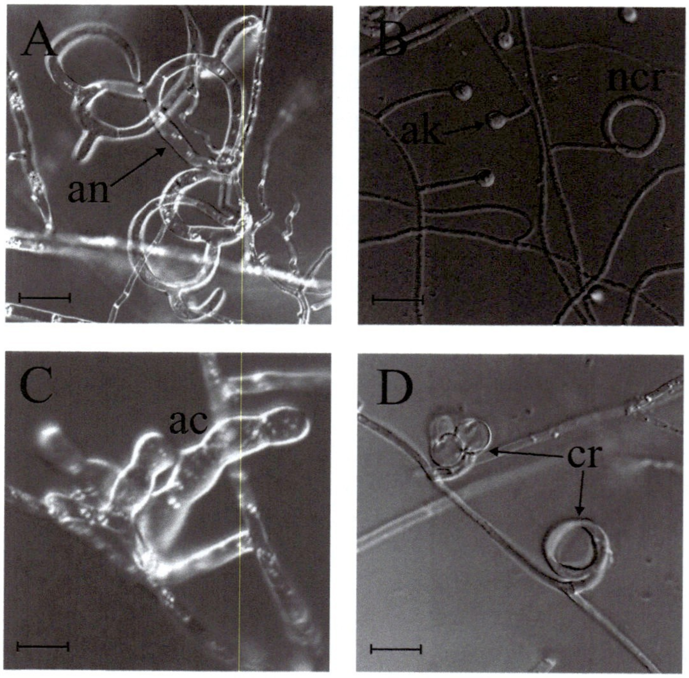

Figure 15.5. Natural nematode-trapping devices. (A) Adhesive network (an), the most widely distributed trap. (B) Adhesive knob (ak) with nonconstricting rings (ncr). (C) Adhesive column (ac) is a short erect branch consisting of a few swollen cells produced on a hypha. (D) Constricting ring (cr), the most sophisticated trapping device, captures prey actively; when a nematode enters a constricting ring, the three ring cells are triggered to swell rapidly inwards and firmly lasso the victim within 1 to 2 seconds. The ring at upper left in panel D has been triggered, that at bottom right is an unsprung trap. Scale bars, 10 μm. Modified from Yang *et al.* (2007) using images kindly supplied by Professor Xingzhong Liu and Dr Ence Yang, Institute of Microbiology, Beijing, China. Original images © (2007) National Academy of Sciences, USA.

Phylogenetic analysis suggests that the trapping structures fall into two lineages: one based on constricting rings and the other using adhesive traps. Only the traps, not the rest of the mycelium, have adhesive layers, and traps have thicker cell walls than vegetative hyphae and their cells contain electron-dense bodies. There is a recognition event between **lectins** (sugar-binding

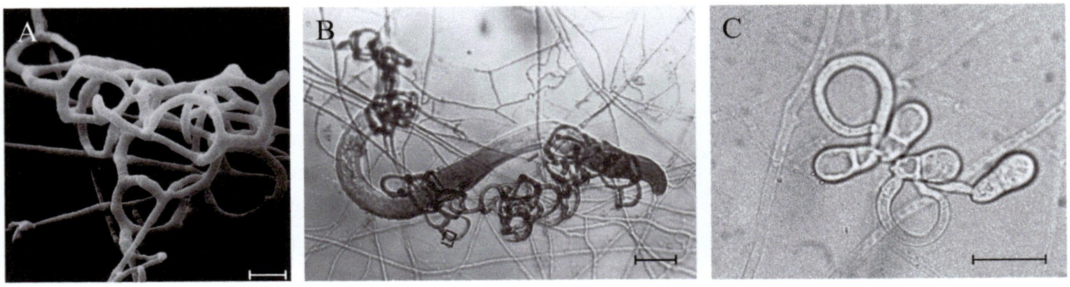

Figure 15.6. Nematode-trapping devices. (A) Scanning electronmicrograph of typical adhesive network trap of *Arthrobotrys oligospora*, bar, 10 μm. (B) Light micrograph of a nematode captured in an adhesive network trap of *Arthrobotrys oligospora*, bar, 20 μm. (C) Light micrograph of conidial traps of *Arthrobotrys oligospora*, induced by the inclusion of peptides in the agar used to germinate the spores, bar, 20 μm. From Nordbring-Hertz (2004); reproduced with permission of Elsevier.

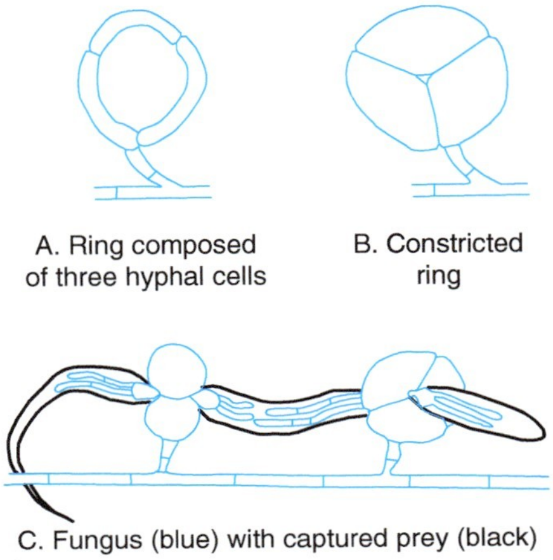

Figure 15.7. Constricting rings of *Drechslerella* snap shut. When a nematode enters a constricting ring the three ring cells (A) are triggered to swell inwards within 1 to 2 seconds and firmly lasso the victim; the cells inflate to maximum size, which is an approximate three-fold increase in cell volume, within 0.1 second, with the swelling of the ring cells being strictly inward (B). The constricted rings clamp the prey firmly (C). Think, for a moment, what that description means in terms of (a) the sensory and signal transduction system that detects the presence of the nematode, (b) the reaction system that generates a burst of metabolic activity to create the osmotic potential and transport the water to inflate the cells and the localised modification of cell wall architecture that directs the morphological expansion to the interior of the loop.

proteins) in the cell wall of the fungus and carbohydrate(s) on the surface of the nematode, among which are the ascarosides. These are lipophilic glycosides of the dideoxy sugar ascarylose, which serve essential functions in regulating nematode development and behaviour, so they are highly specific to the desired prey of the fungus.

Nematophagous fungi, natural predators of soil-dwelling nematodes, can detect and respond to their prey's own ascaroside pheromones (Hsueh *et al.*, 2012). An addendum to this story is that some bacteria can mobilise nematode-trapping fungi to kill nematodes. In their soil habitat, bacteria are consumed by bacterivorous nematodes; however, some of these bacteria release urea, which triggers a lifestyle switch in the fungus *Arthrobotrys oligospora* from saprotrophic to its nematode- predatory, nematode-trapping, form; it seems to be ammonia that the fungus produces from the urea that functions as the signal to form traps. This bacterial defensive mechanism significantly promotes the elimination of nematodes by *Arthrobotrys oligospora* (Wang *et al.*, 2014).

Recognition of the prey results in reorganisation of the **adhesive surface polymer** on the fungus and adhesion of the nematode to the fungus. It also triggers the growth of hyphal branches into the nematode to initiate its digestion. Interaction between predator (e.g. *Arthrobotrys oligospora*) and prey (nematode) shows no species specificity. Hyphae penetrate the nematode within 1 hour of capture. These **hyphae digest the nematode** (Li *et al.*, 2005; Yang *et al.*, 2007; Su *et al.*, 2017).

The capture organs in some predatory nematode-destroying fungi are constitutive, others are inducible. Formation of traps by *Arthrobotrys oligospora* is induced by presence of nematodes or peptides secreted by them; in the absence of nematodes the fungus grows saprotrophically. It is thought that the nematode-trapping habit is a way that saprotrophic soil fungi compensate for the poor nitrogen content of their substrates. Trap formation and nematophagous activity of *Arthrobotrys oligospora* were observed *in vitro* only where conidia were inoculated on nutrient-poor water agar, low-nitrogen medium or a medium containing no amino acids or vitamins.

Trapping devices remain inducible after many years of culture as saprotrophs on artificial media, demonstrating that these highly differentiated trapping structures remain crucial to the survival and virulence of their producer.

Arthrobotrys is characterised by adhesive networks and unstalked adhesive knobs that grow out to form networks; *Dactylellina* by stalked adhesive knobs and nonconstricting rings, and unstalked adhesive knobs that grow out to form loops; and *Drechslerella* by constricting rings.

Predatory fungi appear to have been **derived from non-predatory members of the *Orbiliaceae*** (*Ascomycota*), although this is a lifestyle that has appeared several times and in different fungal phyla. For example, the **oyster mushroom** (*Pleurotus ostreatus*), which is in the *Basidiomycota*, is a **nematode-trapping fungus** that uses adhesive hyphal branches. We recently received this e-mail from Mr John L. Taylor of the North West Fungus Group:

> While I have read about *Pleurotus* species being able to capture nematode worms, I imagined this skill to be limited to hyphae within the woody host material. However, I needed to check a misidentified oyster mushroom, *Pleurotus ostreatus*, so sectioned across cap and 5 gills, finding 15 dead nematodes curled in 2 spaces in the cap tissue. One easily visible had a single cup-shaped hyphal attachment (said to present a nematode toxin), and hyphal growth branching into the nematode from the point of attachment, suggesting ingestion. While this mushroom was structurally sound, it was visually past its edible condition, so those who eat fresh oyster mushroom should not be deterred!

Singh *et al.* (2018) report that *Pleurotus ostreatus* is equally able to immobilise saprotrophic and plant-parasitic nematodes, and so offers the dual prospect of parasitic nematode management in addition to edible mushroom production.

In the orbiliaceous fungi, the adhesive knob is considered to be the **ancestral trapping device** from which constricting rings and networks were derived via two pathways:

- one in which adhesive knobs developed first into adhesive two-dimensional networks, then three-dimensional networks;
- a second pathway in which adhesive was lost and nonconstricting rings developed inflatable cells to form constricting rings (Li *et al.*, 2005; Yang *et al.*, 2007; Su *et al.*, 2017).

Nematode-trapping fungi have been found **fossilised in amber** dated at 100 million years old. The fossil fungi used hyphal rings as trapping devices and are preserved together with their nematode prey. Evidently the predaceous habit of these fungi was well represented in the Cretaceous (the age of the dinosaurs) (Schmidt *et al.*, 2007).

Soil nematodes are very abundant in all soils (commonly millions per square metre) and species diverse (commonly more than 30 taxa). They feed on a wide range of soil organisms within the rhizosphere of agricultural crops and several eelworms are parasitic. Some are important pests of crop plants and farm animals and nematode-trapping fungi may have use in **biological control** (Yeates & Bongers, 1999; Moosavi & Zare, 2012; Moura & Franzener, 2017).

For example, the nematode-trapping fungus *Nematophthora gynophila* can be used to control the cereal nematode, *Heterodera avenae*, which feeds on the host roots damaging the roots and reducing water uptake. Nematodes that are gastrointestinal parasites of farm animals can be controlled with the nematode-trapping *Duddingtonia flagrans*. The resting spores (chlamydospores) can be included in animal feed and survive passage through the animal. Subsequently, the fungus grows in the animal dung where it traps and destroys the parasitic nematode, so reducing pasture infectivity and the worm burden of grazing animals, especially young cattle, sheep and goats. Three months treatment can reduce the worm burden by 90% (Graminha *et al.*, 2005; Larsen, 2006).

15.7 REFERENCES

Aanen, D.K., Eggleton, P., Rouland- Lefèvre, C. *et al.* (2002). The evolution of fungus-growing termites and their mutualistic fungal symbionts. *Proceedings of the National Academy of Sciences of the United States of America*, **99**: 14887–14892. DOI: https://doi.org/10.1073/pnas.222313099.

Aanen, D.K., Ros, V.I.D., de Fine Licht, H.H. *et al.* (2007). Patterns of interaction specificity of fungus-growing termites and *Termitomyces* symbionts in South Africa. *BMC Evolutionary Biology*, **7**: 115. DOI: https://doi.org/10.1186/1471-2148-7-115.

Barke, J., Seipke, R.F., Yu, D.W. & Hutchings, M.I. (2011). A mutualistic microbiome. *Communicative & Integrative Biology*, **4**: 41–43. DOI: https://doi.org/10.4161/cib.13552.

Bass, M. & Cherrett, J.M. (1996). Leaf-cutting ants (Formicidae, Attini) prune their fungus to increase and direct its productivity. *Functional Ecology*, **10**: 55–61. Stable URL: http://www.jstor.org/stable/2390262.

Bignell, D.E. (2000). Ecology of prokaryotic microbes in the guts of wood and litter-feeding termites. In *Termites: Evolution, Sociality, Symbioses, Ecology*, ed. T. Abe, D.E. Bignell & M. Higashi. Dordrecht: Kluwer Academic Publishers, pp. 189–208. ISBN-10: 0792363612, ISBN-13: 9780792363613.

Bobe, R. (2006). The evolution of arid ecosystems in eastern Africa. *Journal of Arid Environments*, **66**: 564–584. DOI: https://doi.org/10.1016/j.jaridenv.2006.01.010.

Bobe, R. & Behrensmeyer, A.K. (2004). The expansion of grassland ecosystems in Africa in relation to mammalian evolution and the origin of the genus *Homo*. *Palaeogeography, Palaeoclimatology, Palaeoecology*, **207**: 399–420. DOI: https://doi.org/10.1016/j.palaeo.2003.09.033.

Bobe, R. & Eck, G.G. (2001). Responses of African bovids to Pliocene climatic change. *Paleobiology*, **27** (Supplement): 1–47. Stable URL: http://www.jstor.org/stable/2666022.

Boulogne, I., Ozier-Lafontaine, H. & Loranger-Merciris, G. (2014). Leaf-cutting ants, biology and control. *Sustainable Agriculture Reviews*, **13**: 1–17. DOI: https://doi.org/10.1007/978-3-319-00915-5_1.

Branstetter, M.G., Ješovnik, A., Sosa-Calvo, J. *et al.* (2017). Dry habitats were crucibles of domestication in the evolution of agriculture in ants. *Proceedings of the Royal Society of London, Series B, Biological Sciences*, **284**: article number 20170095. DOI: https://doi.org/10.1098/rspb.2017.0095.

Bretherton, S., Tordoff, G.M., Jones, T.H. & Boddy, L. (2006). Compensatory growth of *Phanerochaete velutina* mycelial systems grazed by *Folsomia candida* (Collembola). *FEMS Microbiology Ecology*, **58**: 33–40. DOI: https://doi.org/10.1111/j.1574-6941.2006.00149.x.

Cerling, T.E. (1992). Development of grasslands and savannas in East Africa during the Neogene. *Palaeogeography, Palaeoclimatology, Palaeoecology*, **97**: 241–247. DOI: https://doi.org/10.1016/0031-0182(92)90211-M.

Currie, C.R. (2001). A community of ants, fungi, and bacteria: a multilateral approach to studying symbiosis. *Annual Review of Microbiology*, **55**: 357–380. DOI: https://doi.org/10.1146/annurev.micro.55.1.357.

Currie, C.R., Poulsen, M., Mendenhall, J., Boomsma, J.J. & Billen, J. (2006). Coevolved crypts and exocrine glands support mutualistic bacteria in fungus-growing ants. *Science*, **311**: 81–83. DOI: https://doi.org/10.1126/science.1119744.

Currie, C.R. Wong, B. Stuart, A.E. *et al.* (2003). Ancient tripartite coevolution in the attine ant-microbe symbiosis. *Science*, **299**: 386–388. DOI: https://doi.org/10.1126/science.1078155.

Edwards, D., Kenrick, P. & Dolan, L. (2018). History and contemporary significance of the Rhynie cherts – our earliest preserved terrestrial ecosystem. *Philosophical Transactions of the Royal Society of London, Series B*, **373**: article number 20160489. DOI: https://doi.org/10.1098/rstb.2016.0489.

Filipiak, M. (2018). Nutrient dynamics in decomposing dead wood in the context of wood eater requirements: the ecological stoichiometry of saproxylophagous insects. In *Saproxylic Insects, Zoological Monographs*, **Vol. 1**., ed. M. Ulyshen. Cham, Switzerland: Springer Nature Switzerland AG, pp. 429–469. ISBN: 9783319759364. DOI: https://doi.org/10.1007/978-3-319-75937-1_13.

Fisher, P.J., Stradling, D.J. & Pegler, D.N. (1994). *Leucoagaricus* basidiomata from a live nest of the leaf-cutting ant *Atta cephalotes*. *Mycological Research*, **98**: 884–888. DOI: https://doi.org/10.1016/S0953-7562(09)80259-1.

Fliegerova, K., Kaerger, K., Kirk, P. & Voigt, K. (2015). Rumen fungi. In *Rumen Microbiology: From Evolution to Revolution*, ed. A. Puniya, R. Singh & D. Kamra. New Delhi: Springer India, pp. 97–112. DOI: https://doi.org/10.1007/978-81-322-2401-3_7.

Flórez, L.V., Biedermann, P.H.W., Engla, T. & Kaltenpoth, M. (2015). Defensive symbioses of animals with prokaryotic and eukaryotic microorganisms. *Natural Product Reports*, **32**: 904–936. DOI: https://doi.org/10.1039/C5NP00010F.

Franz-Odendaal, T.A., Lee-Thorp, J.A. & Chinsamy, A. (2002). New evidence for the lack of C_4 grassland expansions during the early Pliocene at Laangebaanweg, South Africa. *Paleobiology*, **28**: 378–388. Stable URL: http://www.jstor.org/stable/3595487.

Graminha, É.B.N., Costa, A.J., Oliveira, G.P., Monteiro, A.C. & Palmeira, S.B.S. (2005). Biological control of sheep parasite nematodes by nematode-trapping fungi: *in vitro* activity and after passage through the gastrointestinal tract. *World Journal of Microbiology and Biotechnology*, **21**: 717–722. DOI: https://doi.org/10.1007/s11274-004-4045-8.

Gruninger, R.J., Puniya, A.K., Callaghan, T.M. *et al.* (2014). Anaerobic fungi (Phylum Neocallimastigomycota): advances in understanding their taxonomy, life cycle, ecology, role and biotechnological potential. *FEMS Microbiology Ecology*, **90**: 1–17. DOI: https://doi.org/10.1111/1574-6941.12383.

Hanson, A.M., Hodge, K.T. & Porter, L.M. (2003). Mycophagy among primates. *Mycologist*, 17: 6–10. DOI: https://doi.org/10.1017/S0269915X0300106X.

Hsueh, Y.-P., Mahanti, P., Schroeder, F.C. & Sternberg, P.W. (2012). Nematode-trapping fungi eavesdrop on nematode pheromones. *Current Biology*, **23**: 83–86. DOI: https://doi.org/10.1016/j.cub.2012.11.035.

Hulcr, J. & Dunn, R.R. (2011). The sudden emergence of pathogenicity in insect-fungus symbioses threatens naive forest ecosystems. *Proceedings of the Royal Society of London, Series B, Biological Sciences*, **278**: 2866–2873. DOI: https://doi.org/10.1098/rspb.2011.1130.

Hummel, J., Gee, C., Südekum, K.-H. *et al.* (2008). *In vitro* digestibility of fern and gymnosperm foliage: implications for sauropod feeding ecology and diet selection. *Proceedings of the Royal Society B: Biological Sciences*, **275**: 1015–1021. DOI: https://doi.org/10.1098/rspb.2007.1728.

James, T.Y., Letcher, P.M., Longcore, J.E *et al.*(2006). A molecular phylogeny of the flagellated fungi (Chytridiomycota) and description of a new phylum (Blastocladiomycota). *Mycologia*, **98**: 860–871. URL: http://www.mycologia.org/cgi/content/abstract/98/6/860.

Jami, E., Israel, A., Kotser, A. & Mizrahi, I. (2013). Exploring the bovine rumen bacterial community from birth to adulthood. *The ISME Journal*, **7**: 1069. DOI: https://doi.org/10.1038/ismej.2013.2.

Ješovnik, A., González, V.L. & Schultz, E.R. (2016). Phylogenomics and divergence dating of fungus-farming ants (Hymenoptera: Formicidae) of the genera *Sericomyrmex* and *Apterostigma*. *PLoS ONE*, **11**: article number e0151059. DOI: https://doi.org/10.1371/journal.pone.0151059.

Jiang, X., Xiang, M. & Liu, X. (2017). Nematode-trapping fungi. In *The Fungal Kingdom*, ed. J. Heitman, B. Howlett, P. Crous *et al.* Washington, DC: ASM Press, pp. 963–974. DOI: https://doi.org/10.1128/microbiolspec.FUNK-0022-2016.

Jordal, B.H. & Cognato, A.I. (2012). Molecular phylogeny of bark and ambrosia beetles reveals multiple origins of fungus farming during periods of global warming. *BMC Evolutionary Biology*, **12**: 133. DOI: https://doi.org/10.1186/1471-2148-12-133.

Keskitalo, E.C.H., Strömberg, C., Pettersson, M. *et al.* (2018). Implementing plant health regulations with focus on invasive forest pests and pathogens: examples from Swedish forest nurseries. In *The Human Dimensions of Forest and Tree Health*, ed. J. Urquhart, M. Marzano & C. Potter. Cham, Switzerland: Palgrave Macmillan, an imprint of Springer Nature Switzerland AG, pp. 193–210. ISBN: 9783319769554. DOI: https://doi.org/10.1007/978-3-319-76956-1_8.

King, H., Ocko, S. & Mahadevan, L. (2015). Termite mounds harness diurnal temperature oscillations for ventilation. *Proceedings of the National Academy of Sciences of the United States of America*, **112**: 11589–11593. DOI: https://doi.org/10.1073/pnas.1423242112.

Krings, M., Harper, C.J. & Taylor, E.L. (2018). Fungi and fungal interactions in the Rhynie chert: a review of the evidence, with the description of *Perexiflasca tayloriana* gen. et sp. nov. *Philosophical Transactions of the Royal Society of London, Series B*, **373**: article number 20160500. DOI: https://doi.org/10.1098/rstb.2016.0500.

Kumar, S., Indugu, N., Vecchiarelli, B. & Pitta, D.W. (2015). Associative patterns among anaerobic fungi, methanogenic archaea, and bacterial communities in response to changes in diet and age in the rumen of dairy cows. *Frontiers in Microbiology*, **6**: 781. DOI: https://doi.org/10.3389/fmicb.2015.00781.

Kumpula, J. (2001). Winter grazing of reindeer in woodland lichen pasture: effect of lichen availability on the condition of reindeer. *Small Ruminant Research*, **39**: 121–130. DOI: https://doi.org/10.1016/S0921-4488(00)00179-6.

Larsen, M. (2006). Biological control of nematode parasites in sheep. *Journal of Animal Science*, **84**: E133–E139. DOI: https://doi.org/10.1079/AHRR200350.

Lemons, A., Clay, K. & Rudgers, J.A. (2005). Connecting plant-microbial interactions above and belowground: a fungal endophyte affects decomposition. *Oecologia*, **145**: 595–604. DOI: https://doi.org/10.1007/s00442-005-0163-8.

Li, Y., Hyde, K.D., Jeewon, R. *et al.* (2005). Phylogenetics and evolution of nematode-trapping fungi (Orbiliales) estimated from nuclear and protein coding genes. *Mycologia*, **97**: 1034–1046. DOI: https://doi.org/10.3852/mycologia.97.5.1034.

Mackie, R.I. (2002). Mutualistic fermentative digestion in the gastrointestinal tract: diversity and evolution. *Integrative & Comparative Biology*, **42**: 319–326. DOI: https://doi.org/10.1093/icb/42.2.319.

Mehdiabadi, N.J., Mueller, U.G., Brady, S.G., Himler, A.G. & Schultz, T.R. (2012). Symbiont fidelity and the origin of species in fungus-growing ants. *Nature Communications*, **3**: article number 840. DOI: https://doi.org/10.1038/ncomms1844.

Meirelles, L.A., Solomon, S.E., Bacci, M. *et al.* (2015). Shared *Escovopsis* parasites between leaf-cutting and non-leaf-cutting ants in the higher attine fungus-growing ant symbiosis. *Royal Society Open Science*, **2**: 150257. DOI: https://doi.org/10.1098/rsos.150257.

Moosavi, M.R. & Zare, R. (2012). Fungi as biological control agents of plant-parasitic nematodes. In *Plant Defence: Biological Control. Progress in Biological Control*, **Vol. 12**, ed. J. Mérillon & K. Ramawat. Dordrecht: Springer International Publishing AG, pp. 67–107. ISBN: 9789400719323. DOI: https://doi.org/10.1007/978-94-007-1933-0_4.

Moura, G.S. & Franzener, G. (2017). Biodiversity of nematodes biological indicators of soil quality in the agroecosystems. *Arquivos do Instituto Biológico São Paulo*, **84**: article e0142015. DOI: https://doi.org/10.1590/1808-1657000142015.

Mueller, U.G. & Rabeling, C. (2008). A breakthrough innovation in animal evolution. *Proceedings of the National Academy of Sciences of the United States of America*, **105**: 5287–5288. DOI: https://doi.org/10.1073/pnas.0801464105.

Mueller, U.G., Schultz, T.R., Currie, C.R., Adams, R.M.M. & Malloch, D. (2001). The origin of the attine ant–fungus mutualism. *Quarterly Review of Biology*, **76**: 169–197. URL: http://www.jstor.org/stable/2664003.

Munkacsi, A.B., Pan, J.J., Villesen, P. *et al.* (2004). Convergent coevolution in the domestication of coral mushrooms by fungus-growing ants. *Proceedings of the Royal Society, Series B, Biological Sciences*, **271**: 1777–1782. DOI: https://doi.org/10.1098/rspb.2004.2759.

Ngosong, C., Gabriel, E. & Ruess, L. (2014). Collembola grazing on arbuscular mycorrhiza fungi modulates nutrient allocation in plants. *Pedobiologia*, **57**: 171–179. DOI: https://doi.org/10.1016/j.pedobi.2014.03.002.

Nordbring-Hertz, B. (2004). Morphogenesis in the nematode-trapping fungus *Arthrobotrys oligospora*: an extensive plasticity of infection structures. *Mycologist*, **18**: 125–133. DOI: https://doi.org/10.1017/S0269915XO4003052.

Nygaard, S., Zhang, G., Schiøtt, M. *et al.* (2011). The genome of the leaf-cutting ant *Acromyrmex echinatior* suggests key adaptations to advanced social life and fungus farming. *Genome Research*, **21**: 1339–1348. DOI: https://doi.org/10.1101/gr.121392.111.

Oliveras, I. & Malhi, Y. (2016). Many shades of green: the dynamic tropical forest – savannah transition zones. *Philosophical Transactions of the Royal Society of London, Series B*, **371**: article number 20150308. DOI: https://doi.org/10.1098/rstb.2015.0308.

Orpin, C.G. (1975). Studies on the rumen flagellate *Neocallimastix frontalis*. *Journal of General Microbiology*, **91**: 249–262. DOI: https://doi.org/10.1099/00221287-91-2-249.

Pagnocca, F.C., Bacci, M., Fungaro, M.H. *et al.* (2001). RAPD analysis of the sexual state and sterile mycelium of the fungus cultivated by the leaf-cutting ant *Acromyrmex hispidus fallax*. *Mycological Research*, **105**: 173–176. DOI: https://doi.org/10.1017/S0953756200003191.

Pasanen, H., Junninen, K., Boberg, J. *et al.* (2018). Life after tree death: does restored dead wood host different fungal communities to natural woody substrates? *Forest Ecology and Management*, **409**: 863–871. DOI: https://doi.org/10.1016/j.foreco.2017.12.021.

Poelman, E.H. & Kessler, A. (2016). Keystone herbivores and the evolution of plant defenses. *Trends in Plant Science*, **21**: 477–485. DOI: https://doi.org/10.1016/j.tplants.2016.01.007.

Põldmaa, K., Kaasik, A., Tammaru, T., Kurina, O., Jürgenstein, S. & Teder, T. (2016). Polyphagy on unpredictable resources does not exclude host specialization: insects feeding on mushrooms. *Ecology*, **97**: 2824–2833. DOI: https://doi.org/10.1002/ecy.1526.

Poulsen, M., Hu, H., Li, C. *et al.* (2014). Complementary symbiont contributions to plant decomposition in a fungus-farming termite. *Proceedings of the National Academy of Sciences of the United States of America*, **111**: 14500–14505. DOI: https://doi.org/10.1073/pnas.1319718111.

Puniya, A.K., Singh, R. & Kamra, D.N. (ed.) (2015). *Rumen Microbiology: From Evolution to Revolution.* New Delhi: Springer India. 379 pp. ISBN: 9788132224006. DOI: https://doi.org/10.1007/978-81-322-2401-3.

Rassati, D., Faccoli, M., Haack, R.A. *et al.* (2016). Bark and ambrosia beetles show different invasion patterns in the USA. *PLoS ONE*, **11**: article number e0158519. DOI: https://doi.org/10.1371/journal.pone.0158519.

Rezaeian, M., Beakes, G.W. & Parker, D.S. (2004a). Methods for the isolation, culture and assessment of the status of anaerobic rumen chytrids in both *in vitro* and *in vivo* systems. *Mycological Research*, **108**: 1215–1226. DOI: https://doi.org/10.1017/S0953756204000917.

Rezaeian, M., Beakes, G.W. & Parker, D.S. (2004b). Distribution and estimation of anaerobic zoosporic fungi along the digestive tracts of sheep. *Mycological Research*, **108**: 1227–1233. DOI: https://doi.org/10.1017/S0953756204000929.

Roberts, E.M., Todd, C.N., Aanen, D.K. *et al.* (2016). Oligocene termite nests with *in situ* fungus gardens from the Rukwa Rift Basin, Tanzania, support a Paleogene African origin for insect agriculture. *PLoS ONE*, **11**: article number e0156847. DOI: https://doi.org/10.1371/journal.pone.0156847.

Sánchez-Peña, S. (2005). New view on origin of attine ant-fungus mutualism: exploitation of a preexisting insect-fungus symbiosis (Hymenoptera: Formicidae). *Annals of the Entomological Society of America*, **98**: 151–164. DOI: https://doi.org/10.1603/0013-8746(2005)098[0151:NVOOOA]2.0.CO;2.

Schmidt, A.R., Dörfelt, H. & Perrichot, V. (2007). Carnivorous fungi from Cretaceous amber. *Science*, **318**: 1743. DOI: https://doi.org/10.1126/science.1149947.

Schultz, T.R. & Brady, S.G. (2008). Major evolutionary transitions in ant agriculture. *Proceedings of the National Academy of Sciences of the United States of America*, **105**: 5435–5440. DOI: https://doi.org/10.1073/pnas.0711024105.

Shearer, C.A., Descals, E., Kohlmeyer, B. *et al.* (2007). Fungal biodiversity in aquatic habitats. *Biodiversity and Conservation*, **16**: 49–67. DOI: https://doi.org/10.1007/s10531-006-9120-z.

Singh, R.K., Pandey, S.K., Singh, D. & Masurkar, P. (2018). First report of edible mushroom *Pleurotus ostreatus* from India with potential to kill plant parasitic nematodes. *Indian Phytopathology*, published online December 2018. DOI: http://doi.org/10.1007/s42360-018-0093-0.

Strömberg, C.A.E. (2011). Evolution of grasses and grassland ecosystems. *Annual Review of Earth and Planetary Sciences*, **39**: 517–544. DOI: https://doi.org/10.1146/annurev-earth-040809-152402.

Su, H., Zhao, Y., Zhou, J. *et al.* (2017). Trapping devices of nematode-trapping fungi: formation, evolution, and genomic perspectives. *Biological Reviews*, **92**: 357–368. DOI: https://doi.org/10.1111/brv.12233.

Su, N.-Y. & Scheffrahn, R.H. (1998). A review of subterranean termite control practices and prospects for integrated pest management programmes. *Integrated Pest Management Reviews*, **3**: 1–13. DOI: https://doi.org/10.1023/A:1009684821954.

Suen, G., Teiling, C., Li, L. *et al.* (2011). The genome sequence of the leaf-cutter ant *Atta cephalotes* reveals insights into its obligate symbiotic lifestyle. *PLoS Genetics*, **7**: article number e1002007. DOI: https://doi.org/10.1371/journal.pgen.1002007.

Taylor, T.N., Klavins, S.D., Krings, M. *et al.* (2004). Fungi from the Rhynie chert: a view from the dark side. *Transactions of the Royal Society of Edinburgh: Earth Sciences*, **94**: 457–473.

Taylor, T.N., Krings M. & Taylor, E.L. (2015). *Fossil Fungi.* San Diego, CA: Academic Press, an imprint of Elsevier Inc. 398 pp. ISBN-10: 0123877318, ISBN-13: 978-0123877314. DOI: https://doi.org/10.1016/B978-0-12-387731-4.00014-1.

Trierveiler-Pereira, L., Silva, H.C.S., Funez, L.A. & Baltazar, J.M. (2016). Mycophagy by small mammals: new and interesting observations from Brazil. *Mycosphere*, **7**: 297–304. DOI: https://doi.org/10.5943/mycosphere/7/3/5.

Trinci, A.P.J., Davies, D.R., Gull, K. *et al.* (1994). Anaerobic fungi in herbivorous animals. *Mycological Research*, **98**: 129–152. DOI: https://doi.org/10.1016/S0953-7562(09)80178-0.

Turunen, M., Oksanen, P., Vuojala-Magga, T. *et al.* (2013). Impacts of winter feeding of reindeer on vegetation and soil in the sub-Arctic: insights from a feeding experiment. *Polar Research*, **32**: 18610. DOI: https://doi.org/10.3402/polar.v32i0.18610.

van der Giezen, M. (2002). Strange fungi with even stranger insides. *Mycologist*, **16**: 129–131. DOI: https://doi.org/10.1017/S0269-915X(02)00305-1.

Varma, A., Kolli, B.K., Paul, J., Saxena, S. & König, H. (1994). Lignocellulose degradation by microorganisms from termite hills and termite guts: a survey on the present state of art. *FEMS Microbiology Reviews*, **15**: 9–28. DOI: https://doi.org/10.1111/j.1574-6976.1994.tb00120.x.

Vega, F.E. & Blackwell, M. (2005). *Insect–Fungal Associations: Ecology and Evolution*. Oxford, UK: Oxford University Press. ISBN-10: 0195166523, ISBN-13: 9780195166521.

Wang, X., Li, G.-H., Zou, C.-G *et al.* (2014). Bacteria can mobilize nematode-trapping fungi to kill nematodes. *Nature Communications*, **5**: 5776. DOI: https://doi.org/10.1038/ncomms6776.

Yang, Y., Yang, E., An, Z. & Liu, X. (2007). Evolution of nematode-trapping cells of predatory fungi of the Orbiliaceae based on evidence from rRNA-encoding DNA and multiprotein sequences. *Proceedings of the National Academy of Sciences of the United States of America*, **104**: 8379–8384. DOI: https://doi.org/10.1073/pnas.0702770104.

Yarden, O. (2014). Fungal association with sessile marine invertebrates. *Frontiers in Microbiology*, **5**: article 228. DOI: https://doi.org/10.3389/fmicb.2014.00228.

Yeates, G.W. & Bongers, T. (1999). Nematode diversity in agroecosystems. *Agriculture, Ecosystems and Environment*, **74**: 113–135. DOI: https://doi.org/10.1016/S0167-8809(99)00033-X.

CHAPTER 16

Fungi As Pathogens of Plants

When early humans gave up their nomadic hunter-gatherer existence and turned to agriculture to solve their food problem they would quickly have been challenged by the fungi. Early farmers must have learned very rapidly that crops are very uncertain resources, prone to variations in weather, fire, floods, weeds, insect pests and those troubles that came to be referred to collectively as 'blights' which were due to various sorts of plant disease.

Great plant losses, caused by any of these factors, can be suffered in natural ecosystems but by bringing the crops together into fields in the first place, the early agriculturalist created ideal conditions for the spread of plant disease. And the more selective their farming, the closer their crops came to being true monocultures, the greater the extent of agricultural losses due to any single agency like a particular plant disease, so in this chapter we look at fungi as pathogens of plants.

Fungi are the main disease organisms of plants, being responsible for major losses of world agricultural production. Because of the number that exist, we can only give a few specific examples, so we limit these to the headline crop diseases: the rice blast fungus (*Magnaporthe oryzae*), the bootlace or honey fungus (*Armillaria*), rusts, mildews and smuts (pathogens that produce haustoria that penetrate the plant cells), leaf spot (*Cercospora*), Dutch elm disease (*Ophiostoma*) and black stem rust of wheat (*Puccinia graminis*).

Following these specific examples, we discuss the basics of plant disease; the disease triangle, differences between necrotrophic and biotrophic pathogens of plants, and the effects of pathogens on their hosts. In the final sections of the chapter, we describe how pathogens attack plants, comparing penetration through stomatal openings with direct penetration of the host cell wall, involving both physical and enzymatic penetration of the host. Finally, we discuss the defence mechanisms of plants and the coevolution of disease systems which match the genetic variation of pathogens to that of their hosts.

16.1 FUNGAL DISEASES AND LOSS OF WORLD AGRICULTURAL PRODUCTION

Standing out among the examples of how damaging a crop disease can be is the Irish famine of 1845–1849, which was caused by the failure of the potato crop in Europe because of just one plant disease, the **potato late blight** (caused by a filamentous fungus-like member of the *Oomycota*, *Phytophthora infestans*). This is an astonishing story of how a crop disease affected the structure of our civilisation and our understanding of nature, while causing the deaths of one in eight of the Irish population. It is a story that goes far beyond statistics of number of deaths due to starvation, number of people emigrating, or crop losses and reduction in agricultural yield, and you can read that story in more detail in Chapter 2 of the book *Slayers, Saviors, Servants and Sex. An Exposé of Kingdom Fungi* by David Moore (2000). But it is a piece of our history which we must read about in the knowledge that even today world agriculture suffers significant losses due to plant disease, despite all our scientific advances of the past 150 years. Hopefully, in that time we have learned enough at least to avoid massive calamities like the Irish famine, and today's losses can be reported in terms of monetary losses. But behind each such statistic there must be personal tragedies in which the lives of individuals and families are changed dramatically.

Although **weeds** are the major cause of crop loss on a global scale, major losses are suffered by agricultural crops due to **insect damage** and **plant diseases**. In rounded (approximate) figures, the worldwide annual production tonnage percentage lost to various pests at the start of the twenty-first century have been estimated as follows: losses due to animal pests, 18%; microbial diseases, 16% (and 70–80% of these losses were caused by fungi); weeds, 34%; making a grand total of 68% average annual loss of crop production tonnage (data from Oerke, 2006). Of course, it is not only fungi that cause plant disease (Figure 16.1). There are bacteria, viruses, nematode worms (eelworms), aphids and insects, as well as fungi. Serious plant diseases are caused by all these other pests, but fungi probably cause the most severe losses due to disease around the world. For one thing there are **more plant pathogenic fungi** than there are plant pathogenic bacteria or viruses. One survey made several years ago in the American state of Ohio came up with the estimate that the state had 1,000 diseases of plants caused by fungi, 100 caused by viruses and 50 due to bacteria. Crop protection measures include weed control, which can be managed mechanically or chemically, and the control of animal pests or diseases, which relies heavily on synthetic chemicals. Pesticide use has enabled farmers to modify production systems to increase crop productivity while still maintaining some measure of control over the damaging effect of pests. Unfortunately, despite large increases in **pesticide use**, crop losses have not significantly decreased during the last 40 years.

Crop losses are caused by both biological and physical aspects of the environment that lead to a lower actual yield than the site can be expected to attain (Figures 16.1 and 16.2). The **attainable yield** is the realistic technical maximum under the best achievable growth conditions. It is generally much less than the **yield potential**, which is the theoretical maximum that cannot be reached under practical growth conditions in the field. Crop losses are best expressed as a proportion of attainable yield but sometimes the proportion of the actual yield is calculated. Pests reduce crop productivity in various ways, for example by:

- **reducing the stand** (i.e. the population) of plants (pathogens that kill the host (necrotrophs), like damping-off pathogens that kill seedlings, are examples);
- **reducing photosynthetic rate** (fungal, bacterial, virus diseases);
- accelerating plant senescence (most pathogens);
- shading and 'stealing' light (weeds, some pathogens);
- **depleting assimilates** (nematodes, pathogens, sucking arthropods);
- **consuming tissue** (chewing animals, necrotrophic pathogens);
- **competition** for inorganic nutrients (weeds).

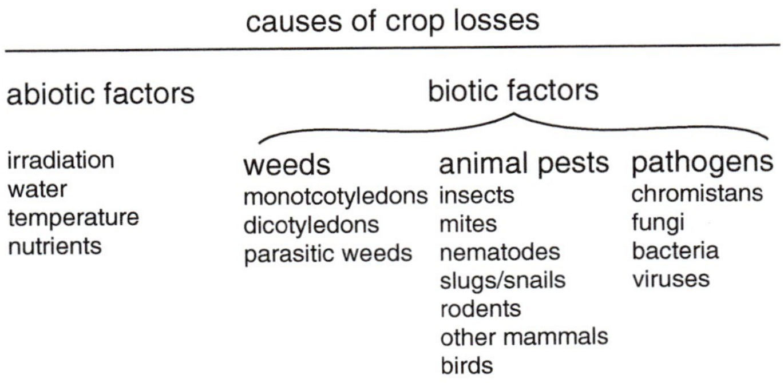

Figure 16.1. Biological and physical aspects of the environment that lead to a lower actual yield than the site can be expected to attain under ideal circumstances. Modified from Oerke (2006).

Crop losses can be quantitative and/or qualitative, and expressed in absolute terms (kg ha^{-1}, or financial loss ha^{-1}, for example) or in relative terms (% loss in production tonnage, for example):

- **quantitative losses** result from reduced productivity giving a lesser yield per unit area;
- **qualitative losses** from pests can result from:
  - reduced content of a normal ingredient(s) of the crop,
  - reduced market quality (for example mis-shaped, blemished fruit and vegetables),
  - reduced storage quality,
  - contamination of the harvested product with pests, parts of pests or toxic products of the pests (for example, mycotoxins).

Agricultural **survey statistics** make it clear that crop losses directly attributable to fungi are very considerable. Of course, it's changing all the time because, at least in part, losses depend on the weather, but it appears that world agriculture sustains average losses (in terms of monetary value) of around 16% annually because of plant diseases. Among crops, the total global **potential loss** due to all pests varied from about 50% in wheat to more than 80% in cotton production. Other estimated **actual losses** are 26–29% for soy bean, wheat and cotton, and 31, 37 and 40% for maize, rice and potatoes, respectively (data from Oerke, 2006).

Crop losses occur at every stage in the food system; in addition to losses during agricultural production, inefficiencies and consumer behaviour and waste all play a role. Alexander *et al.* (2017) found that, due to cumulative losses, the proportion of global agricultural dry biomass production finally consumed as food is only 6% of that actually produced. And we do mean **SIX** per cent; that's not a typographical error. These authors found that the highest rates of loss are associated with livestock production; and although the largest absolute losses of biomass occur **before harvest** (that is, the field losses due to weeds, pests and diseases as indicated above), losses of **harvested** crops were also found to be substantial, with 44% of crop dry matter being lost prior to human consumption. In this survey, over-eating was found to be at least as large a contributor to food system losses as consumer food waste; they conclude 'the findings suggest that influencing consumer behaviour, e.g. to eat less animal products, or to reduce per capita consumption closer to nutrient requirements, offer substantial potential to improve food security for the rising global population in a sustainable manner' (Alexander *et al.*, 2017). It's not just a matter of crop disease.

Efforts to quantify yield losses and identify their causes are still inadequate, and this is especially true for perennial crops, in which this year's disease will have an adverse effect on next year's crop. Cerda *et al.* (2017), researching yield losses caused by pests and diseases in coffee, neatly summarised the situation as follows:

> For some authors, crop loss is the reduction of the crop yield, defined both in terms of quantity and quality, that can occur in the field (pre-harvest) or in the storage (post-harvest) due to biotic or abiotic factors. For others, crop loss

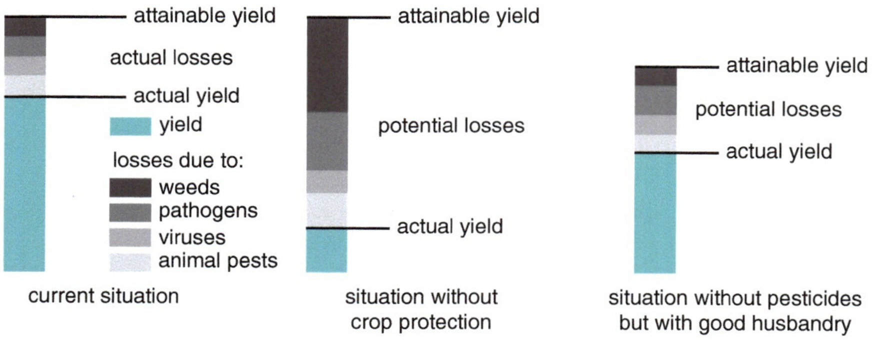

Figure 16.2. Typical crop losses and yield levels estimated with and without various protection regimes. The value of crop protection practices (shown at left as 'current situation') can be calculated as the percentage of potential losses prevented by all the crop protection measures that are employed (compare with centre panel). In contrast, the impact of pesticide uses on crop productivity (right-hand panel) takes into account consequential changes in the agricultural system (for example, use of alternative varieties of the crop, modified crop rotation, reduced fertiliser use), which are provoked by the abandonment of pesticides and which are often accompanied by reduced attainable yield. Redrawn after Oerke (2006).

> also includes the decrease in the value and financial returns of the crop. Furthermore, crop losses comprise primary and secondary losses. Primary crop losses are those caused in the specific year when pest and disease injuries occur; secondary crop losses are those resulting from negative impacts of pests and diseases of the previous year. In annual crops, the inoculum accumulation of pathogens in soil or in seeds and tubers remaining from the previous year can cause secondary losses. These losses however can be avoided by implementing crop rotations or chemical treatments. In perennial crops, premature defoliation or the death of stems and branches caused by leaf injuries lead to loss of vigour and decreased production (secondary losses) in subsequent years. In this case, such secondary losses cannot be avoided since they come from already damaged plants ...

The results of their trials over the years 2013–2015 showed that pests and diseases caused high primary yield losses (26%) and even higher secondary yield losses (38%) (Cerda *et al.*, 2017).

Faced with this overabundance of statistics, it's important to remember that we are talking about 'twenty-first-century agriculture', not some primitive agriculture of the distant past. Today, at this very moment, one in every eight crop plants, on average, will fail to yield because of fungal disease and this includes the positive effect of crop protection policies. Without protection of crops in the developed world, loss of crops would range from 50 to 100% (Figure 16.2). The assessment of crop yield losses is necessary if improvements are to be made in production systems contributing both to the incomes of rural families and worldwide food security and there is increasing interest in the development of pest and disease models to analyse and predict yield losses. Donatelli *et al.* (2017) identified five steps in the simulation of the impacts of plant diseases and pests that need improvement:

- quality and availability of data for input into models;
- quality and availability of data for model evaluation;
- integration with crop models;
- processes for model evaluation;
- develop a community of plant pest and disease modellers.

These authors also point out that climate change is a growing challenge as it is likely to impact on agricultural intensification by altering the ranges, both geographical and host ranges, of pests and pathogens by its effects on the environment (Fones *et al.*, 2017). Further aspects of this discussion are developed in the references cited in Resources Box 16.1.

RESOURCES BOX 16.1 Where to Find More Information About Crop Diseases, Crop Losses and Plant Pathogens

The following papers discuss present and future challenges to agricultural production, including the impact of climate change on the environment.

Bebber, D.P. (2015). Range-expanding pests and pathogens in a warming world. *Annual Review of Phytopathology*, **53**: 335–356. DOI: https://doi.org/10.1146/annurev-phyto-080614-120207.

Beza, E., Silva, J.V., Kooistra, L. & Reidsma, P. (2017). Review of yield gap explaining factors and opportunities for alternative data collection approaches. *European Journal of Agronomy*, **82**: 206–222. DOI: https://doi.org/10.1016/j.eja.2016.06.016.

Fones, H., Fisher, M. & Gurr, S. (2017). Emerging fungal threats to plants and animals challenge agriculture and ecosystem resilience. In *The Fungal Kingdom*, ed. J. Heitman, B.

box continues

Howlett, P. Crous *et al*. Washington, DC: ASM Press, pp. 787–809. DOI: https://doi.org/10.1128/microbiolspec.FUNK-0027-2016.

Juroszek, P. & von Tiedemann, A. (2015). Linking plant disease models to climate change scenarios to project future risks of crop diseases: a review. *Journal of Plant Diseases and Protection*, **122**: 3–15. DOI: https://doi.org/10.1007/BF03356525.

Reidsma, P. & Jeuffroy, M.-H. (2017). Farming systems analysis and design for sustainable intensification: new methods and assessments. *European Journal of Agronomy*, **82**: 203–205. DOI: https://doi.org/10.1016/j.eja.2016.11.007.

Stergiopoulos, I. & Gordon, T.R. (2014). Cryptic fungal infections: the hidden agenda of plant pathogens. *Frontiers in Plant Science*, **5**: 506. DOI: https://doi.org/10.3389/fpls.2014.00506.

van Dijk, M., Morley, T., Jongeneel, R. *et al*. (2017). Disentangling agronomic and economic yield gaps: an integrated framework and application. *Agricultural Systems*, **154**: 90–99. DOI: https://doi.org/10.1016/j.agsy.2017.03.004.

More Information About Plant Pathogens ...

... can be found in the British Society of Plant Pathology's ***Pathogen Profiles***, which are regular features, providing an up-to-date overview of the latest research on a particular pathogen. A list of profile summaries is available at http://www.bspp.org.uk/ (go to <Journals> and then click on <Pathogen Profiles>).

View the Royal Horticultural Society's *Diseases & Disorders* pages at this URL: https://www.rhs.org.uk/advice/plant-problems/diseases-disorders.

Several US universities place their teaching materials online. Among our favourites are:

The ***Master Gardener Resources in Plant Diseases*** at the Department of Plant Pathology of the University of Wisconsin-Madison, USA at this URL: https://pddc.wisc.edu/master-gardener-resources/.

Department of Plant Pathology and Environmental Microbiology at Pennsylvania State University ***Plant Disease Identification and Control*** slide shows, blog posts and extension programmes listed at: https://extension.psu.edu/pests-and-diseases/pest-disease-and-weed-identification/plant-disease-identification-and-control.

The ***Plant Diseases*** pages on EDIS (Electronic Data Information Source) of the University of Florida's Institute of Food and Agricultural Sciences Extension Program at: http://edis.ifas.ufl.edu/topic_plant_diseases.

The ultimate source of ***food and agriculture data*** is the Statistics Division of the Food and Agriculture Organization of the United Nations at: http://www.fao.org/faostat/.

Be warned: there's a LOT of statistical data here. FAOSTAT provides free access to food and agriculture data for over 245 countries and territories and covers all FAO regional groupings with data from 1961 to the most recent year available (which usually means to the year before the current one).

All groups of fungi and fungus-like organisms can cause serious plant diseases and we will expand on some specific examples below. Here, just to indicate the range, we will give you the example of rusts and smuts, which are diseases caused by members of the group of fungi which is the most advanced in evolutionary terms, the *Basidiomycota*; while late blight of potatoes and downy mildew of grapes are diseases caused by the most ancient of fungal-like organisms, belonging to the *Oomycota* in Kingdom *Straminipila*. Diseases such as chestnut blight, peach leaf curl, Dutch elm disease, net blotch of barley, beet leaf spot, apple blotch, maple leaf spot and thousands of others are caused by all those fungi in between these extremes. There is an enormous number of plant disease fungi, so many that there's a rumour about a monastery somewhere in the Himalayas where the monks are listing all the names of plant diseases. When they've entered the last one in their list, the universe ends, and we start all over again. But next time mushrooms rule, OK?

Crop losses due to weeds, pathogens and animal pests can be very substantial (see above) and crop protection measures are needed to safeguard food crops; but not all cultivated crops susceptible to disease are food crops. A disease of the native American chestnut, chestnut blight (caused by an introduced parasite), effectively eliminated a valuable timber and nut-crop tree from the United States. A similar loss happened in the UK when large elm trees were killed by Dutch elm disease (also caused by an introduced parasite but this time the introduction was from the United States and into Europe), although this loss is more difficult to quantify because it is a loss of amenity as much as commercial value. Before the development of fungicides, fungal disease periodically caused massive devastation of crops and consequently mass starvation. The Irish potato famine in the mid-nineteenth century is the prime example; but it was caused by *Phytophthora infestans* (*Oomycota*). So far in this text we have refused to include these fungus-like organisms of Kingdom *Straminipila* as true fungi at all (see section 3.10), but every crop we grow suffers at least one *Phytophthora* disease (Erwin & Ribeiro, 1996) so we cannot ignore these organisms in a chapter about plant diseases. Fungi cause most plant diseases, but they figure in only a minority of animal diseases; these will be discussed in Chapter 17.

16.2 A FEW EXAMPLES OF HEADLINE CROP DISEASES

There are too many plant pathogenic fungi for us to list here, so we will limit our examples to a few that illustrate specific general points that we will develop in more detail later in this chapter. Generally, 'disease' is any physiological departure from the normal health of a plant. Disease may be caused by living (biotic) agents (fungi, bacteria, viruses) or by environmental (abiotic) factors, for example nutrient or water deficiency, anoxia, excessive heat, high ultraviolet radiation or pollution. To protect themselves from damage, plants have developed a wide variety of constitutive and inducible defences. Constitutive defences include many preformed barriers such as cell walls, waxy epidermal cuticles and bark. These substances not only provide the plant with a barrier to invasion, they also give the plant strength and rigidity. In addition to preformed barriers, virtually all living plant cells can detect invading pathogens and respond with inducible defences that include production of toxic chemicals, pathogen-degrading enzymes and deliberate cell suicide. Plants often wait until pathogens are detected before producing such defences because of the high energy costs and nutrient requirements associated with their production and maintenance (Freeman & Beattie, 2008; Faulkner & Robatzek, 2012).

The pages that immediately follow give only a brief description of a mere handful of the numerous fungal diseases of plants. If you would like to get more information on plant diseases we suggest you look at the URLs shown in Resources Box 16.1.

16.3 THE RICE BLAST FUNGUS *MAGNAPORTHE ORYZAE* (*ASCOMYCOTA*)

Magnaporthe oryzae is the most destructive pathogen of rice worldwide, destroying enough rice to feed 60 million people each year. The fungus infects its host by producing an appressorium from the germinating conidium. This has become an experimental system of choice for studying spore attachment, germination and plant surface recognition, the formation of infection structures, and their penetration of host cells and tissues (Knogge, 1998; Martin-Urdiroz *et al.*, 2016; Yan & Talbot, 2016).

Magnaporthe oryzae, which causes the blast disease of rice (*Oryza sativa*), is morphologically indistinguishable from *Magnaporthe grisea*, a species that affects the crabgrass *Digitaria*; but the two species of *Magnaporthe* are taxonomically distinct (Couch & Kohn, 2002). Unfortunately, you will often find **rice blast** referred to as '*Magnaporthe grisea*'; as we do not consider this to be correct we will use the name ***Magnaporthe oryzae*** for the **rice blast fungus**.

Magnaporthe oryzae has also become the main '**model organism**' for studying the molecular aspects of fungal plant disease. The draft sequence of the *Magnaporthe oryzae* genome was published in 2005 (Dean *et al.*, 2005) and subsequently a functional genomics study of pathogenicity revealed many new gene functions required for rice blast disease, improving understanding of the adaptations required by a fungus to cause disease. Research into the molecular genetics of fungal pathogenesis has been stimulated by the availability of increasingly sophisticated molecular analytical methods such as insertional mutagenesis and whole genome, transcriptome, proteome, secretome and metabolome studies (Brown & Holden, 1996; Lorenz, 2002; Jeon *et al.*, 2007; Talbot, 2007).

Jeon *et al.* (2007) used *Agrobacterium tumefaciens*-mediated transformation (AMT; see section 6.10) to create more than 20,000 mutant strains of the rice blast fungus *Magnaporthe oryzae*. This fungus infects plants by forming specialised infection structures, **appressoria**, that penetrate the rice cuticle, leading to the invasion of epidermal and cortical cells and the development of disease symptoms. The insertional mutants (a mutant characterised by the insertion of one or a few nucleotide base pairs to a chromosome) were grown in multiwell plates and analysed for growth, conidiation, appressorium development and pathogenicity in a high-throughput plant infection assay. The resulting data were stored in a relational database and mutants were selected for in-depth phenotyping and molecular characterisation of the mutation; 202 new pathogenicity genes were identified in this study. The fungus is responsible for the most serious disease of rice and is a continuing threat to global food security so there is a large community of researchers trying to develop new disease control strategies. Of special interest is the remarkable ability of the rice blast fungus to invade plant tissue and manipulate the host plant using a battery of secreted **effector proteins** that suppress plant immunity, altering the organisation of the host cells and enabling rapid fungal growth (Martin-Urdiroz *et al.*, 2016; Yan & Talbot, 2016; Zhang *et al.*, 2019).

16.4 *ARMILLARIA* (*BASIDIOMYCOTA*)

Several species of *Armillaria* cause '**honey fungus**' (or 'bootlace fungus') diseases of trees and shrubs. They are extremely aggressive pathogens. One individual clone of *Armillaria ostoyae* has been found in the mixed-conifer forest in the Blue Mountains/Malheur National Forest of northeast Oregon in the United States, which is estimated to be over 1,900 years old and has killed 30% of the Ponderosa pines in its area. This is currently the world's **largest individual organism** and covers 965 ha (9.65 km^2 or 2,384 acres) of forest. The maximum distance between isolates from this 965 ha individual was approximately 3.8 km and use of three estimates of the rate of *Armillaria ostoyae* spread in conifer forests resulted in age estimates for the fungus ranging from a minimum of 1,900 to a maximum of 8,650 years (Ferguson *et al.*, 2003).

Some years earlier, a clone of *Armillaria gallica* had been identified as the largest and oldest living organism (Smith *et al.*, 1992) in a northern Michigan hardwood forest; this individual was 'only' 15 ha (37 acres) in size and 1,500 years old. It is celebrated by the people of Crystal Falls, Michigan in their ***Humongous Fungus Fest*** every August. However, within a few months of the publication of Myron Smith's paper in 1992, Terry Shaw, then in Colorado with the US Forest Service, and Ken Russell, of the Washington State Department of Natural Resources, reported that they had been working on an even larger fungus, *Armillaria ostoyae*, that covered over 600 ha (1,500 acres, 2.5 square miles) south of Mt. Adams in southwestern Washington State (Volk, 2002).

Massive, old fungi are not uncommon and large individuals of *Armillaria* are certainly not limited to the United States. In the Swiss National Park in the Central European Alps *Armillaria ostoyae* individuals averaged 6.8 ha in size (that's 68,000 m^2),

Figure 16.3. Honey fungus or bootlace fungus, *Armillaria mellea*. (A) Mushrooms and rhizomorphs (photograph by David Moore). (B) Rhizomorphs visibly emerging from beneath the bark of a felled log. (C) Close-up of the rhizomorphs shown in (B) (photographs B and C by Elizabeth Moore).

the largest so far reported being approximately 37 ha (Bendel *et al.*, 2006). Basidiomycete mycelia are ubiquitous in forest soils and several studies show that mycelia of many ectomycorrhizal, saprotrophic and pathogenic basidiomycetes can spread vegetatively for considerable distances through forest soil (Cairney, 2005; Bendel *et al.*, 2006).

What enables *Armillaria* to spread rapidly through a stand of trees and bushes, especially in cold and relatively dry conditions, is the combination of its aggressive pathogenicity with its **rhizomorphs**. The multihyphal, highly organised, highly protected structures allow water and nutrients to be transported across the forest floor from the existing food base to support the exploratory, foraging rhizomorph tips as they search for more food bases (Figure 16.3) (see section 10.7). Clearly then, the rhizomorphs contribute enormously to *Armillaria* pathogenicity, and established mycelia of this fungus are essentially permanent, so *Armillaria* root diseases must be ***diseases of the site***. *Armillaria ostoyae* is a native pathogen in North America, with a broad host range, though it is most common and damaging on Douglas fir, grand fir and subalpine fir. Mortality rates are highest in warm, moist habitats but large disease patches can develop even on dry sites, and cold, elevated sites as well. In large areas of northern Idaho and western Montana the range of the pathogen includes most of the best timber-producing sites. The best course of action to minimise crop losses is to select tree species that will survive in *Armillaria*-infested soil (Hagle, 2010).

16.5 PATHOGENS THAT PRODUCE HAUSTORIA (*ASCOMYCOTA* AND *BASIDIOMYCOTA*)

Flax rust (*Melampsora lini, Basidiomycota*) and **barley powdery mildew** (*Blumeria graminis*, also known as *Erysiphe graminis*, *Ascomycota*), share the ability to produce specialised feeding structures called **haustoria** which penetrate the living cells of their host plant. This gives them intimate contact with live host cell membranes through which they can extract nutrients and suppress plant defences. **Smut fungi** (species of *Ustilago*, *Basidiomycota*) are particularly important pathogens of cereal crops. They do not form haustoria but nevertheless make close contact with living host cells through **intracellular hyphae** which invaginate the host cell membrane.

All these pathogens are of major importance to agriculture and comparison of the ways in which they interact with their hosts in terms of cross-recognition and cross-signalling, and the virulence and resistance genes involved in these interactions is seen as a way of improving understanding of the biology of their relationships with their host plants and a way to improve food production (Ellis *et al.*, 2007). Haustoria are the key features of these extremely successful plant pathogens, the obligate biotrophs. The broad phylogenetic range of organisms that produce haustoria suggests that these structures are specific adaptations to close exploitation of their respective host plants that have arisen many times during evolution. Since it became possible to analyse haustorial function at a molecular level a picture began to emerge indicating that haustoria not only function in nutrient uptake but also synthesise and secrete a panel of virulence factors. Proteins called **effector proteins**, which manipulate the physiological and immune responses of host cells, redirecting the host's metabolic flows to the benefit of the pathogen (Voegele & Mendgen, 2003; Garnica *et al.*, 2014).

16.6 *CERCOSPORA* (*ASCOMYCOTA*)

Many species of this genus cause plant diseases, mostly of the **'leaf spot'** variety. Leaf spots are rounded blemishes occurring on the leaves of the infected plants; a typical spot has a defined edge with a darker border and a central zone varying from yellow to brown. Numerous spots can merge together, and the larger areas can be called blights or **blotch** diseases.

Cercospora is not the only cause of **leaf spot diseases** but it causes disease on: alfalfa, asparagus, banana, brassicas, *Cannabis*, carrot, celery, cereals, coffee, cucumber, figs, geraniums, grapes, grasses, hazel, hops, lentil, lettuce, mango, millet, orchids, papaya, peanut, pear, peas, peppers, potato, roses, sorghum, soy bean, spinach, strawberry, sugar beet, sugar cane (the spots merge into stripes; so the disease is called 'black stripe'), sycamore, tobacco, watermelon, and many wild plants and ornamentals.

Cercospora beticola is the most destructive leaf disease of sugar beet worldwide; the disease reduces yield and quality of sugar beet and the need to use fungicide to control leaf spot disease adds significantly to the cost of production. *Cercospora beticola* **overwinters as stromata** in infected crop residues and spores produced on these are the prime source of infection of the leaves of the next season's crop (Khan *et al.*, 2008).

Disease control strategies rely on a combination of fungicide applications (alternating and combining fungicides differing in modes of action to avoid resistance development in the fungus), growing resistant cultivars, and appropriate crop rotation (to avoid the overwintering stromata). Improvement of these integrated pest management (IPM) systems is the main hope, now, for more effective and environmentally sound disease control. Plant varieties with greater genetic resistance is central to this, but pathogen-resistant strains often have lower yields in the absence of the disease, which is unacceptable in commercial practice. Promisingly, yield performance of recently isolated sugar beet varieties with resistance to *Cercospora beticola* have equalled yields of susceptible varieties and should allow reduced fungicide to be use in IPM programmes (Skaracis *et al.*, 2010; Vogel *et al.*, 2018).

16.7 *OPHIOSTOMA (CERATOCYSTIS) NOVO-ULMI* (*ASCOMYCOTA*)

Ophiostoma novo-ulmi (Dutch elm disease or DED) destroyed most (estimated 25 million) mature elm trees in the UK and was responsible for similarly devastating diseases around the world throughout the twentieth century; an epidemic which still rages on today (Anonymous, 2018). The story illustrates several features:

- disease vectors;
- **geographic isolation**, and the creation of 'new' diseases when geographical barriers are breached by accidental introductions through intercontinental trade and commerce; and
- the importance of 'amenity' plants.

This tree disease first came to notice in Holland in 1918 and 1919. Elms died soon after first showing symptoms and mature trees were lost in large numbers. It came to be called **Dutch elm disease** because its true cause was established in the 1920s by Dutch scientists. They found it was caused by a fungus first called *Ceratocystis*. Today we know the fungus as *Ophiostoma* and recognise three species: *Ophiostoma ulmi*, which caused the infection in Europe in 1910 and which reached North America on imported timber in 1928, *Ophiostoma himal-ulmi*, found naturally in the western Himalaya, and the most virulent *Ophiostoma novo-ulmi*, which was first described in Europe and North America in the 1940s and was responsible for devastating infections of elms in both regions from the late 1960s onwards. *Ophiostoma novo-ulmi* could be a hybrid between *Ophiostoma ulmi* and *Ophiostoma himal-ulmi*. *Ophiostoma ulmi* seems to have been brought to Europe accidentally from the Dutch East Indies in South-East Asia during the late nineteenth century.

Symptoms of Dutch elm disease are lack of water and nutrients; described as **wilting**. Leaves first droop and then turn yellow. In a few days to weeks, the leaves are brown and dead, then larger branches begin to die, and most trees are completely dead within 2 years. The reason for this progression is that the plant reacts to the fungus by plugging its own xylem tissue with gum and tyloses (bubble-like plugs in the xylem cavity formed when adjacent parenchyma cells grow into the conducting cells of the xylem) in a vain attempt to stop the spread of the fungus. As more and more channels are plugged the diseased tree is starved of nutrients and water. The rate of progression of Dutch elm disease in the infected tree can be continuously and quantitatively estimated from sap flow measurements using non-invasive monitoring (Urban & Milon, 2014).

But there is more than a fungus involved. Left to itself, the fungus has difficulty passing from tree to tree. In this case, what turns a disease into a raging epidemic is a relationship between the fungus and several species of **elm bark beetles**. Adult females of these insects lay eggs in recently dead elms. Their eggs hatch and young larvae tunnel into the inner bark and outer wood to feed on it. If the tree has been killed by Dutch elm disease, the fungus sporulates in the **beetle tunnels** so the adult beetles that emerge are covered with fungus spores. These are transported to the first tender, young, healthy elm twigs that the young beetles bite into. Elm bark beetles, therefore (*Scolytus multistriatus* in Europe), are vectors of Dutch elm disease. Indeed, the fungal pathogen makes the host trees ***attract*** insect vectors, thereby manipulating its host tree to enhance their appeal to foraging beetles (McLeod *et al.*, 2005; Raffa *et al.*, 2015).

In North America, *Ophiostoma novo-ulmi* relies on both the (accidentally introduced) European elm bark beetle, *Scolytus multistriatus*, and the native elm bark beetle, *Hylurgopinus rufipes*, for transport to new host elms (*Hylurgopinus rufipes* can withstand colder winter temperatures and is the primary vector of Dutch elm disease across the North American prairies). Chinese and Siberian elms are highly resistant to the disease, but those native to Europe and North America are not. The next step in the story is that Dutch elm disease was first found in North America in 1930 in Cincinnati, Ohio.

The evidence indicates that it was introduced on ***elm logs from Europe*** landed at ports on the US eastern seaboard. The American elm had become an important **amenity tree** throughout the continent, being planted in urban sites to such an extent that it's no exaggeration to describe the plantings as urban forests. By 1950, the disease was spreading through 17 states (and into south-eastern Canada). Today, Dutch elm disease occurs wherever American elms grow in North America. Countless millions of trees have been killed, with a corresponding multi-billion-dollar cost of removing and disposing of them and replacing the trees that were lost with new plantings. But the story has yet another twist.

In May 1963 a shipment of ***American elm logs*** was landed in the UK. The logs were infested with both elm bark beetles and *Ophiostoma*, and during the next few years 25 million mature elm trees died in the UK.

There are hopes for **biological control (biocontrol)** of Dutch elm disease (Scheffer *et al.*, 2008) and other fungal pathogens (Massart & Jijakli, 2007). The *Ophiostoma novo-ulmi* genome is providing insights into the phytopathogenicity of the fungus, although the study identified 1,731 proteins potentially involved in pathogen–host interactions, so it might be a long haul (Bernier *et al.*, 2014; Comeau *et al.*, 2015).

Joan Webber, Principal Pathologist at the UK's Forest Research, has pointed out that in Britain 'we probably have more elm trees now than we did before the current epidemic started in the late 1960s. However, they tend to be relatively young elms while magnificent mature elms that graced much of the countryside are more of a rarity' (Webber, 2016).

As there is no cure yet, the key to control of Dutch elm disease is **sanitation**. Dead, dying or weak elm wood must be destroyed to eradicate both fungus and beetle. It goes beyond the aerial parts of the tree, though. The roots of adjacent elms tend to fuse together over time, resulting in a **shared root system** between several trees. This type of root grafting can occur between elms within 15 m of one another. When a single elm tree in such a group becomes infected, the fungus can move down the diseased tree into the roots and then into the next healthy tree through **root grafts**. The sanitation processes must include disruption of these root grafts by digging a trench at least 60 cm deep along a line around the tree and beyond the longest branches of the diseased tree.

To control the Dutch elm disease syndrome there are several potential targets, because the biological control might depend on:

- developing resistant varieties of elm or techniques to induce resistance to disease-causing fungi in susceptible elm; Princeton Elms, bred for resistance to Dutch elm disease in the United States are only mildly resistant, but the 'resista-elm' (sold as *Ulmus* 'New Horizon': http://www.resista-ulmen.com/en/) has a 100% resistance record up to the time of writing;
- organisms (bacteria or fungi) to compete with *Ophiostoma*;
- hyperparasites (that is, parasites of the parasites) of either *Ophiostoma* or the elm bark beetle;
- or processes that prevent successful breeding of the beetles.

If you want to know more about Dutch elm disease you could start by viewing the web pages listed in Resources Box 16.2.

RESOURCES BOX 16.2 Where to Find More Information About Dutch Elm Disease

The following websites will introduce you to the past, present and future challenges presented by Dutch elm disease.

Royal Horticultural Society
https://www.rhs.org.uk/advice/profile?PID=154
Forestry Commission
https://www.forestry.gov.uk/dutchelmdisease
The American Phytopathological Society
https://www.apsnet.org/edcenter/intropp/lessons/fungi/ascomycetes/Pages/DutchElm.aspx
United States Department of Agriculture, National Invasive Species Information Center
https://www.invasivespeciesinfo.gov/microbes/dutchelm.shtml

16.8 BLACK STEM RUST (*PUCCINIA GRAMINIS* F. SP. *TRITICI*) THREATENS GLOBAL WHEAT HARVEST

Wheat stem rust (*Puccinia graminis* f. sp. *tritici*) is historically the most damaging disease of wheat (*Triticum aestivum*). In the worst cases, yield losses of 70% or more are possible. It was at one time the most feared disease of the wheat crop. Feared because what looked like a healthy crop about three weeks before harvest could be reduced by stem rust to a black tangle of broken stems and shrivelled grain by harvest time. The first detailed reports of crop loss to wheat stem rust appeared in Italy in the eighteenth century and two devastating stem rust epidemics in North America in 1904 and 1916 led directly to research that showed that the stem rust pathogen had various races that differed in their ability to infect wheat varieties because the latter carried different combinations of **resistance genes**.

This was followed by a concentration of research on identifying resistance to stem rust and breeding resistant wheat cultivars, and on understanding rust epidemiology and evolution. Another important point is that these efforts also initiated global collaboration between wheat geneticists and plant pathologists. Understanding the biology of the disease revealed the relationship between the common barberry (*Berberis vulgaris*) and *Puccinia graminis*. Barberry is the **alternate host** of *Puccinia* *graminis* and provides the **sexual stage** of the pathogen's life cycle so is the source of new combinations of virulence genes in the pathogen population. In addition, barberry enables the pathogen to **survive through the winter** and therefore serves as the source of new stem rust epidemics on wheat in the spring. Common barberry had been planted widely as an ornamental bush, but **eradication** of the alternate host in North America and Europe was an important step in controlling stem rust epidemics. Between 1918 and 1980, over 500 million barberry bushes were destroyed in the major wheat-producing states in the United States.

The barberry eradication programme was accompanied by efforts to devise **genetic control strategies**. The International Spring Wheat Rust Nursery Program was initiated in 1950 by the United States Department of Agriculture Agricultural Research Service and operated continuously until the mid-1980s. Its objectives were to:

- find new genes or combinations of genes in wheat for field resistance to rusts throughout the world; and
- test new varieties and selections of wheat developed by plant breeders and pathologists for resistance to rusts (Dubin & Brennan, 2009).

About 50 resistance genes in wheat have been catalogued; all but one of which are race-specific, and therefore derive from the **gene-for-gene relationship** between the host plant resistance gene and corresponding pathogen virulence genes (see section 16.17).

All these efforts stabilised the stem rust situation in many countries where modern wheat cultivars were adopted. For example, changes in stem rust races have not been observed in Mexico for almost 40 years and natural infections are non-existent. Similarly, a resistant wheat cultivar released in 1960 in the Indian subcontinent and subsequently grown on millions of hectares, remains resistant to stem rust today. The incidence of stem rust decreased throughout most of the world to almost non-significant levels by the mid-1990s. Unfortunately, as disease incidence declined, there was a corresponding decline in emphasis on research on stem rust and wheat breeding against this disease; and, in 1999, the fungus got ahead of us.

First identified in Uganda during 1999, *Puccinia graminis tritici* **Race Ug99** is the only known race group that has virulence for a wheat resistance gene (*Sr31*, derived from rye, *Secale cereale*) which is incorporated (as a chromosomal translocation) in many of the most widely used wheat cultivars. Race Ug99 (in North American scientific nomenclature, Ug99 is known as race TTKSK) also carries virulence for most of the resistance genes of wheat that have been introduced over the years by selective breeding. This is the headline: Ug99 carries a unique combination of virulence to known and unknown resistance genes of wheat. Consequently, the unique virulence associated with Ug99, or its race group variants, has rendered a large proportion of global wheat cultivars susceptible to disease, irrespective of where they were developed. It is now believed that 80–90% of all wheat varieties grown in developing countries are susceptible to the new race of stem rust and infection can cause yield losses up to 80%.

Since the first detection of race Ug99 in Uganda in 1999, 10 variants in its race group have been detected in 13 countries: Uganda, Kenya, Ethiopia, Sudan, Tanzania, Eritrea, Rwanda, South Africa, Zimbabwe, Mozambique, Yemen, Iran and Egypt (Patpour *et al.*, 2016). Plants recognise pathogens through receptors that identify pathogen effectors secreted by the pathogen into the plant. Wheat genes that confer resistance to Ug99 and other virulent races of *Puccinia graminis* are being identified (Babiker *et al.*, 2016, 2017; Zhang *et al.*, 2017), and two effectors from *Puccinia graminis* f. sp. *tritici* were identified recently that are expected to aid developments of approaches to track the spread of the rusts of wheat and improve understanding of these pathogens (Moscou & van Esse, 2017; and see http://rusttracker.cimmyt.org/ and http://www.fao.org/agriculture/crops/rust/stem/en/).

16.9 PLANT DISEASE BASICS: THE DISEASE TRIANGLE

Plant diseases can be analysed conveniently using the concept called the '**disease triangle**' (Figure 16.4); this arrangement places the three factors that must interact to cause plant disease at the three corners of a triangle. Those three factors are:

- susceptible host;
- disease-causing organism (the pathogen);
- favourable environment for disease.

The **host** is the plant itself; some can fall victim to many diseases, others only suffer particular ones. All plants have a range of susceptibilities to a range of diseases. The **pathogen** is the disease. Diseases of plants are most often caused by fungi but there ***are*** some plant pathogenic bacteria and viruses.

Without the right host in the right conditions, pathogens cannot cause any harm. Some pathogens are specific to only one or a few host plants, others have broad abilities to attack almost everything. The **favourable environment** essentially means the weather conditions needed for a pathogen to thrive (this is an important point; it's 'a favourable environment for disease' and if the pathogen is present and disease results, it's obviously an **unfavourable** environment for the plant).

Disease results only if all of these three things occur simultaneously; if one or more of the factors is not present, then disease does not occur. The disease triangle was probably first recognised at the beginning of the twentieth century and it has become one of the paradigms of plant pathology. It holds a position in plant pathology rather like that held by Ohm's Law (which relates current, resistance and voltage) in electrical and electronic engineering.

It is a paradigm because occurrence of a disease caused by a biological agent **absolutely requires** the interaction of a

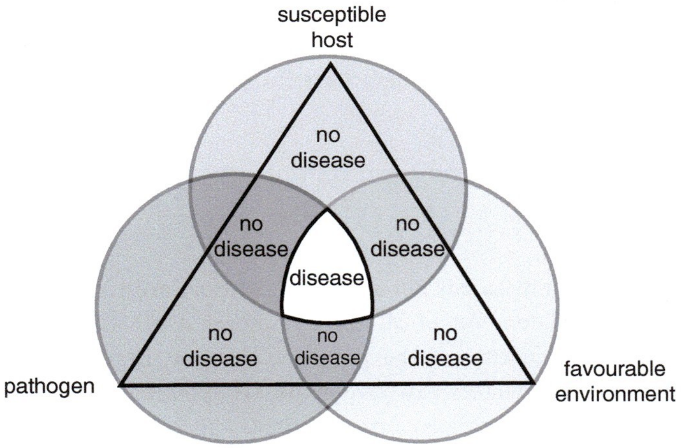

Figure 16.4. The disease triangle illustrating the phenomenon of plant disease as the interior space of a triangle with the three essential factors (susceptible host, favourable environment for disease, and pathogen) at the vertices. Variation in the 'strength' of the contributions of these factors to the relationship ('strength' is indicated by the size of the circles) will quantitatively alter the severity of the disease, which will be shown by a change in the area of the central disease envelope. The diagram is intended to be used dynamically; the static disease triangle allows illustration of the continuum of host reaction from complete susceptibility to immunity, and the degree of pathogen virulence, and the environmental suitability for disease. If any one element is reduced to zero, the triangle transforms into a line and the area occupied by disease collapses to zero. Less dramatic alterations in any factor change the area of the central disease envelope which is an indicator of disease intensity (incidence or severity). For example, a host with some degree of resistance will have a smaller susceptibility circle, and consequently lesser area of overlap and less severe disease. Another example could be a pathogen with greatly increased virulence, which would be shown as a larger 'pathogen circle' and consequently larger area of overlap and more severe disease. Based on a diagram published by the Department of Plant Pathology of the University of Wisconsin-Madison, USA.

susceptible host with a virulent pathogen under environmental conditions favourable for disease development. The mechanisms that contribute to pathogenesis can all be thought of as modifying the disease triangle by reducing or eliminating one of the corners of the triangle. Examples (from among many) include:

- the lack of defences in the host;
- efficient spore dispersal by the pathogen;
- weather conditions favouring spore production, etc.

Methods of disease control (again from among many) include:

- breeding for resistance in the host;
- applying pesticide to hinder the pathogen;
- irrigating to relieve water stress.

It is usually stated that this triangular relationship is unique to plant pathology because the immobility of plants prevents them escaping from inhospitable environments. For example, plants have little thermal storage capacity and are therefore subject to temperature stress much more than animals as even poikilothermic animals can 'bask in the sun' or retire to the shade as appropriate. Further, the immune system of vertebrates arms them with sophisticated mechanisms to recognise and neutralise pathogens. Finally, the predominance of fungi in causing plant diseases is held to reinforce the uniqueness of the plant disease triangle because fungi are also highly dependent on environmental conditions. However, this triangular relationship is only unique to plants if you ignore the fact that **members of Kingdom *Fungi* also suffer disease**, and the severity of that disease also depends on the three essential factors: a susceptible host in an environment favourable for disease challenged by a virulent pathogen.

Some plant pathologists have suggested elaborating the disease triangle by adding additional parameters, such as human activities, disease vectors and time. Humans contribute to the disease triangle because **human activity** in agriculture is pervasive and, if you think about it, impacts on all three factors so far discussed, so can profoundly affect the occurrence and severity of plant diseases. This means that humans are already represented implicitly in the basic triangle configuration and this is the main counterargument against including human activity as a new vertex in a 'disease rectangle'.

Animal and other **vectors** are not essential to all plant diseases even though they play a critical role in many. Vectors are therefore only worth including in those special cases, where the triangular relationship can be modified by placing the vector on the disease triangle side that connects the host and pathogen vertices; this arrangement emphasises the dependence of the pathogen on its vector.

Time is an essential dimension and has been added to the disease triangle by several authors, primarily to convey the idea that disease onset and intensity are affected by the duration that the three prime factors are aligned. Some duration of favourable alignment is necessary for disease to occur; but the length of time depends on your level of analysis. Physiological events in the host that define infection can take place in minutes or hours; disease symptoms in the field can take days or weeks to appear. Showing time as a dimension on the triangle (perhaps converting it into a pyramid) could be a more realistic adaptation of the diagram.

16.10 NECROTROPHIC AND BIOTROPHIC PATHOGENS OF PLANTS

Plant pathogens are often divided into **biotrophs** and **necrotrophs** (and, more recently, **hemibiotrophs**), according to their lifestyles. The definitions of these terms are:

- **biotrophs** derive energy from living cells, they are found on or in living plants, can have very complex nutrient requirements and do not kill host plants rapidly;

- **necrotrophs** derive energy from killed cells; they invade and kill plant tissue rapidly and then live saprotrophically on the dead remains;
- **hemibiotrophs** have an initial period of biotrophy followed by necrotrophy.

This classification suggests a range of generalisations (Table 16.1) that together indicate clearly that the biotroph/necrotroph division is biologically meaningful. The genetic analysis of disease resistance with plants for which the full genomes are available demonstrates that the division is based on how defence against fungal pathogens is controlled.

Defence against biotrophic pathogens is largely due to **programmed cell death** in the host, and to associated activation of defence responses regulated by the salicylic acid-dependent pathway. Necrotrophic pathogens benefit from death of host cells, so they are not limited by this defence, but by responses activated by jasmonate acid and ethylene signalling pathways (see below). Such a 'mode-of-defence' division successfully distinguishes necrotrophs and biotrophs, but it does limit the biotroph category to fungi that produce haustoria (Oliver & Ipcho, 2004; Glazebrook, 2005; Laluk & Mengiste, 2010).

Table 16.2 gives some examples of necrotrophic and biotrophic pathogens and hemibiotrophic pathogens, like *Phytophthora infestans*, that exhibit characteristics of both biotrophs and necrotrophs.

16.11 THE EFFECTS OF PATHOGENS ON THEIR HOSTS

Identifying a disease is not always straightforward and requires careful consideration of all the symptoms. Once the nature, ideally the identity, of the pathogen, its host range and the environmental factors that favour disease symptoms have been established, appropriate control methods can be applied. These might include:

- Using **disease-resistant cultivars** of the plant. Resistance can vary from 'very resistant' (the plant rarely, or never, gets the disease), 'somewhat resistant' (the plant usually does not get the disease, but might in a bad year), 'somewhat susceptible' (the plant often gets the disease) and 'very susceptible' (the plant almost always gets the disease and may be killed by it).

Table 16.1. Comparison of generalised characteristics of biotrophic and necrotrophic plant pathogens

Necrotrophs	Biotrophs
Opportunistic, unspecialised ('non-obligate') pathogens	Specialised ('obligate') pathogens
Host cells killed rapidly	Cause little damage to the host plant; host cells not killed rapidly, but can induce hypersensitive cell death in incompatible interactions
Entry unspecialised via wounds or natural openings	Entry specialised, e.g. direct entry (powdery mildews) or through natural openings (rusts)
Secrete copious cell wall-degrading (lytic) enzymes and toxins	Few if any lytic enzymes or toxins are produced
Appressoria or haustoria not normally produced	Possess appressoria or haustoria
Seldom systemic	Often systemic
Usually attack weak, young or damaged plants	Plants of all ages and vigour attacked
Wide host range	Narrow host range
Easy to culture axenically	Not easily cultured axenically
Survive as competitive saprotrophs	Frequently survive on host or as dormant propagules
Controlled by quantitative resistance genes (example is *Septoria nodorum* blotch caused by *Stagonospora nodorum*)	Controlled by specific (gene-for-gene) resistance genes (for example, tomato leaf mould, the rusts, powdery and downy mildews)
Growth in host intercellular and intracellular through dead cells	Growth intercellular
Controlled by jasmonate- and ethylene-dependent host-defence pathways	Controlled by salicylate-dependent host-defence pathways

Table 16.2. Some common plant pathogens

Necrotrophs		Biotrophs	
Pathogen	**Disease**	**Pathogen**	**Disease**
Botrytis cinerea	Grey mould	*Blumeria (Erysiphe) graminis*	Powdery mildew
Cochliobolus heterostrophus	Corn leaf blight	*Uromyces fabae*	Rust
Pythium ultimum	Damping-off in seedlings	*Ustilago maydis*	Maize smut
Ophiostoma novo- ulmi	Dutch elm disease	*Cladosporium fulvum*	Tomato leaf mould
Fusarium oxysporum	Vascular wilt	*Puccinia graminis*	Black stem rust of cereals
Sclerotinia sclerotiorum	Soft rot	*Phytophthora infestans*	Potato late blight

Hemibiotrophs
Cladosporium fulvum, tomato leaf mould (also called a biotroph).
Colletotrichum lindemuthianum, anthracnose.
Magnaporthe oryzae, rice blast (also called a necrotroph).
Phytophthora infestans, potato late blight (also called a biotroph by some, necrotroph by others).
Mycosphaerella graminicola, septoria leaf blight.

- Another plant-oriented disease control method is to practise **good sanitation**, by removing infected leaves or pruning plants to increase air circulation.
- Similarly, as poorly nourished plants are more susceptible to disease, the **water and nutrient availability** of the soil can be controlled.
- **Biological control methods** offer sustainable approaches to disease management (Maloy, 2005; Massart & Jijakli, 2007) but are not always available or practicable, and ultimately chemical pesticides can be the final resort (Mares *et al.*, 2006).

The symptoms observed in a diseased plant depend on the effect of the pathogen on the physiology of the plant. Photosynthesis is the essential function of plants and any pathogen that interferes with it will cause disease that may appear as **chlorosis** (yellowing) and **necrosis** (browning and death) of the leaves and stems. Even mild impairment of photosynthesis weakens the plant and increases susceptibility to other pests and pathogens.

Pathogens can affect **translocation of water and nutrients** through the vascular system of the host plant. This might be an effect on transpiration through the aerial parts of the plant or poor uptake of nutrients and water through diseased roots. Consequentially sluggish translocation through the vascular system will itself lead to wilting and chlorosis, and possibly necrosis, 'upstream' of the disease focus.

Most, if not all, infectious diseases **increase respiration**, this being a general reaction of the plant to most types of stress. Most of the increase in respiration occurs in the infected host tissue and appears to be a basic response to injury. Consequences of increased oxygen uptake and enhanced activity of respiratory enzymes include: a slight increase in temperature, accumulation of metabolites around points of infection, and even an increase in dry weight of the host tissue.

Changes in plasma membrane and organelle membrane permeability are often the first detectable responses of plant cells to infection by pathogens, and this often leads to **loss of electrolytes** (calcium and potassium ions in particular). The permeability change is a response to toxins produced by the invading fungus. For example, **victorin**, a cyclic peptide, is produced by *Cochliobolus victoriae* in oat leaves. Victorin binds to a membrane protein and changes membrane permeability, eventually causing chlorosis and necrotic stripes ('leaf spots') on the leaves. Victorin is one of the most phytotoxic and selective compounds known, being active against sensitive oats at 10 pg ml^{-1} (13 pM), though it does not affect resistant oats or any other plant even at a million times higher concentration; this is a **host-specific (or host-selective) toxin** (Tada *et al.*, 2005; Gilbert & Wolpert, 2013). Other host-specific toxins are (Dyakov & Ozeretskovskaya, 2007):

- The **HMT-toxin** of *Cochliobolus heterostrophus* Race T, which causes leaf blight in maize. The name of the toxin derives from the old name of the pathogen, which used to be called *Helminthosporium maydis* Race T. This toxin increases membrane permeability of mitochondria to protons by reacting with unique site(s) on the inner mitochondrial membrane of Texas (T) cytoplasmic male-sterile maize.
- The **HC-toxin** produced by Race 1 of *Cochliobolus carbonum* that makes this race exceptionally virulent to certain genotypes of maize is a cyclic tetrapeptide. The toxin is active at 20 ng ml^{-1} and inhibits maize histone deacetylase, consequently affecting histone acetylation, which is implicated in control of chromatin structure, cell cycle progression and gene expression. The peptide is synthesised by HC-toxin synthase 1 (coded for by gene *HTS1*); when all *HTS1* genes are disrupted the fungus is unable to produce HCtoxin and causes only small chlorotic flecks on the

host. Nonspecific resistance of maize to Race 1 is due to a detoxifying enzyme, HC-toxin reductase. Another fungus, *Alternaria jesenskae*, also produces HC-toxin and genomic sequencing shows that the proteins share 75–85% amino acid identity, and the genes for HC-toxin biosynthesis are duplicated in both fungi. The genomic organisation of the genes in the two fungi show a similar but not identical partial clustering arrangement. The genes may have moved by horizontal transfer between the two fungi (see section 17.15), though they may have been present in a common ancestor and lost from other species of *Alternaria* and *Cochliobolus* (Wight *et al.*, 2013).

Fusaric acid (5-butylpyridine-2-carboxylic acid) is a non-host-specific **mycotoxin** with low to moderate toxicity, which is consistently produced by *Fusarium* species pathogenic on several cereals, tomato, banana and tobacco. Fusaric acid may cause many, but not all, of the disease symptoms; it is decarboxylated by tomato plants to the more toxic (100 times) 3-*n*-butyl-pyridone. Fusaric acid is mildly toxic to mice, but has potential value as a clinical pharmaceutical, because it reduces blood pressure by inhibiting dopamine hydroxylase (Deutch & Roth, 2014).

Pathogens also exert influence on transcription and translation in host cells. This is clearly aimed at **changing the host metabolism** in ways that benefit the pathogen. It may lead to a simple adaptation of plant secondary metabolism to enhance the production of chemicals that favour the fungus in some way; for example, by attracting vectors to transport the fungus to other hosts (McLeod *et al.*, 2005). Equally interesting are cases of **excess production of plant hormones** that cause, for example, plant tissue proliferation. Changes in plant hormones can result from the influence of the pathogen on metabolism of the host or from production of a plant growth hormone or its analogue by the pathogen:

- *Albugo candida* (*Oomycota*, Kingdom *Straminipila*) causes **increased indole acetic acid (IAA)** production by infected plants of the genus *Brassica* (which contains more important agricultural and horticultural crops than any other genus: swedes, turnips, kohlrabi, cabbage, sprouts, cauliflower, broccoli, mustard seed, oilseed rape, among others). The disease is called white rust. Indole acetic acid, the original plant auxin, controls plant cell elongation and apical dominance (prevents lateral bud formation), prevents abscission, and promotes continued growth of fruit tissues, cell division in vascular and cork cambium, and formation of lateral and adventitious roots.
- The **gibberellin plant hormones** were actually isolated in the 1930s from rice suffering the bakanae disease (Japanese for 'foolish seedling'), which is caused by *Gibberella fujikuroi* (*Fusarium moniliforme*) (*Ascomycota*). The fungus itself produces a surplus of gibberellic acid, which acts as a growth hormone for the plant. It causes hypertrophy leading to etiolation and chlorosis, and the plants finally collapse and die. In the normal plant gibberellin growth regulators affect cell elongation in stems and leaves, seed germination, dormancy, flowering, enzyme induction, and leaf and fruit senescence. Bakanae still affects rice crops in Asia, Africa and North America. Crop losses in 2003 were estimated at 20–50%; remember that almost half the human population of the world depends on rice (Slavica *et al.*, 2017; and see the International Rice Research Institute's *Rice Knowledge Bank* at http://www.knowledgebank.irri.org/).

16.12 HOW PATHOGENS ATTACK PLANTS

Pathogens need to invade their hosts to cause disease and assure their own survival, so they are equipped with mechanisms for attachment, identification and invasion of suitable hosts. **Entering the host tissues** might involve finding and entering through an existing aperture, such as a stoma or an injury, or direct mechanical attack. How do they find the right place to invade?

A spore that alights on a leaf surface adheres to that surface by **hydrophobic interactions** between hydrophobins in the spore wall and the hydrophobic surface of the leaf cuticle (see section 7.8). The hypha that emerges from the spore senses an appropriate site and when it finds it, the hyphal tip enlarges (differentiating into an **appressorium**) and strengthens the adhesion to the leaf surface. A strong hold on the leaf surface is necessary to support the amount of mechanical force used to penetrate the plant.

Growth of conidial germ tubes of these pathogens on leaves or artificial substrates is usually perpendicular to ridges and furrows on the surface. Ridges or grooves 0.5 μm high by 2.0 μm wide in artificial surfaces induce appressorium formation within four minutes after contact. This is a **thigmotropism**, a directional growth in response to touch or physical contact stimuli with a solid object; a similar mechanism operates in animal pathogens also although there are some subtle differences in the mechanism (Nikawa *et al.*, 1997; Stephenson *et al.*, 2014).

In plant pathogens, the object with which the fungus is interacting is the plant cuticle, which is external to the cell and on the surface of the cell wall of external, usually epidermal, cells. Components of the cuticle are in layers in a matrix of **cutin**. The outer surface is mainly, or only, composed of very **hydrophobic waxes**, with the inner layers, those closest to the epidermal cell wall, being very hydrophilic (cellulose-rich) in composition. This shift from hydrophilic to hydrophobic between inner and outer surface provides the plant with a uniform protective barrier that retards moisture loss from the plant cell surface. The growing hypha senses very minute alterations in the surface texture and the spacing between ridges on the natural leaf surface such as those that occur between epidermal cells and especially around stomata. Extension growth of the germ tube hypha ceases after its apex has grown over a stomatal guard-cell lip (or a depression on an artificial surface *in vitro*) and the appressorium is formed over the stoma (Łaźniewska *et al.*, 2012; Bellincampi *et al.*, 2014; Serrano *et al.*, 2014).

Thigmotropism is not the only **surface cue** involved in plant pathogens; surface hardness and, separately, surface hydrophobicity stimuli are essential for appressorium formation and differentiation in the rice blast fungus *Magnaporthe oryzae*. The pathogen employs a series of receptors and sensors at the plasma membrane to recognise host surface cues and to activate signal transduction pathways required for appressorium formation and pathogenicity. The rice blast fungus senses chemical cues from primary alcohols, which are major components of leaf waxes in grasses, while other fungal sensors recognise hydrophobicity and precursors of cutin molecules on rice leaves. Fungal endocytosis is responsible for internalising the signals and triggering regulators of G-protein signalling cascades, accumulation of cAMP, MAPK pathway proteins, and inducing chitin-deacetylase activity, which is necessary for appressorium formation (Liu *et al.*, 2007; Liu *et al.*, 2011; Kuroki *et al.*, 2017; Li *et al.*, 2017). Additionally, there are indications that these mycelia also respond **chemotropically** to the stomata, presumably to the gases emerging from the substomatal space.

16.13 HOST PENETRATION THROUGH STOMATAL OPENINGS

An appressorium is a highly organised enlarged end of a hypha. When over a stoma, a penetration peg is formed and enters directly through the stoma into the substomatal chamber. From here, the fungal hyphae grow through the middle lamella **between the cells** of the leaf tissue, so they can extract the nutrients they require without killing the host tissue.

Host penetration through stomatal openings occurs in the **rust fungi**, *Uromyces* spp., and a specific example is *Uromyces fabae* (*Basidiomycota*), the bean rust (Voegele, 2006). This species attacks several important crop species: broad bean (*Vicia faba*), pea (*Pisum sativum*) and lentil (*Lens culinaris*), as well as more than 50 other *Vicia* species, and about 20 *Lathyrus* species (sweet peas and grass peas). It is most important in causing rust disease of broad bean (faba bean) on which it can cause up to 50% yield losses. *Uromyces fabae* is a **macrocyclic rust** fungus, having all five types of spore known in the Uredinales (listed in the bullet points below), and as all of its spores are formed on a single host it is also described as **autoecious**. The life cycle of *Uromyces fabae* is as follows:

- diploid **teliospores** that have overwintered on plant debris in the field germinate in spring, forming a **metabasidium**;
- meiosis occurs and the metabasidium forms four haploid **basidiospores** (of two different mating types);
- basidiospores are **ejected** from the metabasidium;
- after landing on a host leaf, basidiospores **germinate** and produce the infection structures;
- in the diseased host the pathogen mycelium produces **pycnia** containing **pycniospores** and receptive hyphae;
- pycniospores are exchanged between pycnia of different mating types and **dikaryotisation** occurs, forming aecial primordia;
- an aecium differentiates and forms dikaryotic **aeciospores;**
- aeciospores **germinate** on the host and form infection structures;
- **uredia** are produced by this dikaryotic infection, and these produce **urediospores;**
- urediospores are the major **aerially dispersed** asexual spore form, being produced in large numbers during the summer in repeated infections of host plants (urediospores are the '**rust**' of rust fungi);
- at the end of summer, uredia differentiate into **telia**, nuclei fuse during sporogenesis to form the single-celled diploid **teliospores** that overwinter on the remains of the host.

Most information about rust infection structures derives from analysis of urediospores (Figure 16.5). Urediospores (Figure 16.5A) germinate with a germ tube which differentiates into a well-defined appressorium that forms a penetration hypha which enters the leaf through the stomatal opening and expands into a vesicle in the stomatal cavity. An infection hypha grows out of the vesicle, and when this contacts a mesophyll cell it differentiates a haustorial mother cell from which a haustorium is formed to penetrate the host cell.

The penetration mechanism of germinating **basidiospores** is completely different, being a direct penetration (detailed below in section 16.14), although the appressorium, vesicle and haustorium infection structures are still formed (Voegele, 2006) (Figure 16.5B).

Urediospores are single-celled, and their walls have spiny, hydrophobic surfaces. Hydrophobic interactions are responsible for **initial adhesion** to the host surface. Initial contact with the host triggers production of an extracellular matrix comprising low molecular weight carbohydrates and glycosylated polypeptides that originates by lysis of the germ pore plug of the spore.

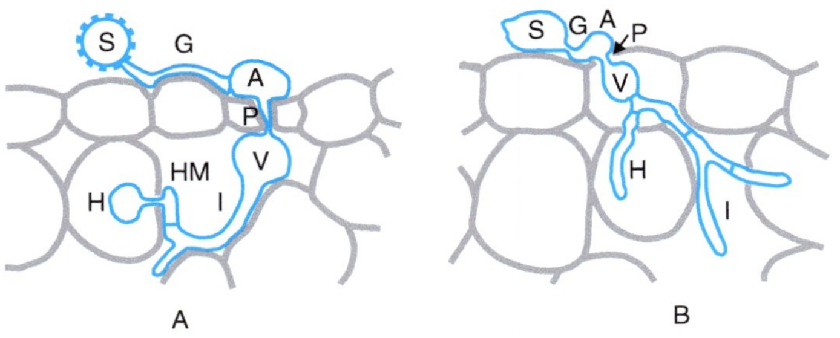

Figure 16.5. Infection structures derived from urediospores (A) and basidiospores (B) of the bean rust, *Uromyces fabae*. S, spore; G, germ tube; A, appressorium; P, penetration hypha; V, vesicle; I, infection hypha, HM, haustorial mother cell; H, haustorium. Redrawn from Voegele (2006).

This contributes to formation of an adhesion pad, to which fungal cutinases and esterases also contribute. A period of at least 40 minutes of darkness is required to induce germination and a temperature in the range 5–26°C (optimum 20°C), but urediospores germinate on almost any surface under these conditions so no signals from the host are needed to induce germination.

Fungal cytoplasm moves into the **germ tube** as it meanders across the surface to which it is attached by the adhesive extracellular matrix material. This is where the topographical signals to which we referred above come into effect. Differentiation of an appressorium in *Uromyces appendiculatus* and *Uromyces vignae* is induced by a ridge 0.4–0.8 μm in height, which corresponds to the height of the stomatal guard-cell lips. On induction, the cytoplasm transfers to the appressorium and the vacuolated germ tube is separated off by a septum. Differentiation of the **appressorium** coincides with the appearance of several (fungal) **lytic enzymes**: cellulases, proteases and chitin-deacetylase (Figure 16.6).

A **penetration hypha** is formed from the base of the appressorium, extending into the stomatal cavity where a substomatal vesicle forms and is separated from the penetration hypha by a septum. The vesicle narrows into an infection hypha and more fungal enzymes appear, which contribute to localised breakdown of the host cell wall: pectin esterases and methylesterases, and cellulases (Figure 16.6). A **haustorial mother cell** differentiates when the infection hypha contacts a mesophyll cell and becomes separated from the infection hypha by a septum. The cytoplasm moves into the haustorial mother cell and earlier structures are vacuolated. Formation of the haustorial mother cell coincides with the onset of polygalacturonate lyase activity (Voegele, 2006) (Figure 16.6).

All the infection structures so far described (which make up the process described as the **penetration phase**) can be generated *in vitro* by germinating spores on plastic membranes. Haustoria, and structures of the **parasitic phase**, and the later **sporulation phase** are only formed in plants (Figure 16.7) because signals from the host are needed to complete differentiation of the haustorium. Formation of the haustorium involves breaching the cell wall of the host cell and invagination of its plasma membrane. Although within the volume of the host cell, the haustorium remains outside the physiological boundary of the host being separated by the plasma membranes of parasite ***and*** host, the fungal cell wall and the **extrahaustorial matrix**. The latter is a mixture of carbohydrates and proteins, which

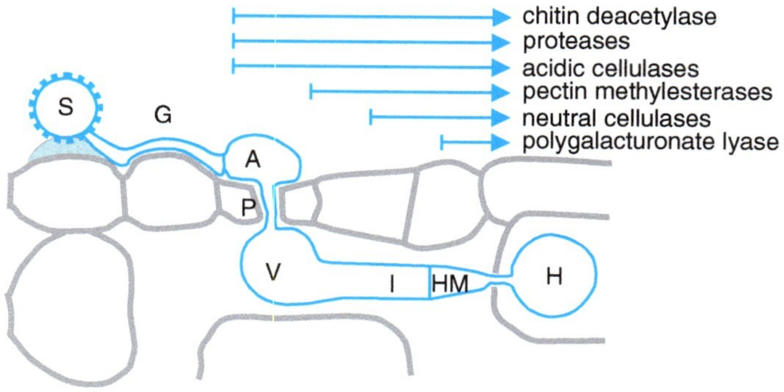

Figure 16.6. A succession of lytic enzyme production occurs during early dikaryotic infection structures of the bean rust, *Uromyces fabae*. Key: S, spore; G, germ tube; A, appressorium; P, penetration hypha; V, vesicle; I, infection hypha; HM, haustorial mother cell; H, haustorium. Redrawn from Voegele (2006).

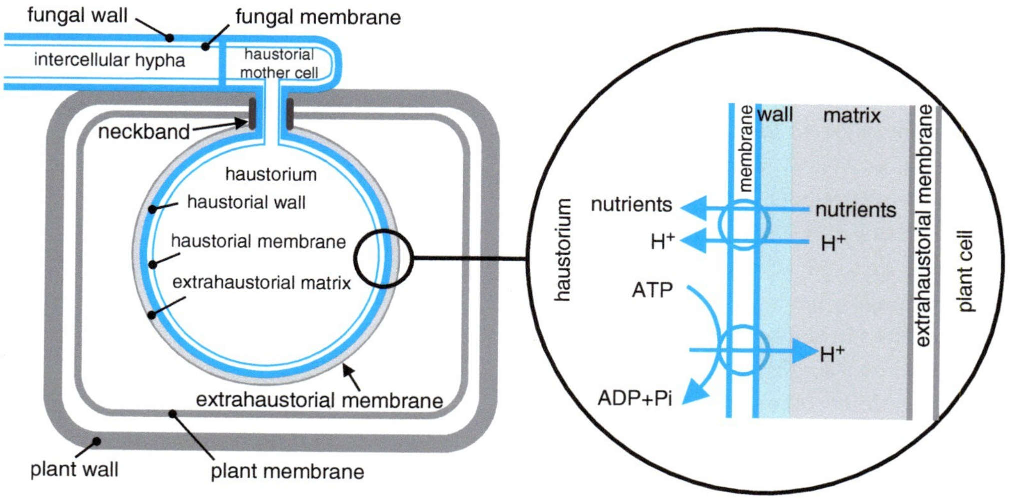

Figure 16.7. Schematic representation of a dikaryotic haustorium of the bean rust, *Uromyces fabae* within a mesophyll cell of its host, *Vicia faba*. This illustration is intended to emphasise the number and arrangement of the barriers (membranes and walls) between the pathogen fungus and its host plant. Structures derived from the fungus are depicted in blue; structures contributed by the plant are shown in grey, including the extrahaustorial matrix (though the fungus also contributes to the matrix) and the extrahaustorial membrane. The magnification shown at right shows how proton pumping by fungal ATPases powers uptake of nutrients from the plant cell by fungal symports (compare with Figures 16.8 and 16.14). Based on illustrations in Voegele (2006).

mainly originates from the host plant and may be a symplastic compartment (the **symplast** is the inner side of the plasma membrane in which water and solutes freely diffuse) or a specialised part of the apoplast (the space outside the plasma membrane in which materials freely diffuse) (Voegele, 2006).

16.14 DIRECT PENETRATION OF THE HOST CELL WALL

Not all appressoria form in response to stomatal ridges or over stomata; some form over the surfaces of intact cells, and this type of appressorium directly penetrates the host cell wall. Direct penetration is not exclusively mechanical, but extremely **high pressures** are developed (Money, 1999). The appressorium develops as a swelling at the tip of the germ tube hypha and is bonded to the host with a strong adhesive. Melanin is deposited within the appressorium wall as it matures, and as this happens the appressorium accumulates metabolites, such as glycerol, which increase the osmotic potential of the cytoplasm. This draws water into the appressorium, generating extremely **high turgor pressure**.

Melanin deposition in the cell wall of the appressorium is essential for maintaining this high turgor pressure; melanin becomes highly cross-linked into the wall to sustain the physical pressure resulting from the high osmotic potential of the cytoplasm.

Next, a thin hypha, called a **penetration peg**, extends from the base of the appressorium and perforates the host surface. The full force of the turgor pressure built up in the appressorium is focused in the minute contact area at the tip of this penetration peg and, in addition, hydrolytic enzymes specific to components of the plant cuticle are secreted from the penetration peg as it passes through the cuticle, being induced by cuticle materials. With penetration achieved, a haustorium is often formed from which hyphae branch and spread through the leaf tissue.

Specific examples of this type of host penetration occur in the **powdery mildews** caused by *Blumeria* (old name: *Erysiphe*) species; and the **rice blast** pathogen *Magnaporthe oryzae*. The grey mould pathogen *Botrytis cinerea* (in the *Ascomycota*), which is also responsible for the 'noble rot' of wine grapes, forms an appressorium that does not generate the same physical pressures and lacks the highly melanised wall and the septum that separates the appressorium from the germ tube.

The powdery mildews form an economically important group of plant pathogens that infect a range of plants, including barley, wheat, pea, apple, sugar beet and grapes. The earliest event in powdery mildew infection is the **secretion of cutinases** and other proteins by the conidia in response to contact with the hydrophobic leaf surface. Within 24 hours from arrival on the plant surface, powdery mildew conidia germinate, develop appressoria, form penetration pegs, penetrate the host outer epidermal cuticle and cell wall, and establish haustorial complexes within epidermal cells. The haustorial complex is the fungal feeding structure that allows the pathogen to draw nutrients from the host cell.

Several transporters, including hexose transporters and amino acid permease, have been identified in the (fungal) plasma membrane of *Blumeria graminis* (wheat powdery mildew) haustoria, demonstrating that glucose and amino acids are transported from wheat to the disease fungus. The flow of nutrients from the host supports continued hyphal growth over the leaf surface (and many more penetration events). Infected host cells enclose fungal haustoria in an **extrahaustorial membrane (EHM)**, which keeps it separate from the host cytoplasm. The EHM is a unique, specialised membrane of host origin: it might be an invagination of the plasma membrane which is subsequently differentiated or be synthesised *de novo* by vesicle trafficking targeted on the invading haustorium. The events involved in a typical powdery mildew infection process are illustrated in Figure 16.8 (Koh *et al.*, 2005).

Approximately 21% of the genes (a total of 2,154) in the genome of *Magnaporthe oryzae* showed **differential expression** (changes in gene expression ≥ two-fold), the majority being upregulated, during spore germination and appressorium formation (Oh *et al.*, 2008). During appressorium formation specifically, 357 genes were differentially expressed (240 being upregulated and 117 downregulated). There was a significant decrease in expression of genes involved in protein synthesis, while expression of genes associated with protein and amino acid degradation, lipid metabolism, secondary metabolism and cellular transportation exhibited a dramatic increase; there being a four-fold enrichment for genes encoding secreted proteins (Oh *et al.*, 2008) (Figure 16.6).

16.15 ENZYMATIC PENETRATION OF THE HOST

Secreted proteins are common components of the descriptions above. Pathogenic fungi share a common lifestyle that involves penetration of the living plant; for biotrophs this also means penetrating plant cell walls, potentially suppression of the defences of the plant, and intimate contact with living host plant cell membranes as a means of acquiring nutrients. All of this requires investment of fungal gene products in the pathogenesis, and we have just referred to the large number of fungal genes that can be shown to be devoted to fungal pathogenicity in the hemibiotroph *Magnaporthe oryzae*. However, comparative analysis of genomes shows that the gene complement for toxin biosynthetic pathways and cell wall-degrading enzymes of necrotrophic fungal pathogens, which kill and consume host cells, is even greater than that found in biotrophs.

As we have seen above, pathogenic fungi generally secrete a blend of **hydrolytic enzymes**, which includes cutinases, cellulases, pectinases and proteases, during germination and host penetration. Such enzymes are also produced by saprotrophs but their activities, substrate specificities and regulation are adapted to the needs of the pathogens and the nature of their host. For

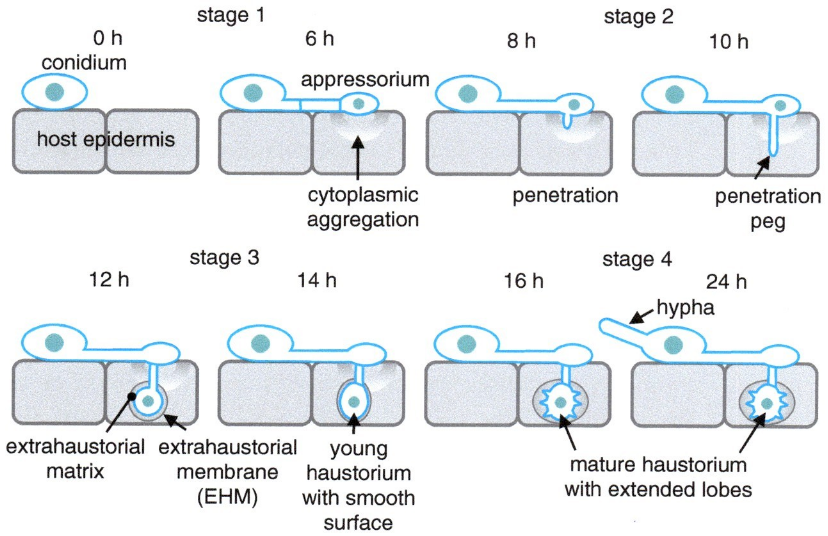

Figure 16.8. 'Comic strip' diagrams of the infection sequence of *Blumeria* (*Erysiphe*) *cichoracearum* on a susceptible *Arabidopsis* host. In stage 1 (0–6 hours after inoculation; all timings are approximate) conidia germinate and develop appressoria, but no penetration into the plant cell occurs. Cytoplasm and organelles of the plant epidermal host cell move towards and accumulate underneath appressoria at this stage. Stage 2 (6–10 h) is the period in which the appressorium developed a long, thin penetration peg, which subsequently penetrates the epidermal cell. Stage 3 (10–14 h) is the stage in which the haustorium first forms at the end of the penetration peg as a swollen, elongated sac with a smooth surface. At this stage, nuclear migration from the hypha on the leaf surface into the haustorium completes, and a septum separates the body from the neck. The haustorial complex, which consists of the fungal haustorium, the gel-like extrahaustorial matrix (which may be of both fungal and host origin) and the host extrahaustorial membrane (EHM) (which is of host origin), develops within 12–18 h. The EHM encases the haustorium, separating it from the host cytoplasm. The last stage of early infection, stage 4 (14–24 h), is characterised by a haustorium with distinct lobes emanating from the body. By stage 4, also, a hypha has emerged from the conidium at the opposite side to the appressorial germ tube. The illustrated sequence of events is representative of other powdery mildews, except that cereal powdery mildews produce a vestigial primary germ tube prior to an appressorial germ tube. Based on illustrations in Koh *et al.* (2005).

example, *Alternaria brassicicola* (*Ascomycota*), which causes black or dark leaf spot on many brassicas but is also found as a common saprotroph in soil, decaying plant material, wood, and on foods, produces **different cutinase isozymes** during saprotrophic and pathogenic stages (Knogge, 1998).

Penetrating a host leaf surface composed of cutin covered with wax is the first challenge to plant pathogens. Infection requires intricate sensing and signalling in both plant and pathogen. The pathogen must choose whether, when and where to germinate, develop an infection structure and/or produce enzymes and metabolites. It does this by sensing the physical and chemical environment, recognising hydrophobic and hydrophilic surfaces, as well as surface hardness and the chemical environment, too (Knogge, 1998; van Kan, 2006).

The grey mould pathogen *Botrytis cinerea* (*Ascomycota*) is a well-studied **necrotroph** which forms an appressorium that is probably not capable of penetrating the host by direct physical pressure. It lacks the highly melanised wall and the septum separating the appressorium from the germ tube, which are characteristic features of appressoria like those of *Blumeria* and *Magnaporthe*. For *Botrytis cinerea* appressoria, **secreted enzymes**, particularly cutinases and lipases, are the key to breaching the plant surface; the genome of *Botrytis cinerea* contains at least five cutinase genes and over a dozen lipase genes.

The tip of the penetration peg that emerges from the appressorium also generates H_2O_2 as a co-substrate for oxidases that degrade cuticular components and aid penetration. Early invasion of epidermal cells also involves pectinases (endopolygalacturonase). A common feature of plants that are most susceptible to *Botrytis cinerea* is a relatively high content of pectin in the cell wall. This host preference of *Botrytis cinerea* reflects an efficient **pectinolytic** machine comprising at least six endopolygalacturonase genes that are differentially regulated according to the nature of the host cell wall.

Botrytis cinerea hastens host cell death in several ways, producing low molecular weight phytotoxic metabolites, the best-studied of which is the sesquiterpene botrydial (Figure 16.9A).

Some strains of *Botrytis cinerea* rely on botrydial alone to kill their host, while others produce additional toxins, such as botcinic acid (Figure 16.9B). *Botrytis cinerea*, and the other common necrotrophic fungus *Sclerotinia sclerotiorum* (*Ascomycota*), also produce the dicarboxylic acid oxalate (Plassard & Fransson, 2009). Oxalate-deficient mutants are non-pathogenic but reversion to oxalate production restores pathogenicity. Oxalic acid seems to serve several functions:

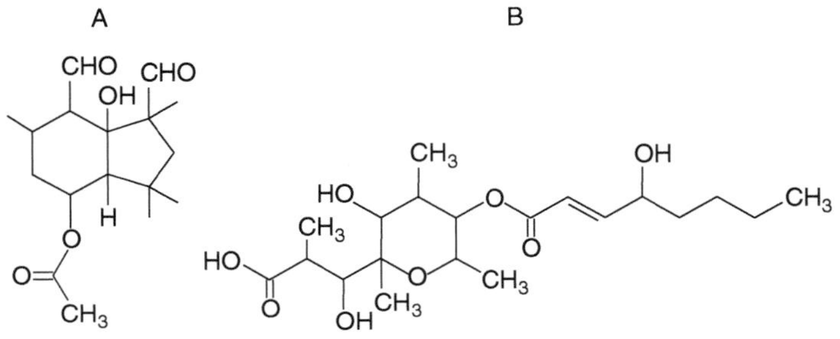

Figure 16.9. Phytotoxic metabolites of *Botrytis cinerea*. (A) The sesquiterpene botrydial. (B) The polyketide called botcinic acid.

- **It is a strong acid** and several enzymes secreted by *Botrytis cinerea* are active in an acidic environment (pectinases, proteases and laccases).
- **Oxalate chelates divalent metal ions**, particularly calcium and copper; most calcium in plant cells is stored in the cell wall, embedded in pectin. Oxalate extracts Ca^{2+} from the pectin after partial hydrolysis by pectinases, and removal of calcium makes pectin more accessible to pectinases for further degradation.
- **Oxalate can reduce plant defence responses** and trigger programmed cell death in plants (van Kan, 2006).

Host-specialised species of *Botrytis*, for example *Botrytis fabae* and *Botrytis elliptica*, produce phytotoxic proteins, known as **host-selective toxins**. These are **necrosis and ethylene inducing proteins (NEPs)**, a family of proteins that induce plant cell death that were originally isolated from culture filtrates of *Fusarium oxysporum* but have since been found in a range of plant pathogenic microorganisms, including bacteria, *Phytophthora* spp. and fungi. Fungal extracellular enzymes are especially important in diseases known as **soft rots** (caused by several bacteria and fungi), to which fruit and some vegetables are prone. The enzymes concerned are pectinases, hemicellulases, cellulases and ligninases (enzymes already discussed in Chapter 10). In commerce, these are most important in relation to potential spoilage of fresh fruits and vegetables in the field, in transit or on displays; there are a few examples indicating that mycotoxins might have a role in the safety of fresh fruit at the retail level (Moss, 2008).

The general pattern coming out of these descriptions is that an infection by *Botrytis* species (and probably other heterotrophs) produces a range of toxic compounds of fungal origin that act as inducers of programmed cell death in the plant cell; in other words, host cell death requires the active participation of the pathogen ***and the host***. Infection induces what is known as an **oxidative burst**. The oxidative burst is an early response to pathogen attack (in plant ***and animal*** cells) leading to the production of **reactive oxygen species** (**ROS** or **superoxide**) including hydrogen peroxide.

The major mechanisms involve either (plant) membrane-bound NADPH oxidases or peroxidases that can occur singly or in combination in different plant species. Most cells produce and detoxify ROS, as they form under normal conditions as by-products of successive one-electron reductions of molecular oxygen. Most cells also have protective mechanisms to maintain the lowest possible intracellular levels of ROS. When the cell is under stress, protective mechanisms are overridden by the rapid production of huge amounts of ROS, and this is '**the oxidative burst**' (a similar reaction process has been known for many years in animal cells, particularly mammalian phagocytes, and is called the '**respiratory burst**'). However, in plants the phenomenon serves:

- to produce **potential protectants** against an invading pathogen: the oxidants themselves are directly protective, they function as substrates for oxidative cross-linking of the plant cell wall, and serve as diffusible signals to induce cell-protection genes in surrounding cells;
- as a central component of a highly amplified and integrated signal system, also involving salicylic acid and changes in cytoplasmic Ca^{2+}, which underlies the expression of the **hypersensitive response** that confers resistance to biotrophic pathogens. This is a programmed cell death reaction involving nuclear condensation and expression of a panel of specific genes and activity of specific proteins. Characteristically, among these are **metacaspases**, which are Arg/Lys-specific equivalents of the Asp-specific caspases that contribute to apoptosis in animals. Metacaspases induce programmed cell death in plants and fungi.

The hypersensitive response protects the plant against biotrophic pathogens by rapidly killing the infected plant cell, and denying it to the invading pathogen before the fungus has had the opportunity to establish the feeding structures it needs to support the germinating spore that has initiated the infection. Sacrifice of one or a few cells of the leaf deprives the pathogens of a supply of food and effectively kills that germinating spore because an obligate biotroph has no saprotrophic ability. Necrotrophic pathogens, such as *Botrytis cinerea* and *Sclerotinia sclerotiorum*, however, can utilise dead tissue.

Indeed, it might be said that the goal of a necrotrophic plant pathogen is to kill its host cell in order to decompose plant biomass and convert it into fungal biomass, so these fungi exploit a host-defence mechanism for their pathogenicity (Govrin & Levine, 2000). *Botrytis cinerea* triggers an oxidative burst during cuticle penetration, using the oxidants to assist penetration, but the infection results in massive accumulation of hydrogen peroxide in the plant plasma membrane, which triggers the hypersensitive response, killing the plant cells and thereby providing plant biomass for the fungus to consume saprotrophically (Knogge, 1998; van Kan, 2006). For this the fungus can produce a full range of digestive enzymes, and a complete expression of digestive metabolism so that the fungus can take full advantage of the plant cell as a nutrient source (Jobic *et al.*, 2007).

16.16 PREFORMED AND INDUCED DEFENCE MECHANISMS IN PLANTS

Plants have evolved many defence mechanisms to protect themselves against pathogens; some are part of the structure and chemistry of the plant that are built in to the healthy plant before infection, which contribute to avoidance or prevention of infection, others are reactions to the presence of the pathogen, though the categories do overlap considerably (Table 16.3).

Pre-existing defence structures comprise:

- surface waxes,
- the chemical structure, thickness, and cross-linking of **epidermal cell walls** and the walls of internal barriers (endodermis);
- the **anatomical position of stomata and lenticels**, and the disposition of epidermal hairs;
- **fungal germination inhibitors** on the plant surface or in surface tissues (for example phenolic compounds, like catechol in onion (*Allium cepa*) bulb scales, or chlorogenic acid in potato epidermis; see Figure 16.10).

Defences formed in response to infection comprise:

- histological defences at the tissue level;
    - formation of cork layers, or abscission layers (if excising a part of the plant will benefit the rest);
    - tyloses which block vascular elements and limit disease spread; and
    - deposition of gums (resinous, viscous substances that dry into brittle solids) on/in the cell wall or plant surface.
- defence tactics at the cellular/subcellular level, including:
    - concentrating cytoplasmic contents around the site of infection;
    - thickening of the cell wall with the β-glucan callose;
    - modification or removal of specific attachment or receptor sites for a pathogen or pathogen-produced toxin;
    - release of inhibitors by the plant into its immediate environment (numerous phenolic compounds can be produced by plants, some as initial defences (Figure 16.10), some as reactive defences and some persisting into the soil as **allelochemicals** (biomolecules released into the environment that influence the growth and development of neighbouring organisms) (Popa *et al.*, 2008);
    - cell wall reinforcement by deposition of cellulose and lignin, and accumulation of hydroxyproline-rich glycoproteins;
    - production of protease inhibitors and lytic enzymes such as chitinase and glucanase (which target the fungal wall);
    - production of **phytoalexins**, which are antimicrobial compounds not found in a healthy plant that increase to antimicrobial concentrations after infection; phytoalexins may be terpenoids, glycosteroids or alkaloids (two examples shown in Figure 16.11), though the definition has stretched over the years and now includes all plant-produced chemicals that contribute to disease and/or pest resistance, so many of the compounds to which we have already referred would come into this category; a simple functional definition recognises phytoalexins as compounds that are synthesised *de novo* and **phytoanticipins** as preformed infection inhibitors, but the distinction is not always obvious (Dixon, 2001). Phytoalexins are toxins to the attacking organism and are highly specific; a phytoalexin active against one pest may not repel a different pest.

The type of induced resistance we have already described, the hypersensitive response, in which the plant produces ROS and hydrogen peroxide to initiate programmed cell death in infected cells so that compromised cells commit suicide to create a physical barrier against the pathogen, is also called the **general short-term response**.

Plants have an alternative long-term resistance response. **Long-term resistance**, also known as **systemic acquired resistance (SAR)**, is a whole-plant resistance response occurring

Table 16.3. Preformed and induced defence mechanisms in plants

Present in plant prior to infection		Induced as a result of infection	
Structural	**Chemical**	**Structural**	**Chemical**
Surface wax	Leaf leachate	Lignification	H_2O_2 (oxidative burst)
Cuticle, bark	Surface pH	Abscission layer	Phytoalexins
Cell walls (thickness)	Toxins (e.g. phenolic compounds)	Tyloses (formed in response to vascular wilt)	Defence response proteins and metabolites
Casparian strip (endodermis)	Enzyme inhibitors (e.g. tannins)	Cork formation	Cork formation
		Hypersensitive response: host cells killed rapidly by triggering programmed cell death	

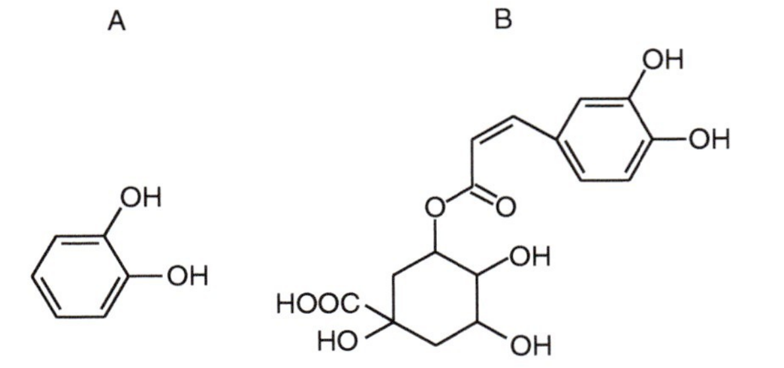

Figure 16.10. Two phenolic compounds which are preformed fungal inhibitors. (A) Catechol in the scales (thin outer tissues) of the onion bulb (*Allium cepa*). (B) Chlorogenic acid (an ester of cinnamic acid) in the epidermis of both the potato tuber and green coffee beans.

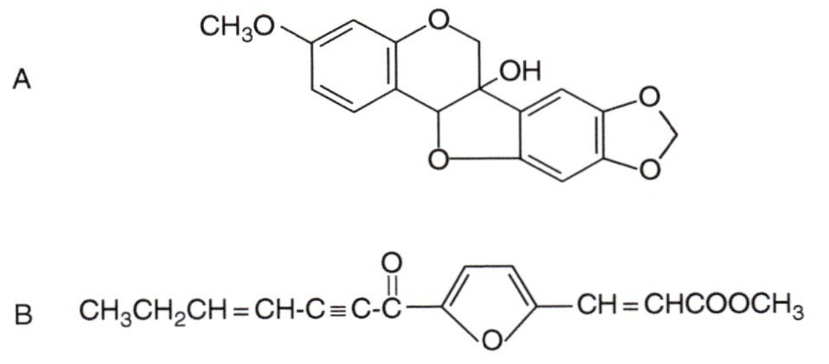

Figure 16.11. Two characteristic phytoalexins. (A) Pisatin, a phytoalexin from pea (*Pisum sativum*), is an isoflavonoid compound; some pathogenic strains of the fungus *Nectria haematococca* produce a pisatin demethylase that detoxifies pisatin. (B) Wyerone is a furanoacetylene phytoalexin produced in broad bean (or faba bean), *Vicia faba*; wyerone is so named because it was discovered at the Plant Growth Substance and Systemic Fungicides Unit at Wye College (University of London).

following a localised exposure to a pathogen. It involves communication with the rest of the plant using the **hormones** jasmonate, ethylene and abscisic acid and the accumulation of endogenous salicylic acid. If the gaseous hormones are released from the injured tissue, it is possible for neighbouring plants to take part in the resistance response as well.

Indeed, even herbivores can detect **aromatic wound response compounds** and view the plant as no longer edible. SAR is effective against a wide range of pathogens and is consequently also called 'broad-spectrum' resistance. The pathogen-induced SAR signal activates a specific molecular signal transduction pathway associated with the induction of a wide range of pathogenesis-related (or 'PR') genes that protect from further pathogen intrusion. These functions include the enzymes needed to synthesise **phytoalexins**.

The signals from the pathogen that cause these active plant responses are called **elicitors**, which are simply fungal products detected by the plant that induce defence reactions. They may be **general elicitors**, which are produced by all members of a pathogen group, or **race or cultivar-specific elicitors** that are produced only by pathogens with a specific genotype. The elicitors are often components, or even parts of components, of the hyphal walls of the pathogen; such as the peptide produced by *Cladosporium fulvum* (*Ascomycota*) that induces phytoalexin formation by tomatoes, and the branched glucans of *Phytophthora megasperma* (*Oomycota*, Kingdom *Straminipila*) that elicit resistance responses when the pathogen infects soya bean (*Glycine max*).

16.17 GENETIC VARIATION IN PATHOGENS AND THEIR HOSTS: COEVOLUTION OF DISEASE SYSTEMS

The genes that determine the varied secreted fungal products essentially constitute the panel of fungal virulence genes, because their products contribute in some way to pathogenicity of the fungus. Equally, the plant genes that determine factors that combat the infection are effectively the plant resistance genes, because their products contribute in some way to combating the infection (Zipfel, 2014).

Ellis *et al.* (2007) described the **coevolution of plant–pathogen disease systems** as involving:

> a complex move-countermove scenario. In step 1, evolution of a plant pathogen from a non-pathogenic ancestor involves the acquisition of molecules called effectors that function to blunt host basal resistance responses that are induced by common microbe-specific molecules, known as pathogen associated molecular patterns (PAMPs). These molecules are recognised by host receptors as 'nonself' and trigger a low-level defence response. Step 2 is evolution by the host plant of 'effector detectors', more commonly known as resistance (R) proteins that specifically recognise pathogen effectors and then trigger strong host defences. When effector proteins are recognised by polymorphic host resistance proteins they are also called avirulence proteins. In step 3, the pathogen's evolutionary countermove is envisaged to be more complex and can involve either modification of effectors to escape R protein recognition but retain virulence function, or loss and replacement of old effector repertoires with new ones. An additional flourish that can be added to step 3 by quintessential pathogens is the evolution of inhibitor proteins that either directly or indirectly block recognition of effectors by R proteins.

Vertical or horizontal resistance categorises the responses of host plants to fungal diseases.

- **Vertical resistance** is generally observed in annual crops which are not vegetatively propagated, and it usually depends on one or a very few pathogen-specific genes in the plant; if the resistance is overcome the crop fails.
- **Horizontal resistance** (probably the most common in nature) is seen in perennial as well as vegetatively propagated annual crops, it is polygenic and tends to be broadly based and nonspecific.

Tolerance of a disease is defined as the ability to produce a crop despite infection. Defence of a plant against pathogens may therefore involve specific resistance conferred by one or a few major genes, which are effective against specific, genetically defined races of a pathogen, or may involve a general polygenic resistance, which is effective against a wide range of pathogens.

These are the defence strategies that the fungal pathogen must overcome to attack the host. We must remember that we are dealing with **fungal pathogens**, so heterokaryosis and the parasexual cycle can contribute to the population genetics of the pathogen. Of course, the plant breeder and, increasingly, the molecular biologist and genetic engineer contribute to the population genetics of the host.

In the mid-twentieth century, combined studies on the inheritance of **specific resistance** in flax and the **virulence factors** of its fungal pathogen, flax rust, introduced the concept that host and parasite genes both played a role in the determination of whether or not a resistance reaction would be observed.

The concept is that the expression of **resistance by the host is dominant** while conversely the expression of **non-virulence by the parasite is dominant** (that is, virulence is a recessive character in the fungus). Specifically, that each individual resistance gene in the host interacted with a corresponding single gene in the pathogen. This is the **gene-for-gene concept**.

In the simplest gene-for-gene interaction the reaction between each pair of resistance and susceptibility alleles in the host with its matching pair of virulence and avirulence alleles in the pathogen is fairly simple, but with several such interactions taking place, the overall relationship between a host and its pathogen is complex. Twenty-nine separate host resistance factors were identified in flax using classic breeding methods, each having a complementary virulence factor in flax rusts, and similar complementary major gene interactions between hosts and pathogens are known or suspected for more than 25 different host-pathogen combinations.

Table 16.4 illustrates the interaction between one gene locus in the host plant and one locus in the pathogen.

A two-factor interaction is shown in Table 16.5, with the array simplified to distinct phenotypes using the dominance relationships. In such a multi-factor interaction, the host is resistant to the pathogen if a matching pair of (host) resistance and (pathogen) virulence alleles occurs at any one of the gene-for-gene loci. Gene-for-gene resistance often causes the hypersensitive response in the host plant to confine the pathogen and limit its proliferation.

The panels of gene-for-gene interactions are interpreted as recognition reactions between the host and the pathogen. The avirulence alleles of the pathogen somehow label it 'pathogen'; the resistance alleles of the host give the plant the ability to recognise and constrain the fungus, but it only needs one of the several potential recognition events to be successful to achieve resistance. For the fungus, future success as a pathogen depends on avoiding recognition; removing all of its labels so it can blend into the background. For the plant, future success depends on more and more sensitive surveillance. Gene-for-gene interactions lock the host and pathogen into a coevolutionary conflict between balanced polymorphisms.

Gene-for-gene interactions are also a target for applied genetics manipulation. Extensive use is made of major genes for resistance in disease control strategies for agricultural crops. The applied approach depends on the reproductive strategies of the pathogen. Pathogens with an exclusively asexual mode of reproduction (clonal pathogens) differ markedly from those with periodic cycles of sexual reproduction. In clonal populations the variety of different pathogen genotypes in the population will be restricted. In contrast, sexual reproduction produces a much wider diversity of pathogen genotypes, including new virulence combinations as a result of recombination.

Disease resistance genes used in breeding agricultural crops originally arose following coevolution of the pathogen (which is now the crop disease) with the ancestors of what is now the

Table 16.4. Single factor gene-for-gene interaction

Fungus pathogen genotype*	Host plant genotype*		
	RR	*Rr*	*rr*
VV	resistant	resistant	disease
Vv	resistant	resistant	disease
vv	disease	disease	disease

*The host alleles are symbolised *R* for resistance and *r* for susceptibility, while the corresponding pathogen alleles are symbolised *V* for avirulence and *v* for virulence. From Moore & Novak Frazer (2002).

Table 16.5. Two-factor gene-for-gene interaction

Fungus pathogen genotype*	Host plant genotype*			
	*R*1-, *R*2-	*R*1-, *r*2*r*2	*r*1*r*1, *R*2-	*r*1*r*1, *r*2*r*2
*V*1-, *V*2-	resistant	resistant	resistant	disease
V 1-, *v*2*v*2	resistant	resistant	disease	disease
*v*1*v*1, *V*2-	resistant	disease	resistant	disease
*v*1*v*1, *v*2*v*2	disease	disease	disease	disease

*The host alleles of factors 1 and 2 are symbolised *R* for resistance and *r* for susceptibility, while the corresponding pathogen alleles are symbolised *V* for avirulence and *v* for virulence. To reduce the size of the table, homozygous and heterozygous dominants (e.g. *R*1*R*1 and *R*1*r*1) are combined (shown as, for example, *R*1-). From Moore and Novak Frazer (2002).

crop plant. The gene-for-gene relationship, particularly in rusts and mildews, is also found in diseases of wild plants in nature. Indeed, wild species related to crops are the commonest sources of the genes used by plant breeders to develop race-specific resistant cultivars of crops. Well-studied natural host–pathogen systems include crown rust (*Puccinia coronata* f. sp. *avenae*) of wild oats (*Avena fatua*) and powdery mildew (*Blumeria fischeri*) of groundsel (*Senecio vulgaris*).

Genetic studies of the interaction between the flax plant (*Linum usitatissimum*) and flax rust (*Melampsora lini*) have identified about 30 flax resistance (*R*) genes, and about 30 'corresponding' flax rust avirulence (*Avr*) genes. In the barley (*Hordeum vulgare*) and barley powdery mildew (*Blumeria graminis hordei*) interaction over 80 resistance genes have been identified. Major gene resistance to smut fungi occurs in wheat and barley and avirulence genes have been genetically mapped in the barley pathogen *Ustilago hordei* but no major gene resistance to smut has been identified in the maize genome (Ellis *et al.*, 2007).

Talbot (2007) described the work of Jeon *et al.* (2007) with the rice blast fungus *Magnaporthe oryzae* as an **industrial-scale study** for using insertional mutagenesis and whole-genome sequence information on such a large scale (see section 16.3). It is inevitable that by the time you read this similarly detailed and comprehensive information will be available about even more of the fungi, the rusts, mildews, blasts, rots, eyespots and wilts, which are responsible for the diseases that afflict so many of our most important crop plants.

Studies of the molecular aspects of virulence have revealed many features contributing to pathogenicity. The only common theme is that no fungus depends on a single molecule for virulence; rather, virulence requires expression of several or even many genes. It has also become clear that microbial symbionts of eukaryotes influence disease resistance in many host-parasite systems. This **symbiont-mediated protection** applies to both bacterial and fungal symbionts, and to microbial and animal parasites. For example, the protection conferred against a fungal pathogen by a vertically transmitted symbiont of an aphid is influenced by **both** host-symbiont and symbiont-pathogen genotype-by-genotype interactions. In this case, variation between symbiont genotypes seemed to be maintained by coevolution with their host and/or their host's parasites (Parker *et al.*, 2017). Microbes that protect their hosts from pathogenic infection are widespread components of the microbiota of both plants and animals. These complex genetic interactions can be genotype-specific and may even motivate the variation in host resistance to pathogenic infection. There being a dynamic coevolutionary association between pathogens and defensive microbes (Kwiatkowski *et al.*, 2012; Ford *et al.*, 2017).

16.18 REFERENCES

Alexander, P., Brown, C., Arneth, A. *et al.* (2017). Losses, inefficiencies and waste in the global food system. *Agricultural Systems*, **153**: 190–200. DOI: https://doi.org/10.1016/j.agsy.2017.01.014.

Anonymous (2018). Aukland (NZ) Council's *Our Aukland* website, 19 February 2018. URL: http://ourauckland.aucklandcouncil.govt.nz/articles/news/2018/2/dutch-elm-disease-found-on-iconic-auckland-trees/.

Babiker, E.M., Gordon, T.C., Chao, S. *et al.* (2016). Genetic mapping of resistance to the Ug99 race group of *Puccinia graminis* f. sp. *tritici* in a spring wheat landrace Cltr 4311. *Theoretical and Applied Genetics*, **129**: 2161–2170. DOI: https://doi.org/10.1007/s00122-016-2764-5.

Babiker, E.M., Gordon, T.C., Chao, S. *et al.* (2017). Molecular mapping of stem rust resistance loci effective against the Ug99 race group of the stem rust pathogen and validation of a single nucleotide polymorphism marker linked to stem rust resistance gene Sr28. *Phytopathology*, **107**: 208–215. DOI: https://doi.org/10.1094/PHYTO-08-16-0294-R.

Bellincampi, D., Cervone, F. & Lionetti, V. (2014). Plant cell wall dynamics and wall-related susceptibility in plant–pathogen interactions. *Frontiers in Plant Science*, **5**: 228. DOI: https://doi.org/10.3389/fpls.2014.00228.

Bendel, M., Kienast, F. & Rigling, D. (2006). Genetic population structure of three *Armillaria* species at the landscape scale: a case study from Swiss *Pinus mugo* forests. *Mycological Research*, **110**: 705–712. DOI: https://doi.org/10.1016/j.mycres.2006.02.002.

Bernier, L., Aoun, M., Bouvet, G.F. *et al.* (2014). Genomics of the Dutch elm disease pathosystem: are we there yet? *iForest – Biogeosciences and Forestry*, **8**: 149–157. DOI: https://doi.org/10.3832/ifor1211-008.

Brown, J.S. & Holden, D.W. (1996). Insertional mutagenesis of pathogenic fungi. *Current Opinion in Microbiology*, **1**: 390–394. DOI: https://doi.org/10.1016/S1369-5274(98)80054-4.

Cairney, J.W.G. (2005). Basidiomycete mycelia in forest soils: dimensions, dynamics and roles in nutrient distribution. *Mycological Research*, **109**: 7–20. DOI: https://doi.org/10.1017/S0953756204001753.

Cerda, R., Avelino, J., Gary, C. *et al.* (2017). Primary and secondary yield losses caused by pests and diseases: assessment and modeling in coffee. *PLoS ONE*, **12**: article number e0169133. DOI: https://doi.org/10.1371/journal.pone.0169133.

Comeau, A., Dufour, J., Bouvet, G. *et al.* (2015). Functional annotation of the *Ophiostoma novo-ulmi* genome: insights into the phytopathogenicity of the fungal agent of Dutch elm disease. *Genome Biology and Evolution*, **7**: 410–430. DOI: https://doi.org/10.1093/gbe/evu281.

Couch, B.C. & Kohn, L.M. (2002). A multilocus gene genealogy concordant with host preference indicates segregation of a new species, *Magnaporthe oryzae, from M. grisea. Mycologia*, **94**: 683–693. DOI: https://doi.org/10.1080/15572536.2003.11833196.

Dean, R.A., Talbot, N.J., Ebbole, D.J. *et al.* (2005). The genome sequence of the rice blast fungus *Magnaporthe grisea. Nature*, **434**: 980–986. DOI: https://doi.org/10.1038/nature03449.

Deutch, A.Y. & Roth, R.H. (2014). Pharmacology and biochemistry of synaptic transmission: classical transmitters. In *From Molecules to Networks: An Introduction to Cellular and Molecular Neuroscience*, 3rd edition, ed. J. Byrne, R. Heidelberger & M.N. Waxham. Boston, MA: Academic Press, pp. 207–237. ISBN: 9780123971791. DOI: https://doi.org/10.1016/B978-0-12-397179-1.00007-5.

Dixon, R.A. (2001). Natural products and plant disease resistance. *Nature*, **411**: 843–847. DOI: https://doi.org/10.1038/35081178.

Donatelli, M., Magarey, R.D., Bregaglio, S. *et al.* (2017). Modelling the impacts of pests and diseases on agricultural systems. *Agricultural Systems*, **155**: 213–224. DOI: https://doi.org/10.1016/j.agsy.2017.01.019.

Dubin, H.J. & Brennan, J.P. (2009). Combating stem and leaf rust of wheat: historical perspective, impacts, and lessons learned. *IFPRI Discussion Paper Number 910*. Washington, DC: International Food Policy Research Institute (IFPRI). URL: http://ebrary.ifpri.org/cdm/ref/collection/p15738coll2/id/17034.

Dyakov Y.T. & Ozeretskovskaya, O.L. (2007). Virulence genes and their products. In *Comprehensive and Molecular Phytopathology*, ed. Y.T. Dyakov, V.G. Dzhavakhiya & T. Korpela. Amsterdam: Elsevier Science. pp. 327–349. ISBN: 978-0-444-52132-3. DOI: https://doi.org/10.1016/B978-044452132-3/50016-X.

Ellis, J.G., Dodds, P.N. & Lawrence, G.J. (2007). The role of secreted proteins in diseases of plants caused by rust, powdery mildew and smut fungi. *Current Opinion in Microbiology*, **10**: 326–331. DOI: https://doi.org/10.1016/j.mib.2007.05.015.

Erwin, D.C. & Ribeiro, O.K. (1996). *Phytophthora Diseases Worldwide*. St Paul, MN: The American Phytopathological Society, APS Press. ISBN: 0-89054-212-0.

Faulkner, C. & Robatzek, S. (2012). Plants and pathogens: putting infection strategies and defence mechanisms on the map. *Current Opinion in Plant Biology*, **15**: 699–707. DOI: https://doi.org/10.1016/j.pbi.2012.08.009.

Ferguson, B.A., Dreisbach, T.A., Parks, C.G., Filip G.M. & Schmitt, C.L. (2003). Coarse-scale population structure of pathogenic *Armillaria* species in a mixed-conifer forest in the Blue Mountains of northeast Oregon. *Canadian Journal of Forest Research*, **33**: 612–623. DOI: https://doi.org/10.1139/x03-065.

Fones, H., Fisher, M. & Gurr, S. (2017). Emerging fungal threats to plants and animals challenge agriculture and ecosystem resilience. In *The Fungal Kingdom*, ed. J. Heitman, B. Howlett, P. Crous *et al.* Washington, DC: ASM Press, pp. 787–809. DOI: https://doi.org/10.1128/microbiolspec. FUNK-0027–2016.

Ford, S.A., Williams, D., Paterson, S. & King, K.C. (2017). Co-evolutionary dynamics between a defensive microbe and a pathogen driven by fluctuating selection. *Molecular Ecology*, **26**: 1778–1789. DOI: https://doi.org/10.1111/mec.13906.

Freeman, B.C. & Beattie, G.A. (2008). An overview of plant defenses against pathogens and herbivores. In *The Plant Health Instructor*. Published online by the American Phytopathological Society. DOI: https://doi.org/10.1094/PHI-I-2008-0226-01.

Garnica, D.P., Nemri, A., Upadhyaya, N.M., Rathjen, J.P. & Dodds, P.N. (2014). The ins and outs of rust haustoria. *PLoS Pathogens*, **10**: article number e1004329. DOI: https://doi.org/10.1371/journal.ppat.1004329.

Gilbert, B.M. & Wolpert, T.J. (2013). Characterization of the LOV1-mediated, Victorin-induced, cell-death response with virus-induced gene silencing. *Molecular Plant–Microbe Interactions*, **26**: 903–917. DOI: https://doi.org/10.1094/MPMI-01-13-0014-R.

Glazebrook, J. (2005). Contrasting mechanisms of defense against biotrophic and necrotrophic pathogens. *Annual Review of Phytopathology*, **43**: 205–227. DOI: https://doi.org/10.1146/annurev.phyto.43.040204.135923.

Govrin, E.M. & Levine, A. (2000). The hypersensitive response facilitates plant infection by the necrotrophic pathogen *Botrytis cinerea*. *Current Biology*, **10**: 751–757. DOI: https://doi.org/10.1016/S0960-9822(00)00560-1.

Hagle, S.K. (2010). *Management Guide for Armillaria Root Disease*. Washington, DC: Forest Health Protection and State Forestry Organizations of the US Forest Service. URL: https://www.fs.usda.gov/Internet/FSE_DOCUMENTS/stelprdb5187208.pdf.

Jeon, J., Park, S.-Y., Chi, M.-H. *et al.* (2007). Genome-wide functional analysis of pathogenicity genes in the rice blast fungus. *Nature Genetics*, **39**: 561–565. DOI: https://doi.org/10.1038/ng2002.

Jobic, C., Boisson, A.-M., Gout, E. *et al.* (2007). Metabolic processes and carbon nutrient exchanges between host and pathogen sustain the disease development during sunflower infection by *Sclerotinia sclerotiorum*. *Planta*, **226**: 251–265. DOI: https://doi.org/10.1007/s00425-006-0470-2.

Juroszek, P. & von Tiedemann, A. (2015). Linking plant disease models to climate change scenarios to project future risks of crop diseases: a review. *Journal of Plant Diseases and Protection*, **122**: 3-15. DOI: https://doi.org/10.1007/BF03356525.

Khan, J., del Rio, L.E., Nelson, R. *et al.* (2008). Survival, dispersal, and primary infection site for *Cercospora beticola* in sugar beet. *Plant Disease*, **92**: 741–745. DOI: https://doi.org/10.1094/PDIS-92-5-0741.

Knogge, W. (1998). Fungal pathogenicity. *Current Opinion in Plant Biology*, **1**: 324–328. DOI: https://doi.org/10.1016/1369-5266(88)80054-2.

Koh, S., André, A., Edwards, H., Ehrhardt, D. & Somerville, S. (2005). *Arabidopsis thaliana* subcellular responses to compatible *Erysiphe cichoracearum* infections. *The Plant Journal*, **44**: 516–529. DOI: https://doi.org/10.1111/j.1365-313X.2005.02545.x.

Kuroki, M., Okauchi, K., Yoshida, S. *et al.* (2017). Chitin-deacetylase activity induces appressorium differentiation in the rice blast fungus *Magnaporthe oryzae*. *Scientific Reports*, **7**: article number 9697. DOI: https://doi.org/10.1038/s41598-017-10322-0.

Kwiatkowski, M., Engelstädter, J. & Vorburger, C. (2012). On genetic specificity in symbiont-mediated host-parasite coevolution. *PLoS Computational Biology*, **8**: article number e1002633. DOI: https://doi.org/10.1371/journal.pcbi.1002633.

Laluk, K. & Mengiste, T. (2010). Necrotroph attacks on plants: wanton destruction or covert extortion? *The Arabidopsis Book (American Society of Plant Biologists)*, **8**: e0136. DOI: https://doi.org/10.1199/tab.0136.

Łaźniewska, J., Macioszek, V.K. & Kononowicz, A.K. (2012). Plant-fungus interface: the role of surface structures in plant resistance and susceptibility to pathogenic fungi. *Physiological and Molecular Plant Pathology*, **78**: 24–30. DOI: https://doi.org/10.1016/j.pmpp.2012.01.004.

Li, X., Gao, C., Li, L. *et al.* (2017). MoEnd3 regulates appressorium formation and virulence through mediating endocytosis in rice blast fungus *Magnaporthe oryzae*. *PLoS Pathogens*, **13**: article number e1006449. DOI: https://doi.org/10.1371/journal.ppat.1006449.

Liu, H., Suresh, A., Willard, F.S. *et al.* (2007). Rgs1 regulates multiple G subunits in *Magnaporthe* pathogenesis, asexual growth and thigmotropism. *The EMBO Journal*, **26**: 690–700. DOI: https://doi.org/10.1038/sj.emboj.7601536.

Liu, W., Zhou, X., Li, G. *et al.* (2011). Multiple plant surface signals are sensed by different mechanisms in the rice blast fungus for appressorium formation. *PLoS Pathogens*, **7**: article number e1001261. DOI: https://doi.org/10.1371/journal.ppat.1001261.

Lorenz, M.C. (2002). Genomic approaches to fungal pathogenicity. *Current Opinion in Microbiology*, **5**: 372–378. DOI: http://www.sciencedirect.com/science/article/B6VS2-468VN2T-2/1/5d06cddd121a56eb8b4f1e3cd94ddc96.

Maloy, O.C. (2005). Plant disease management. In *The Plant Health Instructor*. Washington, DC: American Phytopathological Society. DOI: https://doi.org/10.1094/PHI-I-2005-0202-01.

Mares, D., Romagnoli, C., Andreotti, E. *et al.* (2006). Emerging antifungal azoles and effects on *Magnaporthe grisea*. *Mycological Research*, **110**: 686–696. DOI: https://doi.org/10.1016/j.mycres.2006.03.006.

Martin-Urdiroz, M., Osés-Ruiz, M., Ryder, L.S. & Talbot, N.J. (2016). Investigating the biology of plant infection by the rice blast fungus *Magnaporthe oryzae*. *Fungal Genetics and Biology*, **90**: 61–68. DOI: https://doi.org/10.1016/j.fgb.2015.12.009.

Massart, S. & Jijakli, H.M. (2007). Use of molecular techniques to elucidate the mechanisms of action of fungal biocontrol agents: a review. *Journal of Microbiological Methods*, **69**: 229–241. DOI: https://doi.org/10.1016/j.mimet.2006.09.010.

McLeod, G., Gries, R., von Reuß, S.H. *et al.* (2005). The pathogen causing Dutch elm disease makes host trees attract insect vectors. *Proceedings of the Royal Society of London, Series B, Biological Sciences*, **272**: 2499–2503. DOI: https://doi.org/10.1098/rspb.2005.3202.

Money, N.P. (1999). Biophysics: fungus punches its way in. *Nature*, **401**: 332–333. DOI: https://doi.org/10.1038/43797.

Moore, D. (2000). *Slayers, Saviors, Servants and Sex. An Exposé of Kingdom Fungi*. New York: Springer-Verlag, 176 pp. ISBN-10: 0387951016, ISBN-13: 978-0387951010. See Chapter 2, Blights, rusts, bunts and mycoses. Tales of fungal diseases.

Moore, D. & Novak Frazer, L. (2002). *Essential Fungal Genetics*. New York: Springer-Verlag. ISBN-10: 0387953671, ISBN-13: 978-0387953670.

Moscou, M.J. & van Esse, H.P. (2017). The quest for durable resistance. *Science*, **358**: 1541–1542. DOI: https://doi.org/10.1126/science.aar4797.

Moss, M.O. (2008). Fungi, quality and safety issues in fresh fruits and vegetables. *Journal of Applied Microbiology*, **104**: 1239–1243. DOI: https://doi.org/10.1111/j.1365-2672.2007.03705.x.

Nikawa, H., Nishimura, H., Hamada, T. & Sadamori, S. (1997). Quantification of thigmotropism (contact sensing) of *Candida albicans* and *Candida tropicalis*. *Mycopathologia*, **138**: 13–19. DOI: https://doi.org/10.1023/A:1006849532064.

Oerke, E.-C. (2006). Crop losses to pests. *The Journal of Agricultural Science*, **144**: 31–43. DOI: https://doi.org/10.1017/S0021859605005708.

Oh, Y., Donofrio, N., Pan, H. *et al.* (2008). Transcriptome analysis reveals new insight into appressorium formation and function in the rice blast fungus *Magnaporthe oryzae*. *Genome Biology*, **9**(5): R85. DOI: https://doi.org/10.1186/gb-2008-9-5-r85.

Oliver, R.P. & Ipcho, S.V.S. (2004). *Arabidopsis* pathology breathes new life into the necrotrophs-vs.-biotrophs classification of fungal pathogens. *Molecular Plant Pathology*, **5**: 347–352. DOI: https://doi.org/ 10.1111/J.1364–3703.2004.00228.X.

Parker, B.J., Hrček, J., McLean, A.H.C. & Godfray, H.C.J. (2017). Genotype specificity among hosts, pathogens, and beneficial microbes influences the strength of symbiont-mediated protection. *Evolution*, **71**: 1222–1231. DOI: https://doi.org/10.1111/evo.13216.

Patpour, M., Hovmøller, M.S., Shahin, A.A. *et al.* (2016). First report of the Ug99 race group of wheat stem rust, *Puccinia graminis* f. sp. *tritici*, in Egypt in 2014. *Plant Disease*, **100**: 863–863. DOI: https://doi.org/10.1094/PDIS-08-15-0938-PDN.

Plassard, C. & Fransson, P. (2009). Regulation of low molecular weight organic acid production in fungi. *Fungal Biology Reviews*, **23**: 30–39. DOI: https://doi.org/10.1016/j.fbr.2009.08.002.

Popa, V. I., Dumitru, M., Volf, I. & Anghel, N. (2008). Lignin and polyphenols as allelochemicals. *Industrial Crops and Products*, **27**: 144–149. DOI: https://doi.org/10.1016/j.indcrop.2007.07.019.

Raffa, K.F., Grégoire, J.-C. & Lindgren, B.S. (2015). Natural history and ecology of bark beetles. In *Bark Beetles: Biology and Ecology of Native and Invasive Species*, ed. F.E. Vega & R.W. Hofstetter. San Diego, CA: Academic Press, an imprint of Elsevier Inc., pp. 1–40. ISBN: 9780124171565. DOI: https://doi.org/10.1016/B978-0-12-417156-5.00001-0.

Reidsma, P. & Jeuffroy, M.-H. (2017). Farming systems analysis and design for sustainable intensification: new methods and assessments. *European Journal of Agronomy*, **82**: 203–205. DOI: https://doi.org/10.1016/j.eja.2016.11.007.

Scheffer, R.J., Voeten, J.G.W.F. & Guries, R.P. (2008). Biological control of Dutch elm disease. *Plant Disease*, **92**: 192–200. DOI: https://doi.org/10.1094/PDIS-92-2-0192.

Serrano, M., Coluccia, F., Torres, M., L'Haridon, F. & Métraux, J.-P. (2014). The cuticle and plant defense to pathogens. *Frontiers in Plant Science*, **5**: 274. DOI: https://doi.org/10.3389/fpls.2014.00274.

Skaracis, G.N., Pavli, O.I. & Biancardi, E. (2010). *Cercospora* leaf spot disease of sugar beet. *Sugar Tech*, **12**: 220–228. DOI: https://doi.org/10.1007/s12355-010-0055-z.

Slavica, M., Gullino, M.L. & Spadaro, D. (2017). The puzzle of bakanae disease through interactions between *Fusarium fujikuroi* and rice. *Frontiers in Bioscience, Elite*, **9**: 333–344. DOI: https://doi.org/10.2741/E806.

Smith, M.L., Bruhn J.N. & Anderson, J.B. (1992). The fungus *Armillaria bulbosa* is among the largest and oldest living organisms. *Nature*, **356**: 428–431. DOI: https://doi.org/10.1038/356428a0.

Stephenson, K.S., Gow, N.A.R., Davidson, F.A. & Gadd, G.M. (2014). Regulation of vectorial supply of vesicles to the hyphal tip determines thigmotropism in *Neurospora crassa*. *Fungal Biology*, **118**: 287–294. DOI: https://doi.org/10.1016/j.funbio.2013.12.007.

Stergiopoulos, I. & Gordon, T.R. (2014). Cryptic fungal infections: the hidden agenda of plant pathogens. *Frontiers in Plant Science*, **5**: 506. DOI: https://doi.org/10.3389/fpls.2014.00506.

Tada, Y., Kusaka, K., Betsuyaku, S. *et al.* (2005). Victorin triggers programmed cell death and the defense response via interaction with a cell surface mediator. *Plant and Cell Physiology*, **46**: 1787–1798. DOI: https://doi.org/10.1093/pcp/pci193.

Talbot, N.J. (2007). Fungal genomics goes industrial. *Nature Biotechnology*, **25**: 542–543. DOI: https://doi.org/10.1038/nbt0507-542.

Urban, J. & Milon, D. (2014). Sap flow-based quantitative indication of progression of Dutch elm disease after inoculation with *Ophiostoma novo-ulmi*. *Trees*, **28**: 1599–1605. DOI: https://doi.org/10.1007/s00468-014-1068-0.

van Dijk, M., Morley, T., Jongeneel, R., van Ittersum, M., Reidsma, P. & Ruben, R. (2017). Disentangling agronomic and economic yield gaps: an integrated framework and application. *Agricultural Systems*, **154**: 90–99. DOI: https://doi.org/10.1016/j.agsy.2017.03.004.

van Kan, J.A.L. (2006). Licensed to kill: the lifestyle of a necrotrophic plant pathogen. *Trends in Plant Science*, **11**: 247–253. DOI: https://doi.org/10.1016/j.tplants.2006.03.005.

Voegele, R.T. (2006). *Uromyces fabae*: development, metabolism, and interactions with its host Vicia faba. *FEMS Microbiology Letters*, **259**: 165–173. DOI: https://doi.org/10.1111/j.1574-6968.2006.00248.x.

Voegele, R.T. & Mendgen, K. (2003). Rust haustoria: nutrient uptake and beyond. *New Phytologist*, **159**: 93–100. DOI: https://doi.org/10.1046/j.1469-8137.2003.00761.x.

Vogel, J., Kenter, C., Holst, C. & Märländer, B. (2018). New generation of resistant sugar beet varieties for advanced integrated management of *Cercospora* leaf spot in Central Europe. *Frontiers in Plant Science*, **9**: 222. DOI: https://doi.org/10.3389/fpls.2018.00222.

Volk, T. (2002). The humongous fungus: ten years later. *Inoculum*, **53**: 4–8. URLs: http://www.msafungi.org/wp-content/uploads/Inoculum/53(2).pdf; http://botit.botany.wisc.edu/toms_fungi/apr2002.html.

Webber, J. (2016). The ups and downs of Dutch elm disease. Observatree web blog at http://www.observatree.org.uk/the-ups-and-downs/.

Wight, W.D., Labuda, R. & Walton, J.D. (2013). Conservation of the genes for HC-toxin biosynthesis in *Alternaria jesenskae*. *BMC Microbiology*, **13**: 165. DOI: https://doi.org/10.1186/1471-2180-13-165.

Yan, X. & Talbot, N.J. (2016). Investigating the cell biology of plant infection by the rice blast fungus Magnaporthe oryzae. *Current Opinion in Microbiology*, **34**: 147–153. DOI: https://doi.org/10.1016/j.mib.2016.10.001.

Zhang, L., Zhang, D., Chen, Y. *et al.* (2019). *Magnaporthe oryzae* CK2 accumulates in nuclei, nucleoli, at septal pores and forms a large ring structure in appressoria, and is involved in rice blast pathogenesis. *Frontiers in Cellular and Infection Microbiology*, **9**: article 113. DOI: https://doi.org/10.3389/fcimb.2019.00113.

Zhang, W., Chen, S., Abate, Z. *et al.* (2017). Identification and characterization of Sr13, a tetraploid wheat gene that confers resistance to the Ug99 stem rust race group. *Proceedings of the National Academy of Sciences of the United States of America*, **114**: E9483–E9492. DOI: https://doi.org/10.1073/pnas.1706277114.

Zipfel, C. (2014). Plant pattern-recognition receptors. *Trends in Immunology*, **35**: 345–351. DOI: https://doi.org/10.1016/j.it.2014.05.004.

CHAPTER 17

Fungi As Pathogens of Animals, Including Humans

In this chapter, we study fungi as pathogens of animals, including humans. There are many pathogens of insects among the fungi and fungus-like organisms: *Microsporidia*, trichomycetes, *Laboulbeniales* and entomogenous fungi. Inevitably, discussion of insect disease eventually turns to thoughts of the potential for biological control of arthropod pests. In other animals, cutaneous chytridiomycosis is an **e**merging **i**nfectious **d**isease (**EID**) of amphibians.

In both animals and plants, an unprecedented number of fungal and fungal-like diseases have recently caused some of the most severe die-offs and extinctions ever witnessed in species in the wild, and they are jeopardising our food security. We have already mentioned some EIDs of plants, especially crop plants. We will discuss EIDs further in this chapter. Among animals, fungal EIDs have reduced population abundances in amphibians, bats and even corals, across many species and over large geographical areas, and the most recently recognised fungal disease of snakes may have caused declines in some snake populations in the eastern United States.

Our main concern, though, are the **mycoses** that are the fungus diseases of humans. We describe the clinical groupings set up for human fungal infections, fungi within the home, and their effects on health through production of allergens and toxins. In the penultimate section we attempt a comparison of animal and plant pathogens and briefly discuss the essentials of epidemiology. We finish the chapter with a short discussion of mycoparasitic and fungicolous fungi; that is, fungi that are pathogenic on other fungi.

As with other chapters, we will introduce you to some important and thought-provoking academic publications in this chapter; and here we will direct you to the Editorial published on 25 July 2017 in the journal *Nature Microbiology* (DOI: https://doi.org/10.1038/nmicrobiol.2017.120). If you read no other reference, we suggest you read this. We are so keen on this Editorial because we believe the message of its content should be much more widely appreciated. And we like its title: **'Stop neglecting fungi'**. The following quotation will give you a flavour of its content:

> over 300 million people suffer from serious fungal-related diseases, … fungi collectively kill over 1.6 million people annually, which is more than malaria and similar to the tuberculosis death toll. Fungi and oomycetes destroy a third of all food crops each year, which would be sufficient to feed 600 million people. Furthermore, fungal infestation of amphibians has led to the largest disease-caused loss of biodiversity ever recorded, while fungi also cause mass mortality of bats, bees and other animals, and decimate fruit orchards, pine, elm and chestnut forests …

The report published by the Royal Botanic Gardens, Kew, **'State of the World's Fungi 2018'**, also coined a memorable phrase: 'when looking for nature-based solutions to some of our most critical global challenges, fungi could provide many of the answers' (Willis, 2018).

17.1 PATHOGENS OF INSECTS

Arthropods are the most diverse group of animals on Earth; insects occurring in most terrestrial environments, though only a few species are found in marine habitats, as these are dominated by the crustaceans. Estimates of the number of arthropod species vary between 1,170,000 and 10 million, accounting for over 80% of all known living animal species; with the insects as the most species-rich subgroup in land and freshwater environments. As arthropods were emerging as the (numerically) dominant animals they are today, fungi were also colonising the land. Over the past 400 million years or so, fungi and insects have evolved together in a wide array of intimate associations; including mutualistic endosymbiosis; using fungi as obligate food sources, such as those found in fungus-gardening ants (section 15.2); sexually and behaviourally transmitted parasites, such as *Laboulbeniales* (section 17.4); and with fungi as the most common disease-causing agents of insects, and many entomologists believe that there may be more species of insect ***pathogens*** than there are species of insects (Lovett & St. Leger, 2017).

Insect species are infected by pathogenic viruses, bacteria and protozoa, as well as two groups of organisms that have uncertain relationships, the microsporidia and the trichomycetes; all of these tend to cause infection after being ingested by the insect.

The true fungi that infect insects are **invasive pathogens**; that is, they can produce enzymes and hyphae that can penetrate the insect cuticle. These disease organisms often act as natural control agents by regulating the population size of insect pests. There is consequently considerable interest in harnessing this capability commercially to control insects that are pests because of their adverse effects on our agricultural activities or because they act as vectors of human diseases (Lacey *et al.*, 2015; Butt *et al.*, 2016; and see Chapters 1 & 2 in *Ecofriendly Pest Management for Food Security*, Omkar, 2016; Blackwell, 2017; Lovett & St. Leger, 2017).

17.2 *MICROSPORIDIA*

Microsporidia comprise a phylum of over 1,500 species of fungal-related parasites that can infect nearly all animal hosts. *Microsporidia* are the smallest of eukaryotes: they have genomes in the same size range as bacterial genomes; they have unicellular spores; they lack mitochondria, peroxisomes and centrioles, but have several prokaryotic characteristics, such as 70S ribosomes, and fused 5.8S and 28S rRNAs. Nevertheless, they are highly specialised eukaryotic cells, living only as obligate intracellular parasites of other eukaryotes. Most are important pathogens of insects, but they are also responsible for common diseases of crustaceans and fish, and have been found in most other animal groups, including humans (probably transmitted through contaminated food and/or water) (Weiss & Becnel, 2014; Troemel, 2017).

Microsporidia have been subjected to the most dramatic taxonomic revisions over the years. They were traditionally thought to be a unique phylum of spore-forming protozoa. Then, on the basis of the earliest electron microscopy studies, they were presumed to be one of the most primitive eukaryotic lineages because of a remarkable absence of 'standard' eukaryotic features such as Golgi bodies, peroxisomes, mitochondria and the 9 + 2 arrangement of microtubules. They were placed with other protists lacking mitochondria in a kingdom called Archezoa, and microsporidia were postulated to be direct descendants of a primitive eukaryote that predated mitochondrial endosymbiosis. Finally, genome sequencing and the discovery of microsporidian **mitosomes**, which are highly reduced (in both physical size and biochemical properties) mitochondrial relics, caused the reclassification of microsporidia as fungi. *Microsporidia* share several characteristics with fungi: nuclear division takes place within an intact nuclear membrane, they have the same mRNA capping mechanism, chitin and trehalose in their spores, and some gene structures present only in fungal opisthokonts (Corradi & Keeling, 2009; Capella-Gutiérrez *et al.*, 2012).

While the fungal nature of microsporidia is now accepted, their exact position in the fungal tree of life is still uncertain. They are thought to be related to the *Chytridiomycota* and may be derived from a reduced endoparasitic chytrid (see section 2.8) (Gill & Fast, 2006). In this case, they are now presumed not to be the most primitive eukaryotes, but to be **the most reduced and highly specialised fungi**. Adaptations to an obligate intracellular lifestyle have modified their cell biology and genome by severe **selective reduction** of cell structure, metabolism and gene structure as they separated from the main fungal lineage as a sister clade (see Figure 2.8). Microsporidian genomes, although they are always very small, do have multiple linear chromosomes with telomeres, but in other respects they resemble bacteria by having few introns and transposons, generally short intergenic regions and few duplicated genes. The complement of genes encoded in microsporidian genomes is remarkably similar from one species to another, regardless of genome size, and it is evident that microsporidia have acquired specific genes from unrelated lineages through horizontal gene transfer (see sections 6.5 and 19.15); many of these genes playing a central role in microsporidian evolution (Keeling *et al.*, 2014; Corradi, 2015).

Although their taxonomic classification has evolved through time, understanding which kingdom microsporidia belong to is not an idle academic pastime. *Microsporidia* cause diseases and it is essential for the clinician to understand that the disease organism is a **fungus** and that the disease is likely to be controllable by **antifungal agents** rather than wasting time and resources trying the effects of **antibacterials** or **antiprotozoal** drugs.

The first microsporidian disease to be described, a disease of European silkworms studied in 1857, was shown to be due to a microscopic parasite that was named *Nosema bombycis* and was assigned to a new group of organisms called *Microsporidia* in 1882 (Keeling & Fast, 2002). The microsporidian life cycle ranges from simple to very complex. Reproduction may be asexual, sexual or both, depending on the species, and some species use intermediate hosts. Some may have several cycles of sporulation and several spore types with different functions including reinfection within the same host. Only the spores are viable outside an animal host.

Microsporidian spore morphology typically includes a thick protective outer wall protein and inner wall of **chitin**; but its most obvious morphological feature is the polar filament. This is specialised for host invasion and provides the microsporidian spore with a unique and highly specialised mode of host infection, where the parasite enters its host through a **projectile tube** that is expelled at high velocity. The **polar filament** (or **polar tube**) is a fine hollow tube, which is attached to a plate at one end of the spore, is then coiled within the mature spore, and terminates near the posterior vacuole at the other end of the spore (Figure 17.1) (Keeling & Fast, 2002; Vávra & Larsson, 2014).

The spores are ingested by the host and when the spore germinates in the gut, the polar filament **everts** very rapidly, punctures the membrane of a host cell in the gut wall and effectively

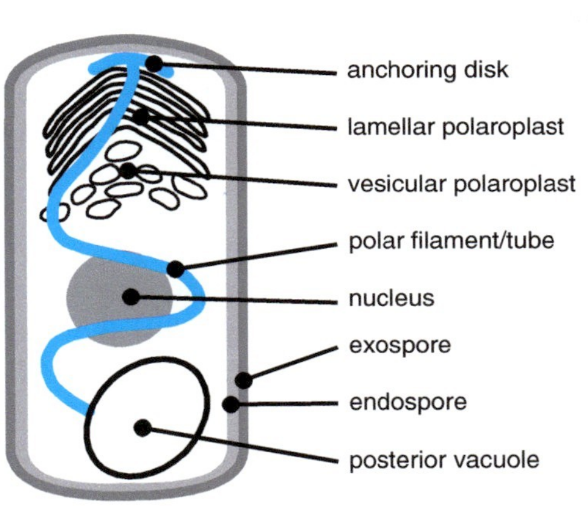

Figure 17.1. The major structures of a typical microsporidian spore. The spore cytoplasm is enclosed by a normal plasma membrane with two rigid extracellular walls. The exospore wall has a dense, proteinaceous matrix; the endospore wall, which thins at the apex, is composed of chitin and proteins. The cytoplasm of the spore (the sporoplasm) is the infectious material of microsporidia. It contains either one nucleus or two closely appressed nuclei (a diplokaryon). The cytoplasm is rich in 70S ribosomes, and is dominated by infection structures: the polaroplast, the polar filament or polar tube, and the posterior vacuole. The membranous polaroplast occupies the anterior part of the spore and is differentiated into the closely stacked membranes of the lamellar polaroplast, and the loosely organised posterior vesicular polaroplast. The polar filament is composed of membrane and glycoprotein layers and ranges from 0.1–0.2 μm in diameter and 50–500 μm in length, being attached at the apex of the spore via an umbrella-shaped anchoring disc. The polar filament is straight for one-third to one-half the length of the spore, and the rest is helically coiled in the sporoplasm. The number of coils, their relative arrangement, and angle of helical tilt are conserved and diagnostic for a particular species. The polar filament terminates at the posterior vacuole, though the point of contact has never been observed. Based on illustrations in Keeling & Fast (2002).

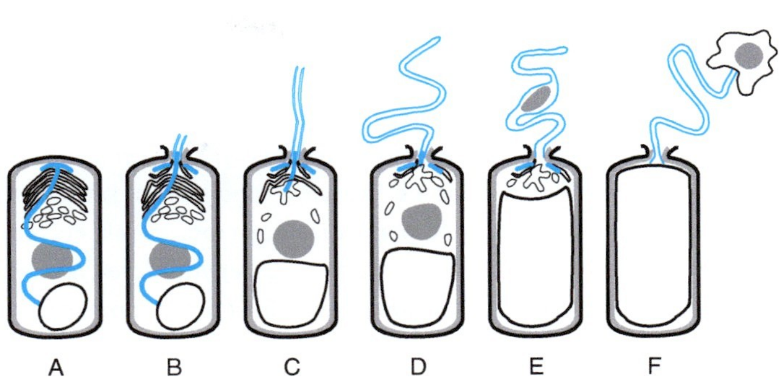

Figure 17.2. Germination of a microsporidian spore depends on eversion of the polar filament or tube and is illustrated in this cartoon strip. (A) Dormant spore, showing polar filament (blue), nucleus (grey), polaroplast and posterior vacuole (as in Figure 17.1). (B) Polaroplast and posterior vacuole begin to swell and anchoring disc ruptures; polar filament begins to emerge, everting as it does so. (C) Polar filament continues to evert. (D) With the polar tube fully everted, the sporoplasm is forced into and then (E) through the polar tube. (F) Sporoplasm emerges from the polar tube bound by new membrane. Based on illustrations in Keeling & Fast (2002).

injects the content of the spore into the cytoplasm of that cell (Figure 17.2). The everted polar filament ranges from 50 to 500 μm in length, which can be up to 100 times the length of the spore. Germination can be complete in less than 2 s and the tip of the polar filament can move at more than 100 μm s^{-1}. When the polar tube is fully extended, the pressure within the spore (most likely generated by the posterior vacuole) forces the sporoplasm through the polar tube (Figure 17.2F). This also occurs rapidly; the sporoplasm is completely injected into the host within 15–500 ms. Following vegetative reproduction in the gut tissues, microsporidia may spread to other tissues.

If this mode of infection is not remarkable enough, a possibly more remarkable feature of microsporidian biology is the ability of the parasite to stimulate **hypertrophic growth** (abnormal enlargement) of the invaded cells of the host animal. The organisation of the host cell structure changes completely and the parasite proliferates inside it, but both are morphologically and physiologically integrated to form a separate entity that develops within the host animal but at the expense of the host animal. It seems to be a **symbiotic co-existence** between the animal cell and its microsporidian parasites to form what is called a **xenoparasitic complex** rather like a tumour. The term **xenoma** is currently used for microsporidian xenoparasitic complex. *Microsporidia* produce xenomas in oligochaetes, crustaceans, insects and poikilothermic vertebrates, mainly fish (and that includes many commercial fish) (Vergneau-Grosset *et al.*, 2016).

Most microsporidia are specific to one or a few closely related host species. More than half of the described species have been isolated from *Lepidoptera* and *Diptera*. Some microsporidia cause acute diseases that result in death of the host, but most are chronic pathogens that reduce longevity and fecundity, prolong larval development, or cause failure to pupate.

Over 1,300 species in 170 genera of microsporidia have been described, and at least 14 species are **pathogens of humans** causing opportunistic infections (generally called **microsporidiosis**) around the world. Human microsporidiosis occurs mainly, but not exclusively, in severely immunocompromised patients, such as those infected with HIV, and transplant patients.

The most common **clinical symptom** is diarrhoea caused by general gastrointestinal infection, but conjunctivitis and other corneal and eye infections occur, as do muscular, respiratory, genitourinary and disseminated infections. Names to look out for as human pathogens are: *Brachiola algerae, rachiola connori, rachiola vesicularum, Encephalitozoon cuniculi* (the most common mammalian microsporidian), *Encephalitozoon hellem, Encephalitozoon intestinalis, Enterocytozoon bieneusi Microsporidium ceylonensis, Microsporidium africanum, Nosema ocularum, Pleistophora* sp., *Trachipleistophora hominis, Trachipleistophora anthropophthera, Vittaforma corneae* (see the US Department of Health & Human

Services, Centers for Disease Control and Prevention web pages at https://www.cdc.gov/dpdx/microsporidiosis/index.html).

Some **domestic and wild animals** carry microsporidian infections, especially *Encephalitozoon cuniculi*, *Encephalitozoon intestinalis* and *Enterocytozoon bieneusi*, and are a source of human infections. Cage birds, particularly parrots, parakeets and budgerigars, can also carry infections of *Encephalitozoon hellem*. *Enterocytozoon bieneusi*, *Nosema* sp. and *Vittaforma corneae* have been identified in storm drains and ditches and can contaminate domestic water supplies (but note that the most common water-borne disease caused by faecal contamination of water supplies is cryptosporidiosis caused by *Cryptosporidium*, which **is** a **protozoan** parasite of mammalian intestines).

Loma salmonae causes microsporidial gill disease of salmon (MGDS) in farmed Pacific salmon, causing severe and extensive inflammation of the gills and consequent respiratory distress and high mortality. It has become an important disease of the Canadian salmon farming industry. The microsporidean parasite *Nosema apis* infects the epithelial cells of the midgut of the honey bee (*Apis mellifera*) and has spread around the world. It is a serious disease of honey bees in the temperate zones causing severe reduction in productive capacity and contributing to unusually high colony losses in recent years (Mitchell & Rodger, 2011; Charbonneau *et al.*, 2016).

17.3 TRICHOMYCETES

Microsporidia are fungi that were long thought to be protozoans and are still called protozoans by many people who should know better. Trichomycetes, on the other hand, includes organisms that are clearly protozoans though they were mistakenly classified as fungi for a long time, and are still called fungi by many people who should know better.

These extremely common organisms live only in the **digestive tracts of insects** and other arthropods, generally as commensals, sometimes as pathogens or symbionts (mutualists), which are associated with, although not penetrating, the cuticle lining the digestive tracts of the host animal. Their hosts include terrestrial, marine and freshwater arthropods, most commonly midges (*Chironomidae*), mosquitoes (*Culicidae*), black flies (*Simuliidae*), beetles (*Coleoptera*), stoneflies (*Plecoptera*) and mayflies (*Ephemeroptera*), as well as several millipedes (*Diplopoda*) and crustaceans.

The traditional taxonomy is based on a few micromorphological characters and the traditional view was to place the Class *Trichomycetes* in the zygomycetes (see section 3.5 for some discussion of the polyphyletic nature of the zygomycetes), the class being divided into four orders:

- *Amoebidiales* (which occur on the external surfaces of freshwater arthropods),
- *Asellariales*,
- *Eccrinales*, and
- *Harpellales*.

Characteristically the trichomycetes develop nonseptate (in the Amoebidiales and Eccrinales) or irregularly septate (in the *Harpellales* and *Asellariales*) vegetative mycelia and asexual sporangia. Further, most genera of the *Harpellales* produce **zygospores** and it is this character which was used to include all the trichomycetes among the zygomycete fungi. The Amoebidiales and the Eccrinales have now been removed from this association; ***they are not fungi, but they are close to the phylogenetically ancient animal-fungal divergence*** (Reynolds *et al.*, 2017).

The Amoebidiales are amoebae-producing organisms that attach to the exoskeleton of freshwater arthropods (*Amoebidium parasiticum* was the first described). Production of amoebae is not otherwise present in Kingdom *Fungi*. More significantly, *Amoebidium parasiticum* has stacked dictyosomes, which do not occur in fungi, and **lacks chitin** in its cell wall. Taken together these features do not make it a good candidate as a fungus. The Eccrinales have unbranched, nonseptate, multinucleate thalli, and produce sporangiospores, which form from the apex downward towards the base of the thallus, a feature found only in Kingdom *Fungi*. These few distinctive morphological characters together with the fact that they share a very specialised ecological niche with genuine fungi like the *Harpellales* was all that classified them within the trichomycetes. Now, sequence analyses have shown that the Eccrinales share a common ancestry with the Amoebidiales and are closely related to members of the protist class ***Mesomycetozoea***, which is positioned at the animal-fungal boundary in the opisthokont lineage (Mendoza *et al.*, 2002; Cafaro, 2005; Reynolds *et al.*, 2017).

Removing these fungus-like protists leaves the trichomycetes consisting of the *Harpellales* and *Asellariales*. These are true fungi having hyphal thalli with cell walls containing chitin fibrils and being regularly septate with incomplete septa having a plugged central pore. No confirmed sexual stage has been reported generally for the *Asellariales*, although conjugation has been reported between cells of the one species *Asellaria ligiae*. Generally, in the *Harpellales*, though, zygospores (biconical and apically thickened when mature, see Figure 3.9) form following conjugation between cells of the same, or different thalli, being borne at the apex of zygosporophores. Phylogenetic analyses confirm that the *Harpellales* and *Asellariales* both belong to Subphylum *Kickxellomycotina* within the traditional zygomycetes, although this is a heterogeneous collection (see our earlier discussion in section 3.5). A major difficulty for study of these organisms is that they are microfungi which are highly specialised for attachment to the gut wall of arthropods and they are **obligate** endosymbionts; consequently, very few of them can be cultured axenically (i.e. in pure culture) although this is needed to prepare fungal macromolecules free of contaminating host molecules. Only 8 of the 38 genera of *Harpellales* have been cultured, but none of the *Asellariales* (Tretter *et al.*, 2014).

The unique **trichospores**, for which the class is named, are the asexual spores. *Harpellales* produce branched or unbranched

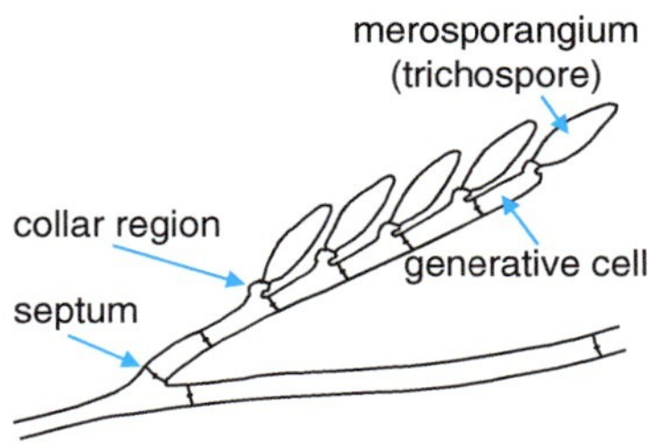

Figure 17.3. Diagram of the general pattern of asexual reproduction in *Smittium* (*Harpellales*). Redrawn after Moss and Young (1978).

thalli, and either the entire thallus or lateral branches of it become regularly septate at maturity to form a series of uninucleate generative cells. From the apical region of each generative cell, a single unisporous merosporangium is produced; this is the trichospore (Figures 3.9 and 17.3). In many genera the merosporangia are borne on short lateral branches, which form the collar region of the generative cell. Members of the *Asellariales* do not produce deciduous merosporangia but the regularly septate branches fragment into single-celled arthrospores. In some species of *Asellariales* the arthrospores germinate by producing a single branch, similar in position and form to a trichospore.

These **insect gut fungi** occur worldwide in all habitats; effectively they have been found whenever they have been looked for. The *Harpellales* are predominantly associated with larval aquatic insects, and, occasionally, with freshwater isopod crustaceans, attached to the midgut or hindgut linings.

They also **attach to the midgut** linings of lower dipterans (*Nematocera*, which includes mosquitoes, crane flies, black flies, gnats and midges), mayflies (*Ephemeroptera*), stoneflies (*Plecoptera*), beetles (*Coleoptera*) and caddisflies (*Trichoptera*).

The *Asellariales* includes species that inhabit terrestrial, freshwater and marine isopods (Isopoda: *Crustacea*) such as woodlice, pill bugs and sea slaters, as well as the hexapod springtails (*Collembola*) that are primitive relatives of insects.

They are most often described as being symbiotic with their hosts, but the nature of the association is not at all clear. Certainly, the fungi are highly specialised to existence within the arthropod guts, so they may be **commensals** (an association in which one species derives some benefit while the other is unaffected). But as fungi they are able to produce many digestive enzymes, so this may be a mutualistic symbiosis in which the fungus provides nutrients to the insect hosts by assisting with **food digestion**. A few species appear to be parasitic at some stage of the host's development. All the evidence suggests that only a fraction of the species of arthropod gut fungi is known.

17.4 LABOULBENIALES

The *Laboulbeniales* is an order of fungi within the Class *Laboulbeniomycetes* of the *Ascomycota* (see section 3.7). The group consists of over 2,000 species that are obligate biotrophs (ectoparasites) of insects (mostly beetles; Figure 17.4), mites and millipedes. They are possibly the most unusual microfungi known because they have no hyphae but cellular thalli (Figure 17.4C). These are formed by enlargement and subsequent cell divisions of the two-celled ascospore, which parasitises the host through a foot cell that penetrates the insect cuticle with a haustorium and permits the fungus to extract nutrients from the insect body tissues. However, the fungal parasites in this group do not cause much damage to the host (Weir & Blackwell, 2001).

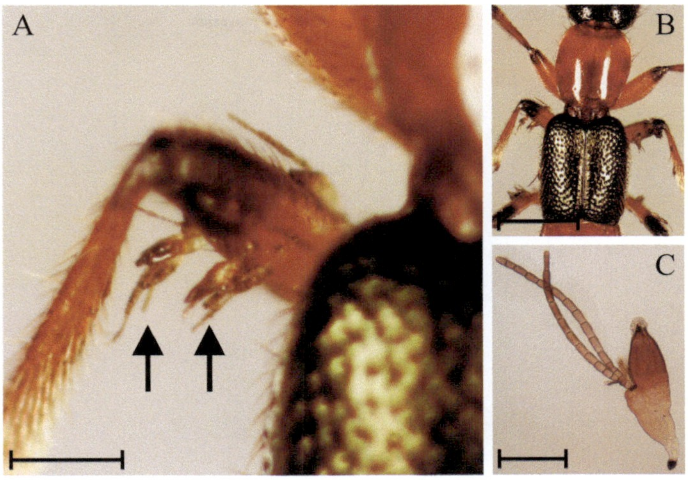

Figure 17.4. *Laboulbenia cristata* on the legs of a rove beetle (*Paederus riparius*). (A) The *Laboulbenia* sporophores (arrowed) *in situ* on one of the legs, and a lower magnification image (B), the scale in relation to the rest of the insect. (C) A *Laboulbenia cristata* sporophore separated from the leg of a rove beetle and mounted in lactophenol for photomicrography. Photographs by Malcolm Storey (http://www.bioimages.org.uk). Images kindly provided by Malcolm Storey.

Although the thallus is morphologically diverse in the group, in each species it consists of a defined number of cells (that is, the thallus shows ***determinate growth***) in a particular arrangement that may include plume-like appendages that help in ascospore release. These appendages, and the small size of the thallus, make the **ectoparasitic fungus** look like bristles on the host (Figures 17.4 and 17.5). So far, the DNA isolations and PCR amplifications that are essential for genomics analyses have been difficult to obtain, mainly because of the small size and determinate growth of the thalli. However, improved isolation techniques that enable efficient and reliable genetic analyses have been published and should soon be applied to these interesting fungi (Haelewaters *et al.*, 2015).

The fungus is transmitted by direct contact between two host individuals, often involved in mating, during which the adhesive ascospores attach to the new host. Ascospores, often discharged in pairs, are spindle shaped and composed of two unequal cells, enclosed in a gelatinous sheath. The longer of the two cells forms the basal portion of the fungal thallus including the foot, through which the fungus attaches to the host and from which the **haustorium** emerges and penetrates into the

Figure 17.5. Scanning electronmicrograph of several thalli of *Rhachomyces philonthinus* (*Laboulbeniales*) on the Rove Beetle *Philonthus*. Image kindly provided by Dr Alex Weir, Faculty of Environmental & Forest Biology, SUNY-ESF, USA. See also http://www.wildaboutbritain.co.uk/pictures/showphoto.php/photo/69703/size/big/cat//ppuser/24039.

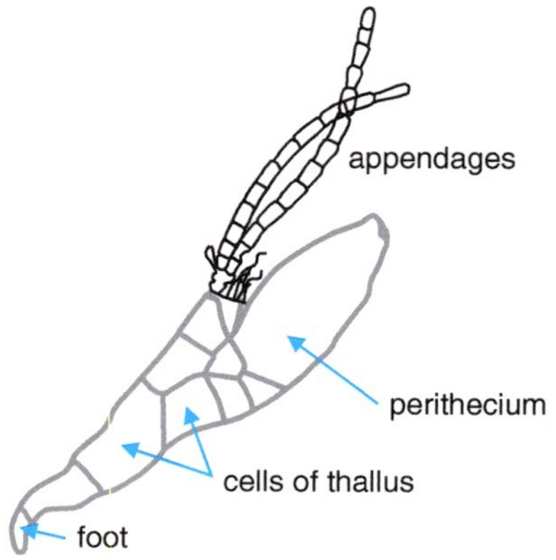

Figure 17.6. Diagram of the main features of the thallus of the *Laboulbeniales*. For more details, see Weir & Beakes (1995).

host tissues (Figure 17.6). The small sporophores (perithecia) are formed as outgrowths from the upper cells of the thallus on the integument of the host following fertilisation by spermatia (microconidia) which may be formed like conventional conidiospores by one or more of the appendages (Figure 17.6).

The *Laboulbeniales* have been recorded from a wide range of **arthropod hosts**: about 80% of *Laboulbeniales* species affect beetles (*Coleoptera*), but in addition cockroaches and allies (*Blattodea*), earwigs (*Dermaptera*), true flies (*Diptera*), bugs (*Heteroptera*), bees and wasps (*Hymenoptera*), ants (*Formicidae*), termites (*Isoptera*), bird lice (*Mallophaga*), crickets and allies (*Orthoptera*), thrips (*Thysanoptera*), millipedes (*Diplopoda*) and mites (*Arachnida*, Subclass *Acari*) are known to carry infections.

Laboulbeniales are the only fungi known to exhibit **extreme host specificity**, which extends to them exhibiting specificity for the parts of the insect on which they will grow; yet the arthropod hosts are considered hyperdiverse, and if there are many fungi on these hosts which are as yet undetected their numbers could considerably inflate the global fungal species total well beyond the range 2.2–3.8 million species of fungi on Earth estimated by Hawksworth and Lücking (2017).

17.5 ENTOMOGENOUS FUNGI

Entomogenous means 'growing on or in the bodies of insects' and strictly speaking applies to all those organisms already discussed in this chapter. However, the word is most often used to describe ***filamentous fungi*** that invade their insect hosts by **penetrating** directly through the cuticle, whereas most organisms in the groups we have already described are either ectoparasites or insect pathogens that generally infect the host when infective spores are ingested into the intestine.

In **entomogenous** (or **'entomopathogenic'**) fungi (Table 17.1), the general infection cycle is that a fungal spore adheres to the cuticle and germinates when conditions are suitable.

The germ tube of the spore then penetrates the host cuticle with a combination of enzymes and physical force and enters the body cavity (haemocoel or haemolymph) of the insect host (diagrammed in Figure 17.7, with an indication of the insect defences) (Butt *et al.*, 2016).

Within the body cavity, the fungus grows vegetatively, often in the form of yeast-like **hyphal bodies** that reproduce by budding, eventually filling the haemocoel and sending hyphae into the solid tissues of the host. As a result, the fungus depletes the tissues of nutrients and the host suffers starvation and may also be poisoned by toxins produced by the fungus. Eventually, the insect dies and the host cuticle ruptures.

The fungus continues to grow **saprotrophically on the cadaver** and the dead insect becomes filled and covered with mycelium; sporophore structures emerge from the cadaver and generate infectious spores (shown in flow chart summary in Figure 17.8 and in the photographs of Figure 17.9) (Castrillo *et al.*, 2005; Labbé *et al.*, 2009).

This final stage is not seen in a few fungal pathogens that rely on dispersal by insect flight or other movement of the infected hosts, for example *Massospora* (*Entomophthorales*), which causes a fungal disease that affects the cicada, and *Strongwellsea* spp. (*Entomophthorales*), which are common fungal pathogens of different types of adult flies. However, an insect cadaver covered by fungal mycelium from which sporophores are emerging and/or around which lays a carpet of released spores is the stage most commonly observed in the field (Figure 17.9C). Because they are easily visible to the naked eye, observation of these final stages allowed early observations of fungal disease in commercially valuable insects like the honey bee and silkworm and helped give birth to invertebrate pathology as a field of study.

Entomogenous fungi have been described from all the major fungal phyla: chytrids, zygomycetes, *Ascomycota* and *Basidiomycota*, and over 700 species of fungi are known to

Table 17.1. Examples of entomogenous fungi

Fungus	Taxonomic position	Host
Coelomomyces psorophorae	*Blastocladiomycota*	Alternates between the larvae of the mosquito (*Culiseta inornata*) and the copepod (*Cyclops vernalis*)
Erynia	*Zoopagomycota*: Order *Entomophthorales*	*Erynia neoaphidis* is a pathogen of aphids; *Erynia radicans* might biocontrol the eastern spruce budworm
Entomophthora muscae	*Zoopagomycota*: Order *Entomophthorales*	Adult *Diptera*, including houseflies
Verticillium lecanii now called *Lecanicillium lecanii*	*Ascomycota*: *Hypocreales*	Aphid, whitefly, used in production of Vertalec (used against glasshouse aphid) and Mycotal (used against glasshouse whitefly)
Beauveria bassiana (the anamorph, asexually reproducing form, of *Cordyceps bassiana*)	*Ascomycota*: *Hypocreales*: *Clavicipitaceae*	Termites, whitefly, different beetles (including potato pest Colorado beetle) and might biocontrol the malaria-transmitting mosquito
Cordyceps militaris	*Ascomycota*: *Hypocreales*: *Clavicipitaceae*	Parasitises and kills moth pupae, including silkworm (*Bombyx mori*)
Nomuraea rileyi	*Ascomycota*: *Hypocreales*: *Clavicipitaceae*	Inhibits larval moult and metamorphosis in the silkworm (*Bombyx mori*)
Metarhizium anisopliae	*Ascomycota*: *Hypocreales*: *Clavicipitaceae*	Infects 200+ insect species, including grasshoppers, termites, thrips and might biocontrol the malaria-transmitting mosquito

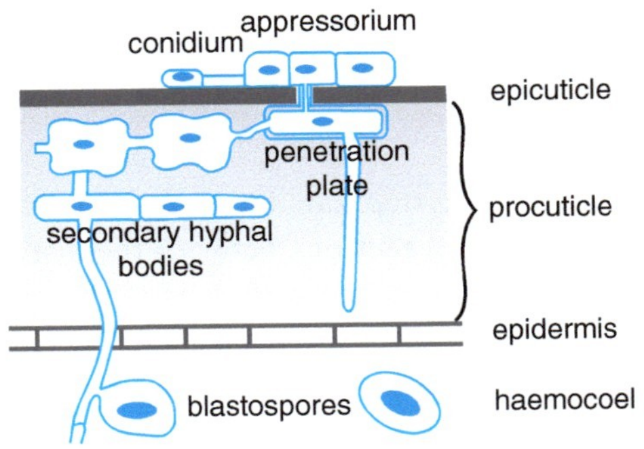

Figure 17.7. Schematic representation of infection structures of *Metarhizium anisopliae* (*Ascomycota*). This is a fungus, which is found in soils throughout the world, that causes disease in over 200 insect species. The illustration shows penetration of the insect cuticle in a diagrammatic cross section. Cuticular resistance barriers may be (a) preformed (including those at the surface, like hydrophobicity, electrostatic charges, other microbes on the surface, low relative humidity and low nutrient availability, toxic lipids and phenols, and/or tanned proteins in the epicuticle; and tanned proteins with crystalline chitin to stiffen and desiccate the procuticle, and protease inhibitors in the procuticle), and/or (b) induced, such as melanisation which cross-links cuticle components to provide a measure of resistance to fungal hydrolytic enzymes.

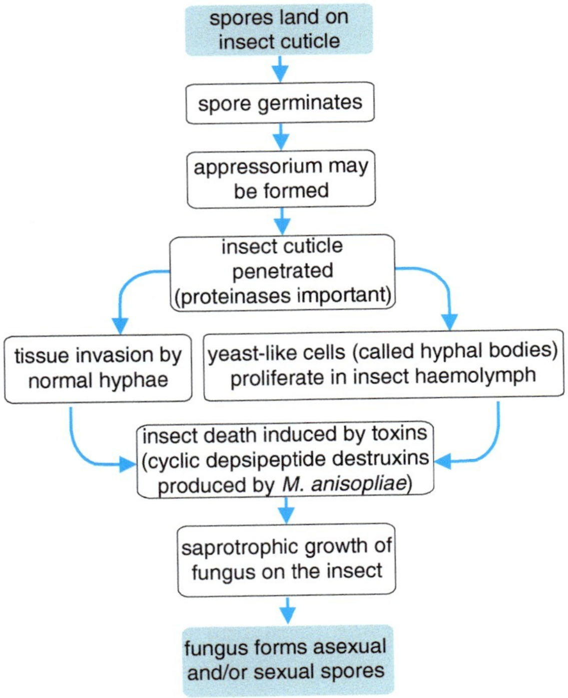

Figure 17.8. Flow chart summarising the generalised infection process of entomogenous fungi.

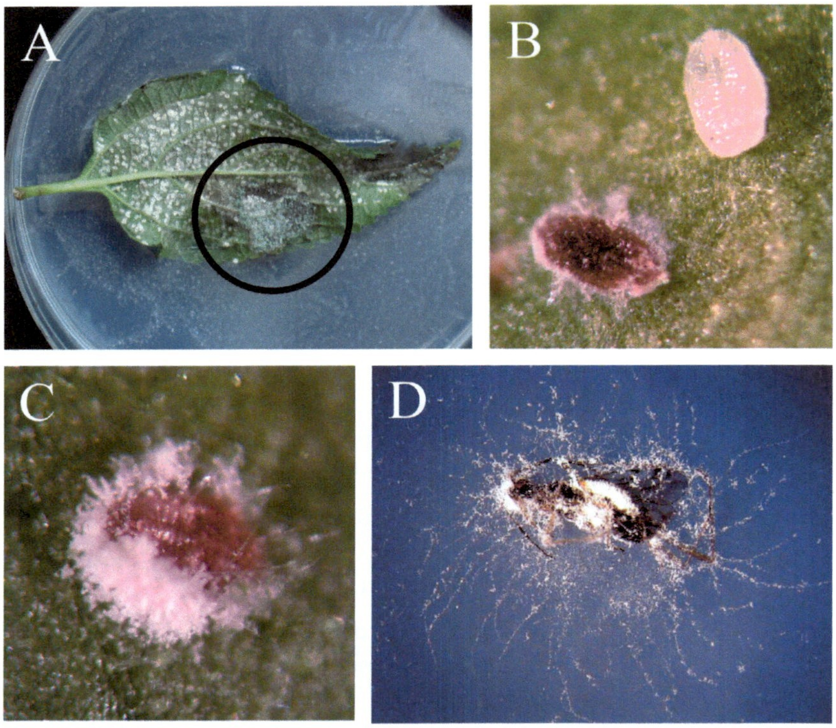

Figure 17.9. The final stages of *Beauveria bassiana* (*Ascomycota*) infections. (A, B and C) show the effect of (B) bassiana infection on whiteflies, *Trialeurodes vaporariorum*. (A) shows the lower surface of a leaf heavily infested with whitefly on which many pupae have been infected by *Beauveria* (circled region in particular). (B) is a magnified view of the leaf surface showing a comparison between an infected whitefly pupa and a healthy one. The infected pupa is red due to the accumulation of large amounts of the red antibiotic oosporein, which is the major secondary metabolite excreted by *Beauveria* and several other fungi. Oosporein is an antifungal agent; is also antagonistic to *Phytophthora infestans* and is a mycotoxin that causes skeletal problems in poultry fed contaminated grain. (C) is another whitefly cadaver showing conidiospore production. (D) is a cadaver of *Dicyphus hesperus*, which is used as a predator in biocontrol of whiteflies and spider mites in greenhouses but is also sensitive to *Beauveria* infection. Photographs by Roselyne Labbé. Modified from Labbé (2005). *Intraguild interactions of the greenhouse whitefly natural enemies, predator* Dicyphus hesperus, *pathogen* Beauveria bassiana *and parasitoid* Encarsia formosa. M.Sc. thesis, Faculté des Sciences de L'Agriculture et de L'Alimentation, Université Laval, Canada, and see Labbé *et al.* (2009). Photographs kindly supplied by Dr Roselyne M. Labbé, Agriculture and Agri-Food Canada.

infest insects. The zygomycetes and *Ascomycota* contain some extremely **common insect pathogens** that are also useful in biocontrol programmes (see section 17.6, below). Although some fungi infect a range of insects (for example, *Verticillium lecanii* (now called *Lecanicillium muscarium*) infects aphids, thrips and whitefly), other species can be extremely specific (*Erynia neoaphidis* only infects aphids). The infection process frequently involves specific fungus–insect host recognition interactions that can even be strain-specific for both fungus and host (Wraight *et al.*, 2007).

For example, two closely related species of *Coelomomyces* (*Blastocladiomycota*), *Coelomomyces dodgei* and *Coelomomyces punctatus*, have sporangia with similar sizes, shapes and general morphology, and their host (mosquito) species ranges and geographic distributions overlap; but they do not interbreed and experimentally forced hybrid zygotes were only partially viable, giving biological support to their classification into two **distinct morphospecies** (Castrillo *et al.*, 2005).

The initial attachment step can be nonspecific, involving **passive hydrophobic interactions** between the insect cuticle and hydrophobin proteins on the surface of the fungal spores (see section 7.8). Spores of the common entomogenous fungi *Beauveria bassiana*, *Metarhizium anisopliae* and *Nomuraea rileyi* are like this and will adhere to both host and non-host insects.

A **mucilaginous coat** also permits passive attachment to surfaces in members of the *Entomophthorales*. In some aquatic entomogenous fungi initial contact between fungal zoospores and mosquito larvae is due to the zoospores having a negative geotaxis that favours collisions with mosquito larvae near the water surface. In contrast, selective attachment depends on the spore (or germ tube) interpreting chemical and physical cues on the host cuticle; this applies to the attachment to mosquito larvae of some species of the chytrid *Coelomomyces*.

In terrestrial filamentous fungi, spore germination results in the emergence of a penetrating germ tube, or a germ tube and appressorium. In either case, a narrow penetration peg then

breaches the insect cuticle using mechanical (turgor pressure) and/or enzymic means, particularly proteases in the latter case because insect cuticle comprises up to 70% protein, but including chitinases and lipases (Charnley, 2003). Many spores also secrete mucilage during the formation of infective structures that enhances adhesion to the host cuticle. **Appressorium formation** and production of cuticle-degrading proteases are triggered by low nutrient levels in *Metarhizium anisopliae*, demonstrating that the fungus senses environmental conditions and host cues at the beginning of the infection process. The enzymes are produced in sequence; protease followed by chitinase is needed to dissolve the insect cuticle. An important point is that cuticle-degrading enzymes not only aid penetration but also provide the fungal germ tube with nutrients (Castrillo *et al.*, 2005).

Entomogenous fungi are **dimorphic** (see section 5.17) and after penetration the fungus changes from filamentous hyphal growth to growing as yeast-like or protoplast hyphal bodies that circulate in the haemolymph and multiply by budding. Later, the fungus changes back to a filamentous mode of growth to invade internal tissues and organs. In *Entomophthora* hyphal growth morphology was induced by low-nutrient conditions, particularly low nitrogen availability (Castrillo *et al.*, 2005). The value of their filamentous morphology to entomogenous fungi is worthy of specific comment. Fungal pathogens of plants and insects have evolved similar methods of breaching the surfaces of their hosts, that is, they both form appressoria from which narrow hyphae penetrate the hosts' surfaces by a combination of mechanical pressure and enzymatic softening. Filamentous fungi are far more successful than unicellular bacteria as pathogens of plants; and the same is true for insects. The reverse applies to their comparative success as pathogens of higher animals.

During growth in the host the fungi also produce a wide variety of **toxic metabolites**, which vary from low molecular weight products of secondary metabolism to complex cyclic peptides and enzymes, some of which are insecticidal. Few compounds have been found in diseased insects in sufficient quantity to account for disease symptoms. An exception is a family of cyclic peptides called the **destruxins**. These are cyclic depsipeptide **toxins** from *Metarhizium* spp. (a depsipeptide is a peptide in which one or more of the amide bonds [-CONHR-] are replaced by ester bonds [-COOR]). Twenty-eight different destruxins have been described with different levels of activity against different insects. The amount of destruxin produced correlates with virulence and host specificity. Destruxins modulate the host immune system and inhibit phagocytes (Charnley, 2003; Liu & Tzeng, 2012).

Many other toxins are produced by entomopathogenic fungi; for example, the entomogenous fungus *Entomophthora muscae*, which infects and kills domestic flies (*Musca domestica*), produces **pheromones** which attract other flies, especially males, towards dead, infected flies producing *Entomophthora* spores: come hither and be infected is the fungus-modified message! Some of the fungal secondary metabolites must be ***neurotoxins*** because changes in insect behaviour are a common feature of fungal infections. The infected host is frequently **driven to climb** up vertical surfaces in the final stages of its life: this might mean climbing up plant stems, rocks, or to the top of walls and windows in domestic pest insects. Being located at altitude is an advantage for spore distribution by the fungus, but it is the dying, infected insect that does the climbing.

Eventually, the fungus emerges through the cuticle and an external mycelium grows over the host and, in appropriate environmental conditions, produces the sporing structures. If circumstances are not suitable, for example dry and/or cold, some fungi form resting structures inside the insect cadaver that produce infective spores when conditions improve.

17.6 BIOLOGICAL CONTROL OF ARTHROPOD PESTS

We have just described a range of arthropod diseases that can govern the development of arthropod populations in nature. They kill specific arthropods and in so doing produce an increased supply of infective agents that can kill more of the same arthropod. Put in those terms, and against the background of our need to produce crops that are subject to attack by numerous arthropod pests and suffer diseases that are transmitted by other such pests, it is not at all surprising that great research effort is devoted to attempting to harness these diseases into **biocontrol agents** that can be used to attack those arthropods that we view as pests. Biological control has been defined as the practice by which the undesirable effects of a pest organism are reduced through the agency of another organism that is not the host, pest or pathogen, or human.

Synthetic chemical **insecticides** and **acaricides** have been the most important element of arthropod control programmes since the discovery of the insecticidal properties of DDT in the late 1930s. They remain the most important element today because they each:

- have a **broad spectrum** of activity that makes them effective against several different types of arthropod;
- are highly efficient;
- are **cheap** and easy to produce;
- are **persistent** and continue to kill pests throughout the season or the field life of the crop.

Chemical pesticides have served us well in the latter half of the twentieth century by reducing the economic damage caused by arthropod pests and by reducing the incidence of many of the diseases of plants and animals that are transmitted by arthropods because of this combination of properties.

But these same properties have also created the problems widely associated with use of chemical pesticides. Because they are highly efficient, cheap and easy to produce, they were rapidly brought into **heavy and widespread use**. Their broad spectrum of activity led to their use throughout the world and in all circumstances in which control of arthropod pests and/or vectors of human and animal diseases was an issue, even if only minor. Their persistence combined with this extensive use resulted in many of these pesticides becoming unnatural components of the natural environment, **contaminating land and food and water** supplies. The varied results included:

- development of **resistance** in many pest populations;
- **emergence of new pests** as old pests were removed;
- elimination of natural enemies of pests;
- disruption of natural ecosystems as **non-pest organisms were also killed**;
- **accumulation of pesticides through food chains** to adversely affect bird and mammalian predators.

All these negatives directed attention towards increased emphasis on **alternative control agents** and strategies, which include:

- **biological control** (biocontrol) agents;
- environmental management;
- use of **pheromones** to control pest mating;
- **genetic modification** to produce arthropod-resistant plants;
- combination of these tactics into **integrated pest management** (IPM) programmes.

Naturally occurring pathogens of arthropods are good candidates as biological control agents and many species are already employed, at least on a small scale, to control arthropod pests in glasshouse crops, orchards, ornamentals, turf and lawn grasses, stored products and forestry, and for moderation of vectors of animal **and** human diseases. There are several theoretical advantages of using **microbial biocontrol agents** although attempts to harness their potential so far have had comparatively minor commercial success. Among the undoubted **advantages of biocontrol** are:

- efficiency and potentially high specificity;
- natural;
- safety for humans and other non-target organisms;
- reduced chemical pesticide use and consequent reduction of residues in food and the environment;
- preservation of natural enemies of the pest;
- maintenance of biodiversity in managed ecosystems.

But there are some negative aspects of biocontrol:

- biocontrol agents can be costly and difficult to produce in quantity;
- they can have short shelf life;
- the pest must be present before the pathogen can be usefully applied (so prophylactic, or preventative, treatment is difficult).

The use of entomogenous fungi for biological control has been disadvantaged by the need for high humidity (80% and above) during the prolonged period required for fungal spores to germinate and then penetrate the surface of insects, because, unlike bacteria, these fungi attack insects through their cuticle, not their digestive tracts. To try to overcome this problem researchers have developed oil-based and other formulations of fungal spores for use in biological control. *Verticillium lecanii* (*Lecanicillium muscarium*) is an effective biocontrol of *Myzus persicae* and other aphids on chrysanthemums because the crop is grown in a glasshouse in which humidity can be controlled. Furthermore, the crop is 'blacked out' with polythene sheeting from mid-afternoon until morning during the summer to control the initiation of flowering (it is a 'short-day' flowering cycle) and this helps to create a high humidity around the plant in which the fungal spores can germinate and infect the aphids. A single spray of spores given just before 'black-out' gives satisfactory control of the aphid within 2–3 weeks. Spores of *Verticillium lecanii* can be used to control whitefly as well as aphids.

It is much more difficult to use entomogenous fungi for biological **control in the field**, although there is promise for biocontrol of pests that have an aquatic stage. Of importance is the fact that recent research has improved the prospect of using insect fungal pathogens for control of diseases such as malaria that depend on insect vectors (Blanford *et al.*, 2005; Scholte *et al.*, 2005; Thomas & Read, 2007). Among 85 genera of entomopathogenic fungi, only six species are commercially available for field application: *Aeschersorzia aleyrodis, Beauveria bassiana, Beauveria brongniarti, Metarhizium anisopliae, Paecilomyces fumosoroseus* and *Verticillium lecanii* (Hussain *et al.*, 2012).

Although several fungi are in use or have promise, the most widely used microbial control agent of arthropod pests now is the bacterium *Bacillus thuringiensis* (Lacey *et al.*, 2015). This is used as the model for biocontrol mechanisms and advances in understanding its molecular biology, mode of action and resistance management contribute to development of fungal control agents (**mycoinsecticides**), especially insights into their pathogenic process, enzymes involved with the penetration of the host cuticle and the role of insecticidal fungal toxins (Charnley, 2003). **Transgenic plants** expressing endotoxin genes from *Bacillus thuringiensis* have been generated to protect crops (tobacco, cotton and maize) against attack from pests. It is possible that the same technology could be applied to toxins produced by entomogenous fungi.

Increased use of microbial control agents as alternatives to broad-spectrum chemical pesticides depends on:

- increased pathogen virulence and speed of kill;
- improved pathogen performance under challenging environmental conditions (cool weather, dry conditions, etc.);
- greater efficiency in their production;
- improvements in formulation that enable ease of application, increased environmental persistence and longer shelf life;
- better understanding of how they will fit into integrated systems and their interaction with the environment and other IPM components;

- greater appreciation of their environmental advantages;
- wider acceptance by growers and the general public.

Future emphasis is likely to be on **IPM programmes**, especially those featuring mixed infections because interaction between pathogens can improve virulence and reproduction of the biocontrol species and improve host–pathogen dynamics (Hussain *et al.*, 2012; Lacey *et al.*, 2015; Ortiz-Urquiza *et al.*, 2015).

17.7 CUTANEOUS CHYTRIDIOMYCOSIS: AN EMERGING INFECTIOUS DISEASE OF AMPHIBIANS

Chytridiomycosis is an infectious disease of the skin that affects amphibians worldwide. It is caused by the chytrid fungus *Batrachochytrium dendrobatidis*, which colonises the epithelium of adult amphibians causing a **fatal inflammatory disease**. It affects over 700 species on all continents where amphibians occur.

The disease was only discovered in 1998, but it has since been linked to dramatic population declines of amphibian species in western North America, Central America, South America, Australia and Africa. The disease impact in the wild is that in some regions **population decline** of native amphibian species, already weakened by the effects of pollution and introduced predators, is so devastating as to amount to potential multiple extinction events. It is certainly the largest disease-caused loss of biodiversity ever recorded, which is not only a loss to the biosphere, but to the pharmaceutical industry too, as amphibians can be an important source of toxins (neurotoxins, vasoconstrictors and pain killers) that could have therapeutic use (Lips, 2016).

The disease was first described from sick and dead adult amphibians collected from montane rain forests in Queensland (Australia) and Panama (Berger *et al.*, 1998), and this was the first report of parasitism of any vertebrate by a chytrid fungus. Chytrids are ubiquitous in aquatic habitats and moist soil, and can degrade cellulose and chitin, and proteins like keratin. Parasitic chytrids are known to infect plants, algae, protists and invertebrates, but *Batrachochytrium dendrobatidis* colonises the keratinised epithelium of adult amphibians. The pathogen has very low species specificity and causes very **high mortality** in susceptible populations, so *Batrachochytrium dendrobatidis* is viewed as a newly emergent pathogen that has reached susceptible amphibian populations in the wild and has probably been distributed around the world by the pet trade for 'exotic' amphibians and freshwater fish (Fisher & Garner, 2007). O'Hanlon *et al.* (2018) used genomics on a panel of more than 200 isolates to trace the source of *Batrachochytrium dendrobatidis* to the Korean peninsula. They dated the emergence of this pathogen to the early twentieth century, coinciding with the global expansion of commercial trade in amphibians, and showed that intercontinental transmission is ongoing.

So far this remains an amphibian disease and remains the **only example of a chytrid parasite of a vertebrate**. Infectious disease is an important component of the population biology of wild animals, as it is one of the balancing factors controlling population size. The occurrence of chytridiomycosis in the wild suggests that *Batrachochytrium* may have coevolved with some amphibian populations that are in balance with the disease by being tolerant or not susceptible; lowland species of a region, for example, remain unaffected although their highland relatives suffer high mortality. Recent disturbances of rain forest habitats may have introduced the parasite into susceptible amphibian populations in the mountain areas.

The clinical signs of amphibian chytridiomycosis are lethargy, abnormal posture and loss of the animal's ability to right itself in the water, and sometimes skin lesions, loss of epidermis and ulceration. Haemorrhages can occur in the skin, muscles or eyes. Death usually occurs a few days after the onset of clinical signs and seems to result from the animal's inflammatory response to the fungus, either to a toxin or to the aggressive digestion of the epidermal structure by the fungus. This causes epidermal hyperplasia (abnormal proliferation of cells within the skin tissue), which in turn **impairs the cutaneous respiration** which is essential to the gas balance and osmoregulation of amphibians.

Once established, *Batrachochytrium* can persist in the keratinised mouthparts of amphibian tadpoles which survive the infection because they are not dependent on cutaneous respiration. This implicates the **larval stage as a reservoir host** for the pathogen that may enable *Batrachochytrium* to persist despite reduced populations of adult amphibians. Persistence may be further enhanced by the ability of the chytrid to grow as a saprotroph. Indeed, this capability may greatly increase the impact of the disease and speed up amphibian population decline, and account for the inability of amphibians to recolonise streams in the worst affected areas.

It may seem odd to consider treatments for a disease of frogs, but when faced with the real possibility of threatened extinction of several or even many species, conservation of those species can become a matter of both disease control and treatment of the last few remaining individuals. Some of the more exotic species under threat have been reduced to populations in the range of a few tens or a few hundred adult animals. In practice, conservation at this late stage cannot be done in the wild, but it should be feasible to prevent mortalities in captive breeding programmes for these most threatened species.

Proper **quarantine procedures** on the part of breeders and aquarists could limit or prevent spread of the infection, and some success in clearing the infection in adults has been claimed for treatments with azole antifungal agents. Another apparent cure is to hold the diseased adults at 37°C for two 8-hour periods, 24 hours apart. The animals have a natural defence in that antimicrobial peptides are produced in glands of the skin and released into skin secretions and the skin is the only organ invaded by the fungus (Rollins-Smith *et al.*, 2002; Zasloff, 2002).

These peptides can kill or inhibit the growth of fungi and are highly effective in inhibiting growth of *Batrachochytrium dendrobatidis* in suspension culture *in vitro*. **Antimicrobial peptides** in the skin secretions could play a role in protection of the skin from initial infection by chytrid zoospores or protect areas of the skin from disease spreading from adjacent infected areas (Rollins-Smith *et al.*, 2002). Although it is not at all clear whether they offer any promise for treatment of diseased animals or intervention in further spread of this disease.

However, the amphibian–chytrid relationship is complex. The response of any amphibian species to chytrid infection depends on the ecology and evolutionary history of the amphibian, the genotype and phenotype of the fungus, and the biological and physical environment. Impacts of chytridiomycosis on amphibians also vary; some species have been driven extinct, populations of others have declined severely and still others do not seem to have been affected. Understanding all these interactions is critical for conservation and management of amphibian populations (Garner *et al.*, 2016; Lips, 2016).

17.8 ASPERGILLOSIS DISEASE OF CORAL

Decline of **coral reef ecosystems** has been a concern since the 1970s. Overall, the loss of coral reefs is thought to be a combined result of global warming, ozone depletion, overfishing, eutrophication, poor land-use practices and other expressions of human activities. There are, however, reports that emphasise the role of coral disease in reef degradation. Many aspects of **coral disease** are poorly understood, but disease syndromes do seem to be increasing, and may parallel a general decline in the marine environment. However, many of the data are purely descriptive and though viruses, bacteria and fungi have all been implicated as disease organisms most reports lack proof that the alleged pathogen is indeed the cause of the disease (Harvell *et al.*, 1999; Work *et al.*, 2008; Woodley *et al.*, 2015).

One disease for which the identity of the pathogen has been proved to a satisfactory level is **aspergillosis of sea fans**. Aspergillosis is a pathogen of the Caribbean sea fan (*Gorgonia* spp.) as well as other gorgonian corals. The impact of disease can range from severe, leading to localised mass mortalities, to mild, leading to partial tissue loss and eventual recovery. *Gorgonia ventalina* sea fans suffered mass mortality on reefs in the Caribbean and the Florida Keys in 1995 and 1996. During the 1995 outbreak, two species of Caribbean sea fans were identified as hosts: *Gorgonia ventalina* and *Gorgoni flabellum*. The pathogen was identified as *Aspergillus sydowii*, a common ***terrestrial*** soil fungus. Symptoms of the disease included expanding areas of the sea fan in which **polyp tissues were destroyed**, exposing the axial skeleton beneath. In some cases, the lesions were so severe that holes appeared in the skeleton. The disease was common over a wide geographical area and though partial destruction of individual *Gorgonia* colonies was most usual, the entire fan was sometimes destroyed. It was demonstrated that tissue-degrading lesions all contained fungal hyphae at their edges and careful analysis demonstrated that sea fan aspergillosis is infectious:

- the same filamentous *Aspergillus* species was isolated from different geographical disease areas;
- the organism could be grown in pure culture *in vitro*;
- mycelium taken from culture and inoculated onto healthy sea fans caused typical disease symptoms;
- the same fungus could be recovered from these experimental disease instances and recultured *in vitro*.

The course of action followed here was established by Robert Koch in the 1870s to identify pathogenic microorganisms (they are called **Koch's postulates**). Such procedures are necessary to demonstrate unambiguously that a presumed disease pathogen is the true cause of a disease (see discussions in Work *et al.*, 2008). This initial study identified the fungus as *Aspergillus fumigatus*, but it was later shown to be *Aspergillus sydowii* (Geiser *et al.*, 1998). The disease persists in the western Atlantic and was probably the cause of mass mortality in sea fans that occurred throughout the Caribbean during the 1980s. *Aspergillus sydowii* is a **common saprotroph** that is found **in both terrestrial and marine** environments. The species is widespread, having been isolated from terrestrial environments as varied as Arctic soils in Alaska and soils in tropical regions, as well as from subtropical marine waters in both coastal regions and oceanic zones as deep as 4,450 m. Although *Aspergillus sydowii* is common and cosmopolitan, it had not previously been recognised as causing disease in plants or animals. However, nonmarine strains of *Aspergillus sydowii* did not cause disease in sea fans so it was concluded that isolates taken from diseased corals have ***acquired* pathogenic potential** not seen in isolates from other sources. Emergence of *Aspergillus sydowii* as a marine pathogen implies that the land–sea boundary is not an effective barrier to disease transmission. Several *Aspergillus* species are known to cause opportunistic infections of animals; generally attacking individuals with weakened immune systems or other impaired defence systems. Sea fans may be infected opportunistically by *Aspergillus sydowii*, the fungus taking advantage of a host weakened by pollution or other environmental factors. Elevated temperature and nutrient pollution are among several factors that can contribute to coral diseases. Elevated temperature alters the host–pathogen interaction and is considered the key driver of coral disease outbreaks (Harvell *et al.*, 1999; Kim & Rypien, 2015).

Aspergillosis in various animal species, and in humans too, has increased over the last few decades. *Aspergillus* species are found worldwide in humans and in almost all domestic animals and birds as well as in many wild species, causing a wide range of diseases from localised infections to fatal disseminated diseases, as well as allergic responses to inhaled conidia. Also, disease-causing isolates may possess a specific pathogenicity factor, perhaps a **mycotoxin** or other secondary metabolite, that is not present in all isolates. Apart from the invasive infections in sea fan corals described in this section, other forms of animal aspergillosis that cause concern are: stonebrood mummification in honey bees, pulmonary and air sac infection in birds, mycotic abortion and mammary gland infections in cattle, guttural pouch mycoses in horses, sinonasal infections in dogs and cats, and invasive pulmonary and cerebral infections in marine mammals and nonhuman primates (Seyedmousavi *et al.*, 2015).

17.9 SNAKE FUNGAL DISEASE

Beginning in 2006, severe skin infections and a rapid population decline were reported in a timber rattlesnake (*Crotalus horridus*) in the north-eastern United States. In 2008, similar infections with a possible fungal cause emerged in Illinois, in an endangered population of Massasauga rattlesnakes (*Sistrurus catenatus*). This infectious disease became known as **snake fungal disease (SFD)**. By 2015, SFD had been documented in most of the eastern United States. SFD is a major conservation concern in North America because it has the potential to cause lethal infections and contribute to extinction of localised snake populations.

SFD is caused by *Ophidiomyces ophiodiicola*, a keratinophilic fungus that is widely distributed in eastern North America. It has a broad host range and is the predominant cause of fungal skin infections in wild snakes. It often causes mild infections in snakes coming out of hibernation and is an emerging pathogen of captive snakes in North America. SFD has been documented in 23 species of snakes, although this is likely to be an underestimate of the number of susceptible species. SFD hosts are both phylogenetically and ecologically very different, which indicates that other species of snakes in the United States might now be infected or susceptible to SFD (Lorch *et al.*, 2016; Burbrink *et al.*, 2017).

Ophidiomyces ophiodiicola has been isolated from captive snakes outside North America, but the pathogen had not been reported from wild snakes elsewhere until carcasses and moulted skins from wild snakes collected during 2010–2016 in Great Britain and the Czech Republic were screened for the presence of skin lesions, and for *Ophidiomyces ophiodiicola* using PCR DNA detection. The fungus was detected in 8.6% of the specimens and further analysis confirmed that *Ophidiomyces* SFD occurs in wild European snakes. Phylogenetic analyses indicated that *Ophidiomyces* isolated from European wild snakes belonged to a clade distinct from the North American isolates; the European and North American diseases involve different strains of *Ophidiomyces ophiodiicola* (Franklinos *et al.*, 2017).

Potentially, therefore, ***Ophidiomyces ophiodiicola* is a global emerging fungal pathogen** of reptiles in the wild. We're talking about reptiles and potential extinctions here, so this is about the right place to remind you about dinosaurs. They are mostly dead. Mostly, because all those birds that are flapping around out there are the remnants of a group of reptiles that once ruled the Earth for several hundred million years; until they went extinct. Taking advantage of the wide-scale extinction events of other organisms is an often-repeated feature of fungal evolution. The period 800–600 million years ago featured successive virtually global glaciations (snowball Earth episodes). The Permian–Triassic (P–Tr) extinction event that occurred approximately 251 million years ago (known as the Great Dying and the Earth's most severe extinction event so far) was studied by Visscher *et al.* (1996) who summed it up with the quotation: 'sedimentary organic matter preserved in latest Permian deposits is characterised by unparalleled abundances of fungal remains, irrespective of depositional environment (marine, lacustrine [= lake sediments], fluviatile [= river/stream deposits]), floral provinciality, and climatic zonation.' Much the same is true for the Cretaceous–Tertiary (K–T) extinction of 65 million years ago, the result of a meteor collision that caused the Chicxulub crater in Mexico, which is blamed for the extinction of the dinosaurs. There was also widespread deforestation right at the end of the Cretaceous, which is assumed to be due to post-impact conditions. However, coincident with all this death and destruction of animal and plant life at the K-T boundary there is a massive proliferation of fungal fossils: 'This fungi-rich interval implies wholesale dieback of photosynthetic vegetation at the K-T boundary in this region. The fungal peak is interpreted to represent a dramatic increase in the available substrates for [saprotrophic] organisms (which are not dependent on photosynthesis) provided by global forest dieback after the Chicxulub impact' (Vajda & McLoughlin, 2004). It is the same story as at the other extinction boundaries: while the rest of the world was dying, the fungi were having a party!

But that Chicxulub meteor might not have had the last word on dinosaur extinction, because the massive increase in the number of fungal spores in the atmosphere of the time may have caused fungal diseases that 'could have contributed to the demise of dinosaurs and the flourishing of mammalian species' (Casadevall, 2005). A reminder, perhaps, that the fungi started the eukaryote journey by spring-cleaning the early Earth, and they've been cleaning up and modifying the planet and its biosphere ever since (Moore, 2013).

17.10 WHITE-NOSE SYNDROME OF BATS

White-nose syndrome (WNS) of bats, caused by the fungal pathogen *Pseudogymnoascus destructans*, has decimated North American hibernating bats since its emergence in 2006 and is predicted to drive several species extinct. WNS was first detected in New York State in 2006 and was detected in Washington State in March 2016. *Pseudogymnoascus destructans* is endemic to Eurasia, where the disease is less severe, and it has a much lesser impact on bat populations.

Mortality from WNS differs substantially between North American species, even when they hibernate at the same sites. Some species declined more than 90% in the first year following WNS detection, whereas population growth rates in other species only decreased 8%.

Research has shown that in areas where the disease has been present for several years, bats first became infected when they returned to their winter quarters (their hibernacula) in the autumn, and both transmission and fungal growth on bats occurs primarily during winter once bats lower their body temperature and begin to hibernate. Despite markedly different mortality rates, most bat populations experience greater than 50% infection incidence; suggesting that variation in mortality is due to variation in response to infection. Mortality is thought to be due to a cascade of physiological disruptions which are reactions to tissue damage from fungal invasion. Deaths occurred approximately 70–120 days after infection. Variation between species in mortality was correlated with fungal loads of individual animals.

The **fungal load** is essentially the amount of fungus present in the animal; it can be measured by measuring the quantity of *Pseudogymnoascus destructans* DNA on bat skin using quantitative PCR on skin swab samples. It had already been established that there is a strong correlation between the abundance of *Pseudogymnoascus destructans* on bat skin measured from a swab and the extent of tissue invasion determined by histology. Although infection incidence was uniformly high in all species, fungal load varied a thousand-fold between species and statistically accounted for 98% of the differences in mortality rates. Fungal loads increased with hibernating roosting temperatures, with bats roosting at warmer temperatures having higher fungal loads and suffering greater WNS impacts. *Pseudogymnoascus destructans* growth increases with temperature across the range of hibernation temperatures commonly used by bats (approximately 1–12°C).

Consequently, differences in behavioural preferences between bat species, in particular their preferred roosting microclimate, determines the impact of a new pathogen. Understanding these behavioural/environmental influences on disease may allow control options that permit minimum intervention. Another aspect of this story is that comparative genomics has demonstrated that the pathogen, *Pseudogymnoascus destructans*, is extremely sensitive to ultraviolet (UV) light, and to DNA alkylating mutagens. The reason for this being that the pathogen's genome has lost a key enzyme in the repair pathway that normally contributes to repair of DNA damage induced by UV light. This feature might be exploited for treatment of bats with WNS (Langwig *et al.*, 2016; Palmer *et al.*, 2018).

17.11 MYCOSES: THE FUNGUS DISEASES OF HUMANS

As we have shown above, there has been an alarming increase in the number of fungal diseases affecting wildlife populations over the past several decades. Although associated primarily with opportunistic infections, fungal diseases of wildlife have caused some of the most important conservation crises in modern times. Studies on the origins of these epidemics are essential to establish how common saprotrophic fungi can gain the ability to cause disease in animals with which they have co-existed for millions of years, and to help us understand how **new diseases emerge**. This is not only important for wildlife populations, but for ourselves, too. We may be next in line (Fisher *et al.*, 2012, 2016; Fones *et al.*, 2017). You might think we don't need more fungal pathogens.

> Fungi cause more than a billion skin infections, more than 100 million mucosal infections, 10 million serious allergies and more than a million deaths each year. Global mortality owing to fungal infections is greater than for malaria and breast cancer and is equivalent to that owing to tuberculosis and HIV … (Gow & Netea, 2016)

The fungus diseases of humans are called mycoses and the majority, perhaps all, are not caused by dedicated pathogens, but rather by fungi common in other situations that take advantage of a particularly beneficial set of environmental conditions or of a host with defences weakened in some way (so-called **opportunistic pathogens**). There are about 135 fungal pathogens that cause diseases of humans and domestic animals, and only about 60 species of fungi cause disease in humans; of these about 30 cause superficial infections of the skin and about 30 cause subcutaneous, lymphatic or systemic infections (Figure 17.10) together with several other species that cause allergic reactions and some commensal fungi that also affect human health and/or disease states (El-Jurdi & Ghannoum, 2017; Köhler *et al.*, 2017).

The fact that there is a relatively short list of mycoses does not mean that fungus diseases of humans are rare. What the human disease fungi lack in diversity, they make up for by being **very widespread**. To quote a recent survey, which expands on the quotation above:

> Nearly a billion people are estimated to have skin, nail and hair fungal infections, many 10's of millions mucosal candidiasis and more than 150 million people have serious

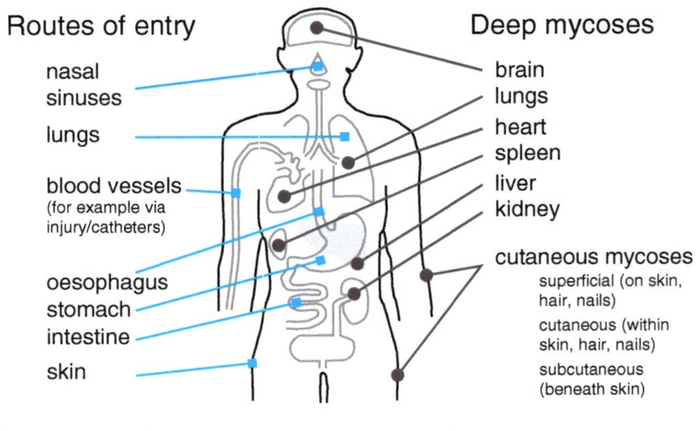

Figure 17.10. Routes of entry and distribution of the fungus diseases of humans. Labels at left indicate routes of entry of pathogenic and opportunistic fungi that cause deep and cutaneous mycoses. At right we indicate the principal tissue sites of deep mycoses in comparison with superficial, cutaneous and subcutaneous mycoses. Based on illustrations produced by Steve Schuenke, University of Texas Medical Branch at Galveston for the book *Medical Microbiology*, 4th edition, by Samuel Baron (1996), ISBN 0-9631172-1-1.

> fungal diseases, which have a major impact on their lives or are fatal. However, severity ranges from asymptomatic-mild mucocutaneous infections to potentially life-threatening systemic infections. Moreover, mortality associated with fungal disease at >1.6 million is similar to that of tuberculosis and >3-fold more than malaria. … HIV/AIDS pandemic, tuberculosis, chronic obstructive pulmonary disease (COPD), asthma and the increasing incidence of cancers are the major drivers of fungal infections in both developed and developing countries globally. Recent global estimates found 3,000,000 cases of chronic pulmonary aspergillosis, about 223,100 cases of cryptococcal meningitis complicating HIV/AIDs, about 700,000 cases of invasive candidiasis, about 500,000 cases of *Pneumocystis jirovecii* pneumonia, about 250,000 cases of invasive aspergillosis, about 100,000 cases of disseminated histoplasmosis, over 10,000,000 cases of fungal asthma and about 1,000,000 cases of fungal keratitis occur ***annually***. (Bongomin *et al.*, 2017)

Many people (perhaps everybody?) suffers from athlete's foot at some time in their life. This is caused by a tropical import called *Trichophyton rubrum*, but you don't have to go to the tropics to collect your share because it so likes warm and moist shoes that it is now distributed throughout the temperate climatic zones and is so common that you can be infected very easily. Athlete's foot is little more than a nuisance though, as it can be successfully treated with over-the-counter remedies. Humans are exceptional among vertebrates in that their living tissue is directly exposed to the outside world; without protective scales, feathers or fur, our skin is our defence against all the opportunistic fungi surrounding us. A consequence of this exposure to the environment is that we are prone to fungal invasion of our skin, nails and hair. Athlete's foot is one example, ringworm is another such infection. This introduces another aspect of fungal biology: 'ringworm' is a family of mycoses because the cause can belong to one of two closely related genera, *Microsporum* or *Trichophyton*. Each fungus is very specific to a specific part of the body. A range of animals can also suffer ringworm diseases of skin and fur, and that range includes farm animals and pets. The fungi spread readily to humans, which introduces another notion, that of a **zoonosis**, being a disease that can be transmitted from other vertebrate animals to humans.

Another remarkable statistic about human mycoses is that it is now unusual for a woman to go through her reproductive years without at least one significant infection by the yeast *Candida albicans*. *Candida albicans* is a normal inhabitant of the human mouth, throat, colon and reproductive organs. Usually, it causes no disease but lives commensally in ecological balance with other microorganisms of the digestive system. However, other factors such as diabetes, old age and pregnancy, but also hormonal changes, can cause *Candida albicans* to grow in a manner that cannot be controlled by the body's defence systems and **candidiasis** results, with symptoms ranging from the irritating to the life threatening. For most people candidiasis, like other superficial infections, is irritating; the fungus remains in the outer layers of the skin because the body's immune defence system prevents the fungus penetrating more deeply.

In patients whose immune systems are compromised in some way, there is no such defence and the infecting fungus becomes deep-seated, systemic and potentially fatal. This category of patient includes transplant patients, where the immune system is pharmaceutically modulated to control rejection, and people with AIDS, where deterioration of the immune system is caused by the decline in $CD4^+$ T cells, the key infection fighters of the immune system, by the **HIV (human immunodeficiency virus)** retrovirus. Infections of those with weakened immune systems are called **opportunistic infections. Candidiasis** is the most common HIV-related fungus infection that can affect the entire body, but there are other opportunistic fungi that typically do not cause disease in healthy people but only those with damaged immune systems; these fungi attack when there is an 'opportunity' to infect and can cause life-threatening disease. The nature of the protective immune response to fungal invaders, and the balance between immune surveillance, disease progression, host invasion and pathology, and other factors that predispose us to infection must be defined before we can effectively manage human fungal infections (Gow & Netea, 2016; Dambuza *et al.*, 2017; de Hoog *et al.*, 2017).

Several filamentous fungi have emerged as new causes of human infections in recent years. There is some uncertainty as to whether such EIDs of humans are caused by newly discovered fungi or are infections caused by fungi that are only now being correctly identified. The truth, as usual, is probably a mixture of all these causes. Recently developed methods of DNA sequence analysis and mass spectrometry have identified several previously unknown species of *Aspergillus* in clinical samples. Strictly speaking, these are not 'new fungi' but rather newly recognised species that the classical morphology-based identification of *Aspergillus* could not distinguish. The other end of the spectrum is illustrated by a case of ***gastrointestinal basidiobolomycosis*** causing bowel perforation in a child. In this case, the 10-year-old patient presented with inflammatory bowel disease (fever, abdominal pain, vomiting) and was first misdiagnosed as suffering intestinal malignancy or schistosomiasis. The patient's condition deteriorated despite anti-schistosomal therapy and radical surgery, and only improved following treatment with **itraconazole** after 'second opinion' histopathology diagnosed infection with *Basidiobolus ranarum*, which was unequivocally identified by PCR. *Basidiobolus ranarum* more typically causes an unusual fungal skin infection that rarely involves the gastrointestinal tract. Once considered rare, there is an increasing number of reported cases of gastrointestinal infection caused by *Basidiobolus* species worldwide, in countries such as the United States, Thailand, Australia, Iran, Egypt and Saudi Arabia (El-Shabrawi *et al.*, 2011; Roilides, 2016; Shaikh *et al.*, 2016). And even 'our favourite mushroom', the ink cap, *Coprinopsis* *cinereus*, can cause disseminated infections in immunocompromised patients, often leading to death. It has been recorded as the cause of endocarditis in 1970 (Spellerya & Maciver, 1971), and an invasive wound infection in a child in 2017 (Correa-Martinez *et al.*, 2018).

17.12 CLINICAL GROUPINGS FOR HUMAN FUNGAL INFECTIONS

There are five types of mycoses to describe, in two main categories:

- skin mycoses
  - superficial mycoses
  - cutaneous mycoses
  - subcutaneous mycoses
- systemic mycoses
  - systemic mycoses due to primary (usually dimorphic) pathogens
  - systemic mycoses due to opportunistic pathogens.

An excellent online source of information about human fungal infections is the US Department of Health & Human Services Centers for Disease Control and Prevention website at this URL: https://www.cdc.gov/fungal/diseases/index.html.

Superficial mycoses (or **tineas**) mostly occur in the tropics and are restricted to the outer surface of the hair and skin. Examples are:

- *Piedraia hortae*, a filamentous member of the *Ascomycota* which causes black piedra, a disease of the hair shaft characterised by brown/black nodules on the scalp hair (actually the ascostromata of the fungus).
- *Trichosporon cutaneum* is a yeast belonging to the *Basidiomycota* that is common in soil, water samples, plants, mammals and birds, as well as being a member of the normal flora of mouth, skin and nails. It causes white piedra, a superficial infection of the skin, and scalp and pubic hair (although it is emerging as an opportunistic pathogen of the immunocompromised) (Taylor & Gurr, 2014).

Cutaneous mycoses. There are three genera of fungi that commonly cause disease in the non-living tissues of skin, hair or nails/claws of people and animals, by growing in a zone just above where the protein keratin is deposited. These three genera are *Microsporum*, *Trichophyton* and *Epidermophyton* (all filamentous *Ascomycota*) and they are often labelled 'dermatophytes' (with the disease being called 'dermatophytosis') although, of course, they are not plants, so they can't be any sort of ' ... phyte' and a better term would be **dermatomycosis**. These fungi can all degrade keratin and grow as non-invasive saprotrophs on skin and its appendages, but their growth causes irritation and inflammation of underlying epithelial cells, this being an allergic reaction that may result in death of these cells (Table 17.2).

Infections of human finger and toenails are so important that they have their own descriptive term: **onychomycosis**,

Table 17.2. Examples of fungi that cause cutaneous mycoses, of which there are about 250,000 cases per year in the UK

Fungus	Disease
Trichophyton rubrum	foot and body ringworm in humans; infections of nail bed (onychomycosis); may spread to groin and hand
Trichophyton mentagrophytes var. *interdigitale*	athlete's foot in humans, mainly in the web of skin between toes and lower surfaces of toes; favoured by humid conditions and common in shoe wearing people; can also affect hand and can cause onychomycosis; common in the fur of rabbits and rodents
Trichophyton verrucosum	infection in the fur of cattle and horses but can be transmitted to humans; can lead to permanent baldness and pitting of scalp skin
Microsporum audouinii	ringworm in children; tends to be restricted to man as host; common in underdeveloped countries; the fungus forms rings in the skin because the mycelium grows like colonies in Petri dish cultures (and the redness is an inflammatory response to proteins secreted by the fungus)
Microsporum canis	common in the fur of dogs and cats where it can cause ringworm; often acquired by children in contact with domestic animal pets
Microsporum gypseum	can be isolated from soil and cause cutaneous mycoses; other 'dermatophytes' are poor competitive saprotrophs and do not survive in soil
Epidermophyton floccosum	man is the primary host of this, the only pathogenic species of *Epidermophyton*, infecting skin, causing tinea corporis, tinea cruris, tinea pedis, and nails, causing onychomycosis
Transmission of dermatophytes: physical contact between hosts, spores become lodged in small wounds, mycelium may persist in sloughed off skin scales or detached hairs, shared facilities such as swimming pools, schools and colleges or 'petting zoos' may act as a focus of an epidemic.	

which is used to refer to nail infections caused by any fungus. Several 'dermatophytes' cause onychomycosis (see Table 17.2), and 'non-dermatophyte' causes include *Scopulariopsis brevicaulis* (filamentous, *Ascomycota*) and *Candida albicans* (yeast, *Ascomycota*). It is not uncommon to have more than one fungus species jointly causing the infection.

Nearly 1.8 million people in the UK suffer from **'nail fungus'** and nail infections can be extremely distressing to the affected individuals. **Prevalence** varies from 2–3% (United States) to 13% (Finland), the disease is twice as frequent among men as women, and it increases with age; rates increasing to about 25% among elderly patients. Prevalence is also high in patients with diabetes and immunosuppressed individuals.

The fungi that cause cutaneous mycoses occur widely in nature and can be:

- **geophilic**, inhabiting the soil as decomposer saprotrophs and causing infections following contact with soil (for example, *Microsporum gypseum*); geophilic fungi can be isolated by baiting soil with 'degreased' hair;
- **zoophilic**, primarily parasitic on animals, but infection is transmitted to humans following contact with the animal host (for example, *Microsporum canis, Trichophyton mentagrophytes* and *Trichophyton verrucosum*);
- **anthropophilic**, primarily parasitic on humans, rarely infecting animals (for example, *Microsporum audouinii, Trichophyton rubrum, Trichophyton tonsurans* and *Epidermophyton floccosum*).

Subcutaneous mycoses are generally caused by fungi that are normally saprotrophic inhabitants of soil, particularly in tropical and subtropical areas of Africa, India and South America, which become infective by being introduced through wounds in the skin. Most infections involve people who normally walk barefoot.

Madurella mycetomatis and *Madurella grisea* (filamentous, *Ascomycota*) cause human **mycetoma** (common name: madura foot), which is a localised infection causing locally invasive tumour-like abscesses, accompanied by chronic inflammation, resulting in swelling, distortion and ulceration of the infected body part. The fungus is introduced through mild wounds in the skin and may grow for several years in the cutaneous and subcutaneous tissues, extending to connective tissues and bones. Mycetomas are usually resistant to chemotherapy, leaving surgery, even amputation, as the only resolution.

Sporothrix schenckii (thermally dimorphic, *Ascomycota*) causes **sporotrichosis**. *Sporothrix* is the anamorph and *Ophiostoma stenoceras* the teleomorph. The fungus occurs in soil worldwide although the disease is localised, with Peru having the highest prevalence of *Sporothrix schenckii* infections. Also called 'rose handler's disease', sporotrichosis starts by entry of the fungus through minor skin injury and can then spread through the lymphatic system. The fungus is dimorphic: forming septate vegetative hyphae, conidiophores and conidia at 25°C, while at 37°C oval to cigar-shaped budding yeast cells are produced. As the yeast form is distributed by the lymphatic system, disseminated sporotrichosis can result in infections of the lungs and bones and joints, endophthalmitis (inflammation of the internal layers of the eye), meningitis and invasive sinusitis.

Systemic mycoses are infections that affect the whole body. We divide these into mycoses due to primary (usually dimorphic) virulent pathogens, and those due to opportunistic pathogens. There is potential for overlap in these two categories. They are all **deep-seated mycoses** that can affect the whole body, but some opportunistic pathogens are dimorphic, some are inhaled. The key difference is in the point that we have already stressed: primary pathogens are **virulent** and generate a disease infection as soon as they encounter the host; opportunistic pathogens are usually saprotrophic or commensal organisms that become pathogenic when the opportunity offered by a weakened immune system is presented.

Systemic mycoses due to primary pathogens are usually acquired by inhalation of virulent spores originating from soil or related substrata, so the disease starts off primarily in the lungs and can spread to other organ systems. The four prime examples are **blastomycosis, coccidioidomycosis, cryptococcosis** and **histoplasmosis.**

Blastomycosis is caused by the dimorphic fungus *Blastomyces dermatitidis* (*Ascomycota*), which is endemic as a saprotroph of decaying woody material in the soil of the Mississippi and Ohio River basins in the United States. Dimorphism is temperature dependent; conversion of the filamentous hyphal growth form to a budding yeast occurs at 37°C. Outbreaks of the disease are associated with occupational or recreational activities around streams and rivers, and wet soils containing rotting organic woody debris. Infection results from inhalation of conidia, which germinate into the yeast form in the lungs. After 30–45 days, an acute **pulmonary disease** indistinguishable from a bacterial pneumonia occurs, although the disseminated disease can affect the skin, genitourinary tract and other organs as well as the lungs. On average there are about 30–60 deaths due to blastomycosis in the United States each year.

Coccidioidomycosis, or **'Valley fever'**, is caused by thermally dimorphic fungal species of the genus *Coccidioides*, which are found (in rodent burrows) in alkaline soil of warm, dry, low altitude areas with high summer temperatures. There are only two species in the genus: *Coccidioides immitis* and *Coccidioides posadasii* which co-exist in the desert south-west of the United States and Mexico, while *Coccidioides immitis* is geographically limited to the San Joaquin valley region of California, and *Coccidioides posadasii* is also found in South America. The division into two species was published in 2002 and before this the two were known as the California and non-California variants of *Coccidioides immitis*. Inhalation of the dry **arthroconidia** of Coccidioides, which are carried by dust storms, is the infection mechanism. Coccidioidomycosis is initially a lung infection; the young and healthy may suffer a slight cough, but it clears up. If the victim is not in the best of health, the infection can spread to skin, bones, joints, lymph nodes, adrenal glands and central nervous system. The disease may prove fatal; it causes about 50–100 deaths per year across North, Central and South America.

Cryptococcosis is caused by *Cryptococcus neoformans* (a basidiomycete encapsulated yeast); the environmental sources of which are soil and other habitats (trees, buildings, etc.) contaminated with **pigeon droppings**. Though it has a worldwide distribution it is predominantly a disease of northern Europe. The infection starts with inhalation of the organism and can remain localised in the lungs but often disseminates throughout the body and the most severe form is probably the non-contagious cryptococcal meningitis when the fungus reaches the brain. Cryptococcosis can be considered an opportunistic infection as it mainly affects **immunosuppressed patients**, being found in about 15% of AIDS patients; the annual incidence before the AIDS epidemic has been estimated at 2–9 cases per million inhabitants; following the emergence of AIDS, the annual incidence of cryptococcosis in the early-to-mid-1990s varied between 17 and 66 cases per ***thousand*** persons living with AIDS (Taylor & Gurr, 2014).

Histoplasmosis exists in two forms; both of which are caused by varieties of *Histoplasma capsulatum*. The most common is the pulmonary disease known as North American histoplasmosis caused by *Histoplasma capsulatum* var. *capsulatum*. The other form has been called African Histoplasmosis, caused by *Histoplasma capsulatum* var. *duboisii*, and is usually a disease of the skin and bones. *Histoplasma* (*Ascomycota*) is a thermally dimorphic anamorph (grows filamentously at 25°C, as a budding yeast at 37°C), which has a teleomorph with the name *Ajellomyces capsulatus*. Soil contaminated with bird or bat droppings is the common natural habitat of *Histoplasma*, which has a worldwide distribution although the disease occurs in tropical areas. Histoplasmosis is endemic in the Tennessee–Ohio–Mississippi River basins of the United States where it causes about 50 deaths per year.

Systemic mycoses due to opportunistic pathogens are infections caused by fungi with low inherent virulence, which includes an almost limitless number of fungi common in any or all environments. Health statistics indicate the sinister fact that mortality from fungal disease has been steadily increasing since the 1980s against a background of steady decline in mortality caused by all other infectious agents. There are several identifiable reasons for this. There has been an increase in clinical awareness leading to **improved diagnosis** of fungal disease (that is, not necessarily more disease, but more of what does occur is being diagnosed as being of fungal origin). To some extent this reflects the introduction of techniques, especially molecular methods, that can rapidly identify fungi. Second, increased availability of international travel has taken more people into the tropics, and tropical regions do seem to harbour more fungal pathogens. Third, and above all, there has been an increase in the number of immunocompromised patients. Drug therapies used to manage the immune system in transplant and cancer patients have the unfortunate side effect of weakening the body's defences against fungal pathogens, and AIDS sufferers have similarly **weakened immune defences** against fungi. The fungi most frequently isolated from immunocompromised patients are the endogenous commensal diploid yeast *Candida* and saprotrophic (that is, environmental) filamentous *Aspergillus* and zygomycete species. As we will point out, the spectrum of fungal species responsible for these diseases is continually, and quite rapidly, changing.

Candidiasis. *Candida* species can colonise any surface of the human body and grow without causing disease; when the normal host-defence mechanisms of an individual are impaired the fungus can become pathogenic. *Candida albicans* is a component of the normal microbiota of the human body. It normally exists in ecological balance with other microbes and you don't know it's there; however, steroid or immunosuppressive drug treatments, and particularly antibacterial antibiotics, reduce the other microbes and consequently increase the opportunities for *Candida* (Table 17.3).

In fact, a decrease in bacterial populations increases the chances of *Candida albicans* adhering to the gut lining; indigenous bacteria suppress adherence of *Candida albicans*, and **adhesion to epithelial surfaces** is an important aspect of *Candida albicans* pathogenicity. *Candida* most commonly forms superficial lesions on the mucous membranes of the mouth and vagina. The disease of the mouth, known as **'thrush'**, occurs on the tongue, gums and buccal mucosa, commonly in infants, and in adults with hormonal or immunological problems. The disease of the vagina, vaginitis, causes itching and smarting of lower vagina and is the second most common infection in the UK. Up to 75% of women suffer at least one episode, with half of them suffering a further episode. It is often associated with pregnancy, when the pH of the vagina is low, and the incidence is increased by use of steroid-based contraceptives.

Mucosal candidiasis occurs in almost **all immunocompromised AIDS sufferers**, and is also common in organ transplantation patients and those undergoing anticancer therapies because of the immunosuppressive effect of the therapies. Defects in

Table 17.3. Isolation of *Candida albicans* from untreated and antibiotic-treated individuals

Patient category	%frequency of isolation from		
	Mouth	Alimentary tract	Vagina
Disease-free individuals	43	15	8
Individuals given antibacterial antibiotics	76	22	15

cell-mediated immunity render AIDS patients particularly susceptible to fungal infections; 60–80% suffer at least one fungal infection and 10–20% will die as a result of fungal infection.

Candida is usually described as a **dimorphic** fungus, although it would be better described as **pleomorphic** with a range of morphologies, from ovoid **yeast** cells at one extreme to **filamentous hyphae** at the other (Odds, 2000). When *Candida albicans* is grown in medium containing serum it forms pseudo-hyphal germ tubes, not buds. This is used as a diagnostic feature. The **hyphal** (or pseudo-hyphal) phase has several **advantages** over the yeast form because it is:

- **more pathogenic** than the yeast phase;
- more efficient at penetrating epithelial layers; and
- **more resistant** to defence systems.

Candida albicans has been regarded as the most common cause of **invasive yeast infection** for a long time. It responds well to treatment with fluconazole and as a result the overall incidence of infections due to *Candida albicans* has decreased. However, since the early 1990s, infections caused by *Candida glabrata* have increased, and this species is less susceptible to fluconazole and has also been associated with oropharyngeal candidiasis in patients receiving radiotherapy for head and neck cancer (Nucci & Marr, 2005).

Candida albicans and *Cryptococcus neoformans* are both capable of rapid phenotypic switching, where genetically identical cells can exist in two distinctive cell forms: white and opaque in the case of *Candida albicans*. The white cells exhibit the classic yeast cell shape, while the opaque cells are elongated and have a distinctive sculptured surface. Each cell type is stably inherited for many generations and switching between the two types of cells occurs rarely and randomly. This phenotypic switch is known as an **epigenetic phenomenon** in the sense that changes in gene expression are heritably maintained without any modification to the primary genomic DNA sequence. A highly interconnected network of sequence-specific DNA-binding proteins control this switch (Whiteway & Oberholzer, 2004; Hernday *et al.*, 2016). The two types of cells interact differently with their mammalian host, with opaque cells being more suited to skin infections, and white cells better suited to bloodstream infections. Odds (2000) suggests that 'switching' could represent a mechanism of accelerated micro-evolution to assure survival of the type that can most rapidly adapt to a new microniche.

Aspergillosis is a disease of the lungs caused by inhaling conidia of *Aspergillus* species. It occurs in mammals and birds and was the first bird mycosis to be identified (at the beginning of the nineteenth century) and can be epidemic in chickens reared under crowded conditions. *Aspergillus* has a worldwide distribution and its saprotrophic lifestyle enables it to be a constant component of our way of life. Aspergillosis has been reported in almost **all domestic mammals and birds** as well as in numerous wild species. The species most commonly associated with aspergillosis are *Aspergillus fumigatus, Aspergillus flavus, Aspergillus niger, Aspergillus nidulans, Aspergillus terreus* and *Aspergillus glaucus*, which are common and widespread, growing on all manner of organic debris.

Aspergillus species can produce enormous numbers of spores and they are common moulds in nature; their spores can heavily contaminate hay, straw and grain. Workers who regularly handle these products (farm and brewery workers) were the groups at greatest risk of 'farmer's lung disease' up to about the middle of the twentieth century when the disease was a rare occupational hazard. Aspergillosis in humans is principally (>90%) caused by *Aspergillus fumigatus* and the greatest risk group is to patients being treated for cancer or with **immunosuppressive therapies** after organ transplantation who suffer neutropenia (reduced neutrophil levels). This type of immune suppression is not usually seen in people with HIV, so aspergillosis is rare among HIV-positive patients. *Aspergillus* spp. cause a spectrum of respiratory diseases as the inhaled spores become lodged in the lungs and bronchi (Figure 17.11).

Allergic aspergillosis is an immediate response to the spores that leads to **asthmatic reactions**. Further damage causes a bronchitis called bronchopulmonary aspergillosis. Ultimately 'colonising aspergillosis' occurs where the fungal hyphae grow into 'fungal balls' (**aspergillomas**) that are formed within the natural lung cavities. Finally, invasive and disseminated aspergillosis spreads from the pulmonary focus to a variety of tissues and organs. The disease was rare before the use of immunosuppressive drugs became common.

Although virtually everyone is exposed to this fungus in their daily environment, it rarely causes disease because healthy respiratory tracts and **healthy immune systems deal effectively with inhaled spores**. It was a problem only in those few occupations that exposed workers to such massive spore concentrations that their healthy respiratory defences were overwhelmed

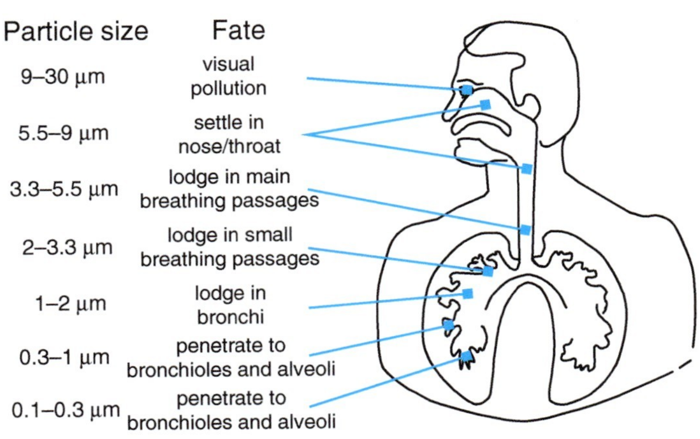

Figure 17.11. Graphic showing how deeply particles can be inhaled into the human respiratory tract. *Aspergillus* conidia are in the size range (2–4 μm) that enables them to penetrate deeply into the human lung. Redrawn using data taken from the US Environmental Protection Agency's website, which acknowledges the original source as Environment Canada.

and respiratory disease resulted. Suppress the immune system and aspergillosis can cause serious lung problems and can also spread to the kidneys, liver, skin, bones and brain. Because these more severe forms of aspergillosis are almost always fatal (Table 17.4), it is important to diagnose and treat this infection quickly (see the *Aspergillus Website* at https://www.aspergillus.org.uk/; the *Support for People with Aspergillosis* website at https://www.nacpatients.org.uk/; and the *National Aspergillosis Centre* at http://www.nationalaspergillosiscentre.org.uk/).

The **epidemiology** of mould infections changed substantially in the 10 years from 1995. The incidence of invasive aspergillosis increased significantly, and infections caused by fungi showing resistance to conventional antifungal agents, such as *Fusarium* species (*Ascomycota*) and zygomycetous species, have also increased. Nucci & Marr (2005) describe this as evolution of the epidemiology of **invasive fungal infection** and show that yeasts other than *Candida albicans* and filamentous fungi other than *Aspergillus fumigatus* are emerging as significant causes of invasive mycoses in severely immunocompromised patients.

For example, in one transplantation clinic *Aspergillus terreus* accounted for 2.1% of cases of invasive aspergillosis in 1996 and 10.2% of cases in 2001. *Aspergillus terreus* is a concern because it shows resistance and poor clinical response to what has been the prime antifungal agent, amphotericin B. Infections caused by newly recognised species of *Aspergillus* ('*Aspergillus lentulus*') have also appeared, isolates of which have low susceptibility to multiple antifungals *in vitro*. The reasons for these changes are a complex mixture of altered prophylactic use of antifungal agents and altered surgical **procedures** which together change the environment that the infecting fungus encounters in organ transplant recipients, premature new-borns and critically ill patients (Nucci & Marr, 2005).

Pneumocystosis is the overall name given to infections caused by *Pneumocystis*, which have four clinical expressions: asymptomatic infections, infantile pneumonia (and may be associated with sudden infant death syndrome (SIDS)), pneumonia in immunocompromised patients, and extrapulmonary infections. Pneumonia due to *Pneumocystis* is frequently written with the acronym PCP (***Pneumocystis*** **p**neumonia) and is one of the most important pneumonias in immunocompromised individuals. *Pneumocystis* was classified as a protozoan until the late 1980s but is now clearly accepted as a **yeast-like fungus** and molecular phylogenies place it in the *Ascomycota*, *Taphrinomycotina* (see section 3.7). The original specific name given to the organism was

Table 17.4. Outcomes for patients treated for invasive aspergillosis

Type of aspergillosis	% overall mortality
Disseminated (disease at multiple sites)	95–100
Cerebral aspergillosis	95

Pneumocystis carinii but isolates causing infections in humans were misidentified and are now referred to *Pneumocystis jirovecii* in honour of the Czech parasitologist **Otto Jirovec**. *Pneumocystis carinii* is still the correct name for this organism when found in hosts other than humans.

Pneumocystis DNA can be detected in air and water, although the organism may not be visible microscopically, suggesting that *Pneumocystis* may not survive in the environment longer than it takes to infect a susceptible host. The organism cannot be cultured *in vitro*, so the only information about the life cycle comes from studies in animals. *Pneumocystis* trophic (vegetative) forms are produced during asexual growth; this is called a **trophozoite**, the terminology surviving from the time the organism was thought to be a protozoan. Trophozoites are variable in morphology and occur clustered together in the host tissue. They are probably haploid and capable of replicating asexually by mitosis, and they produce a diploid zygote by conjugation. Meiosis occurs in the zygote, forming a precyst initially, then an early cyst and, finally, a **mature cyst** (again, terminology surviving from protozoan days). During this maturation process eight intracystic spores or 'daughter cells' are produced, which must be ascospores resulting from meiosis followed by a mitosis. These spores are released when the mature cyst ruptures and germinate into trophic forms. How this cycle relates to release of an infective agent to the environment is unknown. Sources of human infections are patients suffering pneumocystosis or immunocompetent individuals transiently parasitised by *Pneumocystis* (carriers); so *Pneumocystis* is unusual among pathogenic fungal species in that the sufferers somehow transmit the disease.

Pneumocystis is one of the major causes of **opportunistic mycoses** in immunocompromised patients, including those with congenital immunodeficiencies or AIDS, and patients receiving corticosteroid or intensive immunosuppressive therapy for treatment of cancer or prevention of transplant rejection. Following the introduction of highly active antiretroviral therapy for HIV infections in 1996, PCP has been in decline (Sax, 2001; Hammer, 2005), but it is still one of the most significant AIDS-related diseases; extrapulmonary infections, resulting from dissemination of the infection from lungs to lymph nodes, spleen, bone marrow, liver, kidneys, heart, brain, pancreas, skin, and other organs, occur in patients with AIDS.

Zygomycosis is another **emerging disease** that is becoming an increasingly significant infection in transplant recipients. One survey of fungal infections in transplant recipients reported the proportion of cases of zygomycosis increased from 4% to 25% between 2001 and 2003. Analysis of risk factors associated with zygomycosis, compared with those associated with invasive aspergillosis, revealed an association between the use of voriconazole therapy and zygomycosis. Voriconazole is a triazole antifungal used to treat or prevent invasive candidiasis and invasive aspergillosis. Zygomycosis is considered the '**third threat**' of fungal infection in patients who have survived infections caused by *Candida albicans* and *Aspergillus fumigatus* (Nucci & Marr, 2005).

The most common species causing zygomycosis are *Absidia corymbifera*, *Rhizomucor pusillus* and *Rhizopus arrhizus*, which are all **commonly found in soil** but can cause this acute and rapidly developing disease in debilitated patients. The disease is associated with acidotic diabetics, malnourished children, and severely burned patients, and occurs with leukaemia, lymphoma, AIDS and immunosuppressive therapy. The infection typically involves the rhino-facial-cranial area, lungs, gastrointestinal tract, or skin; with the fungi growing in arterial blood vessels, causing embolisms and necrosis of surrounding tissue. Rhinocerebral zygomycosis in acidotic diabetic patients usually results in death within a few days.

Fusariosis in **immunocompromised patients** has also shown a rising trend. The most virulent *Fusarium* species is *Fusarium solani* (filamentous, *Ascomycota*). *Fusarium* is a well-known plant pathogen and is widely distributed on plants, including crop plants, and in the soil. The rate of infection in haematopoietic stem cell transplantation (HSCT) recipients has increased over time. HSCT is now the second most frequent major organ transplant, being used to treat a variety of malignancies and bone marrow failure disorders; an estimated 45,000 such transplants are performed each year in the United States alone, 2,000 in patients under 20 years of age. This population of patients could be at risk of **fusariosis** because of severe **T cell-mediated immunodeficiency** resulting from severe immunosuppression over long periods of time; some cases are diagnosed very late after transplantation (10 years and over) (Nucci & Marr, 2005).

The real worry now is the increasing resistance to the limited collection of antifungal drugs; especially for *Candida* and *Aspergillus* infections, for which the therapeutic options have become limited. The emergence of multiply drug-resistant *Candida* species is a global health threat, and azole-resistant *Aspergillus* has up to 30% prevalence in some European hospitals, which report higher than 90% mortality rates (Fairlamb *et al*., 2016; Meis *et al*., 2016; Robbins *et al*., 2017). Remember that July 2017 editorial in the journal *Nature Microbiology*, **'Stop neglecting fungi'** (Anonymous, 2017), and the Royal Botanic Gardens' report ***'State of the World's Fungi 2018'*** (Willis, 2018).

17.13 FUNGI WITHIN THE HOME AND THEIR EFFECTS ON HEALTH: ALLERGENS AND TOXINS

The effect of fungi within the home is often underestimated. Fungi able to produce spores into the atmosphere are present indoors in a variety of places such as the carpets, wallpaper and other soft furnishings, including sheets and blankets that contain natural fibres, as well as discarded food, pet food, pet cages and other bedding materials. Fungi that cause dry rot of the timber frames and panelling of houses also produce a great many spores (see Figure 14.3). These fungal spores are dispersed around the home by foot traffic and general air movements.

Common mould fungi can also colonise large areas within water-damaged buildings, forming visible indoor mould on a variety of materials but particularly timber and materials containing cellulose such as wallpaper and ceiling tiles. Several studies of the fungal spore content in dust collected from carpet and air samples **within the home** show that about half the houses tested had spore densities exceeding 500 CFU m^{-2} (CFU: colony-forming units). Generally, mould spore concentrations were higher in kitchens than in rooms more likely to be carpeted. The most abundant genera were *Cladosporium*, *Penicillium*, *Aspergillus* and *Stachybotrys* (all filamentous *Ascomycota*). Of these, *Stachybotrys chartarum* is a particular threat.

Patients suffering building-related symptoms, often called 'sick building syndrome', though it now tends to be called **damp building-related illness (DBRI)**, and infants with lung disease (specifically **'infant pulmonary haemorrhage'**) often have histories of living in mouldy, water-damaged buildings. These illnesses have been attributed to **mycotoxins** released by *Stachybotrys chartarum*, inhaled as spores or as mould-covered material. *Stachybotrys* produces trichothecene (see Figure 11.17) mycotoxins known as **satratoxins**, which can lead to pathological changes in animal and human tissues because they are potent inhibitors of DNA, RNA and protein synthesis. *Stachybotrys* also produces the haemolysin, **stachylysin**, which lyses sheep erythrocytes and may contribute to animal disease. Controlled human exposure studies have shown that exposure to volatile organic compounds (VOCs) from indoor mould can cause irritation of the respiratory tract (Pestka *et al*., 2008).

Up to 300 different compounds have been detected in indoor air, including **3-methylfuran** (C_5H_6O; 3-MF), which is a suspected neurotoxin. Because it is a common volatile product of many common fungi, 3-MF can be used as a **marker** for the active growth of fungi in water-damaged buildings; the substance has a characteristic 'musty fungus' smell, as does another common fungal metabolite 2-octen-1-ol ($CH_3(CH_2)_4CH{=}CHCH_2OH$). Pulmonary diseases can be made worse by 3-MF, and it is known to cause acute irritation of the eyes, nose and airways; it could be a contributor to sick building syndrome. Though the degree to which volatile compounds influence human health issues is currently poorly understood, it is evident that the most important sources of volatiles in the domestic environment included cigarette smoking, vehicle-related emissions, building renovation, solvents, household products and pesticides (Wålinder *et al*., 2005; Chin *et al*., 2014).

What is lacking in this area, however, is a **causal link** between the fungus and the reported illness. Also, it is not clear that the attention given to *Stachybotrys* is justified; it is only one of the contaminants inhabiting buildings and, indeed, it is less common and usually occurs in lesser amounts compared to the genera *Aspergillus*, *Penicillium*, *Alternaria* and *Cladosporium*, all of which are isolated more frequently from buildings with sick building syndrome.

Interest was focused on *Stachybotrys* originally by an **alleged** association between this fungus and an outbreak of infant pulmonary haemorrhage in a group of infants in Cleveland, Ohio, from January 1993 to November 1994, but later analysis showed that **the data were misleading** and the 'association' was not statistically significant; other factors such as cigarette smoke exposure emerged as potentially more relevant and the conclusions of the original investigation were retracted. The situation remains that while *Stachybotrys chartarum* produces toxins and can cause disease in animals, it is not ***proved*** to be able to produce symptoms in humans. As *Stachybotrys chartarum* is among the least common of the fungi found in the home, it is likely to be a localised rather than widespread problem.

Most studies, including Ochiai *et al.* (2005), agree that fungi in the home can affect individuals who suffer from **allergies** or who are immunocompromised; whether the fungi cause disease in healthy, immune-competent individuals at the exposure levels encountered in the domestic environment is a matter of debate. Perhaps prevention is better than cure, while the debate continues, particularly for those susceptible to allergy or illness. The home should be kept well ventilated and should be vacuumed regularly. If the environment is damp, a regularly cleaned **dehumidifier** should be used, any water leaks should be repaired immediately, visible mould removed and the area treated with antiseptic (Cheong *et al.*, 2004; Cheong & Neumeister-Kemp, 2005). Allergic reactions are certainly important and severe for the sufferer, as is illustrated by the press report shown in Resources Box 17.1.

RESOURCES BOX 17.1 Mushroom Soup Leak Forces Ryanair Flight Diversion

... on Monday, a passenger on a Ryanair flight from Budapest to Dublin needed medical treatment after a jar of soup leaked in an overhead locker, dripping onto his face.

The man suffered swelling to his neck and struggled to breathe, forcing the aircraft to be diverted to Frankfurt, in Germany.

Yesterday, Ryanair said the leaking jar had contained a 'vegetable oil/mushroom soup type substance' which had caused an allergic reaction.

The Boeing 737 was delayed for two hours while doctors treated the man.

'It is procedure when a passenger requires medical attention to divert to the nearest airport,' a spokeswoman for the airline said. 'The cabin crew and pilot take that decision.'

This is a quotation from a newspaper article extracted from *The Guardian* website at https://www.guardian.co.uk/. Written by James Orr and agencies guardian.co.uk, Thursday, August 28 2008 09:53 BST. It is archived at https://www.theguardian.com/business/2008/aug/28/ryanair.theairlineindustry.

Several studies suggest an association between amounts of fungal spores and visible mould and asthma in children and adolescents. One study predicted that about 5% of individuals will suffer some allergic symptoms from exposure to moulds during their lifetime. Overall, 10–32% of all asthmatics are sensitive to fungi, though this is significantly less than the proportion sensitive to pet or mite allergens (Hossain *et al.*, 2004; Ochiai *et al.*, 2005).

Allergic reactions require prior exposure to the fungal antigens and can then rapidly reoccur on re-exposure to small amounts of the same antigens. A superficially similar reaction is **organic dust toxic syndrome** (**ODTS**), but this follows a single exposure to large amounts of particulates contaminated in some way with fungal material. This syndrome is not mediated by the immune system, being due to the **direct toxicity** of the acute respiratory burden of fungal particulates. It is mainly an **occupational disease**, the risk groups being workers renovating contaminated buildings or those involved in fungal remediation programmes. Evaluating building-related symptoms can involve skin-prick testing with fungal and other likely antigens; this test is based on the proposition that symptoms should occur when the presumed cause is present and abate when the cause is removed. Any sort of organic dust can cause similar health issues; fungi are not the only culprits. There can be a considerable load of bacteria in floor dust, as well as fragments of fungal hyphae and fungal spores. The growth of these microbes is often favoured by characteristics of the dwelling and nearby outdoor environment, but there is still a close dependence on humidity levels (Baumgardner, 2016; Dannemiller *et al.*, 2016).

Almost all episodes of real human illness due to fungal toxins have been caused by ***ingestion*** of the toxin. There are more than 100 toxigenic fungi and over 300 different mycotoxins produced by them, but only two mycotoxins have been shown to be carcinogenic (Bennett & Klich, 2003). We associate drama and tragedy with mushroom poisoning though, in fact, fewer than 50 of the more than 2,000+ species of mushrooms that occur commonly in most countries can be considered poisonous, and only about six are deadly.

Most serious poisonings are due to species of *Amanita* which contain two of the most potent toxins known, cyclic peptides called **amatoxin** and **phallotoxin**. *Amanita phalloides* is aptly named death cap because it accounts for more than 90% of mushroom poisoning fatalities in the United States and western Europe, and the other deadly species is *Amanita virosa* (the common name for which is the destroying angel). Both toxins interfere with basic aspects of normal cell biology.

- **Amatoxins** bind to the RNA polymerase enzyme and inhibit mRNA synthesis, so the cell cannot function properly and eventually dies. Body organs which normally have high rates of protein synthesis are particularly sensitive to this toxin. Liver hepatocytes are in this group and are most commonly involved in mushroom poisoning, but other target organs include kidney, pancreas, testes and white blood cells.

- **Phallotoxin** causes irreversible disruption of the cell membrane and cytoskeleton and cell death inevitably results. Phallotoxin is not absorbed from the human gut, so is not thought to be responsible for the symptoms associated with human poisonings. Amatoxins are readily absorbed from the gut and are then concentrated in the liver where they do their damage.

A patient poisoned with *Amanita* goes through **four characteristic phases**. For the first 6–24 hours the patient is free of symptoms; then there's a 12–24-hour period of severe abdominal cramps, nausea, vomiting and profuse watery diarrhoea. In this stage, the patient is often misdiagnosed as having gastric influenza and, as the patient's gastrointestinal symptoms improve in the third 12–24-hour period, this misdiagnosis may seem to be right as the patient appears to be on the way to recovery. But all through this time the toxin is destroying the liver; killing hepatocytes and literally ripping it apart from within.

Symptoms of **liver damage** only appear 4–8 days after the toxin was consumed, but the liver damage can progress rapidly, accelerating all the time. Death from catastrophic failure of the liver can occur anything from 6–16 days after the fatal meal was consumed. Providing misdiagnosis can be avoided; providing liver function is monitored and providing proper supportive care is made available, the patient has a fair chance of surviving the poisoning. In the long run, though, this may require a liver transplant and even with the best medical care, death still occurs in 20–30% of cases and there is a **mortality rate** of more than 50% in children younger than 10 years old.

Most people who eat a poisonous mushroom suffer an unpleasant but relatively **mild** and short-lived bout of **'food poisoning'**. The quicker that symptoms arise, the less severe is the attack. The toxins of deadly mushrooms are slow acting and do not cause symptoms for 6–12 hours or more after eating. Those causing milder poisonings make people sick in 2 hours or less. After a few hours discomfort, vomiting, nausea and cramps, accompanied by diarrhoea, the victim recovers completely. People at most risk of being poisoned through eating wild mushrooms are toddlers, who are likely to eat anything on an experimental basis; recreational drug users, who think that any trip is worth a try; and immigrants who are in danger of collecting toxic mushrooms that resemble species that are safe to eat in their home regions.

Mushrooms are not the only fungi able to produce toxins that have plagued humanity over the years. Imagine a disease whose victims feel as if they are burning up, literally on fire, or as if ants, or even mice, are crawling about beneath their skin. Imagine that, as the disease develops, they may suffer terrible hallucinations; so bad as to sometimes drive them insane. If they avoid the hallucination and madness an even greater suffering awaits. After a few weeks, limbs become swollen and inflamed, violent burning pains alternate with feelings of deathly cold and gradually the affected parts become numbed, then simply fall off the patient's still-living body. Fingers, toes, hands, feet, arms and legs, all may separate from the body and rot away.

The first record of this terrible disease dates from AD 857; a chronicle from Xanten in the state of North Rhine-Westphalia in the Lower Rhine region of Germany describes a 'great plague of swollen blisters that consumed the people with a loathsome rot, so that their limbs were loosened and fell off before death'. At intervals throughout the following centuries there were epidemics in which an enormous number of victims died. One bad outbreak in France in AD 944 reputedly killed 40,000 people; the last epidemic of any considerable extent was in the 1880s. Records in the eleventh and twelfth centuries refer to this plague as Holy Fire, but it became associated with St. Anthony, as those suffering from the disease started to visit the saint's relics to seek relief and solace in faith. Most cases of what therefore became known as **St. Anthony's fire** occurred in France and its distribution reflected the cultivation of rye and use of the grain to make bread.

The human disease is caused by eating bread prepared from rye which is itself suffering a fungal disease. The plant disease, called **ergot**, is caused by *Claviceps purpurea* which is parasitic on wheat, rye and other grasses. As it is most common on rye it is often called ergot of rye. Spores infect the flower; the spore germinates on the stigma and then grows into the ovary. There it uses the nutrients intended to nourish the seed to produce a curved mass of compacted cells (the sclerotium) in place of the seed (Figure 17.12). When ripe, the sclerotium projects from the head of the rye as a hard purple to black, slightly curved thorn. Ergot is the French name for the spur on the foot of some birds which the sclerotia resemble, hence the common name of the disease. In wild plants, these sclerotia fall to the ground in autumn where they overwinter and germinate in spring to infect the next generation of plants. But on farms the sclerotia are harvested and milled with the grain. The flour then becomes contaminated with the toxins the sclerotia contain.

Figure 17.12. Ergots of *Claviceps purpurea* on wild oat, *Avena fatua*. Photograph by David Moore.

Ergot yields at least three groups of nitrogenous organic compounds ('alkaloids'): **ergotamines, ergobasines** and **ergotoxines** which have now become **medically important** (Tudzynski *et al.*, 2001). The pharmacological effects of the various ergot alkaloids and their derivatives are due to the structural similarity of their tetracyclic ring system to natural neurotransmitters such as noradrenaline, dopamine and serotonin. Ergotamines and ergobasines excite smooth muscles (which control the uterus, blood vessels, stomach and intestines) and are now used to promote contractions of the uterus in childbirth, to stem haemorrhaging and to control migraines. The third group, ergotoxines, have an inhibitory effect on those parts of the nervous system that affect mood and emotional state, and are used to treat psychological disorders such as delirium tremens and hysteria. The use of ergot as a clinical drug has been recorded from the sixteenth century, when the ergots themselves were used by European midwives to induce labour and to terminate unwanted pregnancies.

Ergot poisoning, now known as **ergotism**, causes two groups of symptoms which are manifestations of the same type of poisoning, gangrenous and **convulsive ergotism**. The former was frequent in certain parts of France, the latter occurred mainly in Central Europe. The convulsive form results from ergot poisoning combined with a deficiency of vitamin A, probably itself caused by a lack of dairy produce in the diet. In **gangrenous ergotism**, it is constriction of the blood vessels, especially to the extremities, which eventually causes gangrene and death. In convulsive ergotism, the effects of the ergot alkaloids on the nervous system are most pronounced. Twitching, spasms of the limbs, and strong permanent contractions, particularly of hands and feet. In severe cases, the whole body is subject to sudden, violent, general convulsions, all coupled with hallucinations and visions. Ergotism has been suggested by some to explain the convulsive fits which took hold of eight young girls in Salem, and led to the witch trials of 1692, during which 19 people, mostly women, were pronounced guilty of witchcraft and hanged. Britain has been singularly free from ergotism. There is only one record of typical gangrenous type, in the family of an agricultural labourer from near Bury St. Edmunds, in 1762. The mother and five children all lost one or both feet or legs: the father suffered from numbness of the hands and the loss of finger nails.

Recognition of the disease, improvements in grain preparation and, especially, the introduction of antibiotics in the latter half of this century to control the bacterial infections have all helped to remove human ergotism from our catalogue of miseries. The only UK epidemic occurred in Manchester in 1927 among Jewish immigrants from Central Europe, who lived on bread made from a mixture of wheat and rye. All 200 suffered symptoms of mild convulsive ergotism. At the same time, September 1926 to August 1927, and perhaps because the grain came from the same source, there were 12,000 cases of convulsive ergotism in Russia. The last recorded outbreak, in France in 1951, caused more than 200 people to suffer from hallucinations, and five died. One young victim believed she saw geraniums growing out of her arms. But ergotism in livestock remains a common problem arising from the **ergot infection of fodder grasses**. Sheep suffer from inflammation and ulceration of the tongue. In cattle, the symptoms are lameness and dry gangrene in the feet, and at times the whole foot can be lost.

One final fungal toxin worth a mention, and certainly worth avoiding, is **aflatoxin**. There are several related aflatoxins, which are produced primarily by the moulds *Aspergillus flavus* and *Aspergillus parasiticus*. They are **the most active carcinogenic natural substances** known; the most important of the group is aflatoxin B_1 (AFB_1), which has a range of biological activities, including acute toxicity, teratogenicity, mutagenicity and carcinogenicity. These toxins arise in crops like corn and peanuts, and to a lesser extent rice and soy beans, even before harvest, but particularly when stored under the warm and moist conditions which permit growth of the infesting fungi. This is a problem in developing countries, especially throughout tropical Africa, and the southeast of the United States (Kumar *et al.*, 2016; Umesha *et al.*, 2017).

When the tainted crop is eaten, the aflatoxin is absorbed and metabolically activated by liver enzymes. For, say AFB_1 to exert its effects, it must be converted to its **reactive epoxide** by the action of the mixed function mono-oxygenase enzyme systems (cytochrome P_{450}-dependent) in the tissues, particularly the liver. This epoxide is highly reactive and can form derivatives with several cellular macromolecules, including DNA, RNA and protein. Activated aflatoxins kill hepatocytes, which can cause **hepatitis**, or transformation into cancerous growths. The human liver has a fairly slow metabolism, so susceptibility to acute poisoning is relatively low. Animals are more in danger, particularly poultry. In the 1960s, 100,000 turkeys died on poultry farms in England. The disease was unknown at the time and first became known as turkey X disease. It was eventually found to be due to contaminated poultry feed and this costly event began aflatoxin research. More serious to humans is that prolonged exposure causes primary **liver cancer**, as well as other cancers.

Aflatoxin health risks (collectively referred to as **aflatoxicosis**) differ considerably around the world. Regions with a warm, humid climate generally favourable to mould growth and with poor regulatory and control systems may have as much as 500 times higher incidence of liver cancer than is common in developed countries. The European risk comes mainly from importing **contaminated foods**, often nuts, and animal feeds. Aflatoxin can enter the human food chain from such contaminated feeds. Grains, peanuts, other nuts and cottonseed meal are among the foods on which aflatoxin-producing fungi usually grow. Meat, eggs, milk and other edible products from animals that consume aflatoxin-contaminated feed are also sources of potential exposure to humans.

During January–June 2004, an **aflatoxicosis outbreak** in eastern Kenya resulted in hundreds of cases, nearly half of them fatal. The Kenyan outbreak followed a poor harvest of maize that had been damaged and made susceptible to mould by drought.

Furthermore, to guard against theft of the meagre harvest, people stored the maize in their homes, which were warmer and moister than the granaries where it was usually stored. Over that period, 317 people sought hospital treatment for symptoms of liver failure, and 125 died. Health officials ruled out viral liver diseases; suspecting acute aflatoxin poisoning, they examined maize samples and found aflatoxin B_1 concentrations as high as 4,400 parts per billion (ppb), **220 times the limit** for food and comparable with contents measured in other acute aflatoxicosis outbreaks (Azziz-Baumgartner *et al.*, 2005; Lewis *et al.*, 2005).

Aflatoxins and other mycotoxins contaminate 25% of agricultural crops worldwide and are a source of morbidity and mortality throughout Africa, Asia and Latin America. Control of aflatoxin is restricted to physical separation and destruction of contaminated batches, partial decontamination, and prevention of further contamination by provision of good storage conditions.

Procedures that try to remove aflatoxins are generally inadequate so the best chance of making sure that foods and feeds are free of aflatoxins is by preventing mould infestation and toxin production in the first place. Condensation of moisture on roofs and walls, leaking roofs, seepage of water into warehouses and the like are some of the causes of mould growth during storage. Every place where moisture can get to the product and cause mould growth must be eliminated. Refrigeration and climate control is useful. Reducing the oxygen concentration (to less than 1%) and increasing the carbon dioxide (to above 20%) can also inhibit mould growth. Sadly, all these measures are costly and since huge tonnages of aflatoxin-contaminated produce come from developing countries in the tropics, most of these desirable storage measures are completely impracticable. We must fall back on continued vigilance (Umesha *et al.*, 2017).

To protect against too much exposure, most developed countries measure aflatoxin levels and acceptable levels in human food and animal feeds are controlled by legislative authorities. Still, Americans may consume up to half a microgram of aflatoxins every day, and aflatoxins are among the more than 200 environmental chemicals that have been found in samples of human breast milk. The levels of hazardous chemicals like this in human breast milk are a measure of our progress (or lack of it) in cleaning up our environment. One expert says that 'there is no uncontaminated mothers' milk anywhere in the world ... all mothers carry environmentally-derived chemicals in their bodies'.

17.14 COMPARISON OF ANIMAL AND PLANT PATHOGENS

We have shown in this and earlier chapters that fungi are important pathogens of both plants and animals and you may well have detected some similarities in the concepts and themes that we have described under these topics. It is worth attempting a summary comparison (Sexton & Howlett, 2006). An important difference is that:

- fungi cause more disease in plants than either bacteria or viruses;
- bacteria and viruses cause more disease in animals than fungi.

Epidemiology is the study of disease distribution in populations. As a descriptive or analytic exercise, it examines (often historically) how a disease affects a population; as an activity, it may attempt to control the spread and severity of diseases. There is a human-centric tendency in epidemiology: for the most part animal epidemiology is a branch of medicine concentrated on human diseases, and the epidemiology of plants is concentrated on agricultural and horticultural crops.

Without always labelling it as such, we have already dealt with the epidemiology of many fungal infections. For example, we have already made it clear (section 17.12) that species of *Candida* and *Aspergillus* are the most common causes of invasive **fungal infections** of humans, although several other fungi are emerging as significant opportunistic pathogens (*Pneumocystis*, Zygomycetes, *Fusarium* spp.). Similarly, several examples have been given of factors affecting the spread and management of plant disease in the field and the impact of epidemiological knowledge on **plant disease management** (Chapter 16).

As we have seen in this chapter, fungal pathogenicity in (vertebrate) animals depends very much on the immune status of the host. The decisive significance of the immune system in controlling fungal infection, particularly in humans, is demonstrated by the devastating impact of common saprotrophic fungi as opportunistic fungal pathogens in immunocompromised patients described above. Plants do have defences, but plant cells do not move. A crucial element of the vertebrate immune system is the recruitment (that is, the **movement**) of cells towards the point of challenge. A plant cannot reinforce its defences in this way, so the challenge is between an invading hypha and one defending plant cell. The best defence that a plant can achieve is the programmed death of challenged cells represented in the hypersensitive response that confers resistance to biotrophic pathogens (see section 16.14).

Aside from these biological differences, there are some technical factors that constrain any comparison of fungal pathogenesis in plant and animal hosts:

- plants can be manipulated without the ethical issues associated with animal experimentation;
- experimental study of human pathogenic fungi generally relies on cell lines and experimental animal models, plant pathogens can be studied directly on their hosts;
- because many more fungi infect plants than animals, a larger number and wider range of plant-fungus systems than animal-fungus systems have been studied.

There is a further problem in that the word *virulence* has a slightly different technical meaning to animal and plant pathologists. In plain English, *virulence* refers to the ability of a microbe to cause disease; it is often used interchangeably with the term *pathogenicity* although *virulent* often implies that the pathogen does a ***lot*** of damage to the host. Unfortunately, the 'gene-for-gene' theory of how plants recognise and respond to fungal pathogens suggests that two genes, one in the plant and one in the fungus, are involved and introduced the term ***avirulence gene*** to describe the fungal gene whose product is identified by the plant's resistance gene (see section 16.16). The plant genes act as the plant equivalent of the animal immune system by recognising components of the pathogenic fungus to which they can react to prevent the infection. Animal biologists already have a term for the immune interaction; they call the identifying material on the infective agent an *antigen*.

- for an animal pathologist a virulence gene in an infective agent **produces the antigen**;
- for a plant pathologist a virulence gene is an allele of an avirulence gene and **does not produce** the product that promotes plant resistance.

The complex interactions involved in fungal diseases of plants have been rationalised using the concept called the 'disease triangle' (see section 16.9), which places the three factors which must interact to cause plant disease at the three corners of a triangle. Those three factors being:

- a susceptible host,
- a disease-causing organism (the pathogen),
- a favourable environment for disease.

A similar concept developed for animal pathogens more recently is the 'damage–response' framework described by Casadevall & Pirofski (2003), which emphasises that the outcome of an interaction is determined by the amount of damage incurred by the animal host. This is an attempt to make the concept more realistic by reconciling 'microorganism-centred' and 'host-centred' views of microbial pathogenesis to focus experimental studies around a common principle. The basic tenets of the damage–response framework are that:

- microbial pathogenesis is an outcome of an interaction between a host and a microorganism,
- the host-relevant outcome of the host–microorganism interaction is determined by the amount of damage to the host,
- host damage can result from microbial factors and/or the host response.

Very few fungi can infect both animal and plant hosts, so comparisons must be made between different species. Most is known about fungal pathogens that belong to the *Ascomycota*. Certainly, among vertebrate animal pathogens there is only one known chytrid disease (of amphibians), and relatively few diseases of humans caused by zygomycetes or *Basidiomycota*. Even though, as discussed above, zygomycosis and cryptococcosis are clinically important diseases, they represent too few examples of animal fungal diseases for much worthwhile comparison to be made with the many plant diseases caused by members of these groups. Consequently, we will limit this discussion to *Ascomycota*, and immediately highlight a point of difference in that animal and plant pathogens are concentrated in **different classes among the *Ascomycota*** (see Appendix 1: *Outline Classification of Fungi*):

- the Classes *Leotiomycetes*, *Dothideomycetes*, *Sordariomycetes* and *Taphrinomycetes* are rich in plant pathogens;
- whereas animal pathogens belong to the Class *Eurotiomycetes* (in the Subclasses *Chaetothyriomycetidae* and *Eurotiomycetidae*) that contains few plant pathogens (Berbee, 2001).

There is surprisingly little ***comparative discussion*** of **fungal pathogenesis** in animals and plants, but Sexton & Howlett (2006) and Casadevall (2007) describe parallels in fungal pathogenesis of plant and animal hosts, focusing on the *Ascomycota*. In brief those comparisons are placed in seven stages:

- **Stage 1**, attachment of conidia and/or ascospores, in plant pathogens, or yeast cells, hyphae, or arthrospores in animal pathogens to a surface and recognition of the host.
- **Stage 2**, activation of the infecting fungal propagule.
  - Stage 2a, germination of ascospores (plant pathogens), conidia or arthrospores.
  - Stage 2b, dimorphic switching of animal pathogens from a yeast phase to a pathogenic hyphal stage or from hyphae to a pathogenic yeast phase.
- **Stage 3**, penetration of the host may involve mechanical pressure, such as that produced by appressoria in some plant and some insect pathogens; lytic enzymes, such as proteases; and additional cell wall-degrading enzymes, including cutinases, cellulases, pectinases and xylanases in plant pathogens; protease followed by chitinase is needed to dissolve the insect cuticle. Natural openings in the host, such as stomata in plants or wounds in animals and plants, are also entry points for pathogenic fungi.
  - **Stage 3a**, some animal pathogens (for example *Histoplasma capsulatum*) use receptors on the host cells to bind and facilitate endocytosis as a means of penetrating the host cells.
- **Stage 4**, avoidance of host defences. Pathogenic fungi may detoxify oxidative molecules such as superoxide and antifungal compounds and synthesise protective molecules such as melanin. Animal-pathogenic ascomycetes often avoid or inhibit animal immune system components. Plant pathogens may avoid exposure to fungal wall-degrading enzymes by causing little host cell damage when undergoing intercellular biotrophic growth or by producing inhibitors of these plant enzymes.
- **Stage 5**, colonisation of the host environment, which often results in host cell death, may require specific nutritional mechanisms such as those for iron uptake in animal hosts,

and can produce other changes in host physiology, such as the pH level, to create a more favourable environment for the pathogen.

- **Stage 6**, asexual reproduction in the pathogen.
  - **Stage 6a**, asexual reproduction often results in conidia emerging from lesions on the host surface of plants.
  - **Stage 6b**, spore formation in the host is less common in vertebrate animal pathogens, and direct host-to-host transmission is rare. *Coccidioides immitis* produces endospores in host tissue by numerous mitotic divisions inside a spherule. Remember that copious asexual spore formation is common in invertebrates, particularly following death of the (arthropod) host.
- **Stage 7**, sexual reproduction. Mating and meiotic division produce ascospores during the disease cycle of plant-pathogenic fungi. This can result in recombinant offspring if mating occurs with a genetically different individual (obligatory in heterothallic but not in homothallic fungi). Sexual reproduction has not been reported to occur in animal-pathogenic fungi, with a few possible exceptions, such as *Pneumocystis* species.

Overall, there are clear similarities, particularly in terms of the ability of fungi:

- to adhere to the host;
- to use their filamentous hyphae to penetrate the host; and then to
- use their ability to secrete a comprehensive panel of enzymes to digest and use components of the host as nutrients.

There is one remaining point that we've not addressed so far, namely, that fungi also parasitise other fungi. As fungi and animals are each other's closest relatives, this seems like a reasonable place to deal with this topic.

17.15 MYCOPARASITIC AND FUNGICOLOUS FUNGI

In nature, fungi grow in communities of organisms; those communities include microbes, plants, animals and many other fungi. Fungi in the same substrate will compete with one another for control of that nutrient source and the competition may be so vigorous that the most successfully competitive fungi specialise in being parasitic on other fungi. After all, production of hydrolytic enzymes into the substrate is the lifestyle that enables saprotrophic fungi to exploit dead substrates, and the same capability enables pathogenic fungi to penetrate their animal, plant or even fungal hosts.

Many fungi can secrete glucanases and chitinases, which give them the ability to attack living hyphae of filamentous fungi, causing wall degradation and consequent autolysis, and there are many species of fungi that ***parasitise other fungi***. However, fungi growing with or on other fungi can have many different sorts of relationships, varying from indifferent, saprotrophic, commensal, symbiotic, or a range of intensities of parasitic, and it is not always clear which relationship applies.

Before we go into further detail we should emphasise that some, perhaps many, fungi are able to parasitise organisms from different kingdoms; for example, Figure 17.13 shows *Arthrobotrys oligospora* (*Ascomycota*) acting as a mycoparasite and invading the epidermis of barley roots. We first encountered *Arthrobotrys* in its role as a nematode-trapping fungus (see Figure 15.6), so this species can penetrate the tissues of ***all three eukaryotic kingdoms*** (Nordbring-Hertz, 2004).

Another example occurs in the genus *Metarhizium*, some species of which are host-specific, like the specific locust pathogen *Metarhizium acridum*, while the broad host range *Metarhizium robertsii*, which is an insect pathogen, also colonises plant roots endophytically, and may even be a plant symbiont. Functional

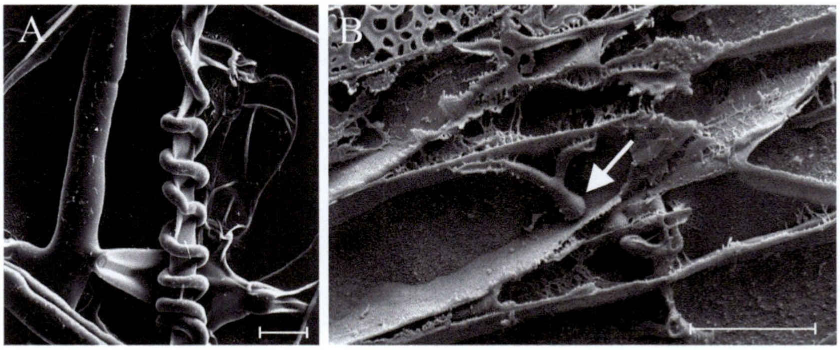

Figure 17.13. Versatility in a fungal pathogen able to attack all three eukaryotic kingdoms. *Arthrobotrys oligospora* is a prime example of a nematode-trapping fungus but here, in (A), it is shown as a mycoparasite coiling around a hypha of *Rhizoctonia solani*. An unaffected hypha of *Rhizoctonia* is shown to the left of the image; scale bar, 10 μm. (B) shows early colonisation of the epidermis of barley roots by *Arthrobotrys oligospora*. Colonisation of plant roots is believed to be endophytic, though it may aid the plant host by making the roots more resistant to plant parasitic nematodes and/or fungal pathogens. See also the nematode-trapping devices of *Arthrobotrys oligospora* shown in Figure 15.6. Bar, 20 μm. From Nordbring-Hertz (2004); reproduced with permission of Elsevier.

genomic approaches confirmed that *Metarhizium robertsii* upregulates different genes in the presence of plants and insects, demonstrating that it has specialist genes for each side of its bifunctional lifestyle. In addition, both *Metarhizium* species secreted more numerous proteins than were secreted by host-specific plant pathogens or non-pathogenic fungi. This implies that in the greater complexity of interactions between *Metarhizium* species and their environments there may be more associations to be found (St. Leger *et al.*, 2011). Certainly, although fungi are major drivers of ecosystem health, they are strongly influenced by environment, and their interactions with bacteria (Araldi-Brondolo *et al.*, 2017).

Various genera of entomopathogenic fungi have displayed the ability to colonise a wide variety of plant species in different families, both naturally and artificially following inoculation, and then confer protection against fungal plant pathogens as well as insect pests. Consequently, there is great research interest now in the possibility of using entomopathogenic fungi as endophytes for dual biological control of both fungal and insect pests (Jaber & Ownley, 2018).

The term **mycoparasitism** refers specifically to ***parasitism*** of one fungus (the host) by another (the mycoparasite). The word hyperparasitism has been used by mycologists for the same phenomenon, but originated as a description of insects that *parasitise* other *parasitic* insects, and is not really appropriate unless the host is itself a parasite. Mycoparasitism describes relationships in which one living fungus acts directly as a nutrient source for another; there is another term, **fungicolous**, which is used for those relationships (in the majority) in which nutrient transfer directly from live host to live parasite has ***not*** been demonstrated. Fungicolous fungi have a constant association with another fungus, but the exact nature of that relationship is difficult to establish (Jeffries, 1995).

By analogy with plant pathogens (see section 16.9) mycoparasitic relationships (Table 17.5) can be described as necrotrophic or biotrophic (Karlsson *et al.*, 2017):

- **Necrotrophs** are destructive parasites that invade and kill their hosts; these fungi often grow well as saprotrophs and include many of the *Ascomycota* that form large numbers of asexual spores (so-called '**mitosporic genera**'), such as species of *Gliodadium* and, especially, *Trichoderma*, as well as zygomycetes like *Dicranophora* and *Spinellus*, and the Oomycete *Pythium*. Although *Pythium* is not a fungus, it rather nicely illustrates the specialisation involved in mycoparasitism because the mycoparasitic species require thiamine for growth *in vitro*, and are unable to utilise inorganic nitrogen sources, these physiological features representing nutritional deficiencies in comparison to plant-pathogenic *Pythium* spp. These have been classified based on the level of interaction between the parasite and its host, which may be through:
  - **hypha-to-hypha** interference (**contact necrotrophs**), an example being hyphae of *Cladosporium*, which cause necrosis and death of basidia of the plant-pathogenic fungus *Exobasidium camelliae* without penetrating the living basidia (Mims *et al.*, 2007);
  - **penetration** of the hyphae of the host by parasitic hyphae (**invasive necrotrophs**).
- **Biotrophs**, where a balanced relationship is established and the parasite grows on the still-living mycelium of the host fungus. Most biotrophic mycoparasites are members of the traditional zygomycetes, for example the genera *Dispira, Dimargaris, Piptocephalis* and *Tieghemomyces*. These can be grown as dual cultures with their hosts, but without the host fungus they grow poorly *in vitro*. These have been classified on the basis of the exact nature of the intimate relationship

Table 17.5. Types of mycoparasite and the nature of the host–parasite interface involved

Type of relationship	Host–parasite interface
Contact necrotroph	Fungi in contact; no penetration of host mycelium by parasitic hyphae; cytoplasm of the host degenerates and hyphal lysis may occur.
Invasive necrotroph	Fungi in contact (Figure 17.13A); hyphae of the parasite penetrate and enter the host; degeneration of host cytoplasm occurs rapidly, often followed by hyphal lysis.
Intracellular biotroph	Complete thallus or mycelium of the mycoparasite enters the hypha of the host; cytoplasm of the host remains healthy.
Haustorial biotroph	Hypha of the host is penetrated by a short haustorial branch from the hypha of the parasite; host cytoplasm remains healthy.
Fusion biotroph	Fungi in intimately close physical contact; parasitic hyphae often coiled around the host hyphae, the two pressed closely together; micropore(s) develop between the closely pressed hyphae, or from a short penetrative hyphal branch of the parasite; host cytoplasm remains healthy.

Table modified from Jeffries (1995).

between the parasite and its host, or the host–parasite interface as it is called (Table 17.5), which may be through:

- entry of the complete thallus of the mycoparasite into the host (**intracellular biotrophs**; mostly chytrids and oomycete organisms that make their way into their host cytoplasm and absorb nutrients directly from it);
- formation of haustoria from the parasitic hyphae that penetrate the host (**haustorial biotrophs**);
- specialised contact cells which accomplish direct cytoplasmic continuity with the host through fine pores in the hyphal walls forming interhyphal channels (**fusion biotroph**).

Mycoparasitism depends on hyphal recognition and interaction, and many specific gene-regulation events in both parasite and host; for example, see the paper by Morissette *et al.* (2008) about gene expression during the interaction of the mycoparasite *Stachybotrys elegans* with its host *Rhizoctonia solani*.

There are several sorts of interaction between fungal hyphae, some of which we have already described in detail. **Intraspecific interactions** involve the same species, though they may come from different mycelia. If the interacting hyphae come from within a single mycelium, hyphal fusions occur readily to improve the interconnections within the mycelium (see Figures 5.12 and 5.13).

Interactions between different mycelia can involve **vegetative compatibility reactions** that determine the individuality of mycelia and are able to regulate the degree of anastomosis and heterokaryosis within a hyphal network (see Figure 7.7). These intraspecific interactions can make the competition between different individual mycelia within a substrate evident when the interaction gives rise to antagonism within the contact zone between opposing mycelia to form, for example, zone lines in infected timber (see Figures 7.8 and 14.5). In extreme cases, where one partner in the interaction is extremely dominant, destruction of one mycelium by the dominant one can result in the transfer of nutrients during these intraspecific interactions, so they can become mycoparasitic. There can also be a test for sexual compatibility, as an essential prelude to plasmogamy and karyogamy as part of the sexual reproductive cycle.

Interspecific interactions may initially depend on some of the same mechanisms. They have been characterised as **neutralistic, mutualistic** or **competitive** (Cooke & Rayner, 1984; Dighton *et al.*, 2005):

- in **neutralistic interactions** the hyphae intermingle without any apparent reaction to one another;
- in **mutualistic interactions** both individuals derive some benefit, which increases the survival value of both;
- **competitive interactions** are detrimental to both species because the antagonism experienced reduces fitness.

Competitive interactions generally result in one of the competing fungi proving to be more **aggressive** than the other and the expression of this is that the more aggressive fungus captures more nutrients than the less aggressive. In the broadest sense, the aggressive fungus is acting as a mycoparasite, but this can be a capture of substrate rather than parasitic absorption of nutrients directly from a host, so this is where the term **fungicolous** is most useful because it covers the full range of associations between fungi that live together whatever the biological nature of the association (Gams *et al.*, 2004).

The most aggressive fungicolous fungi are the true mycoparasites that are divided into the two groups, necrotrophs and biotrophs (Table 17.5), where the **necrotroph** is so strongly aggressive that it dominates and ultimately kills its partner/host. Close direct contact between necrotroph and host, often with the necrotrophic hyphae coiled around hyphae of the host (Figure 17.13A), helps exchange of nutrients without loss to competing third-party microbes. **Biotrophic mycoparasitic** relationships are physiologically balanced; the parasite being highly adapted to co-existing with the host for extended periods of time, and often forming specialised infection structures or host–parasite interfaces (Table 17.5).

Mycoparasites are probably very significant in the natural environment because fungi are major contributors to natural communities and these relationships between fungi must play an important role in development of community structure. **Fungicolous fungi are common**, widespread and numerous. Many fungi occur in nature in obligate associations with the mycelium or sporing structures of other fungi; these may be parasitic relationships, they are certainly interdependent relationships.

Because it is often difficult to obtain clear evidence of mycoparasitism it is frequently inferred as the lifestyle of a fungus for the reason that the mycoparasite causes distinctive growth inhibitions and/or abnormalities in the hyphae of the alleged host. Overall, this makes it difficult to establish the exact number of fungi **known** to grow on other fungi; but there are estimates, and the estimated numbers are large. The 10th edition of the *Dictionary of the Fungi* (Kirk *et al.*, 2008) puts the numbers at 1,100 species of fungi parasitising 2,500 other species of fungi, with probably at least a further 2,000 species of fungi, of all groups, growing specifically on lichens (lichenicolous fungi). Although these numbers are large they are minima; fungicolous fungi occur in **all habitats** and it is probable that mycoparasitism occurs more widely than currently imagined. After all, because fungi are such common, even dominant, members of natural populations in all habitats, the ability to harvest nutrients from either their living or dead hyphae must have very positive selective potential.

Jeffries (1995) details two specific examples of mycoparasitic relationships:

- **Necrotrophic invasion** of spores of arbuscular mycorrhizal fungi. The genera *Acaulospora, Glomus, Gigaspora, Scutellospora* and *Sclerocystis* (*Glomeromycotina*) are the characteristic and ubiquitous arbuscular mycorrhizal fungi. The spores of some of these are the largest known in the fungal kingdom and they can be extracted easily from field soils by simple sieving. In any sample of spores extracted like this there is usually a proportion, often the majority that are parasitised; the spore wall being perforated by many fine radial canals caused by **penetration of the wall by hyphae** of mycoparasitic fungi, or by rhizoids as parasitism of these spores by chytrids is probably widespread in wet soils. Members of the *Oomycota*, including *Spizellomyces* and *Pythium*-like fungi are also frequently seen within arbuscular mycorrhizal spores. Other fungi can often be isolated easily from arbuscular mycorrhizal spores and seem to be facultative parasites, being essentially saprotrophs with some ability to attack live spores. One study of mycoparasites of spores of *Gigaspora gigantea* reported the recovery of 44 species of fungi from surface-disinfected spores; *Acremonium* spp., *Chrysosporium parvum, Exophiala werneckii, Trichoderma* spp. and *Verticillium* sp. (all *Ascomycota*) were most frequently isolated from healthy spores, while *Fusarium* spp., *Gliomastix* spp. (both *Ascomycota*) and *Mortierella ramanniana* (*Mucoromycotina*) were typically isolated from dead or dying spores. Several examples of mycoparasitism of arbuscular mycorrhizas by various aquatic fungi have been described from specimens of the Lower Devonian Rhynie chert; demonstrating the presence of mycoparasitism in a 400 million-year-old ecosystem (Taylor *et al.*, 2015; Krings *et al.*, 2018).

Figure 17.14. Mycoparasitic macrofungi. A sporophore of *Boletus parasiticus* (also known as *Xerocomus parasiticus*) growing on a sporophore of the common earthball, *Scleroderma citrinum*. Collected in the RHS Garden Harlow Carr, Harrogate, Yorkshire, UK. Photograph by David Moore.

- Biotrophic invasion of mucoralean hosts by **haustorial mycoparasites**. Mycoparasites that penetrate their host with haustoria are highly specialised and frequently unable to grow in the absence of the host. Examples, all from the zygomycetes, are parasitic species of *Piptocephalis* (*Zoopagomycotina*) and *Dimargaris* (*Kickxellomycotina*) growing on living saprotrophic hosts from the *Mucoromycotina* (e.g. *Pilobolus, Pilaira* and *Phycomyces*) on the dung of herbivorous animals, especially rabbit pellets. *Piptocephalis* species are also extremely common in many woodland and pasture soils, from which they can be isolated using a baiting technique (spreading the soil sample on a medium already colonised by the host fungus).

These examples involve interactions on a microscopic scale. The impact of parasitic fungi is most obvious, however, in examples where the parasitic fungus produces a macroscopic sporophore from a macroscopic sporophore of its host. Figure 17.14 shows the example of *Boletus parasiticus* (also known as *Xerocomus parasiticus*) growing on a sporophore of *Scleroderma citrinum*.

This bolete species is said to be edible, but its host is poisonous, so it is not recommended for eating! The host, *Scleroderma citrinum*, is mycorrhizal with hardwoods and conifers; it is common and frequently found on the roots of trees growing in mossy, boggy areas in late summer and autumn across the north temperate zone. Although the parasite is described in field guides as 'rare' its host is so common that the parasitism is encountered regularly. There are many other examples of parasitism among our common field fungi (see Michael Kuo's *Key to Mycotrophs* on his MushroomExpert.Com Website at https://www.mushroomexpert.com/mycotrophs.html).

Such **mycotrophism** is not always as obvious as in the case of *Boletus parasiticus* (Figure 17.14); it can go unnoticed because the host mushroom (or other sporophore) can be highly reduced, blackened and almost unrecognisable, or the host may be underground. For example, *Squamanita odorata* is an agaric which is parasitic on other agaric sporophores, but because the host tissues are reduced to unrecognisable galls, molecular techniques were necessary to identify that the host is *Hebeloma mesophaeurn* (Mondiet *et al.*, 2007). Not all of the best examples are basidiomycete 'mushrooms and toadstools'. For example, the ascostromata of the club fungus *Cordyceps* (or *Tolypocladium*; *Ascomycota*) are often found growing out of the subterranean sporophores of the false truffle *Elaphomyces* (*Ascomycota*): see the illustration hyperlinks in the Resources Box 17.2 below. We are more familiar with *Cordyceps* as a widespread parasite of insects (e.g. *Cordyceps militaris* common throughout the northern hemisphere as a parasite of the larvae and pupae of butterflies and moths) so this is another reminder that parasitic genera can have species that specialise on hosts from different kingdoms (compare Figure 17.13 and see Resources Box 17.2).

RESOURCES BOX 17.2

Ascostromata of the club fungus *Cordyceps* (or *Tolypocladium*; *Ascomycota*) are often found growing out of the subterranean sporophores of the false truffle *Elaphomyces* (*Ascomycota*). View the following web pages for illustrations:

Michael Kuo's *Key to 25 mushroom-eating mushrooms and fungi (mycotrophs)* on his MushroomExpert.Com website (URL: http://www.mushroomexpert.com/mycotrophs.html).
Tom Volk's Fungi website (URL: http://botit.botany.wisc.edu/toms_fungi/jan98.html).

It is not known when mycoparasitism evolved, though we have noted above the occurrence of mycoparasites of arbuscular mycorrhizas in 400 million-year-old fossils in the Rhynie chert. Macrofungi are not good candidates for fossilisation, but mycoparasitism of an agaric mushroom in Early Cretaceous (100 million-year-old) amber has been demonstrated (Poinar & Buckley, 2007), this being the **oldest known fossil of an agaric mushroom**. In this specimen, the mycoparasite was itself infected by a **hypermycoparasite**; the specimens demonstrate that complex patterns of mycoparasitism were well developed in the Cretaceous 100 million years ago.

In the present day, a large number of fungi parasitise others of their kind and those that attack commercially grown mushrooms often cause major economic losses to the industry; so, as our final example, we will briefly describe some of the mycoparasites that cause **disease in mushroom farms**, particularly those of the most widely cultivated mushroom, *Agaricus bisporus*. Cultivated mushrooms are prone to bacterial and virus diseases that can cause significant crop losses, but these are outside our current area of interest. The two most common and geographically most widespread mycoparasites of *Agaricus bisporus* are the *Ascomycota Mycogone perniciosa* (cause of the disease known as **wet bubble** or **white mould**) and *Verticillium fungicola* (cause of the disease known as **dry bubble** or **brown spot**). These parasites can cause serious economic losses to the mushroom farmer, and they also cause similar diseases of the paddy straw mushroom, *Volvariella volvacea*, in cultivation.

Vegetative mushroom mycelium is not adversely affected by *Mycogone perniciosa in vitro* until after formation of the strands from which the mushrooms develop. *Verticillium* hyphae grow over the mushroom mycelium *in vitro*, eventually causing severe necrosis of the *Agaricus* mycelium. *Verticillium* causes more mushroom necrosis than *Mycogone perniciosa* and it also kills mushroom mycelium, whereas *Mycogone perniciosa* does not. When *Mycogone perniciosa* grows on a mushroom its growth is thick, velvety and white (which is why it is called white mould), while *Verticillium* produces a fine, greyish-brown felted growth (which is why it is sometimes called a **cobweb mould** though this is usually applied to a *Cladobotryum* infection; see below).

A common result of parasitism by *Mycogone perniciosa* and *Verticillium* is a drastic effect on mushroom development, and **mushroom abnormalities** are the prime disease symptoms. The extent of the symptoms depends on the stage of development reached at the time of infection; the earlier the infection occurs, the greater the deformity caused. The most extreme effect is for a spheroidal mass of undifferentiated tissue to be formed rather than a mushroom; these are the 'bubbles' of these two bubble diseases. The undifferentiated masses caused by *Mycogone perniciosa* can be 5 cm or more across, whereas those caused by *Verticillium* are usually less than 1 cm in diameter.

Infection with *Verticillium* at later stages of development causes developmental deformities, including bulbous stems with vestigial caps and mushrooms in which the tissues are broken and peeled back (Largeteau & Savoie, 2008). Infection of well-developed mushrooms by *Mycogone perniciosa* commonly results in abnormal enlargement of the gills (lamellae) as they become covered by the parasite. Internally, *Mycogone*-infected mushrooms become wet and develop a foetid odour and drops of a clear amber coloured liquid are often extruded from the mushroom (which is why it is called wet bubble disease). *Verticillium*-infected sporophores are dry and shrivelled at first (= dry bubble disease), but in both cases bacteria invade and a **bacterial rot** is the outcome. Although these mycoparasites are certainly able to produce enzymes capable of degrading mushroom hyphal wall polymers most of the rot that occurs in the final stages of these infections is due to secondary (opportunistic) invasion of the necrotic tissues by bacteria and other fungi.

Macrofungi in the wild, particularly agarics, are commonly parasitised by members of the genus *Cladobotryum* (*Ascomycota*), and *Cladobotryum dendroides* is also a **common pathogen** of cultivated mushrooms. The parasite causes general decomposition of the mushroom tissues, characteristic deformities do not occur. *Cladobotryum dendroides* growth over the surface of mushroom beds is conspicuous, giving rise to the common name 'cobweb'. Mushrooms engulfed by the spreading mycelium of the parasite turning pale brown and develop a soft rot, particularly at the base of the stem. Mushroom crops can be damaged by a range of saprotrophic fungi, especially those that favour lignocellulosic substrates, which may be present in poorly-prepared composts or timber containers. These are usually considered to be weed fungi rather than parasites. Many cultivated fungi are not very vigorous (the *Pleurotus* spp. that are cultivated may be an exception to this generalisation), so they tend to come second in any competition for the substrate. *Schizophyllum*, *Stereum*, *Coriolus*, *Coniophora*, *Merulius* (all *Basidiomycota*) and *Hypoxylon* (*Ascomycota*) are among the genera of wood-destroying fungi that have been recorded in mushroom farms using timber containers or in traditional outdoor log cultivation of *Lentinula edodes* (shiitake). The most common weeds on farms that use compost substrates are the smaller ink cap mushrooms, such as *Coprinopsis cinerea* and its relatives, and the extremely common and widespread *Trichoderma* (*Ascomycota*). *Trichoderma* *pleurotum* and *Trichoderma pleuroticola* are highly aggressive parasites causing green mould diseases of the oyster mushroom *Pleurotus ostreatus* that threaten commercial production of oyster mushrooms around the world and may be a major infection risk for other mushroom farms (Komon-Zelazowska *et al.*, 2007).

Trichoderma spp. are common in most soils. They are competitive saprotrophs, opportunistic parasites of other fungi, and possibly symbionts of plants (Harman *et al.*, 2004). *Trichoderma viride* parasitises *Rhizoctonia solani*, which causes the disease of seedlings known as 'damping-off', and *Armillaria* and *Armillariella*, which are important pathogens of trees. At least some strains of *Trichoderma* establish vigorous and long-lasting colonisations of root surfaces by penetrating the epidermis and a few cell layers beneath. This is far from parasitic because the fungus enhances root growth and development and the uptake and use of nutrients and induces resistance responses in the plant that protect it from several plant pathogens as well as abiotic stresses. A very high proportion of agricultural crop loss is due to diseases caused by soil microbes, and strains of *Trichoderma* (and its relative *Gliocladium*) have been developed that can be used for biocontrol of crop pathogens; in other words, *Trichoderma* is being used as a **mycopesticide**. *Trichoderma*, though, is a serious contaminant in the production of mushrooms of the genus *Pleurotus*, which very often causes large crop losses. So, it is interesting to note that acetone-extracts of dried fruit bodies of the polypore *Pycnoporus* inhibited growth of *Trichoderma* strains (isolated from infected substrate in *Pleurotus* production modules) in both agar and wheat straw test cultures (Talavera-Ortiz *et al.*, 2020).

Another example is *Scytalidium parasiticum* (*Ascomycota*; anamorph of *Xylogone* species), which has been reported as a potentially useful necrotrophic mycoparasite of *Ganoderma boninense* causing basal stem rot of oil palm in Malaysia and elsewhere. The mycoparasite was isolated from *Ganoderma*-infected oil palms in plantations and then demonstrated to be environmentally friendly in the sense of having low to very low toxicity to hamster cell lines *in vitro*, and non-toxic when fed to rats. *Scytalidium parasiticum* is therefore seen as a potential candidate for biocontrol of basal stem rot in oil palm caused by *Ganoderma boninense*, there being no other effective treatments (Goh *et al.*, 2016). We will describe more general aspects of the use of parasitic fungi as mycopesticides in more detail in Chapter 19.

17.16 REFERENCES

Anonymous (2017). Stop neglecting fungi. *Nature Microbiology*, **2**: 17120. DOI: https://doi.org/10.1038/nmicrobiol.2017.120.

Araldi-Brondolo, S., Spraker, J., Shaffer, J. *et al.* (2017). Bacterial endosymbionts: master modulators of fungal phenotypes. In *The Fungal Kingdom*, ed. J. Heitman, B. Howlett, P. Crous *et al.* Washington, DC: ASM Press, pp. 981–1004. DOI: https://doi.org/10.1128/microbiolspec.FUNK-0056-2016.

Azziz-Baumgartner, E., Lindblade, K., Gieseker, K. *et al.* (2005). Case-control study of an acute aflatoxicosis outbreak, Kenya, 2004. *Environmental Health Perspectives*, **113**: 1779–1783. DOI: https://doi.org/10.1289/ehp.8384.

Baumgardner, D.J. (2016). Disease-causing fungi in homes and yards in the Midwestern United States. *Journal of Patient-Centered Research and Reviews*, **3**: 99–110. DOI: https://doi.org/10.17294/2330-0698.1053.

Bennett, J.W. & Klich, M. (2003). Mycotoxins. *Clinical Microbiology Reviews*, **16**: 497–516. DOI: https://doi.org/10.1128/CMR.16.3.497-516.2003.

Berbee, M.L. (2001). The phylogeny of plant and animal pathogens in the Ascomycota. *Physiological and Molecular Plant Pathology*, **59**: 165–187. DOI: https://doi.org/10.1006/pmpp.2001.0355.

Berger, L., Speare, R., Daszak, P. *et al.* (1998). Chytridiomycosis causes amphibian mortality associated with population declines in the rain forests of Australia and Central America. *Proceedings of the National Academy of Sciences of the United States of America*, **95**: 9031–9036. DOI: https://doi.org/10.1073/pnas.95.15.9031.

Blackwell, M. (2017). Made for each other: ascomycete yeasts and insects. In *The Fungal Kingdom*, ed. J. Heitman, B. Howlett, P. Crous *et al.* Washington, DC: ASM Press, pp. 945–962. DOI: https://doi.org/10.1128/microbiolspec.FUNK-0081-2016.

Blanford, S., Chan, B.H.K., Jenkins, N. *et al.* (2005). Fungal pathogen reduces potential for malaria transmission. *Science*, **308**: 1638–1641. DOI: https://doi.org/10.1126/science.1108423.

Bongomin, F., Gago, S., Oladele, O.R. & Denning, D.W. (2017). Global and multi-national prevalence of fungal diseases: estimate precision. *Journal of Fungi*, **3**: 57. DOI: https://doi.org/10.3390/jof3040057.

Burbrink, F.T., Lorch, J.M. & Lips, K.R. (2017). Host susceptibility to snake fungal disease is highly dispersed across phylogenetic and functional trait space. *Science Advances*, **3**: article e1701387. DOI: https://doi.org/10.1126/sciadv.1701387.

Butt, T.M., Coates, C.J., Dubovskiy, I.M. *et al.* (2016). Entomopathogenic fungi: new insights into host-pathogen interactions. *Advances in Genetics*, **94**: 307–364. DOI: https://doi.org/10.1016/bs.adgen.2016.01.006.

Cafaro, M.J. (2005). Eccrinales (Trichomycetes) are not fungi, but a clade of protists at the early divergence of animals and fungi. *Molecular Phylogenetics and Evolution*, **35**: 21–34. DOI: https://doi.org/10.1016/j.ympev.2004.12.019.

Capella-Gutiérrez, S., Marcet-Houben, M. & Gabaldón, T. (2012). Phylogenomics supports microsporidia as the earliest diverging clade of sequenced fungi. *BMC Biology*, **10**: 47. DOI: https://doi.org/10.1186/1741-7007-10-47.

Casadevall, A. (2005). Fungal virulence, vertebrate endothermy, and dinosaur extinction: is there a connection? *Fungal Genetics and Biology*, **42**: 98–106. DOI: http://doi.org/10.1016/j".fgb.2004.11.008.

Casadevall, A. (2007). Determinants of virulence in the pathogenic fungi. *Fungal Biology Reviews*, **21**: 130–132. DOI: https://doi.org/10.1016/j.fbr.2007.02.007.

Casadevall, A. & Pirofski, L.-A. (2003). The damage-response framework of microbial pathogenesis. *Nature Reviews Microbiology*, **1**: 17–24. DOI: https://doi.org/10.1038/nrmicro732.

Castrillo, L.A., Roberts, D.W. & Vandenberg, J.D. (2005). The fungal past, present, and future: germination, ramification, and reproduction. *Journal of Invertebrate Pathology (Special SIP Symposium Issue)*, **89**: 46–56. DOI: https://doi.org/10.1016/j.jip.2005.06.005.

Charbonneau, L.R., Hillier, N.K., Rogers, R.E.L., Williams, G.R. & Shutler, D. (2016). Effects of *Nosema apis*, *N. ceranae*, and coinfections on honey bee (*Apis mellifera*) learning and memory. *Scientific Reports*, **6**: article number 22626. DOI: https://doi.org/10.1038/srep22626.

Charnley, A.K. (2003). Fungal pathogens of insects: cuticle degrading enzymes and toxins. *Advances in Botanical Research*, **40**: 241–321. DOI: https://doi.org/10.1016/S0065-2296(05)40006-3.

Cheong, C.D. & Neumeister-Kemp, H.G. (2005). Reducing airborne indoor fungi and fine particulates in carpeted Australian homes using intensive, high efficiency HEPA vacuuming. *Journal of Environmental Health Research*, **4**. Online at https://www.cieh.org/JEHR/reducing_airborne_fungi.html.

Cheong, C.D., Neumeister-Kemp, H.G., Dingle, P.W. & Hardy, G.St J. (2004). Intervention study of airborne fungal spora in homes with portable HEPA filtration units. *Journal of Environmental Monitoring*, **6**: 866–873. DOI: https://doi.org/10.1039/b408135h.

Chin, J.-Y., Godwin, C., Parker, E. *et al.* (2014). Levels and sources of volatile organic compounds in homes of children with asthma. *Indoor Air*, **24**: 403–415. DOI: https://doi.org/10.1111/ina.12086.

Cooke, R.C. & Rayner, A.D.M. (1984). *Ecology of Saprotrophic Fungi*. New York: Longman. ISBN-10: 0582442605, ISBN-13: 9780582442603.

Corradi, N. (2015). Microsporidia: eukaryotic intracellular parasites shaped by gene loss and horizontal gene transfers. *Annual Review of Microbiology*, **69**: 167–183. DOI: https://doi.org/10.1146/annurev-micro-091014-104136.

Corradi, N. & Keeling, P.J. (2009). Microsporidia: a journey through radical taxonomical revisions. *Fungal Biology Reviews*, **23**: 1–8. DOI: https://doi.org/10.1016/j.fbr.2009.05.001.

Correa-Martinez, C., Brentrup, A., Hess, K. *et al.* (2018). First description of a local *Coprinopsis cinerea* skin and soft tissue infection. *New Microbes and New Infections*, **21**: 102–104. DOI: http://doi.org/10.1016/j.nmni.2017.11.008.

Dambuza, I., Levitz, S., Netea, M. & Brown, G. (2017). Fungal recognition and host defense mechanisms. In *The Fungal Kingdom*, ed. J. Heitman, B. Howlett, P. Crous *et al.* Washington, DC: ASM Press, pp. 887–902. DOI: https://doi.org/10.1128/microbiolspec.FUNK-0050-2016.

Dannemiller, K.C., Weschler, C.J. & Peccia, J. (2016). Fungal and bacterial growth in floor dust at elevated relative humidity levels. *Indoor Air*, **27**: 354–363. DOI: https://doi.org/10.1111/ina.12313.

de Hoog, S., Monod, M., Dawson, T. *et al.* (2017). Skin fungi from colonization to infection. In *The Fungal Kingdom*, ed. J. Heitman, B. Howlett, P. Crous *et al.* Washington, DC: ASM Press, pp. 855–871. DOI: https://doi.org/10.1128/microbiolspec.FUNK-0049-2016.

Dighton, J., White, J.F. & Oudemans, P. (2005). *The Fungal Community: Its Organization and Role in the Ecosystem*. Boca Raton, FL: CRC Press, pp. 936. ISBN-10: 0824723554, ISBN-13: 9780824723552.

El-Jurdi, N. & Ghannoum, M. (2017). The Mycobiome: impact on health and disease states. In *The Fungal Kingdom*, ed. J. Heitman, B. Howlett, P. Crous *et al*. Washington, DC: ASM Press, pp. 845–854. DOI: https://doi.org/10.1128/microbiolspec.FUNK-0045-2016.

El-Shabrawi, M.H.F., Kamal, N.M., Jouini, R. *et al*. (2011). Gastrointestinal basidiobolomycosis: an emerging fungal infection causing bowel perforation in a child. *Journal of Medical Microbiology*, **60**: 1395–1402. DOI: https://doi.org/10.1099/jmm.0.028613-0.

Fairlamb, A.H., Gow, N.A.R., Matthews, K.R. & Waters, A.P. (2016). Drug resistance in eukaryotic microorganisms. *Nature Microbiology*, **1**: 16092. DOI: https://doi.org/doi.org/10.1038/nmicrobiol.2016.92.

Fisher, M.C. & Garner, T.W.J. (2007). The relationship between the emergence of *Batrachochytrium dendrobatidis*, the international trade in amphibians and introduced amphibian species. *Fungal Biology Reviews*, **21**: 2–9. DOI: https://doi.org/10.1016/j.fbr.2007.02.002.

Fisher, M.C., Gow, N.A.R. & Gurr, S.J. (2016). Tackling emerging fungal threats to animal health, food security and ecosystem resilience. *Philosophical Transactions of the Royal Society B: Biological Sciences*, **371**: article number 20160332. DOI: https://doi.org/10.1098/rstb.2016.0332.

Fisher, M.C., Henk, D.A., Briggs, C.J. *et al*. (2012). Emerging fungal threats to animal, plant and ecosystem health. *Nature*, **484**: 186–194. DOI: https://doi.org/10.1038/nature10947.

Fones, H., Fisher, M. & Gurr, S. (2017). Emerging fungal threats to plants and animals challenge agriculture and ecosystem resilience. In *The Fungal Kingdom*, ed. J. Heitman, B. Howlett, P. Crous *et al*. Washington, DC: ASM Press, pp. 787–809. DOI: https://doi.org/10.1128/microbiolspec.FUNK-0027-2016.

Franklinos, L.H.V., Lorch, J.M., Bohuski, E. *et al*. (2017). Emerging fungal pathogen *Ophidiomyces ophiodiicola* in wild European snakes. *Scientific Reports*, **7**: article number 3844. DOI: https://doi.org/10.1038/s41598-017-03352-1.

Gams, W., Diederich, P. & Põldmaa, K. (2004). Fungicolous fungi. In *Biodiversity of Fungi, Inventory and Monitoring Methods*, ed. G.M. Mueller, G.F. Bills & M.S. Foster. Burlington, VT: Academic Press, pp. 343–392. ISBN-10: 0125095511, ISBN-13: 9780125095518. DOI: https://doi.org/10.1016/B978-012509551-8/50020-9.

Garner, T.W.J., Schmidt, B.R., Martel, A. *et al*. (2016). Mitigating amphibian chytridiomycoses in nature. *Philosophical Transactions of the Royal Society B: Biological Sciences*, **371**: article number 20160207. DOI: https://doi.org/10.1098/rstb.2016.0207.

Geiser, D.M., Taylor, J.W., Ritchie, K.B. & Smith, G.W. (1998). Cause of sea fan death in the West Indies. *Nature*, **394**: 137–138. DOI: https://doi.org/10.1038/28079.

Gill, E.E. & Fast, N.M. (2006). Assessing the microsporidia-fungi relationship: combined phylogenetic analysis of eight genes. *Gene*, **375**: 103–109. DOI: https://doi.org/10.1016/j.gene.2006.02.023.

Goh, Y.K., Marzuki, N.F., Lim, C.K., Goh, Y.K. & Goh, K.J. (2016). Cytotoxicity and acute oral toxicity of ascomycetous mycoparasitic *Scytalidium parasiticum*. *Transactions on Science and Technology*, **3**: 483–488. URL: https://www.transectscience.org/pdfs/vol3/no3/3_3_483–488.html.

Gow, N.A.R. & Netea, M.G. (2016). Medical mycology and fungal immunology: new research perspectives addressing a major world health challenge. *Philosophical Transactions of the Royal Society B: Biological Sciences*, **371**: article number 20150462. DOI: https://doi.org/10.1098/rstb.2015.0462.

Haelewaters, D., Gorczak, M., Pfliegler, W.P. *et al*. (2015). Bringing Laboulbeniales into the 21st century: enhanced techniques for extraction and PCR amplification of DNA from minute ectoparasitic fungi. *IMA Fungus*, **6**: 363–372. DOI: https://doi.org/10.5598/imafungus.2015.06.02.08.

Hammer, S.M. (2005). Management of newly diagnosed HIV infection. *New England Journal of Medicine*, **353**: 1702–1710. DOI: https://doi.org/10.1056/NEJMcp051203.

Harman, G.E., Howell, C.R., Viterbo, A., Chet, I. & Lorito, M. (2004). *Trichoderma* species: opportunistic, avirulent plant symbionts. *Nature Reviews Microbiology*, **2**: 43–56. DOI: https://doi.org/10.1038/nrmicro797.

Harvell, C.D., Kim, K., Burkholder, J.M. *et al.* (1999). Emerging marine diseases – climate links and anthropogenic factors. *Science*, **285**: 1505–1510. DOI: https://doi.org/10.1126/science.285.5433.1505.

Hawksworth, D.L. & Lücking, R. (2017). Fungal diversity revisited: 2.2 to 3.8 million species. In *The Fungal Kingdom*, ed. J. Heitman, B. Howlett, P. Crous *et al.* Washington, DC: ASM Press, pp. 79–95. DOI: https://doi.org/10.1128/microbiolspec.FUNK-0052-2016.

Hernday, A.D., Lohse, M.B., Nobile, C.J. *et al.* (2016). Ssn6 defines a new level of regulation of white-opaque switching in *Candida albicans* and is required for the stochasticity of the switch. *mBio*, **7**: e01565–15. DOI: https://doi.org/10.1128/mBio.01565-15.

Hossain, M.A., Ahmed, M.S. & Ghannoum, M.A. (2004). Attributes of *Stachybotrys chartarum* and its association with human disease. *Journal of Allergy and Clinical Immunology*, **113**: 200–208. DOI: https://doi.org/10.1016/j.jaci.2003.12.018.

Hussain, A., Tian, M.-Y., Ahmed, S. & Shahid, M. (2012). Current status of entomopathogenic fungi as mycoinecticides and their inexpensive development in liquid cultures. In *Zoology*, ed. M.-D. García. Croatia: InTech, pp. 104–122. ISBN: 9789535103608. URL: https://mts.intechopen.com/books/zoology/current-status-of-entomopathogenic-fungi-as-mycoinecticides-and-their-inexpensive-development-in-liq.

Jaber, L.R. & Ownley, B.H. (2018). Can we use entomopathogenic fungi as endophytes for dual biological control of insect pests and plant pathogens? *Biological Control*, **116**: 36–45. DOI: https://doi.org/doi.org/10.1016/j.biocontrol.2017.01.018.

Jeffries, P. (1995). Biology and ecology of mycoparasitism. *Canadian Journal of Botany* (*Supplement*), **73**: S1284-S1290. DOI: https://doi.org/doi.org/10.1139/b95-389.

Karlsson, M., Atanasova, L., Jensen, D. & Zeilinger, S. (2017). Necrotrophic mycoparasites and their genomes. In *The Fungal Kingdom*, ed. J. Heitman, B. Howlett, P. Crous *et al.* Washington, DC: ASM Press, pp. 1005–1026. DOI: 10.1128/microbiolspec.FUNK-0016-2016.

Keeling, P.J & Fast, N.M. (2002). Microsporidia: biology and evolution of highly reduced intracellular parasites. *Annual Review of Microbiology*, **56**: 93–116. DOI: https://doi.org/10.1146/annurev.micro.56.012302.160854.

Keeling, P.J., Fast, N.M. & Corradi, N. (2014). Microsporidian genome structure and function. In *Microsporidia: Pathogens of Opportunity*, 2nd edition, ed. L.M. Weiss & J.J. Becnel. Hoboken, NJ: Wiley-Blackwell, an imprint of John Wiley & Sons, Inc., pp. 221–229. ISBN: 9781118395226. DOI: https://doi.org/10.1002/9781118395264.ch7.

Kim, K. & Rypien, K. (2015). Aspergillosis of Caribbean sea fan corals, *Gorgonia* spp. In *Diseases of Coral*, ed. C.M. Woodley, C.A. Downs, A.W. Bruckner, J.W. Porter & S.B. Galloway. Hoboken, NJ: John Wiley & Sons, pp. 236–241. DOI: https://doi.org/10.1002/9781118828502.ch16.

Kirk, P.M., Cannon, P.F., Minter, D.W. & Stalpers, J.A. (2008). *Dictionary of the Fungi*, 10th edition. Wallingford: CABI Publishing. ISBN-10: 0851998267, ISBN-13: 978-0851998268.

Köhler, J., Hube, B., Puccia, R., Casadevall, A. & Perfect, J. (2017). Fungi that infect humans. In *The Fungal Kingdom*, ed. J. Heitman, B. Howlett, P. Crous *et al.* Washington, DC: ASM Press, pp. 813–843. DOI: https://doi.org/10.1128/microbiolspec.FUNK-0014-2016.

Komon-Zelazowska, M., Bissett, J., Zafari, D. *et al.* (2007). Genetically closely related but phenotypically divergent *Trichoderma* species cause green mold disease in oyster mushroom farms worldwide. *Applied and Environmental Microbiology*, **73**: 7415–7426. DOI: https://doi.org/10.1128/AEM.01059-07.

Krings, M., Harper, C.J. & Taylor, E.L. (2018). Fungi and fungal interactions in the Rhynie chert: a review of the evidence, with the description of *Perexiflasca tayloriana* gen. et sp. nov. *Philosophical Transactions of the Royal Society of London, Series B*, **373**: article 20160500. DOI: https://doi.org/10.1098/rstb.2016.0500.

Kumar, P., Mahato, D.K., Kamle, M., Mohanta, T.K. & Kang, S.G. (2016). Aflatoxins: a global concern for food safety, human health and their management. *Frontiers in Microbiology*, **7**: 2170. DOI: https://doi.org/10.3389/fmicb.2016.02170.

Labbé, R.M., Gillespie, D., Cloutier, C. & Brodeur, J. (2009). Compatibility of fungus with predator and parasitoid in biological control of greenhouse whitefly. *Biocontrol Science and Technology*, **19**: 429–446. DOI: https://doi.org/10.1080/09583150902803229.

Lacey, L.A., Grzywacz, D., Shapiro-Ilan, D.I. *et al.* (2015). Insect pathogens as biological control agents: back to the future. *Journal of Invertebrate Pathology*, **132**: 1–41. DOI: https://doi.org/10.1016/j.jip.2015.07.009.

Langwig, K.E., Frick, W.F., Hoyt, J.R. *et al.* (2016). Drivers of variation in species impacts for a multi-host fungal disease of bats. *Philosophical Transactions of the Royal Society B: Biological Sciences*, **371**: article number 20150456. DOI: https://doi.org/10.1098/rstb.2015.0456.

Largeteau, M.L. & Savoie, J.M. (2008). Effect of the fungal pathogen *Verticillium fungicola* on fruiting initiation of its host, *Agaricus bisporus*. *Mycological Research*, **112**: 825–828. DOI: https://doi.org/10.1016/j.mycres.2008.01.018.

Lewis, L., Onsongo, M., Njapau, H. *et al.* (2005). Aflatoxin contamination of commercial maize products during an outbreak of acute aflatoxicosis in Eastern and Central Kenya. *Environmental Health Perspectives*, **113**: 1763–1767. DOI: https://doi.org/10.1289/ehp.7998.

Lips, K.R. (2016). Overview of chytrid emergence and impacts on amphibians. *Philosophical Transactions of the Royal Society B: Biological Sciences*, **371**: article number 20150465. DOI: https://doi.org/10.1098/rstb.2015.0465.

Liu, B.-L. & Tzeng, Y.-M. (2012). Development and applications of destruxins: a review. *Biotechnology Advances*, **30**: 1242–1254. DOI: https://doi.org/10.1016/j.biotechadv.2011.10.006.

Lorch, J.M., Knowles, S., Lankton, J.S. *et al.* (2016). Snake fungal disease: an emerging threat to wild snakes. *Philosophical Transactions of the Royal Society B: Biological Sciences*, **371**: article number 20150457. DOI: https://doi.org/10.1098/rstb.2015.0457.

Lovett, B. & St. Leger, R. (2017). The insect pathogens. In *The Fungal Kingdom*, ed. J. Heitman, B. Howlett, P. Crous *et al.* Washington, DC: ASM Press, pp. 925–943. DOI: https://doi.org/10.1128/microbiolspec.FUNK-0001-2016.

Meis, J.F., Chowdhary, A., Rhodes, J.L., Fisher, M.C. & Verweij, P.E. (2016). Clinical implications of globally emerging azole resistance in *Aspergillus fumigatus*. *Philosophical Transactions of the Royal Society B: Biological Sciences*, **371**: article number 20150460. DOI:https://doi.org/10.1098/rstb.2015.0460.

Mendoza, L., Taylor, J.W. & Ajello, L. (2002). The class Mesomycetozoea: a heterogeneous group of microorganisms at the animal-fungal boundary. *Annual Review of Microbiology*, **56**: 315–344. DOI: https://doi.org/10.1146/annurev.micro.56.012302.160950.

Mims, C.W., Hanlin, R.T. & Richardson, E.A. (2007). Light- and electron-microscopic observations of *Cladosporium* sp. growing on basidia of *Exobasidum camelliae* var. *gracilis*. *Canadian Journal of Botany*, **85**: 76–82. DOI: https://doi.org/10.1139/B06-153.

Mitchell, S.O. & Rodger, H.D. (2011). A review of infectious gill disease in marine salmonid fish. *Journal of Fish Diseases*, **34**: 411–432. DOI: https://doi.org/10.1111/j.1365-2761.2011.01251.x.

Mondiet, N., Dubois, M.P. & Selosse, M.A. (2007). The enigmatic *Squamanita odorata* (Agaricales, Basidiomycota) is parasitic on *Hebeloma mesophaeum*. *Mycological Research*, **111**: 599–602. DOI: https://doi.org/10.1016/j.mycres.2007.03.009.

Moore, D. (2013). *Fungal Biology in the Origin and Emergence of Life*. Cambridge, UK: Cambridge University Press. ISBN-10: 1107652774, ISBN-13: 978-1107652774.

Morissette, D.C., Dauch, A., Beech, R. *et al.* (2008). Isolation of mycoparasitic-related transcripts by SSH during interaction of the mycoparasite *Stachybotrys elegans* with its host *Rhizoctonia solani*. *Current Genetics*, **53**: 67–80. DOI: https://doi.org/10.1007/s00294-007-0166-6.

Moss, S.T. & Young, T.W.K. (1978). Phyletic considerations of the Harpellales and Asellariales (Trichomycetes, Zygomycotina) and the Kickxellales (Zygomycetes, Zygomycotina). *Mycologia*, **70**: 944–963. DOI: https://doi.org/10.2307/3759130.

Nordbring-Hertz, B. (2004). Morphogenesis in the nematode-trapping fungus *Arthrobotrys oligospora*: an extensive plasticity of infection structures. *Mycologist*, **18**: 125–133. DOI: https://doi.org/10.1017/S0269915XO4003052.

Nucci, M. & Marr, K.A. (2005). Emerging fungal diseases. *Clinical Infectious Diseases*, **41**: 521–526. DOI: https://doi.org/10.1086/432060.

O'Hanlon, S.J., Rieux, A., Farrer, R.A. *et al.* (2018). Recent Asian origin of chytrid fungi causing global amphibian declines. *Science*, **360**: 621–627. DOI: https://doi.org/10.1126/science.aar1965.

Ochiai, E., Kamei, K., Hiroshima, K. *et al.* (2005). The pathogenicity of *Stachybotrys chartarum*. *Japanese Journal of Medical Mycology*, **46**: 109–117. DOI: https://doi.org/10.3314/jjmm.46.109.

Odds, F.C. (2000). Pathogenic fungi in the 21st century. *Trends in Microbiology*, **8**: 200–201. DOI: https://doi.org/10.1016/S0966-842X(00)01752-2.

Omkar (ed.) (2016). *Ecofriendly Pest Management for Food Security*. San Diego, CA: Academic Press, an imprint of Elsevier Inc. ISBN: 9780128032657.

Ortiz-Urquiza, A., Luo, Z. & Keyhani, N.O. (2015). Improving mycoinsecticides for insect biological control. *Applied Microbiology and Biotechnology*, **99**: 1057–1068. DOI: https://doi.org/10.1007/s00253-014-6270-x.

Palmer, J.M., Drees, K.P., Foster, J.T. & Lindner, D.L. (2018). Extreme sensitivity to ultraviolet light in the fungal pathogen causing white-nose syndrome of bats. *Nature Communications*, **9**: article 35. DOI: https://doi.org/10.1038/s41467-017-02441-z.

Pestka, J.J., Yike, I., Dearborn, D.G., Ward, M.D.W. & Harkema, J.R. (2008). *Stachybotrys chartarum*, trichothecene mycotoxins, and Damp Building-Related Illness: new insights into a public health enigma. *Toxicological Sciences*, **104**: 4–26. DOI: https://doi.org/10.1093/toxsci/kfm284.

Poinar, G.O. & Buckley, R. (2007). Evidence of mycoparasitism and hypermycoparasitism in Early Cretaceous amber. *Mycological Research*, **111**: 503–506. DOI: https://doi.org/10.1016/j.mycres.2007.02.004.

Reynolds, N.K., Smith, M.E., Tretter, E.D. *et al.* (2017). Resolving relationships at the animal-fungal divergence: a molecular phylogenetic study of the protist trichomycetes (Ichthyosporea, Eccrinida). *Molecular Phylogenetics and Evolution*, **109**: 447–464. DOI: https://doi.org/10.1016/j.ympev.2017.02.007.

Robbins, N., Wright, G. & Cowen, L. (2017). Antifungal drugs: the current armamentarium and development of new agents. In *The Fungal Kingdom*, ed. J. Heitman, B. Howlett, P. Crous *et al.* Washington, DC: ASM Press, pp. 903–922. DOI: https://doi.org/10.1128/microbiolspec.FUNK-0002-2016.

Roilides, E. (2016). Editorial. Emerging fungi causing human infection: new or better identified? *Clinical Microbiology and Infection*, **22**: 660–661. DOI: https://doi.org/10.1016/j.cmi.2016.07.023.

Rollins-Smith, L.A., Doersam, J.K., Longcore, J.E. *et al.* (2002). Antimicrobial peptide defenses against pathogens associated with global amphibian declines. *Developmental & Comparative Immunology*, **26**: 63–72. DOI: https://doi.org/10.1016/S0145-305X(01)00041-6.

Sax, P.E. (2001). Opportunistic infections in HIV disease: down but not out. *Infectious Disease Clinics of North America*, **15**: 433–455. DOI: https://doi.org/10.1016/S0891-5520(05)70155-0.

Scholte, E.-J., Ng'habi, K., Kihonda, J. *et al.* (2005). An entomopathogenic fungus for control of adult African malaria mosquitoes. *Science*, **308**: 1641–1642. DOI: https://doi.org/10.1126/science.1108639.

Sexton, A.C. & Howlett, B.J. (2006). Parallels in fungal pathogenesis on plant and animal hosts. *Eukaryotic Cell*, **5**: 1941–1949. DOI: https://doi.org/10.1128/EC.00277-06.

Seyedmousavi, S., Guillot, J., Arné, P. *et al.* (2015). *Aspergillus* and aspergilloses in wild and domestic animals: a global health concern with parallels to human disease. *Medical Mycology*, **53**: 765–797. DOI: https://doi.org/10.1093/mmy/myv067.

Shaikh, N., Hussain, K.A., Petraitiene, R., Schuetz, A.N. & Walsh, T.J. (2016). Entomophthoramycosis: a neglected tropical mycosis. *Clinical Microbiology and Infection*, **22**: 688–694. DOI: https://doi.org/10.1016/j.cmi.2016.04.005.

Spellerya, D. & Maciver, A.J. (1971). Endocarditis caused by a *Coprinus* species: a fungus of the toadstool group. *Journal of Medical Microbiology*, **4**: 370–374. DOI: http://doi.org/10.1099/00222615-4-3-370.

St. Leger, R.J., Wang, C. & Fang, W. (2011). New perspectives on insect pathogens. *Fungal Biology Reviews*, **25**: 84–88. DOI: https://doi.org/10.1016/j.fbr.2011.04.005.

Talavera-Ortiz, A., Chaverri, P., Diaz-Godinez, G., Acosta-Urdapilleta, L., Villegas, E., Tellez-Tellez, M. (2020). Mycelial inhibition of *Trichoderma* spp. isolated from the cultivation of *Pleurotus ostreatus* with an extract of *Pycnoporus* sp. *Acta Botanica Mexicana*, **127**: article number 1537. DOI: https://doi.org/10.21829/abm127.2020.1537.

Taylor, C. & Gurr, S. (2014). Fungal pathogenesis: past, present and future. *Fungal Biology Reviews*, **28**: 24–28. DOI: https://doi.org/10.1016/j.fbr.2014.02.003.

Taylor, T.N., Krings M. & Taylor, E.L. (2015). *Fossil Fungi*. San Diego, CA: Academic Press. 398 pp. ISBN-10: 0123877318, ISBN-13: 978-0123877314. DOI: https://doi.org/10.1016/B978-0-12-387731-4.00014-1.

Thomas, M.B. & Read, A.F. (2007). Can fungal biopesticides control malaria? *Nature Reviews Microbiology*, **5**: 377–383. DOI: https://doi.org/10.1038/nrmicro1638.

Tretter, E.D., Johnson, E.M., Benny, G.L. *et al.* (2014). An eight-gene molecular phylogeny of the Kickxellomycotina, including the first phylogenetic placement of Asellariales. *Mycologia*, **106**: 912–935. DOI: https://doi.org/10.3852/13-253.

Troemel, E. (2017). Host-microsporidia interactions in *Caenorhabditis elegans*, a model nematode host. In *The Fungal Kingdom*, ed. J. Heitman, B. Howlett, P. Crous *et al.* Washington, DC: ASM Press, pp. 975–980. DOI: https://doi.org/10.1128/microbiolspec.FUNK-0003-2016.

Tudzynski, P., Correia, T. & Keller, U. (2001). Biotechnology and genetics of ergot alkaloids. *Applied Microbiology and Biotechnology*, **57**: 593–605. DOI: https://doi.org/10.1007/s002530100801.

Umesha, S., Manukumar, H.M., Chandrasekhar, B. *et al.* (2017). Aflatoxins and food pathogens: impact of biologically active aflatoxins and their control strategies. *Journal of the Science of Food and Agriculture*, **97**: 1698–1707. DOI: https://doi.org/10.1002/jsfa.8144.

Vajda, V. & Mcloughlin, S. (2004). Fungal proliferation at the Cretaceous–Tertiary boundary. *Science*, **303**: 1489. DOI: http://doi.org/10.1126/science.1093807.

Vávra, J. & Larsson, J.I.R. (2014). Structure of Microsporidia. In *Microsporidia: Pathogens of Opportunity*, 2nd edition, ed. L.M. Weiss & J.J. Becnel. Hoboken, NJ: Wiley-Blackwell, an imprint of John Wiley & Sons, Inc., pp. 1–70. ISBN: 9781118395226. DOI: https://doi.org/10.1002/9781118395264.ch1.

Vergneau-Grosset, C., Larrat, S. & Chen, Z.-Y. (2016). Microsporidiosis in vertebrate companion exotic animals. *Journal of Fungi*, **2**: 3. DOI: https://doi.org/10.3390/jof2010003.

Visscher, H., Brinkuis, H., Dilcher, D.L. *et al.* (1996). The terminal Paleozoic fungal event: evidence of terrestrial ecosystem destabilization and collapse. *Proceedings of the National Academy of Sciences of the United States of America*, **93**: 2155–2158. URL: http://www.jstor.org/stable/38482.

Wålinder, R., Ernstgård, L., Johanson, G. *et al.* (2005). Acute effects of a fungal volatile compound. *Environmental Health Perspectives*, **113**: 1775–1778. DOI: https://doi.org/10.1289/ehp.8193.

Weir, A. & Beakes, G.W. (1995). An introduction to the Laboulbeniales: a fascinating group of entomogenous fungi. *Mycologist*, **9**: 6–10. DOI: https://doi.org/10.1016/S0269-915X(09)80238-3.

Weir, A. & Blackwell, M. (2001). Molecular data support the Laboulbeniales as a separate class of Ascomycota, Laboulbeniomycetes. *Mycological Research*, **105**: 1182–1190. DOI: https://doi.org/10.1016/S0953-7562(08)61989-9.

Weiss, L.M. & Becnel, J.J. (ed.) (2014). *Microsporidia: Pathogens of Opportunity*, 2nd edition. Hoboken, NJ: Wiley-Blackwell, an imprint of John Wiley & Sons, Inc. ISBN: 9781118395226.

Whiteway, M. & Oberholzer, U. (2004) *Candida* morphogenesis and host-pathogen interactions. *Current Opinion in Microbiology*, **7**: 350–357. DOI: https://doi.org/10.1016/j.mib.2004.06.005.

Willis, K.J. (ed.) (2018). *State of the World's Fungi 2018*. Report. Richmond, UK: Royal Botanic Gardens, Kew. ISBN: 978-1-84246-678-0. URL: https://stateoftheworldsfungi.org/2018/ (the PDF of the report is a free download).

Woodley, C.M., Downs, C.A., Bruckner, A.W., Porter, J.W. & Galloway, S.B. (2015). *Diseases of Coral*. Hoboken, NJ: John Wiley & Sons, p. 595. ISBN: 9780813824116. DOI: https://doi.org/10.1002/9781118828502.

Work, T.M., Richardson, L.L., Reynolds, T.L. & Willis, B.L. (2008). Biomedical and veterinary science can increase our understanding of coral disease. *Journal of Experimental Marine Biology and Ecology*, **362**: 63–70. DOI: https://doi.org/10.1016/j.jembe.2008.05.011.

Wraight, S.P., Inglis, G.D. & Goettel, M.S. (2007). Fungi. In *Field Manual of Techniques in Invertebrate Pathology*, 2nd edition, ed. L.A. Lacey & H.K. Kaya. Dordrecht, the Netherlands: Springer, pp. 223–248. ISBN: 9781402059315.

Zasloff, M. (2002). Antimicrobial peptides of multicellular organisms. *Nature*, **415**: 389–395. DOI: https://doi.org/10.1038/415389a.

CHAPTER 18

Killing Fungi: Antifungals and Fungicides

In the last two chapters, we have shown how fungi can cause disease in plants and animals, and even other fungi, it is now our intention to discuss how fungal diseases might be controlled and, ideally, cured.

The key to control of any pest or cure of any disease, irrespective of the nature of the disease-causing organism, is a treatment that shows **selective toxicity**, which means that the treatment should inhibit or kill the disease-causing organism with little or no effect on the host. This concept of selective toxicity was established and demonstrated at the end of the nineteenth century by **Paul Ehrlich**, winner of the **Nobel Prize in Physiology or Medicine in 1908**.

We start this chapter with a reminder about selective toxicity, and then go on to examine the molecular effect(s) of agents that disrupt some crucial function of the fungal cell without adversely affecting the host plant or animal. Broadly speaking, this currently involves agents that target either the fungal cell membrane or cell wall. This is because the fungal plasma membrane and cell wall differ from those present in either animals or plants.

A major problem with attempts to control any disease with antibiotics is the occurrence of antibiotic resistance in the disease-causing organism because of the selection pressure exerted by the treatment itself. We will show that clinical control of systemic mycoses in the twenty-first century is still centred on the use of azoles and polyenes, with **combinatorial therapies** providing molecular approaches to management of the development of resistance. Similarly, in the agricultural sector control of fungal disease in the twenty-first century is dominated by the strobilurins, fungicides produced by fungi, and **IPM programmes** use tactics to avoid the emergence of fungicide resistance very similar to those employed in combinatorial therapy.

18.1 AGENTS THAT TARGET FUNGI

Towards the end of the nineteenth century, **Paul Ehrlich** was a physician and scientist working in the fields of haematology, immunology and antimicrobial chemotherapy in the Royal Institute for Experimental Therapy at Goettingen University, Germany. He established the concept that the chemistry of drugs used to control disease must be studied in relation to their modes of action and their affinity for the cells of the organisms against which they were directed. His aim was, as he expressed it, to find chemical substances to serve as '**magic bullets**' which would go straight to the disease organisms at which they were aimed and do minimum damage to the diseased host. Ehrlich was awarded the **Nobel Prize in Physiology or Medicine in 1908** jointly with **Ilya Ilyich Mechnikov** 'in recognition of their work on immunity' (visit: https://www.nobelprize.org/nobel_prizes/medicine/laureates/1908/ehrlich-bio.html).

Ehrlich's team used the now-common approaches of screening many newly synthesised compounds for antimicrobial activity followed by optimisation of the biological activity of a lead compound through systematic chemical modifications. In 1909, the team discovered the organic arsenical antibacterial compound '**Salvarsan**' (arsphenamine) to treat syphilis; it was the 606th compound the team had synthesised for testing and was the **first organic antibiotic**.

Ehrlich established the ***systematic approach*** that has become known as **drug discovery**. Of course, 30 years after this, **penicillin** became the ultimate magic bullet antibacterial as the result of a chance discovery rather than a systematic search (see section 19.15), but penicillin illustrates the selective toxicity principle well. Penicillin has been shown to inhibit bacterial cell wall synthesis, which results in death of the bacterial cell; but, because it targets bacteria-specific proteins and has no effect on human proteins, it is an effective drug to combat many bacterial infections.

Because fungi are themselves eukaryotes, drug discovery programmes aimed at finding antifungal agents with the required specificity must concentrate on targeting features that are unique to fungi. By convention, **fungicides** are used to ***treat fungal diseases of plants***, while **antifungal drugs** are used to ***treat fungal diseases of animals***. Either of these may be **fungistatic**, meaning that the agent stops fungal growth without killing the fungus and thereby 'buys time' for the diseased organism's intrinsic defences to operate; or **fungicidal**, meaning that the agent is fatally toxic to the fungus, without, hopefully, being in the least damaging to the diseased organism.

A **protectant** fungicide or antifungal drug is a chemical that acts outside the plant or animal and protects it from fungal infection. Protectants usually have multiple sites of antifungal activity. Two examples are:

- Milardet's '**Bordeaux mixture**' (a paint-like mixture of calcium hydroxide and copper sulfate), which was originally developed in the Bordeaux region of France in the late nineteenth century to control downy mildew on grape vines but is also used to protect potato plants from *Phytophthora infestans* infection and is an effective protectant of all sorts of plants against all sorts of stem and foliar fungal and fungus-like diseases.
- **Whitfield's ointment** (a greasy mixture of benzoic and salicylic acids) is used to treat athlete's foot and is effective against similar superficial fungal infections of the skin.

A compound like Bordeaux mixture affects a range of sensitive sites in the fungus, but its mode of action at some of these sites may be the same (in Bordeaux mixture it is the copper ion which is fungitoxic). The contrast here is between an antifungal that affects a single fungal site, which hopefully is unique, and an antifungal that affects several fungal sites, with the former exhibiting more selective toxicity than the latter. On the other hand, resistance of the fungus to a multi-site antifungal will develop much more slowly than to a single site antifungal. It's a trade-off between reduction in crop yield, because of lower selective toxicity (with single site antifungals performing better than multi-site ones), and the rate at which resistant strains arise in the fungal population (with multi-site antifungals performing better than single site ones). Of course, selective toxicity is more important for ***clinical antifungal drugs*** than for ***agricultural fungicides***. See Resources Box 18.1 for information about chemical toxicity.

Systemic fungicides and **antifungal drugs** penetrate the plant or animal tissues from the site of application and spread through the tissues of the host, eliminating established infections. Systemic antifungal agents usually have a highly specific (generally single) site of activity, for example:

RESOURCES BOX 18.1 Information About Toxicity

Rapid access to internationally peer-reviewed information on chemicals commonly used throughout the world, including those that occur as contaminants in the environment and in food, is provided by the INCHEM website of the International Programme on Chemical Safety. It consolidates chemical safety information from a number of intergovernmental organisations whose goal it is to assist in the sound management of chemicals.

Now hosted by the World Health Organization, view this URL: https://www.who.int/ipcs/en/

- benzimidazole, targets microtubules of fungal pathogens of plants;
- ketoconazole, targets sterol biosynthesis of fungal pathogens of animals; and
- see Table 18.1.

18.2 ANTIFUNGAL AGENTS THAT TARGET THE FUNGAL MEMBRANE

The basis of the selective toxicity of antifungal agents that target plasma membranes is the relative sterol composition of those membranes (see section 5.12). Sterols influence the fluidity, permeability, microdomain formation and membrane activities, including functionality of membrane-bound proteins. The key fact relevant to antifungal drug discovery is that ***there are three predominant forms of sterol found in eukaryotes***. In plants the most common sterols are the phytosterols; sigmasterol, sitosterol and campesterol. In animal cells and *Oomycota*, cholesterol is the chief sterol in the membrane, while in most fungi the major sterol is ergosterol (the exceptions being the *Chytridiomycota*, where the dominant sterol is again cholesterol). Sterols used in biomembranes are the end products of multistep biosynthetic pathways that all derive from a common initial pathway (acetyl-CoA to squalene epoxide). The three different pathways (fungal, animal, plant) are thought to result from evolution for optimised physiological function in each of the kingdoms. The **ergosterol biosynthesis pathway is fungus-specific** (Figure 18.1) so it is the prime target for treating fungal diseases in animals and plants alike. This difference in the primary sterol component of fungal

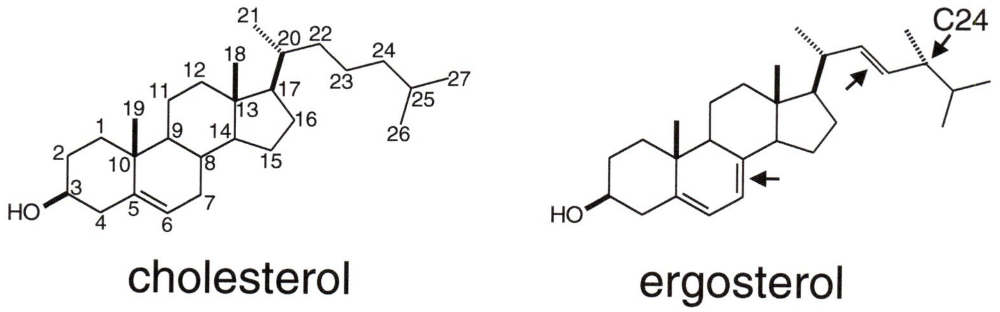

Figure 18.1. Chemical structures of cholesterol and ergosterol. The cholesterol schematic shows the conventional numbering scheme for the carbon atoms. Ergosterol differs from cholesterol by having an extra methyl group at C_{24} and two additional sites of unsaturation (arrows).

Table 18.1. Overview of the sites of action of some systemic antifungal agents

Cellular target	Organism	Agent	Mode of action
Plasma membranes	animals	Polyenes (Figure 18.2)	Bind to ergosterol
	animals & plants	Azoles (Figures 18.5 and 18.6)	Inhibit ergosterol synthesis
	plants	Edifenphos (an organophosphorus fungicide) (Figure 18.11)	Inhibits phosphatidyl choline synthesis
Nucleic acid biosynthesis	plants	Acylalanines (e.g. metalaxyl) (Figure 18.11)	Inhibit RNA polymerase
	animals	5-fluorocytosine	Inhibits RNA synthesis
Mitochondrial respiration	plants	Strobilurins (Figs 18.12 and 18.13)	Bind to cytochrome b, disrupting electron flow and ATP synthesis
Cell wall	plants	Polyoxins, nikkomycins (Figure 18.8)	Inhibit chitin synthesis
	plants	Tricyclazole (Figure 18.11)	inhibits melanin synthesis
	man	Echinocandins (Figure 18.9)	Inhibit β 1–3 glucan synthase
Cytoskeleton	plants	Benzimidazoles (Figure 18.10)	Bind to β-tubulin
	animals	Griseofulvin (Figure 18.10)	Binds to microtubule-associated proteins
Other potential targets include: mannoproteins involved in adhesion, translation elongation factor and proteases.			

and mammalian cells prompted development of two classes of antifungal agents, the polyenes and the azoles (Robson, 1999).

Ergosterol synthesis requires more energy than cholesterol, but ergosterol improves the resistance of membrane lipids to peroxidation and increases the mechanical resistance of the plasma membrane to osmotic stress both of which are caused by transitions between wet and dry conditions. Consequently, it is thought that ergosterol suits the fungi to the terrestrial environment and enables both yeast-form and mycelia to survive cycles of air-drying and wetting (Dupont *et al.* 2012).

Some fungi do not contain ergosterol, including chytrids (which use cholesterol) and the basidiomycete *Puccinia graminis* (which uses fungisterol). The fungus-like *Leptomitales* and *Saprolegniales* (Phylum *Oomycota*, Kingdom *Straminipila*; see section 3.10) use cholesterol and other sterols like demosterol. *Phytophthora cactorum* is also an exception in that it does not ***synthesise*** sterols and must obtain them from its substrate.

The two main classes of antifungals currently available for treating systemic mycoses target ergosterol. **Polyene** antifungals, which include nystatin and amphotericin B, act by selective disruption of membrane structure by hydrophobic binding to ergosterol, forming pores leading to a loss of plasma membrane integrity. These drugs have toxicity and pharmacological difficulties. The **azole** antifungals inhibit the sterol 14-demethylase enzyme step in ergosterol biosynthesis, resulting in ergosterol depletion in the membrane and the accumulation of 14-O′-methylsterols. These changes in composition result in changes in membrane fluidity that adversely affect transport processes and wall biosynthesis and ultimately result in death of the fungus. Use of these agents is being affected adversely by the emergence of stains of fungal pathogens that are azole-resistant.

Polyenes are compounds that contain alternating double and single carbon-carbon bonds. Compounds with two carbon-carbon double bonds alternating with single bonds (−C=C−C=C−) are **conjugated compounds** called **dienes**; those with three such double bonds are **trienes** (−C=C−C=C−C=C−); those with four are tetraenes, etc. Compounds with one or more sequences of many alternating double bonds are the polyenes (Figure 18.2).

Polyenes are terpenes which are derived biosynthetically from units of isoprene (C_5H_8) (see Figure 11.15). These are polyunsaturated organic compounds and the category includes many fatty acids (hence the description 'polyunsaturated fat'). Normally, carbon–carbon double bonds are sufficiently energetic to absorb light in the blue and ultraviolet region of the spectrum, resulting in compounds which are coloured yellow or orange. Consequently, many pigments feature linear polyenes, for example the yellow-orange coloured β-carotene (see Figure 11.20); and polyene antifungals tend to be coloured yellow.

Nystatin (Figure 18.2) was isolated from the bacterium *Streptomyces noursei* in 1950 as a promising antifungal agent, and amphotericin B was isolated from *Streptomyces nodosus* a few years later. In fact, of more than 200 polyene antibiotics so

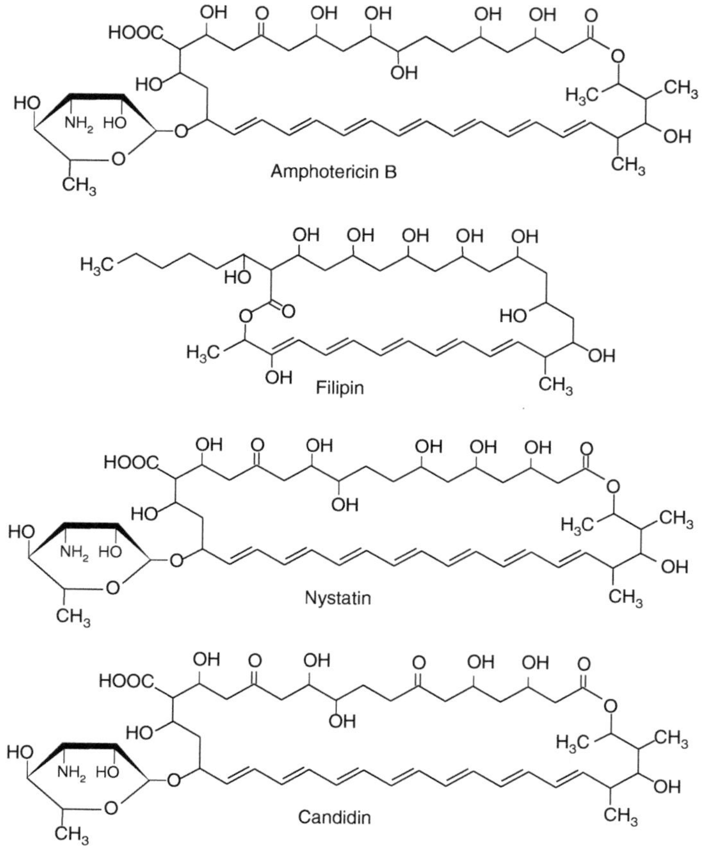

Figure 18.2. Structural formulae of some polyene antibiotics. The top illustration shows Amphotericin B and should be compared with Figure 18.3 as an example of a clinically important polyene antifungal agent. Note the alternating double and single bonds (conjugated double bonds all trans) across the bottom of all the molecules as oriented here and the lactone closure (–C–O–C[=O] –) of the macrolide ring at top right, and the mycosamine amino sugar at bottom left. Filipin lacks both features.

far discovered, the majority are produced by streptomycetes (Table 18.2). Nystatin is used topically (that is, applied to body surfaces; for example, to control vulvovaginal candidiasis) or orally (for example, to control alimentary tract infections) as there is little or no absorption by the alimentary tract, but it is too toxic to be used by intravenous or intramuscular injection (that is, it is too toxic to be administered **parenterally**). **Amphotericin B** is the only polyene which is sufficiently non-toxic to be tolerated parenterally although side effects from prolonged exposure include kidney damage and nausea.

There's more to antifungal polyenes than just the conjugated polyene structure. They are characterised by a **macrolide** ring of carbon atoms (a macrolide is a large macrocyclic structure) which is closed into a ring by formation of a **lactone** (a lactone is a cyclic ester resulting from condensation of an alcohol group and a carboxylic acid group in the same molecule) and to which a 6-carbon deoxyaminosugar is characteristically attached. With one exception, the hexosamine carbohydrate is the deoxyaminosugar **mycosamine** (Figure 18.2).

Table 18.2. Original sources of some polyenes that have been used in experimental and/or clinical settings

Name	Producing organism
Amphotericin B (isolated in 1955)	*Streptomyces nodosus*
Candicidin	*Streptomyces griseus*
Candidin	*Streptomyces viridoflavus*
Etruscomycin (lucensomycin)	*Streptomyces lucensis*
Filipin complex (isolated in 1955)	*Streptomyces filipensis*
Hamycin	*Streptomyces pimprina*
Natamycin (pimaricin)	*Streptomyces natalensis*
Nystatin (isolated in 1950)	*Streptomyces albidus or Streptomyces noursei*
Trichomycin	*Streptomyces hachijoensis* and *Streptomyces abikoensis*

As you can see from Figure 18.2, amphotericin B has seven hydroxyls and the mycosamine sugar on one side of the macrolide ring (which is hydrophilic as a result) and the double-bonded polyene on the other side of ring (and this side is hydrophobic as a result); and as its name implies, the molecule is **amphoteric** (capable of reacting chemically either as an acid or as a base). These structural features are thought to determine the sterol selectivity and biological activity of the molecule. Polyenes have limited solubility in water but readily interact with the sterol molecules or acyl side chains of phospholipids in biological membranes. The basis of the antifungal activity of polyenes is that they interact with membrane sterols, particularly ergosterol, to form polyene–sterol complexes that produce pores (partial or complete) through the membrane that increase membrane permeability.

Amphotericin B forms an **annulus**, which has a pore of 0.8 nm internal diameter, of eight polyene molecules in one leaflet of the membrane bilayer. This is a 'half-pore' but the annuli can form in both leaflets and when two annuli coincide in both opposing leaflets the amphotericin pore spans the lipid bilayer as an ***unregulated ion channel***. The orientation of the aminosugar on the macrolide ring determines sterol preference and specific hydrogen bonding (between 2′-OH of amphotericin and 3β-OH of ergosterol) locks the complex in place so that eight of them enclose the hydrophilic channel (Figure 18.3).

Release of cellular K^+ ions is one of the earliest effects of polyenes and this is associated with uptake of protons and consequent acidification of cytoplasm. Leakage of amino acids, sugars and other metabolites also occurs (Table 18.3).

Polyenes interact with both cholesterol and ergosterol so the key to their value as clinically useful antifungal agents is the fact that some have so much **more affinity for ergosterol** that they can be selectively toxic to the fungus (Figure 18.4 and

Table 18.3. Observable effects of four polyenes that interact with membrane sterols

Feature	Filipin	Nystatin	Amphotericin B	Candicidin
Potassium leakage	+	+	+	+
K^+ or NH_4^+ reversal* of glycolysis inhibition	-	-	+	+
Phosphate leakage	+	+	±	±
Erythrocyte lysis	Rapid	+	+	±
Yeast protoplast lysis	Rapid	+	±	-
Antifungal activity	+	++	+++	++++

*The ammonium ion is physically sufficiently like the potassium ion to serve as a potassium analogue in many chemical reactions. There are many similarities between potassium and ammonium salts.

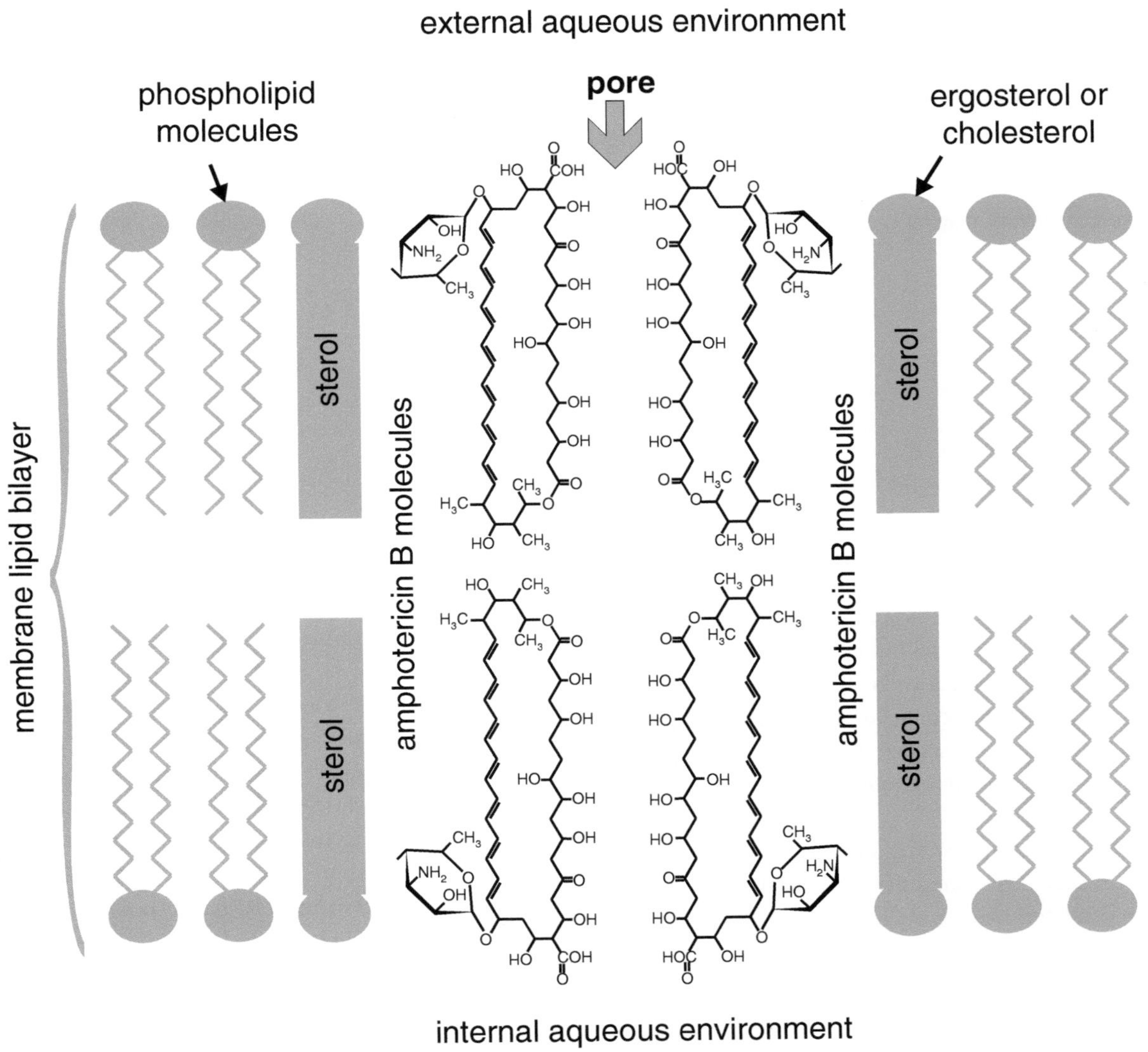

Figure 18.3. Diagrammatic structure of an amphotericin B-sterol pore. The illustration depicts the alignment of two half-pores; the half-pores are assembled from eight molecules of amphotericin in a ring in each leaflet of the lipid bilayer. When two half-pores coincide in both leaflets of the membrane bilayer the hydrophilic channel in the centre forms an unregulated ion channel that spans the membrane.

Table 18.4. Toxicity of polyenes in small mammals[a]

Antibiotic	LD_{50} ($mg\ kg^{-1}$)				Animal
	Intravenous	Intraperitoneal	Subcutaneous	Oral	
Amphotericin B	4–6.6	280–1,640		>8,000	mouse
Candicidin		2.1–7	160–280	98–400	mouse
Nystatin	3	45	2.4	8,000	mouse
Pimaricin (natamycin)	5–10	250	5,000	1,500	rat

[a] Compare these values with the median minimum inhibitory concentrations (MIC) against *Candida albicans* of: Amphotericin B, 0.5 mg l^{-1}; Candicidin, 0.5 mg l^{-1}; Nystatin, 3 mg l^{-1}; and Pimaricin (natamycin), 5 mg l^{-1}.

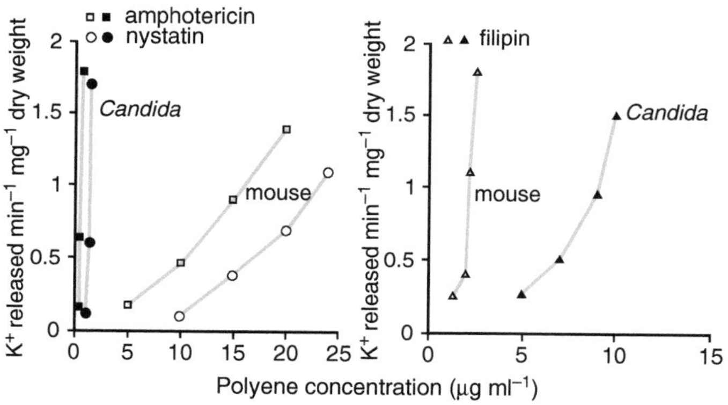

Figure 18.4. Selective toxicity of polyene antifungals against *Candida albicans* illustrated as their effect on release of potassium ions from intact *Candida* cells compared with release from mouse fibroblasts. Left-hand panel compares the methyl ester of amphotericin B and nystatin with mouse fibroblasts (both polyenes have higher affinity for ergosterol than for cholesterol) and the right-hand panel shows the relative effect of the polyene filipin, which lacks an aminosugar in its chemical structure (Figure 18.2) and has much greater affinity for cholesterol than it does for ergosterol.

Table 18.4). Others, like filipin, are too toxic to the host animal to be clinically useful (and see Table 18.4). Filipin does not form ion channels like amphotericin B and nystatin but acts as a membrane disrupter; it is more toxic to animal cells than to fungi (Figure 18.4) because of its affinity for cholesterol. However, this affinity for cholesterol and the fluorescence of the polyene have made filipin an extremely useful ***histochemical stain*** for cholesterol.

Enclosing amphotericin B within **liposomes** enhances drug delivery while reducing toxicity. Liposomes are artificial lipid vesicles (less than 100 nm diameter) made from readily available phospholipids, such as hydrogenated soy bean phosphatidylcholine or egg phosphatidylethanolamine. The phospholipids form into 'bubbles' of membrane bilayers that can be used to encapsulate drugs which are poorly soluble in water. The liposome membranes merge with cell membranes and therefore greatly improve delivery of the insoluble drug. In rabbit and mouse trials, amphotericin B liposomes (called **AmBisome**) were

Table 18.5. Activity of conventional amphotericin B (AmB) and the AmBisome (liposome) preparation

Fungus	Drug preparation	MIC_{90} ($mg\ l^{-1}$)
Candida albicans	AmB	1.25
	AmBisome	0.62
Candida tropicalis	AmB	1.25
	AmBisome	0.62
Aspergillus sp.	AmB	2.50
	AmBisome	1.25
Fusarium sp.	AmB	2.50
	AmBisome	2.50

MIC_{90} = minimum concentration needed to cause 90% inhibition.

more effective against *Candida* and *Aspergillus* than conventional amphotericin B preparations; and, compared to amphotericin B, the animal-toxicity of AmBisome was reduced by two-thirds (Table 18.5); AmBisome is used to treat serious, life-threatening fungal infections of humans.

Resistance to polyenes is **rarely observed clinically** although it can be demonstrated in the laboratory. Combined therapy reduces the risk of drug-resistant strains emerging during antifungal therapy. Amphotericin B can be used in combination with **flucytosine** (5-fluorocytosine or 5-FC) for a beneficial effect. The antifungal medication flucytosine is a fluorinated pyrimidine analogue that targets fungal RNA synthesis and disturbs translation of essential proteins.

Fungicides that **inhibit ergosterol biosynthesis** and thereby reduce or even completely deplete the ergosterol content of the fungal membrane are listed below and their structures are shown in Figure 18.5:

- **pyrimidine analogues** (for example, triarimol),
- **pyridines**, such as buthiobate,

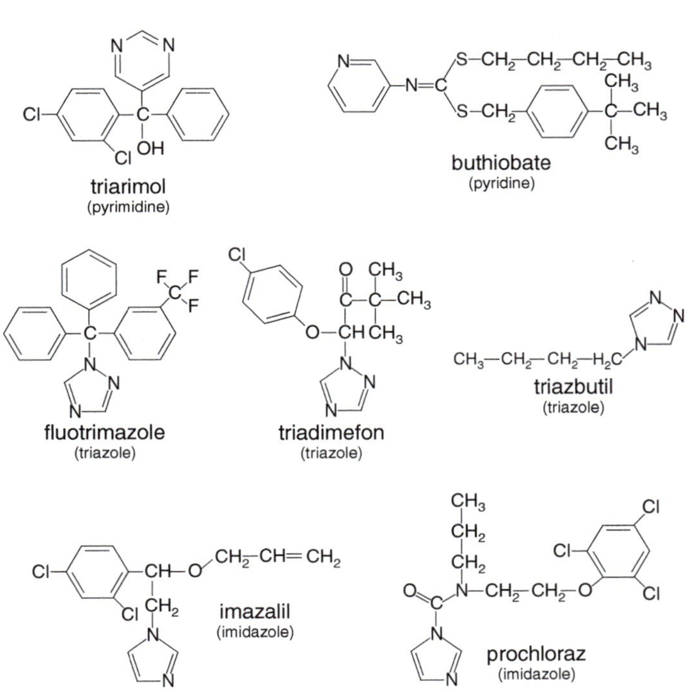

Figure 18.5. Structural formulae of some fungicides that inhibit ergosterol biosynthesis and thereby reduce or even completely deplete the ergosterol content of fungal pests.

- **triazoles** (for example, fluotrimazole, triadimefon and triazbutil), and
- **imidazoles** (for example, imazalil (enilconazole) and prochloraz).

These compounds are used to treat fungal diseases of plants. Similar compounds are used clinically as antifungal drugs (Figure 18.6):

- imidazoles, like **miconazole** (relatively insoluble so used in ointments for topical application), **clotrimazole** and **ketoconazole** (relatively soluble and can be taken by mouth),
- Triazoles, like itraconazole and fluconazole.

The characteristic feature of azoles is a five-membered **heterocyclic ring**; in imidazoles this ring includes two nitrogen atoms. These antifungals were first discovered in the early 1970s and are used to treat **superficial** (miconazole) and **systemic** (ketoconazole) infections of humans. The triazole five-membered ring contains three nitrogen atoms. Triazoles were first used in the 1990s and, compared to the imidazoles, have reduced side effects. Fluconazole and itrasconazole are highly soluble but are excreted renally more rapidly than ketoconazole and have a lower antifungal activity. These compounds affect ergosterol biosynthesis

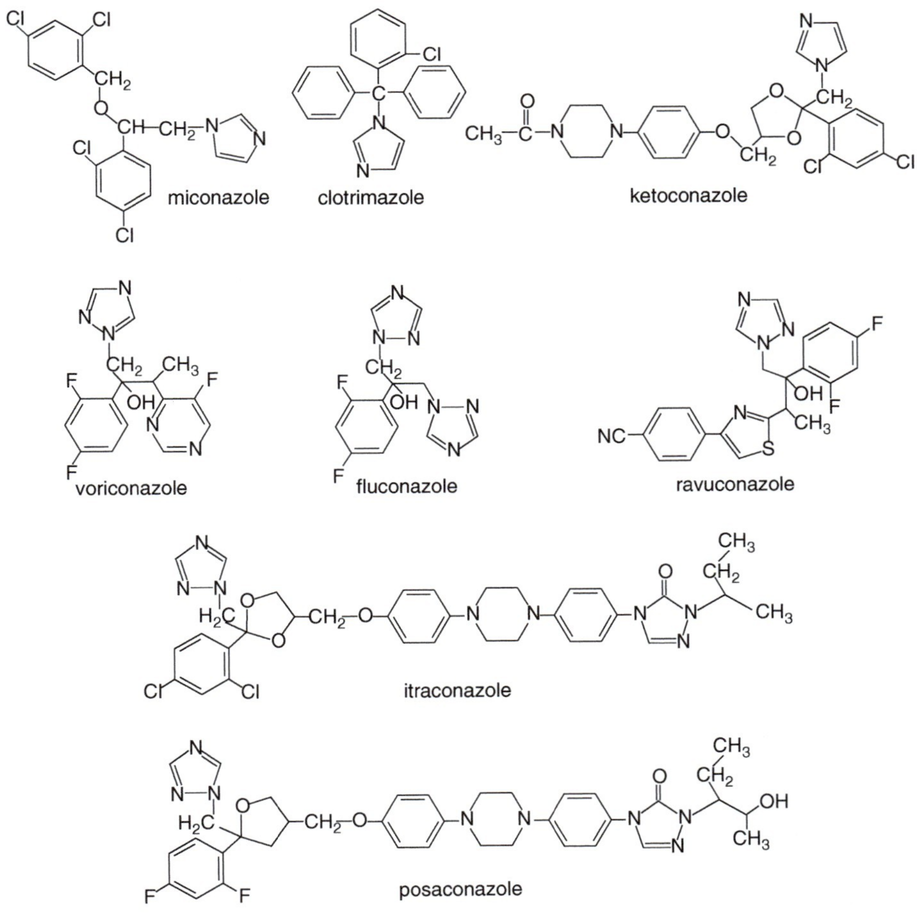

Figure 18.6. Structural formulae of some clinically useful antifungal azoles.

by binding to cytochrome P_{450}. This enzyme is required for the demethylation of lanosterol. Its inhibition hampers membrane formation and leads to accumulation of steroid intermediates which destabilise the membrane and inhibit growth (the agents are fungistatic). Potency of imidazoles is determined by:

- the affinity and geometric orientation of the heterocyclic nitrogen ring to the heme iron atom in cytochrome P_{450};
- protonation of the N-3 or N-4 in the imidazole or triazole ring;
- affinity of the nonligand portion of the drug for the apoprotein of cytochrome P_{450}.

An unprotonated nitrogen atom is needed to bind the heme iron of cytochrome P_{450}. The intracellular pH of the hyphal tip is usually less than 6 and at this pH the triazole ring of itraconazole is unprotonated whereas the imidazole ring of ketoconazole is more than 75% protonated. So **itraconazole** is a **more potent antifungal** drug than ketoconazole. Protracted use of ketoconazole can cause liver damage and fluconazole, which does not affect the liver, has replaced ketoconazole for this reason. However, fluconazole has less antifungal activity than ketoconazole, so in AIDS patients with severe fungal infections and impaired immune systems the risk of ketoconazole liver damage can be worth taking.

When azoles bind to the cytochrome P_{450} demethylase, which is responsible for removal of methyl groups in biosynthesis of ergosterol, the 14-alpha-methyl (lanosterol) sterol content of the membrane increases, and the ergosterol content of the membrane decreases and properties of the membrane consequently deteriorate (Figure 18.7). Ironically, one of the results of membrane deterioration is that chitin synthesis *increases*; but it is an uncoordinated increase and does not benefit the wall.

Sterol biosynthesis is complex, with several branched and potential alternative pathways. The relative importance of each part of the pathway differs between species of fungi and this accounts for differences in their sensitivity to individual azoles (Table 18.6) (Cowen, 2008; Monk & Goffeau, 2008; Howard & Arendrup, 2011).

Resistance to fluconazole can arise due to:

- Increased efflux of a drug from the cell resulting from mutations causing constitutive upregulation of multidrug transporter(s);
- Alteration or amplification of the drug target, which minimises the impact of the drug on the cell;
- Cellular alterations that minimise toxicity of the drug, for example by upregulation of biosynthetic genes producing excess sensitive protein so that the inhibitory agent is effectively titrated out, or by secondary mutations that prevent accumulation of toxic intermediates (Da Silva Ferreira *et al.* 2005; Cowen, 2008). We will discuss below (section 18.5) the potential of combinatorial therapy in lengthening the clinically useful lifetime of established drugs by reducing the evolution of resistance.

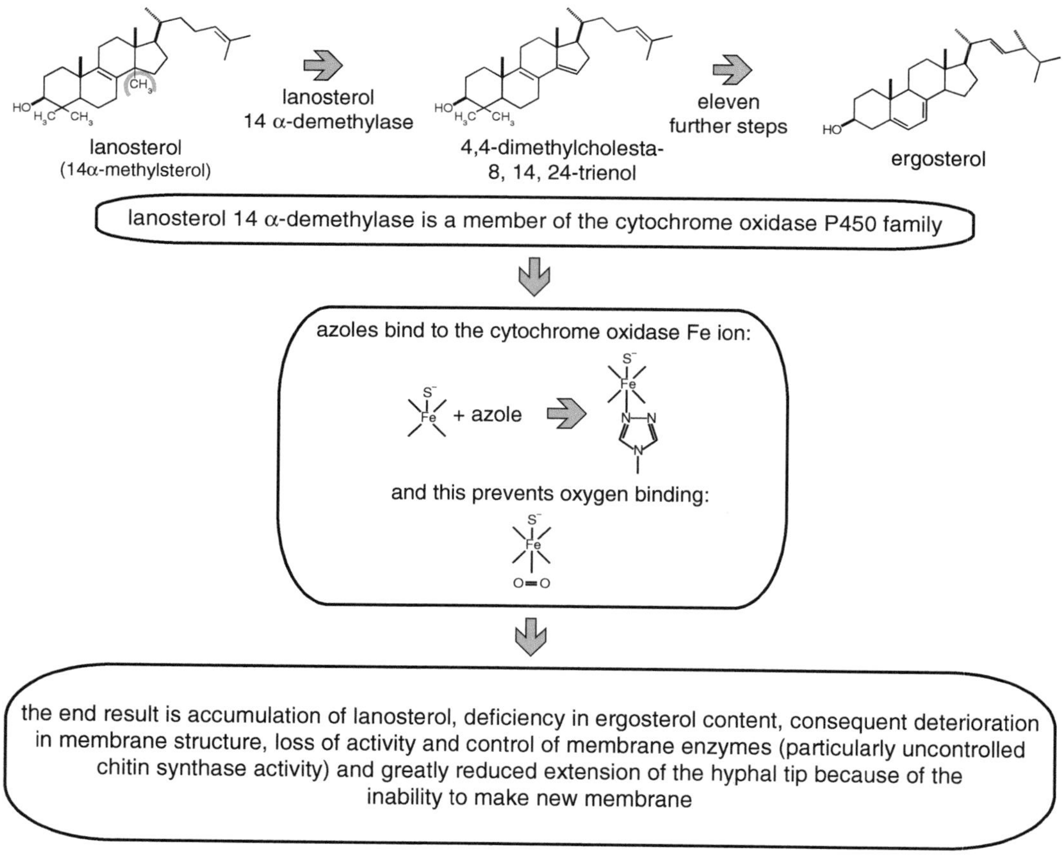

Figure 18.7. Mode of action of ergosterol biosynthesis inhibitors and effects on membrane function and cell proliferation.

Table 18.6. Inhibition of ergosterol synthesis by ketoconazole and itraconazole

Organism	Molar concentration causing 50% inhibition of ergosterol synthesis	
	Ketoconazole	Itraconazole
Candida albicans	2.5×10^{-8} M	5.0×10^{-9} M
Pityrosporum ovale (a lipophilic yeast which is a normal component of human skin flora but can be an opportunistic pathogen)	5.6×10^{-7} M	5.0×10^{-9} M
Aspergillus fumigatus	5.6×10^{-7} M	2.3×10^{-8} M

Voriconazole is the recommended agent for invasive aspergillosis in humans, with lipid amphotericin B or caspofungin as alternating treatments. As the only agents available in oral formulations, azoles are used in chronic infections and often over longer time periods. Unfortunately, in addition to being used in clinical medicine, azoles are employed extensively in agriculture, where they are used to control fungal disease in crop plants and to treat intensively reared farm animals. The compounds are widely deployed for crop protection; for example, over 250 metric tonnes of azoles are used to protect UK crops annually, and the global usage ranges into the thousands of tonnes each year (Kleinkauf *et al.*, 2013).

Aspergillus fumigatus is largely responsible for the increased mortality rates of invasive aspergillosis in immunocompromised patients. *Aspergillus* species are ubiquitous fungal saprotrophs worldwide and azole-resistant *Aspergillus* species can be isolated easily from the environment (Howard & Arendrup, 2011). Azole-resistant aspergillosis in humans was reported in 2007 from the Netherlands (with over 90% of the clinical isolates employing the same resistance mechanism), and since 2007 azole-resistance in clinical isolates of *Aspergillus* has been reported from many other countries. These azole-resistant clinical strains may be a side effect of the use of azole fungicides in agriculture (Verweij *et al.*, 2007, 2009), with such ***field acquired azole resistance*** jeopardising the effectiveness of using azoles to treat aspergillosis and other fungal diseases of humans (Verweij *et al.*, 2016; Garcia-Rubio *et al.*, 2017; Chowdhary & Meis, 2018). These observations serve to emphasise the urgency of the need to identify new fungicides that have targets distinct from those of the currently available azoles. Significantly, the transcription factor that regulates ergosterol biosynthesis in fungi has recently been found to be unique, making it a new potential target for the development of novel antifungals (Yang *et al.*, 2015; Sant *et al.*, 2016; Gumber *et al.*, 2017).

18.3 ANTIFUNGAL AGENTS THAT TARGET THE FUNGAL WALL

As Gow *et al.* (2017) point out, the fungal cell wall 'is composed almost exclusively of molecules that are not represented in the human body yet are important or essential for fungal growth, viability or virulence. As such, the wall is a near ideal target

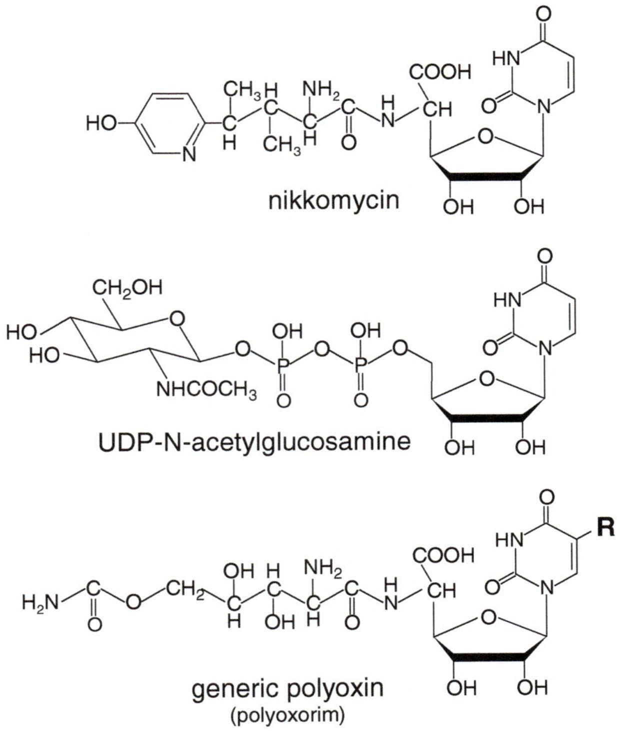

Figure 18.8. Comparison of the structures of the fungicides nikkomycin and polyoxorim (polyoxin) with (centre) the structure of UDP-N-acetylglucosamine which is the normal substrate of chitin synthase. The generic structure of polyoxorim/polyoxin is shown; in polyoxin B, R = CH_2OH; in polyoxin D, R = COOH; and in polyoxin R = CH_3.

for the design of antifungal drugs for clinical use.' We have remarked on these facts earlier in this book (refer to the following: *Fungal cell wall* in section 5.13; *Wall synthesis and remodelling* in section 7.7).

The best-known **chitin synthesis inhibitors** are naturally occurring antibiotics which are analogues of the chitin synthase substrate, UDP-*N*-acetylglucosamine. They are called **nikkomycin** (from the bacterium *Streptomyces tendae*) and **polyoxin** (produced by *Streptomyces cacaoi* var. *asoensis*) (Figure 18.8).

Nikkomycins and polyoxins function as potent and specific competitive inhibitors of chitin synthase, competing as analogues of its normal substrate UDP-*N*-acetylglucosamine (also shown in

Figure 18.8). Zinc salts of **polyoxin D** (the most potent of the 13 different polyoxins which are designated A to M) are used in Japan against rice sheath blight (*Pellicularia sasakii*) and polyoxin B is effective in controlling powdery mildews on fruit trees and grapevines. In the United States, polyoxin D zinc salt was first registered for food use in late 2008. The US Environmental Protection Agency later expanded the tolerance exemption to include polyoxin D zinc salt residues in/on all food commodities. Different registered formulations are effective against *Alternaria* leaf blight and early blight, *Botrytis*, powdery mildew and other diseases on grapes, strawberries, cucurbits, fruiting and leafy vegetables, potatoes, citrus and numerous other crops. Both polyoxin B and D are commercially produced from fermentations of *Streptomyces cacaoi* var. *asoensis*, which was isolated from soil in Japan.

Use of these agents as **agrochemical fungicides** has been hampered by the rapid emergence of resistant fungal strains so it is essential that they are used in rotation with other fungicides for resistance management within **IPM** programmes. Also disappointing is that **clinical treatments** with nikkomycins and polyoxins have not proved effective in controlling human mycoses. This is due to limited uptake of the inhibitors into the cytoplasm of the fungal pathogen and their very rapid elimination from the host animals after intravenous administration. With care, they can be used in conjunction with other antifungal agents in clinical treatment strategies; however, antifungal agents that specifically target the chitin component of the fungal cell wall have found **limited therapeutic use**. Although they often greatly inhibit enzyme activity when assayed *in vitro*, this promise is not realised *in vivo* and consequently, in practice they are not effective antifungal drugs.

The newest family of naturally occurring antibiotics to be developed to treat life-threatening fungal infections are the **echinocandins**, which includes **caspofungin**, micafungin and anidulafungin (Boucher *et al.*, 2004). These are lipopeptide fungal secondary metabolites that feature a cyclic hexapeptide linked to a long-chain fatty acid that is responsible for their antifungal activity and determines species specificity (Figure 18.9). They inhibit β-(1–3) glucan synthase and consequently disrupt cell wall formation and prevent fungal growth. This is a noncompetitive inhibition and although the exact mechanism is not understood yet, echinocandins are known to **bind to the glucan synthase** catalytic subunit. Clinical resistance has been shown to be due to mutations that cause changes in one of two regions in the outer face of the β-(1–3) glucan synthase protein that reduce drug binding sufficiently to permit chitin synthesis in the fungal cell wall.

Echinocandins were discovered during screening of natural products of fungal fermentation for activity specifically against *Candida* species. The first, called papulacandins, were isolated from a strain of *Papularia* (= *Arthrinium*) *sphaerosperma*, a marine species belonging to the *Sordariales* (*Ascomycota*), and led to the discovery of this entirely new group of **broad range antifungals**. Treatment of fungi with echinocandins results in **swelling and lysis** at areas of active cell wall synthesis and the echinocandins have emerged as a **promising therapy** for aspergillosis and candidiasis because of their favourable **safety** profiles; **broad spectrum** of activity; **high potency**; and suitability for **oral administration** (although caspofungin acetate (Cancidas® Figure 18.9) is administered intravenously) (Boucher *et al.*, 2004; Bowman & Free, 2006).

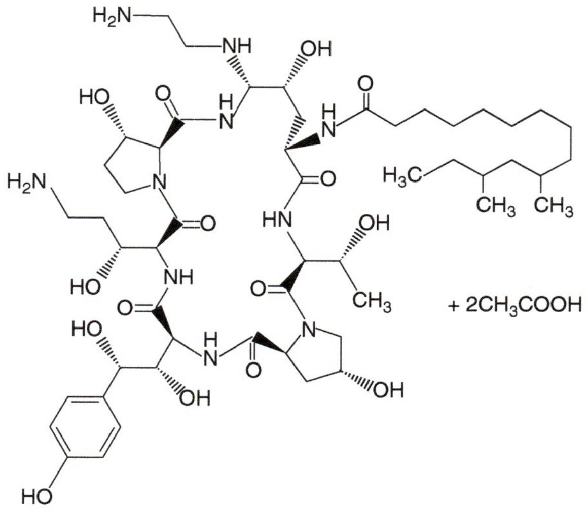

Figure 18.9. The echinocandin antifungal agent known as caspofungin acetate which is used under the brand name Cancidas® worldwide. See https://bnf.nice.org.uk/drug/caspofungin.html. Cancidas® is prepared for intravenous infusion.

The echinocandins are an important class of antifungal agents which are safe and effective. However, most are insoluble or poorly soluble in aqueous solutions; this, and their chemical instability, can restrict their use. Since the introduction of caspofungin in 2001, echinocandins have become increasingly important in treating life-threatening fungal infections. In particular, they are important in treating septicaemia caused by *Candida*, which is responsible for 22% of such infections in American hospitals. **Biafungin**, also called CD101, is a promising echinocandin under development, which importantly has increased chemical stability and increased aqueous solubility. The development of a novel echinocandin which is sufficiently soluble for topical and weekly intravenous administration **and** exhibits prolonged stability in plasma and aqueous solutions up to 40°C is extremely encouraging (Krishnan *et al.*, 2017).

18.4 AGRICULTURAL MYCOCIDES FOR THE TWENTY-FIRST CENTURY: STROBILURINS

Azoles are important agricultural fungicides and we have indicated above some of their uses (Figure 18.5 above). Although the search for new azole fungicides is an active research area, particularly for the more damaging plant pathogens such as the causal agent of rice blast disease, *Magnaporthe oryzae* (Mares *et al.*, 2006), use of clinically useful antimicrobials in agriculture reduces their value in the clinical setting because **broadcast application** (for example by aerial spraying or dusting of crops in the field) exposes enormous populations of microbes to selective pressure for antibiotic resistance. The mechanism of action of all antifungal azoles is based on inhibition of the fungal cytochrome P_{450}, sterol 14α-demethylase, to block biosynthesis of ergosterol. This enzyme activity is present in many other species, including humans, to metabolise the same substrate (lanosterol) to cholesterol, so one of the requirements for the antifungal azoles is to inhibit the fungal pathway but avoid or minimise inhibition of the human enzyme (Mast *et al.*, 2013).

Field acquired resistance (mentioned in section 18.2) is a particular problem for antibacterials where **horizontal transfer of plasmids** carrying resistance can communicate field acquired resistance to commensals and pathogens in the human population. The problem is no less acute in fungi. Horizontal transfer of resistance on plasmids may not occur so readily, but so many of the most troublesome **opportunistic fungal pathogens** are such common soil and environmental fungi that resistance can easily be acquired by some that may later cause infections in humans.

There are alternatives to azoles:

- Melanised walls are important to the survival and functioning of many plant pathogens and the azole fungicide **tricyclazole** (Figure 18.11) inhibits melanin biosynthesis and is effective against fungal plant pathogens that require melanised appressoria to penetrate the plant cell wall (e.g. *Botrytis* and *Magnaporthe*).
- Use of nikkomycin and polyoxorim (polyoxin D) as fungicides that act as competitive inhibitors of chitin synthase is illustrated in Figure 18.8.
- Mitotic inhibitors benomyl and griseofulvin are illustrated in Figure 18.10.
- **Edifenphos** (Figure 18.11) is an organophosphorus ester that is used as a fungicide in agriculture mainly to control rice diseases. Edifenphos is moderately toxic to humans, being rapidly absorbed and able to cause liver, kidney and nerve damage.
- **Metalaxyl** (Figure 18.11), an acylalanine fungicide, inhibits RNA polymerase I in *Oomycota* (Kingdom *Straminipila*) such as the cause of potato late blight, *Phytophthora infestans*. Resistance to this fungicide commonly arises.

Among the most **widely used** and most **effective** classes of agricultural **fungicides** in the world today are a class of chemicals called **strobilurins**, which were first discovered in 1977, although the first useful products were not on sale until 1996. These are **broad-spectrum** fungicides, effective against a wide range of fungal pathogens. Strobilurins work by **inhibiting mitochondrial respiration** by blocking electron transport. This means that the fungus cannot produce energy so can no longer grow and eventually dies. Strobilurins are derived from a natural product so they are environmentally safe because they are rapidly degraded. Refer to Resources Box 18.2 for details of fungicides.

RESOURCES BOX 18.2 Details of Fungicides

We recommend that you visit the American Phytopathological Society (APS) website at this URL: http://www.apsnet.org/publications/apsnetfeatures/Pages/Fungicides.aspx to download the PDF of the feature article entitled *A Short History of Fungicides* by Vince Morton and Theodor Staub (both formerly of Novartis Crop Protection, now Syngenta).

Compendium of Pesticide Common Names

Alan Wood (formerly of CABI) has a website which is well worth a visit at this URL: http://www.alanwood.net/index.html that features a Compendium of Pesticide Common Names and chemical descriptions at http://www.alanwood.net/pesticides/index.html.

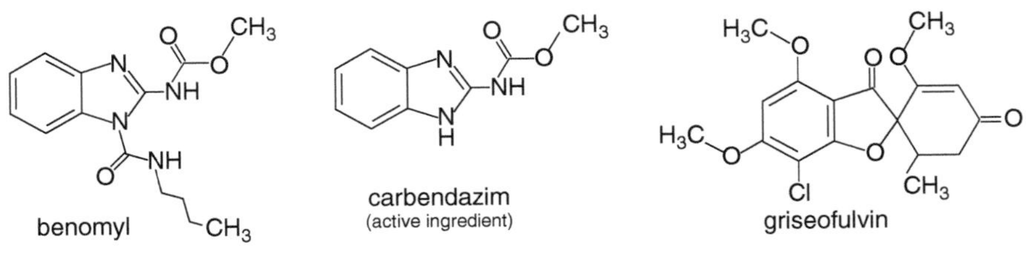

Figure 18.10. Structural formulae of antifungal agents that target microtubule function.

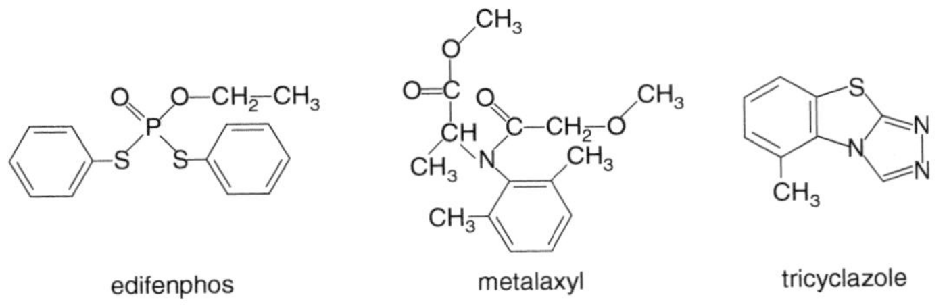

Figure 18.11. Structural formulae of the fungicides edifenphos (an organophosphorus ester), metalaxyl (an acylalanine fungicide, the (-)-isomer is illustrated here), and the melanin-synthesis inhibitor, tricyclazole.

Fungi that produce strobilurins have a modified amino acid sequence in the binding envelope of the coenzyme Q protein that greatly reduces its binding affinity for strobilurin and make the strobilurin-producer strobilurin-resistant. At present there are about eight synthetic strobilurins in the fungicides worldwide market that are used against fungal diseases of a range of agricultural crops. Strobilurins now hold an approximate 20% share of the world **fungicide** market (Balba, 2007). This entire market is based on **fungal toxins** that are toxic to other fungi but have what is described as 'outstanding environmental tolerability' meaning that they have negligible effects on all other organisms. The active ingredients of strobilurin fungicides are synthetic analogues of natural secondary metabolites known as ***p*-methoxyacrylic acids**, specifically: **strobilurin A, oudemansin A**, and **myxothiazol A** (Figure 18.12).

Strobilurin compounds were discovered in the basidiomycete mushroom fungi *Oudemansiella mucida, Pseudohiatula esculenta* var. *tenacellus* (a synonym for *Strobilurus stephanocystis*) and *Strobiluris tenacellus*. It is counterintuitive to expect fungi to produce fungicides; as Balba (2007) puts it 'Many scientists excluded the idea of looking for fungicides in fungus (*sic*)' However, once strobilurin A had been isolated from liquid cultures of the mushroom *Strobilurus tenacellus* the fungicides were found in culture filtrates of many other fungi, and with the benefit of hindsight we can understand why fungicides make a contribution to the biology of the producing organism by contributing to its **competitive ability**.

Strobilurus tenacellus is a saprotrophic wood-rotting fungus which characteristically grows and fruits on the pine cones of *Pinus sylvestris*. The production of strobilurins enables *Strobilurus tenacellus* to predominate in these small substrata despite potential competition from arthropods and other fungi. Strobilurin acts as a feeding deterrent to insects that might compete for the substrate and/or graze the *Strobilurus tenacellus* mycelium. More importantly, strobilurins are fungicidal to all the common fungi that might compete for the resource. This activity identified strobilurins as possible commercial antifungal agents; they are **effective against all major classes** of pathogenic fungi and fungus-like organisms;

- ***Ascomycota*** like powdery mildew and rice blast,
- ***Basidiomycota*** like rusts,
- ***Oomycota*** such as downy mildews.

In 1992, the first artificial strobilurin fungicide was announced and was on the market in 1996. Strobilurins were so successful that by 1999 they accounted for 10% of the global fungicide market, reaching 15% by 2005. Because of their widespread use on cereals and soy beans, strobilurins are now the **second largest group of fungicidal chemicals** in use in agriculture, being second only to the triazoles. Within ten years of their introduction to the market the stobilurins had surpassed all the older fungicide groups in importance excepting only for the triazoles (Sauter *et al.*, 1999; Bartlett *et al.*, 2002; Morton & Staub, 2008; Vincelli, 2012).

Strobilurins **inhibit respiration at a single unique site** which blocks ATP synthesis (see below). This single site specificity raises the problem of potential resistance. Of course, the producing organisms are resistant to their own strobilurins, which itself illustrates that fungi can mutate to strobilurin resistance. Several **fungal pathogens resistant to strobilurins** have been identified in the field, and fungicide resistance affects *Septoria* diseases of wheat in Europe, and of turf in the United States. In response, fungicide production companies are continually researching new strobilurins, developing mixtures with fungicides that have other active principles, and developing new methods of application, such as seed treatments.

Strobilurins act by blocking oxidative phosphorylation in the fungal mitochondrial membrane. The standard pathway for NADH oxidation in mitochondrial respiration involves three protein complexes, I, III and IV. These complexes are involved in energy conservation by linking NADH oxidation and electron/proton translocation (ten protons per two electrons) with ATP synthesis.

Strobilurins bind to the ubiquinone (coenzyme Q) carrier that carries electrons to the cytochrome b-c_1 complex (the most widely occurring electron transfer complex capable of energy transduction). Fungi that produce strobilurins have a modified amino acid sequence in the binding envelope of the coenzyme Q protein that greatly reduces its binding affinity for strobilurin and make the strobilurin-producer strobilurin-resistant (Sauter *et al.*, 1999). They are known as 'Qo inhibitors' (QoI), or **quinone outside inhibitors**, because they act at the quinol **outer** binding site of the cytochrome b-c_1 complex. The QoI group of fungicide families includes the strobilurins and two new families, represented by fenamidone (Bayer Crop Science) and famoxadone (DuPont Crop Protection). The mode of action of QoI fungicides is unique to them. It has made them an important aid in research aimed at understanding the details of mitochondrial respiration, but they are also considered to be the most important recent development made in fungicides by the chemicals industry.

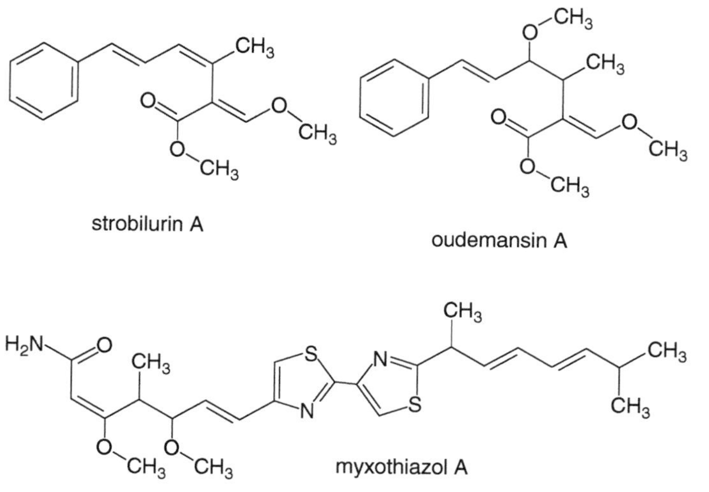

Figure 18.12. The original (natural) strobilurins. Note the characteristic chemical structures: methoxy (CH_3-O-) and acrylate (CH_2=CHCOO$^-$, which features a vinyl group, that is, two carbon atoms double bonded to each other and directly attached to a carbonyl carbon).

The natural strobilurin products are not suitable for agricultural use themselves because they are volatile and unstable in the light. Many agrochemical companies invested in research to produce synthetic analogues of strobilurins, and related compounds (predominantly of strobilurin A due to its simpler structure for chemical synthesis), and a race to patent the first strobilurin fungicide ensued (Sauter *et al.*, 1999; Bartlett *et al.*, 2002). Chemical modification concentrates on improving photostability (diphenyl ethers improve stability) and reducing volatility (by adding benzene rings). Another concern is to improve the systemic properties of the strobilurins, to enhance performance and reduce application rates because of the active redistribution of systemic agents throughout the plant. There is considerable scope for variation in the strobilurin side chains, and this enables steady application of structural changes to improve product functionality. Table 18.7 shows the original strobilurin fungicides and their dates of release (and see Figure 18.13).

The effectiveness of strobilurins is highly dependent on the timing of their application, especially on the stage of growth of the fungus and its respirational demands. **Pyraclostrobin** and **trifloxystrobin** are very effective against *Cercospora beticola* on sugar beet; spore germination is effectively inhibited and spore production by diseased leaves is significantly reduced, both effects being due to the high demand for ATP during spore germination and germ tube development. There is much less inhibition of hyphal growth; this is presumably due to hyphae having different, and perhaps alternative, respiration demands.

A commercial disadvantage of the mode of action of strobilurins is that their target site is in the universal pathway of respiration, yet the fungicides need to be non-toxic to non-target organisms. **Toxicity** to animals varies between the different strobilurins, but strobilurins in general have been classified as having **minimal risk** to human health. Selectivity can be modified by manipulation of the chemical structure; and can be fine-tuned by control of dosage in the field. For example, the activity of *Glomus coronatum* mycorrhiza in 5-week-old maize plants was unaffected by strobilurin treatment to the leaf surfaces at recommended commercial doses.

There is evidence that strobilurin treated plants show **positive physiological side effects** that result in increased crop yields. In cereals, an increase in biomass due to elevated carbohydrate synthesis and consequent increase in photosynthetic capabilities has been observed. Similarly, strobilurin treatment to control early blight in potatoes (caused by *Alternaria solani*) resulted in increased yield, greater tuber size, and improved crop value. These yield improvements following strobilurin treatments are higher than can be explained by their fungicidal activity alone, and the response is not seen when non-strobilurin fungicides, such as azoles, are used.

Development of resistance is an issue with all fungicides and the strobilurins are produced by fungi. Fungi that make the natural products have one of five single nucleotide substitutions in their cytochrome c gene that cause them to be unaffected by the compounds they produce. One of these substitutions is the causative mutation in many resistant pathogen mutants. However, the most damaging **emerging resistance** to strobilurins is associated with an alternative pathway that can bypass the strobilurin target site. **Alternative oxidase** (AOX)

Table 18.7. The original strobilurin fungicides

Fungicide*	Company	Announced	First sales
Azoxystrobin	Syngenta	1992	1996
Kresoxim-methyl	BASF	1992	1996
Metominostrobin	Shionogi	1993	1999
Trifloxystrobin	Bayer	1998	1999
Picoxystrobin	Syngenta	2000	2002
Pyraclostrobin	BASF	2000	2002

*Structural formulae are shown in Figure 18.13. Azoxystrobin was discovered by Imperial Chemical Industries (ICI), the agrochemical interests of which are now part of Syngenta. Trifloxystrobin was discovered by Novartis and sold to Bayer in 2000. Data from Bartlett *et al.* (2002).

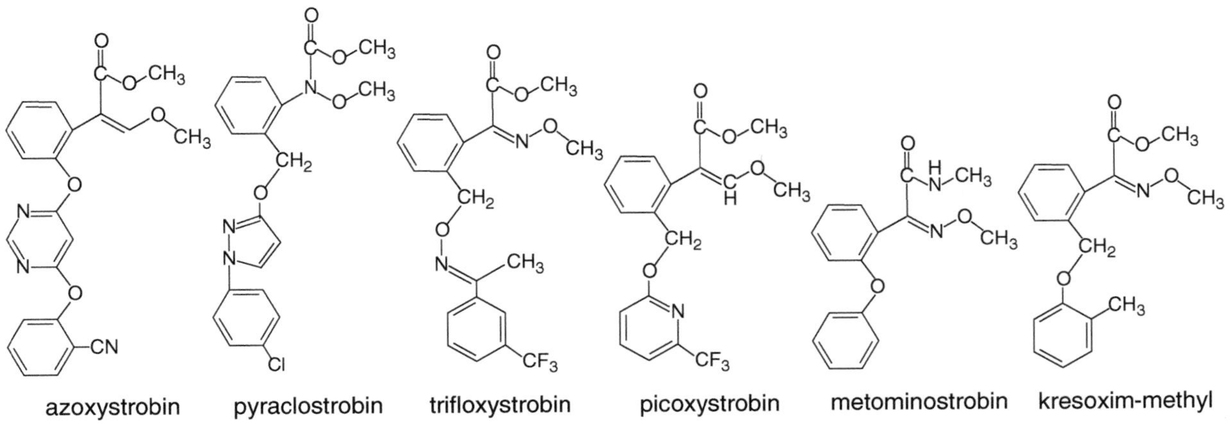

Figure 18.13. Structural formulae of commercial strobilurin fungicides (refer to Table 18.8).

is a strobilurin-insensitive terminal oxidase which is present in many fungi; it accepts electrons directly from ubiquinol and acts as a substitute to the strobilurin target complex. Fungal AOX activity is inhibited by salicyl hydroxamic acid, but this also inhibits the plant AOX and is not a useful selective inhibitor.

Resistance of fungal pathogens to strobilurins does not confer cross resistance to fungicides with different modes of action, like the azoles. But fungi resistant to strobilurins are likely to be cross resistant to other QoI fungicides, famoxadone (a dicarboximide/oxazole fungicide that first went on sale in 1997) and fenamidone (an imidazole fungicide released for sale in 2001).

The emergence of fungicide resistance is avoided (or at least minimised) by alternating fungicides with different modes of action to ensure that specific fungicide chemistries are not overused. There is a **Fungicide Resistance Action Committee (FRAC**; view the URL http://www.frac.info/home) which recommends usage patterns and devises approaches that prolong the effectiveness of fungicides liable to encounter resistance problems and limits crop losses when resistance appears.

The likelihood that resistance will arise can be minimised by restricting the number of applications per season so that no more than two sequential applications are made. Alternating to a non-strobilurin fungicide with a different mode of action is preferable to sequential application of strobilurin. FRAC recommendations are summarised in the following quotation: 'The main resistance management strategies currently recommended are: avoid repetitive and sole use; mix or alternate with an appropriate partner fungicide; limit number and timing of treatments; avoid eradicant use; maintain recommended dose rate; integrate with non-chemical methods. Wherever feasible, several strategies should be used together' (Brent & Hollomon, 2007).

The conceptual similarities between these IPM strategies recommended for control of agricultural crop diseases and the combinatorial antifungal therapies employed to manage human invasive fungal infections will be discussed in the next section.

18.5 CONTROL OF FUNGAL DISEASES FOR THE TWENTY-FIRST CENTURY: INTEGRATED PEST MANAGEMENT AND COMBINATORIAL THERAPY

The antifungal agents that are currently most widely used clinically are the azoles and polyene antibiotics (Table 18.8), both of which target the ergosterol of the fungal plasma membrane. The principal polyene antibiotics, amphotericin B and nystatin, bind selectively to ergosterol and disrupt the fungal membrane making it leaky. The azole antifungal agents, such as fluconazole and ketoconazole, affect the biosynthesis of ergosterol and this affects membrane composition and function.

Other potential targets for antifungal drugs and fungicides include:

- Microtubules; **carbendazim** (benzimidazol-2-yl-carbamate, Figure 18.10) binds to the β-tubulin subunit of microtubules and blocks mitosis. It is used to treat plant diseases caused by ascomycetes, but it is a potential carcinogen. The commercial fungicide is a precursor known as **benomyl** (Figure 18.10), which is hydrolysed within the plant to form the active agent, carbendazim. **Griseofulvin** (Figure 18.10) binds to microtubule-associated proteins, affects the microtubule assembly process and thereby disrupts fungal cell division, blocking mitosis at metaphase. It is used to treat topical skin infections in humans including athlete's foot.
- Dimorphism; the morphological transition between hyphal and yeast-like forms is critical for the pathogenesis and virulence of many fungal pathogens of animals and plants alike. Better understanding of the controls that enable this switch in morphology could reveal potential targets for management of fungal diseases (Gauthier, 2015; Nigg & Bernier, 2016).
- Nucleic acid synthesis is also an established target; **flucytosine** (5-FC) was discovered during searches for antileukaemic drugs, it is used to treat systemic fungal infections of humans and is effective against *Candida* but not *Histoplasma*. **Metalaxyl** (Figure 18.11) an acylalanine fungicide inhibits RNA polymerase I (responsible for ribosomal RNA synthesis) and is used to treat infections in plants caused by *Oomycota* (Kingdom *Straminipila*) such as the cause of potato late blight, *Phytophthora infestans*.
- **Sordarins** selectively inhibit fungal protein synthesis by interacting with translation elongation factor 2 and the large ribosomal subunit. Sordarins have potent fungicidal activity *in vitro* against *Candida albicans* and other *Candida* species, as well as other yeast-like fungi and filamentous moulds. Unfortunately, they have little or no activity against opportunistic filamentous fungi *in vivo*, though they may have **promise for treatment of candidiasis**, histoplasmosis, pneumocystosis and coccidioidomycosis.

Patents for sordarin antifungals were published principally by Glaxo Wellcome and Merck & Co. However, Glaxo Wellcome terminated their development programme for sordarins and there are no indications of any candidates for clinical development. Nevertheless, the sordarins offer potential because of the novelty of their mode of action; more research is needed to find candidates for clinical development (Dominguez *et al.*, 1998; Odds, 2001; Vicente *et al.*, 2009).

A new approach to treatment of **invasive aspergillosis**, yeasts and other fungal infections is provided by the first of the so-called second-generation triazoles, **voriconazole** (Figure 18.6), which can be used in both intravenous and oral formulations (view the website of the US National Library of Medicine at: https://medlineplus.gov/druginfo/meds/a605022.html). Clinical trials have shown that voriconazole is more

Table 18.8. Antifungal agents in common use at the start of the 21st century

Primary target		Detail including nature and names of drugs		
Membrane	Ergosterol inhibitors	Azoles (lanosterol 14-alpha-demethy lase inhibitors)	Imidazoles	Topical: Bifonazole, Clomidazole, Clotrimazole, Croconazole, Econazole, Fenticonazole, Ketoconazole, Isoconazole, Miconazole, Neticonazole, Oxiconazole, Sertaconazole, Sulconazole, Tioconazole
			Triazoles	Topical: Fluconazole, Fosfluconazole
				Systemic: Itraconazole, Posaconazole, Voriconazole
			Benzimidazoles	Topical: Thiabendazole
		Polyene antimycotics (ergosterol binding)	Topical: Natamycin, Nystatin	
			Systemic: Amphotericin B	
		Allylamines (squalene monooxygenase inhibitors)	Topical: Amorolfine, Butenafine, Naftifine, Terbinafine	
			Systemic: Terbinafine	
Wall	β-glucan synthase inhibitors	Echinocandins (anidulafungin, caspofungin, micafungin)		
Intracellular	Pyrimidine analogues/Thymidylate synthase inhibitors	Flucytosine		
	Mitotic inhibitors	Griseofulvin		
Others	Bromochlorosalicylanilide, methylrosaniline, tribromometacresol, undecylenic acid, polynoxylin, chlorophetanol, chlorphenesin, ticlatone, sulbentine, ethyl hydroxybenzoate, haloprogin, salicylic acid, selenium sulfide, ciclopirox, amorolfine, dimazole, tolnaftate, tolciclate, flucytosine			
	Tea tree oil, citronella oil, lemongrass, orange oil, patchouli, lemon myrtle, Whitfield's ointment (a mixture of benzoic and salicylic acid used to treat athlete's foot)			
	Pneumocystis pneumonia treatments: Pentamidine, Dapsone, Atovaquone			
Modified using data published in Wikipedia which provides hyperlinks to further explanations for many of the entries; see https://en.wikipedia.org/wiki/Antifungal and https://en.wikipedia.org/wiki/Template:Antifungals.				

effective than amphotericin B and offers greater survival benefits; making it a new **standard therapy** for invasive aspergillosis. There are safety issues with voriconazole as it has adverse side effects on vision and liver function, as well as a number of drug interactions. **Posaconazole** is effective against difficult to treat infections caused by zygomycetes and **ravuconazole** has broad-spectrum potency *in vitro* and *in vivo* against a wide range of fungal pathogens (Boucher *et al.*, 2004; Cecil & Wenzel, 2009).

Despite the advances offered by new drugs and new techniques the **mortality associated with invasive fungal infections remains unacceptably high**, especially for individuals with severe immunosuppression because of chemotherapy, control of organ transplant rejection or HIV infection.

It remains the case that we have relatively **few effective antifungal agents** and resistance to them is a growing problem. Consequently, there is an urgent need to find new antifungal agents. The search for novel antifungal targets has intensified since the establishment of several partial and complete genome sequences. Gene disruption in fungi can be particularly helpful as gene knockouts can identify essential genes, and a major effort is underway to discover novel antifungal targets for drug screening. These procedures have been described in section 6.10, above.

Despite the **need for new antifungals** to cope with the growing demands of treating systemic mycoses, most of the steps in the **ergosterol biosynthetic pathway** have not been exploited for development of antifungal compounds. Identification of specific enzyme inhibitors is technically difficult on the large scale required in drug discovery programmes because the enzymes are membrane bound and the substrates are difficult to handle. Whole-cell screening methods have been developed and have shown promise in identifying specific, non-azole, inhibitors of enzymes in the ergosterol biosynthesis pathway.

The structure of the **cell wall is unique to the fungi** so there are no homologues in the human genome to ***any*** of the steps involved in fungal cell wall biosynthesis, yet only the echinocandins (Figure 18.13) currently target the fungal cell wall by inhibiting the glucan synthase complex. Other potential targets in the fungal cell wall include the following (Bowman & Free, 2006; Tada *et al.*, 2013; Gow *et al.*, 2017; Mazu *et al.*, 2016):

- The **chitin synthase complex**; nikkomycins and polyoxins are useful competitive inhibitors, but there must be scope for antifungal agents that affect chitin synthase as noncompetitive inhibitors (analogous to the way echinocandins affect glucan synthase).
- **Mannosyltransferases and glycosyltransferases** used in the Golgi apparatus to synthesise the *N*-linked and *O*-linked oligosaccharides attached to cell wall glycoproteins are a target for antifungal drugs because it is known that fungal mutants defective in the biosynthesis of these oligosaccharides are severely affected in both growth and pathogenicity.
- Another target that could be exploited for the development of antifungal drugs is the process of **generating and attaching a GPI anchor to cell wall proteins**. Mutations that affect GPI anchor biosynthesis and attachment are lethal, demonstrating the importance of GPI-anchoring in the fungal cell wall. The core structure of the GPI anchor is conserved between humans and fungi but there are differences in the number and placement of sugar residues onto the core structure that would provide fungal-specific steps in GPI anchor biosynthesis suitable for specific targeting.
- Enzymes that operate in the extracellular space (various **cell wall glycosyltransferases**) to assemble and cross-link cell wall components may offer the best cell wall target for antifungal drugs because they would not have to cross the plasma membrane to get to their site of action and yet would target a critical step in cell wall biosynthesis. The antibacterial penicillins function in a very similar way to inhibit the cross-linking of bacterial cell wall components.

Currently the most promising of the approaches employed in clinical practice is **combinatorial antifungal therapy** which uses antifungal agents with different targets and spectra of activities in combination. Such treatment is expected to broaden antifungal coverage; become more fungicidal because of synergism between the antifungal agents used; minimise the risk of development of resistance; and lastly decrease toxicity. Unfortunately, cumulative evidence supporting the use of combinatorial antifungal therapy is conflicting and controversial. The value of using amphotericin B with flucytosine to reduce mortality in cryptococcal infections has been established, though in many developing countries flucytosine is not available and other, less effective combinations must be used.

The effectiveness of combinatorial antifungal therapy in the treatment of invasive fungal infections, like aspergillosis, is less clear. Very often, treatment outcomes are influenced more by the debilitation of the patient by their underlying condition rather than the combinations of antifungal treatments selected. In many cases combinatorial antifungal therapy is an option of last resort in treating fungal infections that have already shown high intrinsic resistance to current antifungal agents. Further, the patients who develop these invasive fungal infections are usually immunocompromised and because of this have a history of adverse factors that will affect their clinical responses to treatment of their fungal infections. For invasive aspergillosis, combination of voriconazole and anidulafungin has been shown to reduce mortality, so combinatorial antifungal therapy remains a ***potential*** potent option for management of invasive candidiasis, invasive aspergillosis, and rare mould infections (Johnson & Perfect, 2010; Garbati *et al.*, 2012; Belanger *et al.*, 2015).

An alternative combinatorial approach is significant because it illustrates the utility of detailed molecular knowledge well beyond the immediate focus of the drug and its immediate target to develop new methods of treatment based on entirely new concepts. This approach exploits the cellular role of the **heat-shock protein 90 (Hsp90)**, which is a molecular **chaperone** responsible for folding and maturation of client proteins (Panaretou & Zhai, 2008). Client proteins that acquire destabilising mutations can in some cases be stabilised by the molecular chaperone. This makes their activity dependent on Hsp90 activity. But the result is that Hsp90 can allow genetic (that is, mutational) variation to accumulate in a silent state that is revealed only when stress conditions overwhelm Hsp90. In the context of our present discussion 'overwhelming stress' can be taken to mean application of clinical antifungal agent, and Hsp90 certainly **enhances the emergence of antifungal resistance**; but the question is, would **inhibition of Hsp90 *reduce*** resistance to antifungal agents. The answer is yes: inhibition of Hsp90 reduces resistance of *Candida albicans* to azoles and enhances the effectiveness of the echinocandin caspofungin against *Aspergillus* species (Cowen, 2008; Semighini & Heitman, 2009).

In yeast and many other organisms, Hsp90 is expressed in excess of the level required for normal growth, which means that there is plenty of it available to buffer genetic variation and, consequently, influence evolution of the organism and its response to selection. This is biologically interesting but the effect of Hsp90 on emergence of resistance to antifungal agents suggests that Hsp90 could be a novel antifungal target. Several Hsp90 inhibitors (particularly **geldanamycin** and its analogues) have been combined with antifungal drugs with different modes of action against different fungi *in vitro* and in animal pathogenicity models. Geldanamycin is a 1–4-benzoquinone ansamycin antitumor antibiotic (a family of actinomycete secondary metabolites) that inhibits the function of Hsp90 by binding to an unusual ADP/ATP-binding site on the protein.

In all scenarios tested, **inhibition of Hsp90 improved the response to antifungal drugs**. Inhibitors of Hsp90 converted azole's ***fungistatic*** activity against *Candida albicans* into ***fungicidal*** activity and enabled clearance of the fungal infection with **fluconazole** treatment. Similar effects were observed when *Aspergillus fumigatus* was treated with the combination of geldanamycin and **caspofungin** both *in vitro* and in an animal model. A similar strategy is to use genetically **recombinant antibody** targeted against fungal Hsp90 combined with an antifungal drug for treating life-threatening fungal infections (Burnie *et al.*, 2006). Such findings support inhibition of Hsp90 in combination with fungistatic agents as a novel, and effective broad-spectrum therapy against fungal pathogens.

Because of the way that Hsp90 enhances the emergence of antifungal resistance (that is, ***the evolution of resistance***), inhibiting Hsp90 early in the infection has the potential to restrain the range of variation from which resistant isolates might be selected. In Darwinian terms, the ability to produce enough variation to allow adaptive evolution to occur is called **evolvability**. The idea that the evolvability of antifungal resistance can be blocked during treatment is an entirely new concept for a therapeutic approach (Cowen, 2008; Semighini & Heitman, 2009; Li *et al.*, 2015; Lamoth *et al.*, 2016).

18.6 REFERENCES

Balba, H. (2007). Review of strobilurin fungicide chemicals. *Journal of Environmental Science and Health, Part B*, **42**: 441–451. DOI: https://doi.org/10.1080/03601230701316465.

Bartlett, D.W., Clough, J.M., Godwin, J.R. *et al.* (2002). The strobilurin fungicides. *Pest Management Science*, **58**: 649–662. DOI: https://doi.org/10.1002/ps.520.

Belanger, E.S., Yang, E. & Forrest, G.N. (2015). Combination antifungal therapy: when, where, and why. *Current Fungal Infection Reports*, **2**: 67–75. DOI: https://doi.org/10.1007/s40588-015-0017-z.

Boucher, H.W., Groll, A.H., Chiou, C.C. & Walsh, T.J. (2004). Newer systemic antifungal agents: pharmacokinetics, safety and efficacy. *Drugs*, **64**: 1997–2020. DOI: https://doi.org/10.2165/00003495-200464180-00001.

Bowman, S.M. & Free, S.J. (2006). The structure and synthesis of the fungal cell wall. *BioEssays*, **28**: 799–808. DOI: https://doi.org/10.1002/bies.20441.

Brent, K.J. & Hollomon, D.W. (2007). *Fungicide resistance in crop pathogens: how can it be managed?* FRAC Monograph No. 1 (second, revised edition). Brussels, Belgium: Fungicide Resistance Action Committee, a Technical Sub-Group of Croplife International. ISBN: 90-72398-07-6. URL: http://www.frac.info/docs/default-source/publications/monographs/monograph-1.pdf?sfvrsn=769d419a_8.

Burnie, J.P., Carter, T.L., Hodgetts, S.J. & Matthews, R.C. (2006). Fungal heat-shock proteins in human disease. *FEMS Microbiology Reviews*, **30**: 53–88. DOI: https://doi.org/10.1111/j.1574-6976.2005.00001.x.

Cecil, J.A. & Wenzel, R.P. (2009). Voriconazole: a broad-spectrum triazole for the treatment of invasive fungal infections. *Expert Review of Hematology*, **2**: 237–254. DOI: https://doi.org/10.1586/ehm.09.13.

Chowdhary, A. & Meis, J.F. (2018). Emergence of azole resistant *Aspergillus fumigatus* and One Health: time to implement environmental stewardship. *Environmental Microbiology*, **20**: 1299–1301. DOI: https://doi.org/10.1111/1462-2920.14055.

Cowen, L.E. (2008). The evolution of fungal drug resistance: modulating the trajectory from genotype to phenotype. *Nature Reviews Microbiology*, **6**: 187–198. DOI: https://doi.org/10.1038/nrmicro1835.

Da Silva Ferreira, M.E., Colombo, A.L., Paulsen, I. *et al.* (2005). The ergosterol biosynthesis pathway, transporter genes, and azole resistance in *Aspergillus fumigatus*. *Medical Mycology*, **43**: S313–S319. DOI: https://doi.org/10.1080/13693780400029114.

Dominguez, J.M., Kelly, V.A., Kinsman, O.S. *et al.* (1998). Sordarins: a new class of antifungals with selective inhibition of the protein synthesis elongation cycle in yeasts. *Antimicrobial Agents and Chemotherapy*, **42**: 2274–2278. URL: http://aac.asm.org/content/42/9/2274.full.

Dupont, S., Lemetais, G., Ferreira, T. *et al.* (2012). Ergosterol biosynthesis: a fungal pathway for life on land? *Evolution*, **66**: 2961–2968. DOI: https://doi.org/10.1111/j.1558-5646.2012.01667.x.

Garbati, M.A., Alasmari, F.A., Al-Tannir, M.A. & Tleyjeh, I.M. (2012). The role of combination antifungal therapy in the treatment of invasive aspergillosis: a systematic review. *International Journal of Infectious Diseases*, **16**: e76–e81. DOI: https://doi.org/10.1016/j.ijid.2011.10.004.

Garcia-Rubio, R., Cuenca-Estrella, M. & Mellado, E. (2017). Triazole resistance in *Aspergillus* species: an emerging problem. *Drugs*, **77**: 599–613. DOI: https://doi.org/10.1007/s40265-017-0714-4.

Gauthier, G.M. (2015). Dimorphism in fungal pathogens of mammals, plants, and insects. *PLoS Pathogens*, **11**: e1004608 (7 pp.). DOI: https://doi.org/10.1371/journal.ppat.1004608.

Gow, N.A.R., Latgé, J.-P. & Munro, C.A. (2017). The fungal cell wall: structure, biosynthesis, and function. *Microbiology Spectrum*, **5**: FUNK-0035–2016. DOI: https://doi.org/10.1128/microbiolspec.FUNK-0035-2016.

Gumber, K., Sidhu, A. & Sharma, V.K. (2017). *In silico* rationalized novel low molecular weight 1,2,4-triazolyldithiocarbamates: design, synthesis, and mycocidal potential. *Russian Journal of Applied Chemistry*, **90**: 993–1004. DOI: https://doi.org/10.1134/S1070427217060222.

Howard, S.J. & Arendrup, M.C. (2011). Acquired antifungal drug resistance in *Aspergillus fumigatus*: epidemiology and detection. *Medical Mycology*, **49**: S90–S95. DOI: https://doi.org/10.3109/13693786.2010.508469.

Johnson, M.D. & Perfect, J.R. (2010). Use of antifungal combination therapy: agents, order, and timing. *Current Fungal Infection Reports*, **4**: 87–95. DOI: https://doi.org/10.1007/s12281-010-0018-6.

Kleinkauf, N., Verweij, P.E., Arendrup, M. *et al.* (2013). *Risk assessment on the impact of environmental usage of triazoles on the development and spread of resistance to medical triazoles in* Aspergillus *species*. ECDC Technical Report. Stockholm, Sweden: European Centre for Disease Prevention and Control (ECDC). ISBN: 9789291934447. URL: https://library.wur.nl/WebQuery/wurpubs/reports/488729.

Krishnan, B.R., James, K.D., Polowy, K. *et al.* (2017). CD101, a novel echinocandin with exceptional stability properties and enhanced aqueous solubility. *The Journal of Antibiotics*, **70**: 130–135. DOI: https://doi.org/10.1038/ja.2016.89.

Lamoth, F., Juvvadi, P.R. & Steinbach, W.J. (2016). Heat shock protein 90 (Hsp90): a novel antifungal target against *Aspergillus fumigatus*. *Critical Reviews in Microbiology*, **42**: 310–321. DOI: https://doi.org/10.3109/1040841X.2014.947239.

Li, L., An, M., Shen, H. *et al.* (2015). The non-Geldanamycin Hsp90 inhibitors enhanced the antifungal activity of fluconazole. *American Journal of Translational Research*, **7**: 2589–2602. URL: https://www.ncbi.nlm.nih.gov/pmc/articles/PMC4731659/.

Mares, D., Romagnoli, C., Andreotti, E. *et al.* (2006). Emerging antifungal azoles and effects on *Magnaporthe grisea*. *Mycological Research*, **110**: 686–696. DOI: https://doi.org/10.1016/j.mycres.2006.03.006.

Mast, N., Zheng, W., Stout, C.D. & Pikuleva, I.A. (2013). Antifungal azoles: structural insights into undesired tight binding to cholesterol-metabolizing CYP46A1. *Molecular Pharmacology*, **84**: 86–94. DOI: https://doi.org/10.1124/mol.113.085902.

Mazu, T.K., Bricker, B.A., Flores-Rozas, H. & Ablordeppey, S.Y. (2016). The mechanistic targets of antifungal agents: an overview. *Mini-Reviews in Medicinal Chemistry*, **16**: 555–578. DOI: https://doi.org/10.2174/1389557516666160118112103.

Monk, B.C. & Goffeau, A. (2008). Outwitting multidrug resistance to antifungals. *Science*, **321**: 367–369. DOI: https://doi.org/10.1126/science.1159746.

Morton, V. & Staub, T. (2008). *A short history of fungicides*. Washington, DC: American Phytopathological Society APSnet. DOI: https://doi.org/10.1094/APSnetFeature-2008-0308.

Nigg, M. & Bernier, L. (2016). From yeast to hypha: defining transcriptomic signatures of the morphological switch in the dimorphic fungal pathogen *Ophiostoma novo-ulmi*. *BMC Genomics*, **17**: 920 (16 pp). DOI: https://doi.org/10.1186/s12864-016-3251-8.

Odds, F.C. (2001). Sordarin antifungal agents. *Expert Opinion on Therapeutic Patents*, **11**: 283–294. DOI: https://doi.org/10.1517/13543776.11.2.283.

Panaretou, B. & Zhai, C. (2008). The heat shock proteins: their roles as multi-component machines for protein folding. *Fungal Biology Reviews*, **22**: 110–119. DOI: https://doi.org/10.1016/j.fbr.2009.04.002.

Robson, G.D. (1999). Hyphal cell biology. In *Molecular Fungal Biology*, ed. R.P. Oliver & M. Schweizer. Cambridge, UK: Cambridge University Press, pp. 164–184. ISBN-10: 0521561167, ISBN-13: 978-0521561167.

Sant, D.G., Tupe, S.G., Ramana, C.V. & Deshpande, M.V. (2016). Fungal cell membrane: promising drug target for antifungal therapy. *Journal of Applied Microbiology*, **121**: 1498–1510. DOI: https://doi.org/10.1111/jam.13301.

Sauter, H., Steglich, W. & Anke, T. (1999). Strobilurins: evolution of a new class of active substances. *Angewandte Chemie International Edition*, **38**: 1328–1349. DOI: https://doi.org/10.1002/(SICI)1521-3773(19990517)38:10<328::aid-anie1328>3.0.CO;2-1.

Semighini, C.P. & Heitman, J. (2009). Dynamic duo takes down fungal villains. *Proceedings of the National Academy of Sciences of the United States of America*, **106**: 2971–2972. DOI: https://doi.org/10.1073/pnas.0900801106.

Tada, R., Latgé, J.-P. & Aimanianda, V. (2013). Undressing the fungal cell wall/cell membrane – the antifungal drug targets. *Current Pharmaceutical Design*, **19**: 3738–3747. DOI: https://doi.org/10.2174/1381612811319200012.

Verweij, P.E., Lestrade, P.P., Melchers, W.J. & Meis, J.F. (2016). Azole resistance surveillance in *Aspergillus fumigatus*: beneficial or biased? *Journal of Antimicrobial Chemotherapy*, **71**: 2079–2082. DOI: https://doi.org/10.1093/jac/dkw259.

Verweij, P.E., Mellado, E. & Melchers, W.J. (2007). Multiple-triazole-resistant aspergillosis. *New England Journal of Medicine*, **356**: 1481–1483. DOI: https://doi.org/10.1056/NEJMc061720.

Verweij, P.E., Snelders, E., Kema, G.H.J., Mellado, E. & Melchers, W.J.G. (2009). Azole resistance in *Aspergillus fumigatus*: a side-effect of environmental fungicide use? *The Lancet Infectious Diseases*, **9**: 789–795. DOI: https://doi.org/10.1016/S1473-3099(09)70265-8.

Vicente, F., Basilio, A., Platas, G. *et al.* (2009). Distribution of the antifungal agents sordarins across filamentous fungi. *Mycological Research*, **113**: 754–770. DOI: https://doi.org/10.1016/j.mycres.2009.02.011.

Vincelli, P. (2012). *QoI (Strobilurin) fungicides: benefits and risks.* Topics in Plant Pathology Feature Article. Washington, DC: *American Phytopathological Society*. DOI: https://doi.org/10.1094/PHI-I-2002-0809-02.

Yang, H., Tong, J., Lee, C.W. *et al.* (2015). Structural mechanism of ergosterol regulation by fungal sterol transcription factor Upc2. *Nature Communications*, **6**: article 6129. DOI: https://doi.org/10.1038/ncomms7129.

CHAPTER 19

Whole Organism Biotechnology

In this chapter, we examine the biotechnology that uses intact living organisms to produce commercially important products. In the main this means fungal fermentations in submerged liquid cultures, so we describe in detail the essential aspects of cultivating fungi: media, oxygen demand and supply, and fermenter engineering. We describe fungal growth pattern in liquid cultures, fermenter growth kinetics, growth yield, the stationary phase and growth as pellets.

Beyond the batch culture, we discuss fed-batch methods, chemostats and turbidostats. Then we look towards the industrial scene and examine the uses of submerged fermentations, with specific examples: alcoholic fermentations, citric acid biotechnology, penicillin and other pharmaceuticals, enzymes for fabric conditioning and processing, and food processing, steroids and use of fungi to make chemical transformations, the Quorn® fermentation and evolution in fermenters, and the production of spores and other inocula. We hark back to ruminant digestion to consider the 'engineering aspects' of natural digestive fermentations in herbivores and try to work out just how many anaerobic fungi we are cultivating in our livestock.

Many of the most important commercial (particularly food) fermentations take place in the solid state. We look in a little more detail at the digestion of lignocellulosic residues, and then turn to our major foods: bread, cheese and salami manufacture, and soy sauce, tempeh and other food products. We add a few comments about products like chocolate, coffee and even tea, which, though few people realise this, all depend on fermentation processes.

Except for the chytrids (the only true fungi with motile aquatic spores), filamentous fungi have evolved as terrestrial organisms designed to grow and degrade dead or sometimes living plant and animal biomass in terrestrial environments.

However, in biotechnology, fungi are generally grown in submerged culture, in tanks known as **fermenters**. This approach enables high levels of biomass and other products to be formed in a controlled environment. Note that in the United States the last two letters of this word distinguish the ferment**er** (the organism) from the ferment**or** (the cultivation apparatus) but we do not use that convention here.

19.1 FUNGAL FERMENTATIONS IN SUBMERGED LIQUID CULTURES

It was the brewer who invented industrial fermentation, using yeast in a pot to make beer. Today, fermentation engineers manufacture, separate and purify products that serve many different sectors of industry, ranging from chemical feedstocks, pharmaceuticals, foods, and biofuels, as well as beverages. The industry is continuing to refine (and redefine) itself for production of heterologous proteins, vaccines, hormones, urgently needed novel antibiotics, as well as animal cell culture, bioprocessing and post-genomics metabolic engineering. In this chapter, we will give you a good foundation to all this, but if you want to venture further into the topic you need to read these books: *Bioprocess Engineering Principles* (Doran, 2012), *Fermentation and Biochemical Engineering Handbook* (Todaro & Vogel, 2014) and *Principles of Fermentation Technology* (Stanbury *et al.*, 2016).

There are two basic fermenter culture systems: closed ('**batch**') and open ('**continuous**') systems although there are several ways of operating batch cultures, which may be stirred (that is, agitated by being shaken or mixed with an impeller) or unstirred, and there are several ways of controlling continuous systems. Leaving the variations aside for the moment, the main features of these two basic fermenter culture systems are summarised in Table 19.1.

A batch culture is so called because it is a single batch (or quantity) of medium that is inoculated with the organism to produce the new culture. It will be incubated at the best temperature (determined by experiment), but after inoculation **no further change or addition** is made to the culture until it is harvested at the end of the incubation period (at a time chosen by the experimenter). A batch culture may be **stationary**, in which case the microorganism may grow in suspension within the fluid medium or may float and grow as a surface culture. Batch cultures may alternatively be **agitated** in some way, the agitation being intended to improve gas exchange (most usually uptake of **oxygen** into the medium and release of **carbon**

Table 19.1. Comparison of the main features of the two basic fermenter culture systems, batch and continuous

Feature	Type of culture system	
	Closed system Batch culture	Open system Continuous culture
Nutrients	NOT replenished	Replenished
Products of growth	NOT removed	Removed
Environmental conditions	NOT constant	Constant
Exponential growth	Lasts only a few generations	Lasts indefinitely

dioxide from the medium). Small, laboratory-scale cultures are most often cultivated in conical flasks, agitated on shakers that may be **reciprocating** (a back and forth linear motion) or **orbital** (a circular motion that swirls the fluid in the vessel).

Larger cultures, including those on industrial scales of up to hundreds of cubic metres capacity, will be cultivated in what are known as **stirred tank reactors (STRs)** in which the fluid is stirred by a centrally placed **impeller** (an impeller is a bladed rotor). In industrial fermenters the ratio of tank height to diameter averages 1.8 (that is, they tend to be **tall cylinders**); the tanks are filled to about 70% of their absolute volume; impeller rotors extend over about 40% of the tank diameter; agitation power varies between 2 and 6 kW m^{-3}; and impeller tip speed averages 5.5 m s^{-1}. Such impeller speeds fragment mycelia and loosely aggregated masses of mycelia (called **flocs**), which is an advantage overall as it **improves nutrient supply**, including oxygen, to the biomass.

We will discuss variations on the batch process, the various continuous culture processes, and aspects of fermenter design below; for the moment we want to concentrate on the practicalities of batch culture and the behaviour of the fungi, especially filamentous fungi, in laboratory-scale cultures. In the following paragraphs, think in terms of cultures of the sort you might make up yourself in conical flasks for incubation on a shaker in an incubator.

19.2 CULTURING FUNGI

If you **are** planning to prepare a culture, the first consideration is the nature of the **medium**. For *in vitro* cultivations, the medium in which you grow your microorganism must contain all the elements that the organism contains; Table 19.2 can help here, by indicating the **elemental composition** of a typical ascomycete filamentous fungus and a typical bacterium. Table 19.2 provides information about the **type** and **quantities** of nutrients required in a medium; industry does not want to waste nutrients in spent medium and therefore adjusts medium composition accordingly. Heterotrophs use reduced, preformed organic compounds as sources of carbon and energy. There is no naturally occurring organic compound that cannot be used by some microorganisms. Unfortunately, however, many man-made organic compounds (like plastics and pesticides) are degraded slowly or not at all.

During **balanced growth** an increase in biomass is accompanied by a comparable increase of all other properties of the population, such as protein RNA, and DNA content. In general, the chemical composition of the culture remains constant. The term **unbalanced growth** describes instances where the relative concentrations of macromolecules and other components of the biomass becomes altered; for example, when the nitrogen source has been exhausted but the carbon source is present in excess and the various cellular components are synthesised at unequal rates. Most media are designed so that the carbon and energy source, usually glucose, is the first nutrient to become exhausted.

Macro-elements, which are those required in the medium in 'grammes per litre' (g l^{-1}) quantities obviously include carbon, oxygen and hydrogen and also:

- **Nitrogen**; required for synthesis of amino acids, purines, pyrimidines, etc. Many microorganisms can use the nitrogen in amino acids. Other nitrogen sources include NH_4, NO_3.
- **Phosphorus** is present in nucleic acids, phospholipids, nucleotides such as ATP etc. Almost all microorganisms use PO_4^{2-} as a source of phosphorus.
- **Sulfur** is required for synthesis of the amino acids cysteine and methionine and enzyme cofactors like coenzyme A. Many microorganisms use SO_4^{2-} as a source of sulfur.

Minor cations, which are required in 'milligrammes per litre' (mg l^{-1}) quantities in the medium:

- **Calcium** concentration in microbial cells is maintained at extremely low levels by highly specific transport processes. Calcium is an important growth regulator. It is easily released from glassware, which can be a source of contamination for calcium-limited medium.
- **Iron**, (Fe^{2+} (ferrous) and Fe^{3+} (ferric)), is a constituent of cytochromes, haemproteins and many other enzymes. Microorganisms need medium concentrations of 0.36–1.8 × 10^{-3} M iron for growth but, under aerobic conditions at pH 7, Fe^{3+} has a solubility of only 1 × 10^{-17} M. Although solubility of Fe^{2+} is increased under acid conditions, and under anaerobic conditions, Fe^{2+} may attain a concentration of 10^{-1} M, successful uptake of iron in most circumstances depends on biological iron chelators, called **siderophores**,

Table 19.2. Approximate elemental composition of *Fusarium venenatum* and *Escherichia coli* (composition is shown as mg g^{-1} biomass)

Macro-elements, making up about 96% of the biomass			**Minor and trace elements making up about 4% of the biomass**		
Element	***Fusarium venenatum***	***Escherichia coli***	**Element**	***Fusarium venenatum***	***Escherichia coli***
Carbon	447	500	Minor cations		
Oxygen	Not done	200	Potassium	20.0	10
Nitrogen	83	140	Sodium	Not done	10
Hydrogen	69	80	Calcium	0.8	5
Phosphorus	16	31	Magnesium	1.8	5
Sulfur	Not done	10	Iron	0.05	2
			Trace cations		
			Copper	0.04	3% in total, not separately measured
			Manganese	0.12	
			Zinc	0.28	
Cobalt, molybdenum, nickel and a few other metals are important trace elements but were not distinguished in these analyses.					

or materials added to the medium, such as citric acid or ethylenediamine tetra-acetic acid (EDTA).

- **Magnesium** is the most abundant intracellular divalent cation. About 90% of intracellular magnesium is bound to ribosomes and polyanionic cell constituents; the remainder constitutes a relatively constant free concentration of 1–4 mM Mg^{2+}. It has a role in macromolecular synthesis and formation of the energy-rich compound ATP. Enzymes requiring magnesium include superoxide dismutase.
- **Potassium** is required for activity of a number of enzymes and is associated with RNA, but its most significant contribution to fungi is to provide the bulk of the osmotic potential of the hypha.
- **Sodium** is generally regarded as an essential element but a definitive requirement for growth can rarely be demonstrated, though it is certainly required by marine microorganisms.

Trace cations are those required in microgramme (μg) quantities per litre of medium (μg l^{-1}). They include:

- **Cobalt**, which is a component of vitamin B12.
- **Copper** is the prosthetic group of several enzymes, including laccases.
- **Molybdenum, nickel** and **zinc** are also involved in numerous enzymes in several crucial areas of metabolism.

Where there is a deficiency of a trace element, such as Fe^{3+}, the specific growth rate is reduced but biomass yield is largely unaffected (Figure 19.1).

Table 19.3 shows recipes for a few **standard media** and includes preparation notes. During media preparation there are several precautions that are necessary to ensure that the chemistry of the medium does not change when it is sterilised:

- Glucose solutions should be filter sterilised or autoclaved separately from other medium constituents because glucose decomposes (browning caused by caramelisation) if autoclaved in the presence of inorganic salts or organic compounds.
- Ammonium salts should be autoclaved at pH less than 7 to avoid volatilisation of ammonia.
- In contrast, phosphate salts should also be autoclaved separately from other medium components; otherwise insoluble precipitates of magnesium ammonium phosphate, magnesium potassium phosphate or magnesium sodium phosphate can form.

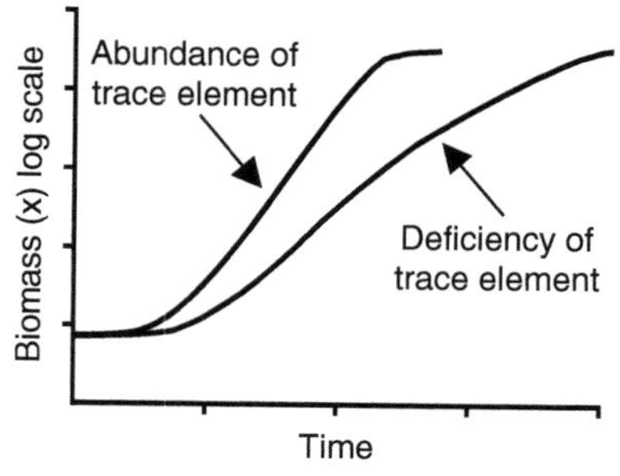

Figure 19.1. Effects of trace element deficiency on growth in batch culture. Growth rate is reduced but biomass yield is largely unaffected.

Table 19.3. Composition and preparation of some widely used culture media

Complex media for fungi

Malt Extract medium (ME) (constituents per litre):

- 20 g malt extract + peptone 1 g
  - add 20 g glucose for Blakeslee's malt extract agar (MEA) medium.
  - add 20 g yeast extract for MYE (malt yeast extract) medium.

Liquid media can be solidified with 12–20 g agar per litre.

Peptones are nutrient extracts obtained by acid or enzyme hydrolysis of natural protein; these extracts are derived from milk proteins (casein), meat, yeasts and plants and provide a source of nitrogen, carbon (as amino acids and peptides) and other nutrients (such as vitamins and minerals) to the cultures. There are numerous peptones and extracts available from commercial suppliers of media (for example: see https://www.sigmaaldrich.com/ and http://www.bd.com/).

Minimal medium for fungi

Vogel's medium (final composition, constituents per litre distilled water; adapted from Vogel, H.L. (1956). A convenient medium for *Neurospora* (medium N). *Microbial Genetics Bulletin*, **13**: 42.)

- Glucose 10 g
- Na_3 citrate 2.5 g
- KH_2PO_4 (anhydrous) 5.0 g
- NH_4NO_3 (anhydrous) 2.0 g
- $MgSO_4.7H_2O$ 0.2 g
- $CaCl_2.2H_2O$ 0.1 g
- Citric acid (monohydrate) 5.26 mg
- $ZnSO_4.7H_2O$ 5.26 mg
- $Fe(NH_4)_2(SO_4)_2.6H_2O$ 1.05 mg
- $CuSO_4.5H_2O$ 0.26 mg
- $MnSO_4.4H_2O$ 0.05 mg
- H_3BO_3 0.05 mg
- $Na_2MoO_4.2H_2O$ 0.05 mg
- Thiamine hydrochloride 0.25 mg (required for phycomycetes and basidiomycetes)
- Biotin 0.05 mg (required for *Neurospora*)

Vogel's Medium is conveniently prepared as four solutions which can be sterilised and mixed aseptically for the liquid medium (agar needs to be added for solid medium).

A. Vogel's stock solution is made up as a 50× concentrated stock solution which is sterilised through membrane filters with 0.2 μm diameter pores. The stock solution contains:

- Na_3 citrate 125 g
- KH_2PO_4 (anhydrous) 250 g
- NH_4NO_3 (anhydrous) 100 g
- $MgSO_4.7H_2O$ 10 g
- $CaCl_2.2H_2O$ 5 g
- 5 ml trace element solution (solution B, below)
- 2.5 ml vitamin solution (solution D, below)
- Distilled water to one litre.

B. Trace element solution:

- Citric acid (monohydrate) 5 g
- $ZnSO_4.7H_2O$ 5 g
- $Fe(NH_4)_2(SO_4)_2.6H_2O$ 1 g
- $CuSO_4.5H_2O$ 0.25 g
- $MnSO_4.4H_2O$ 0.05 g
- H_3BO_3 0.05 g

table continues

- $Na_2MoO_4.2H_2O$ 0.05 g
- Distilled water 95 ml.

C. 5% (w/v) glucose

D. thiamine hydrochloride solution (5 mg ml^{-1}) for phycomycetes or basidiomycetes; or biotin solution (1 mg ml^{-1}) for *Neurospora*.

To prepare 200 ml of Vogel's medium mix the following: 40 ml solution C + 4 ml solution A + 56 ml water + 100 ml 3% (w/v) water agar (for semi-solid medium) or use 156 ml water for liquid medium.

Vogel's medium is a convenient medium for *Neurospora* and other ascomycetes; citric acid is the chelating agent (and weak buffer) in this medium and in the absence of an alternative, some fungi can use citrate as a carbon source, which complicates matters. Autoclaving can have a damaging effect on medium composition: a precipitate is formed that contains all the iron and most of the calcium, manganese and zinc; addition of EDTA to the medium prevents this precipitation during autoclaving.

There are a great many media recipes. Consult the 'media' entry in 10th edition of *Dictionary of the Fungi* (Kirk *et al.*, 2008).

- Iron is a problem as, without **chelating agents**, all the iron is likely to precipitate. To prevent precipitation of iron and other trace elements, and to control their concentration in the medium, it is generally essential to use a chelating agent and citric acid is often employed for this purpose. Several media recipes use other metal-chelating agents, such as EDTA to serve as metal ion buffers (EDTA has greater affinity for Fe^{3+} and least for Ca^{2+} and Mg^{2+}). Unfortunately, EDTA may inhibit growth of some microbes and citric acid may be used as a carbon source following exhaustion of the more readily utilised carbon source in the medium. This can complicate growth studies with such media.

Some components of the medium can be growth factors in the sense that they must be provided for growth to occur. Although most fungi can synthesise most vitamins, in a few cases specific vitamins are required by some fungi. For example, biotin is a required growth factor of *Neurospora crassa* and must be added to media, and thiamine is required by *Coprinopsis cinerea*. We emphasise that these are the characteristics of **wild-type** *Neurospora* and *Coprinopsis*; auxotrophic mutants with deficiencies in vitamin biosynthetic pathways can be obtained just as easily as auxotrophs with deficiencies in other pathways. Where the wild type expresses such a biosynthetic deficiency it presumably means that sufficient supplies of the end product are normally encountered in the natural environment of the fungus for the nutritional deficiency to cause no adverse selection pressure.

19.3 OXYGEN DEMAND AND SUPPLY

A crucial aspect of fermenter operation is the oxygen demand of the culture and the supply of that oxygen to the culture. An estimate of the total amount of oxygen required by a microbe growing on a glucose-mineral salts medium may be obtained from the stoichiometry of glucose oxidation in respiration:

$$C_6H_{12}O_6 + 6O_2 = 6CO_2 + 6H_2O$$

5 g glucose + 5.35 g oxygen = 7.35 g carbon dioxide + 3.0 g water

Table 19.4. The oxygen concentration in medium in equilibrium with air at various temperatures

Temperature (°C)	Oxygen concentration in a typical medium equilibrated with air at one atmosphere pressure (mg l^{-1})
20	9.08
25	8.10
35	6.99

About half of the glucose that may be included in the medium is respired to provide energy for growth. The remaining half (approximately) is used for biosynthesis of the macromolecules and other metabolites of the hypha. The fact that about half the glucose in the medium is used for respiration to generate energy and about half is used for biosynthesis explains why the **growth yield** on glucose is about 0.5; growth yield varies depending on the carbon source in the medium (see section 19.7 below).

Oxygen is **poorly soluble** in water and solubility declines with elevated temperatures (Table 19.4).

Thus, less than 0.2% of the 5.35 g of oxygen which is required for the complete oxidation of 5 g glucose is present in solution in water saturated with air. It is therefore essential to ensure that oxygen is continuously incorporated into the medium as the culture grows; oxygen transfer from the gas to the liquid phase is an important component of this. The simplest concept of the stationary liquid film theory (Figure 19.2), which is simply a visualisation of the movement of oxygen from the bulk atmospheric air, through the air–water interface and into the bulk aqueous medium, sees the process like this:

Oxygen transfer comprises (Figure 19.2):

Bulk gas (air) → stationary liquid film at the fluid surface → bulk liquid of culture

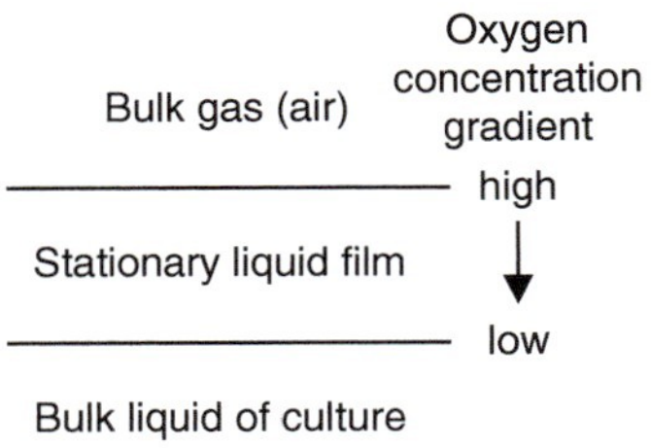

Figure 19.2. Process of oxygen absorption according to the stationary liquid film theory, which envisages a stationary liquid film between bulk atmospheric air and the bulk aqueous medium, with a concentration gradient of dissolved gases through the stationary air–water interface.

The rate of transfer of oxygen from bulk atmospheric gas to bulk liquid medium is given by the equation:

$$R_{oxygen} = K_L a(C^*_{oxygen} - C_{oxygen})$$

where R_{oxygen} = oxygen transfer rate per unit liquid volume, $K_L a$ = mass transfer coefficient (per hour), C^*_{oxygen} = concentration of oxygen which could exist in bulk liquid phase if it were in thermodynamic equilibrium with the gas phase, and C_{oxygen} = actual concentration of oxygen in the bulk liquid phase. When the liquid is saturated with oxygen $C^*_{oxygen} = C_{oxygen}$ and no oxygen transfer occurs.

The value of $K_L a$ is a characteristic of a specific fermenter operated in a particular manner and it has a profound effect on microbial growth in the fermenter (Figure 19.3). Figure 19.3 shows that when biomass is grown under the same cultural conditions (other than $K_L a$), the $K_L a$ of the fermenter affects biomass production; that is, the cultures become oxygen limited at different biomass concentrations. Consequently, fermenters used to grow aerobic organisms, and that category includes most fungi, need to be designed to maximise $K_L a$ values.

When a culture becomes oxygen limited, growth is stoichiometrically related to the quantity of oxygen transferred (shown by the linear part of the growth curve in Figure 19.3). The larger the $K_L a$ the higher the aeration capacity of the fermenter. The oxygen transfer rate ($K_L a$) can be determined by measuring the rate of oxygenation of a 0.5 M solution of sodium sulfite to sodium sulfate in the presence of a catalyst. The **sulfite oxidation rate** is equivalent to the oxygen transfer rate of that specific fermenter operated in that way.

The product $K_L a$ is made up of a constant (K_L) which is effectively a measure of **resistance to oxygen transfer**, and the interface area, a. In a still liquid the interface area is the area of the fluid surface; in an agitated culture this area depends on the number and size of the bubbles of air that the agitation introduces into the fluid medium, and this area can be increased by decreasing bubble size and increasing bubble number; and this is the function of the agitation process (see below).

Factors influencing oxygen transfer rate in shake flask cultures:

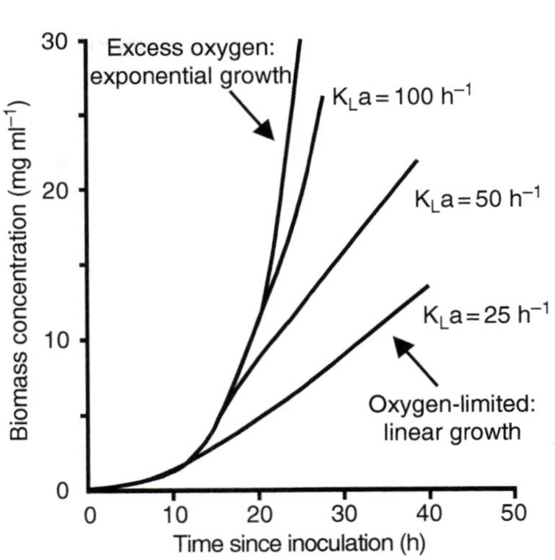

Figure 19.3. Effect of $K_L a$ on microbial growth in batch culture.

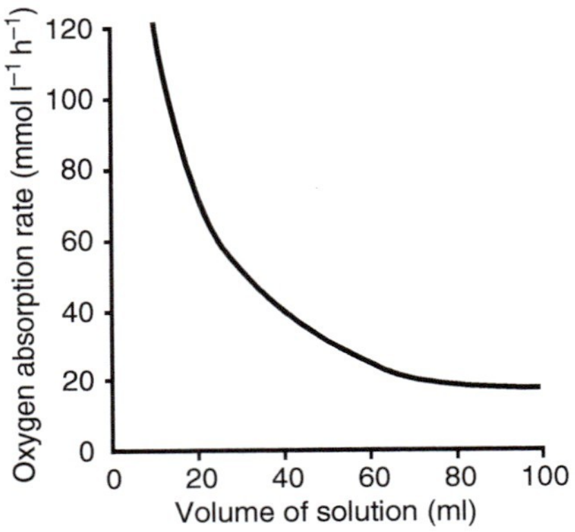

Figure 19.4. Effect of liquid volume on the oxygen transfer rate. The plot shows how the volume of liquid in a 500 ml flask influences oxygen transfer rate when the flask is agitated on an orbital shaker operating at 250 revolutions minute^{-1}.

- effect of volume of liquid in the flask (Figure 19.4);
- effect of the level of agitation, which is measured in terms of the shaker speed and throw (the physical extent of the reciprocation) in shake flask cultures. Increasing speed from 150 to 300 rpm results in 250% increase in oxygen solution rate.

19.4 FERMENTER ENGINEERING

In stirred tank reactors the oxygen transfer rate is affected by the type of stirring (Figure 19.5). In a **vortex STR** rapid rotation of the impeller causes a vortex in the fluid that draws air bubbles into the liquid, which are then dispersed by the impeller. **Vortex STRs** are very rarely used now. It is important to appreciate the contribution made by biochemical engineers to fermentation microbiology in terms of fermenter design and operation. Aeration may be improved further by having the air supply entering the culture just below the impeller, so the incoming air stream is immediately broken up into very small bubbles (a process known as **sparging**). In the **baffled STR**, a baffle on the inner wall of the tank prevents vortex formation but it increases turbulence in the fluid and so increases aeration (Figure 19.5). Baffles can also be used in conical flasks for small-scale cultures. Sometimes four baffles are added to conical flasks by a glass blower indenting the sides of the flask. Flexible metal coils can also be added to conical flasks. Both strategies **increase turbulent shear stress in the fluid**, and hence **aeration**.

Culture viscosity is low in the region of the impeller and this increases the rate of oxygen transfer from the gaseous to liquid phase, so this is a well aerated region in the fermenter. In regions of the fermenter near the walls of the vessel, the culture may suffer poor aeration and be stagnant.

The most efficient type of impeller is a vaned disc, most commonly with 6 or 8 **vertical** vanes or fins (Figure 19.6). Optimum diameter is about 40% of the vessel diameter and optimum vane height is about 17% of the diameter of the impeller. Within the fermenter, the impeller should be one-third to one-half of vessel diameter above the base of vessel and the larger industrial units will have two or three impellers on the same shaft (Figure 19.7).

In terms of oxygen transfer rate, the most effective position for the **outlet of the sparging air** is immediately below the impeller. In practice, cultures are usually sparged at a rate of **one culture-volume of air per minute** (Figure 19.8).

The degree of **agitation** in the STR has a profound effect on K_La because, by forming small bubbles:

- agitation increases the surface area for oxygen transfer, so the rate of stirring affects oxygen transfer rate (Figure 19.8), to be specific, laboratory-scale cultures in 2 l fermenters are

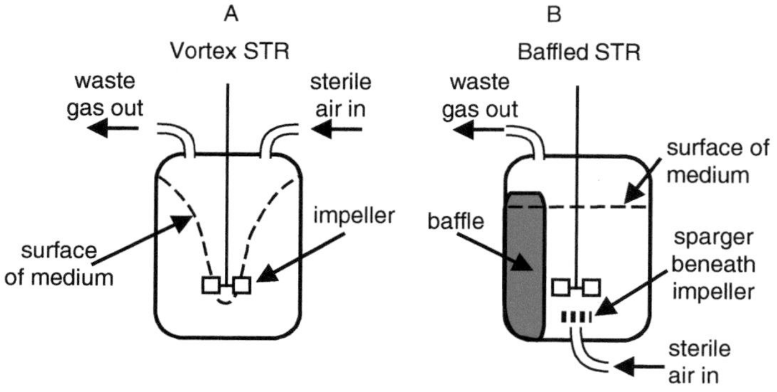

Figure 19.5. Types of stirred tank reactors (STRs). (A) Vortex STR in which the air is entrained by the vortex and dispersed in the liquid by the impeller. (B) Baffled STR in which a baffle on the inner wall of the tank prevents vortex formation and increases turbulence in the fluid and increases aeration that way.

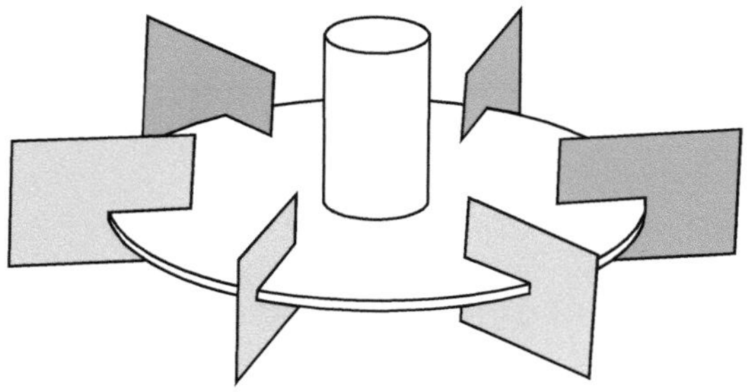

Figure 19.6. Engineering sketch of an efficient fermenter impeller. This pattern is known as a Rushton disc turbine, designed in 1950. The 4- and 6-bladed (shown here) versions of the impeller are most common, but it may have 3, 4, 5, 6, 8 or 12 blades depending on the application. Rotation of this ***radial impeller*** forces the fluid to flow radially at right angles away from the impeller shaft (in contrast to 'marine screw' style ***axial impellers***, which force the fluid to flow along the axis of the shaft).

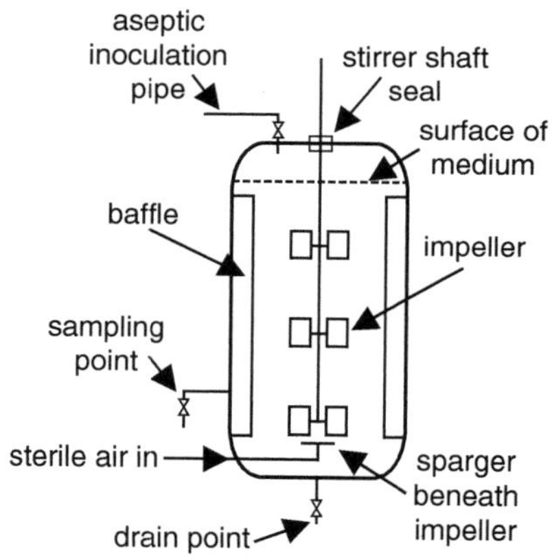

Figure 19.7. Engineering sketch showing, in sectional view, a typical large (100 m³) industrial STR fermenter that would be constructed in stainless steel.

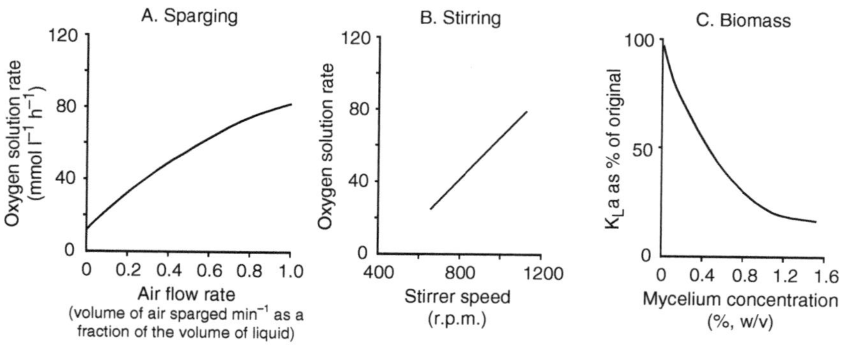

Figure 19.8. Influence of air sparging, stirrer rate and biomass on oxygen transfer rates in stirred tank reactors. (A) Effect of air sparging on oxygen transfer rate. (B) Effect of stirrer rate on oxygen transfer rate in a vortex STR. (C) Effect of biomass (concentration of *Penicillium chrysogenum*) on K_La in a stirred tank reactor.

usually stirred at a rate of 1,000 to 1,200 rpm;
- escape of bubbles from the liquid is delayed, giving a longer period for gas exchange;
- coalescence of bubbles is prevented;
- turbulence decreases the thickness of the stationary liquid film at the gas/liquid interface;
- concentration of biomass also affects oxygen transfer rate because of its effect on the rheology of the culture (Figure 19.8); this depends on morphology; filamentous microorganisms form more viscous cultures than unicellular microorganisms.

It is good practice to **monitor the oxygen content** of the medium during cultivation, so that stirring and sparging can be adjusted if necessary. This can be done with oxygen electrodes.

An important function of the impeller is to **diminish the size of air bubbles**, but there are limits to this because the formation of **foams** must be avoided. A foam is a stable mass of frothy bubbles formed in a matrix of the culture fluid that can accumulate on the surface of the medium. The foam is stabilised by proteins, peptides and other materials that leak into the medium from the biomass. Bubbles trapped in the foam become depleted of oxygen and impede gas exchange because they are isolated from the air flow. Addition of **antifoam** reduces foaming; antifoams are chemical additives that have affinity to the gas-liquid surface where it destabilises the foam by promoting coalescence of the smallest bubbles into larger ones that rupture and release their gas content into the air flow. Examples of commonly used antifoams include: polyalkylene glycols, alkoxylated fatty acid esters, polypropylene glycol (PPG), siloxane polymers, mineral oils and silicates used at concentrations up to 1% of the total volume.

In a stirred tank reactor **heat** is generated by the stirring power needed to aerate the culture and this is additional to the heat generated by the chemical activities of the biomass. To avoid heat build-up a **cooling system** is required. The most usual design is a water jacket made up of water circulating through tubes either on the outside or within the fermenter. Airlift fermenters (Figure 19.9) avoid the heat generated from stirrers. They work because of the density difference between aerated and non-aerated medium (strictly speaking, since the medium is a fluid it is a **specific gravity difference**). The specific gravity of air-depleted culture medium is relatively higher than the specific gravity of air-enriched (that is, **newly sparged**) medium, so the latter tends to **move upwards** and generates the required agitation of the fermenter contents. Furthermore, elevated hydrostatic pressure in the sparging zone increases oxygen transfer into the medium by increasing its solution rate.

19.5 FUNGAL GROWTH IN LIQUID CULTURES

Some fungi grow in liquid cultures in a dispersed, homogenous manner but most will naturally grow in the form of mycelial clumps or pellets:

- moulds which grow as **disperse mycelia** and form homogeneous cultures, e.g. *Geotrichum candidum*, the hyphae of which fragment spontaneously (Figure 19.10A);
- moulds which grow as dispersed mycelia but **aggregate in stationary phase**, forming a large mycelial mass, e.g. *Neurospora crassa*;
- **aggregation** of *Neurospora crassa* is associated with accumulation of galactosaminoglycan in the medium, which helps bind the hyphae together;
- moulds whose spores **aggregate on germination** and result in the formation of numerous macroscopic spherical colonies or pellets, e.g. *Aspergillus nidulans* (Figure 19.10B).

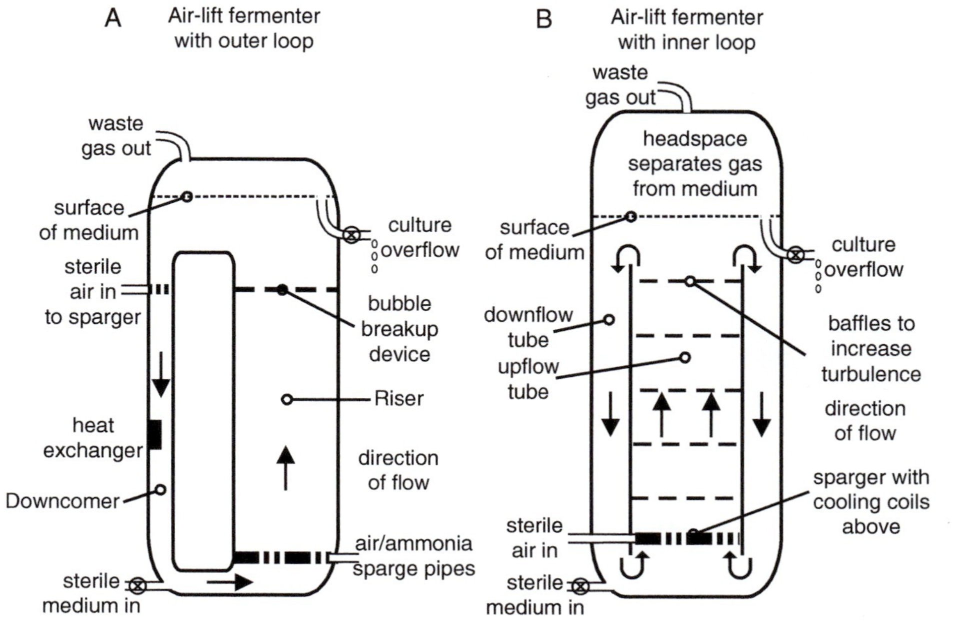

Figure 19.9. Sketch diagrams of typical airlift fermenters. (A) A laboratory-scale (glass) airlift fermenter with outer loop, in which the diameter of the downcomer should be as narrow as possible to minimise hold-up time. (B) An industrial airlift fermenter with inner loop (typical of very large fermenters used to produce single cell protein for animal feed). The rigs can be operated as continuous flow systems when culture overflow and sterile medium input lines are both open; or in batch mode when both are closed after the vessel has been charged with medium and inoculated.

We should not underestimate the impact of these growth patterns on maintaining balanced growth and therefore establishing the kinetic characteristics of the theoretical growth curve in practice. With organisms that grow as single cells (bacteria, yeasts, protozoa and animal cell cultures, unicellular algae) the task is relatively easy, as successive and **representative** samples can be taken from the same culture as it is incubated so that the cell count (i.e. cell density) can be established. It is then just a matter of establishing statistical validity by taking repeated samples from each culture (to ascertain the precision of individual measurements) and running the experiment with a sufficient number of replicate cultures (to determine and measure the experimental variation).

Organisms that are not so easy to deal with are those animal and plant cell cultures that grow as congruent cell layers or calluses and the filamentous fungi; all of which (including those filamentous fungi that grow as pellets) present the experimenter with the problem that **representative** samples cannot be taken from the same culture. In these circumstances to get even a simple measure of growth rate over an extended culture period requires that whole cultures are harvested at each time point and means that many replicate cultures have to be started at the same time to ensure good measurement of precision and variability throughout the incubation.

The type of measurement that is made is also influenced by growth pattern. The object of the experiment is to measure the **live biomass** content of the culture. Where cells cannot be counted, biomass dry weight is the most frequently used alternative, but this raises technical concerns. The harvesting and drying methods are not always straightforward, and at early growth stages there may be so little biomass that accurate weighing can become an issue. Attempts to distinguish between live and dead biomass (which dry weight does not achieve) have led to other measurements being used, including:

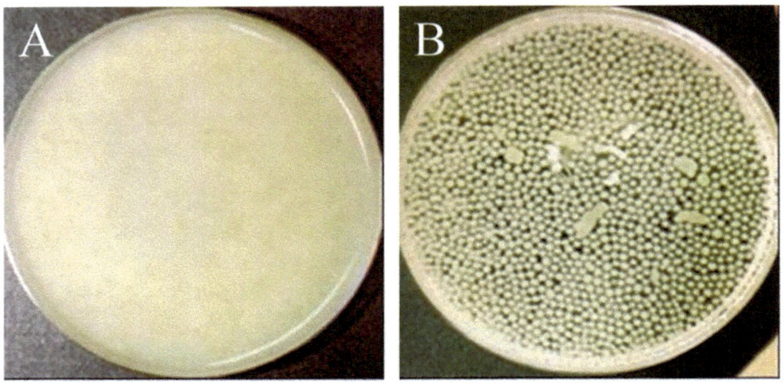

Figure 19.10. Morphology of moulds in submerged culture (shake flask or stirred liquid cultures). Appearance of liquid cultures of filamentous fungi (grown in fermenters and then decanted into 9 cm Petri dishes for photography). (A) Dispersed, homogenous growth of *Geotrichum candidum*. (B) Mycelial pellets as might be formed by fungi such as *Aspergillus nidulans*. Images provided by Dr G.D. Robson.

- content of DNA, which should primarily measure the number of nuclei, and
- ergosterol, which is expected to be a measure of the mass of **fungal** membranes, even if the sample is contaminated with animal and/or plant tissues (for example, when measuring fungal growth in compost, soils or solid tissues, such as mycorrhizas in/on plant roots).

Experiments with the Quorn® fungus, *Fusarium venenatum* described below used image analysis of microscope images to measure total filament length and branching during early stages of growth in liquid cultures. This allowed integration of biomass measurements with hyphal growth parameters, particularly the **hyphal growth unit** (hyphal growth kinetics is discussed in sections 4.4 and 4.9).

Organisms that grow in a dispersed manner have identical growth kinetics to unicellular organisms such as yeast and bacteria: growth curves (Figure 19.11) expressing biomass or cell numbers against time of incubation display a **lag phase, acceleration phase** (caused by the fact that not all cells in the inoculum start growing at the same time), **exponential phase, deceleration phase** (caused by the concentration of the carbon source, the limiting nutrient in the medium, decreasing below that needed to support μ_{max}), and **stationary phase** (caused by exhaustion of the carbon source). This description applies to a medium in which the limiting nutrient is the carbon and energy source, such as glucose. When the limiting nutrient is not the carbon and energy source the situation is more complicated because of unbalanced growth.

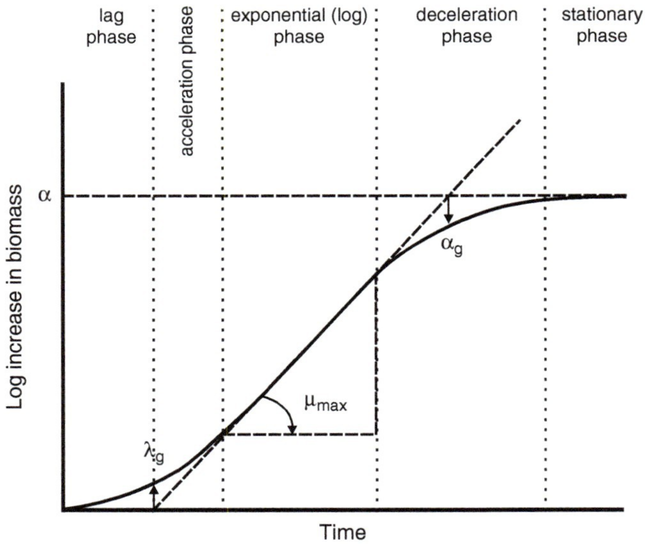

Figure 19.11. The hypothetical growth curve ('classic growth curve') for a cell population grown in submerged liquid culture. Parameters μ_{max} (maximum specific growth rate), λ_g (lag phase), α_g (end of exponential phase) and α (asymptote; which is the straight line that is the limiting value of the curve; λg is determined as the time where the tangent crosses the starting level, α_g is determined as the time where the tangent crosses the asymptote.

The **growth curve** of a filamentous fungus in batch cultures follows essentially the same kinetics as the growth curve of a unicellular organism. Exponential growth is described by the equation:

$$\mu x = \frac{dx}{dt}$$

where, μ = specific growth rate and x = biomass concentration. On integration this is transformed to:

$$\ln x_t = \mu t + \ln x_D$$

You can see the similarity of this to the equation $y = mx + c$, which is the classic linear relationship of the **'straight line graph'**, so you might expect to be able to plot $\log_e$(biomass) on the vertical axis, against time (on the horizontal axis) and obtain the value of μ as the slope of the resultant graph; but this only applies to the exponential phase of a typical growth curve, and the growth curve ***is a curve*** so this straight line is more formally **the tangent of that curve**.

Remember, that in the growth curve of a typical organism in a batch culture there *is* a lag phase before the exponential phase, during which the cells become adapted to the growth conditions and initiate synthesis of RNA, enzymes and other molecules that growth and nuclear divisions require. After the exponential phase nutrients begin to become depleted and growth rate begins to decelerate until nutrient is exhausted and the organism enters stationary phase (Figure 19.11). In practice, most artificial media are designed so that the carbon source is the first nutrient to be exhausted. The exponential phase of the growth curve in Figure 19.11 shows the most significant parameters:

- μ_{max} (maximum specific growth rate), as the slope of the tangent of the exponential curve;
- λ_g (lag phase), is determined as the time where extrapolation of the tangent of the exponential curve intersects the time axis;
- α, which is determined as the extrapolation of the **asymptote**, which is the straight line that the stationary phase part of the curve approaches more and more closely but never quite reaches;
- α_g (the end of exponential phase), which is described as the time where the extrapolation of the tangent of the exponential curve intersects the asymptote.

The slope of the tangent of the exponential curve is the maximum specific growth rate, but another useful parameter that can be calculated using the tangent of the exponential curve is the time it takes to double the biomass (or the number of cells in a unicellular culture), which is called the mean doubling time, often symbolised T_d (Table 19.5).

Table 19.5. Doubling times of some moulds in liquid batch culture

Species	Incubation temperature (°C)	Medium*	Doubling time (T_d, h)
Fusarium venenatum	30	Minimal	2.48
Neurospora crassa	30	Minimal	1.98
Geotrichum candidum	30	Minimal	1.80

*Growth on complex medium is faster than growth on minimal medium because some macromolecules are supplied in the medium and do not have to be synthesised *de novo*.

19.6 FERMENTER GROWTH KINETICS

You will note that Table 19.5 specifies both temperature and nature of the medium. This is necessary because environmental conditions influence the rate of growth of the cells in culture, and therefore the value of any parameter derived from slope of the exponential part of the growth curve. Temperature is an obvious determinant of growth rate, and that's why foodstuffs are refrigerated to keep them fresh; the low temperature slows the growth of microorganisms so much that the food is preserved for longer than its 'shelf life' at normal temperatures. Every other feature of the culture environment also influences growth rate, including factors like pH, gas exchange and, for photosynthetic organisms, illumination too.

However, in most fungal cultures it is the nutrients that matter most (assuming the temperature is adequate). And that means ***all*** nutrients. The maximum specific growth rate (μ_{max}) can only be attained if ***all*** nutrients are present in excess. If supplies of a nutrient are inadequate, that nutrient is described as **limiting**, because the diminished supply limits the growth rate to a suboptimal value. We illustrate this in Figure 19.12, which shows the effect of different initial concentrations of glucose in the medium on the growth rates of fungal batch cultures, but we are using the sugar simply as an example; any other nutrient, whether it be mineral or organic, carbon source, nitrogen source, or sulfur source, could serve in this example although the shape of the curve shown in Figure 19.12 is likely to be different for non-carbon source nutrients because of the complication caused by unbalanced growth.

The curve shown in Figure 19.12 is very similar to the (Michaelis–Menten) single-substrate progress curve that describes the rate of an enzyme reaction (usually symbolised v) as a function of substrate concentration (usually symbolised $[S]$, the square brackets indicating that the term is a *concentration*); by analogy with the Michaelis constant (K_m) of enzyme kinetics, the substrate concentration that supports the half-maximal growth rate is symbolised K_s. This is indicated in Figure 19.12. **Jacques Monod** (in 1942) was the first to realise that when the concentration of a nutrient (such as glucose) decreases, it

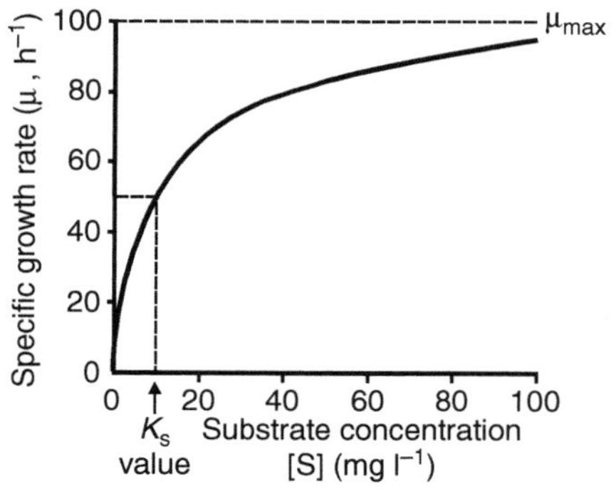

Figure 19.12. Substrate-limited growth of microorganisms. Effect of different initial concentrations of glucose in the medium on the growth rates of fungal batch cultures. Ks is the substrate concentration that supports the half-maximal growth rate. In this graph, specific growth rate (μ) of the culture is plotted as a (hyperbolic) function of substrate concentration.

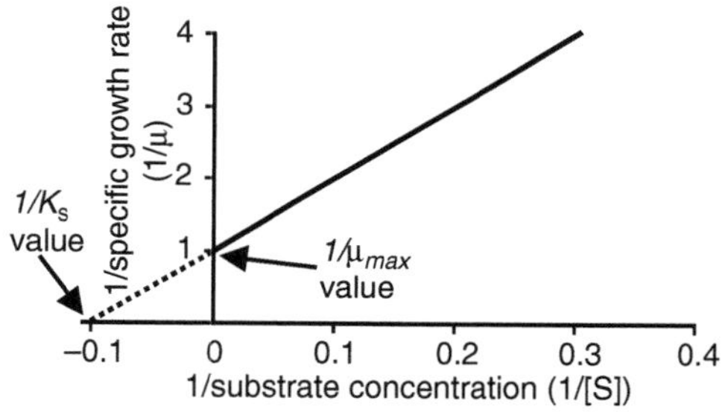

Figure 19.13. A Lineweaver-Burk plot of l/μ against 1/[S].

eventually falls to a level which causes a decrease in specific growth rate of the culture (as shown in Figure 19.12).

Monod suggested that uptake of the limiting nutrient is regulated by an enzyme (a permease) and that the form of the relationship between nutrient concentration and specific growth rate was similar to the effect of substrate concentration upon the rate of an enzyme catalysed reaction. The K_s of a microorganism for a nutrient may reflect the effect of nutrient concentration on the permease protein responsible for its uptake.

The ***Michaelis–Menten relationship*** for the effect of substrate concentration on the rate of an enzyme catalysed reaction is given by the equation:

$$V = \frac{V_{max}S}{(S + K_m)}$$

where V = rate of reaction, V_{max} = maximal rate of reaction, S = substrate concentration, and K_m = substrate concentration at which reaction rate = 0.5 V_{max}.

The analogous ***Monod equation*** for effect of substrate concentration on the specific growth rate of a microorganism is as follows:

$$\mu = \frac{\mu_{max}S}{(S + K_S)}$$

where μ = specific growth rate, μ_{max} = maximum specific growth rate in absence of substrate limitation, S = concentration of growth-limiting substrate, K_s = substrate concentration which allows the organism to grow at 0.5 μ_{max}.

Extending this analogy with enzyme kinetics, where the reaction constants V_{max} and K_m are more effectively obtained from the double-reciprocal (Lineweaver–Burk) plot of $1/v \times 1/[S]$, this Monod equation may be rearranged to give the linear function:

$$\frac{1}{\mu} = \frac{1}{S} \bullet \frac{K_S}{\mu_{max}} + \frac{1}{\mu_{max}}$$

and replotting the data of Figure 19.12 with a vertical axis (ordinate) of 1/μ and horizontal axis (abscissa) of $1/s$ gives a straight line intercepting the abscissa at $1/\mu_{max}$ and the ordinate at $-1/K_s$ (Figure 19.13).

The Michaelis constant, K_m, is a measure of the affinity of an enzyme for its substrate and by analogy the substrate concentration that allows the organism to grow at half-maximal rate, K_s, is taken to be a measure of the affinity of the organism for the substrate. Microorganisms can show high affinities for nutrients. For example, the K_s of *Fusarium venenatum* for glucose measured as described above is 30 μM (= 5.4 mg l^{-1}).

A word of warning is necessary here because though this analogy with enzyme kinetics is a useful way of remembering the **basic kinetics** governing growth of liquid cultures, it has its limitations. In particular it is essential to stay aware that an enzyme reaction rate results from the interaction of a single enzyme with a single substrate whereas the growth rate of a culture, even if you are studying a single substrate, is the outcome of all the integrated metabolic interactions with the substrate and all of its diverse reaction products.

Nevertheless, a low K_s value is an important selective advantage as it allows the organism to grow rapidly even when nutrients like glucose are present in very low concentrations in nature. So the characteristically low K_s values observed indicate that microorganisms have evolved to prosper ecologically at low substrate concentrations. Consequences of low K_s values for such organisms in batch culture are that (a) they grow at μ_{max} until almost all the glucose is used up, and (b) the deceleration phase of growth is over very quickly.

Maximum specific growth rate cannot be attained unless all nutrients (including gaseous oxygen) are provided in excess and it cannot be maintained unless the culture is continually supplied with nutrients to replenish supplies as they are used and unless waste products and other potentially toxic metabolites (including gaseous carbon dioxide) are removed.

Growth rate begins to decline (**'deceleration phase** begins') when a nutrient starts to become depleted, and growth continues to decelerate until the organism enters **stationary phase** when the nutrient becomes exhausted. In most media, the limiting nutrient is usually the carbon source (generally a sugar such as glucose or sucrose) although in industrial-scale fermenters where high biomass densities are favoured to produce maximal quantities of product, oxygen can often be the limiting nutrient. Except for the strictly anaerobic rumen chytrids, fungi are generally aerobic organisms with an absolute requirement for oxygen for respiration and maximal growth.

The onset of stationary phase of a culture in practice may be determined by:

- **nutrient exhaustion** (of, for example, glucose),
- lack of oxygen,
- accumulation of toxic metabolites,
- or any combination of any or all these factors.

Entry into stationary phase may also be caused by a change in the pH of unbuffered media, for example when ammonium sulfate is the nitrogen source.

19.7 GROWTH YIELD

The final **yield** of the culture can be determined when it enters the stationary phase. The yield (X) of a culture is the difference between the initial biomass (X_0) and the maximum biomass at the end of the growth phase (X_{max}):

$$X = X_{max} - X_0$$

The yield can be related to the amount of substrate used (an important consideration in commercial cultures where you must ask how much product you are getting from the expensive substrate you are adding). This is the **yield coefficient** (Y), which is the ratio of the biomass formed (measured in g) to the mass of substrate (e.g. glucose) consumed (S, also in g):

$$\mathrm{Y} = \frac{X}{S}$$

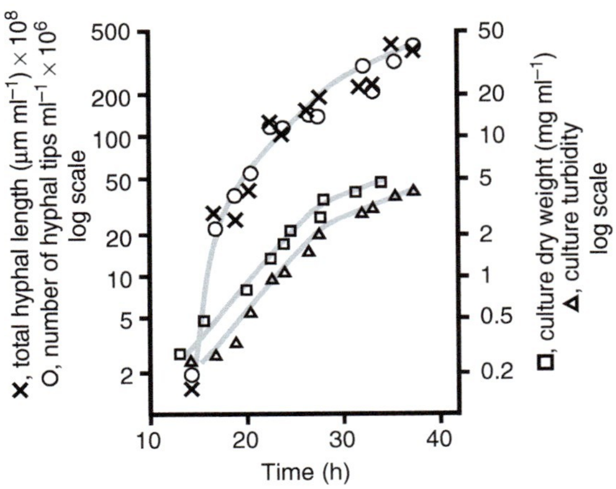

Figure 19.14. Initial growth of the ascomycete *Geotrichum candidum* in shake flask batch cultures. The plots show the increase in dry weight (open square, mg ml^{-1}, log scale), culture tubidity/absorbance (open triangle, log scale), total hyphal length (cross, µm ml^{-1} × 10^8, log scale) and number of hyphal tips ml^{-1} × 10^6 (open circle, log scale) of *Geotrichum candidum* in shake flask culture.

Figure 19.14 shows early growth of the ascomycete *Geotrichum candidum* in shake flask batch cultures; that is, cultures in a defined starting volume of liquid medium that were incubated on a shaking platform in a constant-temperature room. The parameters plotted introduce several features. Note that biomass was quantified both by measuring culture absorbance (or 'turbidity'; open triangles in Figure 19.14) ***and*** by subsequently drying the sample to determine dry weight (open circles in Figure 19.14). The numerical values of these two measurements are obviously different, but the patterns of the curves that result are sufficiently similar for turbidity to be used as a reliable measure of biomass in such cultures of this organism. Because of the morphological difference between unicellular bacteria or yeasts and apically-elongating, non-fragmenting, hyphal fungi, culture absorbance is a less reliable way of assessing the biomass of filamentous fungi than of bacteria or yeasts. Nevertheless, it is non-destructive and has proved to be useful method for most growth experiments with fungi.

The other parameters recorded in Figure 19.14 introduce the features that are characteristic of the filamentous growth form of such cultures; namely, total hyphal length (× symbols in Figure 19.14) and number of hyphal tips (open squares in Figure 19.14) of the biomass; note that these plots also closely follow the form of the dry weight (biomass) growth curve. Again, and as expected, the numerical values represented by these different measurements are different, but the curves described by the observations are so similar as to be reasonably judged identical. Consequently, these observations show that:

- increase in culture dry weight is a reliable measure of increase in total hyphal length;
- culture absorbance is a reliable (but non-destructive) measure of culture dry weight;
- increase in culture absorbance is a reliable measure of increase in the total number of hyphal tips in the culture.

Of course, having measurements of both total hyphal length and number of hyphal tips allows the former to be divided by the latter to calculate the length of the **hyphal growth unit** (G) (Table 19.6; and see section 4.4).

19.8 STATIONARY PHASE

In the stationary phase the biomass does not remain unchanged, rather it becomes senescent. This is associated with a rapid drop in viability as the culture enters stationary phase (Figure 19.15).

Table 19.6. Hyphal growth unit length of Geotrichum candidum in agar and liquid media

Type of medium	Temperature (°C)	Hyphal growth unit length (G, µm)
Liquid	20	112
Agar	25	110

In addition, early stationary phase cultures show signs of **autolysis** (Figure 19.16) following the specific induction of cell features that, in animal systems are diagnostic of **programmed cell death** (Figure 19.17; and see sections 8.5 and 13.15).

The examples shown in Figure 19.17 are:

- Annexin V-FITC: annexins are a family of calcium-dependent phospholipid-binding proteins. Annexin V preferentially binds to phosphatidylserine (PS) which is predominantly located in the cytosol-facing part of the plasma membrane in healthy cells. An early event in apoptosis is the flipping of PS of the plasma membrane from the inside surface to the outside surface when the membrane begins to become disrupted. Annexin V binds specifically to phosphatidylserine and fluorescently labelled annexin V can detect apoptotic cells by detecting PS in the outer membrane leaflet.
- TUNEL staining: terminal deoxynucleotidyl transferase-mediated dUTP-biotin nick end labelling (the TUNEL acronym is derived from those words) detects apoptotic cells by labelling DNA strand breaks ('nicks') revealing the onset of fragmentation of DNA which is a characteristic of apoptotic cells.

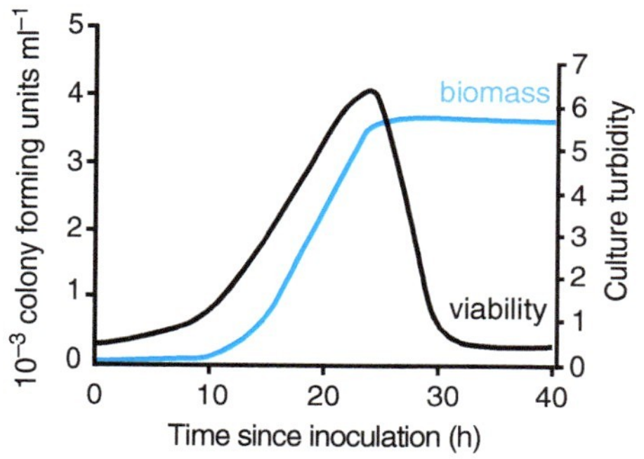

Figure 19.15. Growth and viability of *Aspergillus fumigatus* in batch culture. Growth was followed by using culture turbidity as a measure of biomass, and viability was determined by measuring the colony-forming units ml^{-1}.

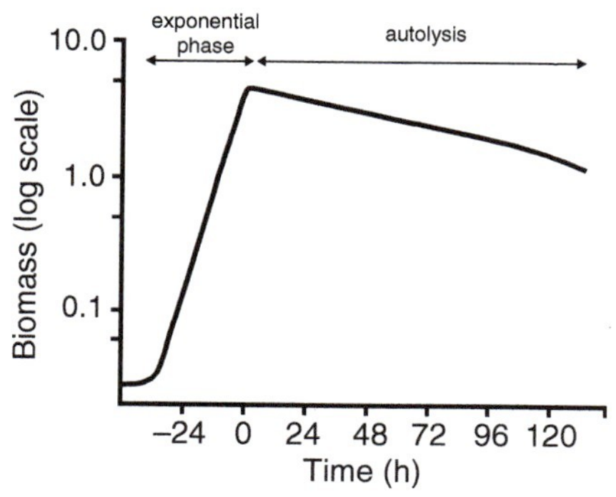

Figure 19.16. Autolysis during stationary phase of batch cultures of *Penicillium chrysogenum*. The deceleration phase is not apparent in this figure because μ_{max} is maintained until almost all the glucose is used up, and because dry weight is a relatively insensitive method of assessing change in biomass concentration.

- Propidium iodide staining (PI): Staining of nuclei with the DNA binding fluorescent (red) dye propidium iodide (PI) also detects fragmented DNA and the intensity of fluorescence correlates with the extent of DNA degradation. Live cells are not permeable to PI, so the stain detects dead cells.

Apoptosis in animals (particularly vertebrates) is a pattern of cell death that maintains cell fragments ***within*** an intact plasma membrane until the dying cell is engulfed by a phagocyte. In fact, the distribution of phosphatidylserine on the outside of the cell membrane (referred to above) identifies the dying cells in such animals as **targets for phagocytosis**. The function of this process is to contain the antigens that might be released by simple cell lysis (necrosis) and protect the animal from autoimmune reactions. This is not an issue in fungi, of course; but the appearance of diagnostic markers of programmed cell death during the stationary phase of a growth curve shows that loss of viability and autolysis in that phase **is an organised process**. Note our comment in section 13.15 that in fungal programmed cell death the cell contents released when the sacrificed cells die are specialised to specific functions. The cells are **killed** as the final step in their growth programme; they don't just die.

During **autolysis**, cell wall and macromolecular components in general are degraded; most are simply returned to the medium as the dead cells undergo necrosis, but some of the breakdown products can be scavenged by surviving hyphal cells in the biomass and used for some regrowth. This secondary growth phase is called **cryptic growth**, this term being coined to remind microbiologists that it is the rule rather than the exception that even during autolysis there is a small (hidden) number of cells that are **still growing**.

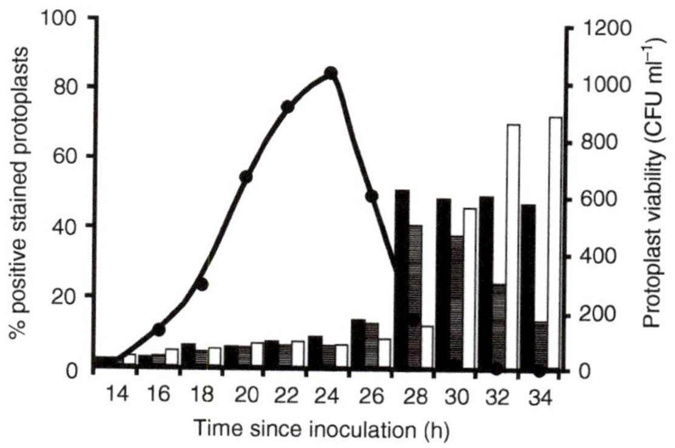

Figure 19.17. Viability and appearance of apoptotic markers during stationary phase of *Aspergillus fumigatus*. Protoplasts were prepared from mycelium at various time points and viability (closed circles) was determined. Onset of characteristics diagnostic of programmed cell death was measured as the percentage staining with annexin V-FITC (black bars) to detect plasma membrane disruption, and TUNEL staining (bars with horizontal hatching) and propidium iodide staining (PI; white bars) for fragmented DNA.

19.9 GROWTH AS PELLETS

Fungi that become aggregated or pelleted in submerged culture display more complex growth kinetics. Initial growth is exponential because during the early stages all the mycelial hyphae of the biomass contribute to growth. However, as small pellets enlarge, nutrients have to diffuse through the pellet and mycelium at **the centre becomes progressively nutrient limited** so that exponential growth becomes limited to the outside of the (spherical) pellet. When large fungal pellets with hollow centres are cut open there is a smell of alcohol, implying that central metabolism has become anaerobic and fermentative. When fungi grow as large pellets the biomass at the pellet centres produce **'toxic'** products, which diffuse outwards into the culture medium. These compounds eventually inhibit growth of the peripheral biomass.

Pellets begin to form when spores initially swell prior to germination and changes to the electrostatic charge of the spore wall leads to aggregation of spores, which following germination, leads to the formation of a mycelial aggregate and subsequently pellets. This is not a numerically trivial matter; cultures of *Aspergillus nidulans* inoculated with 2.3×10^6 conidia cm^{-3} formed only 2.9×10^3 pellets cm^{-3}, a thousand-fold **aggregation factor**.

Several factors influence pellet formation. **Inoculum size is often critical** with high spore concentrations favouring dispersed growth and low concentrations favouring aggregation and pellet formation. Medium composition and impeller agitation of the culture also play a major role in overall morphology in submerged growth and in some cases, a low pH has been shown to favour dispersed (that is, non-pelleted) growth.

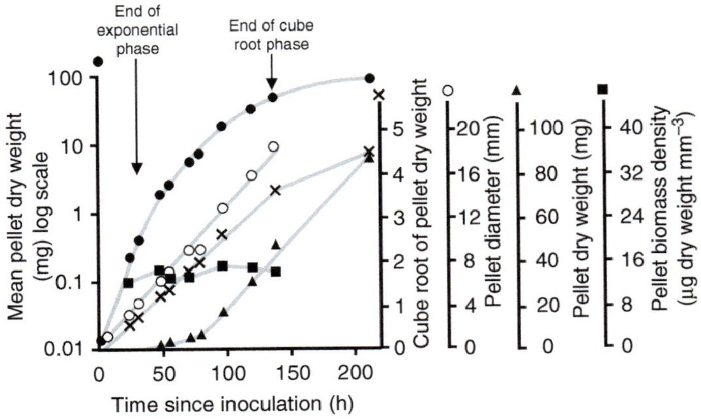

Figure 19.18. Growth of pellets of *Aspergillus nidulans* grown singly at 25°C in submerged batch culture (each flask contained 50 ml of medium and was inoculated with a single pellet). Pellet dry weight is shown plotted on logarithmic (closed circles), cube root (crosses) and arithmetic (triangles) axes. The mean diameter of the pellets (open circles) and biomass density (dry weight per cubic millimetre of pellet; squares) are also recorded. Adapted from Trinci, 1970.

Growth kinetics of a culture in pelleted form leads to biomass increase that is linear when plotted as the cubed root of biomass over time. This is called **cubic growth.** Clearly, growth of moulds that form pellets is based on the growth of a population of spheres and the cubic growth relation emerges due to the spherical nature of the pellet and because the volume of a sphere depends on the cube of its radius:

$$\text{volume of sphere} = \frac{4}{3}\pi r^3$$

Figure 19.18 shows that the increase in dry weight of **single pellets** of *Aspergillus nidulans* grown in shake flask culture is better described by **cube root kinetics**, rather than by exponential kinetics. This is because the pellet increases in radius at a linear rate determined by the width of a peripheral zone of biomass that can continue to grow exponentially; the biomass in regions of the pellet inside this zone is either growing very slowly or has ceased growth.

Three zones have been recognised within pellets of *Aspergillus nidulans* (Figure 19.19). The biomass of a pellet is morphologically and biochemically **heterogeneous**, and although pellets are basically spherical masses of mycelium, the physiology of a fungal pellet is complicated and there is a gradual transition in pellet physiology from its periphery to its centre. The steepness of this transition is influenced by hyphal packing of the biomass (i.e. biomass density), so discrete zonation may be less evident than suggested by Figure 19.19.

Generally, speaking however, the outermost zone grows exponentially and supports radial expansion of the pellet. The width of this **peripheral zone** is determined by the extent to which nutrients (especially oxygen) can be supplied to the biomass in quantities sufficient to support exponential growth. Growth is dependent upon the supply of nutrients to the biomass (by bulk flow and diffusion) and removal by outward diffusion of potentially toxic metabolic excretory products. The **innermost zone** of the pellet is most poorly supplied with nutrients (especially oxygen) so this zone consists of non-growing biomass. Eventually, the mycelium at the centre of the pellet becomes starved of nutrients, stops growing, enters the stationary phase

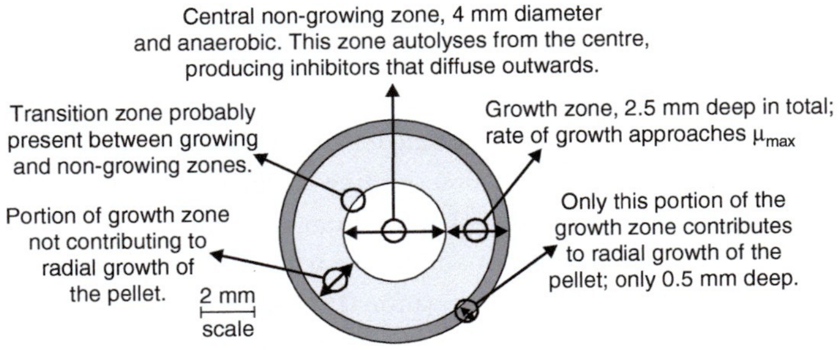

Figure 19.19. Diagrammatic cross section of a 9 mm diameter pellet of *Aspergillus nidulans* showing the growth zones recognised within it. Adapted from Trinci (1970).

and finally starts to autolyse (Trinci, 1970) (Figure 19.19). As the pellet periphery continues to enlarge this leads to a **hollow pellet**. Between the non-growing central biomass and the exponentially-growing peripheral zone is the middle transition zone within which increasingly inadequate nutrient (and oxygen) supply causes growth of biomass to proceed so slowly that it does not contribute to radial expansion of the pellet.

The growth kinetics of pellets are described by these equations:

$$x_t^{\frac{1}{3}} = x_0^{\frac{1}{3}} + kt$$

$$x_t^{\frac{1}{3}} = x_0^{\frac{1}{3}} + k\mu wt$$

where, k = a constant, and x = biomass concentration, μ = specific growth rate, w = width of peripheral growth zone of pellet and t = time.

Transport of nutrients into pellets and secondary metabolites out of pellets varies with diffusibility of the metabolite. Oxygen is so poorly soluble in water (see Table 19.4 and discussion above) that it fails to diffuse into the centre of even small pellets, so the centrally placed cells consequently die.

Exceptionally, this form of growth can be beneficial from a biotechnology perspective. The classic example of this is in the production of the organic acid **citric acid** from *Aspergillus niger*. Citric acid is very widely used as a flavouring agent, preservative and acidity regulator in a range of processed foods and soft drinks. It is also used as a buffering agent in pharmaceuticals and cleaning products, and a metal chelator in soaps and detergents. Originally, citrus fruits, principally lemons, were used as the source of citric acid, a process that was commercialised in Italy in the late 1800s.

However, some strains of *Aspergillus niger* and other fungi were known to secrete large quantities of citric acid under certain conditions and this was developed in the 1920s by Pfizer in the UNITED STATES to produce the **first dedicated filamentous fungal fermentation** (see section 19.14). In this fermentation, high yields of citric acid are critically dependent on the culture growing in the form of pellets; citric acid fermentations start with pellets 0.2–0.5 mm in diameter which grow to a final diameter of 1–2 mm. In addition, efficient conversion of sugar to citrate can only occur if formation of the alternative organic acid, oxalate, is minimised. The key enzyme in the oxalate pathway, oxaloacetate hydrolase, requires the metal manganese and pH values in the range 5–6. Imposition of metal deficiency at very acid pH can completely block oxalate formation, leading to efficient accumulation and secretion of citric acid. The medium is often treated chemically to remove trace metals and fermenter vessels are made from stainless steel to avoid contamination (Plassard & Fransson, 2009).

For the production of most other products, particularly enzymes, high yields are associated with a dispersed culture because this growth morphology maximises the proportion of the biomass that is young, active, and productive. Various techniques have been employed to discourage the formation of pellets. Sometimes this can be achieved by increasing the concentration of the spore inoculum; for example, in *Penicillium chrysogenum*, an inoculum containing fewer than 2–3 × 10^5 conidia ml^{-1} medium forms pellets, whereas an inoculum with more than 2–3 × 10^5 conidia ml^{-1} medium produces filamentous growth.

As pellet formation is often initiated by electrostatic interactions between spores and sporelings, dispersed growth can be favoured by addition of charged **anionic polymers** to the culture medium. Effective materials include *Carbopol®* (carboxypolymethylene), *Junlon*-PW110 (polyacrylic acid) and *Hostacerin®* (sodium polyacrylate), all of which have been used to discourage pellet formation (Jones *et al.*, 1988). Their mode of action seems to be that they coat the surface of the spores with a layer of uniform negative ionic charge and so prevent spore aggregation by electrostatic repulsion.

Culture morphology also has a major impact on the behaviour of the culture in fermenters. When the mycelium is dispersed, culture viscosity increases as the organism grows due to the interaction of the dispersed, branched, mycelial filaments, and at high biomass levels the culture becomes extremely thick and viscous. This has major implications for the fermentation:

- more energy is required to maintain the speed of stirring of the impellers because of the increased resistance of the culture;
- the rate at which oxygen is transferred to the medium is reduced as the viscosity increases.

This latter is the main reason why oxygen is often limiting in industrial fermentations of filamentous fungi. The overall behaviour of the culture under mechanical stress is known as the **rheology** of the culture (Figure 19.20). Rheology is the study of friction between fluids; it is the apparent viscosity, or 'resistance to flow', shown by a complex fluid.

Culture rheology is primarily affected by the **morphology of the mycelium**. When the mycelium is relatively unbranched (high hyphal growth unit; see sections 4.4 and 4.9) the culture viscosity is higher compared to the same concentration of biomass of a highly branched mycelium (low hyphal growth unit) (Figures 19.21 and 19.22). This is because long, sparsely branched hyphae will intertwine and interact more readily than shorter, more highly branched mycelia which behave more as independent entities.

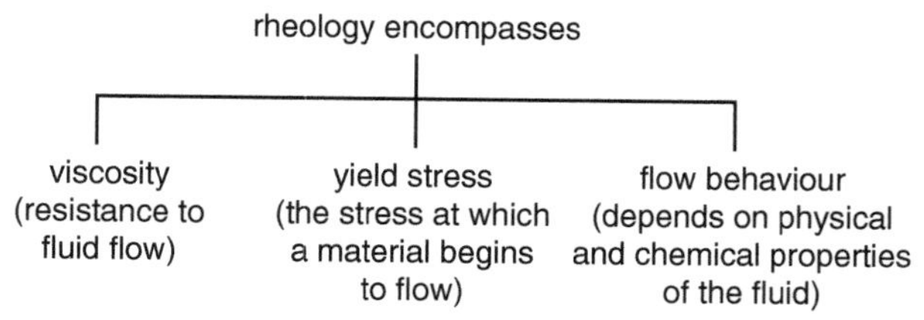

Figure 19.20. The overall behaviour of the culture under mechanical stress is known as the rheology of the culture.

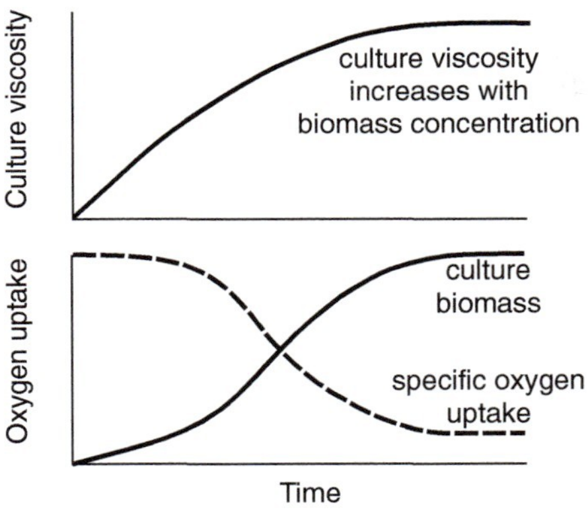

Figure 19.21. Cultures of filamentous fungi have a non-Newtonian behaviour and viscosity is related to biomass concentration and branching frequency. Top panel shows how culture viscosity increases with biomass concentration. Bottom panel shows how oxygen uptake declines with increasing biomass concentration (because the rate of oxygen transfer from the gaseous to the liquid phase decreases with increase in viscosity).

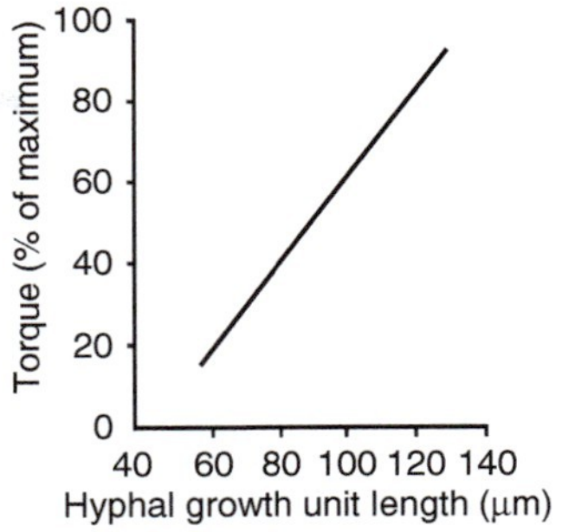

Figure 19.22. Relationship between hyphal branching (expressed as the hyphal growth unit) and viscosity (expressed as the torque required to rotate an impeller) in *Aspergillus oryzae*.

19.10 BEYOND THE BATCH CULTURE

Although we have occasionally mentioned other culture systems in the above, most of the discussion so far has concentrated on batch cultures, so we now want to turn to other types of culture system.

The term **fed-batch culture** (or simply, 'fed-batch') is used to describe a batch culture which is fed with medium either intermittently or continuously. Nothing is removed until final harvest, so the volume of the culture increases with time (Figure 19.23A). Fed-batch is a way of obtaining very slow growth rates resulting from **substrate limitation**; the penicillin fermentation is the classic fed-batch culture, but it is an industrial process that is likely to become important for production of fine chemicals for life sciences; health care, agrochemicals, and nutraceuticals (Blamey *et al.*, 2017).

Although the total biomass in the culture increases with time, biomass concentration remains virtually constant because medium is added, and the volume of the culture increases with time. Providing the dilution rate (D) is less than μ_{max} and Ks is much smaller than substrate concentration in the inflowing medium, a quasi-steady state may be achieved. However, unlike a continuous culture, D (and therefore μ) decreases with time, because the tank has a finite volume and as nothing is removed feeding with medium must stop at some stage in the cultivation (it was stopped at the 6-hour mark in the culture illustrated in Figure 19.23B).

The dilution rate of a fed-batch culture is given by:

$$D = \frac{F}{V_0 + F_t}$$

Where, D = dilution rate, V_0 = original volume of culture and F = flow rate of medium into the culture.

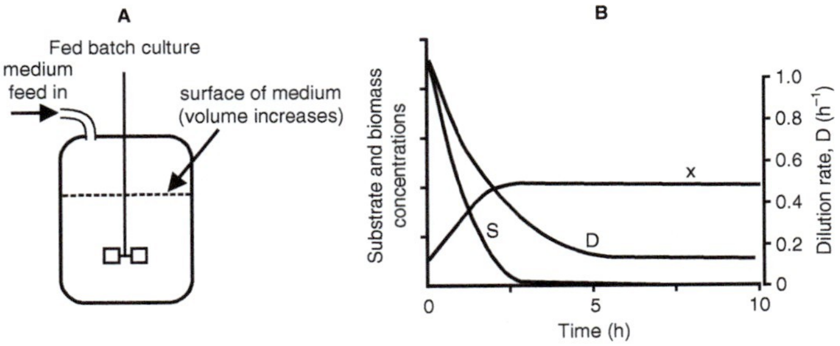

Figure 19.23. (A) Sketch diagram of a fed-batch culture. (B) records the basic kinetics of biomass concentration (x), substrate (glucose) concentration (S) and dilution rate (D) in such a fed-batch culture.

The next development of the fed-batch culture is to remove some of the culture at intervals but to keep feeding the culture with more medium; this is a **repeated fed-batch culture**. In repeated fed-batch cultures the culture volume and consequently the dilution rate and specific growth rate undergo cyclical variations as parts of the culture are removed.

Because it is technically easier to obtain **substrate-limited growth in fed-batch culture** than in chemostat culture, this type of culture is used for many industrial fermentations. Feeding a batch culture with fresh medium can be used to:

- relieve **catabolite repression** when this is an issue in a production process; and
- **avoid toxic effects** of some medium components, that may be required for particular production processes, on the growth of the biomass; for example, sodium phenylacetate, a toxic precursor of penicillin, can be kept at sub-inhibitory concentration by feeding it gradually to the culture.

19.11 CHEMOSTATS AND TURBIDOSTATS

Even a repeated fed-batch culture is a closed culture system in the sense introduced in Table 19.1; the essential open culture system is the **chemostat**. A chemostat consists of a culture:

- into which fresh medium is continuously introduced at a constant rate;
- whose volume is kept constant by continuous removal of culture; and
- in which the supply of a single nutrient controls growth rate.

The fermenter is called a chemostat because the growth rate ***is*** controlled by the availability of a single component of the medium (the **limiting substrate**). Important features are that it is the continuous introduction of fresh medium that feeds the limiting substrate to the culture and the dilution rate (that is, the rate of addition of fresh medium) determines the specific growth rate of the culture (Figure 19.24) (Matteau *et al.*, 2015).

The main features of a chemostat culture are:

- Volume of the culture remains constant.
- The time required to mix a small volume of medium with the culture should be small, that is, it should approach perfect mixing.
- Environmental conditions can be maintained constant (these include nutrient concentration, pH, temperature, antibiotic concentration).
- Specific growth rate can be varied from just above zero to just below μ_{max}.
- An environmental condition (pH, temperature, oxygen tension, etc.) can be varied while maintaining specific growth rate (μ) constant.
- Biomass properties such as macromolecular composition and functional characteristics can be maintained constant.
- Biomass concentration can be set (by altering the concentration of the limiting substrate in inflowing medium) and maintained constant, independently of μ.
- Substrate-limited growth can be maintained indefinitely, and this can offer relief of catabolic repression and induction of secondary metabolism at low substrate (glucose) concentration. When chemostat cultures are operated for a very long time the organism evolves (see below); indeed, chemostat cultures can be used to select mutants with specific physiological characteristics (see Table 19.7).

Dilution rate determines the specific growth rate of the culture,

$$D = \frac{F}{V}$$

Table 19.7. Comparison between turbidostat and chemostat continuous flow cultures

Parameter	Turbidostat	Chemostat
Method of growth rate control	Internal	External
Growth rate of culture	At or close to μ_{max}	From just above zero to just below μ_{max}
Effect of increasing concentration of all nutrients in the medium reservoir	Biomass concentration will only increase if the control settings on the photometer are changed	Increase in biomass concentration
Culture volume	Constant	Constant
Environmental conditions	Constant	Constant
Duration of culture	Indefinite	Indefinite
Types of mutants selected by prolonged culture	μ_{max} mutants and a range of neutral mutants	μ_{max} mutants at high dilution rates, *Ks* mutants at low dilution rates, and a range of neutral mutants

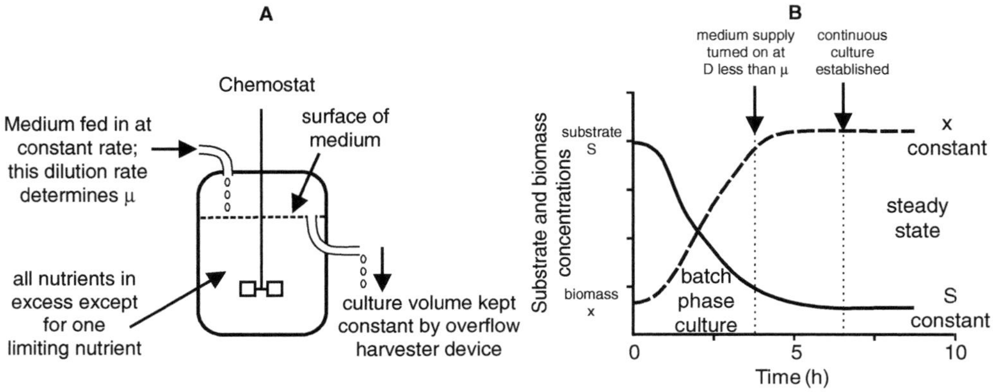

Figure 19.24. A basic chemostat. (A) Sketch diagram of a continuous culture vessel. (B) Record of biomass concentration (x) and substrate (glucose) concentration (S) over the early stages of a cultivation showing the transition from the initial batch phase to the continuous culture phase.

where D = dilution rate, F = medium flow rate and V = culture volume.

Thus, if $F = 1$ l h^{-1} and $V = 5$ l, then $D = 0.2$ h^{-1}. At steady state $\mu = D$. Thus, μ can be varied in a chemostat from just above 0 to just below μ_{max}. At steady state, the biomass concentration and limiting substrate concentration in the culture remain constant (Figure 19.24B). And that link between limiting substrate concentration and biomass concentration is why the rig is called a chemostat.

The biomass concentration in the fermenter vessel at steady state is given by:

$$\bar{x} = Y(s_r - \bar{s})$$

where $\bar{x}$ = biomass concentration in the fermenter vessel, $\bar{s}$ = concentration of growth-limiting substrate in fermenter vessel, Y = yield coefficient for the limiting substrate (Y is about 0.5 for glucose), s_r = concentration of growth-limiting substrate in the inflowing medium.

The effect of dilution rate on biomass concentration and limiting substrate (glucose) concentration is illustrated in Figure 19.25, which shows two scenarios in which the concentrations of glucose in the medium entering the chemostat is either 1.0 or 0.2 g l^{-1}.

As the culture approaches μ_{max} (1.0 h^{-1}) biomass is washed out of the vessel and residual glucose concentration increases. The mean residence time of the organism in fermenter vessels is the reciprocal of the dilution rate (that is, $R = 1/D$, where R = mean residence time of the organism and D = dilution rate).

At dilution rates below what is known as the **critical dilution rate** (D_{crit}), steady state can be maintained; at dilution rates above D_{crit} steady states cannot be maintained because if D is equal to or greater than μ_{max}, a steady state culture cannot be maintained, and the culture will washout (more biomass is lost through the overflow than can be replaced by further growth). The kinetics of washout of biomass from a chemostat culture can be used to calculate μ_{max} since:

$$\mu_{max} = \frac{\ln x_t - \ln x_0}{t} + D$$

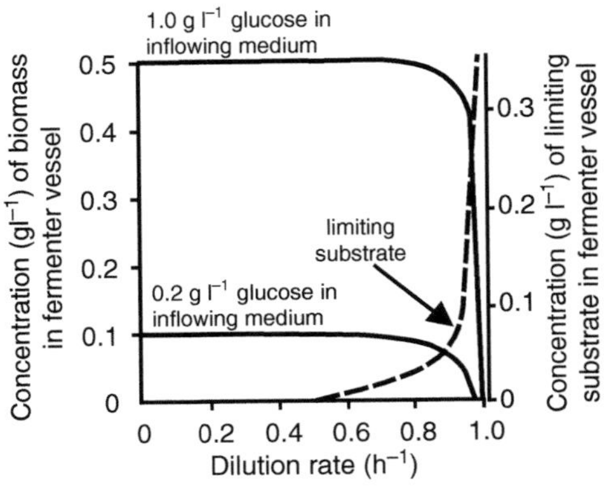

Fig 19.25. The effect of dilution rate on biomass concentration and limiting substrate (glucose) concentration. As the culture approaches μ_{max} (1.0 h^{-1}) biomass is washed out of the vessel and residual glucose concentration increases.

There are two main types of continuous flow cultures, turbidostats and chemostats (Table 19.7). Both must be **perfectly** mixed suspensions of biomass to which the medium is fed at a constant rate and culture is harvested at the same rate. In the chemostat the specific growth rate is controlled **externally** by the concentration of a single nutrient, the limiting substrate. In a glucose-limited chemostat, almost all the glucose is utilised by the culture with the result that biomass concentration reaches a constant value which is essentially proportional to the glucose concentration in the inflowing medium (this is the steady state).

The specific growth rate of a turbidostat culture is at or very close to μ_{max} and is controlled by the rates of **internal** cellular reactions as they are expressed in the optical density of the culture biomass (that is, the turbidity of the culture). This is done by photometers in which incident light is scattered by the culture and the transmitted light is detected and measured by the photometer; when turbidity (that is, transmitted light detected) reaches a certain level, a medium pump is switched on to return the turbidity to the required level (Figure 19.26).

The aim is to hold **culture turbidity** constant by manipulating the rate at which medium is fed. If the turbidity tends to increase, the feed rate is increased to dilute the turbidity back to its set point. When the turbidity tends to fall, the feed rate is lowered so that growth can restore the turbidity to its set point. The optical surfaces of the detectors most often used to measure turbidity are easily fouled by growth of microbial biofilms, foam or precipitates from the medium and the problems presented by this fouling have not been solved. In practice, a **turbidostat** operates well for a brief period, but the control of the turbidity eventually becomes unreliable.

In a culture controlled this way, increasing nutrient concentration will affect biomass concentration but not specific growth rate. As shown above the biomass concentration in the fermenter vessel at steady state is given by $\bar{x} = Y(s_r - \bar{s})$.

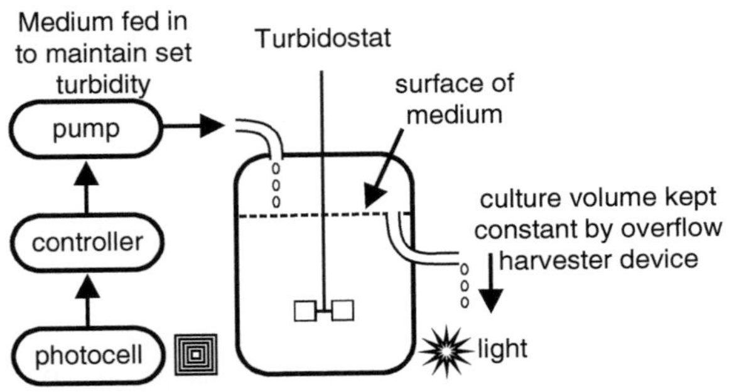

Figure 19.26. Schematic of a turbidostat. Photocells measure the light transmitted through the turbid culture; when turbidity (that is, transmitted light detected) reaches a certain level, a medium pump is switched on to return the turbidity to the required level.

If turbidostats can be maintained for long periods, the system will apply selection pressure for mutants with increased μ_{max} values. Turbidostats have also been used to select mutants with increased resistance to an antibiotic.

Other methods used to monitor biomass and control medium supply include:

- measurement of carbon dioxide in the medium;
- measurement of oxygen in the medium;
- residual nutrient concentration (called a **nutristat** or **glucose-stat**; for Quorn® mycoprotein production, *Fusarium venenatum* is grown in glucose-stat continuous cultures);
- medium pH (called the **pH auxostat**, which controls addition of alkali to the medium);
- dielectric permittivity (permittivity is the ability of a material to transmit ('permit') an electric field), which is affected by changes in the culture as it grows and is measured with a radio-based sensor that controls addition of medium (such chemostats are called **permittistats**).

19.12 USES OF SUBMERGED FERMENTATIONS

The fermentation processes discussed above are used in industry to produce a wide variety of fungal products. **Alcohol** and **citric acid** are the world's most important fungal metabolites in terms of production volume. Although penicillin is still an important antibiotic, most **antibiotics** that we use today actually originate from bacteria (still grown in ways described above). Fungi produce some other crucial pharmaceuticals: for example, the fungal product called **cyclosporine**, which suppresses the **immune response** in transplant patients to avoid organ rejection.

Another natural compound obtained from fungi that has great medical value already and seems to be increasing in clinical significance is called mevinolin. This is produced by the fungus *Aspergillus terreus* and is the basis of the **statins**, which are used to reduce cholesterol levels (risk factors in cardiovascular disease, stroke and several other widespread illnesses). Since their introduction in the late 1980s, statins have become the most widely prescribed cholesterol-lowering drugs in the world. Pfizer's Lipitor (atorvastatin) is the most profitable drug in the history of medicine; at its peak in 2006, yearly revenue from Lipitor exceeded \$US12 billion. Sales of AstraZeneca's Crestor (rosuvastatin) reached \$5.38 billion during 2009, making it one of the company's best-selling medicines. Even though their patents recently expired, revenue for statins is still on track to reach an estimated \$US1 trillion by 2020. Statins are **very** big business (Demasi, 2018).

Fungi also produce compounds like the **ergot alkaloids**, **steroid derivatives**, and **antitumor agents**. Ergot toxins in low, controlled, concentrations are valuable drugs causing vasodilation and a decrease in blood pressure as well as contraction of smooth muscles. Although these alkaloids can be synthesised, strain improvement by mutation and selection of high-yielding strains has been so successful that fermentation remains the most cost-effective means of production.

Most of the **steroids** in clinical use today are modified by fungi and/or fungal enzymes during manufacture. By using fungi and their enzymes to make specific **chemical transformations**, compounds can be made that would otherwise be very difficult, impossible, or just too expensive to produce by direct chemical synthesis. Recently, many fungal products have been found to inhibit the growth of cancers in animal tests. Specificity and safety are the issues that presently limit the medical usefulness of most of these compounds (they may have adverse effects on healthy tissue as well as the tumour). Compounds that work by modifying the activity of the patient's own immune system (**immunomodulators**), making it more active against the cancer cells, seem likely to offer the greatest value. In summary, the fungal products that contribute to our daily lives range through:

- **Primary metabolites**, particularly fatty acids and organic acids, of which citric acid is the main one. Another example is the polyunsaturated fatty acid **arachidonic acid**. This has useful clinical effects in lowering cholesterol and triacylglycerols in plasma, with good effects on arteriosclerosis and other cardiovascular diseases. It is added to infant formula because it is present in breast milk and is important in brain and retina development in new-born infants (Tallima & El Ridi, 2017; Das, 2018);
- **Secondary metabolites**, most of which are used as pharmaceuticals like the antibiotics penicillin and cephalosporin, as well as cyclosporine and the statins (Rodrigues, 2016);
- **Numerous enzymes**, including cellulases, amylases, lipases, proteases, which have a tremendously diverse range of uses in industry (Figure 19.27):
  - **textile processing**: cellulolytic enzymes, particularly in denim treatment (to provide the 'stone-washed' finish) and biofinishing of fabrics with cellulases and proteases to remove loose fibres;
  - **biological detergents**: which must be active under alkaline conditions and therefore favour fungal proteases, lipases, amylases and cellulases, also used for fabric conditioning in colour brightening (removal of broken fibres that scatter light and dull the colour) and fabric restoration;

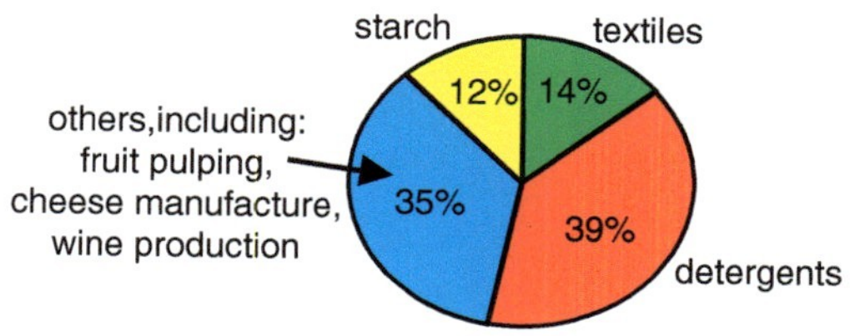

Figure 19.27. Main applications of fungal enzymes.

- **starch hydrolysis** by amylases: for production of high fructose syrups used in confectionery manufacture;
- **cheese making**: fungal proteases have been used as an alternative to bovine chymosin for many years, but increasing use is being made of bovine chymosin expressed in genetically modified yeasts and *Aspergillus niger*;
- **brewing**: fungal α-amylase, glucanases and proteases are used to improve extraction from barley during the 'mashing' process prior to fermentation in beer production;
- **wine and fruit juice conditioning**: fungal pectinases are used to remove pectin to pulp and peel fruits and vegetables and clarify liquors;
- **proteases**: are used to modify protein from a wide variety of sources, for example removal of hairs and bristles from hides during leather production;
- **baking**: xylanases are used to increase dough volume by making more polysaccharides available to the yeast;
- **pharmaceutical production**: extracted fungal enzymes can be used to modify sterols and other precursors during preparation of semisynthetic pharmaceuticals, for example 6-amino penicillanic acid (6-APA) is a semisynthetic penicillin precursor, the side chain of which can be removed by a penicillin-V acylase.

Among the filamentous fungi that are principally used in industry for these production processes are:

- *Aspergillus niger*: glucoamylase, other enzymes, citric acid;
- *Aspergillus oryzae*: α-amylase, other enzymes;
- *Trichoderma reesei*: cellulase;
- *Penicillium chrysogenum*: penicillins;
- *Cephalosporium acremonium*; cephalosporins;
- *Fusarium venenatum*: protein food;
- *Mortierella* sp.: arachidonic acid.

Filamentous fungi are used for enzyme production because many strains have excellent production and **secretion capacity**; for example, strains of *Trichoderma reesei* produce yields of cellulase up to 40 g l^{-1} (50% of which is cellobiohydrolase produced from a single copy gene). These high yields reflect the lifestyle of filamentous fungi, specifically extracellular polymer degradation and the fact that enzyme secretion is linked to hyphal tip growth, so **maximising tip growth maximises protein secretion**. Indeed, genetically modified filamentous fungi offer great promise as **cell factories** for production of heterologous proteins because they:

- express these high production and secretion capacity;
- can be cultivated under established fermenter conditions;
- fungi have similar **post-translational protein modification** processes to mammalian systems; and
- several, if not many, fungi have **'GRAS' (generally recognised as safe)** status because of their long history in food use and/or traditional technologies (Bourdichon *et al.*, 2012, is an authoritative listing of microorganisms with a documented use in food).

We dealt with many of these enzyme production techniques, particularly those that involve genetic modification, in Chapter 18. In this chapter, we intend to concentrate on what might be called the traditional biotechnology industries: those in which the intact fungal organism is carrying out the industrialised process. These include worldwide industries with the greatest commercial values (Table 19.8) (Hutkins, 2006; Montet & Ray, 2015):

- **alcoholic beverage** production (ales, beers, lagers, wines, and spirits);
- **baking** (all leavened bread);
- **cheese** production and cheese finishing;
- production of **high fructose syrup** sweetening agent for the confectionery industries;
- **mushroom cultivation** is the most obvious way in which we exploit fungi. In this worldwide industry, about 30 species are commercially cultivated, 7 being cultivated on what could be described as industrial scale. But mushroom cultivation suits all scales from peasant farms through smallholdings and on to multi-million dollar, highly mechanised farms. However, the mushroom farming industry uses solid-state fermentation processes, rather than submerged liquid cultivations, and we will deal with these towards the end of this chapter.

Table 19.8. Annual commercial values of some fungal products (entries are shown in £ millions)

Product	Totals	Grand totals
Food/drink products		45,970
Alcoholic beverages	24,500	
Cheese	14,500	
Mushrooms	5,000	
High fructose syrup (sweetening agent)	1,050	
Yeast biomass	600	
Citric acid (flavouring agent)	300	
Quorn® mycoprotein	20	
Medicinal products		5,800
Antibiotics	5,500	
Steroids	300	
Industrial products		3,000
Industrial alcohol	2,400	
Enzymes	600	
Grand Total		54,770

19.13 ALCOHOLIC FERMENTATIONS

The most important fermentation worldwide, of course, is the one that is used to make alcoholic drinks (Table 19.8 and see the discussion of fuel ethanol in section 19.22). It's a simple process; starting with something that's rich in sugar, like fruit juices, honey, or cereal grains or roots. Then add yeast (or rely on the yeasts already in the starting material, which is a common procedure in traditional wine fermentation) and allow the mixture to ferment. The final liquid may contain up to 16% alcohol, depending upon the yeast and the process (Walker *et al.*, 2017).

Fermentation is such a simple process that alcohol has been incorporated into the way of life of every civilisation, even the most primitive. The earliest records occur in ancient Egyptian murals and tomb ornaments depicting bread, beer and wine making. Yeast cells can be clearly seen by scanning electron microscopy in beer remains in a tomb pottery vessel from the village where workmen lived who built the tombs in the Valleys of the Kings and Queens, dated 1550–1307 BC (Samuel, 1996). From the biological point of view, it is a remarkable aspect of yeast physiology that the organism responsible for fermentation is invariably the yeast called *Saccharomyces cerevisiae* (not surprisingly known as brewer's yeast) or some closely related variant. Other research has identified the presence of barley, with other components of a beer recipe, in 5,000-year-old archaeological materials from China (Wang *et al.*, 2016).

The chemistry of the ethanol fermentation is summarised in this equation:

$$C_6H_{12}O_6 \rightarrow 2\ C_2H_5OH + 2\ CO_2$$

This equation is telling you one hexose molecule is converted into two ethanol molecules and two molecules of carbon dioxide. Many yeast species can carry out this fermentation, but only in an anaerobic environment; in the presence of oxygen, complete oxidative respiration of sugar to carbon dioxide and water occurs.

However, *Saccharomyces cerevisiae*, and its distant relative *Schizosaccharomyces pombe*, both preferentially use fermentation even in the presence of oxygen and will yield ethanol even under aerobic conditions given appropriate nutrients.

- Yeasts used for making **ales** tend to generate froth and grow on the top of the fermenting liquor, and such top fermenting yeasts are *Saccharomyces cerevisiae* itself.
- **Lager** yeasts do their fermenting at the bottom of the tank and they belong to the related *Saccharomyces carlsbergensis*.

In summary, ales, beers and lagers are generally made from malted barley; other cereal grains are used for specialist beers or in places where barley is not available (Briggs *et al.*, 2004; Hutkins, 2006; Montet & Ray, 2015).

The process starts by encouraging the cereal grains to germinate. In 2–4 days the sprouting seeds start the digestion of their stored starch, producing soluble sugars. Then the sprouted grain is killed by slow heating and **mashed** into hot water with other cereals like maize, wheat or rice. Finally, the sweet mix is boiled with hops to add bitter flavours to the beer. The boiled liquid is cooled and passed to the fermentation vessels for fermenting.

The four main ingredients of beer are:

- **Water**: dissolved salts in the water influence the character of the final beer by affecting extraction of fermentable sugars from the malted grain and the way the yeast behaves during fermentation. Total salts in the water in **Pilsen** (a city in western Bohemia in the Czech Republic) known as the place of origin of Pilsner lager, amounts to around 30 parts per million; whereas in **Burton-upon-Trent** in the UK, which is the home of English pale ale, the local hard water has a salt content of 1,220 ppm. It was calcium sulfate in the local water in Burton-upon-Trent that helped to create the pale ale style of beer; the calcium increases the efficiency of extraction of sugars from the malted grain, while the sulfate enhances the bitterness of the hops. In addition, the divalent Ca^{2+} form salt bridges between suspended polypeptides and polysaccharides and this aggregation helps to clarify the brew.
- **Malt**: which is germinating barley. Starches in the grain are converted into sugars as the seed germinates. When this process is stopped by drying at an optimal point in the process, the grain will contain some sugars plus a quantity of enzymes to aid the further extraction of fermentable sugars. The process of mashing (see below) makes use of these enzymes to do this.
- **Hops**: the flowers of the hop vine contain resins, which provide bitterness to the final product but take at least 15 minutes boiling to extract, and oils, which provide aroma but evaporate quickly if boiled. The typically **English India pale ale** was first brewed in England and exported for the British troops in India during the late 1700s. More hops were added to the recipe and the natural preservatives in hops enabled this ale to travel well over the long sea voyage to the British colonies in India. **British bitter** also contains more hops, but combined with pale malts to differentiate this ale from other brews.
- **Yeast**: some beers make use of wild yeasts, but most modern brewers prefer to control the yeast culture. The top foam is skimmed from the current fermentation and used to start the next beer ferment. The sugar content of the liquid is monitored throughout the fermentation and the process is stopped when the desired alcohol strength is reached.

Preparation for the fermentation starts by crushing the **malted grain**; this is intended to split the grains into smaller pieces to increase their surface area, separating the husks from the grain kernels. The resulting **grist** (made up of grain which is said to be **cracked**) is better able to release the sugars and enzymes.

Mashing is the process of combining the grist with water at 67°C to stimulate the enzymes in the malt to convert starches to sugars. Mashing is done in a vessel called a **mash tun** and is common to all fermentation processes, so the grain is typically malted barley for beers and malt whisky but will be mixed with, or even replaced by, other grains such as maize, sorghum, rye or wheat to prepare for other liquors.

The mash is mixed and sparged for 1.5–2 h to maximise the extraction. Finally, the liquid extracted from the mashing process (now called **wort**) is pumped into a large boiler known as a **copper** and brought to the boil, at which point the first batch of hops is added. These hops provide the bitter taste; volatiles that might contribute to final aroma will be boiled off at this stage. The wort is boiled for 1.5–2 h, then heating is stopped, and aroma hops are added (a process called **dry hopping**).

After a short mixing period, the hot wort is decanted or filtered, rapidly cooled and transferred to the fermentation vessel at 17°C. The yeast is added (or **pitched**) and fermentation starts and is evident in a few hours. After 1–3 weeks, the fresh (or **green**) beer is run off into conditioning tanks cooled below 10°C. After conditioning for a week to several months, the beer is often filtered, and the **bright beer** is ready for serving or packaging.

Wine making is now also a global industry, although Italy, France and Spain still together account for half the world's production, which was estimated at 259 million hectolitres (mhl) in 2016 (source: website of the *International Organisation of Vine and Wine* at http://www.oiv.int/en/). In that year, production in Italy was 48.8 mhl, France 41.9 mhl and Spain 37.8 mhl. For comparison, production of 22.5 mhl was recorded in the United States, and Australian production was 12.5 mhl.

In fact, world production fell by 5% compared to 2015, after a succession of climatic events: severe frosts early in the season, followed later in the year by heatwaves and wildfires, drastically reduced yields in France, Spain and Italy.

The classic wine grape has the scientific name *Vitis vinifera*. The important cultivars include Sauvignon (red Bordeaux), Pinot Noir (the main red Burgundy grape), Riesling and Silvaner (for German white wines), Barbera and Freisa (northern Italian wines) and Palomero, the main sherry grape. The whole of the black grape is crushed to make red wine; it's the grape skin pigment that makes it red. Black or white grapes can be used to make white wines but only the pressed juice is used, and extraction of skin pigments is avoided (Jackson, 2008).

Of course, the quality of the wine depends on the grape used, on production techniques, and on fine points like the type of soil and what the growing season was like. Grape juice is obtained by crushing grapes and usually contains 7–23% pulp, skins, stems and seeds; in the wine industry this grape juice is referred to as **must**.

The most controlled fermentation of the must is achieved with an elliptically shaped variant of *Saccharomyces cerevisiae* called *Saccharomyces ellipsoideus*, which may come from natural sources (the grapes or the preparation machinery) but is more usually added as a starter yeast culture made up in grape juice. Traditional wine makers in Europe prefer to use the **ambient (or wild) yeasts** that occur on the grapes themselves (the **bloom** or blush of the grape) because it is a characteristic of the region. Common genera of such ambient yeasts include *Candida, Pichia, Klöckera* (*Hanseniaspora*), *Metschnikowiaceae* and *Zygosaccharomyces*. High-quality and uniquely flavoured wines can be produced this way, but wild yeasts are often unpredictable and may cause spoilage.

The crucial feature for most wine making is the **controlled fermentation** by that one specifically isolated and cultured yeast, *Saccharomyces ellipsoideus*. Even this exists as several hundred different strains with fermentation characteristics that contribute to the diversity of wine, even when the same grape variety is used.

Fermentation may be carried out in stainless steel tanks, in traditional open wooden vats or wine barrels and can last from 5–14 days for primary fermentation and potentially another 5–10 days secondary fermentation (Hutkins, 2006; Montet & Ray, 2015). The latter, perhaps, inside the wine bottle as in the production of many sparkling wines.

After the yeast fermentation, quality wines take 1–4 years to **age** in wooden casks. For some wines a **bacterial fermentation** is encouraged during aging to mellow the taste by reducing acidity. To make sparkling wines, sugar, a little tannin and a special strain of *Saccharomyces ellipsoideus* that can form granular sediment, are added when the wine is bottled.

A secondary fermentation in the bottle produces carbon dioxide and results in formation of an unstable compound with alcohol (called **ethyl pyrocarbonatene**) which gives the characteristic lingering sparkle of naturally produced sparkling wines (compared with the brief sparkle of those that have only had carbon dioxide injected into them under pressure).

Fortified wines, which have had up to 20% brandy added to them, include **sherry**. This is made from a specific grape and a secondary growth of yeasts in the maturing vats creates the characteristic **sherry** flavour compounds. **Vermouths** are wines flavoured with herbs, like wormwood, and with extra alcohol added. **Port** is a red wine in which the primary fermentation was stopped by adding alcohol or brandy while some sugar remained. The special flavour of **Madeira** results from a heat treatment of the fermented wine before extra alcohol is added.

19.14 CITRIC ACID BIOTECHNOLOGY

Citric acid is a weak 6-carbon organic acid that is a component of the Krebs cycle in which it is generated when the acetyl group from acetyl-CoA is added to the 4-carbon oxaloacetate (see Figure 11.9); it is a naturally occurring component of metabolism in almost all living things.

It gets its name from the fact that it is the principal organic acid in citrus fruits, where it can account for up to 8% of the fruit's dry weight. Its chemical structure (Figure 19.28) features three carboxylic acid groups and a hydroxyl group, which

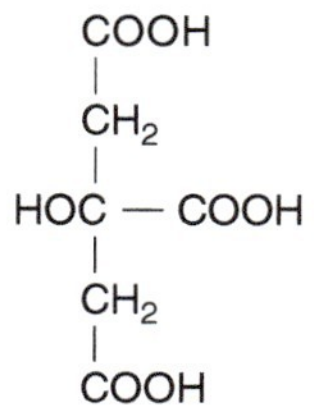

Figure 19.28. Structural formula of citric acid (2-hydroxypropane-1,2,3-tricarboxylic acid).

together confer a variety of properties that make citric acid useful in foods and pharmaceuticals as a pH buffer, acidulant, preservative and/or metal ion chelator.

First isolation of citric acid from lemon juice is credited to **Karl William Scheele** in 1784 (Show *et al.*, 2015). In 1860, the citric acid industry began, isolating the acid from lemon fruits grown in Italy. This method was used until citrus fruit exports from Italy were disrupted during the World War I (1914–1918).

In 1917, **James Currie** found that the common mould fungus *Aspergillus niger* could be used to produce citric acid from sugar in a **surface fermentation** process that became the basis for commercial production employing *Aspergillus* (Show *et al.*, 2015). Surface culture was the original production method employed for large-scale manufacture (introduced in 1919 in Belgium and in 1923 by Pfizer and Co. in the United States).

Although efficient **submerged processes** have been developed (using the pelleted growth morphology, see section 19.9), surface culture is still employed in some production units as it is simple and has low energy costs. The mycelium is grown as a surface mat in shallow trays of 50–100 l capacity that are made of high purity aluminium or stainless steel.

The main **carbohydrate source** of the medium may be refined or crude sucrose, cane syrup or beet molasses (diluted to 15% sugars), and the pH is adjusted to 5–7 and the temperature is maintained at 28–30°C. Fermentation proceeds over 8–12 days and pH decreases to below 2.0.

After completion of fermentation, the fermented medium is separated from the mycelium by being poured out of the trays. Citric acid is **recovered** from the fermentation liquor by precipitation with lime (calcium oxide, CaO); the precipitated **tricalcium citrate** is recovered by filtration and washed several times with water. Treatment of the product cake with sulfuric acid forms insoluble calcium sulfate in a liquor containing citric acid which can be filtered, concentrated and crystallised as **citric acid monohydrate**.

Submerged fermentation can be performed by batch, continuous and fed-batch methods, but batch fermentation is most often used and an aeration system that can maintain a high dissolved oxygen level is essential and under optimal conditions fermentation is complete in 5–10 days. Overall fermentation yields are in the range of 70–75%, and at least **1,600,000 tonnes of citric acid** are produced annually (worldwide) by fermentation. China is the largest player in the citric acid market, accounting for 59% of world production in 2015; it also accounted for 74%, and 12% of world exports and consumption, respectively, in 2015 (source: https://ihsmarkit.com/products/citric-acid-chemical-economics-handbook.html).

19.15 PENICILLIN AND OTHER PHARMACEUTICALS

Citric acid was the **first biotechnology product** and the techniques developed for production of citric acid by surface cultures of a filamentous fungus in batch mode were applied in the 1940s to production of a much more **revolutionary fungal product**, the antibiotic **penicillin**. You have probably heard the story about penicillin being discovered by chance when Alexander Fleming returned from holiday to find that his bacteria had been killed by a contaminating fungus on his Petri dishes. We do not intend to retell that story here because we prefer to emphasise the magnitude of the **change in lifestyle** that was brought about by the availability of antibiotics. However, the 'bare-bones' history of penicillin is as follows:

- 1928, **Alexander Fleming**, in St. Mary's Hospital London, discovered his cultures of *Staphylococcus aureus* were killed by contaminating colonies of *Penicillium notatum*. He is reported to have said some time later: 'I must have had an idea that this was of some importance, for I preserved the original culture plate.'
- 1940, **Howard Florey** and **Ernst Chain**, of Oxford University, as part of a survey of antibacterial substances produced by microorganisms showed the chemotherapeutic properties of penicillin.
- 1941, development of penicillin fermentation in the United States.
- 1945, **Dorothy Hodgkin** and **Barbara Law** (Oxford) used X-ray crystallography to establish the β-lactam structure of penicillin.
- 1956, **John Sheehan** (United States) produced penicillin by chemical synthesis.

Today, we take antibiotics for granted: you have a mildly sore throat so you get an antibiotic; you get a slight injury so you get an antibiotic jab 'in case of complication'. It's all become so trivial. So ordinary. But the changes created by the immediate availability of antibiotics were far from trivial. For the first time in human history, antibiotics brought freedom from fear of death from '**blood poisoning**': a death waiting for everyone for the most trifling of reasons. We can illustrate this with a narrative about one of the first patients to be treated with penicillin.

At the end of December 1940, a man called Albert Alexander was admitted to the Radcliffe Infirmary in Oxford, England. Albert was a 43-year-old police officer. He had been injured about a month before admission to hospital, though the record does not say whether the injury was suffered in the line of duty. The **injury became infected**, the dreaded blood poisoning or

septicaemia. Pathogenic bacteria were spreading through the tissues of his face, head and upper body, growing faster than his immune system could cope with and causing suppurating abscesses all over his face and forehead. His eye became infected and had to be removed on 3 February. Then his lungs became infected. He was close to death and was selected for experimental treatment with partially purified penicillin.

On 12 February, he was injected with 200 mg, followed up with 100 mg doses at three-hourly intervals. The next day his temperature had returned to normal and he was able to sit up in bed. He continued to improve. But so little penicillin was available that it had to be re-extracted from his urine to be re-injected into his veins. Finally, all supplies were exhausted, and his condition worsened rapidly. Despite the early promise of treatment with penicillin, there was not enough of the wonder drug to save Albert Alexander. He died on 15 March. **Killed by an injury that became infected**. And the injury that killed this policeman? Albert Alexander had been ***scratched by the thorn of a garden rose***.

More than any other scientific advance of modern times, the penicillin story can stir interest and imagination because this one fungal product simply revolutionised medicine. It cannot cure everything, but penicillin can be used successfully to treat pneumonia, gangrene, gonorrhoea, septicaemia and osteomyelitis. Diseases that were fatal and widespread, when penicillin first came into use were relegated to medical history. In the 1930s (and before), any injury in which the skin was broken might become infected with soil or airborne bacteria and some of these might grow beyond the level with which the immune system could deal. When that happened, the bacteria were consuming the patient and producing toxins in the blood stream which caused more widespread damage.

An ordinarily active adult might suffer scratches or minor cuts while gardening, walking or climbing. Osteomyelitis was an infection of the bone with *Staphylococcus* which was relatively common in children. Any man who shaved regularly might suffer infection of the inevitable nicks and cuts. This is what killed Lord Caernarvon in 1923. He was a veteran of 19 years in Egypt before he found Tutankhamun's tomb in 1922. And it was not 'The Mummy's Curse' that finally saw an end to the noble lord; it was blood poisoning, septicaemia, caused by an infected shaving cut. Less dramatic infections were painful, debilitating, but above all, common.

Women were even more at risk. Birth is a very messy procedure even today and before the ready availability of antibiotics an astonishing proportion of new mothers suffered, and died, from puerperal fever resulting from internal infection. An equally astonishing number of new-borns were infected during birth with bacteria which were only mildly pathogenic but nevertheless caused blindness, deafness and other life-long disabilities, if the child survived.

Penicillin contributed in a major way to a revolutionary change in medical treatment which in turn changed the human lifestyle to such an extent that diseases that were common causes of death and disability before penicillin are now rarely encountered. This dramatic change to the everyday experience of everyone on the planet is the **most remarkable aspect** of the discovery and introduction of penicillin.

There are several other remarkable aspects of the penicillin story. Supply is one of them. For the first patient, Albert Alexander, there was little more than 1 g available in the whole world and it was not enough to save his life. Today, we produce enough of the totally purified material to give every human on the planet a dose of 5 g! Development of penicillin production on an industrial scale was a triumph in both scientific and technological terms (Table 19.9).

Penicillin was discovered, apparently accidentally **Fleming**. He identified the contaminating fungus as a species of *Penicillium* and named the unknown inhibitory substance penicillin. Fleming studied the material to some extent during the 1930s but was unable to purify or stabilise it. He clearly recognised its potential, suggesting that penicillin might have clinical value if it could be produced on a large scale.

Penicillin was later purified by **Howard Florey, Ernst Chain, Norman Heatley** and other members of a team at Oxford University, which is how the antibiotic became available for experimental use at the Radcliffe Infirmary in Oxford in 1940. The Oxford team developed a purification method using the solvent ether which gave good yields of antibiotic from a meat broth medium on which Fleming's fungus had been grown. The result was a brown powder which was a remarkably potent antibacterial agent even though still not fully purified. The essence of the production process was a **surface fermentation** method and the Oxford group used all types of readily available bottles and dishes, but what proved to be the best of these makeshift containers were hospital bedpans! Later, purpose-made ceramic or glass vessels modelled on this utensil were renamed '**penicillin flasks**'.

The British fermentation industry simply lacked the knowledge and expertise to scale-up production of penicillin effectively. But American academic and industrial scientists had experience of growing fungi in deep fermentation submerged culture and were expert in the selection and development of high-yielding strains. Indeed, the crucial American contribution to industrialising penicillin production grew out of the experience of American scientists during the 1930s, like **Selman Waksman** and **Harold Raistrick**, in the use of fungi for industrial production of chemicals, especially citric acid.

Transfer of penicillin development to the United States enabled the marvel of war-time antibiotic production to be achieved. By the end of World War II, penicillin cost less to produce than the packaging in which it was distributed. It made an enormous contribution to the war effort, too. During World War I, 15% of battle casualties died of infected wounds. When penicillin became widely available in the latter half of World War II, recovery rates from non-mortal wounds were routinely 94–100%;

Table 19.9. Strain development by mutagenesis and strain selection

Species	Strain number and origin	Concentration of penicillin in culture fluid (mM)
Penicillium notatum	Fleming's original strain	0.003–0.006
	Sub-strain isolated from the above	0.035
Penicillium chrysogenum	NRRL* 1951 (new species isolated)	Not recorded
	NRRL 1951.B25	0.25
	NRRL 1951.X-1612	0.85
	NRRL 1951 Wis Q-176 (ancestor of most present-day penicillin producing strains)	2.5
Penicillium chrysogenum	P1 (Panlabs Biologics Inc. San Diego, USA; derived from a strain provided by Toyo Jozo Co. Tokyo, Japan)	22.9
	P8 (Panlabs Biologics Inc. San Diego, USA; derived from a strain provided by Toyo Jozo Co. Tokyo, Japan)	50.0

*NRRL = Northern Regional Research Laboratory (now the National Center For Agricultural Utilization Research, the largest of the US Department of Agriculture Agricultural Research Service research centres) at which the first US national fungal culture collection was opened in 1940.

death from infection of the wound was almost zero. The **Nobel Prize in Physiology or Medicine 1945** was awarded jointly to Sir Alexander Fleming, Ernst Boris Chain and Sir Howard Walter Florey 'for the discovery of penicillin and its curative effect in various infectious diseases' (https://www.nobelprize.org/nobel_prizes/medicine/laureates/1945/).

The microorganism originally chosen for this production process was *Penicillium chrysogenum* (rather than the *Penicillium notatum* that Fleming originally isolated). This choice was made because of improved product accumulation in the culture medium and subsequent cycles of mutation and selection considerably improved penicillin yield (Table 19.9).

Penicillin is a modified **tripeptide** made from aminoadipic acid, cysteine and valine (Figure 19.29). The principle enzyme involved in penicillin biosynthesis is ACV-synthetase, which forms the tripeptide ACV (δ(L-α-aminoadipyl)-L-cysteinyl-D-valine) from the three constituent amino acids:

- α-aminoadipate is an intermediate formed during lysine synthesis, it does not accumulate in exponential phase hyphal cells and is not found in proteins;
- cysteine and valine are both found in exponential phase cells and both commonly occur in polypeptides.

The second step in penicillin biosynthesis is the formation of the β-lactam ring (as penicillin-N) from ACV by the enzyme isopenicillin-N synthetase. Oxygen is involved in the ring closure process but is not incorporated into the molecule. Finally, side chains are added or exchanged by acetyl-CoA: isopenicillin-N acyltransferase in a two-step process:

- release of the isopenicillin-N side chain to yield α-aminoadipate and 6-aminopenicillanic acid (6-APA); then
- addition of a new side chain and release of coenzyme A;
- by providing a supply of one carboxylic acid, the fungus can be persuaded to produce only one type of molecule featuring the supplied carboxylic acid in its side chain. For example, in the naturally occurring penicillin G the side chain is phenoxyacetate; in penicillin-V the side chain is phenylacetate. **Semisynthetic penicillins** are produced by chemical modification of 6-aminopenicillanic acid (6-APA) (Figure 19.30).

The genes coding for enzymes involved in penicillin biosynthesis are clustered together (Figure 19.31) and there is a high degree of homology for these genes in all β-lactam producers: gene pcbAB codes for ACV-synthetase, pcbC for isopenicillin-N synthetase, and penDE for acetyl-CoA: isopenicillin-N acyltransferase. This is known as a **metabolic gene cluster (MGC)**. MGCs encode various functions in fungi; nutrient acquisition, synthesis and/or degradation of metabolites, etc. As well as encoding the enzymes that perform these anabolic or catabolic processes, MGCs may include integrated regulatory sequences and any mechanisms needed to protect their fungal resident from any toxins they may produce (Slot *et al.*, 2017).

This modular, self-contained nature of MGCs contributes to the metabolic and ecological adaptability of fungi and is reminiscent of the **developmental subroutines** we discussed in section 13.14. Being in a cluster enables **pathway amplification by gene duplication**. This seems to be the basis of the

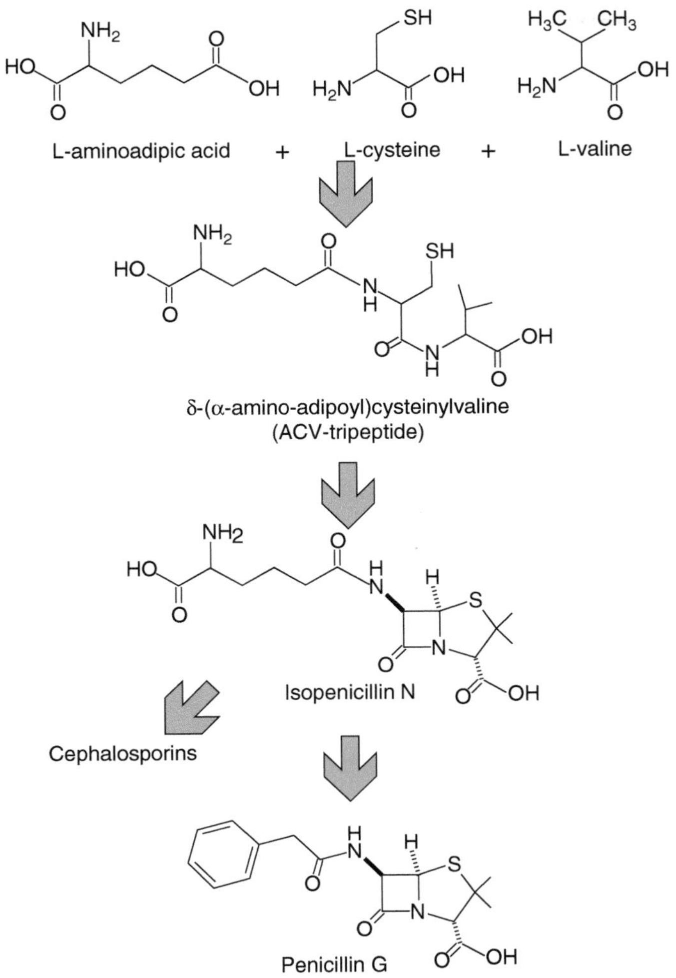

Figure 19.29. Penicillin biosynthesis. Penicillin is a modified tripeptide made from aminoadipic acid, cysteine and valine. The principle enzyme involved in penicillin biosynthesis is ACV-synthetase, which forms the tripeptide ACV (δ-(L-α-aminoadipyl)-L-cysteinyl-D-valine) from the three constituent amino acids. The second step in penicillin biosynthesis is the formation of the β-lactam ring (as penicillin-N) from ACV by the enzyme isopenicillin-N synthetase; oxygen is involved in the ring closure process but is not incorporated into the molecule. Finally, side chains are added or exchanged by acetyl-CoA: isopenicillin-N acyltransferase.

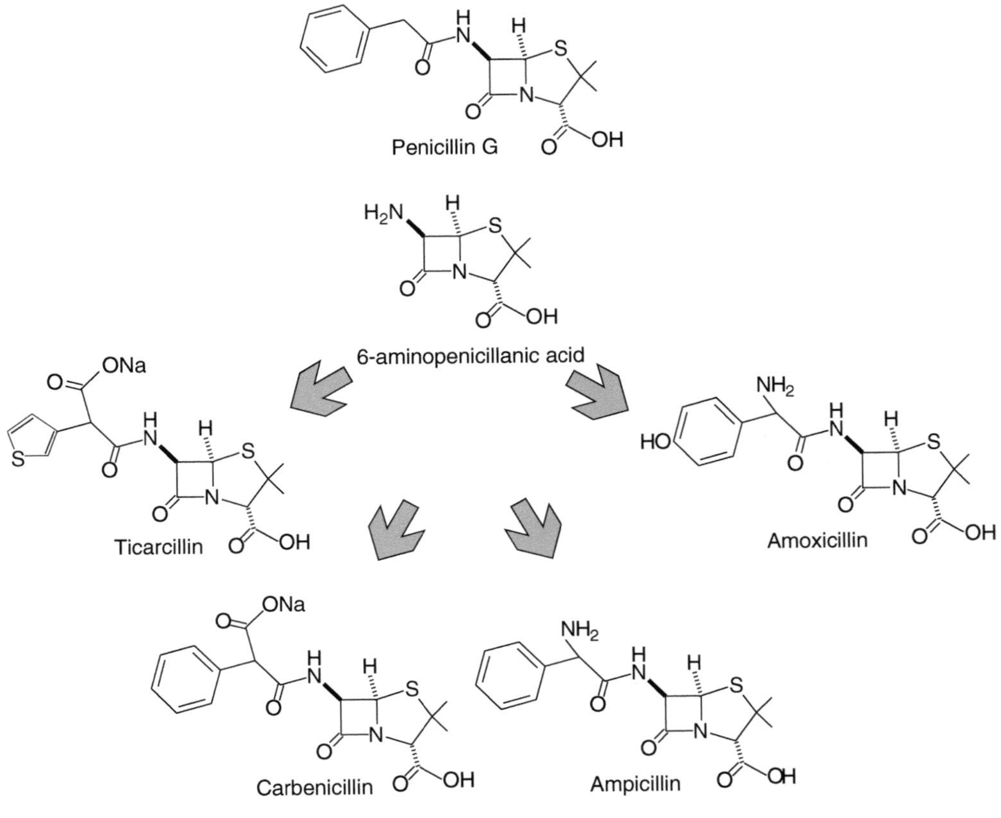

Figure 19.30. Semisynthetic penicillins are produced by chemical modification of 6-aminopenicillanic acid (6-APA).

improvements in yield illustrated in Table 19.9. There are 40–50 times more copies of the genes in present-day industrial strains than in the original *Penicillium chrysogenum* NRRL 1951 strain. It is no surprise that current industrial strains have greatly increased transcription efficiency of genes involved in penicillin synthesis.

The bacterial *Streptomyces* species also produce β-lactam antibiotics and the high degree of homology between ACV-synthetase and isopenicillin-N synthetase produced by *Streptomyces* and those of filamentous fungi suggests that genes involved in penicillin biosynthesis may have originated in bacteria. It is possible that the genes for the penicillin MGC, having first evolved in bacteria, were subsequently transferred to filamentous fungi by a **horizontal gene transfer** event, which phylogenetic analysis dates to about 370 million years ago. Evidence has accumulated suggesting that, through horizontal gene transfer, fungi have been adept at acquiring useful genes from other organisms, particularly bacteria. Horizontal gene transfer is the transmission of genetic material between organisms and across major taxonomic boundaries; that is, from between different species, and even between different domains. The species concept is, of course, difficult to apply to many fungi (see section 3.9), so it's probably best to think of horizontal gene transfer as transmission of genes between reproductively isolated genomes. Note that a consequence of this interpretation is the notion that: 'fungal species are unique evolutionary units that are separated from one another by boundaries that are "porous" under certain conditions ' and this 'fundamentally affects the population genetics of a species, with potentially profound effects on its overall evolution and biology' (Steenkamp *et al.*, 2018). There is evidence for ***many*** horizontal gene transfer events in fungi, that have enhanced both primary and secondary metabolism of ***many*** fungi by expanding their sugar, nitrogen, amino acid, nucleobase and macromolecule metabolism. These transfers have also added significantly to the secretory and metabolite-transporter arrays of fungi, implying that gene transfer has supplemented the basic lifestyle of the fungi concerned (Richards *et al.*, 2011; Slot *et al.*, 2017; see section 6.5).

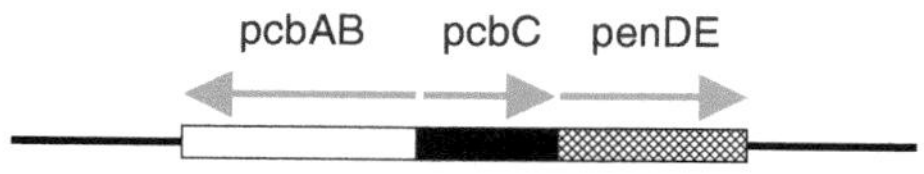

Figure 19.31. Diagram of the cluster of genes that specify enzymes involved in penicillin biosynthesis and their directions of transcription. Gene pcbAB codes for ACV-synthetase, pcbC for isopenicillin-N synthetase, and penDE for acetyl-CoA: isopenicillin-N acyltransferase.

Although original industrial production of penicillin made use of surface batch cultures, most current production uses **fed-batch submerged fermentation** (Figure 19.32). The inoculum for the production stage is amplified through several intermediate seed stages. In general, each of these inoculum development stages provides a 10% (on a volumetric basis) inoculum for the next stage (top panel in Figure 19.32). The production medium may contain glucose, lactose or corn steep liquor (a farm waste product from the wet milling of maize) as nutrient source together with inorganic salts. Addition of penicillin precursors (for example, phenylacetic acid) improves yield but these are toxic, and this is the reason for preference for fed-batch fermentation (because the potentially toxic metabolite can be fed at a concentration below the toxic level).

In the production stage a rich medium supports growth for up to 24 h. Nutrient (glucose) becomes depleted and subsequent growth rate is controlled by addition of glucose, the substrate being carefully monitored to ensure it does not reach a concentration that represses penicillin biosynthesis. As the fermentation continues, other nutrients (nitrogen source, sulfur source, penicillin precursor) are depleted (Figure 19.33) and additional supplies are provided.

Typical production costs as a percentage of total, are: nutrients, 25%; labour, maintenance, steam, electricity, treatment of waste, etc., 25%; fixed costs, depreciation, taxes, insurances, 28%; separation and purification, 22%. Although penicillin can be produced by chemical synthesis, it is much cheaper to produce by fermentation. A 100 m^3 fermenter will yield 2.8 metric tonnes of sodium penicillin G in 7.5 days, which will have a

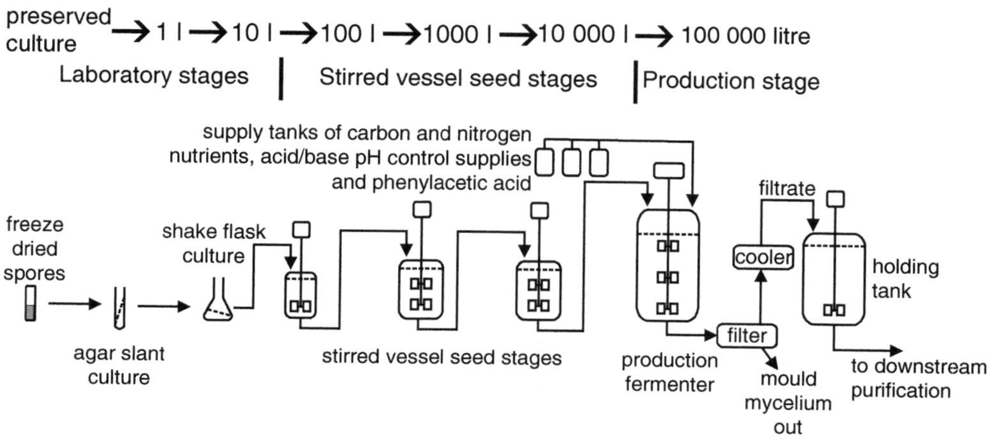

Figure 19.32. Outline diagram of current penicillin production by fed-batch submerged fermentation. Top panel shows the procedure for inoculum development.

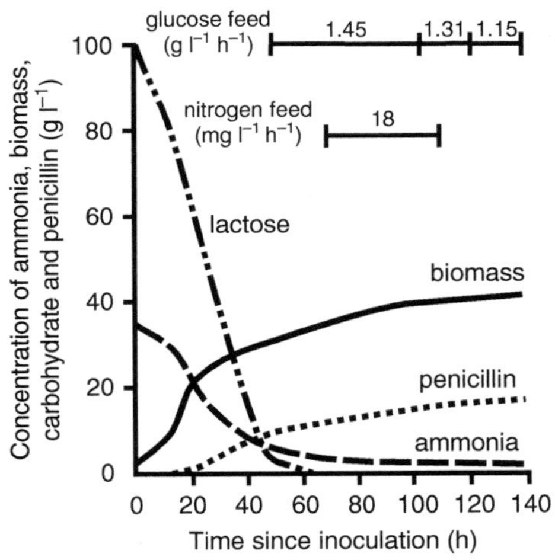

Figure 19.33. Characteristics of a typical penicillin production run.

'factory-gate-value' of about US$70,000. Because penicillin is out of patent it can be produced anywhere in the world and its price varies depending on the number of companies that decide to produce it; but it is always cheap. There are many companies manufacturing antibiotics these days and there are many antibiotics present in the market place. The global antibiotics market was valued at US$39.6 billion in 2013 and penicillin represents about 8% of that total (for comparison, tetracyclines make up about 4% of the overall market, erythromycin 7%, streptomycin 1% and chloramphenicol 1% of the market). It may be useful here to compare the US$39.6 billion market for all antibiotics to the US$1 trillion market for all statins (section 19.12); a comparison that starkly illustrates the reason why 'big pharma' is reluctant to invest in the search for new antibiotics, preferring to leave stimulation of research in this area to national governments.

Arachidonic acid is another important medicine. It is a useful illustration of the use of zygomycete fungi for fermentative production. Arachidonic acid is one of the essential fatty acids required by most mammals. It is a carboxylic acid with a 20-carbon chain and four *cis* double bonds; the first double bond being located at the sixth carbon from the omega end (chemical name: 5,8,11,14-*cis*-eicosatetraenoic acid), and is a major constituent of the membranes of many animal cells, contributing to their fluidity.

Arachidonic acid has a wide variety of physiological functions in animals and, consequently, is used in many ways in medicine, pharmacology, cosmetics, the food industry and agriculture because in metabolism it is a precursor of:

- Prostaglandins, which modulate immune function via the lymphocyte. They are mediators of the vascular phases of inflammation and are potent vasodilators. They increase vascular permeability. Prostacyclin is a vasodilator hormone that works with thromboxane in a homeostasic mechanism in relation to vascular damage.

Table 19.10 Content of lipid and of arachidonic acid in *Mortierella* spp.

Fungus	Fungus lipid as % of dry wt	Arachidonic acid as % of lipid
Mortierella alpina	35 to 40	50 to 65
Mortierella elongata	44	31

- Thromboxane, a powerful vasoconstrictor, which also increases platelet aggregation.
- Leukotrienes, which mediate inflammation, causing vasoconstriction but increased microvascular permeability.

The physiological pathways to which arachidonic acid contributes in the whole animal are major mechanisms for production of pain and inflammation, and control of homeostatic function.

Some mammals lack the ability to convert linoleic acid into arachidonic acid or have a very limited capacity to synthesise arachidonic acid, which makes arachidonic acid an essential part of their diet. Plants contain little or no arachidonic acid so these animals are **obligate carnivores**; the cat family is an example. Humans get their arachidonic acid from both dietary animal sources (meat, eggs and dairy products) or by synthesis from linoleic acid, but dietary supplementation has significant positive effects. Animal liver, fish oil and egg yolk are well-known sources of arachidonic acid but zygomycete fungi of the genus *Mortierella* are also prominent sources (Table 19.10) and production levels can be modified by mutations in genes contributing to fatty acid synthesis in *Mortierella alpina* (Sakuradani *et al.*, 2004).

Conventional fermentation with *Mortierella* is a promising production method for easily-purified arachidonic acid, particularly in a **two-step fed-batch fermentation** process in which impeller speed was reduced after 5 days cultivation from 180 to 40 rpm and aeration rate adjusted from 0.6 to 1.0 fermenter volumes per minute. These adjustments decrease physical damage to the mycelia and extend the stationary phase, which is when most fatty acids synthesis occurs. Feeding the batch culture with 3% (w/v) and 2% (w/v) ethanol after 5 and 7 days cultivation, respectively, enhances production further (Jin *et al.*, 2008).

Arachidonic acid has useful clinical effects in lowering cholesterol and triacylglycerols in plasma, with good effects on **arteriosclerosis** and other **cardiovascular diseases**. It is present in breast milk and is important in brain and retina development in new-born infants and for these reasons is added to **infant feeding** 'formula milk'. Studies have demonstrated that feeding formulae supplemented with arachidonic acid resulted in enhanced growth of infants and provided better developmental outcomes than unsupplemented formulae (Clandinin *et al.*, 2005; Hadley *et al.*, 2016).

19.16 ENZYMES FOR FABRIC CONDITIONING AND PROCESSING, AND FOOD PROCESSING

We will describe the use of fungal enzymes in the dairy industry in section 19.24, but there are many other aspects of both **clothing** and **food manufacture** that depend on enzymes produced by fungal cultures (see Figure 19.27, above). The global industrial enzymes market is projected to reach a value of US$6.30 billion by 2022; proteases accounted for the largest market share due to their wide range of applications in food and beverage, detergent, and biofuel industries. The food and beverages sector had the greatest value overall in 2016, with extensive requirements for carbohydrases in food processing, brewing, baking, and biofuel manufacturing.

The textiles industry is also very large and lucrative, and even with the successful emergence of synthetic fibres such as nylon and polyester, a great many of our fabrics are still made from natural fibres like **cotton** (cellulose) fibres and **wool** and **silk** (protein) fibres which can be processed in various ways with natural enzymes (Vigneswaran *et al.*, 2014).

The fabric processing with which we are individually most familiar is that resulting from the inclusion of fungal enzymes in **detergents** used for clothes washing. Enzymes included in biological detergents must be active under alkaline conditions: they include **proteases, lipases, amylases** (used to remove protein, grease and sugar soiling) and **cellulases** (for colour brightening and fibre restoration). Some of these enzymes are derived from bacterial sources, but many are fungal enzymes.

Enzymes are also used during manufacture of textile products. Emerging areas of development include:

- Enzymatic preparation of natural fibres; **biopreparation** allows the removal of impurities prior to dyeing and saves water, natural fibres can also be modified to produce value-added products (for example the addition of silk peptides to wool), and surface modification of wool fibres to improve the feel of the final fabric. Hydrolytic enzymes including pectinases, cellulases, proteases and xylanases have been applied to cotton processing, resulting in fabrics with good absorbency that feel soft to the touch.
- Enzymatic **bioprocessing of synthetic fibres**, increasing functionality and adding value of polyester, polyamide or acrylic fibres.
- Remediation of textile effluents; **bioremediation** and effluent processing are important aspects of fabric processing, and enzymes can help here, too, for example laccases decolourise waste dyes.

Cellulase treatment of cotton fabrics is an environmentally friendly way of improving the property of the fabrics. Although cellulases were introduced into textile manufacture only in the 1990s, they are now **one of the most frequently used groups of enzymes** for these applications. Cellulases have proved so popular because of their ability to modify cellulosic fibres in a controlled manner that improves the quality of the fabrics (Mojsov, 2014; Shahid *et al.*, 2016).

When **biofinishing cotton fabrics** with cellulases, loose fibres are removed more efficiently and more gently than mechanical methods can achieve. An application is in the preparation of '**stone-washed**' denim by a process known as **biostoning**. The textile industry is, of course, a fashion industry and for some years it has been fashionable for denim clothes to have a worn, faded, look and be soft to the touch. Traditionally, manufacturers of denim jeans have washed their garments with pumice stones to achieve these desirable features, hence stone-washed jeans.

The disadvantages of using stone for washing denim garments included damage to equipment by tumbling stones, difficulty of removing residual pumice from processed clothing, and particulate material clogging drainage channels. The still-fashionable aged look is obtained by combining **cellulase enzyme treatment** with some form of mechanical distressing such as beating or friction. Using cellulase immobilised on a pumice carrier has advantages and can both reduce process costs and raise the activity of the enzyme (Pazarlioğlu *et al.*, 2005; Soares *et al.*, 2011; Vigneswaran *et al.*, 2014).

Many of the treatment and finishing processes in the **wool industry** require chemicals that are environmentally unfriendly, particularly treatments aimed at reducing shrinkage of wool garments. Traditional treatments require strong acidic chlorine and result in polluting wastes. This generates interest in the use of enzymes to achieve the same effects on wool. A range of **proteolytic enzymes** are the candidates, but they penetrate and degrade the internal wool structure during processing, causing unwelcome fibre damage. Natural enzymes can be modified or immobilised to control fibre degradation by enzymes, pointing the way to successful development of enzymatic treatments for achieving a variety of finishing effects for wool-containing products (Shen *et al.*, 2007; Soares *et al.*, 2011; Vigneswaran *et al.*, 2014).

Enzymes are also used extensively in **food biotechnology** (Figure 19.27). Several examples feature elsewhere in this chapter. As an example, we note that the production of **apple juice** is the largest in the juice industry after grapes, and in order to maximise the yield of extracted juice, enzymes are incorporated into the procedure. This is a common feature among a wide variety of sectors in the commercial food processing industry (Table 19.11), as they increase the output of product without additional capital investment.

Processing apples utilises **macerating enzymes**, pectinases, cellulases and hemicellulases (Table 19.11), in two separate steps: extraction and clarification. The enzymes are used in the extraction to liquefy the fruit, increasing the yield of juice collected and enhancing release of other important cell ingredients such as antioxidants and vitamins. In the second step the enzymes are used to remove any precipitates that are formed from insoluble

Table 19.11. Cellulases, hemicellulases and pectinases in food biotechnology

Enzyme	Function	Application
Macerating enzymes (pectinases, cellulases and hemicellulases)	Hydrolysis of soluble pectin and cell wall components; decreasing the viscosity and maintaining the texture of juice from fruits	Improvement in pressing and extraction of juice from fruits and oil from olives; releasing flavour, enzymes, proteins, polysaccharides, starch and agar
Acid and thermostable pectinases with polygalacturonase, pectin esterase and pectin transeliminase	Fast drop in the viscosity of berry and stoned fruits with the breakdown of fruit tissues	Improvement in pressing fruit mashes and high colour extraction
Polygalacturonase with high pro-pectinase and low cellulase	Partial hydrolysis of pro-pectin	Production of high viscosity fruit purees
Polygalacturonase and pectin transeliminase with low pectin esterase and hemicellulase	Partial hydrolysis of pro-pectin and hydrolysis of soluble pectin to medium sized fragments; formation and precipitation of acid moieties; removal of hydrocolloids from cellulose fibres	Production of cloudy vegetable juice of low viscosity
Polygalacturonase, pectin transeliminase and hemicellulase	Complete hydrolysis of pectin, branched polysaccharides and mucous substances	Clarification of fruit juices
Pectinase and β-glucosidase	Infusion of pectinase and glucosidase for easy peeling/firming of fruits and vegetables	Alteration of the sensory properties of fruits and vegetables
Arabinoxylan modifying enzymes (endoxylanases, xylan debranching enzymes)	Modification of cereal arabinoxylan and production of arabinoxylo-oligosaccharides	Improvement in the texture, quality and shelf life of bakery products
Cellulases and hemicellulases	Partial or complete hydrolysis of cell wall polysaccharides and substituted celluloses	Improvement in soaking efficiency; homogeneous water absorption by cereals; the nutritive quality of fermented foods; the rehydrability of dried vegetables and soups; the production of oligosaccharides as functional food ingredients and low-calorie food substituents and biomass conversion
β-Glucanases and mannanases	Solubilisation of fungal and bacterial cell wall	Food safety and preservation
Xylanases and endoglucanases	Hydrolysis of arabinoxylan and starch	Separation and isolation of starch and gluten from wheat flour
Pectin esterase with no polygalacturonase and pectin lyase activities	Fruit processing	Production of high-quality tomato ketchup and fruit pulps
Rhamnogalacturonase, rhamnogalacturonan acetyl esterase and galactanase	Cloud stability	Production of cloud stable apple juice (cloudy juices that do not sediment in storage)
Cellulase and pectinase	Release of antioxidants from pomace (the solid remains of grapes, olives, & other fruit after pressing for juice or oil)	Controlling coronary heart disease and atherosclerosis; reducing food spoilage
Endo-mannanase	Modification of guar-gum (a galactomannan polysaccharide used for thickening and stabilising foods)	Production of water-soluble dietary fibres to enrich the fibre content of foods

Table modified from Bhat (2000).

pectin. Because the enzymes are mixed with the juice, the two need to be separated before packaging and distribution. This is usually achieved by adding gelatine as a **fining agent** which causes the enzymes to clump together in '**flocs**' which can then be removed by filtration. One problem with this process is that the enzymes are used only once. **Immobilisation** of enzymes on polysaccharide gels, nylon filter membrane or particulate carriers enables them to be recycled (Sandri *et al.*, 2011).

The microbial enzymes used for industrial processes are largely produced by liquid fermentation, but solid-state fermentation has great potential for the production of enzymes (Pandey, 2003). For example, enzyme preparations from *Phanerochaete chrysosporium, Aspergillus oryzae, Aspergillus giganteus* and *Trichoderma virens*, produced by solid-state fermentation on cotton seed-coat waste (which serves as both substrate and enzyme inducer) produced enzymes that could be used effectively to degrade impurities in cotton fabrics during biopreparation (Csiszár *et al.*, 2007; Vigneswaran *et al.*, 2014).

The search continues for new enzymes from other organisms by screening microorganisms sampled from new environments or developed by modification of known enzymes using protein engineering or recombinant DNA technology (Olempska-Beer *et al.*, 2006; Anbu *et al.*, 2017).

19.17 STEROIDS AND USE OF FUNGI TO MAKE CHEMICAL TRANSFORMATIONS

Steroid compounds are among the most widely marketed products of the pharmaceutical industry. Today **manufactured steroids** are used for a very wide range of 'over-the-counter' remedies as well as ethical (prescription-only) therapies. Common functions include anti-inflammatory, immunosuppressive, diuretic, anabolic, contraceptive and progesterone analogues. Manufactured steroids are also used in treatments of some forms of cancer, osteoporosis and adrenal insufficiencies, for avoidance of coronary heart disease, as antifungal agents or anti-obesity agents, and, by inhibition of HIV integrase, prevention and treatment of infection by HIV and AIDS. The isolation from natural sources of new steroids and sterols with potential therapeutic applications is an active field of current research (Tong & Dong, 2009; Donova & Egorova, 2012; Meyer *et al.*, 2016).

Research efforts in this topic were prompted in 1949, with the discovery at the Mayo Clinic of the dramatic effect of **cortisone**, an endogenous steroid, in alleviating the symptoms of rheumatoid arthritis. However, steroid molecules are complex and chemical synthesis is difficult; even carrying out the highly specific modification reactions required to produce clinically useful compounds with commercial value is difficult and costly. Consequently, the production of steroid drugs and hormones and growth of the industry in the latter half of the twentieth century is one of the best examples of the successful application of **microbial technology** in large-scale industrial processes.

Soon after the Mayo discovery, the Upjohn Company announced that cultures of the fungus *Rhizopus* are able to introduce (enzymically) a hydroxyl group to a very specific position in the female hormone **progesterone** (namely, 11α-hydroxylation of progesterone), which had been synthesised from the soy bean sterol **stigmasterol** (Hogg, 1992).

This discovery provided a one-step solution to a chemical modification that requires a multistep procedure to achieve by conventional organic chemistry. Indeed, the total synthesis of steroids has only recently been reported (Honma & Nakada, 2007) as an approximately **20-step process** ('approximately' because it depends what you start with and what you want to end with). It is interesting to compare this publication with the equivalent papers of the mid-twentieth century that described the chemical steps needed to complete the transformation by purely chemical means; but we'll leave you to do that (Chamberlin *et al.*, 1951; Peterson & Murray, 1952).

Microbial reactions for the transformation of steroids have proliferated since then, and specific microbial transformation steps have been included in many large-scale production syntheses of new drug and hormone steroids. Modified steroids are favoured over their natural counterparts because of increased potency, longer half-lives in the blood stream, simpler delivery methods and reduced side effects.

The preference for using cultures of intact whole cells over enzymes as the biocatalysts for production of such pharmaceutical derivatives is mostly a matter of the additional cost burden of enzyme isolation, purification and stabilisation.

A wide range of very different microorganisms can be used as **biocatalysts for steroid transformations**:

- various species of **bacteria** (*Pseudomonas, Comamonas, Bacillus, Brevibacterium, Mycobacterium, Streptomyces*, etc.),
- **filamentous fungi** (*Phycomyces, Mucor, Rhizopus, Aspergillus, Penicillium*, etc.),
- **yeasts** (*Hansenula, Pichia, Saccharomyces*, etc.),
- **algae** (*Chlorella, Chlorococcum*, etc.),
- **protozoa** (*Pentatrichomonas, Trichomonas*) (Donova *et al.*, 2005).

The search for novel steroid-like compounds with potential therapeutic applications is extending into wider ranges of natural sources, including **endophytic fungi, corals** and **sponges** (Fernandes *et al.*, 2003).

19.18 THE QUORN® FERMENTATION AND EVOLUTION IN FERMENTERS

In the late 1950s, forecasters predicted a worldwide shortage of protein-rich foods within 30 years (that is, by the 1980s); and it was hoped that **single cell protein (SCP)**, particularly microbes, could provide a means of solving the anticipated world food shortage by industrial production of cheap **protein alternatives to meat** protein, potentially using wastes as substrates.

Fungal SCP must compete with established animal feeds like soya meal, which is a product of conventional farming, so a novel microbial protein that requires costly research and development and/or unusual and expensive production facilities will not be able to compete. The basic requirements for a protein produced for human consumption, as opposed to a feed material for farm animals, are that it should be:

- cheap to manufacture;
- able to be put into large-scale production;
- have a **high protein content**, including essential amino acids.

In 1964, the Rank Hovis McDougall (RHM) Research Centre set out to develop a way of converting starch, which is a waste product from cereal processing, into a protein-rich food. The researchers decided to produce the new food from a filamentous fungus because:

- There is a long history of humans using fungi as food.
- It is relatively easy to harvest fungal mycelia from culture broths.
- It is possible to formulate food products from filamentous fungi which have the appropriate smell, taste and texture, that is the sensory or **organoleptic properties**, of an acceptable food. The market virtues of the material centre on its filamentous structure which enables it to simulate the fibrous nature of meat. Coupled with the inherent nutritional value of fungal biomass, this permits the product to be sold as a low-fat, low-calorie, cholesterol-free health food. Initially, it was deliberately not compared to meat in any way, but sales increased dramatically when marketed as a healthy meat substitute (Table 19.12) which can be efficiently produced (Table 19.13).

Following an extensive screening programme, a strain of *Fusarium venenatum* A3/5, then known as *Fusarium graminearum* A3/5, was selected for evaluation (Figure 19.34). **Mycoprotein** is the term coined by the UK Foods Standards Committee to serve as the generic name for a food product resulting from the fermentation of *Fusarium venenatum*. The material was produced by Marlow Foods Limited, set up in 1984 as a joint venture between RHM, who had developed the product and Imperial Chemical Industries (ICI) who had free fermenter capacity.

A 10-year study (1970–1980) of the product included toxicology testing, involving feeding trials on 11 animal species (including pigs, calves, baboons) that showed no adverse effect on animals or offspring; trials with 2,500 human volunteers, which also showed no ill effects or immunological response; and demonstration that the bacteriological load was like chicken or fish in storage. Eventually a 2 million word, 26-volume report was submitted to the then Ministry of Agriculture, Fisheries and Food for approval, which was granted, and the product was first sold to the public in January 1985. Additional toxicology testing of Quorn® by AstraZeneca was completed in December 1996 for submission to the United States' Food and Drug Administration;

Table 19.12. Comparison of Quorn® mycoprotein with beef

Nutritional content	Braising beef % (w/w)	Quorn® % (w/w)
Protein	30.9	12.2
Dietary fibre	0	5.1
Cholesterol	0.08	0
Fat, total	11.0	2.9
Polyunsaturated/saturated fats ratio	0.1	2.5

Table 19.13. Comparison of protein yields per kg glucose used in production

Producer organism	Protein production (g kg^{-1})
Cattle	14
Pigs	41
Chickens	49
Fusarium venenatum	136

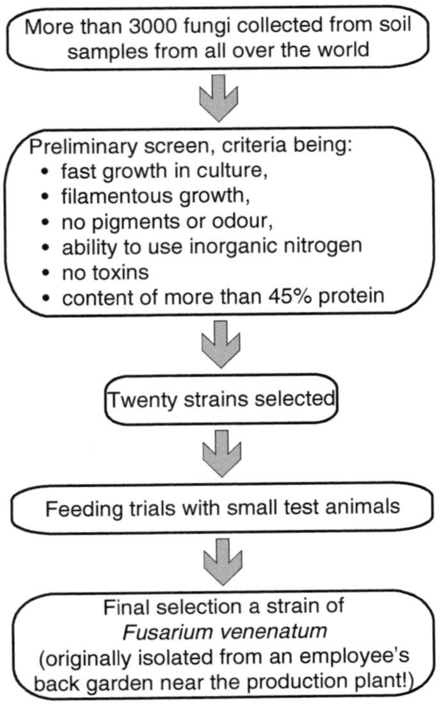

Figure 19.34. Flowchart of the screening programme that was used to find the fungus that eventually became known as the Quorn® fungus.

FDA approval being needed before Quorn® could be sold in the United States. Frozen Quorn® was sold for the first time in the United States in August 2002. No food eaten by humans has been subjected to a more rigorous **safety testing** than Quorn®!

Although mycoprotein was originally conceived as a protein-rich food to supplement what was thought to be a declining world supply of conventional foods, by the early 1980s the predicted global shortage of protein-rich foods had not materialised. Consequently, Marlow Foods decided to sell **Quorn®** mycoprotein as a new healthy food which **lacked animal fats and cholesterol**, is low in calories and saturated fats, and high in dietary fibre (it has more dietary fibre – fungal cell walls – than wholemeal bread). This dramatic change in marketing policy was justified by a survey in 1989, which showed that almost half the UK population was reducing its intake of red meats, while a fifth of young people were vegetarians.

Subsequently, it was decided to market mycoprotein as a **meat analogue**. Today, Quorn® mycoprotein is sold in burgers, sausages and analogues of sliced meats (see Figure 12.12), and as an ingredient in over 50 pre-prepared meals, and in an uncooked, unflavoured form (as chunks, strips or minced) which is used in the home as an ingredient in a wide range of cooked meals. We have discussed the food value of Quorn® in section 12.4; here we will concentrate on the biotechnology of its production (Wiebe, 2004).

All the production systems described earlier were considered:

- Batch and fed-batch because they have the advantages of being:
  - a simple system;
  - commonly used in industrial fermentations;
  - used for manufacture of 'high market value' products.

  Though they have the disadvantage of being too expensive when the product is the biomass, because a new fermentation needs to be set up every few days.
- Continuous flow culture have the advantage of providing:
  - continuous production of biomass;
  - over long periods of time,
  - consequently, comparatively cheap.

  The main disadvantages were that the process is technically very difficult and had not previously been implemented with filamentous fungi on an industrial scale.

A **continuous flow culture system** was chosen for production of *Fusarium venenatum* A3/5 biomass because much higher productivities can be achieved in continuous culture than in batch culture. Up to the beginning of 1994, the airlift fermenter used for mycoprotein production was a fermenter originally built by ICI at Billingham to grow the bacterium, *Methylophilus methylotrophus* for production of an animal feed (Pruteen). This 40 m^3 fermenter (christened Quorn 1), consisted of an elongated loop about 30 m tall and was operated as a glucose-stat at a dilution rate of 0.17–0.20 h^{-1} capable of producing 1,000 tonnes of Quorn® mycoprotein per annum.

In late 1993, Marlow Foods commissioned a new 155 m^3 airlift fermenter (Quorn 2 built on the site of the old Pruteen fermenter) and this was followed by the construction of Quorn 3, the twin of Quorn 2. This has allowed mycoprotein production to be increased to between 10,000 to 14,000 tonnes per annum. The new fermenters are the world's **largest continuous flow culture systems**; each cost £37.5 million to build, is 50 m tall and weighs over 250 tonnes. By 1997, the two new fermenters had enabled sales of Quorn® mycoprotein to be increased to about £74 M per annum.

Each 155 m^3 fermenter is inoculated with 5 l of a batch culture containing 50 g biomass and continuous flow is initiated 4 days later. In an airlift fermenter most of the oxygen transfer takes place near the base of the relatively wide **riser** where sterile air is introduced and where the height of the fermenter creates a high hydrostatic pressure (Figure 19.35). This, together with turbulence (and consequently small bubble size) provides excellent conditions for oxygen transfer from the gaseous to the liquid phase. Thus, the riser contains a two-phase mixture of air and culture flowing together so that air bubbles contribute up to 50% of the volume of the fluid. The rate of transfer of oxygen from the gaseous to the liquid phase decreases as the culture flows to the top of the riser where the gas contains about 10% oxygen.

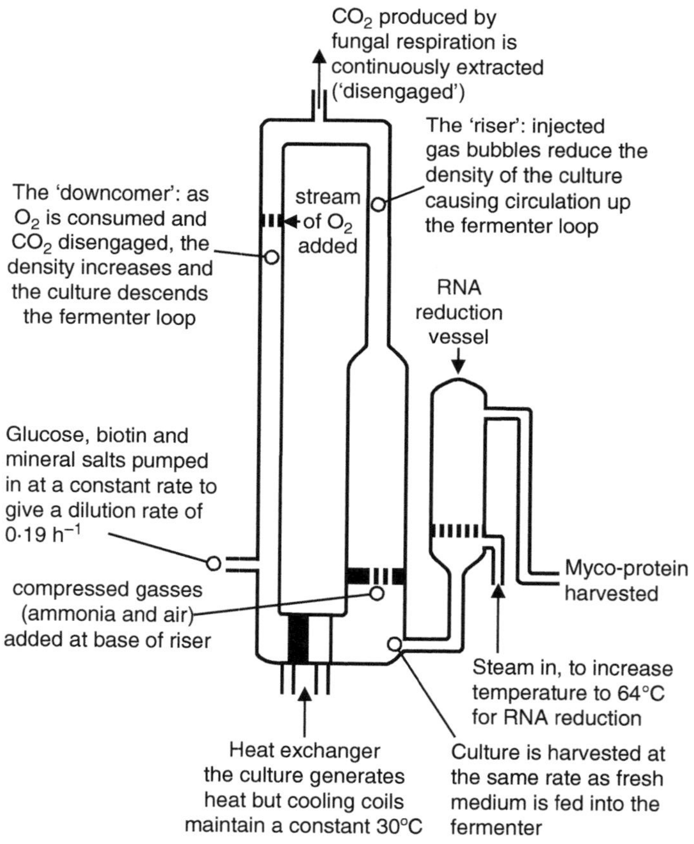

Figure 19.35. Schematic representation of the Quorn® airlift fermenter used for production of mycoprotein in continuous flow culture.

The low-pressure region at the top of the riser causes release of CO_2 and the culture then enters the **downcomer** where, at the bottom, it is directed into the riser and is again charged with air thereby completing the pressure cycle. The culture is maintained at about 30°C by a heat exchanger set into the downcomer. To prevent oxygen-limitation, the downcomer is provided with an oxygen supply. The difference in specific gravity of the relatively aerated culture in the riser and the relatively air-depleted culture in the downcomer (creating a **hydrostatic pressure differential**) ensures that the growing hyphal filaments circulate continuously around the fermenter loop at a rate of about one circulation every 2 minutes for Quorn 2 and 3, compared to every 6 minutes for Quorn 1.

The **nitrogen supply** for growth (which is **ammonia**) is fed into the fermenter with the sterile compressed air, at the base of the riser. The rate of supply of ammonia to the culture is regulated by a pH monitor set to give a culture pH of 6.0. The nutrient solution is fed to the culture to give a dilution rate in the range 0.17–0.20 h^{-1}, and is operated as a glucose-stat, i.e. glucose is always in excess and the fungus always grows at μ_{max} at a biomass concentration of 10 to 15 g l^{-1}.

The fermenter is manned on a 24 h basis. Tests for the presence of mycotoxins are made every 24 h using a method sensitive to two parts per billion; tests made to date have all been negative, as would be expected for the growth conditions employed (the fungus is growing at μ_{max} whereas toxins are secondary metabolites produced by slow-growing or stationary cultures). Product is harvested continuously, and Quorn 2 and 3 each produce an output of 30 tonnes h^{-1}.

The harvested mycelium cannot be used as harvested. If human food contains too much nucleic acid, blood **uric acid** values rise, and the excess accumulates as crystalline deposits in joints and tissues, leading to the disease known as gout, and to kidney stones. This is a peculiarity of human metabolism: in humans, uric acid is produced by the breakdown of nucleic acids whereas in other vertebrates the sparingly soluble uric acid is converted to the highly soluble acid allantoin.

Because of this, the World Health Organization (WHO) laid down a recommendation for human ingestion of RNA from single cell protein (SCP) sources which, for adults, was defined as 2 g RNA per day, with total nucleic acid ingestion from all sources not exceeding 4 g per day. *Fusarium venenatum* biomass cultured at a specific growth rate of 0.19 h^{-1} has a RNA content of 8–9% (w/w) which would limit the ingestion of mycoprotein to not more than 20 g per day. Consequently, a method was developed to **reduce the RNA content of mycoprotein**, while minimising loss of protein and fibrous structure (though, unfortunately, protein loss is still substantial).

In this process, the temperature of the biomass is raised to 68°C for 20–30 minutes to stop growth, disrupt ribosomes, and activate endogenous RNAases which break down cellular RNA to nucleotides which diffuse through the hyphal wall into the culture broth. Importantly, RNAases are more heat resistant than proteases, so **protein loss is minimised**. This method is carried out in the culture broth with no other adjustments, since a pH of 5–6 is optimal for RNAase activity. RNA reduction has a substantial economic penalty attached to it, since as well as removing RNA, other cell constituents are inevitably lost during this process, including perhaps up to 30% of the biomass dry weight and a significant amount of protein, but it is essential to reduce the RNA content. After this treatment, the mycoprotein contains only 1% (w/w) RNA, like that present in animal liver and well within the 2% (w/w) upper limit recommended by WHO.

One of the advantages of using a filamentous fungus rather than a bacterium or yeast for SCP is the comparative ease with which mycelia can be harvested. After RNA reduction, fungal biomass is **harvested by centrifugation** to give a product which contains about 30% (w/w) total solids. Mycoprotein typically contains 44% protein, 18% dietary fibre and only 13% fat (the values for beef are 68%, 0% and 30%, respectively).

Rank Hovis McDougall developed a mechanical process in which the filaments of mycoprotein are aligned so that the required fibrous structure is attained. At this stage, other ingredients may be added to impart colour and flavour to the fungal biomass (for use in pre-prepared meals). Finally, small amounts of egg white are added to the Quorn® mycoprotein as a protein binder which is then 'heat set' to stabilise the alignment of the filaments. Because of this processing mycoprotein has the same 'chewiness' and succulence as meat, but, unlike meat, it retains colourings and flavours even when cooked. Finally, the product is reduced in size and frozen for short- or long-term storage. Quality control checks are carried out at every stage of the process to ensure that the product is of a consistently high quality.

Although *Fusarium venenatum* fermentations were originally intended to run indefinitely, in practice, they are terminated 1,000 h (= 6 weeks) or less after the onset of continuous flow. Although the shear forces experienced by mycelia in stirred tank (the RHM method) and airlift (the Marlow Foods method) fermenters differ, the morphology and growth of mycelia of *Fusarium venenatum* in the two types of fermenter are similar, and highly branched (**colonial**) mutants (Figure 19.36) arise in both systems at approximately the same time after inoculation.

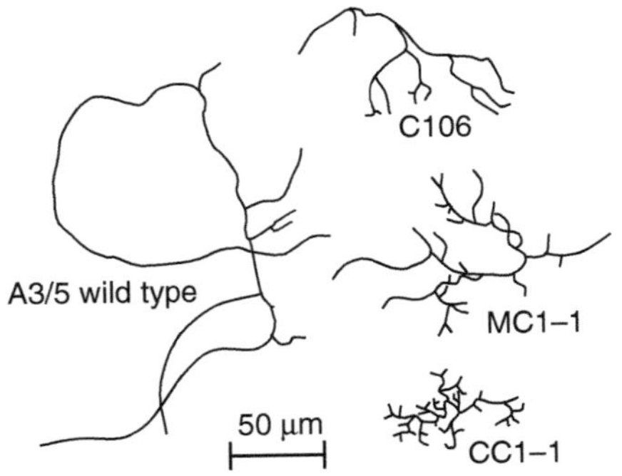

Figure 19.36. Comparison of the morphology of mycelia of *Fusarium venenatum* A3/5 and three highly branched (colonial) mutants (C106, CC1-1 and MC1-1) that arose spontaneously in experimental laboratory chemostat cultures.

Biomass made up of colonial mutants possesses the same chemical composition and is as nutritious as the parental strain, but presence of colonials changes the **texture** of mycoprotein causing it to become more friable, so the filamentous nature of the product cannot be maintained. The culture then needs to be ended and a new one started; appearance of colonials causes premature termination of the fermentations and consequent loss of productivity. When grown in plate culture, highly branched mutants form colonies which are much more compact than those of the parental strain and expand in radius much more slowly than the parental strain (this is why they are called colonial mutants). Despite their reduced radial growth rate, in turbidostat or chemostat culture at a high dilution rate, colonial mutants rapidly supplant the parental strain (Figure 19.37). The colonial mutants appear after about 107 generations (range 99–115). Prevention or delay in the appearance of colonial mutants in mycoprotein fermentations would enhance productivity and decrease the unit cost of the product.

Understanding how this premature termination of the fermentations might be controlled requires an understanding of the evolution of microbial cultures in chemostat culture (Gresham & Hong, 2015). When an organism is grown in a chemostat, the relationship between its specific growth rate (μ) and the concentration of the growth-limiting substrate (*S*) is described by the Monod equation we quoted earlier:

$$\mu = \frac{\mu_{max} S}{(S + K_S)}$$

where μ_{max} is the maximum specific growth rate of the organism and K_s, the saturation constant, is a measure of the organism's affinity for the limiting substrate. *Ks* is the substrate concentration at which the organism grows at half-maximal rate. If cultivation in a chemostat is prolonged, the microbial population adapts to its environment by mutation and natural selection, resulting in the appearance of new, advantageous strains. The competitive advantage of an **advantageous mutant** relative to

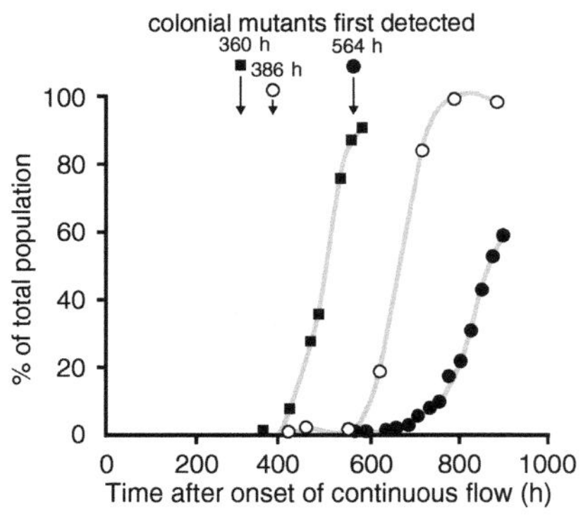

Figure 19.37. Colonial mutant population (expressed as a percentage of the total population) generated during glucose-limited laboratory chemostat cultures of *Fusarium venenatum* grown at 0.19 h^{-1}.

the parental strain can be quantified by calculating the **selection coefficient** (s):

$$s = \frac{\ln[\frac{p_{(t)}}{q_{(t)}}] - \ln[\frac{p_{(0)}}{q_{(0)}}]}{t}$$

where, p_t is the concentration of the mutant at time *t*, q_t is the concentration of the parental strain at time *t*, and p_0 and q_0 are the initial concentrations of each strain. The selection coefficient is a measure of the extent to which selection is acting to ***reduce*** the relative contribution of a given phenotype to the next generation. It is a number between zero and one.

If s = 1, selection against the phenotype is total, and it makes no contribution to the next generation. If s = 0, there is no selection against the phenotype at all (= selectively neutral compared to the favoured phenotype). For example, if the favoured phenotype produces 100 viable progeny, and the alternative phenotype produces only 90, then s = 0.1 and there is adverse selection pressure against the alternative phenotype. An alternative way of expressing this is to describe the **fitness** of the favoured phenotype as 1.0 and that of the alternative phenotype as 0.9.

Selective advantages for mutants appearing in chemostat cultures have been divided into two main categories:

- mutants that have a higher maximum specific growth rate (μ_{max}) than the parental strain; and
- mutants that have a lower saturation constant (*Ks*) for the limiting nutrient than the parental strain.

The former (μ_{max}) mutants are usually selected at high dilution rates in a chemostat or turbidostat, and K_s mutants are usually selected at low dilution rates in a chemostat.

In contrast to advantageous mutations, mutations that confer neither a selective advantage nor disadvantage to the mutant relative to the parental strain (**neutral mutations**) accumulate very slowly in the population (maximally at the forward mutation rate) and never attain high concentrations unless they are linked to an advantageous mutation. Periodic decreases in neutral mutant population have been observed in bacterial cultures maintained in the chemostat for very long periods. This phenomenon is referred to as periodic selection. It is possible to study periodic selection in *Fusarium venenatum* because the fungus produces macroconidia which are formed from uninucleate phialides. Consequently, the nuclei of macroconidia harvested from a culture provide a sample of the nuclei present in the mycelial biomass, and periodic selection can be followed in *Fusarium venenatum* by monitoring neutral mutations occurring in macroconidia (Figure 19.38).

Although neutral mutants accumulate at a linear rate in a chemostat, their concentration decreases when an advantageous mutant arises that does not carry the neutral mutation. This phenomenon of periodic selection provides a means of determining when advantageous mutants appear in a population even when the phenotype of the mutant is not known. For example, the

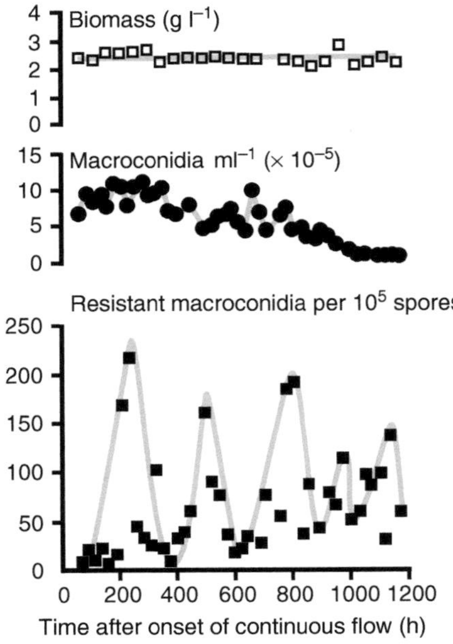

Figure 19.38. Periodic selection in a glucose-limited laboratory chemostat culture of *Fusarium venenatum*. Concentrations of biomass, total macroconidia and macroconidia spontaneously resistant to 250 μM cycloheximide in a glucose-limited chemostat culture of *Fusarium venenatum* A3/5 grown at 25°C and pH 5.8 at a dilution rate of 0.10 h-1 on modified Vogel's medium. The decreases in the concentrations of spontaneous cycloheximide-resistant macroconidia in the population are associated with the appearance of other advantageous mutants of unknown phenotype. This phenomenon is known as periodic selection and similar observations have been made with *Aspergillus* spp.

appearance of advantageous mutants in chemostat populations of *Fusarium venenatum* A3/5 has been determined by monitoring increases and decreases in the levels of chlorate and cycloheximide-resistant macroconidia in the population (an example of the latter is shown in Figure 19.38); at least three advantageous mutants of unknown phenotype appeared in the glucose-limited chemostat population of *Fusarium venenatum* A3/5 shown in Figure 19.38.

In other glucose-limited chemostat cultures of *Fusarium venenatum* A3/5 that were grown at a dilution rate of 0.19 h^{-1}, periodic selection occurred once every 124 h or 34 generations (Wiebe *et al.*, 1993, 1994, 1995). In a glucose-limited chemostat culture of *Fusarium venenatum* A3/5 grown at a dilution rate of 0.05 h^{-1} (doubling time of 13.9 h) the K_s values of some of the populations were determined and found to decrease with evolution of the culture. Thus, growing *Fusarium venenatum* A3/5 in a glucose-limited chemostat at a low dilution rate results in the selection of advantageous mutants of unaltered mycelial morphology but which have a higher affinity for the substrate than the parental strain, which might mean that they have more efficient uptake systems for glucose.

In experiments with glucose-, ammonium- and magnesium-limited chemostat cultures of *Fusarium venenatum* A3/5 grown at dilution rates of 0.18 or 0.19 h^{-1}, colonial mutants were first detected in glucose-limited cultures at 360, 386 and 421 h (99, 106 and 115 generations) after the onset of continuous flow. In ammonium- and magnesium-limited chemostat cultures at 447 h (115 generations) and 260 h (71 generations) respectively after the onset of continuous flow. Whenever the evolution of these cultures could progress, the colonial mutants eventually formed more than 90% of the total population.

Experiment shows that the highly branched phenotype is not responsible for the selective advantage of colonial mutants. Rather, the advantage seems to result from a metabolic alteration which has a pleiotropic effect on branching (Simpson *et al.*, 1998). The development of strategies to prevent or delay the appearance of colonial mutants in industrial mycoprotein fermentations is of considerable economic importance. Three possible strategies have been identified so far:

- operating the fermenter at a low dilution rate;
- periodically changing the selection pressure in the fermenter; and
- isolating, or genetically engineering, sparsely branched strains of *Fusarium venenatum* which are more stable than A3/5, at least as far as mycelial morphology is concerned.

However, in the production process, advantages gained by delaying the appearance of colonial mutants by operating the fermenter at a lower dilution rate would be offset by the decreased rate at which biomass is produced under these conditions. Similarly, periodically changing selection pressure introduces unwelcome complication to the production process with a potential for varying the chemical content of the product. However, it is possible to select variants of *Fusarium venenatum* which are morphologically more stable than A3/5; in laboratory-scale glucose-stat trials with **diploid** strains of A3/5, colonial mutants did not appear in until 1957 and 2028 h after the onset of continuous flow, timings which are equivalent to 540 and 594 generations, compared to the 99 to 115 generations for the haploid A3/5.

Premier Foods bought the entire share capital of RHM in March 2007 and RHM ceased to exist as an independent entity. In January 2011, Exponent Private Equity and Intermediate Capital Group (EPE & ICG) purchased **Quorn® mycoprotein products *and* Cauldron tofu products** from Premier for £205 million. In 2015, **Monde Nissin**, one of the leading food consumer goods companies in the Philippines, agreed to purchase all of Quorn Foods from EPE & ICG for £550 million (https://www.mondenissin.com/). Today, both product ranges are sold in the UK by Quorn Foods under the Marlow Foods Ltd. banner. Quorn Foods has around 800 employees across manufacturing sites in Belasis on Teesside (where Quorn® mycoprotein is produced by fermentation); Stokesley, North Yorkshire (where Quorn® is formed and packed) and Methwold in Norfolk (where cooking and packing of ready-meals takes place). In 2018, Kevin Brennan, chief executive of Quorn Foods, announced plans to build a new HQ at Stokesley on Teesside that will create a 'high tech powerhouse for the meat-free industry', following a 12% jump in like-for-like sales to £112 m compared with 2017. Around the world the meat-free category is growing by between 10% and 20% a year. The chief executive added that Quorn® had recorded 'amazing growth internationally' with "Australian sales up 50% and US sales up 23%."

Quorn® is now in the Philippines (https://www.mondenissin.com/news/title/quorn-world-leader-in-meat-free-food-is-now-in-the-philippines), and the product website is truly international (https://www.quorn.com/).

However, Quorn®'s EU patents, some of which were first filed in 1985, expired in 2010. Although it will be difficult to compete with the present owner's 30-plus years' experience with the fermentation method, fermentation tower and related equipment, a potential competitor company claims to have developed a new biotechnology process to produce mycoprotein cheaper than ever before and with zero waste. **3f bio** (http://www.3fbio.com/) is a technology spinout from the University of Strathclyde. Its own patented technique involves integrating the production of bioethanol with the fermentation of mycoprotein. The current method for producing mycoprotein uses glucose as a feedstock, whereas 3f bio's process uses the cheapest source of a sugar feedstock, wheat hydrolysate, which is produced in the bioethanol refinery process (sources: www.foodmanufacture.co.uk/ (2011); www.theengineer.co.uk/ (2017)).

The expectation of major market success for mycoprotein in the near future is very good news for the whole world. Why this is so derives from the arithmetic that compares the size of the human population (constantly increasing) with the amount of agricultural land available to feed that population (constantly reducing as we build on it, and perhaps reducing further as the climate changes). As it stands, the Earth does not have enough land for all its human inhabitants to enjoy an 'affluent' diet; which is one that includes meat from the billions of animals (see section 19.20) that are farmed each year. This is because so much agricultural land is devoted to feeding those animals with grasses and cereals, and conversion of these crops to meat protein is very inefficient. Assessment of the environmental impacts of food production in general is a growing concern (Poore & Nemecek, 2018; Willett *et al.*, 2019), but meat-free alternatives produced by industrial fermentation, like mycoprotein, would certainly safeguard the Earth's resources while satisfying our taste for meat-like products.

19.19 PRODUCTION OF SPORES AND OTHER INOCULA

Fungi can parasitise organisms from the other eukaryotic kingdoms and have become recognised as biocontrol agents of pests, parasites and pathogens of various sorts. Pathogens of weed plants may be harnessed as **mycoherbicides** (see Chapter 14), and **mycopesticides** can be formulated from pathogens of animals (see Chapter 16), including nematodes (discussed in section 15.6).

We have also mentioned before the many species of fungi that parasitise other fungi and which might serve as biocontrol agents of pathogenic fungi. *Trichoderma* species, which are common in soils are particularly interesting because they are opportunistic parasites of other fungi, especially of common diseases like *Rhizoctonia* damping-off of seedlings and *Armillaria* root diseases of trees, as well as being potential symbionts of plants (section 16.14). Then there are all those mycorrhizal mutualistic associations between fungi and the roots of plants (section 14.8) that could be developed and exploited to improve growth of crop plants without chemical fertilisers. The mycoparasitic and antagonistic fungi are a potential biological alternative to chemical pesticides for the control of weeds and pathogens (Hussain *et al.*, 2012).

In all these cases the potential commercial product is the fungal biomass, in the form of mycelium and/or spores, and fermentation offers a practical method of producing this. For example, liquid fermentation has been used to produce conidia of *Trichoderma harzianum* in bulk to control fruit rot diseases caused by *Botrytis cinerea*. Conidia were produced in a 300-l fermenter operated at 28°C for 60–70 h with an aeration rate of 0.4 volumes of air per volume of medium per minute and a rotor speed of 180 rpm, with constant low-level illumination. Members of the genus *Trichoderma* are used in most of the commercial fungal biocontrol agents of plant diseases (Mendoza-Mendoza *et al.*, 2016).

Similarly, conditions for optimising chlamydospore production, by mycoherbicidal strains of *Fusarium oxysporum* (*Ascomycota*) have been studied with 2.5 l fermenters, and subsequently scaled-up to 20 l, demonstrating that a single step liquid fermentation could produce large numbers of chlamydospores (about 10^7 ml^{-1}) in less than 2 weeks. Chlamydospores are more resistant to desiccation and temperature extremes than other spore types and are therefore easier to formulate into mycoherbicides that have the necessary shelf life for a commercial product.

As a final example we will mention the induction of submerged conidiation of *Ulocladium atrum* (*Ascomycota*), which has potential for controlling grey mould caused by *Botrytis cinerea* in glasshouse and field-grown crops. Yields of maximum total inoculum (mycelial fragments and conidia) were approximately 2×10^7 ml^{-1} after 9 days incubation at 25°C with a 100-rpm rotor. Shelf-life studies showed that the level of viability was comparable to that of aerial spores from 4-week fermentation on oat grains. However, the ability of submerged biomass to germinate in drier conditions declined significantly after 6 months; no such decline was observed in aerial spores formed on solid substrates.

This last example emphasises the fact that the properties of spores produced in **solid-state fermentations** differ distinctly from those obtained in submerged liquid fermentations. Fungal spores used as biocontrol agents are produced preferentially in solid-state fermentations because the spores obtained are of higher quality; meaning specifically that they are more resistant to desiccation and are more stable in a dry commercial preparation, though they also display morphological, functional and biochemical differences compared to those produced in submerged cultures. The best yields of spores are produced by a combination of submerged liquid fermentations (for biomass production in a first 'seed' step) and solid-state fermentation (for subsequent spore production). Substrate types are extremely varied and include all sorts of seeds and cereal grains as well as many waste plant materials. In some fungi (for example *Coniothyrium minitans*, a biocontrol agent active against the plant pathogenic fungus *Sclerotinia sclerotiorum*) spore production only occurs at the surface of the substrate (Reddy, 2013).

Spores for applications in the food industry have often been produced as starter culture by solid-state fermentation because it gives better yields of homogeneous and pure spores. This applies to *Penicillium roqueforti*, *Penicillium camemberti*, used in the production of blue cheese and brie-type cheeses, and *Penicillium nalgiovense* which is used in the production of salami (see section 19.24). Although they can be grown in submerged batch cultures, industrial spore production of these *Penicillium* spp. tends to favour solid substrates (bread, seeds, etc.) that result in yields of up to 2×10^9 spores g^{-1} substrate in a 14-day cultivation, which is approximately 10 times the yield that can be achieved in submerged liquid cultures. We will deal with solid-state fermentation in more detail below (section 19.21) after a brief look at the planet's biggest and most widely distributed submerged liquid fermentation process.

19.20 NATURAL DIGESTIVE FERMENTATIONS IN HERBIVORES

We have already discussed the natural digestive fermentations in herbivores:

- the crucial contribution made by fungi to the herbivore's ability to digest its food (see section 3.3 and Figure 3.3).
- Ruminant biology and the microbial community of the ruminant gastrointestinal tract, which was considered in section 15.5.

What we want to do here, briefly, is add a quantitative view of rumen fermentation; essentially the 'engineering' aspects of ruminant animals which will give you some numbers to excite and interest your friends.

The foregut of ruminants consists of four chambers: rumen, reticulum, omasum and abomasum. The rumen is the largest and, combined with the reticulum, forms a fermentation vessel that has a volume of 100–150 l in cattle and about 10 l in sheep. This fermenter vessel is loaded with:

- solids in the form of the plant biomass on which the animal feeds; and
- fermenter fluid, because the ruminants produce copious quantities of saliva, about 150 l day^{-1} in cattle and 11 l day^{-1} in sheep (that is, about 1 fermenter volume of saliva day^{-1}).

Retention times for liquids and small particles, including microorganisms, in the rumen are in the range 10–24 h, but for large particles, like plant fragments it ranges from 2 to 3 days. The **rumen is a continuous culture system**, but it is not well mixed, and because of the heterogeneous nature of the substrate plant material, it may become stratified. Indeed, the digestive physiology of all animals is quite complicated; the animals depend for their own digestion on the microbial communities in their gastrointestinal tracts and those microbial communities are strongly influenced by diet (Karasov & Douglas, 2013; Dearing & Kohl, 2017).

The ruminant animal relies on microorganisms to convert the plant biomass into nutrients accessible to the animal. The concentrations of these populations in rumen fluid are in the range 10^9–10^{10} ml^{-1} bacteria; 10^5–10^6 ml^{-1} protozoa and about 10^9 ml^{-1} thallus-forming units of anaerobic fungi (Trinci *et al.*, 1994). An **average cow contains about 1.5×10^{14} anaerobic fungi** and an **average sheep contains about 10^{11} anaerobic fungi**.

The total number of dairy cows in the world is 2.8×10^8, but in addition there are about 1.5×10^9 beef cattle worldwide (sources: https://dairy.ahdb.org.uk/ and https://www.drovers.com/). In total, therefore, there are approximately 1.8×10^9 cattle worldwide, which amounts to a grand total of 2.7×10^{11} litres of rumen fluid walking around containing **2.7×10^{23} anaerobic fungi in the world's cattle herd**. Just the cattle herd.

There are about 10^9 sheep in the world, each one a walking 10 l fermentation culture. And then there are all those antelopes, bison, buffalo, camels, deer, goats, oxen, wildebeest ... and we've not even mentioned the hindgut fermenting horses, zebra, rhinoceros, elephants, tapirs, sloths, pigs, peccaries, guinea pigs, chinchillas ... and rabbits!

There are a lot of chytrids doing a lot of fermentation in the world; that's why we call it the **biggest and most widely distributed submerged liquid fermentation process**. You do the maths; and you might like to consider the climatic impact of the so-called 'greenhouse gas' methane released as a by-product of fermentation in herbivores: 'livestock methane emissions, while not the dominant overall source of global methane emissions, may be a major contributor to the observed annual emissions increases over the 2000s to 2010s ' (Wolf *et al.*, 2017).

19.21 SOLID-STATE FERMENTATIONS

In **solid-state fermentations**, microbial growth occurs at the surface of solid substrates and the main difference from submerged liquid fermentation is the quantity of fluid involved. As described above, the biomass grows in suspension in the culture medium in liquid fermentations; whereas a solid-state fermentation has a continuous gas phase between the particles or fragments of solid substrate, with a minimum of visible water.

Most of the water in the system is absorbed within the moist substrate particles and most of the **space between particles is**

filled by the gas phase, although some water droplets may be visible between substrate particles and the particles themselves may be covered in a water film. This type of fermentation is one type of the more general **solid substrate fermentation** which can vary from processes that involve suspending the solid substrate particles in a continuous liquid phase (rather like food digestion in a stomach or rumen) through to processes (such as **trickle filters**) in which the liquid phase flows around and through an immobilised substrate.

Solid-state fermentations can be divided into two groups according to the physical state of the substrate:

- low moisture solids fermented with or without agitation; and
- columns packed with solid substrate which is fermented as liquid is trickled through the column (with or without recirculation).

The substrate usually provides a rich and complex source of nutrients, which may not need to be supplemented. Mixed substrates of this sort are ideal for filamentous fungi able to form mycelia that can grow on and through particulate substrata producing a variety of extracellular enzymes. Traditional substrates for solid-state fermentations are various agricultural products, like cereal straw and other plant litter, and rice, wheat, maize and soy bean seeds. The majority of solid-state fermentations involve filamentous fungi, though yeasts and bacteria can also be cultivated in this way; most are obligate aerobes (though there is no reason why the system should not be operated anaerobically) and the process may feature pure cultures of specific organisms or mixtures of several organisms.

Solid-state fermenters are still the method of choice to produce spores of some (most?) fungi for use as biological control agents. It is difficult to produce good fungal spores for biological control purposes in submerged culture (*Lecanicillium muscarium* (*Verticillium lecanii*) is a notable exception). These features are dealt with in section 19.19.

A major problem with solid-state fermentations is that they are **not easy to control**, and this can make it difficult to meet regulatory requirements. Composting processes can release noxious gases (ammonia and sulfides in particular) and aerial spores produced by the organisms involved may be potential health hazards (as allergens).

Traditional examples of solid-state fermentations are:

- the formation of compost;
- **mushroom cultivation** (and, in a separate process, the production of starter cultures or **mushroom spawn**);
- leavening of bread dough;
- **mould ripening of cheeses**, and other food products, such as **salami** and **soy sauce**.

What are effectively solid-state fermentations also play a role in the **production of chocolate** and **coffee**. In these cases, the fermentation is responsible for the preparation of the 'bean', to separate it from the fruit flesh and mucilage in cocoa; or the cherry and mucilage in coffee and to impart **flavour**.

Cocoa products from the cacao tree, *Theobroma cacao*, are used in the food, chemical, pharmaceutical and cosmetic industries. The plant originated in the Amazon basin but is now an important cultivated crop worldwide in tropical regions. Cacao fruits are oval pods, 20–30 cm long and weighing in the region of 0.5 kg. They grow directly from the trunk and main branches of the tree and are usually harvested by hand. Each pod contains about 40 seeds (the cocoa 'beans', though they are not true beans) embedded in a thick pectinaceous pulp. Most of the initial processing is done on the farm. Harvested pods are cut open with a machete and the contents of pulp and cocoa seeds scooped out by hand. The rind is discarded, but pulp and seeds are piled in heaps for several days. During this time, the seeds and pulp undergo what is called **'sweating'**; this is a natural **microbial fermentation** that liquefies the thick pulp so that it trickles away, leaving the cocoa seeds behind to be collected. Two simultaneous processes occur during their fermentation:

- **microbial activity** in the mucilaginous pulp **produces alcohols and organic acids** as by-products of microbial metabolism, which also liberates heat, **raising the temperature** to about 50°C;
- complex biochemical **reactions occur within the seed cotyledons** due to the diffusion of metabolites from the microorganisms and the rising temperature.

The fermentation features a succession of a **range of yeasts**, lactic acid bacteria and acetic acid bacteria. Rapid growth of yeasts occurs in the first 24 hours (including *Saccharomyces cerevisiae, Klöckera apiculata, Candida bombi, Candida rugopelliculosa, Candida pelliculosa, Candida rugosa, Pichia fermentans, Torulospora pretoriensis, Lodderomyces elongiosporus, Kluyveromyces marxianus* and *Kluyveromyces thermotolerans*). The most important roles of the yeasts are:

- breakdown of citric acid in the pulp leading to a change in pH from 3.5 to 4.2, which allows growth of bacteria;
- ethanol production (substrate for acetic acid bacteria);
- formation of organic acids (oxalic, succinic, malic and acetic);
- production of some organic volatiles which are probably precursors of chocolate flavour in the cocoa 'bean';
- secretion of pectinases to break down the pulp; pectin is the major plant polysaccharide responsible for the viscosity of the pulp and the yeast pectinases are needed to reduce viscosity of the pulp allowing fluid to drain away to increase aeration (required by acetic acid bacteria).

Unfermented cacao seeds do not produce chocolate flavour (their flavour is said to be similar to that of raw potatoes) and it has not been possible to mimic the production of chocolate flavouring by treating fresh cacao seeds with hot alcohol and organic acids. **Flavour development** has an absolute requirement for the complex physical and organic biochemistry that occurs during the fermentation. The fermented beans are sun-dried before transfer to processing plants which roast the beans, then separate cocoa butter (used to make chocolate bars) from the solid matter (powdered to make cocoa for beverages and cooking) (Schwan & Wheals, 2004; Nielsen *et al.*, 2013).

Coffee undergoes a similar fermentation, and for similar reasons. Neither coffee beans nor cocoa beans are really beans. Strictly, 'beans' are the seeds of plants belonging to the legume family (like faba beans, butter beans, etc.). Cocoa beans are cacao seeds, formed in a seed pod made of several carpels, while coffee beans are seeds formed in a fruit made from a single carpel (called a drupe) which has an outer fleshy part surrounding a central shell (the pit or stone) with one or more seeds inside. Common fruits that are drupes include cherry, apricot, damson, nectarine, peach, plum, mango, olive, date, coconut and coffee. Several species of the bushy tree *Coffea* are grown for the beans, but *Coffea arabica* has the best flavour. 'Robusta' beans (*Coffea canephora* var. *robusta*) are usually grown on land unsuitable for *Coffea arabica*. The tree produces red or purple fruits, which are **cherry-like drupes** containing two seeds (the coffee 'beans'). Harvesting is again done by hand; freshly harvested berries are washed and soaked in water, then most of the berry flesh is removed by a machine that has a roller with a roughened surface that scours away the pulp under a stream of water. In the second stage, the coffee 'beans' are **fermented in large water containers**; fermentation dissolves any remaining fruit flesh and removes a parchment-like pectinaceous film (also called 'pergamino') surrounding the coffee seeds. Removing the skin from the coffee bean is called 'coffee hulling'. Fermentation takes approximately 2 days and is important to give the coffee its rich aroma and special flavour. When fermentation is complete, the coffee beans are washed, sun-dried for 5 or 6 days and finally this hulled '**pergamino coffee**' is ready for distribution and/ or export.

According to Wikipedia, 'After water, tea is the most widely-consumed beverage in the world ' (https://en.wikipedia.org/wiki/Tea). **Tea** is made from the processed leaves, leaf buds, and internodes of *Camellia sinensis*:

- The most tender leaves and buds are **plucked** from the bushes by hand.
- The leaves are then **withered** to reduce moisture content by up to 70%.
- Limp (withered) leaves are **rolled** mechanically, which breaks open the leaves and creates the twisted wiry looking tea leaves.
- When rolling is complete, the leaves are spread on tables to provide good aeration to the **released enzymes**. The leaves turn progressively from green, through light brown, to a deep brown, and the oxidation takes from between 30 and 180 minutes at about 26°C.

This enzymatic oxidation, which is generally called **tea fermentation**, is an essential key step in the processing, during which chlorophyll breaks down, tannins are released, and polyphenols in the tea leaf are enzymically oxidised and condensed to form the coloured compounds that contribute to the flavour, colour and strength qualities of the final beverage. The tea industry maintains that the term fermentation is a misnomer, because the enzymes concerned in the oxidation are endogenous to the plant and microbes are not involved. However, we're not convinced; given the prevalence of endophytic and epiphytic fungi (sections 13.19 and 13.20)!

Recently, the solid-state approach has been developed for the commercial production of **extracellular enzymes** and other fungal products, and fungal spores for use as inoculum for biotransformations and, especially, as **mycopesticides**. The main advantage of this approach for microbial pesticide production is that the process can be done by individual farmers or local communes. Local production avoids some of the major problems associated with large-scale commercial production facilities such as poor stability and short shelf life, and associated storage and long-distance shipping problems. The locally produced mycopesticide can be much cheaper and the formulation can be optimised for local environmental conditions (Pandey *et al.*, 1999; Pandey, 2003; Ghosh, 2015).

The fermenters used for large-scale solid-state fermentations are called **bioreactors**. They may be as simple as a plastic bag or an open tray of some sort, including even simple stacks of compost, which may be so large as to require heavy plant for mixing. More 'engineered' equipment might consist of stacked arrangements of trays through which temperature and humidity-controlled air is circulated, or rotary drum type bioreactors possibly including some additional agitation (Mitchell *et al.*, 2006).

A practical generalised approach for production of fungal spores for mycopesticide preparations is a two-stage fermentation procedure in which the fungus is first grown in liquid medium batch culture that is used to seed a solid substrate (usually autoclaved seeds or cereal grains, often rice) for a solid-state fermentation that enables conidiation.

At the end of the solid-state fermentation the conidiated substrate is air dried before extraction of the conidia using a **MycoHarvester**, after which the purified spores are dried, suitably formulated into the final product and packaged.

MycoHarvesters are two-stage devices consisting of a rotating drum agitator for the overgrown substrate and a spore extractor section comprising four or more stainless steel chambers containing '**air cyclones**'. These cyclones are rapidly rotating air vortices driven by a powerful fan that draws air through the equipment. The agitator releases spores mechanically, and substrate debris and spores are sucked into the cyclones.

The **cyclones effectively centrifuge the particles** so that air containing the conidia is drawn out of the machine into a collector, while the large substrate particles fall out of the cyclone to the bottom and sides of the unit (see: http://www.dropdata.net/mycoharvester/MycoHarvester6_brochure.pdf).

19.22 DIGESTION OF LIGNOCELLULOSIC RESIDUES

Lignocelluloses as agricultural, industrial and forest residues account for the majority of the total biomass present in the world. This is a renewable resource with a potential for bioconversion into many useful biological and chemical products. Accumulation of large quantities of this biomass every year can result in both environmental deterioration and loss of potentially valuable materials that might otherwise be processed to yield energy, food, animal feed or fine chemicals.

Agricultural residues and forest materials containing high levels of lignocellulose are particularly abundant because the crop represents such a small proportion of what is grown (see our brief discussion of this in section 11.1). The major component of lignocellulose materials is cellulose, followed by hemicellulose and lignin, intermeshed and chemically bonded by noncovalent forces and by covalent cross-linkages. As fungi, particularly white-rot fungi (*Basidiomycota*), have particularly efficient lignocellulose degradation enzyme machinery they might be attractive components of low-cost bioremediation projects (Kumar *et al.*, 2008; Sánchez, 2009; Huberman *et al.*, 2016; Rouches *et al.*, 2016); and compare the use of laccase for remediation of petroleum-contaminated soil (section 14.6) (Liu *et al.*, 2017).

White-rot basidiomycete fungi, such as *Phanerochaete chrysosporium* (*Sporotrichum pulverulentum*) can **mineralise lignin** completely to CO_2 and water. Many microorganisms are capable of degrading and utilising cellulose and hemicellulose as carbon and energy sources, but only the white-rot fungi can breakdown lignin, the most recalcitrant component of plant cell walls. In nature these wood- and litter-degrading fungi play an important role in the carbon cycle and in addition to lignin, white-rot fungi can degrade a variety of persistent environmental pollutants, such as chlorinated aromatic compounds, heterocyclic aromatic and synthetic high polymers (see section 14.6). These additional capabilities of white-rot fungi are probably due to the strong oxidative activity and the low substrate specificity of their ligninolytic enzymes.

Fungi have two types of extracellular enzymatic systems: the hydrolytic system, which consists of hydrolases responsible for polysaccharide degradation; and a unique oxidative and extracellular ligninolytic system, which **degrades lignin and opens phenyl rings**. An important thought in understanding the mechanism(s) of degradation is that the enzymes concerned are **too large to penetrate** sound, intact wood; this task is performed by hydrogen peroxide and the active oxygen radicals derived from it (section 11.7). The most widely studied white-rot organism is the basidiomycete *Phanerochaete chrysosporium*. *Trichoderma reesei* and its mutants are the most studied ascomycete species and are used for the commercial production of hemicellulases and cellulases. We have already dealt with the enzymes involved in biodegradation of the components of lignocellulose:

- cellulose (section 11.2),
- hemicellulose (section 11.3),
- lignin (section 11.7).

Here, we wish to emphasise a few important points relating to the ***exploitation*** of these natural enzyme systems and illustrate a few of the biotechnologies to which they contribute.

In nature the various enzymes act synergistically to catalyse the release and hydrolysis of cellulose. Several physical factors, like pH, temperature, adsorption onto minerals and chemical factors, such as availability of oxygen, nitrogen and phosphorus and presence of phenolic compounds and other potential inhibitors can influence the **bioconversion** of lignocellulose in both natural and artificial systems.

Phanerochaete chrysosporium simultaneously degrades cellulose, hemicellulose and lignin, whereas other species, such as *Ceriporiopsis subvermispora*, tend to remove lignin in advance of cellulose and hemicellulose (and, of course, brown-rot fungi rapidly digest cellulose, but only slowly modify lignin; this particularly affects softwoods and *Piptoporus betulinus*, *Serpula lacrymans* and *Coniophora puteana* are examples). Consult sections 3.11 and 13.5 (Peralta *et al.*, 2017).

The steady growth of agro-industrial activity has led to the accumulation of a large quantity of **lignocellulosic residues** from agriculture, forestry, municipal solid wastes and various industrial wastes around the world (Table 19.14).

Bioconversion of lignocellulose into useful, higher value, products normally requires multistep processes, the steps including:

- **collection** and mechanical, chemical or biological pretreatment;
- **hydrolysis of polymers** to produce readily metabolised (usually sugar) molecules (Bhattacharya *et al.*, 2015);
- **fermentation of the sugars** to produce a microbial or chemical end product;
- separation, purification, **packaging and marketing**.

Note that Table 19.14 shows official estimates of the ***annual*** production of these agricultural wastes. The amounts are staggering; the total of the entries shown in Table 19.14 is 1.2 ***billion*** metric tonnes and does not include municipal solid wastes like waste paper or garden refuse collected for recycling. Several uses have been suggested for bioprocessed lignocellulosic wastes (Figure 19.39), for example:

- Use as raw material to produce **ethanol**, in the hope that an alternative fuel manufactured using biological methods will have environmental benefits. Ethanol is either used as a chemical feedstock or as an additive enhancer for petrol. Softwood, the dominant source of lignocellulose in the northern hemisphere, has been the subject of interest as a raw material for fuel ethanol production in Sweden, Canada and Western United States (Sánchez, 2009). Brazil and the United States produce ethanol from the fermentation of cane

Table 19.14. Annual production of lignocellulose residues generated by different agricultural sources

Lignocellulosic residues	Millions of tons
Sugar cane bagasse	380
Maize straw	191
Rice shell	188
Wheat straw	185
Soya straw	65
Yuca straw	48
Barley straw	42
Cotton fibre	20
Sorghum straw	18
Banana waste	15
Mani shell	11.1
Sunflower straw	9.0
Bean straw	5.9
Rye straw	5.2
Pine waste	4.6
Coffee straw	1.9
Almond straw	0.49
Hazelnut husk	0.24
Sisal and henequen (*Agave*) straw	0.093

Table modified from Sánchez (2009) and based on FAO and similar official sources.

juice and maize, respectively. It is estimated that 25% of the maize and other cereals grown in the United States (107 million tonnes in 2009) is used to produce fuel ethanol). In the United States, fuel ethanol has been used in gasohol or oxygenated fuels, containing up to 10% ethanol by volume, since the 1980s. It's a very big business; 103 billion litres of fuel ethanol worldwide in 2017, of which 60 billion litres is produced in the United States, 27 billion in Brazil and 5.3 billion in the EU (source: www.statista.com/). Use of ethanol produced from cereals as an oxygenating fuel additive has been criticised claiming use of maize for this purpose has pushed the prices of the crop up to record levels in the commodities markets. This has led to an increase in some food prices because livestock feed produced from maize has increased in price correspondingly. Of course, this criticism would not apply to processes that used **crop wastes** for the fermentation.

- High-value bioproducts such as **organic acids, amino acids, vitamins** and a number of bacterial and fungal polysaccharides, for example xanthans, are produced by fermentation using glucose as the base substrate but theoretically these same products could be manufactured from sugars derived from lignocellulose residues. Based on the known metabolism of *Phanerochaete chrysosporium*, several potential high-value products could be derived from lignin. Rumen microorganisms convert cellulose and other plant carbohydrates in large amounts to acetic, propionic and butyric acids, which ruminant animals then use as energy and carbon sources; these fungi also have promise for commercial bioprocessing of lignocellulose wastes **anaerobically** in liquid digesters.
- Compost making for cultivation of edible mushrooms. Good compost is the essential prerequisite for successful mushroom farming (section 12.6). The basic raw material for mushroom compost in Europe is wheat straw, although straws of other cereals are sometimes used. Ideally, the straw is obtained already mixed with horse manure after being used as stable bedding. On commercial scale, this is not possible and other animal wastes, like chicken manure, are mixed with the straw, together with gypsum (calcium sulfate) and large quantities of water. The excess calcium of gypsum precipitates the mucous and slimy components of manure and so prevents water logging of the compost and generally improves aeration and its mechanical properties that aid thorough mixing. All of this enables the compost to ferment uniformly, which itself results in large crops being grown reliably. Cultivation of edible mushrooms using lignocellulosic residues is a value-addition process to convert these materials into human food. It is one of the most efficient biological ways by which these **residues can be recycled**. Mushrooms can be grown successfully on a wide variety of lignocellulose residues such as cereal straws, banana leaves, sawdust, peanut hulls, coffee pulp, soy bean and cotton stalks, indeed any lignocellulosic substrate that has a substantial cellulose component. Another advantage of this strategy is that the production unit can vary from the small-holder local farmer through to a multi-million-pound mushroom farm.
- Lignocellulose bioconversion can be used to produce **animal feeds**. For example, fungi can be used to improve the nutritional quality of cereals like barley to compensate for the latter's deficiency in the amino acid lysine. Inoculating soaked barley with *Aspergillus oryzae* or *Rhizopus arrhizus* increases the protein content as the fungus grows. The product is used as pig food. With a view to future product developments there is a good deal of research under way on growing fungi such as *Pleurotus* or *Trichoderma* species on cheap lignocellulose residues. Growth of the mycelium of such fungi on agricultural wastes, for example, releases components of lignocellulose residues and converts cellulosic and otherwise non-digestible materials into sugars and glycogen that are readily available to animals and

increases the protein content to the point where the product becomes a nutritious animal feed.

- Although not a 'waste bioconversion' process we finally want to mention the use of fungi to **remodel timber** to increase penetration by wood preservatives, providing environmentally friendly methods for wood protection. Construction timber tends to have high structural strength but low natural durability. Increasing durability by treatment with preservatives is lessened in efficiency by the timber's low permeability because of closure of the pits between cells in the wood. These pits enable lateral permeability when the tree is alive, but they close when the timber is harvested and seasoned. Permeability is improved conventionally by mechanically incising the wood, but selective degradation of pits by white-rot fungus (*Physisporinus vitreus*) is a biotechnological alternative '**bioincising**' process (Schwarze, 2007).

Further development of the bioprocessing potential of lignocellulose biodegradation is likely to depend on understanding the molecular mechanisms the organisms use. For instance, cloning and sequencing of the various cellulolytic genes could make cellulase production more economical.

19.23 BREAD: THE OTHER SIDE OF THE ALCOHOLIC FERMENTATION EQUATION

Leavened bread is another product of the fermentation of sugars from cereal grains by *Saccharomyces cerevisiae*. Known, not surprisingly, as baker's yeast in this industry (though, in the old days, supplied ***to*** bakers ***by*** brewers), the yeast still uses the same chemistry of the ethanol fermentation we have summarised before in this equation:

$$C_6H_{12}O_6 \rightarrow 2\ C_2H_5OH + 2\ CO_2$$

One hexose molecule is converted into two ethanol molecules and two molecules of carbon dioxide. For the brewer the key product is ethanol; for the baker the key product is the **carbon dioxide**. In summary, the fermentation process in bread dough depends on the breakdown of the starches in flour producing carbon dioxide, which makes bubbles in the dough stabilised by gluten proteins from the flour. This causes the dough to expand. A small amount of alcohol is also produced, but this evaporates when the bread is baked. As with alcoholic fermentations, the process depends on the ability of *Saccharomyces cerevisiae* to preferentially use fermentation even in the presence of oxygen, given appropriate nutrients.

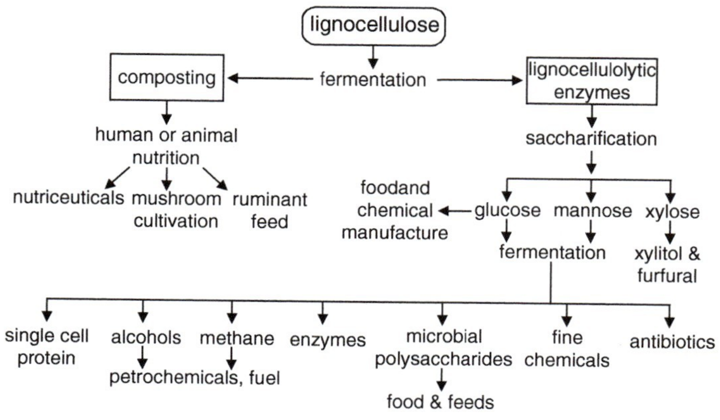

Figure 19.39. Generalised process stages in lignocellulose waste bioconversion and the range of potential products. Adapted from Sánchez (2009).

Those 'appropriate nutrients' are contained in the culture medium for the baker's yeast, which of course is the bread dough; so perhaps we should look at a bread recipe. There are **hundreds of bread recipes**: white, brown, wholemeal and country, French bread, baguettes, Bath buns, Chelsea buns, Scotch baps, barm cakes, granary bread, and so on, and so on (see, for example http://www.cookitsimply.com/category-0020-0e1.html; Hutkins, 2006; Montet & Ray, 2015). The ingredients for a basic white bread recipe are yeast, sugar, water, milk, white flour, salt and butter.

The yeast, sugar, water and milk are mixed and put in a warm place until fermentation makes it frothy, which only takes about 10 minutes. This foaming fermenting mixture is called **barm**, and many years ago it was the practice to take the barm from beer fermentations to leaven bread. The flour and salt are mixed with the butter and the yeast liquid is added and the whole mixed (kneaded) to smooth dough. The dough is left to **rise** in a warm place until it has doubled in size; this takes 60–90 minutes. The dough is then kneaded again for 5 minutes, then shaped into a baking tin and left for the second rising (called **proving**) for about 1 h. Finally, the loaf is baked in an oven at 230°C for 30–40 minutes.

Let's look at the biology of this process. After all the ingredients are mixed together, forming the dough, this doubles in volume over 60–90 minutes in a warm place (remember, this is a bakery we're talking about, so warm places are not hard to find).

The dough does this because the yeast fermentation produces carbon dioxide ***and*** because the wheat flour contains the protein mixture known as gluten. Gluten is a combination of the proteins gliadin and glutenin, which are stored in the endosperm of wheat, rye and barley seeds to nourish the germinating plants along with starch. Together, gliadin and glutenin comprise about 80% of the protein contained in wheat seeds. The gluten they form is an important source of nutrition in foods, including bread, prepared from these cereal grains. But it is **crucial to the structure** of leavened bread, which depends on the sticky and viscous properties of the gluten protein in the dough. Gliadin is a glycoprotein and forms a three-dimensional **network** with glutenin by inter- and intra-molecular sulfur cross-linkages which develop in the dough during the kneading process.

Subsequently, the carbon dioxide gas produced during fermentation is trapped into bubbles by the gluten and as more and more bubbles are formed in the dough they make it 'rise'.

The second **kneading** process redistributes the bubbles and forms more disulfide cross-links, helping more carbon dioxide gas bubbles to form during the proving stage. After this final fermentation period the dough is cooked, the alcohol evaporates and the bubbly structure of the dough is fixed into the open structure of bread.

The first records of any sort of bread are in **ancient Egyptian** hieroglyphs from over 5,000 years ago which show bakeries with dough rising next to bread ovens. Wine making, brewing and baking occurred alongside one another in ancient Egypt. An Egyptian wooden model dating from 2000 BC of a combined **brewery and bakery** can be seen at the British Museum. Optical and scanning electron microscopy of desiccated bread loaves and beer remains preserved in tombs from about 2000–1200 BC (Samuel, 1996) allows us to presume that bread was made with flour from raw grain and with malt and yeast, suggesting that fermenting brewing liquor, what we now call barm, was mixed with bread dough in ancient Egypt for the first leavened bread.

The exact nature of yeast was unknown until Louis Pasteur, in 1859, demonstrated that wine yeast is a living organism and that only active living cells can cause fermentation. There seems always to have been a direct relation between **brewing and baking**. The Faculty of Medicine in seventeenth century Paris debated for months whether bakers should be allowed to use beer barm for their bread. They eventually decided to ban the practice but the bakers didn't take much notice and continued to use barm for the fine light bread that the upper classes liked so much. British cookery books included recipes and instructions for brewing as well as baking as a matter of course until the nineteenth century; beer-making being the only dependable source of baking yeast. Wine-making barm is too bitter to be used in baking. Even beer barm needs to be washed to reduce the bitterness of hops used in brewing ale (Hutkins, 2006; Montet & Ray, 2015).

However, there are different requirements for the yeasts in the two industries. The **baker needs yeast that tolerates higher temperatures** with no particular preference for alcohol production, the **brewer needs yeast that tolerates and produces high alcohol concentrations** with no particular preference for performance at higher temperatures (and strains used for dried yeast production (see below) must be tolerant to the drying process) (Hutkins, 2006; Gibson *et al.*, 2007; Ali *et al.*, 2012; Montet & Ray, 2015).

Today, reliable and highly specialised yeasts are produced commercially and marketed around the world as dried or compressed preparations. Commercial yeast production starts with a pure culture tube or frozen vial of the appropriate yeast strain. This sample is the inoculum for the first of a series of progressively larger cultures that amplify the volume of the yeast suspension. The early stages are grown as batch fermentations, using a medium comprising molasses, phosphate, ammonia and minerals; later stages are conducted as fed-batch fermentations during which molasses and other nutrients are fed to the yeast at a rate that maximises yeast multiplication but prevents the production of alcohol. From the last of these seed fermentations, the vessel contents are pumped to separators that separate the yeast from the spent molasses. After being washed with cold water the yeast cream is held at 1–2°C until used to inoculate the final commercial fermentation tanks.

These commercial fermenters have working volumes up to 250 m^3. Water is first pumped into the fermenter, then this is pitched with yeast cream from the seed store. Aeration (sparging with about one fermenter volume of air per minute), cooling (through an external heat exchanger to maintain culture temperature at 30°C) and nutrient additions initiate a 15–20 h fermentation. At the start of the fermentation, the culture occupies 30–50% of the fermenter volume. The unit is operated as a fed-batch and additions of nutrients at increasing rate during fermentation (to support growth of the increasing cell population) and maintenance of pH in the range of 4.5–5.5 bring the fermenter to its final volume. The number of yeast cells increases about 5–8 times during this fermentation.

At the end of fermentation, yeast is separated from the fermenter broth by centrifuges, washed with water and re-centrifuged to make a yeast cream with a solids concentration of approximately 18%. The yeast cream is cooled to about 7°C and stored in refrigerated stainless steel tanks. Cream yeast can be delivered to customers directly with tanker trucks. Alternatively, the yeast cream can be pumped to a filter press and sufficient water removed to produce a press cake having 30–32% yeast solids content. The press cake yeast can be crumbled into pieces and bagged for distribution to customers in refrigerated trucks. Harvesting and processing wet yeast can take several hours and chilling is needed to store the resultant yeast products; cream and compressed yeast products must be used within 10–28 days, respectively.

Dried yeast manufacture involves drying the yeast with evaporative cooling on fluidised beds (in which air passing through crumbled press cake yeast forms a bed-of-air, which suspends and tumbles the product to dry it). Spray drying is also effective in preserving viable yeast, but spray drying is a high energy-demand operation (in practice requiring 7,500–10,000 J g^{-1} of evaporated water) so it needs to be optimised carefully (Luna-Solano *et al.*, 2005). The final dried product is vacuum packed and can be stored at normal room temperatures for months to years.

19.24 CHEESE AND SALAMI MANUFACTURE

Cheese is a solid or semi-solid protein food product manufactured from milk and before the advent of modern methods of food preservation, like refrigeration, pasteurisation and canning; cheese manufacture was the only method of preserving milk. Although basic cheese making is a **bacterial fermentation**, there are two important processes to which fungi make a crucial contribution. These are the provision of enzymes for **milk coagulation** right at the start of the process, and **mould ripening** to change the flavour and/or consistency of the product.

Cheese production relies on the action of enzymes which coagulate the proteins in milk, forming solid **curds** (from which the cheese is made) and liquid **whey**. Making cheese involves the following basic steps:

- The milk is adjusted to a pH between 5.8 and 6.4 and a source of calcium (for example, calcium chloride) is added.
- The milk-coagulating enzyme is mixed into the milk in a **cheese making vat**, at this stage the milk temperature is about 45°C.
- The enzyme is allowed to react with the casein for sufficient time (around 10 minutes) to start solidifying the milk in the vat.
- The initial coagulum is cut into segments, heated (for example, by direct steam injection or through steam-jacketing of the vat) to about 60°C, and stirred for about 15 min, which causes the fluid whey to start separating from the solid curds. Extraction or expulsion of a liquid from a gel like this is called **syneresis**. Enzymatic digestion of casein molecules enables calcium bonds to form and hydrophobic regions develop that force water molecules to leave the structure. The curd is cooked, or scalded, to expel the whey.
- The curds are cooled and separated from the whey by draining and pressing.
- Cheese curds can then be made into the product the cheese maker requires (Figure 19.40), which is where the lactobacterial fermentation comes into prominence, though other factors (like temperature, pH and additives) can be varied for particular recipes.
- Salt is added to the cheese to slow down bacterial activity and to enhance the taste of cheese. Salt also affects enzymatic activity in cheese during **cheese ripening** (Farkye, 2004).

A preparation of animal enzymes (called **rennet** or **chymosin**) extracted from the stomach membranes of unweaned ruminants has traditionally been the primary coagulant of milk protein in the manufacture of cheese. Rapid expansion of the production and consumption of cheese caused attention to shift to alternative sources of such enzymes. Stomachs from older animals were unsuitable because their high content of pepsin resulted in a less effective coagulation and more proteolysis, causing lower cheese yields and development of off-flavours. Moulds like *Aspergillus* and *Mucor* were identified as potential sources and **aspartyl proteases** from *Rhizomucor pusillus, Rhizomucor miehei* and *Cryphonectria parasitica* were found to be satisfactory for at least some cheeses, so fungal enzymes supplied the market to an extent. Eventually, the gene for calf chymosin (or pro-chymosin) was cloned, first in prokaryotic and then eukaryotic microorganisms and in recent years animal enzymes produced by **genetically modified microbes** have come to dominate the market.

Indeed, **chymosin**, the milk-clotting enzyme used to make cheese and other dairy products, was the **first enzyme from a genetically modified source** to gain approval for use in food manufacture. In March 1990, the US FDA issued the first regulation in the United States for the use in food of any substance produced by recombinant DNA technology by affirming that such chymosin was **'generally recognised as safe'** (**GRAS**). The importance of this was that it exempted the product from the pre-market approval requirements that apply to new food additives.

At that time, the source of the new enzyme was a bovine pro-chymosin gene expressed in the bacterium *Escherichia coli*. Subsequently, chymosin preparations produced commercially from the yeast *Kluyveromyces lactis* and the filamentous *Aspergillus awamori* (both *Ascomycota*) genetically modified to include bovine pro-chymosin were also given GRAS status. The FDA concluded that **fermentation-derived chymosin (FDC)** was purer than traditional calf rennet and was identical to its natural counterpart. It was verified that the yield, texture, and quality of cheese made with FDC was comparable to that made with calf chymosin and that FDC gave superior yield to other coagulants.

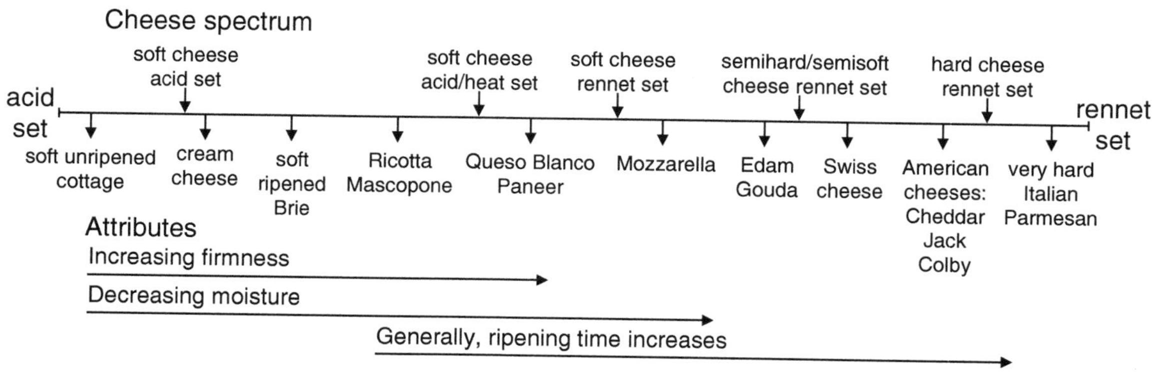

Figure 19.40. The cheese spectrum. There is a wide range of cheeses on the market; they vary from extremely soft to extremely hard, a factor which is controlled by the extent of the proteolysis which is allowed, the cooking temperatures and the pH. They vary in flavour according to the microbes the milk originally contained and/or according to the microbes and other materials introduced into the cheese during fermentation or later processing. They also vary in flavour, of course, according to the origin of their substrate. Milk is not a uniform commodity; cow's milk differs between breeds of cattle, and cheese is made from ewe's, goat's, buffalo's and camel's milk (redrawn after Farkye, 2004).

Microbial rennets have been improved in a number of ways including reducing their nonspecific proteolysis (which improves yield) and thermal stability (which makes it easier to control the elasticity of the product; keeping soft cheeses soft in other words). Today, about 90% of cheese production depends for the coagulation step on enzymes from genetically modified microbes (mainly yeasts but including *Aspergillus awamori*). **Commercial microbial proteases** derived from *Bacillus, Aspergillus* species or from *Rhizomucor niveus*, and nonspecific aminopeptidases from *Aspergillus oryzae* are used for a range of other food modifications including casein and whey protein hydrolysates with reduced allergenicity and rich in bioactive peptides, debittering and flavour generation in protein hydrolysates, and enzyme-accelerated cheese ripening and production of enzyme-modified cheeses. But it is the rennet coagulation that develops the basic structure of cheese, and there are hundreds of cheese varieties ranging from very soft cheeses, such as Camembert to the hard ones, such as Italian Grana cheeses. The rennet enzyme coagulates the casein micelles in milk by reducing repulsion between casein particles. Rennet-altered micelles aggregate into clusters and chains that eventually form a network that surrounds the fat globules. Formation of the rennet gels can be monitored and controlled by the cheese maker (Stepaniak, 2004; Lucey, 2007).

For what may be a **traditional farming industry**, there have been many major biotechnological developments in recent years, but then it ***is*** a big industry, and it grew rapidly across the start of the twenty-first century, though it has faltered more recently as an increase in value of the US dollar together with a sharp decline in the price of milk reduced both cheese market prices and industry revenue figures in dollar terms. Using 2017 statistics, the European Union produced 9.9 million metric tons and the United States 5.6 million (source: https://www.statista.com/), and global revenue was US$99 billion, although the global cheese manufacturing industry suffered an annual 3% decline in the period 2012–2017 (source: www.ibisworld.com/).

Cheese ripening depends on a host of metabolic processes and involves a complex of interrelated events. The biochemical pathways through which lactose, lactate, milk fat and caseins are converted to flavour compounds are now known in general terms. More than 300 different volatile and non-volatile compounds have been implicated in cheese flavours. These flavour compounds originate from biochemical pathways like proteolysis, lipolysis, glycolysis, citrate and lactate metabolism; the relative contribution of these processes depends on the variety of cheese and its characteristic microbial flora (McSweeney, 2004; Stepaniak, 2004). Indeed, the variety of flavours come from a succession of microbes (bacteria, yeasts and fungi) that determine the consistency and flavour of the cheese. Surveys based on 133 isolates provided by producers of cheese and cheese starter cultures, together with 97 isolates from culture collections showed that cheese fungi are found in two classes, the *Eurotiomycetes* with *Penicillium* species (*Eurotiales*) and *Sporendonema casei*/*Sphaerosporium equinum* (*Onygenales*), and the *Sordariomycetes* with *Scopulariopsis* species (*Microascales*) and *Fusarium domesticum* (*Hypocreales*). Some of these fungi, such as *Penicillium camemberti, Fusarium domesticum, Scopulariopsis flava* and *Scopulariopsis casei*, are only known from cheeses and are probably adapted to this specific habitat, which is extremely rich in protein and fat. Other cheese fungi are ubiquitous, such as *Penicillium roqueforti, Scopulariopsis candida* and *Scopulariopsis fusca* (Ropars *et al.*, 2012).

By comparing the genomes of ten *Penicillium* species, Ropars *et al.* (2015) showed that adaptation to the cheese habitat was associated with several recent horizontal transfers of large genomic regions carrying crucial metabolic genes (see section 19.15). Seven horizontally transferred regions, of more than 10 kb each, were identified flanked by specific transposable elements, and displaying nearly 100% identity between distant *Penicillium* species. Two of these regions carried genes with functions involved in the utilisation of cheese nutrients or competition, which were associated with faster growth and greater competitiveness on cheese and contained genes highly expressed in the early stage of cheese maturation. These regions were found to be nearly identical in many strains and species of cheese-associated *Penicillium* fungi, indicating that recent 'selective sweeps' had increased their frequency in the population by natural selection (Ropars *et al.*, 2015).

Mould ripening is a traditional method of **finishing cheeses** which has been in use for at least two thousand years. **Blue cheeses**, like Roquefort, Gorgonzola, Stilton, Danish Blue and Blue Cheshire, all use *Penicillium roqueforti* which is inoculated into the cheese prior to storage at controlled temperature and humidity by having its spores forced into the new cheese on the tines of a metal comb. In some procedures the *Penicillium roqueforti* spores are mixed with the curds during the final whey-removal process and the function of the metal tines is then to introduce air channels through the cheese to encourage the (aerobic) fungus to grow. The inocula come from starter cultures of proprietary strains of *Penicillium roqueforti, Penicillium camemberti* (for cheeses) and *Penicillium nalgiovense* (for salami) produced predominantly by solid-state fermentation because this process gives better yields of homogeneous and pure spores.

After inoculation the fungus grows throughout the cheese and into the voids between curd particles, producing flavour and odour compounds. The holes and tracks made by the inoculation/aeration device are usually evident at point of sale (Figure 19.41).

Camembert and Brie are ripened by a mould called *Penicillium camemberti*, which changes the **texture of the cheese** more than its flavour. This fungus grows on the surface of the cheese extruding enzymes which digest the curds to a softer consistency from the outside towards the centre (Figure 19.42).

Mushrooms are a useful, and acceptable, addition ('meat extender') to meat content in beef sausage recipes (Al-Dalain, 2018), but *Penicillium nalgiovense* is a filamentous fungus (*Ascomycota*) which is the most widely used fungus as a starter

Figure 19.41. Samples of blue cheeses (Gorgonzola on the left, Danish Blue on the right) bought from a Stockport supermarket which show (top) holes in the outer 'rind' made by the tines of the inoculating device when the *Penicillium roqueforti* spores were injected into the newly pressed cheese, and (bottom) spore production within the cheese revealing the tracks made by the tines during inoculation and revealing growth of the fungus into the air voids between the curds. Photographs by David Moore.

culture for cured and **fermented meat products**. Use of moulds in the production of fermented sausages is known from eighteenth century Italy where fermented and air-dried sausage was a popular peasant food because it could be stored safely at room temperature for long periods. Today **salami** is widely produced in southern European countries using a range of meats, including beef, goat, horse, lamb, pork, poultry and/or venison. The chopped meat is mixed with minced animal fat, cereals, herbs and spices, and salt and allowed to ferment for a day. This mixture gives the salami sausage its typical marbled appearance when cut. The mixture is then stuffed into a casing, treated by dipping or spraying with a suspension of the *Penicillium nalgiovense* starter culture at a concentration of about 10^6 spores ml^{-1}, and finally hung to cure.

The fungus does several jobs as it grows over the sausage and into the meat and the rest of the mixture. Basically, it **imparts flavour** and **prevents spoilage** during the curing process, but this is achieved by:

- proteolysis, lactate oxidation, amino acid degradation, lipolysis and fatty acid metabolism during the maturing process creating the desirable flavour;
- with the surface of the sausage colonised by a specific, usually proprietary, mould, the air-exposed sausage is protected against spoilage by other undesirable species of yeasts, moulds and bacteria;

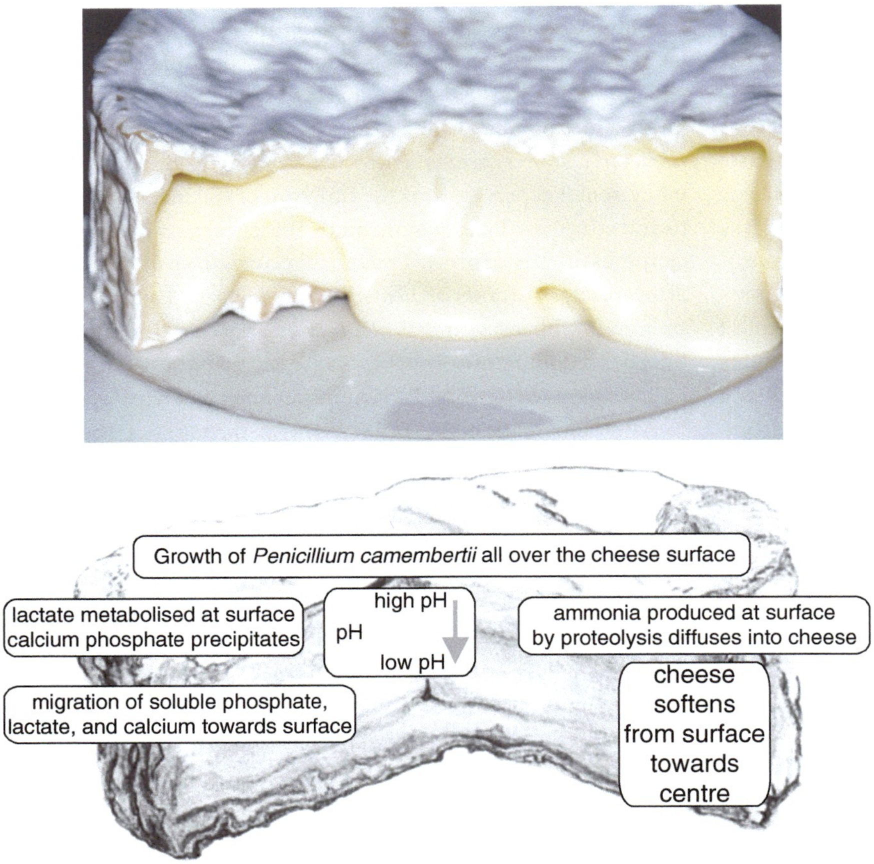

Figure 19.42. A mature Camembert cheese ready for eating (top). The sketch below is a schematic representation of the changes that occur during ripening of such a cheese because of the growth of *Penicillium camemberti* over the surface of the cheese. Photograph by David Moore, schematic modified and redrawn from McSweeney (2004).

- the surface covering of the mould mycelium controls the drying process and ensures a smooth and uniform surface appearance of the product.

19.25 SOY SAUCE, TEMPEH AND OTHER FOOD PRODUCTS

Fungi are used in the processing of several food products that enjoy large markets in Asia. In these, the fungus is mainly responsible for the characteristic odour, flavour or texture and may or may not become part of the final edible product, though many microorganisms may be involved in these microbial ecosystems. Growing fungi on water-soaked seeds of plants is a popular way to produce soy sauce and various other **fermented foods** (Moore & Chiu, 2001; Hutkins, 2006; Montet & Ray, 2015; Wolfe & Dutton, 2015; Joshi, 2016).

Soy sauce has been used in China for more than 2,500 years, so it is one of the world's oldest condiments. It is a seasoning agent with a salty taste and a distinctly meaty aroma, although it is made by a complex fermentation process of a combination of soy beans and wheat in water and salt, in which carbohydrates are fermented to alcohol and lactic acid and proteins broken down to peptides and amino acids. This traditional brewing, or fermentation, method can take 6 months or more to produce the finest soy sauce, which should be a dark brown, transparent, salty liquid with balanced flavour and aroma. The brown colour is a result of sugar caramelisation during the 6–8 month maturation process. There is an alternative non-brewed method: an acid hydrolysis that often produces opaque liquor with a harsher flavour but takes only 2 days to make. We note the alternative but will not discuss it further.

The raw materials are:

- **soy beans** (seeds of *Glycine max*; also known as soya or soja beans) are mashed prior to mixing them with other ingredients;
- **pulverised wheat grains** are mixed with the crushed soy beans (soy sauce produced by chemical hydrolysis generally does not use wheat);
- **salt** (specifically NaCl) is added at the start of second fermentation to 12–18%; this obviously contributes to flavour but it also establishes a selective environment for the lactic acid bacteria and yeasts to complete the fermentation; high salt concentration also serves as a preservative.

Manufacturing through the traditional brewed method occurs in three steps (Luh, 1995):

- **Koji-making**, in which the crushed soy beans and wheat are blended together in water that is boiled until the grains are thoroughly cooked and softened. The mash, as it is known, is allowed to cool to about 27°C and fermented with the filamentous ascomycetes *Aspergillus oryzae* and *Aspergillus sojae*. Depending on the size of the factory, the soy beans may be fermented in fist-sized balls (the traditional method) or on trays. The culture of soy, wheat and mould is known as **koji**. The main function of these moulds is to break down the proteins in the mash; this process takes 3 days and the fungi (in what is known as the **seed starter**) are often closely guarded **proprietary strains** because this step has a vital role in determining the flavour of the final soy sauce, and methods are now being developed to use genomics for strain improvement (Zhong *et al.*, 2018).
- **Brine fermentation**: when the substrate has become overgrown with the mould fungus the koji is transferred to fermentation tanks and mixed with water and salt to produce a mash called **moromi. Lactic acid bacteria** (*Pediococcus halophilus*) and, 30 days later, **yeast** (*Saccharomyces rouxii*), are then added to complete the fermentation. The moromi ferments for several months, during which the soy and wheat paste is digested into a semi-liquid, reddish-brown 'mature mash' which contains over 200 flavour compounds produced as the bacteria and yeast enzymatically digest the protein and other residues to produce amino acids and derivatives.
- **Refinement**: the raw soy sauce is separated from the wheat and soy residue after 6–9 months of moromi fermentation by pressure-filtration through cloth. The filtrate is pasteurised (which forms additional flavour compounds) and bottled ready for sale. Soy sauce press cake is a valuable by-product as an animal feed.

Clearly, three major groups of microorganisms are involved in soy sauce fermentation: *Aspergillus* fungi involved in the koji production, and communities of halotolerant bacteria and yeasts are responsible for the moromi fermentation. Enzymes produced by all these various microbes hydrolyse the raw materials during the complex soy sauce fermentation process. Present-day manufacture is done in large factories using carefully controlled production conditions. Soy beans are now steamed instead of being boiled and temperature and humidity controls are applied during koji production. But in the traditional method in China, moromi is fermented in the open air; and there are significant differences between the products produced in different seasons of the year (Cui *et al.*, 2014).

Soy sauce is the liquid produced as soy beans are fermented, several other traditional soyfoods are made mainly from **whole soy beans** (Moore & Chiu, 2001; Hutkins, 2006). Soy beans are highly nutritious, containing large amounts of protein and other nutrients, and the traditional soyfoods offer real health benefits. Traditional soyfoods are classified into nonfermented and fermented products (Liu, 2008). **Nonfermented soyfoods** include soymilk, **tofu**, soy sprouts, yuba (soymilk film), okara (soy pulp), vegetable soy beans, soynuts and toasted soy flour. **Fermented soyfoods** include the bacterial products **natto** (fermented with the rice straw bacterium *Bacillus natto*), **soy yoghurt** (soymilk fermented by bacteria) and the fungus-dependent products:

- **Sufu**, which is pickled tofu made by fermenting with the zygomycete fungi *Actinomucor elegans, Mucor racemosus* or *Rhizopus* spp.
- **Soy nuggets**: large whole soy beans soaked and steamed and mixed with roasted wheat- or glutinous-rice flour before fermentation with the koji mould *Aspergillus oryzae*. After several days of incubation, the resultant soy bean koji is packed in kegs with salt water and various spices, seeds and/or root ginger slivers (and occasionally rice wine), then aged for several months. The soy nuggets are then sun-dried.
- **Tempeh** is a fermented food made by the controlled fermentation of cooked germinated soy beans with the filamentous zygomycete *Rhizopus oligosporus*. Fermentation by *Rhizopus* binds the soy beans into a protein-rich compact white cake that can be used as a meat substitute. Tempeh has been a favourite food and staple source of protein in Indonesia for several hundred years (https://www.tempeh.info/).
- **Miso** is possibly the **most important fermented food in Japanese cuisine**. It is basically fermented soy bean paste but rice and several cereal grains and even other seeds can be combined into an extremely wide variety of miso that differ according to the combination of ingredients. The basic production process is essentially the same for all recipes: rice, barley or soy beans are steamed, cooled and inoculated with the koji mould *Aspergillus oryzae*. When the koji has become established it is added, as a seed culture, to a mixture of washed, cooked, cooled and crushed soy beans, water and salt. This is placed in vats and allowed to ferment for 12–15 months to allow the proteins in the mixture to be broken down slowly and naturally, forming a paste that is flavoursome and nutritious.

Another fermented product of this sort, **ang-kak** is a bright reddish purple fermented rice product popular in China and the Philippines which is fermented using *Monascus* species (*Ascomycota*, Order *Eurotiales*). In Petri dish cultures, the mycelium is white in early growth stages, however, it rapidly changes to a rich pink and subsequently to a distinctive yellow-orange colour (illustrated in Pattanagul *et al.*, 2007). *Monascus purpureus* produces the characteristic pigments and ethanol which are used for red rice wine and food colouring. The pigments are a mixture of red, yellow and purple polyketides and about 10 times more pigment is obtained from solid-state fermentation than from submerged liquid fermentation.

The rice is first soaked in water until the grains are fully saturated and may then be directly inoculated or steamed prior to inoculation. Inoculation is done by mixing *Monascus purpureus* spores or powdered red yeast rice together with the processed rice. The mix is then incubated at room temperature for 3–6 days. At the end of this time, the rice grains should have turned bright red in the centre and reddish purple on the outside. **Red rice** is sold as a pasteurised wet aggregate in jars, as whole dried grains, or as a ground powder (to be used as a red food colourant) (Pattanagul *et al.*, 2007; Lin *et al.*, 2008; Panda *et al.*, 2010).

19.26 FUNGI AS CELL FACTORIES

We have shown throughout this book how fungi relate to their environment and all the other life forms on the planet, and how we have come to depend on the ways we can manipulate fungi for our benefits in agriculture, medicine and industrial biotechnology. In Chapter 6, we introduced you to a range of techniques that enable the most detailed manipulation of the fungal genome and such molecular tools have already found application in every area of fungal biology (the 'genome-enabled mycology' described by Hibbett *et al.*, 2013; McCluskey & Baker, 2017).

'Fungal cell factories' is a phrase that is often used in those 'blue skies' discussions about what could be done in the future. This is because the metabolic activities of fungi have already been harnessed for so long in applications ranging from food fermentation to pharmaceutical production that they are naturally thought of as indispensable biotechnological tools. The more so because fungal bioprocesses of earlier generations, like those that produce citric acid and penicillin, and those of today's generation producing lovastatin, have had such positive impacts on human society.

We have discussed how fungi are utilised for industrial production processes throughout this book, most notably in Chapters 12 and this, Chapter 19. The growing amount of information that seems to cascade from the various forms of genomic analyses described in section 6.10 is demanding application to these industrial processes for our greater good. The metabolic and enzymatic diversity encoded in the genomes of fungi will continue to be developed for production of new generations of enzymes, pharmaceuticals, chemicals and biofuels. Though there must be many applications which will only emerge with time and further knowledge; there are some which are immediately obvious. Currently, fungal derived enzymes that degrade plant-derived biomass are being utilised for the development of bioprocesses for biofuel and renewable chemical production, particularly the growing demand for sustainable production of biochemicals that substitute for chemicals otherwise obtained from fossil fuels.

Filamentous fungi are of great interests as biocatalysts in biorefineries as they naturally produce and secrete a variety of different organic acids that can be used as building blocks in the chemical industry; ideally, in a lignocellulosic biorefineries, the fungus could be considered in a combined approach where it hydrolyses plant biomass wastes and ferments the resulting sugars into different organic acids. Genomic and metabolomic analyses enable rapid identification of ***novel secondary metabolites*** open to industrial exploitation through the design of high-yielding fungal cell factories (Karagiosis & Baker, 2012; Khan *et al.*, 2014; Nielsen & Nielsen, 2017). There is no shortage of novel methods to obtain new metabolites by engineering fungal secondary metabolism, but increased yield is the key essential and regulation of secondary metabolite biosynthesis is incompletely understood. However, the identification of the *mcrA* gene as a principal regulator of *Aspergillus* secondary metabolism indicates that further advance in this direction is imminent. The *mcrA* gene is conserved, and it encodes a transcription factor that regulates transcription of hundreds of genes including at least 10 secondary metabolite gene clusters in *Aspergillus terreus* and *Penicillium canescens* (Scharf & Brakhage, 2013; Oakley *et al.*, 2016).

Production of recombinant ***proteins*** by filamentous fungi was initially focused on exploiting the extraordinary enzyme synthesis ***and secretion*** ability of fungi to produce single recombinant protein products, especially by industrial strains of *Aspergillus, Trichoderma, Penicillium* and *Rhizopus* species. Two disadvantages of filamentous fungi as hosts for recombinant protein production became apparent immediately: one is their common ability to produce homologous proteases which could degrade the heterologous protein product and the other is that the protein glycosylation patterns in filamentous fungi and in mammals are quite different. Specifically, fungi lack the functionally important terminal sialylation of the glycans that occurs in mammalian cells. Without engineering, filamentous fungi, despite their other advantages, are not the most suitable microbial hosts for production of recombinant human glycoproteins for therapeutic use. Nevertheless, strategies to prevent proteolysis have already met with some success and new scientific information being generated through genomics and proteomics research will extend the biomanufacturing capabilities of recombinant filamentous fungi, enabling them to express genes encoding multiple proteins, making filamentous fungi even better candidates to produce proteins and protein complexes for therapeutic use (Ward, 2012; Fernández & Vega, 2013; Nevalainen & Peterson, 2014).

Most of what we have discussed so far in this section has either stated or implied submerged (liquid) fermentation of fungi, but it is essential to remember that solid-state fermentation is a crucial process for producing enzymes, organic acids, flavour compounds, pharmaceutical agents and food processing (see also Chapter 11; and review by Ghosh, 2015). Of course, it is also the foundation of the mushroom cultivation industry (section 12.6; and see Petre, 2015). This last is especially important in relation to potential improvements in the biotechnological procedures for producing mushrooms as healthy and highly nutritive food in their own right, while at the same time using mushroom farming as a bioremediation tool, by using recalcitrant wastes as substrates for the mushroom farming compost before crop production and/or by using spent mushroom compost for soil remediation after cropping (Purnomo *et al.*, 2011; Camacho-Morales & Sánchez, 2015). *Ganoderma* is mainly farmed for use as a traditional Chinese medicine. Sporophores of the *Ganoderma lucidum* species complex contain many bioactive compounds; indeed, well over 400 secondary metabolites have been isolated from various *Ganoderma* species (Baby *et al.*, 2015; Ahmad, 2019). A mixture of *Ganoderma lucidum* polysaccharides (known as GLP) is the main bioactive component in the water-soluble extracts of this mushroom, and there is some evidence that GLP possesses potential anticancer activity (Sohretoglu & Huang, 2018).

The clinical evidence for antitumor and other medicinal activities of mushroom metabolites comes primarily from some commercialised purified polysaccharides, and polysaccharide preparations can be obtained from medicinal mushrooms cultured in bioreactors. Mushroom polysaccharides do not attack cancer cells directly but produce their antitumor effects by activating various immune responses in the host. Structurally different β-glucans have different affinities towards receptors and thus generate different host responses. Immunomodulating and antitumor activities of these metabolites are related to immune cells such as haematopoietic stem cells, lymphocytes, macrophages, T cells, dendritic cells, and natural killer cells, which are involved in the innate and adaptive immunity, resulting in the therapeutic immune modification (Berovič & Podgornik, 2015; Sudheer *et al.*, 2018). Among the wide range of pharmaceutically interesting metabolites found in extracts of *Ganoderma*, some have been found to be stimulators of neural stem cell proliferation *in vitro* (which could be of value in treatment of neurodegenerative diseases). Other extracts have been assessed for genotoxicity and anti-genotoxicity using comet assays of mouse lymphocytes; no evidence was found for genotoxic chromosomal breakage nor cytotoxic effects by *Ganoderma* extract in the mouse, nor did it protect against the effects of the mutagen ethyl methanesulfonate. This study found no evidence for the extract having any value in protecting against the test mutagen (Chiu *et al.* 2000; Yan *et al.*, 2015). Shah (2012) stresses the importance of genotoxicity testing for pharmaceuticals to ensure compliance with the guideline of the International Conference on Harmonisation of Technical Requirements for Registration of Pharmaceuticals for Human Use (ICH; a unique project that brings together regulatory authorities of Europe, Japan and the United States with pharmaceutical industry representatives).

We mention in section 14.6 that lack of degradability and growing pollution problems on land, in water courses and in the open seas have led to mounting concern about waste plastics. These synthetic polymers are ubiquitous in the modern world but the global environmental problems they pose are caused by their careless disposal. Poly-(ethylene terephthalate) (PET) is one of the most abundantly produced synthetic polymers and is accumulating in the environment at a staggering rate as discarded packaging and textiles. Unfortunately, the properties that make PET so useful to us in our daily lives also endow it with an alarming resistance to biodegradation, with the potential of it lasting for centuries in most natural environments. Most applications that employ PET, such as single-use beverage bottles, clothing, packaging and carpeting employ crystalline PET, which is recalcitrant to catalytic or biological depolymerisation due to the limited accessibility of the ester linkages.

PET can be depolymerised to its constituents if the ester bonds of the polymer can be cleaved. Doing this with available chemical techniques is too costly to be a viable recycling solution. Recently, a newly discovered bacterium isolated from outside a bottle-recycling facility in Japan, *Ideonella sakaiensis*, was shown to exhibit the rare ability to grow on PET as a major carbon and energy source. When grown on PET, this strain produces two enzymes capable of hydrolysing PET and the reaction intermediate, mono(2-hydroxyethyl) terephthalic acid. Both enzymes are required to enzymatically convert PET efficiently into its two environmentally benign monomers, terephthalic acid and ethylene glycol; yielding the monomers for further plastics manufacture (Yoshida *et al.*, 2016; Austin *et al.*, 2018). This paragraph suggests the essentials of a plastics pollution remediation technology; devising it is only a matter of time. What is the prospect of generating recombinant fungi that possess this metabolic activity? First, is it a bacterial technology? Probably not, we think, and to whet your mycological appetite we will mention another study with plastics remediation potential.

Ahuactzin-Pérez *et al.* (2018) discovered that *Pleurotus ostreatus* degrades and uses (as carbon and energy source) high concentrations of di-(2-ethyl hexyl) phthalate (DEHP), and *Fusarium culmorum* has been shown to produce a range of esterase enzymes when challenged with DEHP (Ferrer-Parra *et al.* 2018; Portillo-Ojeda *et al.*, 2018; Ocaña-Romo *et al.*, 2018). Phthalates are plasticisers, primarily used as additives in plastics like polyvinyl chloride (PVC) polymers to make them more flexible. Because they are not chemically bound, phthalates are easily released from plastic articles, through direct release, leaching, and abrasion, and phthalate esters are one of the most frequently detected persistent organic pollutants in the environment (Gao & Wen, 2015). In laboratory animal studies, some phthalates have been associated with developmental and reproductive toxicity and they are generally considered to be toxins that interfere with endocrine systems in mammals (Hauser & Calafat, 2005). If common ascomycete and basidiomycete fungi can produce enzyme systems enabling them to use such pollutants for growth, they provide an opportunity for bioremediation of plastic waste. *Pleurotus ostreatus* degrades lignin efficiently, grows well in both liquid and solid fermentation systems, and is an ideal candidate for genome engineering into a plastic-eating oyster mushroom.

In this book, we have tried to show you what twenty-first century mycology has to offer. If you have read the entire book, well, congratulations on your dedication! But, hopefully, having reached this point you can now appreciate how important fungi are to life on Earth, and particularly, human life on Earth; and that includes ***your everyday life***. You could also take the knowledge to which we guide you and decide to manipulate, control and engineer fungi in ways of which earlier mycologists could only dream in their wildest fantasies. Please also add 'conserving fungi' to that to-do list; we can't imagine life on Earth without fungi and their functions. Indeed, some of our colleagues have claimed that ***even mycologists*** are in danger of needing conservation (Courtecuisse, 2001; Minter, 2001), and others have stridently pointed out the importance of fungi (Anonymous, 2017; Willis, 2018).

We have shown you what we know about fungi. We will end by pointing out some of the topics about which we remain astonishingly ignorant.

- We wonder how **free cell formation** works (section 3.4). How do some fungi and fungal-like organisms accomplish the three-dimensional positioning of wall- and membrane-forming vesicles to subdivide large volumes of cytoplasm to create motile or non-motile spores? They are the only organisms that do this.
- We wonder what mechanisms are used to ensure that the **nuclear membrane remains intact** as the nuclear division progresses (section 5.6). Another unique characteristic of present-day fungi.
- We wonder how the **Spitzenkörper** operates, how it is assembled, regulated, and directed (section 5.9). It is only found in filamentous fungi.
- We wonder what it is that **controls the multinucleated nature of most hyphae** and some specific tissues (sections 5.6 and 5.16). How do fungi (and only fungi) control the synchronicity of mitotic divisions and then the (rapid) migration (section 4.8) and distribution of daughter nuclei within their hyphae?
- We wonder how **hyphal fusion** is managed, and how ***incompatible fusions*** trigger the death of hyphae (section 5.15 and Chapter 8, especially Figure 8.7, and Chapter 9).
- We wonder what regulates the **placement of septa** in hyphae (section 4.12 and section 5.16) and how different states of differentiation can be controlled on the two sides of perforated septa.
- We wonder how the **yeast–hyphal dimorphism** is controlled (section 5.17); it's important in the lifestyle of so many pathogenic fungi and control may be analogous to multicellular growth regulation in plants (Cogo *et al.*, 2018).
- We wonder how **hyphal branching** is controlled in time and space.
- We wonder how multitudes of independent hyphal apices are orchestrated to create tissue layers (like hymenia; section 13.6) and how **determinate-growth controls** are selectively applied to such large hyphal communities.

You will notice that we have not mentioned the fungal plasma membrane (which is unique in using ergosterol) or the fungal wall (which is also exclusive to fungi); this is because, as well as being established targets for selective toxicity, these structures have been investigated exhaustively. It's too easy to try a slightly different agent that targets the wall or membrane; another batch of chemicals, another lot of routine tests. We need now to be more imaginative when trying to identify new antifungal drugs and fungicides. Perhaps we would have more success if we directed our attention to some of the other structures and/

or processes that are unique to fungi. Bearing in mind the bulleted list above we would suggest: (a) vesicle transportation and positioning; (b) nuclear membrane dynamics; (c) Spitzenkörper dynamics; (d) nucleus migration and control of nuclear division; (e) hyphal fusion and how programmed cell death is triggered by the nonself-recognition system; (f) septation dynamics and physiological function; (g) control of dimorphism; (h) dynamics of hyphal branching; (i) selective control of determinate growth of hyphal apices.

This is as far as **we** can go. Now it's up to **you** to decide what will happen in the rest of the twenty-first century. Where are ***you*** taking us from here?

19.27 REFERENCES

Ali, A., Shehzad, A., Khan, M.R., Shabbir, M.A. & Amjid, M.R. (2012). Yeast, its types and role in fermentation during bread making process: a review. *Pakistan Journal of Food Sciences*, **22**: 171–179. URL: https://pdfs.semanticscholar.org/5965/78c82c06e22d0af02f5da33bbee1d72f3a30.pdf.

Ahmad, M.F. (2019). *Ganoderma lucidum*: a macro fungus with phytochemicals and their pharmacological properties. In *Plant and Human Health*, Vol. 2, ed. M. Ozturk & K. Hakeem. Cham, Switzerland: Springer Nature, pp. 491–515. ISBN: 9783030033439. DOI: https://doi.org/10.1007/978-3-030-03344-6_21.

Ahuactzin-Pérez, M., Tlecuitl-Beristain, S., García-Dávila, J. *et al.* (2018). A novel biodegradation pathway of the endocrine-disruptor di(2-ethyl hexyl) phthalate by *Pleurotus ostreatus* based on quantum chemical investigation. *Ecotoxicology and Environmental Safety*, **147**: 494–499. DOI: https://doi.org/10.1016/j.ecoenv.2017.09.004.

Al-Dalain, S.Y.A. (2018). Utilization of mushroom fungi in processing of meat sausage. *Research on Crops*, **19**: 294–299. DOI: https://doi.org/10.5958/2348-7542.2018.00044.X.

Anbu, P., Gopinath, S.C.B., Chaulagain, B.P. & Lakshmipriya, T. (2017). Microbial enzymes and their applications in industries and medicine 2016. *BioMed Research International*, **2017**: article 2195808. DOI: https://doi.org/10.1155/2017/2195808.

Anonymous (2017). Stop neglecting fungi. *Nature Microbiology*, **2**: 17120. DOI:https://doi.org/10.1038/nmicrobiol.2017.120.

Austin, H.P., Allen, M.D., Donohoe, B.S. *et al.* (2018). Characterisation and engineering of a plastic-degrading aromatic polyesterase. *Proceedings of the National Academy of Sciences of the United States of America*, **115**: E4350–E4357. DOI: https://doi.org/10.1073/pnas.1718804115.

Baby, S., Johnson, A.J. & Govindan, B. (2015). Secondary metabolites from *Ganoderma*. *Phytochemistry*, **114**: 66–101. DOI: https://doi.org/10.1016/j.phytochem.2015.03.010.

Berovič, M. & Podgornik, B.B. (2015). Cultivation of medicinal fungi in bioreactors. In *Mushroom Biotechnology: Developments and Applications*, ed. M. Petre. London: Academic Press, an imprint of Elsevier Inc., pp. 155–171. ISBN: 9780128027943. DOI: https://doi.org/10.1016/B978-0-12-802794-3.00009-6.

Bhat, M.K. (2000). Cellulases and related enzymes in biotechnology. *Biotechnology Advances*, **18**: 355–383. DOI: https://doi.org/10.1016/S0734-9750(00)00041-0.

Bhattacharya, A.S., Bhattacharya, A. & Pletschke, B.I. (2015). Synergism of fungal and bacterial cellulases and hemicellulases: a novel perspective for enhanced bio-ethanol production. *Biotechnology Letters*, **37**: 1117–1129. DOI: https://doi.org/10.1007/s10529-015-1779-3.

Blamey, J.M., Fischer, F., Meyer, H.-P., Sarmiento, F. & Zinn, M. (2017). Enzymatic biocatalysis in chemical transformations: a promising and emerging field in green chemistry practice. In *Biotechnology of Microbial Enzymes: Production, Biocatalysis and Industrial Applications*, ed. G. Brahmachari. San Diego, CA: Academic Press, an imprint of Elsevier, pp. 347–403. ISBN: 9780128037256. DOI: https://doi.org/10.1016/B978-0-12-803725-6.00014-5.

Bourdichon, F., Casaregola, S., Choreh, F. *et al.* (2012). Food fermentations: microorganisms with technological beneficial use. *International Journal of Food Microbiology*, **154**: 87–97. DOI: https://doi.org/10.1016/j.ijfoodmicro.2011.12.030.

Briggs, D.E., Brookes, P.A. & Boulton, C.A. (2004). *Brewing: Science and Practice*. Cambridge, UK: Woodhead Publishing Limited. ISBN: 1855734907, ISBN-13: 9781855734906.

Camacho-Morales, R.L. & Sánchez, J.E. (2015). Biotechnological use of fungi for the degradation of recalcitrant agro-pesticides. In *Mushroom Biotechnology: Developments and Applications*, ed. M. Petre. London: Academic Press, an imprint of Elsevier Inc., pp. 203–214. ISBN: 9780128027943. DOI: https://doi.org/10.1016/B978-0-12-802794-3.00012-6.

Chamberlin, E.M., Ruyle, W.V., Erickson, A.E. *et al.* (1951). Synthesis of 11-keto steroids. *Journal of the American Chemical Society*, **73**: 2396–2397. DOI: https://doi.org/10.1021/ja01149a551.

Chiu, S.W., Wang, Z.M., Leung, T.M. & Moore, D. (2000). Nutritional value of *Ganoderma* extract and assessment of its genotoxicity and anti-genotoxicity using comet assays of mouse lymphocytes. *Food and Chemical Toxicology*, **38**: 173–178. DOI: https://doi.org/10.1016/S0278-6915(99)00146-5.

Clandinin, M., Van Aerde, J., Merkel, K. *et al.* (2005). Growth and development of preterm infants fed infant formulas containing docosahexaenoic acid and arachidonic acid. *The Journal of Pediatrics*, **146**: 461–468. DOI: https://doi.org/10.1016/j.jpeds.2004.11.030.

Cogo, A.J.D., Dutra Ferreira, K. dos R., Okorokov, L.A. *et al.* (2018). Spermine modulates fungal morphogenesis and activates plasma membrane H^+-ATPase during yeast to hyphae transition. *Biology Open*, **7**: article bio029660. DOI: https://doi.org/10.1242/bio.029660.

Courtecuisse, R. (2001). Current trends and perspectives for the global conservation of fungi. In *Fungal Conservation: Issues and Solutions*, ed. D. Moore, M.M. Nauta, S.E. Evans & M. Rotheroe. Cambridge, UK: Cambridge University Press, pp. 7–18. ISBN-10: 0521048184, ISBN-13: 978-0521048187.

Csiszár, E., Szakács, G. & Koczka, B. (2007). Biopreparation of cotton fabric with enzymes produced by solid-state fermentation. *Enzyme and Microbial Technology*, **40**: 1765–1771. DOI: https://doi.org/10.1016/j.enzmictec.2006.12.003.

Cui, C., Zhao, M., Li, D., Zhao, H. & Sun, W. (2014). Biochemical changes of traditional Chinese-type soy sauce produced in four seasons during processing. *CyTA – Journal of Food*, **12**: 166–175. DOI: https://doi.org/10.1080/19476337.2013.810673.

Das, U.N. (2018). Arachidonic acid in health and disease with focus on hypertension and diabetes mellitus: a review. *Journal of Advanced Research*, **11**: 43–55. DOI: https://doi.org/10.1016/j.jare.2018.01.002.

Dearing, M.D. & Kohl, K.D. (2017). Beyond fermentation: other important services provided to endothermic herbivores by their gut microbiota. *Integrative and Comparative Biology*, **57**: 723–731. DOI: https://doi.org/10.1093/icb/icx020.

Demasi, M. (2018). Statin wars: have we been misled about the evidence? A narrative review. *British Journal of Sports Medicine*, **52**: 905–909. DOI: https://doi.org/10.1136/bjsports-2017-098497.

Donova, M.V. & Egorova, O.V. (2012). Microbial steroid transformations: current state and prospects. *Applied Microbiology and Biotechnology*, **94**: 1423–1447. DOI: https://doi.org/10.1007/s00253-012-4078-0.

Donova, M.V., Egorova, O.V. & Nikolayeva, V.M. (2005). Steroid 17β-reduction by microorganisms: a review. *Process Biochemistry*, **40**: 2253–2262. DOI: https://doi.org/10.1016/j.procbio.2004.09.025.

Doran, P.M. (2012). *Bioprocess Engineering Principles*, 2nd edition. Oxford, UK: Academic Press, an imprint of Elsevier. ISBN: 978-0122208515.

Farkye, N.Y. (2004). Cheese technology. *International Journal of Dairy Technology*, **57**: 91–98. DOI: https://doi.org/10.1111/j.1471-0307.2004.00146.x.

Fernandes, P., Cruz, A., Angelova, B., Pinheiro, H.M. & Cabral, J.M.S. (2003). Microbial conversion of steroid compounds: recent developments. *Enzyme and Microbial Technology*, **32**: 688–705. DOI: https://doi.org/10.1016/S0141-0229(03)00029-2.

Fernández, F.J. & Vega, M.C. (2013). Technologies to keep an eye on: alternative hosts for protein production in structural biology. *Current Opinion in Structural Biology*, **23**: 365–373. DOI: https://doi.org/10.1016/j.sbi.2013.02.002.

Ferrer-Parra, L., López-Nicolás, D.I., Martínez-Castillo, R. *et al.* (2018). Partial characterization of esterases from *Fusarium culmorum* grown in media supplemented with di-(2-ethyl hexyl phthalate) in solid-state and submerged fermentation. *Mexican Journal of Biotechnology*, **3**: 82–94. DOI: https://doi.org/10.29267/mxjb.2018.3.1.83.

Gao, D. & Wen, Z.-D. (2015). Phthalate esters in the environment: a critical review of their occurrence, biodegradation, and removal during wastewater treatment processes. *The Science of the Total Environment*, **541**: 986–1001. DOI: https://doi.org/10.1016/j.scitotenv.2015.09.148.

Ghosh, J.S. (2015). Solid-state fermentation and food processing: a short review. *Journal of Nutrition & Food Sciences*, **6**: 453 (7 pp.). DOI: https://doi.org/10.4172/2155-9600.1000453.

Gibson, B.R., Lawrence, S.J., Leclaire, J.P.R., Powell, C.D. & Smart, K.A. (2007). Yeast responses to stresses associated with industrial brewery handling. *FEMS Microbiology Reviews*, **31**: 535–569. DOI: https://doi.org/10.1111/j.1574-6976.2007.00076.x.

Gresham, D. & Hong, J. (2015). The functional basis of adaptive evolution in chemostats. *FEMS Microbiology Reviews*, **39**: 2–16. DOI: https://doi.org/10.1111/1574-6976.12082.

Hadley, K.B., Ryan, A.S., Forsyth, S., Gautier, S. & Salem, N. (2016). The essentiality of arachidonic acid in infant development. *Nutrients*, **8**: 216. DOI: https://doi.org/10.3390/nu8040216.

Hauser, R. & Calafat, A.M. (2005). Phthalates and human health. *Occupational and Environmental Medicine*, **62**: 806–818. DOI: https://doi.org/10.1136/oem.2004.017590.

Hibbett, D.S., Stajich, J.E. & Spatafora, J.W. (2013). Toward genome-enabled mycology. *Mycologia*, **105**: 1339–1349. DOI: https://doi.org/10.3852/13-196.

Hogg, J.A. (1992). Steroids, the steroid community, and Upjohn in perspective: a profile of innovation. *Steroids*, **57**: 593–616. DOI: https://doi.org/10.1016/0039-128X(92)90013-Y.

Honma, M. & Nakada, M. (2007). Enantioselective total synthesis of (+)-digitoxigenin. *Tetrahedron Letters*, **48**: 1541–1544. DOI: https://doi.org/10.1016/j.tetlet.2007.01.024.

Huberman, L.B., Liu, J., Qin, L. & Glass, N.L. (2016). Regulation of the lignocellulolytic response in filamentous fungi. *Fungal Biology Reviews*, **30**: 101–111. DOI: https://doi.org/10.1016/j.fbr.2016.06.001.

Hussain, A., Tian, M.-Y., Ahmed, S. & Shahid, M. (2012). Current status of entomopathogenic fungi as mycoinecticides and their inexpensive development in liquid cultures. In *Zoology*, ed. M.-D. García. Croatia: InTech, pp. 104–122. ISBN: 9789535103608. URL: https://mts.intechopen.com/books/zoology/current-status-of-entomopathogenic-fungi-as-mycoinecticides-and-their-inexpensive-development-in-liq.

Hutkins, R.W. (2006). *Microbiology of Fermented Foods: A Modern Approach.* Oxford, UK: Blackwell Publishing. ISBN-10: 0813800188, ISBN-13: 978-0813800189.

Jackson, R.S. (2008). *Wine Science: Principles and Applications*, 3rd edition. London: Academic Press. ISBN: 9780080568744.

Jin, M.-J., Huang, H., Xiao, A.-H. *et al.* (2008). A novel two-step fermentation process for improved arachidonic acid production by *Mortierella alpina*. *Biotechnology Letters*, **30**: 1087–1091. DOI: https://doi.org/10.1007/s10529-008-9661-1.

Jones, P., Shahab, B.A., Trinci, A.P.J. & Moore, D. (1988). Effect of polymeric additives, especially Junlon and Hostacerin, on growth of some basidiomycetes in submerged culture. *Transactions of the British Mycological Society*, **90**: 577–583. DOI: https://doi.org/10.1016/S0007-1536(88)80062-7.

Joshi, V.K. (ed.) (2016). *Indigenous Fermented Foods of South Asia.* Boca Raton, FL: CRC Press, an imprint of Taylor & Francis Group. ISBN: 9781439887905.

Karagiosis, S.A. & Baker, S.E. (2012). Fungal cell factories. In *Food and Industrial Bioproducts and Bioprocessing*, ed. N.T. Dunford. Oxford, UK: Wiley-Blackwell, pp. 205–220. DOI: https://doi.org/10.1002/9781119946083.ch8.

Karasov, W.H. & Douglas, A.E. (2013). Comparative digestive physiology. *Comprehensive Physiology*, **3**: 741–783. DOI: https://doi.org/10.1002/cphy.c110054.

Khan, A.A., Bacha, N., Ahmad, B. *et al.* (2014). Fungi as chemical industries and genetic engineering for the production of biologically active secondary metabolites. *Asian Pacific Journal of Tropical Biomedicine*, **4**: 859–870. DOI: https://doi.org/10.12980/APJTB.4.2014APJTB-2014-0230.

Kirk, P.M., Cannon, P.F., Minter, D.W. & Stalpers, J.A. (2008). *Dictionary of the Fungi*, 10th edition. Wallingford: CABI Publishing. ISBN-10: 0851998267, ISBN-13: 978-0851998268. Kindle edition: ASIN: B00K7ANXJI.

Kumar, R., Singh, S. & Singh, O.V. (2008). Bioconversion of lignocellulosic biomass: biochemical and molecular perspectives. *Journal of Industrial Microbiology and Biotechnology*, **35**: 377–391. DOI: https://doi.org/10.1007/s10295-008-0327-8.

Lin, Y.-L., Wang, T.-H., Lee, M.-H. & Su, N.-W. (2008). Biologically active components and nutraceuticals in the *Monascus*-fermented rice: a review. *Applied Microbiology and Biotechnology*, **77**: 965–973. DOI: https://doi.org/10.1007/s00253-007-1256-6.

Liu, B., Liu, J., Ju, M., Lic, X. & Wang, P. (2017). Bacteria-white-rot fungi joint remediation of petroleum-contaminated soil based on sustained-release of laccase. *RSC Advances*, **7**: 39075–39081. DOI: https://doi.org/10.1039/C7RA06962F.

Liu, K.S. (2008). Food use of whole soybeans. In *Soybeans: Chemistry, Production. Processing and Utilization*, ed. L.A. Johnson, P.J. White & R. Galloway. Urbana, IL: AOCS Press, pp. 441–481. ISBN-10: 1893997642, ISBN-13: 978-1893997646.

Lucey, J.A. (2007). Microstructural approaches to the study and improvement of cheese and yogurt products. In *Understanding and Controlling the Microstructure of Complex Foods*, ed. D.J. McClements. Cambridge, UK: Woodhead Publishing, an imprint of Elsevier Ltd, pp. 600–621. ISBN: 978-1-84569-151-6. DOI: https://doi.org/https://doi.org/10.1533/9781845693671.4.600.

Luh, B.S. (1995). Industrial production of soy sauce. *Journal of Industrial Microbiology and Biotechnology*, **14**: 467–471. DOI: https://doi.org/10.1007/BF01573959.

Luna-Solano, G., Salgado-Cervantes, M.A., Rodriguez-Jimenes, G.C. & Garcia-Alvarado, M.A. (2005). Optimization of brewer's yeast spray drying process. *Journal of Food Engineering*, **68**: 9–18. DOI: https://doi.org/10.1016/j.jfoodeng.2004.05.019.

Matteau, D., Baby, V., Pelletier, S. & Rodrigue, S. (2015). A small-volume, low-cost, and versatile continuous culture device. *PLoS ONE*, **10**: article e0133384. DOI: https://doi.org/10.1371/journal.pone.0133384.

McCluskey, K. & Baker, S.E. (2017). Diverse data supports the transition of filamentous fungal model organisms into the post-genomics era. *Mycology*, **8**: 67–83. DOI: 10.1080/21501203.2017.1281849.

McSweeney, P.L.H. (2004). Biochemistry of cheese ripening. *International Journal of Dairy Technology*, **57**: 127–144. DOI: https://doi.org/10.1111/j.1471-0307.2004.00147.x.

Mendoza-Mendoza, A., Clouston, A., Li, J.-H. *et al.* (2016). Isolation and mass production of *Trichoderma*. In *Microbial-Based Biopesticides. Methods in Molecular Biology*, **Vol. 1477**, ed. T. Glare & M. Moran-Diez. New York: Humana Press, pp. 13–20. ISBN: 978-1-4939-6365-2. DOI: https://doi.org/10.1007/978-1-4939-6367-6_2.

Meyer, V., Andersen, M.R., Brakhage, A.A. *et al.* (2016). Current challenges of research on filamentous fungi in relation to human welfare and a sustainable bio-economy: a white paper. *Fungal Biology and Biotechnology*, **3**: 6. DOI: https://doi.org/10.1186/s40694-016-0024-8.

Minter, D.W. (2001). Fungal conservation in Cuba. In *Fungal Conservation: Issues and Solutions*, ed. D. Moore, M.M. Nauta, S.E. Evans & M. Rotheroe. Cambridge, UK: Cambridge University Press, pp. 182–196. ISBN-10: 0521048184, ISBN-13: 978-0521048187.

Mitchell, D.A., Krieger, N. & Berovič, M. (2006). *Solid-State Fermentation Bioreactors: Fundamentals of Design and Operation*. Berlin: Springer-Verlag. ISBN-10: 3540312854, ISBN-13: 978-3540312857.

Mojsov, K.D. (2014). Trends in bio-processing of textiles: a review. *Advanced Technologies*, **3**: 135–138. URL: http://www.tf.ni.ac.rs/casopis-arhiva/sveska3vol2/c17.pdf.

Montet, D. & Ray, R.C. (ed.) (2015). *Fermented Foods, Part I: Biochemistry and Biotechnology* (*Food Biology* Series). Boca Raton, FL: CRC Press, an imprint of Taylor & Francis Group. ISBN: 978-1498740791.

Moore, D. & Chiu, S.W. (2001). Fungal products as food. Chapter 10 in *Bio-Exploitation of Filamentous Fungi*, ed. S.B. Pointing & K.D. Hyde. Hong Kong; Fungal Diversity Press, pp. 223–251.

Nevalainen, H. & Peterson, R. (2014). Making recombinant proteins in filamentous fungi: are we expecting too much? *Frontiers in Microbiology*, **5**: 75. DOI: https://doi.org/10.3389/fmicb.2014.00075.

Nielsen, D.S., Crafack, M., Jespersen, L. & Jakobsen, M. (2013). The microbiology of cocoa fermentation. In *Chocolate in Health and Nutrition*, ed. R. Watson, V. Preedy & S. Zibadi, **Volume 7** of the *Nutrition and Health* book series. Totowa, NJ: Humana Press, an imprint of Springer Science+Business Media, LLC, pp. 39–60. ISBN: 978-1-61779-802-3. DOI: https://doi.org/10.1007/978-1-61779-803-0_4.

Nielsen, J.C. & Nielsen, J. (2017). Development of fungal cell factories for the production of secondary metabolites: linking genomics and metabolism. *Synthetic and Systems Biotechnology*, **2**: 5–12. DOI: https://doi.org/10.1016/j.synbio.2017.02.002.

Oakley, C.E., Ahuja, M., Sun, W.-W. *et al.* (2016). Discovery of McrA, a master regulator of *Aspergillus* secondary metabolism. *Molecular Microbiology*, **103**: 347–365. DOI: https://doi.org/10.1111/mmi.13562.

Ocaña-Romo, E., Ferrer-Parra, L., López-Nicolás D.I. *et al.* (2018). Partial characterization of esterases from *Fusarium culmorum* grown in media containing di (2-ethyl hexyl phthalate) in solid-state and submerged fermentation. *New Biotechnology*, **44**: Abstracts Supplement, S137:25–11. DOI: https://doi.org/10.1016/j.nbt.2018.05.1099.

Olempska-Beer, Z.S., Merker, R.I., Ditto, M.D. & DiNovi, M.J. (2006). Food-processing enzymes from recombinant microorganisms: a review. *Regulatory Toxicology and Pharmacology*, **45**: 144–158. DOI: https://doi.org/10.1016/j.yrtph.2006.05.001.

Panda, B.P., Javed, S. & Ali, M. (2010). Production of angkak through co-culture of *Monascus purpureus* and *Monascus ruber*. *Brazilian Journal of Microbiology*, **41**: 757–764. DOI: https://doi.org/10.1590/S1517-83822010000300028.

Pandey, A., Selvakumar, P., Soccol, C.R. & Nigam, P. (1999). Solid-state fermentation for the production of industrial enzymes. *Current Science*, **77**: 149–162. URL: http://www.jstor.org/stable/24102923.

Pandey, A. (2003). Solid-state fermentation. *Biochemical Engineering Journal*, **13**: 81–84. DOI: https://doi.org/10.1016/S1369-703X(02)00121-3.

Pattanagul, P., Pinthong, R., Phianmongkhol, A. & Leksawasdi, N. (2007). Review of angkak production (*Monascus purpureus*). *Chiang Mai Journal of Science*, **34**: 319–328. URL: https://pdfs.semanticscholar.org/2e66/64a3d44b4f4cb362ae894b1615907f6e2578.pdf.

Pazarlioğlu, N.K., Sariişik, M. & Telefoncu, A. (2005). Treating denim fabrics with immobilized commercial cellulases. *Process Biochemistry*, **40**: 767–771. DOI: https://doi.org/10.1016/j.procbio.2004.02.003.

Peralta, R.M., da Silva, B.P., Gomes Côrrea, R.C. *et al.* (2017). Enzymes from basidiomycetes: peculiar and efficient tools for biotechnology. In *Biotechnology of Microbial Enzymes: Production, Biocatalysis and Industrial Applications*, ed. G. Brahmachari. Amsterdam: Academic Press, an imprint of Elsevier, pp. 119–149. DOI: https://doi.org/10.1016/B978-0-12-803725-6.00005-4.

Peterson, D.H. & Murray, H.C. (1952). Microbiological oxygenation of steroids at carbon-11. *Journal of the American Chemical Society*, **74**: 1871–1872. DOI: https://doi.org/10.1021/ja01127a531.

Petre, M. (ed) (2015). *Mushroom Biotechnology: Developments and Applications*. London: Academic Press, an imprint of Elsevier Inc. ISBN: 9780128027943.

Plassard, C. & Fransson, P. (2009). Regulation of low molecular weight organic acid production in fungi. *Fungal Biology Reviews*, **23**: 30–39. DOI: https://doi.org/10.1016/j.fbr.2009.08.002.

Poore, J. & Nemecek, T. (2018). Reducing food's environmental impacts through producers and consumers. *Science*, **360**: 987–992. DOI: https://doi.org/10.1126/science.aaq0216.

Portillo-Ojeda, M.L., Arteaga-Mejía, M., González-Márquez, A. & Sánchez, C. (2018). Effect of the pH on growth and esterase activity of *Fusarium culmorum* grown on media supplemented with di (2-ethylhexyl) phthalate in submerged fermentation. *New Biotechnology*, **44**: Abstracts Supplement, S138:25–12. DOI: https://doi.org/10.1016/j.nbt.2018.05.1100.

Purnomo, A.S., Mori, T., Takagi, K. & Kondo, R. (2011). Bioremediation of DDT contaminated soil using brown-rot fungi. *International Biodeterioration & Biodegradation*, **65**: 691–695. DOI: https://doi.org/10.1016/j.ibiod.2011.04.004.

Reddy, P.P. (2013). *Recent Advances in Crop Protection*. New Delhi: Springer India. ISBN: 978-8132207221.

Richards, T.A., Leonard, G., Soanes, D.M. & Talbot, N.J. (2011). Gene transfer into the fungi. *Fungal Biology Reviews*, **25**: 98–110. DOI: https://doi.org/10.1016/j.fbr.2011.04.003.

Rodrigues, A.G. (2016). Secondary metabolism and antimicrobial metabolites of Aspergillus. In *New and Future Developments in Microbial Biotechnology and Bioengineering, Aspergillus System Properties and Applications*, ed. V.K. Gupta. Amsterdam: Elsevier, pp. 81–93. ISBN: 978-0-444-63505-1. DOI: https://doi.org/10.1016/B978-0-444-63505-1.00006-3.

Ropars, J., Cruaud, C., Lacoste, S. & Dupont, J. (2012). A taxonomic and ecological overview of cheese fungi. *International Journal of Food Microbiology*, **155**: 199–210. DOI: https://doi.org/10.1016/j.ijfoodmicro.2012.02.005.

Ropars, J., Rodríguez de la Vega, R.C., López-Villavicencio, M. *et al.* (2015). Adaptive horizontal gene transfers between multiple cheese-associated fungi. *Current Biology*, **25**: 2562–2569. DOI: https://doi.org/10.1016/j.cub.2015.08.025.

Rouches, E., Herpoël-Gimbert, I., Steyer, J.P. & Carrere, H. (2016). Improvement of anaerobic degradation by white-rot fungi pretreatment of lignocellulosic biomass: a review. *Renewable and Sustainable Energy Reviews*, **59**: 179–198. DOI: https://doi.org/10.1016/j.rser.2015.12.317.

Sakuradani, E., Hirano, Y., Kamada, N. *et al.* (2004). Improvement of arachidonic acid production by mutants with lower n-3 desaturation activity derived from *Mortierella alpina* 1S-4. *Applied Microbiology and Biotechnology*, **66**: 243–248. DOI: https://doi.org/10.1007/s00253-004-1682-7.

Samuel, D. (1996). Investigation of ancient Egyptian baking and brewing methods by correlative microscopy. *Science*, **273**: 488–490. DOI: https://doi.org/10.1126/science.273.5274.488.

Sánchez, C. (2009). Lignocellulosic residues: biodegradation and bioconversion by fungi. *Biotechnology Advances*, **27**: 185–194. DOI: https://doi.org/10.1016/j.biotechadv.2008.11.001.

Sandri, I.G., Fontana, R.C., Barfknecht, D.M. & da Silveira, M.M. (2011). Clarification of fruit juices by fungal pectinases. *LWT – Food Science and Technology*, **44**: 2217–2222. DOI: https://doi.org/10.1016/j.lwt.2011.02.008.

Scharf, D.H. & Brakhage, A.A. (2013). Engineering fungal secondary metabolism: a roadmap to novel compounds. *Journal of Biotechnology*, **163**: 179–183. DOI: https://doi.org/10.1016/j.jbiotec.2012.06.027.

Schwan, R.F. & Wheals, A.E. (2004). The microbiology of cocoa fermentation and its role in chocolate quality. *Critical Reviews in Food Science and Nutrition*, **44**: 205–221. DOI: https://doi.org/10.1080/10408690490464104.

Schwarze, F.W.M.R. (2007). Wood decay under the microscope. *Fungal Biology Reviews*, **1**: 133–170. DOI: https://doi.org/10.1016/j.fbr.2007.09.001.

Shah, S.U. (2012). Importance of genotoxicity & S2A guidelines for *IOSR. Journal of Pharmacy and Biological Sciences*, **1**: 43–54. DOI: https://doi.org/10.9790/3008-0124354.

Shahid, M., Mohammad, F., Chen, G., Tang, R.-C. & Xing, T. (2016). Enzymatic processing of natural fibres: white biotechnology for sustainable development. *Green Chemistry*, **18**: 2256–2281. DOI: https://doi.org/10.1039/C6GC00201C.

Shen, J., Rushforth, M., Cavaco-Paulo, A., Guebitz, G. & Lenting, H. (2007). Development and industrialisation of enzymatic shrink-resist process based on modified proteases for wool machine washability. *Enzyme and Microbial Technology*, **40**: 1656–1661. DOI: https://doi.org/10.1016/j.enzmictec.2006.07.034.

Show, P.L., Oladele, K.O., Siew, Q.Y. *et al.* (2015). Overview of citric acid production from *Aspergillus niger*. *Frontiers in Life Science*, **8**: 271–283. DOI: https://doi.org/10.1080/21553769.2015.1033653.

Simpson, D.R., Withers, J.M., Wiebe, M.G. *et al.* (1998). Mutants with general growth rate advantages are the predominant morphological mutants to be isolated from the Quorn production plant. *Mycological Research*, **102**: 221–227. DOI: https://doi.org/10.1017/S0953756297004644.

Slot, J.C., Townsend, J.P. & Wang, Z. (2017). Fungal gene cluster diversity and evolution. *Advances in Genetics*, **100**: 141–178. DOI: https://doi.org/10.1016/bs.adgen.2017.09.005.

Soares, J.C., Moreira, P.R., Queiroga, A.C. *et al.* (2011). Application of immobilized enzyme technologies for the textile industry: a review. *Biocatalysis and Biotransformation*, **29**: 223–237. DOI: https://doi.org/10.3109/10242422.2011.635301.

Sohretoglu, D. & Huang, S. (2018). *Ganoderma lucidum* polysaccharides as an anti-cancer agent. *Anti-Cancer Agents in Medicinal Chemistry*, **18**: 667–674. DOI: https://doi.org/10.2174/1871520617666171113121246.

Stanbury, P.F., Whitaker, A. & Hall, S.J. (2016). *Principles of Fermentation Technology*, 3rd edition. Oxford, UK: Butterworth-Heinemann, an imprint of Elsevier. ISBN: 978-0080999531.

Steenkamp, E.T., Wingfield, M.J., McTaggart, A.R. & Wingfield, B.D. (2018). Fungal species and their boundaries matter: definitions, mechanisms and practical implications. *Fungal Biology Reviews*, **32**: 104–116. DOI: https://doi.org/10.1016/j.fbr.2017.11.002.

Stepaniak, L. (2004). Dairy enzymology. *International Journal of Dairy Technology*, **57**: 153–171. DOI: https://doi.org/10.1111/j.1471-0307.2004.00144.x.

Sudheer, S., Alzorqi, I., Manickam, S. & Ali, A. (2018). Bioactive compounds of the wonder medicinal mushroom '*Ganoderma lucidum*'. In *Bioactive Molecules in Food*, ed. J.M. Mérillon & K. Ramawat. Cham, Switzerland: Springer International Publishing, pp 1–31. ISBN: 978-3-319-54528-8. DOI: https://doi.org/10.1007/978-3-319-54528-8_45-1.

Tallima, H. & El Ridi, R. (2017). Arachidonic acid: physiological roles and potential health benefits – a review. *Journal of Advanced Research*, **11**, 33–41. DOI: https://doi.org/10.1016/j.jare.2017.11.004.

Todaro, C.M. & Vogel, H.C. (ed.) (2014). *Fermentation and Biochemical Engineering Handbook*. Norwich, NY: William Andrew, an imprint of Elsevier. ISBN: 978-1455725533.

Tong, W.-Y. & Dong, X. (2009). Microbial biotransformation: recent developments on steroid drugs. *Recent Patents on Biotechnology*, **3**: 141–153. DOI: https://doi.org/10.2174/187220809788700157.

Trinci, A.P.J. (1970). Kinetics of the growth of mycelial pellets of *Aspergillus nidulans*. *Archives of Microbiology*, **73**: 353–367. DOI: https://doi.org/10.1007/BF00412302.

Trinci, A.P.J., Davies, D.R., Gull, K. *et al.* (1994). Anaerobic fungi in herbivorous animals. *Mycological Research*, **98**: 129–152. DOI: https://doi.org/10.1016/S0953-7562(09)80178-0.

Vigneswaran, C., Ananthasubramanian, M. & Kandhavadivu, P. (2014). *Bioprocessing of Textiles*. Boca Raton, FL: CRC Press/Woodhead Publishing India in Textiles, an imprint of Taylor & Francis Group. ISBN: 9789380308425.

Walker, G.M., Abbas, C., Ingledew, W.M. & Pilgrim, C. (ed.) (2017). *The Alcohol Textbook: A Reference for the Beverage, Fuel and Industrial Alcohol Industries*, 6th edition. Duluth, GA: The Ethanol Technology Institute, Lallemand Biofuels & Distilled Spirits. ISBN: 978-0-692-93088-5.

Ward, O.P. (2012). Production of recombinant proteins by filamentous fungi. *Biotechnology Advances*, **30**: 1119–1139. c10.1016/j.biotechadv.2011.09.012.

Wang, J., Liu, L., Ball, T. *et al.* (2016). Revealing a 5,000-y-old beer recipe in China. *Proceedings of the National Academy of Sciences of the United States of America*, **113**: 6444–6448. DOI: https://doi.org/10.1073/pnas.1601465113.

Wiebe, M.G (2004). Quorn™ Myco-protein: overview of a successful fungal product. *Mycologist*, **18**: 17–20. DOI: https://doi.org/10.1017/S0269-915X(04)00108-9.

Wiebe, M.G., Robson, G.D., Cuncliffe, B., Oliver, S.G. & Trinci, A.P.J. (1993). Periodic selection in long term continuous-flow cultures of the filamentous fungus *Fusarium graminearum*. *Journal of General Microbiology*, **139**: 2811–2817. DOI: https://doi.org/10.1099/00221287-139-11-2811.

Wiebe, M.G., Robson, G.D., Oliver, S.G. & Trinci, A.P.J. (1994). Evolution of *Fusarium graminearum* A3/5 grown in a glucose-limited chemostat culture at a slow dilution rate. *Microbiology*, **140**: 3023–3029. DOI: https://doi.org/10.1099/13500872-140-11-3023.

Wiebe, M.G., Robson, G.D., Oliver, S.G. & Trinci, A.P.J. (1995). Evolution of *Fusarium graminearum* A3/5 grown in a series of glucose-limited chemostat cultures at a high dilution rate. *Mycological Research*, **99**: 173–178. DOI: https://doi.org/10.1016/S0953-7562(09)80883-6.

Willett, W., Rockström, J., Loken, B. *et al.* (2019). Food in the Anthropocene: the EAT-Lancet Commission on healthy diets from sustainable food systems. *Lancet*, published online 16 January 2019. DOI: http://doi.org/10.1016/S0140-6736(18)31788-4. The complete report is available as a free 47-page PDF download, as is the 36-page Appendix.

Willis, K.J. (ed.) (2018). *State of the World's Fungi 2018*. Report. Richmond, UK: Royal Botanic Gardens, Kew. ISBN: 978-1-84246-678-0. URL: https://stateoftheworldsfungi.org/2018/.

Wolf, J., Asrar, G.R. & West, T.O. (2017). Revised methane emissions factors and spatially distributed annual carbon fluxes for global livestock. *Carbon Balance and Management*, **12**: 16. DOI: https://doi.org/10.1186/s13021-017-0084-y.

Wolfe, B.E. & Dutton, R.J. (2015). Fermented foods as experimentally tractable microbial ecosystems. *Cell*, **161**: 49–55. DOI: https://doi.org/10.1016/j.cell.2015.02.034.

Yan, Y.-M., Wang, X.-L., Luo, Q. *et al.* (2015). Metabolites from the mushroom *Ganoderma lingzhi* as stimulators of neural stem cell proliferation. *Phytochemistry*, **114**: 155–162. DOI: https://doi.org/10.1016/j.phytochem.2015.03.013.

Yoshida, S., Hiraga, K., Takehana, T. *et al.* (2016). A bacterium that degrades and assimilates poly(ethylene terephthalate). *Science*, **351**: 1196–1199. DOI: https://doi.org/10.1126/science.aad6359.

Zhong, Y., Lu, X., Xing, L., Ho, S.W.A. & Kwan, H.S. (2018). Genomic and transcriptomic comparison of *Aspergillus oryzae* strains: a case study in soy sauce koji fermentation. *Journal of Industrial Microbiology & Biotechnology*, **45**: 839–853. DOI: https://doi.org/10.1007/s10295-018-2059-8.

APPENDIX 1

Outline Classification of Fungi

A trend in the last two or three decades has been the growing awareness of the enormous and numerous influences that fungi have exerted on the development of the biosphere of this planet. We have indicated in the text of this book that animals and fungi are sister groups; they are each other's closest relatives and share a common ancestor, which is known as the opisthokont clade. The name 'opisthokont' comes from the Greek and means 'posterior flagellum', so the common characteristic that gives them their name is that flagellate cells, when they occur, are propelled by a single posterior flagellum, and this applies as much to chytrid zoospores as to animal sperm. In contrast, other eukaryotic organisms that have motile cells propel them with one or more ***anterior*** flagella (these are the heterokonts). A recognisably fungal grade of organisation has been evident from the very earliest stages in evolution of higher organisms. And perhaps fungal pioneering goes further than this, because the idea that the very first terrestrial eukaryotes might have been fungal is increasingly gaining support. A couple of titles of scholarly articles will illustrate this: 'Terrestrial life: fungal from the start?' (Blackwell, 2000); 'Early cell evolution, eukaryotes, anoxia, sulfide, oxygen, fungi first (?), and a Tree of Genomes revisited' (Martin *et al.*, 2003) and *Fungal Biology in the Origin and Emergence of Life* (Moore, 2013a).

The evolution of fungi is ornamented with some of the most crucial mutualisms, or coevolutions, of the living world. Lichens could be the most ancient; they have been found with certainty in some of the oldest fossils and are currently thought of as being able to flourish in the most extreme environments. Indeed, lichens can also survive 16 days open exposure to the space environment in orbit (Sancho *et al.*, 2007). Fully formed mycorrhizas can also be found in the most ancient plant fossils, and today about 95% of all terrestrial plants depend on this fungal infection of their roots to provide the plant with phosphorus and other nutrients. Fungi may also have an impact on the aerial parts of plants today because endophytes (fungi that live in the spaces within the leaves and stems of living plants) 'are present in any healthy plant tissue' (Sieber, 2007). Animal mutualisms also abound; from the dependence of leaf-cutter ants on their fungal gardens to make the ants the dominant herbivores of their tropical rain forests, to the dependence of even-toed ungulates on the chytrids in their rumens that make them the dominant herbivores of their savanna grasslands.

A possible explanation for fungal success throughout geological time is that their fundamental lifestyle is to recycle the dead remains of other organisms. This means that extinction events of other organisms are just ways of providing additional nutritional resources for fungi. The Permian–Triassic extinction event that occurred approximately 250 million years ago is informally known as the Great Dying. It was the Earth's most severe extinction event (so far!), with about 96% of all marine species and 70% of terrestrial vertebrates becoming extinct. Plants suffered huge extinctions as well as animals, with such massive dieback of vegetation that terrestrial ecosystems destabilised and collapsed throughout the world. This global ecological catastrophe was caused by changes in atmospheric chemistry resulting from the volcanic eruption that formed what are now known as the Siberian Traps flood basalts. This eruption is thought to have covered an area of what is now Siberia equivalent to the current area of Australia. However, the result of all this death and destruction to the flora and fauna is that 'sedimentary organic matter preserved in the latest Permian deposits is characterised by unparalleled abundances of fungal remains, irrespective of depositional environment ... floral provinciality, and climatic zonation' (Visscher *et al.*, 1996). The fungi were having a ball, while the animals and plants were dying in unprecedented numbers.

Today, we humans depend on fungi for a great deal of our daily existence. Obviously, we depend on mycorrhizas to grow our crops, and chytrids to feed our farm animals to provide meat, dairy products, leather and woollen textiles. But we also depend on them for fungicides (strobilurins) to keep our farm crops healthy, enzymes to process our food and our textiles, and wonder drugs to keep ourselves and our animals healthy (penicillins, cephalosporins, cyclosporins, statins); and we've not even mentioned bread, cheese, wine and ale. After such a catalogue of the crucial contributions of fungi to life on Earth (past and present), it is worrying to find that fungi are marginalised or totally ignored in schools around the world. Globally, school curricula call for comparisons only between animals and plants, leaving their pupils not only ignorant of the great Kingdom *Fungi*, but convinced that fungi are bacteria (Moore *et al.*, 2006)! An established spiral of ignorance seems to exist: learn little or nothing about fungi so that when you are old enough to teach, you can teach little or nothing about fungi. If this spiral continues uncorrected, mycologists will doubtless become extinct. At which point, there will likely be an increase in populations of fungi digesting the carcasses of the few remaining previously eminent mycologists.

The taxonomy of the fungi is in continuous development as data about relationships accumulate and opinions change. At the time of writing, fungi are by far the best sampled eukaryotic kingdom, in terms of the number of complete genome sequences that are available. The recent deluge of genomic data available for the fungal kingdom (described as a tsunami by one author) emerging from the JGI's 1,000 Fungal Genomes Project provides genomic sequence data for an accelerating pace of taxonomic revisions. A 1,000 genomes may sound a lot, and indeed sequencing this number of genomes represents several thousand person-years of effort; ***but*** it also represents only a tiny fraction of the 2–4 million species thought to exist on this planet (Hawksworth & Lücking, 2017). It is with a degree of trepidation, therefore, that we present here a reasonably detailed list of taxa of Kingdom *Fungi*. We know that this section will inevitably become outdated, perhaps even by the time you read this. However, we cannot claim to give you a guidebook to the kingdom without giving you some information about the population of that kingdom. ***Please accept the following as a snapshot of Kingdom Fungi, in this book at this time.***

The following classification is adapted from the 9th and 10th editions of the *Dictionary of the Fungi* (Kirk *et al.*, 2001, 2008), but it adopts the phylogenetic arrangement emerging from the AFTOL (*Assembling the Fungal Tree of Life*) project funded by the US National Science Foundation (visit: http://www.aftol.org/; and see Blackwell *et al.*, 2006), as set out by Hibbett *et al.* (2007). AFTOL is a work in progress and uncertainties remain about the exact relationships of many groups. These are indicated in the annotated classification below with '*incertae sedis*', which means 'of uncertain position', this being the standard term for a taxonomic group of currently unknown or undefined relationship. Some of the material contained in this section is sourced from the '*Classification of the fungi*' entry (written by David Moore) in the *Encyclopaedia Britannica* and is used with permission, by courtesy of Encyclopaedia Britannica, Inc.© 2008.

The taxonomic arrangement shown here has been further adapted from that shown in the first edition to reflect the genome analyses of more recent times particularly the detailed reviews of phylogenetic analyses of McCarthy and Fitzpatrick (2017); as well as Choi and Kim (2017), Ren *et al.* (2016), Yarza *et al.* (2017), Zhang *et al.* (2017); and the two monographic volumes constituting the second edition of volume **VII** of *The Mycota* (McLaughlin & Spatafora, ed., 2014, 2015). Other publications (shown immediately below) are referenced in the taxonomic listings that follow. A cladogram (phylogenetic tree) showing the relationships of the 5–7 'Supergroups' into which the global tree of eukaryotes is currently subdivided has been shown above in Chapter 2 (Figure 2.5). The taxonomically formal kingdoms have been assigned to the deliberately informal supergroups by phylogenomic analysis, a method that reconstructs evolutionary histories using large alignments of tens to hundreds of genes, as well as whole-genome sequences. The intention is to maintain flexibility as further work on phylogenomics resolves contentious relationships (Adl *et al.*, 2005, 2012; Burki, 2014; and see https://www.sciencenews.org/article/tree-life-gets-makeover).

Molecular methods are revolutionising our understanding concerning the extent of fungal diversity in nature and the definition of species among entities that are known only from the polynucleotides recovered from nature. Analysis of the phylogenetic and evolutionary ecological relationships among the fungal genomes (and partial genomes) so recovered are threatening to drastically alter the morphology-based Linnaean classification system which is the basis for the classification scheme detailed below. In particular, there is a move towards recognising taxonomic ranks using phylogenetic divergence time as a universally standardised criterion (Zhao *et al.*, 2016; Tedersoo *et al.*, 2018; He *et al.*, 2019).

OUTLINE CLASSIFICATION OF FUNGI

Supergroup: Opisthokonta

Kingdom: *Fungi*

Basal fungi

Phylum: *Cryptomycota*

Also known as Rozellida and Rozellomycota, but *Cryptomycota* is preferred and has been validly published. These organisms were first detected as DNA sequences retrieved from a freshwater pond and were found in samples taken from other freshwater environments, soils and marine sediments. Phylogenetic analysis of these sequences formed a unique terminal clade. The only formally described genera in the clade are *Rozella*, which was previously considered a primitive chytrid, *Nucleophaga* and *Paramicrosporidium*. The existence of related organisms is known from environmental DNA sequences. *Rozella* differs from classical fungi in lacking chitinous cell walls at any stage in their life cycle, although chitin has been detected in the inner layer of resting spores and the organisms possess a fungal-specific chitin synthase gene. *Cryptomycota* are phagotrophic parasites that feed by attaching to, engulfing, or living inside other cells; although most fungi feed by taking in nutrients from outside the cell (osmotrophy) (Jones *et al.*, 2011).

Phylum: *Microsporidia*

No formal subdivision is proposed yet because of the lack of well-sampled multigene phylogenies within the group. *Microsporidia* are unicellular parasites of animals and protists with highly reduced mitochondria. *Microsporidia* may be a sister group of the rest of the Fungi, but this suggestion may have arisen from incomplete sampling (Didier *et al.*, 2014).

TRADITIONAL *CHYTRIDIOMYCOTA*

Phylum: *Chytridiomycota*

Water moulds that live as aquatic saprotrophs or parasites in freshwater and soils, a few are marine. Chytrids produce motile asexual zoospores (with a single posterior flagellum, both a kinetosome and nonfunctional centriole, nine flagellar props, and a microbody-lipid globule complex) in zoosporangia. Golgi apparatus with stacked cisternae; nuclear envelope fenestrated at poles during mitosis. Thallus may be unicellular or filamentous, and holocarpic (where all the thallus is involved in formation of the sporangium) or eucarpic (where only part of the thallus is converted into the sporophore, monocentric, polycentric, or filamentous. Sexual reproduction with zygotic meiosis where known; sometimes produce motile sexual zoogametes. Considered to be the most ancestral group of fungi (Powell & Letcher, 2014). Type: *Chytridium*.

Class: *Chytridiomycetes*

Reproducing asexually by zoospores bearing a single posteriorly directed flagellum; zoospores containing a kinetosome and a non-flagellated centriole; thallus monocentric or rhizomycelial polycentric; sexual reproduction not oogamous. Type: *Chytridium*.

Order: *Chytridiales*

Thallus monocentric or polycentric. No mycelium formed, but produce short absorptive filaments called rhizoids that lack nuclei and may be sufficiently extensive to be described as a rhizomycelium. Zoospores typically with flagellar base containing an electron-opaque plug, microtubules extending from one side of the kinetosome in a parallel array, ribosomes aggregated near the nucleus, kinetosome parallel to non-flagellated centriole and connected to it by fibrous material, nucleus not associated with kinetosome, fenestrated cisterna (rumposome) adjacent to lipid globule. Mostly freshwater saprotrophs or parasites of algae, other fungi and higher plants (e.g. *Synchytrium endobioticum*, cause of potato wart disease). Previously included the plant parasite *Olpidium* but this is now *incertae sedis*. Type: *Chytridium*; example genera: *Chytridium, Chytriomyces, Nowakowskiella*.

Order: *Rhizophydiales*; example genus: *Rhizophydium*

Order: *Spizellomycetales*; example genera: *Spizellomyces, Powellomyces*

Phylum: *Monoblepharidomycota*

Thallus filamentous, either extensive or a simple unbranched thallus, often with a basal holdfast; asexual reproduction by zoospores or autospores; zoospores containing a kinetosome parallel to a non-flagellated centriole, a striated disc partially extending around the kinetosome, microtubules radiating anteriorly from the striated disc, a ribosomal aggregation, and rumposome (fenestrated cisterna) adjacent to a microbody; sexual reproduction oogamous by means of posteriorly uniflagellate antherozoids borne in antheridia and non-flagellate female gametes borne in oogonia (Powell & Letcher, 2014). Type: *Monoblepharis*.

Order: *Monoblepharidales*; example genus: *Monoblepharis*

Phylum: *Neocallimastigomycota*

Thallus monocentric or polycentric; anaerobic, found in digestive system of larger herbivorous mammals and possibly in other terrestrial and aquatic anaerobic environments; lacks mitochondria but contains hydrogenosomes of mitochondrial origin; zoospores posteriorly uniflagellate or polyflagellate, kinetosome present but nonfunctional centriole absent, kinetosome-associated complex composed of a skirt, strut, spur and circumflagellar ring, microtubules extend from spur and radiate around nucleus, forming a posterior fan, flagellar props absent; nuclear envelope remains intact throughout mitosis (Gruninger *et al.*, 2014; Powell & Letcher, 2014; Edwards *et al.*, 2017; Wang *et al.*, 2017).

Class: *Neocallimastigomycetes*

Order: *Neocallimastigales*; example genus: *Neocallimastix*

Phylum: *Blastocladiomycota*

Very like the chytrids, characteristically, the *Blastocladiomycota* have life cycles with what is described as a sporic meiosis; that is, meiosis results in the production of haploid spores that can develop directly into a new, but now haploid, individual. This results in a regular alternation of generations between haploid gametothallus and diploid sporothallus individuals. Members of this phylum were included in *Chytridiomycota* in older textbooks. Saprotrophs as well as parasites of fungi, algae, plants and invertebrates, and may be facultatively anaerobic in oxygen-depleted environments. All members of this phylum have zoospores with a distinct ribosome-filled cap around the nucleus. The thallus may be monocentric or polycentric and becomes mycelial in *Allomyces*. Other representative genera are: *Physoderma, Blastocladiella* and *Coelomomyces*. *Physoderma* spp. are parasitic on higher plants, *Coelomomyces* is an obligate endoparasite of insects with alternating sporangia and gametangia stages in mosquito larvae and copepod hosts, respectively (James *et al.*, 2014).

Class: *Blastocladiomycetes*

Order: *Blastocladiales*

Water moulds with a restricted thallus, characterised by the production of thick-walled, pitted, resistant sporangia; sexual reproduction by isogamous (equal in size and alike in form) or anisogamous (unequal in size but still similar in form) planogametes; *Allomyces* exhibits an alternation of two equal generations; most are saprotrophs, but various species of *Coelomomyces* are parasitic in mosquito larvae; uniquely, their hyphae are devoid of cell walls. More than 50 species; example genera: *Allomyces, Coelomomyces.*

TRADITIONAL 'ZYGOMYCOTAN' (ZYGOMYCETE) FUNGI

The traditional 'Zygomycota', being saprotrophs or parasites (especially of arthropods) that produce non-motile, asexual sporangiospores in sporangia and sexual spores known as zygospores. This grouping includes the common moulds such as *Mucor, Rhizopus,* and *Phycomyces*. At one time the *Chytridiomycota, Oomycota* (see below) and 'Zygomycota' were classified together in Class 'Phycomycetes'. This is no longer valid although the word is often used still as a 'catch-all' phrase covering lower fungi. The problem with 'Zygomycota' is that the group of organisms is polyphyletic, and the name was first published without a Latin diagnosis and is invalid. When relationships among the constituent fungal lineages are resolved the name 'Zygomycota' could be resurrected and validated. It cannot be properly defined in this classification, so some of the zygomycetes are left in an uncertain position (*incertae sedis*). The following listing reflects the consensus cladogram of Spatafora *et al.* (2016), which is based on genome-scale data.

Phylum: *Mucoromycota*

Subphylum: *Glomeromycotina*

Until recently, arbuscular mycorrhizal (AM) fungi have generally been classified in the 'Zygomycota' (being placed in the Order Glomales), but they do not form the zygospores characteristic of zygomycetes, and all 'glomalean' fungi form mutualistic symbioses. Recent molecular studies have suggested separate subphylum status is appropriate for the AM fungi, the *Glomeromycotina*, and this is the position taken by the AFTOL study (Redecker *et al.*, 2013; Redecker & Schüßler, 2014; Spatafora *et al.*, 2016).

Class: *Glomeromycetes*

Order: *Archaeosporales*; example genera: *Archaeospora, Geosiphon*

Order: *Diversisporales*; example genera: *Acaulospora, Diversispora, Pacispora*

Order: *Glomerales*; example genus: *Glomus*

Order: *Paraglomerales*; example genus: *Paraglomus*

Subphylum: *Mortierellomycotina*

Previously included in Subphylum *Mucoromycotina*, Subphylum *Mortierellomycotina* comprises a single order, *Mortierellales* (Hoffmann *et al.*, 2011; Benny *et al.*, 2014).

Order: *Mortierellales*; example genera: *Mortierella, Dissophora, Modicella*

Subphylum: *Mucoromycotina*

Fungi saprotrophs, or rarely gall-forming, nonhaustorial, facultative mycoparasites, or forming ectomycorrhiza. Mycelium branched, coenocytic when young, sometimes producing septa that contain micropores at maturity. Asexual reproduction by sporangia, sporangiola, or merosporangia, or rarely by chlamydospores, arthrospores, or blastospores. Sexual reproduction by more or less globose zygospores formed on opposed or apposed suspensors. This group includes the *Mucorales*, which is the core group of the traditional zygomycetes (Benny *et al.*, 2014).

Order: *Mucorales*; often called the bread moulds; saprotrophic, weakly parasitic on plants, or parasitic on humans and then causing mucormycosis (a pulmonary infection); asexual reproduction by sporangiospores, 1-spored sporangiola (a small deciduous sporangium), or conidia; in the genus *Pilobolus* the heavily cutinised sporangium is forcibly discharged. About 360 species; example genera: *Mucor, Parasitella, Phycomyces, Pilobolus, Rhizopus.*

Order: *Endogonales*; example genera: *Endogone, Peridiospora, Sclerogone, Youngiomyces*

Phylum: *Zoopagomycota*

Subphylum: *Entomophthoromycotina*

Obligate pathogens of animals (primarily arthropods), cryptogamic plants, or saprotrophs; occasionally facultative parasites of vertebrates. Somatic state consisting of a well-defined mycelium, coenocytic or septate, walled or protoplasmic, which may fragment to form multinucleate hyphal bodies; protoplasts either hyphoid or amoeboid and changeable in shape; cystidia or rhizoids formed by some taxa. Such nuclear characters as overall size, location and comparative size of nucleoli, presence or absence of granular heterochromatin in chemically unfixed interphasic nuclei, and mitotic patterns are important at the family level. Conidiophores branched or unbranched. Primary spores true conidia, uni-, pluri-, or multinucleate, forcibly discharged by diverse possible means or passively dispersed; secondary conidia often produced. Resting spores with thick bi-layered walls form as zygospores after conjugations of undifferentiated gametangia from different or the same hyphal bodies or hypha or as azygospores arising without prior gametangial conjugations (Benny *et al.*, 2014).

Order: *Entomophthorales*; insect parasites or saprotrophs, some implicated in animal or human diseases; asexual reproduction by modified sporangia functioning as conidia, forcibly discharged. About 150 species; example genera: *Entomophthora, Ballocephala, Conidiobolus, Entomophaga, Neozygites.*

Subphylum: *Kickxellomycotina*

Fungi saprotrophs, mycoparasites, or obligate symbionts. Thallus arising from a holdfast on other fungi as a haustorial parasite, or branched, septate, subaerial hyphae. Mycelium branched or unbranched, regularly septate. Septa with median, disciform cavities containing plugs. Asexual production by 1- or 2-spored merosporangia, trichospores, or arthrospores. Sexual reproduction by zygospores that are globose, biconical, or allantoid and coiled (Benny *et al.*, 2014).

Order: *Kickxellales*; example genera: *Kickxella, Coemansia, Linderina, Spirodactylon*

Order: *Dimargaritales*; example genera: *Dimargaris, Dispira, Tieghemiomyces*

Order: *Harpellales*; thallus simple or branched, septate; asexual reproduction by trichospores; sexual reproduction zygomycetous; about 35 species. These fungi are commensals (organisms living parasitically on another organism but conferring some benefit in return, or at least not harming the host) with their filamentous thallus attached by a holdfast or basal cell to the digestive tract or external cuticle of living arthropods. Taxa in this order have been referred to as 'Trichomycetes' but this is no longer a useful taxon because it describes a polyphyletic group and is more an ecological rather than a phylogenetic grouping, consequently, the term should not be capitalised, i.e. use as 'trichomycetes'). Example genera: *Harpella, Furculomyces, Legeriomyces, Smittium.*

Order: *Asellariales*; thallus branched, septate, attached by a basal coenocytic cell; asexual reproduction by arthrospores; *Asellariales* are retained in the *Fungi* because of their ultrastructural characteristics. Six species; example genera: *Asellaria, Orchesellaria.*

Subphylum: *Zoopagomycotina*

Endo- or ectoparasites of microanimals and fungi. Vegetative body consisting of a simple, branched or unbranched thallus or more of less extensively branched mycelium. Ectoparasites forming haustoria inside the host. Asexual reproduction by arthrospores, chlamydospores or uni- or multispored sporangiola; sporangiospores of multispored sporangiola formed in simple or branched chains (merosporangia). Sexual reproduction by nearly globose zygospores; sexual hyphae like the vegetative hyphae or more or less enlarged (Benny *et al.*, 2014).

Order: *Zoopagales*; parasitic on amoebas, rotifers, nematodes, or other small animals, which they trap by various specialised mechanisms; asexual reproduction by conidia borne singly or in chains, not forcibly discharged. About 60 species; example genera: *Cochlonema, Rhopalomyces, Piptocephalis, Sigmoideomyces, Syncephalis, Zoopage.*

Subkingdom: *Dikarya*

Unicellular or filamentous Fungi, lacking flagella, often with a dikaryotic state; contains *Ascomycota* and *Basidiomycota*. The name alludes to the putative synapomorphy (being a derived, that is non-ancestral, character shared by the two constituent phyla) of ***dikaryotic hyphae***.

PHYLUM: *ASCOMYCOTA*

This is the largest group of fungi, and the lifestyles adopted cover the complete range from saprotrophs, to symbionts (notably lichens), and to parasites and pathogens (plant pathogens are particularly numerous, but there are many important human pathogens in this group also). The *Ascomycota* are characterised by having sexual spores (ascospores) formed ***endogenously*** within an ascus (indeed the original Latin diagnosis consisted of only two words: '*sporae intracellulares*' and while it is questionable that this description is truly diagnostic for the *Ascomycota*, as a validating diagnosis it is acceptable under the Code). A layered hyphal wall with a thin relatively electron-dense outer layer and a thicker electron-transparent inner layer also appears to be diagnostic. Except for the ascosporogenous yeasts (such as *Saccharomyces* and *Schizosaccharomyces*, which are only distantly related), asci are usually produced in complex fruit bodies (ascomata). The phylum comprises at least 64,000 species in 6,355 genera. In the past, these fungi were grouped based on sporophore shape and ascus arrangement. For example, Hemiascomycetes = no sporophores, asci naked; Euascomycetes = asci contained within ascomata, of which there are three main sorts, cleistothecia being closed, usually globose, but may be rudimentary or may consist of loosely interwoven hyphae, perithecia are flask-shaped, and in apothecia the asci are contained within saucer- or disc-shaped ascomata. The arrangement given below reflects the impact of molecular sequence data, but it is important to emphasise that further data are bound to change this interpretation (Prieto *et al.*, 2013; Wijayawardene *et al.*, 2018). Basic type genus: *Peziza*.

Subphylum: *Taphrinomycotina* (Kurtzman & Sugiyama, 2015)

Class: *Taphrinomycetes*

Order: *Taphrinales*; parasites on vascular plants; asci produced from binucleate ascogenous (ascus-producing) cells formed from the hyphae in the manner of chlamydospores (thick-walled spores). Example genera: *Taphrina, Protomyces.*

Class: *Neolectomycetes*

Order: *Neolectales*; example genus: *Neolecta*

Class: *Pneumocystidomycetes*

Order: *Pneumocystidales*; example genus: *Pneumocystis*

Class: *Schizosaccharomycetes*

Order: *Schizosaccharomycetales*; example genus: *Schizosaccharomyces*

Subphylum: *Saccharomycotina* (Kurtzman & Sugiyama, 2015)

Class: *Saccharomycetes*

Order: *Saccharomycetales*

Growth usually by individual yeast cells, often accompanied by pseudohyphae and/or true hyphae. Cell walls predominately of β-glucan. Ascomata not formed; one to many ascospores formed in asci that often are converted from individual cells or borne on simple ascophores. Mitotic and meiotic nuclear divisions within an intact nuclear membrane. Enveloping membrane system in ascospore delimitation associated independently with post-meiotic nuclei. Asexual reproduction by holoblastic budding, conidia or fission (arthrospores). Example genera: *Saccharomyces, Candida, Dipodascopsis, Metschnikowia*.

Subphylum: *Pezizomycotina* (Pfister, 2015)

Class: *Arthoniomycetes*

Order: *Arthoniales*; example genera: *Arthonia, Dirina, Roccella*

Class: *Dothideomycetes* (Schoch & Grube, 2015)

Subclass: *Dothideomycetidae*

Order: *Capnodiales*; example genera: *Capnodium, Scorias, Mycosphaerella*

Order: *Dothideales*; asci borne in fascicles (clusters) in a locule without sterile elements. About 350 species; example genera: *Dothidea, Dothiora, Sydowia, Stylodothis*.

Order: *Myriangiales*; example genera: *Myriangium, Elsinoe*

Subclass: *Pleosporomycetidae*

Order: *Pleosporales*; asci borne in a basal layer among pseudoparaphyses. More than 4,705 species; example genera: *Pleospora, Phaeosphaeria, Lophiostoma, Sporormiella, Montagnula*.

Dothideomycetes incertae sedis (not placed in any subclass)

Order: *Botryosphaeriales*; example genera: *Botryosphaeria, Guignardia*

Order: *Hysteriales*; stroma boat-shaped, opening by a longitudinal slit, which renders it apothecium-like; asci borne among pseudoparaphyses. About 110 species; example genera: *Hysterium, Hysteropatella*.

Order: *Patellariales*; example genus: *Patellaria*

Order: *Jahnulales*; example genera: *Aliquandostipite, Jahnula, Patescospora*

Class: *Eurotiomycetes*; three subclasses, *Chaetothyriomycetidae, Eurotiomycetidae* and *Mycocaliciomycetidae* are defined to represent the major lineages within *Eurotiomycetes* (Geiser *et al*., 2015).

Subclass: *Chaetothyriomycetidae*

Lichenised, parasitic and saprotrophic ascomycetes with mostly bitunicate/fissitunicate to evanescent asci, produced in perithecial ascomata arranged superficially or immersed in a thallus. Thalli often produced on the surfaces of rocks, lichens, decaying plant material and other substrata. Ascospores variable, from colourless to pigmented, simple to muriform. Hamathecium, when present, consisting of pseudoparaphyses. Pigments, when present, generally related to melanin. Asexual stages with phialidic and annellidic anamorphs observed in non-lichenised taxa.

Order: *Chaetothyriales*; example genera: *Capronia, Ceramothyrium, Chaetothyrium*

Order: *Pyrenulales*; example genera: *Pyrenula, Pyrgillus*

Order: *Verrucariales*; example genera: *Agonimia, Dermatocarpon, Polyblastia, Verrucaria*

Subclass: *Eurotiomycetidae*

Saprotrophic, parasitic and mycorrhizal. Ascomata, when present, usually cleistothecial/gymnothecial, globose, often produced in surrounding stromatic tissue and brightly coloured; hamathecial elements lacking; gametangia usually undifferentiated and consisting of hyphal coils. Asci usually evanescent, sometimes bitunicate, scattered throughout the ascoma, rarely from a hymenium. Ascospores usually single-celled, lenticular, sometimes spherical or elliptical. Anamorphs variable, including phialidic and arthroconidial forms. Type: *Eurotium*.

Order: *Coryneliales*; asci in ascostromata with funnel-shaped ostioles at maturity. About 46 species; example genera: *Corynelia, Caliciopsis*.

Order: *Eurotiales*; asci globose to broadly oval, typically borne at different levels in cleistothecia (completely closed ascoma or sporophore); most of the human and animal dermatophytes belong here, also many saprotrophic soil or coprophilous fungi. Up to 930 species; example genera: *Eurotium, Emericella, Talaromyces, Elaphomyces, Trichocoma, Byssochlamys*.

Order: *Onygenales*; asci formed in a mazaedium (a sporophore consisting of a powdery mass of free spores interspersed with sterile threads, enclosed in a peridium or wall structure), evanescent, and liberating the ascospores as a powdery mass among sterile threads. About 270 species; example genera: *Onygena, Gymnoascus, Arthroderma*.

Subclass: *Mycocaliciomycetidae*

Parasites or commensals on lichens or saprotrophs. Ascomata disciform, stalked or sessile. Excipulum cupulate, and like the stalk hyphae at least in part sclerotised. Spore dispersal active, more rarely passive and ascomata then with a moderately developed mazaedium. Asci unitunicate, cylindrical, mostly with a distinctly thickened apex, 8-spored. Ascospores pale to blackish brown, ellipsoidal or spherical to cuboid, nonseptate or transversely 1–7-septate. Spore wall pigmented, smooth or with an ornamentation formed within the plasmalemma. Vulpinic acid derivatives occur in a few species. A variety of coelomycetous and hyphomycetous anamorphs occur in the group. Type genus: *Mycocalicium*.

Order: *Mycocaliciales*; example genera: *Mycocalicium, Chaenothecopsis, Stenocybe, Sphinctrina*

Class: *Laboulbeniomycetes*

Order: *Laboulbeniales*; minute parasites of insects and arachnids with mycelium represented only by haustoria and stalks. About 2,050 species; example genera: *Laboulbenia, Rickia, Ceratomyces.*

Order: *Pyxidiophorales*; example genus: *Pyxidiophora*

Class: *Lecanoromycetes* (Gueidan *et al.*, 2015)

Subclass: *Acarosporomycetidae*

Order: *Acarosporales*

Lichen-forming ascomycetes with chlorococcoid photobiont. Ascomata immersed or sessile, disciform or perithecioid. True exciple hyaline, annulate. Hymenium non-amyloid. Paraphyses moderately to poorly branched, septate, moderately to poorly anastomosing. Asci functionally unitunicate, lecanoralean, non-amyloid or with slightly amyloid tholi, polyspored, generally with more than 100 ascospores per ascus. Ascospores hyaline, small, nonseptate, non-halonate. The members of this order were formerly classified within the *Lecanorales*, but Acarosporaceae diverged earlier than the *Lecanoromycetidae* and Ostropomycetidae; example genera: *Acarospora, Pleopsidium, Sarcogyne.*

Subclass: *Lecanoromycetidae*

Lichen-forming ascomycetes with green algal or cyanobacterial photobiont. Ascomata immersed, sessile or stalked, usually disciform. True exciple hyaline or pigmented, annulate or cupulate. Hymenium amyloid or non-amyloid. Paraphyses simple or moderately to richly branched, septate, anastomosing or not. Asci bitunicate, functionally unitunicate, or prototunicate, lecanoralean, non-amyloid or amyloid, mostly 8-spored, but varying from 1- to polyspored. Ascospores hyaline or brown, nonseptate, trans-septate or muriform, halonate or non-halonate. This subclass includes the bulk of lichenised discomycetes.

Order: *Lecanorales*; example genera: *Cladonia, Lecanora, Parmelia, Ramalina, Usnea*

Order: *Peltigerales*; example genera: *Coccocarpia, Collema, Nephroma, Pannaria, Peltigera*

Order: *Teloschistales*; example genera: *Caloplaca, Teloschistes, Xanthoria*

Subclass: *Ostropomycetidae*

Order: *Agyriales*; example genera: *Agyrium, Placopsis, Trapelia, Trapeliopsis*

Order: *Baeomycetales*

Lichen-forming ascomycetes with chlorococcoid photobiont. Ascomata sessile or rarely stalked, disciform. True exciple hyaline or pigmented, annulate or cupulate. Hymenium non-amyloid. Paraphyses moderately to richly branched, septate. Asci unitunicate, non-amyloid or with slightly amyloid tholi, 8-spored. Ascospores hyaline, nonseptate or trans-septate, halonate or non-halonate. Type genus: *Baeomyces.*

Order: *Ostropales*; ascoma an apothecium (an open, often cup-like ascoma); asci inoperculate (without a terminal pore); ascospores septate, threadlike. This order includes taxa formerly classified in separate orders, such as Gomphillales, Graphidales, Gyalectales and Trichotheliales; example genera: *Ostropa, Stictis, Gyalecta, Gomphillus, Graphis, Odontotrema, Porina, Thelotrema.*

Order: *Pertusariales*, this order may not be monophyletic as currently circumscribed, with Ochrolechiaceae and some groups of the heterogeneous *Pertusaria* clustering in a separate clade, but without support. Nonetheless, a cluster of taxa in a 'core' group of *Pertusariales* has been strongly supported as monophyletic in phylogenetic analyses. Example genera: *Coccotrema, Icmadophila, Ochrolechia, Pertusaria.*

Lecanoromycetes incertae sedis (not placed in any subclass)

Order: *Candelariales*

Lichen-forming ascomycetes with chlorococcoid photobiont, predominantly nitrophilous. Thallus of various morphology, yellow to orange (pulvinic acid derivatives). Ascomata apothecial, sessile, with or without a distinct margin, yellow to orange. The ascomatal wall formed from densely septate twisted hyphae. Paraphyses mostly simple. Excipulum hyaline, hymenium amyloid. Asci unitunicate of *Candelaria* type with the amyloid lower part of the apical dome and broad apical cushion, often multispored. Ascospores hyaline, aseptate, rarely 1-septate. Type: *Candelaria*. Example genera: *Candelaria, Candelariella.*

Order: *Umbilicariales*

Lichen-forming ascomycetes with chlorococcoid photobiont. Ascomata sessile, or rarely immersed or stalked, mostly black, irregular, disciform. True exciple pigmented, annulate. Hymenium amyloid. Paraphyses simple or slightly branched, septate, apically thickened. Asci unitunicate, with slightly amyloid tholi, 1–8-spored. Ascospores hyaline or brown, nonseptate to muriform. Type: *Umbilicaria*. Example genera: *Lasallia, Umbilicaria.*

Class: *Leotiomycetes* (Zhang & Wang, 2015)

Order: *Cyttariales*; example genus: *Cyttaria*

Order: *Erysiphales*; obligate parasites on flowering plants causing powdery mildews; mycelium white, superficial in most, feeding by means of haustoria sunken into the epidermal cells of the host; one to several asci in a cleistothecium, if more than one, in a basal layer at maturity; asci globose to broadly oval; cleistothecia with appendages. About 150 species; example genera: *Erysiphe, Blumeria, Uncinula.*

Order: *Helotiales*; ascoma an apothecium bearing inoperculate asci exposed from an early stage; some important plant diseases are caused by members of this group. Monophyly of Helotiales *sensu lato* is not well supported by current data. There exists a minimum of five helotialean lineages that are intermixed with other leotiomycetan taxa (e.g. *Cyttariales*,

Erysiphales) and relationships are poorly resolved, thus preventing accurate phylogenetic classification at this time. Example genera: *Mitrula, Hymenoscyphus, Ascocoryne*.

Order: *Rhytismatales*; example genera: *Rhytisma, Lophodermium, Cudonia*

Order: *Thelebolales*; example genera: *Thelebolus, Coprotus, Ascozonus*

Class: *Lichinomycetes*

Order: *Lichinales*; example genera: *Heppia, Lichina, Peltula*

Class: *Orbiliomycetes*

Order: *Orbiliales*; example genera: *Orbilia, Hyalorbilia*

Class: *Pezizomycetes*

Order: *Pezizales*; ascoma an apothecium bearing operculate (with a hinged cap) asci above the ground; apothecia often large, cup- or saucer-shaped, spongy, brain-like, saddle-shaped, etc.; this group includes morels, false morels, saddle fungi and cup fungi among others. About 1,700 species; example genera: *Peziza, Glaziella, Morchella, Pyronema, Tuber*.

Class: *Sordariomycetes* (Zhang & Wang, 2015; Hyde *et al.*, 2017)

Subclass: *Hypocreomycetidae*

Order: *Coronophorales*; asci in ascostromata with irregular or round, never funnel-shaped, openings. About 90 species; example genera: *Nitschkia, Scortechinia, Bertia, Chaetosphaerella*.

Order: *Hypocreales*; perithecia and stromata when present, often brightly coloured; asci in a basal layer among apical paraphyses; perithecia immersed in a stroma formed from a sclerotium (a hard-resting body resistant to unfavourable environmental conditions); asci with a thick apex penetrated by a central canal through which the septate, threadlike ascospores are ejected. The ergot fungus (*Claviceps purpurea*), cause of ergotism in plants, animals and humans, and the original *source of lysergic acid diethylamide (LSD), belongs to* this order; *Cordyceps* spp. parasitise insect larvae. About 2,700 species; example genera: *Hypocrea, Nectria, Cordyceps, Claviceps, Niesslia*.

Order: *Melanosporales*

Ascoma perithecial or secondarily cleistothecial, peridium derived from base of an ascogonial coil, translucent; centrum pseudoparenchymatous, paraphyses absent in development; asci unitunicate, evanescent; ascospores dark, with germ pores at both ends; anamorphs hyphomycetous; often mycoparasitic. Example genus: *Melanospora*.

Order: *Microascales*; example genera: *Microascus, Petriella, Halosphaeria, Lignincola, Nimbospora*

Subclass: *Sordariomycetidae*

Order: *Boliniales*; example genera: *Camarops, Apiocamarops*

Order: *Calosphaeriales*; example genera: *Calosphaeria, Togniniella, Pleurostoma*

Order: *Chaetosphaeriales*; example genera: *Chaetosphaeria, Melanochaeta, Zignoëlla, Striatosphaeria*

Order: *Coniochaetales*; example genera: *Coniochaeta, Coniochaetidium*

Order: *Diaporthales*; perithecia immersed in plant tissue or in a stroma with their long ostioles protruding; ascal stalks gelatinising, freeing the asci from their basal attachment; paraphyses lacking. The chestnut blight fungus (*Endothia parasitica*) belongs here. Close to 1,200 species; example genera: *Diaporthe, Gnomonia, Cryphonectria, Valsa*.

Order: *Ophiostomatales*; example genera: *Ophiostoma, Fragosphaeria*

Order: *Sordariales*; example genera: *Sordaria, Podospora, Neurospora, Lasiosphaeria, Chaetomium*

Subclass: *Xylariomycetidae*

Order: *Xylariales*; perithecia with dark, membranous or carbonous (appearing as black burned wood) walls, with or without a stroma (a compact structure on or in which fructifications are formed); asci persistent, borne in a basal layer among paraphyses (elongate structures resembling asci but sterile), which may ultimately gelatinise and disappear; example genera: *Xylaria, Hypoxylon, Anthostomella, Diatrype, Graphostroma*.

Sordariomycetes incertae sedis (not placed in any subclass)

Order: *Lulworthiales* (this order includes members formerly placed in the *Spathulosporales*). Example genera: *Lulworthia, Lindra*.

Order: *Meliolales*; mycelium dark, superficial on leaves and stems of vascular plants, typically bearing appendages (termed hyphopodia or setae); asci in basal layers in ostiolate perithecia without appendages; mostly tropical fungi. More than 1,900 species; example genus: *Meliola*.

Order: *Phyllachorales*; example genus: *Phyllachora*

Order: *Trichosphaeriales*; example genus: *Trichosphaeria*

***Pezizomycotina incertae sedis*:** (not placed in any class)

Order: *Lahmiales*; example genus: *Lahmia*

Order: *Lahmiales*; example genus: *Medeolaria*

Order: *Triblidiales*; example genera: *Huangshania, Pseudographis, Triblidium*

PHYLUM: *BASIDIOMYCOTA*

Saprotrophic or parasitic on plants or insects; filamentous; hyphae septate, the septa typically inflated (dolipore) and centrally perforated; mycelium of two types, primary (homokaryotic) of uninucleate cells, succeeded by secondary (heterokaryotic), consisting of dikaryotic cells, this often bearing bridge-like clamp connections over the septa; asexual reproduction by fragmentation, oidia (thin-walled, free, hyphal cells behaving as spores), or conidia; sexual reproduction by fusion of hyphae with each other or with hyphal fragments or with germinating spores (somatogamy), resulting in dikaryotic hyphae that eventually give rise to basidia, either singly on the hyphae or in variously shaped basidiomata. The meiospores (sexual spores) are basidiospores borne exogenously on basidia; many are ballistospores that are actively discharged from the small hyphal branches (sterigmata) on which they arise. This is a large phylum of fungi containing the rusts, smuts, jelly fungi, club fungi, coral and shelf (bracket) fungi, mushrooms, puffballs, stinkhorns and bird's-nest fungi. Known *Basidiomycota* comprise about 1,600 genera and 32,000 species. The majority of these are in Subphylum *Agaricomycotina*; about 250 genera (8,400 species) occur in the *Pucciniomycotina*, and 62 genera (1,200 species) are assigned to the *Ustilaginomycotina*. Subdivision of the *Basidiomycota* is based on the form of the basidium and, traditionally, the shape and morphology of the mature sporophore. Developmental comparisons show that apparently similar structures and shapes can arise in different ways; for example, there are several ways of creating the folded spore-bearing surface that we call gills. Since they arise in different ways such features are only superficially similar because they have evolved in different ways (like, for example, the wings of butterflies and birds); they are described as analogous organs (as distinct from homologous organs, which are those that have a shared evolutionary ancestry). Shape and morphology of mature sporophores, even though these are the easiest to find in the field, can be misleading because they do not contribute to a natural classification. The introduction of molecular sequence comparisons is prompting major, and many surprising, revisions of the classification scheme originally constructed based on shape and morphology. Some of these revisions are reflected in the arrangement shown below. But this is far from being the final story (Zhao *et al.*, 2017).

Subphylum: *Pucciniomycotina* (equivalent to the traditional *Urediniomycetes*) (Aime *et al.*, 2014)

Class: *Pucciniomycetes*

Order: *Septobasidiales*; example genera: *Septobasidium, Auriculoscypha*

Order: *Pachnocybales*; example genus: *Pachnocybe*

Order: *Helicobasidiales*; example genera: *Helicobasidium, Tuberculina*

Order: *Platygloeales*; example genera: *Platygloea, Eocronartium*

Order: *Pucciniales*; example genera: *Puccinia, Uromyces*

Class: *Cystobasidiomycetes*

Order: *Cystobasidiales*; example genera: *Cystobasidium, Occultifur, Rhodotorula*

Order: *Erythrobasidiales*; example genera: *Erythrobasidium, Rhodotorula, Sporobolomyces, Bannoa*

Order: *Naohideales*; example genus: *Naohidea*

Class: *Agaricostilbomycetes*

Order: *Agaricostilbales*; example genera: *Agaricostilbum, Chionosphaera*

Order: *Spiculogloeales*; example genera: *Mycogloea, Spiculogloea*

Class: *Microbotryomycetes*

Order: *Heterogastridiales*; example genus: *Heterogastridium*

Order: *Microbotryales*; example genera: *Microbotryum, Ustilentyloma*

Order: *Leucosporidiales*; example genera: *Leucosporidiella, Leucosporidium, Mastigobasidium*

Order: *Sporidiobolales*; example genera: *Sporidiobolus, Rhodosporidium, Rhodotorula*

Class: *Atractiellomycetes*

Order: *Atractiellales*; example genera: *Atractiella, Saccoblastia, Helicogloea, Phleogena*

Class: *Classiculomycetes*

Order: *Classiculales*; example genera: *Classicula, Jaculispora*

Class: *Mixiomycetes*

Order: *Mixiales*; example genus: *Mixia*

Class: *Cryptomycocolacomycetes*

Order: *Cryptomycocolacales*; example genera: *Cryptomycocolax, Colacosiphon*

Subphylum: *Ustilaginomycotina* (equivalent to the traditional *Ustilaginomycetes*) (Begerow *et al.*, 2014; Kijpornyongpan *et al.*, 2018)

Class: *Ustilaginomycetes*

Order: *Urocystales*; example genera: *Urocystic, Ustacystis, Doassansiopsis*

Order: *Ustilaginales*; example genera: *Ustilago, Cintractia*

Class: *Exobasidiomycetes*

Order: *Doassansiales*; example genera: *Doassansia, Rhamphospora, Nannfeldtiomyces*

Order: *Entylomatales*; example genera: *Entyloma, Tilletiopsis*

Order: *Exobasidiales*; example genera: *Exobasidium, Clinoconidium, Dicellomyces*

Order: *Georgefischeriales*; example genera: *Georgefischeria, Phragmotaenium, Tilletiaria, Tilletiopsis*

Order: *Microstromatales*; example genera: *Microstroma, Sympodiomycopsis, Volvocisporium*

Order: *Tilletiales*; example genera: *Tilletia, Conidiosporomyces, Erratomyces*

Ustilaginomycotina incertae sedis: (not placed in any class)

Order: *Malasseziales*; example genus: *Malassezia*

Subphylum: *Agaricomycotina* (equivalent to the traditional Hymenomycetes or Basidiomycetes)

Class: *Tremellomycetes* (Weiss *et al.*, 2014)

Order: *Cystofilobasidiales*; example genera: *Cystofilobasidium, Mrakia, Itersonilia*

Order: *Filobasidiales*; example genera: *Filobasidiella, Cryptococcus*

Order: *Tremellales*; (jelly fungi) sporophores (basidiomata) usually bright-coloured to black gelatinous masses; a few are parasitic on mosses, vascular plants, or insects; most are saprotrophs. About 350 species; example genera: *Tremella, Trichosporon, Christiansenia.*

Class: *Dacrymycetes* (Oberwinkler, 2014)

Order: *Dacrymycetales*; example genera: *Dacrymyces, Calocera, Guepiniopsis*

Class: *Agaricomycetes* (Hibbett *et al.*, 2014)

Subclass: *Agaricomycetidae*

Order: *Agaricales*; example genera: *Agaricus* (Kerrigan, 2016; Zhao *et al.*, 2016), *Coprinopsis* (Moore, 2013b), *Pleurotus*

Order: *Atheliales*; example genera: *Athelia, Piloderma, Tylospora*

Order: *Boletales*; example genera: *Boletus, Scleroderma, Coniophora, Rhizopogon*

Subclass: *Phallomycetidae*

Order: *Geastrales*; example genera: *Geastrum, Radiigera, Sphaerobolus*

Order: *Gomphales*; example genera: *Gomphus, Gautieria, Ramaria*

Order: *Hysterangiales*; example genera: *Hysterangium, Phallogaster, Gallacea, Austrogautieria*

Order: *Phallales*; example genera: *Phallus, Clathrus, Claustula*

Agaricomycetes incertae sedis (not placed in any subclass)

Order: *Auriculariales*; example genera: *Auricularia, Exidia, Bourdotia*

Order: *Cantharellales*; example genera: *Cantharellus, Botryobasidium, Craterellus, Tulasnella*

Order: *Corticiales*; basidiomycetes with effused or discoid (*Cytidia*) basidiomata, a smooth hymenophore, and a monomitic hyphal system with clamped, rarely simple-septate, hyphae. Dendrohyphidia common. Species with or without cystidia. A probasidial resting stage is present in many species. Spores smooth, in masses white to pink. Saprotrophic, parasitic, or lichenicolous. Type: *Corticium*. Example genera: *Corticium, Vuilleminia, Punctularia.*

Order: *Gloeophyllales*; sporophores perennial or annual and long-lived, with hymenium maturing and thickening over time. Stature resupinate, effused-reflexed or dimidiate, with smooth, wrinkled, dentate, lamellate or regularly poroid hymenophore, or pileate-stipitate with lamellae. (Aborted, coralloid or flabelliform sporophores may be formed under conditions of darkness or high carbon dioxide concentration.) Leptocystidia or hyphoid hairs originating in the context and extending into or protruding from the hymenial layer (or lamellar margin in *Neolentinus*) are common; these often with thick brown walls and brownish incrustation. Context brown (but pallid in *Neolentinus*) and generally darkening in potassium hydroxide (the brownish incrustation in *Boreostereum* turning green in potassium hydroxide). Monomitic (if so, with sclerified generative hyphae), dimitic, or trimitic; generative hyphae with or without clamp connections. Basidiospores hyaline, ellipsoid to cylindrical or suballantoid, with thin, smooth walls, and neither amyloid, dextrinoid nor cyanophilous. Where this is known, basidiospores are binucleate and sexuality is heterothallic and bipolar (but tetrapolar in *Vuilleminia berkeleyi*). Type: *Gloeophyllum*. Causing brown rots (*Gloeophyllum, Neolentinus, Veluticeps*) or stringy white rot (*Boreostereum, Donkioporia*) of wood of gymnosperms, monocots and dicots. Occurrence on 'wood in service' (e.g. railway ties, paving blocks, wooden chests) seems to be common (in *Donkioporia, Gloeophyllum, Heliocybe* and *Neolentinus*); often on charred wood (*Boreostereum* and *Veluticeps*). Example genera: *Gloeophyllum, Neolentinus, Veluticeps.*

Order: *Hymenochaetales*; example genera: *Hymenochaete, Phellinus, Trichaptum*

Order: *Polyporales*; example genera: *Polyporus, Fomitopsis, Phanerochaete*

Order: *Russulales*; example genera: *Russula, Aleurodiscus, Bondarzewia, Hericium, Peniophora, Stereum*

Order: *Sebacinales*; example genera: *Sebacina, Tremellodendron, Piriformospora*

Order: *Thelephorales*; example genera: *Thelephora, Bankera, Polyozellus*

Order: *Trechisporales* (basidiomycetes with effused, stipitate or clavarioid basidiomata). Hymenophore smooth, granular (grandinioid), hydnoid or poroid. Hyphal system monomitic, hyphae clamped, subicular hyphae with or without ampullate septa. Cystidia present in some species, mostly lacking. Basidia with four to six sterigmata. Spores smooth or ornamented. On wood or soil. Type: *Trechispora*. Example genera: *Trechispora, Sistotremastrum, Porpomyces.*

Basidiomycota incertae sedis (not placed in any subphylum)

Class: *Wallemiomycetes*

Order: *Wallemiales*; example genus: *Wallemia*

Class: *Entorrhizomycetes*

Order: *Entorrhizales*; example genus: *Entorrhiza*

TAXONOMIC OUTLINE FOR FUNGI AND FUNGUS-LIKE ORGANISMS

Kingdom *Fungi* is a member of one of the six **'Supergroups'** into which the Domain *Eukaryota* is currently subdivided, these being (Adl *et al.*, 2005, 2012):

Excavata: flagellate protozoa; containing free-living, symbiotic forms, and some important parasites of humans. Classified on the basis of flagellar structures and considered to be the basal lineage of flagellated organisms.

Amoebozoa: amoeba-like protists, many with blunt, finger-like, lobed pseudopods and tubular mitochondrial cristae. Contains many of the best-known amoeboid organisms, including the genus *Amoeba* itself, as well as several varieties of slime moulds. Amoebozoa is a monophyletic clade, often shown as the sister group to Opisthokonta.

Opisthokonta: a broad group of eukaryotes, including both Kingdom *Animalia* and Kingdom *Fungi*, together with some eukaryotic microorganisms (choanoflagellates) that are sometimes grouped in Phylum *Choanozoa*.

Rhizaria: a species-rich supergroup of mostly unicellular eukaryotes, defined mainly from ribosomal DNA sequences as they vary considerably in morphology. The three main groups of Rhizaria are Cercozoa (amoebae and flagellates common in soil); Foraminifera (amoeboids with reticulose pseudopods, common as marine benthos); and Radiolaria (amoeboids with axopods, common as marine plankton). Many produce shells or skeletons of complex structure, which form the bulk of protozoan fossils.

Chromalveolata: a varied assemblage divided into four major subgroups, ***Cryptophyta*** (cryptomonad algae), **Haptophyta** (coccoliths and other phytoplankton, some of which form toxic algal blooms), **Stramenopiles** (brown algae, kelps, diatoms and water moulds) and **Alveolata** (includes ciliates, dinoflagellates, photosynthetic protozoa and parasitic protozoa). The emerging consensus is that the Chromalveolata is not monophyletic (Adl *et al.*, 2012; Cavalier-Smith *et al.*, 2015). The four original subgroups fall into at least two categories: one comprises the Stramenopiles and the Alveolata, to which the Rhizaria are now usually added to form the **SAR supergroup** (see below), while the other comprises the *Cryptophyta* and the *Haptophyta*.

Archaeplastida or **Primoplantae** (the latter name avoids confusion with an obsolete name once applied to cyanobacteria). This supergroup comprises red algae (Rhodophyta), green algae, and all land plants, together with a small group of freshwater unicellular algae called glaucophytes. In all these organisms, chloroplasts are surrounded by two membranes, suggesting they developed directly from endosymbiotic cyanobacteria (as opposed to other photosynthetic protists where chloroplasts are surrounded by three or four membranes, suggesting they were acquired secondarily from red or green algae).

Other than Kingdom *Fungi* in the Opisthokonta, mycologists might study organisms assigned to several of the supergroups into which eukaryotes are currently divided.

SLIME MOULDS

Slime moulds are assigned to several supergroups:

Supergroup: Amoebozoa

Superphylum **Mycetozoa** includes:

Class *Myxogastria* of syncytial or plasmodial slime moulds, such as *Physarum*, *Fuligo* and *Stemonitis*

Class *Dictyostelia* of cellular slime moulds like *Dictyostelium* and *Polysphondylium*

Class *Protostelia*, an intermediate group that form much smaller sporophores than the rest, such as *Protostelium*

Supergroup: Rhizaria

This includes the plasmodiophorids, which are obligate intracellular symbionts or parasites of plant, algal or fungal cells living in freshwater or soil habitats. They form multinucleate, unwalled plasmodia and so are traditionally considered slime moulds, because of the plasmodial stage. Previously classified as fungi (and called Plasmodiophoromycota), genetic and ultrastructural studies place them in a diverse group of protists called the *Cercozoa*.

Class *Phytomyxea*, a group of about 15 genera with 50 species of parasites of plants. The genera *Plasmodiophora* and *Spongospora* cause serious plant diseases (cabbage club root disease and powdery scab potato tuber disease, respectively).

Supergroup: Excavata

Includes cellular slime moulds that have a similar lifestyle to dictyostelids, but their amoebae behave differently, having temporary projections called pseudopods (false feet).

Class: *Heterolobosea*

Order: *Acrasida*; example genus *Acrasis*

Supergroup: SAR

Superphylum **Heterokonta** includes a class of marine protists that produce a network of filaments or tubes (slime nets) that form labyrinthine networks of tubes in which amoeba without pseudopods can travel and absorb nutrients. Commonly found as parasites on algae and seagrasses or as decomposers on dead plant material; they also include some parasites of marine invertebrates.

Class *Labyrinthulomycetes* (Botanical Nomenclature) or *Labyrinthulea* (Zoological Nomenclature) in two main groups, labyrinthulids (slime nets) and thraustochytrids. Example genera are *Labyrinthula*, *Pseudoplasmodium* and *Thraustochytrium*. Included in the Kingdom *Straminipila*, Phylum *Labyrinthulomycota*, below; this reflects the borderline nature of these organisms.

Supergroup: Opisthokonta

Fonticula is a genus (with only one species, *Fonticula alba*) of cellular slime mould which forms a sporophore in a volcano shape. It is not closely related to other well-established groups of cellular slime moulds (dictyostelids or acrasids) but molecular phylogenetics found homologies with members of the Opisthokonta as a sister taxon to *Nuclearia*, which is an opisthokont protist thought to be close to the animal-fungus boundary.

Finally, there is a collection of ***'fungus-like' organisms*** or ***water moulds***, currently placed in **Supergroup SAR**, which have a long tradition of being studied by mycologists. These are placed in **Kingdom *Straminipila*** (Beakes *et al.*, 2014).

KINGDOM *STRAMINIPILA*

We follow Beakes *et al.* (2014) in the use of this name ***and*** this spelling. There are alternative spellings (Stramenopila, Straminopila) and arguments for the use of the kingdom name ***Chromista***. A total of approximately 126 genera and 1,040 species; the majority being in the ***Oomycota***. These are common microorganisms and include important plant pathogens such as the cause of potato blight (*Phytophthora*). They have motile spores which swim by means of two flagella and grow as hyphae with cellulose-containing walls.

PHYLUM *OOMYCOTA*

About 1,000 species in 110 genera of cosmopolitan and widespread 'water moulds' occurring in freshwater, soil-water and marine habitats, some being economically ***very*** important pathogens such as *Saprolegnia, Pythium* and *Phytophthora*. *Phytophthora* species are serious pathogens of native vegetation, commercial forests, agricultural and horticultural crops, and cultivated landscapes worldwide. Over 100 species of *Phytophthora* have been described and it has been estimated that there may be up to 500 yet to be discovered. A disturbing recent trend has been the rapid dispersal of new *Phytophthora* species around the world by international movement of nursery stocks and other plant material.

Order: *Leptomitales*; example genera: *Apodachlyella, Ducellieria, Leptolegniella, Leptomitus*

Order: *Myzocytiopsidales*; example genus: *Crypticola*

Order: *Olpidiopsidales*; example genus: *Olpidiopsis*

Order: *Peronosporales*; example genera: *Albugo, Peronospora, Bremia, Plasmopara*

Order: *Pythiales*; example genera *Pythium, Phytophthora, Pythiogeton*

Order: *Rhipidiales*; example genus: *Rhipidium*

Order: *Salilagenidiales*; example genus: *Haliphthoros*

Order: *Saprolegniales*; example genera: *Leptolegnia, Achlya, Saprolegnia*

Order: *Sclerosporales*; example genera: *Sclerospora, Verrucalvus*

Order: *Anisolpidiales*; example genus: *Anisolpidium*

Order: *Lagenismatales*; example genus: *Lagenisma*

Order: *Rozellopsidales*; example genera: *Pseudosphaerita, Rozellopsis*

Order: *Haptoglossales*; example genera: *Haptoglossa, Lagena, Electrogella, Eurychasma, Pontisma, Sirolpidium*

PHYLUM *HYPHOCHYTRIOMYCOTA*

Microscopic organisms that form a small thallus, often with branched rhizoids, which occur as parasites or saprotrophs on algae and fungi in freshwater and in soil. The whole of the thallus is eventually converted into a reproductive structure. Only 24 species (in six genera) are known.

Order: *Hyphochytriales*, example genera: *Hyphochytrium, Rhizidiomyces*

PHYLUM LABYRINTHULOMYCOTA

Feeding stage comprises an ectoplasmic network and spindle-shaped or spherical cells that move within the network by gliding over one another. Occur in both salt- and freshwater in association with algae and other chromists. About 56 species in 12 genera. Included in Superphylum Heterokonta in the discussion of slime moulds above. This reflects the border-line nature of these organisms.

Order: *Labyrinthulales*; example genus: *Labyrinthula*

Order: *Thraustochytriales*; example genus: *Thraustochytrium*

REFERENCES

Adl, S.M., Simpson, A.G.B., Farmer, M.A. *et al.* (2005). The new higher level classification of Eukaryotes with emphasis on the taxonomy of protists. *Journal of Eukaryotic Microbiology*, **52**: 399–451. DOI: https://doi.org/10.1111/j.1550-7408.2005.00053.x.

Adl, S.M., Simpson, A.G.B., Lane, C.E. *et al.* (2012). The revised classification of Eukaryotes. *Journal of Eukaryotic Microbiology*, **59**: 429–493. DOI: https://doi.org/10.1111/j.1550-7408.2012.00644.x.

Aime, M.C., Toome, M. & Mclaughlin, D.J. (2014). Pucciniomycotina. In *The Mycota Systematics and Evolution*, VII part A, 2nd edition, ed. D.J. McLaughlin & J.W. Spatafora. Berlin: Springer-Verlag, pp. 271–294. ISBN: 978-3-642-55317-2. DOI: https://doi.org/10.1007/978-3-642-55318-9_10.

Beakes, G.W., Honda, D. & Thines, M. (2014). Systematics of the Straminipila: Labyrinthulomycota, Hyphochytriomycota, and Oomycota. In *The Mycota Systematics and Evolution*, VII part A, 2nd edition, ed. D.J. McLaughlin & J.W. Spatafora. Berlin: Springer-Verlag, pp. 39–97. ISBN: 978-3-642-55317-2. DOI: https://doi.org/10.1007/978-3-642-55318-9_3.

Begerow, D., Schäfer, A.M., Kellner, R. *et al.* (2014). Ustilaginomycotina. In *The Mycota Systematics and Evolution*, **VII part A**, 2nd edition, ed. D.J. McLaughlin & J.W. Spatafora. Berlin: Springer-Verlag, pp. 295–329. ISBN: 978-3-642-55317-2. DOI: https://doi.org/10.1007/978-3-642-55318-9_11.

Benny, G.L., Humber, R.A. & Voigt, K. (2014). Zygomycetous fungi: Phylum Entomophthoromycota and subphyla Kickxellomycotina, Mortierellomycotina, Mucoromycotina, and Zoopagomycotina. In *The Mycota Systematics and Evolution*, VII part A, 2nd edition, ed. D.J. McLaughlin & J.W. Spatafora. Berlin: Springer-Verlag, pp. 209–250. ISBN: 978-3-642-55317-2. DOI: https://doi.org/10.1007/978-3-642-55318-9_8.

Blackwell, M. (2000). Terrestrial life: fungal from the start? *Science*, **289**: 1884–1885. DOI: https://doi.org/10.1126/science.289.5486.1884.

Blackwell, M., Hibbett, D.S., Taylor, J.W. & Spatafora, J.W. (2006). Research Coordination Networks: a phylogeny for Kingdom Fungi (Deep Hypha). *Mycologia*, **98**: 829–837. DOI: https://doi.org/10.3852/mycologia.98.6.829.

Burki, F. (2014). The eukaryotic tree of life from a global phylogenomic perspective. *Cold Spring Harbor Perspectives in Biology*, **6**: article a016147. DOI: https://doi.org/10.1101/cshperspect.a016147.

Cavalier-Smith, T., Chao, E.E. & Lewis, R. (2015). Multiple origins of Heliozoa from flagellate ancestors: new cryptist Subphylum Corbihelia, Superclass Corbistoma, and monophyly of Haptista, Cryptista, Hacrobia and Chromista. *Molecular Phylogenetics and Evolution*, **93**: 331–362. DOI: https://doi.org/10.1016/j.ympev.2015.07.004.

Choi, J.-J. & Kim, S.-H. (2017). A genome Tree of Life for the fungi kingdom. *Proceedings of the National Academy of Sciences of the United States of America*, **114**: 9391–9396. DOI: https://doi.org/10.1073/pnas.1711939114.

Didier, E.S., Becnel, J.J., Kent, M.L., Sanders, J.L. & Weiss, L.M. (2014). Microsporidia. In *The Mycota Systematics and Evolution*, VII part A, 2nd edition, ed. D.J. McLaughlin & J.W. Spatafora. Berlin: Springer-Verlag, pp. 115–140. ISBN: 978-3-642-55317-2. DOI: https://doi.org/10.1007/978-3-642-55318-9_5.

Edwards, J.E., Forster, R.J., Callaghan, T.M. *et al.* (2017). PCR and omics based techniques to study the diversity, ecology and biology of anaerobic fungi: insights, challenges and opportunities. *Frontiers in Microbiology*, **8**: 1657 (27 pp.). DOI: https://doi.org/10.3389/fmicb.2017.01657.

Geiser, D.M., Lobuglio, K.F. & Gueidan, C. (2015). Pezizomycotina: Eurotiomycetes. In *The Mycota Systematics and Evolution*, VII part A, 2nd edition, ed. D.J. McLaughlin & J.W. Spatafora. Berlin: Springer-Verlag, pp. 121–141. ISBN: 978-3-662-46010-8. DOI: https://doi.org/10.1007/978-3-662-46011-5_5.

Gruninger, R.J., Puniya, A.K., Callaghan, A.M. *et al.* (2014). Anaerobic fungi (Phylum Neocallimastigomycota): advances in understanding their taxonomy, life cycle, ecology, role and biotechnological potential. *FEMS Microbiology Ecology*, **90**: 1–17. DOI: https://doi.org/10.1111/1574-6941.12383.

Gueidan, C., Hill, D.J., Miądlikowska, J. & Lutzoni, F. (2015). Pezizomycotina: Lecanoromycetes. In *The Mycota Systematics and Evolution*, VII part B, 2nd edition, ed. D.J. McLaughlin & J.W. Spatafora. Berlin: Springer-Verlag, pp. 89–120. ISBN: 978-3-662-46010-8. DOI: https://doi.org/10.1007/978-3-662-46011-5_4.

Hawksworth, D.L. & Lücking, R. (2017). Fungal diversity revisited: 2.2 to 3.8 million species. In *The Fungal Kingdom*, ed. J. Heitman, B. Howlett, P. Crous *et al.* Washington, DC: ASM Press, pp. 79–95. DOI: https://doi.org/10.1128/microbiolspec.FUNK-0052-2016.

Hawksworth, D.L., May, T.W. & Redhead, S.A. (2017). Fungal nomenclature evolving: changes adopted by the 19th International Botanical Congress in Shenzhen 2017, and procedures for the Fungal Nomenclature Session at the 11th International Mycological Congress in Puerto Rico 2018. *IMA Fungus*, **8**(2): 211–218. DOI: https://doi.org/10.5598/imafungus.2017.08.02.01.

He, M.-Q., Zhao, R.-L., Hyde, K.D., Begerow, D., Kemler, M. and 65 others. (2019). Notes, outline and divergence times of Basidiomycota. *Fungal Diversity*, **99**: 105–367. DOI: https://doi.org/10.1007/s13225-019-00435-4. [Open Access PDF download].

Hibbett, D.S., Bauer, R., Binder, M. *et al.* (2014). Agaricomycetes. In *The Mycota Systematics and Evolution*, VII part A, 2nd edition, ed. D.J. McLaughlin & J.W. Spatafora. Berlin: Springer-Verlag, pp. 373–429. ISBN: 978-3-642-55317-2. DOI: https://doi.org/10.1007/978-3-642-55318-9_14.

Hibbett, D.S., Bindera, M., Bischoff, J.F. *et al.* (2007). A higher-level phylogenetic classification of the Fungi. *Mycological Research*, **111**: 509–547. DOI: https://doi.org/10.1016/j.mycres.2007.03.004.

Hoffmann, K., Voigt, K. & Kirk, P.M. (2011). Mortierellomycotina subphyl. nov., based on multigene genealogies. *Mycotaxon*, **115**: 353–363. DOI: https://doi.org/10.5248/115.353.

Hyde, K.D., Maharachchikumbura, S.S.N., Hongsanan, S. *et al.* (2017). The ranking of fungi: a tribute to David L. Hawksworth on his 70th birthday. *Fungal Diversity*, **84**: 1–23. DOI: https://doi.org/10.1007/s13225-017-0383-3.

James, T.Y., Porter, T.M. & Martin, W.W. (2014). Blastocladiomycota. In *The Mycota Systematics and Evolution*, VII part A, 2nd edition, ed. D.J. McLaughlin & J.W. Spatafora. Berlin: Springer-Verlag, pp. 177–208. ISBN: 978-3-642-55317-2. DOI: https://doi.org/10.1007/978-3-642-55318-9_7.

Jones, M.D.M., Richards, T.A., Hawksworth, D.L. & Bass, D. (2011). Validation and justification of the phylum name Cryptomycota phyl. nov. *IMA Fungus*, **2**: 173–175. DOI: https://doi.org/10.5598/imafungus.2011.02.02.08.

Kerrigan, R.W. (2016). *Agaricus* of North America. *Memoirs of the New York Botanical Garden*, **Vol. 114**. New York: NYBG Press. ISBN: 978-0-89327-536-5.

Kijpornyongpan, T., Mondo, S.J., Barry, K. *et al.* (2018). Broad genomic sampling reveals a smut pathogenic ancestry of the fungal clade Ustilaginomycotina. *Molecular Biology and Evolution*, msy072, in press. DOI: https://doi.org/10.1093/molbev/msy072.

Kirk, P.M., Cannon, P.F., David, J.C. & Stalpers, J.A. (2001). *Dictionary of the Fungi*, 9th edition. Wallingford: CABI Publishing, ISBN: 085199377X.

Kirk, P.M., Cannon, P.F., Minter, D.W. & Stalpers, J.A. (2008). *Dictionary of the Fungi*, 10th edition. Wallingford: CABI Publishing. ISBN-10: 0851998267, ISBN-13: 978-0851998268. Kindle edition: ASIN: B00K7ANXJI.

Kurtzman, C.P. & Sugiyama, J. (2015). Saccharomycotina and Taphrinomycotina: the yeasts and yeastlike fungi of the Ascomycota. In *The Mycota Systematics and Evolution*, VII part A, 2nd edition, ed. D.J. McLaughlin & J.W. Spatafora. Berlin: Springer-Verlag, pp. 3–33. ISBN: 978-3-662-46010-8. DOI: https://doi.org/10.1007/978-3-662-46011-5_1.

Martin, W., Rotte, C., Hoffmeister, M. *et al.* (2003). Early cell evolution, eukaryotes, anoxia, sulfide, oxygen, fungi first (?), and a Tree of Genomes revisited. *International Union of Biochemistry and Molecular Biology: Life*, **55**: 193–204. DOI: https://doi.org/10.1080/1521654031000141231.

McCarthy, C.G.P. & Fitzpatrick, D.A. (2017). Multiple approaches to phylogenomic reconstruction of the fungal kingdom. *Advances in Genetics*, **100**: 211–266. DOI: https://doi.org/10.1016/bs.adgen.2017.09.006.

McLaughlin, D.J. & Spatafora, J.W. (ed.) (2014). *The Mycota Systematics and Evolution*, VII part A, 2nd edition. Berlin, Heidelberg: Springer-Verlag. ISBN: 978-3-642-55317-2. DOI: https://doi.org/10.1007/978-3-642-55318-9.

McLaughlin, D.J. & Spatafora, J.W. (ed.) (2015). *The Mycota Systematics and Evolution*, VII part B, 2nd edition. Berlin, Heidelberg: Springer-Verlag. ISBN: 978-3-662-46010-8. DOI: https://doi.org/10.1007/978-3-662-46011-5.

Moore, D. (2013a). *Fungal Biology in the Origin and Emergence of Life*. Cambridge, UK: Cambridge University Press. ISBN-10: 1107652774, ISBN-13: 978-1107652774.

Moore, D. (2013b). *Coprinopsis: an autobiography*. London: CreateSpace Independent Publishing Platform. ISBN-10: 1482618974; ISBN-13: 978-1482618976. Download full text: https://www.researchgate.net/publication/321361317_Coprinopsis_an_autobiography.

Moore, D., Pöder, R., Molitoris, H.-P. *et al.* (2006). Crisis in teaching future generations about fungi. *Mycological Research*, **110**: 626–627. DOI: https://doi.org/10.1016/j.mycres.2006.05.005.

Oberwinkler, F. (2014). Dacrymycetes. In *The Mycota Systematics and Evolution*, VII part A, 2nd edition, ed. D.J. McLaughlin & J.W. Spatafora. Berlin: Springer-Verlag, pp. 357–372. ISBN: 978-3-642-55317-2. DOI: https://doi.org/10.1007/978-3-642-55318-9_13.

Pfister, D.H. (2015). Pezizomycotina: Pezizomycetes, Orbiliomycetes. In *The Mycota Systematics and Evolution*, VII part B, 2nd edition, ed. D.J. McLaughlin & J.W. Spatafora. Berlin: Springer-Verlag, pp. 35–55. ISBN: 978-3-662-46010-8. DOI: https://doi.org/10.1007/978-3-662-46011-5_2.

Powell, M.J. & Letcher, P.M. (2014). Chytridiomycota, Monoblepharidomycota, and Neocallimastigomycota. In *The Mycota Systematics and Evolution*, VII part A, 2nd edition, ed. D.J. McLaughlin & J.W. Spatafora. Berlin: Springer-Verlag, pp. 141–176. ISBN: 978-3-642-55317-2. DOI: https://doi.org/10.1007/978-3-642-55318-9_6.

Prieto, M., Baloch, E., Tehler, A. & Wedin, M. (2013). Mazaedium evolution in the Ascomycota (Fungi) and the classification of mazaediate groups of formerly unclear relationship. *Cladistics*, **29**: 296–308. DOI: https://doi.org/10.1111/j.1096-0031.2012.00429.x.

Redecker, D. & Schüßler, A. (2014). Glomeromycota. In *The Mycota Systematics and Evolution*, VII part A, 2nd edition, ed. D.J. McLaughlin & J.W. Spatafora. Berlin: Springer-Verlag, pp. 251–270. ISBN: 978-3-642-55317-2. DOI: https://doi.org/10.1007/978-3-642-55318-9_9.

Redecker, D., Schüßler, A., Stockinger, H. *et al.* (2013). An evidence-based consensus for the classification of arbuscular mycorrhizal fungi (Glomeromycota). *Mycorrhiza*, **23**: 515–531. DOI: https://doi.org/10.1007/s00572-013-0486-y.

Ren, R., Sun, Y., Zhao, Y. *et al.* (2016). Phylogenetic resolution of deep eukaryotic and fungal relationships using highly conserved low-copy nuclear genes. *Genome Biology and Evolution*, **8**: 2683–2701. DOI: https://doi.org/10.1093/gbe/evw196.

Sancho, L.G., de la Torre, R., Horneck, G. *et al.* (2007). Lichens survive in space: results from the 2005 LICHENS experiment. *Astrobiology*, **7**: 443–454. DOI: https://doi.org/10.1089/ast.2006.0046.

Schoch, C. & Grube, M. (2015). Pezizomycotina: Dothideomycetes and Arthoniomycetes. In *The Mycota Systematics and Evolution*, VII part B, 2nd edition, ed. D.J. McLaughlin & J.W. Spatafora. Berlin: Springer-Verlag, pp. 143–176. ISBN: 978-3-662-46010-8. DOI: https://doi.org/10.1007/978-3-662-46011-5_6.

Sieber, T.N. (2007). Endophytic fungi in forest trees: are they mutualists? *Fungal Biology Reviews*, **21**: 75–89. DOI: https://doi.org/10.1016/j.fbr.2007.05.004.

Spatafora, J.W., Chang, Y., Benny, G.L. *et al.* (2016). A phylum-level phylogenetic classification of zygomycete fungi based on genome-scale data. *Mycologia*, **108**: 1028–1046. DOI: https://doi.org/10.3852/16-042.

Tedersoo, L., Sánchez-Ramírez, S., Kõljalg, U. *et al.* (2018). High-level classification of the Fungi and a tool for evolutionary ecological analyses. *Fungal Diversity*, **90**: 135–159. DOI: https://doi.org/10.1007/s13225-018-0401-0.

Visscher, H., Brinkuis, H., Dilcher, D.L. *et al.* (1996). The terminal Paleozoic fungal event: evidence of terrestrial ecosystem destabilization and collapse. *Proceedings of the National Academy of Sciences of the United States of America*, **93**: 2155–2158.

Wang, X., Liu, X. & Groenewald, J.Z. (2017). Phylogeny of anaerobic fungi (Phylum Neocallimastigomycota), with contributions from yak in China. *Antonie van Leeuwenhoek*, **110**: 87–103. DOI: https://doi.org/10.1007/s10482-016-0779-1.

Weiss, M., Bauer, R., Sampaio, J.P. & Oberwinkler, F. (2014). Tremellomycetes and related groups. In *The Mycota Systematics and Evolution*, VII part A, 2nd edition, ed. D.J. McLaughlin & J.W. Spatafora. Berlin: Springer-Verlag, pp. 331–355. ISBN: 978-3-642-55317-2. DOI: https://doi.org/10.1007/978-3-642-55318-9_12.

Wijayawardene, N.N., Hyde, K.D., Lumbsch, H.T. *et al.* (2018). Outline of Ascomycota: 2017. *Fungal Diversity*, **88**: 167–263. DOI: https://doi.org/10.1007/s13225-018-0394-8.

Yarza, P., Yilmaz, P., Panzer, K., Glöckner, F.O. & Reich, M. (2017). A phylogenetic framework for the Kingdom Fungi based on 18S rRNA gene sequences. *Marine Genomics*, **36**: 33–39. DOI: https://doi.org/10.1016/j.margen.2017.05.009.

Zhang, N., Luo, J. & Bhattacharya, D. (2017). Advances in fungal phylogenomics and its impact on fungal systematics. *Advances in Genetics*, **100**: 309–328. DOI: https://doi.org/10.1016/bs.adgen.2017.09.004.

Zhang, N. & Wang, Z. (2015). Pezizomycotina: Sordariomycetes and Leotiomycetes. In *The Mycota Systematics and Evolution*, VII part B, 2nd edition, ed. D.J. McLaughlin & J.W. Spatafora. Berlin: Springer-Verlag, pp. 57–88. ISBN: 978-3-662-46010-8. DOI: https://doi.org/10.1007/978-3-662-46011-5_3.

Zhao, R.-L., Li, G.-J., Sánchez-Ramírez, S. *et al.* (2017). A six-gene phylogenetic overview of Basidiomycota and allied phyla with estimated divergence times of higher taxa and a phyloproteomics perspective. *Fungal Diversity*, **84**: 43–74. DOI: https://doi.org/10.1007/s13225-017-0381-5.

Zhao, R.-L., Zhou, J.-L., Chen, J. *et al.* (2016). Towards standardizing taxonomic ranks using divergence times – a case study for reconstruction of the *Agaricus* taxonomic system. *Fungal Diversity*, **78**: 239–292. DOI: https://doi.org/10.1007/s13225-016-0357-x.

APPENDIX 2

Mycelial and Hyphal Differentiation

MYCELIAL DIFFERENTIATION

The ways in which mycelia differentiate morphologically have been recorded in a very extensive ***historical*** literature, all of which is worth reading (even if you can't find an electronic version). We recommend just a few of these publications in the following list (and offer our apologies to our friends and colleagues for describing them as 'historical').

Boddy, L. (1993). Saprotrophic cord-forming fungi: warfare strategies and other ecological aspects. *Mycological Research*, **97**: 641–655. DOI: https://doi.org/10.1016/S0953-7562(09)80141-X.

Boddy, L., Frankland, J.C. & van West, P. (2007). *Ecology of Saprotrophic Basidiomycetes*. Amsterdam: Elsevier. ISBN-10: 0123741858, ISBN-13: 978-0123741851.

Boddy, L. & Rayner, A.D.M. (1983a). Ecological roles of basidiomycetes forming decay columns in attached oak branches. *New Phytologist*, **93**: 77–88. URL: http://www.jstor.org/stable/2431897.

Boddy, L. & Rayner, A.D.M. (1983b). Mycelial interactions, morphogenesis and ecology of *Phlebia radiata* and *P. rufa* from oak. *Transactions of the British Mycological Society*, **80**: 437–448. DOI: https://doi.org/10.1016/S0007-1536(83)80040-0.

Burnett, J.H. & Trinci, A.P.J. (1979). *Fungal Walls and Hyphal Growth*. Cambridge, UK: Cambridge University Press. ISBN-10: 0521224993, ISBN-13: 978-0521224994.

Dowson, C.G., Rayner, A.D.M. & Boddy, L. (1986). Outgrowth patterns of mycelial cord-forming basidiomycetes from and between woody resource units in soil. *Journal of General Microbiology*, **132**: 203–211. DOI: https://doi.org/10.1099/00221287-132-1-203.

Frankland, J.C., Hedger, J.N. & Smith, M.J. (1982). *Decomposer Basidiomycetes*. Cambridge, UK: Cambridge University Press. ISBN-10: 052110680X, ISBN-13: 978-0521106801.

Jennings, D.H. & Rayner, A.D.M. (1984). *The Ecology and Physiology of the Fungal Mycelium*. Cambridge, UK: Cambridge University Press. ISBN-13: 9780521106269.

Rayner, A.D.M. (1992). Conflicting flows: the dynamics of mycelial territoriality. *McIlvainea*, **10**: 24–35.

Rayner, A.D.M. (1997). *Degrees of Freedom. Living in Dynamic Boundaries*. London: Imperial College Press. ISBN-10: 1860940374, ISBN-13: 978-1860940378.

Rayner, A.D.M. & Boddy, L. (1988). *Fungal Decomposition of Wood*. Chichester, UK: John Wiley & Sons. ISBN-10: 0471103101, ISBN-13: 978-0471103103.

Rayner, A.D.M. & Coates, D. (1987). Regulation of mycelial organisation and responses. In *Evolutionary Biology of the Fungi*, ed. A.D.M. Rayner, C.M. Brasier & D. Moore. Cambridge, UK: Cambridge University Press, pp. 115–136. ISBN-10: 0521330505, ISBN-13: 978-0521330503.

Rayner, A.D.M., Griffith, G.S. & Ainsworth, A.M. (1995a). Mycelial interconnectedness. In *The Growing Fungus*, ed. N.A.R. Gow & G.M. Gadd. London: Chapman & Hall, pp. 21–40. ISBN-10: 0412466007, ISBN-13: 978-0412466007.

Rayner, A.D.M., Griffith, G.S. & Wildman, H.G. (1994). Differential insulation and the generation of mycelial patterns. In *Shape and Form in Plants and Fungi*, ed. D.S. Ingram & A. Hudson. London: Academic Press, pp. 291–310. ISBN-10: 0123710359, ISBN-13: 978-0123710352.

Rayner, A.D.M., Powell, K.A., Thompson, W. & Jennings, D.H. (1985a). Morphogenesis of vegetative organs. In *Developmental Biology of Higher Fungi*, ed. D. Moore, L.A. Casselton, D.A. Wood & J.C. Frankland. Cambridge,

UK: Cambridge University Press, pp. 249–279. ISBN-10: 0521301610, ISBN-13: 978-0521301619.

Rayner, A.D.M., Ramsdale, M. & Watkins, Z.R. (1995b). Origins and significance of genetic and epigenetic instability in mycelial systems. *Canadian Journal of Botany*, **73**: S1241–S1248. DOI: https://doi.org/10.1139/b95-384.

Rayner, A.D.M. & Todd, N.K. (1979) Population and community structure and dynamics of fungi in decaying wood. *Advances in Botanical Research*, **7**: 333–420. DOI: https://doi.org/10.1016/S0065-2296(08)60090-7.

Rayner, A.D.M., Watling, R. & Frankland, J.C. (1985b). Resource relations – an overview. In *Developmental Biology of Higher Fungi*, ed. D. Moore, L.A. Casselton, D.A. Wood & J.C. Frankland. Cambridge, UK: Cambridge University Press, pp. 1–40. ISBN-10: 0521301610, ISBN-13: 978-0521301619.

Rayner, A.D.M. & Webber, J.F. (1984). Interspecific mycelial interactions: an overview. In *The Ecology and Physiology of the Fungal Mycelium*, ed. D.H. Jennings & A.D.M. Rayner. Cambridge, UK: Cambridge University Press, pp. 383–417. ISBN-13: 9780521106269.

Sharland, P.R. & Rayner, A.D.M. (1986) Mycelial interactions in *Daldinia concentrica*. *Transactions of the British Mycological Society*, **86**: 643–649. DOI: https://doi.org/10.1016/S0007-1536(86)80068-7.

Trinci, A.P.J., Wiebe, M.G. & Robson, G.D. (1994) The mycelium as an integrated entity. In *The Mycota*, **Vol. I**, *Growth, Differentiation and Sexuality*, ed. J.G.H. Wessels & F. Meinhardt. Berlin, Heidelberg, New York: Springer-Verlag, pp. 175–193. ISBN-10: 3540577815, ISBN-13: 978-3540577812.

HYPHAL DIFFERENTIATION

Differentiation of hyphae gives rise to an almost infinite variety of hyphal and cell shapes that have (a) attracted the attention of taxonomists over the years, who have used this cellular diversity to define taxa of various ranks, and (b) been given an equally wide variety of descriptive names. The prime source for explanation of names and terms in mycology is the *Dictionary of the Fungi* (Kirk *et al.*, 2008), and a good alternative is the *Illustrated Dictionary of Mycology* (Ulloa & Hanlin, 2012).

In this Appendix we use extracts of the text and illustrations from the *Dictionary of the Fungi, 10th Edition* (with permission of the publisher, CAB International, Wallingford, UK) to show you the usefulness of this reference source by gathering together some entries describing the range of diversity that has been observed in fungal cells. We have arranged the extracts to tell our particular story (in the *Dictionary* they are in alphabetical order). You can find explanations for technical terms not explained in the text by referring to the *Dictionary of the Fungi, 10th Edition*, where you will also find explanations of the abbreviations used in the reference citations in this text, although we have added some URLs for your convenience.

REFERENCES

Kirk, P.M., Cannon, P.F., Minter, D.W. & Stalpers, J.A. (2008). *Dictionary of the Fungi*, 10th edition. Wallingford: CABI Publishing. ISBN-10: 0851998267, ISBN-13: 978-0851998268. Kindle edition: ASIN: B00K7ANXJI.

Ulloa, M. & Hanlin, R.T. (2012). *Illustrated Dictionary of Mycology*, 2nd edition. St Paul, MN: APS Press. ISBN: 978-0-89054-502-7. DOI: https://doi.org/10.1094/9780890545027.

CONIDIA AND CONIDIOGENESIS

Conidial nomenclature. The traditional approach to the nomenclature of the spores of anamorphic fungi is that of Saccardo who differentiated seven morphological types (amero-, dictyo-, didymo-, helico-, phragmo-, scoleco- and staurospores), based on shape and septation, the terminology for which was further qualified according to whether the spores were pigmented (phaeoamerospores, etc.) or not (hyaloamerospores, etc.); see Anamorphic fungi. Kendrick and Nag Raj (in Kendrick, (ed.), *The Whole Fungus* **1**: 43, 1979 (https://doi.org/10.1002/fedr.19800910312)) discussed Saccardo's categories and offered precise definitions. Since Hughes (*CJB* **31**: 577, 1953 (https://doi.org/10.1139/b53-045)) drew attention to the systematic importance of conidiogenous events (q.v.), a nomenclature for conidia has evolved based on the methods by which the conidia develop, giving rise to terms such as annelloconidia, phialoconidia, tretoconidia etc. (cf.). These are not universally accepted, **the preferred usage now being 'conidium' with qualifying adjectives or descriptors**.

Main contributions to conidial terminology have been by the first Kananaskis Workshop–Conference in 1969 (see Kendrick (ed.), *Taxonomy of Fungi Imperfecti*, 1971 (https://trove.nla.gov.au/version/25412018)), Ellis (*Dematiaceous Hyphomycetes*, 1971 (https://www.cabdirect.org/cabdirect/abstract/19721100118)) offered a series of definitions as did Kendrick and Carmichael (in Ainsworth *et al.* (ed.), *The Fungi: A Taxonomic Review with Keys: Ascomycetes and Fungi Imperfecti*, **4A**: 323, 1973 (ISBN-10: 0120456044, ISBN-13: 9780120456048)), Cole (*CJB* **53**: 2983, 1975 (https://doi.org/10.1139/b75-328)), Cole and Samson (*Patterns of Development in Conidial Fungi*, 1979, (ISBN-10: 0273084070; ISBN-13: 978-0273084075)). The descriptive terminology in both conidial morphology and conidiogenous events was initiated by Minter *et al.* (*TBMS* **79**: 75, 1982 (https://doi.org/10.1016/S0007-1536(82)80193–9); **80**: 39, 1983 (https://doi.org/10.1016/S0007-1536(83)80163–6); **81**: 109, 1983 (https://doi.org/10.1016/S0007-1536(83)80210–1)), developed by Sutton (in Reynolds & Taylor (ed.), *The Fungal Holomorph*: 28, 1993 (ISBN-10: 0851988652, ISBN-13: 978-0851988658)) and Hennebert and Sutton, and Sutton and Hennebert (in Hawksworth (ed.), *Ascomycete Systematics*, 65, 77, 1994 (ISBN-13: 9781475792928)), and is used in this edition of the dictionary.

conidiogenous, producing conidia

cell, any cell from or within which a conidium is directly produced

locus, the place on a conidiogenous cell at which a conidium arises; Kendrick (1971: 258 (https://trove.nla.gov.au/version/25412018)).

Conidiogenous events are illustrated in Figures A2.1–A2.14.

conidiophore, a simple or branched hypha (a **fertile hypha**) bearing or consisting of conidiogenous cells from which conidia are produced; sometimes used when describing reduced structures for the conidiogenous cell.

wall-building, descriptive of hyphal growth in which cell wall material is produced by certain ultrastructural secretory bodies in the cytoplasm. Three types of wall-building may be distinguished (Figure A2.1):

- **apical** – in which the bodies are concentrated at the hyphal tip, producing new wall by distal growth, forming a cylindrical hypha in which the youngest wall material is at the tip;
- **ring** – in which the bodies are concentrated adjacent to the cell wall at some point below the tip, in the shape of an imaginary ring, producing new wall by proximal growth, forming a cylindrical hypha in which the youngest wall material is always at the base;
- **diffuse** – in which the bodies occur throughout the cytoplasm at a low concentration, producing lateral growth (i.e. swelling of the cylindrical hypha) by alteration of pre-existing wall.

The terms assume special significance in conidial development where they have been used to clarify the concepts of thallic

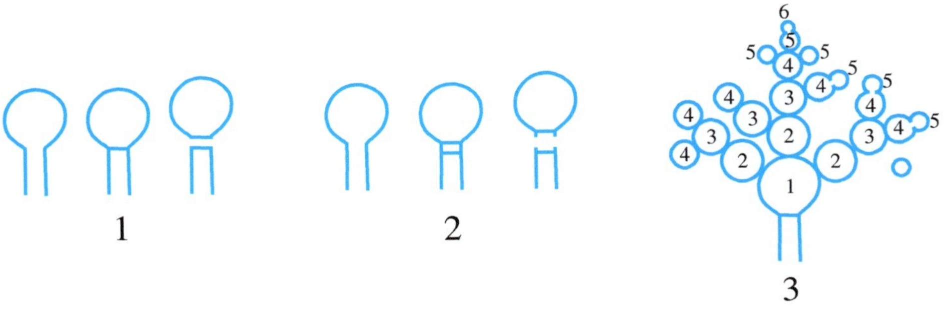

Figure A2.1. Conidiogenous events (cc = conidiogenous cell). **1**, conidial ontogeny holoblastic, one locus per cc, solitary conidia, delimited by one septum, maturation by diffuse wall-building, secession schizolytic, no proliferation of cc; **2**, conidial ontogeny holoblastic, one locus per cc, solitary conidia, delimitation by two septa (or a separating cell), secession rhexolytic or by fracture of the cc, maturation by diffuse wall-building, no proliferation of cc; **3**, conidial ontogeny holoblastic, apical wall-building random at more than one locus per cc and conidia becoming conidiogenous to form connected branched chains, each conidium delimited by one septum, maturation by diffuse wall-building, secession schizolytic, no cc proliferation; … continued in Figure A2.2.

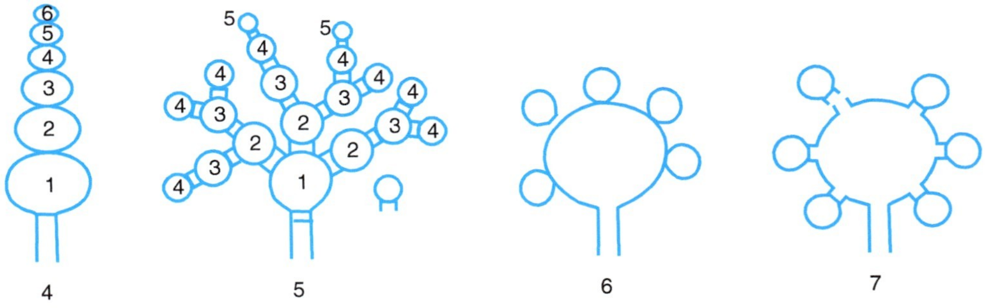

Figure A2.2. Conidiogenous events (continued) (cc = conidiogenous cell). **4**, conidial ontogeny holoblastic, apical wall-building at one locus per cc and each conidium with one locus to form a connected unbranched chain, each conidium delimited by one septum, maturation by diffuse wall-building, secession schizolytic, no proliferation of cc; **5**, conidial ontogeny holoblastic, apical wall-building randomly at more than one locus per cc and conidia becoming conidiogenous to form connected branched chains, each conidium delimited by two septa (or a separating cell), secession rhexolytic or by fracture of the cc, maturation by diffuse wall-building, no cc proliferation; **6**, conidial ontogeny holoblastic, with localised apical wall-building simultaneously at different loci over the whole cc, each locus forming one conidium, delimited by one septum, maturation by diffuse wall-building, secession schizolytic, no cc proliferation; **7**, conidial ontogeny holoblastic, with localised apical wall-building simultaneously at different loci on denticles over the whole cc, each locus forming one conidium, delimited by one septum, maturation by diffuse wall-building, secession by rupture of denticle, no cc proliferation; … continued in Figure A2.3.

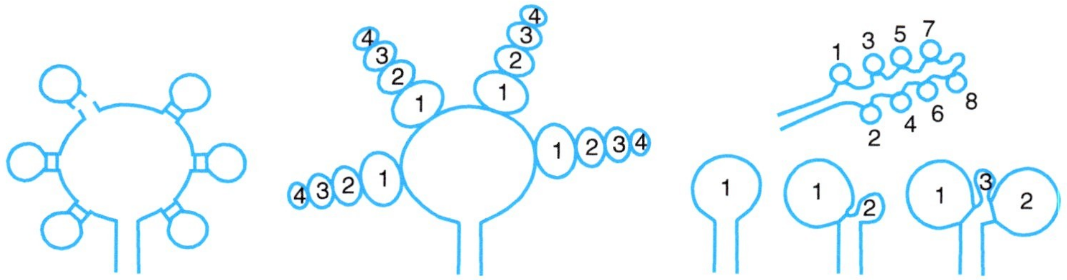

8 9 10

Figure A2.3. Conidiogenous events (continued) (cc = conidiogenous cell). **8**, conidial ontogeny holoblastic, with localised apical wall-building simultaneously at different loci over the whole cc, each conidium delimited by two septa (or a separating cell), secession rhexolytic or by fracture of the cc, each locus forming one conidium, maturation by diffuse wall-building, no cc proliferation; **9**, conidial ontogeny holoblastic, apical wall-building simultaneously at several loci per cc and conidia becoming conidiogenous to form connected branched chains, each conidium delimited by one septum, maturation by diffuse wall-building, secession schizolytic, no cc proliferation; **10**, conidial ontogeny holoblastic, regularly alternating with holoblastic sympodial cc proliferation, maturation by diffuse wall-building, each conidium delimitated by one septum, secession schizolytic; ... continued in Figure A2.4.

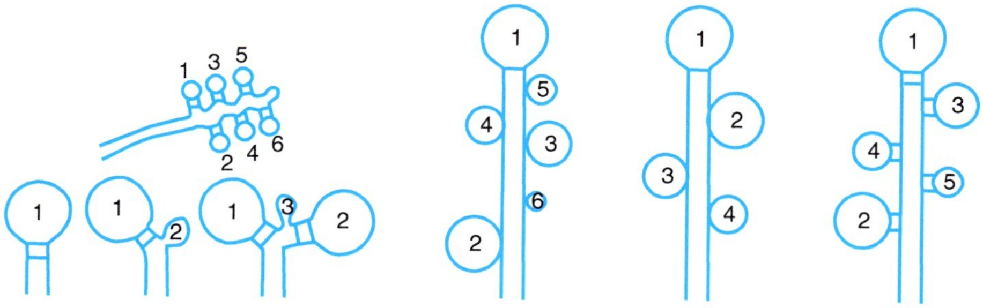

11 12 13 14

Figure A2.4. Conidiogenous events (continued) (cc = conidiogenous cell). **11**, conidial ontogeny holoblastic, regularly alternating with holoblastic sympodial cc proliferation, maturation by diffuse wall-building, each conidium delimited by two septa (or a separating cell), secession rhexolytic or by fracture of the cc; **12**, conidial ontogeny holoblastic, each from apical or lateral loci, delimited by one septum, secession schizolytic, holoblastic cc proliferation sympodial or irregular, maturation by diffuse wall-building; **13**, conidial ontogeny holoblastic, first from an apical locus, delimited by one septum, secession schizolytic, other conidia from lateral loci proceeding down the cc, maturation by diffuse wall-building; **14**, conidial ontogeny holoblastic, first from an apical locus, each conidium delimited by two septa (or a separating cell), secession rhexolytic or by fracture of the cc, other conidia from lateral loci proceeding down the cc, maturation by diffuse wall-building....continued in Figure A2.5.

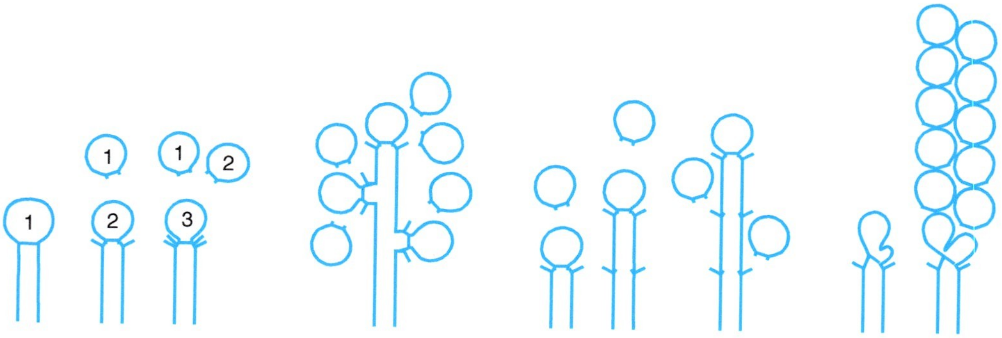

15 16 17 18

Figure A2.5. Conidiogenous events (continued) (cc = conidiogenous cell). **15**, conidial ontogeny holoblastic, delimitation by one septum, schizolytic secession, maturation by diffuse wall-building, percurrent enteroblastic cc proliferation followed by conidial ontogeny by replacement apical wall-building, successive conidia seceding at the same level, sometimes in unconnected chains, collarette variable; **16**, same as 15 but with several random or irregular conidiogenous loci to each cc; **17**, conidial ontogeny holoblastic, delimitation by one septum, schizolytic secession, maturation by diffuse wall-building, percurrent enteroblastic cc proliferation followed by conidial ontogeny by replacement apical wall-building, successive conidia seceding at the same level, collarette variable, conidiogenous activity interspersed periodically with percurrent vegetative proliferation; **18**, conidial ontogeny holoblastic, delimitation by one septum, schizolytic secession, maturation by diffuse wall-building, percurrent and sympodial enteroblastic cc proliferation followed by conidial ontogeny by replacement apical wall-building, successive conidia seceding at the same level, collarette variable; ... continued in Figure A2.6.

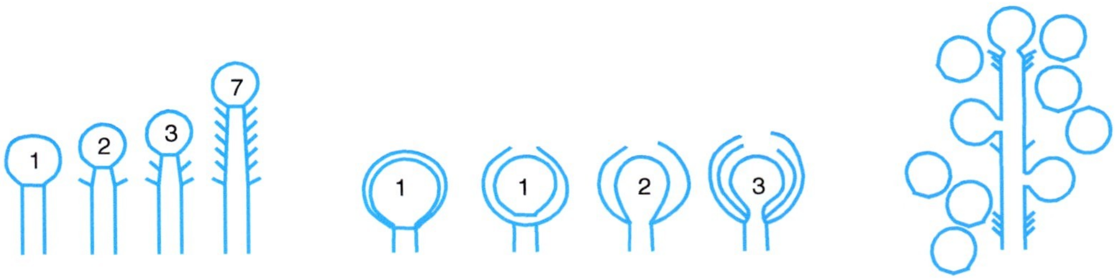

19 20 21

Figure A2.6. Conidiogenous events (continued) (cc = conidiogenous cell). **19**, conidial ontogeny holoblastic, delimitation by one septum, schizolytic secession, maturation by diffuse wall-building, percurrent enteroblastic cc proliferation followed by conidial ontogeny by replacement apical wall-building, successive conidia seceding at progressively higher levels, sometimes in unconnected chains, collarette variable; **20**, conidial ontogeny enteroblastic, delimitation by one septum, schizolytic secession, maturation by diffuse wall-building, outer wall of the cc remaining as a conspicuous collarette, percurrent enteroblastic cc proliferation followed by conidial enteroblastic ontogeny by replacement apical wall-building, successive conidia seceding at the same level, a succession of collarettes formed; **21**, combination of 10, 12 and 19, where the sequences occur at random, irregularly or interchangeably; … continued in Figure A2.7.

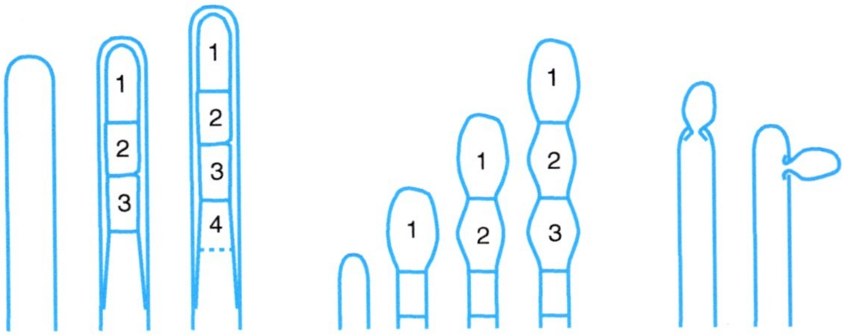

22 23 24 25

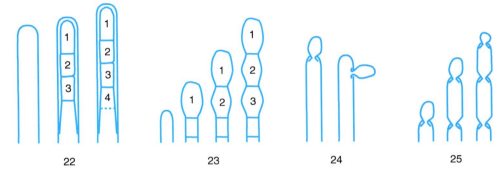

Figure A2.7. Conidiogenous events (continued) (cc = conidiogenous cell). **22**, conidial ontogeny holoblastic with new inner walls constituting the conidia laid down retrogressively by diffuse wall-building, delimitation retrogressive, loss of apical wall-building followed by replacement ring wall-building at the base of the cc adding more retrogressively delimited conidia, the outer (original) cc wall breaks as a connected chain of conidia is formed, collarette variable, one locus per cc, secession schizolytic; **23**, conidial ontogeny holoblastic, one locus per cc, first conidium delimited by one septum, maturation by diffuse wall-building, loss of apical wall-building, replaced by ring wall-building below the delimiting septum which produces conidia in a connected unbranched chain, secession schizolytic, no proliferation of cc; **24**, conidial ontogeny holoblastic, simultaneous with minimal enteroblastic percurrent proliferation at the preformed pore in the outer cc wall, conidia solitary, delimited by one septum, secession schizolytic, maturation by diffuse wall-building, one locus per cc; **25**, conidial ontogeny holoblastic, simultaneous with minimal enteroblastic percurrent proliferation at the preformed pore in the outer cc wall, conidia solitary, delimited by one septum, secession schizolytic, maturation by diffuse wall-building, after one conidium formed extensive enteroblastic percurrent proliferation by apical wall-building occurs until the next apical locus is formed; … continued in Figure A2.8.

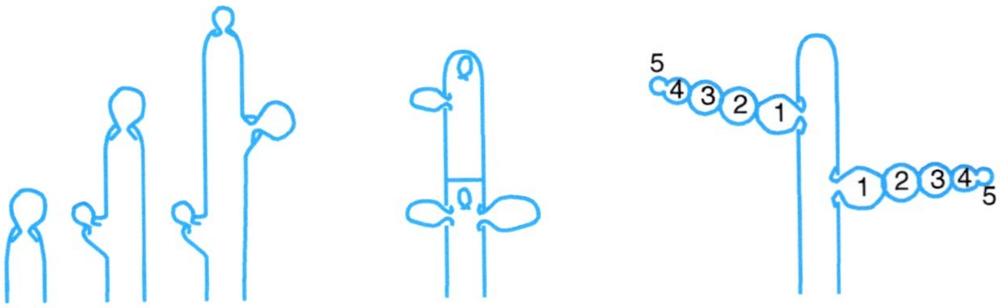

26 27 28 29

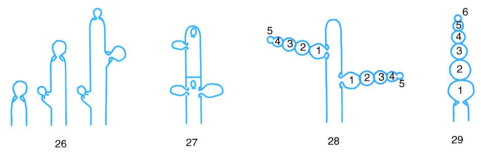

Figure A2.8. Conidiogenous events (continued) (cc = conidiogenous cell). **26**, same as 24 but with holoblastic sympodial proliferation of the cc with conidiogenesis occurring between loci; **27**, same as 24 but with several conidiogenous loci produced in the apical cc and laterally below septa in other ccs constituting the conidiophore; **28**, same as 24 but several loci to each cc and first and subsequent conidia becoming conidiogenous by apical wall-building to form unbranched connected chains; more than one locus to a conidium will produce branched chains; **29**, same as 24 but first conidium becoming conidiogenous by apical wall-building to form an unbranched connected chain; … continued in Figure A2.9.

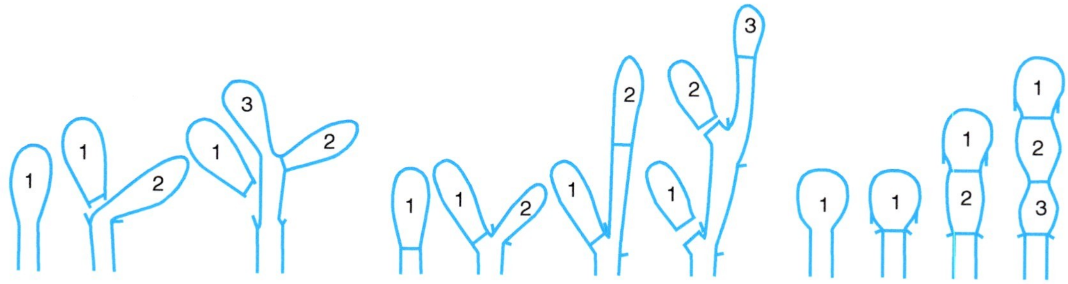

Figure A2.9. Conidiogenous events (continued) (cc = conidiogenous cell). **30**, conidial ontogeny holoblastic, delimitation by one septum, maturation by apical and diffuse wall-building, secession schizolytic and coincident with enteroblastic sympodial cc proliferation below the previous locus; subsequent conidia formed similarly but with holoblastic sympodial cc proliferation; **31**, conidial ontogeny holoblastic, delimitation by one septum, maturation by apical and diffuse wall-building, secession schizolytic and coincident with enteroblastic sympodial cc proliferation below the previous conidiogenous locus, the sequence giving geniculate conidiophores; **32**, conidial ontogeny holoblastic, with new inner walls continuous with all conidia laid down by diffuse wall-building, delimitation by one septum, loss of apical wall-building followed by replacement continuous ring wall-building immediately below delimiting septum, the outer cc wall breaks between the first conidium and the cc to produce a variable collarette, followed by alternation of holoblastic conidial ontogeny by ring wall-building giving connected chains of conidia, maturation by diffuse wall-building, retrogressive delimitation, secession schizolytic; ... continued in Figure A2.10.

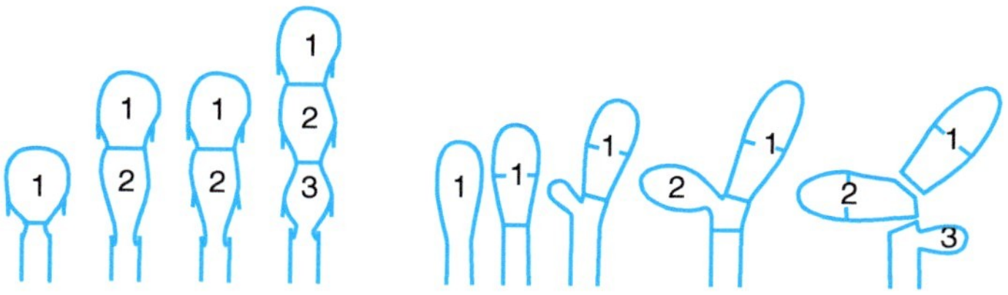

Figure A2.10. Conidiogenous events (continued) (cc = conidiogenous cell). **33**, conidial ontogeny holoblastic with new inner walls laid down by diffuse wall-building, delimitation by one septum, loss of apical wall-building followed by replacement ring wall-building immediately below delimiting septum, the outer cc wall breaks between the first conidium and the cc to produce a variable collarette, subsequent conidia formed by new inner walls for each conidium by ring wall-building giving connected chains of conidia, maturation by diffuse wall-building, retrogressive delimitation, secession schizolytic; **34**, conidial ontogeny holoblastic, delimitation by one septum, secession schizolytic, enteroblastic sympodial cc proliferation below the previous locus and delimiting septum, the second and subsequent conidia formed from proliferations and delimited retrogressively, cc reduced in length with each conidium formed; ... continued in Figure A2.11.

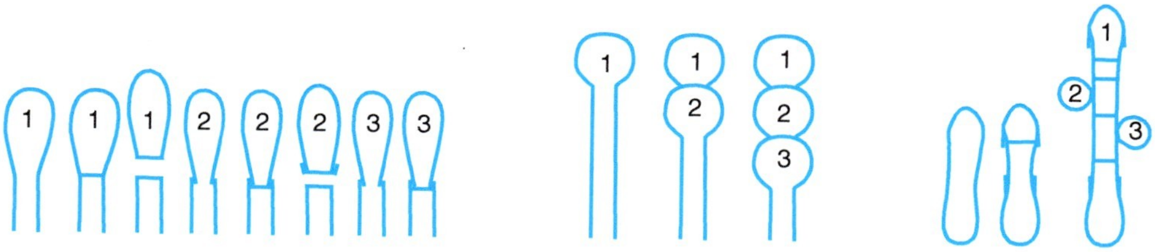

Figure A2.11. Conidiogenous events (continued) (cc = conidiogenous cell). **35**, conidial ontogeny holoblastic, maturation by diffuse wall-building, delimitation by one septum, secession schizolytic, enteroblastic percurrent cc proliferation with retrogressive delimitation of next conidium, producing unconnected chains of conidia, the cc reduced in length with each conidium formed; **36**, conidial ontogeny holoblastic, delimitation by one septum with loss of apical wall-building but replaced by diffuse wall-building below the previous conidium to form the next conidium which is retrogressively delimited giving an unconnected chain of conidia, secession schizolytic, cc reduced in length with each conidium formed; **37**, conidial ontogeny holoblastic, delimitation by one septum with loss of apical wall-building, replaced by ring wall-building below the delimiting septum, outer wall of first conidium and cc breaks, followed by enteroblastic percurrent proliferation by ring wall-building, succeeding conidia holoblastic, delimited laterally and retrogressively, secession schizolytic, several loci per cc; ... continued in Figure A2.12.

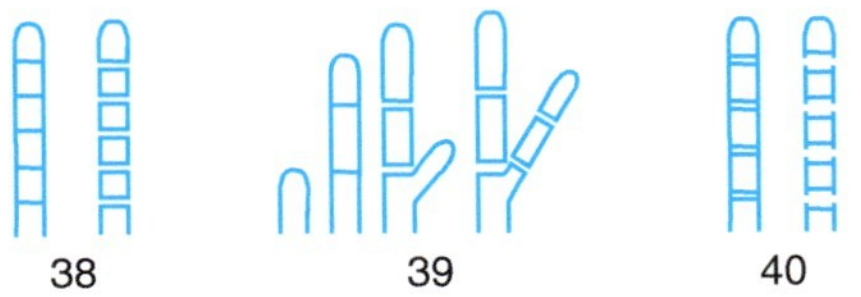

Figure A2.12. Conidiogenous events (continued) (cc = conidiogenous cell). **38**, conidial ontogeny holothallic, ccs formed by apical wall-building coincident with conidial ontogeny, random delimitation by one septum at each end, no maturation during conidiogenesis, secession randomly schizolytic; **39**, conidial ontogeny holothallic, ccs formed by apical wall-building coincident with conidial ontogeny, random delimitation by one septum at each end, no maturation during conidiogenesis, secession randomly schizolytic, cc proliferation holoblastic, irregular or sympodial, constituent cells conidiogenous; **40**, same as 38 but conidial delimitation by two septa or separating cells at each end, secession rhexolytic; ... continued in Figure A2.13.

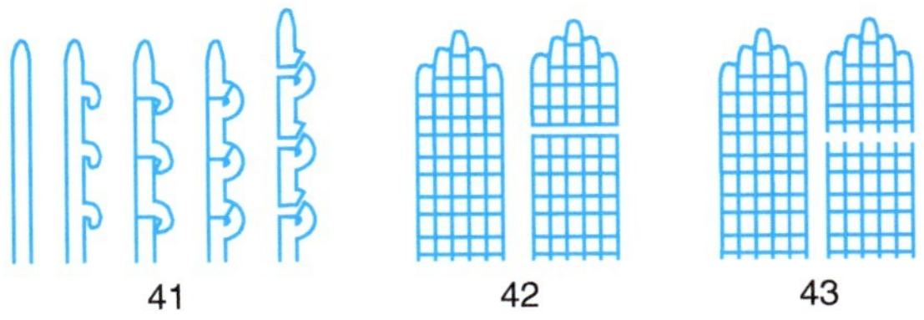

Figure A2.13. Conidiogenous events (continued) (cc = conidiogenous cell). **41**, conidial ontogeny holothallic, ccs formed in association with clamp connexions, random delimitation by septa in cc and the backwardly directed branch in the clamp connexion, maturation by diffuse and localised apical wall-building, secession randomly schizolytic, individual conidia comprised of part of the preceding and following clamp connexions; **42**, conidial ontogeny holoblastic by simultaneous apical wall-building in adjacent cells, delimitation by septa in each of these cells, maturation by diffuse wall-building, secession simultaneous, multicellular, schizolytic, no cc proliferation; **43**, conidial ontogeny holoblastic by simultaneous apical wall-building in adjacent cells, delimitation by septa in each of these cells, maturation by diffuse wall-building, followed by replacement apical wall-building in conidia to form additional conidia in connected chains, secession simultaneous, multicellular, rhexolytic, no cc proliferation.

and blastic. Apical wall-building occurs in *Geniculosporium*, *Cladosporium* and *Scopulariopsis*, and 'phialides' where conidia are produced in gummy masses (e.g. *Trichoderma*) or false chains (e.g. *Mariannaea*). Ring wall-building occurs in 'phialides' with conidia in true chains (e.g. *Penicillium*, *Chalara*), in so-called meristem arthrospores (e.g. *Wallemia*) and in conidiogenous cells of basauxic fungi (e.g. *Arthrinium*). Diffuse wall-building occurs simultaneously with or shortly after apical or ring wall-building in most of the preceding examples, but its occurrence is much delayed or even absent in thallic development (e.g. *Geotrichum*). Wall-building is a preferable term to meristem which implies growth by cell division rather than within a single cell. See Minter *et al.* (*TBMS* **79**: 75, 1982 (https://doi.org/10.1016/S0007-1536(82)80193–9); **80**: 39, 1983 (https://doi.org/10.1016/

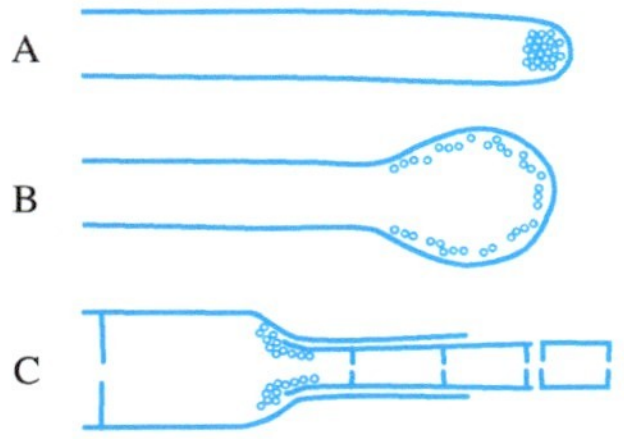

Figure A2.14. Wall-building in relation to conidiogenesis. **(A)** Apical wall-building. **(B)** Diffuse wall-building. **(C)** Ring wall-building.

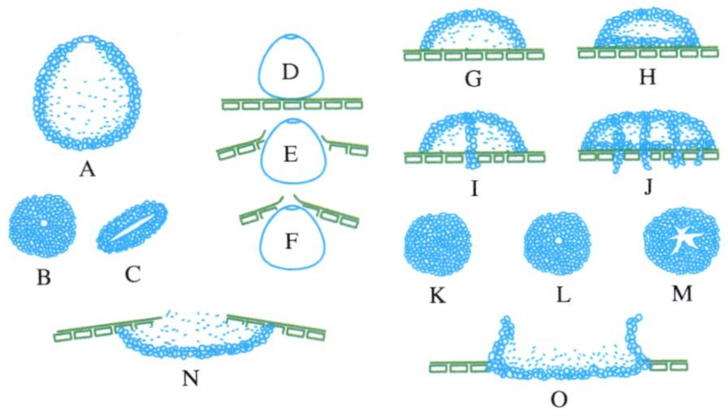

Figure A2.15. Conidiomatal types. **(A–F)**, pycnidial; **(B)** dehiscence by a central circular ostiole; **(C)** dehiscence by a longitudinal ostiole (raphe); **(D)** superficial; **(E)** semi-immersed; **(F)** immersed; **(G–M)**, pycnothyrial; **(G)** with upper wall only; **(H)** with upper and lower walls; **(I)** with a central supporting column; **(J)** multilocular with several supporting columns; **(K)** dehiscence from the margin; **(L)** dehiscence by a central ostiole; **(M)** dehiscence by irregular fissures; **(N)** acervular; **(O)** cupulate; ... continued in Figure A2.16.

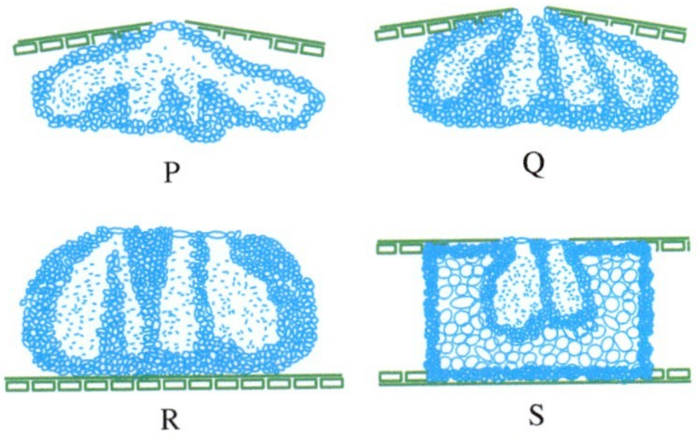

Figure A2.16. Conidiomatal types (continued). **(P)** convoluted immersed; **(Q)** multilocular, immersed; **(R)** multilocular, superficial; **(S)** pseudostromatic.

S0007-1536(83)80163–6); **81**: 109, 1983 (https://doi.org/10.1016/S0007-1536(83)80210–1)). Also see Anamorphic fungi.

conidioma (pl. **-ata**), a specialised multihyphal, conidia-bearing structure (Kendrick & Nag Raj in Kendrick (ed.), *The Whole Fungus* **1**: 51, 1979 (https://doi.org/10.1002/fedr.19800910312)). See acervulus, pycnidium, sporodochium, synnema (all obsolete nouns, but used adjectivally, e.g. acervular conidioma). See Figures A2.15 and A2.16. Cf. conidiophore.

DIFFERENTIATION IN ASCOMATA

ascus (pl. **asci**), term introduced by Nees (Syst. Pilze: 164, 1817) for the typically sac-like cell (first figured in *Pertusaria* by Micheli in 1729; q.v.) characteristic of *Ascomycota* (q.v.), in which (after karyogamy and meiosis) ascospores (generally 8) are produced by 'free cell formation' (this ***Guidebook***, section 3.4). Asci vary considerably in structure, and work in the last two decades has shown previous separation into only 2–3 categories (e.g. **bitunicate, prototunicate, unitunicate**) to be an oversimplification. Sherwood (1981) illustrated nine main types distinguishable by light microscopy: prototunicate, bitunicate, astropalean, annellate, hypodermataceous, pseudoperculate, operculate, lecanoralean and verrucarioid). Eriksson (1981) distinguished seven types of dehiscence in bitunicate asci with an ectotunica and distinct endotunica (see p. 37 of edition 7 of the dictionary). These classifications mask a much wider range of variation; Bellemere (1994) recognised three predehiscence types and 11 dehiscence categories (Figures A2.17 and A2.18). The details of the asci are stressed in ascomycete systematics, especially. in lichen-forming orders where reactions with iodine are emphasised (q.v.) (Hafellner, 1984).

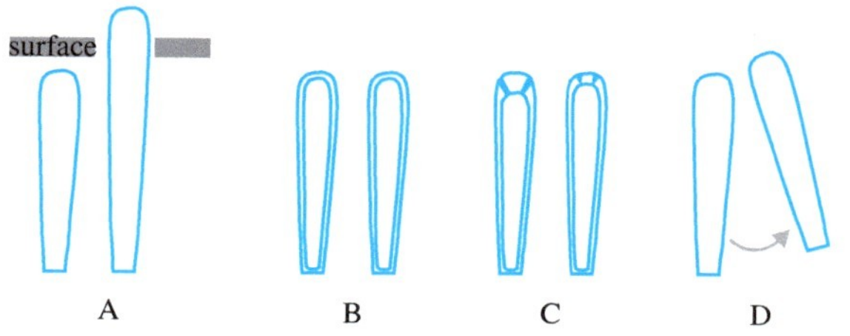

Figure A2.17. Predehiscence stage of asci. **(A)** Protruding ascus; **(B)** ascus wall becoming thinner; **(C)** change in apical structure; **(D)** ascus liberation ... continued in Figure A2.18.

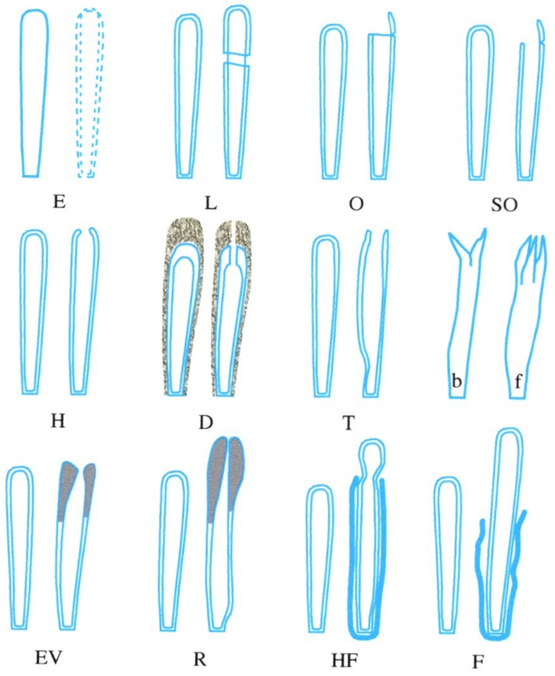

Figure A2.18. (continued); II Dehiscence stage of asci; (**E**) evanescent ascus; (**L**) rupture of lateral wall; subapical rupture (**O**, operculate, and **SO**, suboperculate dehiscence); rupture by apical wall without extrusion (**H**, pore-like dehiscence); **(D)** *Dactylospora*-type; (**T**), *Telochistes*-type = extenditunicate (**b**= bivalve, **f** = fissurate variants); rupture with extrusion (**EV**, eversion; **R**, rostrate; **HF**, hemifissitunicate).

Bitunicate asci with two functional wall layers; those splitting at discharge (**fissitunicate**; 'jack-in-the-box') had been correlated with an ascolocular ontogeny by Luttrell (1951). Reynolds (1989) critically examined this paradigm and found the term to be applied to different ascus types and that an exclusive link to ascostromatic fungi could not be upheld; he also introduced the term **extenditunicate** for asci which extend without any splitting of the wall layers (Reynolds, *Cryptogamie Mycologie* 10: 305, 1989).

Much variation depends on the modifications in the various wall layers, especially the thickness of the walls and the c and d layers, and the details of apical differentiation (Bellemere, 1994) (Figure A2.19). Caution is needed in comparing ascus staining reactions (see iodine) and structures in the absence of ultrastructural data. For terms used to describe the various structures see Figure A2.19.

Also encountered are **ascus crown** (annular thickenings in *Phyllachora*) and **ascus plug** (thickening in the apex through which the spores are forcibly discharged).

tissue types, Korf distinguished the types of hyphal tissues in discomycetes as different **texturas** and this is now applied to all ascomycetes and coelomycetes. Tissue (textura) types (from Korf, *Science Reports of the Yokohama National University*, **II 7**: 13, 1958; which is derived from Starbäck, 1895). See also Dargan (*Nova Hedwigia*, **44**: 489, 1987; *Xylariaceae*), plectenchyma. See Figure A2.20.

DIFFERENTIATION IN BASIDIOMATA

dolipore septum, a septum of a dikaryotic basidiomycete hypha which flares out in the middle portion forming a barrel-shaped

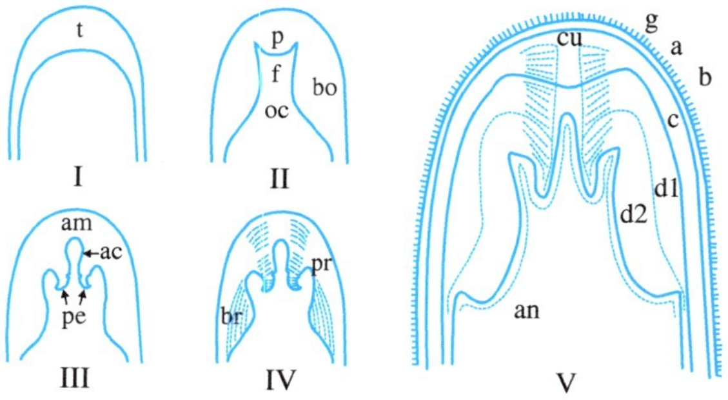

Figure A2.19. I–V. Ascus apex components. **ac** = axial canal; **am** = axial mass; **bo** = bourrelet; **br** = ring in bourrelet; **f** = furrow; **oc** = ocular chamber; **p** = plug; **pe** = pendant; **pr** = rings in the plug and pendant; **t** = tholus; **V**, ascus apex structure. **a** = a layer; **an** = apical nasse; **b** = b layer; **c** = c layer; **d1** and **d2** = sublayers of the d layer.

structure with open ends as shown by electron microscopy; see Markham (*MR* **98**: 1089, 1994; review (https://doi.org/10.1016/S0953-7562(09)80195–0)), Moore (in Hawksworth (ed.), *Identification and Characterization of Pest Organisms*: 249, 1994 (ISBN-10: 0851989047, ISBN-13: 978–0851989044); summary in Figure A2.21), Moore and McAlear (*Am. J. Bot.* **49**: 86, 1962 (https://doi.org/10.2307/2439393)); septal pore swelling; cf. parenthesome.

basidium (pl. **-ia**), (1) the cell or organ, diagnostic for basidiomycetes, from which, after karyogamy and meiosis, basidiospores (generally 4) are produced externally each on an extension (sterigma, q.v.) of its wall (Figure A2.22); (2) a conidiophore or phialide (obsol.).

The confused terminology applied to basidia (sense 1) and their parts has been traced by Clémençon (*Zeitschrift für Mykologie*, **54**: 3, 1988 (https://www.dgfm-ev.de/publikationen/artikelarchiv/die-basidie/download)) and is analysed by Talbot (*TBMS* **61**: 497, 1973 (https://doi.org/10.1016/S0007-1536(73)80119–6)) whose recommended usage (basically that of Donk) and synonymy is adopted in the series of definitions which follow (see Figures A2.23 to A2.25): **pro-**, the morphological part or developmental stage of the basidium in which karyogamy occurs; primary basidial cell; probasidial cyst; **hypo-** (Martin) p.p.; teliospore of Uredinales. **meta-**, (1) the morphological part or developmental stage in which meiosis occurs; hypo- (Martin) p.p.; **epi-** (Martin) p.p.; promycelium of Uredinales. (When the whole meta- includes pro- remnants the distal and functional part may be distinguished as a **pario-** (Talbot, 1973). (2) See proto- below, **holo-**, a basidium (e.g. of *Agaricus*) in which the meta- is not divided by primary septa (see septum) but may become adventitiously septate (see septum) (Talbot, 1968). A holo- may be a **sticho-**, cylindrical, with

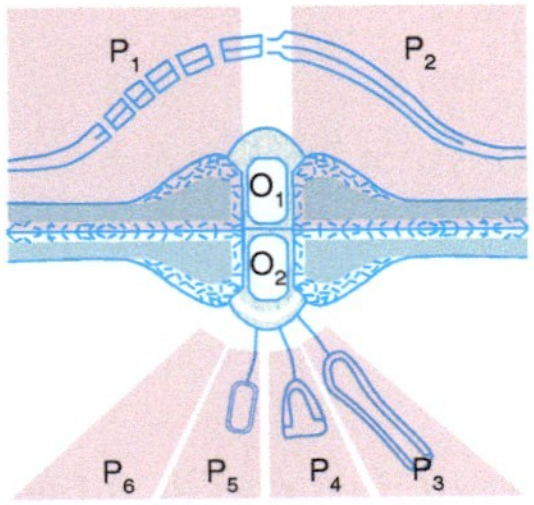

Figure A2.21. Summary diagram of the several forms of the dolipore/parenthesome (**d/p**) septum in the *Basidiomycota*. The outer part of the mushroom-shaped occlusions can appear either as a granule (**O**$_1$) or a striated band (**O**$_2$); the parenthesomes can be regularly perforate (**P**$_1$), imperforate (**P**$_2$), vesiculate (**P**$_{3\text{-}5}$), or absent (**P**$_6$). Particular taxa are generally associated with distinct d/p combinations: O$_1$/P$_1$, Subclass Homobasidiomycetidae; **O**$_1$/**P**$_2$, polypore tribes Phellineae and Hirschioporae, *Heterobasidiomycetidae* Orders Auriculariales, Dacrymycetales and Tulasnellales, and Suborder Exidiineae (note the central thin spot in which the intervening line in the double membrane morphology is absent); **O**$_2$/**P**$_{3,4,6}$), Suborder Tremellineae; **O**$_2$/**P**$_5$), *Wallemia sebi*.

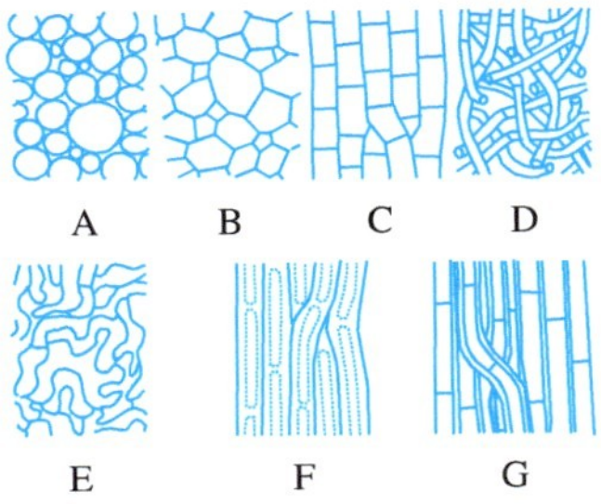

Figure A2.20. Hyphal tissue (**textura**) types. (A) Textura globulosa; (B) textura angularis; (C) textura prismatica; (D) textura intricata; (E) textura epidermoidea; (F) textura oblita; (G) textura porrecta.

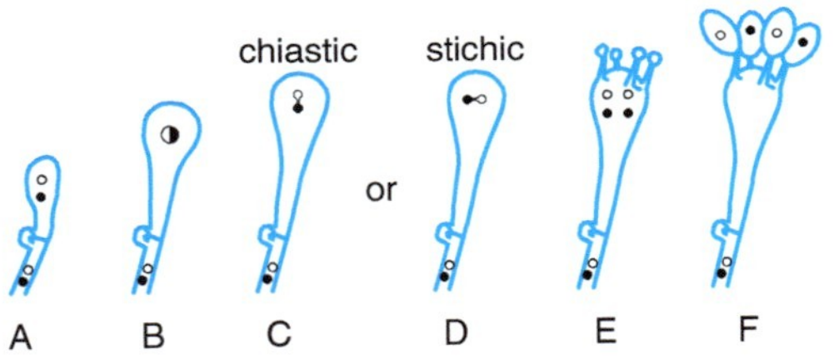

Figure A2.22. Basidiospore development (diagrammatic). (**A-E**) meiosis ((**C**) chiastic and (**D**) stichic). (**E**) Diploid probasidium; (**F**) basidium (metabasidium) with four basidiospores on sterigmata.

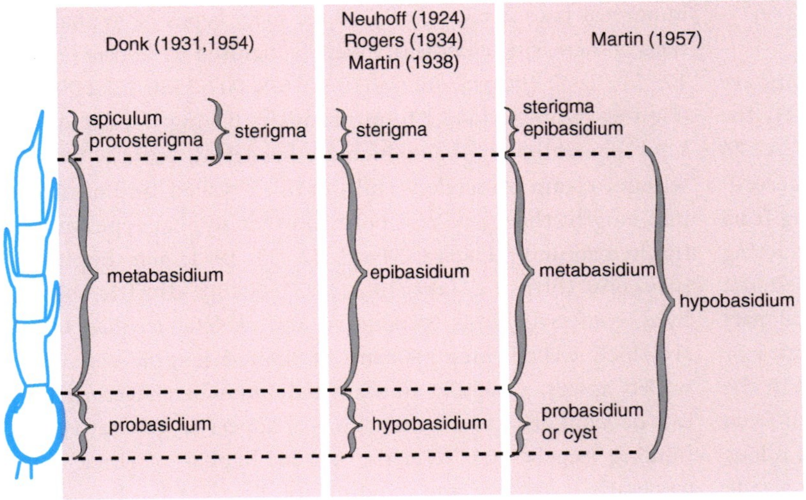

Figure A2.23. Basidium terminology, to compare the terminology of different authors, illustrated with reference to the *Septobasidium*-type (Talbot, 1973 (https://doi.org/10.1016/S0007-1536(73)80119-6)). Note that in this extreme case the metabasidium of Donk coincides with that of Martin.

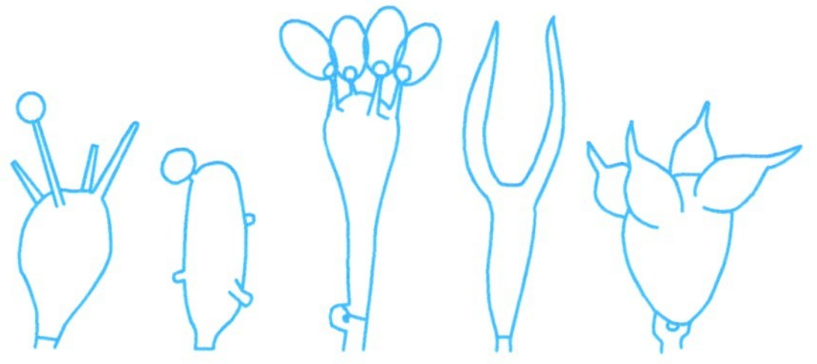

Figure A2.24. Basidial types. **(A–E)**, holobasidial **(A, B** apobasidial; **C, D, E,** autobasidial). **(A)** Lycoperdales; **(B)** Tulostomatales; **(C)** Agaricales; **(D)** Dacrymycetales; **(E)** Tulasnellales ... continued in Figure A2.25.

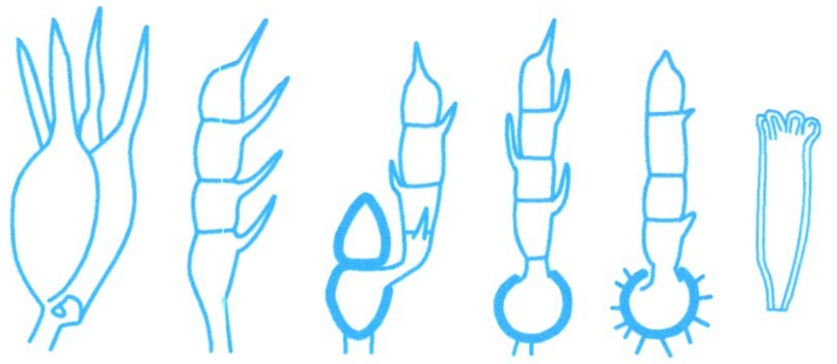

Figure A2.25. Basidial types (continued). **(F–K,)** phragmobasidial **(F, G)**, Basidiomycetes; **(H, I)** Teliomycetes; **(J, K)** Ustomycetes). **(F)** Tremellales; **(G)** Auriculariales; **(H)** Uredinales; **(I)** Septobasidiales; **(J)** Ustilaginales; **(K)** Cryptobasidiales.

nuclear spindles longitudinal and at different levels, or a **chiasto-**, clavate, with nuclear spindles across the basidium and at the same level (see Figure A2.22C, D). **phragmo-**, a basidium in which the meta- is divided by primary septa, usually cruciate (e.g. *Tremella*) or transverse (e.g. *Auricularia*) (Talbot, *Taxon* **17**: 625, 1968 (https://doi.org/10.2307/1218002)).

Among other terms applied to basidia are: **apo-**, one with non-apiculate spores borne symmetrically on the sterigmata and not forcibly discharged (Rogers, *Mycol.* **39**: 558, 1947 (https://doi.org/10.2307/3755195)); **auto-**, one with spores borne asymmetrically and forcibly discharged; **endo-**, one developing within the basidioma, as in gasteromycetes; **epi-**, Martin's term for protosterigma, see sterigma; **hetero-**, a basidium of the Heterobasidiomycetes, usually a phragmo-; **homo-**, a basidium of the Homobasidiomycetes, usually a holo-; **hypo-**, = pro- (Donk), meta- (Martin); (of *Septobasidium*) = basidium (Martin); **pleuro-**, one relatively broad at the base and with bifurcated spreading 'roots', as in *Pleurobasidium* (Donk); **proto-**, a primitive basidium; the opposite of meta- in the sense of changed or degenerate basidium; **repeto-**, see Chadefaud (*Revue de mycologie*, **39**: 173, 1975); **sclero-**, the thick-walled, encysted, gemma-like pro- of the Uredinales (teliospore) and the *Auriculariales* (Janchen, 1923). See also Wells and Wells (*Basidium and Basidiocarp Evolution, Cytology, Function and Development* 1982 (https://doi.org/10.1007/978-1-4612-5677-9), see pp. 9–35).

cystidium (pl. **-ia**), a sterile body, frequently of distinctive shape, occurring at any surface of a basidioma, particularly the hymenium from which it frequently projects (Figures A2.26 and A2.27). Cystidia have been classified and named according to their: (1) *origin*: **hymenial – (tramal)**, originating from hymenial (tramal) hyphae; **pseudo-**, derived from a conducting element, filamentous to fusoid, oily contents, embedded not projecting; **coscino-**, see coscinoid; **skeleto-**, the apical part of a skeletal hypha (frequently ± inflated) projecting into or through the hymenium; false seta; **macro-**, arising deep in the trama in Lactario-Russulae; **hypho-**, hypha-like, derived from generative hyphae. (2) *position* (first by Buller, **2**): on the pileus surface (**pileo-**, **dermato-**, Fayod); at the edge (**cheilo-**), side (**pleuro-**), or within (**endo-**) a lamella; on the stipe (**caulo-**). (3) **form: lepto-**, smooth, thin walled; **lampro-**, thick walled, with or without encrustation (**setiform lampro-**, awl-shaped, wall pigmented; **asteroseta**, a radially branched lampro-; **microsclerid**, a versiform, endolampro-; **lyo-**, cylindrical to conical, very thick walled, apex abruptly thin walled, not encrusted, hyaline, as in *Tubulicrinis* (Donk, 1956); **monilioid gloeo-**, (torulose gloeo- (Bourdot & Galzin, 1928)); monili-form paraphysis (Burt, 1918); pseudophysis; **schizo-** (Nikolayeve, 1956, 1961)), monilioid, frequently with a beaded apex (as in *Hericiaceae* and *Corticiaceae*). (4) *contents*: **gloeo-**, thin walled, usually irregular, contents hyaline or yellowish and highly refractile; **chryso-**, like lepto- but with highly staining contents; **hypo-**, (Larsen & Burdsall, *Memoirs of the New York Botanical Garden*, **28**: 123, 1976); **oleo-**, having an oily resinous exudate; **pseudo-**, see (1) above. See also hyphidium, seta. Reviews of cystidia include Romagnesi (*Revue de Mycologie (Paris)* **9** (suppl.): 4, 1944), Talbot (*Bothalia* **6**: 249, 1954), Lentz (*Bot. Rev.* **20**: 135, 1954 (http://www.jstor.org/stable/4353512)), Smith (in Ainsworth & Sussman (ed.), *The Fungi* **2**: 151, 1966 (http://krishikosh.egranth.ac.in/handle/1/2061852)), Price (*Nova Hedwigia*, **24**: 515, 1975; types in polypores).

Hyphal analysis. A procedure by which the development and structure of basidiomata can be investigated, providing important taxonomic criteria. Three main types of hyphal systems of increasing complexity were recognised by Corner (*TBMS* **17**: 51, 1932 (https://doi.org/10.1016/S0007-1536(32)80026–4)) (Figures A2.28 and A2.29): **monomitic**, having hyphae of one kind (generative hyphae which are branched, septate, with or without clamp connexions, thin to thick walled, and of unlimited length; they give rise both to other hyphal types and to the hymenium) (Teixeira, *Mycol.* **52**: 30, 1961; gen. hyphae of polypores (https://doi.org/10.2307/3756248)); **dimitic**, having hyphae of two kinds (generative and skeletal hyphae which are thick walled, aseptate, and of limited length, with thin-walled apices, generally unbranched but when terminal they can develop arboriform branching or taper) or generative and binding (see below); **trimitic**, having hyphae of three kinds (generative, skeletal and binding (or ligative) hyphae which

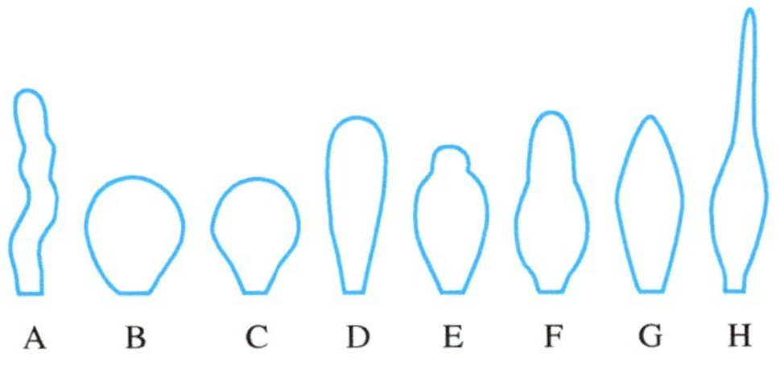

Figure A2.26. Cystidia. (A) hyphoid (*Collybia*); (B) globose (*Agaricus*); (C) pyriform (*Agaricus*); (D) clavate (*Inocybe*); (E) utriform (*Psathyrella*); (F) lageniform (*Pholiota*); (G) fusoid (*Psathyrella*); (H) lanceolate (*Hypholoma*); … continued in Figure A2.27.

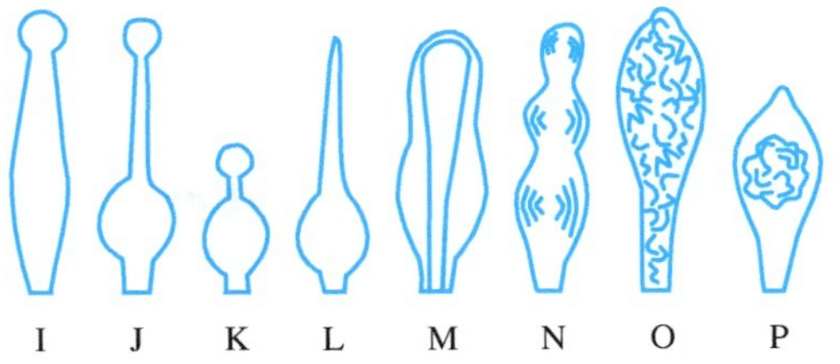

Figure A2.27. Cystidia (continued). (I) capitate (*Hyphoderma*), (J) tibiiform (*Galerina*); (K) lecythiform (*Conocybe*); (L) urticoid (*Naucoria*); (M) metuloid (*Lentinus*), N, gloeocystidium (*Gloeocystidiellum*); (O) macrocystidium (*Russula*); (P) chrysocystidium (*Stropharia*). Not to scale.

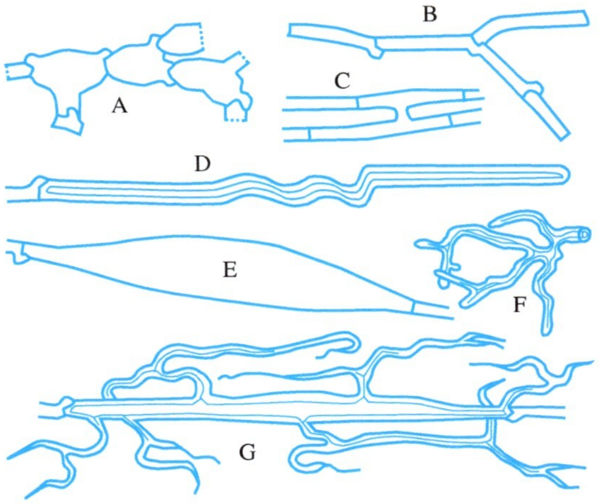

Figure A2.28. Hyphal types. (A) inflated generative hyphae; (B) non-inflated generative hyphae with clamp connexions; (C) generative hyphae without clamp connexions; (D) unbranched skeletal hypha; (E) sarco-hypha; (F) highly branched ligative (binding) hypha; (G) skeleto-ligative hypha. Not to scale.

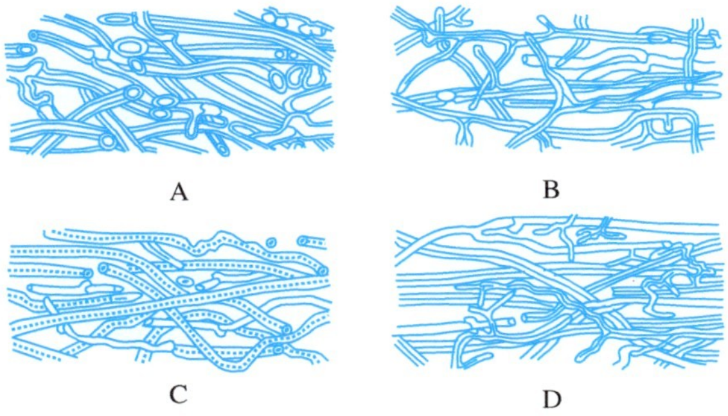

Figure A2.29. Hyphal systems. See Pegler (1973, *Bulletin of the British Mycological Society*, 7 (supplement): pp. 3–43 (https://doi.org/10.1016/S0007-1528(73)80020-3)). (A) monomitic hyphal system, with thick-walled generative hyphae; (B) dimitic hyphal system, with generative and ligative (binding) hyphae; (C) dimitic hyphal system, with generative and skeletal hyphae; (D) trimitic hyphal system, with generative, skeletal and ligative hyphae.

are aseptate, thick walled, much branched, either *Bovista*-type with tapering branches or coralloid; they bind the skeletal and generative hyphae together). In *Polyporaceae* and *Lentinaceae*, intercalary skeletal hyphae can give rise to ligative branching, the entire element being termed a skeleto-binding cell (Corner, 1981) or skeleto-ligative hypha (Pegler, 1973, The polypores: with keys to world genera and British species. *Bulletin of the British Mycological Society*, **7** (supplement): pp. 3–43 (https://doi.org/10.1016/S0007-1528(73)80020–3); 1983, *The Genus* Lentinus*: A World Monograph* (ISBN: 0112426271, 9780112426271)). Corner also recognises **sarcodimitic** (in which the skeletal hyphae are replaced by thick-walled, long, inflating fusiform elements) and **sarcotrimitic** (in which the generative hyphae give both thick-walled inflated elements similar to binding hyphae but septate) types.

Most soft and fleshy basidiomata are monomitic, with hyphae which are generally inflated (most agaricoid and clavarioid fungi). Hard and tough basidiomata may be monomitic with the generative hyphae developing thickened walls, dimitic, with skeletal hyphae (e.g. *Phellinus* spp.) or (especially when perennial) trimitic (e.g. *Fomes, Ganoderma, Microporus xanthopus*). Every species has a well-defined and constant construction, which is maintained regardless of changes in the external morphology of the basidioma due to environmental conditions, hence the importance of hyphal analysis in taxonomy. *Lit.*: Corner (*Annals of Botany*, **46**: 71, 1932, *Phytomorphology* **3**: 152, 1953, *Beihefte zur Nova Hedwigia*, **75**: 13, 1983, **78**: 13, 1984), Cunningham (*New Zealand Journal of Science and Technology*, **28**(A): 238, 1946, *TBMS* **37**: 44, 1954 (https://doi.org/10.1016/S0007-1536(54)80066–0)), Lentz (*Botanical Review*, **20**: 135, 1954), Talbot (*Bothalia* **6**: 1, 1951 (https://doi.org/10.4102/abc.v6i1.1681)).

hyphidium (pl. **-ia**), (paraphysis, pseudoparaphysis, paraphysoid, dikaryoparaphysis, and pseudophysis *sensu* Singer (1962) are syn. or near syn.), a little, or strongly, modified terminal hypha in the hymenium of hymenomycetes (Figures A2.30 and A2.31). Donk (*Persoonia* **3**: 229, 1964 (http://www.repository.naturalis.nl/record/532067)) distinguished; **haplo-**

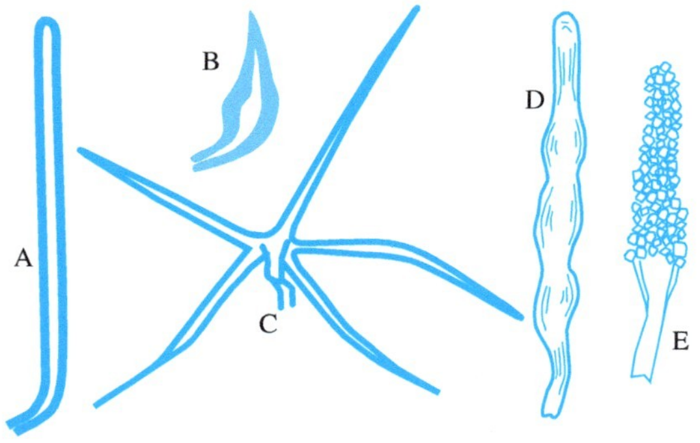

Figure A2.30. Hyphidia. (A) setal hypha (*Phellinus*); (B) seta (*Inonotus*); (C) asteroseta (*Asterostroma*); (D) gloeohypha (*Gloeocystidiellum*); (E) encrusted (*Peniophora*); ... continued in Figure A2.31.

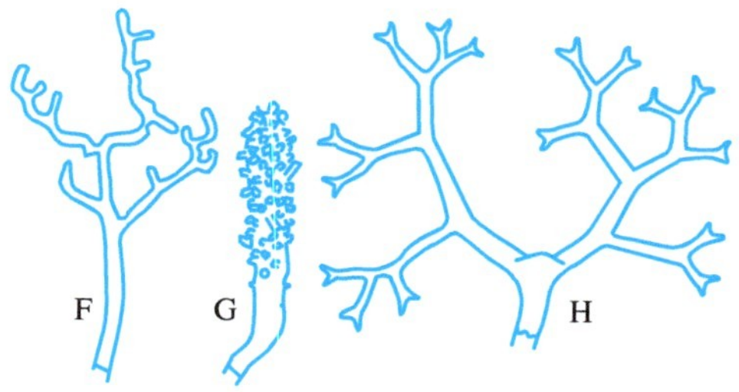

Figure A2.31. Hyphidia (continued). (F) dendrohyphidium (*Cytidia*); (G) acanthohyphidium (*Aleurodiscus*); (H) dendrohyphidium (*Vararia*). Not to scale.

(simple-), unmodified, unbranched or little branched; **dendro-** (dendrophysis), irregularly strongly branched; **dicho-** (dichophysis), repeatedly dichotomously branched; **acantho-** (acanthophysis; bottle-brush paraphysis (Burt, 1918)), having pin-like outgrowths near the apex; in *Corticiaceae* may be botryose, clavate, coralloid, or cylindrical. Cf. cystidium.

How do you know when something's finished? Well, it creeps up on you.

You think there's hours of work to do. And then the brush lifts from the last stroke.

You take two paces back. And there. It's finished.

Your whole life.

[Attributed to Rembrandt van Rijn (1606–1669) by the writer Tim Niel in episode 3 of the 2019 BBC4 TV series *Looking for Rembrandt*]

[View the video clip at https://www.bbc.co.uk/programmes/m0004dds].

INDEX